## Some Physical Constants

| Quantity | Symbol | Value[a] |
|---|---|---|
| Atomic mass unit | u | $1.660\,538\,73\ (13) \times 10^{-27}$ kg |
| | | $931.494\,013\ (37)$ MeV/$c^2$ |
| Avogadro's number | $N_A$ | $6.022\,141\,99\ (47) \times 10^{23}$ particles/mol |
| Bohr magneton | $\mu_B = \dfrac{e\hbar}{2m_e}$ | $9.274\,008\,99\ (37) \times 10^{-24}$ J/T |
| Bohr radius | $a_0 = \dfrac{\hbar^2}{m_e e^2 k_e}$ | $5.291\,772\,083\ (19) \times 10^{-11}$ m |
| Boltzmann's constant | $k_B = \dfrac{R}{N_A}$ | $1.380\,650\,3\ (24) \times 10^{-23}$ J/K |
| Compton wavelength | $\lambda_C = \dfrac{h}{m_e c}$ | $2.426\,310\,215\ (18) \times 10^{-12}$ m |
| Coulomb constant | $k_e = \dfrac{1}{4\pi\epsilon_0}$ | $8.987\,551\,788 \times 10^{9}$ N·m²/C² (exact) |
| Deuteron mass | $m_d$ | $3.343\,583\,09\ (26) \times 10^{-27}$ kg |
| | | $2.013\,553\,212\,71\ (35)$ u |
| Electron mass | $m_e$ | $9.109\,381\,88\ (72) \times 10^{-31}$ kg |
| | | $5.485\,799\,110\ (12) \times 10^{-4}$ u |
| | | $0.510\,998\,902\ (21)$ MeV/$c^2$ |
| Electron volt | eV | $1.602\,176\,462\ (63) \times 10^{-19}$ J |
| Elementary charge | $e$ | $1.602\,176\,462\ (63) \times 10^{-19}$ C |
| Gas constant | $R$ | $8.314\,472\ (15)$ J/mol·K |
| Gravitational constant | $G$ | $6.673\ (10) \times 10^{-11}$ N·m²/kg² |
| Josephson frequency–voltage ratio | $\dfrac{2e}{h}$ | $4.835\,978\,98\ (19) \times 10^{14}$ Hz/V |
| Magnetic flux quantum | $\Phi_0 = \dfrac{h}{2e}$ | $2.067\,833\,636\ (81) \times 10^{-15}$ T·m² |
| Neutron mass | $m_n$ | $1.674\,927\,16\ (13) \times 10^{-27}$ kg |
| | | $1.008\,664\,915\,78\ (55)$ u |
| | | $939.565\,330\ (38)$ MeV/$c^2$ |
| Nuclear magneton | $\mu_n = \dfrac{e\hbar}{2m_p}$ | $5.050\,783\,17\ (20) \times 10^{-27}$ J/T |
| Permeability of free space | $\mu_0$ | $4\pi \times 10^{-7}$ T·m/A (exact) |
| Permittivity of free space | $\epsilon_0 = \dfrac{1}{\mu_0 c^2}$ | $8.854\,187\,817 \times 10^{-12}$ C²/N·m² (exact) |
| Planck's constant | $h$ | $6.626\,068\,76\ (52) \times 10^{-34}$ J·s |
| | $\hbar = \dfrac{h}{2\pi}$ | $1.054\,571\,596\ (82) \times 10^{-34}$ J·s |
| Proton mass | $m_p$ | $1.672\,621\,58\ (13) \times 10^{-27}$ kg |
| | | $1.007\,276\,466\,88\ (13)$ u |
| | | $938.271\,998\ (38)$ MeV/$c^2$ |
| Rydberg constant | $R_H$ | $1.097\,373\,156\,854\,9\ (83) \times 10^{7}$ m$^{-1}$ |
| Speed of light in vacuum | $c$ | $2.997\,924\,58 \times 10^{8}$ m/s (exact) |

*Note:* These constants are the values recommended in 1998 by CODATA, based on a least-squares adjustment of data from different measurements. For a more complete list, see P. J. Mohr and B. N. Taylor, "CODATA recommended values of the fundamental physical constants: 1998." *Rev. Mod. Phys.* 72:351, 2000.

[a] The numbers in parentheses for the values represent the uncertainties of the last two digits.

## Solar System Data

| Body | Mass (kg) | Mean Radius (m) | Period (s) | Distance from the Sun (m) |
|---|---|---|---|---|
| Mercury | $3.18 \times 10^{23}$ | $2.43 \times 10^{6}$ | $7.60 \times 10^{6}$ | $5.79 \times 10^{10}$ |
| Venus | $4.88 \times 10^{24}$ | $6.06 \times 10^{6}$ | $1.94 \times 10^{7}$ | $1.08 \times 10^{11}$ |
| Earth | $5.98 \times 10^{24}$ | $6.37 \times 10^{6}$ | $3.156 \times 10^{7}$ | $1.496 \times 10^{11}$ |
| Mars | $6.42 \times 10^{23}$ | $3.37 \times 10^{6}$ | $5.94 \times 10^{7}$ | $2.28 \times 10^{11}$ |
| Jupiter | $1.90 \times 10^{27}$ | $6.99 \times 10^{7}$ | $3.74 \times 10^{8}$ | $7.78 \times 10^{11}$ |
| Saturn | $5.68 \times 10^{26}$ | $5.85 \times 10^{7}$ | $9.35 \times 10^{8}$ | $1.43 \times 10^{12}$ |
| Uranus | $8.68 \times 10^{25}$ | $2.33 \times 10^{7}$ | $2.64 \times 10^{9}$ | $2.87 \times 10^{12}$ |
| Neptune | $1.03 \times 10^{26}$ | $2.21 \times 10^{7}$ | $5.22 \times 10^{9}$ | $4.50 \times 10^{12}$ |
| Pluto | $\approx 1.4 \times 10^{22}$ | $\approx 1.5 \times 10^{6}$ | $7.82 \times 10^{9}$ | $5.91 \times 10^{12}$ |
| Moon | $7.36 \times 10^{22}$ | $1.74 \times 10^{6}$ | — | — |
| Sun | $1.991 \times 10^{30}$ | $6.96 \times 10^{8}$ | — | — |

## Physical Data Often Used

| | |
|---|---|
| Average Earth–Moon distance | $3.84 \times 10^{8}$ m |
| Average Earth–Sun distance | $1.496 \times 10^{11}$ m |
| Average radius of the Earth | $6.37 \times 10^{6}$ m |
| Density of air (20°C and 1 atm) | 1.20 kg/m$^3$ |
| Density of water (20°C and 1 atm) | $1.00 \times 10^{3}$ kg/m$^3$ |
| Free-fall acceleration | 9.80 m/s$^2$ |
| Mass of the Earth | $5.98 \times 10^{24}$ kg |
| Mass of the Moon | $7.36 \times 10^{22}$ kg |
| Mass of the Sun | $1.99 \times 10^{30}$ kg |
| Standard atmospheric pressure | $1.013 \times 10^{5}$ Pa |

*Note:* These values are the ones used in the text.

## Some Prefixes for Powers of Ten

| Power | Prefix | Abbreviation | Power | Prefix | Abbreviation |
|---|---|---|---|---|---|
| $10^{-24}$ | yocto | y | $10^{1}$ | deka | da |
| $10^{-21}$ | zepto | z | $10^{2}$ | hecto | h |
| $10^{-18}$ | atto | a | $10^{3}$ | kilo | k |
| $10^{-15}$ | femto | f | $10^{6}$ | mega | M |
| $10^{-12}$ | pico | p | $10^{9}$ | giga | G |
| $10^{-9}$ | nano | n | $10^{12}$ | tera | T |
| $10^{-6}$ | micro | $\mu$ | $10^{15}$ | peta | P |
| $10^{-3}$ | milli | m | $10^{18}$ | exa | E |
| $10^{-2}$ | centi | c | $10^{21}$ | zetta | Z |
| $10^{-1}$ | deci | d | $10^{24}$ | yotta | Y |

# PRINCIPLES OF PHYSICS

## A CALCULUS-BASED TEXT FOURTH EDITION
## VOLUME 1

Raymond A. Serway
*Emeritus, James Madison University*

John W. Jewett, Jr.
*California State Polytechnic University—Pomona*

Australia • Canada • Mexico • Singapore • Spain • United Kingdom • United States

*Physics Acquisitions Editor:* Chris Hall
*Publisher:* David Harris
*Editor-in-Chief:* Michelle Julet
*Senior Developmental Editor:* Susan Dust Pashos
*Assistant Editor:* Sarah Lowe
*Editorial Assistant:* Seth Dobrin
*Technology Project Manager:* Sam Subity
*Marketing Managers:* Erik Evans, Julie Conover
*Marketing Assistant:* Leyla Jowza
*Advertising Project Manager:* Stacey Purviance
*Senior Project Manager, Editorial Production:* Teri Hyde
*Print/Media Buyer:* Barbara Britton
*Permissions Editor:* Joohee Lee
*Production Service:* Progressive Publishing Alternatives
*Text Designer:* John Walker
*Art Director:* Rob Hugel
*Photo Researcher:* Dena Digilio Betz
*Copy Editor:* Kathleen Lafferty
*Illustrator:* Progressive Information Technologies
*Cover Designer:* John Walker
*Cover Image:* Transrapid International, Berlin, Germany
*Cover Printer:* Transcontinental-Interglobe
*Compositor:* Progressive Information Technologies
*Printer:* Transcontinental-Interglobe



Printed in Canada
1 2 3 4 5 6 7 09 08 07 06 05



**Library of Congress Control Number: 2004113841**

PRINCIPLES OF PHYSICS, Fourth Edition, Volume 1
ISBN: 0-534-49144-8

**Brooks/Cole — Thomson Learning**
**10 Davis Drive**
**Belmont, CA 94002-3098**
**USA**

**Asia**
Thomson Learning
5 Shenton Way, #01-01
UIC Building
Singapore 068808

**Australia/New Zealand**
Thomson Learning
102 Dodds Street
Southbank, Victoria 3006
Australia

**Canada**
Nelson
1120 Birchmount Road
Toronto, Ontario M1K 5G4
Canada

**Europe/Middle East/Africa**
Thomson Learning
High Holborn House
50/51 Bedford Row
London WC1R 4LR
United Kingdom

**Welcome to PhysicsNow, your fully integrated system for physics tutorials and self-assessment on the web. To get started, just follow these simple instructions.**

### Your first visit to PhysicsNow

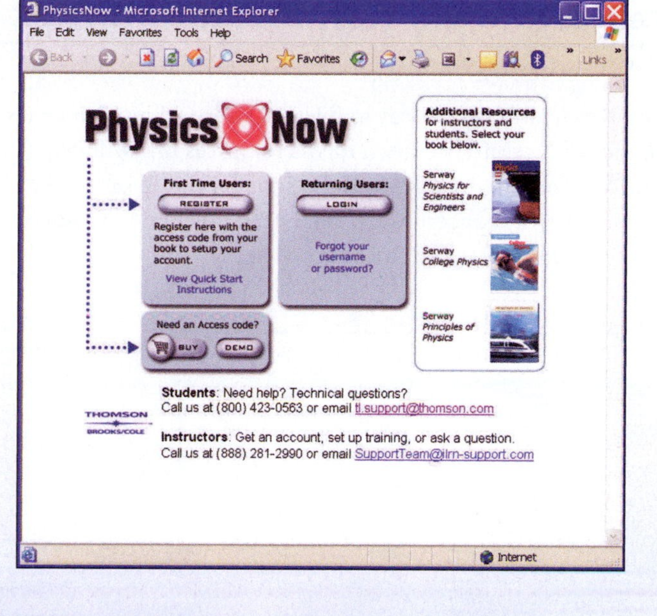


1. Go to **http://www.pop4e.com** and click the **Register** button.
2. The first time you visit, you will be asked to select your school. Choose your state from the drop-down menu, then type in your school's name in the box provided and click **Search.** A list of schools with names similar to what you entered will show on the right. Find your school and click on it.
3. On the next screen, enter the access code from the card that came with your textbook in the "Content or Course Access Code" box*. Enter your email address in the next box and click **Submit.**

   * **PhysicsNow** access codes may be purchased separately. Should you need to purchase an access code, go back to **http://www.pop4e.com** and click the **Buy** button.
4. On the next screen, choose a password and click **Submit.**
5. Lastly, fill out the registration form and click **Register and Enter iLrn.** This information will only be used to contact you if there is a problem with your account.
6. You should now see the **PhysicsNow** homepage. Select a chapter and begin!

**Note:** Your account information will be sent to the email address that you entered in Step 3, so be sure to enter a valid email address. You will use your email address as your username the next time you login.

### Second and later visits

1. Go to **http://www.pop4e.com** and click the **Login** button.
2. Enter your user name (the email address you entered when you registered) and your password and then click **Login.**

**SYSTEM REQUIREMENTS:**
***(Please see the System Requirements link at* www.ilrn.com *for complete list.)***
**PC:** Windows 98 or higher, Internet Explorer 5.5 or higher
**Mac:** OS X or higher, Mozilla browser 1.2.1 or higher

**TECHNICAL SUPPORT:**
For online help, click on **Technical Support** in the upper right corner of the screen, or contact us at:

**1-800-423-0563** Monday–Friday • 8:30 A.M. to 6:00 P.M. EST

**tl.support@thomson.com**

**Turn the page to learn more about PhysicsNow and how it can help you achieve success in your course!**

# What do you need to learn now?

Take charge of your learning with **PhysicsNow**™, a powerful student-learning tool for physics! This interactive resource helps you gauge your unique study needs, then gives you a *Personalized Learning Plan* that will help you focus in on the concepts and problems that will most enhance your understanding. With **PhysicsNow**, you have the resources you need to take charge of your learning!

The access code card included with this new copy of ***Principles of Physics*** is your ticket to all of the resources in **PhysicsNow.** (See the previous page for login instructions.)

## Interact at every turn with the POWER and SIMPLICITY of PhysicsNow!

**PhysicsNow** combines Serway and Jewett's best-selling *Principles of Physics* with carefully crafted media resources that will help you learn. This dynamic resource and the Fourth Edition of the text were developed in concert, to enhance each other and provide you with a seamless, integrated learning system.

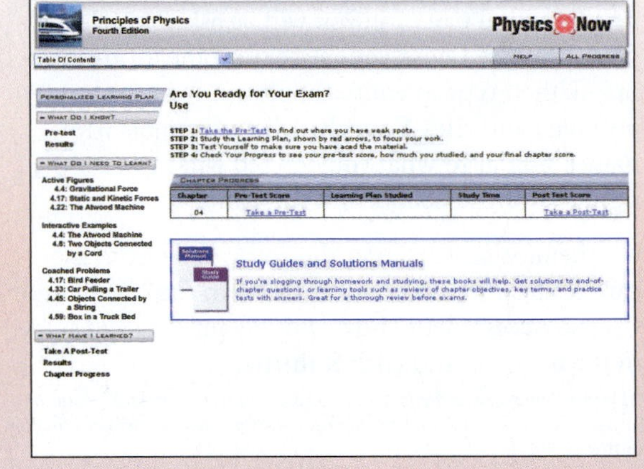


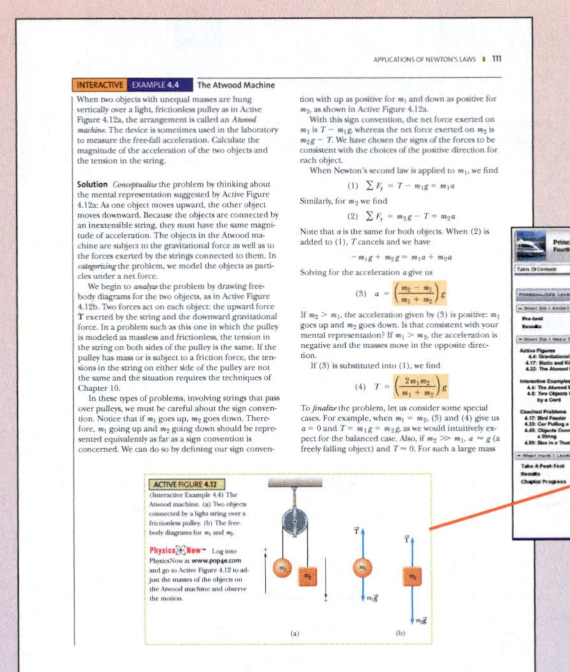

APPLICATIONS OF NEWTON'S LAWS ▮ 111

INTERACTIVE EXAMPLE 4.4 The Atwood Machine

When two objects with unequal masses are hung vertically over a light, frictionless pulley as in Active Figure 4.12a, the arrangement is called an *Atwood machine.* The device is sometimes used in the laboratory to measure the free-fall acceleration. Calculate the magnitude of the acceleration of the two objects and the tension in the string.

**Solution** *Conceptualize* the problem by thinking about the mental representation suggested by Active Figure 4.12a: As one object moves upward, the other object moves downward. Because the objects are connected by an inextensible string, they must have the same magnitude of acceleration. The objects in the Atwood machine are subject to the gravitational force as well as to the forces exerted by the strings connected to them. In *categorizing* the problem, we model the objects as particles under a net force.

We begin to *analyze* the problem by drawing free-body diagrams for the two objects, as in Active Figure 4.12b. Two forces act on each object: the upward force $\vec{T}$ exerted by the string and the downward gravitational force. In a problem such as this one in which the pulley is modeled as massless and frictionless, the tension in the string on both sides of the pulley is the same. If the pulley has mass or is subject to a friction force, the tensions in the string on either side of the pulley are not the same and the situation requires the techniques of Chapter 10.

In these types of problems, involving strings that pass over pulleys, we must be careful about the sign convention. Notice that if $m_1$ goes up, $m_2$ goes down. Therefore, $m_1$ going up and $m_2$ going down should be represented equivalently as far as a sign convention is concerned. We can do so by defining our sign convention with up as positive for $m_1$ and down as positive for $m_2$, as shown in Active Figure 4.12a.

With this sign convention, the net force exerted on $m_1$ is $T - m_1 g$, whereas the net force exerted on $m_2$ is $m_2 g - T$. We have chosen the signs of the forces to be consistent with the choices of the positive direction for each object.

When Newton's second law is applied to $m_1$, we find

$$(1)\quad \sum F_y = T - m_1 g = m_1 a$$

Similarly, for $m_2$ we find

$$(2)\quad \sum F_y = m_2 g - T = m_2 a$$

Note that $a$ is the same for both objects. When (2) is added to (1), $T$ cancels and we have

$$-m_1 g + m_2 g = m_1 a + m_2 a$$

Solving for the acceleration $a$ give us

$$(3)\quad a = \left(\frac{m_2 - m_1}{m_1 + m_2}\right) g$$

If $m_2 > m_1$, the acceleration given by (3) is positive: $m_1$ goes up and $m_2$ goes down. Is that consistent with your mental representation? If $m_1 > m_2$, the acceleration is negative and the masses move in the opposite direction.

If (3) is substituted into (1), we find

$$(4)\quad T = \left(\frac{2 m_1 m_2}{m_1 + m_2}\right) g$$

To *finalize* the problem, let us consider some special cases. For example, when $m_1 = m_2$, (3) and (4) give us $a = 0$ and $T = m_1 g = m_2 g$, as we would intuitively expect for the balanced case. Also, if $m_2 \gg m_1$, $a \approx g$ (a freely falling object) and $T \approx 0$. For such a large mass

ACTIVE FIGURE 4.12

(Interactive Example 4.4) The Atwood machine. (a) Two objects connected by a light string over a frictionless pulley. (b) The free-body diagrams for $m_1$ and $m_2$.

PhysicsNow™ Log into PhysicsNow at **www.pop4e.com** and go to Active Figure 4.12 to adjust the masses of the objects on the Atwood machine and observe the motion.



As you work through the text, you will see notes that direct you to the media-enhanced activities in **PhysicsNow.** This precise page-by-page integration means you'll spend less time flipping through pages or navigating websites looking for useful exercises. These multimedia exercises will make all the difference when you're studying and taking exams . . . after all, it's far easier to understand physics if it's seen in action, and **PhysicsNow** enables you to become a part of the action!

## Begin at http://www.pop4e.com and build your own Personalized Learning Plan now!

Log into **PhysicsNow** at **http://www.pop4e.com** by using the free access code packaged with the text. You'll immediately notice the system's simple, browser-based format. You can build a complete *Personalized Learning Plan* for yourself by taking advantage of all three powerful components found on PhysicsNow:

- **What I Know**
- **What I Need to Learn**
- **What I've Learned**

**The best way to maximize the system and optimize your time is to start by taking the *Pre-Test* ▶▶▶**

## What I Know

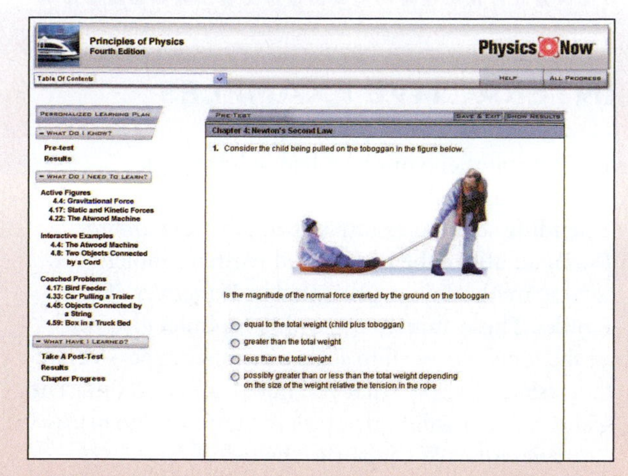


▲ You take a *Pre-Test* to measure your level of comprehension after reading a chapter. Each *Pre-Test* includes approximately 15 questions. The *Pre-Test* is your first step in creating your custom-tailored *Personalized Learning Plan.*

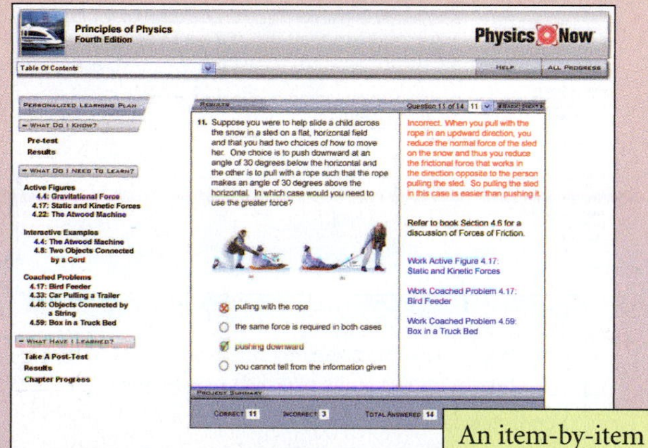


An item-by-item analysis gives you feedback on each of your answers.

▲ Once you've completed the "What I Know" *Pre-Test,* you are presented with a detailed *Personalized Learning Plan,* with text references that outline the elements you need to review in order to master the chapter's most essential concepts. This roadmap to concept mastery guides you to exercises designed to improve skills and to increase your understanding of the basic concepts.

At each stage, the *Personalized Learning Plan* refers to ***Principles of Physics*** to reinforce the connection between text and technology as a powerful learning tool.

## What I Need to Learn

Once you've completed the *Pre-Test,* you're ready to work through tutorials and exercises that will help you master the concepts that are essential to your success in the course.

### ACTIVE FIGURES

A remarkable bank of more than 200 animated figures helps you visualize physics in action. Taken straight from illustrations in the text, these *Active Figures* help you master key concepts from the book. By interacting with the animations and accompanying quiz questions, you come to an even greater understanding of the concepts you need to learn from each chapter. ▼

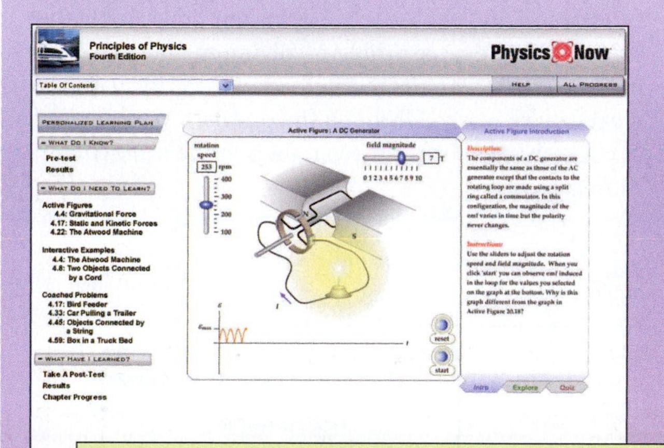


Each figure is titled so you can easily identify the concept you are seeing. The final tab features a *Quiz.* The *Explore* tab guides you through the animation so you understand what you should be seeing and learning.

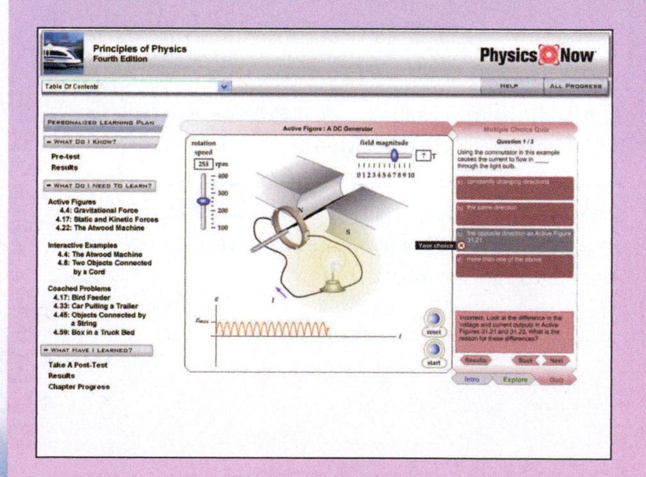


▲ The brief *Quiz* ensures that you mastered the concept played out in the animation—and gives you feedback on each response.

Continued on the next page

## COACHED PROBLEMS

Engaging *Coached Problems* reinforce the lessons in the text by taking a step-by-step approach to problem-solving methodology. Each *Coached Problem* gives you the option of breaking down a problem from the text into steps with feedback to 'coach' you toward the solution. There are approximately five *Coached Problems* per chapter.

You can choose to work through the *Coached Problems* by inputting an answer directly or working in steps with the program. If you choose to work in steps, the problem is solved with the same problem-solving methodology used in ***Principles of Physics*** to reinforce these critical skills. Once you've worked through the problem, you can click **Try Another** to change the variables in the problem for more practice.

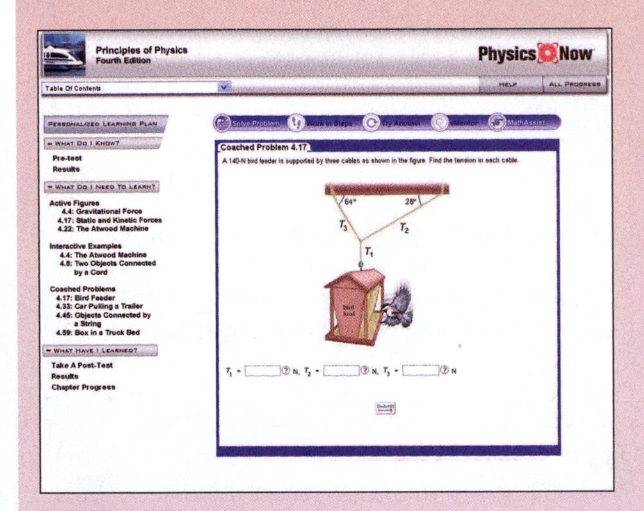


▲ Also built into each *Coached Problem* is a link to Brooks/Cole's exclusive **vMentor**™ web-based tutoring service site that lets you interact directly with a live physics tutor. If you're stuck on math, a *MathAssist* link on each *Coached Problem* launches tutorials on math specific to that problem.

## INTERACTIVE EXAMPLES

You'll strengthen your problem-solving and visualization skills with *Interactive Examples*. Extending selected examples from the text, *Interactive Examples* utilize the proven and trusted problem-solving methodology presented in ***Principles of Physics.*** These animated learning modules give you all the tools you need to solve a problem type—you're then asked to apply what you have learned to different scenarios. You will find approximately two *Interactive Examples* for each chapter of the text.▼

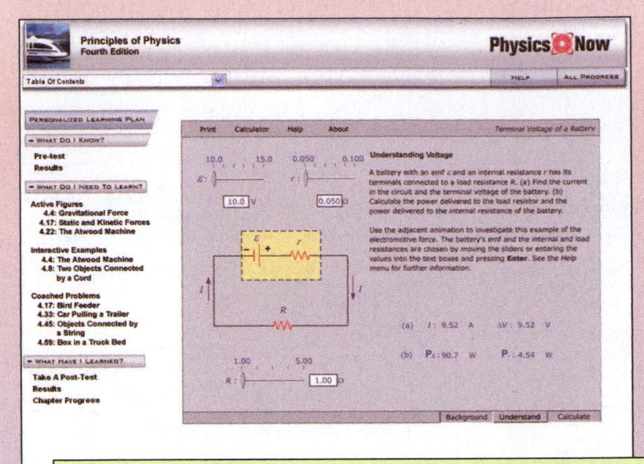


You're guided through the steps to solve the problem and then asked to input an answer in a simulation to see if your result is correct. Feedback is instantaneous.

# What I've Learned

► After working through the problems highlighted in your *Personalized Learning Plan,* you move on to a *Post-Test,* about 15 questions per chapter.

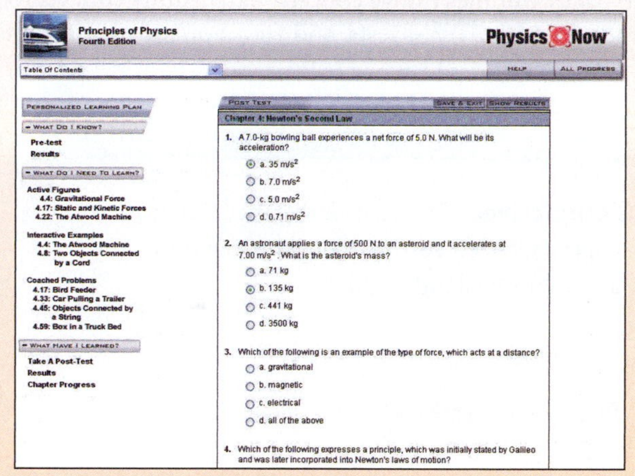


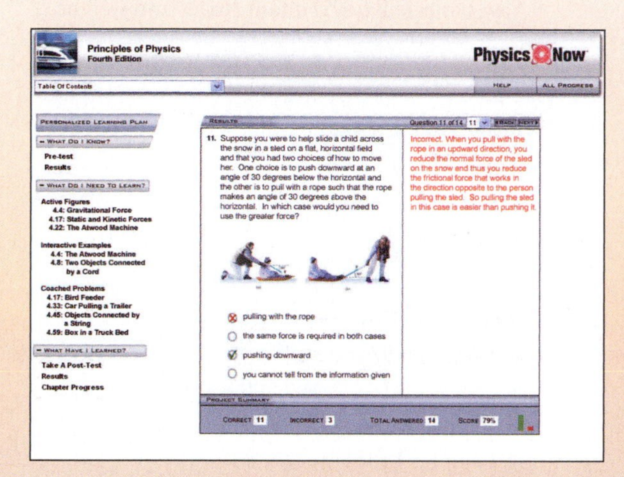


◄ Once you've completed the *Post-Test*, you receive your percentage score and specific feedback on each answer. The *Post-Tests* give you a new set of questions with each attempt, so you can take them over and over as you continue to build your knowledge and skills and master concepts.

## Also available to help you succeed in your course

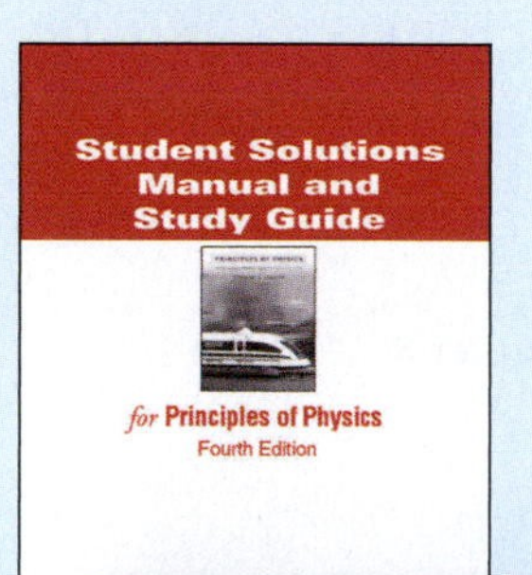


**Student Solutions Manual and Study Guide**

**Volume I (Ch. 1–15) ISBN: 0-534-49145-6**
**Volume II (Ch. 16–31) ISBN: 0-534-49147-2**

These manuals contain detailed solutions to approximately 20-percent of the end-of-chapter problems. These problems are indicated in the textbook with boxed problem numbers. Each manual also features a skills section, important notes from key sections of the text, and a list of important equations and concepts.

**Core Concepts in Physics CD-ROM, Version 2.0**

**ISBN: 0-03-033731-3**

Explore the core of physics with this powerful CD-ROM/workbook program! Content screens provide in-depth coverage of abstract and often difficult principles, building connections between physical concepts and mathematics. The presentation contains more than 350 movies—both animated and live video—including laboratory demonstrations, real-world examples, graphic models, and step-by-step explanations of essential mathematics. An accompanying workbook contains practical physics problems directly related to the presentation, along with worked solutions. Package includes three discs and a workbook.

# Welcome to your MCAT Test Preparation Guide

The **MCAT Test Preparation Guide** makes your copy of *Principles of Physics,* **Fourth Edition,** the most comprehensive MCAT study tool and classroom resource in introductory physics. The grid, which begins below and continues on the next two pages, outlines twelve concept-based **study courses** for the physics part of your MCAT exam. Use it to prepare for the MCAT, class tests, and your homework assignments.

## Vectors

**Skill Objectives:** To calculate distance, calculate angles between vectors, calculate magnitudes, and to understand vectors.

**Review Plan:**

**Distance and Angles:** Chapter 1
- Section 1.6
- Active Figure 1.4
- Chapter Problem 33

**Using Vectors:** Chapter 1
- Sections 1.7–1.9
- Quick Quizzes 1.4–1.8
- Examples 1.6–1.8
- Active Figures 1.9, 1.16
- Chapter Problems 37, 38, 45, 47, 51, 53

## Motion

**Skill Objectives:** To understand motion in two dimensions, to calculate speed and velocity, to calculate centripetal acceleration, and acceleration in free fall problems.

**Review Plan:**

**Motion in 1 Dimension:** Chapter 2
- Sections 2.1, 2.2, 2.4, 2.6, 2.7
- Quick Quizzes 2.3–2.6
- Examples 2.1, 2.2, 2.4–2.10
- Active Figure 2.12
- Chapter Problems 3, 5,13, 19, 21, 29, 31, 33

**Motion in 2 Dimensions:** Chapter 3
- Sections 3.1–3.3
- Quick Quizzes 3.2, 3.3
- Examples 3.1–3.4
- Active Figures 3.4, 3.5, 3.8
- Chapter Problems 1, 7, 15

**Centripetal Acceleration:** Chapter 3
- Sections 3.4, 3.5
- Quick Quizzes 3.4, 3.5
- Example 3.5
- Active Figure 3.12
- Chapter Problems 23, 31

## Force

**Skill Objectives:** To know and understand Newton's Laws, to calculate resultant forces and weight.

**Review Plan:**

**Newton's Laws:** Chapter 4
- Sections 4.1–4.6
- Quick Quizzes 4.1–4.6
- Example 4.1
- Chapter Problem 7

**Resultant Forces:** Chapter 4
- Section 4.7
- Quick Quiz 4.7
- Example 4.6
- Chapter Problems 27, 35

**Gravity:** Chapter 11
- Section 11.1
- Quick Quiz 11.1
- Chapter Problem 3

## Equilibrium

**Skill Objectives:** To calculate momentum and impulse, center of gravity, and torque.

**Review Plan:**

**Momentum:** Chapter 8
- Section 8.1
- Quick Quiz 8.2
- Examples 8.2, 8.3

**Impulse:** Chapter 8
- Sections 8.2, 8.3
- Quick Quizzes 8.3, 8.4
- Examples 8.4, 8.6
- Active Figures 8.8, 8.9
- Chapter Problems 7, 9, 15, 19, 21

**Torque:** Chapter 10
- Sections 10.5, 10.6
- Quick Quiz 10.7
- Example 10.8
- Chapter Problems 21, 27

## Work

**Skill Objectives:** To calculate friction, work, kinetic energy, power, and potential energy.

**Review Plan:**

**Friction:** Chapter 5
- Section 5.1
- Quick Quizzes 5.1, 5.2

**Work:** Chapter 6
- Section 6.2
- Chapter Problems 1, 3

**Kinetic Energy:** Chapter 6
- Section 6.5
- Example 6.4

**Power:** Chapter 6
- Section 6.8
- Chapter Problem 35

**Potential Energy:** Chapter 7
- Sections 7.1, 7.2
- Quick Quizzes 7.1, 7.2
- Chapter Problem 5

## Waves

**Skill Objectives:** To understand interference of waves, to calculate basic properties of waves, properties of springs, and properties of pendulums.

**Review Plan:**

**Wave Properties:** Chapters 12, 13
- Sections 12.1, 12.2, 13.1-13.3
- Quick Quiz 13.1
- Examples 12.1, 13.2
- Active Figures 12.1, 12.2, 12.4, 12.6, 12.10

Chapter 13
- Problem 9

**Pendulum:** Chapter 12
- Sections 12.4, 12.5
- Quick Quizzes 12.3, 12.4
- Examples 12.5, 12.6
- Active Figure 12.11
- Chapter Problem 23

**Interference:** Chapter 14
- Sections 14.1–14.3
- Quick Quiz 14.1
- Active Figures 14.1–14.3

## Matter

**Skill Objectives:** To calculate density, pressure, specific gravity, and flow rates.

**Review Plan:**

**Density:** Chapters 1, 15
- Sections 1.1, 15.2

**Pressure:** Chapter 15
- Sections 15.1–15.4
- Quick Quizzes 15.1–15.4
- Examples 15.1, 15.3
- Chapter Problems 3, 7, 19, 23, 27

**Flow rates:** Chapter 15
- Section 15.6
- Quick Quiz 15.5

## Sound

**Skill Objectives:** To understand interference of waves, calculate properties of waves, the speed of sound, Doppler shifts, and intensity.

**Review Plan:**

**Sound Properties:** Chapters 13, 14
- Sections 13.3, 13.4, 13.7, 13.8, 14.4
- Quick Quizzes 13.2, 13.3, 13.6
- Example 14.3
- Active Figures 13.6–13.8, 13.21, 13.22

Chapter 13
- Problems 3, 17, 23, 29, 35, 37

Chapter 14
- Problem 23

**Interference/Beats:** Chapter 14
- Sections 14.1, 14.2, 14.6
- Quick Quiz 14.6
- Active Figures 14.1–14.3, 14.12
- Chapter Problems 5, 39, 41

## Light

**Skill Objectives:** To understand mirrors and lenses, to calculate the angles of reflection, to use the index of refraction, and to find focal lengths.

**Review Plan:**

**Reflection:** Chapter 25
- Sections 25.1–25.3
- Example 25.1
- Active Figure 25.5

**Refraction:** Chapter 25
- Sections 25.4, 25.5
- Quick Quizzes 25.2–25.5
- Example 25.2
- Chapter Problems 7, 13

**Mirrors and Lenses:** Chapter 26
- Sections 26.1–26.4
- Quick Quizzes 26.1–26.6
- Examples 26.1–26.7
- Active Figures 26.2, 26.24
- Chapter Problems 23, 27, 31, 35

## Electrostatics

**Skill Objectives:** To understand and calculate the electric field, the electrostatic force, and the electric potential.

**Review Plan:**

**Coulomb's Law:** Chapter 19
- Section 19.2–19.4
- Quick Quiz 19.1–19.3
- Examples 19.1, 19.2
- Active Figure 19.7
- Chapter Problems 3, 5

**Electric Field:** Chapter 19
- Sections 19.5, 19.6
- Quick Quizzes 19.4, 19.5
- Active Figures 19.10, 19.19, 19.21

**Potential:** Chapter 20
- Sections 20.1–20.3
- Examples 20.1, 20.2
- Active Figure 20.6
- Chapter Problems 1, 5, 11, 13

## Circuits

**Skill Objectives:** To understand and calculate current, resistance, voltage, and power, and to use circuit analysis.

**Review Plan:**

**Ohm's Law:** Chapter 21
- Sections 21.1, 21.2
- Quick Quizzes 21.1, 21.2
- Examples 21.1, 21.2
- Chapter Problem 7

**Power and energy:** Chapter 21
- Section 21.5
- Quick Quiz 21.4
- Example 21.5
- Active Figure 21.10
- Chapter Problems 17, 19, 23

**Circuits:** Chapter 21
- Section 21.6–21.8
- Quick Quizzes 21.5–21.8
- Example 21.7–21.9
- Active Figures 21.13, 21.14, 21.16
- Chapter Problems 25, 29, 35

## Atoms

**Skill Objectives:** To understand decay processes and nuclear reactions and to calculate half-life.

**Review Plan:**

**Atoms:** Chapter 30
- Sections 30.1
- Quick Quizzes 30.1, 30.2
- Active Figure 30.1

**Decays:** Chapter 30
- Sections 30.3, 30.4
- Quick Quizzes 30.3–30.6
- Examples 30.3–30.6
- Active Figures 30.11–30.14, 30.16, 30.17
- Chapter Problems 13, 19, 23

**Nuclear reactions:** Chapter 30
- Sections 30.5
- Active Figure 30.21
- Chapter Problems 27, 29

## DEDICATION

IN MEMORY OF

**Emily and Fargo Serway**

Two hard working and dedicated parents, for their unforgettable love, vision, and wisdom.

**John W. Jewett**

**Marvin V. Schober**

These fathers and fathers-in-law provided models for hard work, inspiration for creativity, and motivation for excellence.

They are sincerely missed.

# BRIEF CONTENTS

## VOLUME 1

An Invitation to Physics 1

1 Introduction and Vectors 4

CONTEXT 1 Alternative-Fuel Vehicles 34

2 Motion in One Dimension 37

3 Motion in Two Dimensions 69

4 The Laws of Motion 96

5 More Applications of Newton's Laws 125

6 Energy and Energy Transfer 156

7 Potential Energy 188

CONTEXT 1 ■ CONCLUSION: Present and Future Possibilities 220

CONTEXT 2 Mission to Mars 223

8 Momentum and Collisions 226

9 Relativity 259

10 Rotational Motion 291

11 Gravity, Planetary Orbits, and the Hydrogen Atom 337

CONTEXT 2 ■ CONCLUSION: A Successful Mission Plan 367

CONTEXT 3 Earthquakes 371

12 Oscillatory Motion 373

13 Mechanical Waves 400

14 Superposition and Standing Waves 432

CONTEXT 3 ■ CONCLUSION: Minimizing the Risk 459

CONTEXT 4 Search for the *Titanic* 462

15 Fluid Mechanics 464

CONTEXT 4 ■ CONCLUSION: Finding and Visiting the *Titanic* 493

Appendices A.1

Answers to Odd-Numbered Problems A.38

Index I.1

# CONTENTS

## VOLUME 1


**An Invitation to Physics 1**

**1 Introduction and Vectors 4**

1.1 Standards of Length, Mass, and Time 5
1.2 Dimensional Analysis 8
1.3 Conversion of Units 9
1.4 Order-of-Magnitude Calculations 10
1.5 Significant Figures 11
1.6 Coordinate Systems 12
1.7 Vectors and Scalars 14
1.8 Some Properties of Vectors 15
1.9 Components of a Vector and Unit Vectors 17
1.10 Modeling, Alternative Representations, and Problem-Solving Strategy 22

**CONTEXT 1**
Alternative-Fuel Vehicles 34

**2 Motion in One Dimension 37**

2.1 Average Velocity 38
2.2 Instantaneous Velocity 41
2.3 Analysis Models—The Particle Under Constant Velocity 45
2.4 Acceleration 47

2.5 Motion Diagrams 50
2.6 The Particle Under Constant Acceleration 51
2.7 Freely Falling Objects 55
2.8 Context Connection—Acceleration Required by Consumers 59

**3 Motion in Two Dimensions 69**

3.1 The Position, Velocity, and Acceleration Vectors 69
3.2 Two-Dimensional Motion with Constant Acceleration 71
3.3 Projectile Motion 73
3.4 The Particle in Uniform Circular Motion 79
3.5 Tangential and Radial Acceleration 82
3.6 Relative Velocity 83
3.7 Context Connection—Lateral Acceleration of Automobiles 86

**4 The Laws of Motion 96**

4.1 The Concept of Force 97
4.2 Newton's First Law 98
4.3 Mass 100
4.4 Newton's Second Law—The Particle Under a Net Force 101
4.5 The Gravitational Force and Weight 103
4.6 Newton's Third Law 104
4.7 Applications of Newton's Laws 107
4.8 Context Connection—Forces on Automobiles 114

**5 More Applications of Newton's Laws 125**

5.1 Forces of Friction 126
5.2 Newton's Second Law Applied to a Particle in Uniform Circular Motion 132
5.3 Nonuniform Circular Motion 138
5.4 Motion in the Presence of Velocity-Dependent Resistive Forces 140
5.5 The Fundamental Forces of Nature 143
5.6 Context Connection—Drag Coefficients of Automobiles 145

**6 Energy and Energy Transfer 156**

6.1 Systems and Environments 157
6.2 Work Done by a Constant Force 157
6.3 The Scalar Product of Two Vectors 160
6.4 Work Done by a Varying Force 162
6.5 Kinetic Energy and the Work–Kinetic Energy Theorem 166
6.6 The Nonisolated System 169
6.7 Situations Involving Kinetic Friction 173
6.8 Power 177
6.9 Context Connection—Horsepower Ratings of Automobiles 179

**7 Potential Energy 188**

7.1 Potential Energy of a System 188
7.2 The Isolated System 190
7.3 Conservative and Nonconservative Forces 195
7.4 Conservative Forces and Potential Energy 200
7.5 The Nonisolated System in Steady State 202
7.6 Potential Energy for Gravitational and Electric Forces 203

7.7 Energy Diagrams and Stability of Equilibrium 206
7.8 Context Connection—Potential Energy in Fuels 207

CONTEXT 1 ■ CONCLUSION
Present and Future Possibilities 220

CONTEXT 2
Mission to Mars 223

**8 Momentum and Collisions 226**

8.1 Linear Momentum and Its Conservation 227
8.2 Impulse and Momentum 231
8.3 Collisions 233
8.4 Two-Dimensional Collisions 239
8.5 The Center of Mass 242
8.6 Motion of a System of Particles 245
8.7 Context Connection—Rocket Propulsion 248

**9 Relativity 259**
9.1 The Principle of Newtonian Relativity 260
9.2 The Michelson–Morley Experiment 262
9.3 Einstein's Principle of Relativity 263
9.4 Consequences of Special Relativity 264
9.5 The Lorentz Transformation Equations 272
9.6 Relativistic Momentum and the Relativistic Form of Newton's Laws 275
9.7 Relativistic Energy 276
9.8 Mass and Energy 279
9.9 General Relativity 280
9.10 Context Connection—From Mars to the Stars 283

**10 Rotational Motion 291**
10.1 Angular Position, Speed, and Acceleration 292
10.2 Rotational Kinematics: The Rigid Object Under Constant Angular Acceleration 295
10.3 Relations Between Rotational and Translational Quantities 296
10.4 Rotational Kinetic Energy 298
10.5 Torque and the Vector Product 303
10.6 The Rigid Object in Equilibrium 306
10.7 The Rigid Object Under a Net Torque 309
10.8 Angular Momentum 313
10.9 Conservation of Angular Momentum 316
10.10 Precessional Motion of Gyroscopes 319
10.11 Rolling Motion of Rigid Objects 320
10.12 Context Connection—Turning the Spacecraft 323

**11 Gravity, Planetary Orbits, and the Hydrogen Atom 337**
11.1 Newton's Law of Universal Gravitation Revisited 338
11.2 Structural Models 341
11.3 Kepler's Laws 342
11.4 Energy Considerations in Planetary and Satellite Motion 345
11.5 Atomic Spectra and the Bohr Theory of Hydrogen 351
11.6 Context Connection—Changing from a Circular to an Elliptical Orbit 357

CONTEXT 2 ■ CONCLUSION
A Successful Mission Plan 367

CONTEXT 3
Earthquakes 371

**12 Oscillatory Motion 373**
12.1 Motion of a Particle Attached to a Spring 374
12.2 Mathematical Representation of Simple Harmonic Motion 375
12.3 Energy Considerations in Simple Harmonic Motion 381
12.4 The Simple Pendulum 384
12.5 The Physical Pendulum 386
12.6 Damped Oscillations 387
12.7 Forced Oscillations 389
12.8 Context Connection—Resonance in Structures 390

**13 Mechanical Waves 400**
13.1 Propagation of a Disturbance 401
13.2 The Wave Model 403
13.3 The Traveling Wave 405
13.4 The Speed of Transverse Waves of Strings 408
13.5 Reflection and Transmission of Waves 411
13.6 Rate of Energy Transfer by Sinusoidal Waves on Strings 413
13.7 Sound Waves 415
13.8 The Doppler Effect 417
13.9 Context Connection—Seismic Waves 421

**14 Superposition and Standing Waves 432**
14.1 The Principle of Superposition 433
14.2 Interference of Waves 434
14.3 Standing Waves 437
14.4 Standing Waves in Strings 440
14.5 Standing Waves in Air Columns 443
14.6 Beats: Interference in Time 446
14.7 Nonsinusoidal Wave Patterns 448
14.8 Context Connection—Building on Antinodes 450

**CONTEXT 3 ▪ CONCLUSION**
Minimizing the Risk 459

**CONTEXT 4**
Search for the *Titanic* 462

**15 Fluid Mechanics 464**
15.1 Pressure 465
15.2 Variation of Pressure with Depth 466
15.3 Pressure Measurements 470
15.4 Buoyant Forces and Archimedes's Principle 470
15.5 Fluid Dynamics 475
15.6 Streamlines and the Continuity Equation for Fluids 476
15.7 Bernoulli's Equation 478
15.8 Other Applications of Fluid Dynamics 480
15.9 Context Connection—A Near Miss Even Before Leaving Southampton 481

**CONTEXT 4 ▪ CONCLUSION**
Finding and Visiting the *Titanic* 493

**APPENDIX A TABLES A.1**
A.1 Conversion Factors A.1
A.2 Symbols, Dimensions, and Units of Physical Quantities A.2
A.3 Table of Atomic Masses A.4

**APPENDIX B MATHEMATICS REVIEW A.13**
B.1 Scientific Notation A.13
B.2 Algebra A.14
B.3 Geometry A.19
B.4 Trigonometry A.20
B.5 Series Expansions A.22
B.6 Differential Calculus A.22
B.7 Integral Calculus A.24
B.8 Propagation of Uncertainty A.27

**APPENDIX C PERIODIC TABLE OF THE ELEMENTS A.30**

**APPENDIX D SI UNITS A.32**
D.1 SI Base Units A.32
D.2 Some Derived SI Units A.32

**APPENDIX E NOBEL PRIZES A.33**

**ANSWERS TO ODD-NUMBERED PROBLEMS A.38**

**INDEX I.1**

# ABOUT THE AUTHORS

**RAYMOND A. SERWAY** received his doctorate at Illinois Institute of Technology and is Professor Emeritus at James Madison University. In 1990, he received the Madison Scholar Award at James Madison University, where he taught for 17 years. Dr. Serway began his teaching career at Clarkson University, where he conducted research and taught from 1967 to 1980. He was the recipient of the Distinguished Teaching Award at Clarkson University in 1977 and of the Alumni Achievement Award from Utica College in 1985. As Guest Scientist at the IBM Research Laboratory in Zurich, Switzerland, he worked with K. Alex Müller, 1987 Nobel Prize recipient. Dr. Serway also was a visiting scientist at Argonne National Laboratory, where he collaborated with his mentor and friend, Sam Marshall. In addition to earlier editions of this textbook, Dr. Serway is the co-author of *Physics for Scientists and Engineers,* Sixth Edition; *College Physics,* Seventh Edition; and *Modern Physics,* Third Edition. He also is the author of the high-school textbook *Physics,* published by Holt, Rinehart, & Winston. In addition, Dr. Serway has published more than 40 research papers in the field of condensed matter physics and has given more than 70 presentations at professional meetings. Dr. Serway and his wife Elizabeth enjoy traveling, golfing, and spending quality time with their four children and seven grandchildren.

**JOHN W. JEWETT, JR.** earned his doctorate at Ohio State University, specializing in optical and magnetic properties of condensed matter. Dr. Jewett began his academic career at Richard Stockton College of New Jersey, where he taught from 1974 to 1984. He is currently Professor of Physics at California State Polytechnic University, Pomona. Throughout his teaching career, Dr. Jewett has been active in promoting science education. In addition to receiving four National Science Foundation grants, he helped found and direct the Southern California Area Modern Physics Institute (SCAMPI). He also directed Science IMPACT (Institute for Modern Pedagogy and Creative Teaching), which works with teachers and schools to develop effective science curricula. Dr. Jewett's honors include the Stockton Merit Award at Richard Stockton College in 1980, the Outstanding Professor Award at California State Polytechnic University for 1991–1992, and the Excellence in Undergraduate Physics Teaching Award from the American Association of Physics Teachers (AAPT) in 1998. He has given over 80 presentations at professional meetings, including presentations at international conferences in China and Japan. In addition to his work on this textbook, he is co-author of *Physics for Scientists and Engineers,* Sixth Edition with Dr. Serway and author of *The World of Physics . . . Mysteries, Magic, and Myth.* Dr. Jewett enjoys playing keyboard with his all-physicist band, traveling, and collecting antiques that can be used as demonstration apparatus in physics lectures. Most importantly, he relishes spending time with his wife Lisa and their children and grandchildren.

# PREFACE

*Principles of Physics* is designed for a one-year introductory calculus-based physics course for engineering and science students and for premed students taking a rigorous physics course. This fourth edition contains many new pedagogical features—most notably, an integrated Web-based learning system and a structured problem-solving strategy that uses a modeling approach. Based on comments from users of the third edition and reviewers' suggestions, a major effort was made to improve organization, clarity of presentation, precision of language, and accuracy throughout.

This project was conceived because of well-known problems in teaching the introductory calculus-based physics course. The course content (and hence the size of textbooks) continues to grow, while the number of contact hours with students has either dropped or remained unchanged. Furthermore, traditional one-year courses cover little if any physics beyond the 19th century.

In preparing this textbook, we were motivated by the spreading interest in reforming the teaching and learning of physics through physics education research. One effort in this direction was the Introductory University Physics Project (IUPP), sponsored by the American Association of Physics Teachers and the American Institute of Physics. The primary goals and guidelines of this project are to

- Reduce course content following the "less may be more" theme;
- Incorporate contemporary physics naturally into the course;
- Organize the course in the context of one or more "story lines";
- Treat all students equitably.

Recognizing a need for a textbook that could meet these guidelines several years ago, we studied the various proposed IUPP models and the many reports from IUPP committees. Eventually, one of us (RAS) became actively involved in the review and planning of one specific model, initially developed at the U.S. Air Force Academy, entitled "A Particles Approach to Introductory Physics." Part of the summer of 1990 was spent at the Academy working with Colonel James Head and Lt. Col. Rolf Enger, the primary authors of the Particles model, and other members of that department. This most useful collaboration was the starting point of this project.

The other author (JWJ) became involved with the IUPP model called "Physics in Context," developed by John Rigden (American Institute of Physics), David Griffiths (Oregon State University), and Lawrence Coleman (University of Arkansas at Little Rock). This involvement led to the contextual overlay that is used in this book and described in detail later in the Preface.

The combined IUPP approach in this book has the following features:

- It is an evolutionary approach (rather than a revolutionary approach), which should meet the current demands of the physics community.
- It deletes many topics in classical physics (such as alternating current circuits and optical instruments) and places less emphasis on rigid object motion, optics, and thermodynamics.
- Some topics in contemporary physics, such as special relativity, energy quantization, and the Bohr model of the hydrogen atom, are introduced early in the textbook.
- A deliberate attempt is made to show the unity of physics.
- As a motivational tool, the textbook connects physics principles to interesting social issues, natural phenomena, and technological advances.

## OBJECTIVES

This introductory physics textbook has two main objectives: to provide the student with a clear and logical presentation of the basic concepts and principles of physics, and to strengthen an understanding of the concepts and principles through a broad range of interesting applications to the real world. To meet these objectives, we have emphasized sound physical arguments and problem-solving methodology. At the same time, we have attempted to motivate the student through practical examples that demonstrate the role of physics in other disciplines, including engineering, chemistry, and medicine.

## CHANGES IN THE FOURTH EDITION

A number of changes and improvements have been made in the fourth edition of this text. Many of these are in response to recent findings in physics education research and to comments and suggestions provided by the reviewers of the manuscript and instructors using the first three editions. The following represent the major changes in the fourth edition:

**New Context** The context overlay approach is described below under "Text Features." The fourth edition introduces a new Context for Chapters 2–7, "Alternative-Fuel Vehicles." This context addresses the current social issue of the depletion of our supply of petroleum and the efforts being made to develop new fuels and new types of automobiles to respond to this situation.

**Active Figures** Many diagrams from the text have been animated to form **Active Figures,** part of the new ***PhysicsNow*™** integrated Web-based learning system. There are 66 Active Figures in Volume 1, available at **www.pop4e.com.** By visualizing phenomena and processes that cannot be fully represented on a static page, students greatly increase their conceptual understanding. An addition to the figure caption, marked with the **PhysicsNow™** icon, describes briefly the nature and contents of the animation. In addition to viewing animations of the figures, students can change variables to see the effects, conduct suggested explorations of the principles involved in the figure, and take and receive feedback on quizzes related to the figure.

**Interactive Examples** Twenty-eight of the worked examples in Volume 1 have been identified as interactive. As part of the ***PhysicsNow*™** Web-based learning system, students can engage in an extension of the problem solved in the example. This often includes elements of both visualization and calculation, and may also involve prediction and intuition-building. Interactive Examples are available at **www.pop4e.com.**

**Quick Quizzes** Quick Quizzes have been cast in an objective format, including multiple choice, true-false, and ranking. Quick Quizzes provide students with opportunities to test their understanding of the physical concepts presented. The questions require students to make decisions on the basis of sound reasoning, and some of them have been written to help students overcome common misconceptions. Answers to all Quick Quiz questions are found at the end of each chapter. Additional Quick Quizzes that can be used in classroom teaching are available on the instructor's companion Web site. Many instructors choose to use such questions in a "peer instruction" teaching style, but they can be used in standard quiz format as well. To support the use of classroom response systems, we have coded the Quick Quiz questions so that they may be used within the response system of your choice.

**General Problem-Solving Strategy** A general strategy to be followed by the student is outlined at the end of Chapter 1 and provides students with a structured process for solving problems. In the remaining chapters, the steps of the Strategy appear explicitly in one example per chapter so that students are encouraged throughout the course to follow the procedure.

**Line-by-Line Revision** The text has been carefully edited to improve clarity of presentation and precision of language. We hope that the result is a book both accurate and enjoyable to read.

**Problems** In an effort to improve variety, clarity and quality, the end-of-chapter problems were substantially revised. Approximately 15% of the problems (about 300) are new to this edition. The new problems especially are chosen to include interesting applications, notably biological applications. As in previous editions, many problems require students to make order-of-magnitude calculations. More problems now explicitly ask students to design devices and to change among different representations of a situation. All problems have been carefully edited and reworded where necessary. Solutions to approximately 20% of the end-of-chapter problems are included in the *Student Solutions Manual and Study Guide.* Boxed numbers identify these problems. A smaller subset of problems will be available with coached solutions as part of the ***PhysicsNow***™ Web-based learning system and will be accessible to students and instructors using *Principles of Physics.* These coached problems are identified with the **PhysicsNow**™ icon.

**Biomedical Applications** For biology and premed students, icons point the way to various practical and interesting applications of physical principles to biology and medicine. Where possible, an effort was made to include more problems that would be relevant to these disciplines.

## TEXT FEATURES

Most instructors would agree that the textbook selected for a course should be the student's primary guide for understanding and learning the subject matter. Furthermore, the textbook should be easily accessible as well as styled and written to facilitate instruction and learning. With these points in mind, we have included many pedagogical features that are intended to enhance the textbook's usefulness to both students and instructors. These features are as follows:

**Style** To facilitate rapid comprehension, we have attempted to write the book in a clear, logical, and engaging style. The somewhat informal and relaxed writing style is intended to increase reading enjoyment. New terms are carefully defined, and we have tried to avoid the use of jargon.

**Organization** We have incorporated a "context overlay" scheme into the textbook, in response to the "Physics in Context" approach in the IUPP. This feature adds interesting applications of the material to real issues. We have developed this feature to be flexible, so that the instructor who does not wish to follow the contextual approach can simply ignore the additional contextual features without sacrificing complete coverage of the existing material. We believe, though, that the benefits students will gain from this approach will be many.

The context overlay organization divides the text into nine sections, or "Contexts," after Chapter 1, as follows:

| Context Number | Context | Physics Topics | Chapters |
|---|---|---|---|
| 1 | Alternative-Fuel Vehicles | Classical mechanics | 2–7 |
| 2 | Mission to Mars | Classical mechanics | 8–11 |
| 3 | Earthquakes | Vibrations and waves | 12–14 |
| 4 | Search for the *Titanic* | Fluids | 15 |
| 5 | Global Warming | Thermodynamics | 16–18 |
| 6 | Lightning | Electricity | 19–21 |
| 7 | Magnetic Levitation Vehicles | Magnetism | 22–23 |
| 8 | Lasers | Optics | 24–27 |
| 9 | The Cosmic Connection | Modern physics | 28–31 |

Each Context begins with an introduction, leading to a "central question" that motivates study within the Context. The final section of each chapter is a "Context Connection," which discusses how the material in the chapter relates to the Context and to the central question. The final chapter in each Context is followed by a "Context Conclusion." Each Conclusion uses the principles learned in the Context to respond fully to the central question. Each chapter, as well as the Context Conclusions, includes problems related to the context material.

**Pitfall Prevention** These features are placed in the margins of the text and address common student misconceptions and situations in which students often follow unproductive paths. Over 70 Pitfall Preventions are provided in Volume 1 to help students avoid common mistakes and misunderstandings.

**Modeling** A modeling approach, based on four types of models commonly used by physicists, is introduced to help students understand they are solving problems that approximate reality. They must then learn how to test the validity of the model. This approach also helps students see the unity in physics, as a large fraction of problems can be solved with a small number of models. The modeling approach is introduced in Chapter 1.

**Alternative Representations** We emphasize alternative representations of information, including mental, pictorial, graphical, tabular, and mathematical representations. Many problems are easier to solve if the information is presented in alternative ways, to reach the many different methods students use to learn.

**Problem-Solving Strategies** We have included specific strategies for solving the types of problems featured both in the examples and in the end-of-chapter problems. These specific strategies are structured according to the steps in the General Problem-Solving Strategy introduced in Chapter 1. This feature helps students identify necessary steps in solving problems and eliminate any uncertainty they might have.

**Worked Examples** A large number of worked examples of varying difficulty are presented to promote students' understanding of concepts. In many cases, the examples serve as models for solving the end-of-chapter problems. Because of the increased emphasis on understanding physical concepts, many examples are conceptual in nature. The examples are set off in boxes, and the answers to examples with numerical solutions are highlighted with a tan screen.

**Thinking Physics** We have included many Thinking Physics examples throughout each chapter. These questions relate the physics concepts to common experiences or extend the concepts beyond what is discussed in the textual material. Immediately following each of these questions is a "Reasoning" section that responds to the question. Ideally, the student will use these features to better understand physical concepts before being presented with quantitative examples and working homework problems.

**Previews** Most chapters begin with a brief preview that includes a discussion of the particular chapter's objectives and content.

**Important Statements and Equations** Most important statements and definitions are set in boldface type or are highlighted with a blue outline for added emphasis and ease of review. Similarly, important equations are highlighted with a tan background screen to facilitate location.

**Marginal Notes** Comments and notes appearing in the margin can be used to locate important statements, equations, and concepts in the text.

**Illustrations and Tables** The readability and effectiveness of the text material and worked examples are enhanced by the large number of figures, diagrams, photographs, and tables. Full color adds clarity to the artwork and makes illustrations as realistic as possible. For example, vectors are color coded, and curves in graphs are drawn in color. The color photographs have been carefully selected, and their accompanying captions have been written to serve as an added instructional tool.

**Mathematical Level** We have introduced calculus gradually, keeping in mind that students often take introductory courses in calculus and physics concurrently. Most steps are shown when basic equations are developed, and reference is often made to mathematical appendices at the end of the textbook. Vector products are discussed in detail later in the text, where they are needed in physical applications. The dot product is introduced in Chapter 6, which addresses work and energy; the cross product is introduced in Chapter 10, which deals with rotational dynamics.

**Significant Figures** Significant figures in both worked examples and end-of-chapter problems have been handled with care. Most numerical examples and problems are worked out to either two or three significant figures, depending on the accuracy of the data provided.

**Questions** Questions requiring verbal responses are provided at the end of each chapter. Over 260 questions are included in Volume 1. Some questions provide the student with a means of self-testing the concepts presented in the chapter. Others could serve as a basis for initiating classroom discussions. Answers to selected questions are included in the *Student Solutions Manual and Study Guide.*

**Problems** The end-of-chapter problems are more numerous in this edition and more varied (in all, over 930 problems are given throughout Volume 1). For the convenience of both the student and the instructor, about two thirds of the problems are keyed to specific sections of the chapter, including Context Connection sections. The remaining problems, labeled "Additional Problems," are not keyed to specific sections. The icon identifies problems dealing with applications to the life sciences and medicine. One or more problems in each chapter ask students to make an order-of-magnitude calculation based on their own estimated data. Other types of problems are described in more detail below. Answers to odd-numbered problems are provided at the end of the book.

Usually, the problems within a given section are presented so that the straightforward problems (those with black problem numbers) appear first. For ease of identification, the numbers of intermediate-level problems are printed in blue, and those of challenging problems are printed in magenta.

Solutions to approximately 20% of the problems in each chapter are in the *Student Solutions Manual and Study Guide.* Among these, selected problems are identified with PhysicsNow™ icons and have coached solutions available at **www.pop4e.com.**

**Review Problems** Many chapters include review problems requiring the student to relate concepts covered in the chapter to those discussed in previous chapters. These problems can be used by students in preparing for tests and by instructors in routine or special assignments and for classroom discussions.

**Paired Problems** As an aid for students learning to solve problems symbolically, paired numerical and symbolic problems are included in Chapters 1 through 4 and 16 through 21. Paired problems are identified by a common background screen.

**Computer- and Calculator-Based Problems** Many chapters include one or more problems whose solution requires the use of a computer or graphing calculator. Modeling of physical phenomena enables students to obtain graphical representations of variables and to perform numerical analyses.

**Units** The international system of units (SI) is used throughout the text. The U.S. customary system of units is used only to a limited extent in the chapters on mechanics and thermodynamics.

**Summaries** Each chapter contains a summary that reviews the important concepts and equations discussed in that chapter.

**Appendices and Endpapers** Several appendices are provided at the end of the textbook. Most of the appendix material represents a review of mathematical concepts and techniques used in the text, including scientific notation, algebra, geometry, trigonometry, differential calculus, and integral calculus. Reference to these appendices is made throughout the text. Most mathematical review sections in the appendices include worked examples and exercises with answers. In addition to the mathematical reviews, the appendices contain tables of physical data, conversion factors, atomic masses, and the SI units of physical quantities, as well as a periodic table of the elements and a list of Nobel Prize recipients. Other useful information, including fundamental constants and physical data, planetary data, a list of standard prefixes, mathematical symbols, the Greek alphabet, and standard abbreviations of units of measure, appears on the endpapers.

## ANCILLARIES

The ancillary package has been updated substantially and streamlined in response to suggestions from users of the third edition. The most essential parts of the student package are the two-volume *Student Solutions Manual and Study Guide* with a tight focus on problem-solving and the Web-based ***PhysicsNow***™ learning system. Instructors will find increased support for their teaching efforts with new electronic materials.

### Student Ancillaries

***Student Solutions Manual and Study Guide*** by John R. Gordon, Ralph McGrew, and Raymond A. Serway. This two-volume manual features detailed solutions to approximately 20% of the end-of-chapter problems from the textbook. Boxed numbers identify those problems in the textbook whose complete solutions are found in the manual. The manual also features a summary of important chapter notes, a list of important equations and concepts, a short list of important study skills and strategies as well as answers to selected end-of-chapter conceptual questions.

**PhysicsNow**™ Students log into ***PhysicsNow***™ at **www.pop4e.com** by using the free access code packaged with this text.* The ***PhysicsNow***™ system is made up of three interrelated parts:

- How much do you know?
- What do you need to learn?
- What have you learned?

*Free access codes are only available with new copies of *Principles of Physics*, 4th edition.

Students maximize their success by starting with the Pre-Test for the relevant chapter. Each Pre-Test is a mix of conceptual and numerical questions. After completing the Pre-Test, each student is presented with a detailed Learning Plan. The Learning Plan outlines elements to review in the text and Web-based media (Active Figures, Interactive Examples, and Coached Problems) in order to master the chapter's most essential concepts. After working through these materials, students move on to a multiple-choice Post-Test presenting them with questions similar to those that might appear on an exam. Results can be e-mailed to instructors.

**WebTutor™ on WebCT and Blackboard** WebTutor™ offers students real-time access to a full array of study tools, including a glossary of terms and a selection of animations.

**The Brooks/Cole Physics Resource Center** You'll find additional online quizzes, Web links, and animations at **http://physics.brookscole.com.**

## Instructor's Ancillaries

The following ancillaries are available to qualified adopters. Please contact your local Brooks/Cole • Thomson sales representative for details.

***Instructor's Solutions Manual*** by Ralph McGrew. This single manual contains worked solutions to all the problems in the textbook (Volumes 1 and 2) and answers to the end-of-chapter questions. The solutions to problems new to the fourth edition are marked for easy identification by the instructor.

***Test Bank*** by Edward Adelson. Contains approximately 2,000 multiple-choice questions. It is provided in print form for the instructor who does not have access to a computer. The questions in the *Test Bank* are also available in electronic format with complete answers and solutions in iLrn Computerized Testing. The number of conceptual questions has been increased for the 4th edition.

**Multimedia Manager** This easy-to-use multimedia lecture tool allows you to quickly assemble art and database files with notes to create fluid lectures. The CD-ROM set (Volume 1, Chapters 1–15; Volume 2, Chapters 16–31) includes a database of animations, video clips, and digital art from the text as well as PowerPoint lectures and electronic files of the *Instructor's Solutions Manual* and *Test Bank.*

PhysicsNow™ ***PhysicsNow™* Course Management Tools** This extension to the student tutorial environment of *PhysicsNow*™ allows instructors to deliver online assignments in an environment that is familiar to students. This powerful system is your gateway to managing on-line homework, testing, and course administration all in one shell with the proven content to make your course a success. *PhysicsNow*™ is a fully integrated testing, tutorial, and course management software accessible by instructors and students anytime, anywhere. To see a demonstration of this powerful system, contact your Thomson representative or go to **www.pop4e.com**.

PhysicsNow™ ***PhysicsNow™* Homework Management** *PhysicsNow*™ gives you a rich array of problem types and grading options. Its library of assignable questions includes all of the end-of-chapter problems from the text so that you can select the problems you want to include in your online homework assignments. These well-crafted problems are algorithmically generated so that you can assign the same problem with different variables for each student. A flexible grading tolerance feature allows you to specify a percentage range of correct answers so that your students are not penalized for rounding errors. You can give students the option to work an assignment multiple times and record the highest score or limit the times they are able to attempt it. In addition, you can create your own problems to complement the problems from the text. Results flow automatically to an exportable grade book so that instructors are better able to assess student understanding of the material, even prior to class or to an actual test.

**iLrn Computerized Testing** Extend the student experience with **PhysicsNow™** into a testing or quizzing environment. The test item file from the text is included to give you a bank of well-crafted questions that you can deliver online or print out. As with the homework problems, you can use the program's friendly interface to craft your own questions to complement the Serway/Jewett questions. You have complete control over grading, deadlines, and availability and can create multiple tests based on the same material.

**WebTutor™ on WebCT and Blackboard** With **WebTutor™**'s text-specific, preformatted content and total flexibility, instructors can easily create and manage their own personal Web site. **WebTutor™**'s course management tool gives instructors the ability to provide virtual office hours, post syllabi, set up threaded discussions, track student progress with the quizzing material, and much more. **WebTutor™** also provides robust communication tools, such as a course calendar, asynchronous discussion, real-time chat, a whiteboard, and an integrated e-mail system.

## Additional Options for Online Homework

**WebAssign: A Web-Based Homework System** WebAssign is the most utilized homework system in physics. Designed by physicists for physicists, this system is a trusted companion to your teaching. An enhanced version of WebAssign is available for *Principles of Physics.* This enhanced version includes animations with conceptual questions and tutorial problems with feedback and hints to guide student content mastery. Take a look at this new innovation from the most trusted name in physics homework at **www.webassign.net.**

**LON-CAPA: A Computer-Assisted Personalized Approach** LON-CAPA is a Web-based course management system. For more information, visit the LON-CAPA Web site at **www.lon-capa.org.**

**University of Texas Homework Service** With this service, instructors can browse problem banks, select those problems they wish to assign to their students, and then let the Homework Service take over the delivery and grading. Details about and a demonstration of this service are available at **http://hw.ph.utexas.edu/hw.html.**

# TEACHING OPTIONS

Although some topics found in traditional textbooks have been omitted from this textbook, instructors may find that the current text still contains more material than can be covered in a two-semester sequence. For this reason, we would like to offer the following suggestions. If you wish to place more emphasis on contemporary topics in physics, you should consider omitting parts or all of Chapters 15, 16, 17, 18, 24, 25, and 26. On the other hand, if you wish to follow a more traditional approach that places more emphasis on classical physics, you could omit Chapters 9, 11, 28, 29, 30, and 31. Either approach can be used without any loss in continuity. Other teaching options would fall somewhere between these two extremes by choosing to omit some or all of the following sections, which can be considered optional:

- 3.6 Relative Velocity
- 7.7 Energy Diagrams and Stability of Equilibrium
- 9.9 General Relativity
- 10.11 Rolling Motion of Rigid Objects
- 12.6 Damped Oscillations
- 12.7 Forced Oscillations
- 14.7 Nonsinusoidal Wave Patterns
- 15.8 Other Applications of Fluid Dynamics

16.6 Distribution of Molecular Speeds
17.7 Molar Specific Heats of Ideal Gases
17.8 Adiabatic Processes for an Ideal Gas
17.9 Molar Specific Heats and the Equipartition of Energy
20.10 Capacitors with Dielectrics
22.11 Magnetism in Matter
27.9 Diffraction of X-Rays by Crystals
28.13 Tunneling Through a Potential Energy Barrier

## ACKNOWLEDGMENTS

The fourth edition of this textbook was prepared with the guidance and assistance of many professors who reviewed part or all of the manuscript, the pre-revision text, or both. We wish to acknowledge the following scholars and express our sincere appreciation for their suggestions, criticisms, and encouragement:

Anthony Aguirre, *University of California at Santa Cruz*
Royal Albridge, *Vanderbilt University*
Billy E. Bonner, *Rice University*
Richard Cardenas, *St. Mary's University*
Christopher R. Church, *Miami University (Ohio)*
Athula Herat, *Northern Kentucky University*
Huan Z. Huang, *University of California at Los Angeles*
George Igo, *University of California at Los Angeles*
Edwin Lo
Michael J. Longo, *University of Michigan*
Rafael Lopez-Mobilia, *University of Texas at San Antonio*
Ian S. McLean, *University of California at Los Angeles*
Richard Rolleigh, *Hendrix College*
Gregory Severn, *University of San Diego*
Satinder S. Sidhu, *Washington College*
Fiona Waterhouse, *University of California at Berkeley*

*Principles of Physics,* fourth edition was carefully checked for accuracy by James E. Rutledge (University of California at Irvine), Harry W. K. Tom (University of California at Riverside), Gregory Severn (University of San Diego), Bruce Mason (University of Oklahoma at Norman), and Ralf Rapp (Texas A&M University). We thank them for their dedication and vigilance.

We thank the following people for their suggestions and assistance during the preparation of earlier editions of this textbook:

Edward Adelson, *Ohio State University*
Yildirim M. Aktas, *University of North Carolina—Charlotte*
Alfonso M. Albano, *Bryn Mawr College*
Subash Antani, *Edgewood College*
Michael Bass, *University of Central Florida*
Harry Bingham, *University of California, Berkeley*
Anthony Buffa, *California Polytechnic State University, San Luis Obispo*
James Carolan, *University of British Columbia*
Kapila Clara Castoldi, *Oakland University*
Ralph V. Chamberlin, *Arizona State University*
Gary G. DeLeo, *Lehigh University*
Michael Dennin, *University of California, Irvine*
Alan J. DeWeerd, *Creighton University*
Madi Dogariu, *University of Central Florida*
Gordon Emslie, *University of Alabama at Huntsville*
Donald Erbsloe, *United States Air Force Academy*
William Fairbank, *Colorado State University*
Marco Fatuzzo, *University of Arizona*
Philip Fraundorf, *University of Missouri—St. Louis*
Patrick Gleeson, *Delaware State University*
Christopher M. Gould, *University of Southern California*
James D. Gruber, *Harrisburg Area Community College*

John B. Gruber, *San Jose State University*
Todd Hann, *United States Military Academy*
Gail Hanson, *Indiana University*
Gerald Hart, *Moorhead State University*
Dieter H. Hartmann, *Clemson University*
Richard W. Henry, *Bucknell University*
Laurent Hodges, *Iowa State University*
Michael J. Hones, *Villanova University*
Joey Huston, *Michigan State University*
Herb Jaeger, *Miami University*
David Judd, *Broward Community College*
Thomas H. Keil, *Worcester Polytechnic Institute*
V. Gordon Lind, *Utah State University*
Roger M. Mabe, *United States Naval Academy*
David Markowitz, *University of Connecticut*
Thomas P. Marvin, *Southern Oregon University*
Martin S. Mason, *College of the Desert*
Wesley N. Mathews, Jr., *Georgetown University*
John W. McClory, *United States Military Academy*
L. C. McIntyre, Jr., *University of Arizona*
Alan S. Meltzer, *Rensselaer Polytechnic Institute*
Ken Mendelson, *Marquette University*
Roy Middleton, *University of Pennsylvania*
Allen Miller, *Syracuse University*
Clement J. Moses, *Utica College of Syracuse University*
John W. Norbury, *University of Wisconsin—Milwaukee*
Anthony Novaco, *Lafayette College*
Romulo Ochoa, *The College of New Jersey*
Melvyn Oremland, *Pace University*
Desmond Penny, *Southern Utah University*
Steven J. Pollock, *University of Colorado—Boulder*
Prabha Ramakrishnan, *North Carolina State University*
Rex D. Ramsier, *The University of Akron*
Rogers Redding, *University of North Texas*
Charles R. Rhyner, *University of Wisconsin—Green Bay*
Perry Rice, *Miami University*
Dennis Rioux, *University of Wisconsin—Oshkosh*
Janet E. Seger, *Creighton University*
Gregory D. Severn, *University of San Diego*
Antony Simpson, *Dalhousie University*
Harold Slusher, *University of Texas at El Paso*
J. Clinton Sprott, *University of Wisconsin at Madison*
Shirvel Stanislaus, *Valparaiso University*
Randall Tagg, *University of Colorado at Denver*
Cecil Thompson, *University of Texas at Arlington*
Chris Vuille, *Embry–Riddle Aeronautical University*
Robert Watkins, *University of Virginia*
James Whitmore, *Pennsylvania State University*

We are indebted to the developers of the IUPP models, “A Particles Approach to Introductory Physics” and “Physics in Context,” upon which much of the pedagogical approach in this textbook is based.

Ralph McGrew coordinated the end-of-chapter problems. Problems new to this edition were written by Edward Adelson, Michael Browne, Andrew Duffy, Robert Forsythe, Perry Ganas, John Jewett, Randall Jones, Boris Korsunsky, Edwin Lo, Ralph McGrew, Clement Moses, Raymond Serway, and Jerzy Wrobel. Daniel Fernandez, David Tamres, and Kevin Kilty made corrections in problems from the previous edition.

We are grateful to John R. Gordon and Ralph McGrew for writing the *Student Solutions Manual and Study Guide,* to Ralph McGrew for preparing an excellent *Instructor's Solutions Manual,* and to Edward Adelson of Ohio State University for preparing the *Test Bank.* We thank M & N Toscano for the attractive layout of these volumes. During the development of this text, the authors benefited from many useful discussions with colleagues and other physics instructors, including Robert Bauman,

William Beston, Don Chodrow, Jerry Faughn, John R. Gordon, Kevin Giovanetti, Dick Jacobs, Harvey Leff, Clem Moses, Dorn Peterson, Joseph Rudmin, and Gerald Taylor.

Special thanks and recognition go to the professional staff at the Brooks/Cole Publishing Company—in particular, Susan Pashos, Jay Campbell, Sarah Lowe, Seth Dobrin, Teri Hyde, Michelle Julet, David Harris, and Chris Hall—for their fine work during the development and production of this textbook. We are most appreciative of Sam Subity's masterful management of the ***PhysicsNow***™ media program. Julie Conover is our enthusiastic Marketing Manager, and Stacey Purviance coordinates our marketing communications. We recognize the skilled production service provided by Donna King and the staff at Progressive Publishing Alternatives and the dedicated photo research efforts of Dena Betz.

Finally, we are deeply indebted to our wives and children for their love, support, and long-term sacrifices.

**RAYMOND A. SERWAY**
St. Petersburg, Florida

**JOHN W. JEWETT, JR.**
Pomona, California

# TO THE STUDENT

It is appropriate to offer some words of advice that should benefit you, the student. Before doing so, we assume you have read the Preface, which describes the various features of the text that will help you through the course.

## HOW TO STUDY

Very often instructors are asked, "How should I study physics and prepare for examinations?" There is no simple answer to this question, but we would like to offer some suggestions based on our own experiences in learning and teaching over the years.

First and foremost, maintain a positive attitude toward the subject matter, keeping in mind that physics is the most fundamental of all natural sciences. Other science courses that follow will use the same physical principles, so it is important that you understand and are able to apply the various concepts and theories discussed in the text.

The Contexts in the text will help you understand how the physical principles relate to real issues, phenomena, and applications. Be sure to read the Context Introductions, Context Connection sections in each chapter, and Context Conclusions. These will be most helpful in motivating your study of physics.

## CONCEPTS AND PRINCIPLES

It is essential that you understand the basic concepts and principles before attempting to solve assigned problems. You can best accomplish this goal by carefully reading the textbook before you attend your lecture on the covered material. When reading the text, you should jot down those points that are not clear to you. We've purposely left wide margins in the text to give you space for doing this. Also be sure to make a diligent attempt at answering the questions in the Quick Quizzes as you come to them in your reading. We have worked hard to prepare questions that help you judge for yourself how well you understand the material. Pay careful attention to the many Pitfall Preventions throughout the text. These will help you avoid misconceptions, mistakes, and misunderstandings as well as maximize the efficiency of your time by minimizing adventures along fruitless paths. During class, take careful notes and ask questions about those ideas that are unclear to you. Keep in mind that few people are able to absorb the full meaning of scientific material after only one reading.

After class, several readings of the text and your notes may be necessary. Be sure to take advantage of the features available in the ***PhysicsNow***™ learning system, such as the Active Figures, Interactive Examples, and Coached Problems. Your lectures and laboratory work supplement your reading of the textbook and should clarify some of the more difficult material. You should minimize your memorization of material. Successful memorization of passages from the text, equations, and derivations does not necessarily indicate that you understand the material.

Your understanding of the material will be enhanced through a combination of efficient study habits, discussions with other students and with instructors, and your ability to solve the problems presented in the textbook. Ask questions whenever you feel clarification of a concept is necessary.

## STUDY SCHEDULE

It is important for you to set up a regular study schedule, preferably a daily one. Make sure you read the syllabus for the course and adhere to the schedule set by your instructor. The lectures will be much more meaningful if you read the corresponding textual material before attending them. As a general rule, you should devote about two hours of study time for every hour you are in class. If you are having trouble with the course, seek the advice of the instructor or other students who have taken the course. You may find it necessary to seek further instruction from experienced students. Very often, instructors offer review sessions in addition to regular class periods. It is important that you avoid the practice of delaying study until a day or two before an exam. More often than not, this approach has disastrous results. Rather than undertake an all-night study session, briefly review the basic concepts and equations and get a good night's rest. If you feel you need additional help in understanding the concepts, in preparing for exams, or in problem-solving, we suggest that you acquire a copy of the *Student Solutions Manual and Study Guide* that accompanies this textbook; this manual should be available at your college bookstore.

## USE THE FEATURES

You should make full use of the various features of the text discussed in the preface. For example, marginal notes are useful for locating and describing important equations and concepts, and **boldfaced** type indicates important statements and definitions. Many useful tables are contained in the Appendices, but most tables are incorporated in the text where they are most often referenced. Appendix B is a convenient review of mathematical techniques.

Answers to odd-numbered problems are given at the end of the textbook, answers to Quick Quizzes are located at the end of each chapter, and answers to selected end-of-chapter questions are provided in the *Student Solutions Manual and Study Guide*. Problem-Solving Strategies are included in selected chapters throughout the text and give you additional information about how you should solve problems. The Table of Contents provides an overview of the entire text, while the Index enables you to locate specific material quickly. Footnotes sometimes are used to supplement the text or to cite other references on the subject discussed.

After reading a chapter, you should be able to define any new quantities introduced in that chapter and to discuss the principles and assumptions used to arrive at certain key relations. The chapter summaries and the review sections of the *Student Solutions Manual and Study Guide* should help you in this regard. In some cases, it may be necessary for you to refer to the index of the text to locate certain topics. You should be able to correctly associate with each physical quantity the symbol used to represent that quantity and the unit in which the quantity is specified. Furthermore, you should be able to express each important relation in a concise and accurate prose statement.

## PROBLEM-SOLVING

R. P. Feynman, Nobel laureate in physics, once said, "You do not know anything until you have practiced." In keeping with this statement, we strongly advise that you develop the skills necessary to solve a wide range of problems. Your ability to solve problems will be one of the main tests of your knowledge of physics; therefore, you should try to solve as many problems as possible. It is essential that you understand basic concepts and principles before attempting to solve problems. It is good practice to try to find alternative solutions to the same problem. For example, you can solve problems in mechanics using Newton's laws, but very often an alternative method that draws on energy considerations is more direct. You should not deceive yourself into thinking you understand a problem merely because you have seen it solved in class. You must be able to solve the problem and similar problems on your own.

The approach to solving problems should be carefully planned. A systematic plan is especially important when a problem involves several concepts. First, read the problem several times until you are confident you understand what is being asked. Look for any key words that will help you interpret the problem and perhaps allow you to make certain assumptions. Your ability to interpret a question properly is an integral part of problem-solving. Second, you should acquire the habit of writing down the information given in a problem and those quantities that need to be found; for example, you might construct a table listing both the quantities given and the quantities to be found. This procedure is sometimes used in the worked examples of the textbook. After you have decided on the method you feel is appropriate for a given problem, proceed with your solution. Finally, check your results to see if they are reasonable and consistent with your initial understanding of the problem. General problem-solving strategies of this type are included in the text and are set off in their own boxes. We have also developed a General Problem-Solving Strategy, making use of models, to help guide you through complex problems. This strategy is located at the end of Chapter 1. If you follow the steps of this procedure, you will find it easier to come up with a solution and also gain more from your efforts.

Often, students fail to recognize the limitations of certain equations or physical laws in a particular situation. It is very important that you understand and remember the assumptions underlying a particular theory or formalism. For example, certain equations in kinematics apply only to a particle moving with constant acceleration. These equations are not valid for describing motion whose acceleration is not constant, such as the motion of an object connected to a spring or the motion of an object through a fluid.

## EXPERIMENTS

Physics is a science based on experimental observations. In view of this fact, we recommend that you try to supplement the text by performing various types of "hands-on" experiments, either at home or in the laboratory. For example, the common Slinky™ toy is excellent for studying traveling waves; a ball swinging on the end of a long string can be used to investigate pendulum motion; various masses attached to the end of a vertical spring or rubber band can be used to determine their elastic nature; an old pair of Polaroid sunglasses and some discarded lenses and a magnifying glass are the components of various experiments in optics; and the approximate measure of the free-fall acceleration can be determined simply by measuring with a stopwatch the time it takes for a ball to drop from a known height. The list of such experiments is endless. When physical models are not available, be imaginative and try to develop models of your own.

## NEW MEDIA

We strongly encourage you to use the ***PhysicsNow***™ Web-based learning system that accompanies this textbook. It is far easier to understand physics if you see it in action, and these new materials will enable you to become a part of that action. ***PhysicsNow***™ media described in the Preface are accessed at the URL **www.pop4e.com,** and feature a three-step learning process consisting of a Pre-Test, a personalized learning plan, and a Post-Test.

In addition to the Coached Problems identified with icons, ***PhysicsNow***™ includes the following Active Figures and Interactive Examples from Volume 1:

**Chapter 1**
Active Figures 1.4, 1.9, and 1.16
Interactive Example 1.8

**Chapter 2**
Active Figures 2.1, 2.2, 2.8, 2.11, and 2.12
Interactive Examples 2.8 and 2.10

**Chapter 3**
Active Figures 3.4, 3.5, 3.8, and 3.12
Interactive Examples 3.2 and 3.6

**Chapter 4**
Active Figures 4.12 and 4.13
Interactive Examples 4.4 and 4.5

**Chapter 5**
Active Figures 5.1, 5.9, 5.15, and 5.18
Interactive Examples 5.7 and 5.8

**Chapter 6**
Active Figure 6.8
Interactive Examples 6.6 and 6.7

**Chapter 7**
Active Figures 7.3, 7.6, and 7.15
Interactive Examples 7.1 and 7.2

**Chapter 8**
Active Figures 8.8, 8.9, 8.11, 8.13, and 8.14
Interactive Examples 8.2 and 8.8

**Chapter 9**
Active Figures 9.3, 9.5, and 9.8
Interactive Example 9.5

**Chapter 10**
Active Figures 10.4, 10.11, 10.12, 10.21, and 10.28
Interactive Examples 10.5, 10.8, and 10.9

**Chapter 11**
Active Figures 11.1, 11.5, 11.7, 11.19, and 11.20
Interactive Examples 11.1 and 11.3

**Chapter 12**
Active Figures 12.1, 12.2, 12.4, 12.6, 12.9, 12.10, 12.11, and 12.14
Interactive Example 12.1

**Chapter 13**
Active Figures 13.6, 13.7, 13.8, 13.14, 13.15, 13.21, 13.22, and 13.24
Interactive Examples 13.5 and 13.7

**Chapter 14**
Active Figures 14.1, 14.2, 14.3, 14.8, 14.9, 14.12, 14.15, and 14.16
Interactive Examples 14.1 and 14.3

**Chapter 15**
Active Figures 15.9 and 15.10
Interactive Examples 15.4 and 15.7

It is our sincere hope that you too will find physics an exciting and enjoyable experience and that you will profit from this experience, regardless of your chosen profession. Welcome to the exciting world of physics!

*The scientist does not study nature because it is useful; he studies it because he delights in it, and he delights in it because it is beautiful. If nature were not beautiful, it would not be worth knowing, and if nature were not worth knowing, life would not be worth living.*

**Henri Poincaré**

# List of Life Science Applications and Problems in Volume 1


CHAPTER 1: Introduction and Vectors 4

**Example 1.5**
**Problem 1.8**
**Problem 1.64**

CHAPTER 2: Motion in One Dimension 37

**Example 2.5**
**Problem 2.39**
**Problem 2.40**
**Problem 2.41**

CHAPTER 3: Motion in Two Dimensions 69

**Problem 3.6**
**Problem 3.9**
**Problem 3.14**

CHAPTER 4: The Laws of Motion 96

**Problem 4.51**

CHAPTER 5: More Applications of Newton's Laws 125

**Question 5.12**
**Problem 5.4**
**Problem 5.54**

CHAPTER 6: Energy and Energy Transfer 156

**Page 172, bioluminescence**
**Problem 6.38**
**Problem 6.39**
**Problem 6.43**
**Problem 6.44**

CHAPTER 7: Potential Energy 188

**Page 203, the human body as a nonisolated system**
**Question 7.14**
**Problem 7.22**
**Problem 7.45**

CHAPTER 8: Momentum and Collisions 226

**Page 232, advantages of air bags in reducing injury**
**Page 234, glaucoma testing**
**Problem 8.3**
**Problem 8.49**

CHAPTER 9: Relativity 259

**Page 268, varying rates of aging in relativity**
**Example 9.1**
**Problem 9.6**

CHAPTER 10: Rotational Motion 291

**Problem 10.26**
**Problem 10.70**
**Problem 10.71**

CHAPTER 12: Oscillatory Motion 373

**Problem 12.45**

CHAPTER 13: Mechanical Waves 400

**Page 419, Doppler measurements of blood flow**
**Problem 13.24**
**Problem 13.26**
**Problem 13.28**
**Problem 13.34**
**Problem 13.59**

CHAPTER 14: Superposition and Standing Waves 432

**Problem 14.29**
**Problem 14.32**

CHAPTER 15: Fluid Mechanics 464

**Page 466, hypodermic needles**
**Page 468, measuring blood pressure**
**Page 481, vascular flutter**
**Question 15.12**
**Question 15.17**
**Question 15.20**
**Problem 15.8**
**Problem 15.16**
**Problem 15.29**
**Problem 15.45**
**Problem 15.57**

# An Invitation to Physics

(© Stockbyte)

Technicians use electronic devices to test motherboards for computer systems. The principles of physics are involved in the design, manufacturing, and testing of these motherboards. ■

Physics, the most fundamental physical science, is concerned with the basic principles of the universe. It is the foundation on which engineering, technology, and the other sciences—astronomy, biology, chemistry, and geology—are based. The beauty of physics lies in the simplicity of its fundamental theories and in the manner in which just a small number of basic concepts, equations, and assumptions can alter and expand our view of the world around us.

*Classical physics,* developed prior to 1900, includes the theories, concepts, laws, and experiments in classical mechanics, thermodynamics, electromagnetism, and optics. For example, Galileo Galilei (1564–1642) made significant contributions to classical mechanics through his work on the laws of motion with constant acceleration. In the same era, Johannes Kepler (1571–1630) used astronomical observations to develop empirical laws for the motions of planetary bodies.

The most important contributions to classical mechanics, however, were provided by Isaac Newton (1642–1727), who developed classical mechanics as a systematic theory and was one of the originators of calculus as a mathematical tool. Although major developments in classical physics continued in the 18th century, thermodynamics and electromagnetism were not developed until the latter part of the 19th century, principally because the apparatus for controlled experiments was either too crude or unavailable until then. Although many electric and magnetic phenomena had been studied earlier, the work of James Clerk Maxwell (1831–1879) provided a unified theory of electromagnetism. In this text, we shall treat the various disciplines of classical physics in separate sections; we will see, however, that the disciplines of mechanics and electromagnetism are basic to all the branches of physics.

A major revolution in physics, usually referred to as *modern physics,* began near the end of the 19th century. Modern physics developed mainly because many physical phenomena could not be explained by classical physics. The two most important developments in this modern era were the theories of relativity and quantum mechanics. Albert Einstein's theory of relativity completely revolutionized the traditional concepts of space, time, and energy. This theory correctly describes the motion of objects moving at speeds comparable to the speed of light. The theory of relativity also shows that the speed of light is the upper limit of the speed of an object and that mass and energy are related. Quantum mechanics was formulated by a number of distinguished scientists to provide descriptions of physical phenomena at the atomic level.

Scientists continually work at improving our understanding of fundamental laws, and new discoveries are made every day. In many research areas, a great deal of overlap exists among physics, chemistry, and biology. Evidence for this overlap is seen in the names of some subspecialties in science: biophysics, biochemistry, chemical physics, biotechnology, and so on. Numerous technological advances in recent times are the result of the efforts of many scientists, engineers, and technicians. Some of the most notable developments in the latter half of the 20th century were (1) space missions to the Moon and other planets, (2) microcircuitry and high-speed computers, (3) sophisticated imaging techniques used in scientific research and medicine, and (4) several remarkable accomplishments in genetic engineering. The impact of such developments and discoveries on society has indeed been great, and future discoveries and developments will very likely be exciting, challenging, and of great benefit to humanity.

To investigate the impact of physics on developments in our society, we will use a *contextual* approach to the study of the content in this textbook. The book is divided into nine *Contexts,* which relate the physics to social issues, natural phenomena, or technological applications, as outlined here:

| **Chapters** | **Context** |
|---|---|
| 2–7 | Alternative-Fuel Vehicles |
| 8–11 | Mission to Mars |
| 12–14 | Earthquakes |
| 15 | Search for the *Titanic* |
| 16–18 | Global Warming |
| 19–21 | Lightning |
| 22–23 | Magnetic Levitation Vehicles |
| 24–27 | Lasers |
| 28–31 | The Cosmic Connection |

The Contexts provide a story line for each section of the text, which will help provide relevance and motivation for studying the material.

Each Context begins with a discussion of the topic, culminating in a *central question,* which forms the focus for the study of the physics in the Context. The final section of each chapter is a Context Connection, in which the material in the chapter is explored with the central question in mind. At the end of each Context, a

(© David Parker Photo Researchers, Inc.)

A technician works on the H1 detector in the Hadron Electron Accelerator Ring at the Deutsche Elektronen Synchrotron near Hamburg, Germany. Technicians educated in the physical sciences contribute their skills in many areas of modern technology. ■

Context Conclusion brings together all the principles necessary to respond as fully as possible to the central question.

In Chapter 1, we investigate some of the mathematical fundamentals and problem-solving strategies that we will use in our study of physics. The first Context, *Alternative-Fuel Vehicles,* is introduced just before Chapter 2; in this Context, the principles of classical mechanics are applied to the problem of designing, developing, producing, and marketing a vehicle that will help to reduce dependence on foreign oil and emit fewer harmful by-products into the atmosphere than current gasoline engines.

# CHAPTER 1

# Introduction and Vectors

These controls in the cockpit of a commercial aircraft assist the pilot in maintaining control over the *velocity* of the aircraft—how fast it is traveling and in what direction it is traveling—allowing it to land safely. Quantities that are defined by both a magnitude and a direction, such as velocity, are called *vectors*.

(Mark Wagner/Stone/Getty Images)

## CHAPTER OUTLINE


1.1 Standards of Length, Mass, and Time
1.2 Dimensional Analysis
1.3 Conversion of Units
1.4 Order-of-Magnitude Calculations
1.5 Significant Figures
1.6 Coordinate Systems
1.7 Vectors and Scalars
1.8 Some Properties of Vectors
1.9 Components of a Vector and Unit Vectors
1.10 Modeling, Alternative Representations, and Problem-Solving Strategy
SUMMARY


The goal of physics is to provide a quantitative understanding of certain basic phenomena that occur in our Universe. Physics is a science based on experimental observations and mathematical analyses. The main objectives behind such experiments and analyses are to develop theories that explain the phenomenon being studied and to relate those theories to other established theories. Fortunately, it is possible to explain the behavior of various physical systems using relatively few fundamental laws. Analytical procedures require the expression of those laws in the language of mathematics, the tool that provides a bridge between theory and experiment. In this chapter, we shall discuss a few mathematical concepts and techniques that will be used throughout the text. In addition, we will outline an effective problem-solving strategy that should be adopted and used in your problem-solving activities throughout the text.

**PhysicsNow™** This icon throughout the text indicates an opportunity for you to test yourself on key concepts and explore animations and interactions on the PhysicsNow Web site at **http://www.pop4e.com.**

## 1.1 STANDARDS OF LENGTH, MASS, AND TIME

If we measure a certain quantity and wish to describe it to someone, a unit for the quantity must be specified and defined. For example, it would be meaningless for a visitor from another planet to talk to us about a length of 8 "glitches" if we did not know the meaning of the unit glitch. On the other hand, if someone familiar with our system of measurement reports that a wall is 2.0 meters high and our unit of length is defined to be 1.0 meter, we then know that the height of the wall is twice our fundamental unit of length. An international committee has agreed on a system of definitions and standards to describe fundamental physical quantities. It is called the **SI system** (Système International) of units. Its units of length, mass, and time are the meter, kilogram, and second, respectively.

### Length

In A.D. 1120, King Henry I of England decreed that the standard of length in his country would be the yard and that the yard would be precisely equal to the distance from the tip of his nose to the end of his outstretched arm. Similarly, the original standard for the foot adopted by the French was the length of the royal foot of King Louis XIV. This standard prevailed until 1799, when the legal standard of length in France became the **meter,** defined as one ten-millionth of the distance from the equator to the North Pole.

Many other systems have been developed in addition to those just discussed, but the advantages of the French system have caused it to prevail in most countries and in scientific circles everywhere. Until 1960, the length of the meter was defined as the distance between two lines on a specific bar of platinum–iridium alloy stored under controlled conditions. This standard was abandoned for several reasons, a principal one being that the limited accuracy with which the separation between the lines can be determined does not meet the current requirements of science and technology. The definition of the meter was modified to be equal to 1 650 763.73 wavelengths of orange–red light emitted from a krypton-86 lamp. In October 1983, the meter was redefined to be **the distance traveled by light in a vacuum during a time interval of 1/299 792 458 second.** This value arises from the establishment of the speed of light in a vacuum as exactly 299 792 458 meters per second. We will use the standard scientific notation for numbers with more than three digits in which groups of three digits are separated by spaces rather than commas. Therefore, 1 650 763.73 and 299 792 458 in this paragraph are the same as the more popular American cultural notations of 1,650,763.73 and 299,792,458. Similarly, $\pi = 3.14159265$ is written as 3.141 592 65.

■ Definition of the meter

### Mass

Mass represents a measure of the resistance of an object to changes in its motion. The SI unit of mass, the **kilogram,** is defined as **the mass of a specific platinum–iridium alloy cylinder kept at the International Bureau of Weights and Measures at Sèvres, France.** At this point, we should add a word of caution. Many beginning students of physics tend to confuse the physical quantities called *weight* and *mass*. For the present we shall not discuss the distinction between them; they will be clearly defined in later chapters. For now you should note that they are distinctly different quantities.

■ Definition of the kilogram

### Time

Before 1960, the standard of time was defined in terms of the average length of a solar day in the year 1900. (A solar day is the time interval between successive appearances of the Sun at the highest point it reaches in the sky each day.) The basic

(Courtesy of National Institute of Standards and Technology, U.S. Department of Commerce)

**FIGURE 1.1** The nation's primary time standard is a cesium fountain atomic clock developed at the National Institute of Standards and Technology laboratories in Boulder, Colorado. The clock will neither gain nor lose a second in 20 million years.

unit of time, the **second,** was defined to be $(1/60)(1/60)(1/24) = 1/86\,400$ of the average solar day. In 1967, the second was redefined to take advantage of the great precision obtainable with a device known as an atomic clock (Fig. 1.1), which uses the characteristic frequency of the cesium-133 atom as the "reference clock." The second is now defined as **9 192 631 770 times the period of oscillation of radiation from the cesium atom.** It is possible today to purchase clocks and watches that receive radio signals from an atomic clock in Colorado, which the clock or watch uses to continuously reset itself to the correct time.

■ Definition of the second

## Approximate Values for Length, Mass, and Time

Approximate values of various lengths, masses, and time intervals are presented in Tables 1.1, 1.2, and 1.3, respectively. Note the wide range of values for these quantities.[1] You should study the tables and begin to generate an intuition for what is meant by a mass of 100 kilograms, for example, or by a time interval of $3.2 \times 10^7$ seconds.

**PITFALL PREVENTION 1.1**

**Reasonable Values** The generation of intuition about typical values of quantities suggested here is critical. An important step in solving problems is to think about your result at the end of a problem and determine if it seems reasonable. If you are calculating the mass of a housefly and arrive at a value of 100 kg, this value is *unreasonable;* there is an error somewhere. If you are calculating the length of a spacecraft on a launch pad and end up with a value of 10 cm, this value is *unreasonable;* and you should look for a mistake.

Systems of units commonly used in science, commerce, manufacturing, and everyday life are (1) the *SI system,* in which the units of length, mass, and time are the meter (m), kilogram (kg), and second (s), respectively; and (2) the *U.S. customary system,* in which the units of length, mass, and time are the foot (ft), slug, and second, respectively. Throughout most of this text we shall use SI units because they are almost universally accepted in science and industry. We will make limited use of U.S. customary units in the study of classical mechanics.

Some of the most frequently used prefixes for the powers of ten and their abbreviations are listed in Table 1.4. For example, $10^{-3}$ m is equivalent to 1 millimeter (mm), and $10^3$ m is 1 kilometer (km). Likewise, 1 kg is $10^3$ grams (g), and 1 megavolt (MV) is $10^6$ volts (V).

The variables length, time, and mass are examples of *fundamental quantities.* A much larger list of variables contains *derived quantities,* or quantities that can be expressed as a mathematical combination of fundamental quantities. Common examples are *area,* which is a product of two lengths, and *speed,* which is a ratio of a length to a time interval.

[1] If you are unfamiliar with the use of powers of ten (scientific notation), you should review Appendix B.1.

**TABLE 1.1 Approximate Values of Some Measured Lengths**

| | Length (m) |
|---|---|
| Distance from the Earth to the most remote quasar known | $1.4 \times 10^{26}$ |
| Distance from the Earth to the most remote normal galaxies known | $4 \times 10^{25}$ |
| Distance from the Earth to the nearest large galaxy (M 31, the Andromeda galaxy) | $2 \times 10^{22}$ |
| Distance from the Sun to the nearest star (Proxima Centauri) | $4 \times 10^{16}$ |
| One lightyear | $9.46 \times 10^{15}$ |
| Mean orbit radius of the Earth | $1.5 \times 10^{11}$ |
| Mean distance from the Earth to the Moon | $3.8 \times 10^{8}$ |
| Distance from the equator to the North Pole | $1 \times 10^{7}$ |
| Mean radius of the Earth | $6.4 \times 10^{6}$ |
| Typical altitude of an orbiting Earth satellite | $2 \times 10^{5}$ |
| Length of a football field | $9.1 \times 10^{1}$ |
| Length of this textbook | $2.8 \times 10^{-1}$ |
| Length of a housefly | $5 \times 10^{-3}$ |
| Size of smallest visible dust particles | $1 \times 10^{-4}$ |
| Size of cells of most living organisms | $1 \times 10^{-5}$ |
| Diameter of a hydrogen atom | $1 \times 10^{-10}$ |
| Diameter of a uranium nucleus | $1.4 \times 10^{-14}$ |
| Diameter of a proton | $1 \times 10^{-15}$ |

**TABLE 1.2 Masses of Various Objects (Approximate Values)**

| | Mass (kg) |
|---|---|
| Visible Universe | $10^{52}$ |
| Milky Way galaxy | $10^{42}$ |
| Sun | $2 \times 10^{30}$ |
| Earth | $6 \times 10^{24}$ |
| Moon | $7 \times 10^{22}$ |
| Shark | $3 \times 10^{2}$ |
| Human | $7 \times 10^{1}$ |
| Frog | $1 \times 10^{-1}$ |
| Mosquito | $1 \times 10^{-5}$ |
| Bacterium | $1 \times 10^{-15}$ |
| Hydrogen atom | $1.67 \times 10^{-27}$ |
| Electron | $9.11 \times 10^{-31}$ |

Another example of a derived quantity is **density.** The density $\rho$ (Greek letter rho; a table of the letters in the Greek alphabet is provided at the back of the book) of any substance is defined as its *mass per unit volume:*

$$\rho \equiv \frac{m}{V} \qquad [1.1]$$

■ Definition of density

which is a ratio of mass to a product of three lengths. For example, aluminum has a density of $2.70 \times 10^3\ \text{kg/m}^3$, and lead has a density of $11.3 \times 10^3\ \text{kg/m}^3$. An extreme difference in density can be imagined by thinking about holding a 10-centimeter (cm) cube of Styrofoam in one hand and a 10-cm cube of lead in the other.

**TABLE 1.3 Approximate Values of Some Time Intervals**

| | Time Interval (s) |
|---|---|
| Age of the Universe | $5 \times 10^{17}$ |
| Age of the Earth | $1.3 \times 10^{17}$ |
| Time interval since the fall of the Roman empire | $5 \times 10^{12}$ |
| Average age of a college student | $6.3 \times 10^{8}$ |
| One year | $3.2 \times 10^{7}$ |
| One day (time interval for one revolution of the Earth about its axis) | $8.6 \times 10^{4}$ |
| One class period | $3.0 \times 10^{3}$ |
| Time interval between normal heartbeats | $8 \times 10^{-1}$ |
| Period of audible sound waves | $1 \times 10^{-3}$ |
| Period of typical radio waves | $1 \times 10^{-6}$ |
| Period of vibration of an atom in a solid | $1 \times 10^{-13}$ |
| Period of visible light waves | $2 \times 10^{-15}$ |
| Duration of a nuclear collision | $1 \times 10^{-22}$ |
| Time interval for light to cross a proton | $3.3 \times 10^{-24}$ |

**TABLE 1.4 Some Prefixes for Powers of Ten**

| Power | Prefix | Abbreviation |
|---|---|---|
| $10^{-24}$ | yocto | y |
| $10^{-21}$ | zepto | z |
| $10^{-18}$ | atto | a |
| $10^{-15}$ | femto | f |
| $10^{-12}$ | pico | p |
| $10^{-9}$ | nano | n |
| $10^{-6}$ | micro | $\mu$ |
| $10^{-3}$ | milli | m |
| $10^{-2}$ | centi | c |
| $10^{-1}$ | deci | d |
| $10^{3}$ | kilo | k |
| $10^{6}$ | mega | M |
| $10^{9}$ | giga | G |
| $10^{12}$ | tera | T |
| $10^{15}$ | peta | P |
| $10^{18}$ | exa | E |
| $10^{21}$ | zetta | Z |
| $10^{24}$ | yotta | Y |

## 1.2 DIMENSIONAL ANALYSIS

The word *dimension* has a special meaning in physics. It denotes the physical nature of a quantity. Whether a distance is measured in units of feet or meters or miles, it is a distance. We say its dimension is *length*.

The symbols used in this book to specify the dimensions[2] of length, mass, and time are L, M, and T, respectively. We shall often use square brackets [ ] to denote the dimensions of a physical quantity. For example, in this notation the dimensions of velocity $v$ are written $[v] = \mathrm{L/T}$, and the dimensions of area $A$ are $[A] = \mathrm{L}^2$. The dimensions of area, volume, velocity, and acceleration are listed in Table 1.5, along with their units in the two common systems. The dimensions of other quantities, such as force and energy, will be described as they are introduced in the text.

In many situations, you may be faced with having to derive or check a specific equation. Although you may have forgotten the details of the derivation, a useful and powerful procedure called *dimensional analysis* can be used as a consistency check, to assist in the derivation, or to check your final expression. Dimensional analysis makes use of the fact that **dimensions can be treated as algebraic quantities.** For example, quantities can be added or subtracted only if they have the same dimensions. Furthermore, the terms on both sides of an equation must have the same dimensions. By following these simple rules, you can use dimensional analysis to help determine whether an expression has the correct form because the relationship can be correct only if the dimensions on the two sides of the equation are the same.

To illustrate this procedure, suppose you wish to derive an expression for the position $x$ of a car at a time $t$ if the car starts from rest at $t = 0$ and moves with constant acceleration $a$. In Chapter 2, we shall find that the correct expression for this special case is $x = \frac{1}{2}at^2$. Let us check the validity of this expression from a dimensional analysis approach.

The quantity $x$ on the left side has the dimension of length. For the equation to be dimensionally correct, the quantity on the right side must also have the dimension of length. We can perform a dimensional check by substituting the basic dimensions for acceleration, $\mathrm{L/T^2}$ (Table 1.5), and time, T, into the equation $x = \frac{1}{2}at^2$. That is, the dimensional form of the equation $x = \frac{1}{2}at^2$ can be written as

$$[x] = \frac{\mathrm{L}}{\cancel{\mathrm{T}}^2}\cdot\cancel{\mathrm{T}}^2 = \mathrm{L}$$

The dimensions of time cancel as shown, leaving the dimension of length, which is the correct dimension for the position $x$. Notice that the number $\frac{1}{2}$ in the equation has no units, so it does not enter into the dimensional analysis.

**QUICK QUIZ 1.1** True or false: Dimensional analysis can give you the numerical value of constants of proportionality that may appear in an algebraic expression.

**TABLE 1.5 Units of Area, Volume, Velocity, and Acceleration**

| System | Area ($\mathrm{L}^2$) | Volume ($\mathrm{L}^3$) | Velocity (L/T) | Acceleration ($\mathrm{L/T^2}$) |
|---|---|---|---|---|
| SI | $\mathrm{m}^2$ | $\mathrm{m}^3$ | m/s | $\mathrm{m/s^2}$ |
| U.S. customary | $\mathrm{ft}^2$ | $\mathrm{ft}^3$ | ft/s | $\mathrm{ft/s^2}$ |

[2] The *dimensions* of a variable will be symbolized by a capitalized, nonitalic letter, such as, in the case of length, L. The *symbol* for the variable itself will be italicized, such as $L$ for the length of an object or $t$ for time.

**EXAMPLE 1.1** **Analysis of an Equation**

Show that the expression $v_f = v_i + at$ is dimensionally correct, where $v_f$ and $v_i$ represent velocities at two instants of time, $a$ is acceleration, and $t$ is an instant of time.

**Solution** The dimensions of the velocities are

$$[v_f] = [v_i] = \frac{\text{L}}{\text{T}}$$

and the dimensions of acceleration are $\text{L}/\text{T}^2$. Therefore, the dimensions of $at$ are

$$[at] = \frac{\text{L}}{\text{T}^2} \cdot \text{T} = \frac{\text{L}}{\text{T}}$$

and the expression is dimensionally correct. On the other hand, if the expression were given as $v_f = v_i + at^2$, it would be dimensionally *incorrect*. Try it and see!

## 1.3 CONVERSION OF UNITS

Sometimes it is necessary to convert units from one system to another or to convert within a system, for example, from kilometers to meters. Equalities between SI and U.S. customary units of length are as follows:

1 mile (mi) = 1 609 m = 1.609 km    1 ft = 0.304 8 m = 30.48 cm

1 m = 39.37 in. = 3.281 ft    1 inch (in.) = 0.025 4 m = 2.54 cm

A more complete list of equalities can be found in Appendix A.

Units can be treated as algebraic quantities that can cancel each other. To perform a conversion, a quantity can be multiplied by a **conversion factor,** which is a fraction equal to 1, with numerator and denominator having different units, to provide the desired units in the final result. For example, suppose we wish to convert 15.0 in. to centimeters. Because 1 in. = 2.54 cm, we multiply by a conversion factor that is the appropriate ratio of these equal quantities and find that

$$15.0\text{ in.} = (15.0\ \cancel{\text{in.}})\left(\frac{2.54\text{ cm}}{1\ \cancel{\text{in.}}}\right) = 38.1\text{ cm}$$

where the ratio in parentheses is equal to 1. Notice that we put the unit of an inch in the denominator and that it cancels with the unit in the original quantity. The remaining unit is the centimeter, which is our desired result.

**PITFALL PREVENTION 1.2**

**Always include units** When performing calculations, make it a habit to include the units with every quantity and carry the units through the entire calculation. Avoid the temptation to drop the units during the calculation steps and then apply the expected unit to the number that results for an answer. By including the units in every step, you can detect errors if the units for the answer are incorrect.

**QUICK QUIZ 1.2** The distance between two cities is 100 mi. The number of kilometers in the distance between the two cities is **(a)** smaller than 100, **(b)** larger than 100, **(c)** equal to 100.

**EXAMPLE 1.2** **Is He Speeding?**

On an interstate highway in a rural region of Wyoming, a car is traveling at a speed of 38.0 m/s.

**A** Is this car exceeding the speed limit of 75.0 mi/h?

**Solution** We first convert meters to miles:

$$(38.0\ \cancel{\text{m}}/\text{s})\left(\frac{1\text{ mi}}{1\ 609\ \cancel{\text{m}}}\right) = 2.36 \times 10^{-2}\text{ mi/s}$$

Now we convert seconds to hours:

$$(2.36 \times 10^{-2}\text{ mi}/\cancel{\text{s}})\left(\frac{60\ \cancel{\text{s}}}{1\ \cancel{\text{min}}}\right)\left(\frac{60\ \cancel{\text{min}}}{1\text{ h}}\right) = 85.0\text{ mi/h}$$

Therefore, the car is exceeding the speed limit and should slow down.

**B** What is the speed of the car in kilometers per hour?

**Solution** We convert our answer in part A to the appropriate units:

$$(85.0\ \cancel{\text{mi}}/\text{h})\left(\frac{1.609\ \text{km}}{1\ \cancel{\text{mi}}}\right) = 137\ \text{km/h}$$

Figure 1.2 shows the speedometer of an automobile, with speeds in both miles per hour and kilometers per hour. Can you check the conversion we just performed using this photograph?

(Phil Boorman/Getty Images)

**FIGURE 1.2** (Example 1.2) The speedometer of this vehicle shows speeds in both miles per hour and kilometers per hour.

## 1.4 ORDER-OF-MAGNITUDE CALCULATIONS

It is often useful to compute an approximate answer to a given physical problem even when little information is available. This answer can then be used to determine whether a more precise calculation is necessary. Such an approximation is usually based on certain assumptions, which must be modified if greater precision is needed. Therefore, we will sometimes refer to an *order of magnitude* of a certain quantity as the power of ten of the number that describes that quantity. Usually, when an order-of-magnitude calculation is made, the results are reliable to within about a factor of 10. If a quantity increases in value by three orders of magnitude, its value increases by a factor of $10^3 = 1\,000$. We use the symbol $\sim$ for "is on the order of." Therefore,

$$0.008\,6 \sim 10^{-2} \qquad 0.0021 \sim 10^{-3} \qquad 700 \sim 10^3$$

### EXAMPLE 1.3 The Number of Atoms in a Solid

Estimate the number of atoms in 1 $\text{cm}^3$ of a solid.

**Solution** From Table 1.1 we note that the diameter $d$ of an atom is about $10^{-10}$ m. Let us assume that the atoms in the solid are spheres of this diameter. Then the volume of each sphere is about $10^{-30}\ \text{m}^3$ (more precisely, volume $= 4\pi r^3/3 = \pi d^3/6$, where $r = d/2$). Therefore, because $1\ \text{cm}^3 = 10^{-6}\ \text{m}^3$, the number of atoms in the solid is on the order of $10^{-6}/10^{-30} = 10^{24}$ atoms.

A more precise calculation would require additional knowledge that we could find in tables. Our estimate, however, agrees with the more precise calculation to within a factor of 10.

### EXAMPLE 1.4 How Much Gas Do We Use?

Estimate the number of gallons of gasoline used by all cars in the United States each year.

**Solution** Because there are about 280 million people in the United States, an estimate of the number of cars in the country is $7 \times 10^7$ (assuming one car and four people per family). We can also estimate that the average distance traveled per year is $1 \times 10^4$ miles. If we assume gasoline consumption of 0.05 gal/mi (equivalent to 20 miles per gallon), each car uses about $5 \times 10^2$ gal/year. Multiplying this number by the total number of cars in the United States gives an estimated total consumption of about $10^{11}$ gal, which corresponds to a yearly consumer expenditure on the order of $10^2$ billion dollars. This estimate is probably low because we haven't accounted for commercial consumption.

## 1.5 SIGNIFICANT FIGURES

When certain quantities are measured, the measured values are known only to within the limits of the experimental uncertainty. The value of the uncertainty can depend on various factors, such as the quality of the apparatus, the skill of the experimenter, and the number of measurements performed. The number of **significant figures** in a measurement can be used to express something about the uncertainty.

As an example of significant figures, consider the population of New York State, as reported in a published road atlas: 18 976 457. Notice that this number reports the population *to the level of one individual.* We would describe this number as having eight significant figures. Can the population really be this accurate? First of all, is the census process accurate enough to measure the population to one individual? By the time this number was actually published, had the number of births and immigrations into the state balanced the number of deaths and emigrations out of the state, so that the change in the population is exactly zero?

The claim that the population is measured and known to the level of one individual is unjustified. We would describe it by saying that *there are too many significant figures in the measurement.* To account for the inherent uncertainty in the census-taking process and the inevitable changes in population by the time the number is read in the road atlas, it might be better to report the population as something like 19.0 million. This number has three significant figures rather than the eight significant figures in the published population.

Let us look at a more scientific example. Suppose we are asked in a laboratory experiment to measure the area of a rectangular plate using a meter stick as a measuring instrument. Let us assume that the accuracy to which we can measure a particular dimension of the plate is $\pm 0.1$ cm. If the length of the plate is measured to be 16.3 cm, we can claim only that its length lies somewhere between 16.2 cm and 16.4 cm. In this case, we say that the measured value has three significant figures. Likewise, if its width is measured to be 4.5 cm, the actual value lies between 4.4 cm and 4.6 cm. This measured value has only two significant figures. Note that the significant figures include the first estimated digit. Therefore, we could write the measured values as $16.3 \pm 0.1$ cm and $4.5 \pm 0.1$ cm.

Suppose we would now like to find the area of the plate by multiplying the two measured values. If we were to claim that the area is $(16.3 \text{ cm})(4.5 \text{ cm}) = 73.35 \text{ cm}^2$, our answer would be unjustifiable because it contains four significant figures, which is greater than the number of significant figures in either of the measured lengths. The following is a good rule of thumb to use in determining the number of significant figures that can be claimed:

> When multiplying several quantities, the number of significant figures in the final answer is the same as the number of significant figures in the quantity having the lowest number of significant figures. The same rule applies to division.

Applying this rule to the previous multiplication example, we see that the answer for the area can have only two significant figures because the length of 4.5 cm has only two significant figures. Therefore, all we can claim is that the area is $73 \text{ cm}^2$, realizing that the value can range between $(16.2 \text{ cm})(4.4 \text{ cm}) = 71 \text{ cm}^2$ and $(16.4 \text{ cm})(4.6 \text{ cm}) = 75 \text{ cm}^2$.

Zeros may or may not be significant figures. Those used to position the decimal point in such numbers as 0.03 and 0.007 5 are not significant. Therefore, there are one and two significant figures, respectively, in these two values. When the positioning of zeros comes after other digits, however, there is the possibility of misinterpretation. For example, suppose the mass of an object is given as 1 500 g. This value is ambiguous because we do not know whether the two zeros are being used to locate the decimal point or whether they represent significant figures in the measurement. To remove this ambiguity, it is common to use scientific notation to indicate the number of significant figures. In this case, we would express the mass as $1.5 \times 10^3$ g if the measured value has two significant figures, $1.50 \times 10^3$ g if it has three significant figures, and $1.500 \times 10^3$ g if it has four significant figures. Likewise, 0.000 150 should be expressed in scientific notation as $1.5 \times 10^{-4}$ if it has two significant figures or as $1.50 \times 10^{-4}$ if it has three significant figures. The three zeros between the decimal point and the digit 1 in the number 0.000 150 are not counted as significant figures because they are present only to locate the decimal point. In general, a **significant figure** in a measurement is a reliably known digit (other than a zero used to locate the decimal point) or the first estimated digit.

For addition and subtraction, the number of decimal places must be considered when you are determining how many significant figures to report.

> When numbers are added or subtracted, the number of decimal places in the result should equal the smallest number of decimal places of any term in the sum.

For example, if we wish to compute 123 + 5.35, the answer is 128 and not 128.35. If we compute the sum 1.000 1 + 0.000 3 = 1.000 4, the result has the correct number of decimal places; consequently, it has five significant figures even though one of the terms in the sum, 0.000 3, has only one significant figure. Likewise, if we perform the subtraction 1.002 − 0.998 = 0.004, the result has only one significant figure even though one term has four significant figures and the other has three. In this book, **most of the numerical examples and end-of-chapter problems will yield answers having three significant figures.**

If the number of significant figures in the result of an addition or subtraction must be reduced, a general rule for rounding numbers states that the last digit retained is to be increased by 1 if the last digit dropped is greater than 5. If the last digit dropped is less than 5, the last digit retained remains as it is. If the last digit dropped is equal to 5, the last digit retained should be rounded to the nearest even number. (This rule helps avoid accumulation of errors in long arithmetic processes.)

### EXAMPLE 1.5 The Area of a Dish

A biologist is filling a rectangular dish with growth culture and wishes to know the area of the dish. The length of the dish is measured to be 12.71 cm (four significant figures), and the width is measured to be 7.46 cm (three significant figures). Find the area of the dish.

**Solution** If you multiply 12.71 cm by 7.46 cm on your calculator, you will obtain an answer of 94.816 6 $cm^2$. How many of these numbers should you claim? Our rule of thumb for multiplication tells us that you can claim only the number of significant figures in the quantity with the smallest number of significant figures. In this example, that number is three (in the width 7.46 cm), so we should express our final answer as 94.8 $cm^2$.

## 1.6 COORDINATE SYSTEMS

Many aspects of physics deal in some way or another with locations in space. For example, the mathematical description of the motion of an object requires a method for specifying the object's position. Therefore, we first discuss how to describe the position of a point in space by means of coordinates in a graphical representation. A point on a line can be located with one coordinate, a point in a plane is located with two coordinates, and three coordinates are required to locate a point in space.

A coordinate system used to specify locations in space consists of

- A fixed reference point $O$, called the origin
- A set of specified axes or directions with an appropriate scale and labels on the axes
- Instructions that tell us how to label a point in space relative to the origin and axes

One convenient coordinate system that we will use frequently is the *Cartesian coordinate system,* sometimes called the *rectangular coordinate system.* Such a system in two dimensions is illustrated in Figure 1.3. An arbitrary point in this system is labeled with the coordinates $(x, y)$. Positive $x$ is taken to the right of the origin, and positive $y$ is upward from the origin. Negative $x$ is to the left of the origin, and negative $y$ is downward from the origin. For example, the point $P$, which has coordinates (5, 3), may be reached by going first 5 m to the right of the origin and then 3 m above the origin (*or* by going 3 m above the origin and then 5 m to the right). Similarly, the point $Q$ has coordinates $(-3, 4)$, which correspond to going 3 m to the left of the origin and 4 m above the origin.

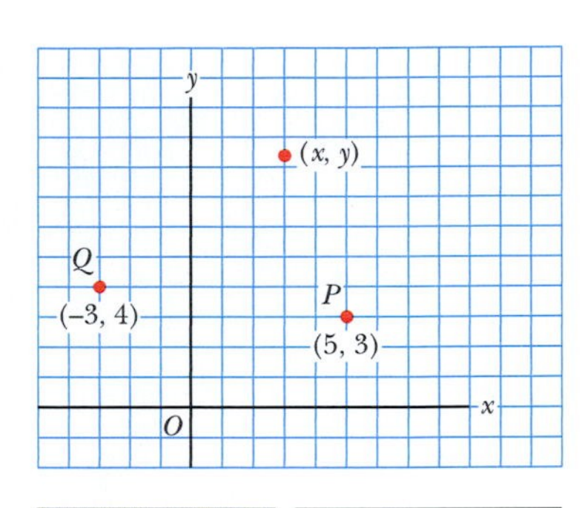


**FIGURE 1.3** Designation of points in a Cartesian coordinate system. Each square in the $xy$ plane is 1 m on a side. Every point is labeled with coordinates $(x,y)$.

Sometimes it is more convenient to represent a point in a plane by its *plane polar coordinates* $(r, \theta)$, as in Active Figure 1.4a. In this coordinate system, $r$ is the length of the line from the origin to the point, and $\theta$ is the angle between that line and a fixed axis, usually the positive $x$ axis, with $\theta$ measured counterclockwise. From the right triangle in Active Figure 1.4b, we find that $\sin\theta = y/r$ and $\cos\theta = x/r$. (A review of trigonometric functions is given in Appendix B.4.) Therefore, starting with plane polar coordinates, one can obtain the Cartesian coordinates through the equations

$$x = r\cos\theta \quad [1.2]$$

$$y = r\sin\theta \quad [1.3]$$

Furthermore, it follows that

$$\tan\theta = \frac{y}{x} \quad [1.4]$$

and

$$r = \sqrt{x^2 + y^2} \quad [1.5]$$

You should note that these expressions relating the coordinates $(x, y)$ to the coordinates $(r, \theta)$ apply only when $\theta$ is defined as in Active Figure 1.4a, where positive $\theta$ is an angle measured *counterclockwise* from the positive $x$ axis. Other choices are

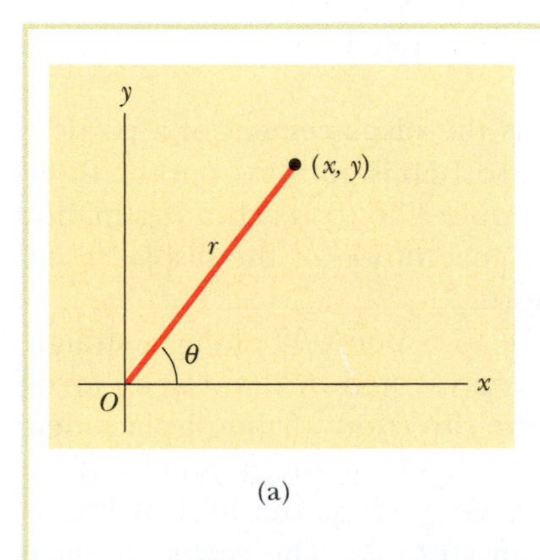


(a)

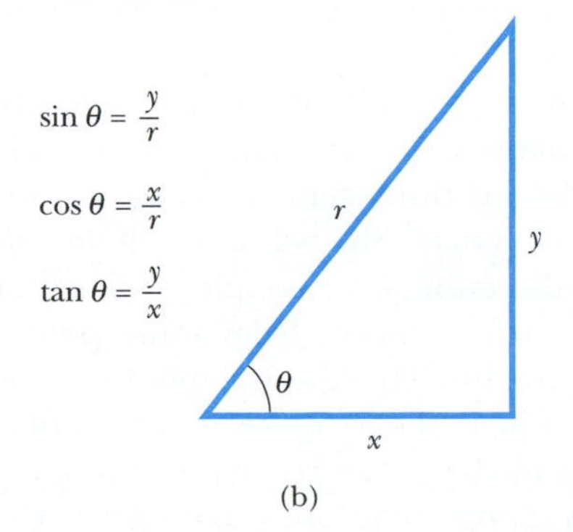


(b)

**ACTIVE FIGURE 1.4** (a) The plane polar coordinates of a point are represented by the distance $r$ and the angle $\theta$, where $\theta$ is measured in a counterclockwise direction from the positive $x$ axis. (b) The right triangle used to relate $(x, y)$ to $(r, \theta)$.

**PhysicsNow™** Log into PhysicsNow at **www.pop4e.com** and go to Active Figure 1.4 to move the point and see the changes to the rectangular and polar coordinates as well as to the sine, cosine, and tangent of angle $\theta$.

(a)

(b)

**FIGURE 1.5** (a) The number of grapes in this bunch is one example of a scalar quantity. Can you think of other examples? (b) This helpful person pointing in the correct direction tells us to travel five blocks north to reach the courthouse. A vector is a physical quantity that is specified by both magnitude and direction.

made in navigation and astronomy. If the reference axis for the polar angle $\theta$ is chosen to be other than the positive $x$ axis or if the sense of increasing $\theta$ is chosen differently, the corresponding expressions relating the two sets of coordinates will change.

## 1.7 VECTORS AND SCALARS

Each of the physical quantities that we shall encounter in this text can be placed in one of two categories, either a scalar or a vector. A **scalar** is a quantity that is completely specified by a positive or negative number with appropriate units. On the other hand, a **vector** is a physical quantity that must be specified by both magnitude and direction.

The number of grapes in a bunch (Fig. 1.5a) is an example of a scalar quantity. If you are told that there are 38 grapes in the bunch, this statement completely specifies the information; no specification of direction is required. Other examples of scalars are temperature, volume, mass, and time intervals. The rules of ordinary arithmetic are used to manipulate scalar quantities; they can be freely added and subtracted (assuming that they have the same units!), multiplied and divided.

Force is an example of a vector quantity. To describe the force on an object completely, we must specify both the direction of the applied force and the magnitude of the force.

Another simple example of a vector quantity is the **displacement** of a particle, defined as its *change in position*. The person in Figure 1.5b is pointing out the direction of your desired displacement vector if you would like to reach a destination such as the courthouse. She will also tell you the magnitude of the displacement along with the direction, for example, "5 blocks north."

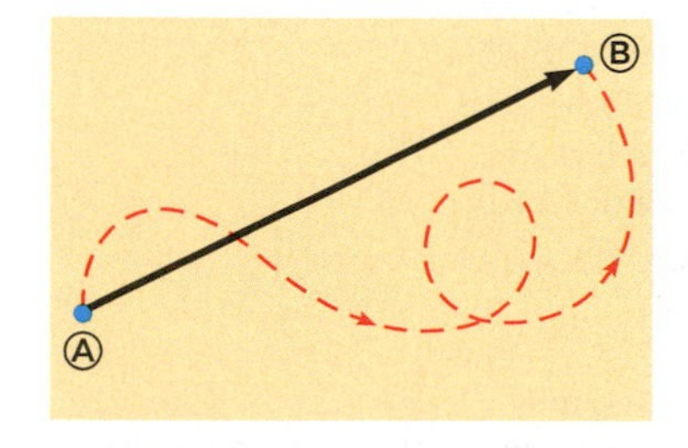


**FIGURE 1.6** After a particle moves from Ⓐ to Ⓑ along an arbitrary path represented by the broken line, its displacement is a vector quantity shown by the arrow drawn from Ⓐ to Ⓑ.

Suppose a particle moves from some point Ⓐ to a point Ⓑ along a straight path, as in Figure 1.6. This displacement can be represented by drawing an arrow from Ⓐ to Ⓑ, where the arrowhead represents the direction of the displacement and the length of the arrow represents the magnitude of the displacement. If the particle travels along some other path from Ⓐ to Ⓑ, such as the broken line in Figure 1.6, its displacement is still the vector from Ⓐ to Ⓑ. The vector displacement along any indirect path from Ⓐ to Ⓑ is defined as being equivalent to the displacement represented by the direct path from Ⓐ to Ⓑ. The magnitude of the displacement is the shortest distance between the end points. Therefore, **the**

**displacement of a particle is completely known if its initial and final coordinates are known.** The path need not be specified. In other words, the **displacement is independent of the path** if the end points of the path are fixed.

■ Distance

Note that the **distance** traveled by a particle is distinctly different from its displacement. The distance traveled (a scalar quantity) is the length of the path, which in general can be much greater than the magnitude of the displacement. In Figure 1.6, the length of the curved red path is much larger than the magnitude of the black displacement vector.

If the particle moves along the $x$ axis from position $x_i$ to position $x_f$, as in Figure 1.7, its displacement is given by $x_f - x_i$. (The indices $i$ and $f$ refer to the initial and final values.) We use the Greek letter delta ($\Delta$) to denote the *change* in a quantity. Therefore, we define the change in the position of the particle (the displacement) as

$$\Delta x \equiv x_f - x_i \qquad [1.6]$$

From this definition we see that $\Delta x$ is positive if $x_f$ is greater than $x_i$ and negative if $x_f$ is less than $x_i$. For example, if a particle changes its position from $x_i = -5$ m to $x_f = 3$ m, its displacement is $\Delta x = +8$ m.

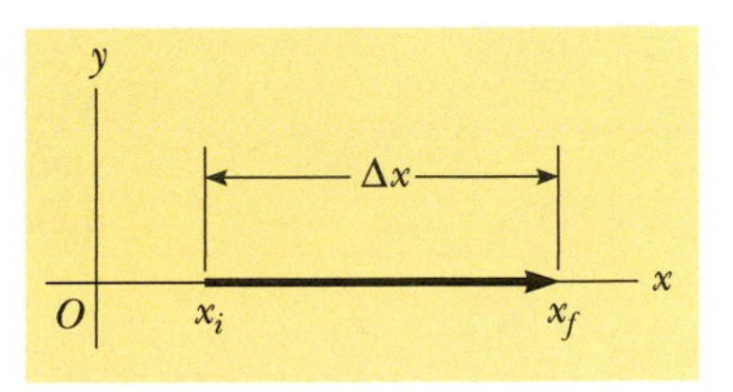


FIGURE 1.7 A particle moving along the $x$ axis from $x_i$ to $x_f$ undergoes a displacement $\Delta x = x_f - x_i$.

Many physical quantities in addition to displacement are vectors. They include velocity, acceleration, force, and momentum, all of which will be defined in later chapters. In this text, we will use boldface letters with an arrow over the letter, such as $\vec{\mathbf{A}}$, to represent vectors. Another common notation for vectors with which you should be familiar is a simple boldface character: **A**.

The magnitude of the vector $\vec{\mathbf{A}}$ is written with an italic letter $A$ or, alternatively, $|\vec{\mathbf{A}}|$. The magnitude of a vector is always positive and carries the units of the quantity that the vector represents, such as meters for displacement or meters per second for velocity. Vectors combine according to special rules, which will be discussed in Sections 1.8 and 1.9.

QUICK QUIZ 1.3 Which of the following are scalar quantities and which are vector quantities? (**a**) your age (**b**) acceleration (**c**) velocity (**d**) speed (**e**) mass

## ■ Thinking Physics 1.1

Consider your commute to work or school in the morning. Which is larger, the distance you travel or the magnitude of the displacement vector?

**Reasoning** Unless you have a very unusual commute, the distance traveled *must* be larger than the magnitude of the displacement vector. The distance includes all the twists and turns you make in following the roads from home to work or school. On the other hand, the magnitude of the displacement vector is the length of a straight line from your home to work or school. This length is often described informally as "the distance as the crow flies." The only way that the distance could be the same as the magnitude of the displacement vector is if your commute is a perfect straight line, which is highly unlikely! The distance could *never* be less than the magnitude of the displacement vector because the shortest distance between two points is a straight line. ■

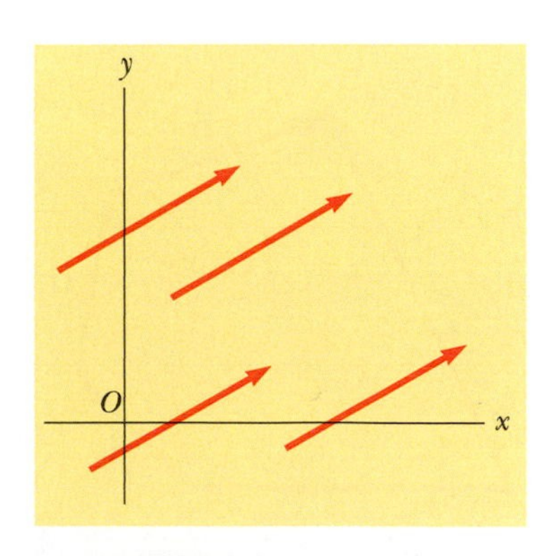


FIGURE 1.8 These four representations of vectors are equal because all four vectors have the same magnitude and point in the same direction.

# 1.8 SOME PROPERTIES OF VECTORS

## Equality of Two Vectors

Two vectors $\vec{\mathbf{A}}$ and $\vec{\mathbf{B}}$ are defined to be equal if they have the same units, the same magnitude, and the same direction. That is, $\vec{\mathbf{A}} = \vec{\mathbf{B}}$ only if $A = B$ *and* $\vec{\mathbf{A}}$ and $\vec{\mathbf{B}}$ point in the same direction. For example, all the vectors in Figure 1.8 are equal even

ACTIVE FIGURE 1.9

(a) When vector $\vec{\mathbf{B}}$ is added to vector $\vec{\mathbf{A}}$, the resultant $\vec{\mathbf{R}}$ is the vector that runs from the tail of $\vec{\mathbf{A}}$ to the tip of $\vec{\mathbf{B}}$. (b) This construction shows that $\vec{\mathbf{A}} + \vec{\mathbf{B}} = \vec{\mathbf{B}} + \vec{\mathbf{A}}$; vector addition is commutative.

**Physics Now™** Log into PhysicsNow at **www.pop4e.com** and go to Active Figure 1.9 to explore the addition of two vectors.

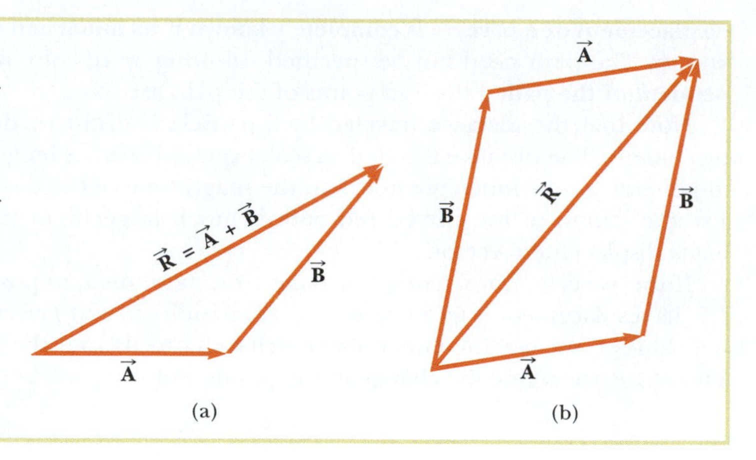


though they have different starting points. This property allows us to translate a vector parallel to itself in a diagram without affecting the vector.

## Addition

When two or more vectors are added together, they must *all* have the same units. For example, it would be meaningless to add a velocity vector to a displacement vector because they are different physical quantities. Scalars obey the same rule. For example, it would be meaningless to add time intervals and temperatures.

The rules for vector sums are conveniently described using geometry. To add vector $\vec{\mathbf{B}}$ to vector $\vec{\mathbf{A}}$, first draw a diagram of vector $\vec{\mathbf{A}}$ on graph paper, with its magnitude represented by a convenient scale, and then draw vector $\vec{\mathbf{B}}$ to the same scale with its tail starting from the tip of $\vec{\mathbf{A}}$, as in Active Figure 1.9a. The *resultant vector* $\vec{\mathbf{R}} = \vec{\mathbf{A}} + \vec{\mathbf{B}}$ is the vector drawn from the tail of $\vec{\mathbf{A}}$ to the tip of $\vec{\mathbf{B}}$. If these vectors are displacements, $\vec{\mathbf{R}}$ is the single displacement that has the same effect as the displacements $\vec{\mathbf{A}}$ and $\vec{\mathbf{B}}$ performed one after the other. This process is known as the *triangle method of addition* because the three vectors can be geometrically modeled as the sides of a triangle.

When vectors are added, the sum is independent of the order of the addition. This independence can be seen for two vectors from the geometric construction in Active Figure 1.9b and is known as the **commutative law of addition:**

$$\vec{\mathbf{A}} + \vec{\mathbf{B}} = \vec{\mathbf{B}} + \vec{\mathbf{A}} \qquad [1.7]$$

If three or more vectors are added, their sum is independent of the way in which they are grouped. A geometric demonstration of this property for three vectors is given in Figure 1.10. It is called the **associative law of addition:**

$$\vec{\mathbf{A}} + (\vec{\mathbf{B}} + \vec{\mathbf{C}}) = (\vec{\mathbf{A}} + \vec{\mathbf{B}}) + \vec{\mathbf{C}} \qquad [1.8]$$

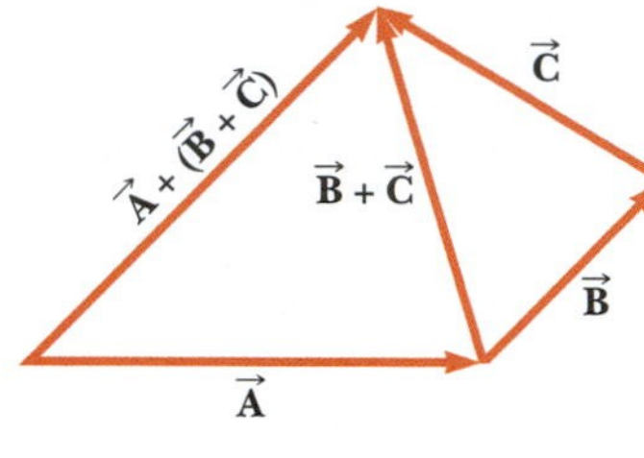


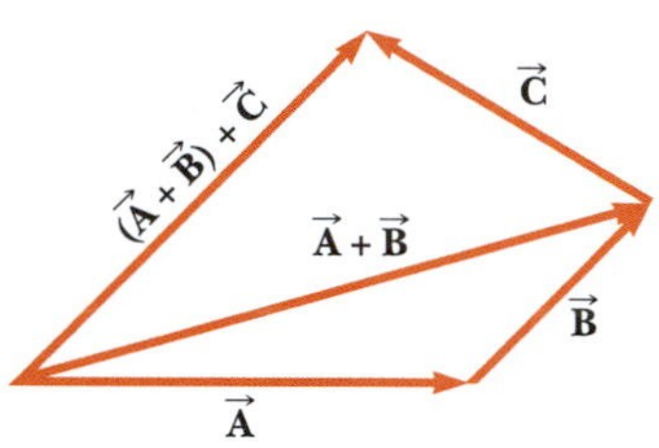


FIGURE 1.10 Geometric constructions for verifying the associative law of addition.

Geometric constructions can also be used to add more than three vectors, as shown in Figure 1.11 for the case of four vectors. The resultant vector $\vec{\mathbf{R}} = \vec{\mathbf{A}} + \vec{\mathbf{B}} + \vec{\mathbf{C}} + \vec{\mathbf{D}}$ is the *vector that closes the polygon formed by the vectors being added.* In other words, $\vec{\mathbf{R}}$ is the *vector drawn from the tail of the first vector to the tip of the last vector.* Again, the order of the summation is unimportant.

We conclude that a vector is a quantity that has both magnitude and direction and also obeys the laws of vector addition described in Figures 1.9 to 1.11.

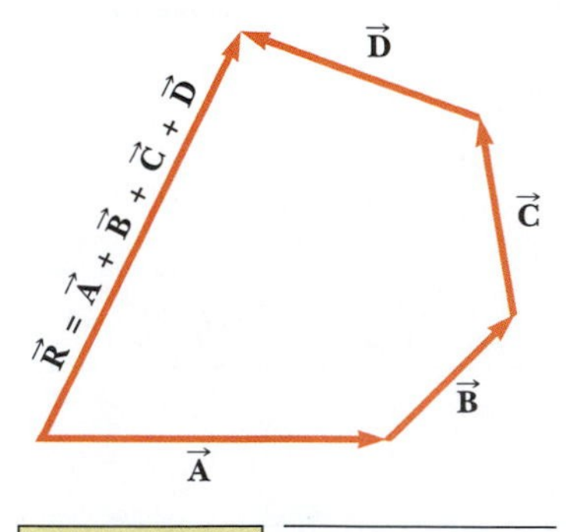


FIGURE 1.11 Geometric construction for summing four vectors. The resultant vector $\vec{\mathbf{R}}$ closes the polygon and points from the tail of the first vector to the tip of the final vector.

## Negative of a Vector

The negative of the vector $\vec{\mathbf{A}}$ is defined as the vector that, when added to $\vec{\mathbf{A}}$, gives zero for the vector sum. That is, $\vec{\mathbf{A}} + (-\vec{\mathbf{A}}) = 0$. The vectors $\vec{\mathbf{A}}$ and $-\vec{\mathbf{A}}$ have the same magnitude but opposite directions.

## Subtraction of Vectors

The operation of vector subtraction makes use of the definition of the negative of a vector. We define the operation $\vec{\mathbf{A}} - \vec{\mathbf{B}}$ as vector $-\vec{\mathbf{B}}$ added to vector $\vec{\mathbf{A}}$:

$$\vec{\mathbf{A}} - \vec{\mathbf{B}} = \vec{\mathbf{A}} + (-\vec{\mathbf{B}}) \quad [1.9]$$

A diagram for subtracting two vectors is shown in Figure 1.12.

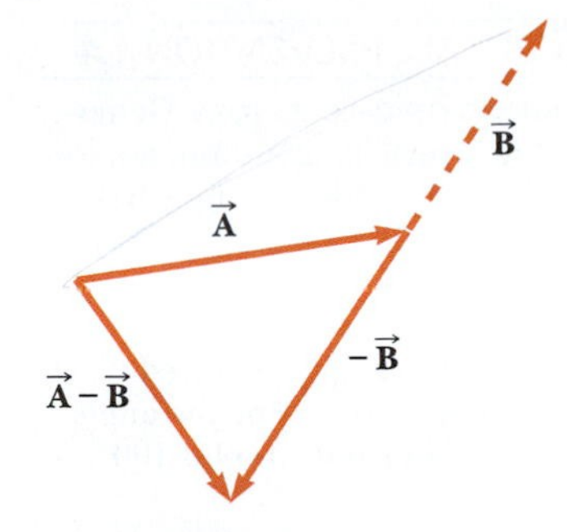


FIGURE 1.12 This construction shows how to subtract vector $\vec{\mathbf{B}}$ from vector $\vec{\mathbf{A}}$: Add the vector $-\vec{\mathbf{B}}$ to vector $\vec{\mathbf{A}}$. The vector $-\vec{\mathbf{B}}$ is equal in magnitude and opposite to the vector $\vec{\mathbf{B}}$.

## Multiplication of a Vector by a Scalar

If a vector $\vec{\mathbf{A}}$ is multiplied by a positive scalar quantity $s$, the product $s\vec{\mathbf{A}}$ is a vector that has the same direction as $\vec{\mathbf{A}}$ and magnitude $sA$. If $s$ is a negative scalar quantity, the vector $s\vec{\mathbf{A}}$ is directed opposite to $\vec{\mathbf{A}}$. For example, the vector $5\vec{\mathbf{A}}$ is five times greater in magnitude than $\vec{\mathbf{A}}$ and has the same direction as $\vec{\mathbf{A}}$. On the other hand, the vector $-\frac{1}{3}\vec{\mathbf{A}}$ has one third the magnitude of $\vec{\mathbf{A}}$ and points in the direction opposite $\vec{\mathbf{A}}$ (because of the negative sign).

## Multiplication of Two Vectors

Two vectors $\vec{\mathbf{A}}$ and $\vec{\mathbf{B}}$ can be multiplied in two different ways to produce either a scalar or a vector quantity. The **scalar product** (or dot product) $\vec{\mathbf{A}}\cdot\vec{\mathbf{B}}$ is a scalar quantity equal to $AB \cos \theta$, where $\theta$ is the angle between $\vec{\mathbf{A}}$ and $\vec{\mathbf{B}}$. The **vector product** (or cross product) $\vec{\mathbf{A}} \times \vec{\mathbf{B}}$ is a vector quantity whose magnitude is equal to $AB \sin \theta$. We shall discuss these products more fully in Chapters 6 and 10, where they are first used.

**PITFALL PREVENTION 1.3**

**VECTOR ADDITION VERSUS SCALAR ADDITION** Keep in mind that $\vec{\mathbf{A}} + \vec{\mathbf{B}} = \vec{\mathbf{C}}$ is very different from $A + B = C$. The first is a vector sum, which must be handled carefully, such as with the graphical method described in Active Figure 1.9. The second is a simple algebraic addition of numbers that is handled with the normal rules of arithmetic.

**QUICK QUIZ 1.4** The magnitudes of two vectors $\vec{\mathbf{A}}$ and $\vec{\mathbf{B}}$ are $A = 12$ units and $B = 8$ units. Which of the following pairs of numbers represents the *largest* and *smallest* possible values for the magnitude of the resultant vector $\vec{\mathbf{R}} = \vec{\mathbf{A}} + \vec{\mathbf{B}}$? (a) 14.4 units, 4 units (b) 12 units, 8 units (c) 20 units, 4 units (d) none of these answers

**QUICK QUIZ 1.5** If vector $\vec{\mathbf{B}}$ is added to vector $\vec{\mathbf{A}}$, under what condition does the resultant vector $\vec{\mathbf{A}} + \vec{\mathbf{B}}$ have magnitude $A + B$? (a) $\vec{\mathbf{A}}$ and $\vec{\mathbf{B}}$ are parallel and in the same direction. (b) $\vec{\mathbf{A}}$ and $\vec{\mathbf{B}}$ are parallel and in opposite directions. (c) $\vec{\mathbf{A}}$ and $\vec{\mathbf{B}}$ are perpendicular.

# 1.9 COMPONENTS OF A VECTOR AND UNIT VECTORS

The geometric method of adding vectors is not the recommended procedure for situations in which great precision is required or in three-dimensional problems because we are forced to represent them on two-dimensional paper. In this section, we describe a method of adding vectors that makes use of the *projections* of a vector along the axes of a rectangular coordinate system.

Consider a vector $\vec{\mathbf{A}}$ lying in the $xy$ plane and making an arbitrary angle $\theta$ with the positive $x$ axis, as in Figure 1.13a. The vector $\vec{\mathbf{A}}$ can be represented by its rectangular **components,** $A_x$ and $A_y$. The component $A_x$ represents the projection of $\vec{\mathbf{A}}$ along the $x$ axis, and $A_y$ represents the projection of $\vec{\mathbf{A}}$ along the $y$ axis. The components of a vector, which are scalar quantities, can be positive or negative. For example, in Figure 1.13a, $A_x$ and $A_y$ are both positive. The absolute values of the components are the magnitudes of the associated **component vectors** $\vec{\mathbf{A}}_x$ and $\vec{\mathbf{A}}_y$.

Figure 1.13b shows the component vectors again, but with the $y$ component vector shifted so that it is added vectorially to the $x$ component vector. This diagram

### PITFALL PREVENTION 1.4

**Tangents on calculators** Generally, the inverse tangent function on calculators provides an angle between $-90°$ and $+90°$. As a consequence, if the vector you are studying lies in the second or third quadrant, the angle measured from the positive $x$ axis will be the angle your calculator returns plus $180°$.

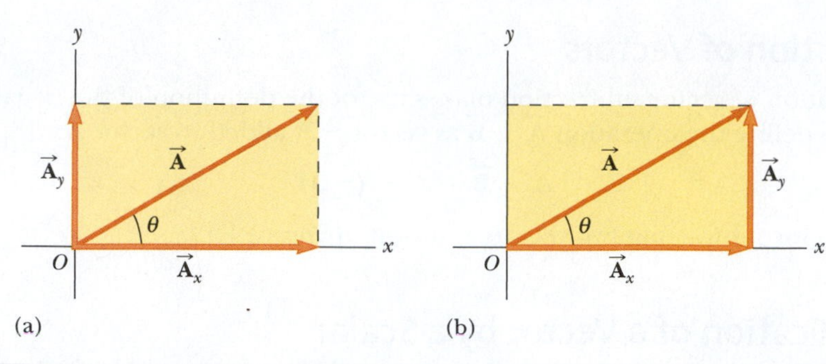


**FIGURE 1.13** (a) A vector $\vec{\mathbf{A}}$ lying in the $xy$ plane can be represented by its component vectors $\vec{\mathbf{A}}_x$ and $\vec{\mathbf{A}}_y$. (b) The $y$ component vector $\vec{\mathbf{A}}_y$ can be moved to the right so that it adds to $\vec{\mathbf{A}}_x$. The vector sum of the component vectors is $\vec{\mathbf{A}}$. These three vectors form a right triangle.

| $A_x$ negative, $A_y$ positive | $A_x$ positive, $A_y$ positive |
|---|---|
| $A_x$ negative, $A_y$ negative | $A_x$ positive, $A_y$ negative |

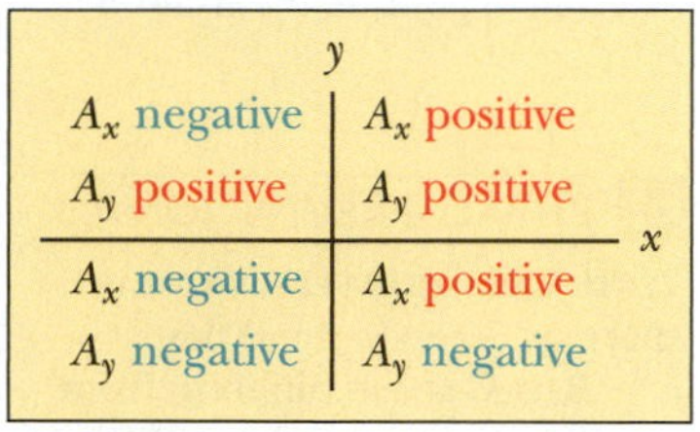

**FIGURE 1.14** The signs of the components of a vector $\vec{\mathbf{A}}$ depend on the quadrant in which the vector is located.

shows us two important features. First, a vector is equal to the sum of its component vectors. Therefore, the combination of the component vectors is a valid substitute for the actual vector. The second feature is that the vector and its component vectors form a right triangle. Therefore, we can let the triangle be a model for the vector and can use right triangle trigonometry to analyze the vector. The legs of the triangle are of lengths proportional to the components (depending on what scale factor you have chosen), and the hypotenuse is of a length proportional to the magnitude of the vector.

From Figure 1.13b and the definition of the sine and cosine of an angle, we see that $\cos\theta = A_x/A$ and $\sin\theta = A_y/A$. Hence, the components of $\vec{\mathbf{A}}$ are given by

$$A_x = A\cos\theta \qquad \text{and} \qquad A_y = A\sin\theta \qquad [1.10]$$

When using these component equations, $\theta$ must be measured counterclockwise from the positive $x$ axis. From our triangle, it follows that the magnitude of $\vec{\mathbf{A}}$ and its direction are related to its components through the Pythagorean theorem and the definition of the tangent function:

■ Magnitude of $\vec{\mathbf{A}}$

$$A = \sqrt{A_x^2 + A_y^2} \qquad [1.11]$$

■ Direction of $\vec{\mathbf{A}}$

$$\tan\theta = \frac{A_y}{A_x} \qquad [1.12]$$

To solve for $\theta$, we can write $\theta = \tan^{-1}(A_y/A_x)$, which is read "$\theta$ equals the angle whose tangent is the ratio $A_y/A_x$." *Note that the signs of the components $A_x$ and $A_y$ depend on the angle $\theta$.* For example, if $\theta = 120°$, $A_x$ is negative and $A_y$ is positive. On the other hand, if $\theta = 225°$, both $A_x$ and $A_y$ are negative. Figure 1.14 summarizes the signs of the components when $\vec{\mathbf{A}}$ lies in the various quadrants.

If you choose reference axes or an angle other than those shown in Figure 1.13, the components of the vector must be modified accordingly. In many applications, it is more convenient to express the components of a vector in a coordinate system having axes that are not horizontal and vertical but are still perpendicular to each other. Suppose a vector $\vec{\mathbf{B}}$ makes an angle $\theta'$ with the $x'$ axis defined in Figure 1.15. The components of $\vec{\mathbf{B}}$ along these axes are given by $B_{x'} = B\cos\theta'$ and $B_{y'} = B\sin\theta'$, as in Equation 1.10. The magnitude and direction of $\vec{\mathbf{B}}$ are obtained from expressions equivalent to Equations 1.11 and 1.12. Therefore, we can express the components of a vector in *any* coordinate system that is convenient for a particular situation.

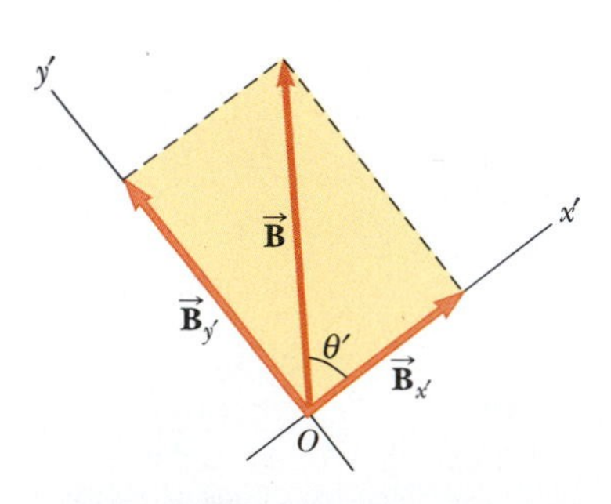


**FIGURE 1.15** The component vectors of vector $\vec{\mathbf{B}}$ in a coordinate system that is tilted.

**QUICK QUIZ 1.6** Choose the correct response to make the sentence true: A component of a vector is **(a)** always, **(b)** never, or **(c)** sometimes larger than the magnitude of the vector.

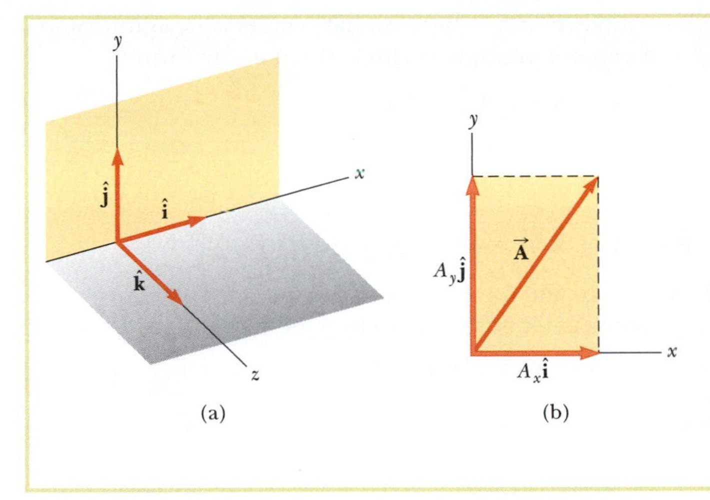


**ACTIVE FIGURE 1.16**

(a) The unit vectors $\hat{\mathbf{i}}$, $\hat{\mathbf{j}}$, and $\hat{\mathbf{k}}$ are directed along the *x*, *y*, and *z* axes, respectively. (b) A vector $\vec{\mathbf{A}}$ lying in the *xy* plane has component vectors $A_x\hat{\mathbf{i}}$ and $A_y\hat{\mathbf{j}}$, where $A_x$ and $A_y$ are the components of $\vec{\mathbf{A}}$.

**PhysicsNow™** Log into PhysicsNow at **www.pop4e.com** and go to Active Figure 1.16 to rotate the coordinate axes in three-dimensional space and view a representation of vector $\vec{\mathbf{A}}$ in three dimensions.

Vector quantities are often expressed in terms of unit vectors. **A unit vector is a dimensionless vector with a magnitude of 1 and is used to specify a given direction.** Unit vectors have no other physical significance. They are used simply as a bookkeeping convenience when describing a direction in space. We will use the symbols $\hat{\mathbf{i}}$, $\hat{\mathbf{j}}$, and $\hat{\mathbf{k}}$ to represent unit vectors pointing in the *x*, *y*, and *z* directions, respectively. The "hat" over the letters is a common notation for a unit vector; for example, $\hat{\mathbf{i}}$ is called "i-hat." The unit vectors $\hat{\mathbf{i}}$, $\hat{\mathbf{j}}$, and $\hat{\mathbf{k}}$ form a set of mutually perpendicular vectors as shown in Active Figure 1.16a, where the magnitude of each unit vector equals 1; that is, $|\hat{\mathbf{i}}| = |\hat{\mathbf{j}}| = |\hat{\mathbf{k}}| = 1$.

Consider a vector $\vec{\mathbf{A}}$ lying in the *xy* plane, as in Active Figure 1.16b. The product of the component $A_x$ and the unit vector $\hat{\mathbf{i}}$ is the component vector $\vec{\mathbf{A}}_x = A_x\hat{\mathbf{i}}$ parallel to the *x* axis with magnitude $A_x$. Likewise, $A_y\hat{\mathbf{j}}$ is a component vector of magnitude $A_y$ parallel to the *y* axis. When using the unit-vector form of a vector, we are simply multiplying a vector (the unit vector) by a scalar (the component). Therefore, the unit-vector notation for the vector $\vec{\mathbf{A}}$ is written

$$\vec{\mathbf{A}} = A_x\hat{\mathbf{i}} + A_y\hat{\mathbf{j}} \qquad [1.13]$$

Now suppose we wish to add vector $\vec{\mathbf{B}}$ to vector $\vec{\mathbf{A}}$, where $\vec{\mathbf{B}}$ has components $B_x$ and $B_y$. The procedure for performing this sum is simply to add the *x* and *y* components separately. The resultant vector $\vec{\mathbf{R}} = \vec{\mathbf{A}} + \vec{\mathbf{B}}$ is therefore

$$\vec{\mathbf{R}} = (A_x + B_x)\hat{\mathbf{i}} + (A_y + B_y)\hat{\mathbf{j}} \qquad [1.14]$$

From this equation, the components of the resultant vector are given by

$$\begin{aligned} R_x &= A_x + B_x \\ R_y &= A_y + B_y \end{aligned} \qquad [1.15]$$

The magnitude of $\vec{\mathbf{R}}$ and the angle it makes with the *x* axis can then be obtained from its components using the relationships

$$R = \sqrt{R_x^2 + R_y^2} = \sqrt{(A_x + B_x)^2 + (A_y + B_y)^2} \qquad [1.16]$$

$$\tan\theta = \frac{R_y}{R_x} = \frac{A_y + B_y}{A_x + B_x} \qquad [1.17]$$

The procedure just described for adding two vectors $\vec{\mathbf{A}}$ and $\vec{\mathbf{B}}$ using the component method can be checked using a diagram like Figure 1.17.

**PITFALL PREVENTION 1.5**

*x* **COMPONENTS** Equation 1.10 for the *x* and *y* components of a vector associates the cosine of the angle with the *x* component and the sine of the angle with the *y* component. This association occurs *solely* because we chose to measure the angle with respect to the *x* axis, so don't memorize these equations. Invariably, you will face a problem in the future in which the angle is measured with respect to the *y* axis, and the equations will be incorrect. It is much better to always think about which side of the triangle containing the components is adjacent to the angle and which side is opposite, and then assign the sine and cosine accordingly.

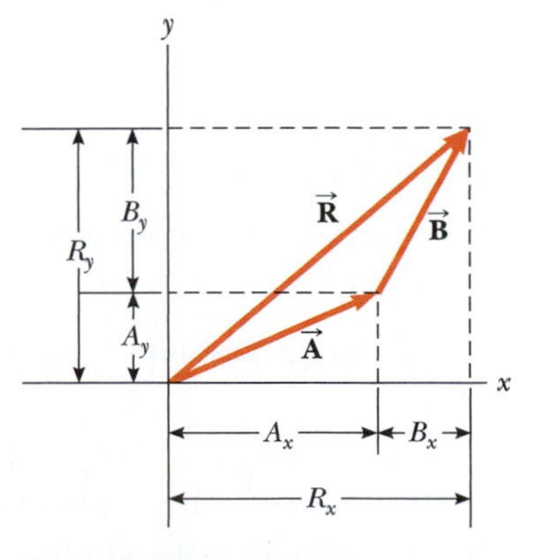


**FIGURE 1.17** A geometric construction showing the relation between the components of the resultant $\vec{\mathbf{R}}$ of two vectors and the individual components.

The extension of these methods to three-dimensional vectors is straightforward. If $\vec{\mathbf{A}}$ and $\vec{\mathbf{B}}$ both have $x$, $y$, and $z$ components, we express them in the form

$$\vec{\mathbf{A}} = A_x\hat{\mathbf{i}} + A_y\hat{\mathbf{j}} + A_z\hat{\mathbf{k}}$$
$$\vec{\mathbf{B}} = B_x\hat{\mathbf{i}} + B_y\hat{\mathbf{j}} + B_z\hat{\mathbf{k}}$$

The sum of $\vec{\mathbf{A}}$ and $\vec{\mathbf{B}}$ is

$$\vec{\mathbf{R}} = \vec{\mathbf{A}} + \vec{\mathbf{B}} = (A_x + B_x)\hat{\mathbf{i}} + (A_y + B_y)\hat{\mathbf{j}} + (A_z + B_z)\hat{\mathbf{k}} \qquad [1.18]$$

The same procedure can be used to add three or more vectors.

If a vector $\vec{\mathbf{R}}$ has $x$, $y$, and $z$ components, the magnitude of the vector is

$$R = \sqrt{R_x^2 + R_y^2 + R_z^2}$$

The angle $\theta_x$ that $\vec{\mathbf{R}}$ makes with the $x$ axis is given by

$$\cos\theta_x = \frac{R_x}{R}$$

with similar expressions for the angles with respect to the $y$ and $z$ axes.

**QUICK QUIZ 1.7** If at least one component of a vector is a positive number, the vector cannot **(a)** have any component that is negative, **(b)** be zero, **(c)** have three dimensions.

**QUICK QUIZ 1.8** If $\vec{\mathbf{A}} + \vec{\mathbf{B}} = 0$, the corresponding components of the two vectors $\vec{\mathbf{A}}$ and $\vec{\mathbf{B}}$ must be **(a)** equal, **(b)** positive, **(c)** negative, **(d)** of opposite sign.

## ■ Thinking Physics 1.2

You may have asked someone directions to a destination in a city and been told something like, "Walk 3 blocks east and then 5 blocks south." If so, are you experienced with vector components?

**Reasoning** Yes, you are! Although you may not have thought of vector component language when you heard these directions, that is exactly what the directions represent. The perpendicular streets of the city reflect an $xy$ coordinate system; we can assign the $x$ axis to the east–west streets, and the $y$ axis to the north–south streets. Thus, the comment of the person giving you directions can be translated as, "Undergo a displacement vector that has an $x$ component of +3 blocks and a $y$ component of −5 blocks." You would arrive at the same destination by undergoing the $y$ component first, followed by the $x$ component, demonstrating the commutative law of addition. ■

### EXAMPLE 1.6 The Sum of Two Vectors

Find the sum of two vectors $\vec{\mathbf{A}}$ and $\vec{\mathbf{B}}$ lying in the $xy$ plane and given by

$$\vec{\mathbf{A}} = 2.00\hat{\mathbf{i}} + 3.00\hat{\mathbf{j}} \quad \text{and} \quad \vec{\mathbf{B}} = 5.00\hat{\mathbf{i}} - 4.00\hat{\mathbf{j}}$$

**Solution** It might be helpful for you to draw a diagram of the vectors to clarify what they look like on the $xy$ plane. Using the rule given by Equation 1.14, we solve this problem mathematically as follows. Note that $A_x = 2.00$, $A_y = 3.00$, $B_x = 5.00$, and $B_y = -4.00$. Therefore, the resultant vector $\vec{\mathbf{R}}$ is

$$\vec{\mathbf{R}} = \vec{\mathbf{A}} + \vec{\mathbf{B}} = (2.00 + 5.00)\hat{\mathbf{i}} + (3.00 - 4.00)\hat{\mathbf{j}}$$
$$= 7.00\,\hat{\mathbf{i}} - 1.00\hat{\mathbf{j}}$$

or

$$R_x = 7.00, \qquad R_y = -1.00$$

The magnitude of $\vec{\mathbf{R}}$ is

$$R = \sqrt{R_x^2 + R_y^2}$$
$$= \sqrt{(7.00)^2 + (-1.00)^2} = \sqrt{50.0} = 7.07$$

**EXAMPLE 1.7** The Resultant Displacement

A particle undergoes three consecutive displacements: $\Delta\vec{\mathbf{r}}_1 = (1.50\hat{\mathbf{i}} + 3.00\hat{\mathbf{j}} - 1.20\hat{\mathbf{k}})$ cm, $\Delta\vec{\mathbf{r}}_2 = (2.30\hat{\mathbf{i}} - 1.40\hat{\mathbf{j}} - 3.60\hat{\mathbf{k}})$ cm, and $\Delta\vec{\mathbf{r}}_3 = (-1.30\hat{\mathbf{i}} + 1.50\hat{\mathbf{j}})$ cm. Find the components of the resultant displacement and its magnitude.

**Solution** We use Equation 1.18 for three vectors:

$$\begin{aligned}\vec{\mathbf{R}} &= \Delta\vec{\mathbf{r}}_1 + \Delta\vec{\mathbf{r}}_2 + \Delta\vec{\mathbf{r}}_3 = (1.50 + 2.30 - 1.30)\hat{\mathbf{i}}\,\text{cm} \\ &\quad + (3.00 - 1.40 + 1.50)\hat{\mathbf{j}}\,\text{cm} \\ &\quad + (-1.20 - 3.60 + 0)\hat{\mathbf{k}}\,\text{cm} \\ &= (2.50\hat{\mathbf{i}} + 3.10\hat{\mathbf{j}} - 4.80\hat{\mathbf{k}})\ \text{cm}\end{aligned}$$

That is, the resultant displacement has components $R_x = 2.50$ cm, $R_y = 3.10$ cm, and $R_z = -4.80$ cm. Its magnitude is

$$\begin{aligned}R &= \sqrt{R_x^2 + R_y^2 + R_z^2} \\ &= \sqrt{(2.50\ \text{cm})^2 + (3.10\ \text{cm})^2 + (-4.80\ \text{cm})^2} \\ &= 6.24\ \text{cm}\end{aligned}$$

**INTERACTIVE EXAMPLE 1.8** Taking a Hike

A hiker begins a two-day trip by first walking 25.0 km due southeast from her car. She stops and sets up her tent for the night. On the second day she walks 40.0 km in a direction 60.0° north of east, at which point she discovers a forest ranger's tower.

**A** Determine the components of the hiker's displacements on the first and second days.

**Solution** If we denote the displacement vectors on the first and second days by $\vec{\mathbf{A}}$ and $\vec{\mathbf{B}}$, respectively, and use the car as the origin of coordinates, we obtain the vectors shown in the diagram in Figure 1.18. Notice that the resultant vector $\vec{\mathbf{R}}$ can be drawn in the diagram to provide an approximation of the final result of the two hikes.

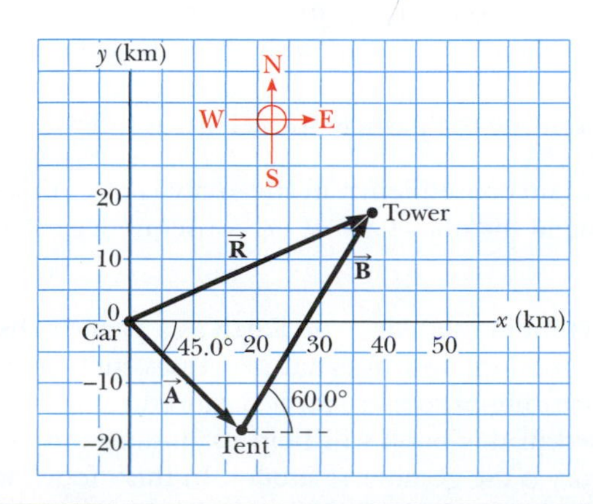


**FIGURE 1.18** (Interactive Example 1.8) The total displacement of the hiker is the vector $\vec{\mathbf{R}} = \vec{\mathbf{A}} + \vec{\mathbf{B}}$.

Displacement $\vec{\mathbf{A}}$ has a magnitude of 25.0 km and is 45.0° southeast. Its components are

$$A_x = A\cos(-45.0°) = (25.0\ \text{km})(0.707) = 17.7\ \text{km}$$

$$A_y = A\sin(-45.0°) = (25.0\ \text{km})(-0.707) = -17.7\ \text{km}$$

The positive value of $A_x$ indicates that the $x$ coordinate increased in this displacement. The negative value of $A_y$ indicates that the $y$ coordinate decreased in this displacement. Notice in the diagram of Figure 1.18 that vector $\vec{\mathbf{A}}$ lies in the fourth quadrant, consistent with the signs of the components we calculated.

The second displacement $\vec{\mathbf{B}}$ has a magnitude of 40.0 km and is 60.0° north of east. Its components are

$$B_x = B\cos 60.0° = (40.0\ \text{km})(0.500) = 20.0\ \text{km}$$

$$B_y = B\sin 60.0° = (40.0\ \text{km})(0.866) = 34.6\ \text{km}$$

**B** Determine the components of the hiker's total displacement for the trip.

**Solution** The resultant displacement vector for the trip, $\vec{\mathbf{R}} = \vec{\mathbf{A}} + \vec{\mathbf{B}}$, has components given by

$$R_x = A_x + B_x = 17.7\ \text{km} + 20.0\ \text{km} = 37.7\ \text{km}$$

$$R_y = A_y + B_y = -17.7\ \text{km} + 34.6\ \text{km} = 16.9\ \text{km}$$

In unit-vector form, we can write the total displacement as

$$\vec{\mathbf{R}} = (37.7\hat{\mathbf{i}} + 16.9\hat{\mathbf{j}})\ \text{km}$$

**PhysicsNow™** Investigate this vector addition situation by logging into PhysicsNow at www.pop4e.com and going to Interactive Example 1.8.

## 1.10 MODELING, ALTERNATIVE REPRESENTATIONS, AND PROBLEM-SOLVING STRATEGY

Most courses in general physics require the student to learn the skills of problem solving, and examinations usually include problems that test such skills. This section describes some useful ideas that will enable you to enhance your understanding of physical concepts, increase your accuracy in solving problems, eliminate initial panic or lack of direction in approaching a problem, and organize your work.

One of the primary problem-solving methods in physics is to form an appropriate **model** of the problem. **A model is a simplified substitute for the real problem that allows us to solve the problem in a relatively simple way.** As long as the predictions of the model agree to our satisfaction with the actual behavior of the real system, the model is valid. If the predictions do not agree, the model must be refined or replaced with another model. The power of modeling is in its ability to reduce a wide variety of very complex problems to a limited number of classes of problems that can be approached in similar ways.

In science, a model is very different from, for example, an architect's scale model of a proposed building, which appears as a smaller version of what it represents. A scientific model is a theoretical construct and may have no visual similarity to the physical problem. A simple application of modeling is presented in Example 1.9, and we shall encounter many more examples of models as the text progresses.

Models are needed because the actual operation of the Universe is extremely complicated. Suppose, for example, we are asked to solve a problem about the Earth's motion around the Sun. The Earth is very complicated, with many processes occurring simultaneously. These processes include weather, seismic activity, and ocean movements as well as the multitude of processes involving human activity. Trying to maintain knowledge and understanding of all these processes is an impossible task.

The modeling approach recognizes that none of these processes affects the motion of the Earth around the Sun to a measurable degree. Therefore, these details are all ignored. In addition, as we shall find in Chapter 11, the size of the Earth does not affect the gravitational force between the Earth and the Sun; only the masses of the Earth and Sun and the distance between them determine this force. In a simplified model, the Earth is imagined to be a particle, an object with mass but zero size. This replacement of an extended object by a particle is called the **particle model,** which is used extensively in physics. By analyzing the motion of a particle with the mass of the Earth in orbit around the Sun, we find that the predictions of the particle's motion are in excellent agreement with the actual motion of the Earth.

The two primary conditions for using the particle model are as follows:

- The size of the actual object is of no consequence in the analysis of its motion.
- Any internal processes occurring in the object are of no consequence in the analysis of its motion.

Both of these conditions are in action in modeling the Earth as a particle. Its radius is not a factor in determining its motion, and internal processes such as thunderstorms, earthquakes, and manufacturing processes can be ignored.

Four categories of models used in this book will help us understand and solve physics problems. The first category is the **geometric model.** In this model, we form a geometric construction that represents the real situation. We then set aside the real problem and perform an analysis of the geometric construction. Consider a popular problem in elementary trigonometry, as in the following example.

### EXAMPLE 1.9 Finding the Height of a Tree

You wish to find the height of a tree but cannot measure it directly. You stand 50.0 m from the tree and determine that a line of sight from the ground to the top of the tree makes an angle of 25.0° with the ground. How tall is the tree?

**Solution** Figure 1.19 shows the tree and a right triangle corresponding to the information in the problem superimposed over it. (We assume that the tree is exactly perpendicular to a perfectly flat ground.) In the triangle, we know the length of the horizontal leg and the angle between the hypotenuse and the horizontal leg. We can find the height of the tree by calculating the length of the vertical leg. We do so with the tangent function:

$$\tan\theta = \frac{\text{opposite side}}{\text{adjacent side}} = \frac{h}{50.0\text{ m}}$$

$$h = (50.0\text{ m})\tan\theta = (50.0\text{ m})\tan 25.0° = 23.3\text{ m}$$

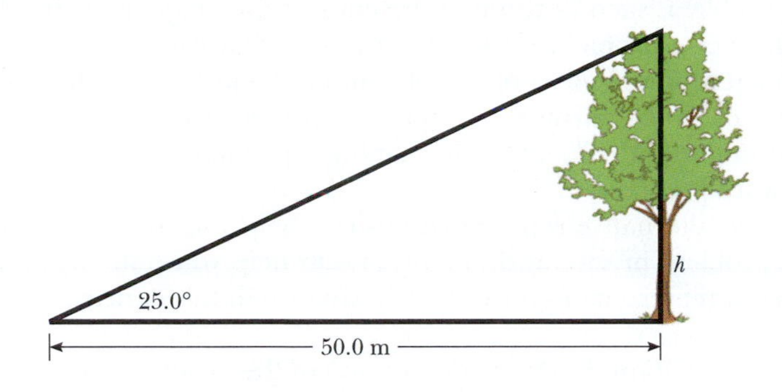


**FIGURE 1.19** (Example 1.9) The height of a tree can be found by measuring the distance from the tree and the angle of sight to the top above the ground. This problem is a simple example of geometrically *modeling* the actual problem.

You may have solved a problem very similar to Example 1.9 but never thought about the notion of modeling. From the modeling approach, however, once we draw the triangle in Figure 1.19, the triangle is a geometric model of the real problem; it is a *substitute.* Until we reach the end of the problem, we do not imagine the problem to be about a tree but to be about a triangle. We use trigonometry to find the vertical leg of the triangle, leading to a value of 23.3 m. Because this leg *represents* the height of the tree, we can now return to the original problem and claim that the height of the tree is 23.3 m.

Other examples of geometric models include modeling the Earth as a perfect sphere, a pizza as a perfect disk, a meter stick as a long rod with no thickness, and an electric wire as a long, straight, cylinder.

The particle model is an example of the second category of models, which we will call **simplification models.** In a simplification model, details that are not significant in determining the outcome of the problem are ignored. When we study rotation in Chapter 10, objects will be modeled as *rigid objects.* All the molecules in a rigid object maintain their exact positions with respect to one another. We adopt this simplification model because a spinning rock is much easier to analyze than a spinning block of gelatin, which is *not* a rigid object. Other simplification models will assume that quantities such as friction forces are negligible, remain constant, or are proportional to some power of the object's speed.

The third category is that of **analysis models,** which are general types of problems that we have solved before. An important technique in problem solving is to cast a new problem into a form similar to one we have already solved and which can be used as a model. As we shall see, there are about two dozen analysis models that can be used to solve most of the problems you will encounter. We will see our first analysis models in Chapter 2, where we will discuss them in more detail.

The fourth category of models is **structural models.** These models are generally used to understand the behavior of a system that is far different in scale from our macroscopic world—either much smaller or much larger—so that we cannot interact with it directly. As an example, the notion of a hydrogen atom as an electron in a circular orbit around a proton is a structural model of the atom. We will discuss this model and structural models in general in Chapter 11.

Intimately related to the notion of modeling is that of forming **alternative representations** of the problem. **A representation is a method of viewing or presenting the information related to the problem.** Scientists must be able to communicate complex ideas to individuals without scientific backgrounds. The best representation to use in conveying the information successfully will vary from one individual to the next. Some will be convinced by a well-drawn graph, and others will require a picture. Physicists are often persuaded to agree with a point of view by examining an equation, but nonphysicists may not be convinced by this mathematical representation of the information.

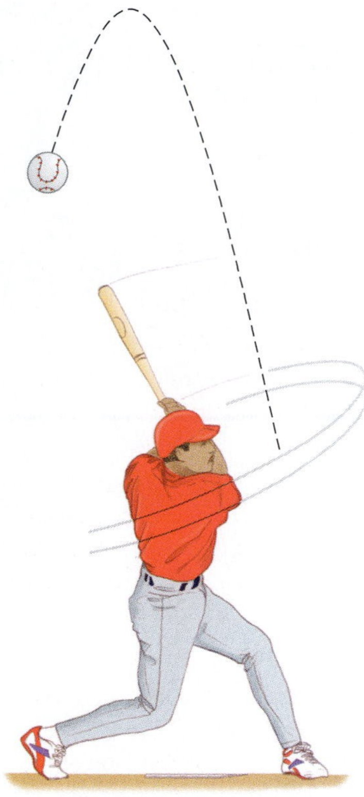

A word problem, such as those at the ends of the chapters in this book, is one representation of a problem. In the "real world" that you will enter after graduation, the initial representation of a problem may be just an existing situation, such as the effects of global warming or a patient in danger of dying. You may have to identify the important data and information, and then cast the situation into an equivalent word problem!

Considering alternative representations can help you think about the information in the problem in several different ways to help you understand and solve it. Several types of representations can be of assistance in this endeavor:

FIGURE 1.20 A pictorial representation of a pop foul being hit by a baseball player.

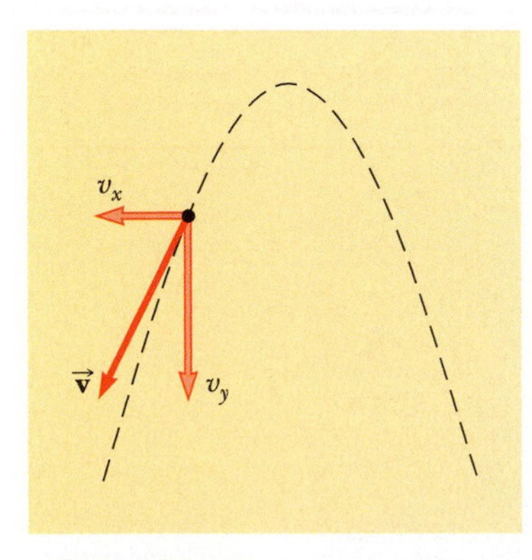


FIGURE 1.21 A simplified pictorial representation for the situation shown in Figure 1.20.

- **Mental representation.** From the description of the problem, imagine a scene that describes what is happening in the word problem, then let time progress so that you understand the situation and can predict what changes will occur in the situation. This step is critical in approaching *every* problem.
- **Pictorial representation.** Drawing a picture of the situation described in the word problem can be of great assistance in understanding the problem. In Example 1.9, the pictorial representation in Figure 1.19 allows us to identify the triangle as a geometric model of the problem. In architecture, a blueprint is a pictorial representation of a proposed building.

  Generally, a pictorial representation describes *what you would see* if you were observing the situation in the problem. For example, Figure 1.20 shows a pictorial representation of a baseball player hitting a short pop foul. Any coordinate axes included in your pictorial representation will be in two dimensions: $x$ and $y$ axes.
- **Simplified pictorial representation.** It is often useful to redraw the pictorial representation without complicating details by applying a simplification model. This process is similar to the discussion of the particle model described earlier. In a pictorial representation of the Earth in orbit around the Sun, you might draw the Earth and the Sun as spheres, with possibly some attempt to draw continents to identify which sphere is the Earth. In the simplified pictorial representation, the Earth and the Sun would be drawn simply as dots, representing particles. Figure 1.21 shows a simplified pictorial representation corresponding to the pictorial representation of the baseball trajectory in Figure 1.20. The notations $v_x$ and $v_y$ refer to the components of the velocity vector for the baseball. We shall use such simplified pictorial representations throughout the book.
- **Graphical representation.** In some problems, drawing a graph that describes the situation can be very helpful. In mechanics, for example, position–time graphs can be of great assistance. Similarly, in thermodynamics, pressure–volume graphs are essential to understanding. Figure 1.22 shows a graphical representation of the position as a function of time of a block on the end of a vertical spring as it oscillates up and down. Such a graph is helpful for understanding simple harmonic motion, which we study in Chapter 12.

  A graphical representation is different from a pictorial representation, which is also a two-dimensional display of information but whose axes, if any, represent *length* coordinates. In a graphical representation, the axes may represent any two related variables. For example, a graphical representation may have axes for temperature and time. Therefore, in comparison to a pictorial representation, a graphical representation is generally *not* something you would see when observing the situation in the problem with your eyes.

- **Tabular representation.** It is sometimes helpful to organize the information in tabular form to help make it clearer. For example, some students find that making tables of known quantities and unknown quantities is helpful. The periodic table is an extremely useful tabular representation of information in chemistry and physics.
- **Mathematical representation.** The ultimate goal in solving a problem is often the mathematical representation. You want to move from the information contained in the word problem, through various representations of the problem that allow you to understand what is happening, to one or more equations that represent the situation in the problem and that can be solved mathematically for the desired result.

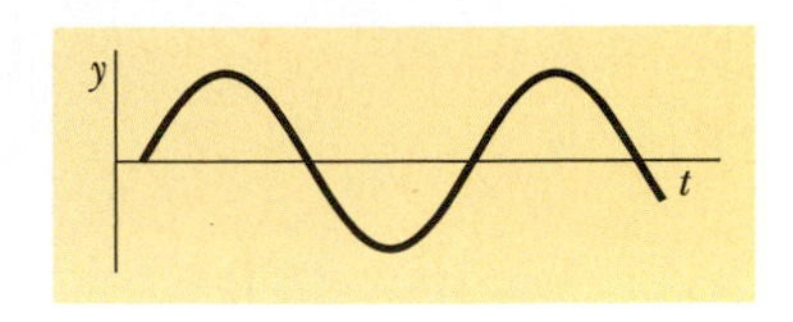


**FIGURE 1.22** A graphical representation of the position as a function of time of a block hanging from a spring and oscillating.

## GENERAL PROBLEM-SOLVING STRATEGY

An important way to become a skilled problem solver is to adopt a problem-solving strategy. This General Problem-Solving Strategy provides useful steps for solving numerical problems.

### Conceptualize

- Read the problem carefully at least twice. Be sure you understand the nature of the problem before proceeding further. Imagine a movie, running in your mind, of what happens in the problem. This step allows you to set up the mental representation of the problem.
- Draw a suitable diagram with appropriate labels and coordinate axes, if needed. This process provides the pictorial representation. If appropriate, generate a graphical representation. If you find it helpful, generate a tabular representation.
- Now focus on the expected result of solving the problem. Exactly what is the question asking? Will the final result be numerical or algebraic? Do you know what units to expect?
- Don't forget to incorporate information from your own experiences and common sense. What should a reasonable answer look like? For example, you wouldn't expect to calculate the speed of an automobile to be $5 \times 10^6$ m/s.

### Categorize

- Once you have a good idea of what the problem is about, you need to *simplify* the problem by drawing a simplified pictorial representation. Use a simplification model to remove additional unnecessary details if the conditions for the model are satisfied. If it helps you solve the problem, identify a useful geometric model from the diagrams.
- Once the problem is simplified, it is important to *categorize* the problem. Is it a simple *plug-in problem,* such that numbers can be simply substituted into a definition? If so, the problem is likely to be finished when this substitution is done. If not, you face an *analysis problem,* and the situation must be analyzed more deeply to reach a solution.
- Once you have eliminated the unnecessary details and have simplified the problem to its fundamental level, identify an analysis model for the problem. (We will see how to identify analysis models as we introduce them throughout the book.)

### Analyze

- Now you must analyze the problem and strive for a mathematical representation of the problem. From the analysis model, identify the basic physical principle or principles that are involved, listing the knowns and unknowns. Select relevant equations that apply to the model.
- Use algebra (and calculus, if necessary) to solve symbolically for the unknown variable in terms of what is given. Substitute in the appropriate numbers, calculate the result, and round it to the proper number of significant figures.

### Finalize

- This final step is the most important part. Examine your numerical answer. Does it have the correct units? Is it of reasonable value? Does it meet your expectations from your conceptualization of the problem? What about the algebraic form of the result, before you substituted numerical values? Does it make sense? Examine the variables in the problem to see whether the answer would change in a physically meaningful way if the variables were drastically increased, decreased, or even became zero. Looking at limiting cases to see whether they yield expected values is a very useful way to make sure that you are obtaining reasonable results.

Although this problem-solving strategy may look complicated, it may not be necessary to perform all the steps for a given problem. Examples in this text focus on how to apply these steps explicitly to help you become an effective problem solver. Many chapters include a section labeled "Problem-Solving Strategy" that should help you through the rough spots. These sections are organized according to the General Problem-Solving Strategy and tailor this strategy to the specific types of problems addressed in individual chapters. Once you have developed an organized system for examining problems and extracting relevant information, you will become a more confident problem solver in physics as well as in other areas.

## SUMMARY

**PhysicsNow™** Take a practice test by logging into PhysicsNow at **www.pop4e.com** and clicking on the Pre-Test link for this chapter.

Mechanical quantities can be expressed in terms of three fundamental quantities—**length, mass,** and **time**—which in the SI system have the units **meters** (m), **kilograms** (kg), and **seconds** (s), respectively. It is often useful to use the method of **dimensional analysis** to check equations and to assist in deriving expressions.

The **density** of a substance is defined as its mass per unit volume:

$$\rho \equiv \frac{m}{V} \qquad [1.1]$$

**Vectors** are quantities that have both magnitude and direction and obey the vector law of addition. **Scalars** are quantities that add algebraically.

Two vectors $\vec{\mathbf{A}}$ and $\vec{\mathbf{B}}$ can be added using the triangle method. In this method (see Fig. 1.9), the vector $\vec{\mathbf{R}} = \vec{\mathbf{A}} + \vec{\mathbf{B}}$ runs from the tail of $\vec{\mathbf{A}}$ to the tip of $\vec{\mathbf{B}}$.

The $x$ component $A_x$ of the vector $\vec{\mathbf{A}}$ is equal to its projection along the $x$ axis of a coordinate system, where $A_x = A\cos\theta$ and where $\theta$ is the angle $\vec{\mathbf{A}}$ makes with the $x$ axis. Likewise, the $y$ component $A_y$ of $\vec{\mathbf{A}}$ is its projection along the $y$ axis, where $A_y = A\sin\theta$.

If a vector $\vec{\mathbf{A}}$ has an $x$ component equal to $A_x$ and a $y$ component equal to $A_y$, the vector can be expressed in unit-vector form as $\vec{\mathbf{A}} = (A_x\hat{\mathbf{i}} + A_y\hat{\mathbf{j}})$. In this notation, $\hat{\mathbf{i}}$ is a unit vector in the positive $x$ direction and $\hat{\mathbf{j}}$ is a unit vector in the positive $y$ direction. Because $\hat{\mathbf{i}}$ and $\hat{\mathbf{j}}$ are unit vectors, $|\hat{\mathbf{i}}| = |\hat{\mathbf{j}}| = 1$. In three dimensions, a vector can be expressed as $\vec{\mathbf{A}} = (A_x\hat{\mathbf{i}} + A_y\hat{\mathbf{j}} + A_z\hat{\mathbf{k}})$, where $\hat{\mathbf{k}}$ is a unit vector in the $z$ direction.

The resultant of two or more vectors can be found by resolving all vectors into their $x$, $y$, and $z$ components and adding their components:

$$\vec{\mathbf{R}} = \vec{\mathbf{A}} + \vec{\mathbf{B}} = (A_x + B_x)\hat{\mathbf{i}} + (A_y + B_y)\hat{\mathbf{j}} + (A_z + B_z)\hat{\mathbf{k}} \qquad [1.18]$$

Problem-solving skills and physical understanding can be improved by **modeling** the problem and by constructing **alternative representations** of the problem. Models helpful in solving problems include **geometric, simplification,** and **analysis models.** Scientists use **structural models** to understand systems larger or smaller in scale than those with which we normally have direct experience. Helpful representations include the **mental, pictorial, simplified pictorial, graphical, tabular,** and **mathematical representations.**

## QUESTIONS

☐ = answer available in the *Student Solutions Manual and Study Guide*

**1.** What types of natural phenomena could serve as time standards?

**2.** Suppose the three fundamental standards of the metric system were length, *density,* and time rather than length, *mass,* and time. The standard of density in this system is to be defined as that of water. What considerations about water would you need to address to make sure that the standard of density is as accurate as possible?

**3.** Express the following quantities using the prefixes given in Table 1.4: (a) $3 \times 10^{-4}$ m, (b) $5 \times 10^{-5}$ s, (c) $72 \times 10^{2}$ g.

**4.** Suppose two quantities $A$ and $B$ have different dimensions. Determine which of the following arithmetic operations *could* be physically meaningful: (a) $A + B$, (b) $A/B$, (c) $B - A$, (d) $AB$.

**5.** If an equation is dimensionally correct, does that mean that the equation must be true? If an equation is not dimensionally correct, does that mean that the equation cannot be true?

**6.** Find the order of magnitude of your age in seconds.

**7.** What level of precision is implied in an order-of-magnitude calculation?

**8.** In reply to a student's question, a guard in a natural history museum says of the fossils near his station, "When I started work here twenty-four years ago, they were eighty million years old, so you can add it up." What should the student conclude about the age of the fossils?

**9.** Can the magnitude of a particle's displacement be greater than the distance traveled? Explain.

**10.** Which of the following are vectors and which are not: force, temperature, the volume of water in a can, the ratings of a TV show, the height of a building, the velocity of a sports car, the age of the Universe?

11. A vector $\vec{A}$ lies in the $xy$ plane. For what orientations of $\vec{A}$ will both of its components be negative? For what orientations will its components have opposite signs?

12. A book is moved once around the perimeter of a tabletop with the dimensions 1.0 m × 2.0 m. If the book ends up at its initial position, what is its displacement? What is the distance traveled?

13. While traveling along a straight interstate highway you notice that the mile marker reads 260. You travel until you reach the 150-mile marker and then retrace your path to the 175-mile marker. What is the magnitude of your resultant displacement from mile marker 260?

14. If the component of vector $\vec{A}$ along the direction of vector $\vec{B}$ is zero, what can you conclude about the two vectors?

15. Can the magnitude of a vector have a negative value? Explain.

16. Under what circumstances would a nonzero vector lying in the $xy$ plane have components that are equal in magnitude?

17. Is it possible to add a vector quantity to a scalar quantity? Explain.

18. In what circumstance is the $x$ component of a vector given by the magnitude of the vector multiplied by the sine of its direction angle?

19. Identify the type of model (geometrical, simplification, or structural) represented by each of the following. (a) In its orbit around the Sun, the Earth is treated as a particle. (b) The distance the Earth travels around the Sun is calculated as $2\pi$ multiplied by the Earth–Sun distance. (c) The atomic structure of a solid material is imagined to consist of small objects (atoms) connected to neighboring identical objects by springs. (d) For an object you drop, air resistance is ignored. (e) The volume of water in a bottle is estimated by calculating the volume of a cylinder. (f) A bat hits a baseball. In studying the motion of the baseball, any distortion of the ball while it is in contact with the bat is not considered. (g) In the early 20th century, the atom was proposed to consist of electrons in orbit around a very small but massive nucleus.

## PROBLEMS

1, 2, 3 = straightforward, intermediate, challenging
☐ = full solution available in the *Student Solutions Manual and Study Guide*
PhysicsNow™ = coached problem with hints available at www.pop4e.com
= computer useful in solving problem
= paired numerical and symbolic problems
= biomedical application

*Note:* Consult the endpapers, appendices, and tables in the text whenever necessary in solving problems. For this chapter, Appendix B.3 and Table 15.1 may be particularly useful. Answers to odd-numbered problems appear in the back of the book.

### Section 1.1 ▪ Standards of Length, Mass, and Time

1. Use information on the endpapers of this book to calculate the average density of the Earth. Where does the value fit among those listed in Table 15.1? Look up the density of a typical surface rock like granite in another source and compare the density of the Earth to it.

2. A major motor company displays a die-cast model of its first automobile, made from 9.35 kg of iron. To celebrate its hundredth year in business, a worker will recast the model in gold from the original dies. What mass of gold is needed to make the new model? The density of iron is $7.86 \times 10^3$ kg/m$^3$, and that of gold is $19.3 \times 10^3$ kg/m$^3$.

3. What mass of a material with density $\rho$ is required to make a hollow spherical shell having inner radius $r_1$ and outer radius $r_2$?

4. Two spheres are cut from a certain uniform rock. One has radius 4.50 cm. The mass of the other is five times greater. Find its radius.

### Section 1.2 ▪ Dimensional Analysis

5. The position of a particle moving under uniform acceleration is some function of time and the acceleration. Suppose we write this position as $x = ka^m t^n$, where $k$ is a dimensionless constant. Show by dimensional analysis that this expression is satisfied if $m = 1$ and $n = 2$. Can this analysis give the value of $k$?

6. Figure P1.6 shows a *frustrum of a cone*. Of the following mensuration (geometrical) expressions, which describes (a) the total circumference of the flat circular faces, (b) the volume, and (c) the area of the curved surface? (i) $\pi(r_1 + r_2)[h^2 + (r_1 - r_2)^2]^{1/2}$ (ii) $2\pi(r_1 + r_2)$ (iii) $\pi h(r_1{}^2 + r_1 r_2 + r_2{}^2)$

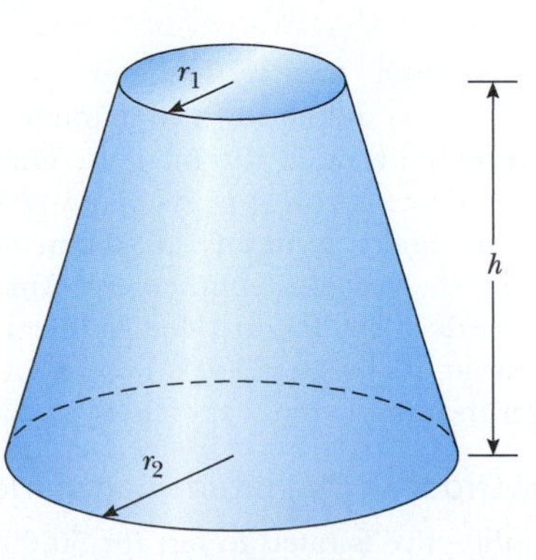


FIGURE P1.6

**7.** Which of the following equations are dimensionally correct? (a) $v_f = v_i + ax$ (b) $y = (2\text{ m})\cos(kx)$, where $k = 2\text{ m}^{-1}$.

## Section 1.3 ▪ Conversion of Units

**8.** Suppose your hair grows at the rate $\frac{1}{32}$ in. per day. Find the rate at which it grows in nanometers (nm) per second. Because the distance between atoms in a molecule is on the order of 0.1 nm, your answer suggests how rapidly layers of atoms are assembled in this protein synthesis.

**9.** Assume it takes 7.00 minutes to fill a 30.0-gal gasoline tank. (a) Calculate the rate at which the tank is filled in gallons per second. (b) Calculate the flow rate of the gasoline in cubic meters per second. (c) Determine the time interval, in hours, required to fill a 1.00-$m^3$ volume at the same rate. (1 U.S. gal = 231 $in.^3$)

**10.** A *section* of land has an area of 1 square mile and contains 640 acres. Determine the number of square meters in 1 acre.

**11.** An ore loader moves 1 200 tons/h from a mine to the surface. Convert this rate to pounds per second, using 1 ton = 2 000 lb.

**12.** At the time of this book's printing, the U.S. national debt is about $7 trillion. (a) If payments were made at the rate of $1 000 per second, how many years would it take to pay off the debt assuming that no interest were charged? (b) A one-dollar bill is about 15.5 cm long. If seven trillion one-dollar bills were laid end to end around the Earth's equator, how many times would they encircle the planet? Take the radius of the Earth at the equator to be 6 378 km. (*Note:* Before doing any of these calculations, try to guess at the answers. You may be very surprised.)

**13.** **Physics Now™** One gallon of paint with a volume of $3.78 \times 10^{-3}\text{ m}^3$ covers an area of 25.0 $m^2$. What is the thickness of the paint on the wall?

**14.** The mass of the Sun is $1.99 \times 10^{30}$ kg, and the mass of an atom of hydrogen, of which the Sun is mostly composed, is $1.67 \times 10^{-27}$ kg. How many atoms are in the Sun?

**15.** **Physics Now™** One cubic meter (1.00 $m^3$) of aluminum has a mass of $2.70 \times 10^3$ kg, and 1.00 $m^3$ of iron has a mass of $7.86 \times 10^3$ kg. Find the radius of a solid aluminum sphere that will balance a solid iron sphere of radius 2.00 cm on an equal-arm balance.

**16.** Let $\rho_{Al}$ represent the density of aluminum and $\rho_{Fe}$ that of iron. Find the radius of a solid aluminum sphere that balances a solid iron sphere of radius $r_{Fe}$ on an equal-arm balance.

**17.** A hydrogen atom has a diameter of approximately $1.06 \times 10^{-10}$ m, as defined by the diameter of the spherical electron cloud around the nucleus. The hydrogen nucleus has a diameter of approximately $2.40 \times 10^{-15}$ m. (a) For a scale model, represent the diameter of the hydrogen atom by the playing length of an American football field (100 yards = 300 ft) and determine the diameter of the nucleus in millimeters. (b) The atom is how many times larger in volume than its nucleus?

## Section 1.4 ▪ Order-of-Magnitude Calculations

**18.** An automobile tire is rated to last for 50 000 miles. To an order of magnitude, through how many revolutions will it turn? In your solution, state the quantities you measure or estimate and the values you take for them.

**19.** **Physics Now™** Estimate the number of Ping-Pong balls that would fit into a typical-size room (without being crushed). In your solution, state the quantities you measure or estimate and the values you take for them.

**20.** Compute the order of magnitude of the mass of a bathtub half full of water. Compute the order of magnitude of the mass of a bathtub half full of pennies. In your solution, list the quantities you take as data and the value you measure or estimate for each.

**21.** To an order of magnitude, how many piano tuners are in New York City? The physicist Enrico Fermi was famous for asking questions like this one on oral Ph.D. qualifying examinations. His own facility in making order-of-magnitude calculations is exemplified in Problem 30.58.

**22.** Soft drinks are commonly sold in aluminum containers. To an order of magnitude, how many such containers are thrown away or recycled each year by U.S. consumers? How many tons of aluminum does this number represent? In your solution, state the quantities you measure or estimate and the values you take for them.

## Section 1.5 ▪ Significant Figures

**23.** How many significant figures are in the following numbers: (a) $78.9 \pm 0.2$, (b) $3.788 \times 10^9$, (c) $2.46 \times 10^{-6}$, (d) 0.005 3?

**24.** Carry out the following arithmetic operations: (a) the sum of the measured values 756, 37.2, 0.83, and 2.5; (b) the product $0.003\,2 \times 356.3$; (c) the product $5.620 \times \pi$.

**25.** The *tropical year,* the time interval from vernal equinox to vernal equinox, is the basis for our calendar. It contains 365.242 199 days. Find the number of seconds in a tropical year.

*Note:* Appendix B.8 on propagation of uncertainty may be useful in solving the next two problems.

**26.** The radius of a sphere is measured to be $(6.50 \pm 0.20)$ cm, and its mass is measured to be $(1.85 \pm 0.02)$ kg. The sphere is solid. Determine its density in kilograms per cubic meter and the uncertainty in the density.

**27.** A sidewalk is to be constructed around a swimming pool that measures $(10.0 \pm 0.1)$ m by $(17.0 \pm 0.1)$ m. If the sidewalk is to measure $(1.00 \pm 0.01)$ m wide by $(9.0 \pm 0.1)$ cm thick, what volume of concrete is needed and what is the approximate uncertainty of this volume?

*Note:* The next four problems call upon mathematical skills that will be useful throughout the course.

**28.** **Review problem.** Prove that one solution of the equation

$$2.00x^4 - 3.00x^3 + 5.00x = 70.0$$

is $x = -2.22$.

**29.** **Review problem.** Find every angle $\theta$ between 0 and 360° for which $\sin\theta$ is equal to $-3.00$ multiplied by $\cos\theta$.

**30.** **Review problem.** A highway curve forms a section of a circle. A car goes around the curve. Its dashboard compass shows that the car is initially heading due east. After it travels 840 m, it is heading 35.0° south of east. Find the radius of curvature of its path.

**31. Review problem.** From the set of equations

$$p = 3q$$

$$pr = qs$$

$$\tfrac{1}{2}pr^2 + \tfrac{1}{2}qs^2 = \tfrac{1}{2}qt^2$$

involving the unknowns $p$, $q$, $r$, $s$, and $t$, find the value of $t/r$.

## Section 1.6 ■ Coordinate Systems

**32.** The polar coordinates of a point are $r = 5.50$ m and $\theta = 240°$. What are the Cartesian coordinates of this point?

**33.** A fly lands on one wall of a room. The lower left corner of the wall is selected as the origin of a two-dimensional Cartesian coordinate system. If the fly is located at the point having coordinates (2.00, 1.00) m, (a) how far is it from the corner of the room? (b) What is its location in polar coordinates?

**34.** Two points in the $xy$ plane have Cartesian coordinates (2.00, −4.00) m and (−3.00, 3.00) m. Determine (a) the distance between these points and (b) their polar coordinates.

**35.** Let the polar coordinates of the point $(x, y)$ be $(r, \theta)$. Determine the polar coordinates for the points (a) $(-x, y)$, (b) $(-2x, -2y)$, and (c) $(3x, -3y)$.

## Section 1.7 ■ Vectors and Scalars

## Section 1.8 ■ Some Properties of Vectors

**36.** A plane flies from base camp to Lake A, 280 km away in the direction 20.0° north of east. After dropping off supplies, it flies to Lake B, which is 190 km at 30.0° west of north from Lake A. Graphically determine the distance and direction from Lake B to the base camp.

**37.** **Physics Now™** A skater glides along a circular path of radius 5.00 m. Assuming he coasts around one half of the circle, find (a) the magnitude of the displacement vector and (b) how far the person skated. (c) What is the magnitude of the displacement if he skates all the way around the circle?

**38.** Each of the displacement vectors $\vec{\mathbf{A}}$ and $\vec{\mathbf{B}}$ shown in Figure P1.38 has a magnitude of 3.00 m. Find graphically (a) $\vec{\mathbf{A}} + \vec{\mathbf{B}}$, (b) $\vec{\mathbf{A}} - \vec{\mathbf{B}}$, (c) $\vec{\mathbf{B}} - \vec{\mathbf{A}}$, and (d) $\vec{\mathbf{A}} - 2\vec{\mathbf{B}}$. Report all angles counterclockwise from the positive $x$ axis.

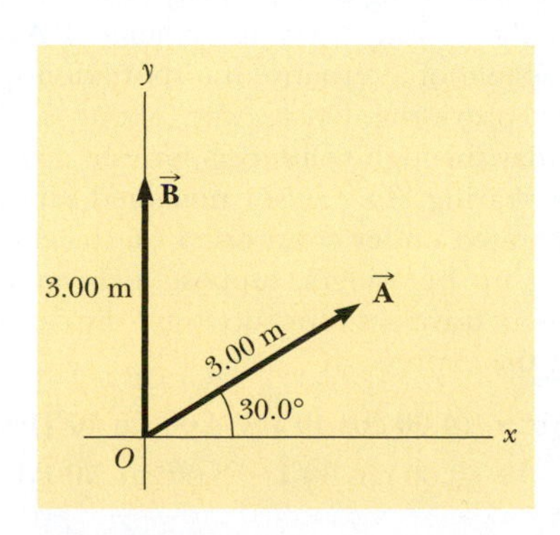


FIGURE **P1.38** Problems 1.38 and 1.48

**39.** A roller coaster car moves 200 ft horizontally and then rises 135 ft at an angle of 30.0° above the horizontal. It then travels 135 ft at an angle of 40.0° downward. What is its displacement from its starting point? Use graphical techniques.

## Section 1.9 ■ Components of a Vector and Unit Vectors

**40.** Find the horizontal and vertical components of the 100-m displacement of a superhero who flies from the top of a tall building following the path shown in Figure P1.40.

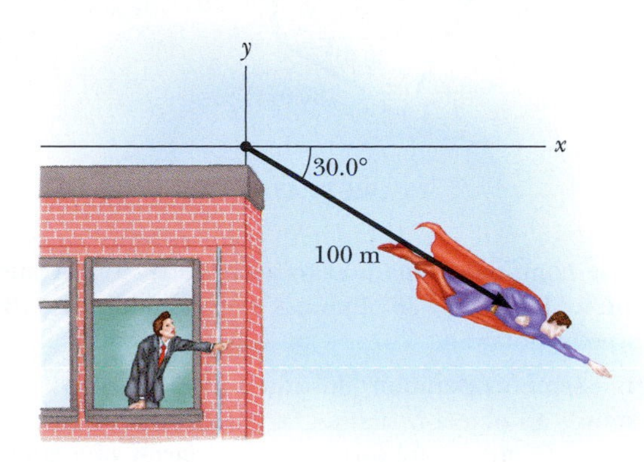


FIGURE **P1.40**

**41.** A vector has an $x$ component of −25.0 units and a $y$ component of 40.0 units. Find the magnitude and direction of this vector.

**42.** For the vectors $\vec{\mathbf{A}} = 2.00\hat{\mathbf{i}} + 6.00\hat{\mathbf{j}}$ and $\vec{\mathbf{B}} = 3.00\hat{\mathbf{i}} - 2.00\hat{\mathbf{j}}$, (a) draw the vector sum $\vec{\mathbf{C}} = \vec{\mathbf{A}} + \vec{\mathbf{B}}$ and the vector difference $\vec{\mathbf{D}} = \vec{\mathbf{A}} - \vec{\mathbf{B}}$. (b) Calculate $\vec{\mathbf{C}}$ and $\vec{\mathbf{D}}$, first in terms of unit vectors and then in terms of polar coordinates, with angles measured with respect to the positive $x$ axis.

**43.** A man pushing a mop across a floor causes it to undergo two displacements. The first has a magnitude of 150 cm and makes an angle of 120° with the positive $x$ axis. The resultant displacement has a magnitude of 140 cm and is directed at an angle of 35.0° to the positive $x$ axis. Find the magnitude and direction of the second displacement.

**44.** Vector $\vec{\mathbf{A}}$ has $x$ and $y$ components of −8.70 cm and 15.0 cm, respectively; vector $\vec{\mathbf{B}}$ has $x$ and $y$ components of 13.2 cm and −6.60 cm, respectively. If $\vec{\mathbf{A}} - \vec{\mathbf{B}} + 3\vec{\mathbf{C}} = 0$, what are the components of $\vec{\mathbf{C}}$?

**45.** Consider the two vectors $\vec{\mathbf{A}} = 3\hat{\mathbf{i}} - 2\hat{\mathbf{j}}$ and $\vec{\mathbf{B}} = -\hat{\mathbf{i}} - 4\hat{\mathbf{j}}$. Calculate (a) $\vec{\mathbf{A}} + \vec{\mathbf{B}}$, (b) $\vec{\mathbf{A}} - \vec{\mathbf{B}}$, (c) $|\vec{\mathbf{A}} + \vec{\mathbf{B}}|$, (d) $|\vec{\mathbf{A}} - \vec{\mathbf{B}}|$, and (e) the directions of $\vec{\mathbf{A}} + \vec{\mathbf{B}}$ and $\vec{\mathbf{A}} - \vec{\mathbf{B}}$.

**46.** Consider the three displacement vectors $\vec{\mathbf{A}} = (3\hat{\mathbf{i}} + 3\hat{\mathbf{j}})$ m, $\vec{\mathbf{B}} = (\hat{\mathbf{i}} - 4\hat{\mathbf{j}})$ m, and $\vec{\mathbf{C}} = (-2\hat{\mathbf{i}} + 5\hat{\mathbf{j}})$ m. Use the component method to determine (a) the magnitude and direction of the vector $\vec{\mathbf{D}} = \vec{\mathbf{A}} + \vec{\mathbf{B}} + \vec{\mathbf{C}}$ and (b) the magnitude and direction of $\vec{\mathbf{E}} = -\vec{\mathbf{A}} - \vec{\mathbf{B}} + \vec{\mathbf{C}}$.

**47.** A person going for a walk follows the path shown in Figure P1.47. The total trip consists of four straight-line paths. At the end of the walk, what is the person's resultant displacement measured from the starting point?

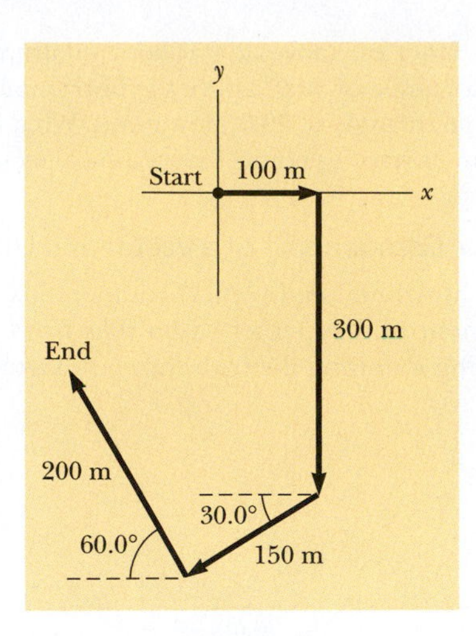


FIGURE **P1.47**

**48.** Use the component method to add the vectors $\vec{\mathbf{A}}$ and $\vec{\mathbf{B}}$ shown in Figure P1.38. Express the resultant $\vec{\mathbf{A}} + \vec{\mathbf{B}}$ in unit-vector notation.

**49.** In an assembly operation illustrated in Figure P1.49, a robot moves an object first straight upward and then also to the east, around an arc forming one quarter of a circle of radius 4.80 cm that lies in an east–west vertical plane. The robot then moves the object upward and to the north, through a quarter of a circle of radius 3.70 cm that lies in a north–south vertical plane. Find (a) the magnitude of the total displacement of the object and (b) the angle the total displacement makes with the vertical.

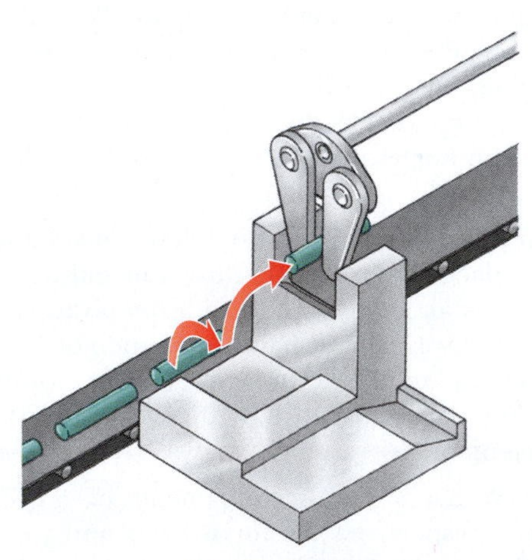

FIGURE **P1.49**

**50.** Vector $\vec{\mathbf{B}}$ has $x$, $y$, and $z$ components of 4.00, 6.00, and 3.00 units, respectively. Calculate the magnitude of $\vec{\mathbf{B}}$ and the angles that $\vec{\mathbf{B}}$ makes with the coordinate axes.

**51.** The vector $\vec{\mathbf{A}}$ has $x$, $y$, and $z$ components of 8.00, 12.0, and $-4.00$ units, respectively. (a) Write a vector expression for $\vec{\mathbf{A}}$ in unit-vector notation. (b) Obtain a unit-vector expression for a vector $\vec{\mathbf{B}}$ one fourth the length of $\vec{\mathbf{A}}$ pointing in the same direction as $\vec{\mathbf{A}}$. (c) Obtain a unit-vector expression for a vector $\vec{\mathbf{C}}$ three times the length of $\vec{\mathbf{A}}$ pointing in the direction opposite the direction of $\vec{\mathbf{A}}$.

**52.** (a) Vector $\vec{\mathbf{E}}$ has magnitude 17.0 cm and is directed 27.0° counterclockwise from the $+x$ axis. Express it in unit-vector notation. (b) Vector $\vec{\mathbf{F}}$ has magnitude 17.0 cm and is directed 27.0° counterclockwise from the $+y$ axis. Express it in unit-vector notation. (c) Vector $\vec{\mathbf{G}}$ has magnitude 17.0 cm and is directed 27.0° clockwise from the $-y$ axis. Express it in unit-vector notation.

**53.** **Physics Now™** Three displacement vectors of a croquet ball are shown in Figure P1.53, where $|\vec{\mathbf{A}}| = 20.0$ units, $|\vec{\mathbf{B}}| = 40.0$ units, and $|\vec{\mathbf{C}}| = 30.0$ units. Find (a) the resultant in unit-vector notation and (b) the magnitude and direction of the resultant displacement.

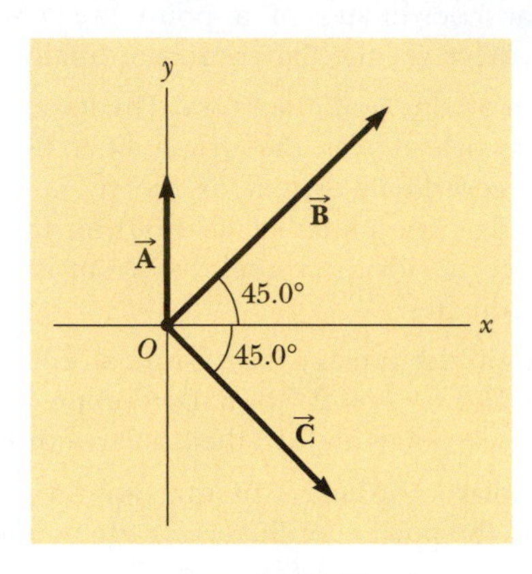


FIGURE **P1.53**

**54.** Taking $\vec{\mathbf{A}} = (6.00\hat{\mathbf{i}} - 8.00\hat{\mathbf{j}})$ units, $\vec{\mathbf{B}} = (-8.00\hat{\mathbf{i}} + 3.00\hat{\mathbf{j}})$ units, and $\vec{\mathbf{C}} = (26.0\hat{\mathbf{i}} + 19.0\hat{\mathbf{j}})$ units, determine $a$ and $b$ such that $a\vec{\mathbf{A}} + b\vec{\mathbf{B}} + \vec{\mathbf{C}} = 0$.

## Section 1.10 ■ Modeling, Alternative Representations, and Problem-Solving Strategy

**55.** A surveyor measures the distance across a straight river by the following method. Starting directly across from a tree on the opposite bank, she walks 100 m along the riverbank to establish a baseline. Then she sights across to the tree. The angle from her baseline to the tree is 35.0°. How wide is the river?

**56.** On December 1, 1955, Rosa Parks stayed seated in her bus seat when a white man demanded it. Police in Montgomery, Alabama, arrested her. On December 5, blacks began refusing to use all city buses. Under the leadership of the Montgomery Improvement Association, an efficient system of alternative transportation sprang up immediately, providing blacks with about 35 000 essential trips per day through volunteers, private taxis, carpooling, and ride sharing. The buses remained empty until they were integrated under court order on December 21, 1956. In picking up her riders, suppose a driver in downtown Montgomery traverses four successive displacements represented by the expression

$$(-6.30\hat{\mathbf{i}})\text{b} - (4.00\cos 40°\hat{\mathbf{i}} + 4.00\sin 40°\hat{\mathbf{j}})\text{b} + (3.00\cos 50°\hat{\mathbf{i}} - 3.00\sin 50°\hat{\mathbf{j}})\text{b} - (5.00\hat{\mathbf{j}})\text{b}$$

Here b represents one city block, a convenient unit of distance of uniform size; $\hat{\mathbf{i}}$ = east and $\hat{\mathbf{j}}$ = north. (a) Draw a map of the successive displacements. (b) What total distance did she travel? (c) Compute the magnitude and direction of her total displacement. The logical structure of this problem and of several problems in later chapters was suggested by Alan Van Heuvelen and David Maloney, *American Journal of Physics* 67(3) (March 1999) 252–256.

**57.** A crystalline solid consists of atoms stacked up in a repeating lattice structure. Consider a crystal as shown in Figure P1.57a. The atoms reside at the corners of cubes of side $L = 0.200$ nm. One piece of evidence for the regular arrangement of atoms comes from the flat surfaces along which a crystal separates, or cleaves, when it is broken. Suppose this crystal cleaves along a face diagonal, as shown in Figure P1.57b. Calculate the spacing $d$ between two adjacent atomic planes that separate when the crystal cleaves.

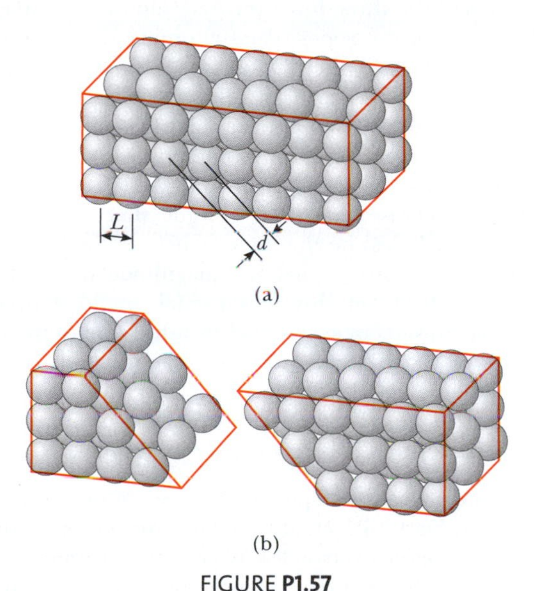


FIGURE **P1.57**

## Additional Problems

**58.** In a situation where data are known to three significant figures, we write 6.379 m = 6.38 m and 6.374 m = 6.37 m. When a number ends in 5, we arbitrarily choose to write 6.375 m = 6.38 m. We could equally well write 6.375 m = 6.37 m, "rounding down" instead of "rounding up," because we would change the number 6.375 by equal increments in both cases. Now consider an order-of-magnitude estimate. Here factors of change, rather than increments, are important. We write 500 m $\sim 10^3$ m because 500 differs from 100 by a factor of 5 whereas it differs from 1 000 by only a factor of 2. We write 437 m $\sim 10^3$ m and 305 m $\sim 10^2$ m. What distance differs from 100 m and from 1 000 m by equal factors, so that we could equally well choose to represent its order of magnitude either as $\sim 10^2$ m or as $\sim 10^3$ m?

**59.** The basic function of the carburetor of an automobile is to "atomize" the gasoline and mix it with air to promote rapid combustion. As an example, assume that 30.0 cm$^3$ of gasoline is atomized into $N$ spherical droplets, each with a radius of $2.00 \times 10^{-5}$ m. What is the total surface area of these $N$ spherical droplets?

**60.** The consumption of natural gas by a company satisfies the empirical equation $V = 1.50t + 0.008\,00t^2$, where $V$ is the volume in millions of cubic feet and $t$ the time in months. Express this equation in units of cubic feet and seconds. Assign proper units to the coefficients. Assume that a month is 30.0 days.

**61.** There are nearly $\pi \times 10^7$ s in one year. Find the percentage error in this approximation, where "percentage error" is defined as

$$\text{Percentage error} = \frac{|\text{assumed value} - \text{true value}|}{\text{true value}} \times 100\%$$

**62.** In physics, it is important to use mathematical approximations. Demonstrate that for small angles (<20°)

$$\tan \alpha \approx \sin \alpha \approx \alpha = \pi\alpha'/180°$$

where $\alpha$ is in radians and $\alpha'$ is in degrees. Use a calculator to find the largest angle for which tan $\alpha$ may be approximated by $\alpha$ with an error less than 10.0%.

**63.** A child loves to watch as you fill a transparent plastic bottle with shampoo. Every horizontal cross-section is a circle, but the diameters of the circles have different values, so the bottle is much wider in some places than others. You pour in bright green shampoo with constant volume flow rate 16.5 cm$^3$/s. At what rate is its level in the bottle rising (a) at a point where the diameter of the bottle is 6.30 cm and (b) at a point where the diameter is 1.35 cm?

**64.** One cubic centimeter of water has a mass of $1.00 \times 10^{-3}$ kg. (a) Determine the mass of 1.00 m$^3$ of water. (b) Biological substances are 98% water. Assume that they have the same density as water to estimate the masses of a cell that has a diameter of 1.00 $\mu$m, a human kidney, and a fly. Model the kidney as a sphere with a radius of 4.00 cm and the fly as a cylinder 4.00 mm long and 2.00 mm in diameter.

**65.** The distance from the Sun to the nearest star is $4 \times 10^{16}$ m. The Milky Way galaxy is roughly a disk of diameter $\sim 10^{21}$ m and thickness $\sim 10^{19}$ m. Find the order of magnitude of the number of stars in the Milky Way. Assume that the distance between the Sun and our nearest neighbor is typical.

**66.** Two vectors $\vec{\mathbf{A}}$ and $\vec{\mathbf{B}}$ have precisely equal magnitudes. For the magnitude of $\vec{\mathbf{A}} + \vec{\mathbf{B}}$ to be larger than the magnitude of $\vec{\mathbf{A}} - \vec{\mathbf{B}}$ by the factor $n$, what must be the angle between them?

**67.** The helicopter view in Figure P1.67 shows two people pulling on a stubborn mule. (a) Find the single force that is equivalent to the two forces shown. The forces are measured in units of newtons (symbolized N). (b) Find the force that a third person would have to exert on the mule to make the resultant force equal to zero.

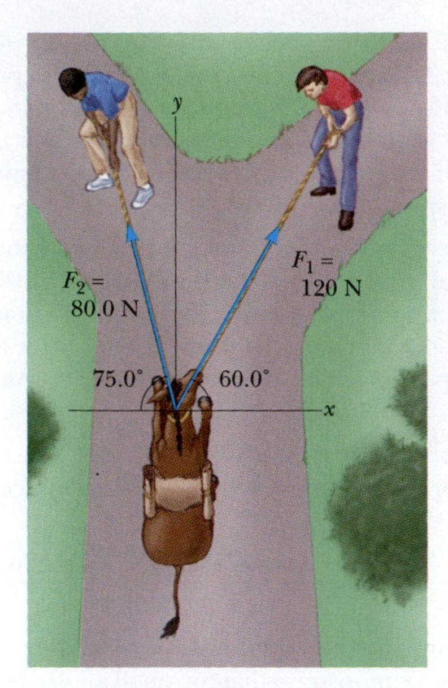


FIGURE P1.67

68. An air-traffic controller observes two aircraft on his radar screen. The first is at altitude 800 m, horizontal distance 19.2 km, and 25.0° south of west. The second aircraft is at altitude 1 100 m, horizontal distance 17.6 km, and 20.0° south of west. What is the distance between the two aircraft? (Place the $x$ axis west, the $y$ axis south, and the $z$ axis vertical.)

69. Long John Silver, a pirate, has buried his treasure on an island with five trees, located at the following points: (30.0 m, − 20.0 m), (60.0 m, 80.0 m), (− 10.0 m, − 10.0 m), (40.0 m, − 30.0 m), and (− 70.0 m, 60.0 m), all measured relative to some origin, as shown in Figure P1.69. His ship's log instructs you to start at tree A and move toward tree B, but to cover only one half of the distance between A and B. Then move toward tree C, covering one third of the distance between your current location and C. Next move toward D, covering one fourth of the distance between where you are and D. Finally, move toward E, covering one fifth of the distance between you and E, stop, and dig. (a) Assume that you have correctly determined the order in which the pirate labeled the trees as A, B, C, D, and E, as shown in the figure. What are the coordinates of the point where his treasure is buried? (b) What if you do not really know the way the pirate labeled the trees? Rearrange the order of the trees [for instance, B(30 m, − 20 m), A(60 m, 80 m), E(− 10 m, − 10 m), C(40 m, − 30 m), and D(− 70 m, 60 m)] and repeat the calculation to show that the answer does not depend on the order in which the trees are labeled.

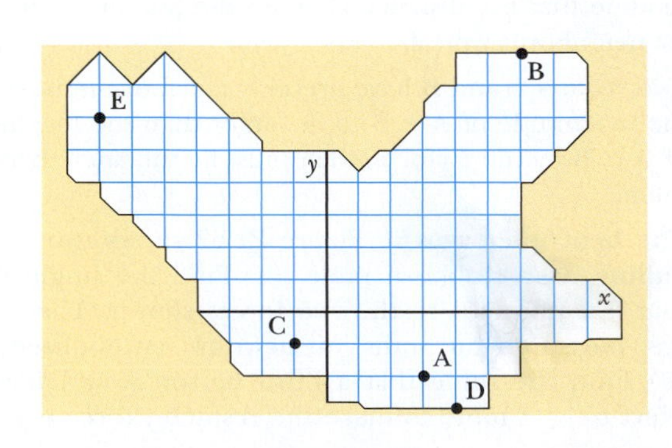


FIGURE P1.69

70. Consider a game in which $N$ children position themselves at equal distances around the circumference of a circle. At the center of the circle is a rubber tire. Each child holds a rope attached to the tire and, at a signal, pulls on his or her rope. All children exert forces of the same magnitude $F$. In the case $N = 2$, it is easy to see that the net force on the tire will be zero because the two oppositely directed force vectors add to zero. Similarly, if $N = 4$, 6, or any even integer, the resultant force on the tire must be zero because the forces exerted by each pair of oppositely positioned children will cancel. When an odd number of children are around the circle, it is not as obvious whether the total force on the central tire will be zero. (a) Calculate the net force on the tire in the case $N = 3$ by adding the components of the three force vectors. Choose the $x$ axis to lie along one of the ropes. (b) Determine the net force for the general case where $N$ is any integer, odd or even, greater than 1. Proceed as follows: Assume that the total force is not zero. Then it must point in some particular direction. Let every child move one position clockwise. Give a reason that the total force must then have a direction turned clockwise by $360°/N$. Argue that the total force must nevertheless be the same as before. Explain that the contradiction proves that the magnitude of the force is zero. This problem illustrates a widely useful technique of proving a result "by symmetry," by using a bit of the mathematics of *group theory*. The particular situation is actually encountered in physics and chemistry when an array of electric charges (ions) exerts electric forces on an atom at a central position in a molecule or in a crystal.

71. A rectangular parallelepiped has dimensions $a$, $b$, and $c$, as shown in Figure P1.71. (a) Obtain a vector expression for the face diagonal vector $\vec{\mathbf{R}}_1$. What is the magnitude of this vector? (b) Obtain a vector expression for the body diagonal vector $\vec{\mathbf{R}}_2$. Note that $\vec{\mathbf{R}}_1$, $c\hat{\mathbf{k}}$, and $\vec{\mathbf{R}}_2$ make a right triangle and prove that the magnitude of $\vec{\mathbf{R}}_2$ is $\sqrt{a^2 + b^2 + c^2}$.

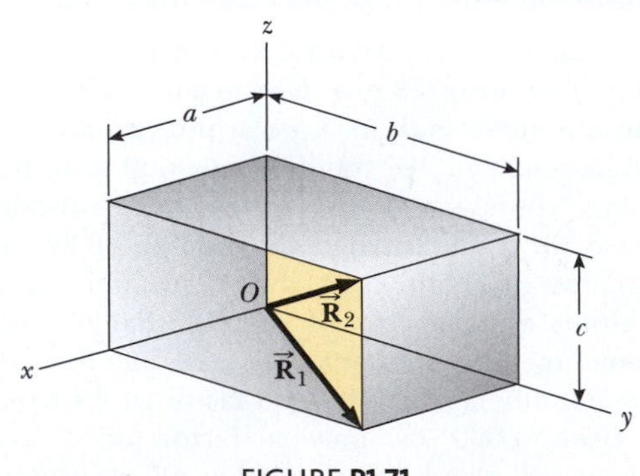


FIGURE P1.71

72. Vectors $\vec{\mathbf{A}}$ and $\vec{\mathbf{B}}$ have equal magnitudes of 5.00. The sum of $\vec{\mathbf{A}}$ and $\vec{\mathbf{B}}$ is the vector $6.00\hat{\mathbf{j}}$. Determine the angle between $\vec{\mathbf{A}}$ and $\vec{\mathbf{B}}$.

## ANSWERS TO QUICK QUIZZES

**1.1** False. Dimensional analysis gives the units of the proportionality constant but provides no information about its numerical value. Determining its numerical value requires either experimental data or mathematical reasoning. For example, in the generation of the equation $x = \frac{1}{2}at^2$, because the factor $\frac{1}{2}$ is dimensionless, there is no way of determining it using dimensional analysis.

**1.2** (b). Because kilometers are shorter than miles, it takes a larger number of kilometers than miles to represent a given distance.

**1.3** Scalars: (a), (d), (e). None of these quantities has a direction. Vectors: (b), (c). For these quantities, the direction is important to completely specify the quantity.

**1.4** (c). The resultant has its maximum magnitude $A + B = 12 + 8 = 20$ units when vector $\vec{\mathbf{A}}$ is oriented in the same direction as vector $\vec{\mathbf{B}}$. The resultant vector has its minimum magnitude $A - B = 12 - 8 = 4$ units when vector $\vec{\mathbf{A}}$ is oriented in the direction opposite vector $\vec{\mathbf{B}}$.

**1.5** (a). The resultant has magnitude $A + B$ when $\vec{\mathbf{A}}$ is oriented in the same direction as $\vec{\mathbf{B}}$.

**1.6** (b). From the Pythagorean theorem, the magnitude of a vector is always larger than the absolute value of each component, unless there is only one nonzero component, in which case the magnitude of the vector is equal to the absolute value of that component.

**1.7** (b). From the Pythagorean theorem, we see that the magnitude of a vector is nonzero if at least one component is nonzero.

**1.8** (d). Each set of components, for example, the two $x$ components $A_x$ and $B_x$, must add to zero, so the components must be of opposite sign.

**THE WIZARD OF ID** By Parker and Hart

# CONTEXT 1

## Alternative-Fuel Vehicles

The idea of self-propelled vehicles has been part of the human imagination for centuries. Leonardo da Vinci drew plans for a vehicle powered by a wound spring in 1478. This vehicle was never built although models have been constructed from his plans and appear in museums. Isaac Newton developed a vehicle in 1680 that operated by ejecting steam out the back, similar to a rocket engine. This invention did not develop into a useful device. Despite these and other attempts, self-propelled vehicles did not succeed; that is, they did not begin to replace the horse as a primary means of transportation until the 19th century.

The history of *successful* self-propelled vehicles begins in 1769 with the invention of a military tractor by Nicolas Joseph Cugnot in France. This vehicle, as well as Cugnot's follow-up vehicles, was powered by a steam engine. During the remainder of the 18th century and for most of the 19th century, additional steam-driven vehicles were developed in France, Great Britain, and the United States.

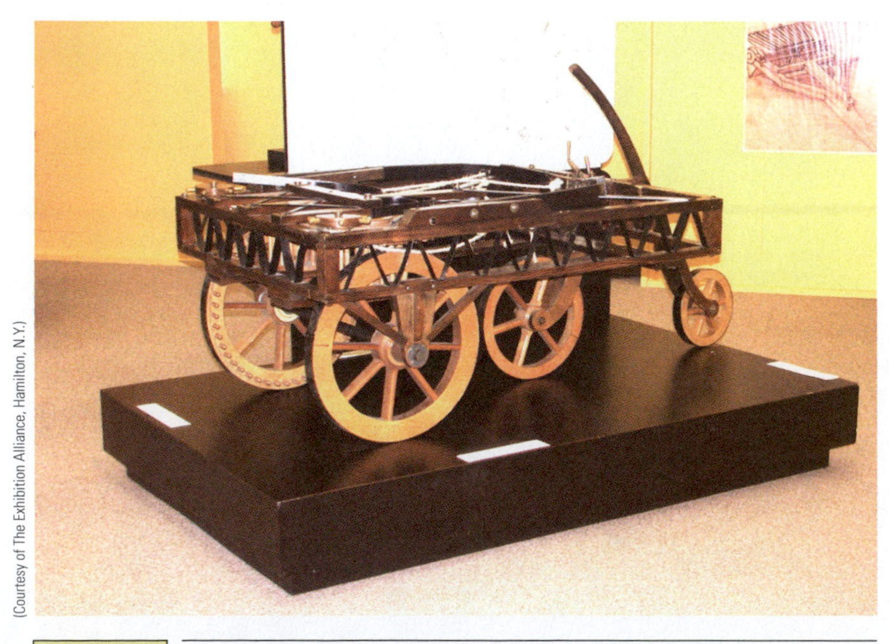
(Courtesy of The Exhibition Alliance, Hamilton, N.Y.)

FIGURE 1 A model of a spring-drive car designed by Leonardo da Vinci.

After the invention of the electric battery by Italian Alessandro Volta at the beginning of the 19th century and its further development over three decades came the invention of early electric vehicles in the 1830s. The development in 1859 of the storage battery, which could be recharged, provided significant impetus to the development of electric vehicles. By the early 20th century, electric cars with a range of about 20 miles and a top speed of 15 miles per hour had been developed.

An internal combustion engine was designed but never built by Dutch physicist Christiaan Huygens in 1680. The invention of modern gasoline-powered internal combustion vehicles is generally credited to Gottlieb Daimler in 1885 and Karl Benz in 1886. Several earlier vehicles, dating back to 1807, however, used internal combustion engines operating on various fuels, including coal gas and primitive gasoline.

At the beginning of the 20th century, steam-powered, gasoline-powered, and electric cars shared the roadways in the United States. Electric cars did not possess the vibration, smell, and noise of gasoline-powered cars and did not suffer from the long start-up time intervals, up to 45 minutes, of steam-powered cars on cold mornings. Electric cars were especially preferred by women, who did not enjoy the difficult task of cranking a gasoline-powered car to start the engine. The limited range of electric cars was not a significant problem because the only roads that existed were in highly populated areas and cars were primarily used for short trips in town.

The end of electric cars in the early 20th century began with the following developments:

(© Bettmann/CORBIS)

FIGURE 2 This magazine advertisement for an electric car is typical of this popular type of car in the early 20th century.

- 1901: A major discovery of crude oil in Texas reduced prices of gasoline to widely affordable levels.
- 1912: The electric starter for gasoline engines was invented, removing the physical task of cranking the engine.
- During the 1910s: Henry Ford successfully introduced mass production of internal combustion vehicles, resulting in a drop in the price of these vehicles to significantly less than that of an electric car.
- By the early 1920s: Roadways in the United States were of much better quality than previously and connected cities, requiring vehicles with a longer range than that of electric cars.

Because of these factors, the roadways were ruled by gasoline-powered cars almost exclusively by the 1920s. Gasoline, however, is a finite and short-lived commodity. We are approaching the end of our ability to use gasoline in transportation; some experts predict that diminishing supplies of crude oil will push the cost of gasoline to prohibitively high levels within two more decades. Furthermore, gasoline and diesel fuel result in serious tailpipe emissions that are harmful to the environment. As we look for a replacement for gasoline, we also want to pursue fuels that will be kinder to the atmosphere. Such fuels will help reduce the effects of global warming, which we will study in Context 5.

What do the steam engine, the electric motor, and the internal combustion engine have in common? That is, what do they each extract from a source, be it a type of fuel or an electric battery? The answer to this question is *energy*. Regardless of the type of automobile, some source of energy must be

(© Martin Bond/Photo Researchers, Inc.)

FIGURE 3 Development of new energy sources requires modifications in the infrastructure to deliver the energy. In this photograph, a bus powered by natural gas is refueled in Bristol, Great Britain.

(© Mehau Kulyk/Photo Researchers, Inc.)

FIGURE 4 Modern electric cars can take advantage of an infrastructure set up in some localities to provide charging stations in parking lots.

provided. Energy is one of the physical concepts that we will investigate in this Context. A fuel such as gasoline contains energy due to its chemical composition and its ability to undergo a combustion process. The battery in an electric car also contains energy, again related to chemical composition, but in this case it is associated with an ability to produce an electric current.

One difficult social aspect of developing a new energy source for automobiles is that there must be a synchronized development of the new automobile along with the infrastructure for delivering the new source of energy. This aspect requires close cooperation between automotive corporations and energy manufacturers and suppliers. For example, electric cars cannot be used to travel long distances unless an infrastructure of charging stations develops in parallel with the development of electric cars.

As we draw near to the time when we run out of gasoline, our central question in this first Context is an important one for our future development:

**What source besides gasoline can be used to provide energy for an automobile while reducing environmentally damaging emissions?**

# CHAPTER 2

# Motion in One Dimension

(Jean Y. Rusznieswki/Getty Images)

One of the physical quantities we will study in this chapter is the velocity of an object moving in a straight line. Downhill skiers can reach velocities with a magnitude greater than 100 km/h.

## CHAPTER OUTLINE


2.1 Average Velocity
2.2 Instantaneous Velocity
2.3 Analysis Models—The Particle Under Constant Velocity
2.4 Acceleration
2.5 Motion Diagrams
2.6 The Particle Under Constant Acceleration
2.7 Freely Falling Objects
2.8 Context Connection—Acceleration Required by Consumers
SUMMARY


To begin our study of motion, it is important to be able to *describe* motion using the concepts of space and time without regard to the causes of the motion. This portion of mechanics is called *kinematics*. In this chapter, we shall consider motion along a straight line, that is, one-dimensional motion. Chapter 3 extends our discussion to two-dimensional motion.

From everyday experience we recognize that motion represents continuous change in the position of an object. For example, if you are driving from your home to a destination, your position on the Earth's surface is changing.

The movement of an object through space (translation) may be accompanied by the rotation or vibration of the object. Such motions can be quite complex. It is often possible to simplify matters, however, by temporarily ignoring rotation and internal motions of the moving object. The result is the simplification model that we call the particle model, discussed in Chapter 1. In many situations, an object can be treated as a particle if the only motion being considered is translation through space. We will use the particle model extensively throughout this book.

## 2.1 AVERAGE VELOCITY

We begin our study of kinematics with the notion of average velocity. You may be familiar with a similar notion, average speed, from experiences with driving. If you drive your car 100 miles according to your odometer and it takes 2.0 hours to do so, your average speed is (100 mi)/(2.0 h) = 50 mi/h. For a particle moving through a distance $d$ in a time interval $\Delta t$, the **average speed** $v_{avg}$ is mathematically defined as

■ Definition of average speed

$$v_{avg} \equiv \frac{d}{\Delta t} \quad [2.1]$$

Speed is not a vector, so there is no direction associated with average speed.

Average velocity may be a little less familiar to you due to its vector nature. Let us start by imagining the motion of a particle, which, through the particle model, can represent the motion of many types of objects. We shall restrict our study at this point to one-dimensional motion along the $x$ axis.

The motion of a particle is completely specified if the position of the particle in space is known at all times. Consider a car moving back and forth along the $x$ axis and imagine that we take data on the position of the car every 10 s. Active Figure 2.1a is a *pictorial* representation of this one-dimensional motion that shows the positions of the car at 10-s intervals. The six data points we have recorded are represented by the letters Ⓐ through Ⓕ. Table 2.1 is a *tabular* representation of the motion. It lists the data as entries for position at each time. The black dots in Active Figure 2.1b show a *graphical* representation of the motion. Such a plot is often called a **position–time graph.** The curved line in Active Figure 2.1b cannot be unambiguously drawn through our six data points because we have no information about what happened between these points. The curved line is, however, a *possible* graphical representation of the position of the car at all instants of time during the 50 s.

**PITFALL PREVENTION 2.1**

**AVERAGE SPEED AND AVERAGE VELOCITY** The magnitude of the average velocity is *not* the average speed. Consider a particle moving from the origin to $x = 10$ m and then back to the origin in a time interval of 4.0 s. The magnitude of the average velocity is zero because the particle ends the time interval at the same position at which it started; the displacement is zero. The average speed, however, is the total distance divided by the time interval: 20 m/4.0 s = 5.0 m/s.

If a particle is moving during a time interval $\Delta t = t_f - t_i$, the displacement of the particle is described as $\Delta \vec{\mathbf{x}} = \vec{\mathbf{x}}_f - \vec{\mathbf{x}}_i = (x_f - x_i)\hat{\mathbf{i}}$. (Recall that displacement is defined as the change in the position of the particle, which is equal to its final position value minus its initial position value.) Because we are considering only one-dimensional motion in this chapter, we shall drop the vector notation at this point and pick it up again in Chapter 3. The direction of a vector in this chapter will be indicated by means of a positive or negative sign.

The **average velocity** $v_{x,\text{avg}}$ of the particle is defined as the ratio of its displacement $\Delta x$ to the time interval $\Delta t$ during which the displacement takes place:

■ Definition of average velocity

$$v_{x,\text{avg}} \equiv \frac{\Delta x}{\Delta t} = \frac{x_f - x_i}{t_f - t_i} \quad [2.2]$$

where the subscript $x$ indicates motion along the $x$ axis. From this definition we see that average velocity has the dimensions of length divided by time: meters per second in SI units and feet per second in U.S. customary units. The average velocity is *independent* of the path taken between the initial and final points. This independence is a major difference from the average speed discussed at the beginning of this section. The average velocity is independent of path because it is proportional

**ACTIVE FIGURE 2.1**

(a) A pictorial representation of the motion of a car. The positions of the car at six instants of time are shown and labeled. (b) A graphical representation, known as a position–time graph, of the car's motion in part (a). The average velocity $v_{x,\text{avg}}$ in the interval $t = 0$ to $t = 10$ s is obtained from the slope of the straight line connecting points Ⓐ and Ⓑ. (c) A velocity–time graph of the motion of the car in part (a).

**Physics Now™** Log into PhysicsNow at **www.pop4e.com** and go to Active Figure 2.1. You can move each of the six points Ⓐ through Ⓕ and observe the car's motion in both a pictorial and a graphical representation as the car follows a smooth path through the six points.

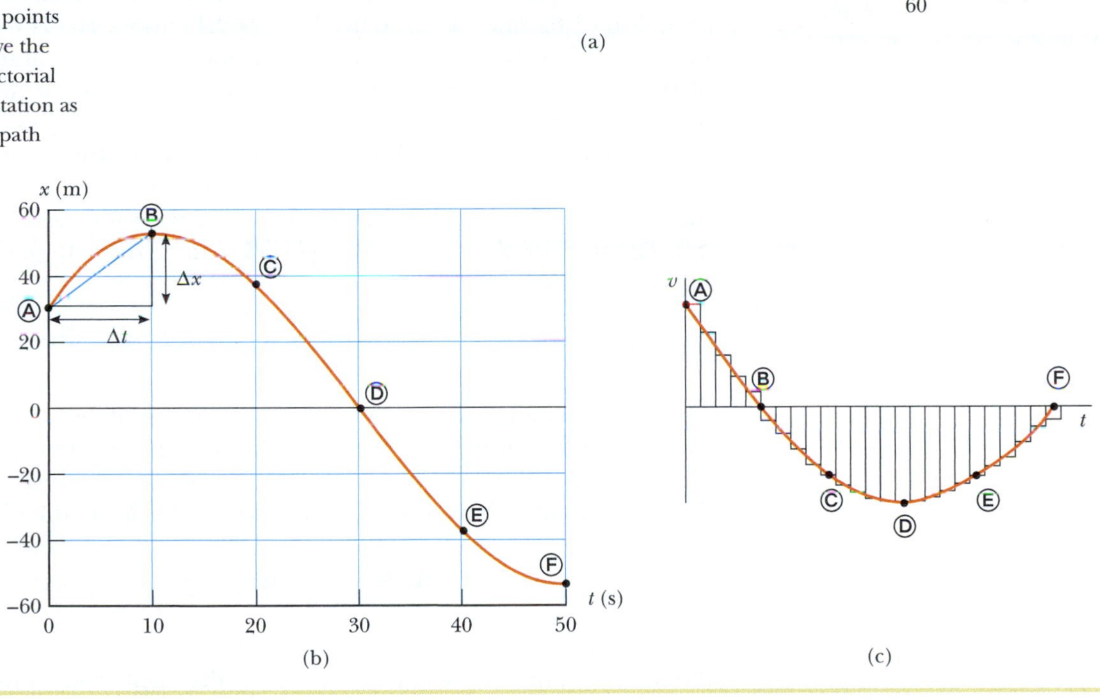


to the displacement $\Delta x$, which depends only on the initial and final coordinates of the particle. Average speed (a scalar) is found by dividing the *distance* traveled by the time interval, whereas average velocity (a vector) is the *displacement* divided by the time interval. Therefore, average velocity gives us no details of the motion; rather, it only gives us the result of the motion. Finally, note that the average velocity in one dimension can be positive or negative, depending on the sign of the displacement. (The time interval $\Delta t$ is always positive.) If the $x$ coordinate of the particle increases during the time interval (i.e., if $x_f > x_i$), $\Delta x$ is positive and $v_{x,\text{avg}}$ is positive, which corresponds to an average velocity in the positive $x$ direction. On the other hand, if the coordinate decreases over time ($x_f < x_i$), $\Delta x$ is negative; hence, $v_{x,\text{avg}}$ is negative, which corresponds to an average velocity in the negative $x$ direction.

**TABLE 2.1 Positions of the Car at Various Times**

| Position | $t$(s) | $x$ (m) |
|---|---|---|
| Ⓐ | 0 | 30 |
| Ⓑ | 10 | 52 |
| Ⓒ | 20 | 38 |
| Ⓓ | 30 | 0 |
| Ⓔ | 40 | −37 |
| Ⓕ | 50 | −53 |

**QUICK QUIZ 2.1** Under which of the following conditions is the magnitude of the average velocity of a particle moving in one dimension smaller than the average speed over some time interval? **(a)** A particle moves in the $+x$ direction without reversing. **(b)** A particle moves in the $-x$ direction without reversing. **(c)** A particle moves in the $+x$ direction and then reverses the direction of its velocity. **(d)** There are no conditions for which it is true.

**PITFALL PREVENTION 2.2**

**Slopes of Graphs** The word *slope* is often used in reference to the graphs of physical data. Regardless of what data are plotted, the word *slope* will represent the ratio of the change in the quantity represented on the vertical axis to the change in the quantity represented on the horizontal axis. Remember that *a slope has units* (unless both axes have the same units). Therefore, the units of the slope in Active Figure 2.1b are m/s, the units of velocity.

The average velocity can also be interpreted geometrically, as seen in the graphical representation in Active Figure 2.1b. A straight line can be drawn between any two points on the curve. Active Figure 2.1b shows such a line drawn between points Ⓐ and Ⓑ. Using a geometric model, this line forms the hypotenuse of a right triangle of height $\Delta x$ and base $\Delta t$. The slope of the hypotenuse is the ratio $\Delta x/\Delta t$. Therefore, we see that **the average velocity of the particle during the time interval $t_i$ to $t_f$ is equal to the slope of the straight line joining the initial and final points on the position–time graph.** For example, the average velocity of the car between points Ⓐ and Ⓑ is $v_{x,\text{avg}} = (52\text{ m} - 30\text{ m})/(10\text{ s} - 0) = 2.2\text{ m/s}$.

We can also identify a geometric interpretation for the total displacement during the time interval. Active Figure 2.1c shows the velocity–time graphical representation of the motion in Active Figures 2.1a and 2.1b. The total time interval for the motion has been divided into small increments of duration $\Delta t_n$. During each of these increments, if we model the velocity as constant during the short increment, the displacement of the particle is given by $\Delta x_n = v_n \Delta t_n$.

Geometrically, the product on the right side of this expression represents the area of a thin rectangle associated with each time increment in Active Figure 2.1c; the height of the rectangle (measured from the time axis) is $v_n$, and the width is $\Delta t_n$. The total displacement of the particle will be the sum of the displacements during each of the increments:

$$\Delta x \approx \sum_n \Delta x_n = \sum_n v_n \Delta t_n$$

This sum is an approximation because we have modeled the velocity as constant in each increment, which is not the case. The term on the right represents the total area of all the thin rectangles. Now let us take the limit of this expression as the time increments shrink to zero, in which case the approximation becomes exact:

$$\Delta x = \lim_{\Delta t_n \to 0} \sum_n \Delta x_n = \lim_{\Delta t_n \to 0} \sum_n v_n \Delta t_n$$

In this limit, the sum of the areas of all the very thin rectangles becomes equal to the total area under the curve. Therefore, **the displacement of a particle during the time interval $t_i$ to $t_f$ is equal to the area under the curve between the initial and final points on the velocity–time graph.** We will make use of this geometric interpretation in Section 2.6.

## EXAMPLE 2.1 Calculate the Average Velocity

A particle moving along the $x$ axis is located at $x_i = 12$ m at $t_i = 1$ s and at $x_f = 4$ m at $t_f = 3$ s. Find its displacement and average velocity during this time interval.

**Solution** First, establish the mental representation. Imagine the particle moving along the axis. Based on the information in the problem, which way is it moving? You may find it useful to draw a pictorial representation, but for this simple example, we will go straight to the mathematical representation. The displacement is

$$\Delta x = x_f - x_i = 4\text{ m} - 12\text{ m} = -8\text{ m}$$

The average velocity is, according to Equation 2.2,

$$v_{x,\text{avg}} = \frac{\Delta x}{\Delta t} = \frac{4\text{ m} - 12\text{ m}}{3\text{ s} - 1\text{ s}} = -4\text{ m/s}$$

Because the displacement is negative for this time interval, we conclude that the particle has moved to the left, toward decreasing values of $x$. Is this conclusion consistent with your mental representation? Keep in mind that it may not have *always* been moving to the left. We only have information about its location at two points in time. After $t_i = 1$ s, it could have moved to the right, turned around, and ended up farther to the left than its original position by the time $t_f = 3$ s. To be completely confident that we know the motion of the particle, we would need to have information about its location at *every* instant of time.

### EXAMPLE 2.2 Motion of a Jogger

A jogger runs in a straight line, with a magnitude of average velocity of 5.00 m/s for 4.00 min and then with a magnitude of average velocity of 4.00 m/s for 3.00 min.

**A** What is the magnitude of the final displacement from her initial position?

**Solution** That this problem involves a jogger is not important; we model the jogger as a particle. We have data for two separate portions of the motion, so we use these data to find the displacement for each portion, using Equation 2.2:

$$v_{x,\,\text{avg}} = \frac{\Delta x}{\Delta t} \quad \rightarrow \quad \Delta x = v_{x,\,\text{avg}}\,\Delta t$$

$$\Delta x_{\text{portion 1}} = (5.00\text{ m/s})(4.00\text{ min})\left(\frac{60\text{ s}}{1\text{ min}}\right) = 1.20 \times 10^3\text{ m}$$

$$\Delta x_{\text{portion 2}} = (4.00\text{ m/s})(3.00\text{ min})\left(\frac{60\text{ s}}{1\text{ min}}\right) = 7.20 \times 10^2\text{ m}$$

We add these two displacements to find the total displacement of $1.92 \times 10^3$ m.

**B** What is the magnitude of her average velocity during this entire time interval of 7.00 min?

**Solution** We now have the data we need to find the average velocity for the entire time interval using Equation 2.2:

$$v_{x,\,\text{avg}} = \frac{\Delta x}{\Delta t} = \frac{1.92 \times 10^3\text{ m}}{7.00\text{ min}}\left(\frac{1\text{ min}}{60\text{ s}}\right) = 4.57\text{ m/s}$$

Notice that the average velocity is *not* calculated as the simple arithmetic mean of the two velocities given in the problem.

## 2.2 INSTANTANEOUS VELOCITY

Suppose you drive your car through a displacement of magnitude 40 miles and it takes exactly 1 hour to do so, from 1:00:00 P.M. to 2:00:00 P.M. Then the magnitude of your average velocity is 40 mi/h for the 1-h interval. How fast, though, were you going at the particular *instant* of time 1:20:00 P.M.? It is likely that your velocity varied during the trip, owing to hills, traffic lights, slow drivers ahead of you, and the like, so that there was not a single velocity maintained during the entire hour of travel. The velocity of a particle at any instant of time is called the **instantaneous velocity.**

Consider again the motion of the car shown in Active Figure 2.1a. Active Figure 2.2a is the graphical representation again, with two blue lines representing average velocities over very different time intervals. One blue line represents the average velocity we calculated earlier over the interval from Ⓐ to Ⓑ. The second blue line represents the average velocity over the much longer interval Ⓐ to Ⓕ. How well does either of these represent the instantaneous velocity at point Ⓐ? In Active Figure 2.1a, the car begins to move to the right, which we identify as a positive velocity. The average velocity from Ⓐ to Ⓕ is *negative* (because the slope of the line from Ⓐ to Ⓕ is negative), so this velocity clearly is not an accurate representation of the instantaneous velocity at Ⓐ. The average velocity from interval Ⓐ to Ⓑ is *positive,* so this velocity at least has the right sign.

In Active Figure 2.2b, we show the result of drawing the lines representing the average velocity of the car as point Ⓑ is brought closer and closer to point Ⓐ. As

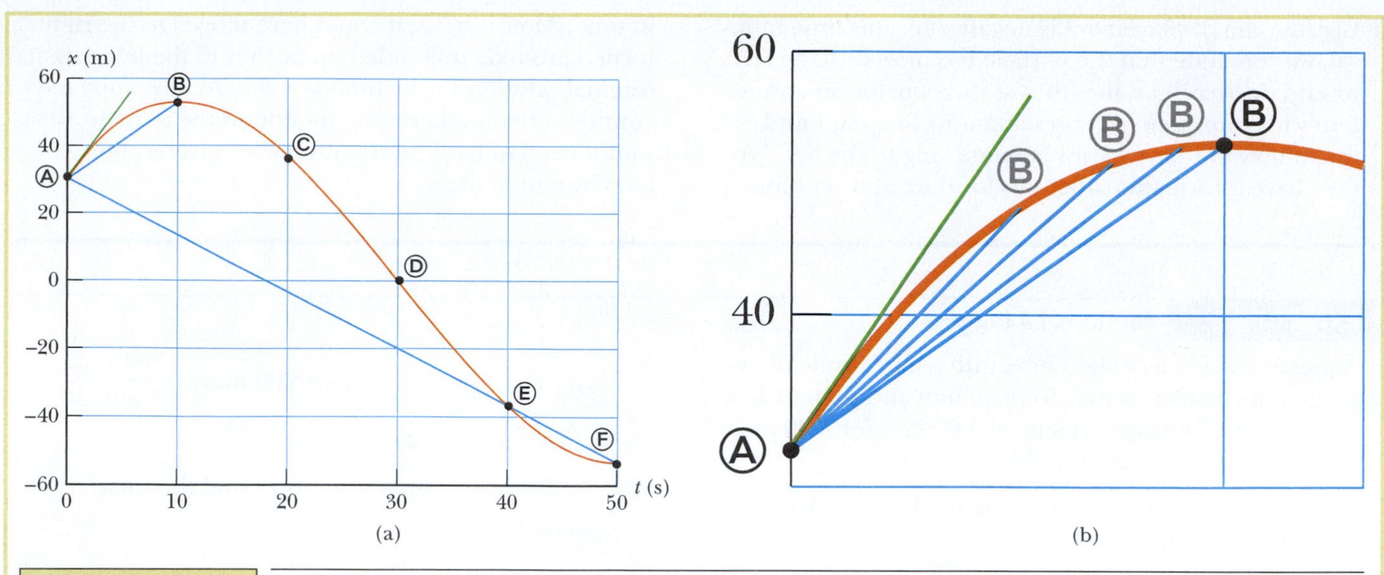


ACTIVE FIGURE 2.2 (a) Position–time graph for the motion of the car in Active Figure 2.1. (b) An enlargement of the upper left-hand corner of the graph in part (a) shows how the blue line between positions Ⓐ and Ⓑ approaches the green tangent line as point Ⓑ is moved closer to point Ⓐ.

**Physics Now™** Log into PhysicsNow at **www.pop4e.com** and go to Active Figure 2.2. You can move point Ⓑ as suggested in part (b) and observe the blue line approaching the green tangent line.

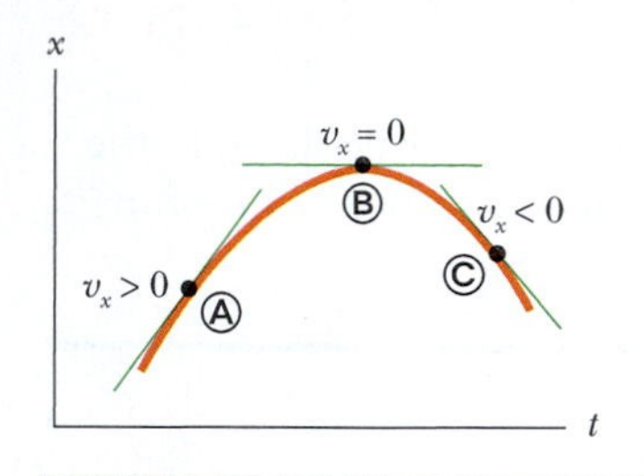


FIGURE 2.3 In the position–time graph shown, the velocity is positive at Ⓐ, where the slope of the tangent line is positive; the velocity is zero at Ⓑ, where the slope of the tangent line is zero; and the velocity is negative at Ⓒ, where the slope of the tangent line is negative.

■ Definition of instantaneous velocity

that occurs, the slope of the blue line approaches that of the green line, which is the line drawn tangent to the curve at point Ⓐ. As Ⓑ approaches Ⓐ, the time interval that includes point Ⓐ becomes infinitesimally small. Therefore, the average velocity over this interval as the interval shrinks to zero can be interpreted as the instantaneous velocity at point Ⓐ. Furthermore, the slope of the line tangent to the curve at Ⓐ is the instantaneous velocity at the time $t_Ⓐ$. In other words, **the instantaneous velocity $v_x$ equals the limiting value of the ratio $\Delta x/\Delta t$ as $\Delta t$ approaches zero:**[1]

$$v_x \equiv \lim_{\Delta t \to 0} \frac{\Delta x}{\Delta t}$$

In calculus notation, this limit is called the *derivative* of $x$ with respect to $t$, written $dx/dt$:

$$v_x \equiv \lim_{\Delta t \to 0} \frac{\Delta x}{\Delta t} = \frac{dx}{dt} \qquad [2.3]$$

The instantaneous velocity can be positive, negative, or zero. When the slope of the position–time graph is positive, such as at point Ⓐ in Figure 2.3, $v_x$ is positive. At point Ⓒ, $v_x$ is negative because the slope is negative. Finally, the instantaneous velocity is zero at the peak Ⓑ (the turning point), where the slope is zero. From here on, we shall usually use the word *velocity* to designate instantaneous velocity.

The **instantaneous speed** of a particle is defined as the magnitude of the instantaneous velocity vector. Hence, by definition, *speed* can never be negative.

[1] Note that the displacement $\Delta x$ also approaches zero as $\Delta t$ approaches zero. As $\Delta x$ and $\Delta t$ become smaller and smaller, however, the ratio $\Delta x/\Delta t$ approaches a value equal to the *true* slope of the line tangent to the $x$ versus $t$ curve.

**QUICK QUIZ 2.2** Are members of the highway patrol more interested in **(a)** your average speed or **(b)** your instantaneous speed as you drive?

If you are familiar with calculus, you should recognize that specific rules exist for taking the derivatives of functions. These rules, which are listed in Appendix B.6, enable us to evaluate derivatives quickly.

Suppose $x$ is proportional to some power of $t$, such as

$$x = At^n$$

where $A$ and $n$ are constants. (This equation is a very common functional form.) The derivative of $x$ with respect to $t$ is

$$\frac{dx}{dt} = nAt^{n-1}$$

For example, if $x = 5t^3$, we see that $dx/dt = 3(5)t^{3-1} = 15t^2$.

## Thinking Physics 2.1

Consider the following motions of an object in one dimension. **(a)** A ball is thrown directly upward, rises to its highest point, and falls back into the thrower's hand. **(b)** A race car starts from rest and speeds up to 100 m/s along a straight line. **(c)** A spacecraft on the way to another star drifts through empty space at constant velocity. Are there any instants of time in the motion of these objects at which the instantaneous velocity at the instant and the average velocity over the entire interval are the same? If so, identify the point(s).

**Reasoning** **(a)** The average velocity over the entire interval for the thrown ball is zero; the ball returns to the starting point at the end of the time interval. There is one point—at the top of the motion—at which the instantaneous velocity is zero. **(b)** The average velocity for the motion of the race car cannot be evaluated unambiguously with the information given, but its magnitude must be some value between 0 and 100 m/s. Because the magnitude of the instantaneous velocity of the car will have every value between 0 and 100 m/s at some time during the interval, there must be some instant at which the instantaneous velocity is equal to the average velocity over the entire interval. **(c)** Because the instantaneous velocity of the spacecraft is constant, its instantaneous velocity at *any* time and its average velocity over *any* time interval are the same. ■

### EXAMPLE 2.3 The Limiting Process

The position of a particle moving along the $x$ axis varies in time according to the expression[2] $x = 3t^2$, where $x$ is in meters and $t$ is in seconds. Find the velocity in terms of $t$ at any time.

**Solution** The position–time graphical representation for this motion is shown in Figure 2.4. We can compute the velocity at any time $t$ by using the definition of the instantaneous velocity. If the initial coordinate of the particle at time $t$ is $x_i = 3t^2$, the coordinate at a later time $t + \Delta t$ is

$$x_f = 3(t + \Delta t)^2 = 3[t^2 + 2t\,\Delta t + (\Delta t)^2]$$
$$= 3t^2 + 6t\,\Delta t + 3(\Delta t)^2$$

Therefore, the displacement in the time interval $\Delta t$ is

$$\Delta x = x_f - x_i = (3t^2 + 6t\,\Delta t + 3(\Delta t)^2) - (3t^2)$$
$$= 6t\,\Delta t + 3(\Delta t)^2$$

[2] Simply to make it easier to read, we write the equation as $x = 3t^2$ rather than as $x = (3.00\ \text{m/s}^2)t^{2.00}$. When an equation summarizes measurements, consider its coefficients to have as many significant digits as other data quoted in a problem. Also consider its coefficients to have the units required for dimensional consistency. When we start our clocks at $t = 0$, we usually do not mean to limit precision to a single digit. Consider any zero value in this book to have as many significant figures as you need.

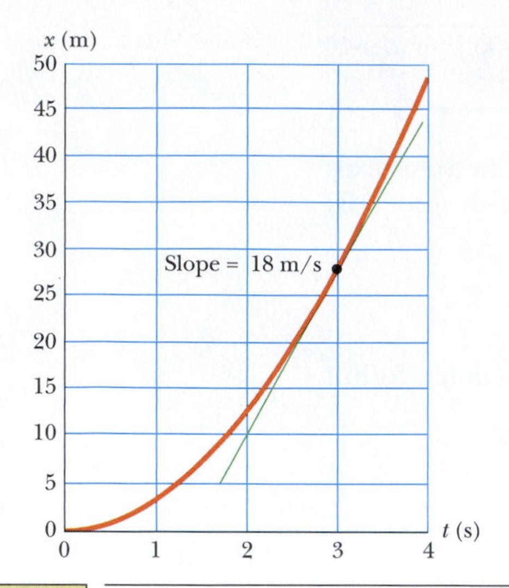


FIGURE 2.4 (Example 2.3) Position–time graph for a particle having an $x$ coordinate that varies in time according to $x = 3t^2$. Note that the instantaneous velocity at $t = 3.0$ s is obtained from the slope of the green line tangent to the curve at this point.

The average velocity in this time interval is

$$v_{x,\,\text{avg}} = \frac{\Delta x}{\Delta t} = \frac{6t\,\Delta t + 3(\Delta t)^2}{\Delta t} = 6t + 3\,\Delta t$$

To find the instantaneous velocity, we take the limit of this expression as $\Delta t$ approaches zero. In doing so, we see that the term $3\,\Delta t$ goes to zero; therefore,

$$v_x = \lim_{\Delta t \to 0} \frac{\Delta x}{\Delta t} = 6t$$

Notice that this expression gives us the velocity at *any* general time $t$. It tells us that $v_x$ is increasing linearly in time. It is then a straightforward matter to find the velocity at some specific time from the expression $v_x = 6t$ by substituting the value of the time. For example, at $t = 3.0$ s, the velocity is $v_x = 6(3) = 18$ m/s. Again, this answer can be checked from the slope at $t = 3.0$ s (the green line in Fig. 2.4).

We can also find $v_x$ by taking the first derivative of $x$ with respect to time, as in Equation 2.3. In this example, $x = 3t^2$, and we see that $v_x = dx/dt = 6t$, in agreement with our result from taking the limit explicitly.

## EXAMPLE 2.4 Average and Instantaneous Velocity

A particle moves along the $x$ axis. Its $x$ coordinate varies with time according to the expression $x = -4t + 2t^2$, where $x$ is in meters and $t$ is in seconds. The position–time graph for this motion is shown in Figure 2.5.

**A** Determine the displacement of the particle in the time intervals $t = 0$ to $t = 1$ s and $t = 1$ s to $t = 3$ s.

**Solution** This problem provides a graphical representation of the motion in Figure 2.5. In your mental representation, note that the particle moves in the negative $x$ direction for the first second of motion, stops instantaneously at $t = 1$ s, and then heads back in the positive $x$ direction for $t > 1$ s. Remember that it is a one-dimensional problem, so the curve in Figure 2.5 does not represent the path the particle follows through space; be sure not to confuse a graphical representation with a pictorial representation of the motion in space (see Active Fig. 2.1 for a comparison). In your mental representation, you should imagine the particle moving to the left and then to the right, with all the motion taking place along a single line.

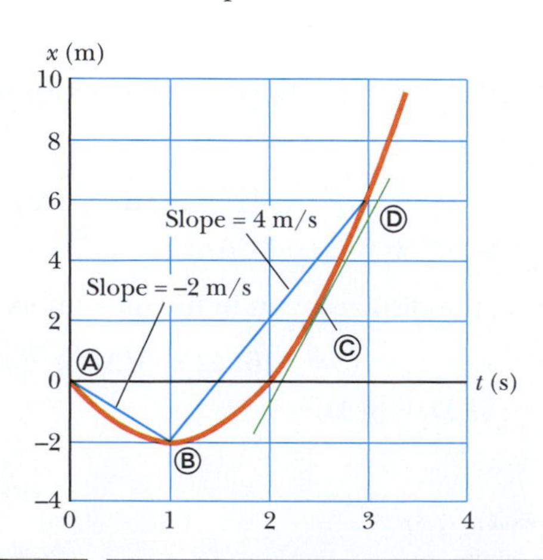


FIGURE 2.5 (Example 2.4) Position–time graph for a particle having an $x$ coordinate that varies in time according to $x = -4t + 2t^2$.

In the first time interval (Ⓐ to Ⓑ), we set $t_i = 0$ and $t_f = 1$ s. Because $x = -4t + 2t^2$, the displacement during the first time interval is

$$\Delta x_{AB} = x_f - x_i = -4(1) + 2(1)^2 - [-4(0) + 2(0)^2] = -2\text{ m}$$

Likewise, in the second time interval (Ⓑ to Ⓓ), we can set $t_i = 1$ s and $t_f = 3$ s. Therefore, the displacement in this interval is

$$\Delta x_{BD} = x_f - x_i = -4(3) + 2(3)^2 - [-4(1) + 2(1)^2] = 8\text{ m}$$

These displacements can also be read directly from the position–time graph (see Fig. 2.5).

**B** Calculate the average velocity in the time intervals $t = 0$ to $t = 1$ s and $t = 1$ s to $t = 3$ s.

**Solution** In the first time interval, $\Delta t = t_f - t_i = 1$ s. Therefore, using Equation 2.2 and the result from part A gives

$$v_{x,\text{avg}} = \frac{\Delta x_{AB}}{\Delta t} = \frac{-2\text{ m}}{1\text{ s}} = -2\text{ m/s}$$

Likewise, in the second time interval, $\Delta t = 2$ s; therefore,

$$v_{x,\text{avg}} = \frac{\Delta x_{BD}}{\Delta t} = \frac{8\text{ m}}{2\text{ s}} = 4\text{ m/s}$$

These values agree with the slopes of the lines joining these points in Figure 2.5.

C Find the instantaneous velocity of the particle at $t = 2.5$ s (point ©).

**Solution** We can find the instantaneous velocity at any time $t$ by taking the first derivative of $x$ with respect to $t$:

$$v_x = \frac{dx}{dt} = \frac{d}{dt}(-4t + 2t^2) = -4 + 4t$$

Therefore, at $t = 2.5$ s, we find that

$$v_x = -4 + 4(2.5) = 6\text{ m/s}$$

We can also obtain this result by measuring the slope of the position–time graph at $t = 2.5$ s. Do you see any symmetry in the motion? For example, are there points at which the speed is the same? Is the velocity the same at these points?

## 2.3 ANALYSIS MODELS—THE PARTICLE UNDER CONSTANT VELOCITY

As mentioned in Section 1.10, the third category of models used in this book is that of **analysis models.** Such models help us analyze the situation in a physics problem and guide us toward the solution. **An analysis model is a problem we have solved before. It is a description of either (1) the behavior of some physical entity or (2) the interaction between that entity and the environment.** When you encounter a new problem, you should identify the fundamental details of the problem and attempt to recognize which, if any, of the types of problems you have already solved might be used as a model for the new problem. For example, suppose an automobile is moving along a straight freeway at a constant speed. Is it important that it is an automobile? Is it important that it is a freeway? If the answers to both questions are "no," we model the situation as a *particle under constant velocity,* which we will discuss in this section.

This method is somewhat similar to the common practice in the legal profession of finding "legal precedents." If a previously resolved case can be found that is very similar legally to the present one, it is offered as a model and an argument is made in court to link them logically. The finding in the previous case can then be used to sway the finding in the present case. We will do something similar in physics. For a given problem, we search for a "physics precedent," a model with which we are already familiar and that can be applied to the present problem.

We shall generate analysis models based on four fundamental simplification models. The first simplification model is the particle model discussed in Chapter 1. We will look at a particle under various behaviors and environmental interactions. Further analysis models are introduced in later chapters based on simplification models of a *system,* a *rigid object,* and a *wave.* Once we have introduced these analysis models, we shall see that they appear over and over again later in the book in different situations.

Let us use Equation 2.2 to build our first analysis model for solving problems. We imagine a particle moving with a constant velocity. The particle under constant velocity model can be applied in *any* situation in which an entity that can be modeled as a particle is moving with constant velocity. This situation occurs frequently, so it is an important model.

If the velocity of a particle is constant, its instantaneous velocity at any instant during a time interval is the same as the average velocity over the interval, $v_x = v_{x,\text{avg}}$. Therefore, we start with Equation 2.2 to generate an equation to be

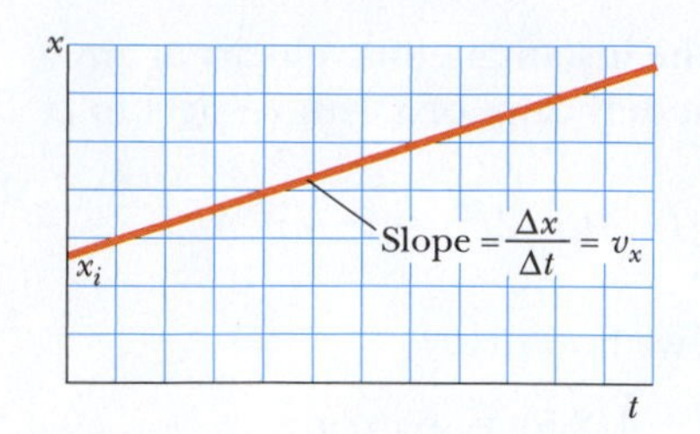


**FIGURE 2.6** Position–time graph for a particle under constant velocity. The value of the constant velocity is the slope of the line.

used in the mathematical representation of this situation:

$$v_x = v_{x,\text{avg}} = \frac{\Delta x}{\Delta t} \qquad [2.4]$$

Remembering that $\Delta x = x_f - x_i$, we see that $v_x = (x_f - x_i)/\Delta t$, or

$$x_f = x_i + v_x \Delta t$$

This equation tells us that the position of the particle is given by the sum of its original position $x_i$ plus the displacement $v_x \Delta t$ that occurs during the time interval $\Delta t$. In practice, we usually choose the time at the beginning of the interval to be $t_i = 0$ and the time at the end of the interval to be $t_f = t$, so our equation becomes

■ Position of a particle under constant velocity

$$x_f = x_i + v_x t \qquad \text{(for constant } v_x\text{)} \qquad [2.5]$$

Equations 2.4 and 2.5 are the primary equations used in the model of a particle under constant velocity. They can be applied to particles or objects that can be modeled as particles.

Figure 2.6 is a graphical representation of the particle under constant velocity. On the position–time graph, the slope of the line representing the motion is constant and equal to the velocity. It is consistent with the mathematical representation, Equation 2.5, which is the equation of a straight line. The slope of the straight line is $v_x$ and the $y$ intercept is $x_i$ in both representations.

**EXAMPLE 2.5 Modeling a Runner as a Particle**

A scientist is studying the biomechanics of the human body. She determines the velocity of an experimental subject while he runs at a constant rate. The scientist starts the stopwatch at the moment the runner passes a given point and stops it at the moment the runner passes another point 20 m away. The time interval indicated on the stopwatch is 4.4 s.

**A** What is the runner's velocity?

**Solution** We model the runner as a particle, as we did in Example 2.2, because the size of the runner and the movement of arms and legs are unnecessary details. This choice, in combination with the velocity being constant, allows us to use Equation 2.4 to find the velocity:

$$v_x = \frac{\Delta x}{\Delta t} = \frac{x_f - x_i}{\Delta t} = \frac{20\text{ m} - 0}{4.4\text{ s}} = 4.5\text{ m/s}$$

**B** What is the position of the runner after 10 s has passed?

**Solution** In this part of the problem, we use Equation 2.5 to find the position of the particle at the time $t = 10$ s. Using the velocity found in part A,

$$x_f = x_i + v_x t = 0 + (4.5\text{ m/s})(10\text{ s}) = 45\text{ m}$$

The mathematical manipulations for the particle under constant velocity stem from Equation 2.4 and its descendent, Equation 2.5. These equations can be used to solve for any variable in the equations that happens to be unknown if the other variables are known. For example, in part B of Example 2.5, we find the position when the velocity and the time are known. Similarly, if we know the velocity and the final position, we could use Equation 2.5 to find the time at which the runner is at this position. We shall present more examples of a particle under constant velocity in Chapter 3.

A particle under constant velocity moves with a constant speed along a straight line. Now consider a particle moving with a constant speed along a curved path. It can be represented with the *particle under constant speed* model. The primary equation for this model is Equation 2.1, with the average speed $v_{\text{avg}}$ replaced by the constant speed $v$. As an example, imagine a particle moving at a constant speed in a circular path. If the speed is 5.00 m/s and the radius of the path is 10.0 m, we can calculate the time interval required to complete one trip around the circle:

$$v = \frac{d}{\Delta t} \quad \rightarrow \quad \Delta t = \frac{d}{v} = \frac{2\pi r}{v} = \frac{2\pi(10.0\text{ m})}{5.00\text{ m/s}} = 12.6\text{ s}$$

## 2.4 ACCELERATION

When the velocity of a particle changes with time, the particle is said to be *accelerating.* For example, the speed of a car increases when you "step on the gas," the car slows down when you apply the brakes, and it changes direction when you turn the wheel; these changes are all accelerations. We will need a precise definition of acceleration for our studies of motion.

Suppose a particle moving along the $x$ axis has a velocity $v_{xi}$ at time $t_i$ and a velocity $v_{xf}$ at time $t_f$. The **average acceleration** $a_{x,\text{avg}}$ of the particle in the time interval $\Delta t = t_f - t_i$ is defined as the ratio $\Delta v_x/\Delta t$, where $\Delta v_x = v_{xf} - v_{xi}$ is the *change* in velocity of the particle in this time interval:

$$a_{x,\text{avg}} \equiv \frac{v_{xf} - v_{xi}}{t_f - t_i} = \frac{\Delta v_x}{\Delta t} \qquad [2.6]$$

■ Definition of average acceleration

Therefore, **acceleration is a measure of how rapidly the velocity is changing.** Acceleration is a vector quantity having dimensions of length divided by $(\text{time})^2$, or $L/T^2$. Some of the common units of acceleration are meters per second per second ($m/s^2$) and feet per second per second ($ft/s^2$). For example, an acceleration of 2 $m/s^2$ means that the velocity changes by 2 m/s during each second of time that passes.

In some situations, the value of the average acceleration may be different for different time intervals. It is therefore useful to define the **instantaneous acceleration** as the limit of the average acceleration as $\Delta t$ approaches zero, analogous to the definition of instantaneous velocity discussed in Section 2.2:

$$a_x \equiv \lim_{\Delta t \to 0} \frac{\Delta v_x}{\Delta t} = \frac{dv_x}{dt} \qquad [2.7]$$

■ Definition of instantaneous acceleration

That is, the instantaneous acceleration equals the derivative of the velocity with respect to time, which by definition is the slope of the velocity–time graph. Note that if $a_x$ is positive, the acceleration is in the positive $x$ direction, whereas negative $a_x$ implies acceleration in the negative $x$ direction. A negative acceleration does not necessarily mean that the particle is *moving* in the negative $x$ direction, a point we shall address in more detail shortly. From now on, we use the term *acceleration* to mean instantaneous acceleration.

Because $v_x = dx/dt$, the acceleration can also be written

$$a_x = \frac{dv_x}{dt} = \frac{d}{dt}\left(\frac{dx}{dt}\right) = \frac{d^2x}{dt^2} \qquad [2.8]$$

This equation shows that the acceleration equals the *second derivative* of the position with respect to time.

Figure 2.7 shows how the acceleration–time curve in a graphical representation can be derived from the velocity–time curve. In these diagrams, the acceleration of a particle at any time is simply the slope of the velocity–time graph at that time. Positive values of the acceleration correspond to those points (between $t_A$ and $t_B$)

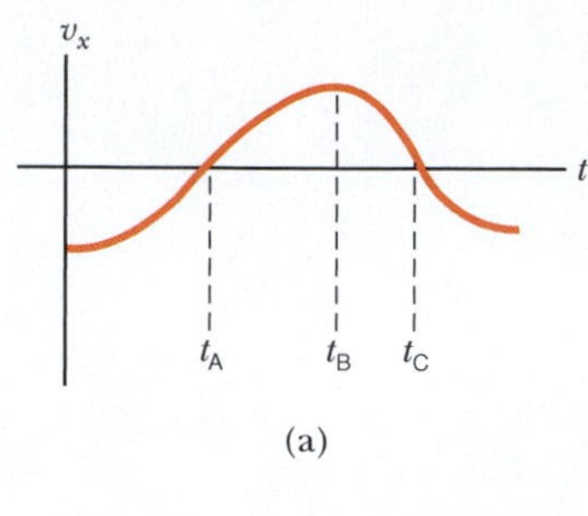


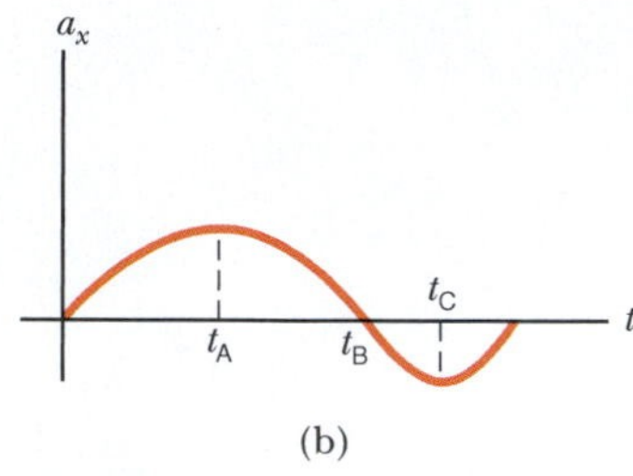


FIGURE 2.7 The instantaneous acceleration can be obtained from the velocity–time graph (a). At each instant the acceleration in the $a_x$ versus $t$ graph (b) equals the slope of the line tangent to the $v_x$ versus $t$ curve.

ACTIVE FIGURE 2.8 (Quick Quiz 2.3) Parts (a), (b), and (c) are velocity–time graphs of objects in one-dimensional motion. The possible acceleration–time graphs of each object are shown in scrambled order in parts (d), (e), and (f).

**Physics Now™** Log into PhysicsNow at **www.pop4e.com** and go to Active Figure 2.8 to practice matching appropriate velocity versus time graphs and acceleration versus time graphs.

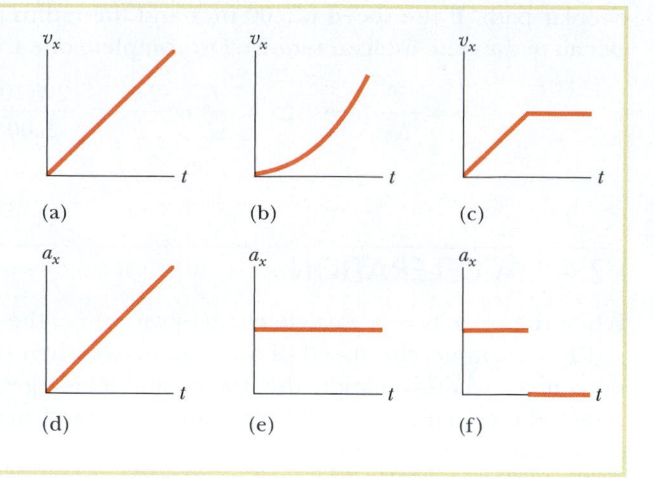


where the velocity in the positive $x$ direction is increasing in magnitude (the particle is speeding up) or those points (between $t = 0$ and $t_A$) where the velocity in the negative $x$ direction is decreasing in magnitude (the particle is slowing down). The acceleration reaches a maximum at time $t_A$, when the slope of the velocity–time graph is a maximum. The acceleration then goes to zero at time $t_B$, when the velocity is a maximum (i.e., when the velocity is momentarily not changing and the slope of the $v$ versus $t$ graph is zero). Finally, the acceleration is negative when the velocity in the positive $x$ direction is decreasing in magnitude (between $t_B$ and $t_C$) or when the velocity in the negative direction is increasing in magnitude (after $t_C$).

**PITFALL PREVENTION 2.3**

**Negative acceleration** Keep in mind that *negative acceleration does not necessarily mean that an object is slowing down.* If the acceleration is negative and the velocity is negative, the object is speeding up!

**QUICK QUIZ 2.3** Using Active Figure 2.8, match each of the velocity–time graphical representations on the top with the acceleration–time graphical representation on the bottom that best describes the motion.

**PITFALL PREVENTION 2.4**

**Deceleration** The word *deceleration* has a common popular connotation as *slowing down.* When combined with the misconception in Pitfall Prevention 2.3 that negative acceleration means slowing down, the situation can be further confused by the use of the word *deceleration.* We will not use this word in this text.

As an example of the computation of acceleration, consider the pictorial representation of a car's motion in Figure 2.9. In this case, the velocity of the car has changed from an initial value of 30 m/s to a final value of 15 m/s in a time interval of 2.0 s. The average acceleration during this time interval is

$$a_{x,\text{avg}} = \frac{15 \text{ m/s} - 30 \text{ m/s}}{2.0 \text{ s}} = -7.5 \text{ m/s}^2$$

The negative sign in this example indicates that the acceleration vector is in the negative $x$ direction (to the left in Figure 2.9). For the case of motion in a straight line, the direction of the velocity of an object and the direction of its acceleration are related as follows. **When the object's velocity and acceleration are in the same direction, the object is speeding up in that direction. On the other hand, when the object's velocity and acceleration are in opposite directions, the speed of the object decreases in time.**

To help with this discussion of the signs of velocity and acceleration, let us take a peek ahead to Chapter 4, where we shall relate the acceleration of an object to the *force* on the object. We will save the details until that later discussion, but for now, let us borrow the notion that **force is proportional to acceleration:**

$$\vec{\mathbf{F}} \propto \vec{\mathbf{a}}$$

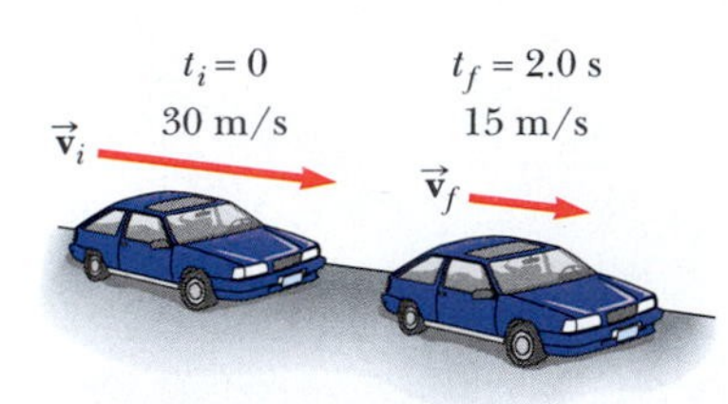


FIGURE 2.9 The velocity of the car decreases from 30 m/s to 15 m/s in a time interval of 2.0 s.

This proportionality indicates that acceleration is caused by force. What's more, as indicated by the vector notation in the proportionality, force and acceleration are in the same direction. Therefore, let us think about the signs of velocity and acceleration by forming a mental representation in which a force is applied to the object to cause the acceleration. Again consider the case in which the velocity and acceleration are in the same direction. This situation is equivalent to an object moving in a given direction and experiencing a force that pulls on it in the same direction. It is clear in this case that the object speeds up! If the velocity and acceleration are in opposite directions, the object moves one way and a force pulls in the opposite direction. In this case, the object slows down! It is very useful to equate the direction of the acceleration in these situations to the direction of a force because it is easier from our everyday experience to think about what effect a force will have on an object than to think only in terms of the direction of the acceleration.

**QUICK QUIZ 2.4** If a car is traveling eastward and slowing down, what is the direction of the force on the car that causes it to slow down? **(a)** eastward **(b)** westward **(c)** neither of these directions

## EXAMPLE 2.6 Average and Instantaneous Acceleration

The velocity of a particle moving along the $x$ axis varies in time according to the expression $v_x = 40 - 5t^2$, where $t$ is in seconds.

**A** Find the average acceleration in the time interval $t = 0$ to $t = 2.0$ s.

**Solution** Build your mental representation from the mathematical expression given for the velocity. For example, which way is the particle moving at $t = 0$? How does the velocity change in the first few seconds? Does it move faster or slower? The velocity–time graphical representation for this function is given in Figure 2.10. The velocities at $t_i = t_A = 0$ and $t_f = t_B = 2.0$ s are found by substituting these values of $t$ into the expression given for the velocity:

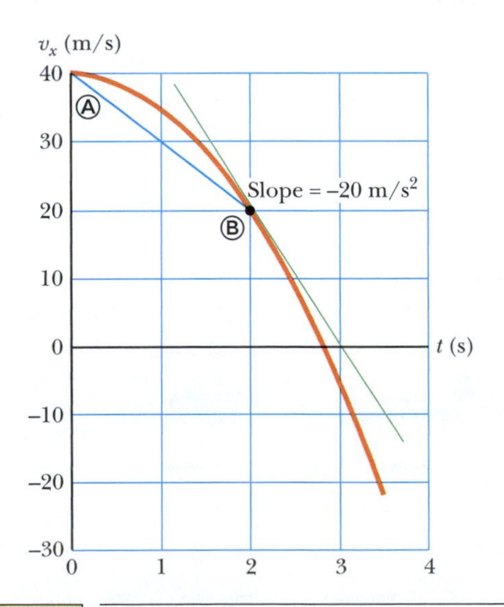


**FIGURE 2.10** (Example 2.6) The velocity–time graph for a particle moving along the $x$ axis according to the relation $v_x = 40 - 5t^2$. The acceleration at $t = 2.0$ s is obtained from the slope of the green tangent line at that time.

$$v_{xA} = 40 - 5t_A{}^2 = 40 - 5(0)^2 = 40 \text{ m/s}$$

$$v_{xB} = 40 - 5t_B{}^2 = 40 - 5(2.0)^2 = 20 \text{ m/s}$$

Therefore, the average acceleration in the specified time interval $\Delta t = t_B - t_A$ is

$$a_{x,\text{avg}} = \frac{20 \text{ m/s} - 40 \text{ m/s}}{2.0 \text{ s}} = -10 \text{ m/s}^2$$

The negative sign is consistent with the negative slope of the line joining the initial and final points on the velocity–time graph.

**B** Determine the acceleration at $t = 2.0$ s.

**Solution** Because this question refers to a specific instant of time, it is asking for an instantaneous acceleration. The velocity at time $t$ is $v_{xi} = 40 - 5t^2$, and the velocity at time $t + \Delta t$ is

$$v_{xf} = 40 - 5(t + \Delta t)^2 = 40 - 5t^2 - 10t\,\Delta t - 5(\Delta t)^2$$

Therefore, the change in velocity over the time interval $\Delta t$ is

$$\Delta v_x = v_{xf} - v_{xi} = -10t\,\Delta t - 5(\Delta t)^2$$

Dividing this expression by $\Delta t$ and taking the limit of the result as $\Delta t$ approaches zero gives the acceleration at *any* time $t$:

$$a_x = \lim_{\Delta t \to 0} \frac{\Delta v_x}{\Delta t} = \lim_{\Delta t \to 0} (-10t - 5\Delta t) = -10t$$

Therefore, at $t = 2.0$ s we find that

$$a_x = (-10)(2.0) \text{ m/s}^2 = -20 \text{ m/s}^2$$

This result can also be obtained by measuring the slope of the velocity–time graph at $t = 2.0$ s (see Fig. 2.10) or by taking the derivative of the velocity expression.

## 2.5 MOTION DIAGRAMS

The concepts of velocity and acceleration are often confused with each other, but in fact they are quite different quantities. It is instructive to make use of the specialized pictorial representation called a **motion diagram** to describe the velocity and acceleration vectors while an object is in motion.

A *stroboscopic photograph* of a moving object shows several images of the object taken as the strobe light flashes at a constant rate. Active Figure 2.11 represents three sets of strobe photographs of cars moving along a straight roadway in a single direction, from left to right. The time intervals between flashes of the stroboscope are equal in each part of the diagram. To distinguish between the two vector quantities, we use red for velocity vectors and violet for acceleration vectors in Active Figure 2.11. The vectors are sketched at several instants during the motion of the object. Let us describe the motion of the car in each diagram.

In Active Figure 2.11a, the images of the car are equally spaced, and the car moves through the same displacement in each time interval. Therefore, the car moves with *constant positive velocity* and has *zero acceleration.* We could model the car as a particle and describe it as a particle under constant velocity.

In Active Figure 2.11b, the images of the car become farther apart as time progresses. In this case, the velocity vector increases in time because the car's displacement between adjacent positions increases as time progresses. Therefore, the car is moving with a *positive velocity* and a *positive acceleration.* The velocity and acceleration are in the same direction. In terms of our earlier force discussion, imagine a force pulling on the car in the same direction it is moving: it speeds up.

In Active Figure 2.11c, we interpret the car as slowing down as it moves to the right because its displacement between adjacent positions decreases as time progresses. In this case, the car moves initially to the right with a *positive velocity* and a *negative acceleration.* The velocity vector decreases in time and eventually reaches zero. (This type of motion is exhibited by a car that skids to a stop after its brakes are applied.) From this diagram we see that the acceleration and velocity vectors are *not* in the same direction. The velocity and acceleration are in opposite directions. In terms of our earlier force discussion, imagine a force pulling on the car opposite to the direction it is moving: it slows down.

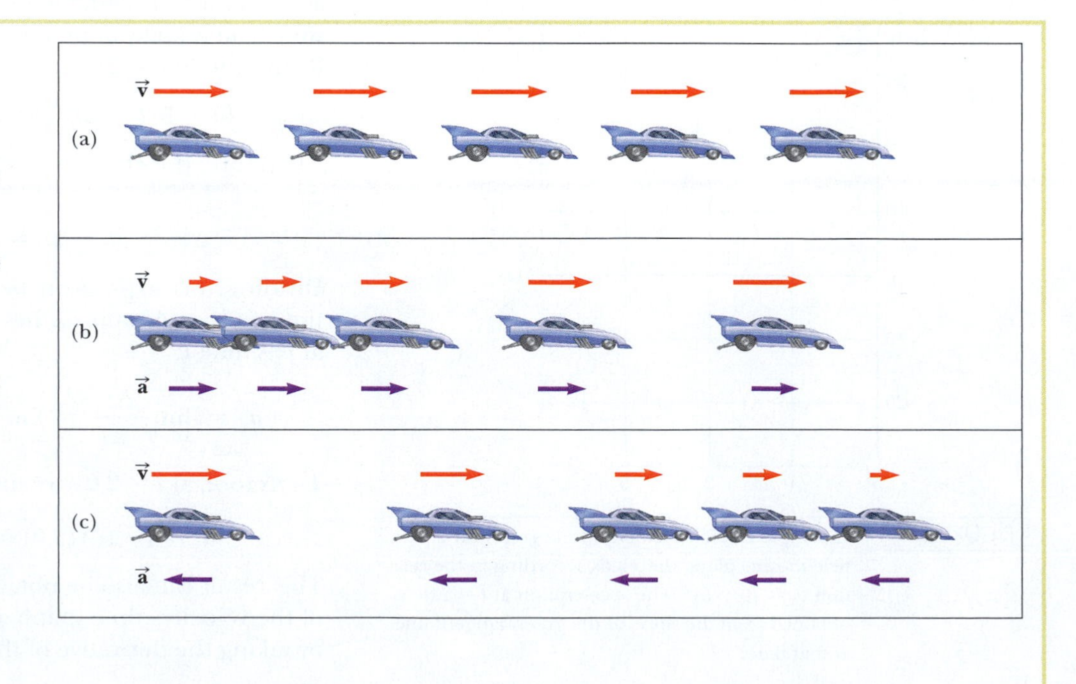


ACTIVE FIGURE 2.11
(a) Motion diagram for a car moving at constant velocity.
(b) Motion diagram for a car whose constant acceleration is in the direction of its velocity. The velocity vector at each instant is indicated by a red arrow, and the constant acceleration vector is indicated by a violet arrow.
(c) Motion diagram for a car whose constant acceleration is in the direction *opposite* the velocity at each instant.

**PhysicsNow**™ Log into PhysicsNow at **www.pop4e.com** and go to Active Figure 2.11 to select the constant acceleration and initial velocity of the car and observe pictorial and graphical representations of its motion.

The violet acceleration vectors in Active Figures 2.11b and 2.11c are all the same length. Therefore, these diagrams represent a motion with constant acceleration. This important type of motion is discussed in the next section.

**QUICK QUIZ 2.5** Which of the following is true? **(a)** If a car is traveling eastward, its acceleration is eastward. **(b)** If a car is slowing down, its acceleration must be negative. **(c)** A particle with constant acceleration can never stop and stay stopped.

## 2.6 THE PARTICLE UNDER CONSTANT ACCELERATION

If the acceleration of a particle varies in time, the motion may be complex and difficult to analyze. A very common and simple type of one-dimensional motion occurs when the acceleration is constant, such as for the motion of the cars in Active Figures 2.11b and 2.11c. In this case, the average acceleration over any time interval equals the instantaneous acceleration at any instant of time within the interval. Consequently, the velocity increases or decreases at the same rate throughout the motion. The *particle under constant acceleration* model is a common analysis model that we can apply to appropriate problems. It is often used to model situations such as falling objects and braking cars.

If we replace $a_{x,\text{avg}}$ with the constant $a_x$ in Equation 2.6, we find that

$$a_x = \frac{v_{xf} - v_{xi}}{t_f - t_i}$$

For convenience, let $t_i = 0$ and $t_f$ be any arbitrary time $t$. With this notation, we can solve for $v_{xf}$:

$$v_{xf} = v_{xi} + a_x t \qquad \text{(for constant } a_x\text{)} \qquad [2.9]$$

▪ Velocity as a function of time for a particle under constant acceleration

This expression enables us to predict the velocity at *any* time $t$ if the initial velocity and constant acceleration are known. It is the first of four equations that can be used to solve problems using the particle under constant acceleration model. A graphical representation of position versus time for this motion is shown in Active Figure 2.12a. The velocity–time graph shown in Active Figure 2.12b is a straight line, the slope of which is the constant acceleration $a_x$. The straight line on this

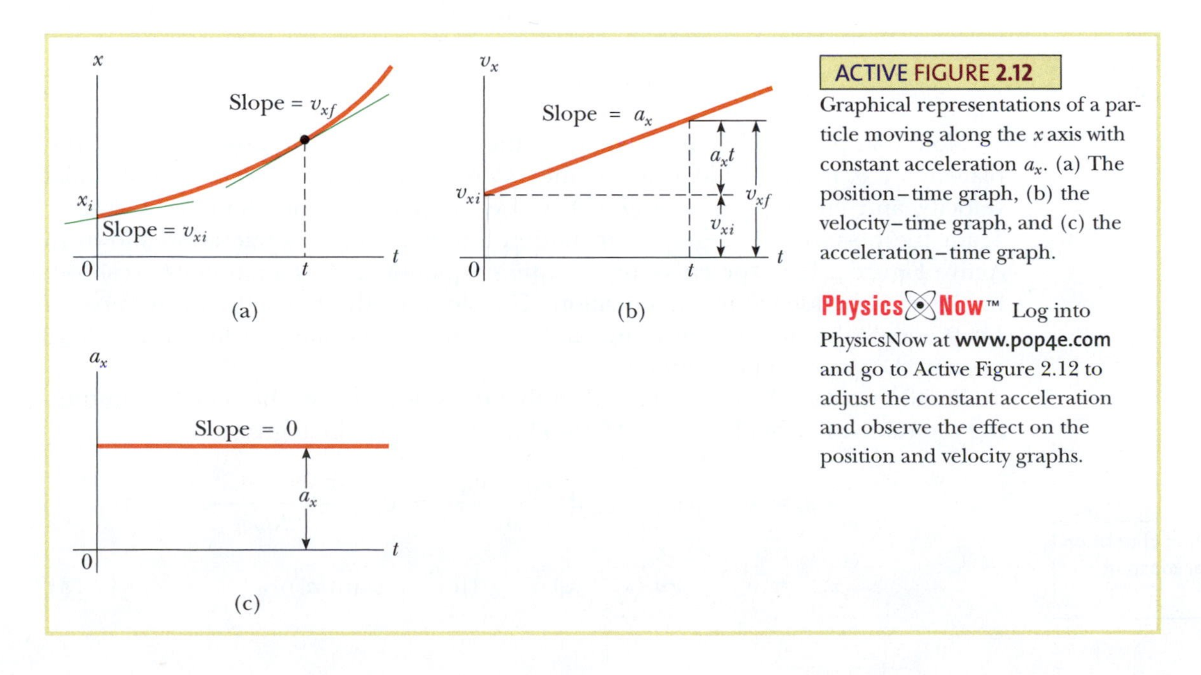


**ACTIVE FIGURE 2.12**
Graphical representations of a particle moving along the $x$ axis with constant acceleration $a_x$. (a) The position–time graph, (b) the velocity–time graph, and (c) the acceleration–time graph.

**Physics Now™** Log into PhysicsNow at **www.pop4e.com** and go to Active Figure 2.12 to adjust the constant acceleration and observe the effect on the position and velocity graphs.

graph is consistent with $a_x = dv_x/dt$ being a constant. From this graph and from Equation 2.9, we see that the velocity at any time $t$ is the sum of the initial velocity $v_{xi}$ and the change in velocity $a_x t$ due to the acceleration. The graph of acceleration versus time (Active Fig. 2.12c) is a straight line with a slope of zero because the acceleration is constant. If the acceleration were negative, the slope of Active Figure 2.12b would be negative and the horizontal line in Active Figure 2.12c would be below the time axis.

We can generate another equation for the particle under constant acceleration model by recalling a result from Section 2.1 that the displacement of a particle is the area under the curve on a velocity–time graph. Because the velocity varies linearly with time (see Active Fig. 2.12b), the area under the curve is the sum of a rectangular area (under the horizontal dashed line in Active Fig. 2.12b) and a triangular area (from the horizontal dashed line upward to the curve). Therefore,

$$\Delta x = v_{xi}\,\Delta t + \tfrac{1}{2}(v_{xf} - v_{xi})\Delta t$$

which can be simplified as follows:

$$\Delta x = (v_{xi} + \tfrac{1}{2}v_{xf} - \tfrac{1}{2}v_{xi})\Delta t = \tfrac{1}{2}(v_{xi} + v_{xf})\Delta t$$

In general, from Equation 2.2, the displacement for a time interval is

$$\Delta x = v_{x,\,\text{avg}}\,\Delta t$$

Comparing these last two equations, we find that the average velocity in any time interval is the arithmetic mean of the initial velocity $v_{xi}$ and the final velocity $v_{xf}$:

■ Average velocity for a particle under constant acceleration

$$v_{x,\,\text{avg}} = \tfrac{1}{2}(v_{xi} + v_{xf}) \qquad \text{(for constant } a_x\text{)} \qquad [2.10]$$

Remember that this expression is valid only when the acceleration is constant, that is, when the velocity varies linearly with time.

We now use Equations 2.2 and 2.10 to obtain the position as a function of time. Again we choose $t_i = 0$, at which time the initial position is $x_i$, which gives

$$\Delta x = v_{x,\,\text{avg}}\,\Delta t = \tfrac{1}{2}(v_{xi} + v_{xf})t$$

■ Position as a function of velocity and time for a particle under constant acceleration

$$x_f = x_i + \tfrac{1}{2}(v_{xi} + v_{xf})t \qquad \text{(for constant } a_x\text{)} \qquad [2.11]$$

We can obtain another useful expression for the position by substituting Equation 2.9 for $v_{xf}$ in Equation 2.11:

$$x_f = x_i + \tfrac{1}{2}[v_{xi} + (v_{xi} + a_x t)]t$$

■ Position as a function of time for a particle under constant acceleration

$$x_f = x_i + v_{xi}t + \tfrac{1}{2}a_x t^2 \qquad \text{(for constant } a_x\text{)} \qquad [2.12]$$

Note that the position at any time $t$ is the sum of the initial position $x_i$, the displacement $v_{xi}t$ that would result if the velocity remained constant at the initial velocity, and the displacement $\tfrac{1}{2}a_x t^2$ because the particle is accelerating. Consider again the position–time graph for motion under constant acceleration shown in Active Figure 2.12a. The curve representing Equation 2.12 is a parabola, as shown by the $t^2$ dependence in the equation. The slope of the tangent to this curve at $t = 0$ equals the initial velocity $v_{xi}$, and the slope of the tangent line at any time $t$ equals the velocity at that time.

Finally, we can obtain an expression that does not contain the time by substituting the value of $t$ from Equation 2.9 into Equation 2.11, which gives

$$x_f = x_i + \tfrac{1}{2}(v_{xi} + v_{xf})\left(\frac{v_{xf} - v_{xi}}{a_x}\right) = x_i + \frac{v_{xf}^2 - v_{xi}^2}{2a_x}$$

■ Velocity as a function of position for a particle under constant acceleration

$$v_{xf}^2 = v_{xi}^2 + 2a_x(x_f - x_i) \qquad \text{(for constant } a_x\text{)} \qquad [2.13]$$

**TABLE 2.2** **Kinematic Equations for Motion in a Straight Line Under Constant Acceleration**

| Equation | Information Given by Equation |
|---|---|
| $v_{xf} = v_{xi} + a_x t$ | Velocity as a function of time |
| $x_f = x_i + \frac{1}{2}(v_{xf} + v_{xi})t$ | Position as a function of velocity and time |
| $x_f = x_i + v_{xi}t + \frac{1}{2}a_x t^2$ | Position as a function of time |
| $v_{xf}^2 = v_{xi}^2 + 2a_x(x_f - x_i)$ | Velocity as a function of position |

*Note:* Motion is along the $x$ axis. At $t = 0$, the position of the particle is $x_i$ and its velocity is $v_{xi}$.

This expression is *not* an independent equation because it arises from combining Equations 2.9 and 2.11. It is useful, however, for those problems in which a value for the time is not involved.

If motion occurs in which the constant value of the acceleration is *zero,* Equations 2.9 and 2.12 become

$$\left.\begin{aligned} v_{xf} &= v_{xi} \\ x_f &= x_i + v_{xi}t \end{aligned}\right\} \quad \text{when } a_x = 0$$

That is, when the acceleration is zero, the velocity remains constant and the position changes linearly with time. In this case, the particle under constant *acceleration* becomes the particle under constant *velocity* (Equation 2.5).

Equations 2.9, 2.11, 2.12, and 2.13 are four **kinematic equations** that may be used to solve any problem in one-dimensional motion of a particle (or an object that can be modeled as a particle) under constant acceleration. Keep in mind that these relationships were derived from the definitions of velocity and acceleration together with some simple algebraic manipulations and the requirement that the acceleration be constant. It is often convenient to choose the initial position of the particle as the origin of the motion so that $x_i = 0$ at $t = 0$. We will see cases, however, in which we must choose the value of $x_i$ to be something other than zero.

The four kinematic equations for the particle under constant acceleration are listed in Table 2.2 for convenience. The choice of which kinematic equation or equations you should use in a given situation depends on what is known beforehand. Sometimes it is necessary to use two of these equations to solve for two unknowns, such as the position and velocity at some instant. You should recognize that the quantities that vary during the motion are velocity $v_{xf}$, position $x_f$, and time $t$. The other quantities—$x_i$, $v_{xi}$, and $a_x$—are *parameters* of the motion and remain constant.

## PROBLEM-SOLVING STRATEGY Particle Under Constant Acceleration

The following procedure is recommended for solving problems that involve an object undergoing a constant acceleration. As mentioned in Chapter 1, individual strategies such as this one will follow the outline of the General Problem-Solving Strategy from Chapter 1, with specific hints regarding the application of the general strategy to the material in the individual chapters.

**1. Conceptualize** Think about what is going on physically in the problem. Establish the mental representation.

**2. Categorize** Simplify the problem as much as possible. Confirm that the problem involves either a particle or an object that can be modeled as a particle and that it is moving with a constant acceleration. Construct an appropriate pictorial representation, such as a motion diagram, or a graphical representation. Make sure all the units in the problem are consistent. That is, if positions are measured in meters, be sure that velocities have units of m/s and accelerations have units of m/s$^2$. Choose a coordinate system to be used throughout the problem.

**3. Analyze** Set up the mathematical representation. Choose an instant to call the "initial" time $t = 0$ and another to call the "final" time $t$. Let your choice be guided by what you know about the particle and what you want to know about it. The initial instant need not be when the particle starts to move, and the final instant will only rarely be when the particle stops moving. Identify all the quantities given in the problem and a separate list of those to be determined. A tabular representation of these quantities may be helpful to you. Select from the list of kinematic equations the one or ones that will enable you to determine the unknowns. Solve the equations.

**4. Finalize** Once you have determined your result, check to see if your answers are consistent with the mental and pictorial representations and that your results are realistic.

### EXAMPLE 2.7 Accelerating an Electron

An electron in the cathode-ray tube of a television set enters a region in which it accelerates uniformly in a straight line from a speed of $3.00 \times 10^4$ m/s to a speed of $5.00 \times 10^6$ m/s in a distance of 2.00 cm. For what time interval is the electron accelerating?

**Solution** For this example, we shall identify the individual steps in the General Problem-Solving Strategy in Chapter 1. In subsequent examples, you should identify the portions of the solution that correspond to each step. For step 1 (*Conceptualize*), think about the electron moving through space. Note that it is moving faster at the end of the interval than before, so imagine it speeding up as it covers the 2.00-cm displacement. In step 2 (*Categorize*), ignore that it is an electron and that it is in a television. The electron is easily modeled as a particle, and the phrase "accelerates uniformly" tells us that it is a particle under constant acceleration. All the parts of Active Figure 2.12 represent the motion of the particle as a function of time, although you may want to graph velocity versus position because no time is given in the problem. Note that all units are metric, although we must convert 2.00 cm to meters to put all units in SI. We make the simple choice of the $x$ axis lying along the straight line mentioned in the text of the problem.

We are now ready to move on to step 3 (*Analyze*), in which we develop the mathematical representation of the problem. Notice that no acceleration is given in the problem and that the time interval is requested, which provides a hint that we should use an equation that does not involve acceleration. We can find the time at which the particle is at the end of the 2.00-cm distance from Equation 2.11:

$$x_f = x_i + \tfrac{1}{2}(v_{xi} + v_{xf})t \quad \rightarrow \quad t = \frac{2(x_f - x_i)}{v_{xi} + v_{xf}}$$

$$t = \frac{2(0.020\,0 \text{ m})}{3.00 \times 10^4 \text{ m/s} + 5.00 \times 10^6 \text{ m/s}}$$
$$= 7.95 \times 10^{-9} \text{ s}$$

Finally, we check if the answer is reasonable (step 4, *Finalize*). The average speed is on the order of $10^6$ m/s. Let us estimate the time interval required to move 1 cm at this speed:

$$\Delta t = \frac{\Delta x}{v} \approx \frac{0.01 \text{ m}}{10^6 \text{ m/s}} \approx 10^{-8} \text{ s} = 10 \times 10^{-9} \text{ s}$$

This result is the same order of magnitude as our answer, providing confidence that our answer is reasonable.

### INTERACTIVE EXAMPLE 2.8 Watch Out for the Speed Limit!

A car traveling at a constant velocity of magnitude 45.0 m/s passes a trooper hidden behind a billboard. One second after the speeding car passes the billboard, the trooper sets out from the billboard to catch it, accelerating at a constant rate of 3.00 m/s$^2$. How long does it take her to overtake the speeding car?

**Solution** We will point out again in this example steps in the General Problem-Solving Strategy. A pictorial representation of the situation is shown in Figure 2.13. Establish the mental representation (*Conceptualize*) of this situation for yourself; in the following solution, we will go straight to the mathematical representation. As you become more proficient at solving physics problems, a quick thought about the mental representation may be enough to allow you to skip pictorial representations and go right to the mathematics. Let us model the speeding car as a particle under constant velocity and the trooper's motorcycle as a particle under constant acceleration (*Categorize*). We shall ignore that they are vehicles and instead will imagine the speeder and the trooper as point particles undergoing the motion described in the problem.

Note that all units are in the same system. To solve this problem algebraically, we will write an expression for the position of each vehicle as a function of time. It is convenient to choose the origin at the position of the billboard and take $t_B = 0$ as the time the trooper

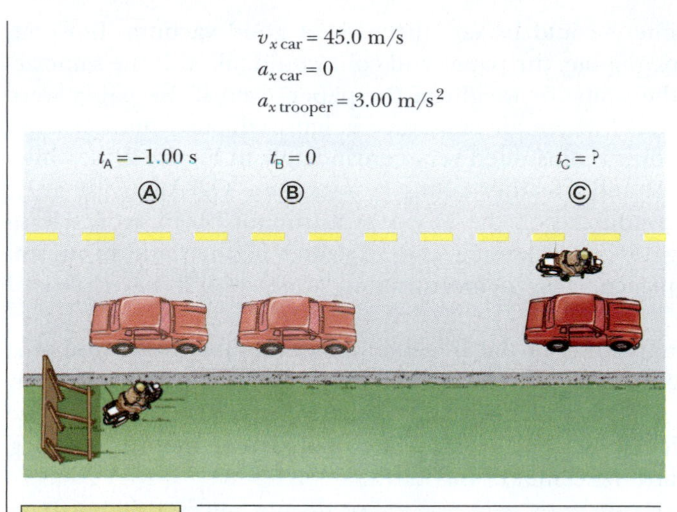


**FIGURE 2.13** (Interactive Example 2.8) A speeding car passes a hidden trooper. The trooper catches up to the car at point Ⓒ.

begins moving. At that instant, the speeding car has already traveled a distance of 45.0 m because it has traveled at a constant speed of $v_x = 45.0$ m/s for 1.00 s; it is at point Ⓑ in Figure 2.13. Therefore, the initial position of the speeding car is $x_i = x_B = 45.0$ m. We do *not* choose $t = 0$ as the time at which the car passes the trooper (point Ⓐ in Fig. 2.13), because then the acceleration of the trooper is not constant during the problem. Her acceleration is $a_x = 0$ for the first second and then 3.00 m/s$^2$ after that. Therefore, we could not model the trooper as a particle under constant acceleration with this choice.

Now we set up the mathematical representation (*Analyze*). Because the car moves with constant velocity, its acceleration is zero, and applying Equation 2.5 gives us

$$x_f = x_B + v_x t \quad \rightarrow \quad x_{\text{car}} = 45.0 \text{ m} + (45.0 \text{ m/s})t$$

Note that at $t = 0$, this expression gives the car's correct initial position, $x_{\text{car}} = 45.0$ m.

For the trooper, who starts from the origin at $t = 0$, we have $x_i = 0$, $v_{xi} = 0$, and $a_x = 3.00$ m/s$^2$. Hence, from Equation 2.12 for a particle under constant acceleration, the position of the trooper as a function of time is

$$x_f = x_i + v_{xi}t + \tfrac{1}{2}a_x t^2 \quad \rightarrow \quad x_{\text{trooper}} = \tfrac{1}{2}a_x t^2 = \tfrac{1}{2}(3.00 \text{ m/s}^2)t^2$$

The trooper overtakes the car at the instant that $x_{\text{trooper}} = x_{\text{car}}$, which is at position Ⓒ in Figure 2.13:

$$\tfrac{1}{2}(3.00 \text{ m/s}^2)t^2 = 45.0 \text{ m} + (45.0 \text{ m/s})t$$

This result gives the quadratic equation (dropping the units)

$$1.50t^2 - 45.0t - 45.0 = 0$$

whose positive solution is $t = 31.0$ s.

From your everyday experience, is this value reasonable (*Finalize*)? (For help in solving quadratic equations, see Appendix B.2.).

**Physics Now™** You can study the motion of the car and the trooper for various velocities of the car by logging into PhysicsNow at www.pop4e.com and going to Interactive Example 2.8.

## 2.7 FREELY FALLING OBJECTS

It is well known that all objects, when dropped, fall toward the Earth with nearly constant acceleration. Legend has it that Galileo Galilei first discovered this fact by observing that two different weights dropped simultaneously from the Leaning Tower of Pisa hit the ground at approximately the same time. (Air resistance plays a role in the falling of an object, but for now we shall model falling objects as if they are falling through a vacuum; this is a simplification model.) Although there is some doubt that this particular experiment was actually carried out, it is well established that Galileo did perform many systematic experiments on objects moving on inclined planes. Through careful measurements of distances and time intervals, he was able to show that the displacement from an origin of an object starting from rest is proportional to the square of the time interval during which the object is in motion. This observation is consistent with one of the kinematic equations we derived for a particle under constant acceleration (Eq. 2.12, with $v_{xi} = 0$). Galileo's achievements in mechanics paved the way for Newton in his development of the laws of motion.

If a coin and a crumpled-up piece of paper are dropped simultaneously from the same height, there will be a small time difference between their arrivals at the

(North Wind Picture Archive)

**Galileo Galilei (1564–1642)**

Italian physicist and astronomer Galileo formulated the laws that govern the motion of objects in free-fall. He also investigated the motion of an object on an inclined plane, established the concept of relative motion, invented the thermometer, and discovered that the motion of a swinging pendulum could be used to measure time intervals. After designing and constructing his own telescope, he discovered four of Jupiter's moons, found that the Moon's surface is rough, discovered sunspots and the phases of Venus, and showed that the Milky Way consists of an enormous number of stars. Galileo publicly defended Nicolaus Copernicus's assertion that the Sun is at the center of the Universe (the heliocentric system). He published *Dialogue Concerning Two New World Systems* to support the Copernican model, a view that the Catholic Church declared to be heretical. After being taken to Rome in 1633 on a charge of heresy, he was sentenced to life imprisonment and later was confined to his villa at Arcetri, near Florence, where he died.

floor. If this same experiment could be conducted in a good vacuum, however, where air friction is truly negligible, the paper and coin would fall with the same acceleration, regardless of the shape or weight of the paper, even if the paper were still flat. In the idealized case, where air resistance is ignored, such motion is referred to as *free-fall.* This point is illustrated very convincingly in Figure 2.14, which is a photograph of an apple and a feather falling in a vacuum. On August 2, 1971, such an experiment was conducted on the Moon by astronaut David Scott. He simultaneously released a geologist's hammer and a falcon's feather, and in unison they fell to the lunar surface. This demonstration surely would have pleased Galileo!

We shall denote the magnitude of the free-fall acceleration with the symbol $g$, representing a vector acceleration $\vec{\mathbf{g}}$. At the surface of the Earth, $g$ is approximately $9.80 \text{ m/s}^2$, or $980 \text{ cm/s}^2$, or $32 \text{ ft/s}^2$. Unless stated otherwise, we shall use the value $9.80 \text{ m/s}^2$ when doing calculations. Furthermore, we shall assume that the vector $\vec{\mathbf{g}}$ is directed downward toward the center of the Earth.

When we use the expression *freely falling object,* we do not necessarily mean an object dropped from rest. A freely falling object is an object moving freely under the influence of gravity alone, regardless of its initial motion. Therefore, objects thrown upward or downward and those released from rest are all freely falling objects once they are released! Because the value of $g$ is constant as long as we are close to the surface of the Earth, we can model a freely falling object as a particle under constant acceleration.

In previous examples in this chapter, the particles were undergoing constant acceleration, as stated in the problem. Therefore, it may have been difficult to understand the need for modeling. We can now begin to see the need for modeling; we are *modeling* a real falling object with an analysis model. Notice that we are (1) ignoring air resistance and (2) assuming that the free-fall acceleration is constant. Therefore, the model of a particle under constant acceleration is a *replacement* for the real problem, which could be more complicated. If air resistance and any variation in $g$ are small, however, the model should make predictions that agree closely with the real situation.

The equations developed in Section 2.6 for objects moving with constant acceleration can be applied to the falling object. The only necessary modification that we need to make in these equations for freely falling objects is to note that the

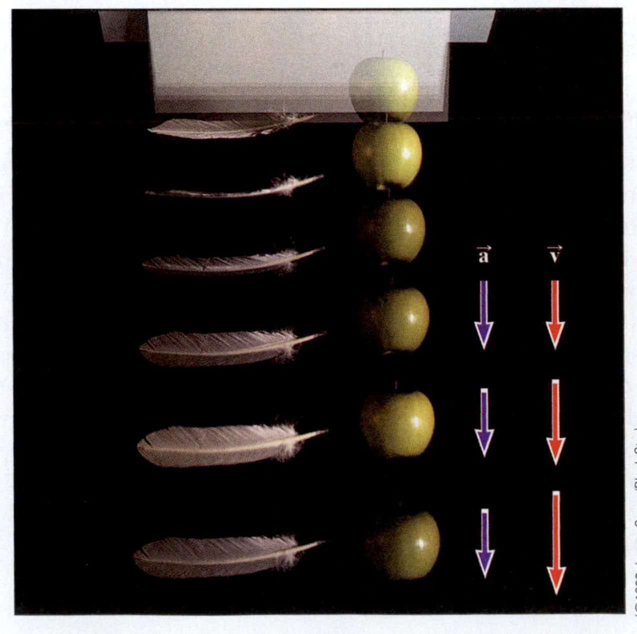

(©1993 James Sugar/Black Star)

**FIGURE 2.14** An apple and a feather, released from rest in a vacuum chamber, fall at the same rate, regardless of their masses. Ignoring air resistance, all objects fall to the Earth with the same acceleration of magnitude $9.80 \text{ m/s}^2$, as indicated by the violet arrows in this multiflash photograph. The velocity of the two objects increases linearly with time, as indicated by the series of red arrows.

motion is in the vertical direction, so we will use $y$ instead of $x$, and that the acceleration is downward and of magnitude $9.80 \text{ m/s}^2$. Therefore, for a freely falling object we commonly take $a_y = -g = -9.80 \text{ m/s}^2$, where the negative sign indicates that the acceleration of the object is downward. The choice of negative for the downward direction is arbitrary, but common.

**PITFALL PREVENTION 2.5**

**ACCELERATION AT THE TOP OF THE MOTION** Imagine throwing a baseball straight up into the air. It is a common misconception that the acceleration of a projectile at the top of its trajectory is zero. This misconception generally arises owing to confusion between velocity and acceleration. Although the velocity at the top of the motion of an object thrown upward momentarily goes to zero, *the acceleration is still that due to gravity* at this point. Remember that acceleration is proportional to force and that the gravitational force still acts at the moment that the object has stopped. If the velocity and acceleration were both zero, the projectile would stay at the top!

**QUICK QUIZ 2.6** A ball is thrown upward. While the ball is in free-fall, does its acceleration **(a)** increase, **(b)** decrease, **(c)** increase and then decrease, **(d)** decrease and then increase, or **(e)** remain constant?

## Thinking Physics 2.2

A sky diver steps out of a stationary helicopter. A few seconds later, another sky diver steps out, so that both sky divers fall along the same vertical line. Ignore air resistance, so that both sky divers fall with the same acceleration, and model the sky divers as particles under constant acceleration. Does the vertical separation distance between them stay the same? Does the difference in their speeds stay the same?

**Reasoning** At any given instant of time, the speeds of the sky divers are definitely different, because one had a head start over the other. In any time interval, however, each sky diver increases his or her speed by the same amount, because they have the same acceleration. Therefore, the difference in speeds remains the same. The first sky diver will always be moving with a higher speed than the second. In a given time interval, then, the first sky diver will have a larger displacement than the second. Therefore, the separation distance between them increases. ■

**PITFALL PREVENTION 2.6**

**THE SIGN OF $g$** Keep in mind that $g$ is a *positive number.* It is tempting to substitute $-9.80 \text{ m/s}^2$ for $g$, but resist the temptation. That the gravitational acceleration is downward is indicated explicitly by stating the acceleration as $a_y = -g$.

### EXAMPLE 2.9 Try to Catch the Dollar

Emily challenges David to catch a dollar bill as follows. She holds the bill vertically, as in Figure 2.15, with the center of the bill between David's index finger and thumb. David must catch the bill after Emily releases it without moving his hand downward. The reaction time of most people is at best about 0.2 s. Who would you bet on?

**Solution** Place your bets on Emily. There is a time delay between the instant Emily releases the bill and the time David reacts and closes his fingers. We model the bill as a particle. When released, the bill will probably flutter downward to the floor due to the effects of the air, but for the very early part of its motion, we will assume that it can be modeled as a particle falling through a vacuum. Because the bill is in free-fall and undergoes a downward acceleration of magnitude $9.80 \text{ m/s}^2$, in 0.2 s it falls a distance of $y = \frac{1}{2}gt^2 \approx 0.2 \text{ m} = 20 \text{ cm}$. This distance is about twice the distance between the center of the bill and its top edge ($\approx 8$ cm). Therefore, David will be unsuccessful.

You might want to try this "trick" on one of your friends.

(George Semple)

**FIGURE 2.15** (Example 2.9)

## INTERACTIVE EXAMPLE 2.10 Not a Bad Throw for a Rookie!

A stone is thrown at point Ⓐ from the top of a building with an initial velocity of 20.0 m/s straight upward. The building is 50.0 m high, and the stone just misses the edge of the roof on its way down, as in the pictorial representation of Figure 2.16.

**A** Determine the time at which the stone reaches its maximum height.

**Solution** Think about the mental representation: the stone rises upward, slowing down. It stops momentarily (point Ⓑ) and then begins to fall downward again. During the entire motion, it is accelerating downward because the gravitational force is always pulling downward on it. Ignoring air resistance, we model the stone as a particle under constant acceleration.

To begin the mathematical representation, we consider the portion of the motion from Ⓐ to Ⓑ and find the time at which the stone reaches the maximum height, point Ⓑ. We use the vertical modification of Equation 2.9, noting that $v_{yf} = 0$ at the maximum height:

$$v_{yf} = v_{yi} + a_y t \quad \rightarrow \quad 0 = 20.0 \text{ m/s} + (-9.80 \text{ m/s}^2)t_{\text{B}}$$

$$t_{\text{B}} = \frac{20.0 \text{ m/s}}{9.80 \text{ m/s}^2} = \boxed{2.04 \text{ s}}$$

**B** Determine the maximum height of the stone above the roof top.

**Solution** The value of time from part A can be substituted into Equation 2.12 to give the maximum height measured from the position of the thrower:

$$\begin{aligned} y_{\max} = y_{\text{B}} &= y_{\text{A}} + v_{y\text{A}}t_{\text{B}} + \tfrac{1}{2}a_y t_{\text{B}}^2 \\ &= 0 + (20.0 \text{ m/s})(2.04 \text{ s}) + \tfrac{1}{2}(-9.80 \text{ m/s}^2)(2.04 \text{ s})^2 \\ &= \boxed{20.4 \text{ m}} \end{aligned}$$

**C** Determine the time at which the stone returns to the level of the thrower.

**Solution** Now we identify the initial point of the motion as Ⓐ and the final point as Ⓒ. When the stone is back at the height of the thrower, the $y$ coordinate is zero. From Equation 2.12, letting $y_f = y_{\text{C}} = 0$, we obtain the expression

$$y_{\text{C}} = y_{\text{A}} + v_{y\text{A}}t_{\text{C}} + \tfrac{1}{2}a_y t_{\text{C}}^2 \quad \rightarrow \quad 0 = 20.0t_{\text{C}} - 4.90t_{\text{C}}^2$$

This result is a quadratic equation and has two solutions for $t_{\text{C}}$. The equation can be factored to give

$$t_{\text{C}}(20.0 - 4.90t_{\text{C}}) = 0$$

One solution is $t_{\text{C}} = 0$, corresponding to the time the stone starts its motion. The other solution—the one we are after—is $\boxed{t_{\text{C}} = 4.08 \text{ s}}$. Note that this value is twice the value for $t_{\text{B}}$. The fall from Ⓑ to Ⓒ is the reverse of the rise from Ⓐ to Ⓑ, and the stone requires exactly the same time interval to undergo each part of the motion.

**D** Determine the velocity of the stone at this instant.

**Solution** The value for $t_{\text{C}}$ found in part C can be inserted into Equation 2.9 to give

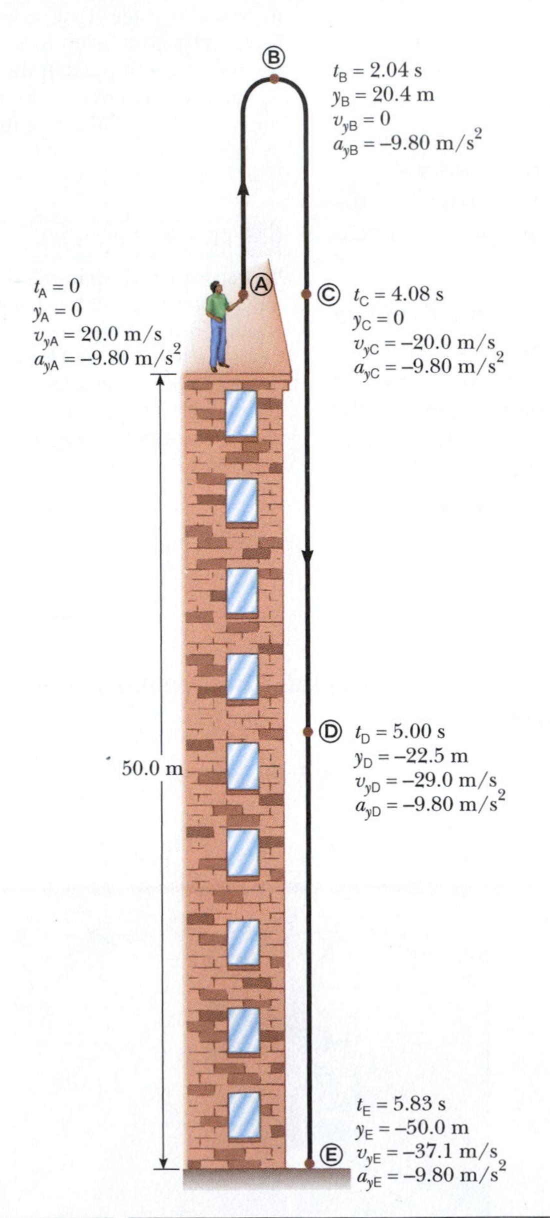


FIGURE 2.16 (Interactive Example 2.10) Position, velocity, and acceleration at various instants of time for a freely falling particle initially thrown upward with a velocity $v_y = 20.0$ m/s.

$$v_{yC} = v_{yA} + a_y t_C$$
$$= 20.0 \text{ m/s} + (-9.80 \text{ m/s}^2)(4.08 \text{ s})$$
$$= -20.0 \text{ m/s}$$

Note that the velocity of the stone when it arrives back at its original height is equal in magnitude to its initial velocity but opposite in direction. This equal magnitude, along with the equal time intervals noted at the end of part C, indicates that the motion to this point is symmetric.

E Determine the velocity and position of the stone at $t = 5.00$ s.

**Solution** For this part of the problem, we analyze the portion of the motion from Ⓐ to Ⓓ. From Equation 2.9, the velocity at Ⓓ after 5.00 s is

$$v_{yD} = v_{yA} + a_y t_D$$
$$= 20.0 \text{ m/s} + (-9.80 \text{ m/s}^2)(5.00 \text{ s})$$
$$= -29.0 \text{ m/s}$$

We can use Equation 2.12 to find the position of the stone at $t = 5.00$ s:

$$y_D = y_A + v_{yA} t_D + \tfrac{1}{2} a_y t_D^2$$
$$= 0 + (20.0 \text{ m/s})(5.00 \text{ s}) + \tfrac{1}{2}(-9.80 \text{ m/s}^2)(5.00 \text{ s})^2$$
$$= -22.5 \text{ m}$$

F Determine the position of the stone at $t = 6.00$ s. How does the model fail this last part of the problem?

**Solution** We use Equation 2.12 again to find the position of the stone at $t = 6.00$ s:

$$y = 0 + (20.0 \text{ m/s})(6.00 \text{ s}) + \tfrac{1}{2}(-9.80 \text{ m/s}^2)(6.00 \text{ s})^2$$
$$= -56.4 \text{ m}$$

The failure of the model is that the building is only 50.0 m high, so the stone cannot be at a position 6.4 m below ground. Our model does not include that the ground exists at $y = -50.0$ m, so the mathematical representation gives us an answer that is not consistent with our expectations in this case.

**PhysicsNow™** You can study the motion of the thrown ball by logging into PhysicsNow at www.pop4e.com and going to Interactive Example 2.10.

## 2.8 ACCELERATION REQUIRED BY CONSUMERS

CONTEXT CONNECTION

We now have our first opportunity to address a Context in a closing section, as we will do in each remaining chapter. Our present Context is *Alternative-Fuel Vehicles,* and our central question is, *What source besides gasoline can be used to provide energy for an automobile while reducing environmentally damaging emissions?*

Consumers have been driving gasoline-powered vehicles for decades and have become used to a certain amount of acceleration, such as that required to enter a freeway on-ramp. This experience raises the question as to what kind of acceleration today's consumer would expect for an alternative-fuel vehicle that might replace a gasoline-powered vehicle. In turn, developers of alternative-fuel vehicles should strive for such an acceleration so as to satisfy consumer expectations and hope to generate a demand for the new vehicle.

If we consider published time intervals for accelerations from 0 to 60 mi/h for a number of automobile models, we find the data shown in the middle column of Table 2.3. The average acceleration of each vehicle is calculated from these data using Equation 2.6. It is clear from the upper part of this table (*Performance vehicles*) that acceleration upward of 10 mi/h · s is very expensive. The highest accelerations are 16.7 mi/h · s and cost either $480,000 for the Ferrari F50 or a bargain at $292,000 for the Lamborghini Diablo GT. For the less affluent driver, the accelerations in the middle part of the table (*Traditional vehicles*) have an average value of 7.8 mi/h · s. This number is typical of consumer-oriented gasoline-powered vehicles and provides an approximate standard for the acceleration desired in an alternative-fuel vehicle.

In the lower part of Table 2.3, we see data for three alternative vehicles. The General Motors EV1 is an electric car that was discontinued in 2001, even though it was a technological success. Notice that its acceleration is similar to those in the

**TABLE 2.3** Accelerations of Various Vehicles, 0–60 mi/h

| Automobile | Model Year | Time Interval, 0–60 mi/h (s) | Average Acceleration (mi/h · s) | Price |
|---|---|---|---|---|
| *Performance vehicles* | | | | |
| Aston Martin DB7 Vantage | 2001 | 5.0 | 12.0 | $170,000 |
| BMW Z8 | 2001 | 4.6 | 13.0 | $134,000 |
| Chevrolet Corvette | 2000 | 4.6 | 13.0 | $46,000 |
| Dodge Viper GTS-R | 1998 | 4.2 | 14.3 | $92,000 |
| Ferrari F50 | 1997 | 3.6 | 16.7 | $480,000 |
| Ferrari 360 Spider F1 | 2000 | 4.6 | 13.0 | $171,000 |
| Lamborghini Diablo GT | 2000 | 3.6 | 16.7 | $292,000 |
| Porsche 911 GT2 | 2002 | 4.0 | 15.0 | $182,000 |
| *Traditional vehicles* | | | | |
| Acura Integra GS | 2000 | 7.9 | 7.6 | $22,000 |
| BMW Mini Cooper S | 2003 | 6.9 | 8.7 | $17,500 |
| Cadillac Escalade (SUV) | 2002 | 8.6 | 7.0 | $51,000 |
| Dodge Stratus | 2002 | 7.5 | 8.0 | $22,000 |
| Lexus ES300 | 1997 | 8.6 | 7.0 | $29,000 |
| Mitsubishi Eclipse GT | 2000 | 7.0 | 8.6 | $23,000 |
| Nissan Maxima | 2000 | 6.7 | 9.0 | $25,000 |
| Pontiac Grand Prix | 2003 | 8.5 | 7.1 | $25,000 |
| Toyota Sienna (SUV) | 2004 | 8.3 | 7.2 | $23,000 |
| Volkswagen Beetle | 1999 | 7.6 | 7.9 | $19,000 |
| *Alternative vehicles* | | | | |
| GM EV1 | 1998 | 7.6 | 7.9 | (lease only) $399/month |
| Toyota Prius | 2004 | 12.7 | 4.7 | $21,000 |
| Honda Insight | 2001 | 11.6 | 5.2 | $21,000 |

*Note:* Data given in this table as well as in similar tables in Chapters 3 through 6 were gathered from a number of websites. Other data, such as the accelerations in this table, were calculated from the raw data.

middle part of Table 2.3. This acceleration is sufficiently large that it satisfies consumer demand for a car with "get-up-and-go."

The Toyota Prius and Honda Insight are *hybrid vehicles*, which we will discuss further in the Context Conclusion. These vehicles combine a gasoline engine and an electric motor. The accelerations for these vehicles are the lowest in the table. The disadvantage of the low acceleration is offset by other factors. These vehicles obtain relatively high gas mileage, have very low emissions, and do not require recharging as does a pure electric vehicle.

(George Lepp/Stone/Getty)

**FIGURE 2.17** In drag racing, acceleration is a highly desired quantity. In a distance of 1/4 mile, speeds of over 320 mi/h are reached, with the entire distance being covered in under 5 s.

In comparison to the vehicles in the upper part of the table, consider the acceleration of an even higher-level "performance vehicle," a typical drag racer, as shown in Figure 2.17. Typical data show that such a vehicle covers a distance of 0.25 mi in 5.0 s, starting from rest. We can find the acceleration from Equation 2.12:

$$x_f = x_i + v_i t + \tfrac{1}{2}a_x t^2 = 0 + 0(t) + \tfrac{1}{2}(a_x)(t)^2 \quad \rightarrow \quad a_x = \frac{2x_f}{t^2}$$

$$= \frac{2(0.25 \text{ mi})}{(5.0 \text{ s})^2} = 0.020 \text{ mi/s}^2 \left(\frac{3\,600 \text{ s}}{1 \text{ h}}\right) = 72 \text{ mi/h} \cdot \text{s}$$

This value is much larger than any accelerations in the table, as would be expected. We can show that the acceleration due to gravity has the following value in units of mi/h · s:

$$g = 9.80 \text{ m/s}^2 = 21.9 \text{ mi/h} \cdot \text{s}$$

Therefore, the drag racer is moving horizontally with 3.3 times as much acceleration as it would move vertically if you pushed it off a cliff! (Of course, the horizontal acceleration can only be maintained for a very short time interval.)

As we investigate two-dimensional motion in the next chapter, we shall consider a different type of acceleration for vehicles, that associated with the vehicle turning in a sharp circle at high speed. ■

## SUMMARY

**PhysicsNow™** Take a practice test by logging into PhysicsNow at **www.pop4e.com** and clicking on the Pre-Test link for this chapter.

The **average speed** of a particle during some time interval is equal to the ratio of the distance $d$ traveled by the particle and the time interval $\Delta t$:

$$v_{\text{avg}} \equiv \frac{d}{\Delta t} \quad [2.1]$$

The **average velocity** of a particle moving in one dimension during some time interval is equal to the ratio of the displacement $\Delta x$ and the time interval $\Delta t$:

$$v_{x,\text{avg}} \equiv \frac{\Delta x}{\Delta t} \quad [2.2]$$

The **instantaneous velocity** of a particle is defined as the limit of the ratio $\Delta x/\Delta t$ as $\Delta t$ approaches zero:

$$v_x \equiv \lim_{\Delta t \to 0} \frac{\Delta x}{\Delta t} = \frac{dx}{dt} \quad [2.3]$$

The **instantaneous speed** of a particle is defined as the magnitude of the instantaneous velocity vector.

If the velocity $v_x$ is constant, the preceding equations can be modified and used to solve problems describing the motion of a *particle under constant velocity:*

$$v_x = \frac{\Delta x}{\Delta t} \quad [2.4]$$

$$x_f = x_i + v_x t \quad [2.5]$$

The **average acceleration** of a particle moving in one dimension during some time interval is defined as the ratio of the change in its velocity $\Delta v_x$ and the time interval $\Delta t$:

$$a_{x,\text{avg}} \equiv \frac{\Delta v_x}{\Delta t} \quad [2.6]$$

The **instantaneous acceleration** is equal to the limit of the ratio $\Delta v_x/\Delta t$ as $\Delta t \to 0$. By definition, this limit equals the derivative of $v_x$ with respect to $t$, or the time rate of change of the velocity:

$$a_x \equiv \lim_{\Delta t \to 0} \frac{\Delta v_x}{\Delta t} = \frac{dv_x}{dt} \quad [2.7]$$

The slope of the tangent to the $x$ versus $t$ curve at any instant gives the instantaneous velocity of the particle.

The slope of the tangent to the $v$ versus $t$ curve gives the instantaneous acceleration of the particle.

The **kinematic equations** for a *particle under constant acceleration* $a_x$ (constant in magnitude and direction) are

$$v_{xf} = v_{xi} + a_x t \quad [2.9]$$

$$x_f = x_i + \tfrac{1}{2}(v_{xi} + v_{xf})t \quad [2.11]$$

$$x_f = x_i + v_{xi}t + \tfrac{1}{2}a_x t^2 \quad [2.12]$$

$$v_{xf}^2 = v_{xi}^2 + 2a_x(x_f - x_i) \quad [2.13]$$

An object falling freely experiences an acceleration directed toward the center of the Earth. If air friction is ignored and if the altitude of the motion is small compared with the Earth's radius, one can assume that the magnitude of the free-fall acceleration $g$ is constant over the range of motion, where $g$ is equal to 9.80 m/s$^2$, or 32 ft/s$^2$. Assuming $y$ to be positive upward, the acceleration is given by $-g$, and the equations of kinematics for an object in free-fall are the same as those already given, with the substitutions $x \to y$ and $a_y \to -g$.

## QUESTIONS

☐ = answer available in the *Student Solutions Manual and Study Guide*

**1.** The speed of sound in air is 331 m/s. During the next thunderstorm, try to estimate your distance from a lightning bolt by measuring the time lag between the flash and the thunderclap. You can ignore the time interval it takes for the light flash to reach you. Why?

**2.** The average velocity of a particle moving in one dimension has a positive value. Is it possible for the instantaneous velocity to have been negative at any time in the interval? Suppose the particle started at the origin $x = 0$. If its average velocity is positive, could the particle ever have been in the $-x$ region of the axis?

**3.** If the average velocity of an object is zero in some time interval, what can you say about the displacement of the object for that interval?

**4.** Can the instantaneous velocity of an object at an instant of time ever be greater in magnitude than the average velocity over a time interval containing the instant? Can it ever be less?

**5.** If an object's average velocity is nonzero over some time interval, does that mean that its instantaneous velocity is never zero during the interval? Explain your answer.

**6.** An object's average velocity is zero over some time interval. Show that its instantaneous velocity must be zero at some time during the interval. It may be useful in your proof to sketch a graph of $x$ versus $t$ and to note that $v_x(t)$ is a continuous function.

**7.** If the velocity of a particle is nonzero, can its acceleration be zero? Explain.

**8.** If the velocity of a particle is zero, can its acceleration be nonzero? Explain.

**9.** Two cars are moving in the same direction in parallel lanes along a highway. At some instant, the velocity of car A exceeds the velocity of car B. Does that mean that the acceleration of A is greater than that of B? Explain.

**10.** Is it possible for the velocity and the acceleration of an object to have opposite signs? If not, state a proof. If so, give an example of such a situation and sketch a velocity–time graph to prove your point.

**11.** Consider the following combinations of signs and values for velocity and acceleration of a particle with respect to a one-dimensional $x$ axis:

| Velocity | Acceleration |
|---|---|
| a. Positive | Positive |
| b. Positive | Negative |
| c. Positive | Zero |
| d. Negative | Positive |
| e. Negative | Negative |
| f. Negative | Zero |
| g. Zero | Positive |
| h. Zero | Negative |

Describe what a particle is doing in each case and give a real-life example for an automobile on an east–west one-dimensional axis, with east considered the positive direction.

**12.** Can the kinematic equations (Eqs. 2.9 through 2.13) be used in a situation where the acceleration varies in time? Can they be used when the acceleration is zero?

**13.** A child throws a marble into the air with an initial speed $v_i$. Another child drops a ball at the same instant. Compare the accelerations of the two objects while they are in flight.

**14.** An object falls freely from height $h$. It is released at time zero and strikes the ground at time $t$. (a) When the object is at height $0.5h$, is the time earlier than $0.5t$, equal to $0.5t$, or later than $0.5t$? (b) When the time is $0.5t$, is the height of the object greater than $0.5h$, equal to $0.5h$, or less than $0.5h$? Give reasons for your answers.

**15.** A student at the top of a building of height $h$ throws one ball upward with a speed of $v_i$ and then throws a second ball downward with the same initial speed. How do the final velocities of the balls compare when they reach the ground?

**16.** You drop a ball from a window on an upper floor of a building. It strikes the ground with speed $v$. You now repeat the drop, but you have a friend down on the street who throws another ball upward at speed $v$. Your friend throws the ball upward at precisely the same time that you drop yours from the window. At some location, the balls pass each other. Is this location *at* the halfway point between window and ground, *above* this point, or *below* this point?

## PROBLEMS

1, 2, 3 = straightforward, intermediate, challenging
□ = full solution available in the *Student Solutions Manual and Study Guide*
**PhysicsNow™** = coached problem with hints available at www.pop4e.com
= computer useful in solving problem
= paired numerical and symbolic problems
= biomedical application

### Section 2.1 ■ Average Velocity

**1.** The position of a pinewood derby car was observed at various times; the results are summarized in the following table. Find the average velocity of the car for (a) the first second, (b) the last 3 s, and (c) the entire period of observation.

| $t$ (s) | 0 | 1.0 | 2.0 | 3.0 | 4.0 | 5.0 |
|---|---|---|---|---|---|---|
| $x$ (m) | 0 | 2.3 | 9.2 | 20.7 | 36.8 | 57.5 |

**2.** A particle moves according to the equation $x = 10t^2$, where $x$ is in meters and $t$ is in seconds. (a) Find the average velocity for the time interval from 2.00 s to 3.00 s. (b) Find the average velocity for the time interval from 2.00 s to 2.10 s.

**3.** The position versus time for a certain particle moving along the $x$ axis is shown in Figure P2.3. Find the average velocity in the time intervals (a) 0 to 2 s, (b) 0 to 4 s, (c) 2 s to 4 s, (d) 4 s to 7 s, and (e) 0 to 8 s.

**4.** A person walks first at a constant speed of 5.00 m/s along a straight line from point $A$ to point $B$ and then back along the line from $B$ to $A$ at a constant speed of 3.00 m/s. (a) What is her average speed over the entire trip? (b) What is her average velocity over the entire trip?

### Section 2.2 ■ Instantaneous Velocity

**5.** **PhysicsNow™** A position–time graph for a particle moving along the $x$ axis is shown in Figure P2.5. (a) Find

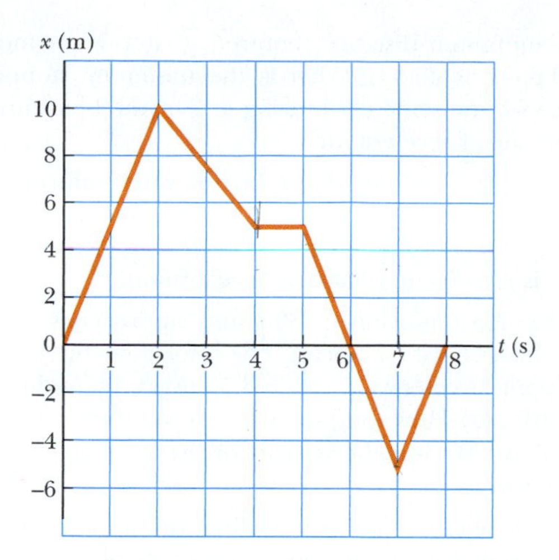

FIGURE P2.3 Problems 2.3 and 2.8.

the average velocity in the time interval $t = 1.50$ s to $t = 4.00$ s. (b) Determine the instantaneous velocity at $t = 2.00$ s by measuring the slope of the tangent line shown in the graph. (c) At what value of $t$ is the velocity zero?

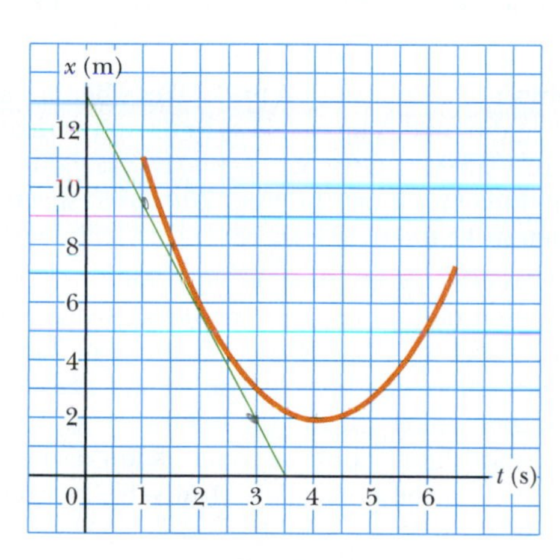

FIGURE P2.5

6. The position of a particle moving along the $x$ axis varies in time according to the expression $x = 3t^2$, where $x$ is in meters and $t$ is in seconds. Evaluate its position (a) at $t = 3.00$ s and (b) at $3.00 \text{ s} + \Delta t$. (c) Evaluate the limit of $\Delta x/\Delta t$ as $\Delta t$ approaches zero to find the velocity at $t = 3.00$ s.

7. (a) Use the data in Problem 2.1 to construct a smooth graph of position versus time. (b) By constructing tangents to the $x(t)$ curve, find the instantaneous velocity of the car at several instants. (c) Plot the instantaneous velocity versus time and, from this information, determine the average acceleration of the car. (d) What was the initial velocity of the car?

8. Find the instantaneous velocity of the particle described in Figure P2.3 at the following times: (a) $t = 1.0$ s, (b) $t = 3.0$ s, (c) $t = 4.5$ s, (d) $t = 7.5$ s.

## Section 2.3 ▪ Analysis Models—The Particle Under Constant Velocity

9. A hare and a tortoise compete in a race over a course 1.00 km long. The tortoise crawls straight and steadily at its maximum speed of 0.200 m/s toward the finish line. The hare runs at its maximum speed of 8.00 m/s toward the goal for 0.800 km and then stops to taunt the tortoise. How close to the goal can the hare let the tortoise approach before resuming the race, which the tortoise wins in a photo finish? Assume that, when moving, both animals move steadily at their respective maximum speeds.

## Section 2.4 ▪ Acceleration

10. A 50.0-g superball traveling at 25.0 m/s bounces off a brick wall and rebounds at 22.0 m/s. A high-speed camera records this event. If the ball is in contact with the wall for 3.50 ms, what is the magnitude of the average acceleration of the ball during this time interval? (*Note:* 1 ms = $10^{-3}$ s.)

11. A particle starts from rest and accelerates as shown in Figure P2.11. Determine (a) the particle's speed at $t = 10.0$ s and at $t = 20.0$ s, and (b) the distance traveled in the first 20.0 s.

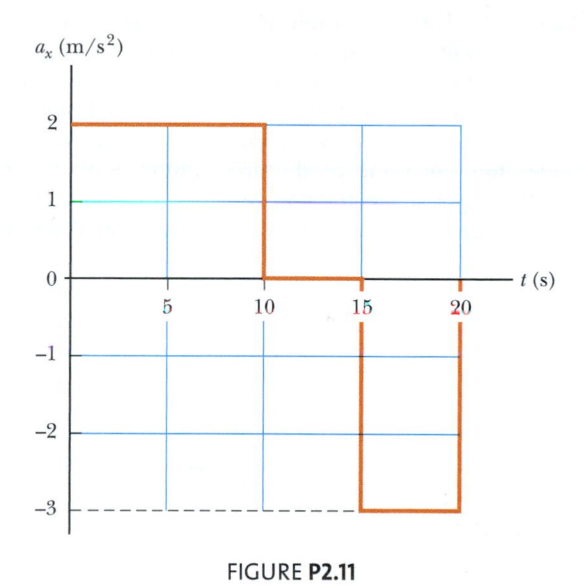

FIGURE P2.11

12. An object moves along the $x$ axis according to the equation $x(t) = (3.00t^2 - 2.00t + 3.00)$m, where $t$ is in seconds. Determine (a) the average speed between $t = 2.00$ s and $t = 3.00$ s, (b) the instantaneous speed at $t = 2.00$ s and at $t = 3.00$ s, (c) the average acceleration between $t = 2.00$ s and $t = 3.00$ s, and (d) the instantaneous acceleration at $t = 2.00$ s and $t = 3.00$ s.

13. **PhysicsNow™** A particle moves along the $x$ axis according to the equation $x = 2.00 + 3.00t - 1.00t^2$, where $x$ is in meters and $t$ is in seconds. At $t = 3.00$ s, find (a) the position of the particle, (b) its velocity, and (c) its acceleration.

14. A student drives a moped along a straight road as described by the velocity versus time graph in Figure P2.14. Sketch this graph in the middle of a sheet of graph paper. (a) Directly above your graph, sketch a graph of the

position versus time, aligning the time coordinates of the two graphs. (b) Sketch a graph of the acceleration versus time directly below the $v_x$-$t$ graph, again aligning the time coordinates. On each graph, show the numerical values of $x$ and $a_x$ for all points of inflection. (c) What is the acceleration at $t = 6$ s? (d) Find the position (relative to the starting point) at $t = 6$ s. (e) What is the moped's final position at $t = 9$ s?

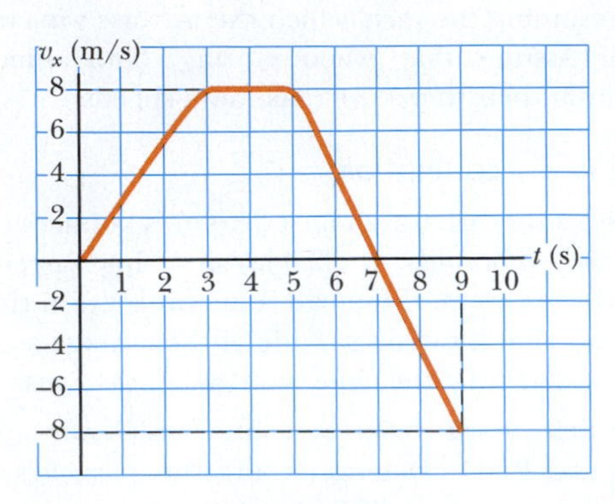


FIGURE P2.14

**15.** Figure P2.15 shows a graph of $v_x$ versus $t$ for the motion of a motorcyclist as he starts from rest and moves along the road in a straight line. (a) Find the average acceleration for the time interval $t = 0$ to $t = 6.00$ s. (b) Estimate the time at which the acceleration has its greatest positive value and the value of the acceleration at that instant. (c) When is the acceleration zero? (d) Estimate the maximum negative value of the acceleration and the time at which it occurs.

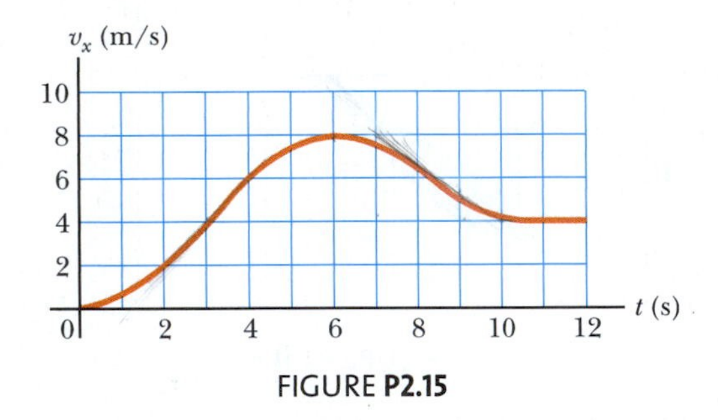


FIGURE P2.15

## Section 2.5 ▪ Motion Diagrams

**16.** Draw motion diagrams for (a) an object moving to the right at constant speed, (b) an object moving to the right and speeding up at a constant rate, (c) an object moving to the right and slowing down at a constant rate, (d) an object moving to the left and speeding up at a constant rate, and (e) an object moving to the left and slowing down at a constant rate. (f) How would your drawings change if the changes in speed were not uniform, that is, if the speed were not changing at a constant rate?

## Section 2.6 ▪ The Particle Under Constant Acceleration

**17.** A truck covers 40.0 m in 8.50 s while smoothly slowing down to a final speed of 2.80 m/s. (a) Find its original speed. (b) Find its acceleration.

**18.** The minimum distance required to stop a car moving at 35.0 mi/h is 40.0 ft. What is the minimum stopping distance for the same car moving at 70.0 mi/h, assuming the same rate of acceleration?

**19.** PhysicsNow™ An object moving with uniform acceleration has a velocity of 12.0 cm/s in the positive $x$ direction when its $x$ coordinate is 3.00 cm. If its $x$ coordinate 2.00 s later is − 5.00 cm, what is its acceleration?

**20.** A speedboat moving at 30.0 m/s approaches a no-wake buoy marker 100 m ahead. The pilot slows the boat with a constant acceleration of $-3.50$ m/s$^2$ by reducing the throttle. (a) How long does it take the boat to reach the buoy? (b) What is the velocity of the boat when it reaches the buoy?

**21.** A jet plane comes in for a landing with a speed of 100 m/s and can accelerate at a maximum rate of $-5.00$ m/s$^2$ as it comes to rest. (a) From the instant the plane touches the runway, what is the minimum time interval needed before it can come to rest? (b) Can this plane land at a small tropical island airport where the runway is 0.800 km long?

**22.** A particle moves along the $x$ axis. Its position is given by the equation $x = 2 + 3t - 4t^2$ with $x$ in meters and $t$ in seconds. Determine (a) its position when it changes direction and (b) its velocity when it returns to the position it had at $t = 0$.

**23.** The driver of a car slams on the brakes when he sees a tree blocking the road. The car slows uniformly with an acceleration of $-5.60$ m/s$^2$ for 4.20 s, making straight skid marks 62.4 m long ending at the tree. With what speed does the car then strike the tree?

**24.** *Help! One of our equations is missing!* We describe constant-acceleration motion with the variables and parameters $v_{xi}$, $v_{xf}$, $a_x$, $t$, and $x_f - x_i$. Of the equations in Table 2.2, the first does not involve $x_f - x_i$. The second does not contain $a_x$, the third omits $v_{xf}$, and the last leaves out $t$. So, to complete the set there should be an equation *not* involving $v_{xi}$. Derive it from the others. Use it to solve Problem 2.23 in one step.

**25.** A truck on a straight road starts from rest, accelerating at 2.00 m/s$^2$ until it reaches a speed of 20.0 m/s. Then the truck travels for 20.0 s at constant speed until the brakes are applied, stopping the truck in a uniform manner in an additional 5.00 s. (a) How long is the truck in motion? (b) What is the average velocity of the truck for the motion described?

**26.** An electron in a cathode-ray tube accelerates uniformly from $2.00 \times 10^4$ m/s to $6.00 \times 10^6$ m/s over 1.50 cm. (a) In what time interval does the electron travel this 1.50 cm? (b) What is its acceleration?

**27.** Speedy Sue, driving at 30.0 m/s, enters a one-lane tunnel. She then observes a slow-moving van 155 m ahead traveling at 5.00 m/s. Sue applies her brakes but can accelerate only at $-2.00$ m/s$^2$ because the road is wet. Will there be a collision? If yes, determine how far into the tunnel and at what time the collision occurs. If no, determine the distance of closest approach between Sue's car and the van.

## Section 2.7 ■ Freely Falling Objects

*Note:* In all problems in this section, ignore the effects of air resistance.

**28.** In a classic clip on *America's Funniest Home Videos,* a sleeping cat rolls gently off the top of a warm TV set. Ignoring air resistance, calculate the position and velocity of the cat after (a) 0.100 s, (b) 0.200 s, and (c) 0.300 s.

**29.** A baseball is hit so that it travels straight upward after being struck by the bat. A fan observes that it takes 3.00 s for the ball to reach its maximum height. Find (a) its initial velocity and (b) the height it reaches.

**30.** *Every morning at seven o'clock*
*There's twenty terriers drilling on the rock.*
*The boss comes around and he says, "Keep still*
*And bear down heavy on the cast-iron drill*
*And drill, ye terriers, drill." And drill, ye terriers, drill.*
*It's work all day for sugar in your tea*
*Down beyond the railway. And drill, ye terriers, drill.*

*The foreman's name was John McAnn.*
*By God, he was a blamed mean man.*
*One day a premature blast went off*
*And a mile in the air went big Jim Goff. And drill . . .*
*Then when next payday came around*
*Jim Goff a dollar short was found.*
*When he asked what for, came this reply:*
*"You were docked for the time you were up in the sky."*
*And drill . . .*
—American folksong

What was Goff's hourly wage? State the assumptions you make in computing it.

**31.** **PhysicsNow™** A student throws a set of keys vertically upward to her sorority sister, who is in a window 4.00 m above. The keys are caught 1.50 s later by the sister's outstretched hand. (a) With what initial velocity were the keys thrown? (b) What was the velocity of the keys just before they were caught?

**32.** A ball is thrown directly downward, with an initial speed of 8.00 m/s, from a height of 30.0 m. After what time interval does the ball strike the ground?

**33.** **PhysicsNow™** A daring ranch hand sitting on a tree limb wishes to drop vertically onto a horse galloping under the tree. The constant speed of the horse is 10.0 m/s, and the distance from the limb to the level of the saddle is 3.00 m. (a) What must be the horizontal distance between the saddle and limb when the ranch hand makes his move? (b) How long is he in the air?

**34.** It is possible to shoot an arrow at a speed as high as 100 m/s. (a) If friction can be ignored, how high would an arrow launched at this speed rise if shot straight up? (b) How long would the arrow be in the air?

## Section 2.8 ■ Context Connection—Acceleration Required by Consumers

**35.** (a) Show that the largest and smallest average accelerations in Table 2.3 are correctly computed from the measured time intervals required for the cars to speed up from 0 to 60 mi/h. (b) Convert both of these accelerations to the standard SI unit. (c) Modeling each acceleration as constant, find the distance traveled by both cars as they speed up. (d) If an automobile were able to maintain an acceleration of magnitude $a = g = 9.80 \text{ m/s}^2$ on a horizontal roadway, what time interval would be required to accelerate from zero to 60.0 mi/h?

**36.** A certain automobile manufacturer claims that its deluxe sports car will accelerate from rest to a speed of 42.0 m/s in 8.00 s. (a) Determine the average acceleration of the car. (b) Assume that the car moves with constant acceleration. Find the distance the car travels in the first 8.00 s. (c) What is the speed of the car 10.0 s after it begins its motion if it can continue to move with the same acceleration?

**37.** A steam catapult launches a jet aircraft from the aircraft carrier *John C. Stennis,* giving it a speed of 175 mi/h in 2.50 s. (a) Find the average acceleration of the plane. (b) Modeling the acceleration as constant, find the distance the plane moves in this time interval.

**38.** *Vroom—vroom!* As soon as a traffic light turns green, a car speeds up from rest to 50.0 mi/h with constant acceleration 9.00 mi/h · s. In the adjoining bike lane, a cyclist speeds up from rest to 20.0 mi/h with constant acceleration 13.0 mi/h · s. Each vehicle maintains constant velocity after reaching its cruising speed. (a) For what time interval is the bicycle ahead of the car? (b) By what maximum distance does the bicycle lead the car?

## Additional Problems

*Note:* The human body can undergo brief accelerations up to 15 times the free-fall acceleration without injury or with only strained ligaments. Acceleration of long duration can do damage by preventing circulation of blood. Acceleration of larger magnitude can cause severe internal injuries, such as by tearing the aorta away from the heart. Problems 2.35, 2.37, and 2.39 through 2.41 deal with variously large accelerations of the human body that you can compare with the $15g$ datum.

**39.** For many years Colonel John P. Stapp, USAF, held the world's land speed record. He participated in studying whether a jet pilot could survive emergency ejection. On March 19, 1954, he rode a rocket-propelled sled that moved down a track at 632 mi/h. He and the sled were safely brought to rest in 1.40 s (Fig. P2.39). Determine (a) the negative acceleration he experienced and (b) the distance he traveled during this negative acceleration, assumed to be constant.

**40.** A woman is reported to have fallen 144 ft from the 17th floor of a building, landing on a metal ventilator box that she crushed to a depth of 18.0 in. She suffered only minor injuries. Ignoring air resistance, calculate (a) the speed of the woman just before she collided with the ventilator and (b) her average acceleration while in contact with the box. (c) Modeling her acceleration as constant, calculate the time interval it took to crush the box.

**41.** Jules Verne in 1865 suggested sending people to the Moon by firing a space capsule from a 220-m-long cannon with a final velocity of 10.97 km/s. What would have been

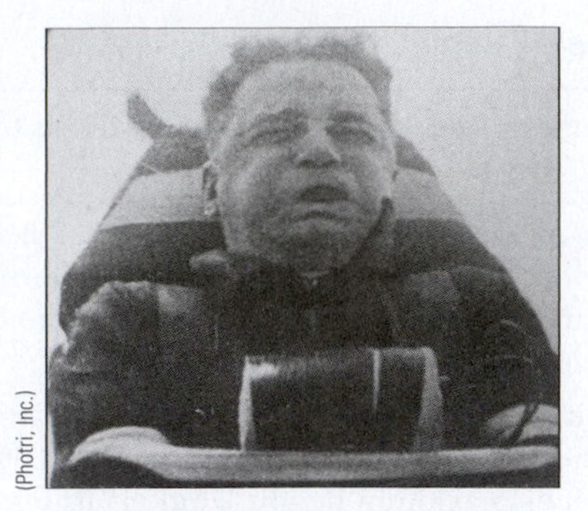


FIGURE P2.39 (*Left*) Col. John Stapp on the rocket sled. (*Right*) Col. Stapp's face is contorted by the stress of rapid negative acceleration.

the unrealistically large acceleration experienced by the space travelers during launch? Compare your answer with the free-fall acceleration 9.80 m/s$^2$.

**42.** **Review problem.** The biggest stuffed animal in the world is a snake 420 m long constructed by Norwegian children. Suppose the snake is laid out in a park as shown in Figure P2.42, forming two straight sides of a 105° angle, with one side 240 m long. Olaf and Inge run a race they invent. Inge runs directly from the tail of the snake to its head and Olaf starts from the same place at the same time but runs along the snake. If both children run steadily at 12.0 km/h, Inge reaches the head of the snake how much earlier than Olaf?

FIGURE P2.42

**43.** A ball starts from rest and accelerates at 0.500 m/s$^2$ while moving down an inclined plane 9.00 m long. When it reaches the bottom, the ball rolls up another plane, where it comes to rest after moving 15.0 m on that plane. (a) What is the speed of the ball at the bottom of the first plane? (b) During what time interval does the ball roll down the first plane? (c) What is the acceleration along the second plane? (d) What is the ball's speed 8.00 m along the second plane?

**44.** A glider on an air track carries a flag of length $\ell$ through a stationary photogate, which measures the time interval $\Delta t_d$ during which the flag blocks a beam of infrared light passing across the photogate. The ratio $v_d = \ell/\Delta t_d$ is the average velocity of the glider over this part of its motion. Assume that the glider moves with constant acceleration. (a) Argue for or against the idea that $v_d$ is equal to the instantaneous velocity of the glider when it is halfway through the photogate in space. (b) Argue for or against the idea that $v_d$ is equal to the instantaneous velocity of the glider when it is halfway through the photogate in time.

**45.** Liz rushes down onto a subway platform to find her train already departing. She stops and watches the cars go by. Each car is 8.60 m long. The first moves past her in 1.50 s and the second in 1.10 s. Find the constant acceleration of the train.

**46.** The Acela is the Porsche of American trains. Shown in Figure P2.46a, the electric train whose name is pronounced ah-SELL-ah is in service on the Washington–New York–Boston run. With two power cars and six coaches, it can carry 304 passengers at 170 mi/h. The carriages tilt as much as 6° from the vertical to prevent passengers from feeling pushed to the side as they go around curves. Its braking mechanism uses electric generators to recover its energy of motion. A velocity–time graph for the Acela is shown in Figure P2.46b. (a) Describe the motion of the train in each successive time interval. (b) Find the peak positive acceleration of the train in the motion graphed. (c) Find the train's displacement in miles between $t = 0$ and $t = 200$ s.

**47.** A test rocket is fired vertically upward from a well. A catapult gives it initial speed 80.0 m/s at ground level. Its engines then fire and it accelerates upward at 4.00 m/s$^2$ until it reaches an altitude of 1 000 m. At that point its engines fail and the rocket goes into free-fall, with an acceleration of $-9.80$ m/s$^2$. (a) How long is the rocket in motion above the ground? (b) What is its maximum altitude? (c) What is its velocity just before it collides with the Earth? (You will need to consider the motion while the engine is operating separately from the free-fall motion.)

**48.** A motorist drives along a straight road at a constant speed of 15.0 m/s. Just as she passes a parked motorcycle police officer, the officer starts to accelerate at 2.00 m/s$^2$ to overtake her. Assuming that the officer maintains this acceleration, (a) determine the time interval required for the

(a)

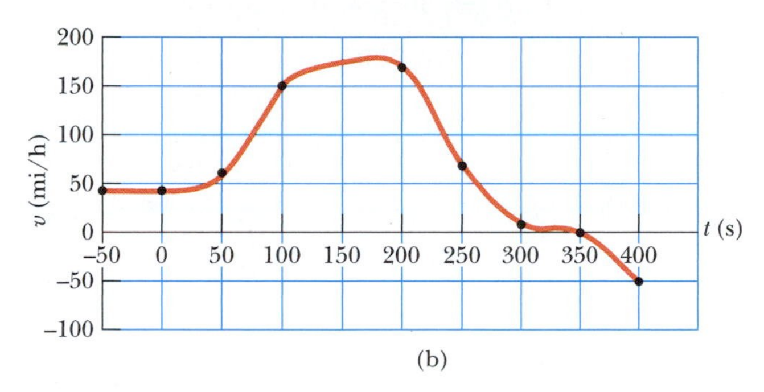


(b)

**FIGURE P2.46** (a) The Acela, 1 171 000 lb of cold steel thundering along at 150 mi/h. (b) Velocity versus time graph for the Acela.

police officer to reach the motorist. Find (b) the speed and (c) the total displacement of the officer as he overtakes the motorist.

**49.** Setting a world record in a 100-m race, Maggie and Judy cross the finish line in a dead heat, both taking 10.2 s. Accelerating uniformly, Maggie took 2.00 s and Judy 3.00 s to attain maximum speed, which they maintained for the rest of the race. (a) What was the acceleration of each sprinter? (b) What were their respective maximum speeds? (c) Which sprinter was ahead at the 6.00-s mark and by how much?

**50.** A commuter train travels between two downtown stations. Because the stations are only 1.00 km apart, the train never reaches its maximum possible cruising speed. During rush hour the engineer minimizes the time interval $\Delta t$ between two stations by accelerating at a rate $a_1 = 0.100\text{ m/s}^2$ for a time interval $\Delta t_1$ and then immediately braking with acceleration $a_2 = -0.500\text{ m/s}^2$ for a time interval $\Delta t_2$. Find the minimum time interval of travel $\Delta t$ and the time interval $\Delta t_1$.

**51.** An inquisitive physics student and mountain climber climbs a 50.0-m cliff that overhangs a calm pool of water. He throws two stones vertically downward, 1.00 s apart, and observes that they cause a single splash. The first stone has an initial speed of 2.00 m/s. (a) How long after release of the first stone do the two stones hit the water? (b) What initial velocity must the second stone have if the two stones are to hit simultaneously? (c) What is the speed of each stone at the instant the two hit the water?

**52.** A hard rubber ball, released at chest height, falls to the pavement and bounces back to nearly the same height. When it is in contact with the pavement, the lower side of the ball is temporarily flattened. Suppose the maximum depth of the dent is on the order of 1 cm. Compute an order-of-magnitude estimate for the maximum acceleration of the ball while it is in contact with the pavement. State your assumptions, the quantities you estimate, and the values you estimate for them.

**53.** To protect his food from hungry bears, a Boy Scout raises his food pack with a rope that is thrown over a tree limb at height $h$ above his hands. He walks away from the vertical rope with constant velocity $v_{boy}$, holding the free end of the rope in his hands (Fig. P2.53). (a) Show that the speed $v$ of the food pack is given by $x(x^2 + h^2)^{-1/2}\, v_{boy}$ where $x$ is the distance he has walked away from the vertical rope. (b) Show that the acceleration $a$ of the food pack is $h^2(x^2 + h^2)^{-3/2} v_{boy}^2$. (c) What values do the acceleration and velocity $v$ have shortly after the boy leaves the point under the pack ($x = 0$)? (d) What values do the pack's velocity and acceleration approach as the distance $x$ continues to increase?

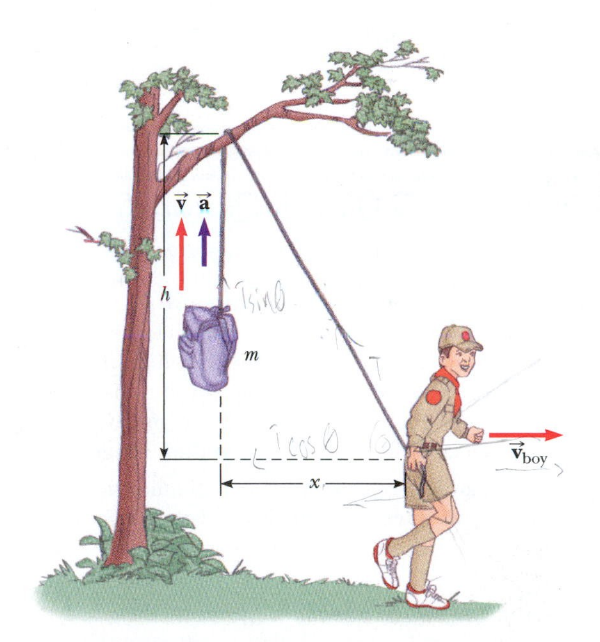


**FIGURE P2.53** Problems 2.53 and 2.54.

**54.** In Problem 2.53, let the height $h$ equal 6.00 m and the speed $v_{boy}$ equal 2.00 m/s. Assume that the food pack starts from rest. (a) Tabulate and graph the speed–time graph. (b) Tabulate and graph the acceleration–time graph. Let the range of time be from 0 s to 5.00 s and the time intervals be 0.500 s.

**55.** A rock is dropped from rest into a well. (a) The sound of the splash is heard 2.40 s after the rock is released from rest. How far below the top of the well is the surface of the water? The speed of sound in air (at the ambient temperature) is 336 m/s. (b) If the travel time for the sound is ignored, what percentage error is introduced when the depth of the well is calculated?

**56.** Astronauts on a distant planet toss a rock into the air. With the aid of a camera that takes pictures at a steady rate, they record the height of the rock as a function of time as given in the Table P2.56. (a) Find the average velocity of the rock in the time interval between each measurement and the next. (b) Using these average velocities to approximate instantaneous velocities at the midpoints of the time intervals, make a graph of velocity as a function of time. Does the rock move with constant acceleration? If so, plot a straight line of best fit on the graph and calculate its slope to find the acceleration.

**TABLE P2.56** Height of a Rock Versus Time

| Time (s) | Height (m) | Time (s) | Height (m) |
|---|---|---|---|
| 0.00 | 5.00 | 2.75 | 7.62 |
| 0.25 | 5.75 | 3.00 | 7.25 |
| 0.50 | 6.40 | 3.25 | 6.77 |
| 0.75 | 6.94 | 3.50 | 6.20 |
| 1.00 | 7.38 | 3.75 | 5.52 |
| 1.25 | 7.72 | 4.00 | 4.73 |
| 1.50 | 7.96 | 4.25 | 3.85 |
| 1.75 | 8.10 | 4.50 | 2.86 |
| 2.00 | 8.13 | 4.75 | 1.77 |
| 2.25 | 8.07 | 5.00 | 0.58 |
| 2.50 | 7.90 | | |

**57.** Two objects, A and B, are connected by a rigid rod that has a length $L$. The objects slide along perpendicular guide rails, as shown in Figure P2.57. If A slides to the left with a constant speed $v$, find the velocity of B when $\alpha = 60.0°$.

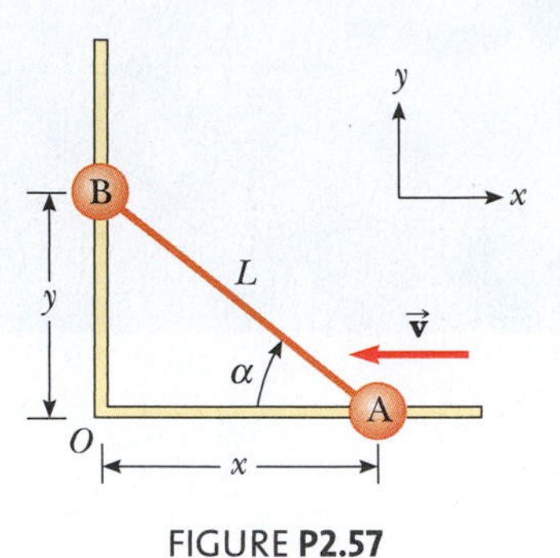


FIGURE **P2.57**

## ANSWERS TO QUICK QUIZZES

**2.1** (c). If the particle moves along a line without changing direction, the displacement and distance over any time interval will be the same. As a result, the magnitude of the average velocity and the average speed will be the same. If the particle reverses direction, however, the displacement will be less than the distance. In turn, the magnitude of the average velocity will be smaller than the average speed.

**2.2** (b). Regardless of your speeds at all other times, if your instantaneous speed at the instant that it is measured is higher than the speed limit, you may receive a speeding ticket.

**2.3** Graph (a) has a constant slope, indicating a constant acceleration; this situation is represented by graph (e). Graph (b) represents a speed that is increasing constantly but not at a uniform rate. Therefore, the acceleration must be increasing, and the graph that best indicates this situation is (d). Graph (c) depicts a velocity that first increases at a constant rate, indicating constant acceleration. Then the velocity stops increasing and becomes constant, indicating zero acceleration. The best match to this situation is graph (f).

**2.4** (b). If the car is slowing down, a force must be acting in the direction opposite to its velocity.

**2.5** (c). If a particle with constant acceleration stops and its acceleration remains constant, it must begin to move again in the opposite direction. If it did not, the acceleration would change from its original constant value to zero. Choice (a) is not correct because the direction of acceleration is independent of the direction of the velocity. Choice (b) is not correct either. For example, a car moving in the negative $x$ direction and slowing down has a positive acceleration.

**2.6** (e). For the entire time interval the ball is in free-fall, the acceleration is that due to gravity.

# CHAPTER 3

# Motion in Two Dimensions

(© Arndt/Premium Stock/PictureQuest)

Lava spews from a volcanic eruption. Notice the parabolic paths of embers projected into the air. We will find in this chapter that all projectiles follow a parabolic path in the absence of air resistance.

## CHAPTER OUTLINE


3.1 The Position, Velocity, and Acceleration Vectors

3.2 Two-Dimensional Motion with Constant Acceleration

3.3 Projectile Motion

3.4 The Particle in Uniform Circular Motion

3.5 Tangential and Radial Acceleration

3.6 Relative Velocity

3.7 Context Connection—Lateral Acceleration of Automobiles

SUMMARY


In this chapter, we shall study the kinematics of an object that can be modeled as a particle moving in a plane. This motion is two dimensional. Some common examples of motion in a plane are the motions of satellites in orbit around the Earth, projectiles such as a thrown baseball, and the motion of electrons in uniform electric fields. We shall also study a particle in uniform circular motion and discuss various aspects of particles moving in curved paths.

## 3.1 THE POSITION, VELOCITY, AND ACCELERATION VECTORS

In Chapter 2, we found that the motion of a particle moving along a straight line is completely specified if its position is known as a function of time. Now let us extend this idea to motion in the $xy$ plane. We will find equations for position and velocity that are the same as those in Chapter 2 except for their vector nature.

We begin by describing the position of a particle with a **position vector** $\vec{\mathbf{r}}$, drawn from the origin of a coordinate system to the location of the particle in the $xy$ plane, as in Figure 3.1. At time $t_i$, the particle is at the point Ⓐ, and at some later time $t_f$, the particle is at Ⓑ, where the subscripts $i$ and $f$ refer to initial and final values. As the particle moves from Ⓐ to Ⓑ in the time interval $\Delta t = t_f - t_i$, the position vector changes from $\vec{\mathbf{r}}_i$ to $\vec{\mathbf{r}}_f$. As we learned in Chapter 2, the displacement of a particle is the difference between its final position and its initial position:

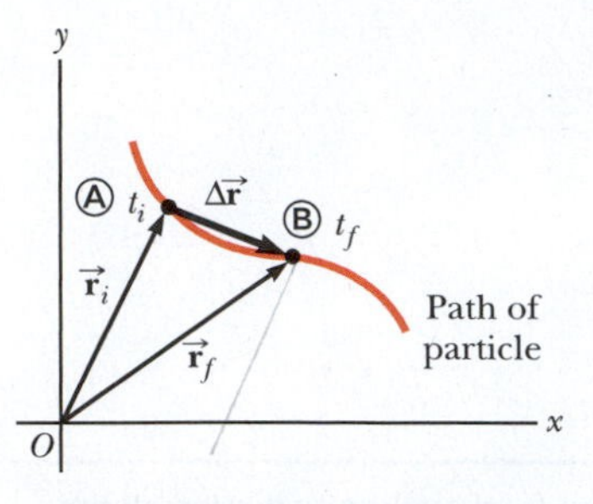


**FIGURE 3.1** A particle moving in the $xy$ plane is located with the position vector $\vec{\mathbf{r}}$ drawn from the origin to the particle. The displacement of the particle as it moves from Ⓐ to Ⓑ in the time interval $\Delta t = t_f - t_i$ is equal to the vector $\Delta\vec{\mathbf{r}} \equiv \vec{\mathbf{r}}_f - \vec{\mathbf{r}}_i$.

$$\Delta\vec{\mathbf{r}} \equiv \vec{\mathbf{r}}_f - \vec{\mathbf{r}}_i \qquad [3.1]$$

The direction of $\Delta\vec{\mathbf{r}}$ is indicated in Figure 3.1.

The **average velocity** $\vec{\mathbf{v}}_{avg}$ of the particle during the time interval $\Delta t$ is defined as the ratio of the displacement to the time interval:

■ Definition of average velocity

$$\vec{\mathbf{v}}_{avg} \equiv \frac{\Delta\vec{\mathbf{r}}}{\Delta t} \qquad [3.2]$$

Because displacement is a vector quantity and the time interval is a scalar quantity, we conclude that the average velocity is a *vector* quantity directed along $\Delta\vec{\mathbf{r}}$. The average velocity between points Ⓐ and Ⓑ is *independent of the path* between the two points. That is because the average velocity is proportional to the displacement, which in turn depends only on the initial and final position vectors and not on the path taken between those two points. As with one-dimensional motion, if a particle starts its motion at some point and returns to this point via any path, its average velocity is zero for this trip because its displacement is zero.

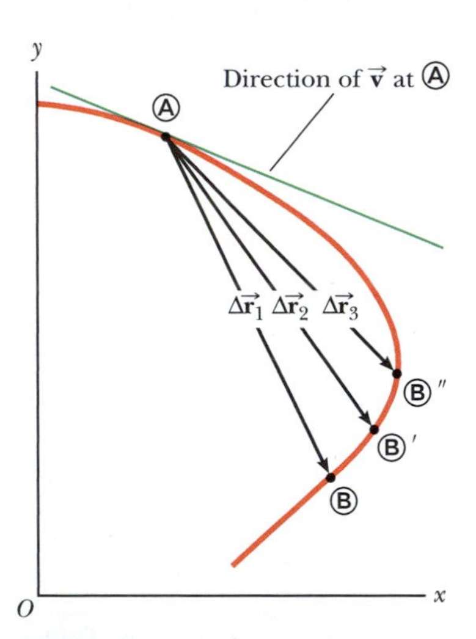


**FIGURE 3.2** As a particle moves between two points, its average velocity is in the direction of the displacement vector $\Delta\vec{\mathbf{r}}$. As the end point of the path is moved from Ⓑ to Ⓑ′ to Ⓑ″, the respective displacements and corresponding time intervals become smaller and smaller. In the limit that the end point approaches Ⓐ, $\Delta t$ approaches zero and the direction of $\Delta\vec{\mathbf{r}}$ approaches that of the line tangent to the curve at Ⓐ. By definition, the instantaneous velocity at Ⓐ is in the direction of this tangent line.

Consider again the motion of a particle between two points in the $xy$ plane, as shown in Figure 3.2. As the time intervals over which we observe the motion become smaller and smaller, the direction of the displacement approaches that of the line tangent to the path at the point Ⓐ.

The **instantaneous velocity** $\vec{\mathbf{v}}$ is defined as the limit of the average velocity $\Delta\vec{\mathbf{r}}/\Delta t$ as $\Delta t$ approaches zero:

$$\vec{\mathbf{v}} \equiv \lim_{\Delta t \to 0} \frac{\Delta\vec{\mathbf{r}}}{\Delta t} = \frac{d\vec{\mathbf{r}}}{dt} \qquad [3.3]$$

That is, the instantaneous velocity equals the derivative of the position vector with respect to time. The direction of the instantaneous velocity vector at any point in a particle's path is along a line that is tangent to the path at that point and in the direction of motion. The magnitude of the instantaneous velocity is called the *speed.*

As a particle moves from point Ⓐ to point Ⓑ along some path as in Figure 3.3, its instantaneous velocity changes from $\vec{\mathbf{v}}_i$ at time $t_i$ to $\vec{\mathbf{v}}_f$ at time $t_f$. The **average acceleration** $\vec{\mathbf{a}}_{avg}$ of a particle over a time interval is defined as the ratio of the change in the instantaneous velocity $\Delta\vec{\mathbf{v}}$ to the time interval $\Delta t$:

■ Definition of average acceleration

$$\vec{\mathbf{a}}_{avg} \equiv \frac{\vec{\mathbf{v}}_f - \vec{\mathbf{v}}_i}{t_f - t_i} = \frac{\Delta\vec{\mathbf{v}}}{\Delta t} \qquad [3.4]$$

Because the average acceleration is the ratio of a vector quantity $\Delta\vec{\mathbf{v}}$ and a scalar quantity $\Delta t$, we conclude that $\vec{\mathbf{a}}_{avg}$ is a vector quantity directed along $\Delta\vec{\mathbf{v}}$. As

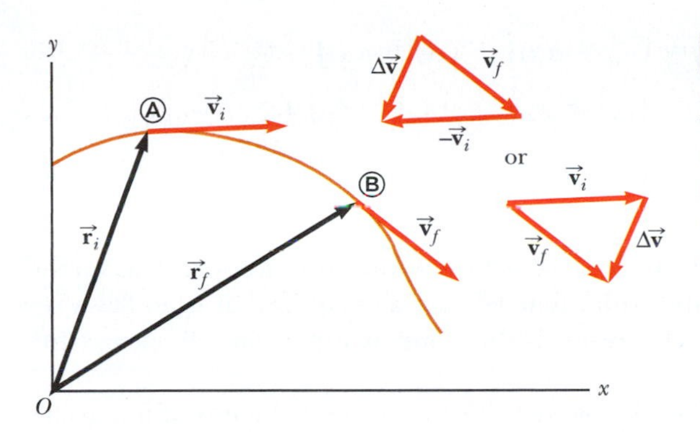


FIGURE 3.3 A particle moves from position Ⓐ to position Ⓑ. Its velocity vector changes from $\vec{v}_i$ at time $t_i$ to $\vec{v}_f$ at time $t_f$. The vector addition diagrams at the upper right show two ways of determining the vector $\Delta\vec{v}$ from the initial and final velocities.

### PITFALL PREVENTION 3.1

**VECTOR ADDITION** The vector addition that was discussed in Chapter 1 involves displacement vectors. Because we are familiar with movements through space in our everyday experience, the addition of displacement vectors can be understood easily. The notion of vector addition can be applied to *any* type of vector quantity. Figure 3.3, for example, shows the addition of velocity vectors using the tip-to-tail approach.

indicated in Figure 3.3, the direction of $\Delta\vec{v}$ is found by adding the vector $-\vec{v}_i$ (the negative of $\vec{v}_i$) to the vector $\vec{v}_f$ because by definition $\Delta\vec{v} = \vec{v}_f - \vec{v}_i$.

The **instantaneous acceleration** $\vec{a}$ is defined as the limiting value of the ratio $\Delta\vec{v}/\Delta t$ as $\Delta t$ approaches zero:

■ Definition of instantaneous acceleration

$$\vec{a} \equiv \lim_{\Delta t \to 0} \frac{\Delta\vec{v}}{\Delta t} = \frac{d\vec{v}}{dt} \qquad [3.5]$$

That is, the instantaneous acceleration equals the derivative of the velocity vector with respect to time.

It is important to recognize that various changes can occur that represent a particle undergoing an acceleration. First, the magnitude of the velocity vector (the speed) may change with time as in straight-line (one-dimensional) motion. Second, the direction of the velocity vector may change with time as its magnitude remains constant. Finally, both the magnitude and the direction of the velocity vector may change.

**QUICK QUIZ 3.1** Consider the following controls in an automobile: gas pedal, brake, steering wheel. The controls in this list that cause an acceleration of the car are (**a**) all three controls, (**b**) the gas pedal and the brake, (**c**) only the brake, or (**d**) only the gas pedal.

## 3.2 TWO-DIMENSIONAL MOTION WITH CONSTANT ACCELERATION

Let us consider two-dimensional motion during which the magnitude and direction of the acceleration remain unchanged. In this situation, we shall investigate motion as a two-dimensional version of the analysis in Section 2.6.

The motion of a particle can be determined if its position vector $\vec{r}$ is known at all times. The position vector for a particle moving in the *xy* plane can be written

$$\vec{r} = x\hat{i} + y\hat{j} \qquad [3.6]$$

where $x$, $y$, and $\vec{r}$ change with time as the particle moves. If the position vector is known, the velocity of the particle can be obtained from Equations 3.3 and 3.6:

$$\vec{v} = \frac{d\vec{r}}{dt} = \frac{dx}{dt}\hat{i} + \frac{dy}{dt}\hat{j} = v_x\hat{i} + v_y\hat{j} \qquad [3.7]$$

Because we are assuming that $\vec{a}$ is constant in this discussion, its components $a_x$ and $a_y$ are also constants. Therefore, we can apply the equations of kinematics to the $x$ and $y$ components of the velocity vector separately. Substituting $v_x = v_{xf} = v_{xi} + a_x t$ and $v_y = v_{yf} = v_{yi} + a_y t$ into Equation 3.7 gives

$$\vec{\mathbf{v}}_f = (v_{xi} + a_x t)\hat{\mathbf{i}} + (v_{yi} + a_y t)\hat{\mathbf{j}}$$
$$= (v_{xi}\hat{\mathbf{i}} + v_{yi}\hat{\mathbf{j}}) + (a_x\hat{\mathbf{i}} + a_y\hat{\mathbf{j}})t$$

■ Velocity vector as a function of time for a particle under constant acceleration

$$\vec{\mathbf{v}}_f = \vec{\mathbf{v}}_i + \vec{\mathbf{a}}t \quad [3.8]$$

This result states that the velocity $\vec{\mathbf{v}}_f$ of a particle at some time $t$ equals the vector sum of its initial velocity $\vec{\mathbf{v}}_i$ and the additional velocity $\vec{\mathbf{a}}t$ acquired at time $t$ as a result of its constant acceleration. This result is the same as Equation 2.9, except for its vector nature.

Similarly, from Equation 2.12 we know that the $x$ and $y$ coordinates of a particle moving with constant acceleration are

$$x_f = x_i + v_{xi}t + \tfrac{1}{2}a_x t^2 \quad \text{and} \quad y_f = y_i + v_{yi}t + \tfrac{1}{2}a_y t^2$$

Substituting these expressions into Equation 3.6 gives

$$\vec{\mathbf{r}}_f = (x_i + v_{xi}t + \tfrac{1}{2}a_x t^2)\hat{\mathbf{i}} + (y_i + v_{yi}t + \tfrac{1}{2}a_y t^2)\hat{\mathbf{j}}$$
$$= (x_i\hat{\mathbf{i}} + y_i\hat{\mathbf{j}}) + (v_{xi}\hat{\mathbf{i}} + v_{yi}\hat{\mathbf{j}})t + \tfrac{1}{2}(a_x\hat{\mathbf{i}} + a_y\hat{\mathbf{j}})t^2$$

■ Position vector as a function of time for a particle under constant acceleration

$$\vec{\mathbf{r}}_f = \vec{\mathbf{r}}_i + \vec{\mathbf{v}}_i t + \tfrac{1}{2}\vec{\mathbf{a}}t^2 \quad [3.9]$$

This equation implies that the final position vector $\vec{\mathbf{r}}_f$ is the vector sum of the initial position vector $\vec{\mathbf{r}}_i$ plus a displacement $\vec{\mathbf{v}}_i t$, arising from the initial velocity of the particle, and a displacement $\frac{1}{2}\vec{\mathbf{a}}t^2$, resulting from the uniform acceleration of the particle. It is the same as Equation 2.12 except for its vector nature.

Pictorial representations of Equations 3.8 and 3.9 are shown in Active Figures 3.4a and 3.4b. Note from Active Figure 3.4b that $\vec{\mathbf{r}}_f$ is generally not along the direction of $\vec{\mathbf{v}}_i$ or $\vec{\mathbf{a}}$ because the relationship between these quantities is a vector expression. For the same reason, from Active Figure 3.4a we see that $\vec{\mathbf{v}}_f$ is generally not along the direction of $\vec{\mathbf{v}}_i$ or $\vec{\mathbf{a}}$. Finally, if we compare the two figures, we see that $\vec{\mathbf{v}}_f$ and $\vec{\mathbf{r}}_f$ are not in the same direction.

Because Equations 3.8 and 3.9 are *vector* expressions, we may also write their $x$ and $y$ component equations:

$$\vec{\mathbf{v}}_f = \vec{\mathbf{v}}_i + \vec{\mathbf{a}}t \quad \rightarrow \quad \begin{cases} v_{xf} = v_{xi} + a_x t \\ v_{yf} = v_{yi} + a_y t \end{cases}$$

$$\vec{\mathbf{r}}_f = \vec{\mathbf{r}}_i + \vec{\mathbf{v}}_i t + \tfrac{1}{2}\vec{\mathbf{a}}t^2 \quad \rightarrow \quad \begin{cases} x_f = x_i + v_{xi}t + \tfrac{1}{2}a_x t^2 \\ y_f = y_i + v_{yi}t + \tfrac{1}{2}a_y t^2 \end{cases}$$

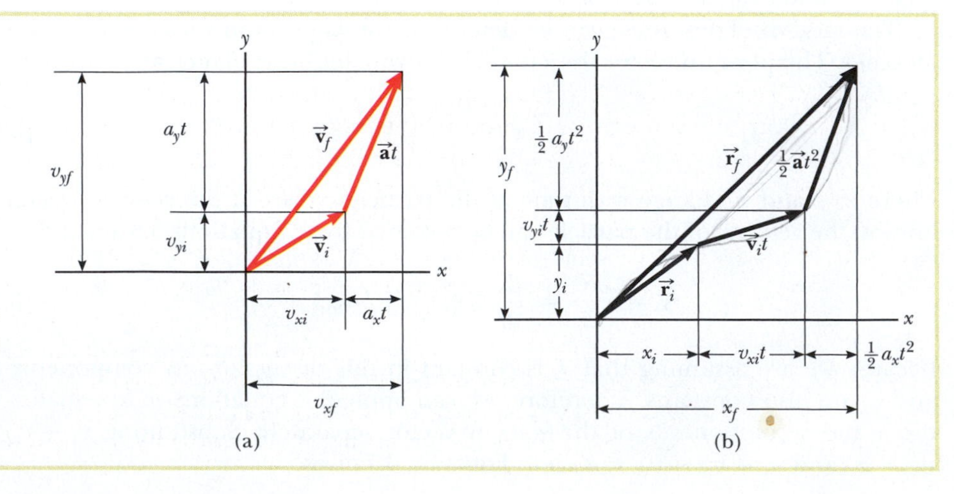


ACTIVE FIGURE 3.4

Vector representations and components of (a) the velocity and (b) the position of a particle under constant acceleration $\vec{\mathbf{a}}$.

**PhysicsNow™** Log into PhysicsNow at **www.pop4e.com** and go to Active Figure 3.4 to investigate the effect of different initial positions and velocities on the final position and velocity (for constant acceleration).

These components are illustrated in Active Figure 3.4. In other words, **two-dimensional motion having constant acceleration is equivalent to two *independent* motions in the $x$ and $y$ directions having constant accelerations $a_x$ and $a_y$.** Motion in the $x$ direction does not affect motion in the $y$ direction and vice versa. Therefore, there is no new model for a particle under two-dimensional constant acceleration; the appropriate model is just the one-dimensional particle under constant acceleration applied twice, in the $x$ and $y$ directions separately!

### EXAMPLE 3.1 Motion in a Plane

A particle moves through the origin of an $xy$ coordinate system at $t = 0$ with initial velocity $\vec{\mathbf{v}}_i = (20\hat{\mathbf{i}} - 15\hat{\mathbf{j}})$ m/s. The particle moves in the $xy$ plane with an acceleration $\vec{\mathbf{a}} = 4.0\hat{\mathbf{i}}$ m/s$^2$.

**A** Determine the components of velocity as a function of time and the total velocity vector at any time.

**Solution** *Conceptualize* by establishing the mental representation and thinking about what the particle is doing. From the given information we see that the particle starts off moving to the right and downward and accelerates only toward the right. What will the particle do under these conditions? It may help if you draw a pictorial representation. To *categorize* consider that because the acceleration is only in the $x$ direction, the moving particle can be modeled as one under constant acceleration in the $x$ direction and one under constant velocity in the $y$ direction.

To *analyze* the situation, we identify $v_{xi} = 20$ m/s and $a_x = 4.0$ m/s$^2$. The equations of kinematics give us, for the $x$ direction,

$$v_{xf} = v_{xi} + a_x t = (20 + 4.0t)$$

Also, with $v_{yi} = -15$ m/s and $a_y = 0$,

$$v_{yf} = v_{yi} + a_y t = -15 \text{ m/s}$$

Therefore, using these results and noting that the velocity vector $\vec{\mathbf{v}}_f$ has two components, we find

$$\vec{\mathbf{v}}_f = v_{xf}\hat{\mathbf{i}} + v_{yf}\hat{\mathbf{j}} = [(20 + 4.0t)\hat{\mathbf{i}} - 15\hat{\mathbf{j}}]$$

Note that only the $x$ component varies in time, reflecting that acceleration occurs only in the $x$ direction.

**B** Calculate the velocity and speed of the particle at $t = 5.0$ s.

**Solution** At $t = 5.0$ s, the velocity expression from part A gives

$$\vec{\mathbf{v}}_f = \{[20 + 4(5.0)]\hat{\mathbf{i}} - 15\hat{\mathbf{j}}\} \text{ m/s} = (40\hat{\mathbf{i}} - 15\hat{\mathbf{j}}) \text{ m/s}$$

That is, at $t = 5.0$ s, $v_{xf} = 40$ m/s and $v_{yf} = -15$ m/s. To determine the angle $\theta$ that $\vec{\mathbf{v}}_f$ makes with the $x$ axis, use $\tan\theta = v_{yf}/v_{xf}$, or

$$\theta = \tan^{-1}\left(\frac{v_{yf}}{v_{xf}}\right) = \tan^{-1}\left(\frac{-15 \text{ m/s}}{40 \text{ m/s}}\right) = -21°$$

The speed is the magnitude of $\vec{\mathbf{v}}_f$:

$$v_f = |\vec{\mathbf{v}}_f| = \sqrt{v_{xf}^2 + v_{yf}^2} = \sqrt{(40)^2 + (-15)^2} \text{ m/s}$$

$$= 43 \text{ m/s}$$

Now we *finalize*. In examining our result, we find that $v_f > v_i$. Does that make sense to you? Is it consistent with your mental representation?

## 3.3 PROJECTILE MOTION

Anyone who has observed a baseball in motion (or, for that matter, any object thrown into the air) has observed projectile motion. The ball moves in a curved path when thrown at some angle with respect to the Earth's surface. This very common form of motion is surprisingly simple to analyze if the following two assumptions are made when building a model for these types of problems: (1) the free-fall acceleration $g$ is constant over the range of motion and is directed downward,[1] and (2) the effect of air resistance is negligible.[2] With these assumptions, the path of a

[1] In effect, this approximation is equivalent to assuming that the Earth is flat within the range of motion considered and that the maximum height of the object is small compared to the radius of the Earth.

[2] This approximation is often *not* justified, especially at high velocities. In addition, the spin of a projectile, such as a baseball, can give rise to some very interesting effects associated with aerodynamic forces (for example, a curve ball thrown by a pitcher).

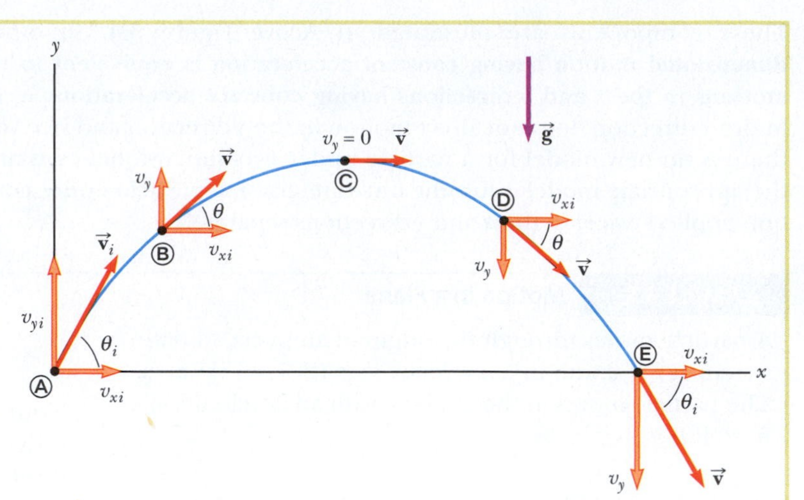


**ACTIVE FIGURE 3.5**

The parabolic path of a projectile that leaves the origin (point Ⓐ) with a velocity $\vec{v}_i$. The velocity vector $\vec{v}$ changes with time in both magnitude and direction. The change in the velocity vector is the result of acceleration in the negative $y$ direction. The $x$ component of velocity remains constant in time because no acceleration occurs in the horizontal direction. The $y$ component of velocity is zero at the peak of the path (point Ⓒ).

**Physics Now™** Log into PhysicsNow at **www.pop4e.com** and go to Active Figure 3.5 to change the launch angle and initial speed. You can also observe the changing components of velocity along the trajectory of the projectile.

projectile, called its *trajectory,* is *always* a parabola. **We shall use a simplification model based on these assumptions throughout this chapter.**

If we choose our reference frame such that the $y$ direction is vertical and positive upward, $a_y = -g$ (as in one-dimensional free-fall) and $a_x = 0$ (because the only possible horizontal acceleration is due to air resistance, and it is ignored). Furthermore, let us assume that at $t = 0$, the projectile leaves the origin (point Ⓐ, $x_i = y_i = 0$) with speed $v_i$, as in Active Figure 3.5. If the vector $\vec{v}_i$ makes an angle $\theta_i$ with the horizontal, we can identify a right triangle in the diagram as a geometric model, and from the definitions of the cosine and sine functions we have

$$\cos \theta_i = \frac{v_{xi}}{v_i} \qquad \text{and} \qquad \sin \theta_i = \frac{v_{yi}}{v_i}$$

Therefore, the initial $x$ and $y$ components of velocity are

$$v_{xi} = v_i \cos \theta_i \qquad \text{and} \qquad v_{yi} = v_i \sin \theta_i$$

Substituting these expressions into Equations 3.8 and 3.9 with $a_x = 0$ and $a_y = -g$ gives the velocity components and position coordinates for the projectile at any time $t$:

$$v_{xf} = v_{xi} = v_i \cos \theta_i = \text{constant} \qquad [3.10]$$

$$v_{yf} = v_{yi} - gt = v_i \sin \theta_i - gt \qquad [3.11]$$

$$x_f = x_i + v_{xi}t = (v_i \cos \theta_i)\, t \qquad [3.12]$$

$$y_f = y_i + v_{yi}t - \tfrac{1}{2}gt^2 = (v_i \sin \theta_i)\, t - \tfrac{1}{2}gt^2 \qquad [3.13]$$

From Equation 3.10 we see that $v_{xf}$ remains constant in time and is equal to $v_{xi}$; there is no horizontal component of acceleration. Therefore, we model the horizontal motion as that of a particle under constant velocity. For the $y$ motion, note that the equations for $v_{yf}$ and $y_f$ are similar to Equations 2.9 and 2.12 for freely falling objects. Therefore, we can apply the model of a particle under constant acceleration to the $y$ component. In fact, *all* the equations of kinematics developed in Chapter 2 are applicable to projectile motion.

A welder cuts holes through a heavy metal construction beam with a hot torch. The sparks generated in the process follow parabolic paths. ■

If we solve for $t$ in Equation 3.12 and substitute this expression for $t$ into Equation 3.13, we find that

$$y_f = (\tan \theta_i)\, x_f - \left(\frac{g}{2v_i^2 \cos^2 \theta_i}\right) x_f^2 \qquad [3.14]$$

which is valid for angles in the range $0 < \theta_i < \pi/2$. This expression is of the form $y = ax - bx^2$, which is the equation of a parabola that passes through the origin. Thus, we have proven that the trajectory of a projectile can be geometrically modeled as a parabola. The trajectory is *completely* specified if $v_i$ and $\theta_i$ are known.

The vector expression for the position of the projectile as a function of time follows directly from Equation 3.9, with $\vec{\mathbf{a}} = \vec{\mathbf{g}}$:

$$\vec{\mathbf{r}}_f = \vec{\mathbf{r}}_i + \vec{\mathbf{v}}_i t + \tfrac{1}{2}\vec{\mathbf{g}} t^2$$

This equation gives the same information as the combination of Equations 3.12 and 3.13 and is plotted in Figure 3.6. Note that this expression for $\vec{\mathbf{r}}_f$ is consistent with Equation 3.13 because the expression for $\vec{\mathbf{r}}_f$ is a vector equation and $\vec{\mathbf{a}} = \vec{\mathbf{g}} = -g\hat{\mathbf{j}}$ when the upward direction is taken to be positive.

The position of a particle can be considered the sum of its original position $\vec{\mathbf{r}}_i$, the term $\vec{\mathbf{v}}_i t$, which would be the displacement if no acceleration were present, and the term $\frac{1}{2}\vec{\mathbf{g}} t^2$, which arises from the acceleration caused by gravity. In other words, if no gravitational acceleration occurred, the particle would continue to move along a straight path in the direction of $\vec{\mathbf{v}}_i$.

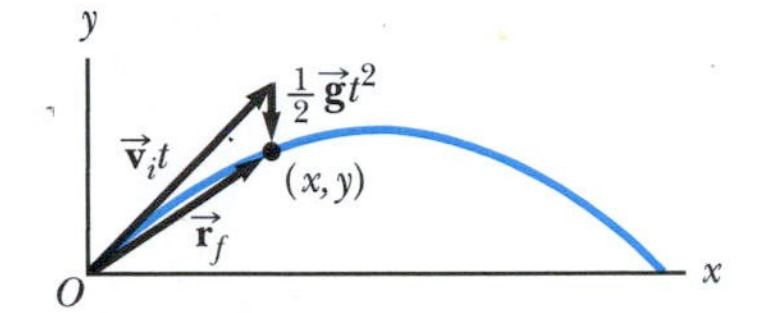


**FIGURE 3.6** The position vector $\vec{\mathbf{r}}_f$ of a projectile whose initial velocity at the origin is $\vec{\mathbf{v}}_i$. The vector $\vec{\mathbf{v}}_i t$ would be the position vector of the projectile if gravity were absent and the vector $\frac{1}{2}\vec{\mathbf{g}}t^2$ is the particle's vertical displacement due to its downward gravitational acceleration.

**QUICK QUIZ 3.2** As a projectile thrown upward moves in its parabolic path (such as in Figure 3.6), at what point along its path are the velocity and acceleration vectors for the projectile perpendicular to each other? **(a)** nowhere **(b)** the highest point **(c)** the launch point At what point are the velocity and acceleration vectors for the projectile parallel to each other? **(d)** nowhere **(e)** the highest point **(f)** the launch point

## Horizontal Range and Maximum Height of a Projectile

Let us assume that a projectile is launched over flat ground from the origin at $t = 0$ with a positive $v_y$ component, as in Figure 3.7. There are two special points that are interesting to analyze: the peak point Ⓐ, which has Cartesian coordinates $(R/2, h)$, and the landing point Ⓑ, having coordinates $(R, 0)$. The distance $R$ is called the *horizontal range* of the projectile, and $h$ is its *maximum height*. Because of the symmetry of the trajectory, the projectile is at the maximum height $h$ when its $x$ position is half the range $R$. Let us find $h$ and $R$ in terms of $v_i$, $\theta_i$, and $g$.

We can determine $h$ by noting that at the peak $v_{yA} = 0$. Therefore, Equation 3.11 can be used to determine the time $t_A$ at which the projectile reaches the peak:

$$t_A = \frac{v_i \sin \theta_i}{g}$$

Substituting this expression for $t_A$ into Equation 3.13 and replacing $y_f$ with $h$ gives $h$ in terms of $v_i$ and $\theta_i$:

$$h = (v_i \sin \theta_i)\frac{v_i \sin \theta_i}{g} - \tfrac{1}{2}g\left(\frac{v_i \sin \theta_i}{g}\right)^2$$

$$h = \frac{v_i^2 \sin^2 \theta_i}{2g} \qquad [3.15]$$

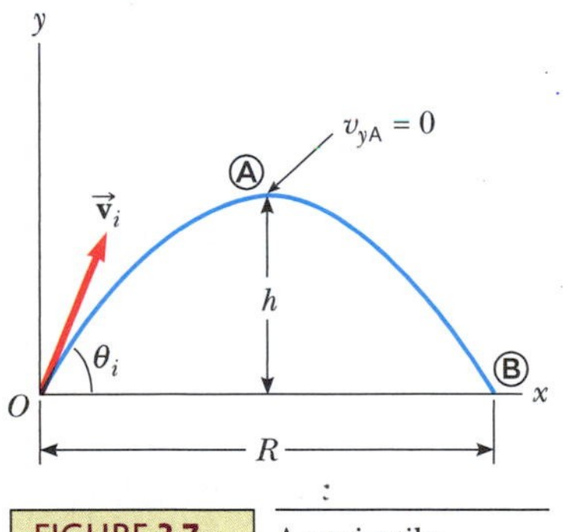


**FIGURE 3.7** A projectile launched from the origin at $t = 0$ with an initial velocity $\vec{\mathbf{v}}_i$. The maximum height of the projectile is $h$, and its horizontal range is $R$. At Ⓐ, the peak of the trajectory, the projectile has coordinates $(R/2, h)$.

Notice from the mathematical representation how you could increase the maximum height $h$: You could launch the projectile with a larger initial velocity, at a higher angle, or at a location with lower free-fall acceleration, such as on the Moon. Is that consistent with your mental representation of this situation?

The range $R$ is the horizontal distance traveled in twice the time interval required to reach the peak. Equivalently, we are seeking the position of the projectile

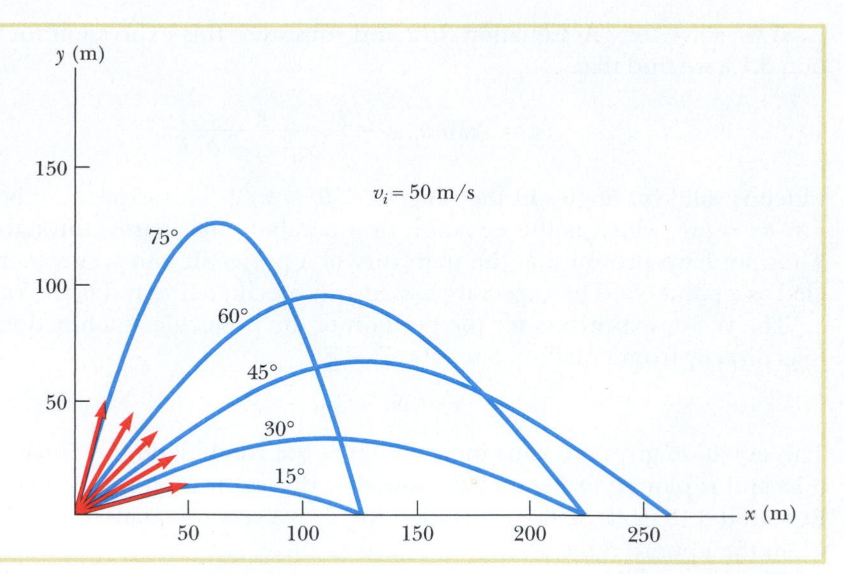


**ACTIVE FIGURE 3.8**
A projectile launched from the origin with an initial speed of 50 m/s at various angles of projection. Note that complementary values of $\theta_i$ will result in the same value of $R$.

**PhysicsNow™** Log into PhysicsNow at **www.pop4e.com** and go to Active Figure 3.8, where you can vary the projection angle to observe the effect on the trajectory and measure the flight time.

at a time $2t_A$. Using Equation 3.12 and noting that $x_f = R$ at $t = 2t_A$, we find that

$$R = (v_i \cos \theta_i)2t_A = (v_i \cos \theta_i)\frac{2v_i \sin \theta_i}{g} = \frac{2v_i^2 \sin \theta_i \cos \theta_i}{g}$$

Because $\sin 2\theta = 2 \sin \theta \cos \theta$, $R$ can be written in the more compact form

$$R = \frac{v_i^2 \sin 2\theta_i}{g} \qquad [3.16]$$

**PITFALL PREVENTION 3.2**

**THE HEIGHT AND RANGE EQUATIONS** Keep in mind that Equations 3.15 and 3.16 are useful for calculating $h$ and $R$ only for a symmetric path, as shown in Figure 3.7. If the path is not symmetric, *do not use these equations*. The general expressions given by Equations 3.10 through 3.13 are the *more important* results because they give the coordinates and velocity components of the projectile at *any* time $t$ for *any* trajectory.

Notice from the mathematical expression how you could increase the range $R$: You could launch the projectile with a larger initial velocity or at a location with lower free-fall acceleration, such as on the Moon. Is that consistent with your mental representation of this situation?

The range also depends on the angle of the initial velocity vector. The maximum possible value of $R$ from Equation 3.16 is given by $R_{max} = v_i^2/g$. This result follows from the maximum value of $\sin 2\theta_i$ being unity, which occurs when $2\theta_i = 90°$. Therefore, $R$ is a maximum when $\theta_i = 45°$.

Active Figure 3.8 illustrates various trajectories for a projectile of a given initial speed. As you can see, the range is a maximum for $\theta_i = 45°$. In addition, for any $\theta_i$ other than 45°, a point with coordinates $(R, 0)$ can be reached by using either one of two complementary values of $\theta_i$, such as 75° and 15°. Of course, the maximum height and the time of flight will be different for these two values of $\theta_i$.

**QUICK QUIZ 3.3** Rank the launch angles for the five paths in Active Figure 3.8 with respect to time of flight, from the shortest time of flight to the longest.

## PROBLEM-SOLVING STRATEGY Projectile Motion

We suggest that you use the following approach when solving projectile motion problems:

**1. Conceptualize** Think about what is going on physically in the problem. Establish the mental representation by imagining the projectile moving along its trajectory.

**2. Categorize** Confirm that the problem involves a particle in free-fall and that air resistance is neglected. Select a coordinate system with $x$ in the horizontal direction and $y$ in the vertical direction.

**3. Analyze** If the initial velocity vector is given, resolve it into $x$ and $y$ components. Treat the horizontal motion and the vertical motion independently. Analyze the horizontal motion of the projectile as a particle under constant velocity. Analyze the vertical motion of the projectile as a particle under constant acceleration.

**4. Finalize** Once you have determined your result, check to see if your answers are consistent with the mental and pictorial representations and that your results are realistic.

## ■ Thinking Physics 3.1

A home run is hit in a baseball game. The ball is hit from home plate into the stands along a parabolic path. What is the acceleration of the ball (**a**) while it is rising, (**b**) at the highest point of the trajectory, and (**c**) while it is descending after reaching the highest point? Ignore air resistance.

**Reasoning** The answers to all three parts are the same: the acceleration is that due to gravity, $a_y = -9.80\ \text{m/s}^2$, because the gravitational force is pulling downward on the ball during the entire motion. During the rising part of the trajectory, the downward acceleration results in the decreasing positive values of the vertical component of the ball's velocity. During the falling part of the trajectory, the downward acceleration results in the increasing negative values of the vertical component of the velocity. ■

## INTERACTIVE EXAMPLE 3.2 That's Quite an Arm

A stone is thrown from the top of a building at an angle of 30.0° to the horizontal and with an initial speed of 20.0 m/s, as in Figure 3.9.

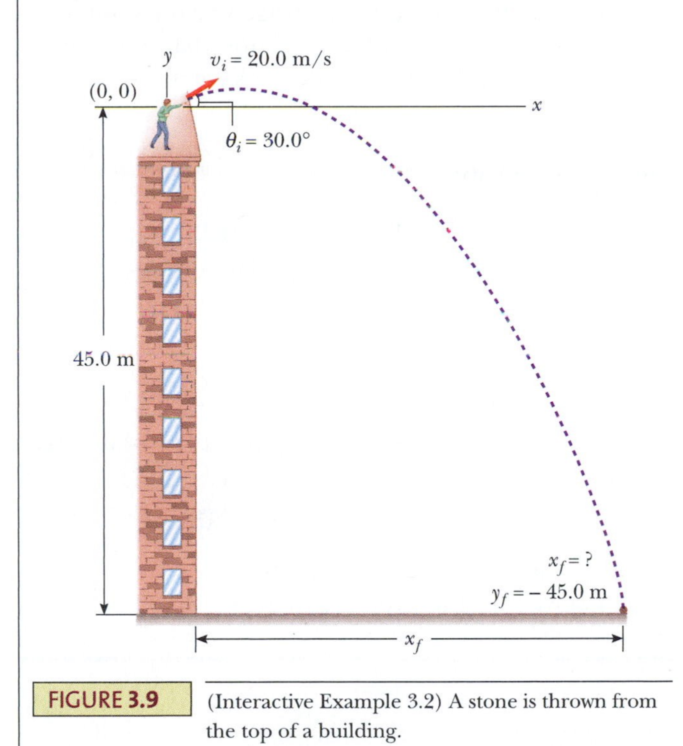


FIGURE 3.9 (Interactive Example 3.2) A stone is thrown from the top of a building.

**A** If the height of the building is 45.0 m, how long is the stone "in flight"?

**Solution** Looking at the pictorial representation in Figure 3.9, it is clear that this trajectory is not symmetric. Therefore, we cannot use Equations 3.15 and 3.16. We use the more general approach described by the Problem-Solving Strategy and represented by Equations 3.10 to 3.13.

The initial $x$ and $y$ components of the velocity are

$$v_{xi} = v_i \cos\theta_i = (20.0\ \text{m/s})(\cos 30.0°) = 17.3\ \text{m/s}$$

$$v_{yi} = v_i \sin\theta_i = (20.0\ \text{m/s})(\sin 30.0°) = 10.0\ \text{m/s}$$

To find $t$, we use the vertical motion, in which we model the stone as a particle under constant acceleration. We use Equation 3.13 with $y_f = -45.0$ m and $v_{yi} = 10.0$ m/s (we have chosen the top of the building as the origin, as in Figure 3.9):

$$y_f = y_i + v_{yi}t - \tfrac{1}{2}gt^2$$

$$-45.0\ \text{m} = 0 + (10.0\ \text{m/s})t - \tfrac{1}{2}(9.80\ \text{m/s}^2)t^2$$

Solving the quadratic equation for $t$ gives, for the positive root, $t = 4.22$ s. Does the negative root have any physical meaning? (Can you think of another way of finding $t$ from the information given?)

**B** What is the speed of the stone just before it strikes the ground?

**Solution** The $y$ component of the velocity just before the stone strikes the ground can be obtained using Equation 3.11, with $t = 4.22$ s:

$$v_{yf} = v_{yi} - gt$$

$$= 10.0\ \text{m/s} - (9.80\ \text{m/s}^2)(4.22\ \text{s}) = -31.4\ \text{m/s}$$

In the horizontal direction, the appropriate model is the particle under constant velocity. Because $v_{xf} = v_{xi} = 17.3$ m/s, the speed as the stone strikes the ground is

$$v_f = \sqrt{v_{xf}^2 + v_{yf}^2}$$

$$= \sqrt{(17.3)^2 + (-31.4)^2}\ \text{m/s} = 35.9\ \text{m/s}$$

**PhysicsNow™** Investigate this projectile situation by logging into PhysicsNow at **www.pop4e.com** and going to Interactive Example 3.2.

### EXAMPLE 3.3 The Stranded Explorers

An Alaskan rescue plane drops a package of emergency rations to a stranded party of explorers, as shown in the pictorial representation in Figure 3.10. If the plane is traveling horizontally at 40.0 m/s at a height of 100 m above the ground, where does the package strike the ground relative to the point at which it is released?

**Solution** We ignore air resistance, so we model this problem as a particle in two-dimensional free-fall, which, as we have seen, is modeled by a combination of a particle under constant velocity in the $x$ direction and a particle under constant acceleration in the $y$ direction. The coordinate system for this problem is selected as shown in Figure 3.10, with the positive $x$ direction to the right and the positive $y$ direction upward.

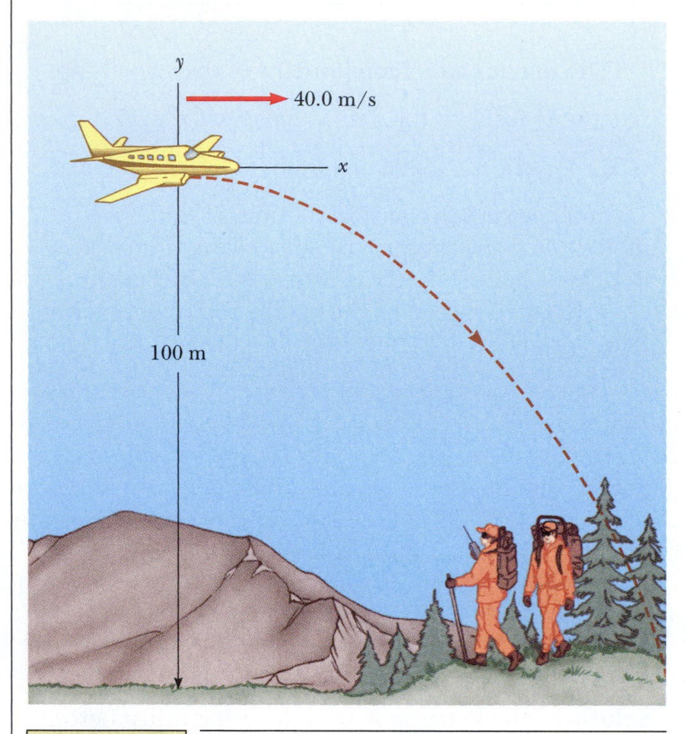


FIGURE 3.10 (Example 3.3) A package of emergency supplies is dropped from a plane to stranded explorers.

Consider first the horizontal motion of the package. From Equation 3.12, the position is given by $x_f = x_i + v_{xi}t$. The initial $x$ component of the package velocity is the same as that of the plane when the package is released, 40.0 m/s. We define the initial position $x_i = 0$ right under the plane at the instant the package is released. Therefore,

$$x_f = x_i + v_{xi}t = 0 + (40.0\text{ m/s})t$$

If we know $t$, the time at which the package strikes the ground, we can determine $x_f$, the final position and therefore the distance traveled by the package in the horizontal direction. At present, however, we have no information about $t$. To find $t$, we turn to the equations for the vertical motion of the package, modeling the package as a particle under constant acceleration. We know that at the instant the package hits the ground, its $y$ coordinate is $-100$ m. We also know that the initial component of velocity $v_{yi}$ of the package in the vertical direction is zero because the package was released with only a horizontal component of velocity. From Equation 3.13, we have

$$y_f = y_i + v_{yi}t - \tfrac{1}{2}gt^2$$

$$-100\text{ m} = 0 + 0 - \tfrac{1}{2}(9.80\text{ m/s}^2)t^2$$

$$t^2 = 20.4\text{ s}^2$$

$$t = 4.52\text{ s}$$

This value for the time at which the package strikes the ground is substituted into the equation for the $x$ coordinate to give us

$$x_f = (40.0\text{ m/s})(4.52\text{ s}) = \boxed{181\text{ m}}$$

The package hits the ground 181 m to the right of the point at which it was dropped in Figure 3.10.

### EXAMPLE 3.4 Javelin Throwing at the Olympics

An athlete throws a javelin a distance of 80.0 m at the Olympics held at the equator, where $g = 9.78\text{ m/s}^2$. Four years later the Olympics are held at the North Pole, where $g = 9.83\text{ m/s}^2$. Assuming that the thrower provides the javelin with exactly the same initial velocity as she did at the equator, how far does the javelin travel at the North Pole?

**Solution** In the absence of any information about how the javelin is affected by moving through the air, we adopt the free-fall model for the javelin. Track and field events are normally held on flat fields. Therefore, we surmise that the javelin returns to the same vertical position from which it was thrown and therefore that the trajectory is symmetric. These assumptions allow us to use Equations 3.15 and 3.16 to analyze the motion. The difference in range is due to the difference in the free-fall acceleration at the two locations.

To solve this problem, we will set up a ratio based on the range of the projectile being mathematically

(Tony Duffy/Getty Images)

A javelin can be thrown over a very long distance by a world class athlete. ■

related to the acceleration due to gravity. This technique of solving by ratios is very powerful and should be studied and understood so that it can be applied in future problem solving. We use Equation 3.16 to express the range of the particle at each of the two locations:

$$R_{\text{North Pole}} = \frac{v_i^2 \sin 2\theta_i}{g_{\text{North Pole}}}$$

$$R_{\text{equator}} = \frac{v_i^2 \sin 2\theta_i}{g_{\text{equator}}}$$

We divide the first equation by the second to establish a relationship between the ratio of the ranges and the ratio of the free-fall accelerations. Because the problem states that the same initial velocity is provided to the javelin at both locations, $v_i$ and $\theta_i$ are the same in the numerator and denominator of the ratio, which gives us

$$\frac{R_{\text{North Pole}}}{R_{\text{equator}}} = \frac{\left(\dfrac{v_i^2 \sin 2\theta_i}{g_{\text{North Pole}}}\right)}{\left(\dfrac{v_i^2 \sin 2\theta_i}{g_{\text{equator}}}\right)} = \frac{g_{\text{equator}}}{g_{\text{North Pole}}}$$

We can now solve this equation for the range at the North Pole and substitute the numerical values:

$$R_{\text{North Pole}} = \frac{g_{\text{equator}}}{g_{\text{North Pole}}} R_{\text{equator}} = \frac{9.78 \text{ m/s}^2}{9.83 \text{ m/s}^2}(80.0 \text{ m})$$

$$= 79.6 \text{ m}$$

Notice one of the advantages of this powerful technique of setting up ratios; we do not need to know the magnitude ($v_i$) nor the direction ($\theta_i$) of the initial velocity. As long as they are the same at both locations, they cancel in the ratio.

## 3.4 THE PARTICLE IN UNIFORM CIRCULAR MOTION

Figure 3.11a shows a car moving in a circular path with *constant speed* $v$. Such motion is called **uniform circular motion** and serves as the basis for a new group of problems we can solve.

It is often surprising to students to find that **even though an object moves at a constant speed in a circular path, it still has an acceleration.** To see why, consider the defining equation for average acceleration, $\vec{\mathbf{a}}_{\text{avg}} = \Delta\vec{\mathbf{v}}/\Delta t$ (Eq. 3.4). The acceleration depends on *the change in the velocity vector.* Because velocity is a vector quantity, an acceleration can be produced in two ways, as mentioned in Section 3.1: by a change in the *magnitude* of the velocity or by a change in the *direction* of the velocity. The latter situation is occurring for an object moving with constant speed in a circular path. The velocity vector is always tangent to the path of the object and perpendicular to the radius of the circular path. We now show that the acceleration vector in uniform circular motion is always perpendicular to the path and always points toward the center of the circle. An acceleration of this nature is called a **centripetal acceleration** (*centripetal* means *center seeking*), and its magnitude is

**PITFALL PREVENTION 3.3**

**ACCELERATION OF A PARTICLE IN UNIFORM CIRCULAR MOTION** Many students have trouble with the notion of a particle moving in a circular path at constant speed and yet having an acceleration because the everyday interpretation of acceleration means *speeding up* or *slowing down.* Remember, though, that acceleration is defined as a change in the *velocity,* not a change in the *speed.* In circular motion, the velocity vector is changing in direction, so there is indeed an acceleration.

■ Magnitude of centripetal acceleration

$$a_c = \frac{v^2}{r} \qquad [3.17]$$

where $r$ is the radius of the circle. The subscript on the acceleration symbol reminds us that the acceleration is centripetal.

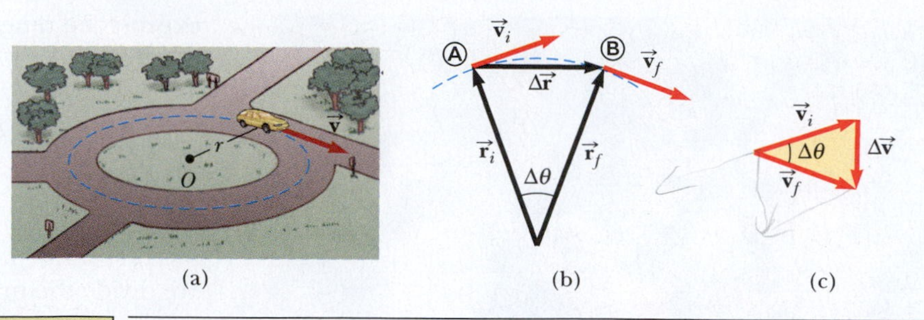


**FIGURE 3.11** (a) A car moving along a circular path at constant speed is in uniform circular motion. (b) As the particle moves from Ⓐ to Ⓑ, its velocity vector changes from $\vec{\mathbf{v}}_i$ to $\vec{\mathbf{v}}_f$. (c) The construction for determining the direction of the change in velocity $\Delta\vec{\mathbf{v}}$, which is toward the center of the circle for small $\Delta\theta$.

Let us first argue conceptually that the acceleration must be perpendicular to the path followed by the particle. If not, there would be a component of the acceleration parallel to the path and therefore parallel to the velocity vector. Such an acceleration component would lead to a change in the speed of the object, which we model as a particle, along the path. This change, however, is inconsistent with our setup of the problem in which the particle moves with constant speed along the path. Therefore, for *uniform* circular motion, the acceleration vector can only have a component perpendicular to the path, which is toward the center of the circle.

To derive Equation 3.17, consider the pictorial representation of the position and velocity vectors in Figure 3.11b. In addition, the figure shows the vector representing the change in position, $\Delta\vec{\mathbf{r}}$. The particle follows a circular path, part of which is shown by the dashed curve. The particle is at Ⓐ at time $t_i$, and its velocity at that time is $\vec{\mathbf{v}}_i$; it is at Ⓑ at some later time $t_f$, and its velocity at that time is $\vec{\mathbf{v}}_f$. Let us also assume that $\vec{\mathbf{v}}_i$ and $\vec{\mathbf{v}}_f$ differ only in direction; their magnitudes are the same (i.e., $v_i = v_f = v$, because it is *uniform* circular motion). To calculate the acceleration of the particle, let us begin with the defining equation for average acceleration (Eq. 3.4):

$$\vec{\mathbf{a}}_{\text{avg}} = \frac{\vec{\mathbf{v}}_f - \vec{\mathbf{v}}_i}{t_f - t_i} = \frac{\Delta\vec{\mathbf{v}}}{\Delta t}$$

In Figure 3.11c, the velocity vectors in Figure 3.11b have been redrawn tail to tail. The vector $\Delta\vec{\mathbf{v}}$ connects the tips of the vectors, representing the vector addition, $\vec{\mathbf{v}}_f = \vec{\mathbf{v}}_i + \Delta\vec{\mathbf{v}}$. In Figures 3.11b and 3.11c, we can identify triangles that can serve as geometric models to help us analyze the motion. The angle $\Delta\theta$ between the two position vectors in Figure 3.11b is the same as the angle between the velocity vectors in Figure 3.11c because the velocity vector $\vec{\mathbf{v}}$ is always perpendicular to the position vector $\vec{\mathbf{r}}$. Therefore, the two triangles are *similar.* (Two triangles are similar if the angle between any two sides is the same for both triangles and if the ratio of the lengths of these sides is the same.) This similarity enables us to write a relationship between the lengths of the sides for the two triangles:

$$\frac{|\Delta\vec{\mathbf{v}}|}{v} = \frac{|\Delta\vec{\mathbf{r}}|}{r}$$

where $v = v_i = v_f$ and $r = r_i = r_f$. This equation can be solved for $|\Delta\vec{\mathbf{v}}|$ and the expression so obtained can be substituted into $\vec{\mathbf{a}}_{\text{avg}} = \Delta\vec{\mathbf{v}}/\Delta t$ (Eq. 3.4) to give the magnitude of the average acceleration over the time interval for the particle to move from Ⓐ to Ⓑ:

$$|\vec{\mathbf{a}}_{\text{avg}}| = \frac{v}{r}\frac{|\Delta\vec{\mathbf{r}}|}{\Delta t}$$

Now imagine that we bring points Ⓐ and Ⓑ in Figure 3.11b very close together. As Ⓐ and Ⓑ approach each other, $\Delta t$ approaches zero and the ratio $|\Delta\vec{\mathbf{r}}|/\Delta t$

approaches the speed $v$. In addition, the average acceleration becomes the instantaneous acceleration at point Ⓐ. Hence, in the limit $\Delta t \rightarrow 0$, the magnitude of the acceleration is

$$a_c = \frac{v^2}{r}$$

Therefore, in uniform circular motion, the acceleration is directed inward toward the center of the circle and has magnitude $v^2/r$.

In many situations, it is convenient to describe the motion of a particle moving with constant speed in a circle of radius $r$ in terms of the **period** $T$, which is defined as the time interval required for one complete revolution. In the time interval $T$, the particle moves a distance of $2\pi r$, which is equal to the circumference of the particle's circular path. Therefore, because its speed is equal to the circumference of the circular path divided by the period, or $v = 2\pi r/T$, it follows that

$$T = \frac{2\pi r}{v} \qquad [3.18]$$

Period of a particle in uniform circular motion

The *particle in uniform circular motion* is a very common physical situation and is useful as an analysis model for problem solving.

### PITFALL PREVENTION 3.4

**CENTRIPETAL ACCELERATION IS NOT CONSTANT** We derived the magnitude of the centripetal acceleration vector and found it to be constant for uniform circular motion, but *the centripetal acceleration vector is not constant.* It always points toward the center of the circle, but it continuously changes direction as the particle moves around the circular path.

**QUICK QUIZ 3.4** Which of the following correctly describes the centripetal acceleration vector for a particle moving in a circular path? **(a)** constant and always perpendicular to the velocity vector for the particle **(b)** constant and always parallel to the velocity vector for the particle **(c)** of constant magnitude and always perpendicular to the velocity vector for the particle **(d)** of constant magnitude and always parallel to the velocity vector for the particle

## Thinking Physics 3.2

An airplane travels from Los Angeles to Sydney, Australia. After cruising altitude is reached, the instruments on the plane indicate that the ground speed holds rock-steady at 700 km/h and that the heading of the airplane does not change. Is the velocity of the airplane constant during the flight?

**Reasoning** The velocity is not constant because of the curvature of the Earth. Even though the speed does not change and the heading is always toward Sydney (is that actually true?), the airplane travels around a significant portion of the Earth's circumference. Therefore, the direction of the velocity vector does indeed change. We could extend this situation by imagining that the airplane passes over Sydney and continues (assuming it has enough fuel!) around the Earth until it arrives at Los Angeles again. It is impossible for an airplane to have a constant velocity (relative to the Universe, not to the Earth's surface) and return to its starting point. ■

### EXAMPLE 3.5 The Centripetal Acceleration of the Earth

What is the centripetal acceleration of the Earth as it moves in its orbit around the Sun?

**Solution** We shall model the Earth as a particle and approximate the Earth's orbit as circular (it's actually slightly elliptical, as we discuss in Chapter 11). Although we don't know the orbital speed of the Earth, with the help of Equation 3.18 we can recast Equation 3.17 in terms of the period of the Earth's orbit, which we know is one year:

$$a_c = \frac{v^2}{r} = \frac{\left(\frac{2\pi r}{T}\right)^2}{r} = \frac{4\pi^2 r}{T^2}$$

$$= \frac{4\pi^2(1.5 \times 10^{11}\text{ m})}{(1\text{ yr})^2}\left(\frac{1\text{ yr}}{3.16 \times 10^7\text{ s}}\right)^2$$

$$= 5.9 \times 10^{-3}\text{ m/s}^2$$

Note that this small acceleration can also be expressed as $6.0 \times 10^{-4}\, g$.

## 3.5 TANGENTIAL AND RADIAL ACCELERATION

Let us consider the motion of a particle along a curved path where the velocity changes both in direction and in magnitude, as described in Active Figure 3.12. In this situation, the velocity vector is always tangent to the path; the acceleration vector $\vec{\mathbf{a}}$, however, is at some angle to the path. At each of three points Ⓐ, Ⓑ, and Ⓒ in Active Figure 3.12, we draw dashed circles that form geometric models of circular paths for the actual path at each point. The radius of the model circle is equal to the radius of curvature of the path at each point.

As the particle moves along the curved path in Active Figure 3.12, the direction of the total acceleration vector $\vec{\mathbf{a}}$ changes from point to point. This vector can be resolved into two components based on an origin at the center of the model circle: a radial component $a_r$ along the radius of the model circle and a tangential component $a_t$ perpendicular to this radius. The *total* acceleration vector $\vec{\mathbf{a}}$ can be written as the vector sum of the component vectors:

$$\vec{\mathbf{a}} = \vec{\mathbf{a}}_r + \vec{\mathbf{a}}_t \quad [3.19]$$

**The tangential acceleration arises from the change in the speed of the particle** and is given by

■ Tangential acceleration

$$a_t = \frac{d|\vec{\mathbf{v}}|}{dt} \quad [3.20]$$

**The radial acceleration is a result of the change in direction of the velocity vector** and is given by

■ Radial acceleration

$$a_r = -a_c = -\frac{v^2}{r}$$

where $r$ is the radius of curvature of the path at the point in question, which is the radius of the model circle. We recognize the radial component of the acceleration as the centripetal acceleration discussed in Section 3.4. The negative sign indicates that the direction of the centripetal acceleration is toward the center of the model circle, opposite the direction of the radial unit vector $\hat{\mathbf{r}}$, which always points away from the center of the circle.

Because $\vec{\mathbf{a}}_r$ and $\vec{\mathbf{a}}_t$ are perpendicular component vectors of $\vec{\mathbf{a}}$, it follows that $a = \sqrt{a_r^2 + a_t^2}$. At a given speed, $a_r$ is large when the radius of curvature is small (as at points Ⓐ and Ⓑ in Active Fig. 3.12) and small when $r$ is large (such as at

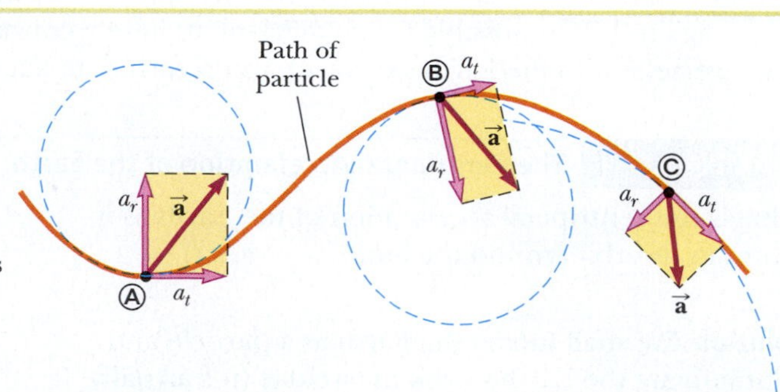


ACTIVE FIGURE 3.12 The motion of a particle along an arbitrary curved path lying in the $xy$ plane. If the velocity vector $\vec{\mathbf{v}}$ (always tangent to the path) changes in direction and magnitude, the acceleration vector $\vec{\mathbf{a}}$ has a tangential component $a_t$ and a radial component $a_r$.

Physics Now™ Log into PhysicsNow at www.pop4e.com and go to Active Figure 3.12 to study the acceleration components of the particle moving on the curved path.

point Ⓒ). The direction of $\vec{a}_t$ is either in the same direction as $\vec{v}$ (if $v$ is increasing) or opposite $\vec{v}$ (if $v$ is decreasing).

In the case of uniform circular motion, where $v$ is constant, $a_t = 0$ and the acceleration is always radial, as described in Section 3.4. In other words, uniform circular motion is a special case of motion along a curved path. Furthermore, if the direction of $\vec{v}$ doesn't change, no radial acceleration occurs and the motion is one dimensional ($a_r = 0$, but $a_t$ may not be zero).

**QUICK QUIZ 3.5** A particle moves along a path and its speed increases with time. **(i)** In which of the following cases are its acceleration and velocity vectors parallel? **(a)** The path is circular. **(b)** The path is straight. **(c)** The path is a parabola. **(d)** Never. **(ii)** From the same choices, in which case are its acceleration and velocity vectors perpendicular everywhere along the path?

## 3.6 RELATIVE VELOCITY

In Section 1.6, we discussed the need for a fixed reference point as the origin of a coordinate system used to locate the position of a point. We have made observations of position, velocity, and acceleration of a particle with respect to this reference point. Now imagine that we have two observers making measurements of a particle located in space and that one of them moves with respect to the other at constant velocity. Each observer can define a coordinate system with an origin fixed with respect to him or her. The origins of the two coordinate systems are in motion with respect to each other. In this section, we explore how we relate the measurements of one observer to that of the other.

As an example, consider two cars, a red one and a blue one, moving on a highway in the same direction, both with speeds of 60 mi/h, as in Figure 3.13. We identify the red car as a particle to be observed, and an observer on the side of the road measures a speed for this car of 60 mi/h. Now consider an observer riding in the blue car. This observer looks out the window and sees that the red car is always in the same position with respect to the blue car. Therefore, this observer measures a speed for the red car of *zero*. This simple example demonstrates that speed measurements differ in different frames of reference. Both observers look at the same particle (the red car) and arrive at different values for its speed. Both are correct; the difference in their measurements is a result of the relative velocity of their frames of reference.

Let us now generate a mathematical representation that will allow us to calculate one observer's measurements from the other's. Consider a particle located at point $P$ in an $xy$ plane, as shown in Figure 3.14. Imagine that the motion of this particle is being observed by two observers. Observer $O$ is in reference frame S. Observer $O'$ is in reference frame S′, which moves with velocity $\vec{v}_{O'O}$ with respect to S,

**FIGURE 3.13** Two observers measure the speed of the red car. Observer $O$ is standing on the ground beside the highway. Observer $O'$ is in the blue car.

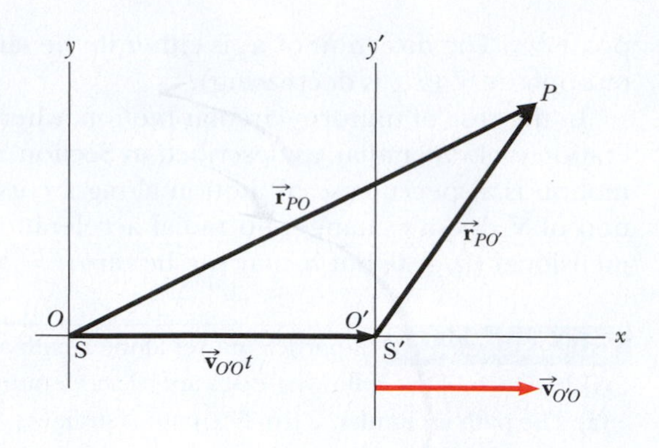


**FIGURE 3.14** Position vectors for an event occurring at point $P$ for two observers. Observer $O'$ is moving to the right at speed $v_{O'O}$ with respect to observer $O$.

where the first subscript describes what is being observed and the second describes who is doing the observing. Therefore, $\vec{v}_{O'O}$ is the velocity of observer $O'$ as measured by observer $O$. At $t = 0$, the origins of the reference frames coincide. Therefore, when modeling the origin of S′ as a particle under constant velocity, the origins of the two reference frames are separated by a displacement $\vec{v}_{O'O}t$ at time $t$. This displacement is shown in Figure 3.14. Also shown in the figure are the position vectors $\vec{r}_{PO}$ and $\vec{r}_{PO'}$ for point $P$ from each of the two origins. They are the position vectors that the two observers would use to describe the location of point $P$, using the same subscript notation. From the diagram, we see that these three vectors form a vector addition triangle:

$$\vec{r}_{PO} = \vec{r}_{PO'} + \vec{v}_{O'O}t$$

Notice the order of subscripts in this expression. The subscripts on the left side are the same as the first and last subscripts on the right. The second and third subscripts on the right are both $O'$. These subscripts are helpful in analyzing these types of situations. On the left, we are looking at the position vector that points directly to $P$ from $O$, as described by the subscripts. On the right, the same point $P$ is located by first going to $P$ from $O'$ and then describing where $O'$ is relative to $O$, again as suggested by the subscripts.

Let us now differentiate this expression with respect to time to find an expression for the velocity of a particle located at point $P$:

$$\frac{d}{dt}(\vec{r}_{PO}) = \frac{d}{dt}(\vec{r}_{PO'} + \vec{v}_{O'O}t) \quad \rightarrow \quad \vec{v}_{PO} = \vec{v}_{PO'} + \vec{v}_{O'O} \qquad [3.21]$$

This expression relates the velocity of the particle as measured by $O$ to that measured by $O'$ and the relative velocity of the two reference frames.

In the one-dimensional case, this equation reduces to

$$v_{PO} = v_{PO'} + v_{O'O}$$

Often, this equation is expressed in terms of the observer $O'$ as

$$v_{PO'} = v_{PO} - v_{O'O} \qquad [3.22]$$

and is called the **relative velocity,** the velocity of a particle as measured by a moving observer (moving with respect to another observer). In our car example, observer $O$ is standing on the side of the road. Observer $O'$ is in the blue car. Both observers are measuring the speed of the red car, which is located at point $P$. Therefore,

$$v_{PO} = 60 \text{ mi/h}$$

$$v_{O'O} = 60 \text{ mi/h}$$

and, using Equation 3.22,

$$v_{PO'} = v_{PO} - v_{O'O} = 60 \text{ mi/h} - 60 \text{ mi/h} = 0$$

The result of our calculation agrees with our previous intuitive discussion. This equation will be used in Chapter 9, when we discuss special relativity. We shall find that this simple expression is valid for low-speed particles but is no longer valid when the particle or observers are moving at speeds close to the speed of light.

## INTERACTIVE EXAMPLE 3.6 A Boat Crossing a River

A boat heading due north crosses a wide river with a speed of 10.0 km/h relative to the water. The river has a current such that the water moves with uniform speed of 5.00 km/h due east relative to the ground.

**A** What is the velocity of the boat relative to a stationary observer on the side of the river?

**Solution** It is often useful to use subscripts other than $O$ and $P$ that make it easy to identify the observers and the object being observed. Observer $O$ is standing on the side of the river. Because he is at rest with respect to the Earth, we will use the subscript E for this observer. Let us identify an imaginary observer $O'$ at rest in the water, floating with the current. Because he is at rest with respect to the water, we will use the subscript w for this observer. Both observers are looking at the boat, denoted by the subscript b. We can identify the velocity of the boat relative to the water as $\vec{\mathbf{v}}_{\text{bw}} = 10.0\hat{\mathbf{j}}$ km/h. The velocity of the water relative to the Earth is that of the current in the river, $\vec{\mathbf{v}}_{\text{wE}} = 5.00\hat{\mathbf{i}}$ km/h.

We are looking for the velocity of the boat relative to the Earth, so, from Equation 3.21,

$$\vec{\mathbf{v}}_{\text{bE}} = \vec{\mathbf{v}}_{\text{bw}} + \vec{\mathbf{v}}_{\text{wE}} = (10.0\hat{\mathbf{j}} + 5.00\hat{\mathbf{i}}) \text{ km/h}$$

This vector addition is shown in Figure 3.15a. The speed of the boat relative to the observer on shore is found from the Pythagorean theorem:

$$v_{\text{bE}} = \sqrt{v_{\text{bw}}^2 + v_{\text{wE}}^2} = \sqrt{(10.0)^2 + (5.00)^2} \text{ km/h}$$
$$= 11.2 \text{ km/h}$$

The direction of the velocity vector can be found with the inverse tangent function:

$$\theta = \tan^{-1}\left(\frac{v_{\text{wE}}}{v_{\text{bw}}}\right) = \tan^{-1}\left(\frac{5.00}{10.0}\right) = 26.6°$$

**B** At what angle should the boat be headed if it is to travel directly north across the river, and what is the speed of the boat relative to the Earth?

**Solution** We now want $\vec{\mathbf{v}}_{\text{bE}}$ to be pointed due north, as shown in Figure 3.15b. From the vector triangle,

$$\theta = \sin^{-1}\left(\frac{v_{\text{wE}}}{v_{\text{bw}}}\right) = \sin^{-1}\left(\frac{5.00 \text{ km/h}}{10.0 \text{ km/h}}\right) = 30.0°$$

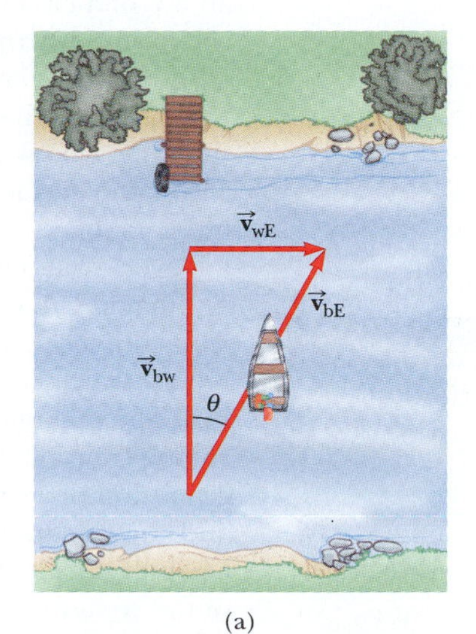


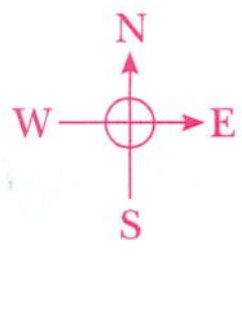


(a)

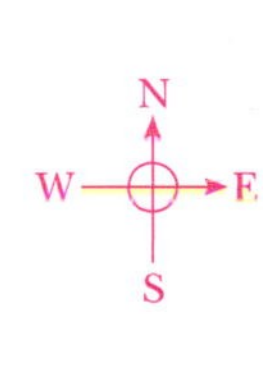


(b)

**FIGURE 3.15** (Interactive Example 3.6) (a) A boat aims directly across a river and ends up downstream. (b) To move directly across the river, the boat must aim upstream.

The speed of the boat relative to the Earth is

$$v_{\text{bE}} = \sqrt{v_{\text{bw}}^2 - v_{\text{wE}}^2} = \sqrt{(10.0)^2 - (5.00)^2} \text{ km/h}$$
$$= 8.66 \text{ km/h}$$

**Physics Now™** Investigate the crossing of the river for various boat speeds and current speeds by logging into PhysicsNow at **www.pop4e.com** and going to Interactive Example 3.6.

## 3.7 LATERAL ACCELERATION OF AUTOMOBILES

CONTEXT CONNECTION

An automobile does not travel in a straight line. It follows a two-dimensional path on a flat Earth surface and a three-dimensional path if there are hills and valleys. Let us restrict our thinking at this point to an automobile traveling in two dimensions on a flat roadway. During a turn, the automobile can be modeled as following an arc of a circular path at each point in its motion. Consequently, the automobile will have a centripetal acceleration.

A desired characteristic of automobiles is that they can negotiate a curve without rolling over. This characteristic depends on the centripetal acceleration. Imagine standing a book upright on a strip of sandpaper. If the sandpaper is moved slowly across the surface of a table with a very small acceleration, the book will stay upright. If the sandpaper is moved with a large acceleration, however, the book will fall over. That is what we would like to avoid in a car.

Imagine that instead of accelerating a book in one dimension we are centripetally accelerating a car in a circular path. The effect is the same. If there is too much centripetal acceleration, the car will "fall over" and will go into a sideways roll. The maximum possible centripetal acceleration that a car can exhibit without rolling over in a turn is called *lateral acceleration.* Two contributions to the lateral acceleration of a car are the height of the center of mass of the car above the ground and the side-to-side distance between the wheels. (We will study center of mass in Chapter 8.) The book in our demonstration has a relatively large ratio of the height of the center of mass to the width of the book upon which it is sitting, so it falls over relatively easily at low accelerations. An automobile has a much lower ratio of the height of the center of mass to the distance between the wheels. Therefore, it can withstand higher accelerations.

**TABLE 3.1** Lateral Accelerations of Various Performance Vehicles

| Automobile | Lateral Acceleration |
|---|---|
| Aston Martin DB7 Vantage | $0.90g$ |
| BMW Z8 | $0.92g$ |
| Chevrolet Corvette | $1.00g$ |
| Dodge Viper GTS-R | $0.98g$ |
| Ferrari F50 | $1.20g$ |
| Ferrari 360 Spider F1 | $0.94g$ |
| Lamborghini Diablo GT | $0.99g$ |
| Porsche 911 GT2 | $0.96g$ |

Consider the documented lateral acceleration of the performance vehicles from Table 2.3 listed in Table 3.1. These values are given as multiples of $g$, the acceleration due to gravity. Notice that all the vehicles have a lateral acceleration close to that due to gravity and that the lateral acceleration of the Ferrari F50 is 20% larger than that due to gravity. The Ferrari is a very stable vehicle!

In contrast, the lateral acceleration of nonperformance cars is lower because they generally are not designed to travel around turns at such a high speed as the performance cars. For example, the Honda Insight has a lateral acceleration of $0.80g$. Sport utility vehicles have lateral accelerations as low as $0.62g$. As a result, they are highly prone to rollovers in emergency maneuvers. ■

## SUMMARY

**PhysicsNow™** Take a practice test by logging into PhysicsNow at www.pop4e.com and clicking on the Pre-Test link for this chapter.

If a particle moves with *constant* acceleration $\vec{\mathbf{a}}$ and has velocity $\vec{\mathbf{v}}_i$ and position $\vec{\mathbf{r}}_i$ at $t = 0$, its velocity and position vectors at some later time $t$ are

$$\vec{\mathbf{v}}_f = \vec{\mathbf{v}}_i + \vec{\mathbf{a}}t \quad [3.8]$$

$$\vec{\mathbf{r}}_f = \vec{\mathbf{r}}_i + \vec{\mathbf{v}}_i t + \tfrac{1}{2}\vec{\mathbf{a}}t^2 \quad [3.9]$$

For two-dimensional motion in the $xy$ plane under constant acceleration, these vector expressions are equivalent to two component expressions, one for the motion along $x$ and one for the motion along $y$.

**Projectile motion** is a special case of two-dimensional motion under constant acceleration, where $a_x = 0$ and $a_y = -g$. In this case, the horizontal components of Equations 3.8 and 3.9 reduce to those of a particle under constant velocity:

$$v_{xf} = v_{xi} = \text{constant} \quad [3.10]$$

$$x_f = x_i + v_{xi}t \quad [3.12]$$

The vertical components of Equations 3.8 and 3.9 are those of a particle under constant acceleration:

$$v_{yf} = v_{yi} - gt \quad [3.11]$$

$$y_f = y_i + v_{yi}t - \tfrac{1}{2}gt^2 \quad [3.13]$$

where $v_{xi} = v_i \cos\theta_i$, $v_{yi} = v_i \sin\theta_i$, $v_i$ is the initial speed of the projectile, and $\theta_i$ is the angle $\vec{\mathbf{v}}_i$ makes with the positive $x$ axis.

A particle moving in a circle of radius $r$ with constant speed $v$ undergoes a **centripetal acceleration** because the direction of $\vec{v}$ changes in time. The magnitude of this acceleration is

$$a_c = \frac{v^2}{r} \quad [3.17]$$

and its direction is always toward the center of the circle.

If a particle moves along a curved path in such a way that the magnitude and direction of $\vec{v}$ change in time, the particle has an acceleration vector that can be described by two components: (1) a radial component $a_r = -a_c$ arising from the change in direction of $\vec{v}$ and (2) a tangential component $a_t$ arising from the change in magnitude of $\vec{v}$.

If an observer $O'$ is moving with velocity $\vec{v}_{O'O}$ with respect to observer $O$, their measurements of the velocity of a particle located at point $P$ are related according to

$$\vec{v}_{PO} = \vec{v}_{PO'} + \vec{v}_{O'O} \quad [3.21]$$

The velocity $\vec{v}_{PO'}$ is called the **relative velocity,** the velocity of a particle as measured by a moving observer (moving at constant velocity with respect to another observer).

## QUESTIONS

☐ = answer available in the *Student Solutions Manual and Study Guide*

1. ☐ If you know the position vectors of a particle at two points along its path and also know the time interval it took to move from one point to the other, can you determine the particle's instantaneous velocity? Its average velocity? Explain.
2. Construct motion diagrams showing the velocity and acceleration of a projectile at several points along its path, assuming that (a) the projectile is launched horizontally and (b) the projectile is launched at an angle $\theta$ with the horizontal.
3. A baseball is thrown such that its initial $x$ and $y$ components of velocity are known. Ignoring air resistance, describe how you would calculate, at the instant the ball reaches the top of its trajectory, (a) its coordinates, (b) its velocity, and (c) its acceleration. How would these results change if air resistance were taken into account?
4. A ball is projected horizontally from the top of a building. One second later another ball is projected horizontally from the same point with the same velocity. At what point in the motion will the balls be closest to each other? Will the first ball always be traveling faster than the second ball? What will be the time interval between the moments when the two balls hit the ground? Can the horizontal projection velocity of the second ball be changed so that the balls arrive at the ground at the same time?
5. ☐ A spacecraft drifts through space at a constant velocity. Suddenly a gas leak in the side of the spacecraft gives it a constant acceleration in a direction perpendicular to the initial velocity. The orientation of the spacecraft does not change, so the acceleration remains perpendicular to the original direction of the velocity. What is the shape of the path followed by the spacecraft in this situation?
6. State which of the following quantities, if any, remain constant as a projectile moves through its parabolic trajectory: (a) speed, (b) acceleration, (c) horizontal component of velocity, (d) vertical component of velocity.
7. ☐ A projectile is launched at some angle to the horizontal with some initial speed $v_i$, and air resistance is negligible. Is the projectile a freely falling body? What is its acceleration in the vertical direction? What is its acceleration in the horizontal direction?
8. The maximum range of a projectile occurs when it is launched at an angle of 45.0° with the horizontal, if air resistance is ignored. If air resistance is not ignored, will the optimum angle be greater or less than 45.0°? Explain.
9. A projectile is launched on the Earth with some initial velocity. Another projectile is launched on the Moon with the same initial velocity. If air resistance can be ignored, which projectile has the greater range? Which reaches the greater altitude? (Note that the free-fall acceleration on the Moon is about 1.6 m/s$^2$.)
10. Correct the following statement: "The racing car rounds the turn at a constant velocity of 90 miles per hour."
11. ☐ Explain whether or not the following particles have an acceleration: (a) a particle moving in a straight line with constant speed, (b) a particle moving around a curve with constant speed.
12. An object moves in a circular path with constant speed $v$. (a) Is the velocity of the object constant? (b) Is its acceleration constant? Explain.
13. Describe how a driver can steer a car traveling at constant speed so that (a) the acceleration is zero or (b) the magnitude of the acceleration remains constant.
14. An ice skater is executing a figure eight, consisting of two equal, tangent circular paths. Throughout the first loop she increases her speed uniformly, and during the second loop she moves at a constant speed. Draw a motion diagram showing her velocity and acceleration vectors at several points along the path of motion.
15. A sailor drops a wrench from the top of a sailboat's mast while the boat is moving rapidly and steadily in a straight line. Where will the wrench hit the deck? (Galileo posed this question.)
16. A ball is thrown upward in the air by a passenger on a train that is moving with constant velocity. (a) Describe the path of the ball as seen by the passenger. Describe the path as seen by an observer standing by the tracks outside the train. (b) How would these observations change if the train were accelerating along the track?

# PROBLEMS

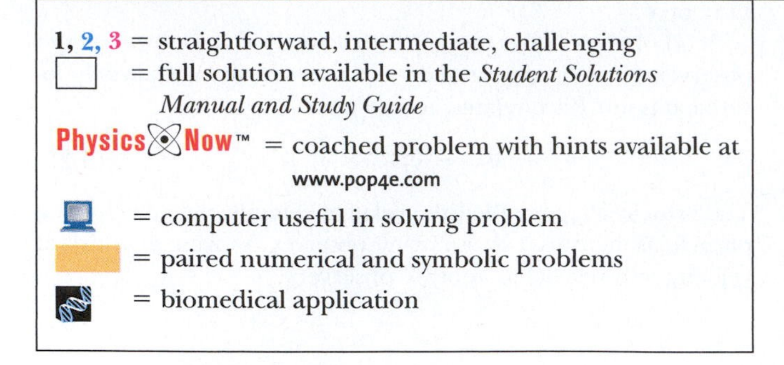


## Section 3.1 ▪ The Position, Velocity, and Acceleration Vectors

1. PhysicsNow™ A motorist drives south at 20.0 m/s for 3.00 min, then turns west and travels at 25.0 m/s for 2.00 min, and finally travels northwest at 30.0 m/s for 1.00 min. For this 6.00-min trip, find (a) the total vector displacement, (b) the average speed, and (c) the average velocity. Let the positive *x* axis point east.

2. Suppose the position vector for a particle is given as a function of time by $\vec{\mathbf{r}}(t) = x(t)\hat{\mathbf{i}} + y(t)\hat{\mathbf{j}}$, with $x(t) = at + b$ and $y(t) = ct^2 + d$, where $a = 1.00$ m/s, $b = 1.00$ m, $c = 0.125$ m/s$^2$, and $d = 1.00$ m. (a) Calculate the average velocity during the time interval from $t = 2.00$ s to $t = 4.00$ s. (b) Determine the velocity and the speed at $t = 2.00$ s.

## Section 3.2 ▪ Two-Dimensional Motion with Constant Acceleration

3. A fish swimming in a horizontal plane has velocity $\vec{\mathbf{v}}_i = (4.00\hat{\mathbf{i}} + 1.00\hat{\mathbf{j}})$ m/s at a point in the ocean where the position relative to a certain rock is $\vec{\mathbf{r}}_i = (10.0\hat{\mathbf{i}} - 4.00\hat{\mathbf{j}})$ m. After the fish swims with constant acceleration for 20.0 s, its velocity is $\vec{\mathbf{v}}_f = (20.0\hat{\mathbf{i}} - 5.00\hat{\mathbf{j}})$ m/s. (a) What are the components of the acceleration? (b) What is the direction of the acceleration with respect to unit vector $\hat{\mathbf{i}}$? (c) If the fish maintains constant acceleration, where is it at $t = 25.0$ s and in what direction is it moving?

4. At $t = 0$, a particle moving in the *xy* plane with constant acceleration has a velocity of $\vec{\mathbf{v}}_i = (3.00\hat{\mathbf{i}} - 2.00\hat{\mathbf{j}})$ m/s and is at the origin. At $t = 3.00$ s, the particle's velocity is $\vec{\mathbf{v}}_f = (9.00\hat{\mathbf{i}} + 7.00\hat{\mathbf{j}})$ m/s. Find (a) the acceleration of the particle and (b) its coordinates at any time $t$.

5. A particle initially located at the origin has an acceleration of $\vec{\mathbf{a}} = 3.00\hat{\mathbf{j}}$ m/s$^2$ and an initial velocity of $\vec{\mathbf{v}}_i = 5.00\hat{\mathbf{i}}$ m/s. Find (a) the vector position and velocity at any time $t$ and (b) the coordinates and speed of the particle at $t = 2.00$ s.

6. It is not possible to see very small objects, such as viruses, using an ordinary light microscope. An electron microscope, however, can view such objects using an electron beam instead of a light beam. Electron microscopy has proved invaluable for investigations of viruses, cell membranes and subcellular structures, bacterial surfaces, visual receptors, chloroplasts, and the contractile properties of muscles. The "lenses" of an electron microscope consist of electric and magnetic fields that control the electron beam. As an example of the manipulation of an electron beam, consider an electron traveling away from the origin along the *x* axis in the *xy* plane with initial velocity $\vec{\mathbf{v}}_i = v_i\hat{\mathbf{i}}$. As it passes through the region $x = 0$ to $x = d$, the electron experiences acceleration $\vec{\mathbf{a}} = a_x\hat{\mathbf{i}} + a_y\hat{\mathbf{j}}$, where $a_x$ and $a_y$ are constants. Taking $v_i = 1.80 \times 10^7$ m/s, $a_x = 8.00 \times 10^{14}$ m/s$^2$, and $a_y = 1.60 \times 10^{15}$ m/s$^2$, determine at $x = d = 0.010\,0$ m (a) the position of the electron, (b) the velocity of the electron, (c) the speed of the electron, and (d) the direction of travel of the electron (i.e., the angle between its velocity and the *x* axis).

## Section 3.3 ▪ Projectile Motion

*Note:* Ignore air resistance in all problems and take $g = 9.80$ m/s$^2$ at the Earth's surface.

7. PhysicsNow™ In a local bar, a customer slides an empty beer mug down the counter for a refill. The bartender is just deciding to go home and rethink his life. He does not see the mug, which slides off the counter and strikes the floor 1.40 m from the base of the counter. If the height of the counter is 0.860 m, (a) with what velocity did the mug leave the counter and (b) what was the direction of the mug's velocity just before it hit the floor?

8. In a local bar, a customer slides an empty beer mug down the counter for a refill. The bartender is momentarily distracted and does not see the mug, which slides off the counter and strikes the floor at distance *d* from the base of the counter. The height of the counter is *h*. (a) With what velocity did the mug leave the counter? (b) What was the direction of the mug's velocity just before it hit the floor?

9. Mayan kings and many school sports teams are named for the puma, cougar, or mountain lion *Felis concolor*, the best jumper among animals. It can jump to a height of 12.0 ft when leaving the ground at an angle of 45.0°. With what speed, in SI units, does it leave the ground to make this leap?

10. An astronaut on a strange planet finds that she can jump a maximum horizontal distance of 15.0 m if her initial speed is 3.00 m/s. What is the free-fall acceleration on the planet?

11. A cannon with a muzzle speed of 1 000 m/s is used to start an avalanche on a mountain slope. The target is 2 000 m from the cannon horizontally and 800 m above the cannon. At what angle, above the horizontal, should the cannon be fired?

12. A ball is tossed from an upper-story window of a building. The ball is given an initial velocity of 8.00 m/s at an angle of 20.0° below the horizontal. It strikes the ground 3.00 s later. (a) How far horizontally from the base of the building does the ball strike the ground? (b) Find the height from which the ball was thrown. (c) How long does it take the ball to reach a point 10.0 m below the level of launching?

13. The speed of a projectile when it reaches its maximum height is one half its speed when it is at half its maximum

height. What is the initial projection angle of the projectile?

14. The small archerfish (length 20 to 25 cm) lives in brackish waters of Southeast Asia from India to the Philippines. This aptly named creature captures its prey by shooting a stream of water drops at an insect, either flying or at rest. The bug falls into the water and the fish gobbles it up. The archerfish has high accuracy at distances of 1.2 m to 1.5 m, and it sometimes makes hits at distances up to 3.5 m. A groove in the roof of its mouth, along with a curled tongue, forms a tube that enables the fish to impart high velocity to the water in its mouth when it suddenly closes its gill flaps. Suppose the archerfish shoots at a target that is 2.00 m away, measured along a line at an angle of 30.0° above the horizontal. With what velocity must the water stream be launched if it is not to drop more than 3.00 cm vertically on its path to the target?

(Frederick McKinney/FPG/Getty)

FIGURE P3.16

15. PhysicsNow™ A placekicker must kick a football from a point 36.0 m (about 40 yards) from the goal, and half the crowd hopes the ball will clear the crossbar, which is 3.05 m high. When kicked, the ball leaves the ground with a speed of 20.0 m/s at an angle of 53.0° to the horizontal. (a) By how much does the ball clear or fall short of clearing the crossbar? (b) Does the ball approach the crossbar while still rising or while falling?

16. A firefighter a distance $d$ from a burning building directs a stream of water from a fire hose at angle $\theta_i$ above the horizontal as shown in Figure P3.16. If the initial speed of the stream is $v_i$, at what height $h$ does the water strike the building?

17. A playground is on the flat roof of a city school, 6.00 m above the street below. The vertical wall of the building is 7.00 m high, forming a 1 m-high railing around the playground. A ball has fallen to the street below, and a passerby returns it by launching it at an angle of 53.0° above the horizontal at a point 24.0 m from the base of the building wall. The ball takes 2.20 s to reach a point vertically above the wall. (a) Find the speed at which the ball was launched. (b) Find the vertical distance by which the ball clears the wall. (c) Find the distance from the wall to the point on the roof where the ball lands.

18. The motion of a human body through space can be precisely modeled as the motion of a particle at the body's center of mass, as we will study in Chapter 8. The components of the displacement of an athlete's center of mass from the beginning to the end of a certain jump are described by the two equations

$$x_f = 0 + (11.2 \text{ m/s})(\cos 18.5°)t$$

$$0.360 \text{ m} = 0.840 \text{ m} + (11.2 \text{ m/s})(\sin 18.5°)t - \tfrac{1}{2}(9.80 \text{ m/s}^2)t^2$$

where $t$ is the time at which the athlete lands after taking off at time $t = 0$. Identify (a) his position and (b) his vector velocity at the takeoff point. (c) The world long jump record is 8.95 m. How far did the athlete in this problem jump? (d) Make a sketch of the motion of his center of mass.

19. A soccer player kicks a rock horizontally off a 40.0-m-high cliff into a pool of water. If the player hears the sound of the splash 3.00 s later, what was the initial speed given to the rock? Assume that the speed of sound in air is 343 m/s.

20. A basketball star covers 2.80 m horizontally in a jump to dunk the ball (Fig. P3.20). His motion through space can be modeled precisely as that of a particle at his *center of mass*, which we will define in Chapter 8. His center of mass is at elevation 1.02 m when he leaves the floor. It reaches a maximum height of 1.85 m above the floor and is at elevation 0.900 m when he touches down again. Determine (a) his time of flight (his "hang time"), (b) his horizontal and (c) vertical velocity components at the instant of takeoff, and (d) his takeoff angle. (e) For comparison, determine the hang time of a whitetail deer making a jump with center of mass elevations $y_i = 1.20$ m, $y_{max} = 2.50$ m, and $y_f = 0.700$ m.

(Top, Jed Jacobsohn/Getty Images; bottom, Bill Lee/Dembinsky Photo Associates)

FIGURE **P3.20**

**21.** A fireworks rocket explodes at height $h$, the peak of its vertical trajectory. It throws out burning fragments in all directions, but all at the same speed $v$. Pellets of solidified metal fall to the ground without air resistance. Find the smallest angle that the final velocity of an impacting fragment makes with the horizontal.

## Section 3.4 ■ The Particle in Uniform Circular Motion

**22.** From information on the endsheets of this book, compute the radial acceleration of a point on the surface of the Earth at the equator owing to the rotation of the Earth about its axis.

**23.** **Physics Now™** The athlete shown in Figure P3.23 rotates a 1.00-kg discus along a circular path of radius 1.06 m. The maximum speed of the discus is 20.0 m/s. Determine the magnitude of the maximum radial acceleration of the discus.

(Sam Sargent/Liaison International)

FIGURE **P3.23**

**24.** Casting of molten metal is important in many industrial processes. *Centrifugal casting* is used for manufacturing pipes, bearings, and many other structures. A variety of sophisticated techniques have been invented, but the basic idea is as illustrated in Figure P3.24. A cylindrical enclosure is rotated rapidly and steadily about a horizontal axis. Molten metal is poured into the rotating cylinder and then cooled, forming the finished product. Turning the cylinder at a high rotation rate forces the solidifying metal strongly to the outside. Any bubbles are displaced toward the axis, so unwanted voids will not be present in the casting. Sometimes it is desirable to form a composite casting, such as for a bearing. Here a strong steel outer surface is poured and then inside it a lining of special low-friction metal. In some applications, a very strong metal is given a coating of corrosion-resistant metal. Centrifugal casting results in strong bonding between the layers.

Suppose a copper sleeve of inner radius 2.10 cm and outer radius 2.20 cm is to be cast. To eliminate bubbles and give high structural integrity, the centripetal acceleration of each bit of metal should be at least $100g$. What rate of rotation is required? State the answer in revolutions per minute.

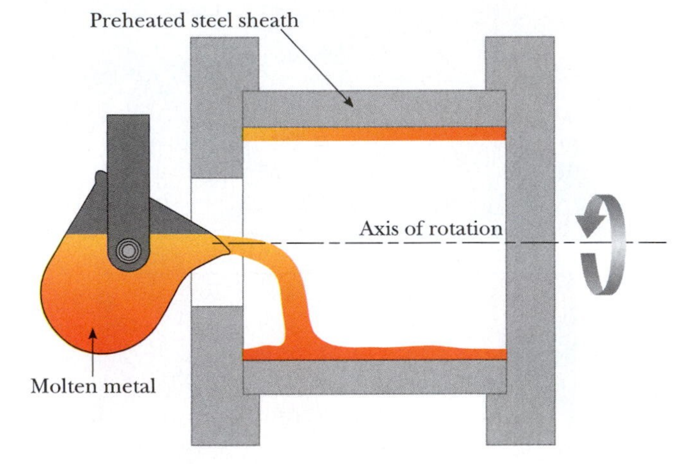


FIGURE **P3.24**

**25.** A tire 0.500 m in radius rotates at a constant rate of 200 rev/min. Find the speed and acceleration of a small stone lodged in the tread of the tire (on its outer edge).

**26.** As their booster rockets separate, Space Shuttle astronauts typically feel accelerations up to $3g$, where $g = 9.80\ \text{m/s}^2$. In their training, astronauts ride in a device where they experience such an acceleration as a centripetal acceleration. Specifically, the astronaut is fastened securely at the end of a mechanical arm, which then turns at constant speed in a horizontal circle. Determine the rotation rate, in revolutions per second, required to give an astronaut a centripetal acceleration of $3.00g$ while in circular motion with radius 9.45 m.

**27.** The astronaut orbiting the Earth in Figure P3.27 is preparing to dock with a Westar VI satellite. The satellite is in a circular orbit 600 km above the Earth's surface, where the free-fall acceleration is $8.21\ \text{m/s}^2$. Take the radius of the Earth as 6 400 km. Determine the speed of the satellite and the time interval required to complete one orbit around the Earth, which is the period of the satellite.

FIGURE P3.27

## Section 3.5 ■ Tangential and Radial Acceleration

**28.** A point on a rotating turntable 20.0 cm from the center accelerates from rest to a final speed of 0.700 m/s in 1.75 s. At $t = 1.25$ s, find the magnitude and direction of (a) the radial acceleration, (b) the tangential acceleration, and (c) the total acceleration of the point.

**29.** A train slows down as it rounds a sharp horizontal turn, slowing from 90.0 km/h to 50.0 km/h in the 15.0 s that it takes to round the bend. The radius of the curve is 150 m. Compute the acceleration at the moment the train speed reaches 50.0 km/h. Assume that it continues to slow down at this time at the same rate.

**30.** A ball swings in a vertical circle at the end of a rope 1.50 m long. When the ball is 36.9° past the lowest point on its way up, its total acceleration is $(-22.5\hat{\mathbf{i}} + 20.2\hat{\mathbf{j}})$ m/s². At that instant, (a) sketch a vector diagram showing the components of its acceleration, (b) determine the magnitude of its radial acceleration, and (c) determine the speed and velocity of the ball.

**31.** Figure P3.31 represents the total acceleration of a particle moving clockwise in a circle of radius 2.50 m at a certain instant of time. At this instant, find (a) the radial acceleration, (b) the speed of the particle, and (c) its tangential acceleration.

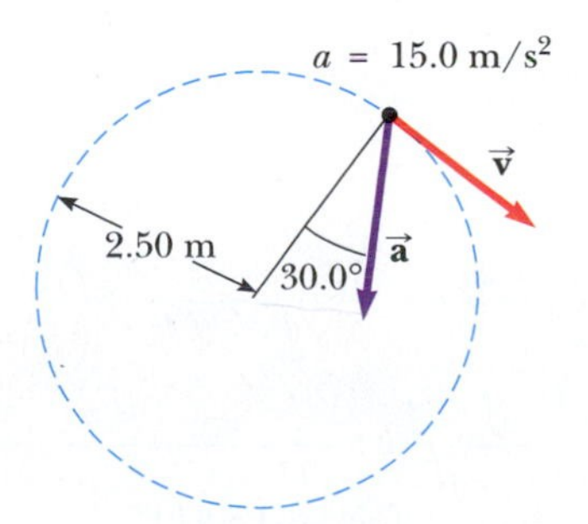


FIGURE P3.31

## Section 3.6 ■ Relative Velocity

**32.** How long does it take an automobile traveling in the left lane at 60.0 km/h to pull alongside a car traveling in the right lane at 40.0 km/h if the cars' front bumpers are initially 100 m apart?

**33.** A river has a steady speed of 0.500 m/s. A student swims upstream a distance of 1.00 km and swims back to the starting point. If the student can swim at a speed of 1.20 m/s in still water, how long does the trip take? Compare this answer with the time interval the trip would take if the water were still.

**34.** A car travels due east with a speed of 50.0 km/h. Raindrops are falling at constant speed vertically with respect to the Earth. The traces of the rain on the side windows of the car make an angle of 60.0° with the vertical. Find the velocity of the rain with respect to (a) the car and (b) the Earth.

**35.** The pilot of an airplane notes that the compass indicates a heading due west. The airplane's speed relative to the air is 150 km/h. The air is moving in a wind at 30.0 km/h toward the north. Find the velocity of the airplane relative to the ground.

**36.** Two swimmers, Alan and Beth, start together at the same point on the bank of a wide stream that flows with a speed $v$. Both move at the same speed $c$ ($c > v$) relative to the water. Alan swims downstream a distance $L$ and then upstream the same distance. Beth swims so that her motion relative to the Earth is perpendicular to the banks of the stream. She swims the distance $L$ and then back the same distance, so that both swimmers return to the starting point. Which swimmer returns first? (*Note:* First, guess the answer.)

**37.** A science student is riding on a flatcar of a train traveling along a straight horizontal track at a constant speed of 10.0 m/s. The student throws a ball into the air along a path that he judges to make an initial angle of 60.0° with the horizontal and to be in line with the track. The student's professor, who is standing on the ground nearby, observes the ball to rise vertically. How high does she see the ball rise?

**38.** A Coast Guard cutter detects an unidentified ship at a distance of 20.0 km in the direction 15.0° east of north. The ship is traveling at 26.0 km/h on a course at 40.0° east of north. The Coast Guard wishes to send a speedboat to intercept the vessel and investigate it. If the speedboat travels 50.0 km/h, in what direction should it head? Express the direction as a compass bearing with respect to due north.

## Section 3.7 ■ Context Connection—Lateral Acceleration of Automobiles

**39.** The cornering performance of an automobile is evaluated on a skid pad, where the maximum speed a car can maintain around a circular path on a dry, flat surface is measured. Then the magnitude of the centripetal acceleration, also called the lateral acceleration, is calculated as a multiple of the free-fall acceleration $g$. Along with the height and width of the car, factors affecting its performance are the tire characteristics and the suspension system. A Dodge Viper GTS-R can negotiate a skid pad of radius 156 m at 139 km/h. Calculate its maximum lateral acceleration from these data to verify the corresponding entry in Table 3.1.

**40.** A certain light truck can go around an unbanked curve having a radius of 150 m with a maximum speed of 32.0 m/s. With what maximum speed can it go around a curve having a radius of 75.0 m?

## Additional Problems

**41.** *The "Vomit Comet."* In zero-gravity astronaut training and equipment testing, NASA flies a KC135A aircraft along a parabolic flight path. As shown in Figure P3.41, the aircraft climbs from 24 000 ft to 31 000 ft, where it enters the zero-$g$ parabola with a velocity of 143 m/s at 45.0° nose high and exits with velocity 143 m/s at 45.0° nose low. During this portion of the flight, the aircraft and objects inside its padded cabin are in free-fall; they have gone ballistic. The aircraft then pulls out of the dive with an upward acceleration of $0.800g$, moving in a vertical circle with radius 4.13 km. (During this portion of the flight, occupants of the plane perceive an acceleration of $1.800g$.) What are the aircraft (a) speed and (b) altitude at the top of the maneuver? (c) What is the time interval spent in zero gravity? (d) What is the speed of the aircraft at the bottom of the flight path?

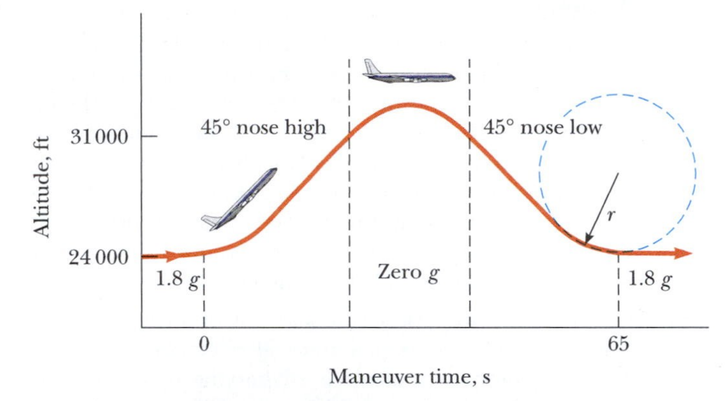

FIGURE **P3.41**

**42.** A landscape architect is planning an artificial waterfall in a city park. Water flowing at 1.70 m/s will leave the end of a horizontal channel at the top of a vertical wall 2.35 m high and from there fall into a pool. (a) Will there be a wide enough space for a walkway on which people can go behind the waterfall? (b) To sell her plan to the city council, the architect wants to build a model to standard scale, one-twelfth actual size. How fast should the water in the channel flow in the model?

**43.** A ball on the end of a string is whirled around in a horizontal circle of radius 0.300 m. The plane of the circle is 1.20 m above the ground. The string breaks and the ball lands 2.00 m (horizontally) away from the point on the ground directly beneath the ball's location when the string breaks. Find the radial acceleration of the ball during its circular motion.

**44.** A projectile is fired up an incline (incline angle $\phi$) with an initial speed $v_i$ at an angle $\theta_i$ with respect to the horizontal ($\theta_i > \phi$), as shown in Figure P3.44. (a) Show that the projectile travels a distance $d$ up the incline, where

$$d = \frac{2v_i^2 \cos\theta_i \sin(\theta_i - \phi)}{g\cos^2\phi}$$

(b) For what value of $\theta_i$ is $d$ a maximum, and what is that maximum value?

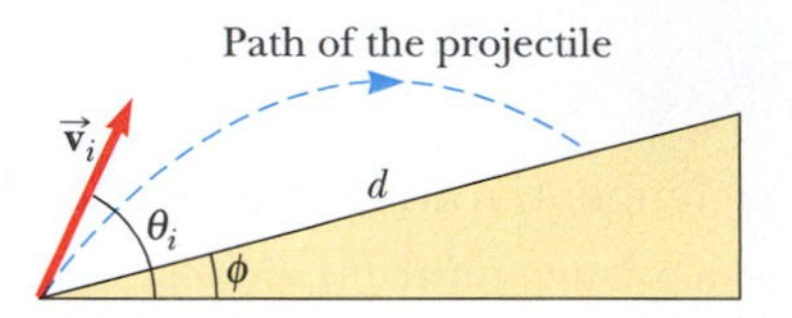


FIGURE **P3.44**

**45.** Barry Bonds hits a home run so that the baseball just clears the top row of bleachers, 21.0 m high, located 130 m from home plate. The ball is hit at an angle of 35.0° to the horizontal, and air resistance is negligible. Find (a) the initial speed of the ball, (b) the time interval that elapses before the ball reaches the top row, and (c) the velocity components and the speed of the ball when it passes over the top row. Assume that the ball is hit at a height of 1.00 m above the ground.

**46.** An astronaut on the surface of the Moon fires a cannon to launch an experiment package, which leaves the barrel moving horizontally. (a) What must be the muzzle speed of the probe so that it travels completely around the Moon and returns to its original location? (b) How long does this trip around the Moon take? Assume that the free-fall acceleration on the Moon is one sixth that on the Earth.

**47.** A basketball player who is 2.00 m tall is standing on the floor 10.0 m from the basket, as shown in Figure P3.47. If he shoots the ball at a 40.0° angle with the horizontal, at

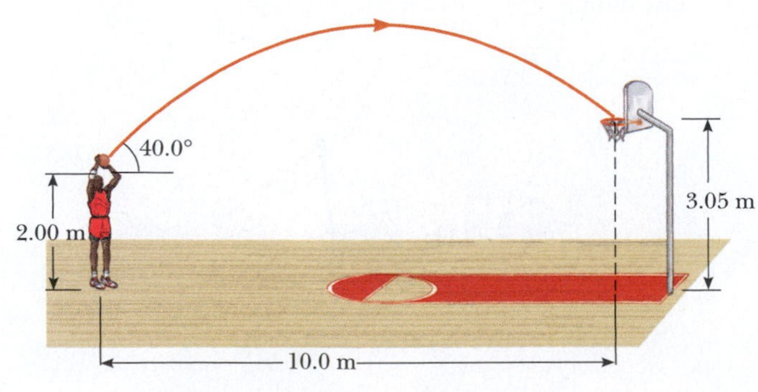


FIGURE **P3.47**

what initial speed must he throw so that it goes through the hoop without striking the backboard? The basket height is 3.05 m.

**48.** When baseball players throw the ball in from the outfield, they usually allow it to take one bounce before it reaches the infield, on the theory the ball arrives sooner that way. Suppose the angle at which a bounced ball leaves the ground is the same as the angle at which the outfielder threw it, as shown in Figure P3.48, but that the ball's speed after the bounce is one half what it was before the bounce. (a) Assume that the ball is always thrown with the same initial speed. At what angle $\theta$ should the fielder throw the ball to make it go the same distance $D$ with one bounce (blue path) as a ball thrown upward at 45.0° with no bounce (green path)? (b) Determine the ratio of the time intervals required for the one-bounce and no-bounce throws.

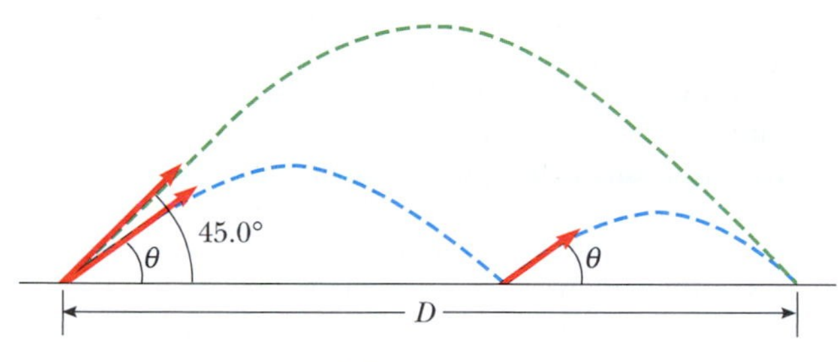


FIGURE **P3.48**

**49.** Your grandfather is copilot of a bomber, flying horizontally over level terrain, with a speed of 275 m/s relative to the ground at an altitude of 3 000 m. (a) The bombardier releases one bomb. How far will the bomb travel horizontally between its release and its impact on the ground? Ignore the effects of air resistance. (b) Firing from the people on the ground suddenly incapacitates the bombardier before he can call, "Bombs away!" Consequently, the pilot maintains the plane's original course, altitude, and speed through a storm of flak. Where will the plane be when the bomb hits the ground? (c) The plane has a telescopic bomb sight set so that the bomb hits the target seen in the sight at the moment of release. At what angle from the vertical was the bomb sight set?

**50.** A person standing at the top of a hemispherical rock of radius $R$ kicks a ball (initially at rest on the top of the rock) to give it horizontal velocity $\vec{\mathbf{v}}_i$ as shown in Figure P3.50.

FIGURE **P3.50**

(a) What must be its minimum initial speed if the ball is never to hit the rock after it is kicked? (b) With this initial speed, how far from the base of the rock does the ball hit the ground?

**51.** **Physics Now™** A car is parked on a steep incline overlooking the ocean, where the incline makes an angle of 37.0° below the horizontal. The negligent driver leaves the car in neutral, and the parking brakes are defective. The car rolls from rest down the incline with a constant acceleration of 4.00 m/s$^2$, traveling 50.0 m to the edge of a vertical cliff. The cliff is 30.0 m above the ocean. Find (a) the speed of the car when it reaches the edge of the cliff and the time interval it takes to get there, (b) the velocity of the car when it lands in the ocean, (c) the total time interval during which the car is in motion, and (d) the position of the car when it lands in the ocean, relative to the base of the cliff.

**52.** A truck loaded with cannonball watermelons stops suddenly to avoid running over the edge of a washed-out bridge (Fig. P3.52). The quick stop causes a number of melons to fly off the truck. One melon rolls over the edge with an initial speed $v_i$ = 10.0 m/s in the horizontal direction. A cross-section of the bank has the shape of the bottom half of a parabola with its vertex at the edge of the road and with the equation $y^2 = 16x$, where $x$ and $y$ are measured in meters. What are the $x$ and $y$ coordinates of the melon when it splatters on the bank?

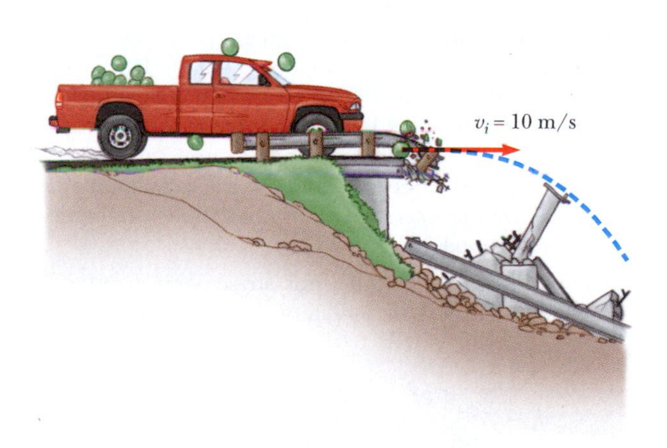


FIGURE **P3.52**

**53.** The determined coyote is out once more in pursuit of the elusive roadrunner. The coyote wears a pair of Acme jet-powered roller skates, which provide a constant horizontal acceleration of 15.0 m/s$^2$ (Fig. P3.53). The coyote starts at rest 70.0 m from the brink of a cliff at the instant the roadrunner zips past him in the direction of the cliff. (a) The roadrunner moves with constant speed. Determine the minimum speed he must have so as to reach the cliff before the coyote. At the edge of the cliff, the roadrunner escapes by making a sudden turn, while the coyote continues straight ahead. The coyote's skates remain horizontal and continue to operate while he is in flight so that his acceleration while in the air is $(15.0\hat{\mathbf{i}} - 9.80\hat{\mathbf{j}})$ m/s$^2$. (b) The cliff is 100 m above the flat floor of a wide canyon. Determine

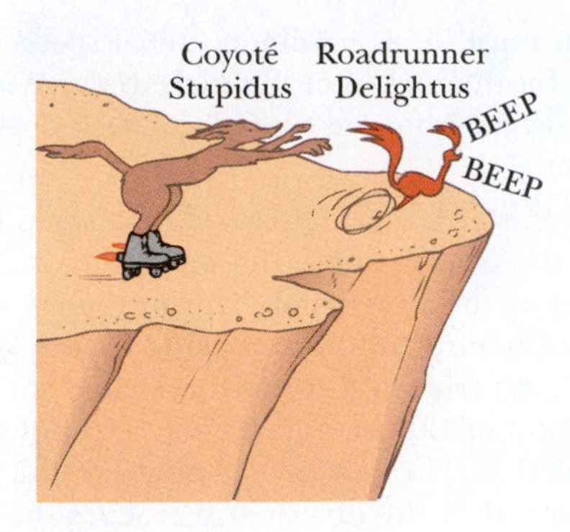


FIGURE **P3.53**

where the coyote lands in the canyon. (c) Determine the components of the coyote's impact velocity.

**54.** A ball is thrown with an initial speed $v_i$ at an angle $\theta_i$ with the horizontal. The horizontal range of the ball is $R$, and the ball reaches a maximum height $R/6$. In terms of $R$ and $g$, find (a) the time interval during which the ball is in motion, (b) the ball's speed at the peak of its path, (c) the initial vertical component of its velocity, (d) its initial speed, and (e) the angle $\theta_i$. (f) Suppose the ball is thrown at the same initial speed found in (d) but at the angle appropriate for reaching the greatest height that it can. Find this height. (g) Suppose the ball is thrown at the same initial speed but at the angle for greatest possible range. Find this maximum horizontal range.

**55.** A catapult launches a rocket at an angle of 53.0° above the horizontal with an initial speed of 100 m/s. The rocket engine immediately starts a burn, and for 3.00 s the rocket moves along its initial line of motion with an acceleration of 30.0 m/s$^2$. Then its engine fails, and the rocket proceeds to move in free-fall. Find (a) the maximum altitude reached by the rocket, (b) its total time of flight, and (c) its horizontal range.

**56.** Do not hurt yourself; do not strike your hand against anything. Within these limitations, describe what you do to give your hand a large acceleration. Compute an order-of-magnitude estimate of this acceleration, stating the quantities you measure or estimate and their values.

FIGURE **P3.57**

**57.** A skier leaves the ramp of a ski jump with a velocity of 10.0 m/s, 15.0° above the horizontal, as shown in Figure P3.57. The slope is inclined at 50.0°, and air resistance is negligible. Find (a) the distance from the ramp to where the jumper lands and (b) the velocity components just before the landing. (How do you think the results might be affected if air resistance were included? Note that jumpers lean forward in the shape of an airfoil, with their hands at their sides, to increase their distance. Why does this method work?)

**58.** In a television picture tube (a cathode-ray tube), electrons are emitted with velocity $\vec{\mathbf{v}}_i$ from a source at the origin of coordinates. The initial velocities of different electrons make different angles $\theta$ with the $x$ axis. As they move a distance $D$ along the $x$ axis, the electrons are acted on by a constant electric field, giving each a constant acceleration $\vec{\mathbf{a}}$ in the $x$ direction. At $x = D$, the electrons pass through a circular aperture, oriented perpendicular to the $x$ axis. At the aperture, the velocity imparted to the electrons by the electric field is much larger than $\vec{\mathbf{v}}_i$ in magnitude. Show that velocities of the electrons going through the aperture radiate from a certain point on the $x$ axis, which is not the origin. Determine the location of this point. This point is called a *virtual source,* and it is important in determining where the electron beam hits the screen of the tube.

**59.** An angler sets out upstream from Metaline Falls on the Pend Oreille River in northwestern Washington State. His small boat, powered by an outboard motor, travels at a constant speed $v$ in still water. The water flows at a lower constant speed $v_W$. He has traveled upstream for 2.00 km when his ice chest falls out of the boat. He notices that the chest is missing only after he has gone upstream for another 15.0 minutes. At that point, he turns around and heads back downstream, all the time traveling at the same speed relative to the water. He catches up with the floating ice chest just as it is about to go over the falls at his starting point. How fast is the river flowing? Solve this problem in two ways. (a) First, use the Earth as a reference frame. With respect to the Earth, the boat travels upstream at speed $v - v_W$ and downstream at $v + v_W$. (b) A second much simpler and more elegant solution is obtained by using the water as the reference frame. This approach has important applications in many more complicated problems, such as calculating the motion of rockets and Earth satellites and analyzing the scattering of subatomic particles from massive targets.

**60.** The water in a river flows uniformly at a constant speed of 2.50 m/s between parallel banks 80.0 m apart. You are to deliver a package directly across the river, but you can swim only at 1.50 m/s. (a) If you choose to minimize the time you spend in the water, in what direction should you head? (b) How far downstream will you be carried? (c) If you choose to minimize the distance downstream that the river carries you, in what direction should you head? (d) How far downstream will you be carried?

**61.** An enemy ship is on the east side of a mountain island, as shown in Figure P3.61. The enemy ship has maneuvered to within 2 500 m of the 1 800-m-high mountain peak and can shoot projectiles with an initial speed of 250 m/s. If the western shoreline is horizontally 300 m from the peak, what are the distances from the western shore at which a ship can be safe from the bombardment of the enemy ship?

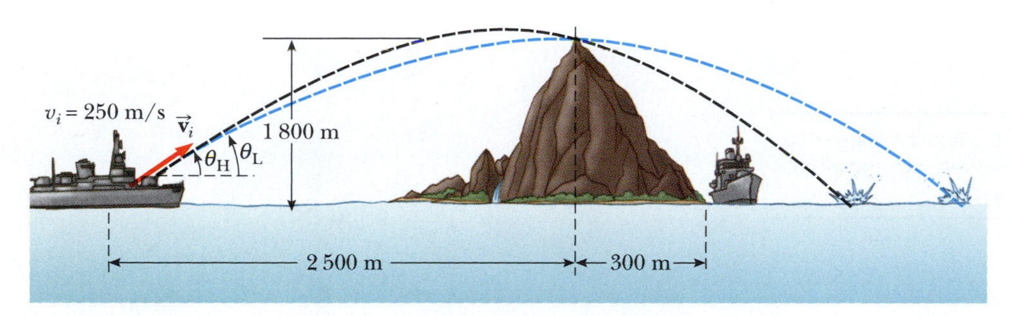


FIGURE **P3.61** View looking south.

## ANSWERS TO QUICK QUIZZES

**3.1** (a) Because acceleration occurs whenever the velocity changes in any way—with an increase or decrease in speed, a change in direction, or both—all three controls are accelerators. The gas pedal causes the car to speed up; the brake pedal causes the car to slow down. The steering wheel changes the direction of the velocity vector.

**3.2** (b), (d). At only one point—the peak of the trajectory—are the velocity and acceleration vectors perpendicular to each other. The velocity vector is horizontal at that point and the acceleration vector is downward. The acceleration vector is always directed downward. The velocity vector is never vertical if the object follows a path such as that in Figure 3.6.

**3.3** 15°, 30°, 45°, 60°, 75°. The greater the maximum height, the longer it takes the projectile to reach that altitude and then fall back down from it. So, as the launch angle increases, the time of flight increases.

**3.4** (c). We cannot choose (a) or (b) because the centripetal acceleration vector is not constant; it continuously changes in direction. Of the remaining choices, only (c) gives the correct perpendicular relationship between $\vec{\mathbf{a}}_c$ and $\vec{\mathbf{v}}$.

**3.5** (i), (b). The velocity vector is tangent to the path. If the acceleration vector is to be parallel to the velocity vector, it must also be tangent to the path. To be tangent requires that the acceleration vector have no component perpendicular to the path. If the path were to change direction, the acceleration vector would have a radial component, perpendicular to the path. Therefore, the path must remain straight. (ii), (d). If the acceleration vector is to be perpendicular to the velocity vector, it must have no component tangent to the path. On the other hand, if the speed is changing, there *must* be a component of the acceleration tangent to the path. Therefore, the velocity and acceleration vectors are never perpendicular in this situation. They can only be perpendicular if there is no change in the speed.

CHAPTER 4

# The Laws of Motion

A small tugboat exerts a force on a large ship, causing it to move. How can such a small boat move such a large object?

(Steve Raymer/CORBIS)

## CHAPTER OUTLINE


4.1 The Concept of Force

4.2 Newton's First Law

4.3 Mass

4.4 Newton's Second Law—The Particle Under a Net Force

4.5 The Gravitational Force and Weight

4.6 Newton's Third Law

4.7 Applications of Newton's Laws

4.8 Context Connection—Forces on Automobiles

SUMMARY


In the preceding two chapters on kinematics, we described the motion of particles based on the definitions of position, velocity, and acceleration. Aside from our discussion of gravity for objects in free-fall, we did not address what causes an object to move as it does. We would like to be able to answer general questions related to the causes of motion, such as "What mechanism causes changes in motion?" and "Why do some objects accelerate at higher rates than others?" In this first chapter on *dynamics,* we shall discuss the causes of the change in motion of particles using the concepts of force and mass. We will discuss the three fundamental laws of motion, which are based on experimental observations and were formulated about three centuries ago by Sir Isaac Newton.

## 4.1 THE CONCEPT OF FORCE

As a result of everyday experiences, everyone has a basic understanding of the concept of force. When you push or pull an object, you exert a force on it. You exert a force when you throw or kick a ball. In these examples, the word *force* is associated with the result of muscular activity and with some change in the state of motion of an object. Forces do not always cause an object to move, however. For example, as you sit reading this book, the gravitational force acts on your body and yet you remain stationary. You can push on a heavy block of stone and yet fail to move it.

This chapter is concerned with the relation between the force on an object and the change in motion of that object. If you pull on a spring, as in Figure 4.1a, the spring stretches. If the spring is calibrated, the distance it stretches can be used to measure the strength of the force. If a child pulls on a wagon, as in Figure 4.1b, the wagon moves. When a football is kicked, as in Figure 4.1c, it is both deformed and set in motion. These examples all show the results of a class of forces called *contact forces.* That is, these forces represent the result of physical contact between two objects.

There exist other forces that do not involve physical contact between two objects. These forces, known as *field forces,* can act through empty space. The gravitational force between two objects that causes the free-fall acceleration described in Chapters 2 and 3 is an example of this type of force and is illustrated in Figure 4.1d. This gravitational force keeps objects bound to the Earth and gives rise to what we commonly call the *weight* of an object. The planets of our solar system are bound to the Sun under the action of gravitational forces. Another common example of a field force is the electric force that one electric charge exerts on another electric

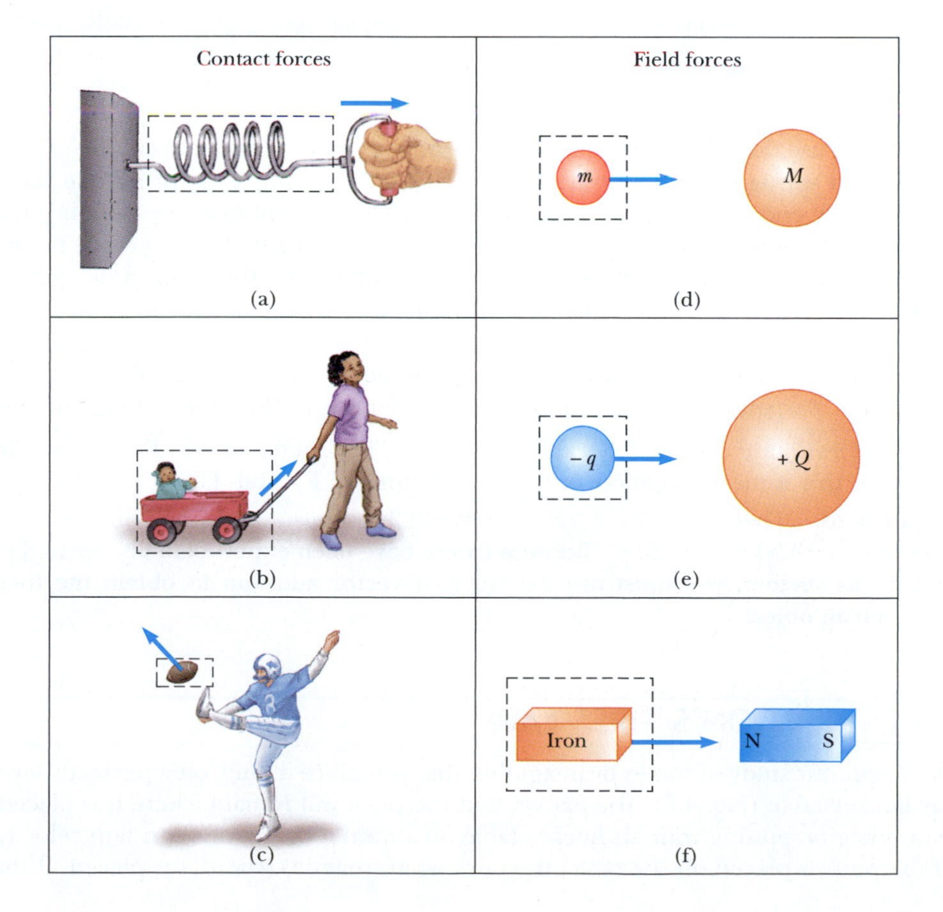


**FIGURE 4.1** Some examples of forces applied to various objects. In each case, a force is exerted on the particle or object within the boxed area. The environment external to the boxed area provides this force.

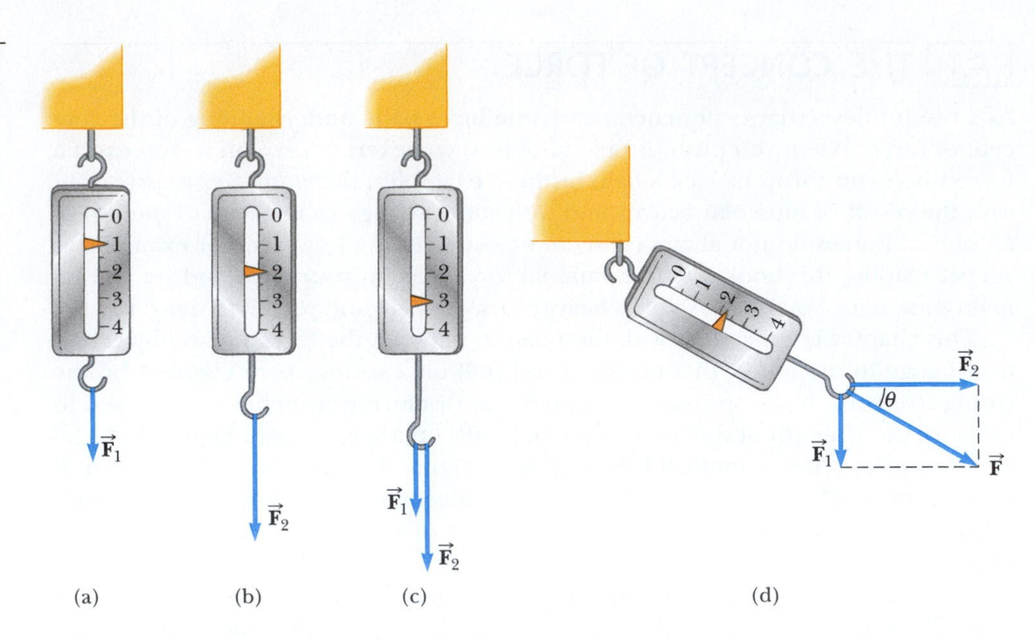


FIGURE 4.2 The vector nature of a force is tested with a spring scale. (a) A downward vertical force $\vec{F}_1$ elongates the spring 1.00 cm. (b) A downward vertical force $\vec{F}_2$ elongates the spring 2.00 cm. (c) When $\vec{F}_1$ and $\vec{F}_2$ are applied simultaneously, the spring elongates by 3.00 cm. (d) When $\vec{F}_1$ is downward and $\vec{F}_2$ is horizontal, the combination of the two forces elongates the spring $\sqrt{(1.00\text{ cm})^2 + (2.00\text{ cm})^2} = \sqrt{5.00}$ cm.

charge, as in Figure 4.1e. These charges might be an electron and proton forming a hydrogen atom. A third example of a field force is the force that a bar magnet exerts on a piece of iron, as shown in Figure 4.1f.

The distinction between contact forces and field forces is not as sharp as you may have been led to believe by the preceding discussion. At the atomic level, all the forces classified as contact forces turn out to be caused by electric (field) forces similar in nature to the attractive electric force illustrated in Figure 4.1e. Nevertheless, in understanding macroscopic phenomena, it is convenient to use both classifications of forces.

We can use the linear deformation of a spring to measure force, as in the case of a common spring scale. Suppose a force is applied vertically to a spring that has a fixed upper end, as in Figure 4.2a. The spring can be calibrated by defining the unit force $\vec{F}_1$ as the force that produces an elongation of 1.00 cm. If a force $\vec{F}_2$, applied as in Figure 4.2b, produces an elongation of 2.00 cm, the magnitude of $\vec{F}_2$ is 2.00 units. If the two forces $\vec{F}_1$ and $\vec{F}_2$ are applied simultaneously, as in Figure 4.2c, the elongation of the spring is 3.00 cm because the forces are applied in the same direction and their magnitudes add. If the two forces $\vec{F}_1$ and $\vec{F}_2$ are applied in perpendicular directions, as in Figure 4.2d, the elongation is $\sqrt{(1.00)^2 + (2.00)^2}\text{ cm} = \sqrt{5.00}\text{ cm} = 2.24\text{ cm}$. The single force $\vec{F}$ that would produce this same elongation is the vector sum of $\vec{F}_1$ and $\vec{F}_2$, as described in Figure 4.2d. That is, $|\vec{F}| = \sqrt{F_1^2 + F_2^2} = 2.24$ units, and its direction is $\theta = \tan^{-1}(-0.500) = -26.6°$. **Because forces have been experimentally verified to behave as vectors, you must use the rules of vector addition to obtain the total force on an object.**

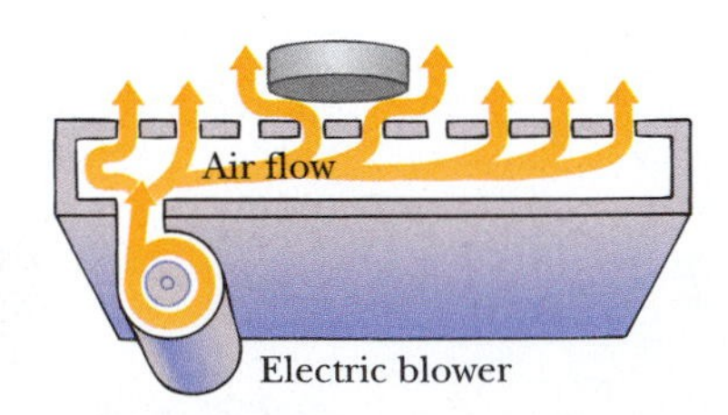


FIGURE 4.3 On an air hockey table, air blown through holes in the surface allows the puck to move almost without friction. If the table is not accelerating, a puck placed on the table will remain at rest with respect to the table if there are no horizontal forces acting on it.

## 4.2 NEWTON'S FIRST LAW

We begin our study of forces by imagining that you place a puck on a perfectly level air hockey table (Fig. 4.3). You expect that the puck will remain where it is placed. Now imagine putting your air hockey table on a train moving with constant velocity. If the puck is placed on the table, the puck again remains where it is placed. If the train were to accelerate, however, the puck would start moving along the table, just as a set of papers on your dashboard falls onto the front seat of your car when you step on the gas.

As we saw in Section 3.6, a moving object can be observed from any number of reference frames. **Newton's first law of motion,** sometimes called the *law of inertia,* defines a special set of reference frames called *inertial frames.* This law can be stated as follows:

■ Newton's first law

> If an object does not interact with other objects, it is possible to identify a reference frame in which the object has zero acceleration.

■ Inertial frame of reference

Such a reference frame is called an **inertial frame of reference.** When the puck is on the air hockey table located on the ground, you are observing it from an inertial reference frame; there are no horizontal interactions of the puck with any other objects, and you observe it to have zero acceleration in the horizontal direction. When you are on the train moving at constant velocity, you are also observing the puck from an inertial reference frame. **Any reference frame that moves with constant velocity relative to an inertial frame is itself an inertial frame.** When the train accelerates, however, you are observing the puck from a **noninertial reference frame** because you and the train are accelerating relative to the inertial reference frame of the surface of the Earth. Although the puck appears to be accelerating according to your observations, we can identify a reference frame in which the puck has zero acceleration. For example, an observer standing outside the train on the ground sees the puck moving with the same velocity as the train had before it started to accelerate (because there is almost no friction to "tie" the puck and the train together). Therefore, Newton's first law is still satisfied even though your observations say otherwise.

(Giraudon/Art Resource)

**ISAAC NEWTON** (1642–1727)

Newton, an English physicist and mathematician, was one of the most brilliant scientists in history. Before the age of 30, he formulated the basic concepts and laws of mechanics, discovered the law of universal gravitation, and invented the mathematical methods of calculus. As a consequence of his theories, Newton was able to explain the motions of the planets, the ebb and flow of the tides, and many special features of the motions of the Moon and the Earth. His contributions to physical theories dominated scientific thought for two centuries and remain important today.

A reference frame that moves with constant velocity relative to the distant stars is the best approximation of an inertial frame, and for our purposes we can consider the Earth as being such a frame. The Earth is not really an inertial frame because of its orbital motion around the Sun and its rotational motion about its own axis, both of which are related to centripetal accelerations. These accelerations, however, are small compared with $g$ and can often be neglected. (This is a simplification model.) For this reason, we assume that the Earth is an inertial frame, as is any other frame attached to it.

Let us assume that we are observing an object from an inertial reference frame. Before about 1600, scientists believed that the natural state of matter was the state of rest. Observations showed that moving objects eventually stopped moving. Galileo was the first to take a different approach to motion and the natural state of matter. He devised thought experiments and concluded that it is not the nature of an object to stop once set in motion; rather, it is its nature to *resist changes in its motion.* In his words, "Any velocity once imparted to a moving body will be rigidly maintained as long as the external causes of retardation are removed."

Given our assumption of observations made from inertial reference frames, we can pose a more practical statement of Newton's first law of motion:

■ Another statement of Newton's first law

> In the absence of external forces, when viewed from an inertial reference frame, an object at rest remains at rest and an object in motion continues in motion with a constant velocity (that is, with a constant speed in a straight line).

In simpler terms, we can say that **when no force acts on an object, the acceleration of the object is zero.** If nothing acts to change the object's motion, its velocity does not change. From the first law, we conclude that any *isolated object* (one that does not interact with its environment) is either at rest or moving with constant

**PITFALL PREVENTION 4.1**

**Newton's first law** Newton's first law does *not* say what happens for an object with *zero net force,* that is, multiple forces that cancel; it says what happens *in the absence of a force.* This subtle but important difference allows us to define force as that which causes a change in the motion. The description of an object under the effect of forces that balance is covered by Newton's second law.

velocity. The tendency of an object to resist any attempt to change its velocity is called **inertia.**

Consider a spacecraft traveling in space, far removed from any planets or other matter. The spacecraft requires some propulsion system to change its velocity. If the propulsion system is turned off when the spacecraft reaches a velocity $\vec{\mathbf{v}}$, however, the spacecraft "coasts" in space with that velocity and the astronauts enjoy a "free ride" (i.e., no propulsion system is required to keep them moving at the velocity $\vec{\mathbf{v}}$).

Finally, recall our discussion in Chapter 2 about the proportionality between force and acceleration:

$$\vec{\mathbf{F}} \propto \vec{\mathbf{a}}$$

Newton's first law tells us that the velocity of an object remains constant if no force acts on an object; the object maintains its state of motion. The preceding proportionality tells us that if a force *does* act, a change does occur in the motion, measured by the acceleration. This notion will form the basis of Newton's second law, and we shall provide more details on this concept shortly.

**QUICK QUIZ 4.1** Which of the following statements is most correct? **(a)** It is possible for an object to have motion in the absence of forces on the object. **(b)** It is possible to have forces on an object in the absence of motion of the object. **(c)** Neither (a) nor (b) is correct. **(d)** Both (a) and (b) are correct.

## 4.3 MASS

Imagine playing catch with either a basketball or a bowling ball. Which ball is more likely to keep moving when you try to catch it? Which ball has the greater tendency to remain motionless when you try to throw it? The bowling ball is more resistant to changes in its velocity than the basketball. How can we quantify this concept?

■ Definition of mass

**Mass** is that property of an object that specifies how much resistance an object exhibits to changes in its velocity, and as we learned in Section 1.1, the SI unit of mass is the kilogram. The greater the mass of an object, the less that object accelerates under the action of a given applied force.

To describe mass quantitatively, we begin by experimentally comparing the accelerations a given force produces on different objects. Suppose a force acting on an object of mass $m_1$ produces an acceleration $\vec{\mathbf{a}}_1$ and the *same force* acting on an object of mass $m_2$ produces an acceleration $\vec{\mathbf{a}}_2$. The ratio of the two masses is defined as the *inverse* ratio of the magnitudes of the accelerations produced by the force:

$$\frac{m_1}{m_2} \equiv \frac{a_2}{a_1} \qquad [4.1]$$

For example, if a given force acting on a 3-kg object produces an acceleration of $4 \text{ m/s}^2$, the same force applied to a 6-kg object produces an acceleration of $2 \text{ m/s}^2$. If one object has a known mass, the mass of the other object can be obtained from acceleration measurements.

**Mass is an inherent property of an object and is independent of the object's surroundings and of the method used to measure it.** Also, **mass is a scalar quantity** and therefore obeys the rules of ordinary arithmetic. That is, several masses can be combined in simple numerical fashion. For example, if you combine a 3-kg mass with a 5-kg mass, their total mass is 8 kg. We can verify this result experimentally by comparing the acceleration that a known force gives to several objects separately with the acceleration that the same force gives to the same objects combined as one unit.

Mass should not be confused with weight. **Mass and weight are two different quantities.** As we shall see later in this chapter, the weight of an object is equal to the magnitude of the gravitational force exerted on the object and varies with location. For example, a person who weighs 180 lb on the Earth weighs only about 30 lb on the Moon. On the other hand, the mass of an object is the same everywhere. An object having a mass of 2 kg on Earth also has a mass of 2 kg on the Moon.

▪ Mass and weight are different quantities

## 4.4 NEWTON'S SECOND LAW—THE PARTICLE UNDER A NET FORCE

Newton's first law explains what happens to an object when no force acts on it: It either remains at rest or moves in a straight line with constant speed. This law allows us to define an inertial frame of reference. It also allows us to identify force as that which changes motion. Newton's second law answers the question of what happens to an object that has a nonzero net force acting on it, based on our discussion of mass in the preceding section.

Imagine you are pushing a block of ice across a frictionless horizontal surface. When you exert some horizontal force $\vec{\mathbf{F}}$, the block moves with some acceleration $\vec{\mathbf{a}}$. Experiments show that if you apply a force twice as large to the same object, the acceleration doubles. If you increase the applied force to $3\vec{\mathbf{F}}$, the original acceleration is tripled, and so on. From such observations, we conclude that **the acceleration of an object is directly proportional to the net force acting on it.** We alluded to this proportionality in our discussion of acceleration in Chapter 2. We are now ready to extend that discussion.

These observations and those in Section 4.3 relating mass and acceleration are summarized in **Newton's second law:**

> The acceleration of an object is directly proportional to the net force acting on it and inversely proportional to its mass.

▪ Newton's second law

We write this law as

$$\vec{\mathbf{a}} \propto \frac{\sum \vec{\mathbf{F}}}{m}$$

where $\Sigma\vec{\mathbf{F}}$ is the **net force,** which is the vector sum of *all* forces acting on the object of mass $m$. If the object consists of a system of individual elements, the net force is the vector sum of all forces *external* to the system. Any *internal* forces—that is, forces between elements of the system—are not included because they do not affect the motion of the entire system. The net force is sometimes called the *resultant force,* the *sum of the forces,* the *total force,* or the *unbalanced force.*

Newton's second law in mathematical form is a statement of this relationship that makes the preceding proportionality an equality:[1]

$$\sum \vec{\mathbf{F}} = m\vec{\mathbf{a}} \qquad [4.2]$$

▪ Mathematical representation of Newton's second law

Note that Equation 4.2 is a *vector* expression and hence is equivalent to the following three component equations:

$$\sum F_x = ma_x \qquad \sum F_y = ma_y \qquad \sum F_z = ma_z \qquad [4.3]$$

▪ Newton's second law in component form

Newton's second law introduces us to a new analysis model, the particle under a net force. If a particle, or an object that can be modeled as a particle, is under the

### PITFALL PREVENTION 4.2

**FORCE IS THE CAUSE OF CHANGES IN MOTION** Be sure that you are clear on the role of force. Many times, students make the mistake of thinking that force is the cause of motion. We can, though, have motion in the absence of forces, as described in Newton's first law. Be sure to understand that force is the cause of *changes* in motion.

### PITFALL PREVENTION 4.3

**$m\vec{\mathbf{a}}$ IS NOT A FORCE** Equation 4.2 does *not* say that the product $m\vec{\mathbf{a}}$ is a force. All forces on an object are added vectorially to generate the net force on the left side of the equation. This net force is then equated to the product of the mass of the object and the acceleration that results from the net force. Do *not* include an "$m\vec{\mathbf{a}}$ force" in your analysis.

[1] Equation 4.2 is valid only when the speed of the object is much less than the speed of light. We will treat the relativistic situation in Chapter 9.

influence of a net force, Equation 4.2, the mathematical statement of Newton's second law, can be used to describe its motion. The acceleration is constant if the net force is constant. Therefore, the particle under a constant net force will have its motion described as a particle under constant acceleration. Of course, not all forces are constant, and when they are not, the particle cannot be modeled as one under constant acceleration. We shall investigate situations in this chapter and the next involving both constant and varying forces.

**QUICK QUIZ 4.2** An object experiences no acceleration. Which of the following *cannot* be true for the object? (**a**) A single force acts on the object. (**b**) No forces act on the object. (**c**) Forces act on the object, but the forces cancel.

**QUICK QUIZ 4.3** You push an object, initially at rest, across a frictionless floor with a constant force for a time interval $\Delta t$, resulting in a final speed of $v$ for the object. You repeat the experiment, but with a force that is twice as large. What time interval is now required to reach the same final speed $v$? (a) $4\,\Delta t$ (b) $2\,\Delta t$ (c) $\Delta t$ (d) $\Delta t/2$ (e) $\Delta t/4$

## Unit of Force

The SI unit of force is the **newton,** which is defined as the force that, when acting on a 1-kg mass, produces an acceleration of 1 m/s$^2$.

From this definition and Newton's second law, we see that the newton can be expressed in terms of the fundamental units of mass, length, and time:

▪ Definition of the newton

$$1\text{ N} \equiv 1\text{ kg}\cdot\text{m/s}^2 \qquad [4.4]$$

The units of mass, acceleration, and force are summarized in Table 4.1. Most of the calculations we shall make in our study of mechanics will be in SI units. Equalities between units in the SI and U.S. customary systems are given in Appendix A.

### ▪ Thinking Physics 4.1

In a train, the cars are connected by *couplers.* The couplers between the cars exert forces on the cars as the train is pulled by the locomotive in the front. Imagine that the train is speeding up in the forward direction. As you imagine moving from the locomotive to the last car, does the force exerted by the couplers *increase, decrease,* or *stay the same*? What if the engineer applies the brakes? How does the force vary from locomotive to last car in this case? (Assume that the only brakes applied are those on the engine.)

**Reasoning** The force *decreases* from the front of the train to the back. The coupler between the locomotive and the first car must apply enough force to accelerate all the remaining cars. As we move back along the train, each coupler is accelerating less mass behind it. The last coupler only has to accelerate the last car, so it exerts the smallest force. If the brakes are applied, the force decreases from front to back of the train also. The first coupler, at the back of the locomotive, must apply a large force to slow down all the remaining cars. The final coupler must only apply a force large enough to slow down the mass of the last car. ▪

**TABLE 4.1 Units of Mass, Acceleration, and Force**

| System of Units | Mass (M) | Acceleration (L/T$^2$) | Force (ML/T$^2$) |
|---|---|---|---|
| SI | kg | m/s$^2$ | N = kg · m/s$^2$ |
| U.S. customary | slug | ft/s$^2$ | lb = slug · ft/s$^2$ |

**EXAMPLE 4.1** An Accelerating Hockey Puck

A 0.30-kg hockey puck slides on the horizontal frictionless surface of an ice rink. It is struck simultaneously by two different hockey sticks. The two constant forces that act on the puck as a result of the hockey sticks are parallel to the ice surface and are shown in the pictorial representation in Figure 4.4. The force $\vec{\mathbf{F}}_1$ has a magnitude of 5.0 N, and $\vec{\mathbf{F}}_2$ has a magnitude of 8.0 N. Determine the acceleration of the puck while it is in contact with the two sticks.

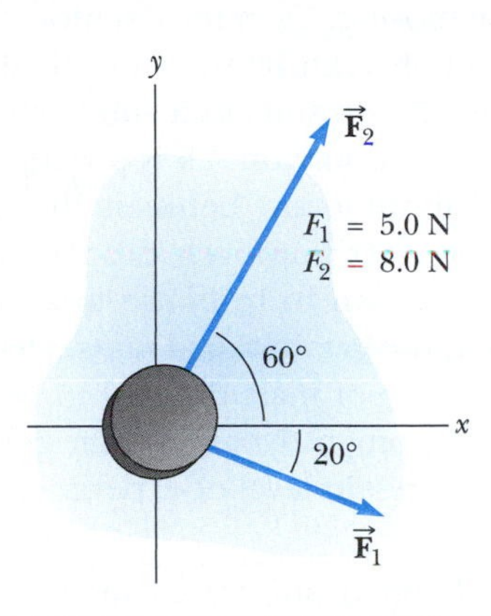


**FIGURE 4.4** (Example 4.1) A hockey puck moving on a frictionless surface accelerates in the direction of the net force, $\Sigma\vec{\mathbf{F}} = \vec{\mathbf{F}}_1 + \vec{\mathbf{F}}_2$.

**Solution** The puck is modeled as a particle under a net force. We first find the components of the net force. The component of the net force in the $x$ direction is

$$\sum F_x = F_{1x} + F_{2x} = F_1 \cos 20° + F_2 \cos 60°$$
$$= (5.0 \text{ N})(0.940) + (8.0 \text{ N})(0.500) = 8.7 \text{ N}$$

The component of the net force in the $y$ direction is

$$\sum F_y = F_{1y} + F_{2y} = -F_1 \sin 20° + F_2 \sin 60°$$
$$= -(5.0 \text{ N})(0.342) + (8.0 \text{ N})(0.866) = 5.2 \text{ N}$$

Now we use Newton's second law in component form to find the $x$ and $y$ components of acceleration:

$$a_x = \frac{\sum F_x}{m} = \frac{8.7 \text{ N}}{0.30 \text{ kg}} = 29 \text{ m/s}^2$$

$$a_y = \frac{\sum F_y}{m} = \frac{5.2 \text{ N}}{0.30 \text{ kg}} = 17 \text{ m/s}^2$$

The acceleration has a magnitude of

$$a = \sqrt{(29 \text{ m/s}^2)^2 + (17 \text{ m/s}^2)^2} = 34 \text{ m/s}^2$$

and its direction is

$$\theta = \tan^{-1}\frac{a_y}{a_x} = \tan^{-1}\frac{17 \text{ m/s}^2}{29 \text{ m/s}^2} = 30°$$

relative to the positive $x$ axis.

## 4.5 THE GRAVITATIONAL FORCE AND WEIGHT

We are well aware that all objects are attracted to the Earth. The force exerted by the Earth on an object is the **gravitational force** $\vec{\mathbf{F}}_g$. This force is directed toward the center of the Earth.[2] The magnitude of the gravitational force is called the **weight** $F_g$ of the object.

We have seen in Chapters 2 and 3 that a freely falling object experiences an acceleration $\vec{\mathbf{g}}$ directed toward the center of the Earth. A freely falling object has only one force on it, the gravitational force, so the net force on the object in this situation is equal to the gravitational force:

$$\sum \vec{\mathbf{F}} = \vec{\mathbf{F}}_g$$

Because the acceleration of a freely falling object is equal to the free-fall acceleration $\vec{\mathbf{g}}$, it follows that

$$\sum \vec{\mathbf{F}} = m\vec{\mathbf{a}} \quad \rightarrow \quad \vec{\mathbf{F}}_g = m\vec{\mathbf{g}}$$

or, in magnitude,

$$F_g = mg \qquad [4.5]$$

■ Relation between mass and weight of an object

[2] This statement represents a simplification model in that it ignores that the mass distribution of the Earth is not perfectly spherical.

Astronaut Edwin E. Aldrin Jr., walking on the Moon after the *Apollo 11* lunar landing. Aldrin's weight on the Moon is less than it is on the Earth, but his mass is the same in both places. ■

Because it depends on $g$, weight varies with location, as we mentioned in Section 4.3. Objects weigh less at higher altitudes than at sea level because $g$ decreases with increasing distance from the center of the Earth. Hence, weight, unlike mass, is not an inherent property of an object. For example, if an object has a mass of 70 kg, its weight in a location where $g = 9.80\ \text{m/s}^2$ is $mg = 686$ N. At the top of a mountain where $g = 9.76\ \text{m/s}^2$, the object's weight would be 683 N. Therefore, if you want to lose weight without going on a diet, climb a mountain or weigh yourself at 30 000 ft during an airplane flight.

Because $F_g = mg$, we can compare the masses of two objects by measuring their weights with a spring scale. At a given location (so that $g$ is fixed) the ratio of the weights of two objects equals the ratio of their masses.

Equation 4.5 quantifies the gravitational force on the object, but notice that this equation does not require the object to be moving. Even for a stationary object, or an object on which several forces act, Equation 4.5 can be used to calculate the magnitude of the gravitational force. This observation results in a subtle shift in the interpretation of $m$ in the equation. The mass $m$ in Equation 4.5 is playing the role of determining the strength of the gravitational attraction between the object and the Earth. This role is completely different from that previously described for mass, that of measuring the resistance to changes in motion in response to an external force. Therefore, we call $m$ in this type of equation the **gravitational mass.** Despite this quantity being different from inertial mass (the type of mass defined in Section 4.3), it is one of the experimental conclusions in Newtonian dynamics that gravitational mass and inertial mass have the same value at the present level of experimental refinement.

**PITFALL PREVENTION 4.4**

**DIFFERENTIATE BETWEEN *g* AND g** Be sure not to confuse the italicized letter *g* that we use for the magnitude of the free-fall acceleration with the abbreviation g that is used for grams.

**QUICK QUIZ 4.4** Suppose you are talking by interplanetary telephone to your friend, who lives on the Moon. He tells you that he has just won a newton of gold in a contest. Excitedly, you tell him that you entered the Earth version of the same contest and also won a newton of gold! Who is richer? **(a)** You are. **(b)** Your friend is. **(c)** You are equally rich.

## 4.6 NEWTON'S THIRD LAW

Newton's third law conveys the notion that forces are always interactions between two objects: **If two objects interact, the force $\vec{\mathbf{F}}_{12}$ exerted by object 1 on object 2 is equal in magnitude but opposite in direction to the force $\vec{\mathbf{F}}_{21}$ exerted by object 2 on object 1:**

■ Newton's third law

$$\vec{\mathbf{F}}_{12} = -\vec{\mathbf{F}}_{21} \qquad [4.6]$$

When it is important to designate forces as interactions between two objects, we will use this subscript notation, where $\vec{\mathbf{F}}_{ab}$ means "the force exerted *by* a *on* b." The third law, illustrated in Figure 4.5a, is equivalent to stating that **forces always occur in pairs** or that **a single isolated force cannot exist.** The force that object 1 exerts on object 2 may be called the *action force,* and the force of object 2 on object 1 may be called the *reaction force.* In reality, either force can be labeled the action or reaction force. **The action force is equal in magnitude to the reaction force and opposite in direction. In all cases, the action and reaction forces act on different objects and must be of the same type.** For example, the force acting on a freely falling projectile is the gravitational force exerted by the Earth on the projectile $\vec{\mathbf{F}}_g = \vec{\mathbf{F}}_{\text{Ep}}$ (E = Earth, p = projectile), and the magnitude of this force is $mg$. The reaction to this force is the gravitational force exerted by the projectile on the Earth $\vec{\mathbf{F}}_{\text{pE}} = -\vec{\mathbf{F}}_{\text{Ep}}$. The reaction force $\vec{\mathbf{F}}_{\text{pE}}$ must accelerate the Earth toward the projectile just as the action force $\vec{\mathbf{F}}_{\text{Ep}}$ accelerates the projectile toward the Earth. Because the Earth has such a large mass, however, its acceleration as a result of this reaction force is negligibly small.

**PITFALL PREVENTION 4.5**

**NEWTON'S THIRD LAW** Newton's third law is such an important and often misunderstood notion that it is repeated here in a Pitfall Prevention. In Newton's third law, action and reaction forces act on *different* objects. Two forces acting on the same object, even if they are equal in magnitude and opposite in direction, *cannot* be an action–reaction pair.

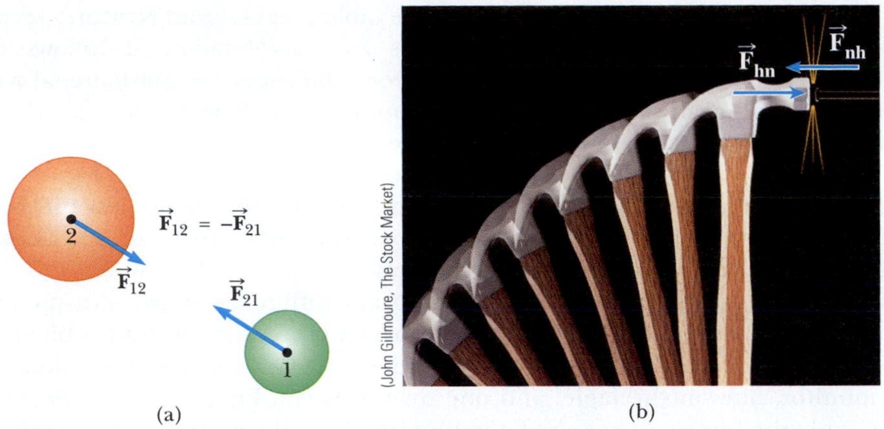

**FIGURE 4.5** Newton's third law. (a) The force $\vec{\mathbf{F}}_{12}$ exerted by object 1 on object 2 is equal in magnitude and opposite in direction to the force $\vec{\mathbf{F}}_{21}$ exerted by object 2 on object 1. (b) The force $\vec{\mathbf{F}}_{hn}$ exerted by the hammer on the nail is equal in magnitude and opposite in direction to the force $\vec{\mathbf{F}}_{nh}$ exerted by the nail on the hammer.

Another example of Newton's third law in action is shown in Figure 4.5b. The force $\vec{\mathbf{F}}_{hn}$ exerted by the hammer on the nail (the action) is equal in magnitude and opposite the force $\vec{\mathbf{F}}_{nh}$ exerted by the nail on the hammer (the reaction). This latter force stops the forward motion of the hammer when it strikes the nail.

The Earth exerts a gravitational force $\vec{\mathbf{F}}_g$ on any object. If the object is a computer monitor at rest on a table, as in the pictorial representation in Figure 4.6a, the reaction force to $\vec{\mathbf{F}}_g = \vec{\mathbf{F}}_{Em}$ is the force exerted by the monitor on the Earth $\vec{\mathbf{F}}_{mE} = -\vec{\mathbf{F}}_{Em}$. The monitor does not accelerate because it is held up by the table. The table exerts on the monitor an upward force $\vec{\mathbf{n}} = \vec{\mathbf{F}}_{tm}$, called the **normal force.**[3] This force prevents the monitor from falling through the table; it can have

■ Normal force

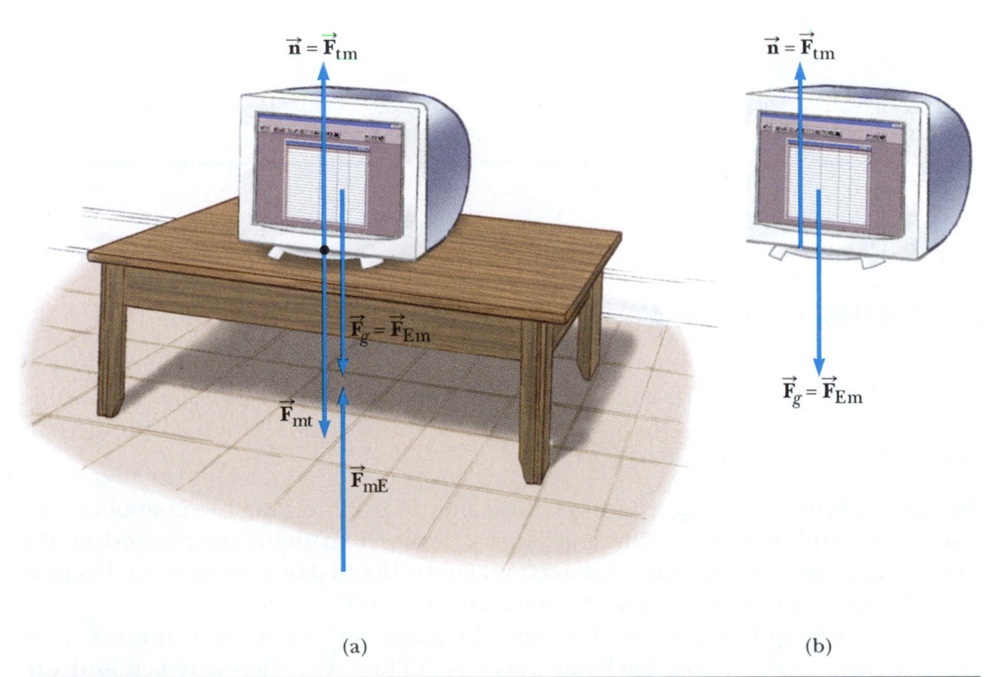


**FIGURE 4.6** (a) When a computer monitor is sitting on a table, several forces are acting. (b) The free-body diagram for the monitor. The forces acting on the monitor are the normal force $\vec{\mathbf{n}} = \vec{\mathbf{F}}_{tm}$ and the gravitational force $\vec{\mathbf{F}}_g = \vec{\mathbf{F}}_{Em}$.

[3] The word *normal* is used because the direction of $\vec{\mathbf{n}}$ is always *perpendicular* to the surface.

### PITFALL PREVENTION 4.6

***n* DOES NOT ALWAYS EQUAL *mg*** In the situation shown in Figure 4.6, we find that $n = mg$. There are many situations in which the normal force has the same magnitude as the gravitational force, but *do not* adopt this equality as a general rule (a common student pitfall). If the problem involves an object on an incline, if there are applied forces with vertical components, or if there is a vertical acceleration of the system, $n \neq mg$. *Always* apply Newton's second law to find the relationship between $n$ and $mg$.

### PITFALL PREVENTION 4.7

**FREE-BODY DIAGRAMS** The *most important* step in solving a problem using Newton's laws is to draw a proper simplified pictorial representation, the free-body diagram. Be sure to draw only those forces that act on the object you are isolating. Be sure to draw *all* forces acting on the object, including any field forces, such as the gravitational force. Do not include velocity, position, or acceleration vectors. Do not include a vector for the net force or for $m\vec{\mathbf{a}}$.

any value needed, up to the point at which the table breaks. From Newton's second law we see that, because the monitor has zero acceleration, it follows that $\Sigma\vec{\mathbf{F}} = \vec{\mathbf{n}} - m\vec{\mathbf{g}} = 0$, or $n = mg$. The normal force balances the gravitational force on the monitor, so the net force on the monitor is zero. The reaction to **n** is the force exerted by the monitor downward on the table, $\vec{\mathbf{F}}_{\text{mt}} = -\vec{\mathbf{F}}_{\text{tm}}$.

Note that the forces acting on the monitor are $\vec{\mathbf{F}}_g$ and $\vec{\mathbf{n}}$, as shown in Figure 4.6b. The two reaction forces $\vec{\mathbf{F}}_{\text{mE}}$ and $\vec{\mathbf{F}}_{\text{mt}}$ are exerted by the monitor on the Earth and the table, respectively. Remember that the two forces in an action–reaction pair always act on two different objects.

Figure 4.6 illustrates an extremely important difference between a pictorial representation and a simplified pictorial representation for solving problems involving forces. Figure 4.6a shows many of the forces in the situation: those on the monitor, one on the table, and one on the Earth. Figure 4.6b, by contrast, shows only the forces on *one object,* the monitor. This illustration is a critical simplified pictorial representation called a **free-body diagram.** When analyzing a particle under a net force, we are interested in the net force on one object, an object of mass $m$, which we will model as a particle. Therefore, a free-body diagram helps us isolate only those forces on the object and eliminate the other forces from our analysis. The free-body diagram can be simplified further, if you wish, by representing the object, such as the monitor in this case, as a particle by simply drawing a dot.

**QUICK QUIZ 4.5** If a fly collides with the windshield of a fast-moving bus, which experiences an impact force with a larger magnitude? **(a)** The fly does. **(b)** The bus does. **(c)** The same force is experienced by both. Which experiences the greater acceleration? **(d)** The fly does. **(e)** The bus does. **(f)** The same acceleration is experienced by both.

**QUICK QUIZ 4.6** Which of the following is the reaction force to the gravitational force acting on your body as you sit in your desk chair? **(a)** the normal force from the chair **(b)** the force you apply downward on the seat of the chair **(c)** neither of these forces

## Thinking Physics 4.2

A horse pulls on a sled with a horizontal force, causing the sled to accelerate as in Figure 4.7a. Newton's third law says that the sled exerts a force of equal magnitude and opposite direction on the horse. In view of this situation, how can the sled accelerate? Don't these forces cancel?

**Reasoning** When applying Newton's third law, it is important to remember that the forces involved act on different objects. Notice that the force exerted by the horse acts *on the sled,* whereas the force exerted by the sled acts *on the horse.* Because these forces act on different objects, they cannot cancel.

The horizontal forces exerted on the *sled* alone are the forward force $\vec{\mathbf{F}}_{\text{hs}}$ exerted by the horse and the backward force of friction $\vec{\mathbf{f}}_{\text{sled}}$ between sled and surface (Fig. 4.7b). When $\vec{\mathbf{F}}_{\text{hs}}$ exceeds $\vec{\mathbf{f}}_{\text{sled}}$, the sled accelerates to the right.

The horizontal forces exerted on the *horse* alone are the forward friction force $\vec{\mathbf{f}}_{\text{horse}}$ from the ground and the backward force $\vec{\mathbf{F}}_{\text{sh}}$ exerted by the sled (Fig. 4.7c). The resultant of these two forces causes the horse to accelerate. When $\vec{\mathbf{f}}_{\text{horse}}$ exceeds $\vec{\mathbf{F}}_{\text{sh}}$, the horse accelerates to the right. ■

FIGURE 4.7 (Thinking Physics 4.2) (a) A horse pulls a sled through the snow. (b) The forces on the sled. (c) The forces on the horse.

## 4.7 APPLICATIONS OF NEWTON'S LAWS

In this section, we present some simple applications of Newton's laws to objects that are either in equilibrium ($\vec{a} = 0$) or are accelerating under the action of constant external forces. We shall assume that the objects behave as particles so that we need not worry about rotational motion or other complications. In this section, we also apply some additional simplification models. We ignore the effects of friction for those problems involving motion, which is equivalent to stating that the surfaces are *frictionless.* We usually ignore the masses of any ropes or strings involved. In this approximation, the magnitude of the force exerted at any point along a string is the same. In problem statements, the terms *light* and *of negligible mass* are used to indicate that a mass is to be ignored when you work the problem. These two terms are synonymous in this context.

When an object such as a block is being pulled by a rope or string attached to it, the rope exerts a force $\vec{T}$ on the object. Its direction is along the rope, away from the object. The magnitude $T$ of this force is called the **tension** in the rope.

■ Tension

Consider a crate being pulled to the right on a frictionless, horizontal surface, as in Figure 4.8a. Suppose you are asked to find the acceleration of the crate and the force the floor exerts on it. Note that the horizontal force $\vec{T}$ being applied to the crate acts through the rope.

Because we are interested only in the motion of the crate, we must be able to *identify any and all external forces acting on it.* These forces are illustrated in the free-body diagram in Figure 4.8b. In addition to the force $\vec{T}$, the free-body diagram for the crate includes the gravitational force $\vec{F}_g$ and the normal force **n** exerted by the floor on the crate. The *reactions* to the forces we have listed—namely, the force exerted by the crate on the rope, the force exerted by the crate on the Earth, and the force exerted by the crate on the floor—are not included in the free-body diagram because they act on *other* objects and not on the crate.

Now let us apply Newton's second law to the crate. First, we must choose an appropriate coordinate system. In this case, it is convenient to use the coordinate system shown in Figure 4.8b, with the $x$ axis horizontal and the $y$ axis vertical. We can apply Newton's second law in the $x$ direction, $y$ direction, or both, depending on what we are asked to find in the problem. In addition, we may be able to use the equations of motion for the particle under constant acceleration that we discussed in Chapter 2. You should use these equations only when the acceleration is constant, however, which is the case if the net force is constant. For example, if the force $\vec{T}$ in Figure 4.8 is constant, the acceleration in the $x$ direction is also constant because $\vec{a} = \vec{T}/m$.

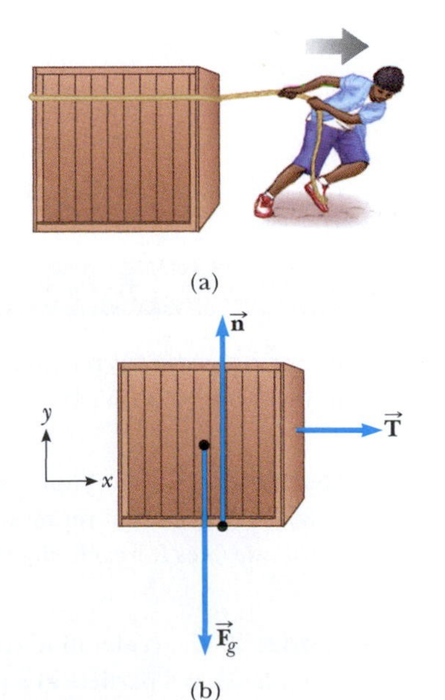


FIGURE 4.8 (a) A crate being pulled to the right on a frictionless surface. (b) The free-body diagram that represents the external forces on the crate.

### The Particle in Equilibrium

Objects that are either at rest or moving with constant velocity are said to be in **equilibrium.** From Newton's second law with $\vec{a} = 0$, this condition of equilibrium can be expressed as

$$\sum \vec{\mathbf{F}} = 0 \qquad [4.7]$$

This statement signifies that the vector sum of all the forces (the net force) acting on an object in equilibrium is zero.[4] If a particle is subject to forces but exhibits an acceleration of zero, we use Equation 4.7 to analyze the situation, as we shall see in some of the following examples.

Usually, the problems we encounter in our study of equilibrium are easier to solve if we work with Equation 4.7 in terms of the components of the external forces acting on an object. In other words, in a two-dimensional problem, the sum of all the external forces in the $x$ and $y$ directions must separately equal zero; that is,

$$\sum F_x = 0 \qquad \sum F_y = 0 \qquad [4.8]$$

The extension of Equations 4.8 to a three-dimensional situation can be made by adding a third component equation, $\Sigma F_z = 0$.

In a given situation, we may have balanced forces on an object in one direction but unbalanced forces in the other. Therefore, for a given problem, we may need to model the object as a particle in equilibrium for one component and a particle under a net force for the other.

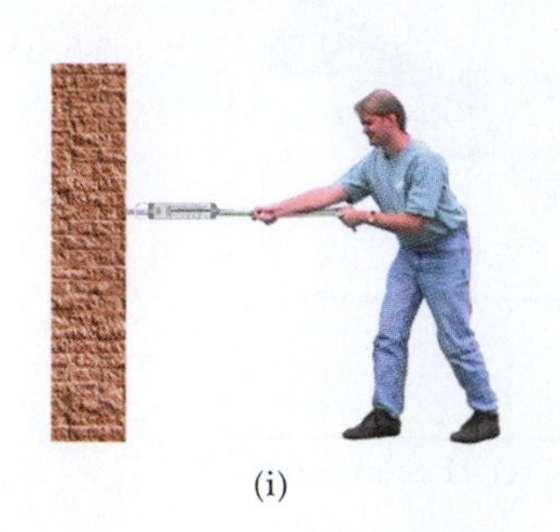

(i)

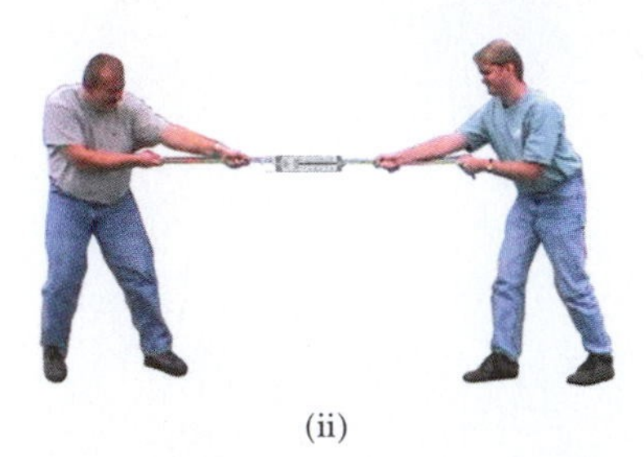

(ii)

**FIGURE 4.9** (Quick Quiz 4.7) (i) An individual pulls with a force of magnitude $F$ on a spring scale attached to a wall. (ii) Two individuals pull with forces of magnitude $F$ in opposite directions on a spring scale attached between two ropes.

**QUICK QUIZ 4.7** Consider the two situations shown in Figure 4.9, in which no acceleration occurs. In both cases, all individuals pull with a force of magnitude $F$ on a rope attached to a spring scale. Is the reading on the spring scale in part (i) of the figure (**a**) greater than, (**b**) less than, or (**c**) equal to the reading in part (ii)?

## The Particle Under a Net Force

In a situation in which a nonzero net force is acting on an object, the object is accelerating. We use Newton's second law to determine the features of the motion:

$$\sum \vec{\mathbf{F}} = m\vec{\mathbf{a}}$$

In practice, this equation is broken into components so that two (or three) equations can be handled independently. The representative suggestions and problems that follow should help you solve problems of this kind.

### PROBLEM-SOLVING STRATEGY Applying Newton's Laws

The following procedure is recommended when dealing with problems involving Newton's law.

**1. Conceptualize** Draw a simple, neat diagram of the system to help establish the mental representation. Establish convenient coordinate axes for each object in the system.

**2. Categorize** If an acceleration component for an object is zero, it is modeled as a particle in equilibrium in this direction and $\Sigma F = 0$. If not, the object is modeled as a particle under a net force in this direction and $\Sigma F = ma$.

**3. Analyze** Isolate the object whose motion is being analyzed. Draw a free-body diagram for this object. For systems containing more than one object, draw *separate* free-body diagrams for each object. *Do not* include in the free-body diagram forces exerted by the object on its surroundings.

Find the components of the forces along the coordinate axes. Apply Newton's second law, $\Sigma\vec{\mathbf{F}} = m\vec{\mathbf{a}}$, in component

[4] This statement is only one condition of equilibrium for an object. An object that can be modeled as a particle moving through space is said to be in translational motion. If the object is spinning, it is said to be in rotational motion. A second condition of equilibrium is a statement of rotational equilibrium. This condition will be discussed in Chapter 10 when we discuss spinning objects. Equation 4.7 is sufficient for analyzing objects in translational motion, which are those of interest to us at this point.

form. Check your dimensions to make sure all terms have units of force.

Solve the component equations for the unknowns. Remember that to obtain a complete solution, you must have as many independent equations as you have unknowns.

**4. Finalize** Make sure your results are consistent with the free-body diagram. Also check the predictions of your solutions for extreme values of the variables. By doing so, you can often detect errors in your results.

We now embark on a series of examples that demonstrate how to solve problems involving a particle in equilibrium or a particle under a net force. You should read and study these examples very carefully.

## EXAMPLE 4.2 A Traffic Light at Rest

A traffic light weighing 122 N hangs from a cable tied to two other cables fastened to a support, as in Figure 4.10a. The upper cables make angles of 37.0° and 53.0° with the horizontal. These upper cables are not as strong as the vertical cable and will break if the tension in them exceeds 100 N. Does the traffic light remain in this situation, or will one of the cables break?

**Solution** Let us assume that the cables do not break, so no acceleration of any sort occurs in any direction. Therefore, we use the model of a particle in equilibrium for both $x$ and $y$ components for any part of the system. We shall construct two free-body diagrams. The first is for the traffic light, shown in Figure 4.10b; the second is for the knot that holds the three cables together, as in Figure 4.10c. The knot is a convenient point to choose because all the forces in which we are interested act through this point. Because the acceleration of the system is zero, we can use the equilibrium conditions that the net force on the light is zero and that the net force on the knot is zero.

Considering Figure 4.10b, we apply the equilibrium condition in the $y$ direction, $\Sigma F_y = 0 \rightarrow T_3 - F_g = 0$, which leads to $T_3 = F_g = 122$ N. Thus, the force $\vec{\mathbf{T}}_3$ exerted by the vertical cable balances the weight of the light.

Considering the knot next, we choose the coordinate axes as shown in Figure 4.10c and resolve the forces into their $x$ and $y$ components, as shown in the following tabular representation:

| Force | $x$ component | $y$ component |
|---|---|---|
| $\vec{\mathbf{T}}_1$ | $-T_1 \cos 37.0°$ | $T_1 \sin 37.0°$ |
| $\vec{\mathbf{T}}_2$ | $T_2 \cos 53.0°$ | $T_2 \sin 53.0°$ |
| $\vec{\mathbf{T}}_3$ | 0 | $-122$ N |

Equations 4.8 give us

$$(1)\ \sum F_x = T_2 \cos 53.0° - T_1 \cos 37.0° = 0$$

$$(2)\ \sum F_y = T_1 \sin 37.0° + T_2 \sin 53.0° - 122\ \text{N} = 0$$

We solve (1) for $T_2$ in terms of $T_1$ to give

$$T_2 = T_1\left(\frac{\cos 37.0°}{\cos 53.0°}\right) = 1.33T_1$$

This value for $T_2$ is substituted into (2) to give

$$T_1 \sin 37.0° + (1.33T_1)(\sin 53.0°) - 122\ \text{N} = 0$$

$$T_1 = 73.4\ \text{N}$$

We then calculate $T_2$:

$$T_2 = 1.33T_1 = 97.4\ \text{N}$$

Both of these values are less than 100 N (just barely for $T_2$!), so the cables do not break.

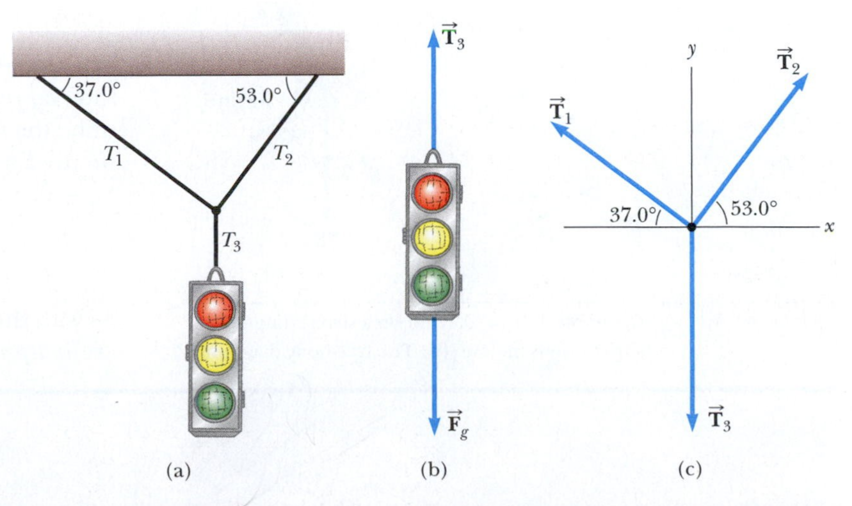


**FIGURE 4.10** (Example 4.2) (a) A traffic light suspended by cables. (b) The free-body diagram for the traffic light. (c) The free-body diagram for the knot in the cable.

## EXAMPLE 4.3 A Sled on Frictionless Snow

A child on a sled is released on a frictionless hill of angle $\theta$, as in Figure 4.11a.

**A** Determine the acceleration of the sled after it is released.

**Solution** We identify the combination of the sled and the child as our object of interest. We model the object as a particle of mass $m$. Newton's second law can be used to determine the acceleration of the particle. First, we construct the free-body diagram for the particle as in Figure 4.11b. The only forces on the particle are the normal force $\vec{\mathbf{n}}$ acting perpendicularly to the incline and the gravitational force $m\vec{\mathbf{g}}$ acting vertically downward. For problems of this type involving inclines, it is convenient to choose the coordinate axes with $x$ *along* the incline and $y$ *perpendicular* to it. Then, we replace $m\vec{\mathbf{g}}$ by a combination of a component vector of magnitude $mg \sin \theta$ along the *positive* $x$ axis (down the incline) and one of magnitude $mg \cos \theta$ in the *negative* $y$ direction.

Applying Newton's second law in component form to the particle and noting that $a_y = 0$ gives

$$(1) \quad \sum F_x = mg \sin \theta = ma_x$$

$$(2) \quad \sum F_y = n - mg \cos \theta = 0$$

From (1) we see that the acceleration along the incline is provided by the component of the gravitational force parallel to the incline, which gives us

$$(3) \quad a_x = g \sin \theta$$

Note that the acceleration given by (3) is *independent* of the mass of the particle; it depends only on the angle of inclination and on $g$. From (2) we conclude that the component of the gravitational force perpendicular to the incline is *balanced* by the normal force; that is, $n = mg \cos \theta$. (Notice, as pointed out in Pitfall Prevention 4.6, that $n$ does not equal $mg$ in this case.)

***Special Cases*** When $\theta = 90°$, (3) gives us $a_x = g$ and (2) gives us $n = 0$. This case corresponds to the particle in free-fall. (For our choice of coordinate system, positive $x$ is in the downward direction when $\theta = 90°$; hence, the acceleration is $+g$ rather than $-g$.) When $\theta = 0°$, $a_x = 0$ and $n = mg$ (its maximum value), which corresponds to the situation in which the particle is on a level surface and not accelerating.

This technique of looking at special cases of limiting situations is often useful in checking an answer. In this situation, if the angle $\theta$ goes to 90°, we know intuitively that the object should be falling parallel to the surface of the incline. That (3) mathematically reduces to $a_x = g$ when $\theta = 90°$ gives us confidence in our answer. It doesn't prove that the answer is correct, but if the acceleration does not reduce to $g$, it would tell us that the answer is incorrect.

**B** Suppose the sled is released from rest at the top of the hill and the distance from the front of the sled to the bottom of the hill is $d$. How long does it take the front of the sled to reach the bottom, and what is its speed just as it arrives at that point?

**Solution** In part A, we found $a_x = g \sin \theta$, which is constant. Hence, we can model the system as a particle under constant acceleration for the motion parallel to the incline. We use Equation 2.12, $x_f = x_i + v_{xi}t + \frac{1}{2}a_x t^2$, to describe the position of the sled's front edge. We define the initial position as $x_i = 0$ and the final position as $x_f = d$. Because the sled starts sliding from rest, $v_{xi} = 0$. With these values, Equation 2.12 becomes simply $d = \frac{1}{2}a_x t^2$, or

$$t = \sqrt{\frac{2d}{a_x}} = \sqrt{\frac{2d}{g \sin \theta}}$$

This equation answers the first question as to the time interval required to reach the bottom. Now, to determine the speed when the sled arrives at the bottom, we use Equation 2.13, $v_{xf}^2 = v_{xi}^2 + 2a_x(x_f - x_i)$ with $v_{xi} = 0$, and we find that $v_{xf}^2 = 2a_x d$, or

$$v_{xf} = \sqrt{2a_x d} = \sqrt{2gd \sin \theta}$$

As with the acceleration parallel to the incline, $t$ and $v_{xf}$ are *independent* of the mass of the sled and child.

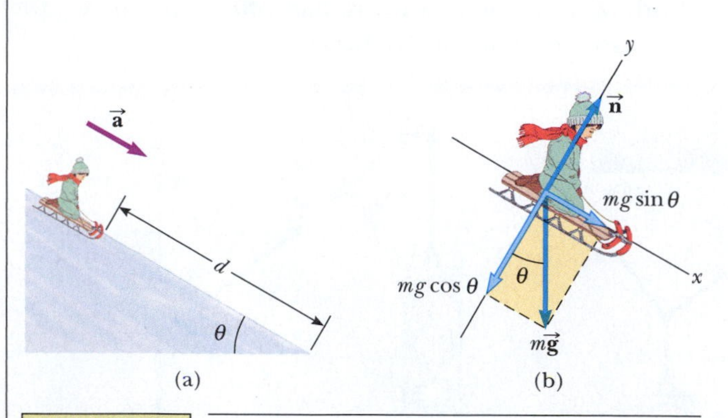


FIGURE 4.11 (Example 4.3) (a) A child on a sled sliding down a frictionless incline. (b) The free-body diagram.

## INTERACTIVE EXAMPLE 4.4 The Atwood Machine

When two objects with unequal masses are hung vertically over a light, frictionless pulley as in Active Figure 4.12a, the arrangement is called an *Atwood machine.* The device is sometimes used in the laboratory to measure the free-fall acceleration. Calculate the magnitude of the acceleration of the two objects and the tension in the string.

**Solution** *Conceptualize* the problem by thinking about the mental representation suggested by Active Figure 4.12a: As one object moves upward, the other object moves downward. Because the objects are connected by an inextensible string, they must have the same magnitude of acceleration. The objects in the Atwood machine are subject to the gravitational force as well as to the forces exerted by the strings connected to them. In *categorizing* the problem, we model the objects as particles under a net force.

We begin to *analyze* the problem by drawing free-body diagrams for the two objects, as in Active Figure 4.12b. Two forces act on each object: the upward force $\vec{T}$ exerted by the string and the downward gravitational force. In a problem such as this one in which the pulley is modeled as massless and frictionless, the tension in the string on both sides of the pulley is the same. If the pulley has mass or is subject to a friction force, the tensions in the string on either side of the pulley are not the same and the situation requires the techniques of Chapter 10.

In these types of problems, involving strings that pass over pulleys, we must be careful about the sign convention. Notice that if $m_1$ goes up, $m_2$ goes down. Therefore, $m_1$ going up and $m_2$ going down should be represented equivalently as far as a sign convention is concerned. We can do so by defining our sign convention with up as positive for $m_1$ and down as positive for $m_2$, as shown in Active Figure 4.12a.

With this sign convention, the net force exerted on $m_1$ is $T - m_1 g$, whereas the net force exerted on $m_2$ is $m_2 g - T$. We have chosen the signs of the forces to be consistent with the choices of the positive direction for each object.

When Newton's second law is applied to $m_1$, we find

$$(1) \quad \sum F_y = T - m_1 g = m_1 a$$

Similarly, for $m_2$ we find

$$(2) \quad \sum F_y = m_2 g - T = m_2 a$$

Note that $a$ is the same for both objects. When (2) is added to (1), $T$ cancels and we have

$$-m_1 g + m_2 g = m_1 a + m_2 a$$

Solving for the acceleration $a$ give us

$$(3) \quad a = \left(\frac{m_2 - m_1}{m_1 + m_2}\right) g$$

If $m_2 > m_1$, the acceleration given by (3) is positive: $m_1$ goes up and $m_2$ goes down. Is that consistent with your mental representation? If $m_1 > m_2$, the acceleration is negative and the masses move in the opposite direction.

If (3) is substituted into (1), we find

$$(4) \quad T = \left(\frac{2m_1 m_2}{m_1 + m_2}\right) g$$

To *finalize* the problem, let us consider some special cases. For example, when $m_1 = m_2$, (3) and (4) give us $a = 0$ and $T = m_1 g = m_2 g$, as we would intuitively expect for the balanced case. Also, if $m_2 \gg m_1$, $a \approx g$ (a freely falling object) and $T \approx 0$. For such a large mass

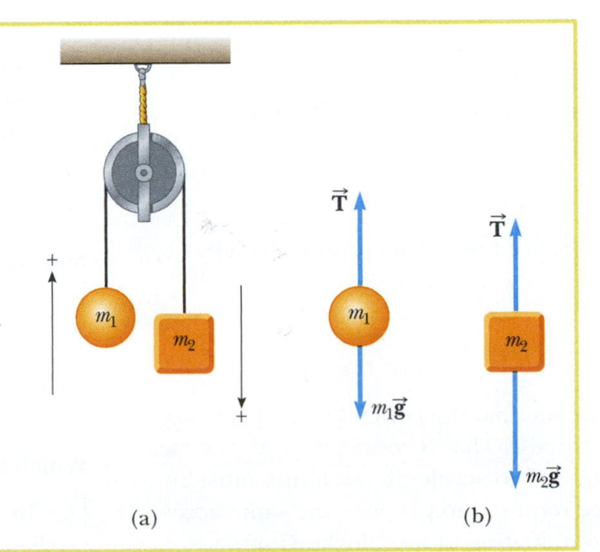


ACTIVE FIGURE 4.12
(Interactive Example 4.4) The Atwood machine. (a) Two objects connected by a light string over a frictionless pulley. (b) The free-body diagrams for $m_1$ and $m_2$.

**PhysicsNow™** Log into PhysicsNow at **www.pop4e.com** and go to Active Figure 4.12 to adjust the masses of the objects on the Atwood machine and observe the motion.

$m_2$, we would expect $m_1$ to have little effect so that $m_2$ is simply falling. Our results are consistent with our intuitive predictions in both of these limiting situations.

**PhysicsNow™** Investigate the motion of the Atwood machine for different masses by logging into PhysicsNow at **www.pop4e.com** and going to Interactive Example 4.4.

## INTERACTIVE EXAMPLE 4.5 One Block Pushes Another

Two blocks of masses $m_1$ and $m_2$, with $m_1 > m_2$, are placed in contact with each other on a frictionless, horizontal surface, as in Active Figure 4.13a. A constant horizontal force $\vec{\mathbf{F}}$ is applied to $m_1$ as shown.

**A** Find the magnitude of the acceleration of the system of two blocks.

**Solution** Both blocks must experience the *same* acceleration because they are in contact with each other and remain in contact with each other. We model the combination of both blocks as a particle under a net force. Because $\vec{\mathbf{F}}$ is the only horizontal force exerted on the particle, we have

$$\sum F_x \text{ (system)} = F = (m_1 + m_2)a$$

$$(1) \quad a = \frac{F}{m_1 + m_2}$$

**B** Determine the magnitude of the contact force between the two blocks.

**Solution** The contact force is internal to the combination of two blocks. Therefore, we cannot find this force by modeling the combination as a single particle. We now need to treat each of the two blocks individually as a particle under a net force. We first construct a free-body diagram for each block, as shown in Active Figures 4.13b and 4.13c, where the contact force is denoted by $\vec{\mathbf{P}}$. From Active Figure 4.13c we see that the only horizontal force acting on $m_2$ is the contact force $\vec{\mathbf{P}}_{12}$ (the force exerted by $m_1$ on $m_2$), which is directed to the right. Applying Newton's second law to $m_2$ gives

$$(2) \quad \sum F_x = P_{12} = m_2 a$$

Substituting the value of the acceleration $a$ given by (1) into (2) gives

$$(3) \quad P_{12} = m_2 a = \left(\frac{m_2}{m_1 + m_2}\right)F$$

From this result we see that the contact force $P_{12}$ is *less* than the applied force $F$. That is consistent with the fact that the force required to accelerate $m_2$ alone must be less than the force required to produce the same acceleration for the combination of two blocks. Compare this result with the forces in the couplers in the train of Thinking Physics 4.1.

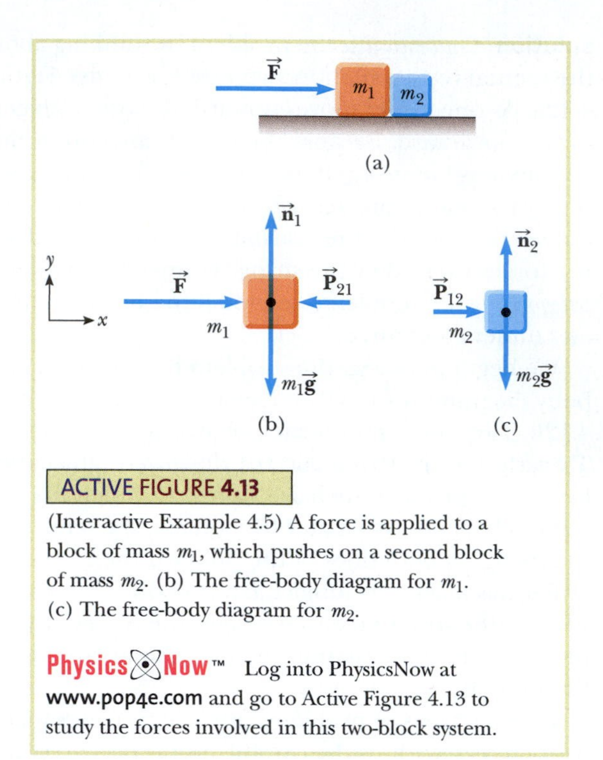


**ACTIVE FIGURE 4.13**
(Interactive Example 4.5) A force is applied to a block of mass $m_1$, which pushes on a second block of mass $m_2$. (b) The free-body diagram for $m_1$. (c) The free-body diagram for $m_2$.

**PhysicsNow™** Log into PhysicsNow at **www.pop4e.com** and go to Active Figure 4.13 to study the forces involved in this two-block system.

It is instructive to check this expression for $P_{12}$ by considering the forces acting on $m_1$, shown in Active Figure 4.13b. The horizontal forces acting on $m_1$ are the applied force $\vec{\mathbf{F}}$ to the right and the contact force $\vec{\mathbf{P}}_{21}$ to the left (the force exerted by $m_2$ on $m_1$). From Newton's third law, $\vec{\mathbf{P}}_{21}$ is the reaction to $\vec{\mathbf{P}}_{12}$, so $P_{21} = P_{12}$. Applying Newton's second law to $m_1$ gives

$$(4) \quad \sum F_x = F - P_{21} = F - P_{12} = m_1 a$$

Solving for $P_{12}$ and substituting the value of $a$ from (1) into (4) gives

$$P_{12} = F - m_1 a$$

$$= F - m_1\left(\frac{F}{m_1 + m_2}\right) = \left(\frac{m_2}{m_1 + m_2}\right)F$$

Which agrees with (3), as it must.

**C** Imagine that the force $\vec{\mathbf{F}}$ in Active Figure 4.13 is applied toward the left on the right-hand block of mass

$m_2$. Is the magnitude of the force $\vec{\mathbf{P}}_{12}$ the same as it was when the force was applied toward the right on $m_1$?

**Solution** With the force applied toward the left on $m_2$, the contact force must accelerate $m_1$. In the original situation, the contact force accelerates $m_2$. Because $m_1 > m_2$, more force is required, so the magnitude of $\vec{\mathbf{P}}_{12}$ is greater.

**Physics Now™** Investigate the motion of the blocks for different mass combinations and applied forces by logging into PhysicsNow at **www.pop4e.com** and going to Interactive Example 4.5.

## EXAMPLE 4.6 Weighing a Fish in an Elevator

A person weighs a fish on a spring scale attached to the ceiling of an elevator, as shown in Figure 4.14. Show that if the elevator accelerates, the spring scale reads an apparent weight different from the fish's true weight.

**Solution** An observer on the accelerating elevator is not in an inertial frame. We need to analyze this situation in an inertial frame, so let us imagine observing it from the stationary ground. We model the fish as a particle under a net force. The external forces acting on the fish are the downward gravitational force $\vec{\mathbf{F}}_g$ and the upward force $\vec{\mathbf{T}}$ exerted on it by the hook hanging from the bottom of the scale. (It might be more fruitful in your mental representation to imagine that the hook is a string connecting the fish to the spring in the scale.) Because the tension is the same everywhere in the hook supporting the fish, the hook pulls downward with a force of magnitude $T$ on the spring scale. Therefore, the tension $T$ in the hook is also the reading of the spring scale.

If the elevator is either at rest or moves at constant velocity, the fish is not accelerating and is a particle in equilibrium, which gives us $\Sigma F_y = T - mg = 0$

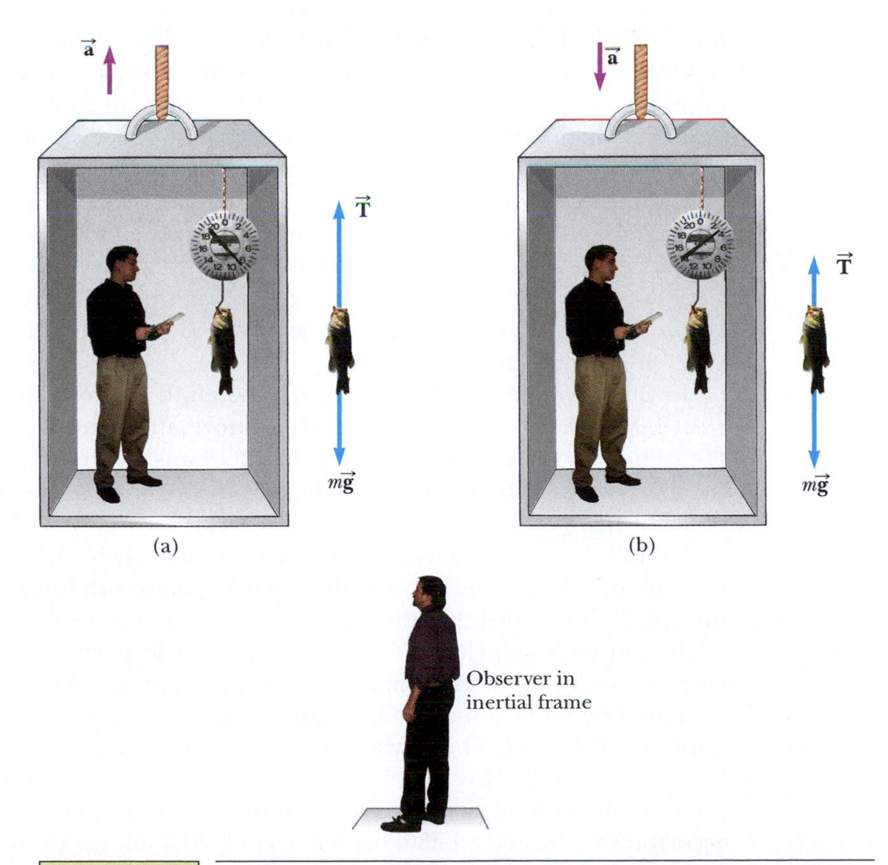


FIGURE 4.14 (Example 4.6) (a) When the elevator accelerates *upward*, the spring scale reads a value *greater* than the fish's true weight. (b) When the elevator accelerates *downward*, the spring scale reads a value *less* than the fish's true weight.

→ $T = mg$. If the elevator accelerates either up or down, however, the tension is no longer equal to the weight of the fish because $T - mg$ does not equal zero.

If the elevator accelerates with an acceleration $\vec{\mathbf{a}}$ relative to an observer in an inertial frame outside the elevator, Newton's second law applied to the fish in the vertical direction gives us

$$\sum F_y = T - mg = ma_y$$

which leads to

$$(1)\quad T = mg + ma_y$$

We conclude from (1) that the scale reading $T$ is greater than the weight $mg$ if $\vec{\mathbf{a}}$ is upward as in Figure 4.14a. Furthermore, we see that $T$ is less than $mg$ if $\vec{\mathbf{a}}$ is downward as in Figure 4.14b. For example, if the weight of the fish is 40.0 N and $\vec{\mathbf{a}}$ is upward with $a_y = 2.00\ \text{m/s}^2$, the scale reading is

$$T = mg + ma_y = mg\left(1 + \frac{a_y}{g}\right)$$

$$= (40.0\ \text{N})\left(1 + \frac{2.00\ \text{m/s}^2}{9.80\ \text{m/s}^2}\right) = 48.2\ \text{N}$$

If $a_y = -2.00\ \text{m/s}^2$ so that $\vec{\mathbf{a}}$ is downward,

$$T = mg\left(1 + \frac{a_y}{g}\right)$$

$$= (40.0\ \text{N})\left(1 + \frac{-2.00\ \text{m/s}^2}{9.80\ \text{m/s}^2}\right) = 31.8\ \text{N}$$

Hence, if you buy a fish in an elevator, make sure the fish is weighed while the elevator is at rest or is accelerating downward!

***Special Case*** If the cable breaks, the elevator falls freely so that $a_y = -g$, and from (1) we see that the tension $T$ is zero; that is, the fish appears to be weightless.

## 4.8 FORCES ON AUTOMOBILES

CONTEXT CONNECTION

In the Context Connections of Chapters 2 and 3, we focused on two types of acceleration exhibited by a number of vehicles. In this chapter, we learned how the acceleration of an object is related to the force on the object. Let us apply this understanding to an investigation of the forces that are applied to automobiles when they are exhibiting their maximum acceleration in speeding up from rest to 60 mi/h.

The force that accelerates an automobile is the friction force from the ground. (We will study friction forces in detail in Chapter 5.) The engine applies a force to the wheels, attempting to rotate them so that the bottoms of the tires apply forces backward on the road surface. By Newton's third law, the road surface applies forces in the forward direction on the tires, causing the car to move forward. If we ignore air resistance, this force can be modeled as the net force on the automobile in the horizontal direction.

In Chapter 2, we investigated the 0 to 60 mi/h acceleration of a number of vehicles. Table 4.2 repeats this acceleration information and also shows the weight of the vehicle in pounds and the mass in kilograms. With both the acceleration and the mass, we can find the force driving the car forward, as shown in the last column of Table 4.2.

We can see some interesting results in Table 4.2. Notice that the forces in the performance vehicle section are all large compared with forces in the other parts of the table. Notice also that the masses of performance vehicles are similar to those of the non-SUV vehicles in the traditional vehicle portion of the table. Thus, the large forces for the performance vehicles translate into the very large accelerations exhibited by these vehicles. One standout in this portion of the table is the Lamborghini Diablo GT. The driving force on it is 15% larger than the next largest, the Porsche 911 GT2. This vehicle is not the most massive in the group, so the large force results in the largest acceleration in the group. The other car with the same acceleration, the Ferrari F50, has a mass only 81% of that of the Lamborghini. Consequently, although the force on the Ferrari is higher than the average in the group, it is only the fourth largest.

As expected, the forces exerted on the traditional vehicles are smaller than those of the performance vehicles, corresponding to the smaller accelerations of

**TABLE 4.2 Driving Forces on Various Vehicles**

| Automobile | Model Year | Acceleration (mi/h · s) | Weight (lb) | Mass (kg) | Force (N) |
|---|---|---|---|---|---|
| *Performance vehicles:* | | | | | |
| Aston Martin DB7 Vantage | 2001 | 12.0 | 3 285 | 1 493 | $8.01 \times 10^3$ |
| BMW Z8 | 2001 | 13.0 | 3 215 | 1 461 | $8.52 \times 10^3$ |
| Chevrolet Corvette | 2000 | 13.0 | 3 115 | 1 416 | $8.25 \times 10^3$ |
| Dodge Viper GTS-R | 1998 | 14.3 | 2 865 | 1 302 | $8.32 \times 10^3$ |
| Ferrari F50 | 1997 | 16.7 | 2 655 | 1 207 | $8.99 \times 10^3$ |
| Ferrari 360 Spider F1 | 2000 | 13.0 | 3 400 | 1 545 | $9.01 \times 10^3$ |
| Lamborghini Diablo GT | 2000 | 16.7 | 3 285 | 1 493 | $11.12 \times 10^3$ |
| Porsche 911 GT2 | 2002 | 15.0 | 3 175 | 1 443 | $9.68 \times 10^3$ |
| *Traditional vehicles:* | | | | | |
| Acura Integra GS | 2000 | 7.6 | 2 725 | 1 239 | $4.20 \times 10^3$ |
| BMW Mini Cooper S | 2003 | 8.7 | 2 678 | 1 217 | $4.73 \times 10^3$ |
| Cadillac Escalade (SUV) | 2002 | 7.0 | 5 542 | 2 519 | $7.86 \times 10^3$ |
| Dodge Stratus | 2002 | 8.0 | 3 192 | 1 451 | $5.19 \times 10^3$ |
| Lexus ES300 | 1997 | 7.0 | 3 296 | 1 498 | $4.67 \times 10^3$ |
| Mitsubishi Eclipse GT | 2000 | 8.6 | 3 186 | 1 448 | $5.55 \times 10^3$ |
| Nissan Maxima | 2000 | 9.0 | 3 221 | 1 464 | $5.86 \times 10^3$ |
| Pontiac Grand Prix | 2003 | 7.1 | 3 384 | 1 538 | $4.85 \times 10^3$ |
| Toyota Sienna (SUV) | 2004 | 7.2 | 3 912 | 1 778 | $5.74 \times 10^3$ |
| Volkswagen Beetle | 1999 | 7.9 | 2 771 | 1 260 | $4.44 \times 10^3$ |
| *Alternative vehicles:* | | | | | |
| GM EV1 | 1998 | 7.9 | 2 970 | 1 350 | $4.76 \times 10^3$ |
| Toyota Prius | 2004 | 4.7 | 2 765 | 1 257 | $2.65 \times 10^3$ |
| Honda Insight | 2001 | 5.2 | 1 967 | 894 | $2.07 \times 10^3$ |

this group. Notice, however, that the forces for the two SUVs are large. Because these two vehicles have accelerations that are somewhat similar to those of the other vehicles in this portion of the table, we can identify these large forces as being required to accelerate the larger mass of the SUVs.

Also as expected, the forces driving the two hybrid vehicles, the Toyota Prius and the Honda Insight, are the lowest in the table. This finding is consistent with the accelerations of these vehicles being much lower than those elsewhere in the table. ■

## SUMMARY

**Physics Now™** Take a practice test by logging into PhysicsNow at www.pop4e.com and clicking on the Pre-Test link for this chapter.

**Newton's first law** states that if an object does not interact with other objects, it is possible to identify a reference frame in which the object has zero acceleration. Thus, if we observe an object from such a frame and no force is exerted on the object, an object at rest remains at rest and an object in uniform motion in a straight line maintains that motion.

Newton's first law defines an **inertial frame of reference,** which is a frame in which Newton's first law is valid.

**Newton's second law** states that the acceleration of an object is directly proportional to the net force acting on the object and inversely proportional to the object's mass. Therefore, the net force on an object equals the product of the mass of the object and its acceleration, or

$$\sum \vec{\mathbf{F}} = m\vec{\mathbf{a}} \qquad [4.2]$$

The **weight** of an object is equal to the product of its mass (a scalar quantity) and the magnitude of the free-fall acceleration, or

$$F_g = mg \qquad [4.5]$$

If the acceleration of an object is zero, the object is modeled as a particle in equilibrium, with the appropriate equations being

$$\sum F_x = 0 \qquad \sum F_y = 0 \qquad [4.8]$$

**Newton's third law** states that if two objects interact, the force exerted by object 1 on object 2 is equal in magnitude but opposite in direction to the force exerted by object 2 on object 1. Therefore, an isolated force cannot exist in nature.

# QUESTIONS

☐ = answer available in the *Student Solutions Manual and Study Guide*

1. A ball is held in a person's hand. (a) Identify all the external forces acting on the ball and the reaction to each. (b) If the ball is dropped, what force is exerted on it while it is falling? Identify the reaction force in this case.
2. What is wrong with the statement, "Because the car is at rest, there are no forces acting on it"? How would you correct this sentence?
3. In the motion picture *It Happened One Night* (Columbia Pictures, 1934), Clark Gable is standing inside a stationary bus in front of Claudette Colbert, who is seated. The bus suddenly starts moving forward and Clark falls into Claudette's lap. Why did that happen?
4. As you sit in a chair, the chair pushes up on you with a normal force. The force is equal to your weight and in the opposite direction. Is this force the Newton's third law reaction to your weight?
5. A passenger sitting in the rear of a bus claims that she was injured as the driver slammed on the brakes, causing a suitcase to come flying toward her from the front of the bus. If you were the judge in this case, what disposition would you make? Why?
6. A space explorer is moving through space in a space ship far from any planet or star. She notices a large rock, taken as a specimen from an alien planet, floating around the cabin of the ship. Should she push it gently or kick it toward the storage compartment? Why?
7. A rubber ball is dropped onto the floor. What force causes the ball to bounce?
8. While a football is in flight, what forces act on it? What are the action–reaction pairs while the football is being kicked and while it is in flight?
9. If gold were sold by weight, would you rather buy it in Denver or in Death Valley? If it were sold by mass, at which of the two locations would you prefer to buy it? Why?
10. If you hold a horizontal metal bar several centimeters above the ground and move it through grass, each leaf of grass bends out of the way. If you increase the speed of the bar, each leaf of grass will bend more quickly. How then does a rotary power lawn mower manage to cut grass? How can it exert enough force on a leaf of grass to shear it off?
11. A weightlifter stands on a bathroom scale. He pumps a barbell up and down. What happens to the reading on the bathroom scale as he does so? What if he is strong enough to actually *throw* the barbell upward? How does the reading on the scale vary now?
12. The mayor of a city decides to fire some city employees because they will not remove the obvious sags from the cables that support the city traffic lights. If you were a lawyer, what defense would you give on behalf of the employees? Who do you think would win the case in court?
13. Suppose a truck loaded with sand accelerates along a highway. If the driving force on the truck remains constant, what happens to the truck's acceleration if its trailer leaks sand at a constant rate through a hole in its bottom?
14. As a rocket is fired from a launching pad, its speed and acceleration increase with time as its engines continue to operate. Explain why that occurs even though the thrust of the engines remains constant.
15. Twenty people participate in a tug-of-war. The two teams of ten people are so evenly matched that neither team wins. After the game they notice that one person's car is mired in mud. They attach the tug-of-war rope to the bumper of the car, and all the people pull on the rope. The heavy car has just moved a couple of decimeters when the rope breaks. Why did the rope break in this situation when it did not break when the same twenty people pulled on it in a tug-of-war?
16. "When the locomotive in Figure Q4.16 broke through the wall of the train station, the force exerted by the locomotive on the wall was greater than the force the wall could exert on the locomotive." Is this statement true or in need of correction? Explain your answer.

(Roger Viollet, Mill Valley, CA, University Science Books, 1982)

FIGURE Q4.16

17. An athlete grips a light rope that passes over a low-friction pulley attached to the ceiling of a gym. A sack of sand precisely equal in weight to the athlete is tied to the rope's other end. Both the sand and the athlete are initially at rest. The athlete climbs the rope, sometimes speeding up and slowing down as he does so. What happens to the sack of sand? Explain.
18. If action and reaction forces are always equal in magnitude and opposite in direction to each other, doesn't the net vector force on any object necessarily add up to zero? Explain your answer.
19. Can an object exert a force on itself? Argue for your answer.

## PROBLEMS

1, 2, 3 = straightforward, intermediate, challenging
☐ = full solution available in the *Student Solutions Manual and Study Guide*
Physics Now™ = coached problem with hints available at www.pop4e.com
= computer useful in solving problem
= paired numerical and symbolic problems
= biomedical application

### Section 4.3 ▪ Mass

1. A force $\vec{F}$ applied to an object of mass $m_1$ produces an acceleration of 3.00 m/s$^2$. The same force applied to a second object of mass $m_2$ produces an acceleration of 1.00 m/s$^2$. (a) What is the value of the ratio $m_1/m_2$? (b) If $m_1$ and $m_2$ are combined, find their acceleration under the action of the force $\vec{F}$.

2. (a) A car with a mass of 850 kg is moving to the right with a constant speed of 1.44 m/s. What is the total force on the car? (b) What is the total force on the car if it is moving to the left?

### Section 4.4 ▪ Newton's Second Law—The Particle Under a Net Force

3. A 3.00-kg object undergoes an acceleration given by $\vec{a} = (2.00\hat{i} + 5.00\hat{j})$ m/s$^2$. Find the resultant force acting on it and the magnitude of the resultant force.

4. Two forces, $\vec{F}_1 = (-6\hat{i} - 4\hat{j})$ N and $\vec{F}_2 = (-3\hat{i} + 7\hat{j})$ N, act on a particle of mass 2.00 kg that is initially at rest at coordinates (−2.00 m, +4.00 m). (a) What are the components of the particle's velocity at $t = 10.0$ s? (b) In what direction is the particle moving at $t = 10.0$ s? (c) What displacement does the particle undergo during the first 10.0 s? (d) What are the coordinates of the particle at $t = 10.0$ s?

5. Physics Now™ To model a spacecraft, a toy rocket engine is securely fastened to a large puck that can glide with negligible friction over a horizontal surface, taken as the *xy* plane. The 4.00-kg puck has a velocity of $3.00\hat{i}$ m/s at one instant. Eight seconds later, its velocity is to be $(8.00\hat{i} + 10.0\hat{j})$ m/s. Assuming that the rocket engine exerts a constant horizontal force, find (a) the components of the force and (b) its magnitude.

6. A 3.00-kg object is moving in a plane, with its $x$ and $y$ coordinates given by $x = 5t^2 - 1$ and $y = 3t^3 + 2$, where $x$ and $y$ are in meters and $t$ is in seconds. Find the magnitude of the net force acting on this object at $t = 2.00$ s.

7. Two forces $\vec{F}_1$ and $\vec{F}_2$ act on a 5.00-kg object. If $F_1 = 20.0$ N and $F_2 = 15.0$ N, find the accelerations in (a) and (b) of Figure P4.7.

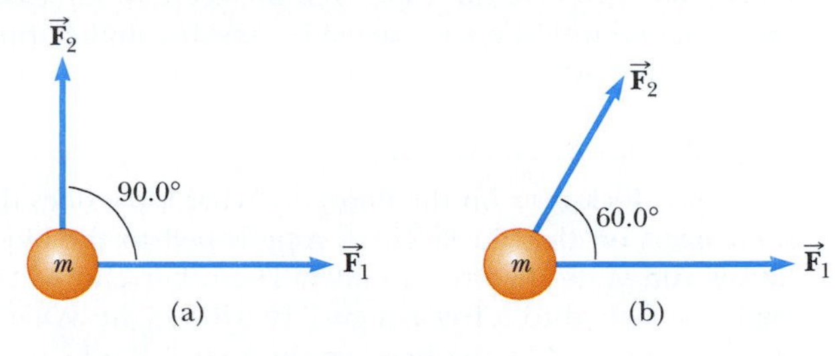


FIGURE P4.7

8. Three forces, given by $\vec{F}_1 = (-2.00\hat{i} + 2.00\hat{j})$ N, $\vec{F}_2 = (5.00\hat{i} - 3.00\hat{j})$ N, and $\vec{F}_3 = (-45.0\hat{i})$ N, act on an object to give it an acceleration of magnitude 3.75 m/s$^2$. (a) What is the direction of the acceleration? (b) What is the mass of the object? (c) If the object is initially at rest, what is its speed after 10.0 s? (d) What are the velocity components of the object after 10.0 s?

### Section 4.5 ▪ The Gravitational Force and Weight

9. A woman weighs 120 lb. Determine (a) her weight in newtons and (b) her mass in kilograms.

10. If a man weighs 900 N on the Earth, what would he weigh on Jupiter, where the free-fall acceleration is 25.9 m/s$^2$?

11. The distinction between mass and weight was discovered after Jean Richer transported pendulum clocks from Paris, France, to Cayenne, French Guiana in 1671. He found that they quite systematically ran slower in Cayenne than in Paris. The effect was reversed when the clocks returned to Paris. How much weight would you personally lose in traveling from Paris, where $g = 9.809\,5$ m/s$^2$, to Cayenne, where $g = 9.780\,8$ m/s$^2$? (We will consider how the free-fall acceleration influences the period of a pendulum in Section 12.4.)

12. The gravitational force on a baseball is $-F_g\hat{j}$. A pitcher throws the baseball with velocity $v\hat{i}$ by uniformly accelerating it straight forward horizontally for a time interval $\Delta t = t - 0 = t$. If the ball starts from rest, (a) through what distance does it accelerate before its release? (b) What force does the pitcher exert on the ball?

13. An electron of mass $9.11 \times 10^{-31}$ kg has an initial speed of $3.00 \times 10^5$ m/s. It travels in a straight line, and its speed increases to $7.00 \times 10^5$ m/s in a distance of 5.00 cm. Assuming that its acceleration is constant, (a) determine the net force exerted on the electron and (b) compare this force with the weight of the electron.

14. Besides its weight, a 2.80-kg object is subjected to one other constant force. The object starts from rest and in 1.20 s experiences a displacement of $(4.20\hat{i} - 3.30\hat{j})$ m, where the direction of $\hat{j}$ is the upward vertical direction. Determine the other force.

### Section 4.6 ▪ Newton's Third Law

15. You stand on the seat of a chair and then hop off. (a) During the time you are in flight down to the floor, the Earth is lurching up toward you with an acceleration of what order of magnitude? In your solution, explain your logic. Model the Earth as a perfectly solid object. (b) The Earth moves up through a distance of what order of magnitude?

**16.** The average speed of a nitrogen molecule in air is about $6.70 \times 10^2$ m/s, and its mass is $4.68 \times 10^{-26}$ kg. (a) If it takes $3.00 \times 10^{-13}$ s for a nitrogen molecule to hit a wall and rebound with the same speed but moving in the opposite direction, what is the average acceleration of the molecule during this time interval? (b) What average force does the molecule exert on the wall?

**17.** A 15.0-lb block rests on the floor. (a) What force does the floor exert on the block? (b) A rope is tied to the block and is run vertically over a pulley. The other end of the rope is attached to a free-hanging 10.0-lb weight. What is the force exerted by the floor on the 15.0-lb block? (c) If we replace the 10.0-lb weight in part (b) with a 20.0-lb weight, what is the force exerted by the floor on the 15.0-lb block?

## Section 4.7 ■ Applications of Newton's Laws

**18.** A bag of cement of weight 325 N hangs in equilibrium from three wires as suggested in Figure P4.18. Two of the wires make angles $\theta_1 = 60.0°$ and $\theta_2 = 25.0°$ with the horizontal. Find the tensions $T_1$, $T_2$, and $T_3$ in the wires.

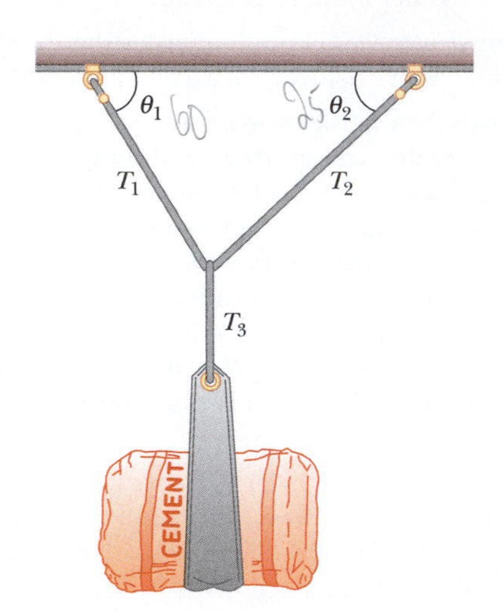


FIGURE **P4.18** Problems 4.18 and 4.19.

**19.** A bag of cement of weight $F_g$ hangs in equilibrium from three wires as shown in Figure P4.18. Two of the wires make angles $\theta_1$ and $\theta_2$ with the horizontal. Show that the tension in the left-hand wire is

$$T_1 = \frac{F_g \cos \theta_2}{\sin (\theta_1 + \theta_2)}$$

**20.** Figure P4.20 shows a worker poling a boat—a very efficient mode of transportation—across a shallow lake. He pushes parallel to the length of the light pole, exerting on the bottom of the lake a force of 240 N. The pole lies in the vertical plane containing the keel of the boat. At one moment the pole makes an angle of 35.0° with the vertical and the water exerts a horizontal drag force of 47.5 N on the boat, opposite to its forward motion at 0.857 m/s. The mass of the boat including its cargo and the worker is 370 kg. (a) The water exerts a buoyant force vertically upward on the boat. Find the magnitude of this force. (b) Model the forces as constant over a short interval of time to find the velocity of the boat 0.450 s after the moment described.

(© Tony Arruza/CORBIS)

FIGURE **P4.20**

**21.** You are a judge in a children's kite-flying contest, and two children will win prizes for the kites that pull most strongly and least strongly on their strings. To measure string tensions, you borrow a weight hanger, some slotted weights, and a protractor from your physics teacher, and you use the following protocol, illustrated in Figure P4.21: Wait for a child to get her kite well controlled, hook the hanger onto the kite string about 30 cm from her hand, pile on weight until that section of string is horizontal, record the mass required, and record the angle between the horizontal and the string running up to the kite. (a) Explain how this method works. As you construct your explanation, imagine that the children's parents ask you about your method, that they might make false assumptions about your ability without concrete evidence, and that your

FIGURE **P4.21**

explanation is an opportunity to give them confidence in your evaluation technique. (b) Find the string tension assuming that the mass is 132 g and the angle of the kite string is 46.3°.

**22.** The systems shown in Figure P4.22 are in equilibrium. If the spring scales are calibrated in newtons, what do they read? (Ignore the masses of the pulleys and strings, and assume that the incline is frictionless.)

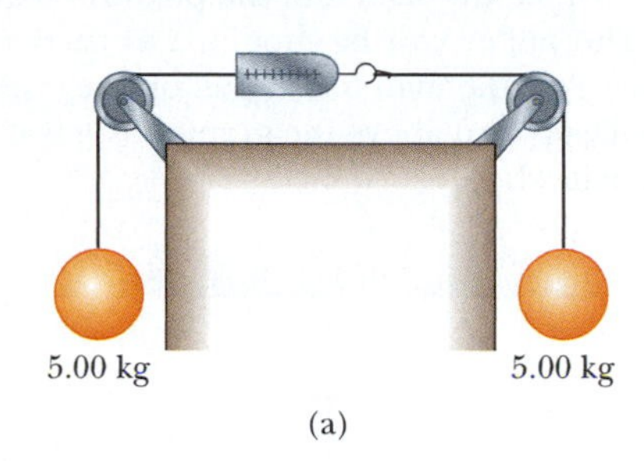


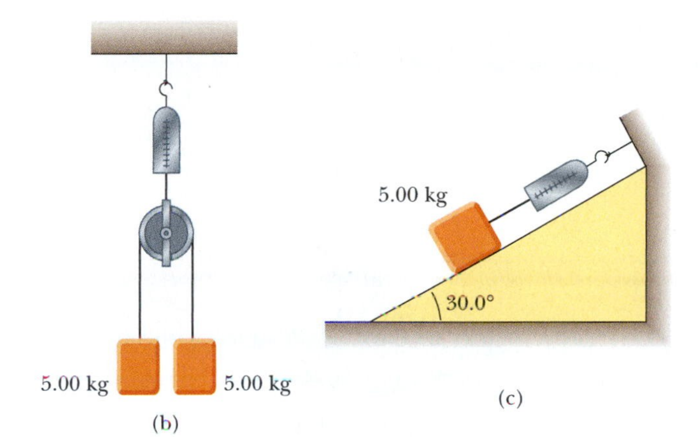


FIGURE **P4.22**

**23.** A simple accelerometer is constructed inside a car by suspending an object of mass $m$ from a string of length $L$ that is tied to the car's ceiling. As the car accelerates the string–object system makes a constant angle of $\theta$ with the vertical. (a) Assuming that the string mass is negligible compared with $m$, derive an expression for the car's acceleration in terms of $\theta$ and show that it is independent of the mass $m$ and the length $L$. (b) Determine the acceleration of the car when $\theta = 23.0°$.

**24.** Figure P4.24 shows loads hanging from the ceiling of an elevator that is moving at constant velocity. Find the tension in each of the three strands of cord supporting each load.

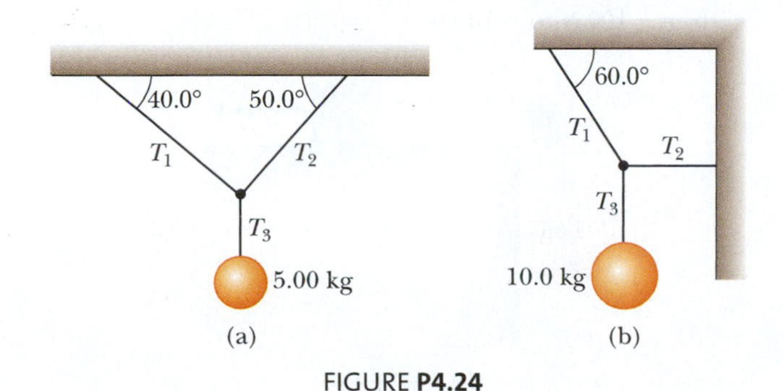


FIGURE **P4.24**

**25.** Two people pull as hard as they can on horizontal ropes attached to a boat that has a mass of 200 kg. If they pull in the same direction, the boat has an acceleration of 1.52 m/s² to the right. If they pull in opposite directions, the boat has an acceleration of 0.518 m/s² to the left. What is the magnitude of the force each person exerts on the boat? Disregard any other horizontal forces on the boat.

**26.** Draw a free-body diagram of a block that slides down a frictionless plane having an inclination of $\theta = 15.0°$ (Fig. P4.26). Assuming that the block starts from rest at the top and that the length of the incline is 2.00 m, find (a) the acceleration of the block and (b) its speed when it reaches the bottom of the incline.

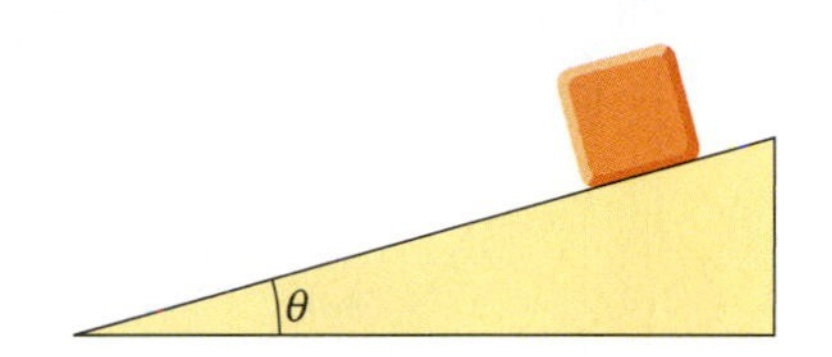


FIGURE **P4.26** Problems 4.26, 4.29, and 4.46.

**27.** **Physics Now™** A 1.00-kg object is observed to accelerate at 10.0 m/s² in a direction 30.0° north of east (Fig. P4.27). The force $\vec{\mathbf{F}}_2$ acting on the object has magnitude 5.00 N and is directed north. Determine the magnitude and direction of the force $\vec{\mathbf{F}}_1$ acting on the object.

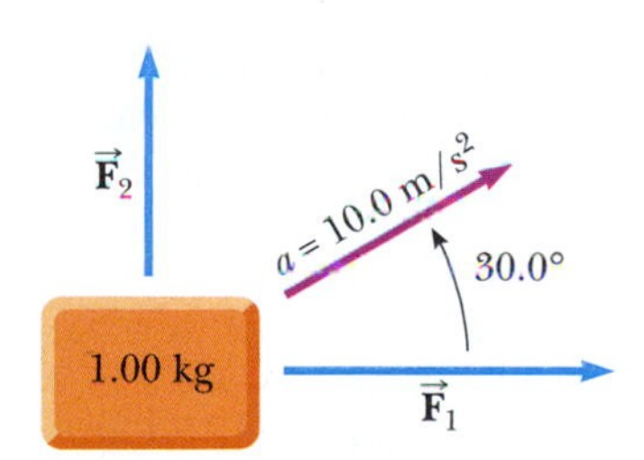


FIGURE **P4.27**

**28.** A 5.00-kg object placed on a frictionless, horizontal table is connected to a cable that passes over a pulley and then is fastened to a hanging 9.00-kg object as shown in Figure P4.28. Draw free-body diagrams of both objects. Find the acceleration of the two objects and the tension in the string.

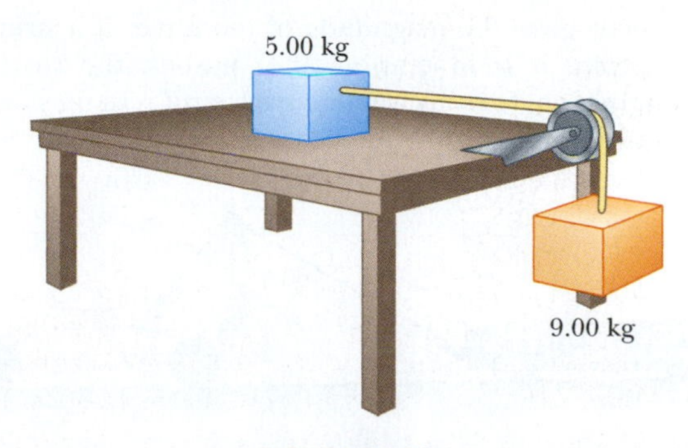


FIGURE **P4.28**

**29.** **Physics Now™** A block is given an initial velocity of 5.00 m/s up a frictionless 20.0° incline (Fig. P4.26). How far up the incline does the block slide before coming to rest?

**30.** Two objects are connected by a light string that passes over a frictionless pulley as shown in Figure P4.30. Draw free-body diagrams of both objects. The incline is frictionless, and $m_1 = 2.00$ kg, $m_2 = 6.00$ kg, and $\theta = 55.0°$. Find (a) the accelerations of the objects, (b) the tension in the string, and (c) the speed of each of the objects 2.00 s after they are released simultaneously from rest.

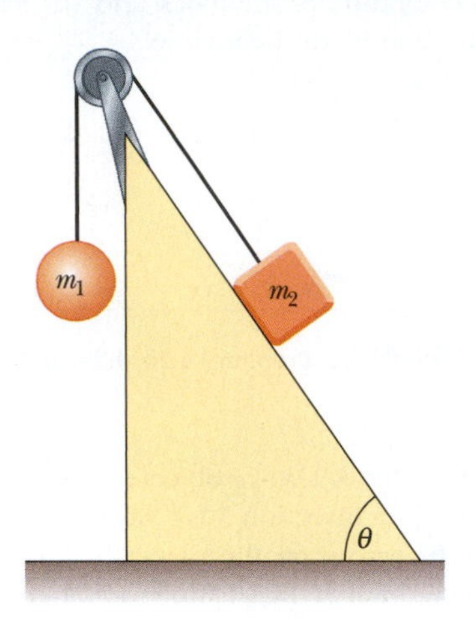


FIGURE **P4.30**

**31.** A car is stuck in the mud. A tow truck pulls on the car with a force of 2 500 N as shown in Fig. P4.31. The tow cable is under tension and therefore pulls downward and to the left on the pin at its upper end. The light pin is held in equilibrium by forces exerted by the two bars A and B. Each bar is a *strut;* that is, each is a bar whose weight is small compared to the forces it exerts and which exerts forces only through hinge pins at its ends. Each strut exerts a force directed parallel to its length. Determine the force of tension or compression in each strut. Proceed as follows. Make a guess as to which way (pushing or pulling) each force acts on the top pin. Draw a free-body diagram of the pin. Use the condition for equilibrium of the pin to translate the free-body diagram into equations. From the equations calculate the forces exerted by struts A and B. If you obtain a positive answer, you correctly guessed the direction of the force. A negative answer means that the direction should be reversed, but the absolute value correctly gives the magnitude of the force. If a strut pulls on a pin, it is in tension. If it pushes, the strut is in compression. Identify whether each strut is in tension or in compression.

FIGURE **P4.31**

**32.** Two objects with masses of 3.00 kg and 5.00 kg are connected by a light string that passes over a light frictionless pulley to form an Atwood machine as shown in Active Figure 4.12a. Determine (a) the tension in the string, (b) the acceleration of each object, and (c) the distance each object will move in the first second of motion if they start from rest.

**33.** In Figure P4.33, the man and the platform together weigh 950 N. The pulley can be modeled as frictionless. Determine how hard the man has to pull on the rope to lift himself steadily upward above the ground. (Or is it impossible? If so, explain why.)

FIGURE **P4.33**

**34.** In the Atwood machine shown in Active Figure 4.12a, $m_1 = 2.00$ kg and $m_2 = 7.00$ kg. The masses of the pulley and string are negligible by comparison. The pulley turns without friction, and the string does not stretch. The lighter object is released with a sharp push that sets it into motion at $v_i = 2.40$ m/s downward. (a) How far will $m_1$ descend below its initial level? (b) Find the velocity of $m_1$ after 1.80 s.

**35.** **Physics Now™** In the system shown in Figure P4.35, a horizontal force $\vec{\mathbf{F}}_x$ acts on the 8.00-kg object. The horizontal surface is frictionless. (a) For what values of $F_x$ does the 2.00-kg object accelerate upward? (b) For what values of $F_x$ is the tension in the cord zero? (c) Plot the acceleration of the 8.00-kg object versus $F_x$. Include values of $F_x$ from $-100$ N to $+100$ N.

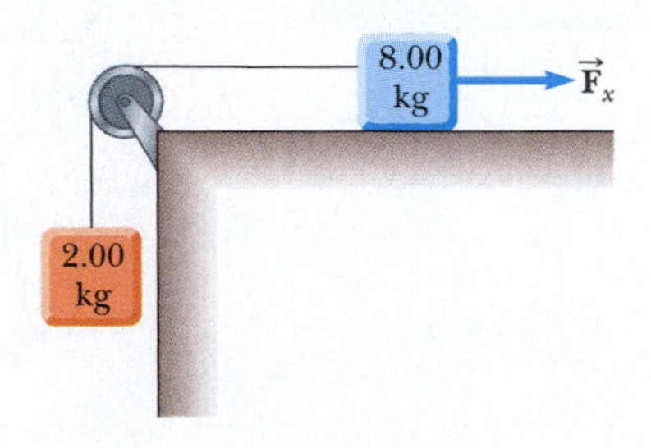


FIGURE **P4.35**

**36.** A frictionless plane is 10.0 m long and inclined at 35.0°. A sled starts at the bottom with an initial speed of 5.00 m/s up the incline. When it reaches the point at which it momentarily stops, a second sled is released from the top of this incline with an initial speed $v_i$. Both sleds reach the bottom of the incline at the same moment. (a) Determine the distance that the first sled traveled up the incline. (b) Determine the initial speed of the second sled.

**37.** A 72.0-kg man stands on a spring scale in an elevator. Starting from rest, the elevator ascends, attaining its maximum speed of 1.20 m/s in 0.800 s. It travels with this constant speed for the next 5.00 s. The elevator then undergoes a uniform acceleration in the negative $y$ direction for 1.50 s and comes to rest. What does the spring scale register (a) before the elevator starts to move, (b) during the first 0.800 s, (c) while the elevator is traveling at constant speed, and (d) during the time it is slowing down?

**38.** An object of mass $m_1$ on a frictionless horizontal table is connected to an object of mass $m_2$ through a very light pulley $P_1$ and a light fixed pulley $P_2$ as shown in Figure P4.38. (a) If $a_1$ and $a_2$ are the accelerations of $m_1$ and $m_2$, respectively, what is the relation between these accelerations? Express (b) the tensions in the strings and (c) the accelerations $a_1$ and $a_2$ in terms of $g$ and the masses $m_1$ and $m_2$.

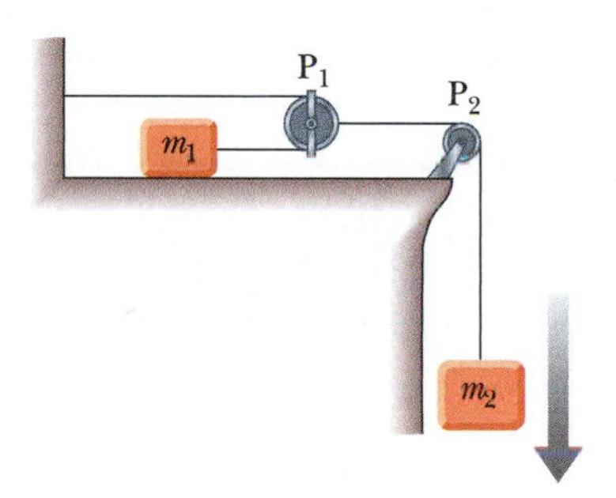


FIGURE **P4.38**

## Section 4.8 ▪ Context Connection—Forces on Automobiles

**39.** A young woman buys an inexpensive used car for stock car racing. It can attain highway speed with an acceleration of 8.40 mi/h·s. By making changes to its engine, she can increase the net horizontal force on the car by 24.0%. With much less expense, she can remove material from the body of the car to decrease its mass by 24.0%. (a) Which of these two changes, if either, will result in the greater increase in the car's acceleration? (b) If she makes both changes, what acceleration can she attain?

**40.** A 1 000-kg car is pulling a 300-kg trailer. Together the car and trailer move forward with an acceleration of 2.15 m/s$^2$. Ignore any force of air drag on the car and all frictional forces on the trailer. Determine (a) the net force on the car, (b) the net force on the trailer, (c) the force exerted by the trailer on the car, and (d) the resultant force exerted by the car on the road.

## Additional Problems

**41.** An inventive child named Pat wants to reach an apple in a tree without climbing the tree. While sitting in a chair connected to a rope that passes over a frictionless pulley (Fig. P4.41), Pat pulls on the loose end of the rope with such a force that the spring scale reads 250 N. Pat's true weight is 320 N, and the chair weighs 160 N. (a) Draw free-body diagrams for Pat and the chair considered as separate systems and another diagram for Pat and the chair considered as one system. (b) Show that the acceleration of the system is *upward* and find its magnitude. (c) Find the force Pat exerts on the chair.

FIGURE **P4.41** Problems 4.41 and 4.42.

**42.** In the situation described in Problem 4.41 and Figure P4.41, the masses of the rope, spring balance, and pulley are negligible. Pat's feet are not touching the ground. (a) Assume that Pat is momentarily at rest when he stops pulling down on the rope and passes the end of the rope to another child, of weight 440 N, who is standing on the ground next to him. The rope does not break. Describe the ensuing motion. (b) Instead, assume that Pat is momentarily at rest when he ties the rope to a strong hook projecting from the tree trunk. Explain why this action can make the rope break.

**43.** Three blocks are in contact with one another on a frictionless, horizontal surface as shown in Figure P4.43. A horizontal force $\vec{\mathbf{F}}$ is applied to $m_1$. Taking $m_1 = 2.00$ kg, $m_2 = 3.00$ kg, $m_3 = 4.00$ kg, and $F = 18.0$ N, draw a separate free-body diagram for each block and find (a) the acceleration of the blocks, (b) the *resultant* force on each block, and (c) the magnitudes of the contact forces between the blocks. (d) You are working on a construction project. A coworker is nailing up plasterboard on one side of a light partition, and you are on the opposite side, providing "backing" by leaning against the wall with your back pushing on it. Every hammer blow makes your back sting.

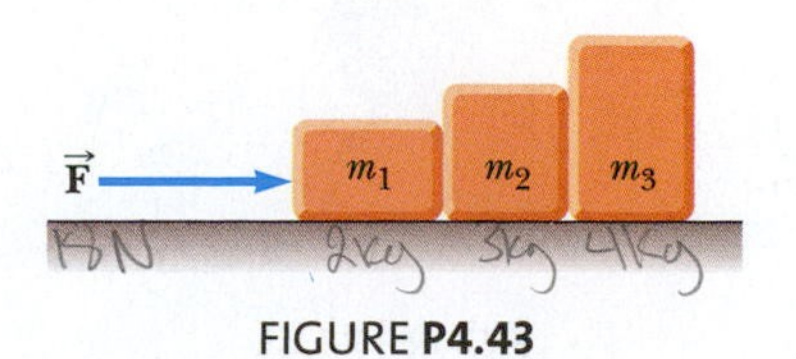


FIGURE **P4.43**

The supervisor helps you put a heavy block of wood between the wall and your back. Using the situation analyzed in parts (a), (b), and (c) as a model, explain how this change works to make your job more comfortable.

**44.** **Review problem.** A block of mass $m = 2.00$ kg is released from rest at $h = 0.500$ m above the surface of a table, at the top of a $\theta = 30.0°$ incline as shown in Figure P4.44. The frictionless incline is fixed on a table of height $H = 2.00$ m. (a) Determine the acceleration of the block as it slides down the incline. (b) What is the velocity of the block as it leaves the incline? (c) How far from the table will the block hit the floor? (d) What time interval elapses between when the block is released and when it hits the floor? (e) Does the mass of the block affect any of the above calculations?

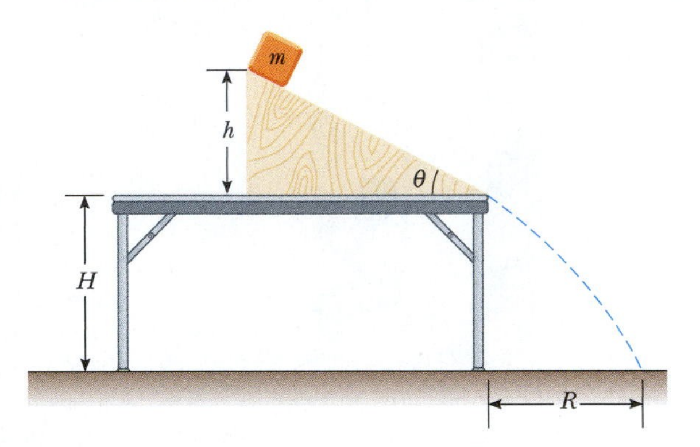


FIGURE **P4.44** Problems 4.44 and 4.55.

**45.** Physics Now™ An object of mass $M$ is held in place by an applied force $\vec{\mathbf{F}}$ and a pulley system as shown in Figure P4.45. The pulleys are massless and frictionless. Find (a) the tension in each section of rope, $T_1$, $T_2$, $T_3$, $T_4$, and $T_5$ and (b) the magnitude of $\vec{\mathbf{F}}$. *Suggestion:* Draw a free-body diagram for each pulley.

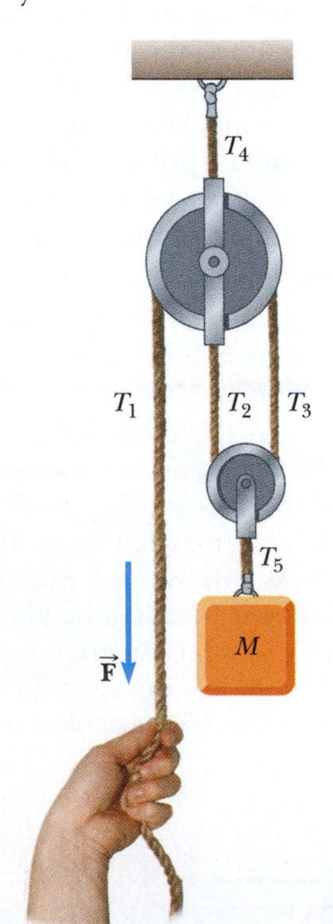


FIGURE **P4.45**

**46.** A student is asked to measure the acceleration of a cart on a "frictionless" inclined plane as shown in Figure P4.26 and analyzed in Example 4.3, using an air track, a stopwatch, and a meter stick. The height of the incline is measured to be 1.774 cm, and the total length of the incline is measured to be $d = 127.1$ cm. Hence, the angle of inclination $\theta$ is determined from the relation $\sin\theta = 1.774/127.1$. The cart is released from rest at the top of the incline, and its position $x$ along the incline is measured as a function of time, where $x = 0$ refers to the initial position of the cart. For $x$ values of 10.0 cm, 20.0 cm, 35.0 cm, 50.0 cm, 75.0 cm, and 100 cm, the measured times at which these positions are reached (averaged over five runs) are 1.02 s, 1.53 s, 2.01 s, 2.64 s, 3.30 s, and 3.75 s, respectively. Construct a graph of $x$ versus $t^2$ and perform a linear least-squares fit to the data. Determine the acceleration of the cart from the slope of this graph and compare it with the value you would get using $a = g \sin\theta$, where $g = 9.80$ m/s$^2$.

**47.** What horizontal force must be applied to the cart shown in Figure P4.47 so that the blocks remain stationary relative to the cart? Assume that all surfaces, wheels, and pulley are frictionless. (*Suggestion:* Note that the force exerted by the string accelerates $m_1$.)

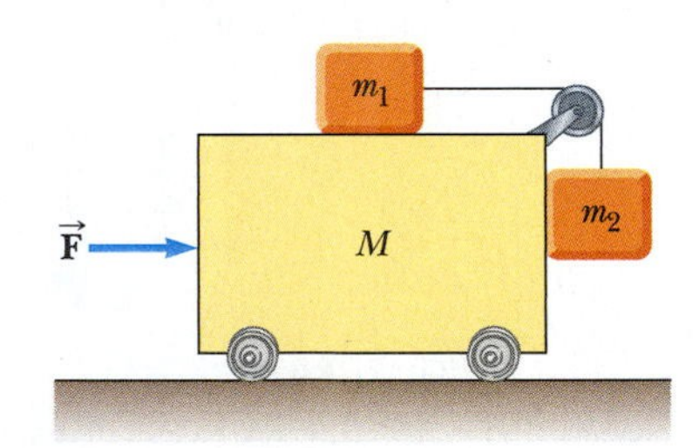


FIGURE **P4.47** Problems 4.47 and 4.48.

**48.** Initially, the system of objects shown in Figure P4.47 is held motionless. The pulley and all surfaces and wheels are frictionless. Let the force $\vec{\mathbf{F}}$ be zero and assume that $m_2$ can move only vertically. At the instant after the system of objects is released, find (a) the tension $T$ in the string, (b) the acceleration of $m_2$, (c) the acceleration of $M$, and (d) the acceleration of $m_1$. (*Note:* The pulley accelerates along with the cart.)

**49.** A 1.00-kg glider on a horizontal air track is pulled by a string at an angle $\theta$. The taut string runs over a pulley and is attached to a hanging object of mass 0.500 kg as shown in Figure P4.49. (a) Show that the speed $v_x$ of the glider and the speed $v_y$ of the hanging object are related by $v_x = uv_y$, where $u = z(z^2 - h_0{}^2)^{-1/2}$. (b) The glider is released from rest. Show that at that instant the acceleration $a_x$ of the glider and the acceleration $a_y$ of the hanging object are related by $a_x = ua_y$. (c) Find the tension in the string at the instant the glider is released for $h_0 = 80.0$ cm and $\theta = 30.0°$.

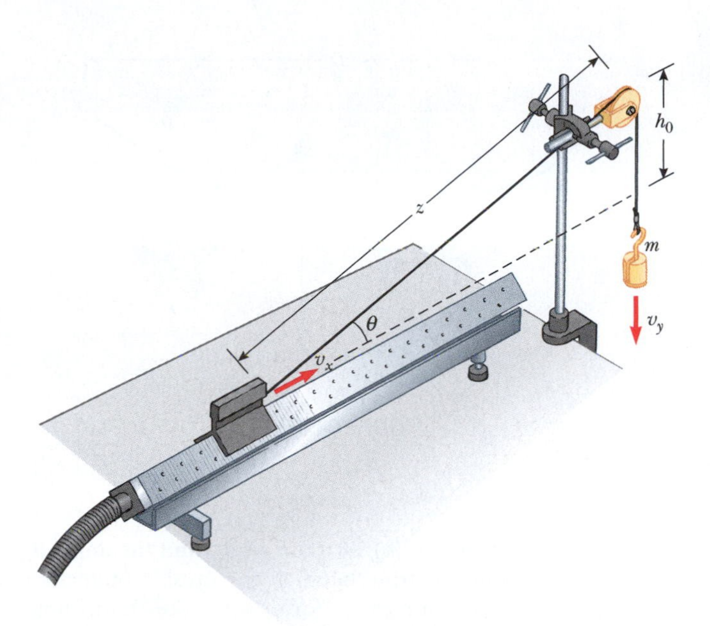


FIGURE **P4.49**

**50.** Cam mechanisms are used in many machines. For example, cams open and close the valves in your car engine to admit gasoline vapor to each cylinder and to allow the escape of exhaust. The principle is illustrated in Figure P4.50, showing a follower rod (also called a pushrod) of mass $m$ resting on a wedge of mass $M$. The sliding wedge duplicates the function of a rotating eccentric disk on a car's camshaft. Assume that there is no friction between the wedge and the base, between the pushrod and the wedge, or between the rod and the guide through which it slides. When the wedge is pushed to the left by the force $\vec{\mathbf{F}}$, the rod moves upward and does something such as opening a valve. By varying the shape of the wedge, the motion of the follower rod could be made quite complex, but assume that the wedge makes a constant angle of $\theta = 15.0°$. Suppose you want the wedge and the rod to start from rest and move with constant acceleration, with the rod moving upward 1.00 mm in 8.00 ms. Take $m = 0.250$ kg and $M = 0.500$ kg. What force $F$ must be applied to the wedge?

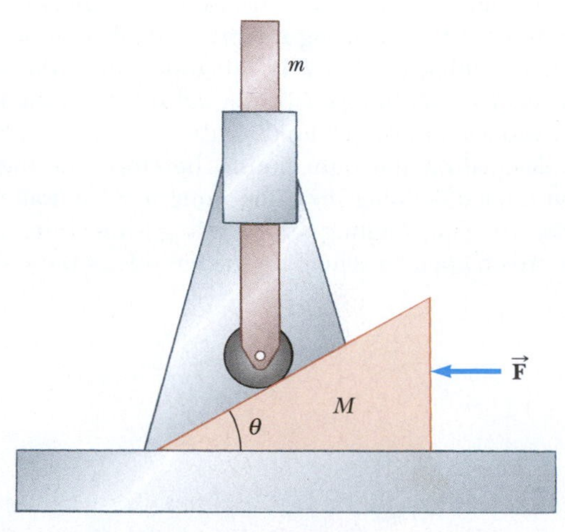


FIGURE **P4.50**

**51.** If you jump from a desktop and land stiff-legged on a concrete floor, you run a significant risk that you will break a leg. To see how that happens, consider the average force stopping your body when you drop from rest from a height of 1.00 m and stop in a much shorter distance $d$. Your leg is likely to break at the point where the cross-sectional area of the bone (the tibia) is smallest. This point is just above the ankle, where the cross-sectional area of one bone is about 1.60 cm$^2$. A bone will fracture when the compressive stress on it exceeds about $1.60 \times 10^8$ N/m$^2$. If you land on both legs, the maximum force that your ankles can safely exert on the rest of your body is then about

$$2(1.60 \times 10^8 \text{ N/m}^2)(1.60 \times 10^{-4} \text{ m}^2) = 5.12 \times 10^4 \text{ N.}$$

Calculate the minimum stopping distance $d$ that will not result in a broken leg if your mass is 60.0 kg. Don't try it! Bend your knees!

**52.** Any device that allows you to increase the force you exert is a kind of *machine.* Some machines, such as the prybar or the inclined plane, are very simple. Some machines do not even look like machines. For example, your car is stuck in the mud and you can't pull hard enough to get it out. You do, however, have a long cable that you connect taut between your front bumper and the trunk of a stout tree. You now pull sideways on the cable at its midpoint, exerting a force $f$. Each half of the cable is displaced through a small angle $\theta$ from the straight line between the ends of the cable. (a) Deduce an expression for the force acting on the car. (b) Evaluate the cable tension for the case where $\theta = 7.00°$ and $f = 100$ N.

**53.** A van accelerates down a hill (Fig. P4.53), going from rest to 30.0 m/s in 6.00 s. During the acceleration, a toy ($m = 0.100$ kg) hangs by a string from the van's ceiling. The acceleration is such that the string remains perpendicular to the ceiling. Determine (a) the angle $\theta$ and (b) the tension in the string.

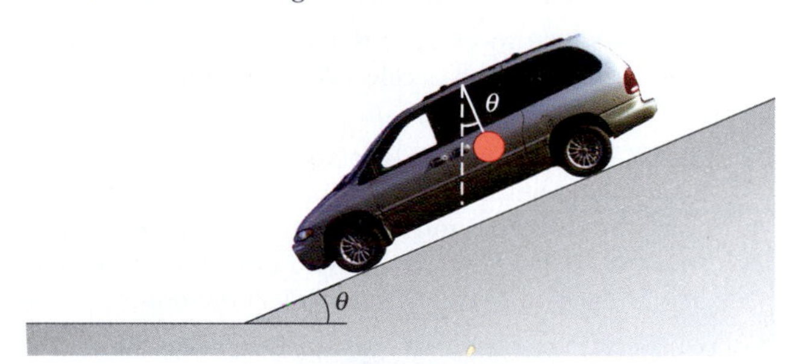


FIGURE **P4.53**

**54.** Two blocks of mass 3.50 kg and 8.00 kg are connected by a massless string that passes over a frictionless pulley (Fig. P4.54). The inclines are frictionless. Find (a) the

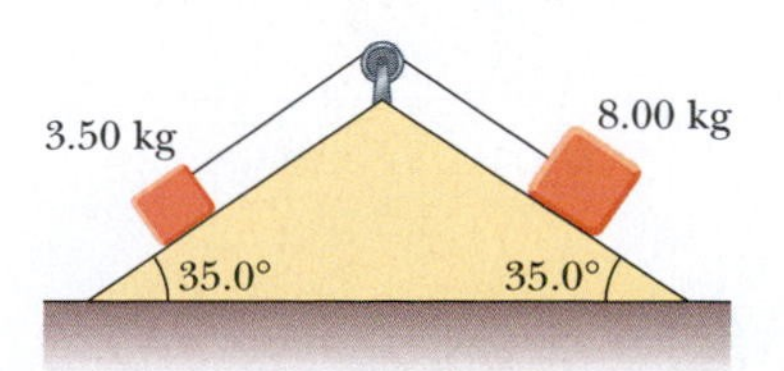


FIGURE **P4.54** Problems 4.54 and 5.41.

magnitude of the acceleration of each block and (b) the tension in the string.

55. In Figure P4.44, the incline has mass $M$ and is fastened to the stationary horizontal tabletop. The block of mass $m$ is placed near the bottom of the incline and is released with a quick push that sets it sliding upward. It stops near the top of the incline, as shown in the figure, and then slides down again, always without friction. Find the force that the tabletop exerts on the incline throughout this motion.

56. An 8.40-kg object slides down a fixed, frictionless inclined plane. Use a computer to determine and tabulate the normal force exerted on the object and its acceleration for a series of incline angles (measured from the horizontal) ranging from 0° to 90° in 5° increments. Plot a graph of the normal force and the acceleration as functions of the incline angle. In the limiting cases of 0° and 90°, are your results consistent with the known behavior?

57. A mobile is formed by supporting four metal butterflies of equal mass $m$ from a string of length $L$. The points of support are evenly spaced a distance $\ell$ apart as shown in Figure P4.57. The string forms an angle $\theta_1$ with the ceiling at each end point. The center section of string is horizontal. (a) Find the tension in each section of string in terms of $\theta_1$, $m$, and $g$. (b) In terms of $\theta_1$, find the angle $\theta_2$ that the sections of string between the outside butterflies and the inside butterflies form with the horizontal. (c) Show that the distance $D$ between the end points of the string is

$$D = \frac{L}{5}\left(2\cos\theta_1 + 2\cos\left[\tan^{-1}\left(\tfrac{1}{2}\tan\theta_1\right)\right] + 1\right)$$

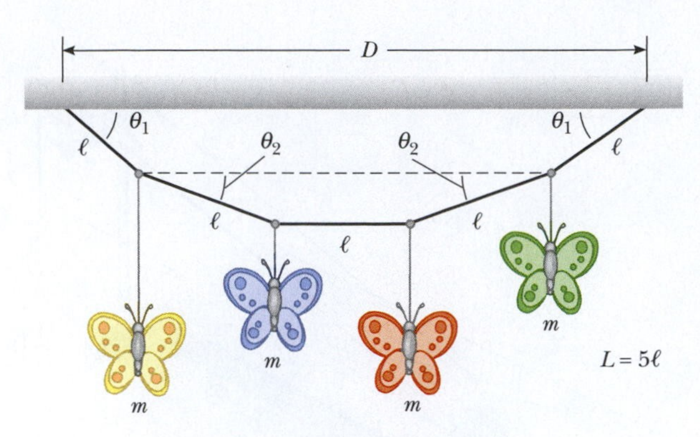


FIGURE **P4.57**

## ANSWERS TO QUICK QUIZZES

**4.1** (d). Choice (a) is true. Newton's first law tells us that motion requires no force: An object in motion continues to move at constant velocity in the absence of external forces. Choice (b) is also true: A stationary object can have several forces acting on it, but if the vector sum of all these external forces is zero, there is no net force and the object remains stationary.

**4.2** (a). If a single force acts, this force constitutes the net force and there is an acceleration according to Newton's second law.

**4.3** (d). With twice the force, the object will experience twice the acceleration. Because the force is constant, the acceleration is constant, and the speed of the object, starting from rest, is given by $v = at$. With twice the acceleration, the object will arrive at speed $v$ at half the time.

**4.4** (b). Because the value of $g$ is smaller on the Moon than on the Earth, more mass of gold would be required to represent 1 N of weight on the Moon. Therefore, your friend on the Moon is richer, by about a factor of 6!

**4.5** (c), (d). In accordance with Newton's third law, the fly and the bus experience forces that are equal in magnitude but opposite in direction. Because the fly has such a small mass, Newton's second law tells us that it undergoes a very large acceleration. The huge mass of the bus means that it more effectively resists any change in its motion and exhibits a small acceleration.

**4.6** (c). The reaction force to the gravitational force on you is an upward gravitational force on the Earth caused by you.

**4.7** (c). The scale is in equilibrium in both situations, so it experiences a net force of zero. Because each individual pulls with a force $F$ and there is no acceleration, each individual is in equilibrium. Therefore, the tension in the ropes must be equal to $F$. In case (i), the individual pulls with force $F$ on a spring mounted rigidly to a brick wall. The resulting tension $F$ in the rope causes the scale to read a force $F$. In case (ii), the individual on the left can be modeled as simply holding the rope tightly while the individual on the right pulls. Therefore, the individual on the left is doing the same thing that the wall does in case (i). The resulting scale reading is the same whether a wall or a person is holding the left side of the scale.

# More Applications of Newton's Laws

(© Paul Hardy/CORBIS)

The London Eye, a ride on the River Thames in downtown London. Riders travel in a large vertical circle for a breathtaking view of the city. In this chapter, we will study the forces involved in circular motion.

## CHAPTER OUTLINE


5.1 Forces of Friction

5.2 Newton's Second Law Applied to a Particle in Uniform Circular Motion

5.3 Nonuniform Circular Motion

5.4 Motion in the Presence of Velocity-Dependent Resistive Forces

5.5 The Fundamental Forces of Nature

5.6 Context Connection—Drag Coefficients of Automobiles

SUMMARY


In Chapter 4, we introduced Newton's laws of motion and applied them to situations in which we ignored friction. In this chapter, we shall expand our investigation to objects moving in the presence of friction, which will allow us to model situations more realistically. Such objects include those sliding on rough surfaces and those moving through viscous media such as liquids and air. We also apply Newton's laws to the dynamics of circular motion so that we can understand more about objects moving in circular paths under the influence of various types of forces.

## 5.1 FORCES OF FRICTION

When an object moves either on a surface or through a viscous medium such as air or water, there is resistance to the motion because the object interacts with its surroundings. We call such resistance a **force of friction.** Forces of friction are very important in our everyday lives. They allow us to walk or run and are necessary for the motion of wheeled vehicles.

Imagine you are working in your garden and have filled a trash can with yard clippings. You then try to drag the trash can across the surface of your concrete patio as in Active Figure 5.1a. The patio surface is *real,* not an idealized, frictionless surface in a simplification model. If we apply an external horizontal force $\vec{\mathbf{F}}$ to the trash can, acting to the right, the trash can remains stationary if $\vec{\mathbf{F}}$ is small. The force that counteracts $\vec{\mathbf{F}}$ and keeps the trash can from moving acts to the left and is called the **force of static friction** $\vec{\mathbf{f}}_s$. As long as the trash can is not moving, it is modeled as a particle in equilibrium and $f_s = F$. Therefore, if $\vec{\mathbf{F}}$ is increased in magnitude, the magnitude of $\vec{\mathbf{f}}_s$ also increases. Likewise, if $\vec{\mathbf{F}}$ decreases, $\vec{\mathbf{f}}_s$ also decreases. Experiments show that the friction force arises from the nature of the two surfaces; because of their roughness, contact is made only at a few points, as shown in the magnified surface view in Active Figure 5.1a.

If we increase the magnitude of $\vec{\mathbf{F}}$, as in Active Figure 5.1b, the trash can eventually slips. When the trash can is on the verge of slipping, $f_s$ is a maximum as shown in Active Figure 5.1c. If $F$ exceeds $f_{s,\max}$, the trash can moves and accelerates to the right. While the trash can is in motion, the friction force is less than $f_{s,\max}$ (Active Fig. 5.1c). We call the friction force for an object in motion the **force of**

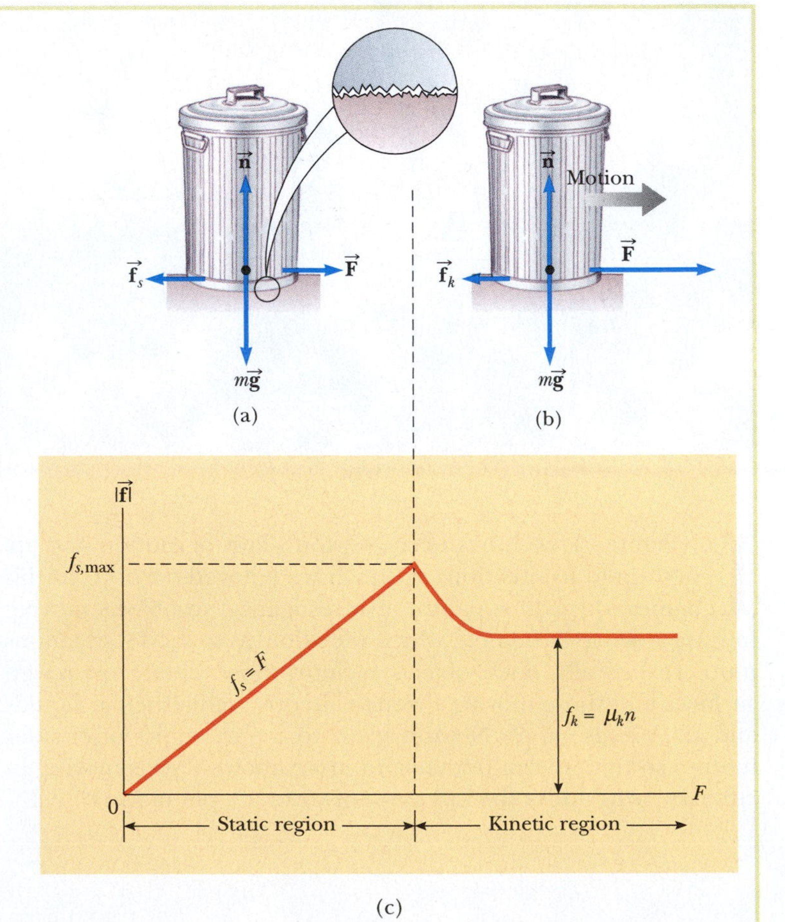


**ACTIVE FIGURE 5.1**

(a) The force of static friction $\vec{\mathbf{f}}_s$ between a trash can and a concrete patio is opposite the applied force $\vec{\mathbf{F}}$. The magnitude of the force of static friction equals that of the applied force. (b) When the magnitude of the applied force exceeds the magnitude of the force of kinetic friction $\vec{\mathbf{f}}_k$, the trash can accelerates to the right. (c) A graph of the magnitude of the friction force versus that of the applied force. In our model, the force of kinetic friction is independent of the applied force and the relative speed of the surfaces. Note that $f_{s,\max} > f_k$.

**PhysicsNow™** You can vary the load in the trash can and practice sliding it on surfaces of varying roughness by logging into PhysicsNow at **www.pop4e.com** and going to Active Figure 5.1. Note the effect on the trash can's motion and the corresponding behavior of the graph in (c).

**kinetic friction** $\vec{\mathbf{f}}_k$. The net force $F - f_k$ in the $x$ direction produces an acceleration to the right, according to Newton's second law. If we reduce the magnitude of $\vec{\mathbf{F}}$ so that $F = f_k$, the acceleration is zero and the trash can moves to the right with constant speed. If the applied force is removed, the friction force acting to the left provides an acceleration of the trash can in the $-x$ direction and eventually brings it to rest.

Experimentally, one finds that, to a good approximation, both $f_{s,\max}$ and $f_k$ for an object on a surface are proportional to the normal force exerted by the surface on the object; thus, we adopt a simplification model in which this approximation is assumed to be exact. The assumptions in this simplification model can be summarized as follows:

- The magnitude of the force of static friction between any two surfaces in contact can have the values

$$f_s \leq \mu_s n \qquad [5.1]$$

■ Force of static friction

where the dimensionless constant $\mu_s$ is called the **coefficient of static friction** and $n$ is the magnitude of the normal force. The equality in Equation 5.1 holds when the surfaces are on the verge of slipping, that is, when $f_s = f_{s,\max} \equiv \mu_s n$. This situation is called *impending motion.* The inequality holds when the component of the applied force parallel to the surfaces is less than this value.
- The magnitude of the force of kinetic friction acting between two surfaces is

$$f_k = \mu_k n \qquad [5.2]$$

■ Force of kinetic friction

where $\mu_k$ is the **coefficient of kinetic friction.** In our simplification model, this coefficient is independent of the relative speed of the surfaces.
- The values of $\mu_k$ and $\mu_s$ depend on the nature of the surfaces, but $\mu_k$ is generally less than $\mu_s$. Table 5.1 lists some measured values.
- The direction of the friction force on an object is opposite to the actual motion (kinetic friction) or the impending motion (static friction) of the object relative to the surface with which it is in contact.

The approximate nature of Equations 5.1 and 5.2 is easily demonstrated by trying to arrange for an object to slide down an incline at constant speed. Especially at low speeds, the motion is likely to be characterized by alternate stick and slip episodes. The simplification model described in the bulleted list above has been developed so that we can solve problems involving friction in a relatively straightforward way.

### PITFALL PREVENTION 5.1

**THE EQUAL SIGN IS USED IN LIMITED SITUATIONS** In Equation 5.1, the equal sign is used *only* when the surfaces are just about to break free and begin sliding. Do not fall into the common trap of using $f_s = \mu_s n$ in *any* static situation.

### PITFALL PREVENTION 5.2

**THE DIRECTION OF THE FRICTION FORCE** Sometimes, an incorrect statement about the friction force between an object and a surface is made—"The friction force on an object is opposite to its motion or impending motion"—rather than the correct phrasing, "The friction force on an object is opposite to its motion or impending motion *relative to the surface.*" Think carefully about Quick Quiz 5.2.

**TABLE 5.1 Coefficients of Friction**

| | $\mu_s$ | $\mu_k$ |
|---|---|---|
| Steel on steel | 0.74 | 0.57 |
| Aluminum on steel | 0.61 | 0.47 |
| Copper on steel | 0.53 | 0.36 |
| Rubber on concrete | 1.0 | 0.8 |
| Wood on wood | 0.25–0.5 | 0.2 |
| Glass on glass | 0.94 | 0.4 |
| Waxed wood on wet snow | 0.14 | 0.1 |
| Waxed wood on dry snow | — | 0.04 |
| Metal on metal (lubricated) | 0.15 | 0.06 |
| Ice on ice | 0.1 | 0.03 |
| Teflon on Teflon | 0.04 | 0.04 |
| Synovial joints in humans | 0.01 | 0.003 |

*Note:* All values are approximate.

Now that we have identified the characteristics of the friction force, we can include the friction force in the net force on an object in the model of a particle under a net force.

**QUICK QUIZ 5.1** You press your physics textbook flat against a vertical wall with your hand, which applies a normal force perpendicular to the book. What is the direction of the friction force on the book due to the wall? **(a)** downward **(b)** upward **(c)** out from the wall **(d)** into the wall

**QUICK QUIZ 5.2** A crate is located in the center of a flatbed truck. The truck accelerates to the east and the crate moves with it, not sliding at all. What is the direction of the friction force exerted by the truck on the crate? **(a)** It is to the west. **(b)** It is to the east. **(c)** No friction force exists because the crate is not sliding.

**QUICK QUIZ 5.3** You are playing with your daughter in the snow. She sits on a sled and asks you to slide her across a flat, horizontal field. You have a choice of **(a)** pushing her from behind, by applying a force downward on her shoulders at 30° below the horizontal (Fig. 5.2a) or **(b)** attaching a rope to the front of the sled and pulling with a force at 30° above the horizontal (Fig 5.2b). Which would require less force for a given acceleration of the daughter?

## Thinking Physics 5.1

In the motion picture *The Abyss* (Twentieth Century Fox, 1989), an underwater oil exploration rig is located at the ocean bottom in very deep water. It is connected to a ship on the ocean surface by a cable called an "umbilical cord" as suggested in Figure 5.3a. On the ship, the umbilical cord is attached to a gantry. During a hurricane, the gantry structure breaks loose from the ship, falls into the water, and sinks to the bottom, passing over the edge of an extremely deep abyss. As a result, the rig is dragged by the umbilical cord along the ocean bottom as described in Figure 5.3b. As the rig approaches the edge of the abyss, however, it is not pulled over the edge but rather, stops just short of the edge as shown in Figure 5.3c. Is this scenario purely a cinematic edge-of-the-seat situation, or do the principles of physics suggest why the moving rig does not topple over the edge?

**Reasoning** Physics can explain this phenomenon. While the rig is being pulled across the ocean floor (Fig. 5.3b), it is pulled by the section of the umbilical cord that is almost horizontal and therefore almost parallel to the ocean floor. Therefore, the rig is subject to two horizontal forces: the tension in the umbilical cord

**FIGURE 5.2** (Quick Quiz 5.3) A father tries to slide his daughter on a sled over snow by (a) pushing downward on her shoulders or (b) pulling upward on a rope attached to the sled. Which is easier?

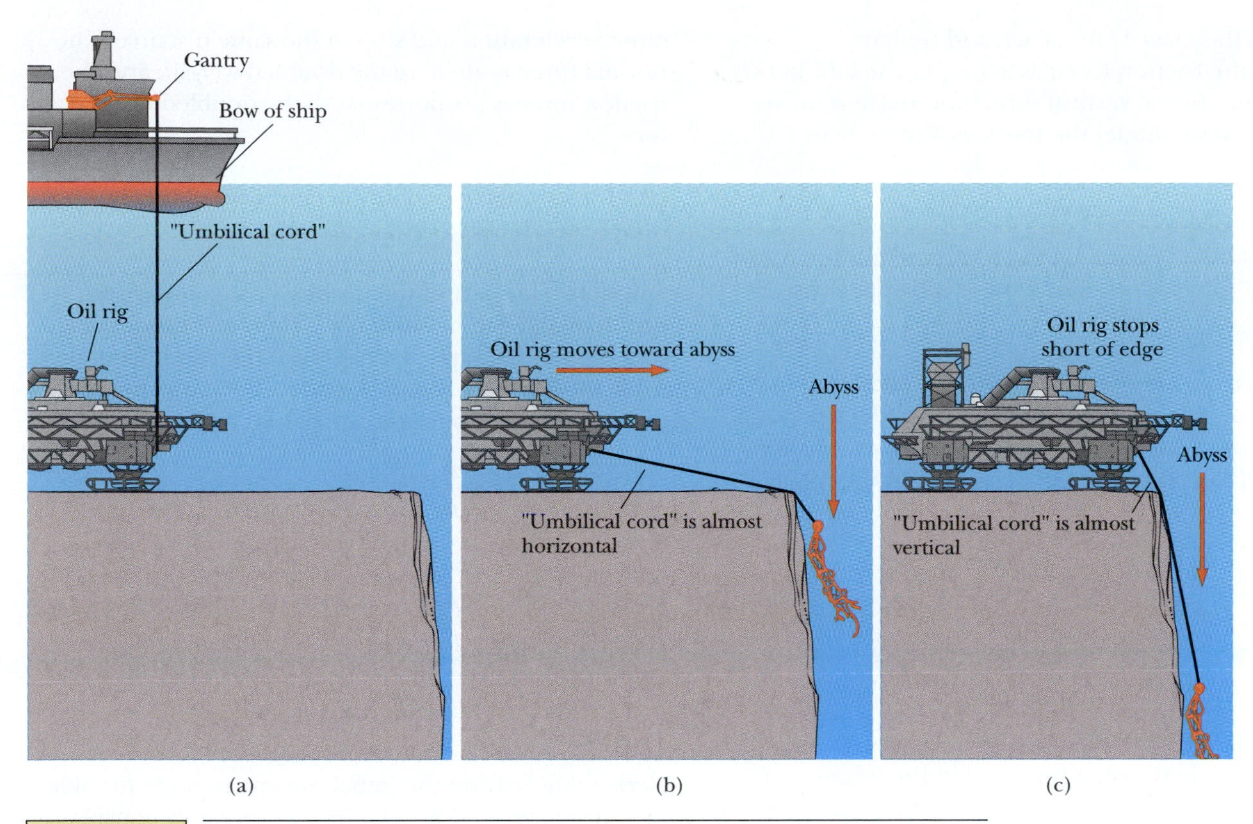


**FIGURE 5.3** (Thinking Physics 5.1) An oil rig at the bottom of the ocean is dragged by a cable.

pulling it forward and friction with the ocean floor pulling back. Let us assume that these forces are equal in magnitude so that the rig moves with constant speed. As the rig nears the edge of the abyss, the angle the umbilical cord makes with the horizontal increases. As a result, the component of the force from the cord parallel to the ocean floor decreases and the downward vertical component increases. As a result of the increased vertical force, the rig is pulled downward more strongly to the ocean floor, increasing the normal force on it and, in turn, increasing the friction force between the rig and the ocean floor. Therefore, with less force pulling it forward (from the umbilical cord) and more force opposing the motion (as a result of friction), the rig slows down. By the time the rig reaches the edge of the abyss, the force from the umbilical cord is almost *straight down* (Fig. 5.3c), resulting in little forward force. Furthermore, this large downward force pulls the rig into the ocean floor, resulting in a very large friction force that stops the rig. ■

## EXAMPLE 5.1 The Skidding Truck

The driver of an empty speeding truck slams on the brakes and skids to a stop through a distance $d$.

**A** If the truck carries a heavy load such that the moving mass is doubled, what would be its skidding distance if it starts from the same initial speed?

**Solution** Figure 5.4 shows a free-body diagram for the skidding truck. The only force in the horizontal direction is the friction force, which is assumed to be independent of speed in our simplification model for friction. Therefore, from Newton's second law,

$$\sum F_x = -f_k = ma$$

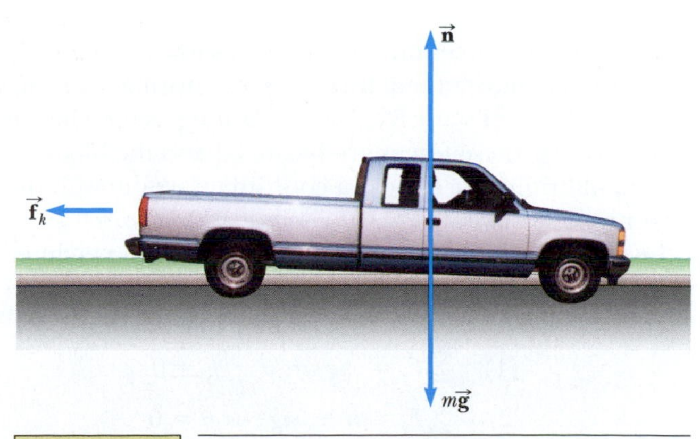


**FIGURE 5.4** (Example 5.1) A truck skids to a stop.

where $m$ is the mass of the truck and we have expressed the friction force as acting to the left, in the $-x$ direction. In the vertical direction, there is no acceleration, so we model the truck as a particle in equilibrium:

$$\sum F_y = n - mg = 0 \quad \rightarrow \quad n = mg$$

Finally, from the relation between the friction force and the normal force, we combine these two equations:

$$f_k = \mu_k n \quad \rightarrow \quad -\mu_k(mg) = ma \quad \rightarrow \quad a = -\mu_k g$$

Because both $\mu_k$ and $g$ are constant, the acceleration of the truck is constant. We therefore model the truck as a particle under constant acceleration. We use Equation 2.13 to find the position of the truck when the velocity is zero:

$$v_{xf}^2 = v_{xi}^2 + 2a_x(x_f - x_i)$$

$$0 = v_{xi}^2 - 2(\mu_k g)(x_f - 0)$$

$$x_f = d = \frac{v_{xi}^2}{2\mu_k g}$$

We can argue from the mathematical representation as follows. The expression for the skidding distance $d$ does not include the mass. Therefore, the truck skids the same distance regardless of the mass of the load. Conceptually, we can argue that the truck with twice the mass requires twice the friction force to exhibit the same acceleration and stop in the same distance. The normal force is equal to the doubled weight, and the friction force is proportional to the doubled normal force!

B If the initial speed of the empty truck is halved, what would be the skidding distance?

**Solution** This part of the problem is a comparison problem and can be solved by a ratio technique such as that used in Example 3.4. We write the result from part A for the skidding distance $d$ twice, once for the original situation and once for the halved initial velocity:

$$d_1 = \frac{v_{1xi}^2}{2\mu_k g}$$

$$d_2 = \frac{v_{2xi}^2}{2\mu_k g} = \frac{\left(\frac{1}{2}v_{1xi}\right)^2}{2\mu_k g} = \frac{1}{4}\frac{v_{1xi}^2}{2\mu_k g}$$

Dividing the first equation by the second, we have

$$\frac{d_1}{d_2} = 4 \quad \rightarrow \quad d_2 = \frac{1}{4}d_1$$

Notice that halving the initial velocity reduces the skidding distance by 75%! This important safety consideration is associated with the possibility of an accident when driving at high speed.

## EXAMPLE 5.2 Experimental Determination of $\mu_s$ and $\mu_k$

The following is a simple method of measuring coefficients of friction. Suppose a block is placed on a rough surface inclined relative to the horizontal, as shown in Figure 5.5. The incline angle $\theta$ is increased until the block starts to move.

A How is the coefficient of static friction related to the critical angle $\theta_c$ at which the block begins to move?

**Solution** The forces on the block, as shown in Figure 5.5, are the gravitational force $m\vec{\mathbf{g}}$, the normal force $\vec{\mathbf{n}}$, and the force of static friction $\vec{\mathbf{f}}_s$. As long as the block is not moving, these forces are balanced and the block is in equilibrium. We choose a coordinate system with the positive $x$ axis parallel to the incline and downhill and the positive $y$ axis upward perpendicular to the incline. Applying Newton's second law in component form to the block gives

$$(1) \quad \sum F_x = mg\sin\theta - f_s = 0$$

$$(2) \quad \sum F_y = n - mg\cos\theta = 0$$

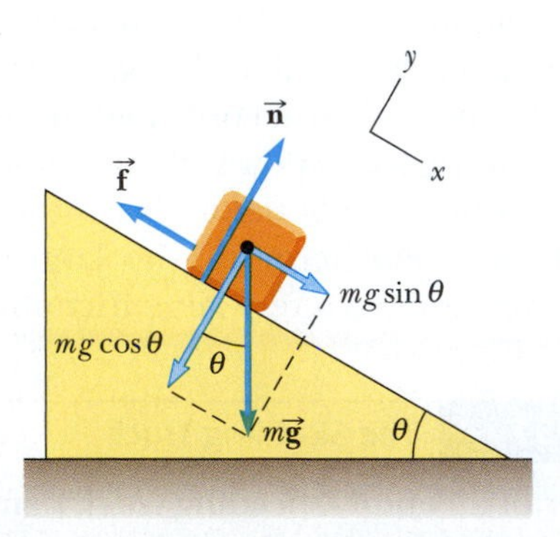


FIGURE 5.5 (Example 5.2) A block on an adjustable incline is used to determine the coefficients of friction.

These equations are valid for any angle of inclination $\theta$. At the critical angle $\theta_c$ at which the block is on the verge of slipping, the friction force has its maximum magnitude $\mu_s n$, so we rewrite (1) and (2) for this condition as

$$(3) \quad mg\sin\theta_c = \mu_s n$$

$$(4) \quad mg\cos\theta_c = n$$

Dividing (3) by (4), we have

$$\tan \theta_c = \mu_s$$

Therefore, the coefficient of static friction is equal to the tangent of the angle of the incline at which the block begins to slide.

B How could we find the coefficient of kinetic friction?

**Solution** Once the block begins to move, the magnitude of the friction force is the kinetic value $\mu_k n$, which is smaller than that of the force of static friction. As a result, if the angle is maintained at the critical angle, the block accelerates down the incline. To restore the equilibrium situation in Equation (1), with $f_s$ replaced by $f_k$, the angle must be reduced to a value $\theta_c'$ such that the block slides down the incline at *constant speed.* In this situation, Equations (3) and (4), with $\theta_c$ replaced by $\theta_c'$ and $\mu_s$ by $\mu_k$, give us

$$\tan \theta_c' = \mu_k$$

## EXAMPLE 5.3 Connected Objects

A ball and a cube are connected by a light string that passes over a frictionless light pulley, as in Figure 5.6a. The coefficient of kinetic friction between the cube and the surface is 0.30. Find the acceleration of the two objects and the tension in the string.

**Solution** To *conceptualize* the problem, imagine the ball moving downward and the cube sliding to the right, both accelerating from rest. We recognize that there are two objects that are accelerating, so we *categorize* this problem as one involving particles under a net force, where one of the forces to be included is the friction force. To begin to *analyze* the problem, we set up a simplified pictorial representation by drawing the free-body diagrams for the two objects as in Figures 5.6b and 5.6c. For the ball, no forces are exerted in the horizontal direction, and we apply Newton's second law in the vertical direction. For the cube, the acceleration is horizontal, so we know the cube is in equilibrium in the vertical direction. We use the fact that the magnitude of the force of kinetic friction acting on the cube is proportional to the normal force according to $f_k = \mu_k n$. Because the pulley is light (massless) and frictionless, the tension in the string is the same on both sides of the pulley. Because the tension acts on both objects, it is the common quantity that applies to both objects and allows us to combine separate equations for the two objects into one equation.

Let us address the cube of mass $m_1$ first. Newton's second law applied to the cube in component form, with the positive $x$ direction to the right, gives

$$\sum F_x = m_1 a \quad \rightarrow \quad T - f_k = m_1 a$$

$$\sum F_y = 0 \quad \rightarrow \quad n - m_1 g = 0$$

where $T$ is the tension in the string. Because $f_k = \mu_k n$ and $n = m_1 g$ from the equilibrium equation for the $y$ direction, we have $f_k = \mu_k m_1 g$. Therefore, from the equation for the $x$ direction,

$$(1) \quad T = \mu_k m_1 g + m_1 a$$

Now we apply Newton's second law to the ball moving in the vertical direction. Because the ball moves downward when the cube moves to the right, we choose the positive direction downward for the ball:

$$(2) \quad \sum F_y = m_2 a \quad \rightarrow \quad m_2 g - T = m_2 a$$

Substituting the expression for $T$ from (1) into (2) gives us

$$m_2 g - (\mu_k m_1 g + m_1 a) = m_2 a$$

$$a = \frac{m_2 - \mu_k m_1}{m_1 + m_2} g$$

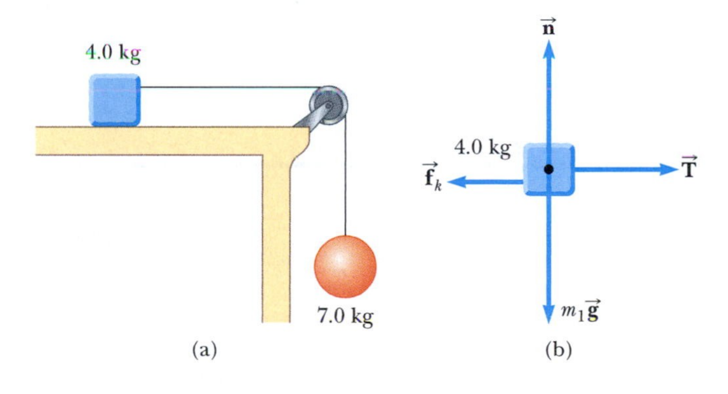


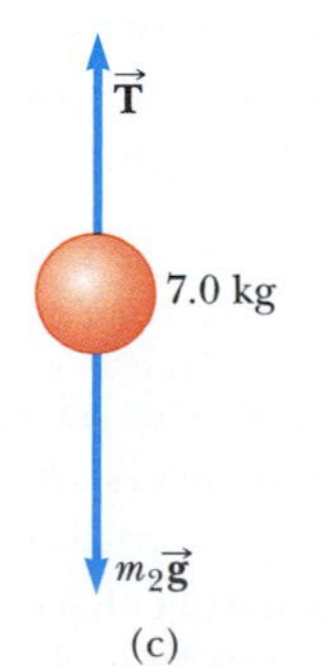


FIGURE 5.6 (Example 5.3) (a) Two objects connected by a light string that passes over a frictionless pulley. (b) Free-body diagram for the sliding cube. (c) Free-body diagram for the hanging ball.

Now, substituting the known values,

$$a = \frac{7.0\text{ kg} - 0.30(4.0\text{ kg})}{7.0\text{ kg} + 4.0\text{ kg}}(9.80\text{ m/s}^2) = 5.2\text{ m/s}^2$$

which is the *magnitude* of the acceleration of each of the two objects. For the ball, the acceleration *vector* is downward and the vector is toward the right for the cube. When the magnitude of the acceleration is substituted into (1), we find the tension:

$$T = 0.30(4.0\text{ kg})(9.80\text{ m/s}^2) + (4.0\text{ kg})(5.2\text{ m/s}^2) = 33\text{ N}$$

To *finalize* the problem, note that the acceleration is smaller than that due to gravity. That does not tell us that the answer is correct, but if the acceleration were larger than $g$, it would tell is that we have made an error. Note also that the tension in the string is smaller than $m_2 g = (7.0\text{ kg})(9.80\text{ m/s}^2) = 69\text{ N}$, which is consistent with $m_2$ accelerating downward.

### EXAMPLE 5.4 The Sliding Crate

A warehouse worker places a crate on a sloped surface that is inclined at 30.0° with respect to the horizontal (Fig. 5.7a). If the crate slides down the incline with an acceleration of magnitude $g/3$, determine the coefficient of kinetic friction between the crate and the surface.

**Solution** Figure 5.7b shows the forces acting on the crate. The $x$ axis is chosen parallel to the incline and the $y$ axis perpendicular. From Newton's second law,

$$(1)\quad \sum F_x = ma \quad\rightarrow\quad mg\sin\theta - f_k = ma$$

$$(2)\quad \sum F_y = 0 \quad\rightarrow\quad n - mg\cos\theta = 0$$

The kinetic friction force is $f_k = \mu_k n$ and, from (2), we find that $n = mg\cos\theta$. Therefore, the friction force can be expressed as $f_k = \mu_k mg\cos\theta$. Substituting into (1) gives us

$$mg\sin\theta - \mu_k mg\cos\theta = ma \quad\rightarrow\quad \mu_k = \frac{g\sin\theta - a}{g\cos\theta}$$

Substituting the known values, we have

$$\mu_k = \frac{\cancel{g}\sin 30.0° - \frac{1}{3}\cancel{g}}{\cancel{g}\cos 30.0°} = \frac{(0.500 - 0.333)}{0.867} = 0.192$$

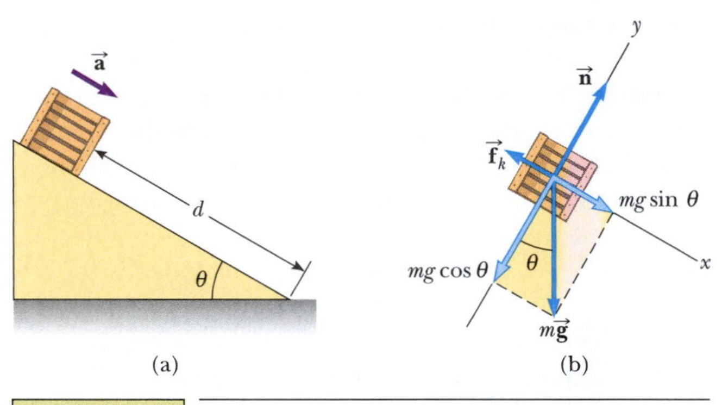


FIGURE 5.7 (Example 5.4) (a) A crate of mass $m$ slides down an incline. (b) Free-body diagram for the sliding crate.

## 5.2 NEWTON'S SECOND LAW APPLIED TO A PARTICLE IN UNIFORM CIRCULAR MOTION

Solving problems involving friction is just one of many applications of Newton's second law. Let us now consider another common situation, associated with a particle in uniform circular motion. In Chapter 3, we found that a particle moving in a circular path of radius $r$ with uniform speed $v$ experiences a centripetal acceleration of magnitude

■ Centripetal acceleration

$$a_c = \frac{v^2}{r}$$

The acceleration vector with this magnitude is directed toward the center of the circle and is *always* perpendicular to $\vec{\mathbf{v}}$.

According to Newton's second law, if an acceleration occurs, a net force must be causing it. Because the acceleration is toward the center of the circle, the net force must be toward the center of the circle. Therefore, when a particle travels in a circular path, a force must be acting *inward* on the particle that causes the circular motion. We investigate the forces causing this type of acceleration in this section.

Consider an object of mass $m$ tied to a string of length $r$ and being whirled in a horizontal circular path on a frictionless table top as in the overhead view in

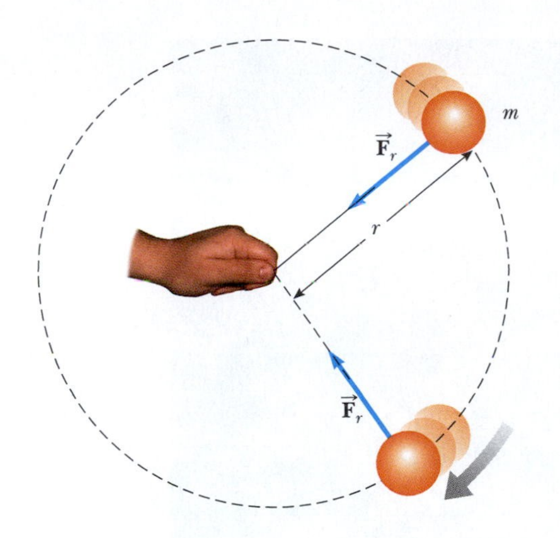


FIGURE 5.8 Overhead view of a ball moving in a circular path in a horizontal plane. A force $\vec{\mathbf{F}}_r$ directed toward the center of the circle keeps the ball moving in its circular path.

Figure 5.8. Let us assume that the object moves with constant speed. The natural tendency of the object is to move in a straight-line path, according to Newton's first law; the string, however, prevents this motion along a straight line by exerting a radial force $\vec{\mathbf{F}}_r$ on the object to make it follow a circular path. This force, whose magnitude is the tension in the string, is directed along the length of the string toward the center of the circle as shown in Figure 5.8.

In this discussion, the tension in the string causes the circular motion. Other forces also cause objects to move in circular paths. For example, friction forces cause automobiles to travel around curved roadways and the gravitational force causes a planet to orbit the Sun.

Regardless of the nature of the force acting on the particle in circular motion, we can apply Newton's second law to the particle along the radial direction:

$$\sum F = ma_c = m\frac{v^2}{r} \qquad [5.3]$$

In general, an object can move in a circular path under the influence of various types of forces, or a *combination* of forces, as we shall see in some of the examples that follow.

If the force acting on an object vanishes, the object no longer moves in its circular path; instead, it moves along a straight-line path tangent to the circle. This idea is illustrated in Active Figure 5.9 for the case of the ball whirling in a circle at the

### PITFALL PREVENTION 5.3

**CENTRIPETAL FORCE** The force causing centripetal acceleration is called *centripetal force* in some textbooks. Giving the force causing circular motion a name leads many students to consider it as a new *kind* of force rather than a new *role* for force. A common mistake is to draw the forces in a free-body diagram and then add another vector for the centripetal force. Yet it is not a separate force; it is one of our familiar forces *acting in the role of causing a circular motion.* For the motion of the Earth around the Sun, for example, the "centripetal force" is *gravity.* For a rock whirled on the end of a string, the "centripetal force" is the *tension* in the string. After this discussion, we shall no longer use the phrase *centripetal force.*

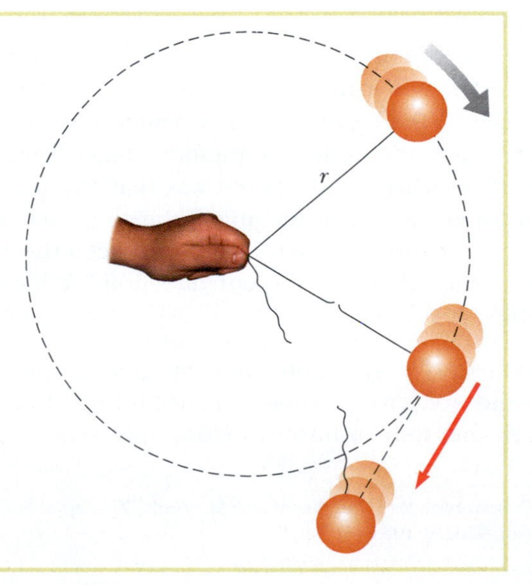


ACTIVE FIGURE 5.9 An overhead view of a ball moving in a circular path in a horizontal plane. When the string breaks, the ball moves in the direction tangent to the circular path.

**PhysicsNow™** Log into PhysicsNow at **www.pop4e.com** and go to Active Figure 5.9 to "break" the string yourself and observe the effect on the ball's motion.

### PITFALL PREVENTION 5.4

**DIRECTION OF TRAVEL WHEN THE STRING IS CUT** Study Active Figure 5.9 carefully. Many students have a misconception that the ball moves *radially* away from the center of the circle when the string is cut. The velocity of the ball is *tangent* to the circle. By Newton's first law, the ball simply continues to move in the direction that it is moving just as the force from the string disappears.

The cars of a corkscrew roller coaster must travel in tight loops. The normal force from the track contributes to the centripetal acceleration. The gravitational force, because it remains constant in direction, is sometimes in the same direction as the normal force, but is sometimes in the opposite direction. ■

(Robin Smith/Getty Images)

### PITFALL PREVENTION 5.5

**CENTRIFUGAL FORCE** The commonly heard phrase "centrifugal force" is described as a force pulling *outward* on an object moving in a circular path. If you are experiencing a "centrifugal force" on a rotating carnival ride, what is the other object with which you are interacting? You cannot identify another object because it is a fictitious force that occurs as a result of your being in a noninertial reference frame.

end of a string. If the string breaks at some instant, the ball moves along the straight-line path tangent to the circle at the point on the circle at which the ball is located at that instant.

**QUICK QUIZ 5.4** You are riding on a Ferris wheel (Fig. 5.10) that is rotating with constant speed. The car in which you are riding always maintains its correct upward orientation; it does not invert. **(i)** What is the direction of the normal force on you from the seat when you are at the top of the wheel? **(a)** upward **(b)** downward **(c)** impossible to determine. **(ii)** What is the direction of the net force on you when you are at the top of the wheel? **(a)** upward **(b)** downward **(c)** impossible to determine

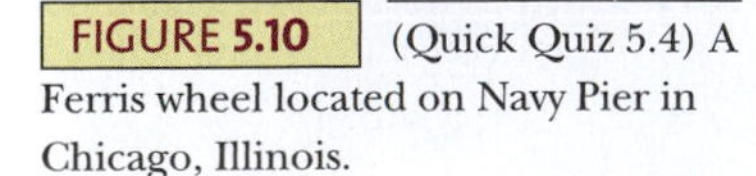

(© Tom Carroll/Index Stock Imagery/PictureQuest)

**FIGURE 5.10** (Quick Quiz 5.4) A Ferris wheel located on Navy Pier in Chicago, Illinois.

## ■ Thinking Physics 5.2

The Copernican theory of the solar system is a structural model in which the planets are assumed to travel around the Sun in circular orbits. Historically, this theory was a break from the Ptolemaic theory, a structural model in which the Earth was at the center. When the Copernican theory was proposed, a natural question arose: What keeps the Earth and other planets moving in their paths around the Sun? An interesting response to this question comes from Richard Feynman[1]: "In those days, one of the theories proposed was that the planets went around because behind them there were invisible angels, beating their wings and driving the planets forward. . . . It turns out that in order to keep the planets going around, the invisible angels must fly in a different direction." What did Feynman mean by this statement?

**Reasoning** The question asked by those at the time of Copernicus indicates that they did not have a proper understanding of inertia as described by Newton's first law. At that time in history, before Galileo and Newton, the interpretation was that

[1] R. P. Feynman, R. B. Leighton, and M. Sands, *The Feynman Lectures on Physics,* Vol. 1, (Reading, MA: Addison-Wesley, 1963), p. 7-2.

*motion* was caused by force. This interpretation is different from our current understanding that *changes in motion* are caused by force. Therefore, it was natural for Copernicus's contemporaries to ask what force propelled a planet in its orbit. According to our current understanding, it is equally natural for us to realize that no force tangent to the orbit is necessary, that the motion simply continues owing to inertia.

Therefore, in Feynman's imagery, the angels do not have to push the planet *from behind.* The angels must push *inward,* to provide the centripetal acceleration associated with the orbital motion of the planet. Of course, the angels are not real from a scientific point of view, but are a metaphor for the *gravitational force.* ■

## EXAMPLE 5.5 How Fast Can It Spin?

An object of mass 0.500 kg is attached to the end of a cord whose length is 1.50 m. The object is whirled in a horizontal circle as in Figure 5.8. If the cord can withstand a maximum tension of 50.0 N, what is the maximum speed the object can have before the cord breaks?

**Solution** Because the magnitude of the force that provides the centripetal acceleration of the object in this case is the tension $T$ exerted by the cord on the object, Newton's second law gives us for the inward radial direction

$$\sum F_r = ma_c \quad \rightarrow \quad T = m\frac{v^2}{r}$$

Solving for the speed $v$, we have

$$v = \sqrt{\frac{Tr}{m}}$$

The maximum speed that the object can have corresponds to the maximum value of the tension. Hence, we find

$$v_{\text{max}} = \sqrt{\frac{T_{\text{max}}r}{m}} = \sqrt{\frac{(50.0 \text{ N})(1.50 \text{ m})}{0.500 \text{ kg}}} = 12.2 \text{ m/s}$$

## EXAMPLE 5.6 The Conical Pendulum

A small object of mass $m$ is suspended from a string of length $L$. The object revolves in a horizontal circle of radius $r$ with constant speed $v$, as in Figure 5.11a. (Because the string sweeps out the surface of a cone, the system is known as a *conical pendulum.*)

**A** Find the speed of the object.

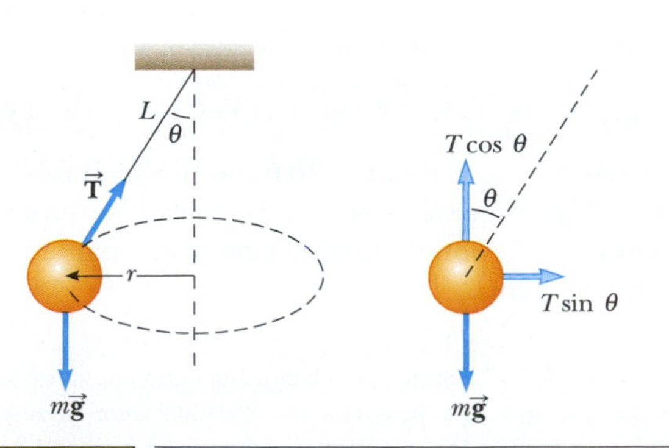


**FIGURE 5.11** (Example 5.6) The conical pendulum and its free-body diagram.

**Solution** The free-body diagram for the object of mass $m$ is shown in Figure 5.11b, where the force $\vec{\mathbf{T}}$ exerted by the string has been resolved into a vertical component $T\cos\theta$ and a horizontal component $T\sin\theta$ acting toward the center of rotation. Because the object does not accelerate in the vertical direction, we model it as a particle in equilibrium in the vertical direction:

$$\sum F_y = 0 \quad \rightarrow \quad T\cos\theta - mg = 0$$

$$(1) \quad T\cos\theta = mg$$

In the horizontal direction, we have a centripetal acceleration so we model the object as a particle under a net force. Because the force that provides the centripetal acceleration in this example is the component $T\sin\theta$, from Newton's second law we have

$$(2) \quad \sum F_r = ma_c \quad \rightarrow \quad T\sin\theta = m\frac{v^2}{r}$$

By dividing (2) by (1), we eliminate $T$ and find that

$$\tan\theta = \frac{v^2}{rg} \quad \rightarrow \quad v = \sqrt{rg\tan\theta}$$

From a triangle we can construct in the pictorial representation in Figure 5.11a, we note that $r = L \sin \theta$; therefore,

$$v = \sqrt{Lg \sin \theta \tan \theta}$$

**B** Find the period of revolution, defined as the time interval required to complete one revolution.

**Solution** The object is traveling at constant speed around its circular path. Because the object travels a distance of $2\pi r$ (the circumference of the circular path) in a time interval $\Delta t$ equal to the period of revolution, we find

$$(3) \quad \Delta t = \frac{2\pi r}{v} = \frac{2\pi r}{\sqrt{rg \tan \theta}} = 2\pi \sqrt{\frac{L \cos \theta}{g}}$$

The intermediate algebraic steps used in obtaining (3) are left to the reader. Note that the period is independent of the mass $m$!

## INTERACTIVE EXAMPLE 5.7 What Is the Maximum Speed of the Car?

A 1 500-kg car moving on a flat, horizontal road negotiates a curve whose radius is 35.0 m (Fig. 5.12a). If the coefficient of static friction between the tires and the dry pavement is 0.523, find the maximum speed the car can have to make the turn successfully.

**Solution** In the rolling motion of each tire, the bit of rubber meeting the road is instantaneously at rest relative to the road. It is prevented from skidding radially outward by a static friction force that acts radially inward, enabling the car to move in its circular path. The car is an extended object with four friction forces acting on it, one on each wheel, but we shall model it as a particle with only one net friction force. Figure 5.12b shows a free-body diagram for the car. From Newton's second law in the horizontal direction, we have

$$(1) \quad \sum F_x = ma \quad \rightarrow \quad f_s = m\frac{v^2}{r}$$

The maximum speed that the car can have around the curve corresponds to the speed at which it is on the verge of skidding toward the side of the road. At this point, the friction force has its maximum value

$$f_{s,\max} = \mu_s n$$

In the vertical direction, no acceleration occurs, so

$$\sum F_y = 0 \quad \rightarrow \quad n - mg = 0$$

Therefore, the magnitude of the normal force equals the weight in this case, and we find

$$f_{s,\max} = \mu_s mg$$

Substituting this expression into (1), we find the maximum speed:

$$\mu_s mg = m\frac{v_{\max}^2}{r} \quad \rightarrow \quad v_{\max} = \sqrt{\mu_s gr}$$

Substituting the numerical values gives us

$$v_{\max} = \sqrt{(0.523)(9.80 \text{ m/s}^2)(35.0 \text{ m})} = 13.4 \text{ m/s}$$

This result is equivalent to 30.0 mi/h, which is less than a typical nonfreeway speed of 35 mi/h. Therefore, this roadway could benefit greatly from some banking, as in the next example!

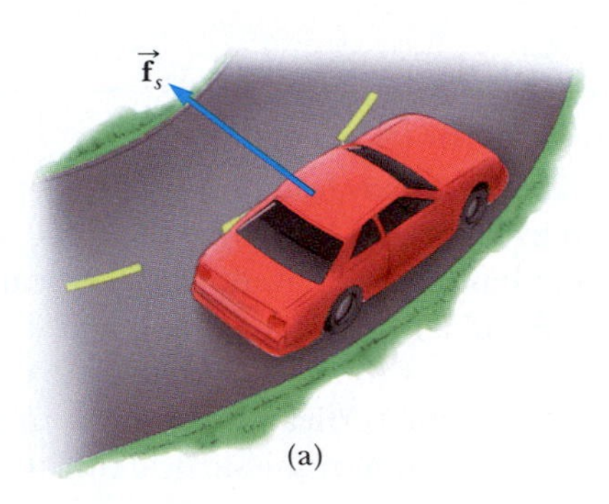


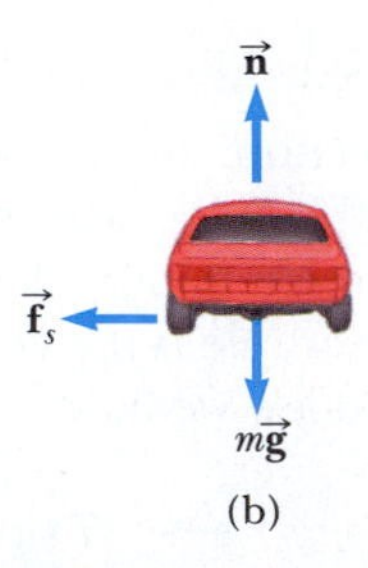


FIGURE 5.12 (Interactive Example 5.7) (a) The force of static friction directed toward the center of the curve keeps the car moving in a circular path. (b) Free-body diagram for the car.

PhysicsNow™ Study the relationship between the car's speed, radius of the turn, and the coefficient of static friction between road and tires by logging into PhysicsNow at www.pop4e.com and going to Interactive Example 5.7.

## INTERACTIVE EXAMPLE 5.8 The Banked Roadway

A civil engineer wishes to redesign the curved roadway in Interactive Example 5.7 in such a way that a car will not have to rely on friction to round the curve without skidding. In other words, a car moving at the designated speed can negotiate the curve even when the road is covered with ice. Such a curve is usually *banked,* meaning that the roadway is tilted toward the inside of the curve. Suppose the designated speed for the curve is to be 13.4 m/s (30.0 mi/h) and the radius of the curve is 35.0 m. At what angle should the curve be banked?

**Solution** On a level (unbanked) road, the force that causes the centripetal acceleration is the force of static friction between car and road, as we saw in the previous example. If the road is banked at an angle $\theta$, however, as in Figure 5.13, the normal force $\vec{\mathbf{n}}$ has a horizontal component $n_x = n \sin\theta$ pointing toward the center of the curve. Because the curve is to be designed so that the force of static friction is zero, only the component $n \sin\theta$ causes the centripetal acceleration. Hence, Newton's second law for the radial direction gives

$$(1) \quad \sum F_r = n \sin\theta = \frac{mv^2}{r}$$

The car is in equilibrium in the vertical direction. Therefore, from $\sum F_y = 0$ we have

$$(2) \quad n \cos\theta = mg$$

Dividing (1) by (2) gives

$$(3) \quad \tan\theta = \frac{v^2}{rg}$$

$$\theta = \tan^{-1}\left(\frac{(13.4\ \text{m/s})^2}{(35.0\ \text{m})(9.80\ \text{m/s}^2)}\right) = 27.6°$$

If a car rounds the curve at a speed less than 13.4 m/s, friction is needed to keep it from sliding down the bank (to the left in Fig. 5.13). A driver who attempts to negotiate the curve at a speed greater than 13.4 m/s has to depend on friction to keep from sliding up the bank (to the right in Fig. 5.13). The banking angle is independent of the mass of the vehicle negotiating the curve.

**PhysicsNow™** Adjust the turn radius and the speed to see the effect on the banking angle by logging into PhysicsNow at www.pop4e.com and going to Interactive Example 5.8.

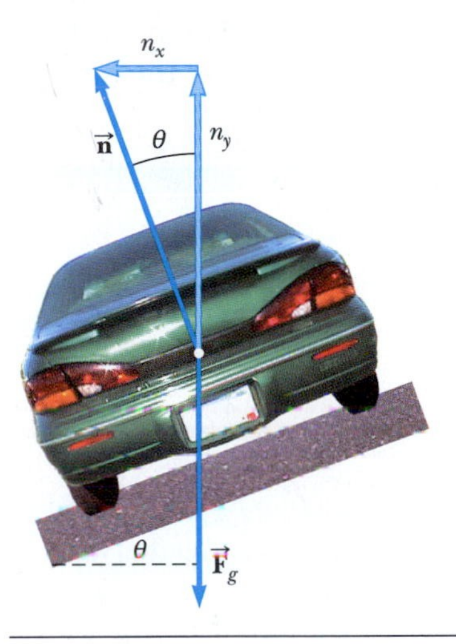


FIGURE 5.13 (Interactive Example 5.8) A car rounding a curve on a road banked at an angle $\theta$ to the horizontal. In the absence of friction the force that causes the centripetal acceleration and keeps the car moving in its circular path is the horizontal component of the normal force.

## EXAMPLE 5.9 Let's Go Loop-the-Loop

A pilot of mass $m$ in a jet aircraft executes a "loop-the-loop" maneuver as illustrated in Figure 5.14a. The aircraft moves in a vertical circle of radius 2.70 km at a *constant speed* of 225 m/s.

**A** Determine the force exerted by the seat on the pilot at the bottom of the loop. Express the answer in terms of the weight $mg$ of the pilot.

**Solution** This example is the first numerical one we have seen in which the force causing the centripetal acceleration is a *combination* of forces rather than a single force. We shall model the pilot as a particle under a net force and analyze the situation at the bottom and top of the circular path.

The free-body diagram for the pilot at the bottom of the loop is shown in Figure 5.14b. The forces acting on the pilot are the downward gravitational force $m\vec{\mathbf{g}}$ and the upward normal force $\vec{\mathbf{n}}_{\text{bot}}$ exerted by the seat on the pilot. Because the net upward force at the bottom that provides the centripetal acceleration has a magnitude $n_{\text{bot}} - mg$, Newton's second law for the radial (upward) direction gives

$$\sum F_y = ma \quad \longrightarrow \quad n_{\text{bot}} - mg = m\frac{v^2}{r}$$

$$n_{\text{bot}} = mg + m\frac{v^2}{r} = mg\left[1 + \frac{v^2}{rg}\right]$$

Substituting the values given for the speed and radius gives

$$n_{\text{bot}} = mg\left[1 + \frac{(225\text{ m/s})^2}{(2.70 \times 10^3\text{ m})(9.80\text{ m/s}^2)}\right]$$

$$= 2.91\,mg$$

Therefore, the force exerted by the seat on the pilot at the bottom of the loop is *greater* than the pilot's weight by a factor of 2.91.

**B** Determine the force exerted by the seat on the pilot at the top of the loop. Express the answer in terms of the weight $mg$ of the pilot.

**Solution** The free-body diagram for the pilot at the top of the loop is shown in Figure 5.14c. At this point, both the gravitational force and the force $\vec{\mathbf{n}}_{\text{top}}$ exerted by the seat on the pilot act *downward,* so the net force downward that provides the centripetal acceleration has a magnitude $n_{\text{top}} + mg$. Applying Newton's second law gives

$$\sum F_y = ma \quad \rightarrow \quad n_{\text{top}} + mg = m\frac{v^2}{r}$$

$$n_{\text{top}} = m\frac{v^2}{r} - mg = mg\left[\frac{v^2}{rg} - 1\right]$$

$$n_{\text{top}} = mg\left[\frac{(225\text{ m/s})^2}{(2.70 \times 10^3\text{ m})(9.80\text{ m/s}^2)} - 1\right]$$

$$= 0.911\,mg$$

In this case, the force exerted by the seat on the pilot is *less* than the weight by a factor of 0.911. Therefore, the pilot feels lighter at the top of the loop.

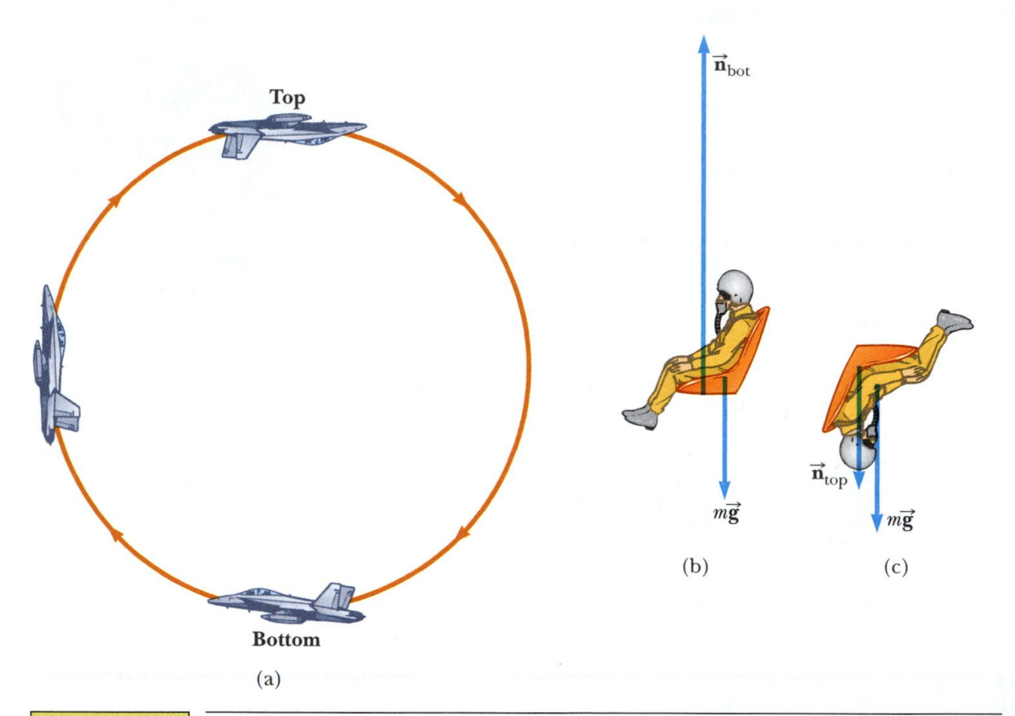


FIGURE 5.14 (Example 5.9) (a) An aircraft executes a loop-the-loop maneuver as it moves in a vertical circle at constant speed. (b) Free-body diagram for the pilot at the bottom of the loop. In this position, the pilot experiences a force from the seat that is larger than his weight. (c) Free-body diagram for the pilot at the top of the loop. Here the force from the seat could be smaller than his weight or larger, depending on the speed of the aircraft.

## 5.3 NONUNIFORM CIRCULAR MOTION

In Chapter 3, we found that if a particle moves with varying speed in a circular path, there is, in addition to the radial component of acceleration, a tangential component of magnitude $dv/dt$. Therefore, the net force acting on the particle must also have a radial and a tangential component as shown in Active Figure 5.15.

That is, because the total acceleration is $\vec{\mathbf{a}} = \vec{\mathbf{a}}_r + \vec{\mathbf{a}}_t$, the total force exerted on the particle is $\sum \vec{\mathbf{F}} = \sum \vec{\mathbf{F}}_r + \sum \vec{\mathbf{F}}_t$. The component vector $\sum \vec{\mathbf{F}}_r$ is directed toward the center of the circle and is responsible for the centripetal acceleration. The component vector $\sum \vec{\mathbf{F}}_t$ tangent to the circle is responsible for the tangential acceleration, which causes the speed of the particle to change with time.

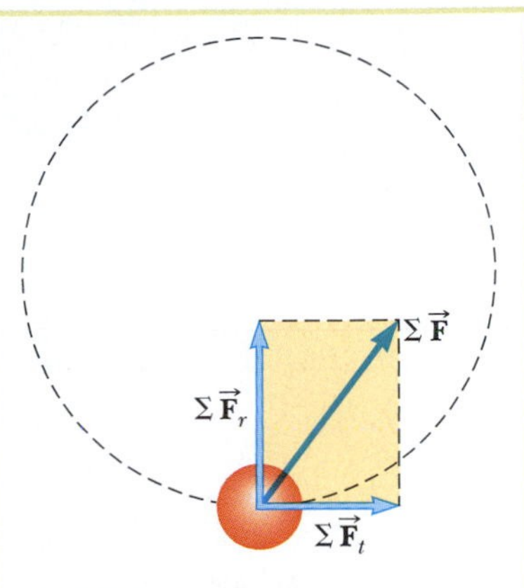


**ACTIVE FIGURE 5.15**
When the net force acting on a particle moving in a circular path has a tangential component vector $\sum \vec{\mathbf{F}}_t$, its speed changes. The total force on the particle also has a component vector $\sum \vec{\mathbf{F}}_r$ directed toward the center of the circular path. Therefore, the total force is $\sum \vec{\mathbf{F}} = \sum \vec{\mathbf{F}}_r + \sum \vec{\mathbf{F}}_t$.

**Physics Now™** Log into PhysicsNow at **www.pop4e.com** and go to Active Figure 5.15 to adjust the initial position of the particle. Compare the component forces acting on the particle to those for a child swinging on a swing set.

**QUICK QUIZ 5.5** Which of the following is *impossible* for a car moving in a circular path? Assume that the car is never at rest. **(a)** The car has tangential acceleration but no centripetal acceleration. **(b)** The car has centripetal acceleration but no tangential acceleration. **(c)** The car has both centripetal acceleration and tangential acceleration.

**QUICK QUIZ 5.6** A bead slides freely along a *horizontal,* curved wire at constant speed, as shown in Figure 5.16. **(a)** Draw the vectors representing the force exerted by the wire on the bead at points Ⓐ, Ⓑ, and Ⓒ. **(b)** Suppose the bead in Figure 5.16 speeds up with constant tangential acceleration as it moves toward the right. Draw the vectors representing the force on the bead at points Ⓐ, Ⓑ, and Ⓒ.

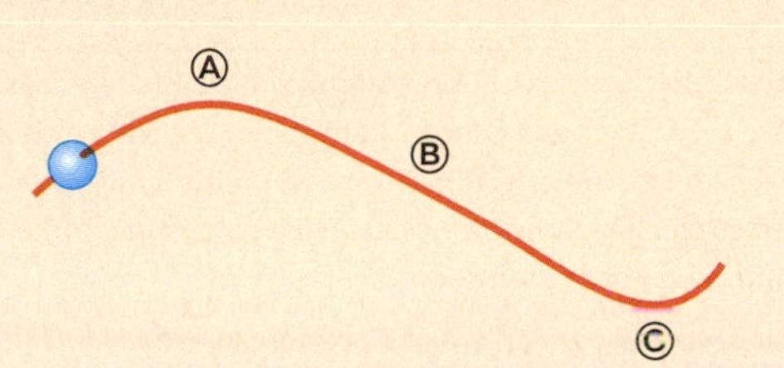


**FIGURE 5.16** (Quick Quiz 5.6) A bead slides along a curved wire.

## EXAMPLE 5.10 Follow the Rotating Ball

A small sphere of mass $m$ is attached to the end of a cord of length $R$, which rotates under the influence of the gravitational force and the force exerted by the cord in a *vertical* circle about a fixed point $O$, as in Figure 5.17a. Let us determine the tension in the cord at any instant when the speed of the sphere is $v$ and the cord makes an angle $\theta$ with the vertical.

**Solution** First, note that the speed is *not* uniform because a tangential component of acceleration arises from the gravitational force on the sphere. Although this example is similar to Example 5.9, it is not identical. From the free-body diagram in Figure 5.17a, we see that the only forces acting on the sphere are the gravitational force $m\vec{\mathbf{g}}$ and the force $\vec{\mathbf{T}}$ exerted by the cord.

We resolve $m\vec{\mathbf{g}}$ into a tangential component $mg\sin\theta$ and a radial component $mg\cos\theta$. Applying Newton's second law for the tangential direction gives

$$\sum F_t = ma_t \quad \rightarrow \quad mg\sin\theta = ma_t$$

$$a_t = g\sin\theta$$

This component causes $v$ to change in time because $a_t = dv/dt$.

Applying Newton's second law to the forces in the radial direction (for which the outward direction is positive), we find

$$\sum F_r = ma_r \quad \rightarrow \quad mg\cos\theta - T = -m\frac{v^2}{r}$$

$$T = m\left(\frac{v^2}{R} + g\cos\theta\right)$$

At the bottom of the path, where $\cos\theta = \cos 0 = 1$, we see that

$$T_{\text{bot}} = m\left(\frac{v_{\text{bot}}^2}{R} + g\right)$$

which is the *maximum* value of $T$; as the sphere passes through the bottom point, the string is under the most tension. This property is of interest to trapeze artists because their support wires must withstand this largest tension at the bottom of the swing as well as to Tarzan when he chooses a nice, strong vine on which to swing to withstand this force.

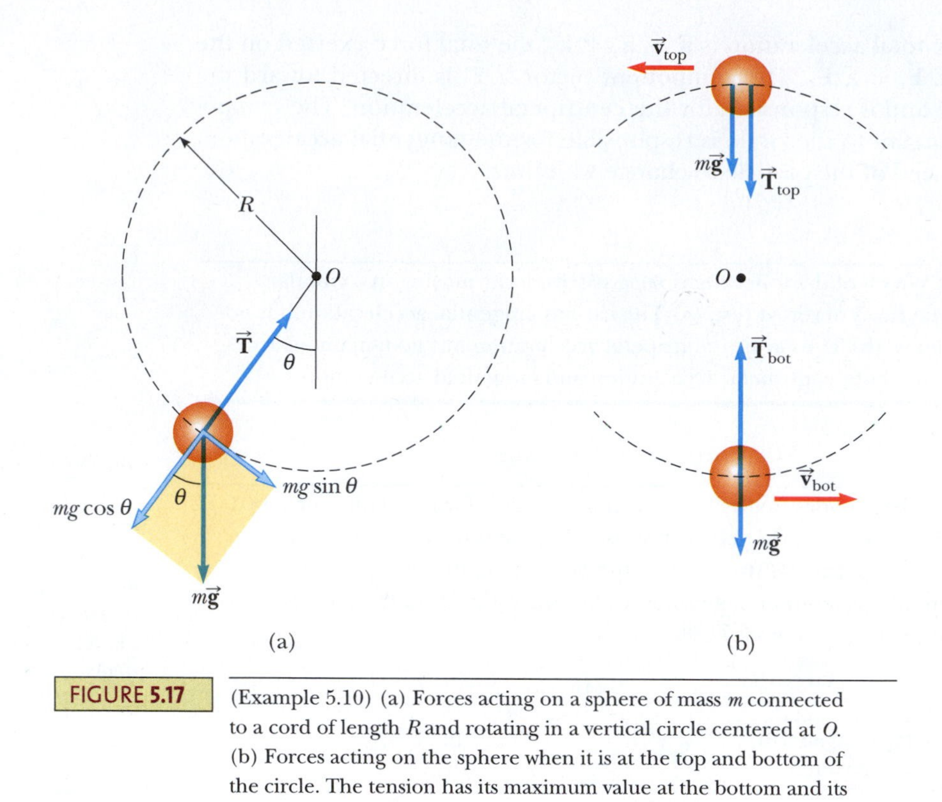


FIGURE 5.17 (Example 5.10) (a) Forces acting on a sphere of mass $m$ connected to a cord of length $R$ and rotating in a vertical circle centered at $O$. (b) Forces acting on the sphere when it is at the top and bottom of the circle. The tension has its maximum value at the bottom and its minimum value at the top.

## 5.4 MOTION IN THE PRESENCE OF VELOCITY-DEPENDENT RESISTIVE FORCES

Earlier, we described the friction force between a moving object and the surface along which it moves. So far, we have ignored any interaction between the object and the *medium* through which it moves. Let us now consider the effect of a medium such as a liquid or gas. The medium exerts a **resistive force $\vec{R}$** on the object moving through it. You feel this force if you ride in a car at high speed with your hand out the window; the force you feel pushing your hand backward is the resistive force of the air rushing past the car. The magnitude of this force depends on the relative speed between the object and the medium, and the direction of $\vec{R}$ on the object is always opposite the direction of the object's motion relative to the medium. Some examples are the air resistance associated with moving vehicles (sometimes called air drag), the force of the wind on the sails of a sailboat, and the viscous forces that act on objects sinking through a liquid.

Generally, the magnitude of the resistive force increases with increasing speed. The resistive force can have a complicated speed dependence. In the following discussions, we consider two simplification models that allow us to analyze these situations. The first model assumes that the resistive force is proportional to the velocity, which is approximately the case for objects that fall through a liquid with low speed and for very small objects, such as dust particles, that move through air. The second model treats situations for which we assume that the magnitude of the resistive force is proportional to the square of the speed of the object. Large objects, such as a sky diver moving through air in free-fall, experience such a force.

### Model 1: Resistive Force Proportional to Object Velocity

At low speeds, the resistive force acting on an object that is moving through a viscous medium is effectively modeled as being proportional to the object's velocity.

The mathematical representation of the resistive force can be expressed as

$$\vec{\mathbf{R}} = -b\vec{\mathbf{v}} \qquad [5.4]$$

where $\vec{\mathbf{v}}$ is the velocity of the object relative to the medium and $b$ is a constant that depends on the properties of the medium and on the shape and dimensions of the object. The negative sign represents that the resistive force is opposite the velocity of the object relative to the medium.

Consider a sphere of mass $m$ released from rest in a liquid, as in Active Figure 5.18a. We assume that the only forces acting on the sphere are the resistive force $\vec{\mathbf{R}}$ and the weight $m\vec{\mathbf{g}}$, and we describe its motion using Newton's second law.[2] Considering the vertical motion and choosing the downward direction to be positive, we have

$$\sum F_y = ma_y \quad \rightarrow \quad mg - bv = m\frac{dv}{dt}$$

Dividing this equation by the mass $m$ gives

$$\frac{dv}{dt} = g - \frac{b}{m}v \qquad [5.5]$$

Equation 5.5 is called a *differential equation;* it includes both the speed $v$ and the derivative of the speed. The methods of solving such an equation may not be familiar to you as yet. Note, however, that if we define $t = 0$ when $v = 0$, the resistive force is zero at this time and the acceleration $dv/dt$ is simply $g$. As $t$ increases, the speed increases, the resistive force increases, and the acceleration decreases. Thus, this problem is one in which neither the velocity nor the acceleration of the particle is constant.

The acceleration becomes zero when the increasing resistive force eventually balances the weight. At this point, the object reaches its **terminal speed** $v_T$ and from then on it continues to move with zero acceleration. After this point, the motion is that of a particle under constant velocity. The terminal speed can be obtained from Equation 5.5 by setting $a = dv/dt = 0$, which gives

$$mg - bv_T = 0 \quad \rightarrow \quad v_T = \frac{mg}{b}$$

The expression for $v$ that satisfies Equation 5.5 with $v = 0$ at $t = 0$ is

$$v = \frac{mg}{b}(1 - e^{-bt/m}) = v_T(1 - e^{-t/\tau}) \qquad [5.6]$$

where $v_T = mg/b$, $\tau = m/b$, and $e = 2.718\,28$ is the base of the natural logarithm. This expression for $v$ can be verified by substituting it back into Equation 5.5. (Try it!) This function is plotted in Active Figure 5.18b.

The mathematical representation of the motion (Eq. 5.6) indicates that the terminal speed is never reached because the exponential function is never exactly equal to zero. For all practical purposes, however, when the exponential function is very small at large values of $t$, the speed of the particle can be approximated as being constant and equal to the terminal speed.

We cannot compare different objects by means of the time interval required to reach terminal speed because, as we have just discussed, this time interval is infinite for all objects! We need some means to compare these exponential behaviors for different objects. We do so with a parameter called the **time constant.** The time constant $\tau = m/b$ that appears in Equation 5.6 is the time interval required for the factor in parentheses in Equation 5.6 to become equal to $1 - e^{-1} = 0.632$. Therefore, the time constant represents the time interval required for the object to reach 63.2% of its terminal speed (Active Fig. 5.18b).

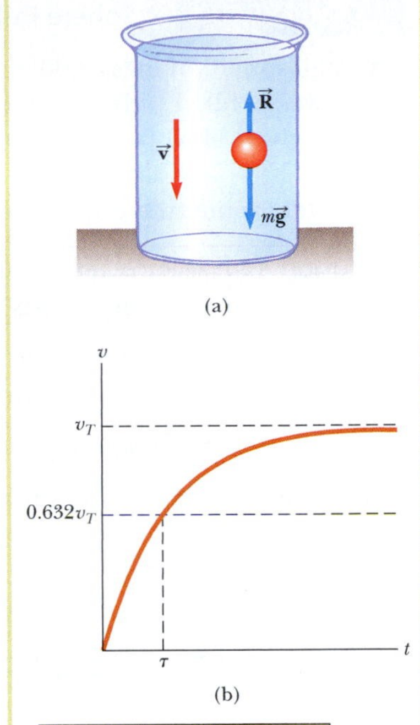


**ACTIVE FIGURE 5.18**
(a) A small sphere falling through a viscous fluid. (b) The speed–time graph for an object falling through a viscous medium. The object approaches a terminal speed $v_T$, and the time constant $\tau$ is the time interval required to reach $0.632v_T$.

**PhysicsNow™** Log into PhysicsNow at www.pop4e.com and go to Active Figure 5.18 to vary the size and mass of the sphere and the viscosity (resistance to flow) of the surrounding medium. Observe the effects on the sphere's motion and its speed–time graph.

[2] A *buoyant* force also acts on any object surrounded by a fluid. This force is constant and equal to the weight of the displaced fluid, as will be discussed in Chapter 15. The effect of this force can be modeled by changing the apparent weight of the sphere by a constant factor, so we can ignore it here.

## EXAMPLE 5.11 A Sphere Falling in Oil

A small sphere of mass 2.00 g is released from rest in a large vessel filled with oil. The sphere approaches a terminal speed of 5.00 cm/s.

A Determine the time constant $\tau$.

**Solution** Because the terminal speed is given by $v_T = mg/b$, the coefficient $b$ is

$$b = \frac{mg}{v_T} = \frac{(2.00 \times 10^{-3}\ \text{kg})(9.80\ \text{m/s}^2)}{5.00 \times 10^{-2}\ \text{m/s}}$$
$$= 0.392\ \text{N}\cdot\text{s/m}$$

Therefore, the time constant $\tau$ is

$$\tau = \frac{m}{b} = \frac{2.00 \times 10^{-3}\ \text{kg}}{0.392\ \text{N}\cdot\text{s/m}} = 5.1 \times 10^{-3}\ \text{s}$$

B Determine the time interval required for the sphere to reach 90.0% of its terminal speed.

**Solution** The speed of the sphere as a function of time is given by Equation 5.6. To find the time $t$ at which the sphere is traveling at a speed of $0.900v_T$, we set $v = 0.900v_T$, substitute into Equation 5.6, and solve for $t$:

$$0.900v_T = v_T(1 - e^{-t/\tau})$$
$$1 - e^{-t/\tau} = 0.900$$
$$e^{-t/\tau} = 0.100$$
$$-\frac{t}{\tau} = \ln 0.100 = -2.30$$
$$t = 2.30\tau = 2.30(5.10 \times 10^{-3}\ \text{s})$$
$$= 11.7 \times 10^{-3}\ \text{s} = 11.7\ \text{ms}$$

## Model 2: Resistive Force Proportional to Object Speed Squared

For large objects moving at high speeds through air, such as airplanes, sky divers, and baseballs, the magnitude of the resistive force is modeled as being proportional to the square of the speed:

$$R = \tfrac{1}{2}D\rho Av^2 \qquad [5.7]$$

where $\rho$ is the density of air, $A$ is the cross-sectional area of the moving object measured in a plane perpendicular to its velocity, and $D$ is a dimensionless empirical quantity called the *drag coefficient*. The drag coefficient has a value of about 0.5 for spherical objects moving through air but can be as high as 2 for irregularly shaped objects.

Consider an airplane in flight that experiences such a resistive force. Equation 5.7 shows that the force is proportional to the density of air and hence decreases with decreasing air density. Because air density decreases with increasing altitude, the resistive force on a jet airplane flying at a given speed will decrease with increasing altitude. Therefore, airplanes tend to fly at very high altitudes to take advantage of this reduced resistive force, which allows them to fly faster for a given engine thrust. Of course, this higher speed *increases* the resistive force, in proportion to the square of the speed, so a balance is struck between fuel economy and higher speed.

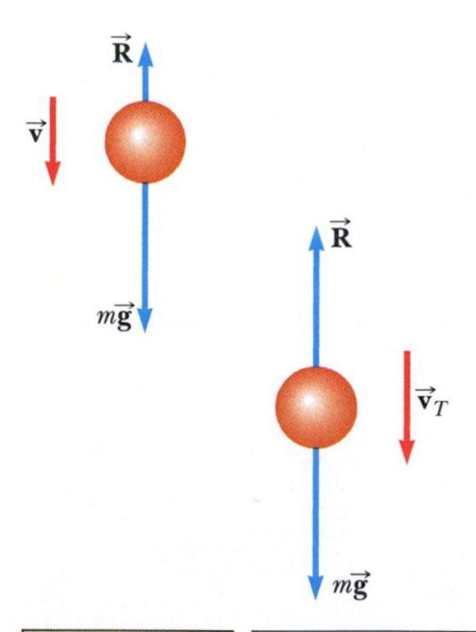


FIGURE 5.19 An object falling through air experiences a resistive drag force $\vec{R}$ and a gravitational force $\vec{F}_g = m\vec{g}$. The object reaches terminal speed (on the right) when the net force acting on it is zero, that is, when $\vec{R} = -\vec{F}_g$, or $R = mg$. Before that occurs, the acceleration varies with speed according to Equation 5.9.

Now let us analyze the motion of a falling object subject to an upward air resistive force whose magnitude is given by Equation 5.7. Suppose an object of mass $m$ is released from rest, as in Figure 5.19, from the position $y = 0$. The object experiences two external forces: the downward gravitational force $m\vec{g}$ and the upward resistive force $\vec{R}$. Hence, using Newton's second law,

$$\sum F = ma \quad \rightarrow \quad mg - \tfrac{1}{2}D\rho Av^2 = ma \qquad [5.8]$$

Solving for $a$, we find that the object has a downward acceleration of magnitude

$$a = g - \left(\frac{D\rho A}{2m}\right)v^2 \qquad [5.9]$$

Because $a = dv/dt$, Equation 5.9 is another differential equation that provides us with the speed as a function of time.

**TABLE 5.2** Terminal Speeds for Various Objects Falling Through Air

| Object | Mass (kg) | Cross-sectional Area ($m^2$) | $v_T$ (m/s)[a] |
|---|---|---|---|
| Sky diver | 75 | 0.70 | 60 |
| Baseball (radius 3.7 cm) | 0.145 | $4.2 \times 10^{-3}$ | 33 |
| Golf ball (radius 2.1 cm) | 0.046 | $1.4 \times 10^{-3}$ | 32 |
| Hailstone (radius 0.50 cm) | $4.8 \times 10^{-4}$ | $7.9 \times 10^{-5}$ | 14 |
| Raindrop (radius 0.20 cm) | $3.4 \times 10^{-5}$ | $1.3 \times 10^{-5}$ | 9.0 |

[a]The drag coefficient $D$ is assumed to be 0.5 in each case.

Again, we can calculate the terminal speed $v_T$ because when the gravitational force is balanced by the resistive force, the net force is zero and therefore the acceleration is zero. Setting $a = 0$ in Equation 5.9 gives

$$g - \left(\frac{D\rho A}{2m}\right)v_T^2 = 0$$

$$v_T = \sqrt{\frac{2mg}{D\rho A}} \qquad [5.10]$$

Table 5.2 lists the terminal speeds for several objects falling through air, all computed on the assumption that the drag coefficient is 0.5.

**QUICK QUIZ 5.7** Consider a sky surfer falling through air, as in Figure 5.20, before reaching her terminal speed. As the speed of the sky surfer increases, the magnitude of her acceleration (**a**) remains constant, (**b**) decreases until it reaches a constant nonzero value, or (**c**) decreases until it reaches zero.

(Jump Run Productions/Image Bank)

**FIGURE 5.20** (Quick Quiz 5.7) A sky surfer takes advantage of the upward force of the air on her board.

## 5.5 THE FUNDAMENTAL FORCES OF NATURE

We have described a variety of forces experienced in our everyday activities, such as the gravitational force acting on all objects at or near the Earth's surface and the force of friction as one surface slides over another. Newton's second law tells us how to relate the forces to the object's or particle's acceleration.

In addition to these familiar macroscopic forces in nature, forces also act in the atomic and subatomic world. For example, atomic forces within the atom are responsible for holding its constituents together and nuclear forces act on different parts of the nucleus to keep its parts from separating.

Until recently, physicists believed that there were four fundamental forces in nature: the gravitational force, the electromagnetic force, the strong force, and the weak force. We shall discuss these forces individually and then consider the current view of fundamental forces.

### The Gravitational Force

The **gravitational force** is the mutual force of attraction between any two objects in the Universe. It is interesting and rather curious that although the gravitational force can be very strong between macroscopic objects, it is inherently the weakest of all the fundamental forces. For example, the gravitational force between the electron and proton in the hydrogen atom has a magnitude on the order of $10^{-47}$ N, whereas the electromagnetic force between these same two particles is on the order of $10^{-7}$ N.

■ Newton's law of universal gravitation

In addition to his contributions to the understanding of motion, Newton studied gravity extensively. **Newton's law of universal gravitation** states that every particle in

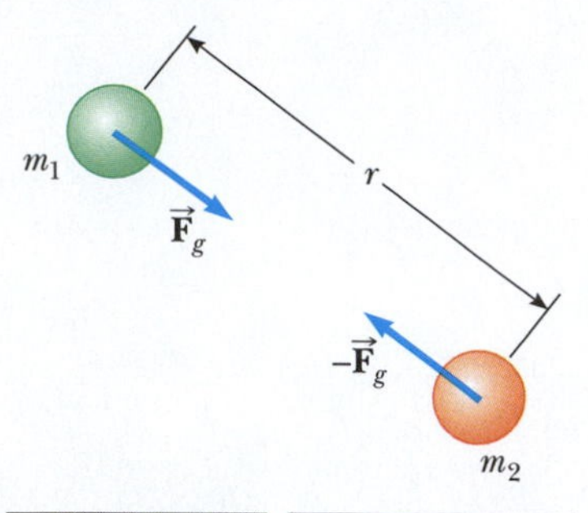


**FIGURE 5.21** Two particles with masses $m_1$ and $m_2$ attract each other with a force of magnitude $Gm_1m_2/r^2$.

the Universe attracts every other particle with a force that is directly proportional to the product of the masses of the particles and inversely proportional to the square of the distance between them. If the particles have masses $m_1$ and $m_2$ and are separated by a distance $r$, as in Figure 5.21, the magnitude of the gravitational force is

$$F_g = G\frac{m_1 m_2}{r^2} \qquad [5.11]$$

where $G = 6.67 \times 10^{-11}\ \mathrm{N \cdot m^2/kg^2}$ is the **universal gravitational constant.** More detail on the gravitational force will be provided in Chapter 11.

## The Electromagnetic Force

The **electromagnetic force** is the force that binds atoms and molecules in compounds to form ordinary matter. It is much stronger than the gravitational force. The force that causes a rubbed comb to attract bits of paper and the force that a magnet exerts on an iron nail are electromagnetic forces. Essentially all forces at work in our macroscopic world, apart from the gravitational force, are manifestations of the electromagnetic force. For example, friction forces, contact forces, tension forces, and forces in elongated springs are consequences of electromagnetic forces between charged particles in proximity.

The electromagnetic force involves two types of particles: those with positive charge and those with negative charge. (More information on these two types of charge is provided in Chapter 19.) Unlike the gravitational force, which is always an attractive interaction, the electromagnetic force can be either attractive or repulsive, depending on the charges on the particles.

■ Coulomb's law

**Coulomb's law** expresses the magnitude of the *electrostatic force*[3] $F_e$ between two charged particles separated by a distance $r$:

$$F_e = k_e\frac{q_1 q_2}{r^2} \qquad [5.12]$$

where $q_1$ and $q_2$ are the charges on the two particles, measured in units called *coulombs* (C), and $k_e$ $(= 8.99 \times 10^9\ \mathrm{N \cdot m^2/C^2})$ is the **Coulomb constant.** Note that the electrostatic force has the same mathematical form as Newton's law of universal gravitation (see Eq. 5.11), with charge playing the mathematical role of mass and the Coulomb constant being used in place of the universal gravitational constant. The electrostatic force is attractive if the two charges have opposite signs and is repulsive if the two charges have the same sign, as indicated in Figure 5.22.

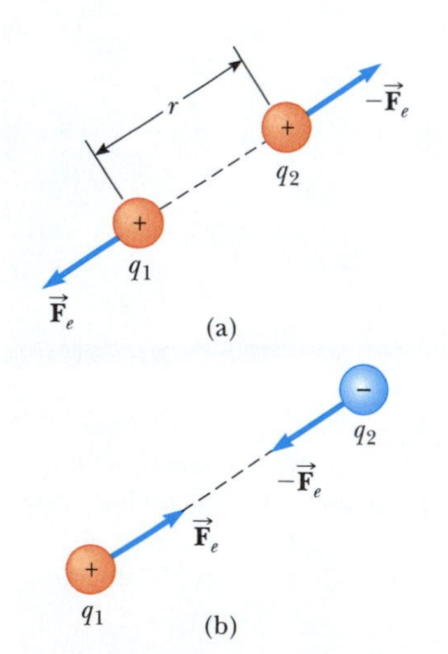


**FIGURE 5.22** Two point charges separated by a distance $r$ exert an electrostatic force on each other given by Coulomb's law. (a) When the charges are of the same sign, the charges repel each other. (b) When the charges are of opposite sign, the charges attract each other.

The smallest amount of isolated charge found in nature (so far) is the charge on an electron or proton. This fundamental unit of charge is given the symbol $e$ and has the magnitude $e = 1.60 \times 10^{-19}$ C. An electron has charge $-e$, whereas a proton has charge $+e$. Theories developed in the latter half of the 20th century propose that protons and neutrons are made up of smaller particles called **quarks,** which have charges of either $\frac{2}{3}e$ or $-\frac{1}{3}e$ (discussed further in Chapter 31). Although experimental evidence has been found for such particles inside nuclear matter, free quarks have never been detected.

## The Strong Force

An atom, as we currently model it, consists of an extremely dense positively charged nucleus surrounded by a cloud of negatively charged electrons, with the electrons attracted to the nucleus by the electric force. All nuclei except those of hydrogen are combinations of positively charged protons and neutral neutrons (collectively

[3]The electrostatic force is the electromagnetic force between two electric charges that are at rest. If the charges are moving, magnetic forces are also present; these forces will be studied in Chapter 22.

called nucleons), yet why does the repulsive electrostatic force between the protons not cause nuclei to break apart? Clearly, there must be an attractive force that counteracts the strong electrostatic repulsive force and is responsible for the stability of nuclei. This force that binds the nucleons to form a nucleus is called the **nuclear force.** It is one manifestation of the **strong force,** which is the force between particles formed from quarks, which we will discuss in Chapter 31. Unlike the gravitational and electromagnetic forces, which depend on distance in an inverse-square fashion, the nuclear force is extremely short range; its strength decreases very rapidly outside the nucleus and is negligible for separations greater than approximately $10^{-14}$ m.

## The Weak Force

The **weak force** is a short-range force that tends to produce instability in certain nuclei. It was first observed in naturally occurring radioactive substances and was later found to play a key role in most radioactive decay reactions. The weak force is about $10^{36}$ times stronger than the gravitational force and about $10^{3}$ times weaker than the electromagnetic force.

## The Current View of Fundamental Forces

For years, physicists have searched for a simplification scheme that would reduce the number of fundamental forces needed to describe physical phenomena. In 1967, physicists predicted that the electromagnetic force and the weak force, originally thought to be independent of each other and both fundamental, are in fact manifestations of one force, now called the **electroweak** force. This prediction was confirmed experimentally in 1984. We shall discuss it more fully in Chapter 31.

We also now know that protons and neutrons are not fundamental particles; current models of protons and neutrons theorize that they are composed of simpler particles called quarks, as mentioned previously. The quark model has led to a modification of our understanding of the nuclear force. Scientists now define the strong force as the force that binds the quarks to one another in a nucleon (proton or neutron). This force is also referred to as a **color force,** in reference to a property of quarks called "color," which we shall investigate in Chapter 31. The previously defined nuclear force, the force that acts between nucleons, is now interpreted as a secondary effect of the strong force between the quarks.

Scientists believe that the fundamental forces of nature are closely related to the origin of the Universe. The Big Bang theory states that the Universe began with a cataclysmic explosion about 14 billion years ago. According to this theory, the first moments after the Big Bang saw such extremes of energy that all the fundamental forces were unified into one force. Physicists are continuing their search for connections among the known fundamental forces, connections that could eventually prove that the forces are all merely different forms of a single superforce. This fascinating search continues to be at the forefront of physics.

# 5.6 DRAG COEFFICIENTS OF AUTOMOBILES

CONTEXT CONNECTION

In the Context Connection of Chapter 4, we ignored air resistance and assumed that the driving force on the tires was the only force on the vehicle in the horizontal direction. Given our understanding of velocity-dependent forces from Section 5.4, we should understand now that air resistance could be a significant factor in the design of an automobile.

Table 5.3 shows the drag coefficients for the vehicles that we have investigated in previous chapters. Notice that the coefficients for the performance and traditional vehicles vary from 0.30 to 0.43, with the average coefficient in the two portions of the table almost the same. A look at the lower part of the table shows that this parameter

**TABLE 5.3 Drag Coefficients of Various Vehicles**

| Automobile | Model Year | Drag Coefficient |
|---|---|---|
| *Performance vehicles* | | |
| Aston Martin DB7 Vantage | 2001 | 0.31 |
| BMW Z8 | 2001 | 0.43 |
| Chevrolet Corvette | 2000 | 0.29 |
| Dodge Viper GTS-R | 1998 | 0.40 |
| Ferrari F50 | 1997 | 0.37 |
| Ferrari 360 Spider F1 | 2000 | 0.33 |
| Lamborghini Diablo GT | 2000 | 0.31 |
| Porsche 911 GT2 | 2002 | 0.34 |
| *Traditional vehicles* | | |
| Acura Integra GS | 2000 | 0.34 |
| BMW Mini Cooper S | 2003 | 0.35 |
| Cadillac Escalade (SUV) | 2002 | 0.42 |
| Dodge Stratus | 2002 | 0.34 |
| Lexus ES300 | 1997 | 0.32 |
| Mitsubishi Eclipse GT | 2000 | 0.30 |
| Nissan Maxima | 2000 | 0.31 |
| Pontiac Grand Prix | 2003 | 0.31 |
| Toyota Sienna (SUV) | 2004 | 0.31 |
| Volkswagen Beetle | 1999 | 0.36 |
| *Alternative vehicles* | | |
| GM EV1 | 1998 | 0.19 |
| Toyota Prius | 2004 | 0.26 |
| Honda Insight | 2001 | 0.25 |

is where the alternative vehicles shine. All three vehicles have drag coefficients lower than all others in the table, and the GM EV1 has a remarkable coefficient of just 0.19.

Designers of alternative-fuel vehicles try to squeeze every last mile of travel out of the energy that is stored in the vehicle in the form of fuel or an electric battery. A significant method of doing so is to reduce the force of air resistance so that the net force driving the car forward is as large as possible.

A number of techniques can be used to reduce the drag coefficient. Two factors that help are a small frontal area and smooth curves from the front of the vehicle to the back. For example, the Chevrolet Corvette shown in Figure 5.23a exhibits a streamlined shape that contributes to its low drag coefficient. As a comparison, consider a large, boxy vehicle, such as the Hummer H2 in Figure 5.23b. The drag coefficient for this vehicle is 0.57. Another factor includes elimination or minimization of

(a)

(b)

(a, Courtesy of GM; b, Courtesy of GM–Hummer)

**FIGURE 5.23** (a) The Chevrolet Corvette has a streamlined shape that contributes to its low drag coefficient of 0.29. (b) The Hummer H2 is not streamlined like the Corvette and consequently has a much higher drag coefficient of 0.57.

as many irregularities in the surfaces as possible, including door handles that project from the body, windshield wipers, wheel wells, and rough surfaces on headlamps and grills. An important consideration is the underside of the carriage. As air rushes beneath the car, there are many irregular surfaces associated with brakes, drive trains, suspension components, and so on. The drag coefficient can be made lower by assuring that the overall surface of the car's undercarriage is as smooth as possible. ■

## SUMMARY

PhysicsNow™ Take a practice test by logging into PhysicsNow at www.pop4e.com and clicking on the Pre-Test link for this chapter.

Forces of friction are complicated, but we design a simplification model for friction that allows us to analyze motion that includes the effects of friction. The **maximum force of static friction** $f_{s,\max}$ between two surfaces is proportional to the normal force between the surfaces. This maximum force occurs when the surfaces are on the verge of slipping. In general, $f_s \leq \mu_s n$, where $\mu_s$ is the **coefficient of static friction** and $n$ is the magnitude of the normal force. When an object slides over a rough surface, the **force of kinetic friction** $\vec{\mathbf{f}}_k$ is opposite the direction of the velocity of the object relative to the surface and its magnitude is proportional to the magnitude of the normal force on the object. The magnitude is given by $f_k = \mu_k n$, where $\mu_k$ is the **coefficient of kinetic friction.** Usually, $\mu_k < \mu_s$.

Newton's second law, applied to a particle moving in uniform circular motion, states that the net force in the inward radial direction must equal the product of the mass and the centripetal acceleration:

$$\sum F = ma_c = m\frac{v^2}{r} \qquad [5.3]$$

An object moving through a liquid or gas experiences a **resistive force** that is velocity dependent. This resistive force, which is opposite the velocity of the object relative to the medium, generally increases with speed. The force depends on the object's shape and on the properties of the medium through which the object is moving. In the limiting case for a falling object, when the resistive force balances the weight $(a = 0)$, the object reaches its **terminal speed.**

The fundamental forces existing in nature can be expressed as the following four: the gravitational force, the electromagnetic force, the strong force, and the weak force.

## QUESTIONS

☐ = answer available in the *Student Solutions Manual and Study Guide.*

1. Draw a free-body diagram for each of the following objects: (a) a projectile in motion in the presence of air resistance, (b) a rocket leaving the launch pad with its engines operating, (c) an athlete running along a horizontal track.
2. What force causes (a) an automobile, (b) a propeller-driven airplane, and (c) a rowboat to move?
3. Identify the action-reaction pairs in the following situations: a man takes a step, a snowball hits a girl in the back, a baseball player catches a ball, a gust of wind strikes a window.
4. In a contest of National Football League behemoths, teams from the Rams and the 49ers engage in a tug-of-war, pulling in opposite directions on a strong rope. The Rams exert a force of 9 200 N and they are winning, making the center of the light rope move steadily toward themselves. Is it possible to know the tension in the rope from the information stated? Is it larger or smaller than 9 200 N? How hard are the 49ers pulling on the rope? Would it change your answer if the 49ers were winning or if the contest were even? The stronger team wins by exerting a larger force, on what? Explain your answers.
5. Suppose you are driving a classic car. Why should you avoid slamming on your brakes when you want to stop in the shortest possible distance? (Many cars have antilock brakes that avoid this problem.)
6. A book is given a brief push to make it slide up a rough incline. It comes to a stop and slides back down to the starting point. Does it take the same time interval to go up as to come down? What if the incline is frictionless?
7. Describe a few examples in which the force of friction exerted on an object is in the direction of motion of the object.
8. An object executes circular motion with constant speed whenever a net force of constant magnitude acts perpendicular to the velocity. What happens to the speed if the force is not perpendicular to the velocity?
9. What causes a rotary lawn sprinkler to turn?
10. It has been suggested that rotating cylinders about 10 miles in length and 5 miles in diameter be placed in space and used as colonies. The purpose of the rotation is to simulate gravity for the inhabitants. Explain this concept for producing an effective imitation of gravity.
11. A pail of water can be whirled in a vertical path such that none is spilled. Why does the water stay in the pail, even when the pail is upside down above your head?
12. Why does a pilot tend to black out when pulling out of a steep dive?
13. If someone told you that astronauts are weightless in orbit because they are beyond the pull of gravity, would you accept the statement? Explain.

14. A falling sky diver reaches terminal speed with her parachute closed. After the parachute is opened, what parameters change to decrease this terminal speed?

15. On long journeys, jet aircraft usually fly at high altitudes of about 30 000 ft. What is the main advantage from an economic viewpoint of flying at these altitudes?

16. Consider a small raindrop and a large raindrop falling through the atmosphere. Compare their terminal speeds. What are their accelerations when they reach terminal speed?

17. "If the current position and velocity of every particle in the Universe were known, together with the laws describing the forces that particles exert on one another, then the whole future of the Universe could be calculated. The future is determinate and preordained. Free will is an illusion." Do you agree with this thesis? Argue for or against it.

# PROBLEMS

1, 2, 3 = straightforward, intermediate, challenging
☐ = full solution available in the *Student Solutions Manual and Study Guide*
PhysicsNow™ = coached problem with hints available at www.pop4e.com
= computer useful in solving problem
= paired numerical and symbolic problems
= biomedical application

## Section 5.1 ■ Forces of Friction

1. A 25.0-kg block is initially at rest on a horizontal surface. A horizontal force of 75.0 N is required to set the block in motion. After it is in motion, a horizontal force of 60.0 N is required to keep the block moving with constant speed. Find the coefficients of static and kinetic friction from this information.

2. A car is traveling at 50.0 mi/h on a horizontal highway. (a) If the coefficient of static friction between road and tires on a rainy day is 0.100, what is the minimum distance in which the car will stop? (b) What is the stopping distance when the surface is dry and $\mu_s = 0.600$?

3. Before 1960, it was believed that the maximum attainable coefficient of static friction for an automobile tire was less than 1. Then around 1962, three companies independently developed racing tires with coefficients of 1.6. Since then, tires have improved, as illustrated in this problem. According to the 1990 *Guinness Book of Records,* the shortest time interval in which a piston-engine car initially at rest has covered a distance of one-quarter mile is 4.96 s. This record was set by Shirley Muldowney in September 1989. (a) Assume that, as shown in Figure P5.3, the rear wheels lifted the front wheels off the pavement. What minimum value of $\mu_s$ is necessary to achieve the record time? (b) Suppose Muldowney were able to double her engine power, keeping other things equal. How would this change affect the elapsed time?

(Mike Powell/Getty Images)

FIGURE **P5.3**

4. The person in Figure P5.4 weighs 170 lb. As seen from the front, each light crutch makes an angle of 22.0° with the vertical. Half of the person's weight is supported by the crutches. The other half is supported by the vertical forces of the ground on the person's feet. Assuming that the person is moving with constant velocity and the force exerted by the ground on the crutches acts along the crutches, determine (a) the smallest possible coefficient of friction between crutches and ground and (b) the magnitude of the compression force in each crutch.

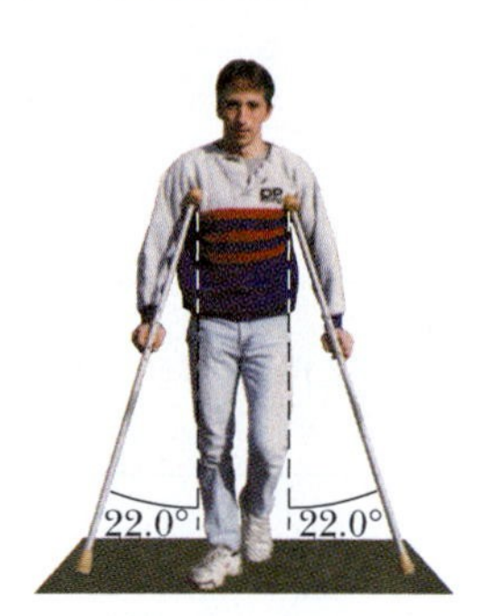


FIGURE **P5.4**

5. To meet a U.S. Postal Service requirement, footwear must have a coefficient of static friction of 0.5 or more on a specified tile surface. A typical athletic shoe has a coefficient of 0.800. In an emergency, what is the minimum time interval in which a person starting from rest can move 3.00 m on a tile surface if she is wearing (a) footwear meeting the Postal Service minimum and (b) a typical athletic shoe?

6. Consider a large truck carrying a heavy load, such as steel beams. A significant hazard for the driver is that the load may slide forward, crushing the cab, if the truck stops suddenly in an accident or even in braking. Assume, for example, that a 10 000-kg load sits on the flatbed of a 20 000-kg truck moving at 12.0 m/s. Assume that the load is not tied down to the truck, but has a coefficient of friction of 0.500 with the flatbed of the truck. (a) Calculate the minimum stopping distance for which the load will not slide forward relative to the truck. (b) Is any piece of data unnecessary for the solution?

7. To determine the coefficients of friction between rubber and various surfaces, a student uses a rubber eraser and an

incline. In one experiment, the eraser begins to slip down the incline when the angle of inclination is 36.0° and then moves down the incline with constant speed when the angle is reduced to 30.0°. From these data, determine the coefficients of static and kinetic friction for this experiment.

8. A woman at an airport is towing her 20.0-kg suitcase at constant speed by pulling on a strap at an angle $\theta$ above the horizontal (Fig. P5.8). She pulls on the strap with a 35.0-N force, and the friction force on the suitcase is 20.0 N. Draw a free-body diagram of the suitcase. (a) What angle does the strap make with the horizontal? (b) What normal force does the ground exert on the suitcase?

FIGURE P5.8

9. Physics Now™ A 3.00-kg block starts from rest at the top of a 30.0° incline and slides a distance of 2.00 m down the incline in 1.50 s. Find (a) the magnitude of the acceleration of the block, (b) the coefficient of kinetic friction between block and plane, (c) the friction force acting on the block, and (d) the speed of the block after it has slid 2.00 m.

10. A 9.00-kg hanging block is connected by a string over a pulley to a 5.00-kg block that is sliding on a flat table (Fig. P5.10). The string is light and does not stretch; the pulley is light and turns without friction. The coefficient of kinetic friction between the sliding block and the table is 0.200. Find the tension in the string.

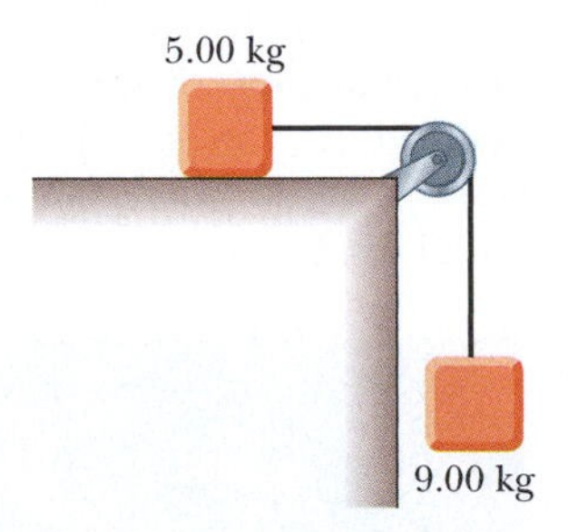


FIGURE P5.10

11. Two blocks connected by a rope of negligible mass are being dragged by a horizontal force $\vec{\mathbf{F}}$ (Fig. P5.11). Suppose $F = 68.0$ N, $m_1 = 12.0$ kg, $m_2 = 18.0$ kg, and the coefficient of kinetic friction between each block and the surface is 0.100. (a) Draw a free-body diagram for each block. (b) Determine the tension $T$ and the magnitude of the acceleration of the system.

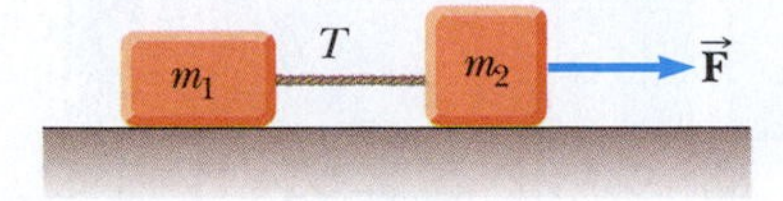


FIGURE P5.11

12. Three objects are connected on the table as shown in Figure P5.12. The table is rough and has a coefficient of kinetic friction of 0.350. The objects have masses 4.00 kg, 1.00 kg, and 2.00 kg, as shown, and the pulleys are frictionless. Draw a free-body diagram for each object. (a) Determine the acceleration of each object and their directions. (b) Determine the tensions in the two cords.

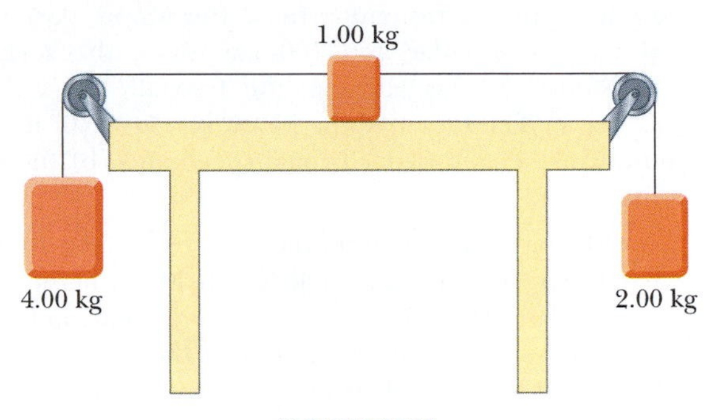


FIGURE P5.12

13. A block of mass 3.00 kg is pushed up against a wall by a force $\vec{\mathbf{P}}$ that makes a 50.0° angle with the horizontal as shown in Figure P5.13. The coefficient of static friction between the block and the wall is 0.250. Determine the possible values for the magnitude of $\vec{\mathbf{P}}$ that allow the block to remain stationary.

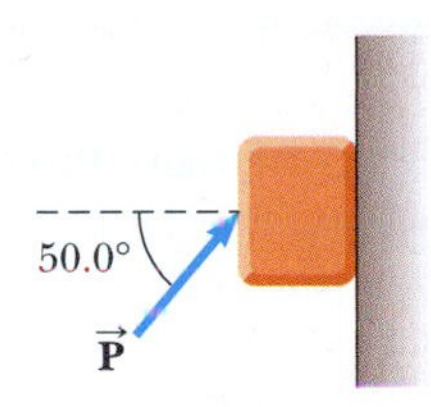


FIGURE P5.13

14. **Review problem.** One side of the roof of a building slopes up at 37.0°. A student throws a Frisbee onto the roof. It strikes with a speed of 15.0 m/s and does not bounce, but instead slides straight up the incline. The coefficient of kinetic friction between the plastic Frisbee and the roof is 0.400. The Frisbee slides 10.0 m up the roof to its peak, where it goes into free-fall, following a parabolic trajectory with negligible air resistance. Determine the maximum height the Frisbee reaches above the point where it struck the roof.

## Section 5.2 ▪ Newton's Second Law Applied to a Particle in Uniform Circular Motion

15. A light string can support a stationary hanging load of 25.0 kg before breaking. A 3.00-kg object attached to the string rotates on a horizontal, frictionless table in a circle of radius 0.800 m, and the other end of the string is held fixed. What range of speeds can the object have before the string breaks?

16. In the Bohr model of the hydrogen atom, the speed of the electron is approximately $2.20 \times 10^6$ m/s. Find (a) the force acting on the electron as it revolves in a circular orbit of radius $0.530 \times 10^{-10}$ m and (b) the centripetal acceleration of the electron.

17. A crate of eggs is located in the middle of the flatbed of a pickup truck as the truck negotiates an unbanked curve in the road. The curve may be regarded as an arc of a circle of radius 35.0 m. If the coefficient of static friction between crate and truck is 0.600, how fast can the truck be moving without the crate sliding?

18. Whenever two *Apollo* astronauts were on the surface of the Moon, a third astronaut orbited the Moon. Assume the orbit to be circular and 100 km above the surface of the Moon. At this altitude, the free-fall acceleration is 1.52 m/s$^2$. The radius of the Moon is $1.70 \times 10^6$ m. Determine (a) the astronaut's orbital speed and (b) the period of the orbit.

19. Consider a conical pendulum with an 80.0-kg bob on a 10.0-m wire making an angle $\theta = 5.00°$ with the vertical (Fig. P5.19). Determine (a) the horizontal and vertical components of the force exerted by the wire on the pendulum and (b) the radial acceleration of the bob.

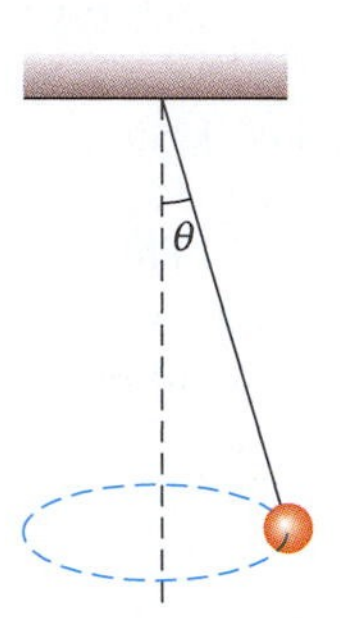


FIGURE **P5.19**

20. A 4.00-kg object is attached to a vertical rod by two strings as shown in Figure P5.20. The object rotates in a horizontal circle at constant speed 6.00 m/s. Find the tension in (a) the upper string and (b) the lower string.

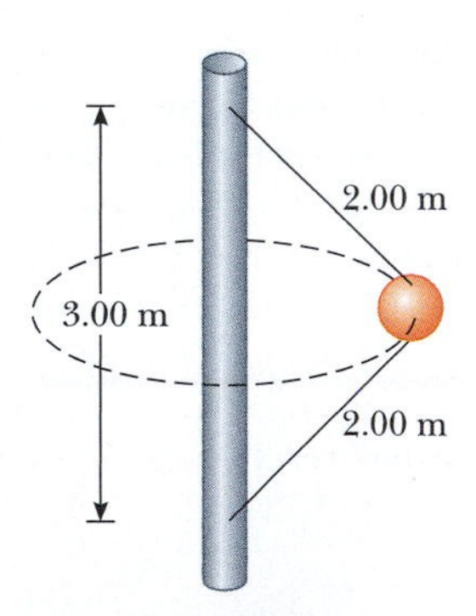


FIGURE **P5.20**

## Section 5.3 ■ Nonuniform Circular Motion

21. **Physics Now™** Tarzan ($m$ = 85.0 kg) tries to cross a river by swinging from a vine. The vine is 10.0 m long, and his speed at the bottom of the swing (as he just clears the water) will be 8.00 m/s. Tarzan doesn't know that the vine has a breaking strength of 1 000 N. Does he make it safely across the river?

22. We will study the most important work of Nobel laureate Arthur Compton in Chapter 28. Disturbed by speeding cars outside the physics building at Washington University in St. Louis, he designed a speed bump and had it installed. Suppose a car of mass $m$ passes over a bump in a road that follows the arc of a circle of radius $R$ as shown in Figure P5.22. (a) What force does the road exert on the car as the car passes the highest point of the bump if the car travels at a speed $v$? (b) What is the maximum speed the car can have as it passes this highest point without losing contact with the road?

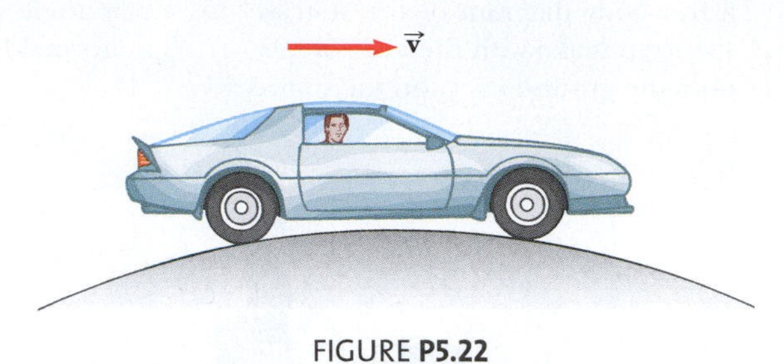


FIGURE **P5.22**

23. **Physics Now™** A pail of water is rotated in a vertical circle of radius 1.00 m. What is the minimum speed of the pail, upside down at the top of the circle, if no water is to spill out?

24. A roller coaster at Six Flags Great America amusement park in Gurnee, Illinois, incorporates some clever design technology and some basic physics. Each vertical loop, instead of being circular, is shaped like a teardrop (Fig. P5.24). The cars ride on the inside of the loop at the top, and the speeds are high enough to ensure that the cars remain on the track. The biggest loop is 40.0 m high, with a maximum speed of 31.0 m/s (nearly 70 mi/h) at the bottom. Suppose the speed at the top is 13.0 m/s and the corresponding centripetal acceleration is $2g$. (a) What is the radius of the arc of the teardrop at the top? (b) If the total mass of a car plus the riders is $M$, what force does the rail exert on the car at the top? (c) Suppose the roller coaster had a circular loop of radius 20.0 m. If the cars have the same speed, 13.0 m/s at the top, what is the centripetal acceleration at the top? Comment on the normal force at the top in this situation.

(Frank Cezus/FPG International)

FIGURE **P5.24**

## Section 5.4 ▪ Motion in the Presence of Velocity-Dependent Resistive Forces

**25.** A small piece of Styrofoam packing material is dropped from a height of 2.00 m above the ground. Until it reaches terminal speed, the magnitude of its acceleration is given by $a = g - bv$. After falling 0.500 m, the Styrofoam effectively reaches terminal speed and then takes 5.00 s more to reach the ground. (a) What is the value of the constant $b$? (b) What is the acceleration at $t = 0$? (c) What is the acceleration when the speed is 0.150 m/s?

**26.** (a) Calculate the terminal speed of a wooden sphere (density 0.830 g/cm$^3$) falling through air if its radius is 8.00 cm and its drag coefficient is 0.500. (b) From what height would a freely falling object reach this speed in the absence of air resistance?

**27.** A small, spherical bead of mass 3.00 g is released from rest at $t = 0$ in a bottle of liquid shampoo. The terminal speed is observed to be $v_T = 2.00$ cm/s. Find (a) the value of the constant $b$ in Equation 5.4, (b) the time $\tau$ at which it reaches $0.632v_T$, and (c) the value of the resistive force when the bead reaches terminal speed.

**28.** A 9.00-kg object starting from rest falls through a viscous medium and experiences a resistive force $\vec{\mathbf{R}} = -b\vec{\mathbf{v}}$, where $\vec{\mathbf{v}}$ is the velocity of the object. The object reaches one-half its terminal speed in 5.54 s. (a) Determine the terminal speed. (b) At what time is the speed of the object three-fourths the terminal speed? (c) How far has the object traveled in the first 5.54 s of motion?

**29.** **PhysicsNow™** A motorboat cuts its engine when its speed is 10.0 m/s and coasts to rest. The equation describing the motion of the motorboat during this period is $v = v_i e^{-ct}$, where $v$ is the speed at time $t$, $v_i$ is the initial speed, and $c$ is a constant. At $t = 20.0$ s, the speed is 5.00 m/s. (a) Find the constant $c$. (b) What is the speed at $t = 40.0$ s? (c) Differentiate the expression for $v(t)$ and thus show that the acceleration of the boat is proportional to the speed at any time.

**30.** Consider an object on which the net force is a resistive force proportional to the square of its speed. For example, assume that the resistive force acting on a speed skater is $f = -kmv^2$, where $k$ is a constant and $m$ is the skater's mass. The skater crosses the finish line of a straight-line race with speed $v_0$ and then slows down by coasting on his skates. Show that the skater's speed at any time $t$ after crossing the finish line is $v(t) = v_0/(1 + ktv_0)$.

## Section 5.5 ▪ The Fundamental Forces of Nature

**31.** Two identical isolated particles, each of mass 2.00 kg, are separated by a distance of 30.0 cm. What is the magnitude of the gravitational force exerted by one particle on the other?

**32.** Find the order of magnitude of the gravitational force that you exert on another person 2 m away. In your solution, state the quantities you measure or estimate and their values.

**33.** When a falling meteor is at a distance above the Earth's surface of 3.00 times the Earth's radius, what is its free-fall acceleration caused by the gravitational force exerted on it?

**34.** In a thundercloud, there may be electric charges of +40.0 C near the top of the cloud and −40.0 C near the bottom of the cloud. These charges are separated by 2.00 km. What is the electric force on the top charge?

## Section 5.6 ▪ Context Connection—Drag Coefficients of Automobiles

**35.** The mass of a sports car is 1 200 kg. The shape of the body is such that the aerodynamic drag coefficient is 0.250 and the frontal area is 2.20 m$^2$. Ignoring all other sources of friction, calculate the initial acceleration of the car assuming that it has been traveling at 100 km/h and is now shifted into neutral and allowed to coast.

**36.** Consider a 1 300-kg car presenting front-end area 2.60 m$^2$ and having drag coefficient 0.340. It can achieve instantaneous acceleration 3.00 m/s$^2$ when its speed is 10.0 m/s. Ignore any force of rolling resistance. Assume that the only horizontal forces on the car are static friction forward exerted by the road on the drive wheels and resistance exerted by the surrounding air, with density 1.20 kg/m$^3$. (a) Find the friction force exerted by the road. (b) Suppose the car body could be redesigned to have a drag coefficient of 0.200. If nothing else changes, what will be the car's acceleration? (c) Assume that the force exerted by the road remains constant. Then what maximum speed could the car attain with $D = 0.340$? (d) With $D = 0.200$?

## Additional Problems

**37.** Consider the three connected objects shown in Figure P5.37. Assume first that the inclined plane is frictionless and that the system is in equilibrium. In terms of $m$, $g$, and $\theta$, find (a) the mass $M$ and (b) the tensions $T_1$ and $T_2$. Now assume that the value of $M$ is double the value found in part (a). Find (c) the acceleration of each object and (d) the tensions $T_1$ and $T_2$. Next, assume that the coefficient of static friction between $m$ and $2m$ and the inclined plane is $\mu_s$ and that the system is in equilibrium. Find (e) the maximum value of $M$ and (f) the minimum value of $M$. (g) Compare the values of $T_2$ when $M$ has its minimum and maximum values.

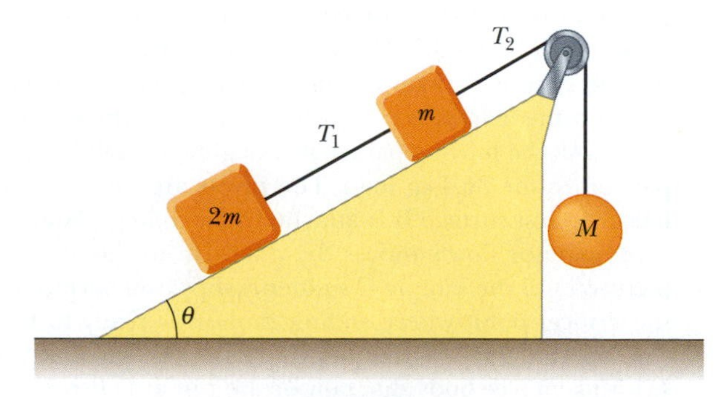


FIGURE **P5.37**

**38.** A 2.00-kg aluminum block and a 6.00-kg copper block are connected by a light string over a frictionless pulley. They sit on a steel surface, as shown in Figure P5.38, where $\theta = 30.0°$. When they are released from rest, will they start to move? If so, determine (a) their acceleration and (b) the tension in the string. If not, determine the sum of the magnitudes of the forces of friction acting on the blocks.

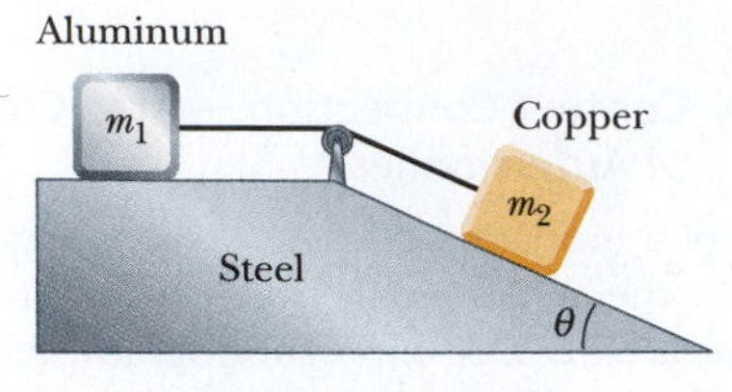


FIGURE **P5.38**

**39.** A crate of weight $F_g$ is pushed by a force $\vec{\mathbf{P}}$ on a horizontal floor. (a) Assuming that the coefficient of static friction is $\mu_s$ and that $\vec{\mathbf{P}}$ is directed at angle $\theta$ below the horizontal, show that the minimum value of $P$ that will move the crate is given by

$$P = \frac{\mu_s F_g \sec \theta}{1 - \mu_s \tan \theta}$$

(b) Find the minimum value of $P$ that can produce motion when $\mu_s = 0.400$, $F_g = 100$ N, and $\theta = 0°$, 15.0°, 30.0°, 45.0°, and 60.0°.

**40.** A 1.30-kg toaster is not plugged in. The coefficient of static friction between the toaster and a horizontal countertop is 0.350. To make the toaster start moving you carelessly pull on its electric cord. (a) For the cord tension to be as small as possible, you should pull at what angle above the horizontal? (b) With this angle, how large must the tension be?

**41.** The system shown in Figure P4.54 (Chapter 4) has an acceleration of magnitude 1.50 m/s². Assume that the coefficient of kinetic friction between block and incline is the same for both inclines. Find (a) the coefficient of kinetic friction and (b) the tension in the string.

**42.** Materials such as automobile tire rubber and shoe soles are tested for coefficients of static friction with an apparatus called a James tester. The pair of surfaces for which $\mu_s$ is to be measured are labeled B and C in Figure P5.42. Sample C is attached to a foot D at the lower end of a pivoting arm E that makes angle $\theta$ with the vertical. The upper end of the arm is hinged at F to a vertical rod G that slides freely in a guide H fixed to the frame of the apparatus and supports a load I of mass 36.4 kg. The hinge pin at F is also the axle of a wheel that can roll vertically on the frame. All the moving parts have weights negligible in comparison to the 36.4-kg load. The pivots are nearly frictionless. The test surface B is attached to a rolling platform A. The operator slowly moves the platform to the left in the picture until the sample C suddenly slips over surface B. At the critical point where sliding motion is ready to begin, the operator notes the angle $\theta_s$ of the pivoting arm. (a) Make a free-body diagram of the pin at F. It is in equilibrium under three forces: the weight of the load I, a horizontal normal force exerted by the frame, and a force of compression directed upward along the arm E. (b) Draw a free-body diagram of the foot D and sample C, considered as one system. (c) Determine the normal force that the test surface B exerts on the sample for any angle $\theta$. (d) Show that $\mu_s = \tan \theta_s$. (e) The protractor on the tester can record angles as large as 50.2°. What is the greatest coefficient of friction it can measure?

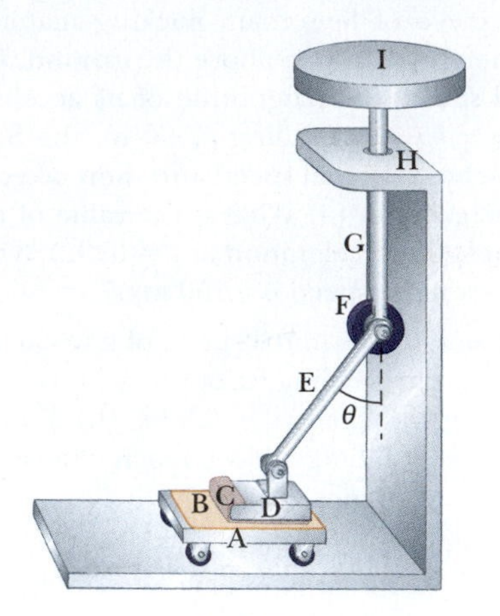


FIGURE **P5.42**

**43.** A block of mass $m = 2.00$ kg rests on the left edge of a block of mass $M = 8.00$ kg. The coefficient of kinetic friction between the two blocks is 0.300, and the surface on which the 8.00-kg block rests is frictionless. A constant horizontal force of magnitude $F = 10.0$ N is applied to the 2.00-kg block, setting it in motion as shown in Figure P5.43a. If the distance $L$ that the leading edge of the smaller block travels on the larger block is 3.00 m, (a) in what time interval will the smaller block make it to the right side of the 8.00-kg block as shown in Figure P5.43b? (*Note:* Both blocks are set into motion when $\vec{\mathbf{F}}$ is applied.) (b) How far does the 8.00-kg block move in the process?

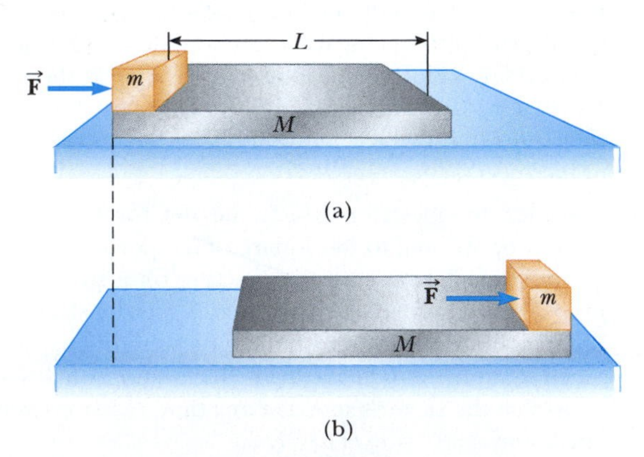


FIGURE **P5.43**

**44.** A 5.00-kg block is placed on top of a 10.0-kg block (Fig. P5.44). A horizontal force of 45.0 N is applied to the 10-kg block, and the 5-kg block is tied to the wall. The coefficient of kinetic friction between all moving surfaces is 0.200. (a) Draw a free-body diagram for each block and identify the action-reaction forces between the blocks. (b) Determine the tension in the string and the magnitude of the acceleration of the 10-kg block.

**45.** A car rounds a banked curve as in Figure 5.13. The radius of curvature of the road is $R$, the banking angle is $\theta$, and the coefficient of static friction is $\mu_s$. (a) Determine the range of speeds the car can have without slipping up or down the road. (b) Find the minimum value for $\mu_s$ such

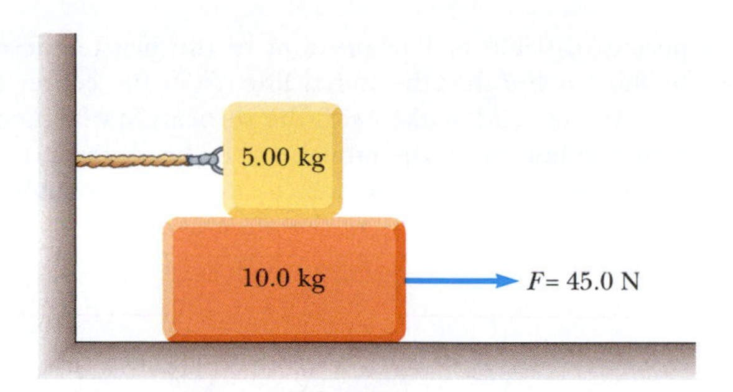


FIGURE **P5.44**

that the minimum speed is zero. (c) What is the range of speeds possible if $R = 100$ m, $\theta = 10.0°$, and $\mu_s = 0.100$ (slippery conditions)?

**46.** The following equations describe the motion of a system of two objects.

$$n - (6.50 \text{ kg})(9.80 \text{ m/s}^2) \cos 13.0° = 0$$

$$f_k = 0.360n$$

$$T + (6.50 \text{ kg})(9.80 \text{ m/s}^2) \sin 13.0° - f_k = (6.50 \text{ kg})a$$

$$-T + (3.80 \text{ kg})(9.80 \text{ m/s}^2) = (3.80 \text{ kg})a$$

(a) Solve the equations for $a$ and $T$. (b) Describe a situation to which these equations apply. Draw free-body diagrams for both objects.

**47.** In a home laundry dryer, a cylindrical tub containing wet clothes is rotated steadily about a horizontal axis as shown in Figure P5.47. The clothes are made to tumble so that they will dry uniformly. The rate of rotation of the smooth-walled tub is chosen so that a small piece of cloth will lose contact with the tub when the cloth is at an angle of 68.0° above the horizontal. If the radius of the tub is 0.330 m, what rate of revolution is needed?

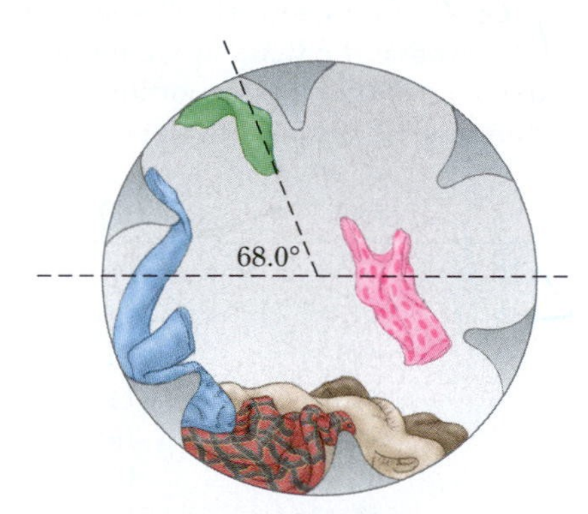


FIGURE **P5.47**

**48.** A student builds and calibrates an accelerometer and uses it to determine the speed of her car around a certain unbanked highway curve. The accelerometer is a plumb bob with a protractor that she attaches to the roof of her car. A friend riding in the car with the student observes that the plumb bob hangs at an angle of 15.0° from the vertical when the car has a speed of 23.0 m/s. (a) What is the centripetal acceleration of the car rounding the curve? (b) What is the radius of the curve? (c) What is speed of the car if the plumb bob deflection is 9.00° while rounding the same curve?

**49.** **Physics Now™** Because the Earth rotates about its axis, a point on the equator experiences a centripetal acceleration of 0.033 7 m/s$^2$, whereas a point at the poles experiences no centripetal acceleration. (a) Show that at the equator the gravitational force on an object must exceed the normal force required to support the object. That is, show that the object's true weight exceeds its apparent weight. (b) What is the apparent weight at the equator and at the poles of a person having a mass of 75.0 kg? (Assume that the Earth is a uniform sphere and take $g = 9.800$ m/s$^2$.)

**50.** An air puck of mass $m_1$ is tied to a string and allowed to revolve in a circle of radius $R$ on a frictionless horizontal table. The other end of the string passes through a hole in the center of the table, and a counterweight of mass $m_2$ is tied to it (Fig. P5.50). The suspended object remains in equilibrium while the puck on the tabletop revolves. What are (a) the tension in the string, (b) the radial force acting on the puck, and (c) the speed of the puck?

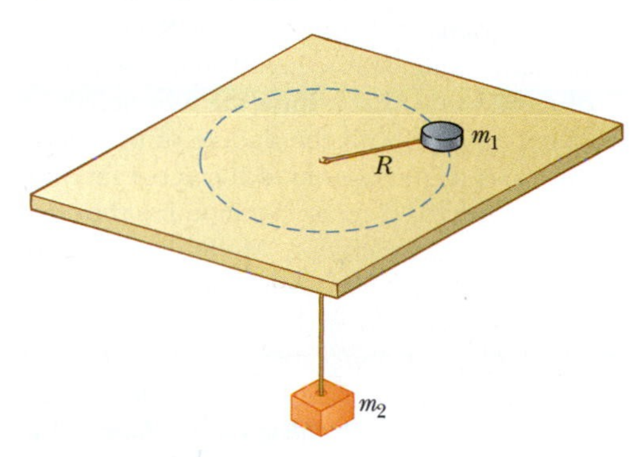


FIGURE **P5.50**

**51.** A Ferris wheel rotates four times each minute. It carries each car around a circle of diameter of 18.0 m. (a) What is the centripetal acceleration of a rider? (b) What force does the seat exert on a 40.0-kg rider at the lowest point of the ride? (c) At the highest point of the ride? (d) What force (magnitude and direction) does the seat exert on a rider when the rider is halfway between top and bottom?

**52.** An amusement park ride consists of a rotating circular platform 8.00 m in diameter from which 10.0-kg seats are suspended at the end of 2.50-m massless chains (Fig. P5.52). When the system rotates, the chains make an

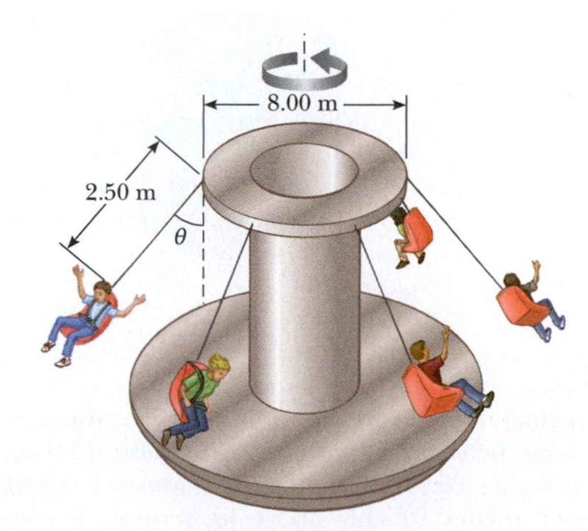


FIGURE **P5.52**

angle $\theta = 28.0°$ with the vertical. (a) What is the speed of each seat? (b) Draw a free-body diagram of a 40.0-kg child riding in a seat and find the tension in the chain.

53. A space station, in the form of a wheel 120 m in diameter, rotates to provide an "artificial gravity" of 3.00 m/s$^2$ for persons who walk around on the inner wall of the outer rim. Find the rate of rotation of the wheel (in revolutions per minute) that will produce this effect.

54. *Sedimentation and centrifugation.* According to Stokes's law, water exerts on a slowly moving immersed spherical object a resistive force described by

$$\vec{\mathbf{R}} = -(0.018\,8\text{ N}\cdot\text{s/m}^2)r\vec{\mathbf{v}}$$

where $r$ is the radius of the sphere and $\vec{\mathbf{v}}$ is its velocity. (a) Consider a spherical grain of gold dust with density $19.3 \times 10^3$ kg/m$^3$ and radius 0.500 $\mu$m. Ignore the buoyant force on the grain. Find the terminal speed at which the grain falls in water. (b) Over what time interval will all such suspended grains settle out of a tube 8.00 cm high? (c) The sedimentation rate can be greatly increased by the use of a centrifuge. Assume that it spins the tube at 3 000 rev/min in a horizontal plane, with the middle of the tube at 9.00 cm from the axis of rotation. Find the acceleration of the middle of the tube. (d) This acceleration has the effect of an enhanced free-fall acceleration. Model it as uniform over the length of the tube. Over what time interval will all the suspended grains of gold settle out of the water in this situation? In biological applications, such as separating blood cells from plasma, the suspended particles also feel a significant buoyant force, as we will study in Chapter 15.

55. An amusement park ride consists of a large vertical cylinder that spins about its axis sufficiently fast that any person inside is held up against the wall when the floor drops away (Fig. P5.55). The coefficient of static friction between person and wall is $\mu_s$, and the radius of the cylinder is $R$. (a) Show that the maximum period of revolution necessary to keep the person from falling is $T = (4\pi^2 R\mu_s/g)^{1/2}$. (b) Obtain a numerical value for $T$ assuming that $R =$ 4.00 m and $\mu_s = 0.400$. How many revolutions per minute does the cylinder make?

FIGURE **P5.55**

56. A single bead can slide with negligible friction on a wire that is bent into a circular loop of radius 15.0 cm as shown in Figure P5.56. (a) The circle is always in a vertical plane and rotates steadily about its vertical diameter with a period of 0.450 s. The position of the bead is described by the angle $\theta$ that the radial line, from the center of the loop to the bead, makes with the vertical. At what angle up from the bottom of the circle can the bead stay motionless relative to the turning circle? (b) Repeat the problem taking the period of the circle's rotation as 0.850 s.

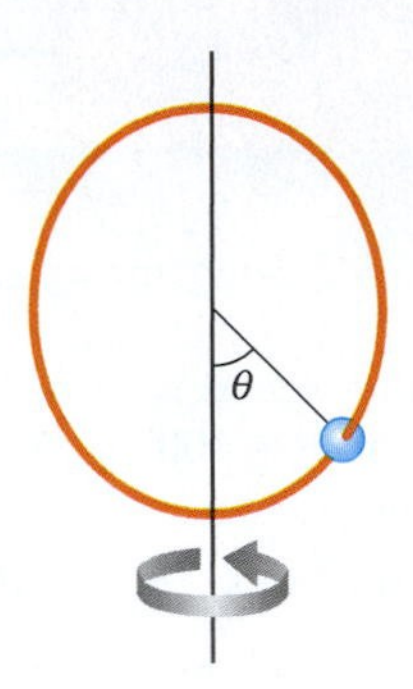


FIGURE **P5.56**

57. The expression $F = arv + br^2v^2$ gives the magnitude of the resistive force (in newtons) exerted on a sphere of radius $r$ (in meters) by a stream of air moving at speed $v$ (in meters per second), where $a$ and $b$ are constants with appropriate SI units. Their numerical values are $a = 3.10 \times 10^{-4}$ and $b = 0.870$. Using this expression, find the terminal speed for water droplets falling under their own weight in air, taking the following values for the drop radii: (a) 10.0 $\mu$m, (b) 100 $\mu$m, (c) 1.00 mm. Note that for (a) and (c) you can obtain accurate answers without solving a quadratic equation by considering which of the two contributions to the air resistance is dominant and ignoring the lesser contribution.

58. Members of a skydiving club were given the following data to use in planning their jumps. In the table, $d$ is the distance fallen from rest by a sky diver in a "free-fall stable spread position" versus the time of fall $t$. (a) Convert the distances in feet into meters. (b) Graph $d$ (in meters) versus $t$. (c) Determine the value of the terminal speed $v_T$ by finding the slope of the straight portion of the curve. Use a least-squares fit to determine this slope.

| $t$ (s) | $d$ (ft) | $t$ (s) | $d$ (ft) |
|---|---|---|---|
| 0 | 0 | 11 | 1 309 |
| 1 | 16 | 12 | 1 483 |
| 2 | 62 | 13 | 1 657 |
| 3 | 138 | 14 | 1 831 |
| 4 | 242 | 15 | 2 005 |
| 5 | 366 | 16 | 2 179 |
| 6 | 504 | 17 | 2 353 |
| 7 | 652 | 18 | 2 527 |
| 8 | 808 | 19 | 2 701 |
| 9 | 971 | 20 | 2 875 |
| 10 | 1 138 | | |

59. A model airplane of mass 0.750 kg flies in a horizontal circle at the end of a 60.0-m control wire with a speed of 35.0 m/s. Compute the tension in the wire assuming that it makes a constant angle of 20.0° with the horizontal. The forces exerted on the airplane are the pull of the control wire, the gravitational force, and aerodynamic lift, which acts at 20.0° inward from the vertical as shown in Figure P5.59.

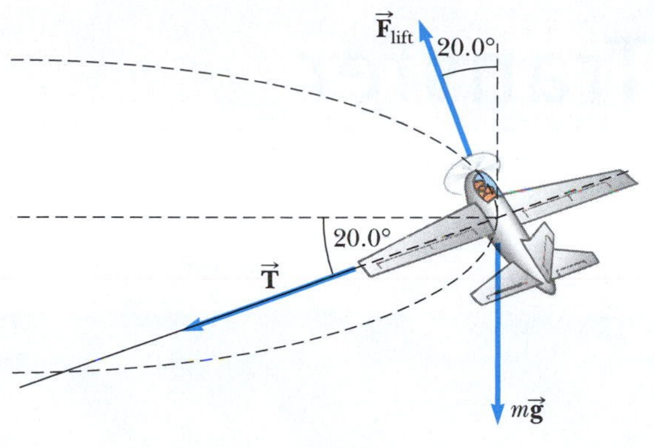


FIGURE **P5.59**

**60.** If a single constant force acts on an object that moves on a straight line, the object's velocity is a linear function of time. The equation $v = v_i + at$ gives its velocity $v$ as a function of time, where $a$ is its constant acceleration. What if velocity is instead a linear function of position? Assume that as a particular object moves through a resistive medium, its speed decreases as described by the equation $v = v_i - kx$, where $k$ is a constant coefficient and $x$ is the position of the object. Find the law describing the total force acting on this object.

## ANSWERS TO QUICK QUIZZES

**5.1** (b). The friction force acts opposite to the weight of the book to keep the book in equilibrium. Because the weight is downward, the friction force must be upward.

**5.2** (b). The crate accelerates to the east. Because the only horizontal force acting on it is the force of static friction between its bottom surface and the truck bed, that force must also be directed to the east.

**5.3** (b). When pulling with the rope, there is a component of your applied force that is upward, which reduces the normal force between the sled and the snow. In turn, the friction force between the sled and the snow is reduced, making the sled easier to move. If you push from behind, with a force with a downward component, the normal force is larger, the friction force is larger, and the sled is harder to move.

**5.4** (i), (a). The normal force is always perpendicular to the surface that applies the force. Because your car maintains its orientation at all points on the ride, the normal force is always upward. (ii), (b). Your centripetal acceleration is downward toward the center of the circle, so the net force on you must be downward.

**5.5** (a). If the car is moving in a circular path, it must have centripetal acceleration given by Equation 3.17.

**5.6** (a) Because the speed is constant, the only direction the force can have is that of the centripetal acceleration. The force is larger at Ⓒ than at Ⓐ because the radius at Ⓒ is smaller. There is no force at Ⓑ because the wire is straight. (b) In addition to the forces in the centripetal direction in (a), there are now tangential forces to provide the tangential acceleration. The tangential force is the same at all three points because the tangential acceleration is constant.

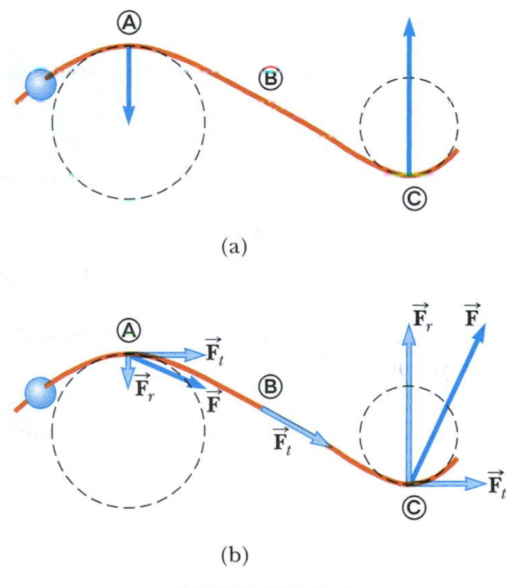


FIGURE **QQA.5.6**

**5.7** (c). When the downward gravitational force $m\vec{g}$ and the upward force of air resistance $\vec{R}$ have the same magnitude, she reaches terminal speed and her acceleration is zero.

CHAPTER 6

# Energy and Energy Transfer

On a wind farm, a technician inspects one of the windmills. Moving air does work on the blades of the windmills, causing the blades and the rotor of an electrical generator to rotate. Energy is transferred out of the system of the windmill by means of electricity.

(Billy Hustace/Getty Images)

## CHAPTER OUTLINE


6.1 Systems and Environments
6.2 Work Done by a Constant Force
6.3 The Scalar Product of Two Vectors
6.4 Work Done by a Varying Force
6.5 Kinetic Energy and the Work–Kinetic Energy Theorem
6.6 The Nonisolated System
6.7 Situations Involving Kinetic Friction
6.8 Power
6.9 Context Connection—Horsepower Ratings of Automobiles

SUMMARY


In the preceding chapters, we analyzed the motion of an object using quantities such as *position, velocity, acceleration,* and *force,* with which you are familiar from everyday life. We developed a number of models using these notions that allow us to solve a variety of problems. Some problems that, in theory, could be solved with these models are very difficult to solve in practice, but they can be made much simpler with a different approach. In this and the following two chapters, we shall investigate this new approach, which will introduce us to new analysis models for problem solving. This approach includes definitions of quantities that may not be familiar to you. You may be familiar with some quantities, but they may have more specific meanings in physics than in everyday life. We begin this discussion by exploring *energy.*

Energy is present in the Universe in various forms. ***Every* physical process in the Universe involves energy and energy transfers**

**or transformations.** Therefore, energy is an extremely important concept to understand. Unfortunately, despite its importance, it cannot be easily defined. The variables in previous chapters were relatively concrete; we have everyday experience with velocities and forces, for example. Although the *notion* of energy is more abstract, we do have *experiences* with energy, such as running out of gasoline or losing our electrical service if we forget to pay the bill.

The concept of energy can be applied to the dynamics of a mechanical system without resorting to Newton's laws. This "energy approach" to describing motion is especially useful when the force acting on a particle is not constant; in such a case, the acceleration is not constant and we cannot apply the particle under constant acceleration model we developed in Chapter 2. Particles in nature are often subject to forces that vary with the particles' positions. These forces include gravitational forces and the force exerted on an object attached to a spring. We will develop a global approach to problems involving energy and energy transfer. This approach extends well beyond physics and can be applied to biological organisms, technological systems, and engineering situations.

## 6.1 SYSTEMS AND ENVIRONMENTS

All our analysis models in the earlier chapters were based on the motion of a particle or an object modeled as a particle. We begin our study of our new approach by identifying a **system.** A system is a simplification model in that we focus our attention on a small region of the Universe—the system—and ignore details of the rest of the Universe outside the system. A critical skill in applying the energy approach to problems in the next three chapters is *correctly identifying the system.* A system may

■ A system

- be a single object or particle
- be a collection of objects or particles
- be a region of space (e.g., the interior of an automobile engine combustion cylinder)
- vary in size and shape (e.g., a rubber ball that deforms upon striking a wall)

A **system boundary,** which is an imaginary surface (often but not necessarily coinciding with a physical surface), divides the Universe between the system and the **environment** of the system.

As an example, imagine a force applied to an object in empty space. We can define the object as the system as in the first item in the bulleted list above. The force applied to it is an influence on the system from the environment and acts across the system boundary. We will see how to analyze this situation using a system approach in a subsequent section of this chapter.

Another example occurs in Example 5.3. Here the system can be defined as the combination of the ball, the cube, and the string, consistent with the second item of the bulleted list. The influences from the environment include the gravitational forces on the ball and the cube, the normal and friction forces on the cube, and the force of the pulley on the string. The forces exerted by the string on the ball and the cube are internal to the system and therefore are not included as influences from the environment.

### PITFALL PREVENTION 6.1

**IDENTIFY THE SYSTEM** One of the most important steps to take in solving a problem using the energy approach is to identify the system of interest correctly. Be sure this step is the *first* step you take in solving a problem.

## 6.2 WORK DONE BY A CONSTANT FORCE

Let us begin our analysis of systems by introducing a term whose meaning in physics is distinctly different from its everyday meaning. This new term is **work.** Imagine that you are trying to push a heavy sofa across your living room floor. If you push on the sofa *and* it moves through a displacement, you have done work on the sofa.

Consider a particle, which we identify as the system, that undergoes a displacement $\Delta\vec{\mathbf{r}}$ along a straight line while acted on by a constant force $\vec{\mathbf{F}}$ that makes

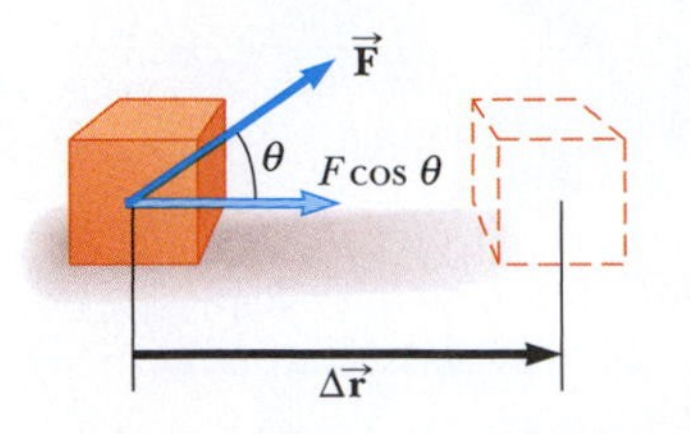


FIGURE 6.1 If an object undergoes a displacement $\Delta\vec{\mathbf{r}}$, the work done by the constant force $\vec{\mathbf{F}}$ on the object is $(F\cos\theta)\Delta r$.

an angle $\theta$ with $\Delta\vec{\mathbf{r}}$, as in Figure 6.1. The force has accomplished something—it has moved the particle—so we say that work was done by the force on the particle.

Notice that we know only the force and the displacement given in the description of the situation. We have no information about how long it took for this displacement to occur, nor any information about velocities or accelerations. The absence of this information provides a hint of the power of the energy approach as well as a hint of how different it will be from our approach in previous chapters. We do not need this information to find the work done. Let us now formally define the work done on a system if the force is constant:

The **work** $W$ done on a system by an external agent exerting a constant force on the system is the product of the magnitude $F$ of the force, the magnitude $\Delta r$ of the displacement of the point of application of the force, and $\cos\theta$, where $\theta$ is the angle between the force and displacement vectors:

▪ Work done by a constant force

$$W \equiv F\Delta r\cos\theta \qquad [6.1]$$

Work is a scalar quantity; no direction is associated with it. Its units are those of force multiplied by length; therefore, the SI unit of work is the **newton · meter** (N · m). The newton · meter, when it refers to work or energy, is called the **joule** (J).

From the definition in Equation 6.1, we see that a force does no work on a system if the point of application of the force does not move. In the mathematical representation, if $\Delta r = 0$, Equation 6.1 gives $W = 0$. In the mental representation, imagine pushing on the sofa mentioned earlier. If it doesn't move, no work has been done on the sofa. Of course, the work is also zero if the applied force is zero. If you don't push on the sofa, no work is done on it!

Also note from Equation 6.1 that the work done by a force is zero when the force is perpendicular to the displacement. That is, if $\theta = 90°$, then $\cos 90° = 0$ and $W = 0$. For example, consider the free-body diagram for a block moving across a frictionless surface in Figure 6.2. The work done by the normal force and the gravitational force on the block during its horizontal displacement are both zero for the same reason: they are both perpendicular to the displacement.

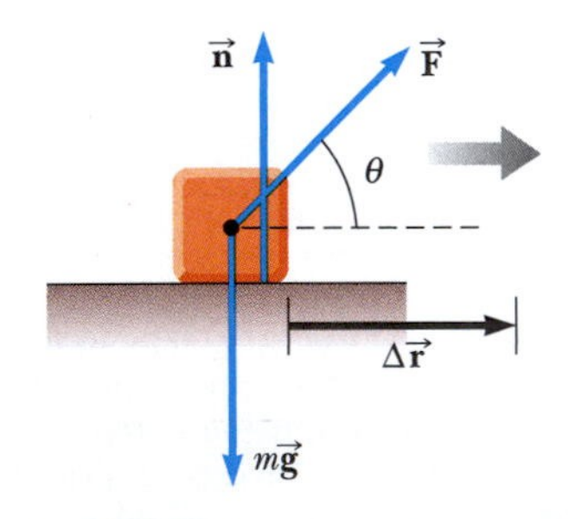


FIGURE 6.2 When an object is displaced horizontally on a flat table, the normal force $\vec{\mathbf{n}}$ and the gravitational force $m\vec{\mathbf{g}}$ do no work.

For now, we restrict our attention to systems consisting of a single particle or a small number of particles. In the case of a force applied to a particle, the displacement of the point of application of the force is necessarily the same as the displacement of the particle. In Chapter 17, we will consider work done in compressing a gas, which is modeled as a system consisting of a large number of particles. In this process, the displacement of the point of application of the force is very different from the displacement of the system.

In general, a particle may be moving under the influence of several forces. In that case, because work is a scalar quantity, the total work done as the particle undergoes some displacement is the algebraic sum of the work done by each of the forces.

The sign of the work depends on the direction of $\vec{\mathbf{F}}$ relative to $\Delta\vec{\mathbf{r}}$. The work done by the applied force is positive when the vector component of magnitude $F\cos\theta$ is in the *same direction* as the displacement. For example, when an object is lifted, the work done by the lifting force on the object is positive because the lifting force is upward, that is, in the same direction as the displacement. When the vector component of magnitude $F\cos\theta$ is in the direction *opposite* the displacement, $W$ is *negative.* In the case of the object being lifted, for instance, the work done by the gravitational force on the object is negative.

If a constant applied force $\vec{\mathbf{F}}$ acts parallel to the direction of the displacement, $\theta = 0$ and $\cos 0 = 1$. In this case, Equation 6.1 gives

$$W = F\Delta r \qquad [6.2]$$

Both Equations 6.1 and 6.2 are special cases of a more generalized definition of work. Both equations assume a constant force, and Equation 6.2 assumes that the force is parallel to the displacement. In the next two sections, we shall consider the situation in which a force is not parallel to the displacement and the more general case of a varying force.

**PITFALL PREVENTION 6.2**

**Work is done by . . . on . . .** Not only must you identify the system, you must also identify the interaction of the system with the environment. When discussing work, always use the phrase, "the work done by . . . on . . ." After "by" insert the part of the environment that is interacting directly with the system. After "on" insert the system. For example, "the work done by the hammer on the nail" identifies the nail as the system and the force from the hammer represents the interaction with the environment. This wording is similar to our use in Chapter 4 of "the force exerted by . . . on . . . ."

**QUICK QUIZ 6.1** Figure 6.3 shows four situations in which a force is applied to an object. In all four cases, the force has the same magnitude and the displacement of the object is to the right and of the same magnitude. Rank the situations in order of the work done by the force on the object, from most positive to most negative.

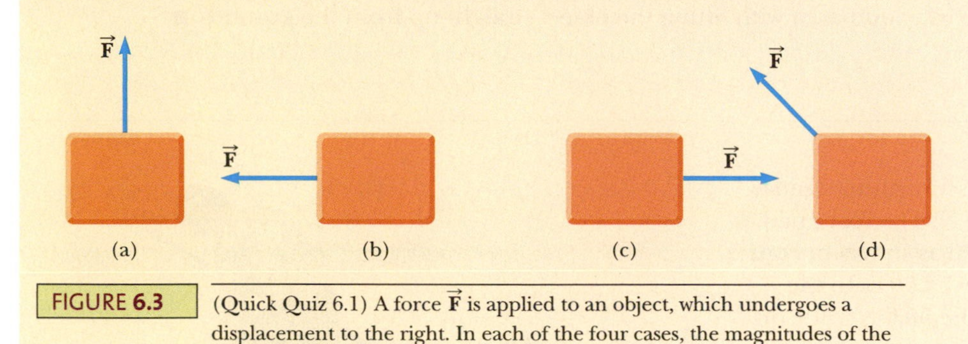


FIGURE 6.3 (Quick Quiz 6.1) A force $\vec{F}$ is applied to an object, which undergoes a displacement to the right. In each of the four cases, the magnitudes of the force and displacement are the same.

**PITFALL PREVENTION 6.3**

**Cause of the displacement** We can calculate the work done by a force on an object, but that force is *not* necessarily the cause of the object's displacement. For example, if you lift an object, work is done by the gravitational force, although gravity is not the cause of the object moving upward!

## Thinking Physics 6.1

A person slowly lifts a heavy box of mass $m$ a vertical height $h$ and then walks horizontally at constant velocity a distance $d$ while holding the box as in Figure 6.4. Determine the work done **(a)** by the person and **(b)** by the gravitational force on the box in this process.

**Reasoning** **(a)** Assume that the person lifts the box with a force of magnitude equal to the weight of the box $mg$. In this case, the work done by the person on the box during the vertical displacement is $W = F\,\Delta r = (mg)(h) = mgh$, which is positive because the lifting force is in the same direction as the displacement. For the horizontal displacement, we assume that the acceleration of the box is approximately zero. As a result, the work done by the person on the box during the horizontal displacement of the box is zero because the horizontal force is approximately zero, and the force supporting the box's weight in this process is perpendicular to the displacement. Therefore, the net work done by the person on the box during the complete process is $mgh$.

**(b)** The work done by the gravitational force on the box during the vertical displacement of the box is $-mgh$, which is negative because this force is opposite the displacement. The work done by the gravitational force is zero during the horizontal displacement because this force is perpendicular to the displacement. Hence, the net work done by the gravitational force for the complete process is $-mgh$. The net work done by all forces on the box is zero, because $+mgh + (-mgh) = 0$. ■

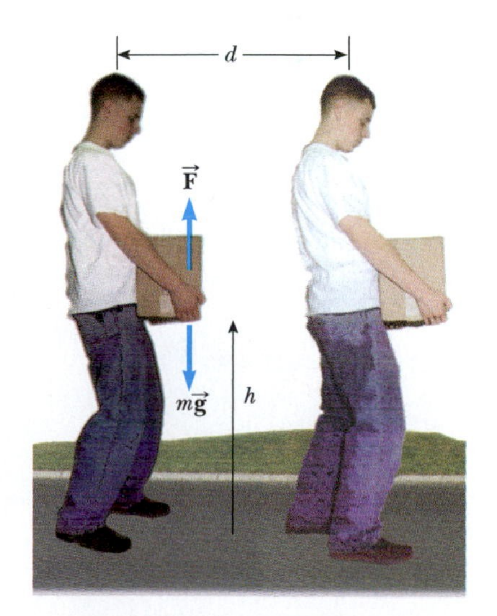


FIGURE 6.4 (Thinking Physics 6.1) A person lifts a heavy box of mass $m$ a vertical distance $h$ and then walks horizontally at constant velocity a distance $d$.

## Thinking Physics 6.2

Roads going up mountains are formed into *switchbacks,* with the road weaving back and forth along the face of the slope so that any portion of the roadway has only a gentle rise. Do switchbacks require that an automobile climbing the mountain do any less work than if it were driving on a roadway that runs straight up the slope? Why are the switchbacks used?

**Reasoning** If we ignore the effects of rolling friction on the tires of the car, the same amount of work would be done in driving up the switchbacks and driving straight up the mountain because the weight of the car is moved upward against the gravitational force by the same vertical distance in each case. So why do we use the switchbacks? The answer lies in the force required, not the work. The force needed from the engine to follow a gentle rise is much less than that required to drive straight up the hill. Roadways running straight uphill would require redesigning engines so as to enable them to apply much larger forces. This situation is similar to the ease with which a heavy object can be rolled up a ramp into a moving van truck, compared with lifting the object straight up from the ground. ■

**EXAMPLE 6.1** **Mr. Clean**

A man cleaning his apartment pulls a vacuum cleaner with a force of magnitude $F = 50.0$ N. The force makes an angle of 30.0° with the horizontal as shown in Figure 6.5. The vacuum cleaner is displaced 3.00 m to the right. Calculate the work done by the 50.0-N force on the vacuum cleaner.

**Solution** Using the definition of work (Equation 6.1), we have

$$W = (F \cos \theta)\, \Delta r = (50.0 \text{ N})(\cos 30.0°)(3.00 \text{ m})$$

$$= 130 \text{ N}\cdot\text{m} = \boxed{130 \text{ J}}$$

Note that the normal force $\vec{\mathbf{n}}$, the gravitational force $m\vec{\mathbf{g}}$, and the upward component of the applied force do *no* work because they are perpendicular to the displacement.

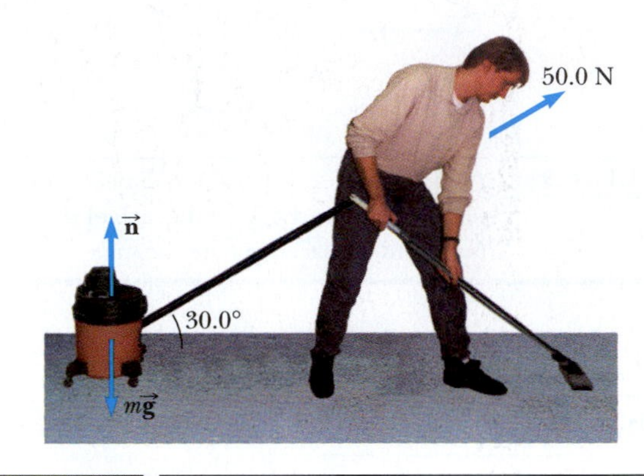


FIGURE 6.5 (Example 6.1) A vacuum cleaner being pulled at an angle of 30.0° with the horizontal.

## 6.3 THE SCALAR PRODUCT OF TWO VECTORS

Based on Equation 6.1, it is convenient to express the definition of work in terms of a **scalar product** of the two vectors $\vec{\mathbf{F}}$ and $\Delta\vec{\mathbf{r}}$. The scalar product was introduced briefly in Section 1.8. We formally provide its definition here:

The scalar product of any two vectors $\vec{\mathbf{A}}$ and $\vec{\mathbf{B}}$ is a scalar quantity equal to the product of the magnitudes of the two vectors and the cosine of the angle $\theta$ between them:

$$\vec{\mathbf{A}} \cdot \vec{\mathbf{B}} \equiv AB \cos \theta \qquad [6.3]$$

where $\theta$ is the angle between $\vec{\mathbf{A}}$ and $\vec{\mathbf{B}}$ as in Figure 6.6.

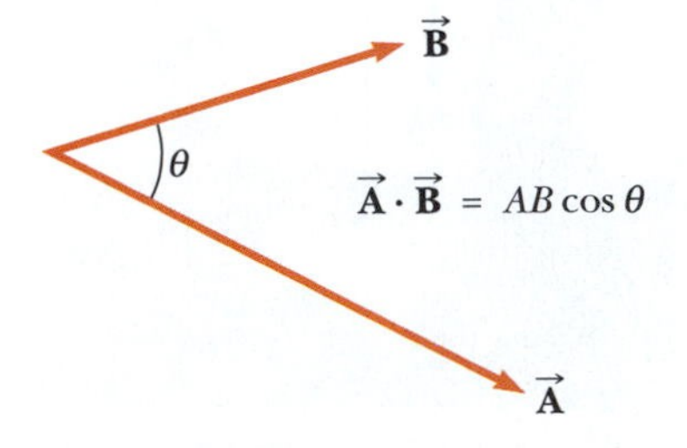


FIGURE 6.6 The scalar product $\vec{\mathbf{A}} \cdot \vec{\mathbf{B}}$ equals the magnitude of $\vec{\mathbf{A}}$ multiplied by the magnitude of $\vec{\mathbf{B}}$ and the cosine of the angle between $\vec{\mathbf{A}}$ and $\vec{\mathbf{B}}$.

Note that $\vec{\mathbf{A}}$ and $\vec{\mathbf{B}}$ need not have the same units. The units of the scalar product are simply the product of the units of the two vectors. Because of the dot symbol, the scalar product is often called the *dot product.*

Notice that the right-hand side of Equation 6.3 has the same mathematical structure as the right-hand side of Equation 6.1. Consequently, we can write the definition of work as the scalar product $\vec{\mathbf{F}} \cdot \Delta\vec{\mathbf{r}}$. Therefore, we can express Equation 6.1 as

$$W = \vec{\mathbf{F}} \cdot \Delta\vec{\mathbf{r}} = F\Delta r \cos\theta \qquad [6.4]$$

■ Work expressed as a scalar product

Before continuing with our discussion of work, let us investigate some properties of the scalar product because we will need to use it later in the book as well. From Equation 6.3 we see that the scalar product is *commutative.* That is,

$$\vec{\mathbf{A}} \cdot \vec{\mathbf{B}} = \vec{\mathbf{B}} \cdot \vec{\mathbf{A}} \qquad [6.5]$$

In addition, the scalar product obeys the *distributive law of multiplication,* so that

$$\vec{\mathbf{A}} \cdot (\vec{\mathbf{B}} + \vec{\mathbf{C}}) = \vec{\mathbf{A}} \cdot \vec{\mathbf{B}} + \vec{\mathbf{A}} \cdot \vec{\mathbf{C}} \qquad [6.6]$$

The scalar product is simple to evaluate from Equation 6.3 when $\vec{\mathbf{A}}$ is either perpendicular or parallel to $\vec{\mathbf{B}}$. If $\vec{\mathbf{A}}$ is perpendicular to $\vec{\mathbf{B}}$ ($\theta = 90°$), then $\vec{\mathbf{A}} \cdot \vec{\mathbf{B}} = 0$. (The equality $\vec{\mathbf{A}} \cdot \vec{\mathbf{B}} = 0$ also holds in the more trivial case when either $\vec{\mathbf{A}}$ or $\vec{\mathbf{B}}$ is zero.) If $\vec{\mathbf{A}}$ and $\vec{\mathbf{B}}$ point in the same direction ($\theta = 0$), then $\vec{\mathbf{A}} \cdot \vec{\mathbf{B}} = AB$. If $\vec{\mathbf{A}}$ and $\vec{\mathbf{B}}$ point in opposite directions ($\theta = 180°$), then $\vec{\mathbf{A}} \cdot \vec{\mathbf{B}} = -AB$. The scalar product is negative when $90° < \theta < 180°$.

The unit vectors $\hat{\mathbf{i}}$, $\hat{\mathbf{j}}$, and $\hat{\mathbf{k}}$, which were defined in Chapter 1, lie in the positive $x$, $y$, and $z$ directions, respectively, of a right-handed coordinate system. Therefore, it follows from the definition of $\vec{\mathbf{A}} \cdot \vec{\mathbf{B}}$ that the scalar products of these unit vectors are given by

$$\hat{\mathbf{i}} \cdot \hat{\mathbf{i}} = \hat{\mathbf{j}} \cdot \hat{\mathbf{j}} = \hat{\mathbf{k}} \cdot \hat{\mathbf{k}} = 1 \qquad [6.7]$$

$$\hat{\mathbf{i}} \cdot \hat{\mathbf{j}} = \hat{\mathbf{i}} \cdot \hat{\mathbf{k}} = \hat{\mathbf{j}} \cdot \hat{\mathbf{k}} = 0 \qquad [6.8]$$

■ Scalar products of unit vectors

Two vectors $\vec{\mathbf{A}}$ and $\vec{\mathbf{B}}$ can be expressed in component form as

$$\vec{\mathbf{A}} = A_x\hat{\mathbf{i}} + A_y\hat{\mathbf{j}} + A_z\hat{\mathbf{k}} \qquad \vec{\mathbf{B}} = B_x\hat{\mathbf{i}} + B_y\hat{\mathbf{j}} + B_z\hat{\mathbf{k}}$$

Therefore, using these expressions, Equations 6.7 and 6.8 reduce the scalar product of $\vec{\mathbf{A}}$ and $\vec{\mathbf{B}}$ to

$$\vec{\mathbf{A}} \cdot \vec{\mathbf{B}} = A_xB_x + A_yB_y + A_zB_z \qquad [6.9]$$

where we have used the distributive law (Eq. 6.6) to simplify the result. This equation and Equation 6.3 are alternative but equivalent expressions for the scalar product. Equation 6.3 is useful if you know the magnitudes and directions of the vectors, and Equation 6.9 is useful if you know the components of the vectors. In the special case where $\vec{\mathbf{A}} = \vec{\mathbf{B}}$, we see that

$$\vec{\mathbf{A}} \cdot \vec{\mathbf{A}} = A_x^2 + A_y^2 + A_z^2 = A^2$$

**PITFALL PREVENTION 6.4**

**WORK IS A SCALAR** Although Equation 6.4 defines the work in terms of two vectors, *work is a scalar;* there is no direction associated with it. *All* types of energy and energy transfer are scalars. This property is a major advantage of the energy approach because we don't need vector calculations!

**QUICK QUIZ 6.2** Which of the following statements is true about the relationship between the scalar product of two vectors and the product of the magnitudes of the vectors? **(a)** $\vec{\mathbf{A}} \cdot \vec{\mathbf{B}}$ is larger than $AB$. **(b)** $\vec{\mathbf{A}} \cdot \vec{\mathbf{B}}$ is smaller than $AB$. **(c)** $\vec{\mathbf{A}} \cdot \vec{\mathbf{B}}$ could be larger or smaller than $AB$, depending on the angle between the vectors. **(d)** $\vec{\mathbf{A}} \cdot \vec{\mathbf{B}}$ could be equal to $AB$.

**EXAMPLE 6.2** The Scalar Product

The vectors $\vec{\mathbf{A}}$ and $\vec{\mathbf{B}}$ are given by $\vec{\mathbf{A}} = 2\hat{\mathbf{i}} + 3\hat{\mathbf{j}}$ and $\vec{\mathbf{B}} = -\hat{\mathbf{i}} + 2\hat{\mathbf{j}}$.

**A** Determine the scalar product $\vec{\mathbf{A}} \cdot \vec{\mathbf{B}}$.

**Solution** We can evaluate the scalar product directly using the unit vector notation:

$$\begin{aligned}\vec{\mathbf{A}} \cdot \vec{\mathbf{B}} &= (2\hat{\mathbf{i}} + 3\hat{\mathbf{j}}) \cdot (-\hat{\mathbf{i}} + 2\hat{\mathbf{j}}) \\ &= -2\hat{\mathbf{i}} \cdot \hat{\mathbf{i}} + 2\hat{\mathbf{i}} \cdot 2\hat{\mathbf{j}} - 3\hat{\mathbf{j}} \cdot \hat{\mathbf{i}} + 3\hat{\mathbf{j}} \cdot 2\hat{\mathbf{j}} \\ &= -2 + 6 = 4\end{aligned}$$

where we have used that $\hat{\mathbf{i}} \cdot \hat{\mathbf{i}} = \hat{\mathbf{j}} \cdot \hat{\mathbf{j}} = 1$ and $\hat{\mathbf{i}} \cdot \hat{\mathbf{j}} = \hat{\mathbf{j}} \cdot \hat{\mathbf{i}} = 0$. The same result is obtained using Equation 6.9 directly, where $A_x = 2$, $A_y = 3$, $B_x = -1$, and $B_y = 2$. Note that the result has no units because no units were specified on the original vectors $\vec{\mathbf{A}}$ and $\vec{\mathbf{B}}$.

**B** Find the angle $\theta$ between $\vec{\mathbf{A}}$ and $\vec{\mathbf{B}}$.

**Solution** The magnitudes of $\vec{\mathbf{A}}$ and $\vec{\mathbf{B}}$ are given by

$$A = \sqrt{A_x^2 + A_y^2} = \sqrt{(2)^2 + (3)^2} = \sqrt{13}$$

$$B = \sqrt{B_x^2 + B_y^2} = \sqrt{(-1)^2 + (2)^2} = \sqrt{5}$$

Using Equation 6.3 and the result from part A gives

$$\cos\theta = \frac{\vec{\mathbf{A}} \cdot \vec{\mathbf{B}}}{AB} = \frac{4}{\sqrt{13}\sqrt{5}} = \frac{4}{\sqrt{65}} = 0.496$$

$$\theta = \cos^{-1}(0.496) = 60.3°$$

## 6.4 WORK DONE BY A VARYING FORCE

Consider a particle being displaced along the $x$ axis under the action of a force with an $x$ component $F_x$ that varies with position, as in the graphical representation in Figure 6.7. The particle is displaced in the direction of increasing $x$ from $x = x_i$ to $x = x_f$. In such a situation, we cannot use Equation 6.1 to calculate the work done by the force because this relationship applies only when $\vec{\mathbf{F}}$ is constant in magnitude and direction. As seen in Figure 6.7, we do not have a *single* value of the force to substitute into Equation 6.1. If, however, we imagine that the point of application of the force undergoes a *small* displacement in the $x$ direction so that $\Delta r = \Delta x$, as shown in Figure 6.7a, the $x$ component $F_x$ of the force is approximately constant over this interval. We can then approximate the work done by the force on the particle for this small displacement as

$$W_1 \approx F_x \,\Delta x \qquad [6.10]$$

This quantity is just the area of the shaded geometric model rectangle in Figure 6.7a. If we imagine that the curve described by $F_x$ versus $x$ is divided into a large number of such intervals, the total work done for the displacement from $x_i$ to $x_f$ is approximately equal to the sum of a large number of such terms:

$$W \approx \sum_{x_i}^{x_f} F_x \,\Delta x$$

If the displacements $\Delta x$ are allowed to approach zero, the number of terms in the sum increases without limit, but the value of the sum approaches a definite value equal to the area under the curve bounded by $F_x$ and the $x$ axis in Figure 6.7b. As you probably have learned in calculus, this limit of the sum is called an *integral* and is represented by

$$\lim_{\Delta x \to 0} \sum_{x_i}^{x_f} F_x \,\Delta x = \int_{x_i}^{x_f} F_x \, dx$$

The limits on the integral $x = x_i$ to $x = x_f$ define what is called a **definite integral.** (An *indefinite integral* is the limit of a sum over an unspecified interval. Appendix B.7 gives a brief description of integration.) This definite integral is numerically

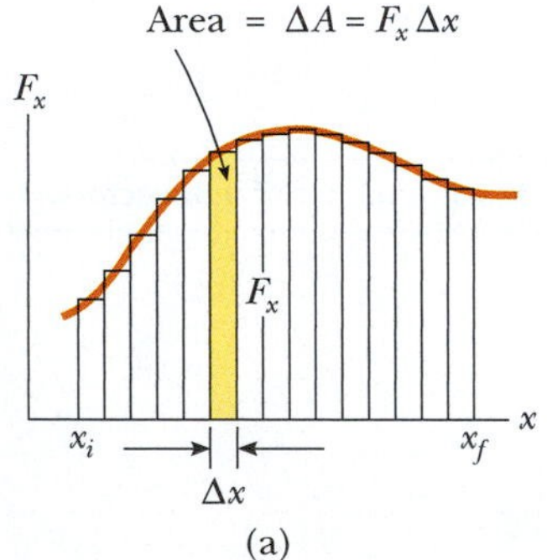


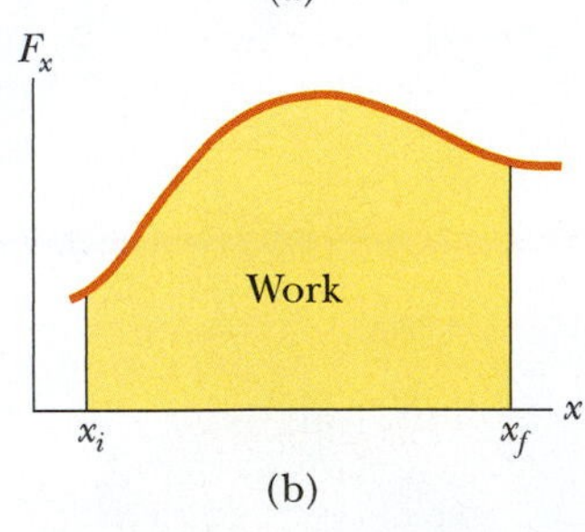


**FIGURE 6.7** (a) The work done by a force of magnitude $F_x$ for the small displacement $\Delta x$ is $F_x \Delta x$, which equals the area of the shaded rectangle. The total work done for the displacement from $x_i$ to $x_f$ is approximately equal to the sum of the areas of all the rectangles. (b) The work done by the variable force $F_x$ as the particle moves from $x_i$ to $x_f$ is *exactly* equal to the area under this curve.

equal to the area under the curve of $F_x$ versus $x$ between $x_i$ and $x_f$. Therefore, we can express the work done by $F_x$ for the displacement from $x_i$ to $x_f$ as

$$W = \int_{x_i}^{x_f} F_x \, dx \qquad [6.11]$$

This equation reduces to Equation 6.1 when $F_x = F \cos \theta$ is constant and $x_f - x_i = \Delta x$.

If more than one force acts on a system *and the system can be modeled as a particle,* the total work done on the system is just the work done by the net force. If we express the $x$ component of the net force as $\sum F_x$, the total work, or *net work,* done on the particle as it moves from $x_i$ to $x_f$ is

$$\sum W = W_{\text{net}} = \int_{x_i}^{x_f} \left( \sum F_x \right) dx$$

For the general case of a particle moving along an arbitrary path while acted on by a net force $\sum \vec{\mathbf{F}}$, we use the scalar product:

■ Work done by a variable net force

$$\sum W = W_{\text{net}} = \int \left( \sum \vec{\mathbf{F}} \right) \cdot d\vec{\mathbf{r}} \qquad [6.12]$$

where the integral is calculated over the path that the particle takes through space.

If the system cannot be modeled as a particle (for example, if the system consists of multiple particles that can move with respect to each other), we cannot use Equation 6.12 because different forces on the system may move through different displacements. In that case, we must evaluate the work done by each force separately and then add the works algebraically.

## Work Done by a Spring

A common physical system for which the force varies with position is shown in Active Figure 6.8. A block on a horizontal, frictionless surface is connected to a spring. If the block is located at a position $x$ relative to its equilibrium position $x = 0$, the stretched or compressed spring exerts a force on the block given by

■ Hooke's law

$$F_s = -kx \qquad [6.13]$$

where $k$ is a positive constant called the *force constant* (or *spring constant* or *stiffness constant*) of the spring. This force law for springs is known as **Hooke's law.** For many springs, Hooke's law can describe the behavior very accurately provided that the displacement from equilibrium is not too large. The value of $k$ is a measure of the stiffness of the spring. Stiff springs have larger $k$ values, and weak springs have smaller $k$ values. We shall employ a simplification model in which all springs obey Hooke's law unless specified otherwise.

The negative sign in Equation 6.13 signifies that the force exerted by the spring on the block is always directed *opposite* the displacement from the equilibrium position $x = 0$. For example, when $x > 0$, such that the block is pulled to the right and the spring is stretched as in Active Figure 6.8a, the spring force is to the left, or negative. When $x < 0$, and the spring is compressed as in Active Figure 6.8c, the spring force is to the right, or positive. Of course, when $x = 0$, as in Active Figure 6.8b, the spring is unstretched and $F_s = 0$. Because the spring force always acts toward the equilibrium position, it is sometimes called a *restoring force.*

If the block is displaced to a position $-x_{\text{max}}$ and then released, it moves from $-x_{\text{max}}$ through zero to $+x_{\text{max}}$ (assuming a frictionless surface) and then turns around and returns to $-x_{\text{max}}$. The details of this oscillating motion will be

**ACTIVE FIGURE 6.8**
The force exerted by a spring on a block varies with the block's displacement from the equilibrium position $x = 0$. (a) When $x$ is positive (stretched spring), the spring force is to the left. (b) When $x$ is zero (natural length of the spring), the spring force is zero. (c) When $x$ is negative (compressed spring), the spring force is to the right. (d) Graph of $F_s$ versus $x$ for the block–spring system. The work done by the spring force as the block moves from $-x_{max}$ to 0 is the area of the shaded triangle, $\frac{1}{2}kx_{max}^2$.

**Physics Now™** Observe the block's motion for various maximum displacements and spring constants by logging into PhysicsNow at **www.pop4e.com** and going to Active Figure 6.8.

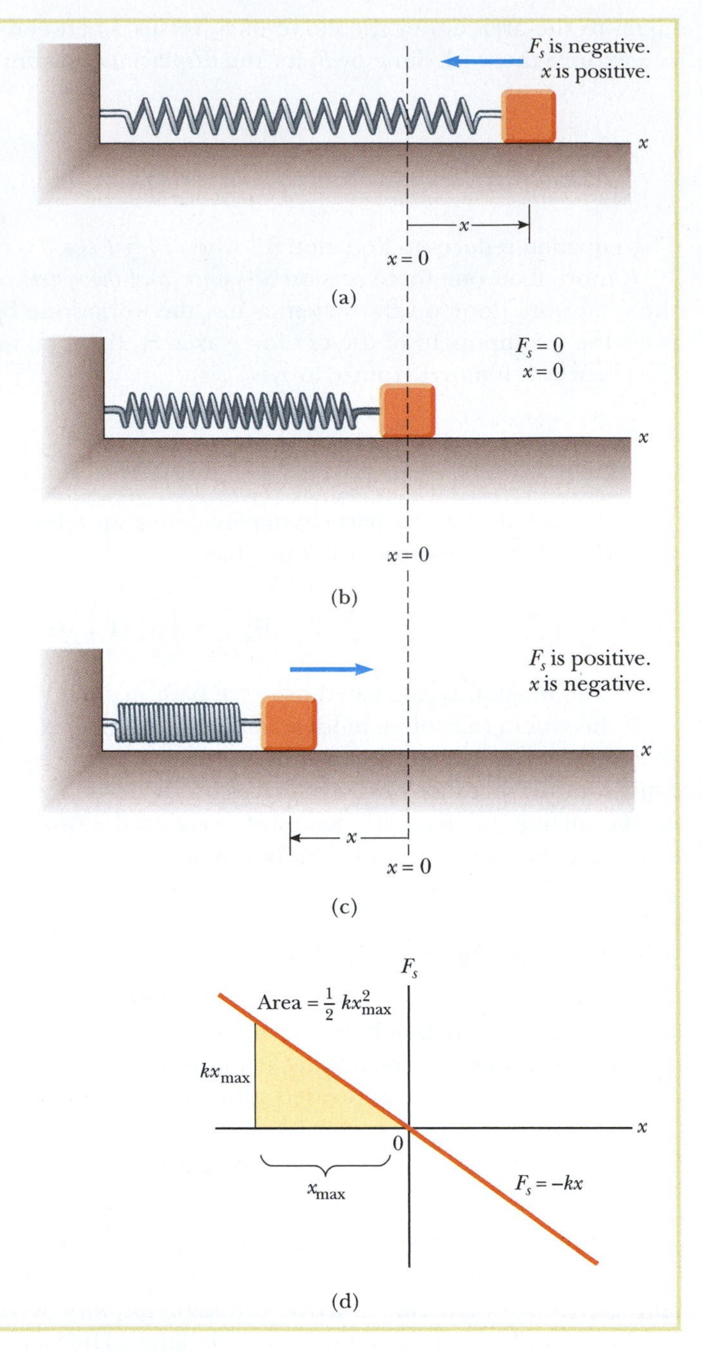


discussed in Chapter 12. For our purposes here, let us calculate the work done by the spring force on the block as the block moves from $x_i = -x_{max}$ to $x_f = 0$. Applying the particle model to the block and using Equation 6.11, we have

$$W_s = \int_{x_i}^{x_f} F_s\,dx = \int_{-x_{max}}^{0} (-kx)\,dx = \tfrac{1}{2}kx_{max}^2 \qquad [6.14]$$

The work done by the spring force on the block is positive because the spring force is in the same direction as the displacement (both are to the right).

If we consider the work done by the spring force on the block as the block continues to move from $x_i = 0$ to $x_f = x_{max}$, we find that $W_s = -\frac{1}{2}kx_{max}^2$. This work is negative because for this part of the motion the displacement is to the right and the spring force is to the left. Therefore, the *net* work done by the spring force on the block as it moves from $x_i = -x_{max}$ to $x_f = x_{max}$ is *zero.*

If we plot $F_s$ versus $x$, as in Active Figure 6.8d, we arrive at the same results. The work calculated in Equation 6.14 is equal to the area of the shaded triangle in Active Figure 6.8d, with base $x_{max}$ and height $kx_{max}$. This area is $\frac{1}{2}kx_{max}^2$.

If the block undergoes an *arbitrary* displacement from $x = x_i$ to $x = x_f$, the work done by the spring force is

$$W_s = \int_{x_i}^{x_f} (-kx)\, dx = \tfrac{1}{2}kx_i^2 - \tfrac{1}{2}kx_f^2 \qquad [6.15]$$

■ Work done by a spring

From this equation we see that the work done by the spring force on the block is zero for any motion that ends where it began ($x_i = x_f$). We shall make use of this important result in Chapter 7, where we describe the motion of this system in more detail. Equation 6.15 also shows that the work done by the spring force is zero when the block moves between any two symmetric locations, $x_i = -x_f$. Consider the curve representing the spring force in Active Figure 6.8d; if the block moves from $x = -x_{max}$ to $x = +x_{max}$, the total work is zero because we are adding a positive area (for $-x_{max} < x < 0$) to a negative area (for $0 < x < +x_{max}$) of equal magnitude.

Equations 6.14 and 6.15 describe the work done by the spring force on the block. Now consider the work done by an *external agent* on the block as the agent applies a force to the spring and stretches it *very slowly* from $x_i = -x_{max}$ to $x_f = 0$ as in Figure 6.9. This work can be easily calculated by noting that the applied force $\vec{\mathbf{F}}_{app}$ is of equal magnitude and opposite direction to the spring force $\vec{\mathbf{F}}_s$ at any value of the position (because the block is not accelerating), so that $F_{app} = -(-kx) = +kx$. The work done by this applied force (the external agent) on the block is therefore

$$W_{F_{app}} = \int_{-x_{max}}^{0} F_{app}\, dx = \int_{-x_{max}}^{0} kx\, dx = -\tfrac{1}{2}kx_{max}^2$$

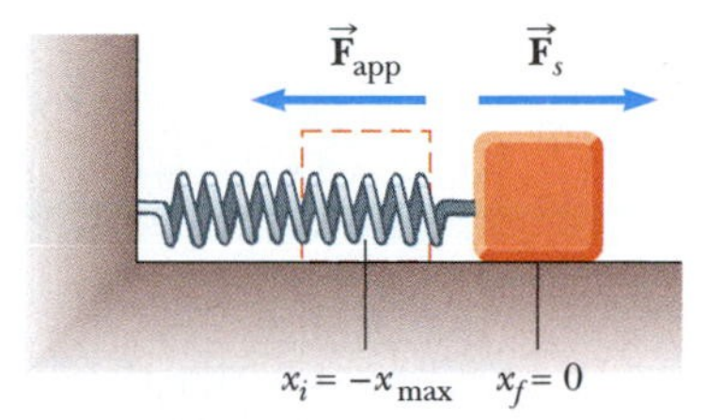


FIGURE 6.9 A block moves from $x_i = -x_{max}$ to $x_f = 0$ on a frictionless surface as a force $\vec{\mathbf{F}}_{app}$ is applied to the block. If the process is carried out very slowly, the applied force is equal in magnitude and opposite in direction to the spring force at all times.

Note that this work is equal to the negative of the work done by the spring force on the block for this displacement (Eq. 6.14). The work is negative because the external agent must push to the left on the spring in Figure 6.9 to prevent it from expanding, and this direction is opposite the direction of the displacement as the block moves from $-x_{max}$ to 0.

**QUICK QUIZ 6.3** A dart is loaded into a spring-loaded toy dart gun by pushing the spring in by a distance *d*. For the next loading, the spring is compressed a distance 2*d*. How much work is required to load the second dart compared to that required to load the first? **(a)** four times as much **(b)** two times as much **(c)** the same **(d)** half as much **(e)** one-fourth as much

## EXAMPLE 6.3 Work Required to Stretch a Spring

One end of a horizontal spring ($k = 80$ N/m) is held fixed while an external force is applied to the free end, stretching it slowly from $x_A = 0$ to $x_B = 4.0$ cm.

**A** Find the work done by the external force on the spring.

**Solution** Because we have not been told otherwise, we assume that the spring obeys Hooke's law. We place the zero reference of the coordinate axis at the free end of the unstretched spring. The applied force is $F_{app} = kx = (80\text{ N/m})(x)$. The work done by $F_{app}$ is the area of the triangle from 0 to 4.0 cm in Figure 6.10:

$$W = \tfrac{1}{2}kx_B^2 = \tfrac{1}{2}(80\text{ N/m})(0.040\text{ m})^2 = \boxed{0.064\text{ J}}$$

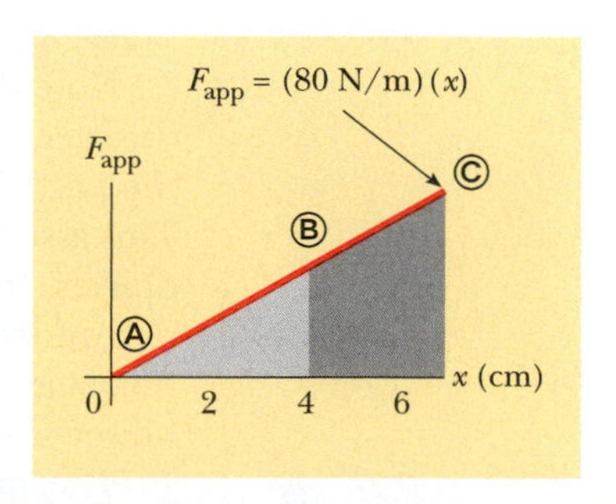


FIGURE 6.10 (Example 6.3) A graph of the applied force required to stretch a spring that obeys Hooke's law versus the elongation of the spring.

Note that the work is positive because the applied force and the displacement are in the same direction.

**B** Find the additional work done in stretching the spring from $x_B = 4.0$ cm to $x_C = 7.0$ cm.

**Solution** The work done in stretching the spring the additional amount is the darker shaded area between these limits in Figure 6.10. Geometrically, it is the difference in area between the large and small triangles:

$$W = \tfrac{1}{2}kx_C^2 - \tfrac{1}{2}kx_B^2$$
$$= \tfrac{1}{2}(80 \text{ N/m}) \, [(0.070 \text{ m})^2 - (0.040 \text{ m})^2] = \boxed{0.13 \text{ J}}$$

Using calculus, we find the same result:

$$W = \int_{x_B}^{x_C} F_{\text{app}} \, dx = \int_{0.040 \text{ m}}^{0.070 \text{ m}} (80 \text{ N/m}) \, x \, dx$$
$$= \tfrac{1}{2}(80 \text{ N/m}) \, (x^2) \Big|_{0.040 \text{ m}}^{0.070 \text{ m}}$$
$$W = \tfrac{1}{2}(80 \text{ N/m}) \, [(0.070 \text{ m})^2 - (0.040 \text{ m})^2] = \boxed{0.13 \text{ J}}$$

## 6.5 KINETIC ENERGY AND THE WORK–KINETIC ENERGY THEOREM

Now that we have explored various means of evaluating the work done by a force on a system, let us explore the significance and benefits of the energy approach. As we shall see in this section, if the work done by the net force on a particle can be calculated for a given displacement, the change in the particle's speed is easy to evaluate. Let's see how it is done.

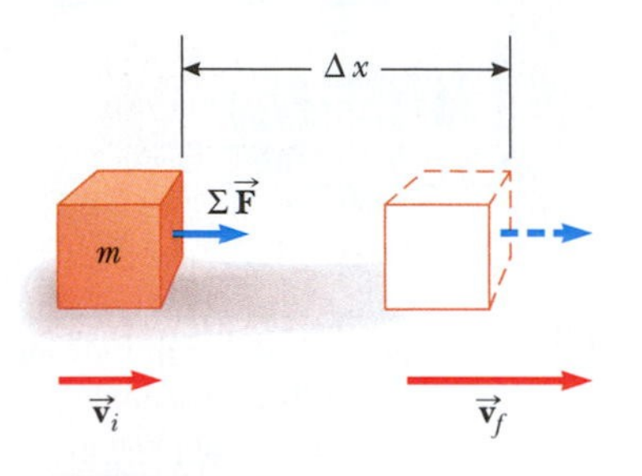


**FIGURE 6.11** An object modeled as a particle undergoes a displacement of magnitude $\Delta x$ and a change in speed under the action of a constant net force $\sum \vec{\mathbf{F}}$.

Figure 6.11 shows an object modeled as a particle of mass $m$ moving to the right along the $x$ axis under the action of a net force $\sum \vec{\mathbf{F}}$, also to the right. If the point of application of the force moves through a displacement $\Delta x = x_f - x_i$, the work done by the force $\sum \vec{\mathbf{F}}$ on the particle is

$$W_{\text{net}} = \int_{x_i}^{x_f} \sum F \, dx \qquad [6.16]$$

Using Newton's second law, we can substitute for the magnitude of the net force $\sum F = ma$ and then perform the following chain-rule manipulations on the integrand:

$$W_{\text{net}} = \int_{x_i}^{x_f} ma \, dx = \int_{x_i}^{x_f} m \frac{dv}{dt} \, dx = \int_{x_i}^{x_f} m \frac{dv}{dx} \frac{dx}{dt} \, dx = \int_{v_i}^{v_f} mv \, dv$$

$$W_{\text{net}} = \tfrac{1}{2}mv_f^2 - \tfrac{1}{2}mv_i^2 \qquad [6.17]$$

This equation was generated for the specific situation of one-dimensional motion, but it can also be used for two- or three-dimensional motion. It tells us that the work done by the net force on a particle of mass $m$ is equal to the difference between the initial and final values of a quantity $\frac{1}{2}mv^2$.

Note that in deriving Equation 6.17, the $dx$ we used to calculate the work was the displacement of the particle. In other words, we assumed that the displacement of the particle is the same as the displacement of the point of application of the force. This assumption is necessarily true for particles, but it may not be true for extended objects. It will only be true if the object is perfectly rigid, so that all parts of the object undergo the same displacement. Most of the situations that we will consider in this chapter and the next will satisfy this requirement. One important exception, however—objects subject to kinetic friction—will be explored in Section 6.7.

The quantity $\frac{1}{2}mv^2$ in Equation 6.17 is so important that we give it a special name. The **kinetic energy** $K$ of an object of mass $m$ moving with a speed $v$ is defined as

■ Kinetic energy of an object

$$K \equiv \tfrac{1}{2}mv^2 \qquad [6.18]$$

Kinetic energy is a scalar quantity and has the same units as work. For example, an object of mass 2.0 kg moving with a speed of 4.0 m/s has a kinetic energy of 16 J.

It is often convenient to write Equation 6.17 in the form

$$W_{\text{net}} = K_f - K_i = \Delta K \qquad [6.19]$$

■ The work–kinetic energy theorem

Equation 6.19 is an important result known as the **work–kinetic energy theorem:**

> When work is done on a system and the only change in the system is in its speed, the work done by the net force equals the change in kinetic energy of the system.

The work–kinetic energy theorem indicates that the speed of a particle increases if the net work done on it is positive because the final kinetic energy will be greater than the initial kinetic energy. The speed decreases if the net work is negative because the final kinetic energy will be less than the initial kinetic energy.

The work–kinetic energy theorem will clarify some results we saw earlier in this chapter that may have seemed odd. In Thinking Physics 6.1, a person lifts a block and moves it horizontally. At the end of the Reasoning, we mentioned that the net work done by all forces on the block is zero. That may seem strange, but it is correct. If we choose the block as the system, the net force on the system is zero because the upward lifting force is modeled as being equal in magnitude to the gravitational force. Therefore, the net force is zero and zero net work is done, which is consistent because the kinetic energy of the block does not change. It may seem incorrect that no work was done because something changed—the block was lifted—but that is correct *because we chose the block as the system.* If we had chosen the block and the Earth as the system, we would have a different result because the work done on this system is not zero. We will explore this idea in the next chapter.

In Section 6.4, we also saw a result of zero work done, when a block on a spring moved from $x_i = -x_{\text{max}}$ to $x_f = x_{\text{max}}$. The work is zero here for a different reason from that for lifting the block. It is the result of the combination of positive work and an equal amount of negative work done by the *same* force. It is also different from the lifting example in that the speed of the block on the spring is continually changing. The work–kinetic energy theorem refers only to the initial and final points for the speeds; it does not depend on details of the path followed between these points. We shall use this concept often in the remainder of this chapter and in the next chapter.

### PITFALL PREVENTION 6.5

**Conditions for the Work–Kinetic Energy Theorem** Always remember the special conditions for the work–kinetic energy theorem. We will see many situations in which other changes occur in the system besides its speed, and there are other interactions with the environment besides work. The work–kinetic energy theorem is important, but it is limited in its application and is not a general principle. We shall present a general principle involving energy in Section 6.6.

**QUICK QUIZ 6.4** A dart is loaded into a spring-loaded toy dart gun by pushing the spring in by a distance $d$. For the next loading, the spring is compressed a distance $2d$. How much faster does the second dart leave the gun compared with the first? **(a)** four times as fast **(b)** two times as fast **(c)** the same **(d)** half as fast **(e)** one-fourth as fast

### EXAMPLE 6.4 A Block Pulled on a Frictionless Surface

A 6.00-kg block initially at rest is pulled to the right along a horizontal frictionless surface by a constant, horizontal force $\vec{\mathbf{F}}$ of magnitude 12.0 N as in Figure 6.12. Find the speed of the block after it has moved 3.00 m.

**Solution** The block is the system, and three external forces interact with it. Neither the gravitational force nor the normal force does work on the block because these forces are vertical and the displacement of the block is horizontal. There is no friction, so the only

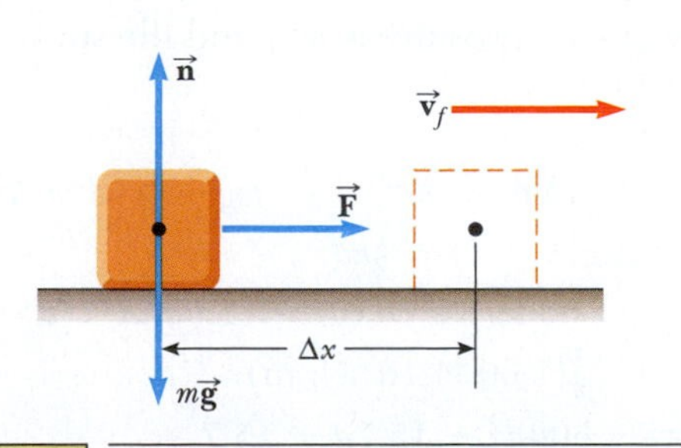


**FIGURE 6.12** (Example 6.4) A block on a frictionless surface is pulled to the right by a constant horizontal force.

external force that we must consider in the calculation is the 12.0-N force.

The work done by the 12.0-N force is

$$W = F\Delta x = (12.0 \text{ N})(3.00 \text{ m}) = 36.0 \text{ N}\cdot\text{m} = 36.0 \text{ J}$$

Using the work–kinetic energy theorem and noting that the initial kinetic energy is zero, we find

$$W = K_f - K_i = \tfrac{1}{2}mv_f^2 - 0$$

$$v_f = \sqrt{\frac{2W}{m}} = \sqrt{\frac{2(36.0 \text{ J})}{6.00 \text{ kg}}} = 3.46 \text{ m/s}$$

Notice that an energy calculation such as this one gives only the speed of the particle, not the velocity. In many cases, that is all you need. If you want to find the direction of the velocity vector, you may need to analyze the pictorial representation or perform other calculations. In this example, it is clear that $\vec{\mathbf{v}}_f$ is directed to the right.

## EXAMPLE 6.5 Dropping a Block onto a Spring

A massless spring that has a force constant of $1.00 \times 10^3$ N/m is placed on a table in a vertical position as in Figure 6.13. A block of mass 1.60 kg is held 1.00 m above the free end of the spring. The block is dropped from rest so that it falls vertically onto the spring. By what maximum distance does the spring compress?

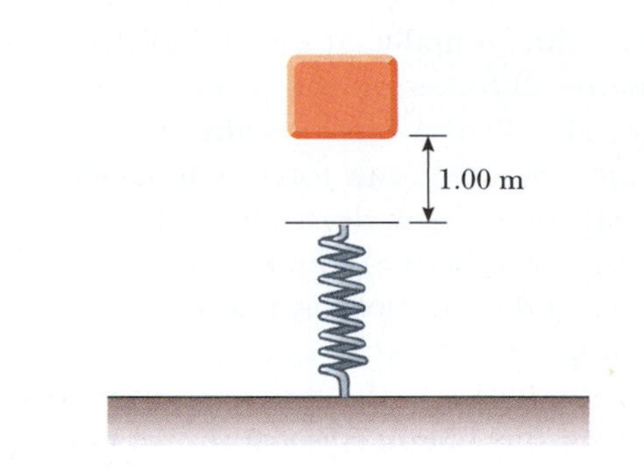


**FIGURE 6.13** (Example 6.5) A block is dropped onto a vertical spring, causing the spring to compress.

**Solution** Conceptualize the problem by imagining the block dropping on the spring and compressing the spring by some distance. The block is at rest momentarily before the compressed spring begins to move the block upward again. We want to focus on that instant of time at which the block is at rest. We identify the block as the system. We identify the initial condition as the release of the block from the height $y_i = h = 1.00$ m above the free end of the spring. The final condition occurs when the block is momentarily at rest with the spring compressed its maximum distance. For this condition, the block is located at $y_f = -d$, where $d$ is the maximum distance by which the spring is compressed. Because both the gravitational force and the spring force are doing work on the block, we categorize the problem as one that can be addressed with the work–kinetic energy theorem. To analyze the problem, we determine that the net work done on the block during its displacement between the initial and final positions by gravity (positive work) and the spring force (negative work) is

$$\begin{aligned} W_{\text{net}} &= \vec{\mathbf{F}}_g \cdot \Delta\vec{\mathbf{r}} - \tfrac{1}{2}kd^2 = (-mg)\hat{\mathbf{j}} \cdot (-d - h)\hat{\mathbf{j}} - \tfrac{1}{2}kd^2 \\ &= mg(h + d) - \tfrac{1}{2}kd^2 \\ &= (1.60 \text{ kg})(9.80 \text{ m/s}^2)(1.00 \text{ m} + d) \\ &\quad -\tfrac{1}{2}(1.00 \times 10^3 \text{ N/m})d^2 \\ &= -500d^2 + 15.7d + 15.7 \end{aligned}$$

The change in kinetic energy of the block is zero because it is at rest at both the initial and final conditions. Therefore, from the work–kinetic energy theorem, the work done by the net force must be equal to zero:

$$-500d^2 + 15.7d + 15.7 = 0$$

This quadratic equation can be solved, and the solutions are $d = 0.19$ m and $d = -0.16$ m. Because we have chosen the value of $d$ as a positive number by claiming that $y = -d$ is *below* the initial position of the end of the spring, we must choose the positive root, $d = 0.19$ m.

To finalize the problem, let us be sure that we can interpret the negative root. The negative root gives the position for the final condition as $y = -d = -(-0.16 \text{ m}) = +0.16$ m, which is the position above the *initial* position $y = 0$ at which the block again comes to rest in its oscillation, assuming that the block remains attached to the spring. These two positions are symmetric around $y = -0.016$ m, which is where the block would rest in equilibrium on the spring, according to Hooke's law.

## 6.6 THE NONISOLATED SYSTEM

We have seen a number of examples in which an object, modeled as a particle, is acted on by various forces, with the result that there is a change in its kinetic energy. This very simple situation is the first example of the **nonisolated system,** which is an important new analysis model for us. Physical problems for which this model is appropriate involve systems that interact with or are influenced by their environment, causing some kind of change in the system.

The work–kinetic energy theorem is our first introduction to the nonisolated system. The interaction is the work done by the external force and the quantity related to the system that changes is its kinetic energy. Because the energy of the system changes, we conceptualize work as a means of **energy transfer; work has the effect of transferring energy between the system and the environment.** If positive work is done on the system, energy is transferred to the system, whereas negative work indicates that energy is transferred from the system to the environment.

So far, we have discussed kinetic energy as the only type of energy in a system. We now argue the existence of a second type of energy. Consider a situation in which an object slides along a surface with friction. Clearly, work is done by the friction force because there is a force and a displacement of the object on which the force acts. Keep in mind, however, that our equations for work involve the displacement of the *point of application of the force.* If an object is perfectly rigid, the displacement of the point of application of the force is the same as the displacement of the object. For a nonrigid object, however, these displacements are not the same. Imagine, for example, a block of gelatin sitting on a plate. Suppose the block is pushed with a horizontal force applied to a vertical side so that the block deforms but does not slide on the plate. There has been a displacement of the object because most of the particles in the object, except for those along the stationary bottom edge, have moved horizontally through various displacements. The displacement of the point of application of the friction force between the block and the plate is zero, however, because the bottom of the block has not moved.

On a microscopic scale, real objects are deformable; it is the deformation and interaction of the surfaces in contact that cause the friction force. In general, the displacement of the point of application of the friction force (assuming that we could calculate it!) is not the same as the displacement of the object.[1]

Let us imagine the book in Figure 6.14 sliding to the right on the surface of a heavy table and slowing down as a result of the friction force. Suppose the *surface* is the system. The sliding book exerts a friction force to the right on the surface. As a result, many atoms on the surface move slightly to the right under the influence of this force. Consequently, the points of application of the friction force move to the right and the friction force does positive work on the surface. The surface, however, is not moving after the book has stopped. Positive work has been done on the surface, yet the kinetic energy of the surface does not increase. Is this situation a violation of the work–kinetic energy theorem?

It is not so much a violation as a misapplication because this situation does not fit the description of the conditions given for the work–kinetic energy theorem. The theorem requires that the only change in the system is in its speed, which is not the case here. Work is done on the system of the surface by the book, but the result of that work is *not* an increase in kinetic energy. From your everyday experience with sliding over surfaces with friction, you can probably guess that the surface will be *warmer* after the book slides over it (rub your hands together briskly to experience that!). Therefore, the work done has gone into warming the surface rather than causing it to increase in speed. We use the phrase **internal energy** $E_{\text{int}}$ for the

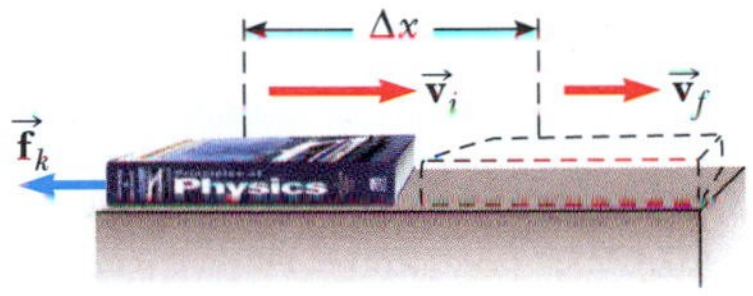

**FIGURE 6.14** A book sliding to the right on a horizontal surface slows down in the presence of a force of kinetic friction acting to the left. The initial velocity of the book is $\vec{\mathbf{v}}_i$, and its final velocity is $\vec{\mathbf{v}}_f$. The normal force and gravitational force are not included in the diagram because they are perpendicular to the direction of motion and therefore do not influence the speed of the book.

[1]For more details on energy transfer situations involving forces of kinetic friction, see B. A. Sherwood and W. H. Bernard, *American Journal of Physics* 52:1001, 1984; and R. P. Bauman, *The Physics Teacher* 30:264, 1992.

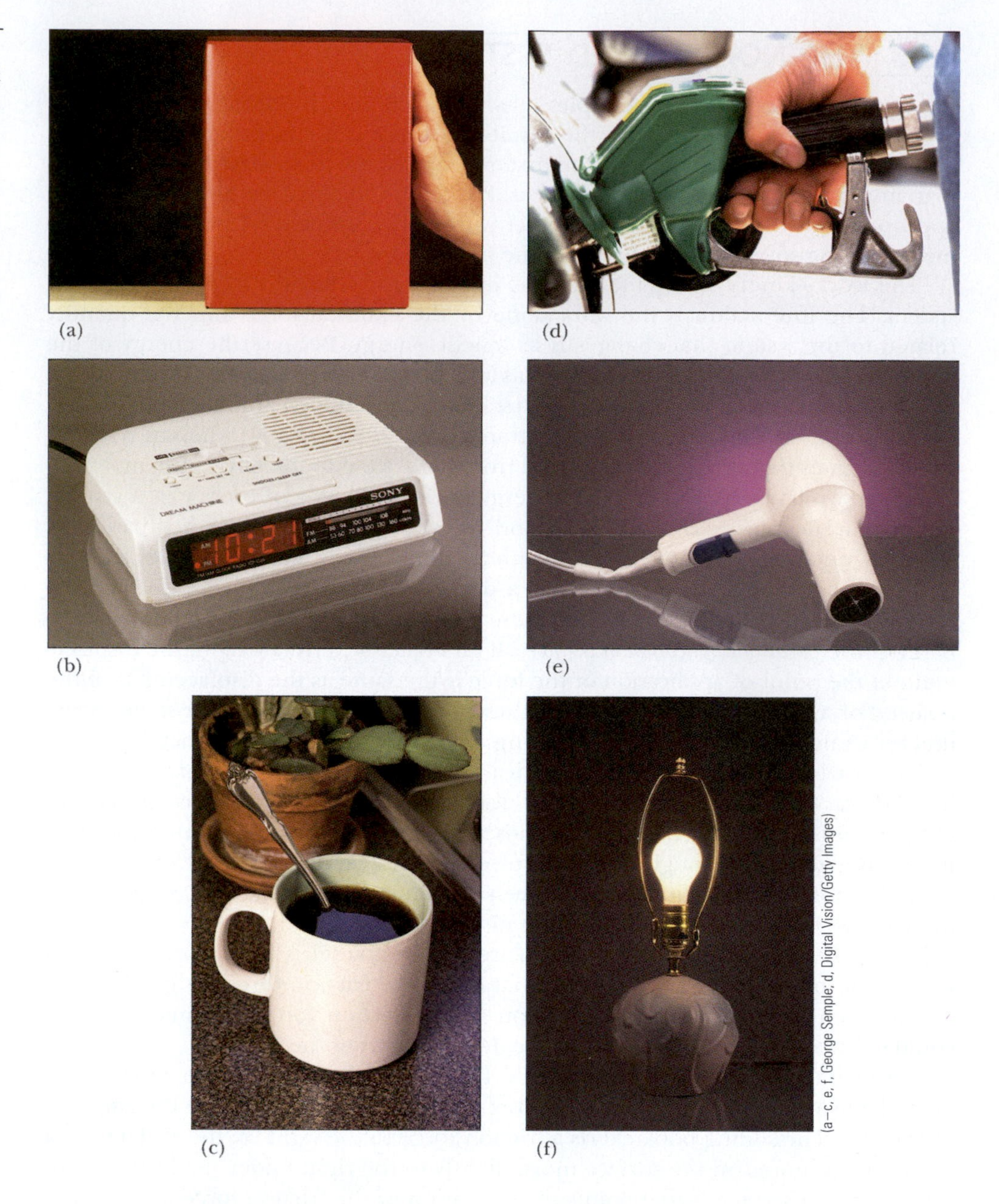


**FIGURE 6.15** Energy transfer mechanisms. (a) Energy is transferred to the block by *work,* (b) energy leaves the radio by *mechanical waves,* (c) energy transfers up the handle of the spoon by *heat,* (d) energy enters the automobile gas tank by *matter transfer,* (e) energy enters the hair dryer by *electrical transmission,* and (f) energy leaves the light bulb by *electromagnetic radiation.*

(a–c, e, f, George Semple; d, Digital Vision/Getty Images)

energy associated with an object's temperature. (We will see a more general definition for internal energy in Chapter 17.) In this case, the work done by the book on the surface does indeed represent energy transferred into the system, but it appears in the system as internal energy rather than kinetic energy.

We have now seen two methods of storing energy in a system: kinetic energy, related to motion of the system, and internal energy, related to its temperature. We have seen only one way to transfer energy into the system so far: work. Next, we introduce a few other ways to transfer energy into or out of a system, which will be studied in detail in other sections of the book. We will focus on the following six methods (Fig. 6.15) for transferring energy between the environment and the system.

■ Methods of energy transfer

**Work** (this chapter) is a method of transferring energy to a system by the application of a force to the system and a displacement of the point of application of the force, as we have seen in the previous sections (Fig. 6.15a).

**Mechanical waves** (Chapter 13) are a means of transferring energy by allowing a disturbance to propagate through air or another medium. This method is the one by which energy leaves a radio (Fig. 6.15b) through the loudspeaker—sound—and by which energy enters your ears to stimulate the hearing process. Mechanical waves also include seismic waves and ocean waves.

**Heat** (Chapter 17) is a method of transferring energy by means of microscopic collisions; for example, the end of a metal spoon in a cup of coffee becomes hot because fast-moving electrons and atoms in the bowl of the spoon bump into slower ones in the nearby part of the handle (Fig. 6.15c). These particles move faster because of the collisions and bump into the next group of slow particles. Therefore, the internal energy of the handle end of the spoon rises from energy transfer as a result of this bumping process. This process, also called *thermal conduction,* is caused by a temperature difference between two regions in space.[2]

In **matter transfer** (Chapter 17), matter physically crosses the boundary of the system, carrying energy with it. Examples include filling the system of your automobile tank with gasoline (Fig. 6.15d) and carrying energy to the rooms of your home by means of circulating warm air from the furnace. Matter transfer occurs in several situations and is introduced in Chapter 17 by means of one example, *convection.*

**Electrical transmission** (Chapter 21) involves energy transfer by means of electric currents. That is how energy transfers into your stereo system or any other electrical device such as a hair dryer (Fig. 6.15e).

**Electromagnetic radiation** (Chapter 24) refers to electromagnetic waves such as light, microwaves, and radio waves (Fig. 6.15f). Examples of this method of transfer include energy going into your baked potato in your microwave oven and light energy traveling from the Sun to the Earth through space.[3]

### PITFALL PREVENTION 6.6

**HEAT IS NOT A FORM OF ENERGY** The word *heat* is one of the most misused words in our popular language. In this text, heat is a method of *transferring* energy across a system boundary, *not* a form of stored energy. Therefore, phrases such as "heat content," "the heat of the summer," and "the heat escaped" all represent uses of this word that are inconsistent with our physics definition. See Chapter 17.

The central feature of the energy approach is the notion that **we can neither create nor destroy energy; energy is *conserved.*** Therefore, **if the amount of energy in a system changes, it can *only* be because energy has crossed the boundary by a transfer mechanism such as those listed above.** This general statement of the principle of **conservation of energy** can be described mathematically as follows:

$$\Delta E_{\text{system}} = \sum T \qquad [6.20]$$

■ Conservation of energy: the continuity equation for energy

where $E_{\text{system}}$ is the total energy of the system, including all methods of energy storage (kinetic, internal, and another to be discussed in Chapter 7) and $T$ (for *transfer*) is the amount of energy transferred across the system boundary by a transfer mechanism. Two of our transfer mechanisms have well-established symbolic notations. For work, $T_{\text{work}} = W$, as we have seen in this chapter, and for heat, $T_{\text{heat}} = Q$, which we will see in detail in Chapter 17. The other four members of our list do not have established symbols, so we will call them $T_{\text{MW}}$ (mechanical waves), $T_{\text{MT}}$ (matter transfer), $T_{\text{ET}}$ (electrical transmission), and $T_{\text{ER}}$ (electromagnetic radiation).

In this chapter, we have seen how to calculate work. The other types of transfers will be discussed in subsequent chapters. Equation 6.20 is called the **continuity**

[2]Many textbooks use the term *heat* to include *conduction, convection,* and *radiation.* Conduction is the only one of these three processes driven by a temperature difference alone, so we will restrict heat to this process in this book. Convection and radiation are included in other types of energy transfer in our list of six.

[3]Electromagnetic radiation and work done by field forces are the only energy transfer mechanisms that do not require molecules of the environment to be available at the system boundary. Therefore, systems surrounded by a vacuum (such as planets) can only exchange energy with the environment by means of these two possibilities.

**equation for energy.** A continuity equation arises in any situation in which the change in a quantity in a system occurs solely because of transfers across the boundary (because the quantity is conserved), several examples of which occur in various areas of physics, as we shall see.

The full expansion of Equation 6.20, with kinetic and internal energy as the storage mechanisms, is

$$\Delta K + \Delta E_{\text{int}} = W + T_{\text{MW}} + Q + T_{\text{MT}} + T_{\text{ET}} + T_{\text{ER}}$$

This equation is the primary mathematical representation of the energy analysis of the nonisolated system. In most cases, it reduces to a much simpler equation because some of the terms are zero. If, for a given system, all terms on the right side of the continuity equation for energy are zero, the system is an *isolated system,* which we study in the next chapter.

The concept described by Equation 6.20 is no more complicated in theory than is that of balancing your checking account statement. If your account is the system, the change in the account balance for a given month is the sum of all the transfers: deposits, withdrawals, fees, interest, and checks written. It may be useful for you to think of energy as the *currency of nature*!

Suppose a force is applied to a nonisolated system and the point of application of the force moves through a displacement. Further, suppose the only effect on the system is to increase its speed. Then the only transfer mechanism is work (so that $\sum T$ in Equation 6.20 reduces to just $W$) and the only kind of energy in the system that changes is the kinetic energy (so that $\Delta E_{\text{system}}$ reduces to just $\Delta K$). Equation 6.20 then becomes

$$\Delta K = W$$

which is the work–kinetic energy theorem, Equation 6.19. This theorem is a special case of the more general continuity equation for energy. In future chapters, we shall see several more examples of other special cases of the continuity equation for energy.

Bioluminescence

Equation 6.20 is not restricted to phenomena commonly described as belonging to the area of physics. For example, Figure 6.16 shows a glow worm whose last three segments of the abdomen glow with *bioluminescence.* In this process, chemical energy in the worm is transformed such that energy leaves the worm by electromagnetic radiation in the form of visible light. For this process, Equation

FIGURE 6.16 The glow worm *Lampyris noctiluca* is found in Great Britain and parts of continental Europe. It exhibits the phenomenon of *bioluminescence.* The light leaving the last three segments of its abdomen represents a transfer of energy out of the system of the worm.

(Sinclair Stammers/Science Photo Library/Photo Researchers, Inc.)

6.20 can be written

$$\Delta E_{\text{chem}} = T_{\text{ER}}$$

Chemical energy is a form of potential energy, which we will study in Chapter 7. Chemical energy is stored in any organism by means of food ingested by the organism. Therefore, the source of the light leaving the worm in Figure 6.16 is food ingested earlier by the worm.

**QUICK QUIZ 6.5** By what transfer mechanisms does energy enter and leave **(a)** your television set, **(b)** Your gasoline-powered lawn mower, and **(c)** your hand-cranked pencil sharpener?

**QUICK QUIZ 6.6** Consider a block sliding over a horizontal surface with friction. Ignore any sound the sliding might make. If we consider the system to be the *block,* this system is **(a)** isolated or **(b)** nonisolated. If we consider the system to be the *surface,* this system is **(c)** isolated or **(d)** nonisolated. If we consider the system to be the *block and the surface,* this system is **(e)** isolated or **(f)** nonisolated.

### Thinking Physics 6.3

A toaster is turned on. Discuss the forms of energy and energy transfer occurring in the coils of the toaster.

**Reasoning** We identify the coils as the system. The energy that changes in the system is *internal energy* because the temperature of the coils rises. The energy transfer mechanism for energy coming into the coils is *electrical transmission* through the wire plugged into the wall. Energy is transferring out of the coils by *electromagnetic radiation* because the coils are hot and glowing. Some transfer of energy also occurs by *heat* from the hot surfaces of the coils into the air. We could express this process in terms of the continuity equation for energy as

$$\Delta E_{\text{int}} = Q + T_{\text{ET}} + T_{\text{ER}}$$

After a short warm-up period, the temperature of the coils reaches a constant value and the internal energy will no longer change. In this situation, the energy input and output are balanced:

$$0 = Q + T_{\text{ET}} + T_{\text{ER}} \quad \rightarrow \quad -T_{\text{ET}} = Q + T_{\text{ER}}$$

Note that $Q$ and $T_{\text{ER}}$ are both negative because they represent energy leaving the system; $T_{\text{ET}}$ is positive because energy continues to enter the system by electrical transmission. ■

## 6.7 SITUATIONS INVOLVING KINETIC FRICTION

In the preceding section, we discussed the nature of the friction force and the situation with deformable objects. Let us see how to handle problems with friction forces such as that on our block in Figure 6.11 sliding on the surface.

Consider a situation in which forces, including friction, are applied to the block as it follows an arbitrary path in space and let us follow a similar procedure to that in generating Equation 6.17. We start by writing Equation 6.12 for all forces other than friction:

$$\sum W_{\text{other forces}} = \int \left( \sum \vec{\mathbf{F}}_{\text{other forces}} \right) \cdot d\vec{\mathbf{r}} \qquad [6.21]$$

The $d\vec{\mathbf{r}}$ in this equation is the displacement of the object because for forces other than friction, under the assumption that these forces do not deform the object, this displacement is the same as that of the point of application of the forces. To each side of Equation 6.21 let us add the integral of the scalar product of the force of kinetic friction and $d\vec{\mathbf{r}}$:

$$\sum W_{\text{other forces}} + \int \vec{\mathbf{f}}_k \cdot d\vec{\mathbf{r}} = \int \left( \sum \vec{\mathbf{F}}_{\text{other forces}} \right) \cdot d\vec{\mathbf{r}} + \int \vec{\mathbf{f}}_k \cdot d\vec{\mathbf{r}}$$

$$= \int \left( \sum \vec{\mathbf{F}}_{\text{other forces}} + \vec{\mathbf{f}}_k \right) \cdot d\vec{\mathbf{r}}$$

The integrand on the right side of this equation is the net force $\Sigma \vec{\mathbf{F}}$, so,

$$\sum W_{\text{other forces}} + \int \vec{\mathbf{f}}_k \cdot d\vec{\mathbf{r}} = \int \sum \vec{\mathbf{F}} \cdot d\vec{\mathbf{r}}$$

Incorporating Newton's second law $\sum \vec{\mathbf{F}} = m\vec{\mathbf{a}}$, gives us

$$\sum W_{\text{other forces}} + \int \vec{\mathbf{f}}_k \cdot d\vec{\mathbf{r}} = \int m\vec{\mathbf{a}} \cdot d\vec{\mathbf{r}} = \int m \frac{d\vec{\mathbf{v}}}{dt} \cdot d\vec{\mathbf{r}} = \int_{t_i}^{t_f} m \frac{d\vec{\mathbf{v}}}{dt} \cdot \vec{\mathbf{v}}\, dt \qquad [6.22]$$

where we have used Equation 3.5 to rewrite $d\vec{\mathbf{r}}$ as $\vec{\mathbf{v}}\,dt$. The scalar product obeys the product rule for differentiation (See Eq. B.30 in Appendix B.6), so the derivative of the scalar product of $\vec{\mathbf{v}}$ with itself can be written

$$\frac{d}{dt}(\vec{\mathbf{v}} \cdot \vec{\mathbf{v}}) = \frac{d\vec{\mathbf{v}}}{dt} \cdot \vec{\mathbf{v}} + \vec{\mathbf{v}} \cdot \frac{d\vec{\mathbf{v}}}{dt} = 2 \frac{d\vec{\mathbf{v}}}{dt} \cdot \vec{\mathbf{v}}$$

where we have used the commutative property of the scalar product to justify the final expression in this equation. Consequently,

$$\frac{d\vec{\mathbf{v}}}{dt} \cdot \vec{\mathbf{v}} = \frac{1}{2} \frac{d}{dt}(\vec{\mathbf{v}} \cdot \vec{\mathbf{v}}) = \frac{1}{2} \frac{dv^2}{dt}$$

Substituting this result into Equation 6.22, we find that

$$\sum W_{\text{other forces}} + \int \vec{\mathbf{f}}_k \cdot d\vec{\mathbf{r}} = \int_{t_i}^{t_f} m \left( \frac{1}{2} \frac{dv^2}{dt} \right) dt = \frac{1}{2} m \int_{v_i}^{v_f} d(v^2)$$

$$= \frac{1}{2} m v_f^2 - \frac{1}{2} m v_i^2 = \Delta K$$

Looking at the left side of this equation, we realize that in the inertial frame of the surface, $\vec{\mathbf{f}}_k$ and $d\vec{\mathbf{r}}$ will be in opposite directions for every increment $d\vec{\mathbf{r}}$ of the path followed by the object. Therefore, $\vec{\mathbf{f}}_k \cdot d\vec{\mathbf{r}} = -f_k\,dr$. The previous expression now becomes

$$\sum W_{\text{other forces}} - \int f_k\, dr = \Delta K$$

If the kinetic friction force is constant, $f_k$ can be brought out of the integral. The remaining integral $\int dr$ is simply the sum of increments of length along the path, which is the total path length $d$. Therefore,

■ The change in kinetic energy of an object due to friction and other forces

$$\sum W_{\text{other forces}} - f_k d = \Delta K \qquad [6.23]$$

This equation can be considered to be a modification of the work–kinetic energy theorem to be used when a constant friction force acts on an object. The change in kinetic energy is equal to the work done by all forces other than friction minus a term $f_k d$ associated with the friction force.

Now consider the larger system consisting of the block *and* the surface as the block slows down under the influence of a friction force alone. No work is done across the boundary of this system; the system does not interact with the environment, so there is no work done by other forces beside friction. In this case, Equation 6.23 becomes $-f_k d = \Delta K$. For this situation, Equation 6.20 becomes

$$\Delta K + \Delta E_{\text{int}} = 0$$

The change in kinetic energy of this system is the same as the change in kinetic energy of the system of the block because the block is the only part of the block–surface system that is moving. Therefore,

$$-f_k d + \Delta E_{\text{int}} = 0$$

$$\Delta E_{\text{int}} = f_k d \qquad [6.24]$$

▪ The increase in internal energy of a system due to friction

Therefore, the increase in internal energy of the system is equal to the product of the friction force and the path length through which the block moves. In summary, **a friction force transforms kinetic energy in a system to internal energy, and for a system in which the friction force alone acts, the increase in internal energy of the system is equal to its decrease in kinetic energy.**

**QUICK QUIZ 6.7** You are traveling along a freeway at 65 mi/h. You suddenly skid to a stop because of congestion in traffic. Where is the energy that your car once had as kinetic energy before you stopped? **(a)** It is all in internal energy in the road. **(b)** It is all in internal energy in the tires. **(c)** Some of it has transformed to internal energy and some of it transferred away by mechanical waves. **(d)** It all transferred away from your car by various mechanisms.

## INTERACTIVE EXAMPLE 6.6 A Block Pulled on a Rough Surface

A block of mass 6.00 kg initially at rest is pulled to the right by a constant horizontal force with magnitude $F = 12.0$ N (Fig. 6.17a). The coefficient of kinetic friction between the block and the surface is 0.150.

**A** Find the speed of the block after it has moved 3.00 m. (This question is Example 6.4 modified so that the surface is no longer frictionless.)

**Solution** We define the system as the block. Because the block moves in a straight line without reversing direction, the displacement $\Delta x$ of the block and the distance $d$ through which it moves are equal. We apply Equation 6.23:

$$\Delta K = -f_k d + \sum W_{\text{other forces}} = -\mu_k n d + Fd$$

The block is modeled as a particle in equilibrium in the vertical direction so that $n = mg$. Therefore,

$$\Delta K = -\mu_k mgd + Fd$$

Evaluating $\Delta K$, we have

$$K = -(0.150)(6.00 \text{ kg})(9.80 \text{ m/s}^2)(3.00 \text{ m}) + (12.0 \text{ N})(3.00 \text{ m}) = 9.54 \text{ J}$$

Now, we find $v_f$

$$\Delta K = \tfrac{1}{2}mv_f^2 - \tfrac{1}{2}mv_i^2$$

$$v_f = \sqrt{\left(\frac{2}{m}\right)\left(\Delta K + \tfrac{1}{2}mv_i^2\right)}$$

Substituting the numerical values, we find

$$v_f = \sqrt{\left(\frac{2}{6.00 \text{ kg}}\right)(9.54 \text{ J} + 0)} = 1.78 \text{ m/s}$$

Notice that this value is less than that calculated in Example 6.4 because of the effect of the friction force.

**B** Suppose the force $\vec{\mathbf{F}}$ is applied at an angle $\theta$ as shown in Figure 6.17b. At what angle should the force be applied to achieve the largest possible speed after the block has moved 3.00 m to the right?

**Solution** At first, we might guess that $\theta = 0$ is the optimal angle to transfer the maximum energy to the

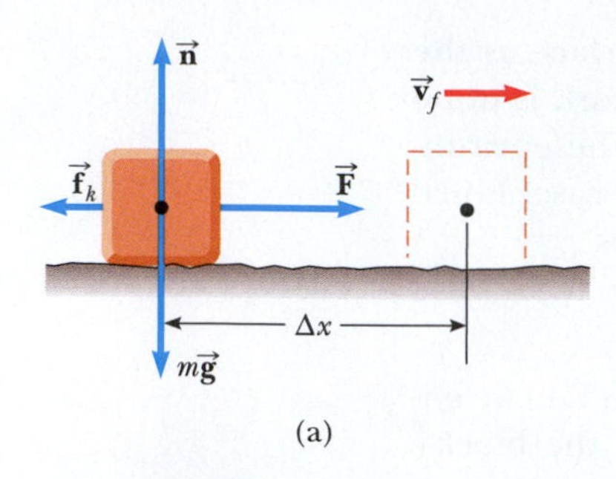

(a)

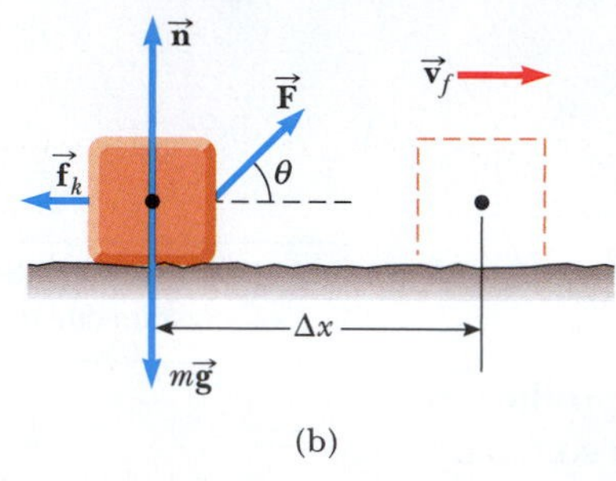

(b)

**FIGURE 6.17** (Interactive Example 6.6) (a) A block is pulled to the right by a constant horizontal force on a surface with friction. (b) The applied force is at an angle $\theta$ to the horizontal.

block. That would indeed be the case when pulling the block on a frictionless surface. With friction, however, pulling the block at some angle $\theta \neq 0$ reduces the normal force on the block, which in turn reduces the friction force. As a result, more energy can be transferred by work by pulling at some nonzero angle. For a nonzero angle $\theta$, the work done by the applied force is

$$W = F\Delta x \cos\theta = Fd\cos\theta$$

The block is in equilibrium in the vertical direction, so

$$\sum F_y = n + F\sin\theta - mg = 0$$

and

$$n = mg - F\sin\theta$$

Because $K_i = 0$, Equation 6.23 can be written as

$$\begin{aligned} K_f &= -f_k d + \sum W_{\text{other forces}} \\ &= -\mu_k n d + Fd\cos\theta \\ &= -\mu_k(mg - F\sin\theta)\,d + Fd\cos\theta \end{aligned}$$

Maximizing the speed is equivalent to maximizing the final kinetic energy. Consequently, we differentiate $K_f$ with respect to $\theta$ and set the result equal to zero:

$$\frac{d(K_f)}{d\theta} = -\mu_k(0 - F\cos\theta)\,d - Fd\sin\theta = 0$$

$$\mu_k\cos\theta - \sin\theta = 0$$

$$\tan\theta = \mu_k$$

For $\mu_k = 0.150$, we have

$$\theta = \tan^{-1}(\mu_k) = \tan^{-1}(0.150) = \boxed{8.53°}$$

If we test this result by examining the second derivative of $K_f$, we find indeed that this angle gives a maximum value.

**PhysicsNow™** Try out the effects of pulling the block at various angles by logging into PhysicsNow at www.pop4e.com and going to Interactive Example 6.6.

## INTERACTIVE EXAMPLE 6.7 A Block–Spring System

A block of mass 1.6 kg is attached to a horizontal spring that has a force constant of $1.0 \times 10^3$ N/m as shown in Active Figure 6.8. The spring is compressed 2.0 cm and is then released from rest.

**A** Calculate the speed of the block as it passes through the equilibrium position $x = 0$ if the surface is frictionless.

**Solution** In this situation, the block starts with $v_i = 0$ at $x_i = -2.0$ cm and we want to find $v_f$ at $x_f = 0$. We use Equation 6.14 to find the work done by the spring with $x_{\max} = x_i = -2.0 \text{ cm} = -2.0 \times 10^{-2}$ m:

$$\begin{aligned} W_s &= \tfrac{1}{2}kx_{\max}^2 = \tfrac{1}{2}(1.0 \times 10^3 \text{ N/m})(-2.0 \times 10^{-2}\text{ m})^2 \\ &= 0.20 \text{ J} \end{aligned}$$

Using the work–kinetic energy theorem with $v_i = 0$, we obtain the change in kinetic energy of the block as a result of the work done on it by the spring:

$$W_s = \tfrac{1}{2}mv_f^2 - \tfrac{1}{2}mv_i^2$$

$$v_f = \sqrt{v_i^2 + \frac{2}{m}W_s}$$

$$= \sqrt{0 + \frac{2}{1.6 \text{ kg}}(0.20 \text{ J})}$$

$$= \boxed{0.50 \text{ m/s}}$$

**B** Calculate the speed of the block as it passes through the equilibrium position if a constant friction force of 4.0 N retards the block's motion from the moment it is released.

**Solution** Certainly, the answer has to be less than what we found in part A because the friction force retards the motion. We use Equation 6.23:

$$\sum W_{\text{other forces}} - f_k d = \Delta K$$

$$W_s - f_k d = \tfrac{1}{2}mv_f^2 - \tfrac{1}{2}mv_i^2$$

$$v_f = \sqrt{v_i^2 + \frac{2}{m}(W_s - f_k d)}$$

Substituting the numerical values, we find

$$v_f = \sqrt{0 + \frac{2}{1.6\text{ kg}}[0.20\text{ J} - (4.0\text{ N})(2.0 \times 10^{-2}\text{ m})]}$$

$$= 0.39\text{ m/s}$$

As expected, this value is somewhat less than the 0.50 m/s we found in part A.

**PhysicsNow™** Investigate the role of the spring constant, amount of spring compression, and surface friction by logging into PhysicsNow at www.pop4e.com and going to Interactive Example 6.7.

## 6.8 POWER

We discussed transfers of energy across the boundary of a system by a number of methods. From a practical viewpoint, it is interesting to know not only the amount of energy transferred to a system but also the *rate* at which the energy is transferred. The time rate of energy transfer is called **power.**

We shall focus on work as our particular energy transfer method in this discussion, but keep in mind that the notion of power is valid for *any* means of energy transfer. If an external force is applied to an object (for which we will adopt the particle model) and if the work done by this force is $W$ in the time interval $\Delta t$, the **average power** during this interval is defined as

$$\mathcal{P}_{\text{avg}} \equiv \frac{W}{\Delta t} \tag{6.25}$$

The **instantaneous power** $\mathcal{P}$ at a particular point in time is the limiting value of the average power as $\Delta t$ approaches zero:

$$\mathcal{P} \equiv \lim_{\Delta t \to 0} \frac{W}{\Delta t} = \frac{dW}{dt} \tag{6.26}$$

where we represent the infinitesimal value of the work done by $dW$. We know from Equation 6.4 that we can write the infinitesimal amount of work done over a displacement $d\vec{\mathbf{r}}$ as $dW = \vec{\mathbf{F}} \cdot d\vec{\mathbf{r}}$. Therefore, the instantaneous power can be written

$$\mathcal{P} = \frac{dW}{dt} = \vec{\mathbf{F}} \cdot \frac{d\vec{\mathbf{r}}}{dt} = \vec{\mathbf{F}} \cdot \vec{\mathbf{v}} \tag{6.27}$$

where we have used $\vec{\mathbf{v}} = d\vec{\mathbf{r}}/dt$.

In general, power is defined for any type of energy transfer. The most general expression for power is therefore

$$\mathcal{P} = \frac{dE}{dt} \tag{6.28}$$

■ General expression for power

where $dE/dt$ is the rate at which energy is crossing the boundary of the system by transfer mechanisms.

The SI unit of power is joules per second (J/s), also called a **watt** (W) (after James Watt):

$$1\text{ W} = 1\text{ J/s} = 1\text{ kg}\cdot\text{m}^2/\text{s}^3$$

The unit of power in the U.S. customary system is the **horsepower** (hp):

$$1\text{ hp} \equiv 550\text{ ft}\cdot\text{lb/s} = 746\text{ W}$$

**PITFALL PREVENTION 6.7**

**Be careful with power** Do not confuse the symbol W for the watt with the italic symbol $W$ for work. Also, remember that the watt already represents a rate of energy transfer, so we do not want to say something like "watts per second" for power. The watt is *the same as* a joule per second.

A new unit of energy can now be defined in terms of the unit of power. One **kilowatt-hour** (kWh) is the energy transferred in a time interval of 1 h at the constant rate of 1 kW. The numerical value of 1 kWh of energy is

$$1 \text{ kWh} = (10^3 \text{ W})(3\,600 \text{ s}) = 3.60 \times 10^6 \text{ J}$$

It is important to realize that a kilowatt-hour is a unit of energy, not power. When you pay your electric bill, you are buying energy, and the amount of energy transferred by electrical transmission into a home during the period represented by the electric bill is usually expressed in kilowatt-hours. For example, your bill may state that you used 900 kWh of energy during a month and that you are being charged a rate of 10¢ per kWh. Your obligation is then $90 for this amount of energy. As another example, suppose an electric bulb is rated at 100 W. In 1.00 h of operation, it will have energy transferred to it by electrical transmission in the amount of $(0.100 \text{ kW})(1.00 \text{ h}) = 0.100 \text{ kWh} = 3.60 \times 10^5 \text{ J}$.

## EXAMPLE 6.8 Power Delivered by an Elevator Motor

A 1 000-kg elevator carries a maximum load of 800 kg. A constant friction force of 4 000 N retards its motion upward as in Figure 6.18.

**A** What is the minimum power delivered by the motor to lift the elevator at a constant speed of 3.00 m/s?

**Solution** We use two analysis models for the elevator. First, we model it as a particle in equilibrium because it moves at constant speed. The motor must supply the force $\vec{\mathbf{T}}$ that results in the tension in the cable that pulls the elevator upward. From Newton's second law and from $a = 0$ because $v$ is constant, we have

$$T - f - Mg = 0$$

where $M$ is the *total* mass (elevator plus load), equal to 1 800 kg. Therefore,

$$\begin{aligned} T &= f + Mg \\ &= 4.00 \times 10^3 \text{ N} + (1.80 \times 10^3 \text{ kg})(9.80 \text{ m/s}^2) \\ &= 2.16 \times 10^4 \text{ N} \end{aligned}$$

We now model the elevator as a nonisolated system. Work is being done on it by the tension force (as well as other forces). We can use Equation 6.27 to evaluate the power delivered by the motor, which is the rate at which work is done on the elevator by the tension force. Because $\vec{\mathbf{T}}$ is in the same direction as $\vec{\mathbf{v}}$, we have

$$\begin{aligned} \mathcal{P} &= \vec{\mathbf{T}} \cdot \vec{\mathbf{v}} = Tv \\ &= (2.16 \times 10^4 \text{ N})(3.00 \text{ m/s}) = 6.48 \times 10^4 \text{ W} \\ &= 64.8 \text{ kW} \end{aligned}$$

Because $\vec{\mathbf{T}}$ is the force the motor applies to the cable, the preceding result represents the rate at which energy is being transferred out of the motor by doing work on the cable.

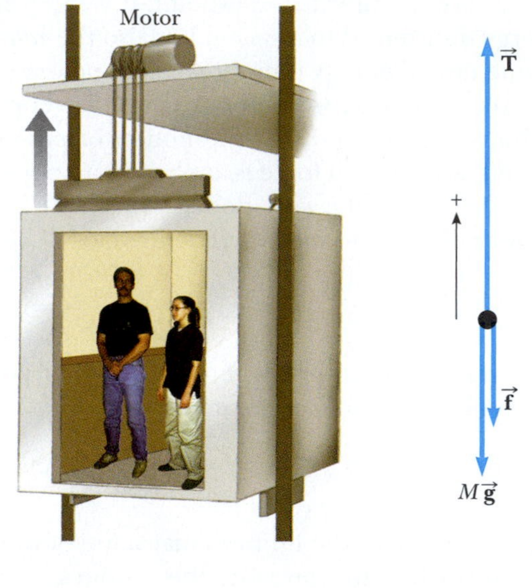


**FIGURE 6.18** (Example 6.8) (a) A motor lifts an elevator car. (b) Free-body diagram for the elevator. The motor exerts an upward force $\vec{\mathbf{T}}$ on the supporting cables. The magnitude of this force is $T$, the tension in the cables, which is applied in the upward direction on the elevator. The downward forces on the elevator are the friction force $\vec{\mathbf{f}}$ and the gravitational force $\vec{\mathbf{F}}_g = M\vec{\mathbf{g}}$.

**B** What power must the motor deliver at any instant if it is designed to provide an upward acceleration of 1.00 m/s²?

**Solution** In this case, we expect the tension to be larger than in part A because the cable must now cause an upward acceleration of the elevator. Modeling the elevator as a particle under a net force, we apply Newton's second law, which gives

$$T - f - Mg = Ma$$

$$T = M(a + g) + f$$
$$= (1.80 \times 10^3 \text{ kg})(1.00 \text{ m/s}^2 + 9.80 \text{ m/s}^2) + 4.00 \times 10^3 \text{ N}$$
$$= 2.34 \times 10^4 \text{ N}$$

Therefore, using Equation 6.27, we have for the required power

$$\mathcal{P} = Tv = (2.34 \times 10^4 \, v)$$

where $v$ is the instantaneous speed of the elevator in meters per second. Hence, the power required increases with increasing speed.

## 6.9 HORSEPOWER RATINGS OF AUTOMOBILES

CONTEXT CONNECTION

As discussed in Section 4.8, an automobile moves because of Newton's third law. The engine attempts to rotate the wheels in such a direction as to push the Earth toward the back of the car because of the friction force between the wheels and the roadway. By Newton's third law, the Earth pushes in the opposite direction on the wheels, which is toward the front of the car. Because the Earth is much more massive than the car, the Earth remains stationary while the car moves forward.

This principle is the same one humans use for walking. By pushing your leg backward while your foot is on the ground, you apply a friction force backward on the surface of the Earth. By Newton's third law, the surface applies a forward friction force on you, which causes your body to move forward.

The strength of the friction force $\vec{\mathbf{f}}$ exerted on a car by the roadway is related to the rate at which energy is transferred to the wheels to set them into rotation, which is the power of the engine:

$$\mathcal{P}_{\text{avg}} = \frac{\Delta E}{\Delta t} = \frac{f\Delta x}{\Delta t} = fv \quad \rightarrow \quad \mathcal{P} \leftrightarrow f$$

where the symbol $\leftrightarrow$ implies a relationship between the variables that is not necessarily an exact proportionality. In turn, the magnitude of the driving force is related to the acceleration of the car owing to Newton's second law:

$$f = ma \quad \rightarrow \quad f \propto a$$

Consequently, there should be a close relationship between the power rating of a vehicle and the possible acceleration of the vehicle:

$$\mathcal{P} \leftrightarrow a$$

Let us see if this relationship exists for actual data. For automobiles, a common unit for power is the *horsepower* (hp), defined in Section 6.8. Table 6.1 shows the gasoline-powered automobiles we have studied in the preceding chapters. The fourth column provides the published horsepower rating of each vehicle. The final column shows the ratio of the horsepower rating to the acceleration. Consider first the *Performance vehicles* section of the table. The ratio of power to acceleration is similar for all these vehicles, demonstrating the relationship between power and acceleration that we proposed.

In the second part of the table, under *Traditional vehicles,* there is a wider range of ratios of power to acceleration. This range is correlated to the range of vehicle masses in this listing. Notice that the BMW Mini Cooper S, Acura Integra GS, and Volkswagen Beetle have relatively low ratios and are cars with relatively small masses. It takes less power to accelerate this much mass to 60 mi/h than for a heavier car. Conversely, the two SUVs in this listing, the Cadillac Escalade and the Toyota Sienna, have the highest ratios of power to acceleration in this part of the table, 49 hp/mi/h · s and 32 hp/mi/h · s, respectively.

**TABLE 6.1 Horsepower Ratings and Accelerations of Various Vehicles**

| Automobile | Time Interval, 0 to 60 mi/h (s) | Acceleration (mi/h · s) | Horsepower Rating (hp) | Ratio of Horsepower Rating to Acceleration (hp/mi/h · s) |
|---|---|---|---|---|
| *Performance vehicles* | | | | |
| Aston Martin DB7 Vantage | 5.0 | 12.0 | 414 | 35 |
| BMW Z8 | 4.6 | 13.0 | 394 | 30 |
| Chevrolet Corvette | 4.6 | 13.0 | 385 | 30 |
| Dodge Viper GTS-R | 4.2 | 14.3 | 460 | 32 |
| Ferrari F50 | 3.6 | 16.7 | 513 | 31 |
| Ferrari 360 Spider F1 | 4.6 | 13.0 | 395 | 30 |
| Lamborghini Diablo GT | 3.6 | 16.7 | 567 | 34 |
| Porsche 911 GT2 | 4.0 | 15.0 | 456 | 30 |
| *Traditional vehicles* | | | | |
| Acura Integra GS | 7.9 | 7.6 | 140 | 18 |
| BMW Mini Cooper S | 6.9 | 8.7 | 163 | 19 |
| Cadillac Escalade (SUV) | 8.6 | 7.0 | 345 | 49 |
| Dodge Stratus | 7.5 | 8.0 | 200 | 25 |
| Lexus ES300 | 8.6 | 7.0 | 200 | 29 |
| Mitsubishi Eclipse GT | 7.0 | 8.6 | 205 | 24 |
| Nissan Maxima | 6.7 | 9.0 | 222 | 25 |
| Pontiac Grand Prix | 8.5 | 7.1 | 200 | 28 |
| Toyota Sienna (SUV) | 8.3 | 7.2 | 230 | 32 |
| Volkswagen Beetle | 7.6 | 7.9 | 150 | 19 |

# SUMMARY

**Physics Now™** Take a practice test by logging intoPhysics-Now at www.pop4e.com and clicking on the Pre-Test link for this chapter.

A **system** can be a single particle, a collection of particles, or a region of space. A **system boundary** separates the system from the **environment.** Many physics problems can be solved by considering the interaction of a system with its environment.

The **work** done by a *constant* force $\vec{\mathbf{F}}$ on a particle is defined as the product of the magnitude $F$ of the force, the magnitude $\Delta r$ of the displacement of the point of application of the force, and $\cos\theta$, where $\theta$ is the angle between the force vector and the displacement vector $\Delta\vec{\mathbf{r}}$:

$$W \equiv F\,\Delta r\cos\theta \qquad [6.1]$$

The **scalar** or **dot product** of any two vectors $\vec{\mathbf{A}}$ and $\vec{\mathbf{B}}$ is defined by the relationship

$$\vec{\mathbf{A}}\cdot\vec{\mathbf{B}} \equiv AB\cos\theta \qquad [6.3]$$

where the result is a scalar quantity and $\theta$ is the angle between the directions of the two vectors. The scalar product obeys the commutative and distributive laws.

The scalar product allows us to write the work done by a constant force $\vec{\mathbf{F}}$ on a particle as

$$W = \vec{\mathbf{F}}\cdot\Delta\vec{\mathbf{r}} \qquad [6.4]$$

The work done by a *varying* force acting on a particle moving along the $x$ axis from $x_i$ to $x_f$ is

$$W = \int_{x_i}^{x_f} F_x\,dx \qquad [6.11]$$

where $F_x$ is the component of force in the $x$ direction. If several forces act on the particle, the net work done by all forces is the sum of the individual amounts of work done by each force.

The **kinetic energy** of a particle of mass $m$ moving with a speed $v$ is

$$K \equiv \tfrac{1}{2}mv^2 \qquad [6.18]$$

The **work–kinetic energy theorem** states that when work is done on a system and the only change in the system is in its speed, the net work done on the system by external forces equals the change in kinetic energy of the system:

$$W_{\text{net}} = K_f - K_i = \Delta K \qquad [6.19]$$

For a nonisolated system, we can equate the change in the total energy stored in the system to the sum of all the transfers of energy across the system boundary:

$$\Delta E_{\text{system}} = \sum T \qquad [6.20]$$

which is the **continuity equation for energy.** Methods of energy transfer ($T$) include **work** ($T = W$), **mechanical waves** ($T_{MW}$), **heat** ($T = Q$), **matter transfer** ($T_{MT}$), **electrical transmission** ($T_{ET}$), and **electromagnetic radiation** ($T_{ER}$). Storage mechanisms ($E_{system}$) seen in this chapter include **kinetic energy** $K$ and **internal energy** $E_{int}$. The continuity equation arises because **energy is conserved;** we can neither create nor destroy energy. The work–kinetic energy theorem is a special case of the continuity equation for energy in situations in which work is the only transfer mechanism and kinetic energy is the only type of energy storage in the system.

In the case of an object sliding through a distance $d$ over a surface with friction, the change in kinetic energy of the system is found from

$$\sum W_{\text{other forces}} - f_k d = \Delta K \qquad [6.23]$$

where $f_k$ is the force of kinetic friction and $\sum W_{\text{other forces}}$ is the work done by all forces other than friction.

**Average power** is the time rate of energy transfer. If we use work as the energy transfer mechanism,

$$\mathcal{P}_{\text{avg}} \equiv \frac{W}{\Delta t} \qquad [6.25]$$

If an agent applies a force $\vec{\mathbf{F}}$ to an object moving with a velocity $\vec{\mathbf{v}}$, the **instantaneous power** delivered by that agent is

$$\mathcal{P} \equiv \frac{dW}{dt} = \vec{\mathbf{F}} \cdot \vec{\mathbf{v}} \qquad [6.27]$$

Because power is defined for any type of energy transfer, the general expression for power is

$$\mathcal{P} = \frac{dE}{dt} \qquad [6.28]$$

# QUESTIONS

☐ = answer available in the *Student Solutions Manual and Study Guide*

1. When a particle rotates in a circle, a force acts on it directed toward the center of rotation. Why is it that this force does no work on the particle?
2. When a punter kicks a football, is he doing any work on the ball while his toe is in contact with it? Is he doing any work on the ball after it loses contact with his toe? Are any forces doing work on the ball while it is in flight?
3. Cite two examples in which a force is exerted on an object without doing any work on the object.
4. Discuss the work done by a pitcher throwing a baseball. What is the approximate distance through which the force acts as the ball is thrown?
5. [5.] As a simple pendulum swings back and forth, the forces acting on the suspended object are the gravitational force, the tension in the supporting cord, and air resistance. (a) Which of these forces, if any, does no work on the pendulum? (b) Which of these forces does negative work at all times during the pendulum's motion? (c) Describe the work done by the gravitational force while the pendulum is swinging.
6. If the scalar product of two vectors is positive, does that imply that the vectors must have positive rectangular components?
7. For what values of $\theta$ is the scalar product (a) positive and (b) negative?
8. A certain uniform spring has spring constant $k$. Now the spring is cut in half. What is the relationship between $k$ and the spring constant $k'$ of each resulting smaller spring? Explain your reasoning.
9. [9.] Can kinetic energy be negative? Explain.
10. Two sharpshooters fire 0.30-caliber rifles using identical shells. A force exerted by expanding gases in the barrels accelerates the bullets. The barrel of rifle A is 2.00 cm longer than the barrel of rifle B. Which rifle will have the higher muzzle speed?
11. [11.] One bullet has twice the mass of a second bullet. If both are fired so that they have the same speed, which has more kinetic energy? What is the ratio of the kinetic energies of the two bullets?
12. You are reshelving books in a library. You lift a book from the floor to the top shelf. The kinetic energy of the book on the floor was zero and the kinetic energy of the book sitting on the top shelf is zero, so no change occurs in the kinetic energy. Yet you did some work in lifting the book. Is the work–kinetic energy theorem violated?
13. [13.] (a) If the speed of a particle is doubled, what happens to its kinetic energy? (b) What can be said about the speed of a particle if the net work done on it is zero?
14. A car salesperson claims that a souped-up 300-hp engine is a necessary option in a compact car in place of the conventional 130-hp engine. Suppose you intend to drive the car within speed limits ($\leq$ 65 mi/h) on flat terrain. How would you counter this sales pitch?
15. Can the average power over a time interval ever be equal to the instantaneous power at an instant within the interval? Explain.
16. Words given quantitative definitions in physics are sometimes used in popular literature in interesting ways. For example, a rock falling from the top of a cliff is said to be "gathering force as it falls to the beach below." What does the phrase "gathering force" mean, and can you repair this phrase?
17. In most circumstances, the normal force acting on an object and the force of static friction do zero work on the object. The reason that the work is zero is different for the two cases, however. Explain why each does zero work.
18. "A level air track can do no work." Argue for or against this statement.

## PROBLEMS

1, 2, 3 = straightforward, intermediate, challenging
☐ = full solution available in the *Student Solutions Manual and Study Guide*
PhysicsNow™ = coached problem with hints available at www.pop4e.com
= computer useful in solving problem
= paired numerical and symbolic problems
= biomedical application

### Section 6.2 ■ Work Done by a Constant Force

1. A block of mass 2.50 kg is pushed 2.20 m along a frictionless horizontal table by a constant 16.0-N force directed 25.0° below the horizontal. Determine the work done on the block by (a) the applied force, (b) the normal force exerted by the table, and (c) the gravitational force. (d) Determine the total work done on the block.

2. A shopper in a supermarket pushes a cart with a force of 35.0 N directed at an angle of 25.0° downward from the horizontal. Find the work done by the shopper on the cart as he moves down an aisle 50.0 m long.

3. **PhysicsNow™** Batman, whose mass is 80.0 kg, is dangling on the free end of a 12.0-m rope, the other end of which is fixed to a tree limb above. He is able to get the rope in motion as only Batman knows how, eventually getting it to swing enough that he can reach a ledge when the rope makes a 60.0° angle with the vertical. How much work was done by the gravitational force on Batman in this maneuver?

4. A raindrop of mass $3.35 \times 10^{-5}$ kg falls vertically at constant speed under the influence of gravity and air resistance. Model the drop as a particle. As it falls 100 m, what is the work done on the raindrop (a) by the gravitational force and (b) by air resistance?

### Section 6.3 ■ The Scalar Product of Two Vectors

In Problems 6.5 through 6.9, calculate numerical answers to three significant figures as usual.

5. Find the scalar product of the vectors in Figure P6.5.

6. For any two vectors $\vec{\mathbf{A}}$ and $\vec{\mathbf{B}}$, show that $\vec{\mathbf{A}} \cdot \vec{\mathbf{B}} = A_xB_x + A_yB_y + A_zB_z$. (*Suggestion:* Write $\vec{\mathbf{A}}$ and $\vec{\mathbf{B}}$ in unit vector form and use Equations 6.7 and 6.8.)

7. **PhysicsNow™** A force $\vec{\mathbf{F}} = (6\hat{\mathbf{i}} - 2\hat{\mathbf{j}})$ N acts on a particle that undergoes a displacement $\Delta\vec{\mathbf{r}} = (3\hat{\mathbf{i}} + \hat{\mathbf{j}})$ m. Find (a) the work done by the force on the particle and (b) the angle between $\vec{\mathbf{F}}$ and $\Delta\vec{\mathbf{r}}$.

8. For $\vec{\mathbf{A}} = 3\hat{\mathbf{i}} + \hat{\mathbf{j}} - \hat{\mathbf{k}}$, $\vec{\mathbf{B}} = -\hat{\mathbf{i}} + 2\hat{\mathbf{j}} + 5\hat{\mathbf{k}}$, and $\vec{\mathbf{C}} = 2\hat{\mathbf{j}} - 3\hat{\mathbf{k}}$, find $\vec{\mathbf{C}} \cdot (\vec{\mathbf{A}} - \vec{\mathbf{B}})$.

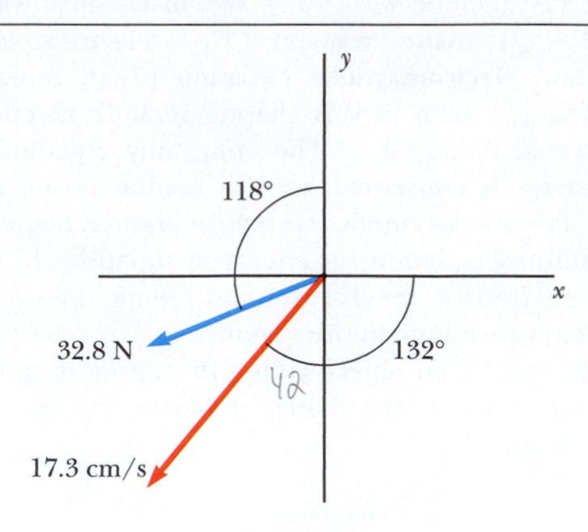


FIGURE P6.5

9. Using the definition of the scalar product, find the angles between (a) $\vec{\mathbf{A}} = 3\hat{\mathbf{i}} - 2\hat{\mathbf{j}}$ and $\vec{\mathbf{B}} = 4\hat{\mathbf{i}} - 4\hat{\mathbf{j}}$ (b) $\vec{\mathbf{A}} = -2\hat{\mathbf{i}} + 4\hat{\mathbf{j}}$ and $\vec{\mathbf{B}} = 3\hat{\mathbf{i}} - 4\hat{\mathbf{j}} + 2\hat{\mathbf{k}}$, and (c) $\vec{\mathbf{A}} = \hat{\mathbf{i}} - 2\hat{\mathbf{j}} + 2\hat{\mathbf{k}}$ and $\vec{\mathbf{B}} = 3\hat{\mathbf{j}} + 4\hat{\mathbf{k}}$.

### Section 6.4 ■ Work Done by a Varying Force

10. The force acting on a particle is $F_x = (8x - 16)$ N, where $x$ is in meters. (a) Make a plot of this force versus $x$ from $x = 0$ to $x = 3.00$ m. (b) From your graph, find the net work done by this force on the particle as it moves from $x = 0$ to $x = 3.00$ m.

11. **PhysicsNow™** A particle is subject to a force $F_x$ that varies with position as shown in Figure P6.11. Find the work done by the force on the particle as it moves (a) from $x = 0$ to $x = 5.00$ m, (b) from $x = 5.00$ m to $x = 10.0$ m, and (c) from $x = 10.0$ m to $x = 15.0$ m. (d) What is the total work done by the force over the distance $x = 0$ to $x = 15.0$ m?

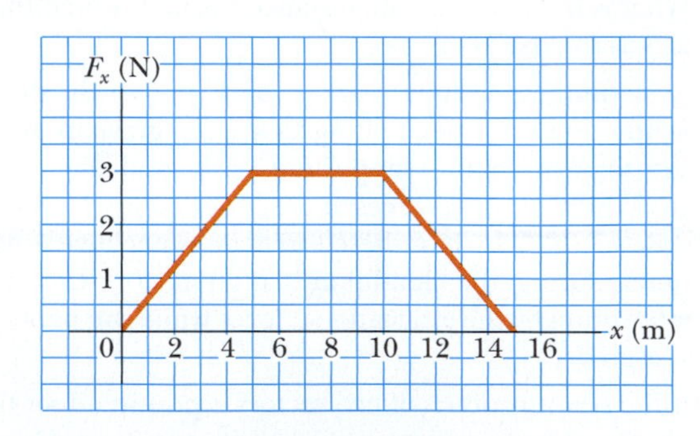


FIGURE P6.11 Problems 11 and 24.

12. A 6 000-kg freight car rolls along rails with negligible friction. The car is brought to rest by a combination of two coiled springs as illustrated in Figure P6.12. Both springs obey Hooke's law with $k_1 = 1\,600$ N/m and $k_2 = 3\,400$ N/m. After the first spring compresses a distance of 30.0 cm, the second spring acts with the first to increase the force as additional compression occurs as shown in the graph. The car comes to rest 50.0 cm after first contacting the two-spring system. Find the car's initial speed.

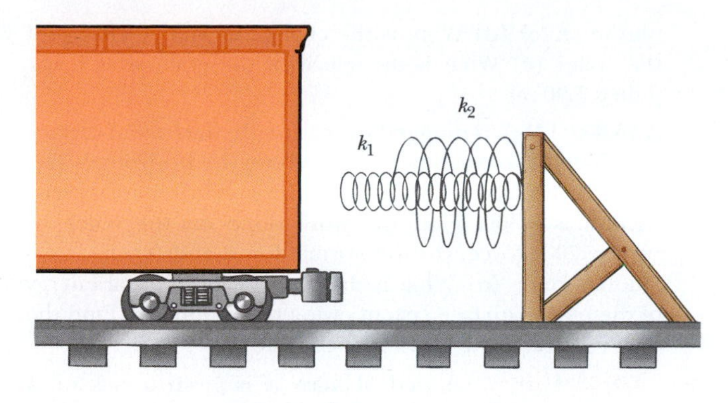


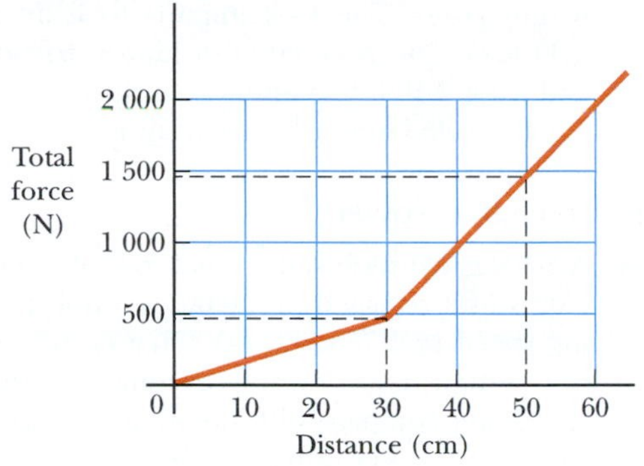


FIGURE **P6.12**

**13.** When a 4.00-kg object is hung vertically on a certain light spring that obeys Hooke's law, the spring stretches 2.50 cm. If the 4.00-kg object is removed, (a) how far will the spring stretch if a 1.50-kg block is hung on it and (b) how much work must an external agent do to stretch the same spring 4.00 cm from its unstretched position?

**14.** A force $\vec{\mathbf{F}} = (4x\hat{\mathbf{i}} + 3y\hat{\mathbf{j}})$ N acts on an object as the object moves in the $x$ direction from the origin to $x = 5.00$ m. Find the work $W = \int \vec{\mathbf{F}} \cdot d\vec{\mathbf{r}}$ done on the object by the force.

**15.** An archer pulls her bowstring back 0.400 m by exerting a force that increases uniformly from zero to 230 N. (a) What is the equivalent spring constant of the bow? (b) How much work does the archer do in drawing the bow?

**16.** A 100-g bullet is fired from a rifle having a barrel 0.600 m long. Choose the origin to be at the location where the bullet begins to move. Then the force (in newtons) exerted by the expanding gas on the bullet is $15\,000 + 10\,000x - 25\,000x^2$, where $x$ is in meters. (a) Determine the work done by the gas on the bullet as the bullet travels the length of the barrel. (b) If the barrel is 1.00 m long, how much work is done and how does this value compare to the work calculated in (a)?

**17.** It takes 4.00 J of work to stretch a Hooke's-law spring 10.0 cm from its unstressed length. Determine the extra work required to stretch it an additional 10.0 cm.

**18.** A cafeteria tray dispenser supports a stack of trays on a shelf that hangs from four identical spiral springs under tension, one near each corner of the shelf. Each tray is rectangular, 45.3 cm by 35.6 cm, is 0.450 cm thick, and has mass 580 g. Demonstrate that the top tray in the stack can always be at the same height above the floor, however many trays are in the dispenser. Find the spring constant each spring should have for the dispenser to function in this convenient way. Is any piece of data unnecessary for this determination?

**19.** A small particle of mass $m$ moves at constant speed as it is pulled to the top of a frictionless half-cylinder (of radius $R$) by a cord that passes over the top of the cylinder as illustrated in Figure P6.19. (a) Show that $F = mg\cos\theta$. (*Note:* If the particle moves at constant speed, the component of its acceleration tangent to the cylinder must be zero at all times.) (b) By directly integrating $W = \int \vec{\mathbf{F}} \cdot d\vec{\mathbf{r}}$, find the work done by the force in moving the particle at constant speed from the bottom to the top of the half-cylinder.

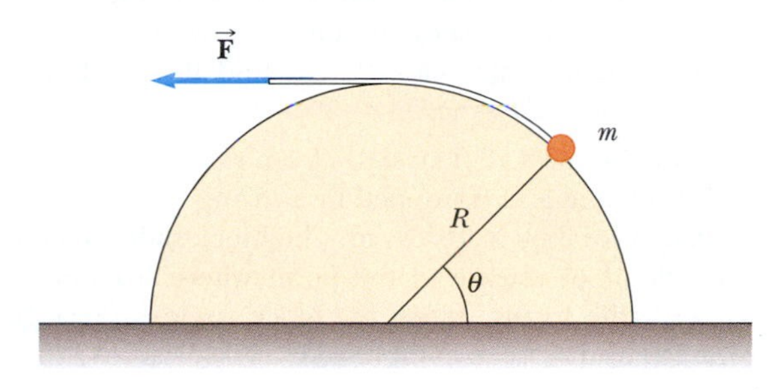


FIGURE **P6.19**

**20.** A light spring with spring constant $k_1$ is hung from an elevated support. From its lower end a second light spring that has spring constant $k_2$ is hung. An object of mass $m$ is hung at rest from the lower end of the second spring. (a) Find the total extension distance of the pair of springs. (b) Find the effective spring constant of the pair of springs as a system. We describe these springs as *in series.*

## Section 6.5 ▪ Kinetic Energy and the Work–Kinetic Energy Theorem

## Section 6.6 ▪ The Nonisolated System

**21.** A 0.600-kg particle has a speed of 2.00 m/s at point Ⓐ and kinetic energy of 7.50 J at point Ⓑ. (a) What is its kinetic energy at Ⓐ? (b) What is its speed at Ⓑ? (c) What is the total work done on the particle as it moves from Ⓐ to Ⓑ?

**22.** A 0.300-kg ball has a speed of 15.0 m/s. (a) What is its kinetic energy? (b) If its speed were doubled, what would be its kinetic energy?

**23.** A 3.00-kg object has an initial velocity $(6.00\hat{\mathbf{i}} - 2.00\hat{\mathbf{j}})$ m/s. (a) What is its kinetic energy at this time? (b) Find the total work done on the object as its velocity changes to $(8.00\hat{\mathbf{i}} + 4.00\hat{\mathbf{j}})$ m/s. (*Note:* From the definition of the scalar product, $v^2 = \vec{\mathbf{v}} \cdot \vec{\mathbf{v}}$.)

**24.** A 4.00-kg particle is subject to a total force that varies with position as shown in Figure P6.11. The particle starts from rest at $x = 0$. What is its speed at (a) $x = 5.00$ m, (b) $x = 10.0$ m, and (c) $x = 15.0$ m?

**25.** A 2 100-kg pile driver is used to drive a steel I-beam into the ground. The pile driver falls 5.00 m before coming into contact with the top of the beam, and it drives the beam 12.0 cm farther into the ground before coming to rest. Using energy considerations, calculate the average force the beam exerts on the pile driver while the pile driver is brought to rest.

**26.** You can think of the work–kinetic energy theorem as a second theory of motion, parallel to Newton's laws in describing how outside influences affect the motion of an object. In this problem, do parts (a) and (b) separately from parts (c) and (d) to compare the predictions of the two theories. In a rifle barrel, a 15.0-g bullet is accelerated from rest to a speed of 780 m/s. (a) Find the work that is done on the bullet. (b) Assuming that the rifle barrel is 72.0 cm long, find the magnitude of the average total force that acted on it as $F = W/(\Delta r \cos \theta)$. (c) Find the constant acceleration of a bullet that starts from rest and gains a speed of 780 m/s over a distance of 72.0 cm. (d) Assuming that the bullet has mass 15.0 g, find the total force that acted on it as $\Sigma F = ma$.

**27.** A block of mass 12.0 kg slides from rest down a frictionless 35.0° incline and is stopped by a strong spring with a force constant of $3.00 \times 10^4$ N/m. The block slides 3.00 m from the point of release to the point where it comes to rest against the spring. When the block comes to rest, how far has the spring been compressed?

**28.** In the neck of the picture tube of a certain black-and-white television set, an electron gun contains two charged metallic plates 2.80 cm apart. An electric force accelerates each electron in the beam from rest to 9.60% of the speed of light over this distance. (a) Determine the kinetic energy of the electron as it leaves the electron gun. Electrons carry this energy to a phosphorescent material on the inner surface of the television screen, making it glow. For an electron passing between the plates in the electron gun, determine (b) the magnitude of the constant electric force acting on the electron, (c) the acceleration, and (d) the time of flight.

## Section 6.7 ▪ Situations Involving Kinetic Friction

**29.** A 40.0-kg box initially at rest is pushed 5.00 m along a rough, horizontal floor with a constant applied horizontal force of 130 N. The coefficient of friction between box and floor is 0.300. Find (a) the work done by the applied force, (b) the increase in internal energy in the box–floor system as a result of friction, (c) the work done by the normal force, (d) the work done by the gravitational force, (e) the change in kinetic energy of the box, and (f) the final speed of the box.

**30.** A 2.00-kg block is attached to a spring of force constant 500 N/m as shown in Active Figure 6.8. The block is pulled 5.00 cm to the right of equilibrium and released from rest. Find the speed the block has as it passes through equilibrium if (a) the horizontal surface is frictionless and (b) the coefficient of friction between block and surface is 0.350.

**31.** A crate of mass 10.0 kg is pulled up a rough incline with an initial speed of 1.50 m/s. The pulling force is 100 N parallel to the incline, which makes an angle of 20.0° with the horizontal. The coefficient of kinetic friction is 0.400, and the crate is pulled 5.00 m. (a) How much work is done by the gravitational force on the crate? (b) Determine the increase in internal energy of the crate–incline system owing to friction. (c) How much work is done by the 100-N force on the crate? (d) What is the change in kinetic energy of the crate? (e) What is the speed of the crate after being pulled 5.00 m?

**32.** A 15.0-kg block is dragged over a rough, horizontal surface by a 70.0-N force acting at 20.0° above the horizontal. The block is displaced 5.00 m, and the coefficient of kinetic friction is 0.300. Find the work done on the block by (a) the 70-N force, (b) the normal force, and (c) the gravitational force. (d) What is the increase in internal energy of the block–surface system owing to friction? (e) Find the total change in the block's kinetic energy.

**33.** **PhysicsNow™** A sled of mass $m$ is given a kick on a frozen pond. The kick imparts to it an initial speed of 2.00 m/s. The coefficient of kinetic friction between sled and ice is 0.100. Use energy considerations to find the distance the sled moves before it stops.

## Section 6.8 ▪ Power

**34.** A 650-kg elevator starts from rest. It moves upward for 3.00 s with constant acceleration until it reaches its cruising speed of 1.75 m/s. (a) What is the average power of the elevator motor during this time interval? (b) How does this power compare with the motor power when the elevator moves at its cruising speed?

**35.** **PhysicsNow™** A 700-N Marine in basic training climbs a 10.0-m vertical rope at a constant speed in 8.00 s. What is his power output?

**36.** A skier of mass 70.0 kg is pulled up a slope by a motor-driven cable. (a) How much work is required to pull the skier a distance of 60.0 m up a 30.0° slope (assumed frictionless) at a constant speed of 2.00 m/s? (b) A motor of what power is required to perform this task?

**37.** An energy-efficient lightbulb, taking in 28.0 W of power, can produce the same level of brightness as a conventional lightbulb operating at power 100 W. The lifetime of the energy-efficient bulb is 10 000 h and its purchase price is $17.0, whereas the conventional bulb has lifetime 750 h and costs $0.420 per bulb. Determine the total savings obtained by using one energy-efficient bulb over its lifetime as opposed to using conventional bulbs over the same time interval. Assume an energy cost of $0.080 0 per kilowatt-hour.

**38.** Energy is conventionally measured in Calories as well as in joules. One Calorie in nutrition is one kilocalorie, defined as 1 kcal = 4 186 J. Metabolizing 1 g of fat can release 9.00 kcal. A student decides to try to lose weight by exercising. She plans to run up and down the stairs in a football stadium as fast as she can and as many times as necessary. Is this plan in itself a practical way to lose weight? To evaluate the program, suppose she runs up a flight of 80 steps, each 0.150 m high, in 65.0 s. For simplicity, ignore the energy she uses in coming down (which is small). Assume that a typical efficiency for human muscles is 20.0%. Therefore when your body converts 100 J from metabolizing fat, 20 J goes into doing mechanical work (here, climbing stairs) and the remainder goes into extra internal energy. Assume that the student's mass is 50.0 kg. (a) How many times must she run the flight of stairs to lose 1 lb of fat? (b) What is her average power output, in watts and in horsepower, as she is running up the stairs?

**39.** For saving energy, bicycling and walking are far more efficient means of transportation than is travel by automobile. For example, when riding at 10.0 mi/h, a cyclist uses food energy at a rate of about 400 kcal/h above what he would use if merely sitting still. (In exercise physiology, power is often measured in kcal/h rather than in watts. Here 1 kcal = 1 nutritionist's Calorie = 4 186 J.) Walking at 3.00 mi/h requires about 220 kcal/h. It is interesting to compare these values with the energy consumption required for travel by car. Gasoline yields about $1.30 \times 10^8$ J/gal. Find the fuel economy in equivalent miles per gallon for a person (a) walking and (b) bicycling.

## Section 6.9 ■ Context Connection—Horsepower Ratings of Automobiles

**40.** Make an order-of-magnitude estimate of the output power a car engine contributes to speeding the car up to highway speed. For concreteness, consider your own car, if you use one, and make the calculation as precise as you wish. In your solution, state the physical quantities you take as data and the values you measure or estimate for them. The mass of the vehicle is given in the owner's manual. If you do not wish to estimate for a car, consider a bus or truck that you specify.

**41.** A certain automobile engine delivers $2.24 \times 10^4$ W (30.0 hp) to its wheels when moving at a constant speed of 27.0 m/s ($\approx$ 60 mi/h). What is the resistive force acting on the automobile at that speed?

## Additional Problems

**42.** A baseball outfielder throws a 0.150-kg baseball at a speed of 40.0 m/s and an initial angle of 30.0°. What is the kinetic energy of the baseball at the highest point of its trajectory?

**43.** While running, a person transforms about 0.600 J of chemical energy to mechanical energy per step per kilogram of body mass. If a 60.0-kg runner transforms energy at a rate of 70.0 W during a race, how fast is the person running? Assume that a running step is 1.50 m long.

**44.** In bicycling for aerobic exercise, a woman wants her heart rate to be between 136 and 166 beats per minute. Assume that her heart rate is directly proportional to her mechanical power output within the range relevant here. Ignore all forces on the woman-plus-bicycle system except for static friction forward on the drive wheel of the bicycle and an air resistance force proportional to the square of her speed. When her speed is 22.0 km/h, her heart rate is 90.0 beats per minute. In what range should her speed be so that her heart rate will be in the range she wants?

**45.** A 4.00-kg particle moves along the $x$ axis. Its position varies with time according to $x = t + 2.0t^3$, where $x$ is in meters and $t$ is in seconds. Find (a) the kinetic energy at any time $t$, (b) the acceleration of the particle and the force acting on it at time $t$, (c) the power being delivered to the particle at time $t$, and (d) the work done on the particle in the interval $t = 0$ to $t = 2.00$ s.

**46.** A bead at the bottom of a bowl is one example of an object in a stable equilibrium position. When a physical system is displaced by an amount $x$ from stable equilibrium, a restoring force acts on it, tending to return the system to its equilibrium configuration. The magnitude of the restoring force can be a complicated function of $x$. For example, when an ion in a crystal is displaced from its lattice site, the restoring force may not be a simple function of $x$. In such cases, we can generally imagine the function $F(x)$ to be expressed as a power series in $x$ as $F(x) = -(k_1x + k_2x^2 + k_3x^3 + \cdots)$. The first term here is just Hooke's law, which describes the force exerted by a simple spring for small displacements. For small excursions from equilibrium we generally ignore the higher-order terms, but in some cases it may be desirable to keep the second term as well. If we model the restoring force as $F = -(k_1x + k_2x^2)$, how much work is done in displacing the system from $x = 0$ to $x = x_{max}$ by an applied force equal in magnitude to the restoring force?

**47.** A traveler at an airport takes an escalator up one floor as shown in Figure P6.47. The moving staircase would itself carry him upward with vertical velocity component $v$ between entry and exit points separated by height $h$. While the escalator is moving, however, the hurried traveler climbs the steps of the escalator at a rate of $n$ steps/s. Assume that the height of each step is $h_s$. (a) Determine the amount of chemical energy converted into mechanical energy by the traveler's leg muscles during his escalator ride given that his mass is $m$. (b) Determine the work the escalator motor does on this person.

(Ron Chapple/FPG/Getty)

FIGURE **P6.47**

**48.** A 5.00-kg steel ball is dropped onto a copper plate from a height of 10.0 m. If the ball leaves a dent 3.20 mm deep, what is the average force exerted by the plate on the ball during the impact?

**49.** In a control system, an accelerometer consists of a 4.70-g object sliding on a horizontal rail. A low-mass spring attaches the object to a flange at one end of the rail. Grease on the rail makes static friction negligible, but rapidly damps out vibrations of the sliding object. When subject to a steady acceleration of $0.800g$, the object is to assume a location 0.500 cm away from its equilibrium position. Find the force constant required for the spring.

**50.** A light spring with force constant 3.85 N/m is compressed by 8.00 cm as it is held between a 0.250-kg block on the left and a 0.500-kg block on the right, both resting on a horizontal surface. The spring exerts a force on each block, tending to push the blocks apart. The blocks are simultaneously released from rest. Find the acceleration with which each block starts to move if the coefficient of kinetic friction between each block and the surface is (a) 0, (b) 0.100, and (c) 0.462.

**51.** A single constant force $\vec{\mathbf{F}}$ acts on a particle of mass $m$. The particle starts at rest at $t = 0$. (a) Show that the instantaneous power delivered by the force at any time $t$ is $(F^2/m)t$. (b) If $F = 20.0$ N and $m = 5.00$ kg, what is the power delivered at $t = 3.00$ s?

**52.** A particle is attached between two identical springs on a horizontal frictionless table. Both springs have spring constant $k$ and are initially unstressed. (a) The particle is pulled a distance $x$ along a direction perpendicular to the initial configuration of the springs as shown in Figure P6.52. Show that the force exerted by the springs on the particle is

$$\vec{\mathbf{F}} = -2kx\left(1 - \frac{L}{\sqrt{x^2 + L^2}}\right)\hat{\mathbf{i}}$$

(b) Determine the amount of work done by this force in moving the particle from $x = A$ to $x = 0$.

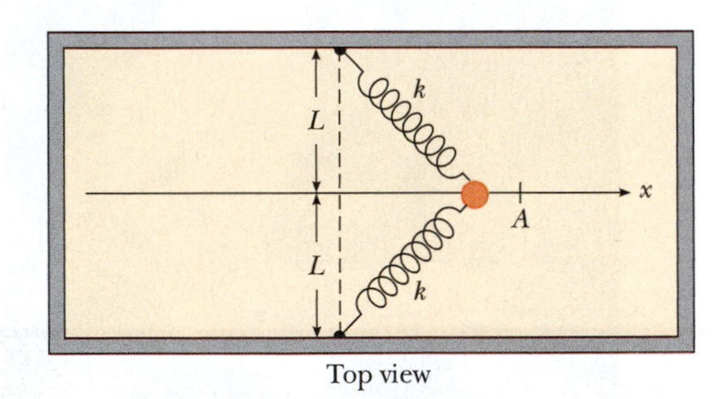


FIGURE **P6.52**

**53.** A 200-g block is pressed against a spring of force constant 1.40 kN/m until the block compresses the spring 10.0 cm. The spring rests at the bottom of a ramp inclined at 60.0° to the horizontal. Using energy considerations, determine how far up the incline the block moves before it stops (a) if there is no friction between the block and the ramp and (b) if the coefficient of kinetic friction is 0.400.

**54.** **Review problem.** Two constant forces act on a 5.00-kg object moving in the $xy$ plane as shown in Figure P6.54. Force $\vec{\mathbf{F}}_1$ is 25.0 N at 35.0°, whereas $\vec{\mathbf{F}}_2$ is 42.0 N at 150°. At time $t = 0$, the object is at the origin and has velocity $(4.00\hat{\mathbf{i}} + 2.50\hat{\mathbf{j}})$ m/s. (a) Express the two forces in unit-vector notation. Use unit-vector notation for your other answers. (b) Find the total force on the object. (c) Find the object's acceleration. Now, considering the instant $t = 3.00$ s, (d) find the object's velocity, (e) its location, (f) its kinetic energy from $\frac{1}{2}mv_f^2$, and (g) its kinetic energy from $\frac{1}{2}mv_i^2 + \sum \vec{\mathbf{F}} \cdot \Delta\vec{\mathbf{r}}$.

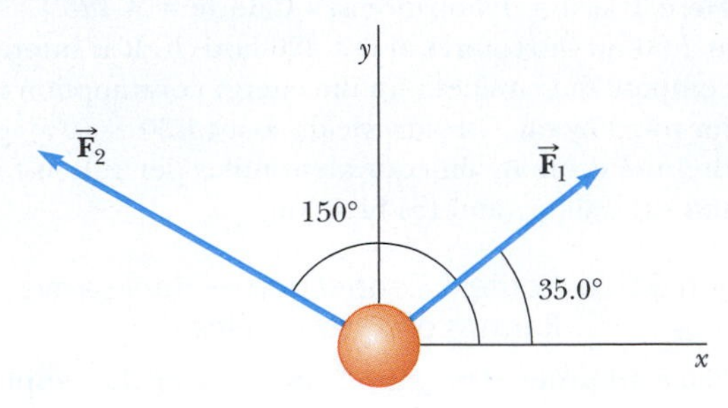


FIGURE **P6.54**

**55.** The ball launcher in a classic pinball machine has a spring that has a force constant of 1.20 N/cm (Fig. P6.55). The surface on which the ball moves is inclined 10.0° with respect to the horizontal. The spring is initially compressed 5.00 cm. Find the launching speed of a 100-g ball when the plunger is released. Friction and the mass of the plunger are negligible.

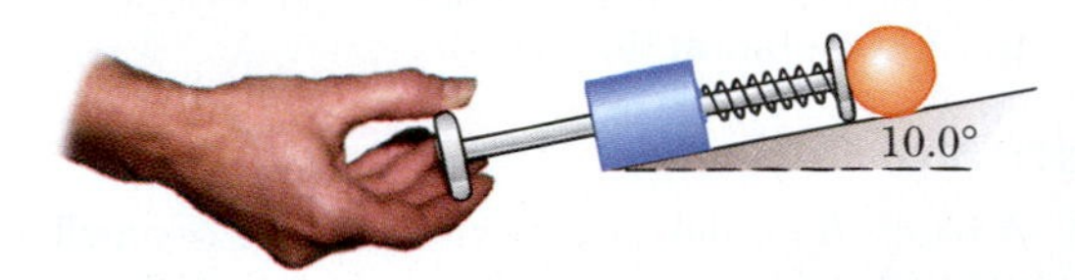


FIGURE **P6.55**

**56.** When objects with different weights are hung on a spring, the spring stretches to different lengths as shown in the following table. (a) Make a graph of the applied force versus the extension of the spring. By least-squares fitting, determine the straight line that best fits the data. (You may not want to use all the data points.) (b) From the slope of the best-fit line, find the spring constant $k$. (c) If the spring is extended to 105 mm, what force does it exert on the object it suspends?

| *F* (N) | *L* (mm) | *F* (N) | *L* (mm) |
|---|---|---|---|
| 2.0 | 15 | 14 | 112 |
| 4.0 | 32 | 16 | 126 |
| 6.0 | 49 | 18 | 149 |
| 8.0 | 64 | 20 | 175 |
| 10 | 79 | 22 | 190 |
| 12 | 98 | | |

**57.** In diatomic molecules, the constituent atoms exert attractive forces on each other at large distances and repulsive forces at short distances. For many molecules, the Lennard–Jones law is a good approximation to the magnitude of these forces:

$$F = F_0\left[2\left(\frac{\sigma}{r}\right)^{13} - \left(\frac{\sigma}{r}\right)^7\right]$$

where $r$ is the center-to-center distance between the atoms in the molecule, $\sigma$ is a length parameter, and $F_0$ is the force when $r = \sigma$. For an oxygen molecule, $F_0 = 9.60 \times 10^{-11}$ N and $\sigma = 3.50 \times 10^{-10}$ m. Determine the work done by this force as the atoms are pulled apart from $r = 4.00 \times 10^{-10}$ m to $r = 9.00 \times 10^{-10}$ m.

**58.** A 0.400-kg particle slides around a horizontal track. The track has a smooth vertical outer wall forming a circle with a radius of 1.50 m. The particle is given an initial speed of 8.00 m/s. After one revolution, its speed has dropped to 6.00 m/s because of friction with the rough floor of the track. (a) Find the energy transformed from mechanical to internal in the system owing to friction in one revolution. (b) Calculate the coefficient of kinetic friction. (c) What is the total number of revolutions the particle makes before stopping?

**59.** A particle moves along the $x$ axis from $x = 12.8$ m to $x = 23.7$ m under the influence of a force

$$F = \frac{375}{x^3 + 3.75x}$$

where $F$ is in newtons and $x$ is in meters. Using numerical integration, determine the total work done by this force on the particle during this displacement. Your result should be accurate to within 2%.

**60.** As it plows a parking lot, a snowplow pushes an ever-growing pile of snow in front of it. Suppose a car moving through the air is similarly modeled as a cylinder pushing a growing plug of air in front of it. The originally stationary air is set into motion at the constant speed $v$ of the cylinder as shown in Figure P6.60. In a time interval $\Delta t$, a new disk of air of mass $\Delta m$ must be moved a distance $v\,\Delta t$

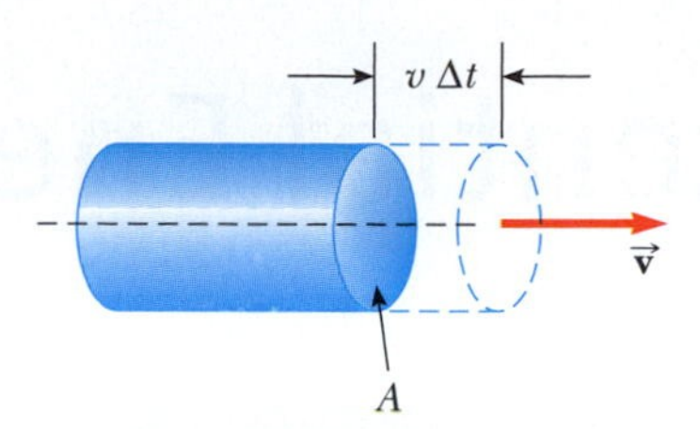


FIGURE **P6.60**

and hence must be given a kinetic energy $\frac{1}{2}(\Delta m)v^2$. Using this model, show that the automobile's power loss owing to air resistance is $\frac{1}{2}\rho A v^3$ and that the resistive force acting on the car is $\frac{1}{2}\rho A v^2$, where $\rho$ is the density of air. Compare this result with the empirical expression $\frac{1}{2}D\rho A v^2$ for the resistive force.

**61.** A windmill, such as that in the opening photograph of this chapter, turns in response to a force of high-speed air resistance, $R = \frac{1}{2}D\rho A v^2$. The power available is $\mathcal{P} = Rv = \frac{1}{2}D\rho\pi r^2 v^3$, where $v$ is the wind speed and we have assumed a circular face for the windmill of radius $r$. Take the drag coefficient as $D = 1.00$ and the density of air from the front endpaper. For a home windmill with $r = 1.50$ m, calculate the power available if (a) $v = 8.00$ m/s and (b) $v = 24.0$ m/s. The power delivered to the generator is limited by the efficiency of the system, about 25%. For comparison, a typical home needs about 3 kW of electric power.

**62.** Consider the block–spring–surface system in part (b) of Interactive Example 6.7. (a) At what position $x$ of the block is its speed a maximum? (b) Explore the effect of an increased friction force of 10.0 N. At what position of the block does its maximum speed occur in this situation?

## ANSWERS TO QUICK QUIZZES

**6.1** c, a, d, b. The work in (c) is positive and of the largest possible value because the angle between the force and the displacement is zero. The work done in (a) is zero because the force is perpendicular to the displacement. In (d) and (b), negative work is done by the applied force because in neither case is there a component of the force in the direction of the displacement. Situation (b) is the most negative value because the angle between the force and the displacement is 180°.

**6.2** (d). Because of the range of values of the cosine function, $\vec{\mathbf{A}} \cdot \vec{\mathbf{B}}$ has values that range from $AB$ to $-AB$.

**6.3** (a). Because the work done in compressing a spring is proportional to the square of the compression distance $x$, doubling the value of $x$ causes the work to increase fourfold.

**6.4** (b). Because the work is proportional to the square of the compression distance $x$ and the kinetic energy is proportional to the square of the speed $v$, doubling the compression distance doubles the speed.

**6.5** (a) For the television set, energy enters by electrical transmission (through the power cord) and electromagnetic radiation (the television signal). Energy leaves by heat (from hot surfaces into the air), mechanical waves (sound from the speaker), and electromagnetic radiation (from the screen). (b) For the gasoline-powered lawn mower, energy enters by matter transfer (gasoline). Energy leaves by work (on the blades of grass), mechanical waves (sound), and heat (from hot surfaces into the air). (c) For the hand-cranked pencil sharpener, energy enters by work (from your hand turning the crank). Energy leaves by work (done on the pencil), mechanical waves (sound), and heat resulting from the temperature increase from friction.

**6.6** (b), (d), (e). For the block, the friction force from the surface represents an interaction with the environment. For the surface, the friction force from the block represents an interaction with the environment. For the block and the surface, the friction force is internal to the system, so there are no interactions with the environment.

**6.7** (c). The brakes and the roadway are warmer, so their internal energy has increased. In addition, the sound of the skid represents transfer of energy away by mechanical waves.

# CHAPTER 7

# Potential Energy

A strobe photograph of a pole vaulter. In the system of the pole vaulter and the Earth, several types of energy transformations occur during this process.



## CHAPTER OUTLINE


7.1 Potential Energy of a System

7.2 The Isolated System

7.3 Conservative and Nonconservative Forces

7.4 Conservative Forces and Potential Energy

7.5 The Nonisolated System in Steady State

7.6 Potential Energy for Gravitational and Electric Forces

7.7 Energy Diagrams and Stability of Equilibrium

7.8 Context Connection—Potential Energy in Fuels

SUMMARY


In Chapter 6, we introduced the concepts of kinetic energy, which is associated with the motion of an object or a particle, and internal energy, which is associated with the temperature of a system. In this chapter, we introduce another form of energy for a system, called *potential energy,* which is associated with the configuration of a system of two or more interacting objects or particles. This new type of energy will provide us with a powerful and universal fundamental principle for an isolated system.

## 7.1 POTENTIAL ENERGY OF A SYSTEM

In Chapter 6, we defined a system in general but focused our attention on single particles under the influence of an external force. In this chapter, we consider systems of two or more objects or particles interacting via a force that is *internal* to the system. The kinetic energy of such a system is the algebraic sum of the kinetic energies of all members of the system. In some systems, however, one object may be so massive that it can be modeled as stationary and its kinetic energy can be ignored. For example, if we consider a ball–Earth system as the ball falls to the ground, the kinetic energy of the system can be considered as only the kinetic energy of the ball. The Earth moves toward the ball so slowly in this process that we can ignore its kinetic energy. (We will justify this claim in Chapter 8.) On the other hand, the kinetic energy of a system of two electrons must include the kinetic energies of both particles.

Let us imagine a system consisting of a book and the Earth, interacting via the gravitational force. We do positive work on the system by lifting the book slowly through a height $\Delta y = y_b - y_a$ as in Figure 7.1. According to the continuity equation for energy (Eq. 6.20) introduced in Chapter 6, this work on the system must appear as an increase in energy of the system. The book is at rest before we perform the work and is at rest after we perform the work; therefore, the kinetic energy of the system does not change. There is no reason to suspect that the temperature of the book or of the Earth should change, so the internal energy of the system experiences no change.

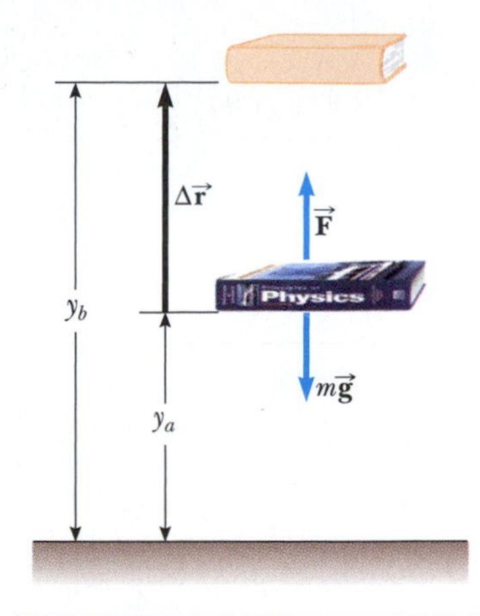


**FIGURE 7.1** The work done by an external agent on the system of the book and the Earth as the book is lifted from $y_a$ to $y_b$ is equal to $mgy_b - mgy_a$.

### PITFALL PREVENTION 7.1

**POTENTIAL ENERGY BELONGS TO A SYSTEM** Keep in mind that potential energy is always associated with a *system* of two or more interacting objects. In the gravitational case, in which a small object moves near the surface of the Earth, we may sometimes refer to the potential energy "associated with the object" rather than the more proper "associated with the system" because the Earth does not move significantly. We will not, however, refer to the potential energy "of the object" because this wording clearly ignores the role of the Earth in the potential energy.

Because the energy change of the system is not in the form of kinetic energy or internal energy, it must appear as some other form of energy storage. After lifting the book, suppose we release it and let it fall to the ground. Notice that the book (and therefore the system) now has kinetic energy, and its origin is in the work that was done in lifting the book. While the book is at the highest point, the energy of the system has the *potential* to become kinetic energy, but does not do so until the book is allowed to fall. Therefore, we call the energy storage mechanism before we release the book **potential energy.** We will find that a potential energy can be associated with a number of types of forces. In this particular case, we are discussing **gravitational potential energy.**

Let us now derive an expression for the gravitational potential energy associated with an object at a given location above the Earth's surface. To do so, consider an external agent lifting an object of mass $m$ from an initial height $y_a$ above the ground to a final height $y_b$ as in Figure 7.1. We assume that the lifting is done slowly, with no acceleration, so that the lifting force can be modeled as equal to the weight of the object; the object is in equilibrium and moving at constant velocity. The work done by the external agent on the system (the object and the Earth) as the object undergoes this upward displacement is given by the product of the upward force $\vec{\mathbf{F}} = -m\vec{\mathbf{g}}$ and the displacement $\Delta\vec{\mathbf{r}} = \Delta y\hat{\mathbf{j}}$:

$$W = (-m\vec{\mathbf{g}})\cdot\Delta\vec{\mathbf{r}} = [-m(-g\hat{\mathbf{j}})]\cdot[(y_b - y_a)\hat{\mathbf{j}}] = mgy_b - mgy_a \qquad [7.1]$$

We have discussed in Chapter 6 that work is a means of transferring energy into a system. Consequently, the expression on the right in Equation 7.1 must represent a change in the energy of the system, equal to the amount of work done on the system. Notice how similar Equation 7.1 is to Equation 6.17 in the preceding chapter. In each equation, the work done on a system equals a difference between the final and initial values of a quantity. Of course, both equations are nothing more than special cases of the continuity equation for energy, Equation 6.20. In Equation 6.17, the work represents a transfer of energy into the system and the increase in energy of the system is kinetic in form. In Equation 7.1, the work represents a transfer of energy into the system and the system energy appears in a different form, which we call gravitational potential energy.

Therefore, we can represent the quantity $mgy$ to be the gravitational potential energy $U_g$ of the object–Earth system:

$$U_g \equiv mgy \qquad [7.2]$$

▪ Gravitational potential energy

The units of gravitational potential energy are joules, the same as those of work and kinetic energy. Potential energy, like work and kinetic energy, is a scalar quantity. Note that Equation 7.2 is valid only for objects near the surface of the Earth, where $g$ is approximately constant.

Using our definition of gravitational potential energy, we can now rewrite Equation 7.1 as

$$W = \Delta U_g$$

which mathematically describes that the work done on the system by the external agent in this situation appears as a change in the gravitational potential energy of the system.

The gravitational potential energy depends only on the vertical height of the object above the Earth's surface. Therefore, the same amount of work is done on an object–Earth system whether the object is lifted vertically from the Earth or whether it starts at the same point and is pushed up a frictionless incline, ending up at the same height. This concept can be shown in a mathematical representation by reperforming the work calculation in Equation 7.1 with a displacement having both vertical and horizontal components:

$$W = (-m\vec{\mathbf{g}}) \cdot \Delta\vec{\mathbf{r}} = [-m(-g\hat{\mathbf{j}})] \cdot [(x_b - x_a)\hat{\mathbf{i}} + (y_b - y_a)\hat{\mathbf{j}}] = mgy_b - mgy_a$$

Note that no term involving $x$ appears in the final result because $\hat{\mathbf{j}} \cdot \hat{\mathbf{i}} = 0$.

In solving problems, it is necessary to choose a reference configuration for which to set the gravitational potential energy equal to some reference value, which is normally zero. The choice of this configuration is completely arbitrary because the important quantity is the *difference* in potential energy, and this difference is independent of the choice of reference configuration.

It is often convenient to choose an object located at the surface of the Earth as the reference configuration for zero gravitational potential energy, but this choice is not essential. Often, the statement of the problem suggests a convenient configuration to use.

**QUICK QUIZ 7.1** Choose the correct answer. The gravitational potential energy of a system **(a)** is always positive, **(b)** is always negative, or **(c)** can be negative or positive.

**QUICK QUIZ 7.2** An object falls off a table to the floor. We wish to analyze the situation in terms of kinetic and potential energy. In discussing the potential energy of the system, we identify the system as **(a)** both the object and the Earth, **(b)** only the object, or **(c)** only the Earth.

## 7.2 THE ISOLATED SYSTEM

The introduction of potential energy allows us to generate a powerful and universally applicable principle for solving problems that are difficult to solve with Newton's laws. Let us develop this new principle by thinking about the book–Earth system in Figure 7.1 again. After we have lifted the book, there is gravitational potential energy stored in the system, which we can calculate from the work done by the external agent on the system using $W = \Delta U_g$.

Let us now shift our focus to the book alone as the system and let the book fall (Fig. 7.2). As the book falls from $y_b$ to $y_a$, the work done by the gravitational force on the book is

$$W_{\text{on book}} = (m\vec{\mathbf{g}})\cdot\Delta\vec{\mathbf{r}} = (-mg\hat{\mathbf{j}})\cdot(y_a - y_b)\hat{\mathbf{j}} = mgy_b - mgy_a \qquad [7.3]$$

From the work–kinetic energy theorem of Chapter 6, the work done on the book is also

$$W_{\text{on book}} = \Delta K_{\text{book}}$$

Therefore, equating these two expressions for the work done on the book gives

$$\Delta K_{\text{book}} = mgy_b - mgy_a \qquad [7.4]$$

Now, let us relate each side of this equation to the *system* of the book and the Earth. For the right-hand side,

$$mgy_b - mgy_a = U_{gi} - U_{gf} = -\Delta U_g$$

where $U_g$ is the gravitational potential energy of the system. Because the book is the only part of the system that is moving, the left-hand side of Equation 7.4 becomes

$$\Delta K_{\text{book}} = \Delta K$$

where $K$ is the kinetic energy of the system. Therefore, by replacing each side of Equation 7.4 with its system equivalent, the equation becomes

$$\Delta K = -\Delta U_g \qquad [7.5]$$

This equation can be manipulated to provide a very important result for a new analysis model. First, we bring the change in potential energy to the left side of the equation:

$$\Delta K + \Delta U_g = 0 \qquad [7.6]$$

Notice that this equation is in the form of the continuity equation for energy, Equation 6.20. On the left, we have a sum of changes of the energy stored in the system. The right-hand side in the continuity equation is the sum of the transfers across the boundary of the system. This sum is equal to zero in this case because our book–Earth system is isolated from the environment.

Let us now write the changes in energy in Equation 7.6 explicitly:

$$(K_f - K_i) + (U_{gf} - U_{gi}) = 0 \quad \rightarrow \quad K_f + U_{gf} = K_i + U_{gi} \qquad [7.7]$$

In general, we define the sum of kinetic and potential energies of a system as the **total mechanical energy of the system.** Therefore, Equation 7.7 is a statement of **conservation of mechanical energy** for an **isolated system.** An isolated system is one for which no energy transfers occur across the boundary. Therefore, the energy in the system is conserved and the sum of the kinetic and potential energies remains constant. **Equation 7.7 is only true when no friction acts between members of the system.** In Section 7.3, we shall see how this equation must be modified to include the effects of friction.

For the falling book situation that we are describing in this discussion, Equation 7.7 can be written as

$$\tfrac{1}{2}mv_f^2 + mgy_f = \tfrac{1}{2}mv_i^2 + mgy_i$$

As the book falls to the Earth, the book–Earth system loses potential energy and gains kinetic energy such that the total of the two types of energy always remains constant. The transformation of one type of energy to another is the result of the process of work done by the gravitational force on the book. Note that this

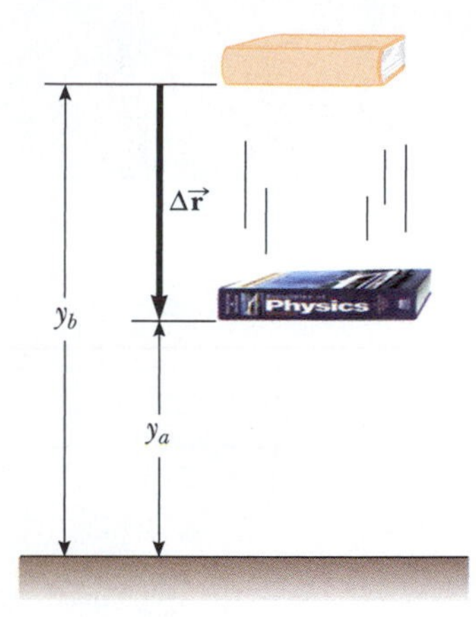


**FIGURE 7.2** The work done by the gravitational force on the book as the book falls from $y_b$ to $y_a$ is equal to $mgy_b - mgy_a$.

**PITFALL PREVENTION 7.2**

**ISOLATED SYSTEMS** The isolated system model goes far beyond Equation 7.7. This equation is only the *mechanical energy* version of this model. We will see shortly how to include internal energy. In later chapters, we will see other isolated systems and generate new versions (and associated equations) related to such quantities as momentum, angular momentum, and electric charge.

▪ Conservation of mechanical energy for an isolated system

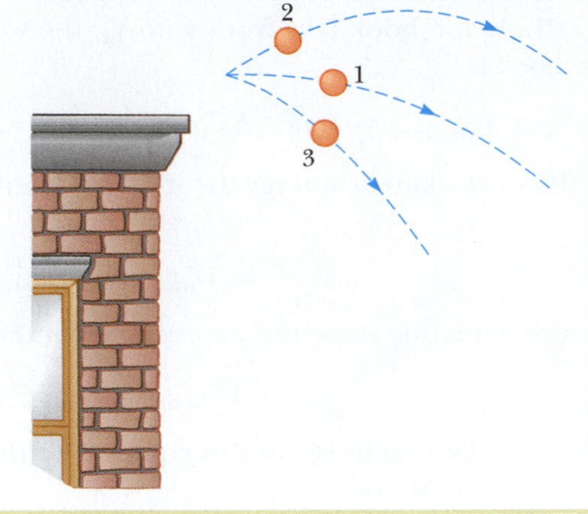


ACTIVE FIGURE 7.3

(Quick Quiz 7.3) Three identical balls are thrown with the same initial speed from the top of a building.

**PhysicsNow™** Log into PhysicsNow at **www.pop4e.com** and go to Active Figure 7.3 to change the angle of projection and observe the trajectory of the ball and the changes in energy of the ball–Earth system.

work is *internal* to the system; it is not work done *on* the system from the environment.

We will see other types of potential energy besides gravitational, so we can write the general form of the definition for mechanical energy as

$$E_{\text{mech}} \equiv K + U \quad [7.8]$$

where $U$ without a subscript refers to the total potential energy of the system, including all types. In addition, $K$ in general refers to the sum of the kinetic energies of all particles in the system.

**QUICK QUIZ 7.3** Three identical balls are thrown from the top of a building, all with the same initial speed. The first is thrown horizontally, the second at some angle above the horizontal, and the third at some angle below the horizontal, as shown in Active Figure 7.3. Neglecting air resistance, rank the speeds of the balls at the instant each hits the ground.

## Thinking Physics 7.1

You have graduated from college and are designing roller coasters for a living. You design a roller coaster in which a car is pulled to the top of a hill of height $h$ and then, starting from a momentary rest, rolls freely down the hill and upward toward the peak of the next hill, which is at height $1.1h$. Will you have a long career in this business?

**Reasoning** Your career will probably not be long because this roller coaster will not work! At the top of the first hill, the roller coaster car has no kinetic energy and the gravitational potential energy of the car–Earth system is that associated with a height for the car of $h$. If the car were to reach the top of the next hill, the system would have higher potential energy, that associated with height $1.1h$. This situation would violate the principle of conservation of mechanical energy. If this coaster were actually built, the car would move upward on the second hill to a height $h$ (ignoring the effects of friction), stop short of the peak, and then start rolling backward, becoming trapped between the two hills. ■

## INTERACTIVE EXAMPLE 7.1 Ball in Free-Fall

A ball of mass $m$ is dropped from rest at a height $h$ above the ground as in Figure 7.4. Ignore air resistance.

**A** Determine the speed of the ball when it is at a height $y$ above the ground.

**Solution** The ball and the Earth do not experience any forces from the environment because we ignore air resistance. The ball–Earth system is isolated and we use the principle of conservation of mechanical energy. Note that the system has potential energy and no kinetic energy at the beginning of our time interval of interest. As the ball falls, the total mechanical energy of the system (the sum of kinetic and potential energies) remains constant and equal to its initial potential energy. The potential energy of the system decreases, and the kinetic energy of the system (which is due only to the ball) increases.

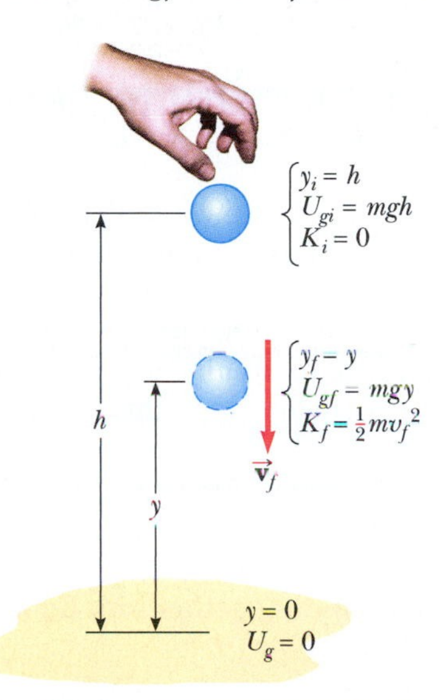


FIGURE 7.4 (Interactive Example 7.1) A ball is dropped from rest at a height $h$ above the ground. Initially, the total energy of the ball–Earth system is gravitational potential energy, equal to $mgh$ when $h = 0$ is at the ground. When the ball is at elevation $y$, the total system energy is the sum of kinetic and potential energies.

Before the ball is released from rest at a height $h$ above the ground, the kinetic energy of the system is $K_i = 0$ and the potential energy is $U_i = mgh$, where the $y$ coordinate is measured from ground level. When the ball is at an arbitrary position $y$ above the ground, its kinetic energy is $K_f = \frac{1}{2}mv_f^2$ and the potential energy of the system is $U_f = mgy$. Applying Equation 7.7, we have

$$K_f + U_{gf} = K_i + U_{gi}$$
$$\tfrac{1}{2}mv_f^2 + mgy = 0 + mgh$$
$$v_f^2 = 2g(h - y)$$
$$v_f = \sqrt{2g(h - y)}$$

**B** Determine the speed of the ball at $y$ if it is given an initial speed $v_i$ at the initial altitude $h$.

**Solution** In this case, the initial energy includes kinetic energy of the ball equal to $\frac{1}{2}mv_i^2$ and Equation 7.7 gives

$$\tfrac{1}{2}mv_f^2 + mgy = \tfrac{1}{2}mv_i^2 + mgh$$
$$v_f^2 = v_i^2 + 2g(h - y)$$
$$v_f = \sqrt{v_i^2 + 2g(h - y)}$$

Note that this result is consistent with Equation 2.13 (Chapter 2), $v_{yf}^2 = v_{yi}^2 - 2g(y_f - y_i)$, for a particle under constant acceleration, where $y_i = h$. Furthermore, this result is valid even if the initial velocity is at an angle to the horizontal (the projectile situation), as discussed in Quick Quiz 7.3.

**PhysicsNow™** Compare the effect of upward, downward, and zero initial velocities by logging into PhysicsNow at **www.pop4e.com** and going to Interactive Example 7.1.

## INTERACTIVE EXAMPLE 7.2 A Grand Entrance

You are designing apparatus to support an actor of mass 65 kg who is to "fly" down to the stage during the performance of a play. You attach the actor's harness to a 130-kg sandbag by means of a lightweight steel cable running smoothly over two frictionless pulleys as in Figure 7.5a. You need 3.0 m of cable between the harness and the nearest pulley so that the pulley can be hidden behind a curtain. For the apparatus to work successfully, the sandbag must never lift above the floor as the actor swings from above the stage to the floor.

Let us identify the angle $\theta$ as the angle that the actor's cable makes with the vertical when he begins his motion from rest. What is the maximum value $\theta$ can have such that the sandbag does not lift off the floor during the actor's swing?

**Solution** We must draw on several concepts to solve this problem. To conceptualize the problem, imagine what happens as the actor approaches the bottom of the swing. At the bottom, the rope is vertical and must

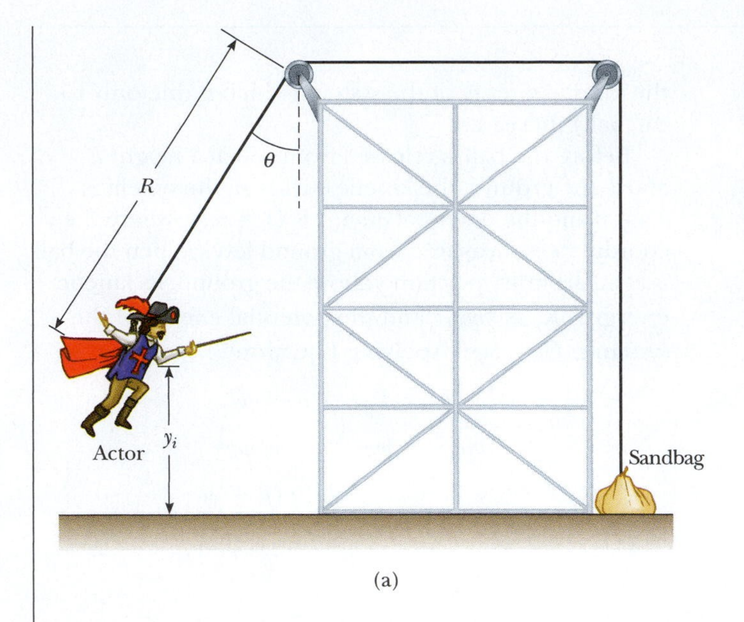


(a)

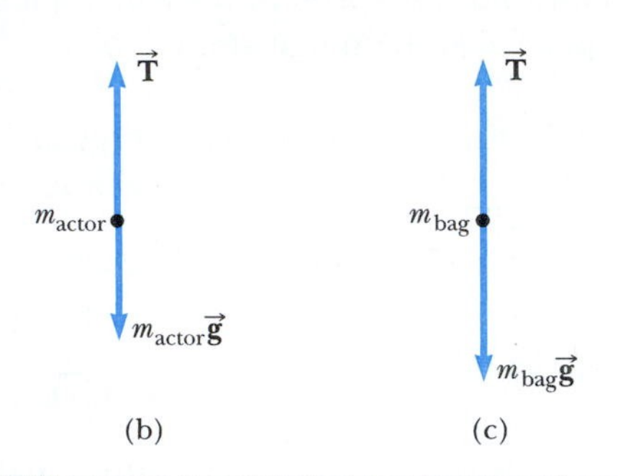


(b) (c)

**FIGURE 7.5** (Interactive Example 7.2) (a) An actor uses some clever staging to make his entrance. (b) Free-body diagram for the actor at the bottom of the circular path. (c) Free-body diagram for the sandbag when it is just lifted from the floor.

support his weight as well as provide centripetal acceleration of his body in the upward direction. At this point, the tension in the rope is the highest and the sandbag is most likely to lift off the floor. Looking first at the swinging of the actor from the initial point to the lowest point, we categorize this problem as an energy problem involving an isolated system, the actor and the Earth. To analyze this part of the problem we use the principle of conservation of mechanical energy for the system to find the actor's speed as he arrives at the floor as a function of the initial angle $\theta$ and the radius $R$ of the circular path through which he swings.

Applying Equation 7.7 to the actor–Earth system gives

$$K_f + U_{gf} = K_i + U_{gi}$$

$$(1)\quad \tfrac{1}{2}m_{\text{actor}}v_f^2 + 0 = 0 + m_{\text{actor}}gy_i$$

where $y_i$ is the initial height of the actor above the floor and $v_f$ is the speed of the actor at the instant before he lands. (Note that $K_i = 0$ because he starts from rest and that $U_f = 0$ because we define the configuration of the actor at the floor as having a gravitational potential energy of zero.) From the geometry in Figure 7.5a, we see that $y_i = R - R\cos\theta = R(1 - \cos\theta)$. Using this relationship in Equation (1), we obtain

$$(2)\quad v_f^2 = 2gR(1 - \cos\theta)$$

Next, we focus on the instant the actor is at the lowest point. Because the tension in the cable is transferred to the sandbag by means of force, we categorize the actor at this instant as a particle under a net force and a particle in uniform circular motion. To analyze, we apply Newton's second law to the actor at the bottom of his path, using the free-body diagram in Figure 7.5b as a guide:

$$\sum F_y = T - m_{\text{actor}}g = m_{\text{actor}}\frac{v_f^2}{R}$$

$$(3)\quad T = m_{\text{actor}}g + m_{\text{actor}}\frac{v_f^2}{R}$$

Finally, we note that the sandbag lifts off the floor when the upward force exerted on it by the cable exceeds the gravitational force acting on it; the normal force is zero when that happens. Because we do not want the sandbag to lift off the floor, we categorize the sandbag as a particle in equilibrium. A force $T$ of the magnitude given by (3) is transmitted by the cord to the sandbag. If the sandbag is to be on the verge of being lifted off the floor, the normal force on it becomes zero and we require that $T = m_{\text{bag}}g$ as in Figure 7.5c. Using this condition together with Equations (2) and (3), we find that

$$m_{\text{bag}}g = m_{\text{actor}}g + m_{\text{actor}}\frac{2gR(1 - \cos\theta)}{R}$$

Solving for $\cos\theta$ and substituting in the given parameters, we obtain

$$\cos\theta = \frac{3m_{\text{actor}} - m_{\text{bag}}}{2m_{\text{actor}}} = \frac{3(65\text{ kg}) - 130\text{ kg}}{2(65\text{ kg})} = \frac{1}{2}$$

$$\theta = 60°$$

To finalize the problem, note that we had to combine techniques from different areas of our study: energy and Newton's second law. Furthermore, the length $R$ of the cable from the actor's harness to the leftmost pulley did not appear in the final algebraic equation. Therefore, the final answer is independent of $R$.

**PhysicsNow™** Let the actor fly or crash without injury by logging into PhysicsNow at **www.pop4e.com** and going to Interactive Example 7.2.

## 7.3 CONSERVATIVE AND NONCONSERVATIVE FORCES

In the preceding section, we showed that the mechanical energy of a system is conserved in a process in which the force between members of the system is the gravitational force. The gravitational force is one example of a category of forces for which the mechanical energy of a system is conserved. These forces are called **conservative forces.** The other possibility for energy storage in a system besides kinetic and potential is internal energy. Therefore, a conservative force, for our purposes in mechanics, is a **force between members of a system that causes no transformation of mechanical energy to internal energy within the system.** If no energy is transformed to internal energy, the mechanical energy of the system is conserved, as described by Equation 7.7.

If a force is conservative, the work done by such a force has a special property as the members of the system move in response either to the force itself or to an external force: **The work done by a conservative force is independent of the path followed by the members of the system and depends only on the initial and final configurations of the system.**

■ A conservative force

From this property, it follows that **the work done by a conservative force when a member of the system is moved through a closed path is equal to zero.**

These statements can be mathematically demonstrated and serve as general mathematical definitions of conservative forces. Both statements can be seen for the gravitational force from Equation 7.3. The work done is expressed only in terms of the initial and final heights, with no indication of what path is followed. If the path is closed, the initial and final heights are the same in Equation 7.3 and the work is equal to zero.

Another example of a conservative force is the force of a spring on an object attached to the spring, where the spring force is given by Hooke's law, $F_s = -kx$. As we learned in Chapter 6 (Eq. 6.15), the work done by the spring force is

$$W_s = \tfrac{1}{2}kx_i^2 - \tfrac{1}{2}kx_f^2$$

where the initial and final positions of the object are measured from its equilibrium position $x = 0$, at which the spring is unstretched. Again we see that $W_s$ depends only on the initial and final coordinates of the object and is zero for any closed path. Hence, the spring force is conservative.

In Section 7.1, we discussed the notion of an external agent lifting a book and storing energy as potential energy in the book–Earth system. In Section 6.4, we discussed an external agent pulling a block attached to a spring from $x = 0$ to $x = x_{\text{max}}$ and calculated the work done on the system as $\frac{1}{2}kx_{\text{max}}^2$. This situation is another one, like that of the book in Section 7.1, in which work is done on a system but there is no change in kinetic energy of the system. Therefore, the energy must be stored in the block–spring system as potential energy. The **elastic potential energy** associated with the spring force is defined by

$$U_s \equiv \tfrac{1}{2}kx^2 \qquad [7.9]$$

■ Elastic potential energy

The elastic potential energy can be considered as the energy stored in the deformed spring (one that is either compressed or stretched from its equilibrium position). The elastic potential energy stored in the spring is zero whenever the spring is undeformed ($x = 0$). Because elastic potential energy is proportional to $x^2$, we see that $U_s$ is always positive in a deformed spring.

Consider Active Figure 7.6a, which shows an undeformed spring on a frictionless, horizontal surface. When the block is pushed against the spring (Active Fig. 7.6b), compressing the spring a distance $x$, the elastic potential energy stored in the spring is $\frac{1}{2}kx^2$. When the block is released, the spring returns to its original

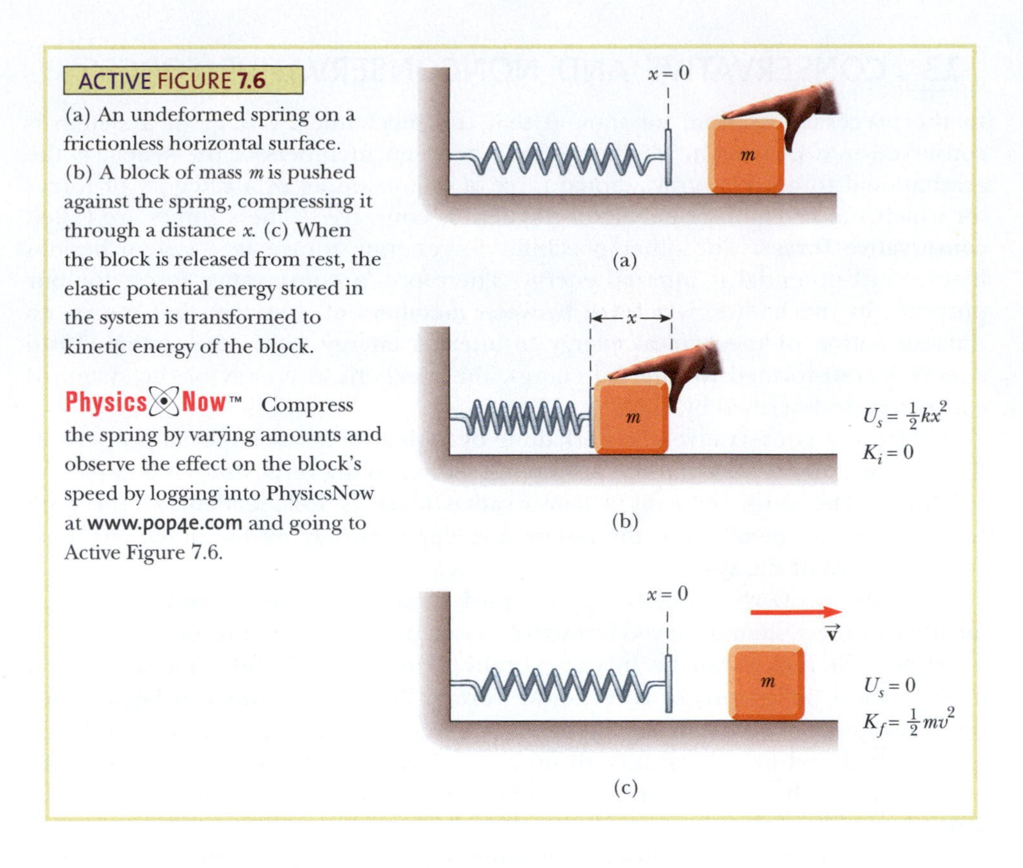


**ACTIVE FIGURE 7.6**

(a) An undeformed spring on a frictionless horizontal surface. (b) A block of mass $m$ is pushed against the spring, compressing it through a distance $x$. (c) When the block is released from rest, the elastic potential energy stored in the system is transformed to kinetic energy of the block.

**PhysicsNow™** Compress the spring by varying amounts and observe the effect on the block's speed by logging into PhysicsNow at **www.pop4e.com** and going to Active Figure 7.6.

length, applying a force to the block. This force does work on the block, resulting in kinetic energy of the block (Active Fig. 7.6c).

▮ A nonconservative force

In comparison to a conservative force, a **nonconservative force** in mechanics is **a force between members of a system that is not conservative;** that is, it causes transformation of mechanical energy to internal energy within the system. A common nonconservative force in mechanics is the friction force. If we consider a system consisting of a block and a surface and imagine an initially sliding block coming to rest because of friction, we see the result of a nonconservative force. Initially, the system has kinetic energy (of the block). Afterward, nothing is moving so the final kinetic energy is zero. The friction force between the block and the surface transforms the mechanical energy into internal energy; the block and surface are both slightly warmer than before.

Let us return to the notion of the work done over a path. The two statements claimed for conservative forces are not true for nonconservative forces. For nonconservative forces, the work done depends on the path taken between the initial and final configurations, and the work done over a closed path is not zero. As an example, consider Figure 7.7. Suppose you displace a book between two points on a table. If the book is displaced in a straight line along the blue path between points Ⓐ and Ⓑ in Figure 7.7, you do a certain amount of work against the kinetic friction force to keep the book moving at a constant speed. Now, imagine that you push the book along the brown semicircular path in Figure 7.7. You perform more work against friction along this longer path than along the straight path. The work done depends on the path, so the friction force cannot be conservative.

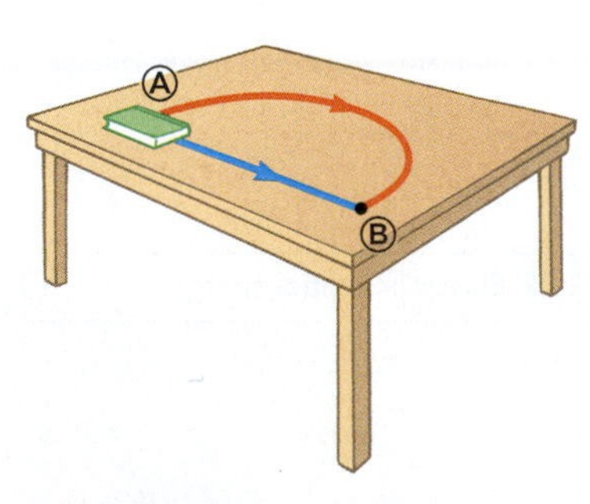


**FIGURE 7.7** The work done against the force of friction depends on the path taken as the book is moved from Ⓐ to Ⓑ; hence, friction is a nonconservative force. The work required is greater along the brown path than the blue path.

In Chapter 6, we discussed that we cannot calculate the work done by friction on an object because the displacement of the point of application of the friction force is not the same as the displacement of the object. In the case of an object subject to a force of kinetic friction, the particle model is not valid. Another example of a situation in which we cannot use the particle model is seen for deformable objects. For example, suppose a rubber ball is flattened against a brick wall. The ball deforms during the pushing process. If the ball is suddenly released, it jumps away from the wall.

The force causing the ball to accelerate is the normal force of the wall on the ball. The point of application of this force, however, does not move through space; it stays fixed at the point of contact between the ball and the wall. Therefore, no work is done on the ball. Yet the ball has kinetic energy afterward. The work–kinetic energy theorem does not describe this situation correctly. It is more valuable to apply the isolated system model to this situation. With the ball as the system, there is no transfer of energy across the boundary as the ball springs off the wall. Rather, there is a *transformation* of energy, from elastic potential energy (stored in the ball when it was flattened) to kinetic energy. In the same way, if a skateboarder pushes off a wall to start rolling, no work is done by the force from the wall; the kinetic energy is transformed within the system from potential energy stored in the body from previous meals.

We also discussed in Chapter 6 the difficulties associated with the nonconservative force of friction in energy calculations. Recall that the result of a friction force is to transform kinetic energy in a system to internal energy and that the increase in internal energy is equal to the decrease in kinetic energy. If a potential energy is associated with the system, the decrease in *mechanical* energy, equals the increase in internal energy in the isolated system. Therefore, for a constant friction force,

$$-f_k d = \Delta K + \Delta U = \Delta E_{\text{mech}} = -\Delta E_{\text{int}} \qquad [7.10]$$

We can recast this expression by putting the changes in all forms of energy storage on one side of the equation:

$$\Delta K + \Delta U + \Delta E_{\text{int}} = \Delta E_{\text{system}} = 0 \qquad [7.11]$$

This gives us the most general expression of the continuity equation for energy for an isolated system. Note that $\Delta K$ may represent more than one term if two or more parts of the system are moving. Also, $\Delta U$ may represent more than one term if different types of potential energy (e.g., gravitational and elastic) are associated with the system. Equation 7.11 is equivalent to

$$K + U + E_{\text{int}} = \text{constant} \qquad [7.12]$$

■ Conservation of energy for an isolated system

which tells us that **the total energy (kinetic, potential, and internal) of an isolated system is conserved, regardless of whether the forces acting within the system are conservative or nonconservative. No violation of this critical conservation principle has ever been observed.** If we consider the Universe as an isolated system, this statement claims that there is a fixed amount of energy in our Universe and that all processes within the Universe represent transformations of energy from one type to another.

**PITFALL PREVENTION 7.3**

**COMPLICATED SYSTEMS** For simplicity, our discussion leading to Equation 7.10 assumes that only one object in the system is sliding on a surface. If there are two or more objects sliding on surfaces in the system, a term $-f_k d$ must be included for each object, with $d$ representing the distance the object slides relative to the surface with which it is in contact.

**QUICK QUIZ 7.4** A ball is connected to a light spring suspended vertically. When displaced downward from its equilibrium position and released, the ball oscillates up and down. **(i)** In the system of *the ball, the spring, and the Earth,* what forms of energy are there during the motion? **(ii)** In the system of *the ball and the spring,* what forms of energy are there during the motion? **(a)** kinetic and elastic potential **(b)** kinetic and gravitational potential **(c)** kinetic, elastic potential, and gravitational potential **(d)** elastic potential and gravitational potential

## PROBLEM-SOLVING STRATEGY Isolated Systems

Many problems in physics can be solved using the principle of conservation of energy for an isolated system. The following procedure should be used when you apply this principle:

1. **Conceptualize** Define your system, which may consist of more than one object and may or may not include springs or other possibilities for storage of potential energy. Choose configurations to represent the initial and final conditions of the system.
2. **Categorize** Determine if any energy transfers occur across the boundary of your system. If so, use the nonisolated system model, $\Delta E_{\text{system}} = \Sigma T$. If not, use the isolated system model, $\Delta E_{\text{system}} = 0$.

Determine whether any nonconservative forces are present. Remember that if friction or air resistance is present, *mechanical* energy is not conserved but the *total* energy of an isolated system is.

3. **Analyze** For each object that changes elevation, select a reference position for the object that will define the zero configuration of gravitational potential energy for the system. For a spring, the zero configuration for elastic potential energy is when the spring is neither compressed nor extended from its equilibrium position. If there is more than one conservative force, write an expression for the potential energy associated with each force.

   If mechanical energy is conserved, write the total initial mechanical energy $E_i$ of the system for some configuration as the sum of the kinetic and potential energy associated with the configuration. Then write a similar expression for the total mechanical energy $E_f$ of the system for the final configuration that is of interest. Because mechanical energy is conserved, equate the two total energies and solve for the quantity that is unknown.

   If nonconservative forces are present (and therefore mechanical energy is not conserved), first write expressions for the total initial and total final mechanical energies. In this case, the difference between the total final mechanical energy and the total initial mechanical energy equals the energy transformed to or from internal energy by the nonconservative forces.

4. **Finalize** Make sure your results are consistent with your mental representation. Also make sure that the values of your results are reasonable and consistent with connections to everyday experience.

## EXAMPLE 7.3 Crate Sliding Down a Ramp

A 3.00-kg crate slides down a ramp at a loading dock. The ramp is 1.00 m in length and is inclined at an angle of 30.0° as shown in Figure 7.8. The crate starts from rest at the top and experiences a constant friction force of magnitude 5.00 N. Use energy methods to determine the speed of the crate when it reaches the bottom of the ramp.

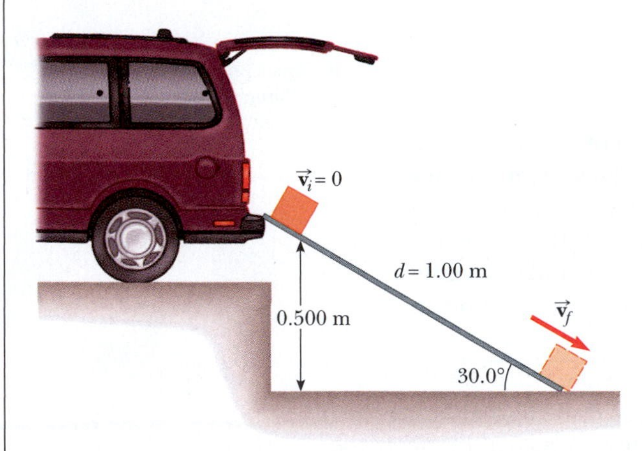


FIGURE 7.8 (Example 7.3) A crate slides down a ramp under the influence of gravity. The potential energy of the crate–Earth system decreases, whereas the kinetic energy of the crate increases.

**Solution** We define the system as the crate, the Earth, and the ramp. This system is isolated. If we had chosen the crate and the Earth as the system, we would need to use the nonisolated system model because the friction force between the crate and the ramp is an external influence. There would be work done across the boundary as well as flow of energy by heat between the crate and the ramp. This problem would be difficult to solve. In general, if a friction force acts, it is easiest to define the system so that the friction force is an internal force.

Because $v_i = 0$ for the crate, the initial kinetic energy of the system is zero. If the $y$ coordinate is measured from the bottom of the ramp, $y_i = (1.00 \text{ m}) \sin 30° = 0.500$ m for the crate. The total mechanical energy of the crate–Earth–ramp system when the crate is at the top is therefore the gravitational potential energy:

$$E_i = U_i = mgy_i$$

When the crate reaches the bottom, the gravitational potential energy of the system is *zero* because the elevation of the crate is $y_f = 0$. The total mechanical energy when the crate is at the bottom is therefore kinetic energy:

$$E_f = K_f = \tfrac{1}{2}mv_f^2$$

We cannot say that $E_f = E_i$ in this case, however, because a nonconservative force—the force of friction—reduces the mechanical energy of the system. In this case, the change in mechanical energy for the system is $\Delta E_{\text{mech}} = -f_k d$, where $d = 1.00$ m. Because $\Delta E_{\text{mech}} = \Delta K + \Delta U = \frac{1}{2}mv_f^2 - mgy_i$ in this situation, Equation 7.10 gives

$$-f_k d = \tfrac{1}{2}mv_f^2 - mgy_i$$

$$v_f = \sqrt{2gy_i - 2\frac{f_k d}{m}}$$

$$= \sqrt{2(9.80 \text{ m/s}^2)(0.500 \text{ m}) - 2\frac{(5.00 \text{ N})(1.00 \text{ m})}{(3.00 \text{ kg})}}$$

$$= 2.54 \text{ m/s}$$

## EXAMPLE 7.4 Motion on a Curved Track

A child of mass $m$ takes a ride on an irregularly curved slide of height $h = 2.00$ m as in Figure 7.9. The child starts from rest at the top.

**A** Determine the speed of the child at the bottom, assuming that no friction is present.

**Solution** We will define the system as the child and the Earth and will model the child as a particle. The normal force $\vec{\mathbf{n}}$ does no work on the system because this force is always perpendicular to each element of the displacement. Furthermore, because no friction is present, no work is done by friction across the boundary of the system. Therefore, we use the isolated system model with no friction forces, for which mechanical energy is conserved; that is, $K + U =$ constant.

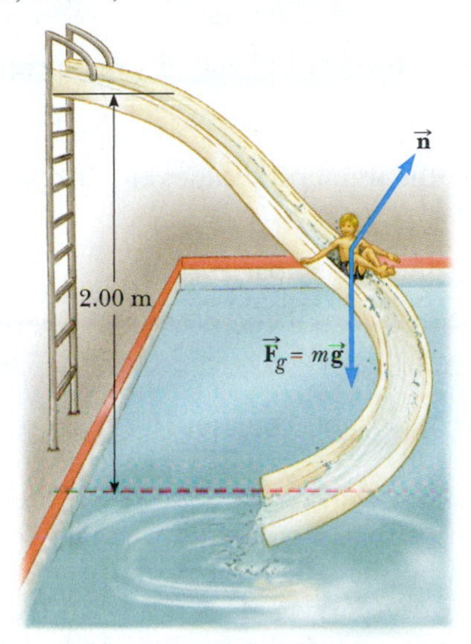


FIGURE 7.9 (Example 7.4) If the slide is frictionless, the speed of the child at the bottom depends only on the height of the slide.

If we measure the $y$ coordinate for the child from the bottom of the slide, $y_i = h$, $y_f = 0$, and we have for the system

$$K_f + U_f = K_i + U_i$$

$$\tfrac{1}{2}mv_f^2 + 0 = 0 + mgh$$

$$v_f = \sqrt{2gh}$$

The result for the speed is the same as if the child simply fell vertically through a distance $h$! In this example, $h = 2.00$ m, giving

$$v_f = \sqrt{2gh} = \sqrt{2(9.80 \text{ m/s}^2)(2.00 \text{ m})} = 6.26 \text{ m/s}$$

**B** If a friction force acts on the 20.0-kg child and he arrives at the bottom of the slide with a speed $v_f = 3.00$ m/s, by how much does the mechanical energy of the system decrease as a result of this force?

**Solution** We define the system as the child, the Earth, and the slide. In this case, a nonconservative force acts within the system and mechanical energy is *not* conserved. We can find the change in mechanical energy as a result of friction, given that the final speed at the bottom is known:

$$\Delta E_{\text{mech}} = K_f + U_f - K_i - U_i = \tfrac{1}{2}mv_f^2 + 0 - 0 - mgh$$

$$\Delta E_{\text{mech}} = \tfrac{1}{2}(20.0 \text{ kg})(3.00 \text{ m/s})^2 - (20.0 \text{ kg})(9.80 \text{ m/s}^2)(2.00 \text{ m})$$

$$= -302 \text{ J}$$

The change in mechanical energy $\Delta E_{\text{mech}}$ is negative because friction reduces the mechanical energy of the system. The change in internal energy in the system is $+302$ J.

## EXAMPLE 7.5 Block–Spring Collision

A block of mass 0.800 kg is given an initial velocity $v_{\text{A}} = 1.20$ m/s to the right and collides with a light spring of force constant $k = 50.0$ N/m as in Figure 7.10.

**A** If the surface is frictionless, calculate the maximum compression of the spring after the collision.

**Solution** We define the system as the block and the spring. No transfers of energy occur across the boundary of this system, so we use the isolated system model. Before the collision, when the block is at Ⓐ, for example, the system has kinetic energy due to the moving block and the spring is uncompressed, so the potential energy stored in the system is zero. Therefore, the total energy of the system before the collision is $\frac{1}{2}mv_{\text{A}}^2$. After the collision, and when the spring is fully compressed at point Ⓒ, the block is momentarily at rest and has zero kinetic energy, whereas the potential energy stored in the spring has its maximum value $\frac{1}{2}kx_{\text{max}}^2$. The total mechanical energy of the system is conserved because no nonconservative forces act within the system.

Because the mechanical energy of the system is conserved,

$$\tfrac{1}{2}mv_{\text{A}}^2 + 0 = 0 + \tfrac{1}{2}kx_{\text{max}}^2$$

$$x_{\text{max}} = \sqrt{\frac{m}{k}}\,v_{\text{A}} = \sqrt{\frac{0.800 \text{ kg}}{50.0 \text{ N/m}}}(1.20 \text{ m/s}) = 0.152 \text{ m}$$

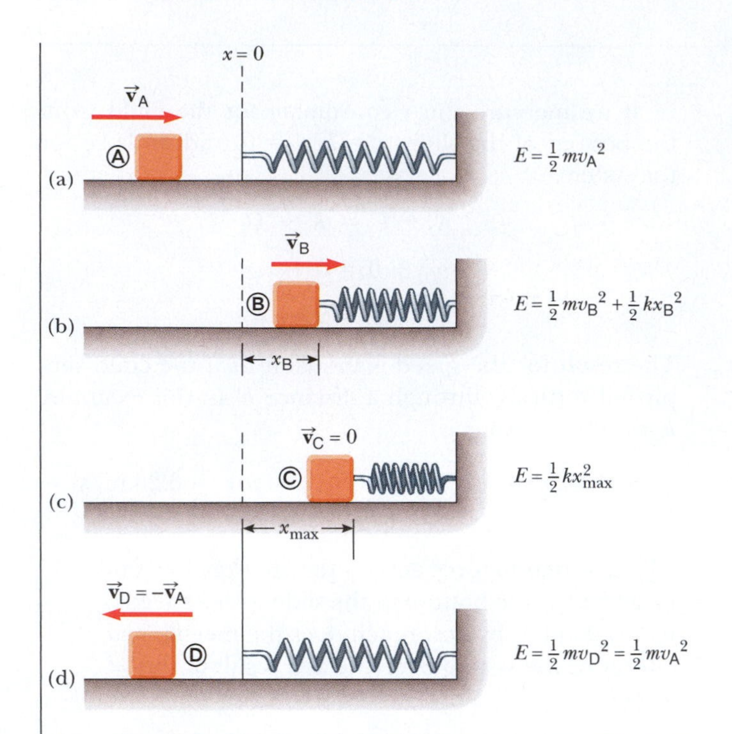


**FIGURE 7.10** (Example 7.5) A block sliding on a smooth, horizontal surface collides with a light spring. (a) Initially, the mechanical energy of the system is all kinetic energy. (b) The mechanical energy is the sum of the kinetic energy of the block and the elastic potential energy in the spring. (c) The mechanical energy is entirely potential energy. (d) The mechanical energy is transformed back to the kinetic energy of the block. The total energy of the block–spring system remains constant throughout the motion.

**B** If a constant force of kinetic friction acts between the block and the surface with $\mu_k = 0.500$ and if the speed of the block just as it collides with the spring is $v_A = 1.20$ m/s, what is the maximum compression in the spring?

**Solution** We define the system as the block, the spring, and the surface. In this case, mechanical energy of the system is *not* conserved because a friction force acts between members of the system. The magnitude of the friction force is

$$f_k = \mu_k n = \mu_k mg = 0.500(0.800 \text{ kg})(9.80 \text{ m/s}^2) = 3.92 \text{ N}$$

where we have used $n = mg$ from Newton's second law in the vertical direction. Therefore, the decrease in mechanical energy due to friction as the block is displaced through a straight line from $x_i = 0$ to the point $x_f = x_{max}$ at which the block stops is

$$\Delta E_{mech} = -f_k x_{max} = -3.92 x_{max}$$

The change in mechanical energy can be expressed as

$$\Delta E_{mech} = E_f - E_i = (0 + \tfrac{1}{2}kx_{max}^2) - (\tfrac{1}{2}mv_A^2 + 0)$$

Substituting the numerical values and dropping the units, we have

$$-3.92x_{max} = \frac{50.0}{2}x_{max}^2 - \tfrac{1}{2}(0.800)(1.20)^2$$

$$25.0x_{max}^2 + 3.92x_{max} - 0.576 = 0$$

Solving the quadratic equation for $x_{max}$ gives $x_{max} = 0.092\ 4$ m and $x_{max} = -0.249$ m. We choose the positive root $x_{max} = 0.092\ 4$ m because the block must be to the right of the origin when it comes to rest. Note that 0.092 4 m is less than the distance obtained in the frictionless case (part A). This result is what we expect because friction retards the motion of the system.

## 7.4 CONSERVATIVE FORCES AND POTENTIAL ENERGY

Let us return to the falling book discussed in Section 7.2. We found that the work done within the book–Earth system by the gravitational force on the book can be expressed as the negative of the difference between two quantities that we called the initial and final potential energies of the system:

$$W_{on\ book} = mgy_b - mgy_a = -\Delta U \quad [7.13]$$

This expression is the hallmark of a conservative force: we can identify a potential energy function such that the work done by the force on a member of the system in which the force acts depends only on the difference in the function's initial and final values. Such a function does not exist for a nonconservative force because the work done depends on the particular path followed between the initial and final points.

For a conservative force, this notion allows us to generate a mathematical relationship between a force and its potential energy function. From the definition of work, we can write Equation 7.13 for a general force in the $x$ direction as

$$W = \int_{x_i}^{x_f} F_x \, dx = -\Delta U = -(U_f - U_i) = -U_f + U_i \quad [7.14]$$

Therefore, the potential energy function can be written as

$$U_f = -\int_{x_i}^{x_f} F_x \, dx + U_i \qquad [7.15]$$

■ **Finding the potential energy of a system associated with a force between members of the system**

This expression allows us to calculate the potential energy function associated with a conservative force if we know the force function. The value of $U_i$ is often taken to be zero at some arbitrary reference point. It really doesn't matter what value we assign to $U_i$ because any value simply shifts $U_f$ by a constant, and it is the *change* in potential energy that is physically meaningful.

As an example, let us calculate the potential energy function for the spring force. We model the spring as obeying Hooke's law, so the force the spring exerts is $F_s = -kx$. The potential energy stored in a block-spring system is

$$U_f = -\int_{x_i}^{x_f} (-kx)\, dx + U_i = \tfrac{1}{2}kx_f^2 - \tfrac{1}{2}kx_i^2 + U_i$$

As mentioned earlier, we can choose the configuration representing the zero of potential energy arbitrarily. Let us choose $U_i = 0$ when the block is at the position $x_i = 0$. Then,

$$U_f = \tfrac{1}{2}kx_f^2 - \tfrac{1}{2}kx_i^2 + U_i = \tfrac{1}{2}kx_f^2 - 0 + 0 \quad \rightarrow \quad U_f = U_s = \tfrac{1}{2}kx^2$$

which is the potential energy function we have already recognized (see Eq. 7.9) for a spring that obeys Hooke's law.

In the preceding discussion, we have seen how to find a potential energy function if we know the force function. Let us now turn this process around. Suppose we know the potential energy function. Can we find the force function? We start from the basic definition of work done by a conservative force for an infinitesimal displacement $d\vec{\mathbf{r}} = dx\hat{\mathbf{i}}$ in the $x$ direction:

$$dW = \vec{\mathbf{F}} \cdot d\vec{\mathbf{r}} = \vec{\mathbf{F}} \cdot dx\hat{\mathbf{i}} = F_x \, dx = -dU$$

This equation can be rewritten as

$$F_x = -\frac{dU}{dx} \qquad [7.16]$$

■ **Finding the force between members of the system from the potential energy of the system**

In general, **the conservative force acting between parts of a system equals the negative derivative of the potential energy associated with that system.**[1]

In the case of an object located a distance $y$ above some reference point, the gravitational potential energy function is given by $U_g = mgy$, and it follows from Equation 7.16 that (considering the $y$ direction rather than $x$)

$$F_y = -\frac{dU_g}{dy} = -\frac{d}{dy}(mgy) = -mg$$

which is the correct expression for the vertical component of the gravitational force.

---

[1]In three dimensions, the appropriate expression is

$$\vec{\mathbf{F}} = -\hat{\mathbf{i}}\frac{\partial U}{\partial x} - \hat{\mathbf{j}}\frac{\partial U}{\partial y} - \hat{\mathbf{k}}\frac{\partial U}{\partial z}$$

where $\partial U/\partial x$ and so on are *partial derivatives.* In the language of vector calculus, $\vec{\mathbf{F}}$ equals the negative of the *gradient* of the scalar potential energy function $U(x, y, z)$.

## 7.5 THE NONISOLATED SYSTEM IN STEADY STATE

We have seen two approaches related to systems so far. In a nonisolated system, the energy stored in the system changes due to transfers across the boundaries of the system. Therefore, nonzero terms occur on both sides of the continuity equation for energy, $\Delta E_{\text{system}} = \sum T$. For an isolated system, no energy transfer takes place across the boundary, so the right-hand side of the continuity equation is zero; that is, $\Delta E_{\text{system}} = 0$.

Another possibility exists that we have not yet addressed. It is possible for no change to occur in the energy of the system even though nonzero terms are present on the right-hand side of the continuity equation, $0 = \sum T$. This situation can only occur if the rate at which energy is entering the system is equal to the rate at which it is leaving. In this case, the system is in steady state under the effects of two or more competing transfers, which we describe as a **nonisolated system in steady state.** The system is nonisolated because it is interacting with the environment, but it is in steady state because the system energy remains constant.

We could identify a number of examples of this type of situation. First, consider your home as a nonisolated system. Ideally, you would like to keep the temperature of your home constant for the comfort of the occupants. Therefore, your goal is to keep the internal energy in the home fixed.

The energy transfer mechanisms for the home are numerous, as we can see in Figure 7.11. Solar electromagnetic radiation is absorbed by the roof and walls of the home and enters the home through the windows. Energy enters by electrical transmission to operate electrical devices. Leaks in the walls, windows, and doors allow warm or cold air to enter and leave, carrying energy across the boundary of the system by matter transfer. Matter transfer also occurs if any devices in the home operate from natural gas because energy is carried in with the gas. Energy transfer by heat occurs through the walls, windows, floor, and roof as a result of temperature differences between the inside and outside of the home. Therefore, we have a variety of transfers, but the energy in the home remains constant in the idealized case. In reality, the home is a system in *quasi-steady state* because some small temperature variations actually occur over a 24-h period, but we can imagine an idealized situation that conforms to the nonisolated system in steady-state model.

As a second example, consider the Earth and its atmosphere as a system. Because this system is located in the vacuum of space, the only possible types of

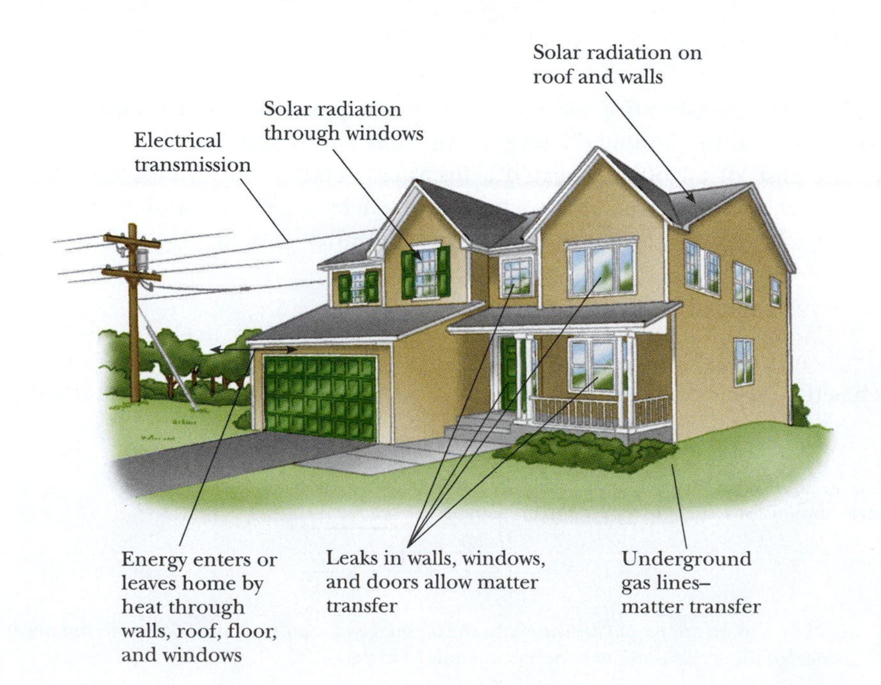


**FIGURE 7.11** Energy enters and leaves a home by several mechanisms. The home can be modeled as a nonisolated system in steady state.

energy transfers are those that involve no contact between the system and external molecules in the environment. As mentioned in the footnote on page 172, only two types of transfer do not depend on contact with molecules: work done by field forces and electromagnetic radiation. The Earth–atmosphere system exchanges energy with the rest of the Universe only by means of electromagnetic radiation (ignoring work done by field forces and ignoring some small matter transfer as a result of cosmic ray particles and meteoroids entering the system and spacecraft leaving the system!). The primary input radiation is that from the Sun, and the output radiation is primarily infrared radiation emitted from the atmosphere and the ground. Ideally, these transfers are balanced so that the Earth maintains a constant temperature. In reality, however, the transfers are not *exactly* balanced, so the Earth is in quasi-steady state; measurements of the temperature show that it does appear to be changing. The change in temperature is very gradual and currently appears to be in the positive direction. This change is the essence of the social issue of global warming. (See Context 5, beginning on page 497.)

If we consider a time interval of several days, the human body can be modeled as another nonisolated system in steady state. If the body is at rest at the beginning and end of the time interval, there is no change in kinetic energy. Assuming that no major weight gain or loss occurs during this time interval, the amount of potential energy stored in the body as food in the stomach and fat remains constant on the average. If no fevers are experienced during this time interval, the internal energy of the body remains constant. Therefore, the change in the energy of the system is zero. Energy transfer methods during this time interval include work (you apply forces on objects which move), heat (your body is warmer than the surrounding air), matter transfer (breathing, eating), mechanical waves (you speak and hear), and electromagnetic radiation (you see, as well as absorb and emit radiation from your skin).

The human body as a nonisolated system

## 7.6 POTENTIAL ENERGY FOR GRAVITATIONAL AND ELECTRIC FORCES

Earlier in this chapter we introduced the concept of gravitational potential energy, that is, the energy associated with a system of objects interacting via the gravitational force. We emphasized that the gravitational potential energy function, Equation 7.2, is valid only when the object of mass $m$ is near the Earth's surface. We would like to find a more general expression for the gravitational potential energy that is valid for all separation distances. Because the free-fall acceleration $g$ varies as $1/r^2$, it follows that the general dependence of the potential energy function of the system on separation distance is more complicated than our simple expression, Equation 7.2.

Consider a particle of mass $m$ moving between two points Ⓐ and Ⓑ above the Earth's surface as in Figure 7.12. The gravitational force on the particle due to the Earth, first introduced in Section 5.6, can be written in vector form as

$$\vec{\mathbf{F}}_g = -\frac{GM_E m}{r^2}\hat{\mathbf{r}} \qquad [7.17]$$

where $\hat{\mathbf{r}}$ is a unit vector directed from the Earth toward the particle and the negative sign indicates that the force is downward toward the Earth. This expression shows that the gravitational force depends on the radial coordinate $r$. Furthermore, the gravitational force is conservative. Equation 7.15 gives

$$U_f = -\int_{r_i}^{r_f} F(r)\,dr + U_i = GM_E m\int_{r_i}^{r_f}\frac{dr}{r^2} + U_i = GM_E m\left(-\frac{1}{r}\right)\Bigg|_{r_i}^{r_f} + U_i$$

or

$$U_f = -GM_E m\left(\frac{1}{r_f} - \frac{1}{r_i}\right) + U_i \qquad [7.18]$$

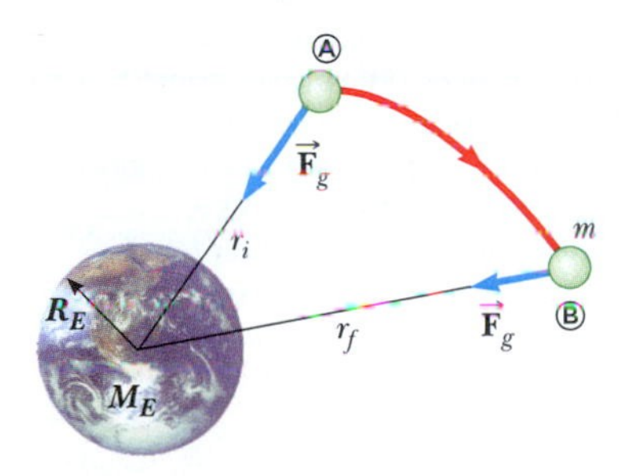


**FIGURE 7.12** As a particle of mass $m$ moves from Ⓐ to Ⓑ above the Earth's surface, the potential energy of the particle–Earth system, given by Equation 7.19, changes because of the change in the particle–Earth separation distance $r$ from $r_i$ to $r_f$.

### PITFALL PREVENTION 7.4

**What is $r$?** In Section 5.5, we discussed the gravitational force between two *particles*. In Equation 7.17, we present the gravitational force between a particle and an extended object, the Earth. We could also express the gravitational force between two extended objects, such as the Earth and the Sun. In these kinds of situations, remember that $r$ is measured *between the centers* of the objects. Be sure *not* to measure $r$ from the surface of the Earth.

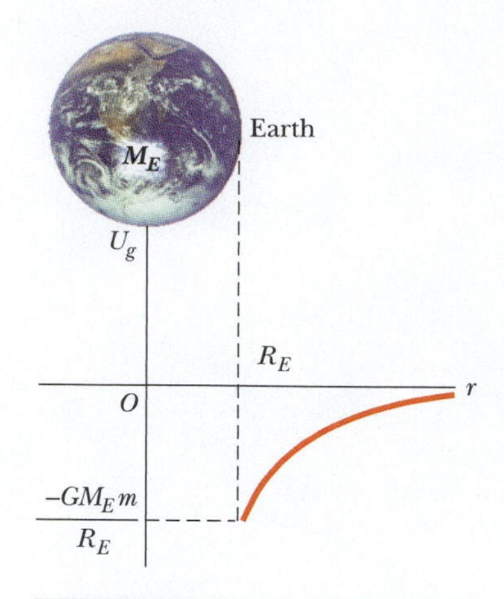


FIGURE 7.13 Graph of the gravitational potential energy $U_g$ versus $r$ for a particle above the Earth's surface. The potential energy of the system goes to zero as $r$ approaches infinity.

As always, the choice of a reference point for the potential energy is completely arbitrary. It is customary to locate the reference point where the force is zero. Letting $U_i \rightarrow 0$ as $r_i \rightarrow \infty$, we obtain the important result

$$U_g = -\frac{GM_E m}{r} \qquad [7.19]$$

for separation distances $r > R_E$, the radius of the Earth. Because of our choice of the reference point for zero potential energy, the function $U_g$ is always negative (Fig. 7.13).

Although Equation 7.19 was derived for the particle–Earth system, it can be applied to *any* two particles. For *any pair* of particles of masses $m_1$ and $m_2$ separated by a distance $r$, the gravitational force of attraction is given by Equation 5.11 and the gravitational potential energy of the system of two particles is

$$U_g = -\frac{Gm_1 m_2}{r} \qquad [7.20]$$

This expression also applies to larger objects *if their mass distributions are spherically symmetric,* as first shown by Newton. In this case, $r$ is measured between the centers of the spherical objects.

Equation 7.20 shows that the gravitational potential energy for any pair of particles varies as $1/r$ (whereas the force between them varies as $1/r^2$). Furthermore, the potential energy is *negative* because the force is attractive and we have chosen the potential energy to be zero when the particle separation is infinity. Because the force between the particles is attractive, we know that an external agent must do positive work to increase the separation between the two particles. The work done by the external agent produces an increase in the potential energy as the two particles are separated. That is, $U_g$ becomes less negative as $r$ increases.

### PITFALL PREVENTION 7.5

**GRAVITATIONAL POTENTIAL ENERGY** Be careful! Equation 7.20 looks similar to Equation 5.14 for the gravitational force, but there are two major differences. The gravitational force is a vector, whereas the gravitational potential energy is a scalar. The gravitational force varies as the *inverse square* of the separation distance, whereas the gravitational potential energy varies as the simple *inverse* of the separation distance.

We can extend this concept to three or more particles. In this case, the total potential energy of the system is the sum over all *pairs* of particles. Each pair contributes a term of the form given by Equation 7.20. For example, if the system contains three particles, as in Figure 7.14, we find that

$$U_{\text{total}} = U_{12} + U_{13} + U_{23} = -G\left(\frac{m_1 m_2}{r_{12}} + \frac{m_1 m_3}{r_{13}} + \frac{m_2 m_3}{r_{23}}\right) \qquad [7.21]$$

The absolute value of $U_{\text{total}}$ represents the work needed to separate all three particles by an infinite distance.

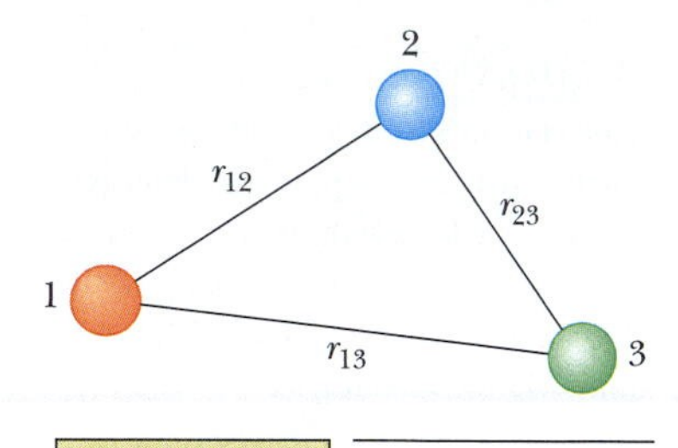


FIGURE 7.14 Three interacting particles.

## Thinking Physics 7.2

Why is the Sun hot?

**Reasoning** The Sun was formed when a cloud of gas and dust coalesced, because of gravitational attraction, into a massive astronomical object. Let us define this cloud as our system and model the gas and dust as particles. Initially, the particles of the system were widely scattered, representing a large amount of gravitational potential energy. As the particles moved together to form the Sun, the gravitational potential energy of the system decreased. According to the isolated system model, this potential energy was transformed to kinetic energy as the particles fell toward the center. As the speeds of the particles increased, many collisions occurred between particles, randomizing their motion and transforming the kinetic energy to internal energy, which represented an increase in temperature. As the particles came together, the temperature rose to a point at which nuclear reactions occurred. These reactions release huge amounts of energy that maintain the high temperature of the Sun. This process has occurred for every star in the Universe. ■

**EXAMPLE 7.6** **The Change in Potential Energy**

A particle of mass $m$ is displaced through a small vertical distance $\Delta y$ near the Earth's surface. Show that the general expression for the change in gravitational potential energy reduces to the familiar relationship $\Delta U_g = mg\,\Delta y$.

**Solution** We can express Equation 7.18 in the form

$$\Delta U_g = -GM_E m\left(\frac{1}{r_f} - \frac{1}{r_i}\right) = GM_E m\left(\frac{r_f - r_i}{r_i r_f}\right)$$

If both the initial and final positions of the particle are close to the Earth's surface, $r_f - r_i = \Delta y$ and $r_i r_f \approx R_E^2$. (Recall that $r$ is measured from the center of the Earth.) The change in potential energy therefore becomes

$$\Delta U_g \approx \frac{GM_E m}{R_E^2}\,\Delta y = F_g \Delta y = mg\,\Delta y$$

where we have used Equation 7.17 to express $GM_E m/R_E^2$ as the magnitude of the gravitational force $F_g$ on an object of mass $m$ at the Earth's surface and then Equation 4.5 to express $F_g$ as $mg$.

In Chapter 5, we discussed the electrostatic force between two point particles, which is given by Coulomb's law,

$$F_e = k_e \frac{q_1 q_2}{r^2} \qquad [7.22]$$

Because this expression looks so similar to Newton's law of universal gravitation, we would expect that the generation of a potential energy function for this force would proceed in a similar way. That is indeed the case, and this procedure results in the **electric potential energy** function,

$$U_e = k_e \frac{q_1 q_2}{r} \qquad [7.23]$$

As with the gravitational potential energy, the electric potential energy is defined as zero when the charges are infinitely far apart. Comparing this expression with that for the gravitational potential energy, we see the obvious differences in the constants and the use of charges instead of masses, but there is one more difference. The gravitational expression has a negative sign, but the electrical expression doesn't. For systems of objects that experience an attractive force, the potential energy decreases as the objects are brought closer together. Because we have defined zero potential energy at infinite separation, all real separations are finite and the energy must decrease from a value of zero. Therefore, all potential energies for systems of objects that attract must be negative. In the gravitational case, attraction is the only possibility. The constant, the masses, and the separation distance are all positive, so the negative sign must be included explicitly, as it is in Equation 7.20.

The electric force can be either attractive or repulsive. Attraction occurs between charges of opposite sign. Therefore, for the two charges in Equation 7.23, one is positive and one is negative if the force is attractive. The product of the charges provides the negative sign for the potential energy mathematically, and we do not need an explicit negative sign in the potential energy expression. In the case of charges with the same sign, either a product of two negative charges or two positive charges will be positive, leading to a positive potential energy. This conclusion is reasonable because to cause repelling particles to move together from infinite separation requires work to be done on the system, so the potential energy increases.

## 7.7 ENERGY DIAGRAMS AND STABILITY OF EQUILIBRIUM

The motion of a system can often be understood qualitatively by analyzing a graphical representation of the system's potential energy curve. An **energy diagram** shows the potential energy of the system as a function of the position of one of the members of the system (or as a function of the separation distance between two members of the system). Consider the potential energy function for the block–spring system, given by $U_s = \frac{1}{2}kx^2$. This function is plotted versus $x$ in Active Figure 7.15a.

The spring force is related to $U_s$ through Equation 7.16:

$$F_s = -\frac{dU_s}{dx} = -kx$$

That is, the force is equal to the negative of the *slope* of the $U_s$ versus $x$ curve. When the block is placed at rest at the equilibrium position ($x = 0$), where $F_s = 0$, it will remain there unless some external force acts on it. If the spring in Active Figure 7.15b is stretched to the right from equilibrium, $x$ is positive and the slope $dU_s/dx$ is positive; therefore, $F_s$ is negative and the block accelerates back toward $x = 0$. If the spring is compressed, $x$ is negative and the slope is negative; therefore, $F_s$ is positive and again the block accelerates toward $x = 0$.

From this analysis, we conclude that the $x = 0$ position is one of **stable equilibrium.** That is, any movement away from this position results in a force directed back toward $x = 0$. (We described this type of force in Chapter 6 as a *restoring force.*) In general, **positions of stable equilibrium correspond to those values of $x$ for which $U(x)$ has a relative minimum value on an energy diagram.**

From Active Figure 7.15 we see that if the block is given an initial displacement $x_{max}$ and is released from rest, the total initial energy of the system is the potential energy stored in the spring, given by $\frac{1}{2}kx_{max}^2$. As motion commences, the system acquires kinetic energy and loses an equal amount of potential energy. From an energy viewpoint, the energy of the system cannot exceed $\frac{1}{2}kx_{max}^2$; therefore, the block must stop at the points $x = \pm x_{max}$ and, because of the spring force, accelerate toward $x = 0$. The block oscillates between the two points $x = \pm x_{max}$, called the *turning points.* The block cannot be farther from equilibrium than $\pm x_{max}$ because the potential energy of the system beyond these points would be larger than the total energy, an

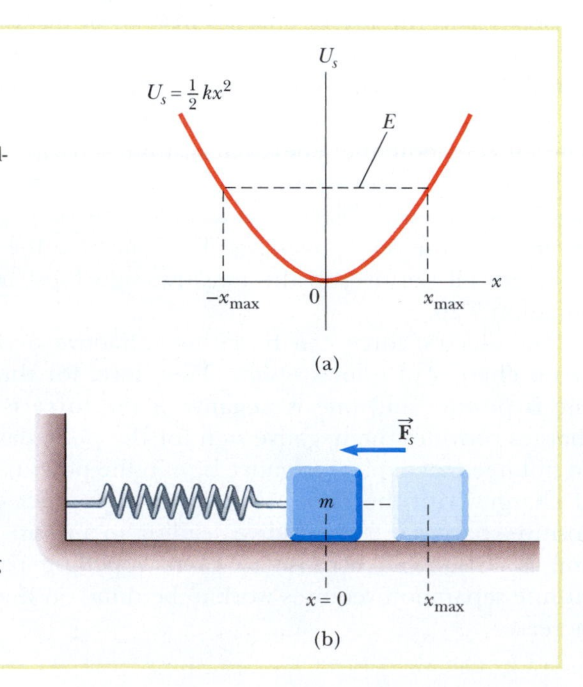


ACTIVE FIGURE 7.15
(a) Potential energy as a function of $x$ for the block–spring system shown in part (b). The block oscillates between the turning points, which have the coordinates $x = \pm x_{max}$. The restoring force exerted by the spring always acts toward $x = 0$, the position of stable equilibrium.

**PhysicsNow™** Log into PhysicsNow at www.pop4e.com and go to Active Figure 7.15 to observe the block oscillate between its turning points and trace the corresponding points on the potential energy curve for varying values of $k$.

impossible situation in classical physics. Because there is no transformation of mechanical energy to internal energy (no friction), the block oscillates between $-x_{max}$ and $+x_{max}$ forever. (We shall discuss these oscillations further in Chapter 12.)

Now consider an example in which the curve of $U$ versus $x$ is as shown in Figure 7.16. In this case, $F_x = 0$ at $x = 0$, and so the particle is in equilibrium at this point. This point, however, is a position of **unstable equilibrium** for the following reason. Suppose the particle is displaced to the *right* of the origin. Because the slope is negative for $x > 0$, $F_x = -dU/dx$ is positive and the particle accelerates away from $x = 0$. Now suppose the particle is displaced to the *left* of the origin. In this case, the force is negative because the slope is positive for $x < 0$, and the particle again accelerates away from the equilibrium position. The $x = 0$ position in this situation is called a position of unstable equilibrium because, for any displacement from this point, the force pushes the particle farther away from equilibrium. In fact, the force pushes the particle toward a position representing lower potential energy of the system. A ball placed on the top of an inverted spherical bowl is in a position of unstable equilibrium. If the ball is displaced slightly from the top and released, it will roll off the bowl. In general, **positions of unstable equilibrium correspond to those values of *x* for which *U*(*x*) has a relative maximum value on an energy diagram.**[2]

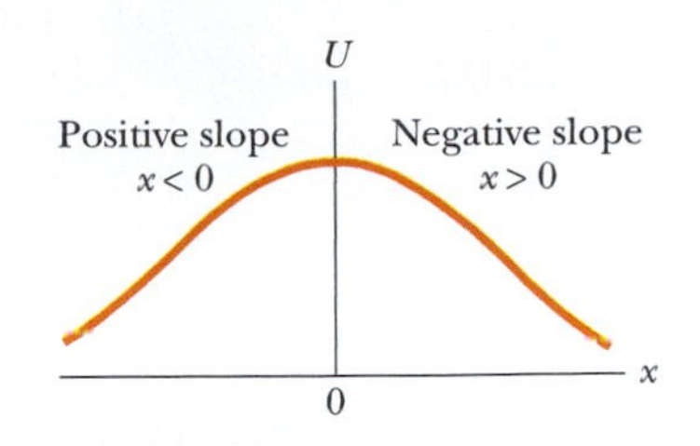


FIGURE 7.16 A plot of $U$ versus $x$ for a particle that has a position of unstable equilibrium, located at $x = 0$. For any finite displacement of the particle, the force on the particle is directed away from $x = 0$.

Finally, a situation may arise in which $U$ is constant over some region, and hence $F = 0$. A point in this region is called a position of **neutral equilibrium.** Small displacements from this position produce neither restoring nor disrupting forces. A ball lying on a flat horizontal surface is an example of an object in neutral equilibrium.

## 7.8 POTENTIAL ENERGY IN FUELS

CONTEXT CONNECTION

Fuel represents a storage mechanism for potential energy to be used to make a vehicle move. The standard fuel for automobiles for several decades has been *gasoline.* Gasoline is refined from crude oil that is present in the Earth. This oil represents the decay products of plant life that existed on the Earth, primarily from 100 to 600 million years ago. The source of energy in crude oil is hydrocarbons produced from molecules in the ancient plants.

The primary chemical reactions occurring in an internal combustion engine involve the oxidation of carbon and hydrogen:

$$C + O_2 \rightarrow CO_2$$

$$4H + O_2 \rightarrow 2H_2O$$

Both reactions release energy that is used to operate the automobile.

Notice the final products in these reactions. One is water, which is not harmful to the environment. Carbon dioxide, however, contributes to the greenhouse effect, which leads to global warming, which we will study in Context 5. The incomplete combustion of carbon and oxygen can form CO, carbon monoxide, which is a poisonous gas. Because air contains other elements besides oxygen, other harmful emission products, such as oxides of nitrogen, exist.

The amount of potential energy stored in a fuel and available from the fuel is typically called the *heat of combustion,* even though this term is a misuse of the word *heat.* For automotive gasoline, this value is about 44 MJ/kg. Because the efficiency of the engine is not 100%, only part of this energy eventually finds its way into kinetic energy of the car. We will study efficiencies of engines in Context 5.

Another common fuel is *diesel fuel.* The heat of combustion for diesel fuel is 42.5 MJ/kg, slightly lower than that for gasoline. Diesel engines, however, operate at a higher efficiency than gasoline engines, so they can extract a larger percentage of the available energy.

[2] You can mathematically test whether an extreme of $U$ represents stable or unstable equilibrium by examining the sign of $d^2U/dx^2$.

A number of additional fuels have been developed to operate internal combustion engines with minimal modifications. They are described briefly below.

## Ethanol

Ethanol is the most widely used alternative fuel and is used primarily on commercial fleet vehicles. It is an alcohol made from such crops as corn, wheat, and barley. Because these crops can be grown, ethanol is renewable. The use of ethanol reduces carbon monoxide and carbon dioxide emissions compared with the use of normal gasoline.

Ethanol is mixed with gasoline to form the following mixtures:

E10: 10% ethanol, 90% gasoline

E85: 85% ethanol, 15% gasoline

The energy content of E85 is about 70% of that for gasoline, so the miles per gallon ratio will be lower than that for a vehicle powered by straight gasoline. On the other hand, the renewable nature of ethanol counteracts this disadvantage significantly.

## Biodiesel

Biodiesel fuel is formed by a chemical reaction between alcohol and oils from field crops as well as vegetable oil, fat, and grease from commercial sources. Pacific Biodiesel in Hawaii makes biodiesel from used restaurant cooking oil, providing a usable fuel as well as diverting this used oil from landfills.

Biodiesel is available in the following forms:

B20: 20% biodiesel, 80% gasoline

B100: 100% biodiesel

B100 is nontoxic and biodegradable. The use of biodiesel reduces environmentally harmful tailpipe emissions significantly. Furthermore, tests have shown that the emission of cancer-causing particulate matter is reduced by 94% with the use of pure biodiesel.

The energy content of B100 is about 90% of that for conventional diesel. As with ethanol, the renewable nature of biodiesel counteracts this disadvantage significantly.

## Natural Gas

Natural gas is a fossil fuel, originating from gas wells or as a by-product of the refining process for crude oil. It is primarily methane ($CH_4$), with smaller amounts of nitrogen, ethane, propane, and other gases. It burns cleanly and generates much lower amounts of harmful tailpipe emissions than gasoline. Natural gas vehicles are used in many fleets of buses, delivery trucks, and refuse haulers.

Although ethanol and biodiesel mixtures can be used in conventional engines with minimal modifications, a natural gas engine is much more heavily modified. In addition, the gas must be carried on board the vehicle in one of two ways that require higher-level technology than a simple fuel tank. One possibility is to liquefy the gas, requiring a well-insulated storage container to keep the gas at $-190°C$. The other possibility is to compress the gas to about 200 times atmospheric pressure and carry it in the vehicle in a high-pressure storage tank.

The energy content of natural gas is 48 MJ/kg, a bit higher than that for gasoline. Note that natural gas, like gasoline, is *not* a renewable source.

## Propane

Propane is available commercially as liquefied petroleum gas, which is actually a mixture of propane, propylene, butane, and butylenes. It is a by-product of natural gas processing and refining of crude oil. Propane is the most widely accessible alternative fuel, with fueling facilities in all states of the United States.

Tailpipe emissions for propane-fueled vehicles are significantly lower than those for gasoline-powered vehicles. Tests show that carbon monoxide is reduced by 30% to 90%.

As with natural gas, high-pressure tanks are necessary to carry the fuel. In addition propane is a nonrenewable resource. The energy content of propane is 46 MJ/kg, slightly higher than that of gasoline.

## Electric Vehicles

In the Context introduction before Chapter 2, we discussed the electric cars that were on the roadways in the early part of the twentieth century. As mentioned, these electric cars virtually disappeared around the 1920s due to several factors. One was that oil was plentiful during the twentieth century and there was little incentive to operate vehicles on anything other than gasoline or diesel.

In the early 1970s, difficulties arose with regard to the availability of oil from the Middle East, leading to shortages at gas stations. At this time, interest arose anew in electric-powered vehicles. An early attempt to market a new electric vehicle was the Electrovette, an electric version of the Chevrolet Chevette.

Although the oil crisis eased somewhat, political instabilities in the Middle East created uncertainty in the availability of oil and interest in electric cars continued, albeit on a small scale. In the late 1980s, General Motors developed a prototype called the Impact, an electric car that could accelerate from 0 to 60 in 8 s and had a drag coefficient of 0.19, much lower than that of traditional cars. The Impact was the hit of the 1990 Los Angeles Auto Show. In the 1990s, the Impact became commercially available as the EV1.

Although the EV1 was a very successful electric car in terms of quality and performance, it was difficult to convince consumers that oil was in short supply and not many consumers chose to drive the car. A few other manufacturers also developed electric cars, and consumer response was similar. Two major disadvantages of electric cars were the limited range, 70 to 100 mi, on a single charging of the batteries and the several hours of time required to recharge the batteries. These difficulties, as well as a federal court ruling that relaxed emissions standards, led General Motors to cancel the EV1 program in 2001. An additional contribution to the demise of contemporary electric cars is the development of hybrid electric vehicles, which will be discussed in the Context Conclusion.

## SUMMARY

Take a practice test by logging into PhysicsNow at www.pop4e.com and clicking on the Pre-Test link for this chapter.

If a particle of mass $m$ is elevated a distance $y$ from a reference point $y = 0$ near the Earth's surface, the **gravitational potential energy** of the particle–Earth system can be defined as

$$U_g \equiv mgy \qquad [7.2]$$

The **total mechanical energy of a system** is defined as the sum of the kinetic energy and potential energy:

$$E_{\text{mech}} \equiv K + U \qquad [7.8]$$

If no energy transfers occur across the boundary of the system, the system is modeled as an **isolated system.** In this model, the principle of **conservation of mechanical energy** states that the total mechanical energy of the system is constant if all of the forces in the system are conservative. For example, if a system involves gravitational forces,

$$K_f + U_{gf} = K_i + U_{gi} \qquad [7.7]$$

A force is **conservative** if the work it does on a particle is independent of the path the particle takes between two given points. A conservative force in mechanics does not cause a transformation of mechanical energy to internal energy. A force that does not meet these criteria is said to be **nonconservative.**

The **elastic potential energy** stored in a spring of force constant $k$ is

$$U_s \equiv \tfrac{1}{2}kx^2 \qquad [7.9]$$

If some of the forces acting within a system are not conservative, the mechanical energy of the system does not remain constant. In the case of a common nonconservative force, a constant force of friction, the change in mechanical energy of the system when an object in the system moves is equal to the product of the kinetic friction force and the distance through which the object moves:

$$-f_k d = \Delta K + \Delta U \qquad [7.10]$$

This decrease in mechanical energy in the system is equal to the increase in internal energy:

$$\Delta E_{\text{int}} = f_k d \qquad [7.10]$$

A potential energy function $U$ can be associated only with a conservative force. If a conservative force $\vec{\mathbf{F}}$ acts within a system on a particle that moves along the $x$ axis from $x_i$ to $x_f$, the potential energy function can be written

$$U_f = -\int_{x_i}^{x_f} F_x \, dx + U_i \qquad [7.15]$$

If we know the potential energy function, the component of a conservative force is given by the negative of the derivative of the potential energy function:

$$F_x = -\frac{dU}{dx} \qquad [7.16]$$

In some situations, a system may have energy crossing the boundary with no change in the energy stored in the system. In such a case, the energy input in any time interval equals the energy output, and we describe this system as a **nonisolated system in steady state.**

The **gravitational potential energy** associated with a system of two particles or uniform spherical distributions of mass separated by a distance $r$ is

$$U_g = -\frac{Gm_1 m_2}{r} \qquad [7.20]$$

where $U_g$ is taken to approach zero as $r \rightarrow \infty$.

The **electric potential energy** associated with two charged particles separated by a distance $r$ is

$$U_e = k_e \frac{q_1 q_2}{r} \qquad [7.23]$$

where $U_e$ is taken to approach zero as $r \rightarrow \infty$.

In an energy diagram, a point of **stable equilibrium** is one at which the potential energy is a minimum. A point of **unstable equilibrium** is one at which the potential energy is a maximum. **Neutral equilibrium** exists if the potential energy function is constant.

## QUESTIONS

☐ = answer available in the *Student Solutions Manual and Study Guide.*

1. If the height of a playground slide is kept constant, will the length of the slide or the presence of bumps make any difference in the final speed of children playing on it? Assume that the slide is slick enough to be considered frictionless. Repeat this question assuming that friction is present.
2. Explain why the total energy of a system can be either positive or negative, whereas the kinetic energy is always positive.
3. One person drops a ball from the top of a building, while another person at the bottom observes its motion. Will these two people agree on the value of the gravitational potential energy of the ball–Earth system? On the change in potential energy? On the kinetic energy?
4. Discuss the changes in mechanical energy of an object–Earth system in (a) lifting the object, (b) holding the object at a fixed position, and (c) lowering the object slowly. Include the muscles in your discussion.
5. In Chapter 6, the work–kinetic energy theorem, $W = \Delta K$, was introduced. This equation states that work done on a system appears as a change in kinetic energy. It is a special-case equation, valid if there are no changes in any other type of energy, such as potential or internal. Give some examples in which work is done on a system but the change in energy of the system is not that of kinetic energy.
6. If three conservative forces and one nonconservative force act within a system, how many potential energy terms appear in the equation that describes the system?
7. If only one external force acts on a particle, (a) does it necessarily change the particle's kinetic energy? (b) Does it change the particle's velocity?
8. A driver brings an automobile to a stop. If the brakes lock so that the car skids, where is the original kinetic energy of the car and in what form is it after the car stops? Answer the same question for the case in which the brakes do not lock but the wheels continue to turn.
9. You ride a bicycle. In what sense is your bicycle solar-powered?
10. In an earthquake, a large amount of energy is "released" and spreads outward, potentially causing severe damage. In what form does this energy exist before the earthquake, and by what energy transfer mechanism does it travel?
11. A bowling ball is suspended from the ceiling of a lecture hall by a strong cord. The ball is drawn away from its equilibrium position and released from rest at the tip of the demonstrator's nose as shown in Figure Q7.11. Assuming

FIGURE Q7.11

that the demonstrator remains stationary, explain why the ball does not strike her on its return swing. Would this demonstrator be safe if the ball were given a push from its starting position at her nose?

**12.** A ball is thrown straight up into the air. At what position is its kinetic energy a maximum? At what position is the gravitational potential energy of the ball–Earth system a maximum?

**13.** A pile driver is a device used to drive objects into the Earth by repeatedly dropping a heavy weight on them. By how much does the energy of the pile driver–Earth system increase when the weight it drops is doubled? Assume that the weight is dropped from the same height each time.

**14.** Our body muscles exert forces when we lift, push, run, jump, and so forth. Are these forces conservative?

**15.** A block is connected to a spring that is suspended from the ceiling. Assuming that the block is set in motion and that air resistance can be ignored, describe the energy transformations that occur within the system consisting of the block, Earth, and spring.

**16.** Discuss the energy transformations that occur during the operation of an automobile.

**17.** What would the curve of $U$ versus $x$ look like if a particle were in a region of neutral equilibrium?

**18.** A ball rolls on a horizontal surface. Is the ball in stable, unstable, or neutral equilibrium?

## PROBLEMS

1, 2, 3 = straightforward, intermediate, challenging
☐ = full solution available in the *Student Solutions Manual and Study Guide*
Physics Now™ = coached problem with hints available at www.pop4e.com
= computer useful in solving problem
= paired numerical and symbolic problems
= biomedical application

### Section 7.1 ■ Potential Energy of a System

**1.** A 1 000-kg roller coaster train is initially at the top of a rise, at point Ⓐ. It then moves 135 ft, at an angle of 40.0° below the horizontal, to a lower point Ⓑ. (a) Choose the train at point Ⓑ to be the zero configuration for gravitational potential energy. Find the potential energy of the roller coaster–Earth system at points Ⓐ and Ⓑ, and the change in potential energy as the coaster moves. (b) Repeat part (a), setting the zero configuration when the train is at point Ⓐ.

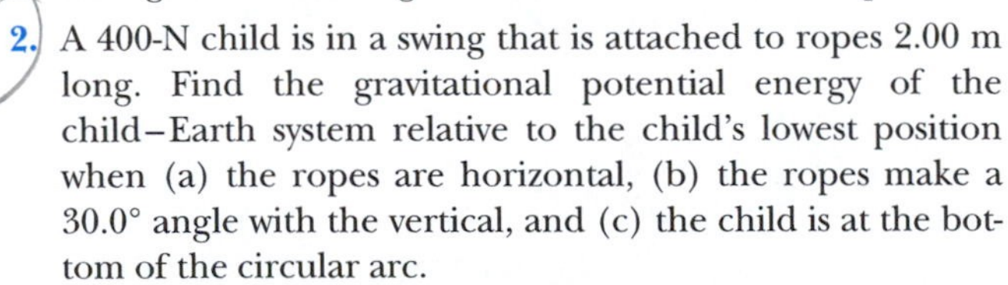

**2.** A 400-N child is in a swing that is attached to ropes 2.00 m long. Find the gravitational potential energy of the child–Earth system relative to the child's lowest position when (a) the ropes are horizontal, (b) the ropes make a 30.0° angle with the vertical, and (c) the child is at the bottom of the circular arc.

**3.** A person with a remote mountain cabin plans to install her own hydroelectric plant. A nearby stream is 3.00 m wide and 0.500 m deep. Water flows at 1.20 m/s over the brink of a waterfall 5.00 m high. The manufacturer promises only 25.0% efficiency in converting the potential energy of the water–Earth system into electric energy. Find the power she can generate. (Large-scale hydroelectric plants, with a much larger drop, can be more efficient.)

### Section 7.2 ■ The Isolated System

**4.** At 11:00 A.M. on September 7, 2001, more than one million British school children jumped up and down for one minute. The curriculum focus of the "Giant Jump" was on earthquakes, but it was integrated with many other topics, such as exercise, geography, cooperation, testing hypotheses, and setting world records. Children built their own seismographs that registered local effects. (a) Find the mechanical energy released in the experiment. Assume that 1 050 000 children of average mass 36.0 kg jump 12 times each, raising their centers of mass by 25.0 cm each time and briefly resting between one jump and the next. The free-fall acceleration in Britain is 9.81 m/s$^2$. (b) Most of the energy is converted very rapidly into internal energy within the bodies of the children and the floors of the school buildings. Of the energy that propagates into the ground, most produces high frequency "microtremor" vibrations that are rapidly damped and cannot travel far. Assume that 0.01% of the energy is carried away by a long-range seismic wave. The magnitude of an earthquake on the Richter scale is given by

$$M = \frac{\log E - 4.8}{1.5}$$

where $E$ is the seismic wave energy in joules. According to this model, what is the magnitude of the demonstration quake? It did not register above background noise overseas or on the seismograph of the Wolverton Seismic Vault, Hampshire.

**5.** A bead slides without friction around a loop-the-loop (Fig. P7.5). The bead is released from a height $h = 3.50R$. (a) What is its speed at point Ⓐ? (b) How large is the normal force on it if its mass is 5.00 g?

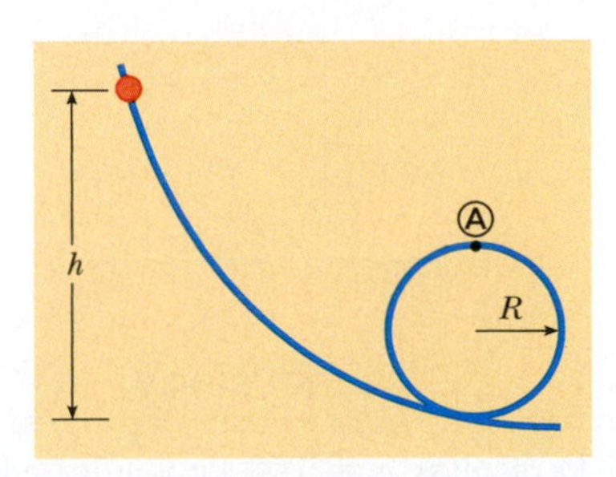


FIGURE P7.5

6. **Review problem.** A particle of mass 0.500 kg is shot from $P$ as shown in Figure P7.6. The particle has an initial velocity $\vec{v}_i$ with a horizontal component of 30.0 m/s. The particle rises to a maximum height of 20.0 m above $P$. Using the law of conservation of energy, determine (a) the vertical component of $\vec{v}_i$, (b) the work done by the gravitational force on the particle during its motion from $P$ to $B$, and (c) the horizontal and the vertical components of the velocity vector when the particle reaches $B$.

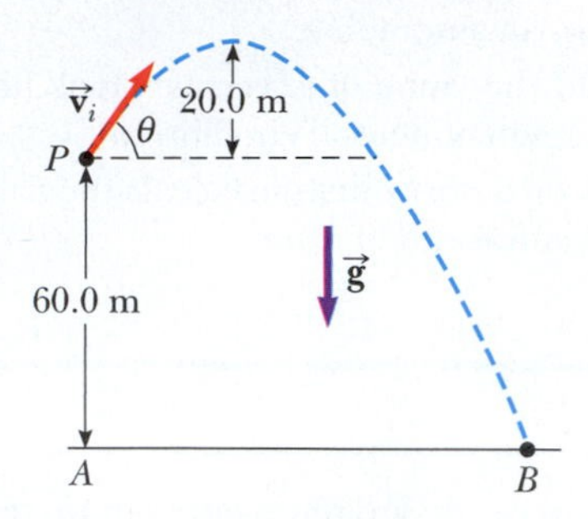


FIGURE P7.6

7. Dave Johnson, the bronze medallist at the 1992 Olympic decathlon in Barcelona, leaves the ground at the high jump with vertical velocity component 6.00 m/s. How far does his center of mass move up as he makes the jump?

8. A simple pendulum, which you will consider in detail in Chapter 12, consists of an object suspended by a string. The object is assumed to be a particle. The string, with its top end fixed, has negligible mass and does not stretch. In the absence of air friction, the system oscillates by swinging back and forth in a vertical plane. The string is 2.00 m long and makes an initial angle of 30.0° with the vertical. Calculate the speed of the particle (a) at the lowest point in its trajectory and (b) when the angle is 15.0°.

9. Two objects are connected by a light string passing over a light frictionless pulley as shown in Figure P7.9. The 5.00-kg object is released from rest. Using the principle of conservation of energy, (a) determine the speed of the 3.00-kg object just as the 5.00-kg object hits the ground, and (b) find the maximum height to which the 3.00-kg object rises.

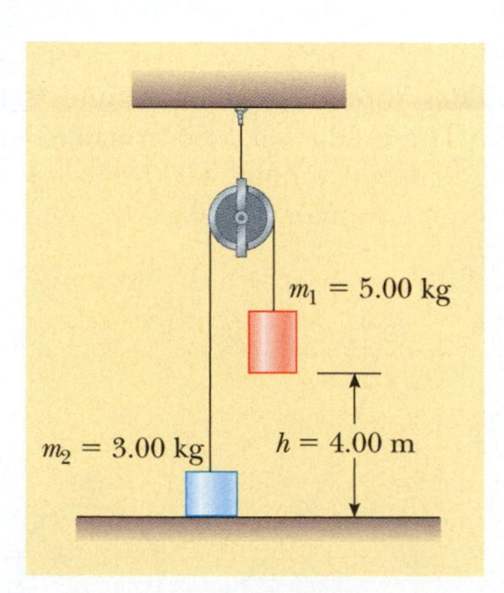


FIGURE P7.9

10. A particle of mass $m$ = 5.00 kg is released from point Ⓐ and slides on the frictionless track shown in Figure P7.10. Determine (a) the particle's speed at points Ⓑ and Ⓒ and (b) the net work done by the gravitational force as the particle moves from Ⓐ to Ⓒ.

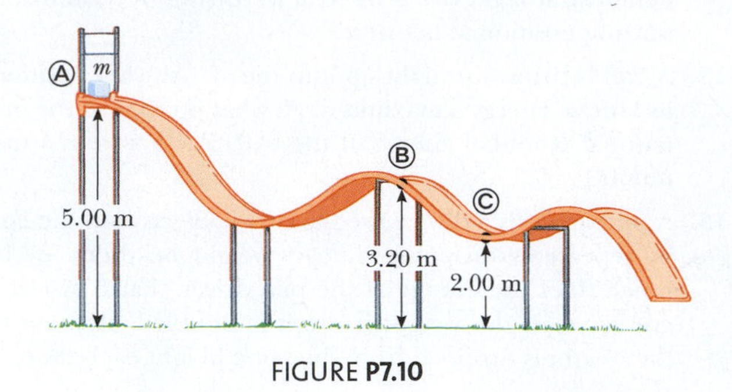


FIGURE P7.10

11. A circus trapeze consists of a bar suspended by two parallel ropes, each of length $\ell$, allowing performers to swing in a vertical circular arc (Fig. P7.11). Suppose a performer with mass $m$ holds the bar and steps off an elevated platform, starting from rest with the ropes at an angle $\theta_i$ with respect to the vertical. Suppose the size of the performer's body is small compared to the length $\ell$, she does not pump the trapeze to swing higher, and air resistance is negligible. (a) Show that when the ropes make an angle $\theta$ with the vertical, the performer must exert a force

$$mg(3 \cos \theta - 2 \cos \theta_i)$$

so as to hang on. (b) Determine the angle $\theta_i$ for which the force needed to hang on at the bottom of the swing is twice the performer's weight.

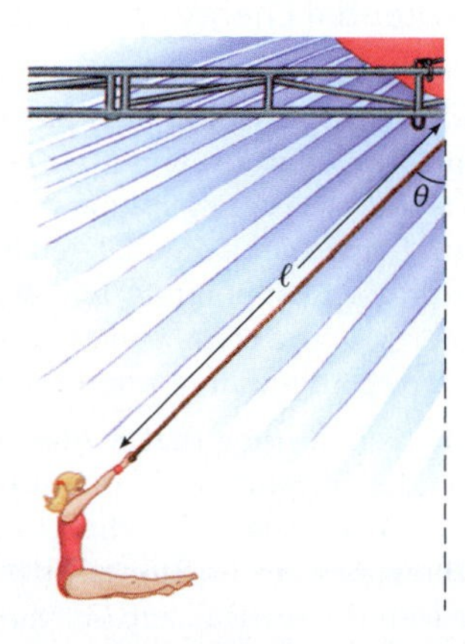


FIGURE P7.11

12. A light rigid rod is 77.0 cm long. Its top end is pivoted on a low-friction horizontal axle. The rod hangs straight down at rest, with a small massive ball attached to its bottom end. You strike the ball, suddenly giving it a horizontal velocity so that it swings around in a full circle. What minimum speed at the bottom is required to make the ball go over the top of the circle?

13. Columnist Dave Barry poked fun at the name "The Grand Cities" adopted by Grand Forks, North Dakota, and East Grand Forks, Minnesota. Residents of the prairie towns then named a sewage pumping station for him. At the Dave Barry Lift Station No. 16, untreated sewage is raised vertically by 5.49 m, in the amount 1 890 000 L each day. The waste has density 1 050 kg/m$^3$. It enters and leaves the

pump at atmospheric pressure, through pipes of equal diameter. (a) Find the output power of the lift station. (b) Assume that an electric motor continuously operating with average power 5.90 kW runs the pump. Find its efficiency. Barry attended the outdoor January dedication of the lift station and a festive potluck supper to which the residents of the different Grand Forks sewer districts brought casseroles, Jell-O salads, and "bars" (desserts).

## Section 7.3 ■ Conservative and Nonconservative Forces

14. (a) Suppose a constant force acts on an object. The force does not vary with time nor with the position or the velocity of the object. Start with the general definition for work done by a force

$$W = \int_i^f \vec{\mathbf{F}} \cdot d\vec{\mathbf{r}}$$

and show that the force is conservative. (b) As a special case, suppose the force $\vec{\mathbf{F}} = (3\hat{\mathbf{i}} + 4\hat{\mathbf{j}})$N acts on a particle that moves from $O$ to $C$ in Figure P7.14. Calculate the work the force $\vec{\mathbf{F}}$ does on the particle as it moves along each one of the three paths $OAC$, $OBC$, and $OC$. (Your three answers should be identical.)

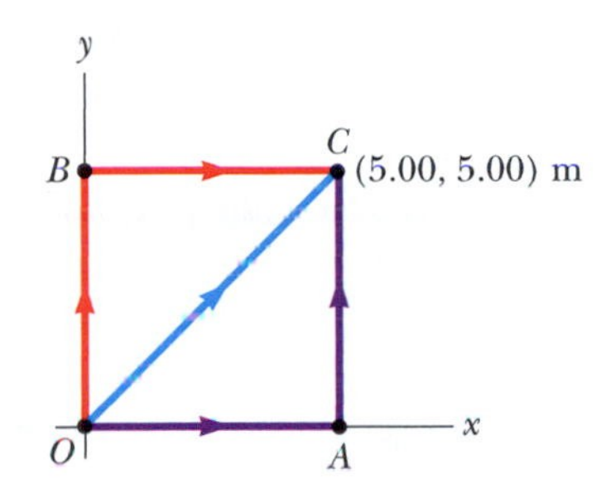


FIGURE P7.14 Problems 7.14 and 7.15.

15. A force acting on a particle moving in the $xy$ plane is given by $\vec{\mathbf{F}} = (2y\hat{\mathbf{i}} + x^2\hat{\mathbf{j}})$N, where $x$ and $y$ are in meters. The particle moves from the origin to a final position having coordinates $x = 5.00$ m and $y = 5.00$ m as shown in Figure P7.14. Calculate the work done by $\vec{\mathbf{F}}$ on the particle as it moves along (a) $OAC$, (b) $OBC$, and (c) $OC$. (d) Is $\vec{\mathbf{F}}$ conservative or nonconservative? Explain.

16. An object of mass $m$ starts from rest and slides a distance $d$ down a frictionless incline of angle $\theta$. While sliding, it contacts an unstressed spring of negligible mass as shown in Figure P7.16. The object slides an additional distance $x$ as it is brought momentarily to rest by compression of the spring (of force constant $k$). Find the initial separation $d$ between the object and the spring.

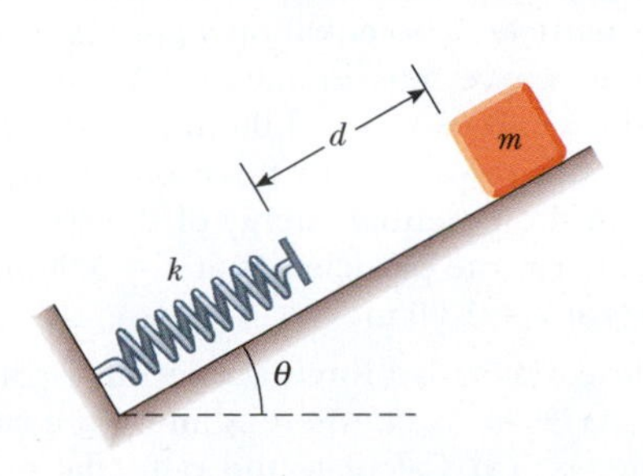


FIGURE P7.16

17. A block of mass 0.250 kg is placed on top of a light vertical spring of force constant 5 000 N/m and pushed downward so that the spring is compressed by 0.100 m. After the block is released from rest it travels upward and then leaves the spring. To what maximum height above the point of release does it rise?

18. A daredevil plans to bungee jump from a balloon 65.0 m above a carnival midway (Fig. P7.18). He will use a uniform elastic cord, tied to a harness around his body, to stop his fall at a point 10.0 m above the ground. Model his body as a particle. Assume that the cord has negligible mass and is described by Hooke's force law. In a preliminary test, hanging at rest from a 5.00-m length of the cord, he finds that his body weight stretches it by 1.50 m. He will drop from rest at the point where the top end of a longer section of the cord is attached to the stationary balloon. (a) What length of cord should he use? (b) What maximum acceleration will he experience?

FIGURE P7.18 Problems 7.18 and 7.64.

19. At time $t_i$, the kinetic energy of a particle is 30.0 J and the potential energy of the system to which it belongs is 10.0 J. At some later time $t_f$, the kinetic energy of the particle is 18.0 J. (a) If only conservative forces act on the particle, what are the potential energy and the total energy of the system at time $t_f$? (b) If the potential energy of the system at time $t_f$ is 5.00 J, are there any nonconservative forces acting on the particle? Explain.

20. Heedless of danger, a child leaps onto a pile of old mattresses to use them as a trampoline. His motion between two particular points is described by the energy conservation equation

$$\tfrac{1}{2}(46.0\text{ kg})(2.40\text{ m/s})^2 + (46.0\text{ kg})(9.80\text{ m/s}^2)(2.80\text{ m} + x) = \tfrac{1}{2}(1.94 \times 10^4\text{ N/m})x^2$$

(a) Solve the equation for $x$. (b) Compose the statement of a problem, including data, for which this equation gives the solution. Identify the physical meaning of the value of $x$.

21. In her hand, a softball pitcher swings a ball of mass 0.250 kg around a vertical circular path of radius 60.0 cm before releasing it from her hand. The pitcher maintains a component of force on the ball of constant magnitude 30.0 N in the direction of motion around the complete path. The ball's speed at the top of the circle is 15.0 m/s. If she releases the ball at the bottom of the circle, what is its speed upon release?

22. In a needle biopsy, a narrow strip of tissue is extracted from a patient using a hollow needle. Rather than being

pushed by hand, to ensure a clean cut the needle can be fired into the patient's body by a spring. Assume that the needle has mass 5.60 g, the light spring has force constant 375 N/m, and the spring is originally compressed 8.10 cm to project the needle horizontally without friction. After the needle leaves the spring, the tip of the needle moves through 2.40 cm of skin and soft tissue, which exerts on it a resistive force of 7.60 N. Next, the needle cuts 3.50 cm into an organ, which exerts on it a backward force of 9.20 N. Find (a) the maximum speed of the needle and (b) the speed at which a flange on the back end of the needle runs into a stop that is set to limit the penetration to 5.90 cm.

**23.** PhysicsNow™ The coefficient of friction between the 3.00-kg block and the surface in Figure P7.23 is 0.400. The system starts from rest. What is the speed of the 5.00-kg ball when it has fallen 1.50 m?

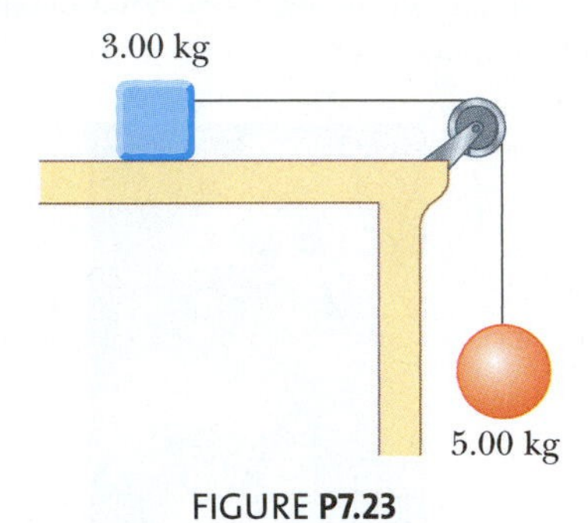


FIGURE P7.23

**24.** A boy in a wheelchair (total mass 47.0 kg) wins a race with a skateboarder. He has speed 1.40 m/s at the crest of a slope 2.60 m high and 12.4 m long. At the bottom of the slope his speed is 6.20 m/s. Assuming that air resistance and rolling resistance can be modeled as a constant friction force of 41.0 N, find the work he did in pushing forward on his wheels during the downhill ride.

**25.** A 5.00-kg block is set into motion up an inclined plane with an initial speed of 8.00 m/s (Fig. P7.25). The block comes to rest after traveling 3.00 m along the plane, which is inclined at an angle of 30.0° to the horizontal. For this motion, determine (a) the change in the block's kinetic energy, (b) the change in the potential energy of the block–Earth system, and (c) the friction force exerted on the block (assumed to be constant). (d) What is the coefficient of kinetic friction?

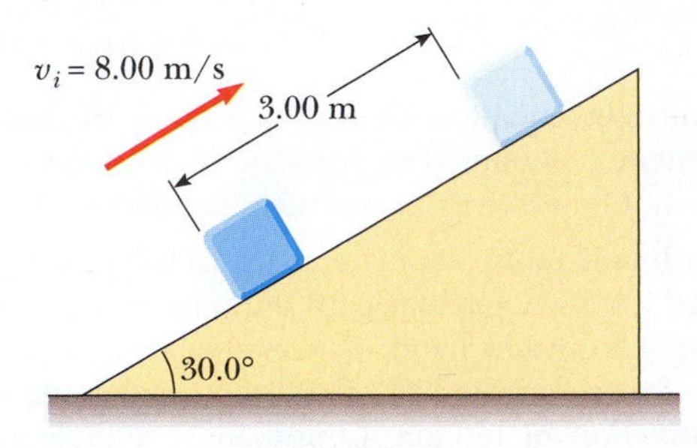


FIGURE P7.25

**26.** An 80.0-kg sky diver jumps out of a balloon at an altitude of 1 000 m and opens the parachute at an altitude of 200 m. (a) Assuming that the total retarding force on the diver is constant at 50.0 N with the parachute closed and constant at 3 600 N with the parachute open, find the speed of the sky diver when he lands on the ground. (b) Do you think the sky diver will be injured? Explain. (c) At what height should the parachute be opened so that the final speed of the sky diver when he hits the ground is 5.00 m/s? (d) How realistic is the assumption that the total retarding force is constant? Explain.

**27.** A toy cannon uses a spring to project a 5.30-g soft rubber ball. The spring is originally compressed by 5.00 cm and has a force constant of 8.00 N/m. When it is fired, the ball moves 15.0 cm through the horizontal barrel of the cannon and the barrel exerts a constant friction force of 0.032 0 N on the ball. (a) With what speed does the projectile leave the barrel of the cannon? (b) At what point does the ball have maximum speed? (c) What is this maximum speed?

**28.** A 50.0-kg block and 100-kg block are connected by a string as shown in Figure P7.28. The pulley is frictionless and of negligible mass. The coefficient of kinetic friction between the 50-kg block and incline is 0.250. Determine the change in the kinetic energy of the 50-kg block as it moves from Ⓐ to Ⓑ, a distance of 20.0 m.

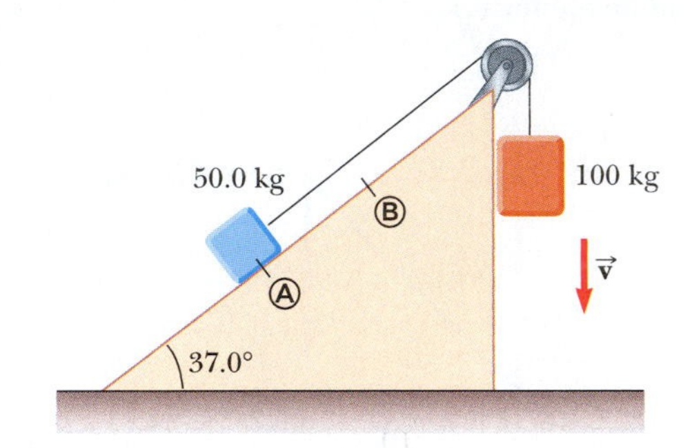


FIGURE P7.28

**29.** A 1.50-kg object is held 1.20 m above a relaxed massless vertical spring with a force constant of 320 N/m. The object is dropped onto the spring. (a) How far does it compress the spring? (b) How far does it compress the spring if the same experiment is performed on the Moon, where $g = 1.63$ m/s$^2$? (c) Repeat part (a), but now assume that a constant air-resistance force of 0.700 N acts on the object during its motion.

**30.** A 75.0-kg sky surfer is falling straight down with terminal speed 60.0 m/s. Determine the rate at which the sky surfer–Earth system is losing mechanical energy.

## Section 7.4 ■ Conservative Forces and Potential Energy

**31.** PhysicsNow™ A single conservative force acts on a 5.00-kg particle. The equation $F_x = (2x + 4)$ N describes the force, where $x$ is in meters. As the particle moves along the $x$ axis from $x = 1.00$ m to $x = 5.00$ m, calculate (a) the work done by this force on the particle, (b) the change in the potential energy of the system, and (c) the kinetic energy the particle has at $x = 5.00$ m if its speed is 3.00 m/s at $x = 1.00$ m.

**32.** A single conservative force acting on a particle varies as $\vec{\mathbf{F}} = (-Ax + Bx^2)\hat{\mathbf{i}}$ N, where $A$ and $B$ are constants and $x$ is in meters. (a) Calculate the potential energy function

$U(x)$ associated with this force, taking $U = 0$ at $x = 0$. (b) Find the change in potential energy of the system and the change in kinetic energy of the particle as it moves from $x = 2.00$ m to $x = 3.00$ m.

**33.** PhysicsNow™ The potential energy of a system of two particles separated by a distance $r$ is given by $U(r) = A/r$, where $A$ is a constant. Find the radial force $\vec{F}$ that each particle exerts on the other.

**34.** A potential energy function for a two-dimensional force is of the form $U = 3x^3y - 7x$. Find the force that acts at the point $(x, y)$.

## Section 7.6 ■ Potential Energy for Gravitational and Electric Forces

**35.** A satellite of the Earth has a mass of 100 kg and is at an altitude of $2.00 \times 10^6$ m. (a) What is the potential energy of the satellite–Earth system? (b) What is the magnitude of the gravitational force exerted by the Earth on the satellite? (c) What force does the satellite exert on the Earth?

**36.** How much energy is required to move a 1 000-kg object from the Earth's surface to an altitude twice the Earth's radius?

**37.** At the Earth's surface, a projectile is launched straight up at a speed of 10.0 km/s. To what height will it rise? Ignore air resistance.

**38.** A system consists of three particles, each of mass 5.00 g, located at the corners of an equilateral triangle with sides of 30.0 cm. (a) Calculate the potential energy describing the gravitational interactions internal to the system. (b) If the particles are released simultaneously, where will they collide?

## Section 7.7 ■ Energy Diagrams and Stability of Equilibrium

**39.** For the potential energy curve shown in Figure P7.39, (a) determine whether the force $F_x$ is positive, negative, or zero at the five points indicated. (b) Indicate points of stable, unstable, and neutral equilibrium. (c) Sketch the curve for $F_x$ versus $x$ from $x = 0$ to $x = 9.5$ m.

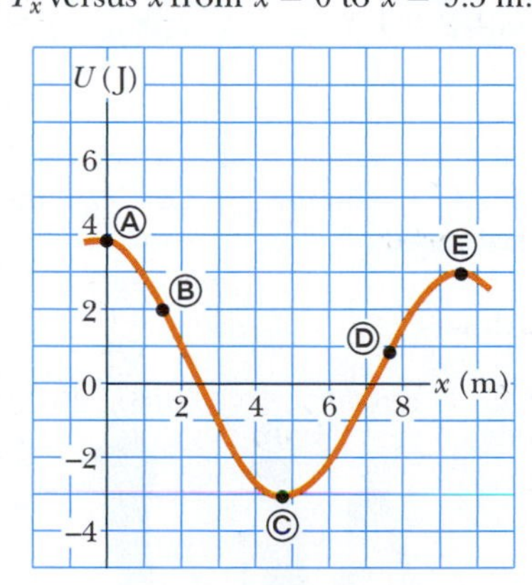


FIGURE **P7.39**

**40.** A particle moves along a line where the potential energy of its system depends on its position $r$ as graphed in Figure P7.40. In the limit as $r$ increases without bound, $U(r)$ approaches +1 J. (a) Identify each equilibrium position for this particle. Indicate whether each is a point of stable, unstable, or neutral equilibrium. (b) The particle will be bound if the total energy of the system is in what range? Now suppose the system has energy −3 J. Determine (c) the range of positions where the particle can be found, (d) its maximum kinetic energy, (e) the location where it has maximum kinetic energy, and (f) the *binding energy* of the system, that is, the additional energy that it would have to be given for the particle to move out to $r \to \infty$.

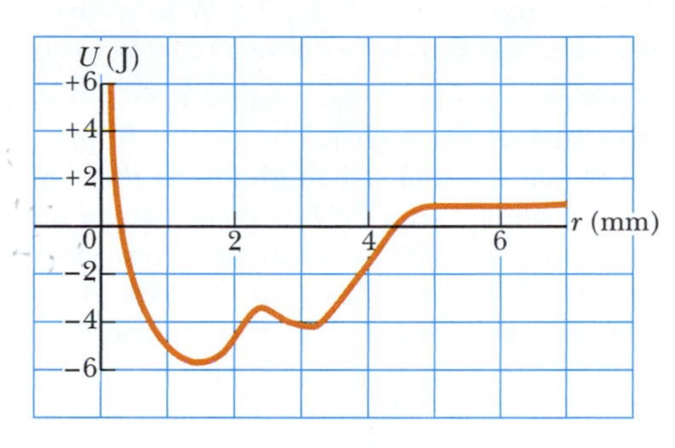


FIGURE **P7.40**

**41.** A particle of mass 1.18 kg is attached between two identical springs on a horizontal, frictionless tabletop. The springs have force constant $k$ and each is initially unstressed. (a) The particle is pulled a distance $x$ along a direction perpendicular to the initial configuration of the springs as shown in Figure P7.41. Show that the potential energy of the system is

$$U(x) = kx^2 + 2kL(L - \sqrt{x^2 + L^2})$$

(*Suggestion:* See Problem 6.52 in Chapter 6.) (b) Make a plot of $U(x)$ versus $x$ and identify all equilibrium points. Assume that $L = 1.20$ m and $k = 40.0$ N/m. (c) If the particle is pulled 0.500 m to the right and then released, what is its speed when it reaches the equilibrium point $x = 0$?

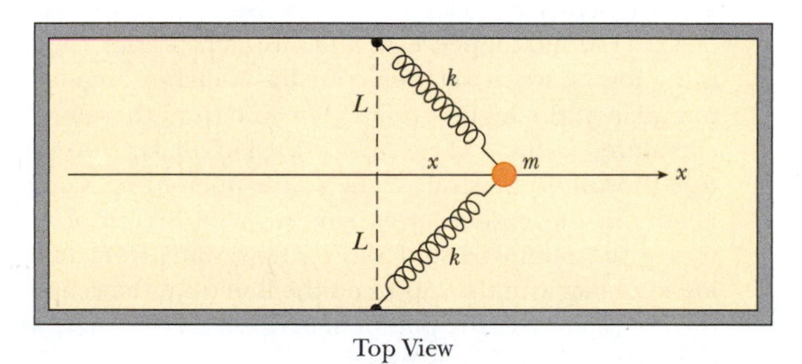


Top View

FIGURE **P7.41**

## Section 7.8 ■ Context Connection—Potential Energy in Fuels

**42.** **Review problem.** The mass of a car is 1 500 kg. The shape of the body is such that its aerodynamic drag coefficient is $D = 0.330$ and the frontal area is 2.50 m². Assuming that the drag force is proportional to $v^2$ and ignoring other sources of friction, calculate the power required to maintain a speed of 100 km/h as the car climbs a long hill sloping at 3.20°.

**43.** In considering the energy supply for an automobile, the energy per unit mass of the energy source is an important parameter. As the chapter text points out, the "heat of combustion" or stored energy per mass is quite similar for gasoline, ethanol, diesel fuel, cooking oil, methane, and propane. For a broader perspective, compare the energy per mass in joules per kilogram for gasoline, lead-acid batteries, hydrogen, and hay. Rank the four in order of increasing energy density and state the factor of increase between each one and the next. Hydrogen has "heat of combustion" 142 MJ/kg. For wood, hay, and dry vegetable matter in general, this parameter is 17 MJ/kg. A fully charged 16.0-kg lead-acid battery can deliver power 1 200 W for 1.0 hr.

**44.** The power of sunlight reaching each square meter of the Earth's surface on a clear day in the tropics is close to 1 000 W. On a winter day in Manitoba the power concentration of sunlight can be 100 W/m$^2$. Many human activities are described by a power-per-footprint-area on the order of $10^2$ W/m$^2$ or less. (a) Consider, for example, a family of four paying $80 to the electric company every 30 days for 600 kWh of energy carried by electrical transmission to their house, which has floor area 13.0 m by 9.50 m. Compute the power-per-area measure of this energy use. (b) Consider a car 2.10 m wide and 4.90 m long traveling at 55.0 mi/h using gasoline having "heat of combustion" 44.0 MJ/kg with fuel economy 25.0 mi/gal. One gallon of gasoline has a mass of 2.54 kg. Find the power-per-area measure of the car's energy use. It can be similar to that of a steel mill where rocks are melted in blast furnaces. (c) Explain why direct use of solar energy is not practical for a conventional automobile.

## Additional Problems

**45.** Make an order-of-magnitude estimate of your power output as you climb stairs. In your solution, state the physical quantities you take as data and the values you measure or estimate for them. Do you consider your peak power or your sustainable power?

**46.** Assume that you attend a state university that was founded as an agricultural college. Close to the center of the campus is a tall silo topped with a hemispherical cap. The cap is frictionless when wet. Someone has somehow balanced a pumpkin at the highest point. The line from the center of curvature of the cap to the pumpkin makes an angle $\theta_i = 0°$ with the vertical. While you happen to be standing nearby in the middle of a rainy night, a breath of wind makes the pumpkin start sliding downward from rest. It loses contact with the cap when the line from the center of the hemisphere to the pumpkin makes a certain angle with the vertical. What is this angle?

**47.** **Review problem.** The system shown in Figure P7.47 consists of a light inextensible cord, light frictionless pulleys, and blocks of equal mass. It is initially held at rest so that the blocks are at the same height above the ground. The blocks are then released. Find the speed of block A at the moment when the vertical separation of the blocks is $h$.

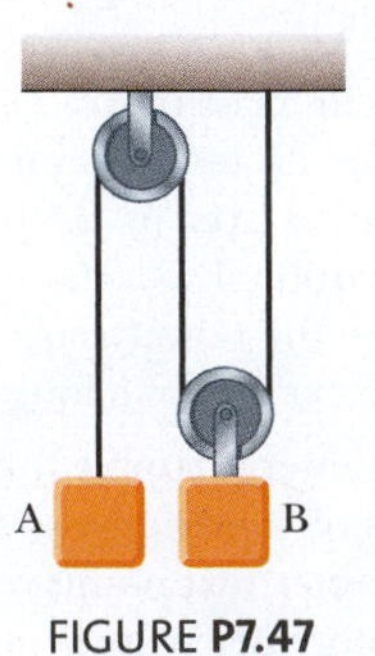


FIGURE **P7.47**

**48.** A 200-g particle is released from rest at point Ⓐ along the horizontal diameter on the inside of a frictionless, hemispherical bowl of radius $R = 30.0$ cm (Fig. P7.48). Calculate (a) the gravitational potential energy of the particle–Earth system when the particle is at point Ⓐ relative to point Ⓑ, (b) the kinetic energy of the particle at point Ⓑ, (c) its speed at point Ⓑ, and (d) its kinetic energy and the potential energy when the particle is at point Ⓒ.

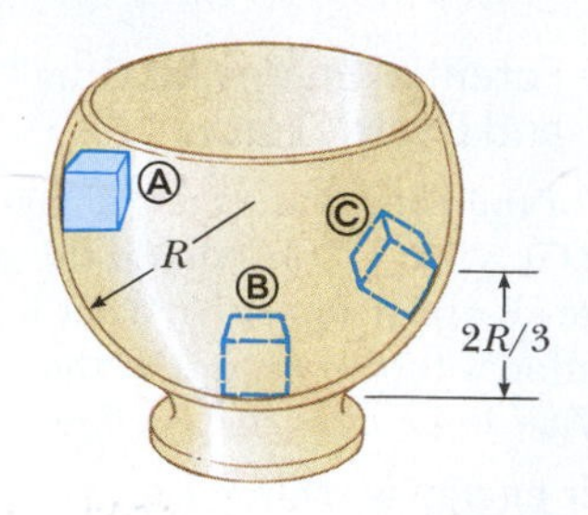


FIGURE **P7.48** Problems 7.48 and 7.49.

**49.** PhysicsNow™ The particle described in Problem 7.48 (Fig. P7.48) is released from rest at Ⓐ, and the surface of the bowl is rough. The speed of the particle at Ⓑ is 1.50 m/s. (a) What is its kinetic energy at Ⓑ? (b) How much mechanical energy is transformed into internal energy as the particle moves from Ⓐ to Ⓑ? (c) Is it possible to determine the coefficient of friction from these results in any simple manner? Explain.

**50.** A child's pogo stick (Fig. P7.50) stores energy in a spring with a force constant of $2.50 \times 10^4$ N/m. At position Ⓐ ($x_A = -0.100$ m), the spring compression is a maximum and the child is momentarily at rest. At position Ⓑ ($x_B = 0$), the spring is relaxed and the child is moving upward. At position Ⓒ, the child is again momentarily at rest at the top of the jump. The combined mass of child and pogo stick is 25.0 kg. (a) Calculate the total energy of the child–stick–Earth system, taking both gravitational and elastic potential energies as zero for $x = 0$. (b) Determine $x_C$. (c) Calculate the speed of the child at $x = 0$. (d) Determine the value of $x$ for which the kinetic energy of the system is a maximum. (e) Calculate the child's maximum upward speed.

FIGURE **P7.50**

51. A 10.0-kg block is released from point Ⓐ in Figure P7.51. The track is frictionless except for the portion between points Ⓑ and Ⓒ, which has a length of 6.00 m. The block travels down the track, hits a spring of force constant 2 250 N/m, and compresses the spring 0.300 m from its equilibrium position before coming to rest momentarily. Determine the coefficient of kinetic friction between the block and the rough surface between Ⓑ and Ⓒ.

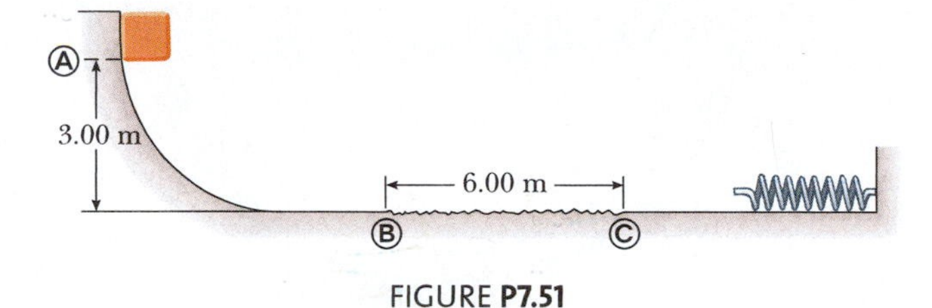


FIGURE **P7.51**

52. The potential energy function for a system is given by $U(x) = -x^3 + 2x^2 + 3x$. (a) Determine the force $F_x$ as a function of $x$. (b) For what values of $x$ is the force equal to zero? (c) Plot $U(x)$ versus $x$ and $F_x$ versus $x$, and indicate points of stable and unstable equilibrium.

53. A 20.0-kg block is connected to a 30.0-kg block by a string that passes over a light frictionless pulley. The 30.0-kg block is connected to a spring that has negligible mass and a force constant of 250 N/m as shown in Figure P7.53. The spring is unstretched when the system is as shown in the figure, and the incline is frictionless. The 20.0-kg block is pulled 20.0 cm down the incline (so that the 30.0-kg block is 40.0 cm above the floor) and released from rest. Find the speed of each block when the 30.0-kg block is 20.0 cm above the floor (that is, when the spring is unstretched).

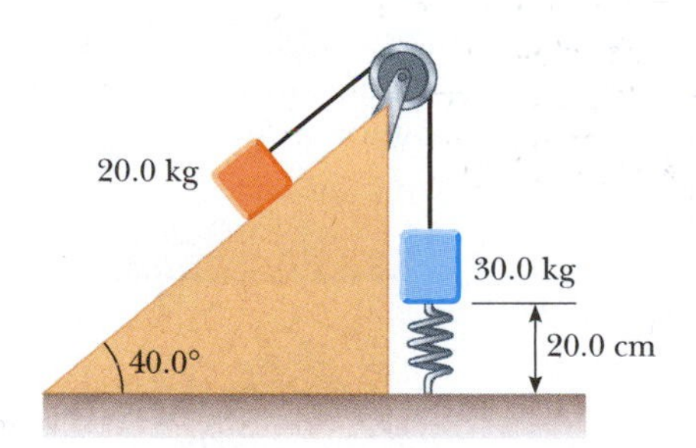


FIGURE **P7.53**

54. A 1.00-kg object slides to the right on a surface having a coefficient of kinetic friction 0.250 (Fig. P7.54). The object has a speed of $v_i = 3.00$ m/s when it makes contact with a light spring that has a force constant of 50.0 N/m. The object comes to rest after the spring has been compressed a distance $d$. The object is then forced toward the left by the spring and continues to move in that direction beyond the spring's unstretched position. The object finally comes to rest a distance $D$ to the left of the unstretched spring. Find (a) the distance of compression $d$, (b) the speed $v$ at the unstretched position when the object is moving to the left, and (c) the distance $D$ where the object comes to rest.

55. **PhysicsNow™** A block of mass 0.500 kg is pushed against a horizontal spring of negligible mass until the spring is compressed a distance $x$ (Fig. P7.55). The force constant of the spring is 450 N/m. When it is released, the block travels along a frictionless, horizontal surface to point $B$, the bottom of a vertical circular track of radius $R = 1.00$ m, and continues to move up the track. The speed of the block at the bottom of the track is $v_B = 12.0$ m/s, and the block experiences an average friction force of 7.00 N while sliding up the track. (a) What is $x$? (b) What speed do you predict for the block at the top of the track? (c) Does the block actually reach the top of the track, or does it fall off before reaching the top?

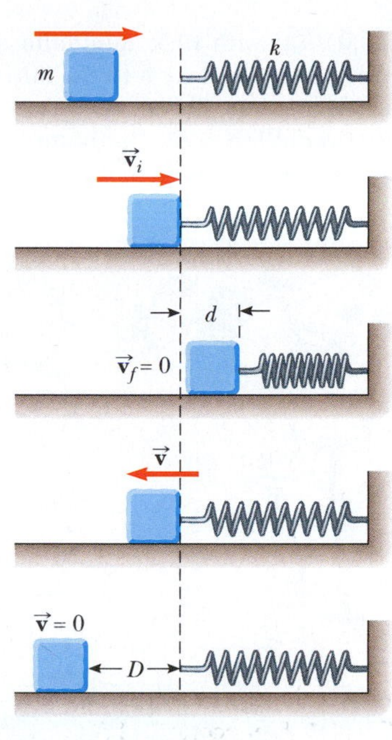


FIGURE **P7.54**

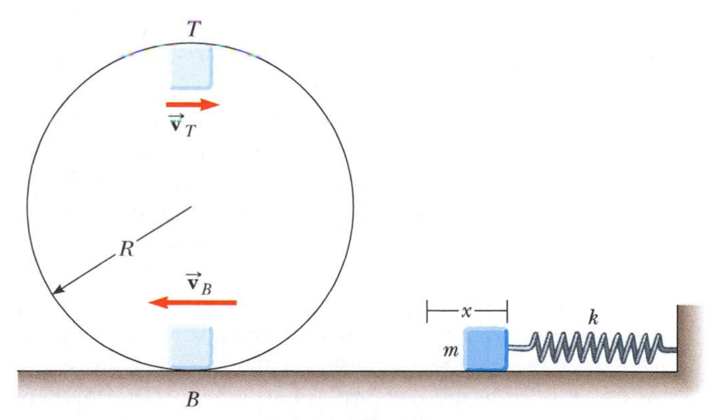


FIGURE **P7.55**

56. A uniform chain of length 8.00 m initially lies stretched out on a horizontal table. (a) Assuming that the coefficient of static friction between chain and table is 0.600, show that the chain will begin to slide off the table if at least 3.00 m of it hangs over the edge of the table. (b) Determine the speed of the chain as it all leaves the table, given that the coefficient of kinetic friction between the chain and the table is 0.400.

57. Jane, whose mass is 50.0 kg, needs to swing across a river (having width $D$) filled with person-eating crocodiles to save Tarzan from danger. She must swing into a wind exerting constant horizontal force $\vec{\mathbf{F}}$, on a vine having length $L$ and initially making an angle $\theta$ with the vertical (Fig. P7.57). Taking $D = 50.0$ m, $F = 110$ N, $L = 40.0$ m,

and $\theta = 50.0°$, (a) with what minimum speed must Jane begin her swing to just make it to the other side? (b) Once the rescue is complete, Tarzan and Jane must swing back across the river. With what minimum speed must they begin their swing? Assume that Tarzan has a mass of 80.0 kg.

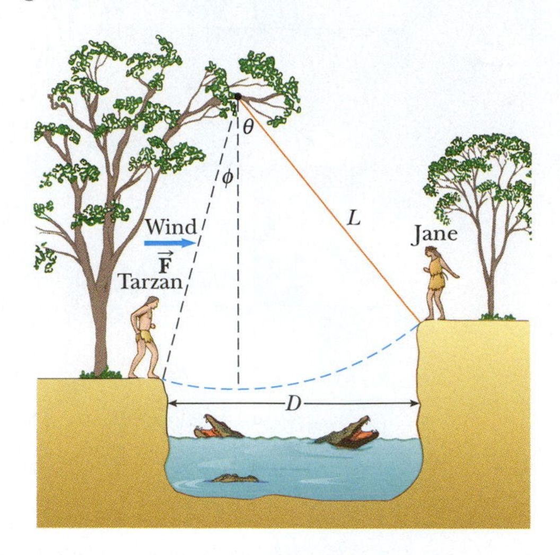


FIGURE **P7.57**

**58.** A 5.00-kg block free to move on a horizontal, frictionless surface is attached to one end of a light horizontal spring. The other end of the spring is held fixed. The spring is compressed 0.100 m from equilibrium and released. The speed of the block is 1.20 m/s when it passes the equilibrium position of the spring. The same experiment is now repeated with the frictionless surface replaced by a surface for which the coefficient of kinetic friction is 0.300. Determine the speed of the block at the equilibrium position of the spring.

**59.** A skateboarder with his board can be modeled as a particle of mass 76.0 kg, located at his center of mass (which we will study in Chapter 8). As shown in Figure P7.59, the skateboarder starts from rest in a crouching position at one lip of a half-pipe (point Ⓐ). The half-pipe is a dry water channel, forming one half of a cylinder of radius 6.80 m with its axis horizontal. On his descent, the skateboarder moves without friction so that his center of mass moves through one quarter of a circle of radius 6.30 m. (a) Find his speed at the bottom of the half-pipe (point Ⓑ). (b) Find his centripetal acceleration. (c) Find the normal force $n_B$ acting on the skateboarder at point Ⓑ. Immediately after passing point Ⓑ, he stands up and raises his arms, lifting his center of mass from 0.500 m to 0.950 m above the concrete (point Ⓒ). To account for the conversion of chemical into mechanical energy, model his legs as doing work by pushing him vertically up with a constant force equal to the normal force $n_B$ over a distance of 0.450 m. (You will be able to solve this problem with a more accurate model in Chapter 10.) (d) What is the work done on the skateboarder's body in this process? Next, the skateboarder glides upward with his center of mass moving in a quarter circle of radius 5.85 m. His body is horizontal when he passes point Ⓓ, the far lip of the half-pipe. (e) Find his speed at this location. At last he goes ballistic, twisting around while his center of mass moves vertically. (f) How high above point Ⓓ does he rise? (g) Over what time interval is he airborne before he touches down, 2.34 m below the level of point Ⓓ? (*Caution:* Do not try this yourself without the required skill and protective equipment or in a drainage channel to which you do not have legal access.)

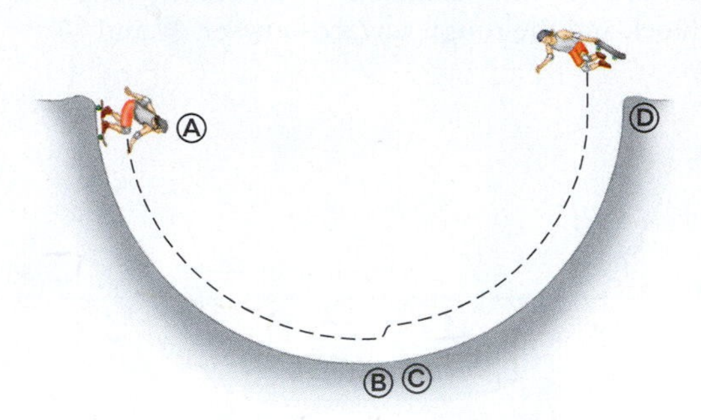


FIGURE **P7.59**

**60.** A block of mass $M$ rests on a table. It is fastened to the lower end of a light vertical spring. The upper end of the spring is fastened to a block of mass $m$. The upper block is pushed down by an additional force $3mg$ so that the spring compression is $4mg/k$. In this configuration, the upper block is released from rest. The spring lifts the lower block off the table. In terms of $m$, what is the greatest possible value for $M$?

**61.** A pendulum, comprising a light string of length $L$ and a small sphere, swings in a vertical plane. The string hits a peg located a distance $d$ below the point of suspension (Fig. P7.61). (a) Show that if the sphere is released from a height below that of the peg, it will return to this height after the string strikes the peg. (b) Show that if the pendulum is released from the horizontal position ($\theta = 90°$) and is to swing in a complete circle centered on the peg, the minimum value of $d$ must be $3L/5$.

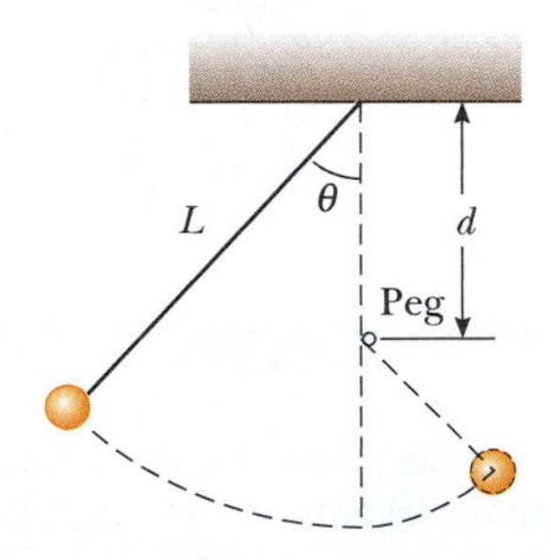


FIGURE **P7.61**

**62.** A roller coaster car is released from rest at the top of the first rise and then moves freely with negligible friction. The roller coaster shown in Figure P7.62 has a circular loop of radius $R$ in a vertical plane. (a) First, suppose the car barely makes it around the loop; at the top of the loop the riders are upside down and feel weightless. Find the required height of the release point above the bottom of the loop, in terms of $R$. (b) Now assume that the release point is at or above the minimum required height. Show that the normal force on the car at the bottom of the loop exceeds the normal force at the top of the loop by six times the weight of the car. The normal force on each rider follows the same rule. Such a large normal force is dangerous and very uncomfortable for the riders. Roller coasters are therefore not built with circular loops in vertical planes. Figure P5.24 and the photograph on page 134 show two actual designs.

FIGURE **P7.62**

**63.** **Review problem.** In 1887 in Bridgeport, Connecticut, C. J. Belknap built the water slide shown in Figure P7.63. A rider on a small sled, of total mass 80.0 kg, pushed off to start at the top of the slide (point Ⓐ) with a speed of 2.50 m/s. The chute was 9.76 m high at the top, 54.3 m long, and 0.51 m wide. Along its length, 725 wheels made friction negligible. Upon leaving the chute horizontally at its bottom end (point Ⓒ), the rider skimmed across the water of Long Island Sound for as much as 50 m, "skipping along like a flat pebble," before at last coming to rest and swimming ashore, pulling his sled after him. According to *Scientific American,* "The facial expression of novices taking their first adventurous slide is quite remarkable, and the sensations felt are correspondingly novel and peculiar." (a) Find the speed of the sled and rider at point Ⓒ. (b) Model the force of water friction as a constant retarding force acting on a particle. Find the work done by water friction in stopping the sled and rider. (c) Find the magnitude of the force the water exerts on the sled. (d) Find the magnitude of the force the chute exerts on the sled at point Ⓑ. (e) At point Ⓒ the chute is horizontal but curving in the vertical plane. Assume that its radius of curvature is 20.0 m. Find the force the chute exerts on the sled at point Ⓒ.

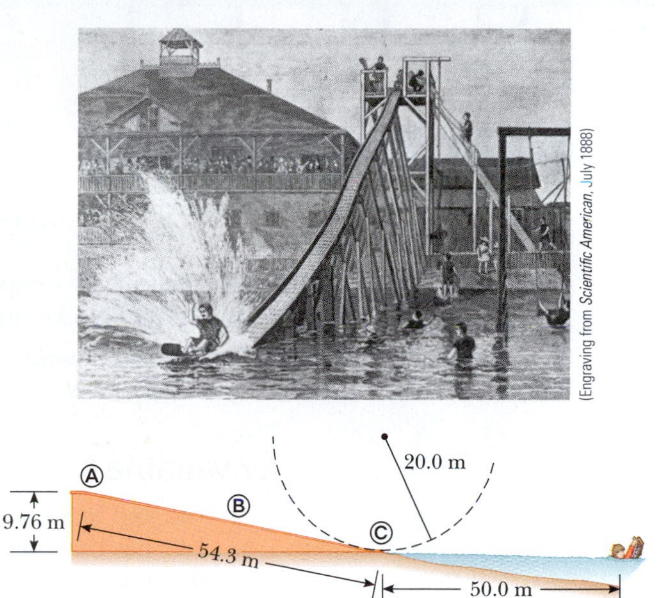


FIGURE **P7.63**

**64.** Starting from rest, a 64.0-kg person bungee jumps from a tethered balloon 65.0 m above the ground (Fig. P7.18). The bungee cord has negligible mass and unstretched length 25.8 m. One end is tied to the basket of the balloon and the other end to a harness around the person's body. The cord is described by Hooke's law with a spring constant of 81.0 N/m. The balloon does not move. (a) Model the person's body as a particle. Express the gravitational potential energy of the person–Earth system as a function of the person's variable height $y$ above the ground. (b) Express the elastic potential energy of the cord as a function of $y$. (c) Express the total potential energy of the person–cord–Earth system as a function of $y$. (d) Plot a graph of the gravitational, elastic, and total potential energies as functions of $y$. (e) Assume that air resistance is negligible. Determine the minimum height of the person above the ground during his plunge. (f) Does the potential energy graph show any equilibrium position or positions? If so, at what elevations? Are they stable or unstable? (g) Determine the jumper's maximum speed.

## ANSWERS TO QUICK QUIZZES

**7.1** (c). The sign of the gravitational potential energy depends on your choice of zero configuration. If the two objects in the system are closer together than in the zero configuration, the potential energy is negative. If they are farther apart, the potential energy is positive.

**7.2** (a). We must include the Earth if we are going to work with gravitational potential energy.

**7.3** $v_1 = v_2 = v_3$. The first and third balls speed up after they are thrown, whereas the second ball initially slows down but then speeds up after reaching its peak. The paths of all three balls are parabolas, and the balls take different time intervals to reach the ground because they have different initial velocities. All three balls, however, have the same speed at the moment they hit the ground because all start with the same kinetic energy and because the ball–Earth system undergoes the same change in gravitational potential energy in all three cases.

**7.4** (i), (c). This system exhibits changes in kinetic energy as well as in both types of potential energy. (ii), (a). Because the Earth is not included in the system, there is no gravitational potential energy associated with the system.

# Present and Future Possibilities

Now that we have explored some fundamental principles of classical mechanics, let us return to our central question for the *Alternative-Fuel Vehicles* Context:

> ***What source besides gasoline can be used to provide energy for an automobile while reducing environmentally damaging emissions?***

## Available Now—The Hybrid Electric Vehicle

As discussed in Section 7.8, electric vehicles such as the GM EV1 have not been successfully marketed and are falling by the wayside. Currently taking their place are a growing number of **hybrid electric vehicles.** In these automobiles, a gasoline engine and an electric motor are combined to increase the fuel economy of the vehicle and reduce its emissions. Currently available models include the Toyota Prius and Honda Insight, which are originally designed hybrid vehicles, as well as other existing models that have been modified with a hybrid drive system, such as the Honda Civic.

Two major categories of hybrid vehicles are the **parallel hybrid** and the **series hybrid.** In a parallel hybrid, both the engine and the motor are connected to the transmission, so either one can provide propulsion energy for the car. In a series hybrid, the gasoline engine does not provide propulsion energy to the transmission directly. The engine turns a generator, which in turn either charges the batteries or powers the electric motor. Only the electric motor is connected directly to the transmission to propel the car.

(Courtesy of Honda Motor Co., Inc.)

FIGURE 1 The Honda Insight.

The Honda Insight (Fig. 1) is a parallel hybrid. Both the engine and the motor provide power to the transmission, and the engine is running at all times while the car is moving. The goal of the development of this hybrid is maximum mileage, which is achieved through a number of design features. Because the engine is small, the Insight has lower emissions than a traditional gasoline-powered vehicle. Because the engine is running at all vehicle speeds, however, its emissions are not as low as those of the Toyota Prius.

Figure 2 shows the engine compartment of the Toyota Prius. In this parallel hybrid, power to the wheels can come from either the gasoline engine or the electric motor. The vehicle has some aspects of a series hybrid, however, in that the electric motor alone accelerates the vehicle from rest until it is moving at a speed of about 15 mph (24 kph). During this acceleration period, the engine is not running, so gasoline is not used and there is no emission. As a result, the average tailpipe emissions are lower than those of the Insight, although the gasoline mileage is not quite as high.

When a hybrid vehicle brakes, the motor acts as a generator and returns some of the kinetic energy of the vehicle back to the battery as electric potential energy. In a normal vehicle, this kinetic energy is not recoverable because it is transformed to internal energy in the brakes and roadway.

Gas mileage for hybrid vehicles is in the range of 45 to 60 mi/gal and emissions are far below those of a standard gasoline engine. A hybrid vehicle does not need to be charged like a purely electric vehicle. The battery that drives the electric motor is charged while the gasoline engine is running. Consequently, even though the hybrid vehicle has an electric motor like a pure electric vehicle, it can simply be filled at a gas station like a normal vehicle.

Hybrid electric vehicles are not strictly alternative-fuel vehicles because they use the same fuel as normal vehicles, gasoline. They do, however, represent an important step toward more efficient cars with lower emissions, and the increased mileage helps conserve crude oil.

(Photo by Brent Romans/www.Edmunds.com)

FIGURE 2 The engine compartment of the Toyota Prius.

## In the Future — The Fuel Cell Vehicle

In an internal combustion engine, the chemical potential energy in the fuel is transformed to internal energy during an explosion initiated by a spark plug. The resulting expanding gases do work on pistons, directing energy to the wheels of the vehicle. In current development is the **fuel cell,** in which the conversion of the energy in the fuel to internal energy is not required. The fuel (hydrogen) is oxidized, and energy leaves the fuel cell by electrical transmission. The energy is used by an electric motor to drive the vehicle.

The advantages of this type of vehicle are many. There is no internal combustion engine to generate harmful emissions, so the vehicle is emission-free. Other than the energy used to power the vehicle, the only by-products are internal energy and water. The fuel is hydrogen, which is the most abundant element in the universe. The efficiency of a fuel cell is much higher than that of an internal combustion engine, so more of the potential energy in the fuel can be extracted.

That is all good news. The bad news is that fuel cell vehicles are still only in the early prototype stage (Fig. 3). It will be many years before fuel cell vehicles are available to consumers. During these years, fuel cells must be perfected to operate in weather extremes, manufacturing infrastructure must be set up to supply the hydrogen, and a fueling infrastructure must be established to allow transfer of hydrogen into individual vehicles.

(© Adam Hart-Davis/Photo Researchers, Inc.)

FIGURE 3 A hydrogen fuel cell in Europe's first hydrogen-powered taxi. Fuel cells convert the energy from a chemical reaction into electricity.

### Problems

1. When a conventional car brakes to a stop, all (100%) its kinetic energy is converted into internal energy. None of this energy is available to get the car moving again. Consider a hybrid electric car of mass 1 300 kg moving at 22.0 m/s. (a) Calculate its kinetic energy. (b) The car uses its regenerative braking system to come to a stop at a red light. Assume that the motor-generator converts 70.0% of the car's kinetic energy into energy delivered to the battery by electrical transmission. The other 30.0% becomes internal energy. Compute the amount of energy charging up the battery. (c) Assume that the battery can give back 85.0% of the energy chemically stored in it. Compute the amount of this energy. The other 15.0% becomes internal energy. (d) When the light turns green, the car's motor-generator runs as a motor to convert 68.0% of the energy from the battery into kinetic energy of the car. Compute the amount of this energy and (e) the speed at which the car will be set moving with no other energy input. (f) Compute the overall efficiency of the braking-and-starting process. (g) Compute the net amount of internal energy produced.

**2.** In both a conventional car and a hybrid electric car, the gasoline engine is the original source of all the energy the car uses to push through the air and against rolling resistance of the road. In city traffic, a conventional gasoline engine must run at a wide variety of rotation rates and fuel inputs. That is, it must run at a wide variety of tachometer and throttle settings. It is almost never running at its maximum-efficiency point. In a hybrid electric car, on the other hand, the gasoline engine can run at maximum efficiency whenever it is on. A simple model can reveal the distinction numerically. Assume that the two cars both do 66.0 MJ of "useful" work in making the same trip to the drugstore. Let the conventional car run at 7.00% efficiency as it puts out useful energy 33.0 MJ and let it run at 30.0% efficiency as it puts out 33.0 MJ. Let the hybrid car run at 30.0% efficiency all the time. Compute (a) the required energy input for each car and (b) the overall efficiency of each.

# CONTEXT 2

## Mission to Mars

In this Context, we shall investigate the physics necessary to send a spacecraft from Earth to Mars. If the two planets were sitting still in space, millions of kilometers apart, it would be a difficult enough proposition, but keep in mind that we are launching the spacecraft from a moving object, the Earth, and are aiming at a moving target, Mars. Furthermore, the spacecraft's motion is influenced by gravitational forces from the Earth, the Sun, and Mars as well as from any other massive objects in the vicinity. Despite these apparent difficulties, we can use the principles of physics to plan a successful mission.

Travel in space began in the early 1960s, with the launch of human-occupied spacecraft in both the United States and the Soviet Union. The first human to ride into space was Yury Gagarin, who made a one-orbit trip in 1961 in the Soviet spacecraft *Vostok.* Competition between the two countries resulted in a "space race," which led to the successful landing of American astronauts on the Moon in 1969.

In the 1970s, the Viking Project landed spacecraft on Mars to analyze the soil for signs of life. These tests were inconclusive.

U.S. efforts in the 1980s focused on the development and implementation of the space shuttle system, a reusable space transportation system. The shuttle has been used extensively in moving supplies and personnel to the International Space Station, which was begun in 1998 and continues to develop. It has also been an important means of performing scientific experiments in space and delivering satellites into orbit.

The United States returned to Mars in the 1990s with the Mars Global Surveyor, designed to perform careful mapping of the Martian surface, and Mars Pathfinder, which landed on Mars and deployed a roving robot to analyze rocks and soil. Not all trips have been successful. In 1999, Mars Polar Lander was launched to land near the polar ice cap and search for water. As it entered the Martian atmosphere, it sent its last data and was never heard from again. Mars Climate Orbiter was also lost in 1999 due to communication errors between the builder of the spacecraft and the mission control team.

In late 2003 and early 2004, arrivals of spacecraft at Mars were expected by three space agencies, the National Aerodynamics and Space Administration (NASA) in the United States, the European Space Agency (ESA) in Europe, and the Japanese Aerospace Exploration Agency (JAXA) in Japan. The extreme difficulties associated with

(Japanese Aerospace Exploration Agency (JAXA))

GURE 1 The *Nozomi* is the first Mars orbiter to be launched by Japan. This photo shows its launch on July 4, 1998 from Kagoshima Space Center. Unfortunately, the *Nozomi* mission was unsuccessful because of technical difficulties, and the spacecraft did not achieve orbit around Mars.

(Courtesy of NASA/JPL)

FIGURE 2 The Mars rover *Spirit* is tested in a clean room at the Jet Propulsion Laboratory in Pasadena, California.

such an endeavor can be appreciated by examining the results of these simultaneous missions. The Japanese mission ended in failure when a stuck valve and electrical circuit problems affected a critical midcourse correction, resulting in the inability of the spacecraft, named *Nozomi,* to achieve an orbit around Mars. It passed about 1 000 km above the Martian surface on December 14, 2003, and then left the planet to continue its orbit around the Sun.

(Courtesy of NASA/JPL/Cornell)

FIGURE 3 An image from a camera on the Mars rover *Opportunity* shows a rock called the "Berry Bowl." The "berries" are sphere-like grains containing hematite, which scientists used to confirm the earlier presence of water on the surface. The circular area on the rock is the result of using the rover's rock abrasion tool to remove a layer of dust. In this way, a clean surface of the rock was available for spectral analysis by the rover's spectrometers.

The European effort resulted in a successful injection of their *Mars Express* spacecraft into an orbit around Mars. A lander, named *Beagle 2,* descended to the surface. Unfortunately, no signals from the lander have been detected and it is presumed lost. The *Mars Express* orbiter continues to send data and is equipped to perform scientific analyses from orbit.

The NASA effort was the most successful of the three missions, with the *Spirit* rover landing successfully on the surface of Mars on January 4, 2004. Its twin, *Opportunity,* also landed successfully, on January 24, 2004, on the opposite side of the planet from *Spirit.* Amazingly, *Opportunity* landed inside a crater, providing scientists with a

(Pierre Mion/National Geographic Image Collection)

FIGURE **4** In this Context, we shall investigate the details of the challenging task of sending a spacecraft from the Earth to Mars.

wonderful opportunity to study the geology of an impact crater. Aside from a computer glitch that was successfully repaired, both rovers performed excellently and sent back very high-quality photographs of the Martian surface as well as large amounts of data including verification of water that once existed on the surface.

Many individuals dream of one day establishing colonies on Mars. This dream is far in the future; we are still learning much about Mars today and have yet taken only a handful of trips to the planet. Travel to Mars is still not an everyday occurrence, although we learn more from each mission. In this Context, we address the central question,

**How can we undertake a successful transfer of a spacecraft from Earth to Mars?**

CHAPTER 8

# Momentum and Collisions

A golf ball is struck by a club and begins to leave the tee. Note the deformation of the ball as a result of the large force exerted on it by the club.



## CHAPTER OUTLINE

8.1 Linear Momentum and Its Conservation
8.2 Impulse and Momentum
8.3 Collisions
8.4 Two-Dimensional Collisions
8.5 The Center of Mass
8.6 Motion of a System of Particles
8.7 Context Connection — Rocket Propulsion
SUMMARY

Consider what happens when a golf ball is struck by a club as in the opening photograph for this chapter. The ball changes its motion from being at rest to having a very large velocity as a result of the collision; consequently, it is able to travel a large distance through the air. Because the ball experiences this change in velocity over a very short time interval, the average force on it during the collision is very large. By Newton's third law, the club experiences a reaction force equal in magnitude and opposite to the force on the ball. This reaction force produces a change in the velocity of the club. Because the club is much more massive than the ball, however, the change in the club's velocity is much less than the change in the ball's velocity.

One main objective of this chapter is to enable you to understand and analyze such events. As a first step, we shall introduce the concept of *momentum,* a term used to describe objects in motion. The concept of momentum leads us to a new conservation law and momentum approaches for treating isolated and nonisolated systems. This conservation law is especially useful for treating problems that involve collisions between objects.

## 8.1 LINEAR MOMENTUM AND ITS CONSERVATION

In the preceding two chapters, we studied situations that are difficult to analyze with Newton's laws. We were able to solve problems involving these situations by applying a conservation principle, conservation of energy. Consider another situation. A 60-kg archer stands on frictionless ice and fires a 0.50-kg arrow horizontally at 50 m/s. From Newton's third law, we know that the force that the bow exerts on the arrow will be matched by a force in the opposite direction on the bow (and the archer). This force will cause the archer to begin to slide backward on the ice. But with what speed? We cannot answer this question using *either* Newton's second law or an energy approach because there is not enough information.

Despite our inability to solve the archer problem using our techniques learned so far, this problem is very simple to solve if we introduce a new quantity that describes motion. To motivate this new quantity, let us apply the General Problem-Solving Strategy from Chapter 1 and *conceptualize* an isolated system of two particles (Fig. 8.1) with masses $m_1$ and $m_2$ and moving with velocities $\vec{\mathbf{v}}_1$ and $\vec{\mathbf{v}}_2$ at an instant of time. Because the system is isolated, the only force on one particle is that from the other particle, and we can *categorize* this situation as one in which Newton's laws can be applied. If a force from particle 1 (for example, a gravitational force) acts on particle 2, there must be a second force—equal in magnitude but opposite in direction—that particle 2 exerts on particle 1. That is, the forces form a Newton's third law action–reaction pair so that $\vec{\mathbf{F}}_{12} = -\vec{\mathbf{F}}_{21}$. We can express this condition as a statement about the *system* of two particles as follows:

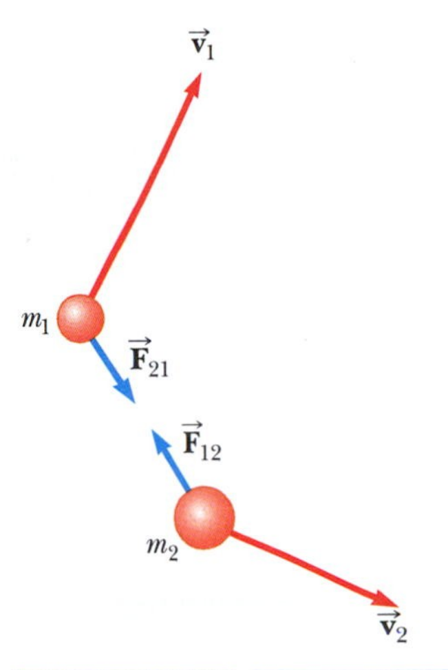


**FIGURE 8.1** Two particles interact with each other. According to Newton's third law, we must have $\vec{\mathbf{F}}_{12} = -\vec{\mathbf{F}}_{21}$.

$$\vec{\mathbf{F}}_{21} + \vec{\mathbf{F}}_{12} = 0$$

Let us further *analyze* this situation by incorporating Newton's second law. Over some time interval, the interacting particles in the system will accelerate. Therefore, replacing each force with $m\vec{\mathbf{a}}$ gives

$$m_1\vec{\mathbf{a}}_1 + m_2\vec{\mathbf{a}}_2 = 0$$

Now we replace the acceleration with its definition from Equation 3.5:

$$m_1\frac{d\vec{\mathbf{v}}_1}{dt} + m_2\frac{d\vec{\mathbf{v}}_2}{dt} = 0$$

If the masses $m_1$ and $m_2$ are constant, we can bring them into the derivatives, which gives

$$\frac{d(m_1\vec{\mathbf{v}}_1)}{dt} + \frac{d(m_2\vec{\mathbf{v}}_2)}{dt} = 0$$

$$\frac{d}{dt}(m_1\vec{\mathbf{v}}_1 + m_2\vec{\mathbf{v}}_2) = 0 \qquad [8.1]$$

To *finalize* this discussion, note that the derivative of the sum $m_1\vec{\mathbf{v}}_1 + m_2\vec{\mathbf{v}}_2$ with respect to time is zero. Consequently, this sum must be constant. We learn from this discussion that the quantity $m\vec{\mathbf{v}}$ for a particle is important in that the sum of the values of this quantity for the particles in an isolated system is conserved. We call this quantity *linear momentum:*

The **linear momentum** $\vec{\mathbf{p}}$ of a particle or an object that can be modeled as a particle of mass $m$ moving with a velocity $\vec{\mathbf{v}}$ is defined to be the product of the mass and velocity:[1]

■ Definition of linear momentum of a particle

$$\vec{\mathbf{p}} \equiv m\vec{\mathbf{v}} \qquad [8.2]$$

Because momentum equals the product of a scalar $m$ and a vector $\vec{\mathbf{v}}$, it is a vector quantity. Its direction is the same as that for $\vec{\mathbf{v}}$, and it has dimensions ML/T. In the SI system, momentum has the units kg · m/s.

If an object is moving in an arbitrary direction in three-dimensional space, $\vec{\mathbf{p}}$ has three components and Equation 8.2 is equivalent to the component equations

$$p_x = mv_x \qquad p_y = mv_y \qquad p_z = mv_z \qquad [8.3]$$

As you can see from its definition, the concept of momentum provides a quantitative distinction between objects of different masses moving at the same velocity. For example, the momentum of a truck moving at 2 m/s is much greater in magnitude than that of a Ping-Pong ball moving at the same speed. Newton called the product $m\vec{\mathbf{v}}$ the *quantity of motion,* perhaps a more graphic description than *momentum,* which comes from the Latin word for movement.

**QUICK QUIZ 8.1** Two objects have equal kinetic energies. How do the magnitudes of their momenta compare? **(a)** $p_1 < p_2$ **(b)** $p_1 = p_2$ **(c)** $p_1 > p_2$ **(d)** not enough information to determine the answer

**QUICK QUIZ 8.2** Your physical education teacher throws a baseball to you at a certain speed and you catch it. The teacher is next going to throw you a medicine ball whose mass is ten times the mass of the baseball. You are given the following choices. You can have the medicine ball thrown with **(a)** the same speed as the baseball, **(b)** the same momentum, or **(c)** the same kinetic energy. Rank these choices from easiest to hardest to catch.

Let us use the particle model for an object in motion. By using Newton's second law of motion, we can relate the linear momentum of a particle to the net force acting on the particle. In Chapter 4, we learned that Newton's second law can be written as $\sum \vec{\mathbf{F}} = m\vec{\mathbf{a}}$. This form applies only when the mass of the particle remains constant, however. In situations where the mass is changing with time, one must use an alternative statement of Newton's second law: **The time rate of change of momentum of a particle is equal to the net force acting on the particle,** or

■ Newton's second law for a particle

$$\sum \vec{\mathbf{F}} = \frac{d\vec{\mathbf{p}}}{dt} \qquad [8.4]$$

If the mass of the particle is constant, the preceding equation reduces to our previous expression for Newton's second law:

$$\sum \vec{\mathbf{F}} = \frac{d\vec{\mathbf{p}}}{dt} = \frac{d(m\vec{\mathbf{v}})}{dt} = m\frac{d\vec{\mathbf{v}}}{dt} = m\vec{\mathbf{a}}$$

It is difficult to imagine a particle whose mass is changing, but if we consider objects, a number of examples emerge. These examples include a rocket that is ejecting its

[1]This expression is nonrelativistic and is valid only when $v \ll c$, where $c$ is the speed of light. In the next chapter, we discuss momentum for high-speed particles.

fuel as it operates, a snowball rolling down a hill and picking up additional snow, and a watertight pickup truck whose bed is collecting water as it moves in the rain.

From Equation 8.4 we see that if the net force on an object is zero, the time derivative of the momentum is zero and therefore the momentum of the object must be constant. This conclusion should sound familiar because it is the case of a particle in equilibrium, expressed in terms of momentum. Of course, if the particle is *isolated* (that is, if it does not interact with its environment), no forces act on it and $\vec{\mathbf{p}}$ remains unchanged, which is Newton's first law.

## Momentum and Isolated Systems

Using the definition of momentum, Equation 8.1 can be written as

$$\frac{d}{dt}(\vec{\mathbf{p}}_1 + \vec{\mathbf{p}}_2) = 0$$

Because the time derivative of the total system momentum $\vec{\mathbf{p}}_{\text{tot}} = \vec{\mathbf{p}}_1 + \vec{\mathbf{p}}_2$ is *zero,* we conclude that the *total* momentum $\vec{\mathbf{p}}_{\text{tot}}$ must remain constant:

$$\vec{\mathbf{p}}_{\text{tot}} = \text{constant} \qquad [8.5]$$

■ Conservation of momentum for an isolated system

or, equivalently,

$$\vec{\mathbf{p}}_{1i} + \vec{\mathbf{p}}_{2i} = \vec{\mathbf{p}}_{1f} + \vec{\mathbf{p}}_{2f} \qquad [8.6]$$

where $\vec{\mathbf{p}}_{1i}$ and $\vec{\mathbf{p}}_{2i}$ are initial values and $\vec{\mathbf{p}}_{1f}$ and $\vec{\mathbf{p}}_{2f}$ are final values of the momentum during a period over which the particles interact. Equation 8.6 in component form states that the momentum components of the isolated system in the $x$, $y$, and $z$ directions are all *independently constant;* that is,

$$\sum_{\text{system}} p_{ix} = \sum_{\text{system}} p_{fx} \qquad \sum_{\text{system}} p_{iy} = \sum_{\text{system}} p_{fy} \qquad \sum_{\text{system}} p_{iz} = \sum_{\text{system}} p_{fz} \qquad [8.7]$$

This result, known as the law of **conservation of linear momentum,** is the mathematical representation of the momentum version of the isolated system model. It is considered one of the most important laws of mechanics. We have generated this law for a system of two interacting particles, but it can be shown to be true for a system of any number of particles. We can state it as follows: **The total momentum of an isolated system remains constant.**

Notice that we have made no statement concerning the nature of the forces acting between members of the system. The only requirement is that the forces must be *internal* to the system. Therefore, momentum is conserved for an isolated system *regardless* of the nature of the internal forces, *even if the forces are nonconservative.*

**PITFALL PREVENTION 8.1**

**MOMENTUM OF A SYSTEM IS CONSERVED** Remember that the momentum of an isolated *system* is conserved. The momentum of one particle within an isolated system is not necessarily conserved because other particles in the system may be interacting with it. Always apply conservation of momentum to an isolated *system.*

### EXAMPLE 8.1 Can We Really Ignore the Kinetic Energy of the Earth?

In Section 7.1, we claimed that we can ignore the kinetic energy of the Earth when considering the energy of a system consisting of the Earth and a dropped ball. Verify this claim.

**Solution** We will verify this claim by setting up a ratio of the kinetic energy of the Earth to that of the ball:

$$(1) \quad \frac{K_E}{K_b} = \frac{\frac{1}{2}m_E v_E^{\,2}}{\frac{1}{2}m_b v_b^{\,2}} = \left(\frac{m_E}{m_b}\right)\left(\frac{v_E}{v_b}\right)^2$$

where $v_E$ and $v_b$ are the speeds of the Earth and the ball, respectively, after the ball has fallen through some distance. Now we find a relationship between these two speeds by considering conservation of momentum in the vertical direction for the system of the ball and the Earth. The initial momentum of the system is zero, so the final momentum must also be zero:

$$p_i = p_f \quad \rightarrow \quad 0 = m_b v_b + m_E v_E$$

$$\rightarrow \quad \frac{v_E}{v_b} = -\frac{m_b}{m_E}$$

Substituting for $v_E/v_b$ in (1), we have

$$\frac{K_E}{K_b} = \left(\frac{m_E}{m_b}\right)\left(-\frac{m_b}{m_E}\right)^2 = \frac{m_b}{m_E}$$

Substituting order-of-magnitude numbers for the masses, this ratio becomes

$$\frac{K_E}{K_b} = \frac{m_b}{m_E} \sim \frac{1 \text{ kg}}{10^{25} \text{ kg}} \sim 10^{-25}$$

The kinetic energy of the Earth is a very small fraction of the kinetic energy of the ball, so we are justified in ignoring it in the kinetic energy of the system.

## INTERACTIVE EXAMPLE 8.2 The Archer

Let us consider the situation proposed at the beginning of this section. A 60-kg archer stands at rest on frictionless ice and fires a 0.50-kg arrow horizontally at 50 m/s (Fig. 8.2). With what velocity does the archer move across the ice after firing the arrow?

**FIGURE 8.2** (Interactive Example 8.2) An archer fires an arrow horizontally. Because he is standing on frictionless ice, he will begin to slide across the ice.

**Solution** We *cannot* solve this problem using Newton's second law, $\sum \vec{\mathbf{F}} = m\vec{\mathbf{a}}$, because we have no information about the force on the arrow or its acceleration. We *cannot* solve this problem using an energy approach because we do not know how much work is done in pulling the bow back or how much potential energy is stored in the bow. We *can,* however, solve this problem very easily with conservation of momentum because momentum does not depend on any of these quantities that we do not know.

Let us take the system to consist of the archer (including the bow) and the arrow. The system is not isolated because the gravitational force and the normal force act on the system. These forces, however, are vertical and perpendicular to the motion of the system. Therefore, there are no external forces in the horizontal direction, and we can consider the system to be isolated in terms of momentum components in this direction.

The total horizontal momentum of the system before the arrow is fired is zero ($m_1\vec{\mathbf{v}}_{1i} + m_2\vec{\mathbf{v}}_{2i} = 0$), where the archer is particle 1 and the arrow is particle 2. Therefore, the total horizontal momentum of the system after the arrow is fired must be zero; that is,

$$m_1\vec{\mathbf{v}}_{1f} + m_2\vec{\mathbf{v}}_{2f} = 0$$

We choose the direction of firing of the arrow as the positive *x* direction. With $m_1 = 60$ kg, $m_2 = 0.50$ kg, and $\vec{\mathbf{v}}_{2f} = 50\hat{\mathbf{i}}$ m/s, solving for $\vec{\mathbf{v}}_{1f}$ we find the recoil velocity of the archer to be

$$\vec{\mathbf{v}}_{1f} = -\frac{m_2}{m_1}\vec{\mathbf{v}}_{2f} = -\left(\frac{0.50 \text{ kg}}{60 \text{ kg}}\right)(50\hat{\mathbf{i}} \text{ m/s})$$
$$= -0.42\hat{\mathbf{i}} \text{ m/s}$$

The negative sign for $\vec{\mathbf{v}}_{1f}$ indicates that the archer is moving to the left after the arrow is fired, in the direction opposite the direction of the arrow's motion, in accordance with Newton's third law. Because the archer is much more massive than the arrow, his acceleration and consequent velocity are much smaller than the arrow's acceleration and velocity.

**PhysicsNow™** Log into PhysicsNow at **www.pop4e.com** and go to Interactive Example 8.2 to change the arrow's speed and the masses of the archer and the arrow.

## EXAMPLE 8.3 Decay of the Kaon at Rest

One type of nuclear particle, called the *neutral kaon* ($K^0$), decays into a pair of other particles called *pions* ($\pi^+$ and $\pi^-$), which are oppositely charged but equal in mass, as in Figure 8.3. Assuming that the kaon is initially at rest, prove that the two pions must have momenta that are equal in magnitude and opposite in direction.

**Solution** The isolated system is the kaon before the decay and the two pions afterward. The decay of the kaon, represented in Figure 8.3, can be written

$$K^0 \rightarrow \pi^+ + \pi^-$$

If we let $\vec{\mathbf{p}}^+$ be the momentum of the positive pion and $\vec{\mathbf{p}}^-$ be the momentum of the negative pion after

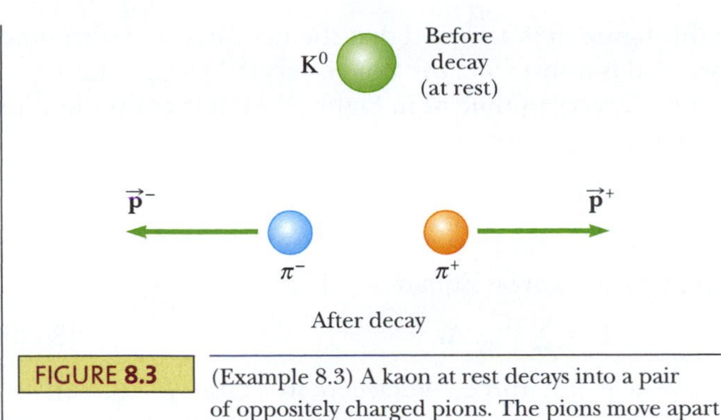


FIGURE 8.3 (Example 8.3) A kaon at rest decays into a pair of oppositely charged pions. The pions move apart with momenta of equal magnitudes but opposite directions.

the decay, the final momentum $\vec{\mathbf{p}}_f$ of the isolated system of two pions can be written

$$\vec{\mathbf{p}}_f = \vec{\mathbf{p}}^{\,+} + \vec{\mathbf{p}}^{\,-}$$

Because the kaon is at rest before the decay, we know that the initial system momentum $\vec{\mathbf{p}}_i = 0$. Furthermore, because the momentum of the isolated system is conserved, $\vec{\mathbf{p}}_i = \vec{\mathbf{p}}_f = 0$, so that $\vec{\mathbf{p}}^{\,+} + \vec{\mathbf{p}}^{\,-} = 0$ or

$$\vec{\mathbf{p}}^{\,+} = -\vec{\mathbf{p}}^{\,-}$$

Therefore, we see that the two momentum vectors of the pions are equal in magnitude and opposite in direction.

## 8.2 IMPULSE AND MOMENTUM

As described by Equation 8.4, the momentum of a particle changes if a net force acts on the particle. Let us assume that a net force $\sum \vec{\mathbf{F}}$ acts on a particle and that this force may vary with time. According to Equation 8.4,

$$d\vec{\mathbf{p}} = \sum \vec{\mathbf{F}}\, dt \qquad [8.8]$$

We can integrate this expression to find the change in the momentum of a particle during the time interval $\Delta t = t_f - t_i$. Integrating Equation 8.8 gives

$$\Delta\vec{\mathbf{p}} = \vec{\mathbf{p}}_f - \vec{\mathbf{p}}_i = \int_{t_i}^{t_f} \sum \vec{\mathbf{F}}\, dt \qquad [8.9]$$

The integral of a force over the time interval during which it acts is called the **impulse** of the force. The impulse of the net force $\sum \vec{\mathbf{F}}$ is a vector defined by

■ Impulse of a net force

$$\vec{\mathbf{I}} \equiv \int_{t_i}^{t_f} \sum \vec{\mathbf{F}}\, dt \qquad [8.10]$$

The direction of the impulse vector is the same as the direction of the change in momentum. Impulse has the dimensions of momentum, ML/T.

Based on this definition, Equation 8.9 tells us that the total impulse of the net force $\sum \vec{\mathbf{F}}$ on a particle equals the change in the momentum of the particle: $\vec{\mathbf{I}} = \Delta\vec{\mathbf{p}}$. This statement, known as the **impulse–momentum theorem,** is equivalent to Newton's second law. It also applies to a system of particles for which the net external force on the system causes a change in the total momentum of the system:

■ Impulse–momentum theorem

$$\vec{\mathbf{I}} \equiv \int_{t_i}^{t_f} \sum \vec{\mathbf{F}}_{\text{ext}}\, dt = \Delta\vec{\mathbf{p}}_{\text{tot}} \qquad [8.11]$$

Impulse is an interaction between the system and its environment. As a result of this interaction, the momentum of the system changes. This idea is an analog to the continuity equation for energy, which relates an interaction with the environment to the change in the energy of the system. Therefore, when we say that an impulse is given to a system, we imply that momentum is transferred from an external agent to that system. In many situations, the system can be modeled as a particle, so Equation 8.10 can be used rather than the more general Equation 8.11.

From the definition, we see that impulse is a vector quantity having a magnitude equal to the area under the curve of the magnitude of the net force versus time, as

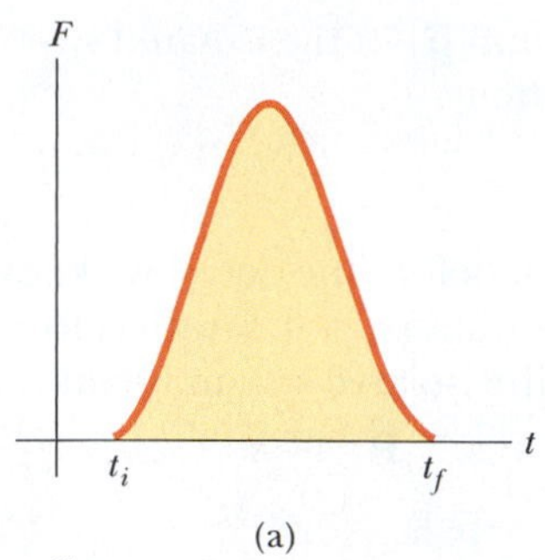


(a)

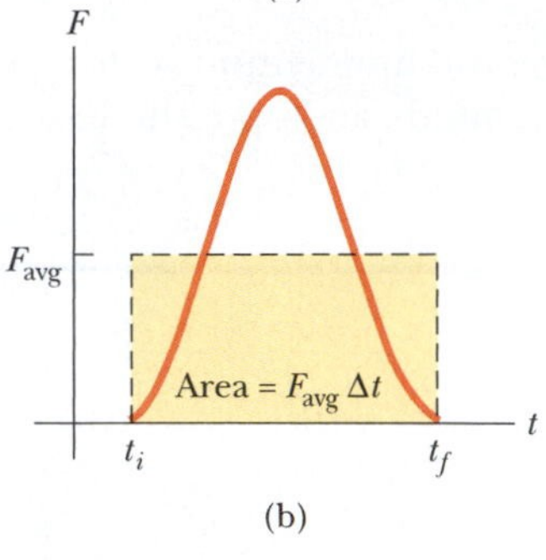


(b)

**FIGURE 8.4** (a) A net force acting on a particle may vary in time. The impulse is the area under the curve of the magnitude of the net force versus time. (b) The average force (*horizontal dashed line*) gives the same impulse to the particle in the time interval $\Delta t$ as the time-varying force described in part (a). The area of the rectangle is the same as the area under the curve.

illustrated in Figure 8.4. In this figure, it is assumed that the net force varies in time in the general manner shown and is nonzero in the time interval $\Delta t = t_f - t_i$.

Because the force can generally vary in time as in Figure 8.4a, it is convenient to define a time-averaged net force $\sum \vec{\mathbf{F}}_{avg}$ given by

$$\sum \vec{\mathbf{F}}_{avg} \equiv \frac{1}{\Delta t}\int_{t_i}^{t_f} \sum \vec{\mathbf{F}}\,dt \qquad [8.12]$$

where $\Delta t = t_f - t_i$. Therefore, we can express Equation 8.10 as

$$\vec{\mathbf{I}} = \sum \vec{\mathbf{F}}_{avg}\,\Delta t \qquad [8.13]$$

The magnitude of this average net force, described in Figure 8.4b, can be thought of as the magnitude of the constant net force that would give the same impulse to the particle in the time interval $\Delta t$ as the actual time-varying net force gives over this same interval.

In principle, if $\sum \vec{\mathbf{F}}$ is known as a function of time, the impulse can be calculated from Equation 8.10. The calculation becomes especially simple if the net force acting on the particle is constant. In this case, $\sum \vec{\mathbf{F}}_{avg}$ over a time interval is the same as the constant $\sum \vec{\mathbf{F}}$ at any instant within the interval, and Equation 8.13 becomes

$$\vec{\mathbf{I}} = \Delta\vec{\mathbf{p}} = \sum \vec{\mathbf{F}}\,\Delta t \qquad [8.14]$$

In many physical situations, we shall use what is called the **impulse approximation: We assume that one of the forces exerted on a particle acts for a short time but is much greater than any other force present.** This simplification model allows us to ignore the effects of other forces because these effects are small for the short time interval during which the large force acts. This approximation is especially useful in treating collisions in which the duration of the collision is very short. When this approximation is made, we refer to the force that is greater as an *impulsive force.* For example, when a baseball is struck with a bat, the duration of the collision is about 0.01 s and the average force the bat exerts on the ball during this time interval is typically several thousand newtons. This average force is much greater than the gravitational force, so we ignore any change in velocity related to the gravitational force during the collision. It is important to remember that $\vec{\mathbf{p}}_i$ and $\vec{\mathbf{p}}_f$ represent the momenta *immediately* before and after the collision, respectively. Therefore in the impulse approximation, very little motion of the particle takes place during the collision.

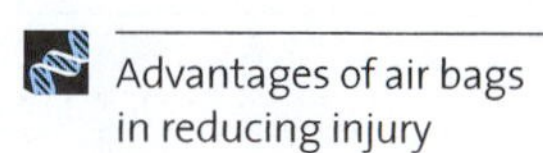


The concept of impulse helps us understand the value of air bags in stopping a passenger in an automobile accident (Fig. 8.5). The passenger experiences the same change in momentum and therefore the same impulse in a collision whether the car has air bags or not. The air bag allows the passenger to experience that change in momentum over a longer time interval, however, reducing the peak force on the passenger and increasing the chances of escaping without injury. Without the air bag, the passenger's head could move forward and be brought to rest in a short time interval by the steering wheel or the dashboard. In this case, the passenger undergoes the same change in momentum, but the short time interval results in a very large force that could cause severe head injury. Such injuries often result in spinal cord nerve damage where the nerves enter the base of the brain.

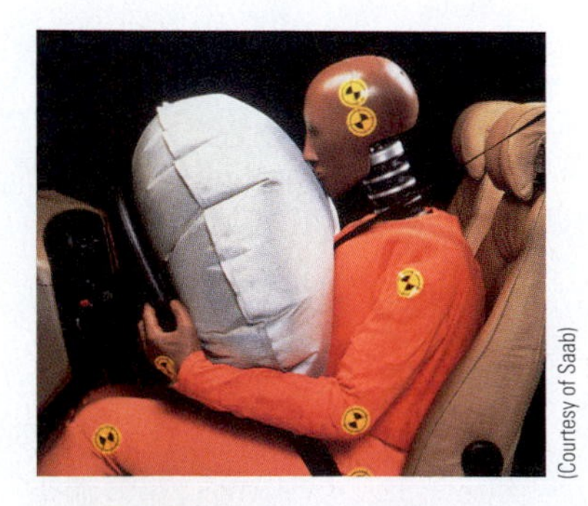

(Courtesy of Saab)

**FIGURE 8.5** A test dummy is brought to rest by an air bag in an automobile.

**QUICK QUIZ 8.3** Two objects are at rest on a frictionless surface. Object 1 has a greater mass than object 2. (i) When a constant force is applied to object 1, it accelerates through a distance $d$. The force is removed from object 1 and is applied to object 2. At the moment when object 2 has accelerated through the same distance $d$, which statements are true? **(a)** $p_1 < p_2$ **(b)** $p_1 = p_2$ **(c)** $p_1 > p_2$ **(d)** $K_1 < K_2$ **(e)** $K_1 = K_2$ **(f)** $K_1 > K_2$ **(ii)** When a constant force is applied to object 1, it accelerates for a time interval $\Delta t$. The force is removed from object 1 and is applied to object 2. After object 2 has accelerated for the same time interval $\Delta t$, which statements are true? **(a)** $p_1 < p_2$ **(b)** $p_1 = p_2$ **(c)** $p_1 > p_2$ **(d)** $K_1 < K_2$ **(e)** $K_1 = K_2$ **(f)** $K_1 > K_2$

**EXAMPLE 8.4** How Good Are the Bumpers?

In a crash test, an automobile of mass 1 500 kg collides with a wall as in Figure 8.6. The initial and final velocities of the automobile are $\vec{\mathbf{v}}_i = -15.0\hat{\mathbf{i}}$ m/s and $\vec{\mathbf{v}}_f = 2.60\hat{\mathbf{i}}$ m/s. If the collision lasts for 0.150 s, find the impulse due to the collision and the average force exerted on the automobile.

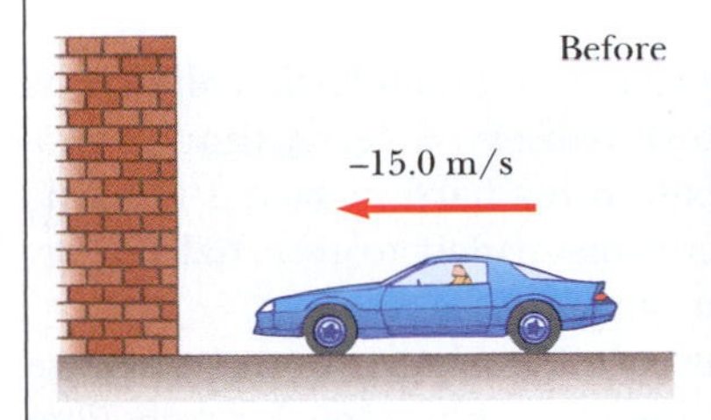


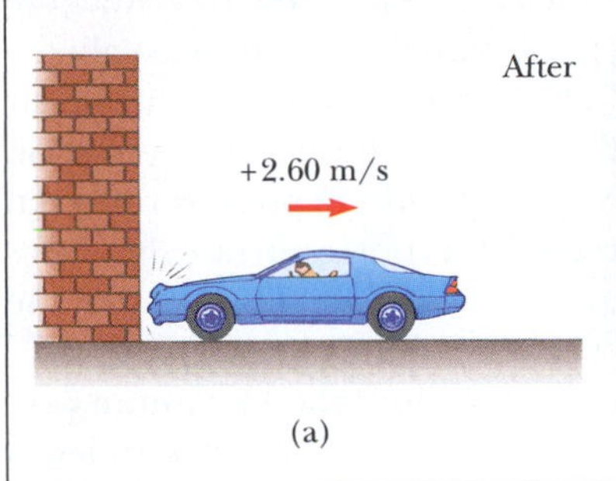


(a)

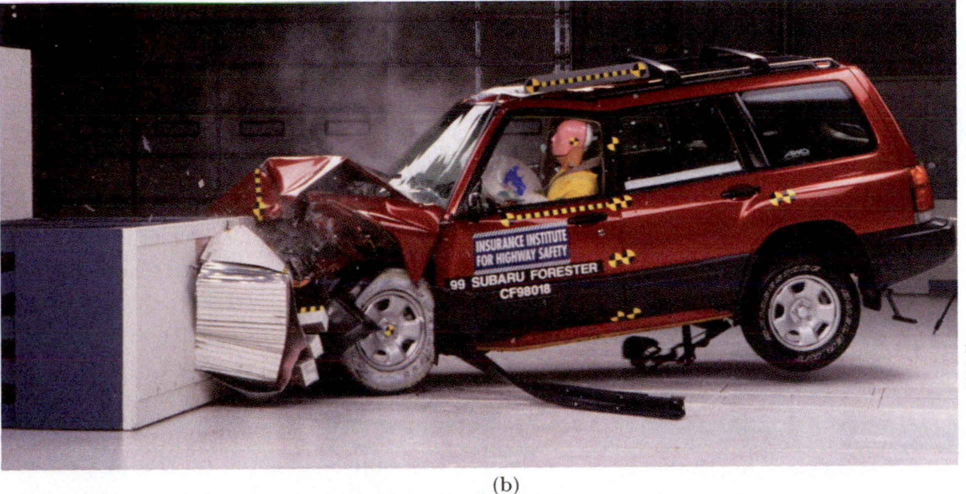


(b)

(Tim Wright/CORBIS)

**FIGURE 8.6** (Example 8.4) (a) The car's momentum changes as a result of its collision with the wall. (b) In a crash test, the large force exerted by the wall on the car produces extensive damage to the car's front end.

**Solution** We identify the automobile as the system. The initial and final momenta of the automobile are

$$\vec{\mathbf{p}}_i = m\vec{\mathbf{v}}_i = (1\,500 \text{ kg})(-15.0\hat{\mathbf{i}} \text{ m/s})$$
$$= -2.25 \times 10^4\hat{\mathbf{i}} \text{ kg}\cdot\text{m/s}$$
$$\vec{\mathbf{p}}_f = m\vec{\mathbf{v}}_f = (1\,500 \text{ kg})(2.60\hat{\mathbf{i}} \text{ m/s})$$
$$= 0.390 \times 10^4\hat{\mathbf{i}} \text{ kg}\cdot\text{m/s}$$

Hence, the impulse is

$$\vec{\mathbf{I}} = \Delta\vec{\mathbf{p}} = \vec{\mathbf{p}}_f - \vec{\mathbf{p}}_i$$
$$= 0.390 \times 10^4\hat{\mathbf{i}} \text{ kg}\cdot\text{m/s} - (-2.25 \times 10^4\hat{\mathbf{i}} \text{ kg}\cdot\text{m/s})$$
$$\vec{\mathbf{I}} = 2.64 \times 10^4\hat{\mathbf{i}} \text{ kg}\cdot\text{m/s}$$

The average force exerted on the automobile is

$$\vec{\mathbf{F}}_{\text{avg}} = \frac{\Delta\vec{\mathbf{p}}}{\Delta t} = \frac{2.64 \times 10^4\hat{\mathbf{i}} \text{ kg}\cdot\text{m/s}}{0.150 \text{ s}} = 1.76 \times 10^5\hat{\mathbf{i}} \text{ N}$$

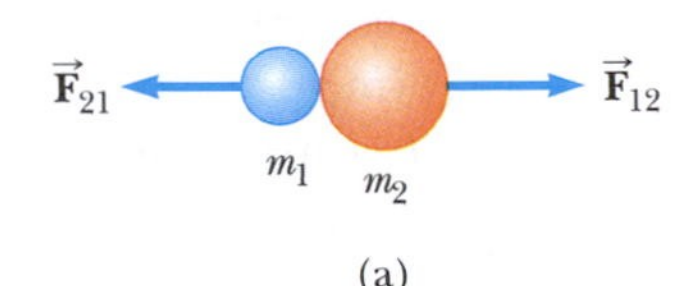


(a)

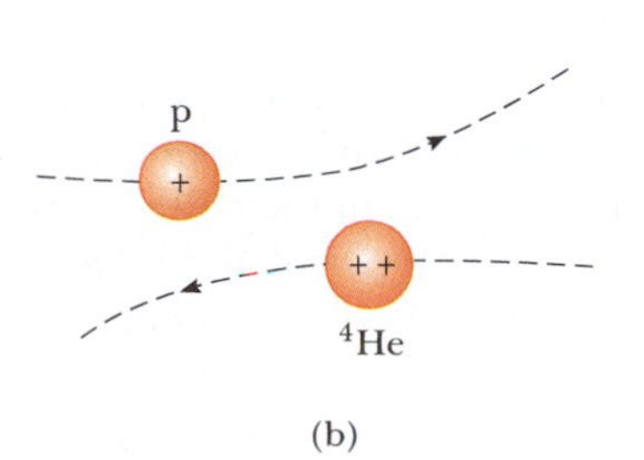


(b)

**FIGURE 8.7** (a) A collision between two objects as the result of direct contact. (b) A "collision" between two charged particles that do not make contact.

## 8.3 COLLISIONS

In this section, we use the law of conservation of momentum to describe what happens when two objects collide. The forces due to the collision are assumed to be much larger than any external forces present, so we use the simplification model we call the impulse approximation. The general goal in collision problems is to relate the final conditions of the system to the initial conditions.

A collision may be the result of physical contact between two objects, as described in Figure 8.7a. This observation is common when two macroscopic objects collide, such as two billiard balls or a baseball and a bat.

The notion of what we mean by *collision* must be generalized because "contact" on a microscopic scale is ill defined. To understand the distinction between macroscopic and microscopic collisions, consider the collision of a proton with an alpha particle (the nucleus of the helium atom), illustrated in Figure 8.7b. Because the two particles are positively charged, they repel each other. A collision has occurred, but the colliding particles were never in "contact."

When two particles of masses $m_1$ and $m_2$ collide, the collision forces may vary in time in a complicated way, as seen in Figure 8.4. As a result, an analysis of the situation with Newton's second law could be very complicated. We find, however, that the momentum concept is similar to the energy concept in Chapters 6 and 7 in that it provides us with a much easier method to solve problems involving isolated systems.

According to Equation 8.5, the momentum of an isolated system is conserved during some interaction event, such as a collision. The kinetic energy of the system, however, is generally *not* conserved in a collision. We define an **inelastic collision** as one in which the kinetic energy of the system is not conserved (even though momentum is conserved). The collision of a rubber ball with a hard surface is inelastic because some of the kinetic energy of the ball is transformed to internal energy when the ball is deformed while in contact with the surface.

Glaucoma testing

A practical example of an inelastic collision is used to detect glaucoma, a disease in which the pressure inside the eye builds up and leads to blindness by damaging the cells of the retina. In this application, medical professionals use a device called a *tonometer* to measure the pressure inside the eye. This device releases a puff of air against the outer surface of the eye and measures the speed of the air after reflection from the eye. At normal pressure, the eye is slightly spongy and the pulse is reflected at low speed. As the pressure inside the eye increases, the outer surface becomes more rigid and the speed of the reflected pulse increases. Therefore, the speed of the reflected puff of air is used to measure the internal pressure of the eye.

**PITFALL PREVENTION 8.2**

**Perfectly inelastic collisions**
Keep in mind the distinction between inelastic and perfectly inelastic collisions. If the colliding particles stick together, the collision is perfectly inelastic. If they bounce off each other (and kinetic energy is not conserved), the collision is inelastic. Generally, inelastic collisions are hard to analyze unless additional information is provided. This difficulty appears in the mathematical representation as having more unknowns than equations.

When two objects collide and stick together after a collision, the maximum possible fraction of the initial kinetic energy is transformed; this collision is called a **perfectly inelastic collision.** For example, if two vehicles collide and become entangled, they move with some common velocity after the perfectly inelastic collision. If a meteorite collides with the Earth, it becomes buried in the ground and the collision is perfectly inelastic.

An **elastic collision** is defined as one in which the kinetic energy of the system is conserved (as well as momentum). Real collisions in the macroscopic world, such as those between billiard balls, are only approximately elastic because some transformation of kinetic energy takes place and some energy leaves the system by mechanical waves, sound. Imagine a billiard game with truly elastic collisions. The opening break would be completely silent! Truly elastic collisions do occur between atomic and subatomic particles. Elastic and perfectly inelastic collisions are *limiting* cases; a large number of collisions fall in the range between them.

In the remainder of this section, we treat collisions in one dimension and consider the two extreme cases: perfectly inelastic collisions and elastic collisions. The important distinction between these two types of collisions is that **the momentum of the system is conserved in all cases, but the kinetic energy is conserved only in elastic collisions.** When analyzing one-dimensional collisions, we can drop the vector notation and use positive and negative signs for velocities to denote directions, as we did in Chapter 2.

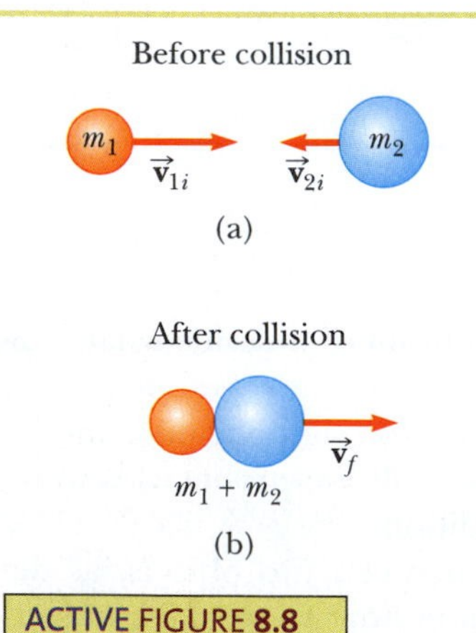


ACTIVE FIGURE **8.8**
A perfectly inelastic head-on collision between two particles: (a) before the collision and (b) after the collision.

**Physics Now™** Log into PhysicsNow at **www.pop4e.com** and go to Active Figure 8.8 to adjust the masses and velocities of the colliding objects and see the effect on the final velocity.

## One-Dimensional Perfectly Inelastic Collisions

Consider two objects of masses $m_1$ and $m_2$ moving with initial velocities $v_{1i}$ and $v_{2i}$ along a straight line as in Active Figure 8.8. If the two objects collide head-on, stick together, and move with some common velocity $v_f$ after the collision, the collision is perfectly inelastic. Because the total momentum of the two-object isolated system before the collision equals the total momentum of the combined-object system after the collision, we have

$$m_1 v_{1i} + m_2 v_{2i} = (m_1 + m_2) v_f \qquad [8.15]$$

$$v_f = \frac{m_1 v_{1i} + m_2 v_{2i}}{m_1 + m_2} \qquad [8.16]$$

Therefore, if we know the initial velocities of the two objects, we can use this single equation to determine the final common velocity.

## One-Dimensional Elastic Collisions

Now consider two objects that undergo an elastic head-on collision (Active Fig. 8.9) in one dimension. In this collision, both momentum and kinetic energy are conserved; therefore, we can write[2]

$$m_1 v_{1i} + m_2 v_{2i} = m_1 v_{1f} + m_2 v_{2f} \qquad [8.17]$$

$$\tfrac{1}{2} m_1 v_{1i}^2 + \tfrac{1}{2} m_2 v_{2i}^2 = \tfrac{1}{2} m_1 v_{1f}^2 + \tfrac{1}{2} m_2 v_{2f}^2 \qquad [8.18]$$

In a typical problem involving elastic collisions, two unknown quantities occur (such as $v_{1f}$ and $v_{2f}$), and Equations 8.17 and 8.18 can be solved simultaneously to find them. An alternative approach, employing a little mathematical manipulation of Equation 8.18, often simplifies this process. Let us cancel the factor of $\frac{1}{2}$ in Equation 8.18 and rewrite the equation as

$$m_1(v_{1i}^2 - v_{1f}^2) = m_2(v_{2f}^2 - v_{2i}^2)$$

Here we have moved the terms containing $m_1$ to one side of the equation and those containing $m_2$ to the other. Next, let us factor both sides:

$$m_1(v_{1i} - v_{1f})(v_{1i} + v_{1f}) = m_2(v_{2f} - v_{2i})(v_{2f} + v_{2i}) \qquad [8.19]$$

We now separate the terms containing $m_1$ and $m_2$ in the equation for conservation of momentum (Eq. 8.17) to obtain

$$m_1(v_{1i} - v_{1f}) = m_2(v_{2f} - v_{2i}) \qquad [8.20]$$

To obtain our final result, we divide Equation 8.19 by Equation 8.20 and obtain

$$v_{1i} + v_{1f} = v_{2f} + v_{2i}$$

or, gathering initial and final values on opposite sides of the equation,

$$v_{1i} - v_{2i} = -(v_{1f} - v_{2f}) \qquad [8.21]$$

This equation, in combination with the condition for conservation of momentum, Equation 8.17, can be used to solve problems dealing with one-dimensional elastic collisions between two objects. According to Equation 8.21, the relative speed[3] $v_{1i} - v_{2i}$ of the two objects before the collision equals the negative of their relative speed after the collision, $-(v_{1f} - v_{2f})$.

Suppose the masses and the initial velocities of both objects are known. Equations 8.17 and 8.21 can be solved for the final velocities in terms of the initial values because we have two equations and two unknowns:

$$v_{1f} = \left(\frac{m_1 - m_2}{m_1 + m_2}\right) v_{1i} + \left(\frac{2m_2}{m_1 + m_2}\right) v_{2i} \qquad [8.22]$$

$$v_{2f} = \left(\frac{2m_1}{m_1 + m_2}\right) v_{1i} + \left(\frac{m_2 - m_1}{m_1 + m_2}\right) v_{2i} \qquad [8.23]$$

It is important to remember that the appropriate signs for the velocities $v_{1i}$ and $v_{2i}$ must be included in Equations 8.22 and 8.23. For example, if $m_2$ is moving to the left initially, as in Active Figure 8.9a, $v_{2i}$ is negative.

Let us consider some special cases. If $m_1 = m_2$, Equations 8.22 and 8.23 show us that $v_{1f} = v_{2i}$ and $v_{2f} = v_{1i}$. That is, the objects exchange speeds if they have equal

### PITFALL PREVENTION 8.3

**Momentum and Kinetic Energy in Collisions** Linear momentum of an isolated system is conserved in *all* collisions. Kinetic energy of an isolated system is conserved *only* in elastic collisions. These statements are true because kinetic energy can be transformed into several types of energy or can be transferred out of the system (so that the system may *not* be isolated in terms of energy during the collision), but there is only one type of linear momentum.

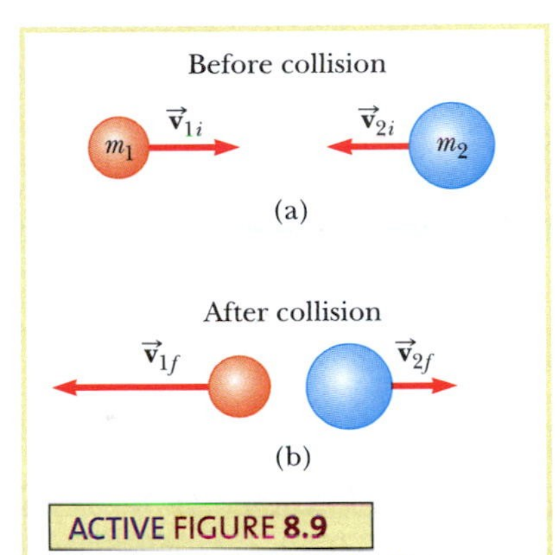


**ACTIVE FIGURE 8.9**
An elastic head-on collision between two particles: (a) before the collision and (b) after the collision.

**PhysicsNow™** Log into PhysicsNow at www.pop4e.com and go to Active Figure 8.9 to adjust the masses and velocities of the colliding objects and see the effect on the final velocities.

### PITFALL PREVENTION 8.4

**Not a General Equation** We have spent some effort on deriving Equation 8.21, but remember that it can be used only in a very *specific* situation: a one-dimensional, elastic collision between two objects. The *general* concept is conservation of momentum (and conservation of kinetic energy if the collision is elastic) for an isolated system.

[2]Notice that the kinetic energy of the system is the sum of the kinetic energies of the two particles. In our energy conservation examples in Chapter 7 involving a falling object and the Earth, we ignored the kinetic energy of the Earth because it is so small. Therefore, the kinetic energy of the *system* is just the kinetic energy of the falling *object*. That is a special case in which the mass of one of the objects (the Earth) is so immense that ignoring its kinetic energy introduces no measurable error. For problems such as those described here, however, and for the particle decay problems we will see in Chapters 30 and 31, we need to include the kinetic energies of *all* particles in the system.

[3]See Section 3.6 for a review of relative speed.

masses. That is what one observes in head-on billiard ball collisions, assuming there is no spin on the ball: The initially moving ball stops and the initially stationary ball moves away with approximately the same speed.

If $m_2$ is initially at rest, $v_{2i} = 0$ and Equations 8.22 and 8.23 become

Elastic collision in one dimension: particle 2 initially at rest

$$v_{1f} = \left(\frac{m_1 - m_2}{m_1 + m_2}\right) v_{1i} \quad [8.24]$$

$$v_{2f} = \left(\frac{2m_1}{m_1 + m_2}\right) v_{1i} \quad [8.25]$$

If $m_1$ is very large compared with $m_2$, we see from Equations 8.24 and 8.25 that $v_{1f} \approx v_{1i}$ and $v_{2f} \approx 2v_{1i}$. That is, when a very heavy object collides head-on with a very light one initially at rest, the heavy object continues its motion unaltered after the collision but the light object rebounds with a speed equal to about twice the initial speed of the heavy object. An example of such a collision is that of a moving heavy atom, such as uranium, with a light atom, such as hydrogen.

If $m_2$ is much larger than $m_1$ and if $m_2$ is initially at rest, we find from Equations 8.24 and 8.25 that $v_{1f} \approx -v_{1i}$ and $v_{2f} \approx 0$. That is, when a very light object collides head-on with a very heavy object initially at rest, the velocity of the light object is reversed and the heavy object remains approximately at rest. For example, imagine what happens when a marble hits a stationary bowling ball.

**QUICK QUIZ 8.4** A Ping-Pong ball is thrown at a stationary bowling ball hanging from a wire. The Ping-Pong ball makes a one-dimensional elastic collision and bounces back along the same line. After the collision, the Ping-Pong ball has, compared with the bowling ball, **(a)** a larger magnitude of momentum and more kinetic energy, **(b)** a smaller magnitude of momentum and more kinetic energy, **(c)** a larger magnitude of momentum and less kinetic energy, **(d)** a smaller magnitude of momentum and less kinetic energy, or **(e)** the same magnitude of momentum and the same kinetic energy

## PROBLEM-SOLVING STRATEGY One-Dimensional Collisions

We suggest that you use the following approach when solving collision problems in one dimension:

1. **Conceptualize** Establish the mental representation by imagining the collision occurring in your mind. Draw simple diagrams of the particles before and after the collision with appropriate velocity vectors. You may have to guess for now at the directions of final velocity vectors.
2. **Categorize** Is the system of particles truly isolated? If so, categorize the collision as elastic, inelastic, or perfectly inelastic.
3. **Analyze** Set up the appropriate mathematical representation for the problem. If the collision is perfectly inelastic, use Equation 8.15. If the collision is elastic, use Equations 8.17 and 8.21. If the collision is inelastic, use Equation 8.17. To find the final velocities in this case, you will need some additional piece of information.
4. **Finalize** Once you have determined your result, check to see that your answers are consistent with the mental and pictorial representations and that your results are realistic.

## EXAMPLE 8.5 Kinetic Energy in a Perfectly Inelastic Collision

We claimed that the maximum amount of kinetic energy was transformed to other forms in a perfectly inelastic collision. Prove this statement mathematically for a one-dimensional two-particle collision.

**Solution** We will assume that the maximum kinetic energy is transformed and prove that the collision must be perfectly inelastic. We set up the fraction $f$ of the final kinetic energy after the collision to the initial kinetic energy:

$$f = \frac{K_f}{K_i} = \frac{\frac{1}{2}m_1 v_{1f}^2 + \frac{1}{2}m_2 v_{2f}^2}{\frac{1}{2}m_1 v_{1i}^2 + \frac{1}{2}m_2 v_{2i}^2} = \frac{m_1 v_{1f}^2 + m_2 v_{2f}^2}{m_1 v_{1i}^2 + m_2 v_{2i}^2}$$

The *maximum* amount of energy transformed to other forms corresponds to the *minimum* value of $f$. For fixed initial conditions, we imagine that the final velocities $v_{1f}$ and $v_{2f}$ are variables. We minimize the fraction $f$ by taking the derivative of $f$ with respect to $v_{1f}$ and setting the result equal to zero:

$$\frac{df}{dv_{1f}} = \frac{d}{dv_{1f}}\left(\frac{m_1 v_{1f}^2 + m_2 v_{2f}^2}{m_1 v_{1i}^2 + m_2 v_{2i}^2}\right)$$

$$= \frac{2m_1 v_{1f} + 2m_2 v_{2f}\dfrac{dv_{2f}}{dv_{1f}}}{m_1 v_{1i}^2 + m_2 v_{2i}^2} = 0$$

$$(1) \quad \rightarrow \quad m_1 v_{1f} + m_2 v_{2f}\frac{dv_{2f}}{dv_{1f}} = 0$$

From the conservation of momentum condition, we can evaluate the derivative in (1). We differentiate Equation 8.17 with respect to $v_{1f}$:

$$\frac{d}{dv_{1f}}(m_1 v_{1i} + m_2 v_{2i}) = \frac{d}{dv_{1f}}(m_1 v_{1f} + m_2 v_{2f})$$

$$\rightarrow \quad 0 = m_1 + m_2\frac{dv_{2f}}{dv_{1f}} \quad \rightarrow \quad \frac{dv_{2f}}{dv_{1f}} = -\frac{m_1}{m_2}$$

Substituting this expression for the derivative into (1), we find

$$m_1 v_{1f} - m_2 v_{2f}\frac{m_1}{m_2} = 0 \quad \rightarrow \quad v_{1f} = v_{2f}$$

If the particles come out of the collision with the same velocities, they are joined together and it is a perfectly inelastic collision, which is what we set out to prove.

### EXAMPLE 8.6 Carry Collision Insurance

An 1 800-kg car stopped at a traffic light is struck from the rear by a 900-kg car and the two become entangled. If the smaller car was moving at 20.0 m/s before the collision, what is the speed of the entangled cars after the collision?

**Solution** The total momentum of the system (the two cars) before the collision equals the total momentum of the system after the collision because the system is isolated (in the impulse approximation). Notice that we ignore friction with the road in the impulse approximation. Therefore, the result we obtain for the final speed will only be approximately true just after the collision. For longer time intervals after the collision, we would use Newton's second law to describe the slowing down of the system as a result of friction. Because the cars "become entangled," it is a perfectly inelastic collision.

The magnitude of the total momentum of the system before the collision is equal to that of only the smaller car because the larger car is initially at rest:

$$p_i = m_1 v_i = (900 \text{ kg})(20.0 \text{ m/s}) = 1.80 \times 10^4 \text{ kg}\cdot\text{m/s}$$

After the collision, the mass that moves is the sum of the masses of the cars. The magnitude of the momentum of the combination is

$$p_f = (m_1 + m_2)v_f = (2\,700 \text{ kg})v_f$$

Equating the initial momentum to the final momentum and solving for $v_f$, the speed of the entangled cars, we have

$$v_f = \frac{p_f}{m_1 + m_2} = \frac{p_i}{m_1 + m_2} = \frac{1.80 \times 10^4 \text{ kg}\cdot\text{m/s}}{2\,700 \text{ kg}}$$

$$= 6.67 \text{ m/s}$$

### EXAMPLE 8.7 Slowing Down Neutrons by Collisions

In a nuclear reactor, neutrons are produced when ${}^{235}_{92}\text{U}$ atoms split in a process called *fission*. These neutrons are moving at about $10^7$ m/s and must be slowed down to about $10^3$ m/s before they take part in another fission event. They are slowed down by being passed through a solid or liquid material called a *moderator*. The slowing-down process involves elastic collisions. Let us show that a neutron can lose most of its kinetic energy if it collides elastically with a moderator containing light nuclei, such as deuterium (in "heavy water," $D_2O$).

**Solution** We identify the system as the neutron and a moderator nucleus. Because the momentum and kinetic energy of this system are conserved in an elastic collision, Equations 8.24 and 8.25 can be applied to a one-dimensional collision of these two particles.

Let us assume that the moderator nucleus of mass $m_m$ is at rest initially and that the neutron of mass $m_n$ and initial speed $v_{ni}$ collides head-on with it. The initial kinetic energy of the neutron is

$$K_{ni} = \tfrac{1}{2}m_n v_{ni}^2$$

After the collision, the neutron has kinetic energy $\frac{1}{2}m_n v_{nf}^2$, where $v_{nf}$ is given by Equation 8.24:

$$K_{nf} = \frac{1}{2}m_n v_{nf}^2 = \frac{1}{2}m_n\left(\frac{m_n - m_m}{m_n + m_m}\right)^2 v_{ni}^2$$

Therefore, the fraction of the total kinetic energy possessed by the neutron after the collision is

$$(1)\quad f_n = \frac{K_{nf}}{K_{ni}} = \frac{\frac{1}{2}m_n\left(\frac{m_n - m_m}{m_n + m_m}\right)^2 v_{ni}^2}{\frac{1}{2}m_n v_{ni}^2} = \left(\frac{m_n - m_m}{m_n + m_m}\right)^2$$

From this result, we see that the final kinetic energy of the neutron is small when $m_m$ is close to $m_n$ and is zero when $m_m = m_n$.

We can calculate the kinetic energy of the moderator nucleus after the collision using Equation 8.25:

$$K_{mf} = \frac{1}{2}m_m v_{mf}^2 = \frac{2m_n^2 m_m}{(m_n + m_m)^2}\, v_{ni}^2$$

Hence, the fraction of the total kinetic energy transferred to the moderator nucleus is

$$(2)\quad f_{\text{trans}} = \frac{K_{mf}}{K_{ni}} = \frac{\frac{2m_n^2 m_m}{(m_n + m_m)^2}\, v_{ni}^2}{\frac{1}{2}m_n v_{ni}^2} = \frac{4m_n m_m}{(m_n + m_m)^2}$$

If $m_m \approx m_n$, we see that $f_{\text{trans}} \approx 1 = 100\%$. Because the system's kinetic energy is conserved, (2) can also be obtained from (1) with the condition that $f_n + f_m = 1$, so that $f_m = 1 - f_n$.

For collisions of the neutrons with deuterium nuclei in $D_2O$ ($m_m = 2m_n$), $f_n = 1/9$ and $f_{\text{trans}} = 8/9$. That is, 89% of the neutron's kinetic energy is transferred to the deuterium nucleus. In practice, the moderator efficiency is reduced because head-on collisions are very unlikely to occur.

## INTERACTIVE EXAMPLE 8.8 Two Blocks and a Spring

A block of mass $m_1 = 1.60$ kg, initially moving to the right with a speed of 4.00 m/s on a frictionless horizontal track, collides with a massless spring attached to a second block of mass $m_2 = 2.10$ kg, moving to the left with a speed of 2.50 m/s as in Figure. 8.10a. The spring has a spring constant of 600 N/m.

**A** At the instant when $m_1$ is moving to the right with a speed of 3.00 m/s as in Figure 8.10b, determine the speed of $m_2$.

**Solution** Figure 8.10 helps conceptualize the problem. Because the blocks move along a frictionless straight track, we categorize this problem as one involving a one-dimensional collision between objects forming an isolated system. We identify the system as the two blocks and the spring and identify the collision as elastic because the force from the spring is conservative. Because the total momentum of the isolated system is conserved, we analyze the problem by recognizing that

$$m_1 v_{1i} + m_2 v_{2i} = m_1 v_{1f} + m_2 v_{2f}$$

$$(1.60\text{ kg})(4.00\text{ m/s}) + (2.10\text{ kg})(-2.50\text{ m/s}) = (1.60\text{ kg})(3.00\text{ m/s}) + (2.10\text{ kg})v_{2f}$$

$$v_{2f} = \boxed{-1.74\text{ m/s}}$$

Note that the initial velocity of $m_2$ is $-2.50$ m/s because its direction is to the left. The negative value for $v_{2f}$ means that $m_2$ is still moving to the left at the instant we are considering.

**B** Determine the distance the spring is compressed at that instant.

**Solution** Because the system is isolated, we can also analyze this problem from the energy version of the

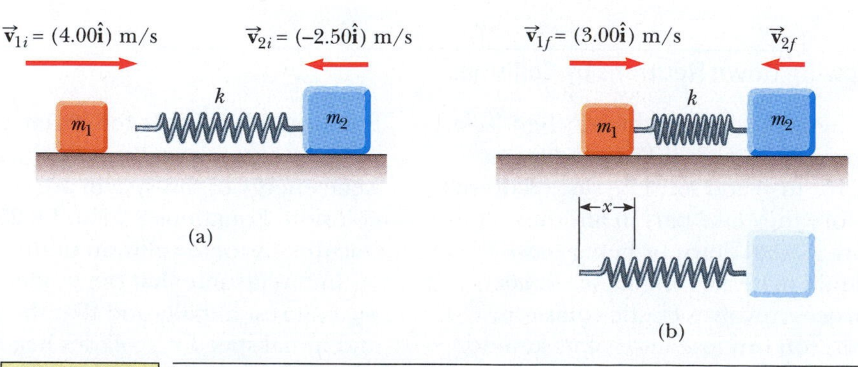


**FIGURE 8.10** (Interactive Example 8.8) A moving block collides with another moving block with a spring attached: (a) before the collision and (b) at one instant during the collision.

isolated system model to determine the compression $x$ in the spring shown in Figure 8.10b. No nonconservative forces are acting within the system, so the mechanical energy of the system is conserved:

$$E_i = E_f$$

$$\tfrac{1}{2}m_1v_{1i}^2 + \tfrac{1}{2}m_2v_{2i}^2 = \tfrac{1}{2}m_1v_{1f}^2 + \tfrac{1}{2}m_2v_{2f}^2 + \tfrac{1}{2}kx^2$$

Substituting the given values and the result to part A into this expression gives

$$x = \boxed{0.173 \text{ m}}$$

C Determine the maximum distance by which the spring is compressed during the collision.

**Solution** The maximum compression of the spring occurs when the two blocks are not moving relative to each other. For their relative velocity to be zero they must be moving with the same velocity in our reference frame as we watch the collision. Therefore, we can model the collision *up to this point* as a perfectly inelastic collision:

$$m_1v_{1i} + m_2v_{2i} = (m_1 + m_2)v_f$$

$$(1.60 \text{ kg})(4.00 \text{ m/s}) + (2.10 \text{ kg})(-2.50 \text{ m/s}) = (1.60 \text{ kg} + 2.10 \text{ kg})v_f$$

$$v_f = 0.311 \text{ m/s}$$

As in part B, mechanical energy is conserved, so we set up a conservation of mechanical energy expression:

$$E_i = E_f$$

$$\tfrac{1}{2}m_1v_{1i}^2 + \tfrac{1}{2}m_2v_{2i}^2 = \tfrac{1}{2}(m_1 + m_2)v_f^2 + \tfrac{1}{2}kx^2$$

Substituting the values into this expression gives

$$x = \boxed{0.253 \text{ m}}$$

To finalize the problem, note that the value for $x$ in part C is larger than that in part B. We can argue that this result is consistent with our mental representation. In part C, the blocks are moving at the same speed so that their relative speed is zero. In parts A and B, note that the blocks are moving with speeds $v_{1f} = 3.00$ m/s and $v_{2f} = -1.74$ m/s at the instant of interest. Therefore, the blocks are moving toward each other with a relative speed of 4.74 m/s. As a result, the spring will continue to compress, and the ultimate maximum value of $x$ will be larger than that value found in part B.

**PhysicsNow™** Investigate this situation with a variety of masses and initial speeds of the blocks by logging into PhysicsNow at **www.pop4e.com** and going to Interactive Example 8.8.

## 8.4 TWO-DIMENSIONAL COLLISIONS

In Section 8.1, we showed that the total momentum of a system is conserved when the system is isolated (i.e., when no external forces act on the system). For a general collision of two objects in three-dimensional space, the principle of conservation of momentum implies that the total momentum in each direction is conserved. An important subset of collisions takes place in a plane. The game of billiards is a familiar example involving multiple collisions of objects moving on a two-dimensional surface. Let us restrict our attention to a single two-dimensional collision between two objects that takes place in a plane. For such collisions, we obtain two component equations for the conservation of momentum:

$$m_1v_{1ix} + m_2v_{2ix} = m_1v_{1fx} + m_2v_{2fx}$$

$$m_1v_{1iy} + m_2v_{2iy} = m_1v_{1fy} + m_2v_{2fy}$$

where we use three subscripts in this general equation to represent, respectively, (1) the identification of the object, (2) initial and final values, and (3) the velocity component in the $x$ or $y$ direction.

Consider a two-dimensional problem in which an object of mass $m_1$ collides with an object of mass $m_2$ that is initially at rest as in Active Figure 8.11. After the collision, $m_1$ moves at an angle $\theta$ with respect to the horizontal and $m_2$ moves at an angle $\phi$ with respect to the horizontal. This collision is called a *glancing* collision. Applying the law of conservation of momentum in component form and noting that the initial $y$ component of momentum is zero, we have

$$x \text{ component:} \quad m_1v_{1i} + 0 = m_1v_{1f}\cos\theta + m_2v_{2f}\cos\phi \qquad [8.26]$$

$$y \text{ component:} \quad 0 + 0 = m_1v_{1f}\sin\theta - m_2v_{2f}\sin\phi \qquad [8.27]$$

**ACTIVE FIGURE 8.11**
A glancing collision between two particles.

**Physics Now™** Log into PhysicsNow at **www.pop4e.com** and go to Active Figure 8.11 to adjust the speed and position of the blue particle and the masses of both particles to see the effects.

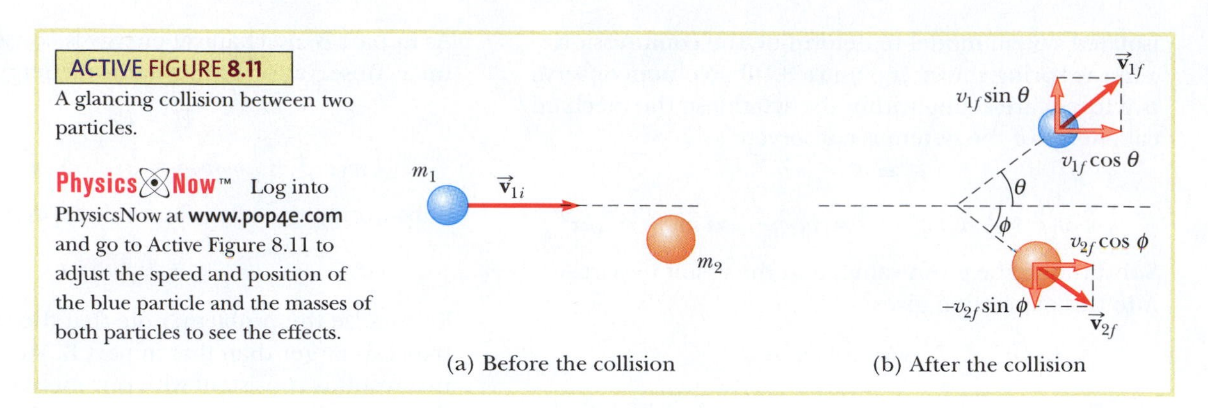

(a) Before the collision (b) After the collision

If the collision is elastic, we can write a third equation for conservation of kinetic energy in the form

$$\frac{1}{2}m_1 v_{1i}^2 = \frac{1}{2}m_1 v_{1f}^2 + \frac{1}{2}m_2 v_{2f}^2 \qquad [8.28]$$

If we know the initial velocity $v_{1i}$ and the masses, we are left with four unknowns ($v_{1f}$, $v_{2f}$, $\theta$, and $\phi$). Because we have only three equations, one of the four remaining quantities must be given to determine the motion after the collision from conservation principles alone.

If the collision is inelastic, kinetic energy is *not* conserved and Equation 8.28 does *not* apply.

## PROBLEM-SOLVING STRATEGY Two-Dimensional Collisions

The following procedure is recommended when dealing with problems involving collisions between two objects.

1. **Conceptualize** Imagine the collisions occurring in your mind and predict the approximate directions in which the particles will move after the collision. Set up a coordinate system and define your velocities with respect to that system. It is convenient to have the $x$ axis coincide with one of the initial velocities. Sketch the coordinate system, draw and label all velocity vectors, and include all the given information.
2. **Categorize** Is the system of particles truly isolated? If so, categorize the collision as elastic, inelastic, or perfectly inelastic.
3. **Analyze** Write expressions for the $x$ and $y$ components of the momentum of each object before and after the collision. Remember to include the appropriate signs for the components of the velocity vectors. It is essential that you pay careful attention to signs.

   Write expressions for the *total* momentum in the $x$ direction *before* and *after* the collision and equate the two. Repeat this procedure for the total momentum in the $y$ direction.

   Proceed to solve the momentum equations for the unknown quantities. If the collision is inelastic, kinetic energy is *not* conserved and additional information is probably required. If the collision is perfectly inelastic, the final velocities of the two objects are equal.

   If the collision is elastic, kinetic energy is conserved and you can equate the total kinetic energy before the collision to the total kinetic energy after the collision. This step provides an additional relationship between the velocity magnitudes.
4. **Finalize** Once you have determined your result, check to see that your answers are consistent with the mental and pictorial representations and that your results are realistic.

## EXAMPLE 8.9 Proton–Proton Collision

A proton collides elastically with another proton that is initially at rest. The incoming proton has an initial speed of $3.5 \times 10^5$ m/s and makes a glancing collision with the second proton as in Active Figure 8.11. (At close separations, the protons exert a repulsive electrostatic force on each other.) After the collision, one proton moves off at an angle of 37° to the original direction of motion and the second deflects at an angle of $\phi$ to the same axis. Find the final speeds of the two protons and the angle $\phi$.

**Solution** The isolated system is the pair of protons. Both momentum and kinetic energy of the system are conserved in this glancing elastic collision. Because

$m_1 = m_2$, $\theta = 37°$, and we are given that $v_{1i} = 3.5 \times 10^5$ m/s, Equations 8.26, 8.27, and 8.28 become

$$(1) \qquad v_{1f}\cos 37° + v_{2f}\cos\phi = 3.5 \times 10^5 \text{ m/s}$$

$$(2) \qquad v_{1f}\sin 37° - v_{2f}\sin\phi = 0$$

$$(3) \quad v_{1f}^2 + v_{2f}^2 = (3.5 \times 10^5 \text{ m/s})^2 = 1.2 \times 10^{11} \text{ m}^2/\text{s}^2$$

We rewrite (1) and (2) as follows:

$$v_{2f}\cos\phi = 3.5 \times 10^5 \text{ m/s} - v_{1f}\cos 37°$$

$$v_{2f}\sin\phi = v_{1f}\sin 37°$$

Now we square these two equations and add them:

$$\begin{aligned} v_{2f}^2\cos^2\phi + v_{2f}^2\sin^2\phi = {} & 1.2 \times 10^{11} \text{ m}^2/\text{s}^2 \\ & - (7.0 \times 10^5 \text{ m/s})v_{1f}\cos 37° \\ & + v_{1f}^2\cos^2 37° + v_{1f}^2\sin^2 37° \end{aligned}$$

$$\rightarrow \quad v_{2f}^2 = 1.2 \times 10^{11} - (5.6 \times 10^5)v_{1f} + v_{1f}^2$$

Substituting this expression into (3) gives

$$v_{1f}^2 + [1.2 \times 10^{11} - (5.6 \times 10^5)v_{1f} + v_{1f}^2] = 1.2 \times 10^{11}$$

$$\rightarrow \quad 2v_{1f}^2 - (5.6 \times 10^5)v_{1f} = (2v_{1f} - 5.6 \times 10^5)v_{1f} = 0$$

One possibility for the solution of this equation is $v_{1f} = 0$, which corresponds to a head-on collision; the first proton stops and the second continues with the same speed in the same direction. This result is not what we want. The other possibility is

$$2v_{1f} - 5.6 \times 10^5 = 0 \quad \rightarrow \quad v_{1f} = \boxed{2.8 \times 10^5 \text{ m/s}}$$

From (3),

$$v_{2f} = \sqrt{1.2 \times 10^{11} - v_{1f}^2} = \sqrt{1.2 \times 10^{11} - (2.8 \times 10^5)^2}$$

$$= \boxed{2.1 \times 10^5 \text{ m/s}}$$

and from (2),

$$\phi = \sin^{-1}\left(\frac{v_{1f}\sin 37°}{v_{2f}}\right) = \sin^{-1}\left(\frac{(2.8 \times 10^5)\sin 37°}{2.1 \times 10^5}\right)$$

$$= \boxed{53°}$$

It is interesting that $\theta + \phi = 90°$. This result is *not* accidental. Whenever two objects of equal mass collide elastically in a glancing collision and one of them is initially at rest, their final velocities are at right angles to each other.

## EXAMPLE 8.10 Collision at an Intersection

A 1 500-kg car traveling east with a speed of 25.0 m/s collides at an intersection with a 2 500-kg van traveling north at a speed of 20.0 m/s as shown in Figure 8.12. Find the direction and magnitude of the velocity of the wreckage after the collision, assuming that the vehicles undergo a perfectly inelastic collision (i.e., they stick together).

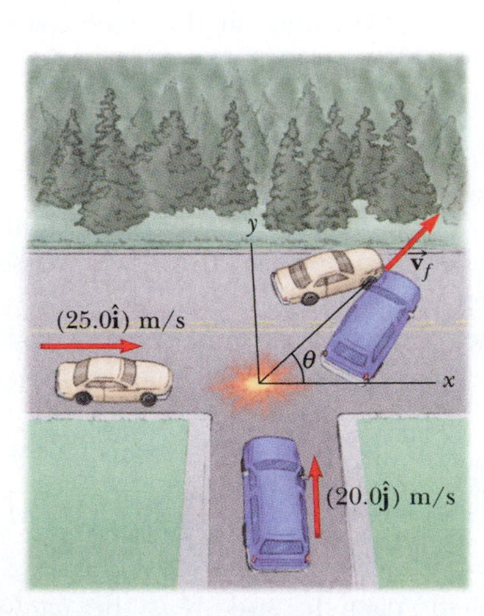


**FIGURE 8.12** (Example 8.10) An eastbound car colliding with a northbound van.

**Solution** Let us choose east to be along the positive $x$ direction and north to be along the positive $y$ direction as in Figure 8.12. Before the collision, the only object having momentum in the $x$ direction is the car. Therefore, the magnitude of the total initial momentum of the system (car plus van) in the $x$ direction is

$$\sum p_{xi} = (1\,500 \text{ kg})(25.0 \text{ m/s}) = 3.75 \times 10^4 \text{ kg}\cdot\text{m/s}$$

The wreckage moves at an angle $\theta$ and speed $v_f$ after the collision. The magnitude of the total momentum in the $x$ direction after the collision is

$$\sum p_{xf} = (4\,000 \text{ kg})v_f\cos\theta$$

Because the total momentum in the $x$ direction is conserved, we can equate these two equations to obtain

$$(1) \quad 3.75 \times 10^4 \text{ kg}\cdot\text{m/s} = (4\,000 \text{ kg})\ v_f\cos\theta$$

Similarly, the total initial momentum of the system in the $y$ direction is that of the van, whose magnitude is equal to (2 500 kg)(20.0 m/s). Applying conservation of momentum to the $y$ direction, we have

$$\sum p_{yi} = \sum p_{yf}$$

$$(2\,500 \text{ kg})(20.0 \text{ m/s}) = (4\,000 \text{ kg})v_f\sin\theta$$

$$(2) \qquad 5.00 \times 10^4 \text{ kg}\cdot\text{m/s} = (4\,000 \text{ kg})v_f\sin\theta$$

If we divide (2) by (1), we find that

$$\tan\theta = \frac{5.00 \times 10^4}{3.75 \times 10^4} = 1.33$$

$$\theta = 53.1°$$

When this angle is substituted into (2), the value of $v_f$ is

$$v_f = \frac{5.00 \times 10^4 \text{ kg}\cdot\text{m/s}}{(4\,000 \text{ kg}) \sin 53.1°} = 15.6 \text{ m/s}$$

## 8.5 THE CENTER OF MASS

In this section, we describe the overall motion of a system of particles in terms of a very special point called the **center of mass** of the system. This notion gives us confidence in the particle model because we will see that the center of mass accelerates as if all the system's mass were concentrated at that point and all external forces act there.

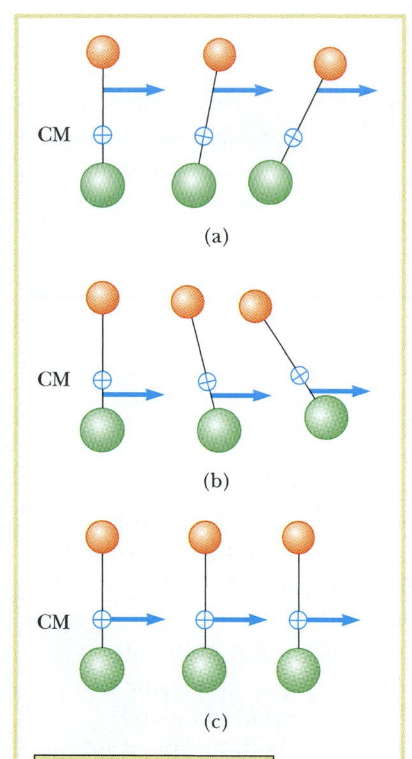


ACTIVE FIGURE 8.13
Two particles of unequal mass are connected by a light, rigid rod. (a) The system rotates clockwise when a force is applied between the less massive particle and the center of mass. (b) The system rotates counterclockwise when a force is applied between the more massive particle and the center of mass. (c) The system moves in the direction of the force without rotating when a force is applied at the center of mass.

**Physics Now™** Log into PhysicsNow at **www.pop4e.com** and go to Active Figure 8.13 to choose the point at which to apply the force.

Consider a system consisting of a pair of particles connected by a light, rigid rod (Active Fig. 8.13). The center of mass as indicated in the figure is located on the rod and is closer to the larger mass in the figure; we will see why soon. If a single force is applied at some point on the rod that is above the center of mass, the system rotates clockwise (Active Fig. 8.13a) as it translates through space. If the force is applied at a point on the rod below the center of mass, the system rotates counterclockwise (Active Fig. 8.13b). If the force is applied exactly at the center of mass, the system moves in the direction of $\vec{\mathbf{F}}$ without rotating (Active Fig. 8.13c) as if the system is behaving as a particle. Therefore, in theory, the center of mass can be located with this experiment.

If we were to analyze the motion in Active Figure 8.13c, we would find that the system moves as if all its mass were concentrated at the center of mass. Furthermore, if the external net force on the system is $\sum\vec{\mathbf{F}}$ and the total mass of the system is $M$, the center of mass moves with an acceleration given by $\vec{\mathbf{a}} = \sum\vec{\mathbf{F}}/M$. That is, the system moves as if the resultant external force were applied to a single particle of mass $M$ located at the center of mass, which justifies our particle model for extended objects. We have ignored all rotational effects for extended objects so far, implicitly assuming that forces were provided at just the right position so as to cause no rotation. We will study rotational motion in Chapter 10, where we will apply forces that do not pass through the center of mass.

The position of the center of mass of a system can be described as being the *average position* of the system's mass. For example, the center of mass of the pair of particles described in Active Figure 8.14 is located on the $x$ axis, somewhere between the particles. The $x$ coordinate of the center of mass in this case is

$$x_{CM} = \frac{m_1x_1 + m_2x_2}{m_1 + m_2} \qquad [8.29]$$

For example, if $x_1 = 0$, $x_2 = d$, and $m_2 = 2m_1$, we find that $x_{CM} = \frac{2}{3}d$. That is, the center of mass lies closer to the more massive particle. If the two masses are equal, the center of mass lies midway between the particles.

We can extend the concept of center of mass to a system of many particles in three dimensions. The $x$ coordinate of the center of mass of $n$ particles is defined to be

$$x_{CM} \equiv \frac{m_1x_1 + m_2x_2 + m_3x_3 + \cdots + m_nx_n}{m_1 + m_2 + m_3 + \cdots + m_n} = \frac{\sum_i m_ix_i}{\sum_i m_i} = \frac{\sum_i m_ix_i}{M} \qquad [8.30]$$

where $x_i$ is the $x$ coordinate of the $i$th particle and $M$ is the *total mass* of the system. The $y$ and $z$ coordinates of the center of mass are similarly defined by the equations

$$y_{CM} \equiv \frac{\sum_i m_i y_i}{M} \quad \text{and} \quad z_{CM} \equiv \frac{\sum_i m_i z_i}{M} \qquad [8.31]$$

The center of mass can also be located by its position vector, $\vec{\mathbf{r}}_{CM}$. The rectangular coordinates of this vector are $x_{CM}$, $y_{CM}$, and $z_{CM}$, defined in Equations 8.30 and 8.31. Therefore,

$$\vec{\mathbf{r}}_{CM} = x_{CM}\hat{\mathbf{i}} + y_{CM}\hat{\mathbf{j}} + z_{CM}\hat{\mathbf{k}} = \frac{\sum_i m_i x_i \hat{\mathbf{i}} + \sum_i m_i y_i \hat{\mathbf{j}} + \sum_i m_i z_i \hat{\mathbf{k}}}{M}$$

$$\vec{\mathbf{r}}_{CM} = \frac{\sum_i m_i \vec{\mathbf{r}}_i}{M} \qquad [8.32]$$

where $\vec{\mathbf{r}}_i$ is the position vector of the $i$th particle, defined by

$$\vec{\mathbf{r}}_i \equiv x_i\hat{\mathbf{i}} + y_i\hat{\mathbf{j}} + z_i\hat{\mathbf{k}}$$

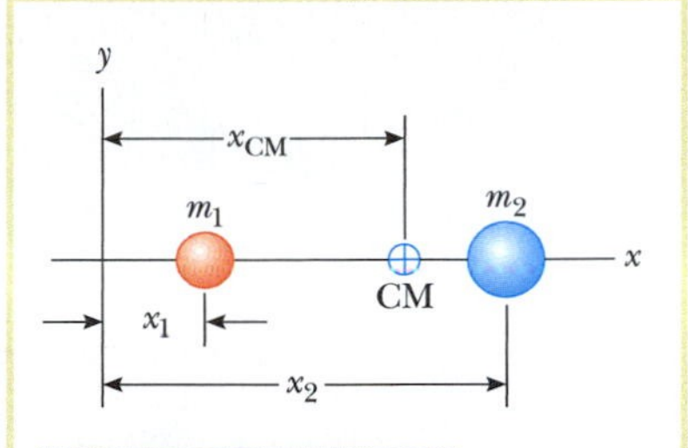


ACTIVE FIGURE 8.14
The center of mass of two particles having unequal mass is located on the $x$ axis at $x_{CM}$, a point between the particles, closer to the one having the larger mass.

**PhysicsNow™** Log into PhysicsNow at **www.pop4e.com** and go to Active Figure 8.14 to adjust the masses of the particles and see the effect on the location of the center of mass.

Although locating the center of mass for an extended object is somewhat more cumbersome than locating the center of mass of a system of particles, this location is based on the same fundamental ideas. We can model the extended object as a system containing a large number of elements (Fig. 8.15). Each element is modeled as a particle of mass $\Delta m_i$, with coordinates $x_i$, $y_i$, $z_i$. The particle separation is very small, so this model is a good representation of the continuous mass distribution of the object. The $x$ coordinate of the center of mass of the particles representing the object, and therefore of the approximate center of mass of the object, is

$$x_{CM} \approx \frac{\sum_i x_i \Delta m_i}{M}$$

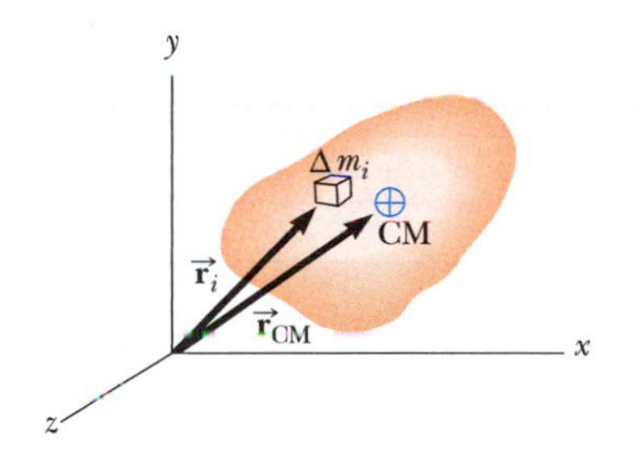


FIGURE 8.15 An extended object can be modeled as a distribution of small elements of mass $\Delta m_i$. The center of mass of the object is located at the vector position $\vec{\mathbf{r}}_{CM}$, which has coordinates $x_{CM}$, $y_{CM}$, and $z_{CM}$.

with similar expressions for $y_{CM}$ and $z_{CM}$. If we let the number of elements approach infinity (and, as a consequence, the size and mass of each element approach zero), the model becomes indistinguishable from the continuous mass distribution and $x_{CM}$ is given precisely. In this limit, we replace the sum by an integral and $\Delta m_i$ by the differential element $dm$:

$$x_{CM} = \lim_{\Delta m_i \to 0} \frac{\sum_i x_i \Delta m_i}{M} = \frac{1}{M}\int x\, dm \qquad [8.33]$$

where the integration is over the length of the object in the $x$ direction. Likewise, for $y_{CM}$ and $z_{CM}$ we obtain

$$y_{CM} = \frac{1}{M}\int y\, dm \quad \text{and} \quad z_{CM} = \frac{1}{M}\int z\, dm \qquad [8.34]$$

We can express the vector position of the center of mass of an extended object as

$$\vec{\mathbf{r}}_{CM} = \frac{1}{M}\int \vec{\mathbf{r}}\, dm \qquad [8.35]$$

■ Center of mass of a continuous mass distribution

which is equivalent to the three expressions in Equations 8.33 and 8.34.

**The center of mass of a homogeneous, symmetric object must lie on an axis of symmetry.** For example, the center of mass of a homogeneous rod must lie midway between the ends of the rod. The center of mass of a homogeneous sphere or a homogeneous cube must lie at the geometric center of the object.

The center of mass of a system is often confused with the **center of gravity** of a system. Each portion of a system is acted on by the gravitational force. The net effect of all these forces is equivalent to the effect of a single force $M\vec{\mathbf{g}}$ acting at a

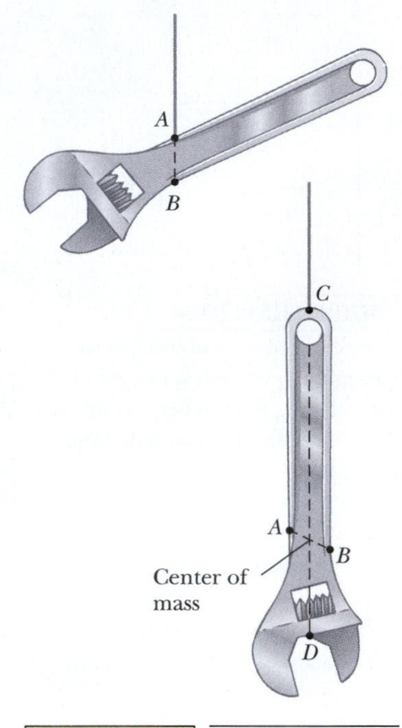


FIGURE 8.16 An experimental technique for determining the center of mass of a wrench. The wrench is hung freely from two different pivots, $A$ and $C$. The intersection of the two vertical lines $AB$ and $CD$ locates the center of mass.

special point called the center of gravity. The center of gravity is the average position of the gravitational forces on all parts of the object. If $\vec{\mathbf{g}}$ is uniform over the system, the center of gravity coincides with the center of mass. If the gravitational field over the system is not uniform, the center of gravity and the center of mass are different. In most cases, for objects or systems of reasonable size, the two points can be considered to be coincident.

One can experimentally determine the center of gravity of an irregularly shaped object, such as a wrench, by suspending the wrench from two different points (Fig. 8.16). An object of this size has virtually no variation in the gravitational field over its dimensions, so this method also locates the center of mass. The wrench is first hung from point $A$, and a vertical line $AB$ is drawn (which can be established with a plumb bob) when the wrench is in equilibrium. The wrench is then hung from point $C$, and a second vertical line $CD$ is drawn. The center of mass coincides with the intersection of these two lines. In fact, if the wrench is hung freely from any point, the vertical line through that point will pass through the center of mass.

**QUICK QUIZ 8.5** A baseball bat is made of wood of uniform density. The bat is cut at the location of its center of mass as shown in Figure 8.17. Which piece has the smaller mass? **(a)** the piece on the right **(b)** the piece on the left **(c)** both pieces have the same mass **(d)** impossible to determine

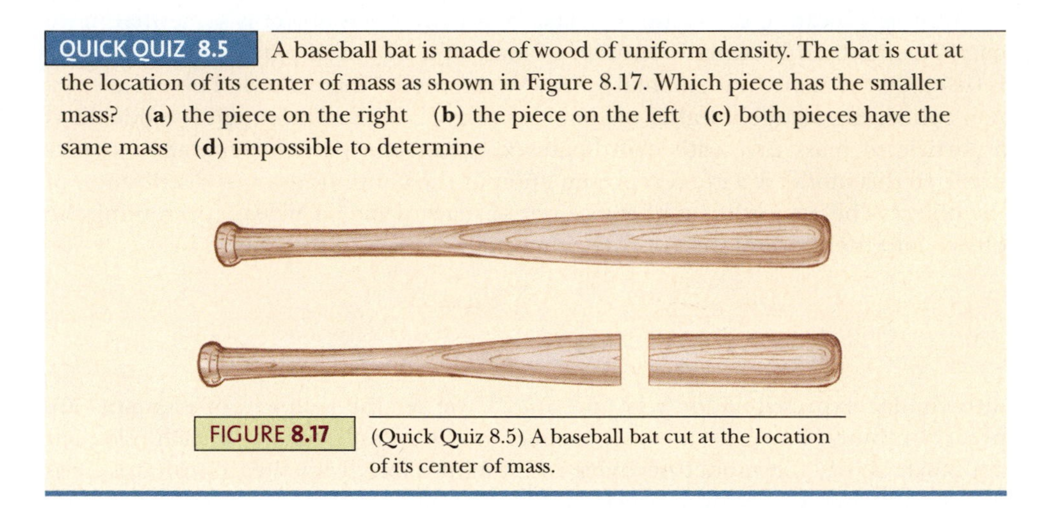

FIGURE 8.17 (Quick Quiz 8.5) A baseball bat cut at the location of its center of mass.

## EXAMPLE 8.11 The Center of Mass of Three Particles

A system consists of three particles located at the corners of a right triangle as in Figure 8.18. Find the center of mass of the system.

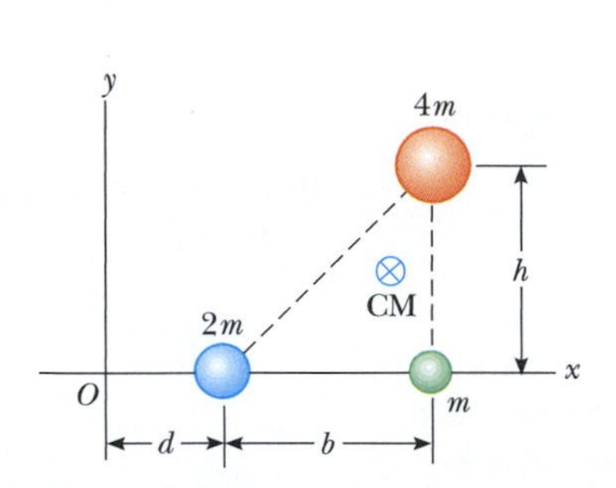


FIGURE 8.18 (Example 8.11) Locating the center of mass for a system of three particles.

**Solution** Using the basic defining equations for the coordinates of the center of mass and noting that $z_{\text{CM}} = 0$, we have

$$x_{\text{CM}} = \frac{\sum_i m_i x_i}{M} = \frac{2md + m(d+b) + 4m(d+b)}{7m}$$
$$= d + \tfrac{5}{7}b$$

$$y_{\text{CM}} = \frac{\sum_i m_i y_i}{M} = \frac{2m(0) + m(0) + 4mh}{7m} = \tfrac{4}{7}h$$

Therefore, we can express the position vector to the center of mass measured from the origin as

$$\vec{\mathbf{r}}_{\text{CM}} = x_{\text{CM}}\hat{\mathbf{i}} + y_{\text{CM}}\hat{\mathbf{j}} + z_{\text{CM}}\hat{\mathbf{k}} = (d + \tfrac{5}{7}b)\hat{\mathbf{i}} + \tfrac{4}{7}h\hat{\mathbf{j}}$$

**EXAMPLE 8.12** **The Center of Mass of a Right Triangle**

You have been asked to hang a metal sign from a single vertical wire. The sign is of the triangular shape shown in Figure 8.19a. The bottom of the sign is to be parallel to the ground. At what distance from the left end of the sign should you attach the wire?

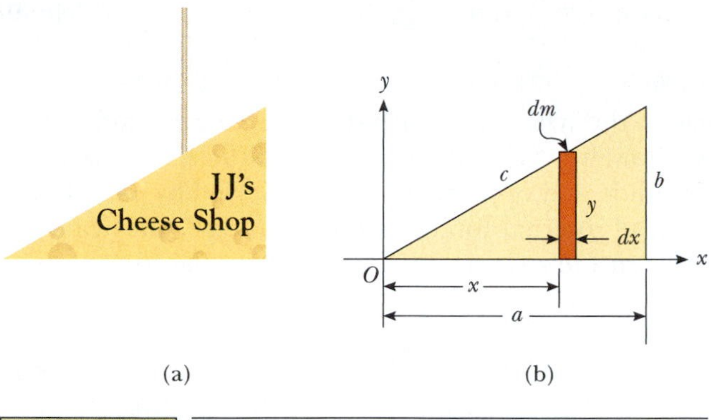


**FIGURE 8.19** (Example 8.12) (a) A triangular sign to be hung from a single wire. (b) Geometric construction for locating the center of mass.

**Solution** We will need to attach the wire at a point directly above the center of gravity of the sign, which is the same as its center of mass because it is in a uniform gravitational field. We model the sign as a perfect triangle. We assume that the sign has a uniform density and total mass $M$. Because the sign is a continuous distribution of mass, we will need to use the integral expression in Equation 8.33 to find the $x$ coordinate of the center of mass.

We divide the triangle into narrow strips of width $dx$ and height $y$ as shown in Figure 8.19b, where $y$ is the height to the hypotenuse of the triangle above the $x$ axis for a given value of $x$. The mass of each strip is the product of the volume of the strip and the density $\rho$ of the material from which the sign is made: $dm = \rho yt\, dx$, where $t$ is the thickness of the metal sign. The density of the material is the total mass of the sign divided by its total volume (area of the triangle times thickness), so

$$dm = \rho yt\, dx = \left(\frac{M}{\frac{1}{2}abt}\right) yt\, dx = \frac{2My}{ab}\, dx$$

Using Equation 8.33 to find the $x$ coordinate of the center of mass gives

$$x_{\text{CM}} = \frac{1}{M}\int x\, dm = \frac{1}{M}\int_0^a x\, \frac{2My}{ab}\, dx = \frac{2}{ab}\int_0^a xy\, dx$$

To proceed further and evaluate the integral, we must express $y$ in terms of $x$. The line representing the hypotenuse of the triangle in Figure 8.19b has a slope of $b/a$ and passes through the origin, so the equation of this line is $y = (b/a)x$. With this substitution for $y$ in the integral, we have

$$x_{\text{CM}} = \frac{2}{ab}\int_0^a x\left(\frac{b}{a}x\right) dx = \frac{2}{a^2}\int_0^a x^2\, dx = \frac{2}{a^2}\left[\frac{x^3}{3}\right]_0^a$$
$$= \frac{2}{3}a$$

Therefore, the wire must be attached to the sign at a distance two thirds of the length of the bottom edge from the left end. We could also find the $y$ coordinate of the center of mass of the sign, but that is not needed to determine where the wire should be attached.

## 8.6 MOTION OF A SYSTEM OF PARTICLES

We can begin to understand the physical significance and utility of the center of mass concept by taking the time derivative of the position vector $\vec{\mathbf{r}}_{\text{CM}}$ of the center of mass, given by Equation 8.32. Assuming that $M$ remains constant—that is, no particles enter or leave the system—we find the following expression for the **velocity of the center of mass:**

Velocity of the center of mass for a system of particles

$$\vec{\mathbf{v}}_{\text{CM}} = \frac{d\vec{\mathbf{r}}_{\text{CM}}}{dt} = \frac{1}{M}\sum_i m_i \frac{d\vec{\mathbf{r}}_i}{dt} = \frac{1}{M}\sum_i m_i \vec{\mathbf{v}}_i \qquad [8.36]$$

where $\vec{\mathbf{v}}_i$ is the velocity of the $i$th particle. Rearranging Equation 8.36 gives

$$M\vec{\mathbf{v}}_{\text{CM}} = \sum_i m_i \vec{\mathbf{v}}_i = \sum_i \vec{\mathbf{p}}_i = \vec{\mathbf{p}}_{\text{tot}} \qquad [8.37]$$

This result tells us that **the total momentum of the system equals its total mass multiplied by the velocity of its center of mass.** In other words, the total momentum of the system is equal to the momentum of a single particle of mass $M$ moving with a velocity $\vec{\mathbf{v}}_{\text{CM}}$; this is the particle model.

If we now differentiate Equation 8.36 with respect to time, we find the **acceleration of the center of mass:**

■ Acceleration of the center of mass for a system of particles

$$\vec{\mathbf{a}}_{\mathrm{CM}} = \frac{d\vec{\mathbf{v}}_{\mathrm{CM}}}{dt} = \frac{1}{M}\sum_i m_i \frac{d\vec{\mathbf{v}}_i}{dt} = \frac{1}{M}\sum_i m_i \vec{\mathbf{a}}_i \qquad [8.38]$$

Rearranging this expression and using Newton's second law, we have

$$M\vec{\mathbf{a}}_{\mathrm{CM}} = \sum_i m_i \vec{\mathbf{a}}_i = \sum_i \vec{\mathbf{F}}_i \qquad [8.39]$$

where $\vec{\mathbf{F}}_i$ is the force on particle $i$.

The forces on any particle in the system may include both external and internal forces. By Newton's third law, however, the force exerted by particle 1 on particle 2, for example, is equal in magnitude and opposite the force exerted by particle 2 on particle 1. When we sum over all internal forces in Equation 8.39, they cancel in pairs. Therefore, the net force on the system is due *only* to external forces and we can write Equation 8.39 in the form

■ Newton's second law for a system of particles

$$\sum \vec{\mathbf{F}}_{\mathrm{ext}} = M\vec{\mathbf{a}}_{\mathrm{CM}} = \frac{d\vec{\mathbf{p}}_{\mathrm{tot}}}{dt} \qquad [8.40]$$

That is, the external net force on the system of particles equals the total mass of the system multiplied by the acceleration of the center of mass, or the time rate of change of the momentum of the system. If we compare this statement to Newton's second law for a single particle, we see that the center of mass moves like an imaginary particle of mass $M$ under the influence of the external net force on the system. In the absence of external forces, the center of mass moves with uniform velocity as in the case of the translating and rotating wrench in Figure 8.20. If the net force acts along a line through the center of mass of an extended object such as the wrench, the object is accelerated without rotation. If the net force does not act through the center of mass, the object will undergo rotation in addition to translation. The linear acceleration of the center of mass is the same in either case, as given by Equation 8.40.

Finally, we see that if the external net force is zero, from Equation 8.40 it follows that

$$\frac{d\vec{\mathbf{p}}_{\mathrm{tot}}}{dt} = M\vec{\mathbf{a}}_{\mathrm{CM}} = 0$$

so that

$$\vec{\mathbf{p}}_{\mathrm{tot}} = M\vec{\mathbf{v}}_{\mathrm{CM}} = \text{constant} \qquad \left(\text{when } \sum \vec{\mathbf{F}}_{\mathrm{ext}} = 0\right) \qquad [8.41]$$

That is, the total linear momentum of a system of particles is constant if no external forces act on the system. It follows that, for an *isolated* system of particles, the total momentum is conserved. The law of conservation of momentum that was derived in Section 8.1 for a two-particle system is thus generalized to a many-particle system.

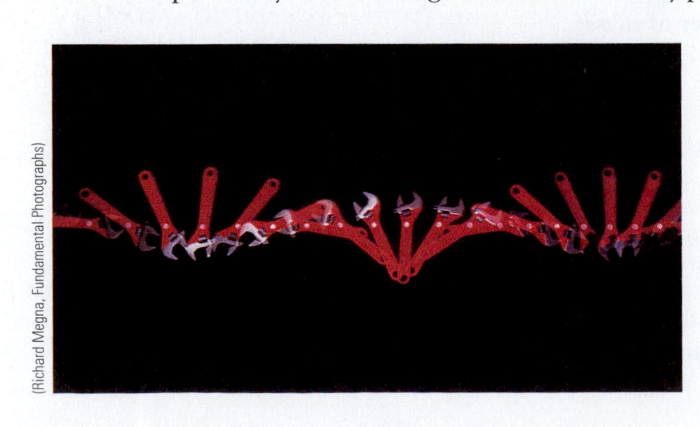

(Richard Megna, Fundamental Photographs)

**FIGURE 8.20** Strobe photograph showing an overhead view of a wrench moving on a horizontal surface. The center of mass of the wrench (marked with a white dot) moves in a straight line as the wrench rotates about this point. The wrench moves from left to right in the photograph and is slowing down due to friction between the wrench and the supporting surface. (*Note* The decreasing distance between the white dots.)

FIGURE 8.21 (Thinking Physics 8.1) A boy takes a step in a canoe. What happens to the canoe?

## Thinking Physics 8.1

A boy stands at one end of a canoe that is stationary relative to the shore (Fig. 8.21). He then walks to the opposite end of the canoe, away from the shore. Does the canoe move?

**Reasoning** Yes, the canoe moves toward the shore. Ignoring friction between the canoe and water, no horizontal force acts on the system consisting of the boy and canoe. The center of mass of the system therefore remains fixed relative to the shore (or any stationary point). As the boy moves away from the shore, the canoe must move toward the shore such that the center of mass of the system remains fixed in position. ■

**QUICK QUIZ 8.6** The vacationers on a cruise ship are eager to arrive at their next destination. They decide to try to speed up the cruise ship by gathering at the bow (the front) and running all at once toward the stern (the back) of the ship. **(i)** While they are running toward the stern, what is the speed of the ship? **(a)** higher than it was before **(b)** unchanged **(c)** lower than it was before **(d)** impossible to determine **(ii)** The vacationers stop running when they reach the stern of the ship. After they have all stopped running, what is the speed of the ship? **(a)** higher than it was before they started running **(b)** unchanged from what it was before they started running **(c)** lower than it was before they started running **(d)** impossible to determine

### EXAMPLE 8.13 An Exploding Projectile

A projectile is fired into the air and suddenly explodes into several fragments (Fig. 8.22). What can be said about the motion of the center of mass of the system made up of all the fragments after the explosion?

**Solution** Neglecting air resistance, the only external force on the projectile is the gravitational force. Therefore, if the projectile did not explode, it would continue to move along the parabolic path indicated by the dashed line in Figure 8.22. Because the forces caused by the explosion are internal, they do not affect the motion of the center of mass of the system (the fragments). Therefore, after the explosion, the center of mass follows the same parabolic path the projectile would have followed if there had been no explosion.

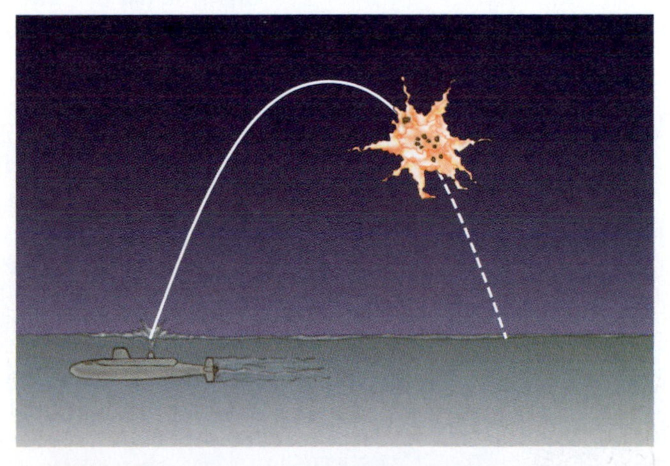

FIGURE 8.22 (Example 8.13) When a projectile explodes into several fragments, where does the center of mass of the fragments land?

## 8.7 ROCKET PROPULSION

CONTEXT CONNECTION

On our trip to Mars, we will need to control our spacecraft by firing the rocket engines. When ordinary vehicles, such as the automobiles in Context 1, are propelled, the driving force for the motion is the friction force exerted by the road on the car. A rocket moving in space, however, has no road to "push" against. The source of the propulsion of a rocket must therefore be different. **The operation of a rocket depends on the law of conservation of momentum as applied to a system, where the system is the rocket plus its ejected fuel.**

The propulsion of a rocket can be understood by first considering the archer on ice in Interactive Example 8.2. As an arrow is fired from the bow, the arrow receives momentum $m\vec{\mathbf{v}}$ in one direction and the archer receives a momentum of equal magnitude in the opposite direction. As additional arrows are fired, the archer moves faster, so a large velocity of the archer can be established by firing many arrows.

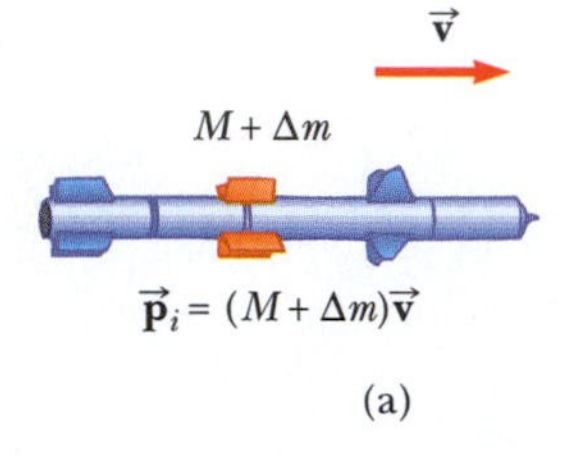


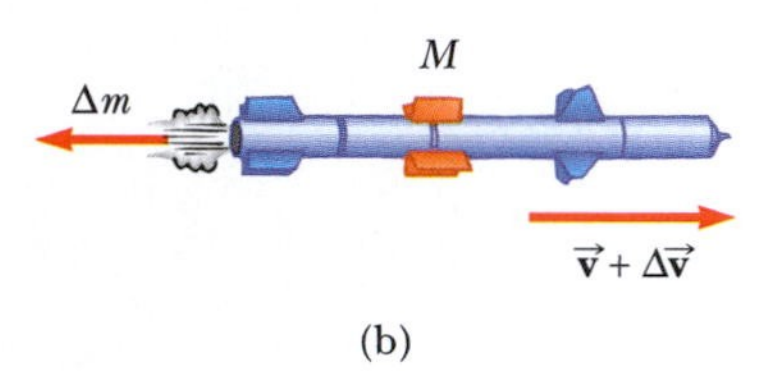


FIGURE 8.23 Rocket propulsion. (a) The initial mass of the rocket and fuel is $M + \Delta m$ at a time $t$, and its speed is $v$. (b) At a time $t + \Delta t$, the rocket's mass has been reduced to $M$, and an amount of fuel $\Delta m$ has been ejected. The rocket's speed increases by an amount $\Delta v$.

In a similar manner, as a rocket moves in free space (a vacuum), its momentum changes when some of its mass is released in the form of ejected gases. Because the ejected gases acquire some momentum, the rocket receives a compensating momentum in the opposite direction. The rocket therefore is accelerated as a result of the "push," or thrust, from the exhaust gases. Note that the rocket represents the *inverse* of an inelastic collision; that is, momentum is conserved, but the kinetic energy of the system is *increased* (at the expense of energy stored in the fuel of the rocket).

Suppose at some time $t$ the magnitude of the momentum of the rocket plus the fuel is $(M + \Delta m)v$ (Fig. 8.23a). During a short time interval $\Delta t$, the rocket ejects fuel of mass $\Delta m$ and the rocket's speed therefore increases to $v + \Delta v$ (Fig. 8.22b). If the fuel is ejected with velocity $\vec{\mathbf{v}}_e$ *relative to the rocket,* the speed of the fuel relative to a stationary frame of reference is $v - v_e$ according to our discussion of relative velocity in Section 3.6. Therefore, if we equate the total initial momentum of the system with the total final momentum, we have

$$(M + \Delta m)v = M(v + \Delta v) + \Delta m(v - v_e)$$

Simplifying this expression gives

$$M\,\Delta v = \Delta m(v_e)$$

If we now take the limit as $\Delta t$ goes to zero, $\Delta v \rightarrow dv$ and $\Delta m \rightarrow dm$. Furthermore, the increase $dm$ in the exhaust mass corresponds to an equal decrease in the rocket mass, so $dm = -dM$. Note that the negative sign is introduced into the equation because $dM$ represents a decrease in mass. Using this fact, we have

$$M\,dv = -v_e\,dM \qquad [8.42]$$

Integrating this equation and taking the initial mass of the rocket plus fuel to be $M_i$ and the final mass of the rocket plus its remaining fuel to be $M_f$, we have

$$\int_{v_i}^{v_f} dv = -v_e \int_{M_i}^{M_f} \frac{dM}{M}$$

■ Velocity change in rocket propulsion

$$v_f - v_i = v_e \ln\left(\frac{M_i}{M_f}\right) \qquad [8.43]$$

which is the basic expression for rocket propulsion. It tells us that the increase in speed is proportional to the exhaust speed $v_e$. The exhaust speed should therefore be very high.

The **thrust** on the rocket is the force exerted on the rocket by the ejected exhaust gases. We can obtain an expression for the instantaneous thrust from Equation 8.42:

■ Rocket thrust

$$\text{Instantaneous thrust} = Ma = M\frac{dv}{dt} = \left|v_e \frac{dM}{dt}\right| \qquad [8.44]$$

Here we see that the thrust increases as the exhaust speed increases and as the rate of change of mass (burn rate) increases.

We can now determine the amount of fuel needed to set us on our journey to Mars. The fuel requirements are well within the capabilities of current technology, as evidenced by the several missions to Mars that have already been accomplished. What if we wanted to visit another *star,* however, rather than another *planet?* This question raises many new technological challenges, including the requirement to consider the effects of relativity, which we investigate in the next chapter.

## ■ Thinking Physics 8.2

When Robert Goddard proposed the possibility of rocket-propelled vehicles, the *New York Times* agreed that such vehicles would be useful and successful within the Earth's atmosphere ("Topics of the Times," *New York Times,* January 13, 1920, p. 12). The *Times,* however, balked at the idea of using such a rocket in the vacuum of space, noting that "its flight would be neither accelerated nor maintained by the explosion of the charges it then might have left. To claim that it would be is to deny a fundamental law of dynamics, and only Dr. Einstein and his chosen dozen, so few and fit, are licensed to do that. . . . That Professor Goddard, with his 'chair' in Clark College and the countenancing of the Smithsonian Institution, does not know the relation of action to reaction, and of the need to have something better than a vacuum against which to react—to say that would be absurd. Of course, he only seems to lack the knowledge ladled out daily in high schools." What did the writer of this passage overlook?

**Reasoning** The writer of this passage was making a common mistake in believing that a rocket works by expelling gases that push on something, propelling the rocket forward. With this belief, it is impossible to see how a rocket fired in empty space would work.

Gases do not need to push on anything; it is the act itself of expelling the gases that pushes the rocket forward. This point can be argued from Newton's third law: The rocket pushes the gases backward, resulting in the gases pushing the rocket forward. It can also be argued from conservation of momentum: As the gases gain momentum in one direction, the rocket must gain momentum in the opposite direction to conserve the original momentum of the rocket–gas system.

The *New York Times* did publish a retraction 49 years later ("A Correction," *New York Times,* July 17, 1969, p. 43) while the *Apollo 11* astronauts were on their way to the Moon. It appeared on a page with two other articles entitled "Fundamentals of Space Travel" and "Spacecraft, Like Squid, Maneuver by 'Squirts'" and contained the following passages: "an editorial feature of the *New York Times* dismissed the notion that a rocket could function in a vacuum and commented on the ideas of Robert H. Goddard. . . . Further investigation and experimentation have confirmed the findings of Isaac Newton in the 17th century, and it is now definitely established that a rocket can function in a vacuum as well as in an atmosphere. The *Times* regrets the error." ■

### EXAMPLE 8.14 A Rocket in Space

A rocket in free space has a speed of $3.0 \times 10^3$ m/s relative to the Earth. Its engines are turned on, and fuel is ejected in a direction opposite the rocket's motion at a speed of $5.0 \times 10^3$ m/s relative to the rocket.

**A** What is the speed of the rocket relative to the Earth once its mass is reduced to one half its mass before ignition?

**Solution** Applying Equation 8.43, we have

$$v_f = v_i + v_e \ln\left(\frac{M_i}{M_f}\right)$$

$$= 3.0 \times 10^3 \text{ m/s} + (5.0 \times 10^3 \text{ m/s}) \ln\left(\frac{M_i}{0.5M_i}\right)$$

$$= 6.5 \times 10^3 \text{ m/s}$$

B What is the thrust on the rocket if it burns fuel at the rate of 50 kg/s?

**Solution** Using Equation 8.44, we have

$$\text{Thrust} = \left| v_e \frac{dM}{dt} \right| = (5.0 \times 10^3 \text{ m/s})(50 \text{ kg/s})$$

$$= 2.5 \times 10^5 \text{ N}$$

## SUMMARY

 Take a practice test by logging into PhysicsNow at www.pop4e.com and clicking on the Pre-Test link for this chapter.

The linear momentum of any object of mass $m$ moving with a velocity $\vec{\mathbf{v}}$ is

$$\vec{\mathbf{p}} \equiv m\vec{\mathbf{v}} \quad [8.2]$$

**Conservation of linear momentum** applied to two interacting objects states that if the two objects form an isolated system, the total momentum of the system at all times equals its initial total momentum:

$$\vec{\mathbf{p}}_{1i} + \vec{\mathbf{p}}_{2i} = \vec{\mathbf{p}}_{1f} + \vec{\mathbf{p}}_{2f} \quad [8.6]$$

The **impulse** of a net force $\Sigma\vec{\mathbf{F}}$ is defined as the integral of the force over the time interval during which it acts. The total impulse on any system is equal to the change in the momentum of the system and is given by

$$\vec{\mathbf{I}} = \int_{t_i}^{t_f} \sum \vec{\mathbf{F}}_{\text{ext}}\, dt = \Delta\vec{\mathbf{p}}_{\text{tot}} \quad [8.11]$$

This is known as the **impulse–momentum theorem.**

When two objects collide, the total momentum of the isolated system before the collision always equals the total momentum after the collision, regardless of the nature of the collision. An **inelastic collision** is one in which kinetic energy is not conserved. A **perfectly inelastic collision** is one in which the colliding objects stick together after the collision. An **elastic collision** is one in which both momentum and kinetic energy are conserved.

In a two- or three-dimensional collision, the components of momentum in each of the directions are conserved independently.

The vector position of the center of mass of a system of particles is defined as

$$\vec{\mathbf{r}}_{\text{CM}} = \frac{\sum_i m_i \vec{\mathbf{r}}_i}{M} \quad [8.32]$$

where $M$ is the total mass of the system and $\vec{\mathbf{r}}_i$ is the position vector of the $i$th particle.

The **velocity of the center of mass for a system of particles** is

$$\vec{\mathbf{v}}_{\text{CM}} = \frac{1}{M} \sum_i m_i \vec{\mathbf{v}}_i \quad [8.36]$$

The total momentum of a system of particles equals the total mass multiplied by the velocity of the center of mass; that is, $\vec{\mathbf{p}}_{\text{tot}} = M\vec{\mathbf{v}}_{\text{CM}}$.

Newton's second law applied to a system of particles is

$$\sum \vec{\mathbf{F}}_{\text{ext}} = M\vec{\mathbf{a}}_{\text{CM}} = \frac{d\vec{\mathbf{p}}_{\text{tot}}}{dt} \quad [8.40]$$

where $\vec{\mathbf{a}}_{\text{CM}}$ is the acceleration of the center of mass and the sum is over all external forces. The center of mass therefore moves like an imaginary particle of mass $M$ under the influence of the resultant external force on the system.

## QUESTIONS

☐ = answer available in the *Student Solutions Manual and Study Guide*

1. Does a large force always produce a larger impulse on an object than a smaller force does? Explain.
2. If the speed of a particle is doubled, by what factor is its momentum changed? By what factor is its kinetic energy changed?
3. If two particles have equal kinetic energies, are their momenta necessarily equal? Explain.
4. While in motion, a pitched baseball carries kinetic energy and momentum. (a) Can we say that it carries a force that it can exert on any object it strikes? (b) Can the baseball deliver more kinetic energy to the object it strikes than the ball carries initially? (c) Can the baseball deliver to the object it strikes more momentum than the ball carries initially? Explain your answers.
5. You are watching a movie about a superhero and notice that the superhero hovers in the air and throws a piano at some villains while remaining stationary in the air. What is wrong with this scenario?
6. If two objects collide and one is initially at rest, is it possible for both to be at rest after the collision? Is it possible for one to be at rest after the collision? Explain.
7. Explain how linear momentum is conserved when a ball bounces from a floor.

8. A bomb, initially at rest, explodes into several pieces. (a) Is linear momentum of the system conserved? (b) Is kinetic energy of the system conserved? Explain.

9. You are standing perfectly still and then you take a step forward. Before the step your momentum was zero, but afterward you have some momentum. Is the principle of conservation of momentum violated in this case?

10. Consider a perfectly inelastic collision between a car and a large truck. Which vehicle experiences a larger change in kinetic energy as a result of the collision?

11. A sharpshooter fires a rifle while standing with the butt of the gun against her shoulder. If the forward momentum of a bullet is the same as the backward momentum of the gun, why isn't it as dangerous to be hit by the gun as by the bullet?

12. A pole-vaulter falls from a height of 6.0 m onto a foam rubber pad. Can you calculate his speed just before he reaches the pad? Can you calculate the force exerted on him by the pad? Explain.

13. Firefighters must apply large forces to hold a fire hose steady (Fig. Q8.13). What factors related to the projection of the water determine the magnitude of the force needed to keep the end of the fire hose stationary?

(© Bill Stormont/The Stock Market)

FIGURE Q8.13 Firefighters attack a burning house with a hose line.

14. A large bedsheet is held vertically by two students. A third student, who happens to be the star pitcher on the baseball team, throws a raw egg at the sheet. Explain why the egg does not break when it hits the sheet, regardless of its initial speed. (If you try this demonstration, make sure the pitcher hits the sheet near its center, and do not allow the egg to fall on the floor after being caught.)

15. NASA often uses a planet's gravity to "slingshot" a probe on its way to a more distant planet. The interaction of the planet and the spacecraft is a collision in which the objects do not touch. How can the probe have its speed increased in this manner?

16. Can the center of mass of an object be located at a position at which there is no mass? If so, give examples.

17. A juggler juggles three balls in a continuous cycle. Any one ball is in contact with his hands for one fifth of the time. Describe the motion of the center of mass of the three balls. What average force does the juggler exert on one ball while he is touching it?

18. Explain how you could use a balloon to demonstrate the mechanism responsible for rocket propulsion.

19. Does the center of mass of a rocket in free space accelerate? Explain. Can the speed of a rocket exceed the exhaust speed of the fuel? Explain.

20. On the subject of the following positions, state your own view and argue to support it. (a) The best theory of motion is that force causes acceleration. (b) The true measure of a force's effectiveness is the work it does, and the best theory of motion is that work on an object changes its energy. (c) The true measure of a force's effect is impulse, and the best theory of motion is that impulse imparted to an object changes its momentum.

## PROBLEMS

1, 2, 3 = straightforward, intermediate, challenging
□ = full solution available in the *Student Solutions Manual and Study Guide*
Physics Now™ = coached problem with hints available at www.pop4e.com
= computer useful in solving problem
= paired numerical and symbolic problems
= biomedical application

### Section 8.1 ■ Linear Momentum and Its Conservation

1. A 3.00-kg particle has a velocity of $(3.00\hat{\mathbf{i}} - 4.00\hat{\mathbf{j}})$ m/s. (a) Find its $x$ and $y$ components of momentum. (b) Find the magnitude and direction of its momentum.

2. How fast can you set the Earth moving? In particular, when you jump straight up as high as you can, what is the order of magnitude of the maximum recoil speed that you give to the Earth? Model the Earth as a perfectly solid object. In your solution, state the physical quantities you take as data and the values you measure or estimate for them.

3. In research in cardiology and exercise physiology, it is often important to know the mass of blood pumped by a person's heart in one stroke. This information can be obtained by means of a *ballistocardiograph.* The instrument works as follows. The subject lies on a horizontal pallet floating on a film of air. Friction on the pallet is negligible. Initially, the momentum of the system is zero. When the heart beats, it expels a mass $m$ of blood into the aorta with speed $v$, and the body and platform move in the opposite direction with speed $V$. The blood velocity can be determined independently (e.g., by observing the Doppler shift of ultrasound). Assume that it is 50.0 cm/s in one typical trial. The mass of the subject plus the pallet is 54.0 kg. The pallet moves $6.00 \times 10^{-5}$ m in 0.160 s after one heartbeat. Calculate the mass of blood that leaves the heart. Assume

that the mass of blood is negligible compared with the total mass of the person. (This simplified example illustrates the principle of ballistocardiography, but in practice a more sophisticated model of heart function is used.)

**4.** (a) A particle of mass $m$ moves with momentum $p$. Show that the kinetic energy of the particle is given by $K = p^2/2m$. (b) Express the magnitude of the particle's momentum in terms of its kinetic energy and mass.

**5.** Two blocks with masses $M$ and $3M$ are placed on a horizontal, frictionless surface. A light spring is attached to one of them, and the blocks are pushed together with the spring between them (Fig. P8.5). A cord initially holding the blocks together is burned; after this, the block of mass $3M$ moves to the right with a speed of 2.00 m/s. (a) What is the speed of the block of mass $M$? (b) Find the original elastic potential energy in the spring, taking $M = 0.350$ kg.

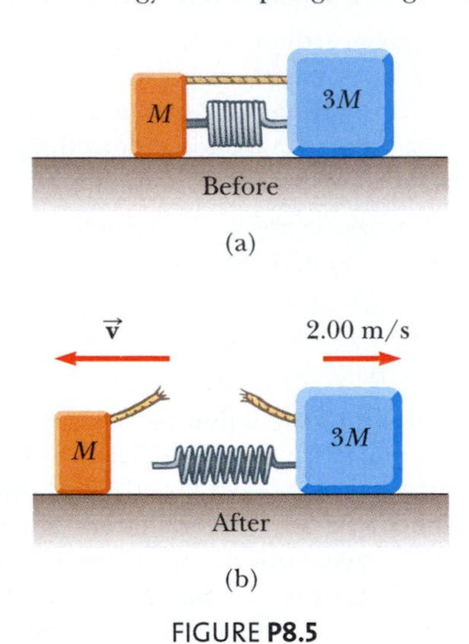


FIGURE **P8.5**

## Section 8.2 ▪ Impulse and Momentum

**6.** A friend claims that as long as he has his seat belt on, he can hold on to a 12.0-kg child in a 60.0 mi/h head-on collision with a brick wall in which the car passenger compartment comes to a stop in 0.050 0 s. Show that the violent force during the collision will tear the child from his arms. (A child should always be in a toddler seat secured with a seat belt in the back seat of a car.)

**7.** An estimated force–time curve for a baseball struck by a bat is shown in Figure P8.7. From this curve, determine (a) the impulse delivered to the ball, (b) the average force exerted on the ball, and (c) the peak force exerted on the ball.

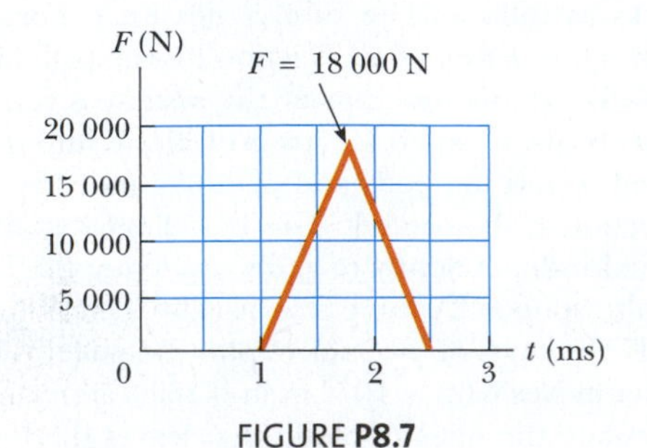


FIGURE **P8.7**

**8.** A tennis player receives a shot with the ball (0.060 0 kg) traveling horizontally at 50.0 m/s and returns the shot with the ball traveling horizontally at 40.0 m/s in the opposite direction. (a) What is the impulse delivered to the ball by the racquet? (b) What work does the racquet do on the ball?

**9.** **PhysicsNow™** A 3.00-kg steel ball strikes a wall with a speed of 10.0 m/s at an angle of 60.0° with the surface. It bounces off with the same speed and angle (Fig. P8.9). If the ball is in contact with the wall for 0.200 s, what is the average force exerted on the ball by the wall?

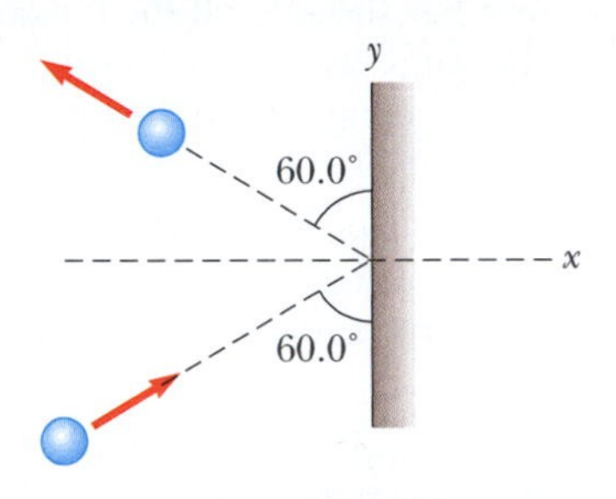


FIGURE **P8.9**

**10.** In a slow-pitch softball game, a 0.200-kg softball crosses the plate at 15.0 m/s at an angle of 45.0° below the horizontal. The batter hits the ball toward center field, giving it a velocity of 40.0 m/s at 30.0° above the horizontal. (a) Determine the impulse delivered to the ball. (b) If the force on the ball increases linearly for 4.00 ms, holds constant for 20.0 ms, and then decreases linearly to zero in another 4.00 ms, what is the maximum force on the ball?

**11.** A garden hose is held as shown in Figure P8.11. The hose is originally full of motionless water. What additional force is necessary to hold the nozzle stationary after the water flow is turned on if the discharge rate is 0.600 kg/s with a speed of 25.0 m/s?

FIGURE **P8.11**

**12.** A glider of mass $m$ is free to slide along a horizontal air track. It is pushed against a launcher at one end of the track. Model the launcher as a light spring of force constant $k$, compressed by a distance $x$. The glider is released from rest. (a) Show that the glider attains a speed $v = x(k/m)^{1/2}$. (b) Does a glider of large or of small mass attain a greater speed? (c) Show that the impulse imparted to the glider is given by the expression $x(km)^{1/2}$. (d) Is a greater impulse imparted to a large or a small mass? (e) Is more work done on a large or a small mass?

## Section 8.3 ▪ Collisions

**13.** A railroad car of mass $2.50 \times 10^4$ kg is moving with a speed of 4.00 m/s. It collides and couples with three other coupled railroad cars, each of the same mass as the single car and moving in the same direction with an initial speed of 2.00 m/s. (a) What is the speed of the four cars after the collision? (b) How much mechanical energy is lost in the collision?

**14.** Four railroad cars, each of mass $2.50 \times 10^4$ kg, are coupled together and coasting along horizontal tracks at speed $v_i$ toward the south. A very strong but foolish movie actor, riding on the second car, uncouples the front car and gives it a big push, increasing its speed to 4.00 m/s southward. The remaining three cars continue moving south, now at 2.00 m/s. (a) Find the initial speed of the four cars. (b) How much work did the actor do? (c) State the relationship between the process described here and the process in Problem 8.13.

**15.** A 45.0-kg girl is standing on a plank that has a mass of 150 kg. The plank, originally at rest, is free to slide on a frozen lake, which is a flat, frictionless supporting surface. The girl begins to walk along the plank at a constant speed of 1.50 m/s relative to the plank. (a) What is her speed relative to the ice surface? (b) What is the speed of the plank relative to the ice surface?

**16.** Two blocks are free to slide along the frictionless wooden track $ABC$ shown in Figure P8.16. A block of mass $m_1 = 5.00$ kg is released from $A$. Protruding from its front end is the north pole of a strong magnet, repelling the north pole of an identical magnet embedded in the back end of the block of mass $m_2 = 10.0$ kg, initially at rest. The two blocks never touch. Calculate the maximum height to which $m_1$ rises after the elastic collision.

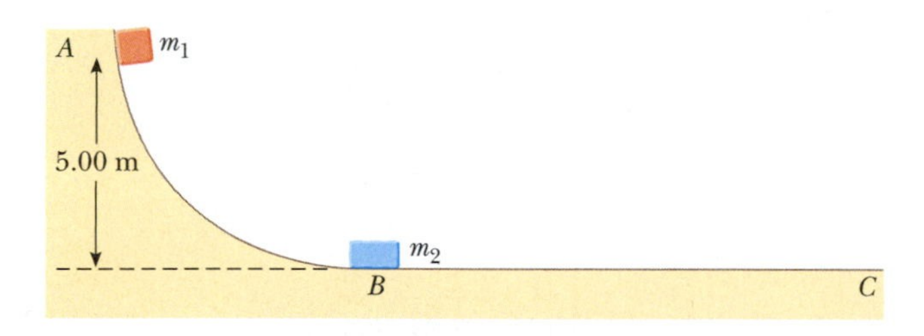


FIGURE **P8.16**

**17.** Most of us know intuitively that in a head-on collision between a large dump truck and a subcompact car, you are better off being in the truck than in the car. Why? Many people imagine that the collision force exerted on the car is much greater than that experienced by the truck. To substantiate this view, they point out that the car is crushed, whereas the truck is only dented. This idea of unequal forces, of course, is false. Newton's third law tells us that both objects experience forces of the same magnitude. The truck suffers less damage because it is made of stronger metal. What about the two drivers? Do they experience the same forces? To answer this question, suppose each vehicle is initially moving at 8.00 m/s and they undergo a perfectly inelastic head-on collision. Each driver has mass 80.0 kg. Including the drivers, the total vehicle masses are 800 kg for the car and 4 000 kg for the truck. If the collision time is 0.120 s, what force does the seat belt exert on each driver?

**18.** As shown in Figure P8.18, a bullet of mass $m$ and speed $v$ passes completely through a pendulum bob of mass $M$. The bullet emerges with a speed of $v/2$. The pendulum bob is suspended by a stiff rod of length $\ell$ and negligible mass. What is the minimum value of $v$ such that the pendulum bob will barely swing through a complete vertical circle?

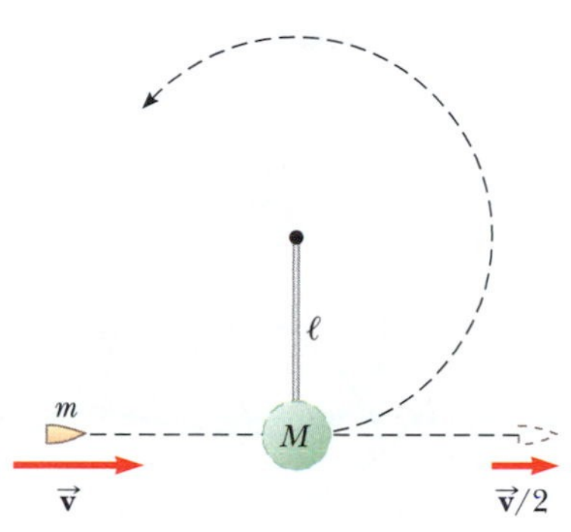


FIGURE **P8.18**

**19.** Physics Now™ A neutron in a nuclear reactor makes an elastic head-on collision with the nucleus of a carbon atom initially at rest. (a) What fraction of the neutron's kinetic energy is transferred to the carbon nucleus? (b) Assume that the initial kinetic energy of the neutron is $1.60 \times 10^{-13}$ J. Find its final kinetic energy and the kinetic energy of the carbon nucleus after the collision. (The mass of the carbon nucleus is nearly 12.0 times the mass of the neutron.)

**20.** A 7.00-g bullet, when fired from a gun into a 1.00-kg block of wood held in a vise, penetrates the block to a depth of 8.00 cm. This block of wood is next placed on a frictionless horizontal surface, and a second 7.00-g bullet is fired from the gun into the block. To what depth will the bullet penetrate the block in this case?

**21.** Physics Now™ A 12.0-g wad of sticky clay is hurled horizontally at a 100-g wooden block initially at rest on a horizontal surface. The clay sticks to the block. After impact, the block slides 7.50 m before coming to rest. If the coefficient of friction between the block and the surface is 0.650, what was the speed of the clay immediately before impact?

**22.** (a) Three carts of masses 4.00 kg, 10.0 kg, and 3.00 kg move on a frictionless, horizontal track with speeds of 5.00 m/s, 3.00 m/s, and 4.00 m/s, respectively, as shown in Figure P8.22. Velcro couplers make the carts stick together after colliding. Find the final velocity of the train of three

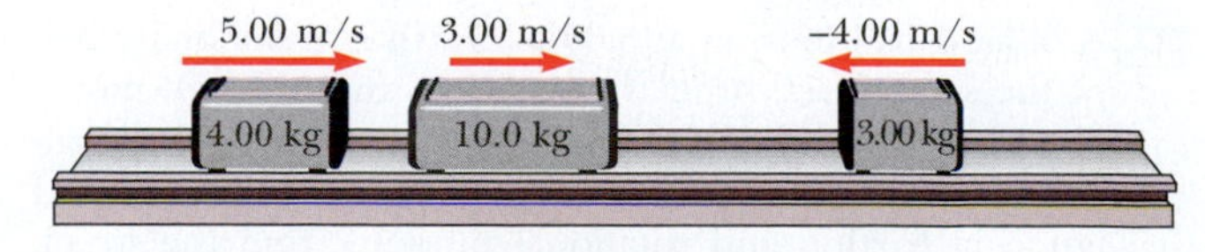


FIGURE **P8.22**

carts. (b) Does your answer require that all the carts collide and stick together at the same time? What if they collide in a different order?

**23.** A tennis ball of mass 57.0 g is held just above a basketball of mass 590 g. With their centers vertically aligned, both are released from rest at the same time, to fall through a distance of 1.20 m, as shown in Figure P8.23. (a) Find the magnitude of the downward velocity with which the basketball reaches the ground. Assume that an elastic collision with the ground instantaneously reverses the velocity of the basketball while the tennis ball is still moving down. Next, the two balls meet in an elastic collision. (b) To what height does the tennis ball rebound?

FIGURE **P8.23**

## Section 8.4 ▪ Two-Dimensional Collisions

**24.** A 90.0-kg fullback running east with a speed of 5.00 m/s is tackled by a 95.0-kg opponent running north with a speed of 3.00 m/s. Noting that the collision is perfectly inelastic, (a) calculate the speed and direction of the players just after the tackle and (b) determine the mechanical energy lost as a result of the collision. Account for the missing energy.

**25.** Two shuffleboard disks of equal mass, one orange and the other yellow, are involved in an elastic, glancing collision. The yellow disk is initially at rest and is struck by the orange disk moving with a speed $v_i$. After the collision, the orange disk moves along a direction that makes an angle $\theta$ with its initial direction of motion. The velocities of the two disks are perpendicular after the collision. Determine the final speed of each disk.

**26.** Two automobiles of equal mass approach an intersection. One vehicle is traveling with velocity 13.0 m/s toward the east, and the other is traveling north with speed $v_{2i}$. Neither driver sees the other. The vehicles collide in the intersection and stick together, leaving parallel skid marks at an angle of 55.0° north of east. The speed limit for both roads is 35 mi/h, and the driver of the northward-moving vehicle claims that he was within the speed limit when the collision occurred. Is he telling the truth?

**27.** A billiard ball moving at 5.00 m/s strikes a stationary ball of the same mass. After the collision, the first ball moves at 4.33 m/s, at an angle of 30.0° with respect to the original line of motion. Assuming an elastic collision (and ignoring friction and rotational motion), find the struck ball's velocity.

**28.** A proton, moving with a velocity of $v_i\hat{\mathbf{i}}$, collides elastically with another proton that is initially at rest. Assuming that the two protons have equal speeds after the collision, find (a) the speed of each proton after the collision in terms of $v_i$ and (b) the direction of the velocity vectors after the collision.

**29.** An object of mass 3.00 kg, with an initial velocity of $5.00\hat{\mathbf{i}}$ m/s, collides with and sticks to an object of mass 2.00 kg, with an initial velocity of $-3.00\hat{\mathbf{j}}$ m/s. Find the final velocity of the composite object.

**30.** A 0.300-kg puck, initially at rest on a horizontal, frictionless surface, is struck by a 0.200-kg puck moving initially along the $x$ axis with a speed of 2.00 m/s. After the collision, the 0.200-kg puck has a speed of 1.00 m/s at an angle of $\theta = 53.0°$ to the positive $x$ axis (see Active Fig. 8.11). (a) Determine the velocity of the 0.300-kg puck after the collision. (b) Find the fraction of kinetic energy lost in the collision.

**31.** **PhysicsNow™** An unstable atomic nucleus of mass $17.0 \times 10^{-27}$ kg initially at rest disintegrates into three particles. One of the particles, of mass $5.00 \times 10^{-27}$ kg, moves along the $y$ axis with a velocity of $6.00 \times 10^{6}$ m/s. Another particle, of mass $8.40 \times 10^{-27}$ kg, moves along the $x$ axis with a speed of $4.00 \times 10^{6}$ m/s. Find (a) the velocity of the third particle and (b) the total kinetic energy increase in the process.

## Section 8.5 ▪ The Center of Mass

**32.** Four objects are situated along the $y$ axis as follows: a 2.00-kg object is at +3.00 m, a 3.00-kg object is at +2.50 m, a 2.50-kg object is at the origin, and a 4.00-kg object is at −0.500 m. Where is the center of mass of these objects?

**33.** A uniform piece of sheet steel is shaped as shown in Figure P8.33. Compute the $x$ and $y$ coordinates of the center of mass of the piece.

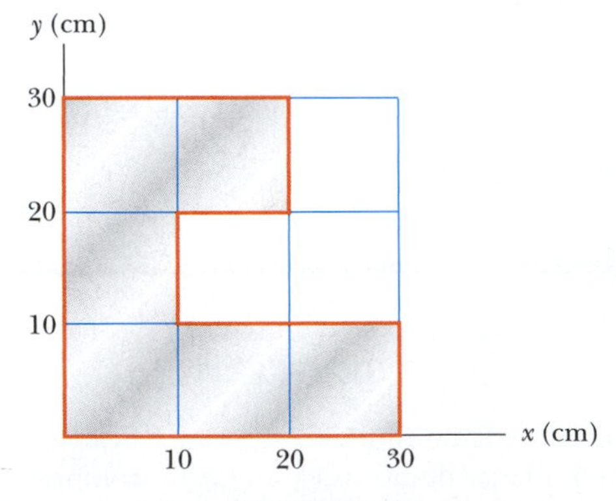


FIGURE **P8.33**

**34.** A water molecule consists of an oxygen atom with two hydrogen atoms bound to it (Fig. P8.34). The angle between the two bonds is 106°. If the bonds are 0.100 nm long, where is the center of mass of the molecule?

**35.** (a) Consider an extended object whose different portions have different elevations. Assume that the free fall acceleration is uniform over the object. Prove that the gravitational potential energy of the object–Earth system is given by $U_g = Mgy_{CM}$, where $M$ is the total mass of the object and

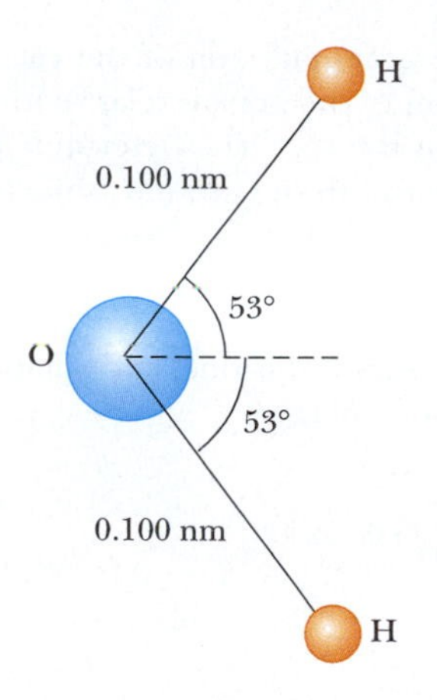


FIGURE P8.34

$y_{CM}$ is the elevation of its center of mass above the chosen reference level. (b) Calculate the gravitational potential energy associated with a ramp constructed on level ground with stone with density 3 800 kg/m$^3$ and everywhere 3.60 m wide (Fig. P8.35). In a side view, the ramp appears as a right triangle with height 15.7 m at the top end and base 64.8 m.

FIGURE P8.35

**36.** A rod of length 30.0 cm has linear density (mass-per-length) given by

$$\lambda = 50.0 \text{ g/m} + 20.0x \text{ g/m}^2$$

where $x$ is the distance from one end, measured in meters. (a) What is the mass of the rod? (b) How far from the $x = 0$ end is its center of mass?

## Section 8.6 ■ Motion of a System of Particles

**37.** A 2.00-kg particle has a velocity $(2.00\hat{\mathbf{i}} - 3.00\hat{\mathbf{j}})$ m/s, and a 3.00-kg particle has a velocity $(1.00\hat{\mathbf{i}} + 6.00\hat{\mathbf{j}})$ m/s. Find (a) the velocity of the center of mass and (b) the total momentum of the system.

**38.** Consider a system of two particles in the $xy$ plane: $m_1 = 2.00$ kg is at the location $\vec{\mathbf{r}}_1 = (1.00\hat{\mathbf{i}} + 2.00\hat{\mathbf{j}})$ m and has a velocity of $(3.00\hat{\mathbf{i}} + 0.500\hat{\mathbf{j}})$ m/s; $m_2 = 3.00$ kg is at $\vec{\mathbf{r}}_2 = (-4.00\hat{\mathbf{i}} - 3.00\hat{\mathbf{j}})$ m and has velocity $(3.00\hat{\mathbf{i}} - 2.00\hat{\mathbf{j}})$ m/s. (a) Plot these particles on a grid or graph paper. Draw their position vectors and show their velocities. (b) Find the position of the center of mass of the system and mark it on the grid. (c) Determine the velocity of the center of mass and also show it on the diagram. (d) What is the total linear momentum of the system?

**39.** Romeo (77.0 kg) entertains Juliet (55.0 kg) by playing his guitar from the rear of their boat at rest in still water, 2.70 m away from Juliet who is in the front of the boat. After the serenade, Juliet carefully moves to the rear of the boat (away from shore) to plant a kiss on Romeo's cheek. How far does the 80.0-kg boat move toward the shore it is facing?

**40.** A ball of mass 0.200 kg has a velocity of $1.50\hat{\mathbf{i}}$ m/s; a ball of mass 0.300 kg has a velocity of $-0.400\hat{\mathbf{i}}$ m/s. They meet in a head-on elastic collision. (a) Find their velocities after the collision. (b) Find the velocity of their center of mass before and after the collision.

## Section 8.7 ■ Context Connection—Rocket Propulsion

**41.** **PhysicsNow™** The first stage of a *Saturn V* space vehicle consumed fuel and oxidizer at the rate of $1.50 \times 10^4$ kg/s, with an exhaust speed of $2.60 \times 10^3$ m/s. (a) Calculate the thrust produced by these engines. (b) Find the acceleration of the vehicle just as it lifted off the launch pad on the Earth, taking the vehicle's initial mass as $3.00 \times 10^6$ kg. You must include the gravitational force to solve part (b).

**42.** Model rocket engines are sized by thrust, thrust duration, and total impulse, among other characteristics. A size C5 model rocket engine has an average thrust of 5.26 N, a fuel mass of 12.7 g, and an initial mass of 25.5 g. The duration of its burn is 1.90 s. (a) What is the average exhaust speed of the engine? (b) If this engine is placed in a rocket body of mass 53.5 g, what is the final velocity of the rocket if it is fired in outer space? Assume that the fuel burns at a constant rate.

**43.** A rocket for use in deep space is to be capable of boosting a total load (payload plus rocket frame and engine) of 3.00 metric tons to a speed of 10 000 m/s. (a) It has an engine and fuel designed to produce an exhaust speed of 2 000 m/s. How much fuel plus oxidizer is required? (b) If a different fuel and engine design could give an exhaust speed of 5 000 m/s, what amount of fuel and oxidizer would be required for the same task?

**44.** *Rocket science.* A rocket has total mass $M_i = 360$ kg, including 330 kg of fuel and oxidizer. In interstellar space, it starts from rest at the position $x = 0$, turns on its engine at time $t = 0$, and puts out exhaust with relative speed $v_e =$ 1 500 m/s at the constant rate $k = 2.50$ kg/s. The fuel will last for an actual burn time of 330 kg/(2.5 kg/s) = 132 s, but define a "projected depletion time" as $T_p = M_i/k =$ 360 kg/(2.5 kg/s) = 144 s (which would be the burn time if the rocket could use its payload and fuel tanks, and even the walls of the combustion chamber, as fuel.) (a) Show that during the burn the velocity of the rocket is given as a function of time by

$$v(t) = -v_e \ln\left(1 - \frac{t}{T_p}\right)$$

(b) Make a graph of the velocity of the rocket as a function of time for times running from 0 to 132 s. (c) Show that the acceleration of the rocket is

$$a(t) = \frac{v_e}{T_p - t}$$

(d) Graph the acceleration as a function of time. (e) Show that the position of the rocket is

$$x(t) = v_e(T_p - t)\ln\left(1 - \frac{t}{T_p}\right) + v_e t$$

(f) Graph the position during the burn as a function of time.

**45.** An orbiting spacecraft is described not as a "zero-$g$" but rather as a "microgravity" environment for its occupants and for onboard experiments. Astronauts experience slight lurches due to the motions of equipment and other astronauts and as a result of venting of materials from the craft. Assume that a 3 500-kg spacecraft undergoes an acceleration of 2.50 $\mu g = 2.45 \times 10^{-5}$ m/s$^2$ due to a leak from one of its hydraulic control systems. The fluid is known to escape with a speed of 70.0 m/s into the vacuum of space. How much fluid will be lost in 1.00 h if the leak is not stopped?

## Additional Problems

**46.** Two gliders are set in motion on an air track. A spring of force constant $k$ is attached to the near side of one glider. The first glider of mass $m_1$ has velocity $\vec{\mathbf{v}}_1$, and the second glider of mass $m_2$ moves more slowly, with velocity $\vec{\mathbf{v}}_2$, as shown in Figure P8.46. When $m_1$ collides with the spring attached to $m_2$ and compresses the spring to its maximum compression $x_{max}$, the velocity of the gliders is $\vec{\mathbf{v}}$. In terms of $\vec{\mathbf{v}}_1$, $\vec{\mathbf{v}}_2$, $m_1$, $m_2$, and $k$, find (a) the velocity $\vec{\mathbf{v}}$ at maximum compression, (b) the maximum compression $x_{max}$, and (c) the velocity of each glider after $m_1$ has lost contact with the spring.

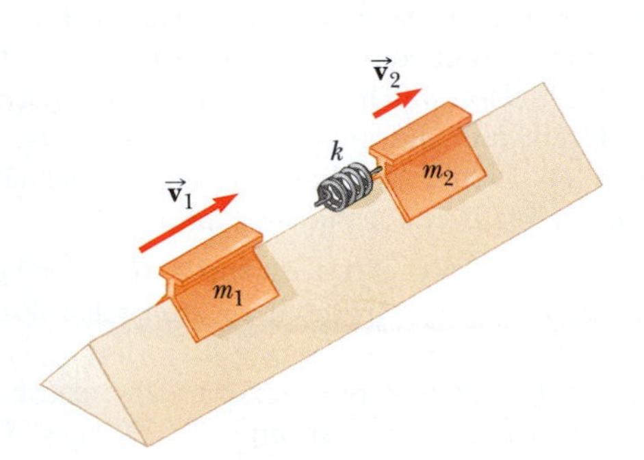


FIGURE **P8.46**

**47.** **Review problem.** A 60.0-kg person running at an initial speed of 4.00 m/s jumps onto a 120-kg cart initially at rest (Fig. P8.47). The person slides on the cart's top surface and finally comes to rest relative to the cart. The coefficient of kinetic friction between the person and the cart is 0.400. Friction between the cart and ground can be ignored. (a) Find the final velocity of the person and cart relative to the ground. (b) Find the friction force acting on the person while he is sliding across the top surface of the cart. (c) How long does the friction force act on the person? (d) Find the change in momentum of the person and the change in momentum of the cart. (e) Determine the displacement of the person relative to the ground while he is sliding on the cart. (f) Determine the displacement of the cart relative to the ground while the person is sliding. (g) Find the change in kinetic energy of the person. (h) Find the change in kinetic energy of the cart. (i) Explain why the answers to (g) and (h) differ. (What kind of collision is this one, and what accounts for the loss of mechanical energy?)

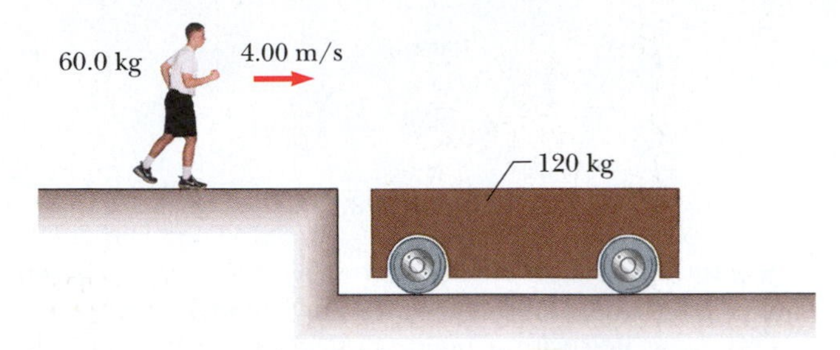


FIGURE **P8.47**

**48.** A bullet of mass $m$ is fired into a block of mass $M$ initially at rest at the edge of a frictionless table of height $h$ (Fig. P8.48). The bullet remains in the block, and after impact the block lands a distance $d$ from the bottom of the table. Determine the initial speed of the bullet.

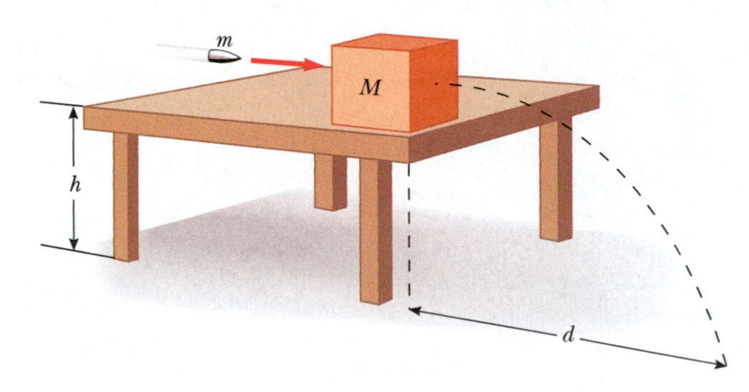


FIGURE **P8.48**

**49.** When it is threatened, a squid can escape by expelling a jet of water, sometimes colored with camouflaging ink. Consider a squid originally at rest in ocean water of constant density 1 030 kg/m$^3$. Its original mass is 90.0 kg, of which a significant fraction is water inside its mantle. It expels this water through its siphon, a circular opening of diameter 3.00 cm, at a speed of 16.0 m/s. (a) As the squid is just starting to move, the surrounding water exerts no drag force on it. Find the squid's initial acceleration. (b) To estimate the maximum speed of the escaping squid, model the drag force of the surrounding water as described by Equation 5.7. Assume that the squid has a drag coefficient of 0.300 and a cross-sectional area of 800 cm$^2$. Find the speed at which the drag force counterbalances the thrust of its jet.

**50.** Pursued by ferocious wolves, you are in a sleigh with no horses, gliding without friction across an ice-covered lake. You take an action described by these equations:

$$(270 \text{ kg})(7.50 \text{ m/s})\hat{\mathbf{i}} = (15.0 \text{ kg})(-v_{1f}\hat{\mathbf{i}}) + (255 \text{ kg})(v_{2f}\hat{\mathbf{i}})$$

$$v_{1f} + v_{2f} = 8.00 \text{ m/s}$$

(a) Complete the statement of the problem, giving the data and identifying the unknowns. (b) Find the values of $v_{1f}$ and $v_{2f}$. (c) Find the work you do.

**51.** A small block of mass $m_1 = 0.500$ kg is released from rest at the top of a curve-shaped, frictionless wedge of mass $m_2 = 3.00$ kg, which sits on a frictionless, horizontal surface as shown in Figure P8.51a. When the block leaves the wedge, its velocity is measured to be 4.00 m/s to the right as shown in Figure P8.51b. (a) What is the velocity of the wedge after the block reaches the horizontal surface? (b) What is the height $h$ of the wedge?

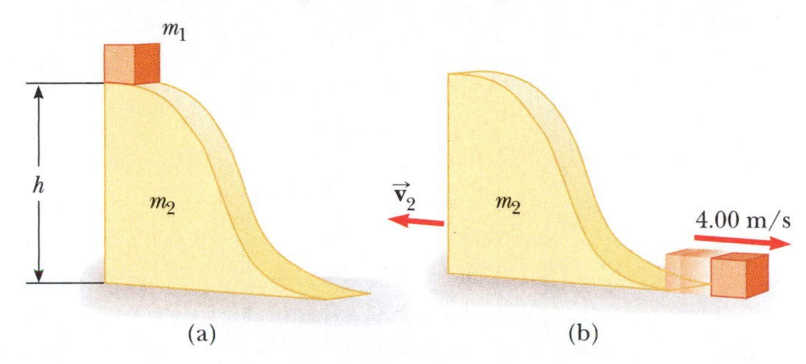


FIGURE **P8.51**

**52.** A jet aircraft is traveling at 500 mi/h (223 m/s) in horizontal flight. The engine takes in air at a rate of 80.0 kg/s and burns fuel at a rate of 3.00 kg/s. The exhaust gases are ejected at 600 m/s relative to the aircraft. Find the thrust of the jet engine and the delivered power.

**53.** **Review problem.** A light spring of force constant 3.85 N/m is compressed by 8.00 cm and held between a 0.250-kg block on the left and a 0.500-kg block on the right. Both blocks are at rest on a horizontal surface. The blocks are released simultaneously so that the spring tends to push them apart. Find the maximum velocity each block attains if the coefficient of kinetic friction between each block and the surface is (a) 0, (b) 0.100, and (c) 0.462. Assume that the coefficient of static friction is larger than that for kinetic friction.

**54.** **Review problem.** There are (one can say) three coequal theories of motion: Newton's second law, stating that the total force on an object causes its acceleration; the work–kinetic energy theorem, stating that the total work on an object causes its change in kinetic energy; and the impulse–momentum theorem, stating that the total impulse on an object causes its change in momentum. In this problem, you compare predictions of the three theories in one particular case. A 3.00-kg object has velocity $7.00\hat{\mathbf{j}}$ m/s. Then, a total force $12.0\hat{\mathbf{i}}$ N acts on the object for 5.00 s. (a) Calculate the object's final velocity, using the impulse–momentum theorem. (b) Calculate its acceleration from $\vec{\mathbf{a}} = (\vec{\mathbf{v}}_f - \vec{\mathbf{v}}_i)/\Delta t$. (c) Calculate its acceleration from $\vec{\mathbf{a}} = \sum\vec{\mathbf{F}}/m$. (d) Find the object's vector displacement from $\Delta\vec{\mathbf{r}} = \vec{\mathbf{v}}_i t + \frac{1}{2}\vec{\mathbf{a}}t^2$. (e) Find the work done on the object from $W = \vec{\mathbf{F}}\cdot\Delta\vec{\mathbf{r}}$. (f) Find the final kinetic energy from $\frac{1}{2}mv_f^2 = \frac{1}{2}m\vec{\mathbf{v}}_f\cdot\vec{\mathbf{v}}_f$. (g) Find the final kinetic energy from $\frac{1}{2}mv_i^2 + W$.

**55.** Two particles with masses $m$ and $3m$ are moving toward each other along the $x$ axis with the same initial speeds $v_i$. The particle with mass $m$ is traveling to the left, and particle $3m$ is traveling to the right. They undergo a head-on elastic collision and each rebounds along the same line as it approached. Find the final speeds of the particles.

**56.** Two particles with masses $m$ and $3m$ are moving toward each other along the $x$ axis with the same initial speeds $v_i$. Particle $m$ is traveling to the left, and particle $3m$ is traveling to the right. They undergo an elastic glancing collision such that particle $m$ is moving downward after the collision at a right angle to its initial direction. (a) Find the final speeds of the two particles. (b) What is the angle $\theta$ at which the particle $3m$ is scattered?

**57.** George of the Jungle, with mass $m$, swings on a light vine hanging from a stationary tree branch. A second vine of equal length hangs from the same point, and a gorilla of larger mass $M$ swings in the opposite direction on it. Both vines are horizontal when the primates start from rest at the same moment. George and the gorilla meet at the lowest point of their swings. Each is afraid that one vine will break, so they grab each other and hang on. They swing upward together, reaching a point where the vines make an angle of 35.0° with the vertical. (a) Find the value of the ratio $m/M$. (b) Try this experiment at home. Tie a small magnet and a steel screw to opposite ends of a string. Hold the center of the string fixed to represent the tree branch and reproduce a model of the motions of George and the gorilla. What changes in your analysis will make it apply to this situation? Assume next that the magnet is strong so that it noticeably attracts the screw over a distance of a few centimeters. Then the screw will be moving faster just before it sticks to the magnet. Does this change make a difference?

**58.** A cannon is rigidly attached to a carriage, which can move along horizontal rails but is connected to a post by a large spring, initially unstretched and with force constant $k = 2.00 \times 10^4$ N/m, as shown in Figure P8.58. The cannon fires a 200-kg projectile at a velocity of 125 m/s directed 45.0° above the horizontal. (a) Assuming that the mass of the cannon and its carriage is 5 000 kg, find the recoil speed of the cannon. (b) Determine the maximum extension of the spring. (c) Find the maximum force the spring exerts on the carriage. (d) Consider the system consisting of the cannon, carriage, and projectile. Is the momentum of this system conserved during the firing? Why or why not?

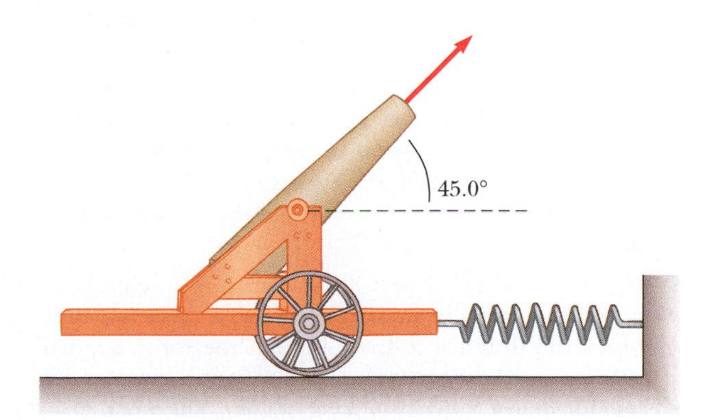


FIGURE **P8.58**

**59.** Sand from a stationary hopper falls onto a moving conveyor belt at the rate of 5.00 kg/s as shown in Figure P8.59. The conveyor belt is supported by frictionless rollers and moves at a constant speed of 0.750 m/s under the action of a constant horizontal external force $\vec{\mathbf{F}}_{\text{ext}}$ supplied by the

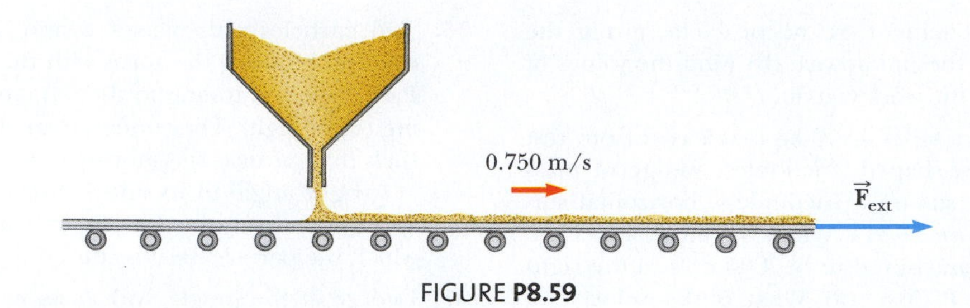


FIGURE **P8.59**

motor that drives the belt. Find (a) the sand's rate of change of momentum in the horizontal direction, (b) the force of friction exerted by the belt on the sand, (c) the external force $\vec{\mathbf{F}}_{ext}$, (d) the work done by $\vec{\mathbf{F}}_{ext}$ in 1 s, and (e) the kinetic energy acquired by the falling sand each second due to the change in its horizontal motion. (f) Why are the answers to (d) and (e) different?

**60.** A chain of length $L$ and total mass $M$ is released from rest with its lower end just touching the top of a table as shown in Figure P8.60a. Find the force exerted by the table on the chain after the chain has fallen through a distance $x$ as shown in Figure P8.60b. (Assume that each link comes to rest the instant it reaches the table.)

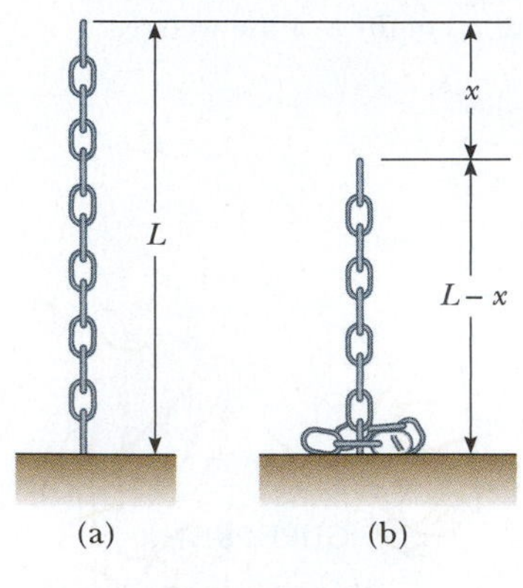


FIGURE **P8.60**

## ANSWERS TO QUICK QUIZZES

**8.1** (d). Two identical objects ($m_1 = m_2$) traveling at the same speed ($v_1 = v_2$) have the same kinetic energies and the same magnitudes of momentum. It also is possible, however, for particular combinations of masses and velocities to satisfy $K_1 = K_2$ but not $p_1 = p_2$. For example, a 1-kg object moving at 2 m/s has the same kinetic energy as a 4-kg object moving at 1 m/s, but the two clearly do not have the same momenta. Because we have no information about masses and speeds, we cannot choose among (a), (b), or (c).

**8.2** (b), (c), (a). The slower the ball, the easier it is to catch. If the momentum of the medicine ball is the same as the momentum of the baseball, the speed of the medicine ball must be one-tenth the speed of the baseball because the medicine ball has ten times the mass. If the kinetic energies are the same, the speed of the medicine ball must be $1/\sqrt{10}$ the speed of the baseball because of the squared speed term in the equation for $K$. The medicine ball is hardest to catch when it has the same speed as the baseball.

**8.3** (i), (c), and (e). Object 2 has a greater acceleration because of its smaller mass. Therefore, it takes less time to travel the distance $d$. Even though the force applied to objects 1 and 2 is the same, the change in momentum is less for object 2 because $\Delta t$ is smaller. The work $W = Fd$ done on both objects is the same, because both $F$ and $d$ are the same in the two cases. Therefore, $K_1 = K_2$. (ii), (b) and (d). The same impulse is applied to both objects, so they experience the same change in momentum. Object 2 has a larger acceleration because of its smaller mass. Therefore, the distance that object 2 covers in the time interval $\Delta t$ is larger than that for object 1. As a result, more work is done on object 2 and $K_2 > K_1$.

**8.4** (b). Because momentum of the two-ball system is conserved, $\vec{\mathbf{p}}_{Pi} + 0 = \vec{\mathbf{p}}_{Pf} + \vec{\mathbf{p}}_B$. Because the Ping-Pong ball bounces back from the much more massive bowling ball with approximately the same speed, $\vec{\mathbf{p}}_{Pf} = -\vec{\mathbf{p}}_{Pi}$. As a consequence, $\vec{\mathbf{p}}_B = 2\vec{\mathbf{p}}_{Pi}$. Kinetic energy can be expressed as $K = p^2/2m$. Because of the much larger mass of the bowling ball, its kinetic energy is much smaller than that of the Ping-Pong ball.

**8.5** (b). The piece with the handle will have less mass than the piece made up of the end of the bat. To see why, take the origin of coordinates as the center of mass before the bat was cut. Replace each cut piece by a small sphere located at the center of mass for each piece. The sphere representing the handle piece is farther from the origin, but the product of less mass and greater distance balances the product of greater mass and less distance for the end piece, as shown.

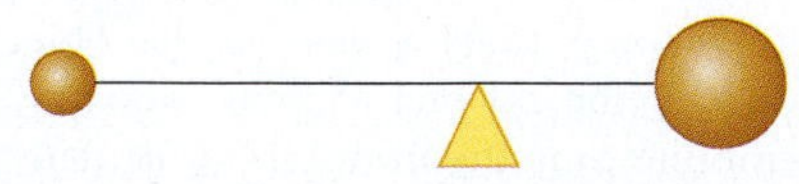

**8.6** (i), (a). The vessel–passengers system is isolated. If the passengers all start running one way, the speed of the vessel increases (a *small* amount!) the other way, so the speed of the center of mass of the system remains constant. (ii), (b). Once they stop running, the momentum of the system is the same as it was before they started running; you cannot change the momentum of an isolated system by means of internal forces. In case you are thinking that the passengers could run from bow to stern over and over to take advantage of the speed increase *while* they are running, remember that they will slow the ship down every time they return to the bow!

# CHAPTER 9

# Relativity

(Emily Serway)

*Standing on the shoulders of a giant.* David Serway, son of one of the authors, watches over his children, Nathan and Kaitlyn, as they frolic in the arms of Albert Einstein at the Einstein memorial in Washington, D.C. It is well known that Einstein, the principal architect of relativity, was very fond of children.

## CHAPTER OUTLINE


9.1 The Principle of Newtonian Relativity
9.2 The Michelson–Morley Experiment
9.3 Einstein's Principle of Relativity
9.4 Consequences of Special Relativity
9.5 The Lorentz Transformation Equations
9.6 Relativistic Momentum and the Relativistic Form of Newton's Laws
9.7 Relativistic Energy
9.8 Mass and Energy
9.9 General Relativity
9.10 Context Connection—From Mars to the Stars
SUMMARY


Our everyday experiences and observations are associated with objects that move at speeds much less than that of light in a vacuum, $c = 3.00 \times 10^8$ m/s. Models based on Newtonian mechanics and early concepts of space and time were formulated to describe the motion of such objects. This formalism is very successful in describing a wide range of phenomena that occur at low speeds, as we have seen in previous chapters. It fails, however, when applied to objects whose speeds approach that of light. Experimentally, the predictions of Newtonian theory can be tested by accelerating electrons or other particles to very high speeds. For example, it is possible to accelerate an electron to a speed of $0.99c$. According to the Newtonian definition of kinetic energy, if the energy transferred to such an electron were increased by a factor of 4, the electron speed should double to $1.98c$. Relativistic calculations, however, show that the speed of the electron—as well as the speeds of all other objects in the Universe—remains less than the speed of light. Because it places no upper limit on speed, Newtonian mechanics is contrary to modern theoretical predictions and experimental results, and the Newtonian models that we have developed are limited to objects moving much slower than the speed of light. Because Newtonian mechanics does not correctly predict the results of experiments carried out on objects moving at high speeds, we need a new formalism that is valid for these objects.

In 1905, at the age of only 26, Albert Einstein published his *special theory of relativity*, which is the subject of most of this chapter. Regarding the theory, Einstein wrote:

> The relativity theory arose from necessity, from serious and deep contradictions in the old theory from which there seemed no escape. The strength of the new theory lies in the consistency and simplicity with which it solves all these difficulties, using only a few very convincing assumptions.[1]

Although Einstein made many important contributions to science, special relativity alone represents one of the greatest intellectual achievements of the 20th century. With special relativity, experimental observations can be correctly predicted for objects over the range of all possible speeds, from rest to speeds approaching the speed of light. This chapter gives an introduction to special relativity, with emphasis on some of its consequences.

## 9.1 THE PRINCIPLE OF NEWTONIAN RELATIVITY

We will begin by considering the notion of relativity at low speeds. This discussion was actually begun in Section 3.6 when we discussed relative velocity. At that time, we discussed the importance of the observer. In a similar way here, we will generate equations that allow us to express one observer's measurements in terms of the other's. This process will lead to some rather unexpected and startling results about our understanding of space and time.

As we have mentioned previously, it is necessary to establish a frame of reference when describing a physical event. You should recall from Chapter 4 that an inertial frame is one in which an object is measured to have no acceleration if no forces act on it. Furthermore, any frame moving with constant velocity with respect to an inertial frame must also be an inertial frame. The laws predicting the results of an experiment performed in a vehicle moving with uniform velocity will be identical for the driver of the vehicle and a hitchhiker on the side of the road. The formal statement of this result is called the **principle of Newtonian relativity:**

**Principle of Newtonian relativity**

The laws of mechanics must be the same in all inertial frames of reference.

The following observation illustrates the equivalence of the laws of mechanics in different inertial frames. Consider a pickup truck moving with a constant velocity as in Figure 9.1a. If a passenger in the truck throws a ball straight up in the air, the passenger observes that the ball moves in a vertical path (ignoring air resistance). The motion of the ball appears to be precisely the same as if the ball were thrown by a person at rest on the Earth and observed by that person. The kinematic equations of Chapter 3 describe the results correctly whether the truck is at rest or in uniform motion. Now consider the ball thrown in the truck as viewed by an observer at rest on the Earth. This observer sees the path of the ball as a parabola as in Figure 9.1b. Furthermore, according to this observer, the ball has a horizontal component of velocity equal to the speed of the truck. Although the two observers measure different velocities and see different paths of the ball, they see the same forces on the ball and agree on the validity of Newton's laws as well as classical principles such as conservation of energy and conservation of momentum. Their measurements differ, but the measurements they make satisfy the same laws. All differences between the two views stem from the relative motion of one frame with respect to the other.

[1]A. Einstein and L. Infeld, *The Evolution of Physics* (New York, Simon and Schuster, 1966), p. 192.

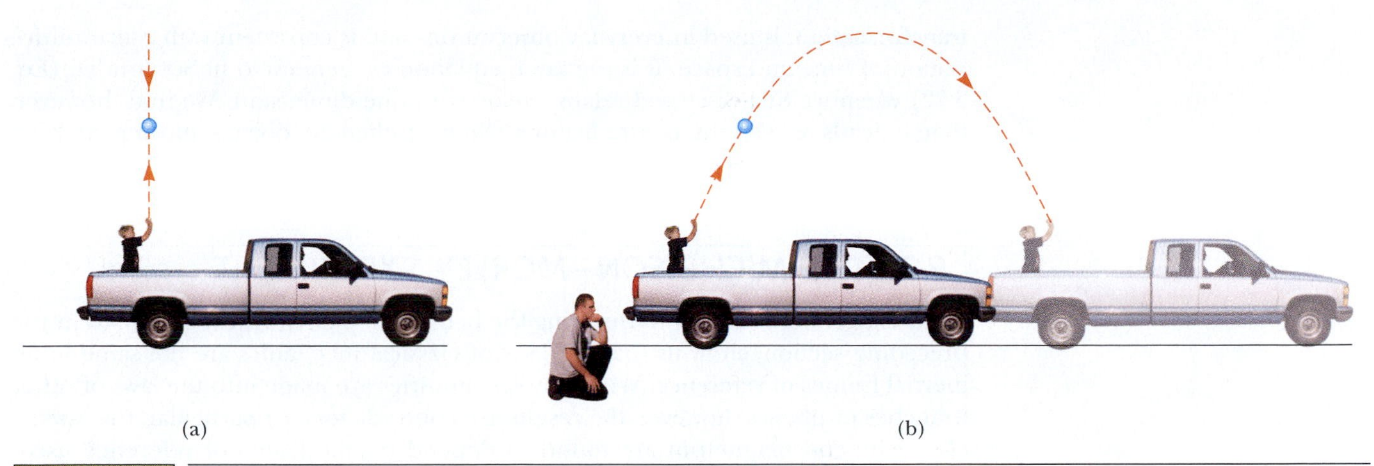


**FIGURE 9.1** (a) The observer in the truck sees the ball move in a vertical path when thrown upward. (b) The Earth observer sees the path of the ball to be a parabola.

Suppose some physical phenomenon, which we call an **event,** occurs. The event's location in space and time of occurrence can be specified by an observer with the coordinates $(x, y, z, t)$. We would like to be able to transform these coordinates from one inertial frame to another moving with uniform relative velocity, which will allow us to express one observer's measurements in terms of the other's.

Consider two inertial frames S and S′ (Fig. 9.2). The frame S′ moves with a constant velocity $\vec{\mathbf{v}}$ along the common $x$ and $x'$ axes, where $\vec{\mathbf{v}}$ is measured relative to S. We assume that the origins of S and S′ coincide at $t = 0$. Therefore, at time $t$, the origin of frame S′ is to the right of the origin of S by a distance $vt$. An event occurs at point $P$. An observer in S describes the event with space–time coordinates $(x, y, z, t)$, and an observer in S′ describes the same event with coordinates $(x', y', z', t')$. As we can see from Figure 9.2, a simple geometric argument shows that the space coordinates are related by the equations

$$x' = x - vt \qquad y' = y \qquad z' = z \tag{9.1}$$

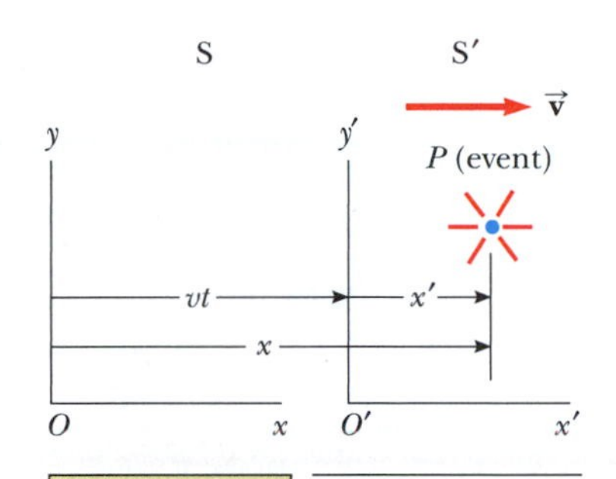


**FIGURE 9.2** An event occurs at a point $P$. The event is seen by two observers $O$ and $O'$ in inertial frames S and S′, where S′ moves with a velocity $\vec{\mathbf{v}}$ relative to S.

Time is assumed to be the same in both inertial frames. That is, within the framework of classical mechanics, all clocks run at the same rate, regardless of their velocity, so that the time at which an event occurs for an observer in S is the same as the time for the same event in S′:

$$t' = t \tag{9.2}$$

Equations 9.1 and 9.2 constitute what is known as the **Galilean transformation of coordinates.**

Now suppose a particle moves through a displacement $dx$ in a time interval $dt$ as measured by an observer in S. It follows from the first of Equations 9.1 that the corresponding displacement $dx'$ measured by an observer in S′ is $dx' = dx - v dt$. Because $dt = dt'$ (Eq. 9.2), we find that

$$\frac{dx'}{dt'} = \frac{dx}{dt} - v$$

or

$$u'_x = u_x - v \tag{9.3}$$

where $u_x$ and $u'_x$ are the instantaneous $x$ components of velocity of the particle[2] relative to S and S′, respectively. This result, which is called the **Galilean velocity**

### PITFALL PREVENTION 9.1

**THE RELATIONSHIP BETWEEN THE S AND S′ FRAMES** Keep in mind the relationship between the S and S′ frames. Otherwise, many of the mathematical representations in this chapter could be misinterpreted. We choose the time $t = 0$ to be the instant at which the origins of the two coordinate systems coincide. The $x$ and $x'$ axes coincide except that their origins are different at all times other than $t = 0$. The $y$ and $y'$ axes (and the $z$ and $z'$ axes) are parallel, but they do not coincide for $t \neq 0$ because of the displacement of the origin of S′ with respect to that of S. If the S′ frame is moving in the positive $x$ direction relative to S, $v$ is positive; otherwise, it is negative.

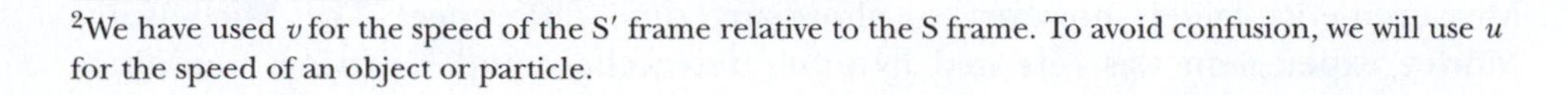

[2]We have used $v$ for the speed of the S′ frame relative to the S frame. To avoid confusion, we will use $u$ for the speed of an object or particle.

**transformation**, is used in everyday observations and is consistent with our intuitive notion of time and space. It is the same equation we generated in Section 3.6 (Eq. 3.22) when we first discussed relative velocity in one dimension. We find, however, that it leads to serious contradictions when applied to objects moving at high speeds.

## 9.2 THE MICHELSON–MORLEY EXPERIMENT

Many experiments similar to throwing the ball in the pickup truck, described in the preceding section, show us that the laws of classical mechanics are the same in all inertial frames of reference. When similar inquiries are made into the laws of other branches of physics, however, the results are contradictory. In particular, the laws of electricity and magnetism are found to depend on the frame of reference used. It might be argued that these laws are wrong, but that is difficult to accept because the laws are in total agreement with known experimental results. The Michelson–Morley experiment was one of many attempts to investigate this dilemma.

The experiment stemmed from a misconception early physicists had concerning the manner in which light propagates. The properties of mechanical waves, such as water and sound waves, were well known, and all these waves require a *medium* to support the propagation of the disturbance, as we shall discuss in Chapter 13. For sound from your stereo system, the medium is the air, and for ocean waves, the medium is the water surface. In the 19th century, physicists subscribed to a model for light in which electromagnetic waves also require a medium through which to propagate. They proposed that such a medium exists, filling all space, and they named it the **luminiferous ether.** The ether would define an **absolute frame of reference** in which the speed of light is $c$.

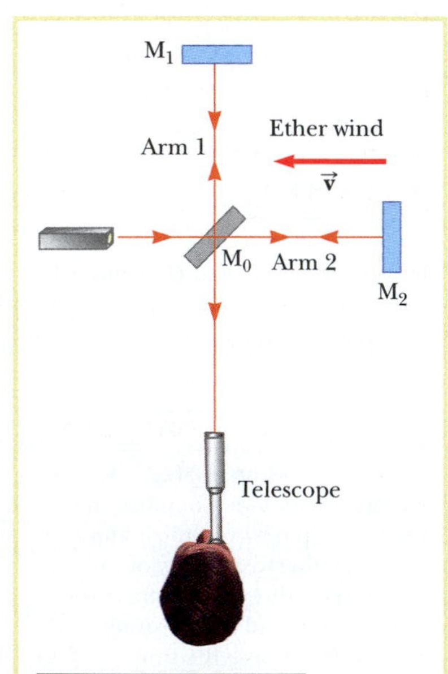


ACTIVE FIGURE 9.3
In the Michelson interferometer, the ether theory claims that the time of travel for a light beam traveling from the beam splitter to mirror $M_1$ and back will be different from that for a light beam traveling from the beam splitter to mirror $M_2$ and back. The interferometer is sufficiently sensitive to detect this time difference.

**PhysicsNow™** Log into PhysicsNow at **www.pop4e.com** and go to Active Figure 9.3 to adjust the speed of the ether wind and see the effect on the light beams if there were an ether.

The most famous experiment designed to show the presence of the ether was performed in 1887 by A. A. Michelson (1852–1931) and E. W. Morley (1838–1923). The objective was to determine the speed of the Earth through space with respect to the ether, and the experimental tool used was a device called the *interferometer,* shown schematically in Active Figure 9.3.

Light from the source at the left encounters a beam splitter $M_0$, which is a partially silvered mirror. Part of the light passes through toward mirror $M_2$, and the other part is reflected upward toward mirror $M_1$. Both mirrors are the same distance from the beam splitter. After reflecting from these mirrors, the light returns to the beam splitter, and part of each light beam propagates toward the observer at the bottom.

Suppose one arm of the interferometer (Arm 2, in Active Fig. 9.3) is aligned along the direction of the velocity $\vec{v}$ of the Earth through space and therefore through the ether. The "ether wind" blowing in the direction opposite the Earth's motion should cause the speed of light, as measured in the Earth's frame of reference, to be $c - v$ as the light approaches mirror $M_2$ in Active Figure 9.3 and $c + v$ after reflection.

The other arm (Arm 1) is perpendicular to the ether wind. For light to travel in this direction, the vector $\vec{c}$ must be aimed "upstream" so that the vector addition of $\vec{c}$ and $\vec{v}$ gives the speed of the light perpendicular to the ether wind as $\sqrt{c^2 - v^2}$. This situation is similar to Example 3.6, in which a boat crosses a river with a current. The boat is a model for the light beam in the Michelson–Morley experiment, and the river current is a model for the ether wind.

Because they travel in perpendicular directions with different speeds, light beams leaving the beam splitter simultaneously will arrive back at the beam splitter at different times. The interferometer is designed to detect this time difference. Measurements failed, however, to show any time difference! The Michelson–Morley experiment was repeated by other researchers under varying conditions and at different locations, but the results were always the same: *No time difference of the magnitude required was ever observed.*[3]

The negative result of the Michelson–Morley experiment not only contradicted the ether hypothesis, but it also meant that it was impossible to measure the absolute speed of the Earth with respect to the ether frame. From a theoretical viewpoint, it was impossible to find the absolute frame. As we shall see in the next section, however, Einstein offered a postulate that places a different interpretation on the negative result. In later years, when more was known about the nature of light, the idea of an ether that permeates all space was abandoned. **Light is now understood to be an electromagnetic wave that requires no medium for its propagation.** As a result, an ether through which light travels is an unnecessary construct.

Modern versions of the Michelson–Morley experiment have placed an upper limit of about 5 cm/s = 0.05 m/s on ether wind velocity. We can show that the speed of the Earth in its orbit around the Sun is $2.97 \times 10^4$ m/s, six orders of magnitude larger than the upper limit of ether wind velocity! These results have shown quite conclusively that the motion of the Earth has no effect on the measured speed of light.

(AIP Emilio Segrè Visual Archives, Michelson Collection)

**ALBERT A. MICHELSON** (1852–1931)

Michelson was born in Prussia in a town that later became part of Poland. He moved to the United States as a small child and spent much of his adult life making accurate measurements of the speed of light. In 1907, he was the first American to be awarded the Nobel Prize in Physics, which he received for his work in optics. His most famous experiment, conducted with Edward Morley in 1887, indicated that it was impossible to measure the absolute velocity of the Earth with respect to the ether.

## 9.3 EINSTEIN'S PRINCIPLE OF RELATIVITY

In the preceding section, we noted the failure of experiments to measure the relative speed between the ether and the Earth. Einstein proposed a theory that boldly removed these difficulties and at the same time completely altered our notion of space and time.[4] He based his relativity theory on two postulates:

1. **The principle of relativity:** All the laws of physics are the same in all inertial reference frames.
2. **The constancy of the speed of light:** The speed of light in vacuum has the same value in all inertial frames, regardless of the velocity of the observer or the velocity of the source emitting the light.

These postulates form the basis of **special relativity,** which is the relativity theory applied to observers moving with constant velocity. The first postulate asserts that *all* the laws of physics—those dealing with mechanics, electricity and magnetism, optics, thermodynamics, and so on—are the same in all reference frames moving with constant velocity relative to each other. This postulate is a sweeping generalization of the principle of Newtonian relativity that only refers to the laws of mechanics. From an experimental point of view, Einstein's principle of relativity means that any kind of experiment performed in a laboratory at rest must agree with the same laws of physics as when performed in a laboratory moving at constant velocity relative to the first one. Hence, no preferred inertial reference frame exists and it is impossible to detect absolute motion.

Note that postulate 2, the principle of the constancy of the speed of light, is required by postulate 1: If the speed of light were not the same in all inertial frames, it would be possible to experimentally distinguish between inertial frames and a

[3]From an Earth observer's point of view, changes in the Earth's speed and direction of motion in the course of a year are viewed as ether wind shifts. Even if the speed of the Earth with respect to the ether were zero at some time, six months later the Earth is moving in the opposite direction, the speed of the Earth with respect to the ether would be nonzero, and a clear time difference should be detected. None has ever been observed, however.

[4]A. Einstein, "On the Electrodynamics of Moving Bodies," *Ann. Physik* 17:891, 1905. For an English translation of this article and other publications by Einstein, see the book by H. Lorentz, A. Einstein, H. Minkowski, and H. Weyl, *The Principle of Relativity* (New York: Dover, 1958).

(AIP Niels Bohr Library)

**ALBERT EINSTEIN** (1879–1955)

Einstein, one of the greatest physicists of all times, was born in Ulm, Germany. He left Germany in 1932 for the United States and became a U.S. citizen in 1940. In 1905, at the age of 26, he published four scientific papers that revolutionized physics. Two of these papers were concerned with what is now considered his most important contribution of all, the special theory of relativity.

In 1916, Einstein published his work on the general theory of relativity. The most dramatic prediction of this theory is the degree to which light is deflected by a gravitational field. Measurements made by astronomers on bright stars in the vicinity of the eclipsed sun in 1919 confirmed Einstein's prediction, and Einstein suddenly became a world celebrity.

Einstein was deeply disturbed by the development of quantum mechanics in the 1920s despite his own role as a scientific revolutionary. In particular, he could never accept the probabilistic view of events in nature that is a central feature of quantum theory. The last few decades of his life were devoted to an unsuccessful search for a unified theory that would combine gravitation and electromagnetism into one picture.

preferred, absolute frame in which the speed of light is $c$, in contradiction to postulate 1. Postulate 2 also eliminates the problem of measuring the speed of the ether by denying the existence of the ether and boldly asserting that light always moves with speed $c$ relative to all inertial observers.

## 9.4 CONSEQUENCES OF SPECIAL RELATIVITY

If we accept the postulates of special relativity, we must conclude that relative motion is unimportant when measuring the speed of light, which is the lesson of the Michelson–Morley experiment. At the same time, we must alter our commonsense notion of space and time and be prepared for some very unexpected consequences, as we shall see now.

### Simultaneity and the Relativity of Time

A basic premise of Newtonian mechanics is that a universal time scale exists that is the same for all observers. In fact, Newton wrote, "Absolute, true, and mathematical time, of itself, and from its own nature, flows equably without relation to anything external." Thus, Newton and his followers simply took simultaneity for granted. In his development of special relativity, Einstein abandoned the notion that two events that appear simultaneous to one observer appear simultaneous to all observers. According to Einstein, **a time measurement depends on the reference frame in which the measurement is made.**

Einstein devised the following thought experiment to illustrate this point. A boxcar moves with uniform velocity and two lightning bolts strike its ends, as in Figure 9.4a, leaving marks on the boxcar and on the ground. The marks on the boxcar are labeled $A'$ and $B'$, and those on the ground are labeled $A$ and $B$. An observer at $O'$ moving with the boxcar is midway between $A'$ and $B'$, and a ground observer at $O$ is midway between $A$ and $B$. The events recorded by the observers are the arrivals of light signals from the lightning bolts.

The two light signals reach observer $O$ at the same time as indicated in Figure 9.4b. As a result, $O$ concludes that the events at $A$ and $B$ occurred simultaneously. Now consider the same events as viewed by the observer on the boxcar at $O'$. From our frame of reference, at rest with respect to the tracks in Figure 9.4, we see the lightning strikes occur as $A'$ passes $A$, $O'$ passes $O$, and $B'$ passes $B$. By the time the light has reached observer $O$, observer $O'$ has moved as indicated in Figure 9.4b. Therefore, the light signal from $B'$ has already swept past $O'$ because it had less distance to travel, but the light from $A'$ has not yet reached $O'$. According to Einstein, **observer $O'$ must find that light travels at the same speed as that measured by observer $O$.** Observer $O'$ therefore concludes that the lightning struck the front of the boxcar before it struck the back. This thought experiment clearly demonstrates that the two events, which appear to be simultaneous to observer $O$, do not appear to be simultaneous to observer $O'$. In general, two events separated in space and observed to be simultaneous in one reference frame are not observed to be simultaneous in a second frame moving relative to the first. That is, simultaneity is not an absolute concept but one that depends on the state of motion of the observer.

Einstein's thought experiment demonstrates that two observers can disagree on the simultaneity of two events. **This disagreement, however, depends on the transit time of light to the observers and therefore does *not* demonstrate the deeper meaning of relativity.** In relativistic analyses of high-speed situations, relativity shows that **simultaneity is relative even when the transit time is subtracted out.** In fact, all the relativistic effects that we will discuss from here on will assume that we are ignoring differences caused by the transit time of light to the observers.

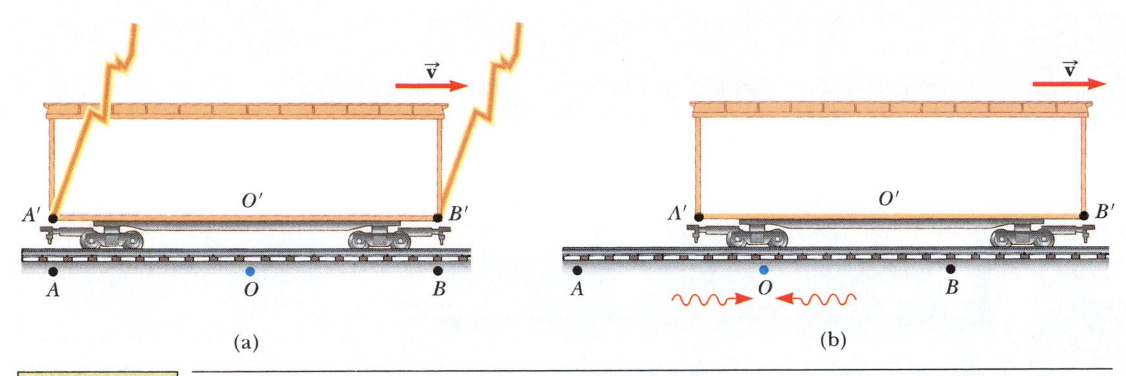


**FIGURE 9.4** (a) Two lightning bolts strike the ends of a moving boxcar. (b) The events appear to be simultaneous to the observer at $O$, who is standing on the ground midway between $A$ and $B$. The events do not appear to be simultaneous to the observer $O'$ riding on the boxcar, who claims that the front of the car is struck before the rear. Note that the leftward-traveling light signal from $B'$ has already passed observer $O'$, but the rightward-traveling light signal from $A'$ has not yet reached $O'$.

## Time Dilation

According to the preceding paragraph, observers in different inertial frames measure different time intervals between a pair of events, independent of the transit time of the light. This situation can be illustrated by considering a vehicle moving to the right with a speed $v$ as in the pictorial representation in Active Figure 9.5a. A mirror is fixed to the ceiling of the vehicle, and observer $O'$, at rest in a frame attached to the vehicle, holds a flashlight a distance $d$ below the mirror. At some instant, the flashlight is turned on momentarily and emits a pulse of light (event 1) directed toward the mirror. At some later time after reflecting from the mirror, the pulse arrives back at the flashlight (event 2). Observer $O'$ carries a clock that she uses to measure the time interval $\Delta t_p$ between these two events. (The subscript $p$ stands for "proper," as will be discussed shortly.) Because the light pulse has a constant speed $c$, the time interval required for the pulse to travel from $O'$ to the mirror and back to $O'$ (a distance of $2d$) can be found by modeling the light pulse as a particle under constant speed as discussed in Chapter 2:

$$\Delta t_p = \frac{2d}{c} \qquad [9.4]$$

This time interval $\Delta t_p$ is measured by $O'$, for whom the two events occur at the same spatial position.

Now consider the same pair of events as viewed by observer $O$ at rest with respect to a second frame attached to the ground as in Active Figure 9.5b. According to this observer, the mirror and flashlight are moving to the right with a speed $v$. The geometry appears to be entirely different as viewed by this observer. By the time the light from the flashlight reaches the mirror, the mirror has moved horizontally a distance $v\,\Delta t/2$, where $\Delta t$ is the time interval required for the light to travel from the flashlight to the mirror and back to the flashlight as measured by observer $O$. In other words, the second observer concludes that because of the motion of the vehicle, if the light is to hit the mirror, it must leave the flashlight at an angle with respect to the vertical direction. Comparing Active Figures 9.5a and 9.5b, we see that the light must travel farther when observed in the second frame than in the first frame.

According to the second postulate of special relativity, both observers must measure $c$ for the speed of light. Because the light travels farther in the second frame but at the same speed, it follows that the time interval $\Delta t$ measured by the observer in the second frame is longer than the time interval $\Delta t_p$ measured by the observer in the first frame. To obtain a relationship between these two time intervals, it is

**PITFALL PREVENTION 9.2**

**WHO'S RIGHT?** At this point, you might wonder which observer in Figure 9.4 is correct concerning the two events. *Both are correct* because the principle of relativity states that *no inertial frame of reference is preferred.* Although the two observers reach different conclusions, both are correct in their own reference frame because the concept of simultaneity is not absolute. In fact, the central point of relativity is that any uniformly moving frame of reference can be used to describe events and do physics.

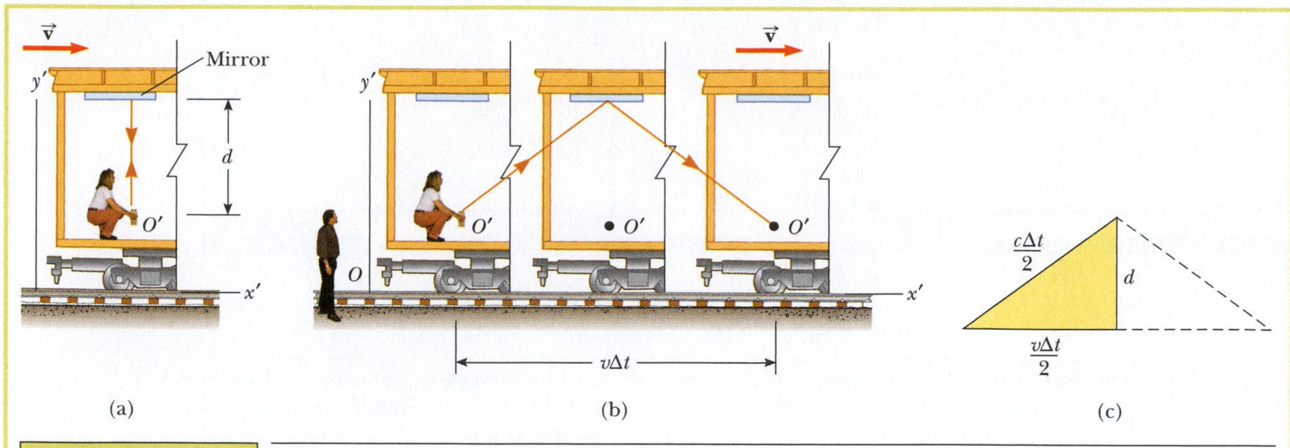


**ACTIVE FIGURE 9.5** A mirror is fixed to a moving vehicle, and two observers measure the time interval between two events: the leaving of a light pulse from a flashlight and the arrival of the reflected light pulse back at the flashlight. (a) Observer $O'$, riding on the vehicle, sees the light pulse travel a total distance of $2d$ and measures a time interval between the events of $\Delta t_p$. (b) Observer $O$ is standing on the Earth and sees the mirror and $O'$ move to the right with speed $v$. Observer $O$ measures the distance that the light pulse travels to be greater than $2d$ and measures a time interval between the events of $\Delta t$. (c) The right triangle for calculating the relationship between $\Delta t$ and $\Delta t_p$.

**PhysicsNow™** By logging into PhysicsNow at **www.pop4e.com** and going to Active Figure 9.5 you can observe the bouncing of the light pulse for various speeds of the train.

convenient to use the right triangle geometric model shown in Active Figure 9.5c. The Pythagorean theorem applied to the triangle gives

$$\left(\frac{c\,\Delta t}{2}\right)^2 = \left(\frac{v\,\Delta t}{2}\right)^2 + d^2$$

Solving for $\Delta t$ gives

$$\Delta t = \frac{2d}{\sqrt{c^2 - v^2}} = \frac{2d}{c\sqrt{1 - \frac{v^2}{c^2}}} \qquad [9.5]$$

Because $\Delta t_p = 2d/c$, we can express Equation 9.6 as

$$\Delta t = \frac{\Delta t_p}{\sqrt{1 - \frac{v^2}{c^2}}} = \gamma\,\Delta t_p \qquad [9.6]$$

where $\gamma = (1 - v^2/c^2)^{-1/2}$. This result says that **the time interval $\Delta t$ measured by $O$ is longer than the time interval $\Delta t_p$ measured by $O'$** because $\gamma$ is always greater than one. That is, $\Delta t > \Delta t_p$. This effect is known as **time dilation.**

We can see that time dilation is not observed in our everyday lives by considering the factor $\gamma$. This factor deviates significantly from a value of 1 only for very high speeds, as shown in Table 9.1. For example, for a speed of $0.1c$, the value of $\gamma$ is 1.005. Therefore, a time dilation of only 0.5% occurs at one-tenth the speed of light. Speeds we encounter on an everyday basis are far slower than that, so we do not see time dilation in normal situations.

The time interval $\Delta t_p$ in Equation 9.6 is called the **proper time interval.** In general, the proper time interval is defined as **the time interval between two events as measured by an observer for whom the events occur at the same point in space.** In our case, observer $O'$ measures the proper time interval. For us to be able to use Equation 9.6, the events must occur at the same spatial position in *some* inertial

**TABLE 9.1**

**Approximate Values for $\gamma$ at Various Speeds**

| $v/c$ | $\gamma$ |
|---|---|
| 0.001 0 | 1.000 000 5 |
| 0.010 | 1.000 05 |
| 0.10 | 1.005 |
| 0.20 | 1.021 |
| 0.30 | 1.048 |
| 0.40 | 1.091 |
| 0.50 | 1.155 |
| 0.60 | 1.250 |
| 0.70 | 1.400 |
| 0.80 | 1.667 |
| 0.90 | 2.294 |
| 0.92 | 2.552 |
| 0.94 | 2.931 |
| 0.96 | 3.571 |
| 0.98 | 5.025 |
| 0.99 | 7.089 |
| 0.995 | 10.01 |
| 0.999 | 22.37 |

frame. Therefore, for instance, this equation cannot be used to relate the measurements made by the two observers in the lightning example described at the beginning of this section because the lightning strikes occur at different positions for both observers.

If a clock is moving with respect to you, the time interval between ticks of the moving clock is measured to be longer than the time interval between ticks of an identical clock in your reference frame. Therefore, it is often said that a moving clock is measured to run more slowly than a clock in your reference frame by a factor $\gamma$. That is true for mechanical clocks as well as for the light clock just described. We can generalize this result by stating that all physical processes, including chemical and biological ones, slow down relative to a stationary clock when those processes occur in a frame moving with respect to the clock. For example, the heartbeat of an astronaut moving through space would keep time with a clock inside the spaceship. Both the astronaut's clock and heartbeat would be measured to be slowed down according to an observer on the Earth comparing time intervals with his own clock at rest with respect to him (although the astronaut would have no sensation of life slowing down in the spaceship).

Time dilation is a verifiable phenomenon; let us look at one situation in which the effects of time dilation can be observed and that served as an important historical confirmation of the predictions of relativity. Muons are unstable elementary particles that have a charge equal to that of the electron and a mass 207 times that of the electron. Muons can be produced as a result of collisions of cosmic radiation with atoms high in the atmosphere. Slow-moving muons in the laboratory have a lifetime measured to be the proper time interval $\Delta t_p = 2.2\ \mu\text{s}$. If we assume that the speed of atmospheric muons is close to the speed of light, we find that these particles can travel a distance during their lifetime of approximately $(3.0 \times 10^8\ \text{m/s})(2.2 \times 10^{-6}\ \text{s}) \approx 6.6 \times 10^2\ \text{m}$ before they decay (Fig. 9.6a). Hence, they are unlikely to reach the surface of the Earth from high in the atmosphere where they are produced; nonetheless, experiments show that a large number of muons *do* reach the surface. The phenomenon of time dilation explains this effect. As measured by an observer on the Earth, the muons have a dilated lifetime equal to $\gamma\Delta t_p$. For example, for $v = 0.99c$, $\gamma \approx 7.1$ and $\gamma\,\Delta t_p \approx 16\ \mu\text{s}$. Hence, the average distance traveled by the muons in this time interval as measured by an observer on the Earth is approximately $(3.0 \times 10^8\ \text{m/s})(16 \times 10^{-6}\ \text{s}) \approx 4.8 \times 10^3\ \text{m}$, as shown in Figure 9.6b.

The results of an experiment reported by J. C. Hafele and R. E. Keating provided direct evidence of time dilation.[5] The experiment involved the use of very stable atomic clocks. Time intervals measured with four such clocks in jet flight were compared with time intervals measured by reference clocks located at the U.S. Naval Observatory. Their results were in good agreement with the predictions of special relativity and can be explained in terms of the relative motion between the Earth's rotation and the jet aircraft. In their paper, Hafele and Keating report the following: "Relative to the atomic time scale of the U.S. Naval Observatory, the flying clocks lost 59 ± 10 ns during the eastward trip and gained 273 ± 7 ns during the westward trip."

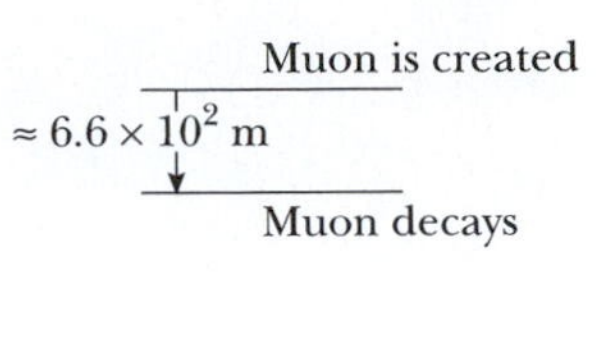


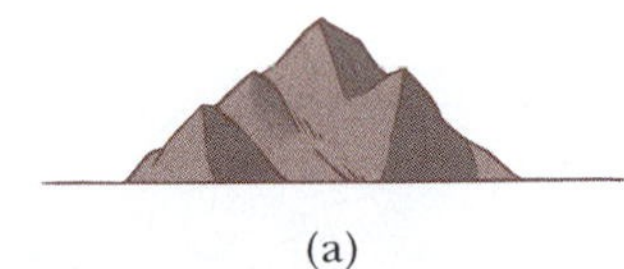

(a)

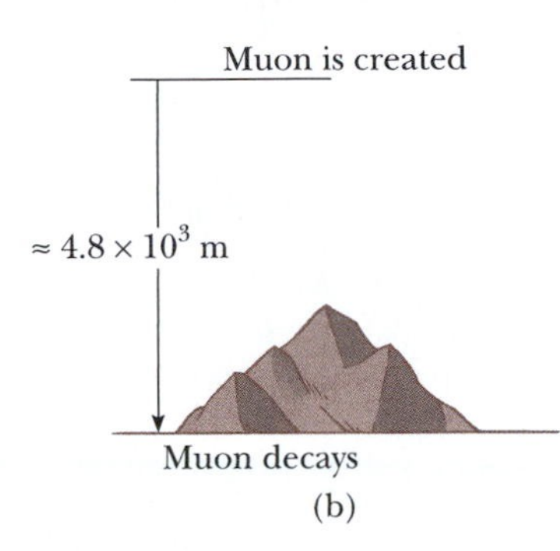


(b)

**FIGURE 9.6** (a) Without relativistic considerations, muons created in the atmosphere and traveling downward with a speed of $0.99c$ would travel only about $6.6 \times 10^2$ m before decaying with an average lifetime of 2.2 $\mu$s. Therefore, very few muons would reach the surface of the Earth. (b) With relativistic considerations, the muon's lifetime is dilated according to an observer on Earth. As a result, according to this observer, the muon can travel about $4.8 \times 10^3$ m before decaying, which results in many of them arriving at the surface.

**QUICK QUIZ 9.1** Suppose the observer $O'$ on the train in Active Figure 9.5 aims her flashlight at the far wall of the boxcar and turns it on and off, sending a pulse of light toward the far wall. Both $O'$ and $O$ measure the time interval between when the pulse leaves the flashlight and when it hits the far wall. Which observer measures the proper time interval between these two events? **(a)** $O'$ **(b)** $O$ **(c)** both observers **(d)** neither observer

[5] J. C. Hafele and R. E. Keating, "Around the World Atomic Clocks: Relativistic Time Gains Observed," *Science*, July 14, 1972, p. 168.

**QUICK QUIZ 9.2** A crew watches a movie that is 2 h long in a spacecraft that is moving at high speed through space. Will an Earth-bound observer, who is watching the movie through a powerful telescope, measure the duration of the movie to be (**a**) longer than, (**b**) shorter than, or (**c**) equal to 2 h?

## The Twin Paradox

An intriguing consequence of time dilation is the so-called twin paradox (Fig. 9.7). Consider an experiment involving a set of twins named Speedo and Goslo. At age 20, Speedo, the more adventuresome of the two, sets out on an epic journey to Planet X, located 20 lightyears (ly) from the Earth. (Note that 1 ly is the distance light travels through free space in 1 year. It is equal to $9.46 \times 10^{15}$ m.) Furthermore, his spaceship is capable of reaching a speed of $0.95c$ relative to the inertial frame of his twin brother back home. After reaching Planet X, Speedo becomes homesick and immediately returns to the Earth at the same speed $0.95c$. Upon his return, Speedo is shocked to discover that Goslo has aged 42 yr and is now 62 yr old. Speedo, on the other hand, has aged only 13 yr.

At this point, it is fair to raise the following question: Which twin is the traveler and which is really younger as a result of this experiment? From Goslo's frame of reference, he was at rest while his brother traveled at a high speed away from him and then came back. According to Speedo, however, he himself remained stationary while Goslo and the Earth raced away from him and then headed back. There is an apparent contradiction due to the apparent symmetry of the observations. Which twin has developed signs of excess aging?

Varying rates of aging in relativity

The situation in this problem is actually not symmetrical. To resolve this apparent paradox, recall that the special theory of relativity describes observations made in inertial frames of reference moving relative to each other. Speedo, the space traveler, must experience a series of accelerations during his journey because he must fire his rocket engines to slow down and start moving back toward the Earth. As a result, his speed is not always uniform, and consequently he is not always in an inertial frame. Therefore, there is no paradox because only Goslo, who is always in a single inertial frame, can make correct predictions based on special relativity. During each passing year noted by Goslo, slightly less than 4 months elapses for Speedo.

Only Goslo, who is in a single inertial frame, can apply the simple time dilation formula to Speedo's trip. Therefore, Goslo finds that instead of aging 42 yr, Speedo ages only $(1 - v^2/c^2)^{1/2}(42 \text{ yr}) = 13$ yr. According to both twins, Speedo spends 6.5 yr traveling to Planet X and 6.5 yr returning, for a total travel time of 13 yr.

**FIGURE 9.7** (a) As Speedo leaves his twin brother, Goslo, on the Earth, both are the same age. (b) When Speedo returns from his journey to Planet X, he is younger than Goslo.

**QUICK QUIZ 9.3** Suppose astronauts are paid according to the amount of time they spend traveling in space. After a long voyage traveling at a speed approaching $c$, would a crew rather be paid according to (**a**) an Earth-based clock, (**b**) their spacecraft's clock, or (**c**) either clock?

## Thinking Physics 9.1

Suppose a student explains time dilation with the following argument: If I start running away from a clock at 12:00 at a speed very close to the speed of light, I would not see the time change, because the light from the clock representing 12:01 would never reach me. What is the flaw in this argument?

**Reasoning** The implication in this argument is that the velocity of light relative to the runner is approximately *zero* because "the light . . . would never reach me." In this Galilean point of view, the relative velocity is a simple subtraction of running velocity from the light velocity. From the point of view of special relativity, one of the fundamental postulates is that the speed of light is the same for all observers, *including one running away from the light source at the speed of light.* Therefore, the light from 12:01 will move toward the runner at the speed of light, as measured by all observers, including the runner. ■

### EXAMPLE 9.1 What Is the Heart Rate of the Astronaut?

An astronaut at rest on the Earth has a heartbeat rate of 70 beats/min.

**A** When the astronaut is traveling on a spacecraft at $0.95c$, what will this rate be as measured by another observer on the spacecraft?

**Solution** The proper time interval between beats of the heart is the period $T_p = 60\text{ s}/70\text{ beats} = 0.86\text{ s/beat}$. The observer in the spacecraft will measure this time interval because two successive beats of the heart take place at the same position according to this observer.

**B** When the astronaut is traveling on a spacecraft at $0.95c$, what will this rate be as measured by an observer at rest on the Earth?

**Solution** For the time interval measured by the observer on the Earth, Equation 9.6 gives

$$T = \gamma T_p = \frac{1}{\sqrt{1 - \dfrac{(0.95c)^2}{c^2}}}\, T_p$$

$$= (3.2)(0.86\text{ s}) = 2.7\text{ s}$$

Therefore, the heart rate as measured by this observer is

$$\text{Heart rate} = \frac{1}{2.7\text{ s/beat}} = 0.36\text{ beat/s}\left(\frac{60\text{ s}}{1\text{ min}}\right)$$

$$= 22\text{ beats/min}$$

That is, measurements by this observer show that the heart rate is measured to slow down compared with that measured by an observer on the spacecraft.

## Length Contraction

The measured distance between two points also depends on the frame of reference. The **proper length** of an object is defined as **the distance in space between the end points of the object measured by someone who is at rest relative to the object.** An observer in a reference frame that is moving with respect to the object will measure a length along the direction of the velocity that is always less than the proper length. This effect is known as **length contraction.** Although we have introduced this effect through the mental representation of an object, the object is not

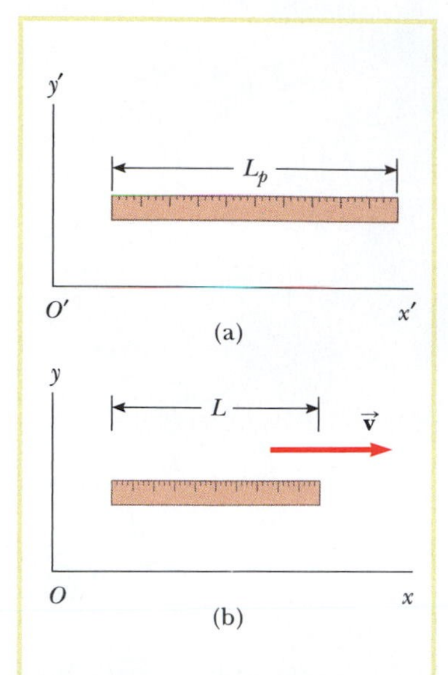


ACTIVE FIGURE 9.8
(a) According to an observer in a frame attached to the meter stick (that is, both the stick and the frame have the same velocity), the stick is measured to have its proper length $L_p$. (b) According to an observer in a frame in which the meter stick has a velocity $\vec{v}$ relative to the frame, the stick is measured to be *shorter* than the proper length $L_p$ by a factor $(1 - v^2/c^2)^{1/2}$.

**Physics Now™** By logging into PhysicsNow at **www.pop4e.com** and going to Active Figure 9.8 you can view the meter stick from the points of view of two observers to compare the measured length of the stick.

necessary. **The distance between *any* two points in space is measured by an observer to be contracted along the direction of the velocity of the observer relative to the points.**

Consider a spacecraft traveling with a speed $v$ from one star to another. We will consider the time interval between two events: (1) the leaving of the spacecraft from the first star and (2) the arrival of the spacecraft at the second star. There are two observers: one on the Earth and the other in the spacecraft. The observer at rest on the Earth (and also at rest with respect to the two stars) measures the distance between the stars to be $L_p$, the proper length. According to this observer, the time interval required for the spacecraft to complete the voyage is $\Delta t = L_p/v$. What does an observer in the moving spacecraft measure for the distance between the stars? This observer measures the proper time interval because the passage of each of the two stars by his spacecraft occurs at the same position in his reference frame, at his spacecraft. Therefore, because of time dilation, the time interval required to travel between the stars as measured by the space traveler will be smaller than that for the Earth-bound observer. Using the time dilation expression, the proper time interval between events is $\Delta t_p = \Delta t/\gamma$. The space traveler claims to be at rest and sees the destination star moving toward the spacecraft with speed $v$. Because the space traveler reaches the star in the time interval $\Delta t_p < \Delta t$, he concludes that the distance $L$ between the stars is shorter than $L_p$. This distance measured by the space traveler is

$$L = v\,\Delta t_p = v\frac{\Delta t}{\gamma}$$

Because $L_p = v\,\Delta t$, we see that

$$L = \frac{L_p}{\gamma} = L_p\left(1 - \frac{v^2}{c^2}\right)^{1/2} \qquad [9.7]$$

Because $(1 - v^2/c^2)^{1/2}$ is less than 1, the space traveler measures a length that is shorter than the proper length. Therefore, an observer in motion with respect to two points in space measures the length $L$ between the points (along the direction of motion) to be shorter than the length $L_p$ measured by an observer at rest with respect to the points (the proper length).

Note that **length contraction takes place only along the direction of motion.** For example, suppose a meter stick moves past an Earth observer with speed $v$ as in Active Figure 9.8. The length of the meter stick as measured by an observer in a frame attached to the stick is the proper length $L_p$ as in Active Figure 9.8a. The length $L$ of the stick measured by the Earth observer is shorter than $L_p$ by the factor $(1 - v^2/c^2)^{1/2}$, but the width is the same. Furthermore, length contraction is a symmetric effect. If the stick is at rest on the Earth, an observer in the moving frame would also measure its length to be shorter by the same factor $(1 - v^2/c^2)^{1/2}$.

It is important to emphasize that the proper length and proper time interval are defined differently. The proper length is measured by an observer at rest with respect to the end points of the length. The proper time interval between two events is measured by someone for whom the events occur at the same position. Often, the proper time interval and the proper length are not measured by the same observer. As an example, let us return to the decaying muons moving at speeds close to the speed of light. An observer in the muon's reference frame would measure the proper lifetime, and an Earth-based observer would measure the proper length (the distance from creation to decay in Fig. 9.6). In the muon's reference frame, no time dilation occurs, but the distance of travel is observed to be shorter when measured in this frame. Likewise, in the Earth observer's reference frame, a time dilation does occur, but the distance of travel is measured to be the proper length. Therefore, when calculations on the muon are performed in both frames, the outcome of the experiment in one frame is the same as the outcome in the other frame: More muons reach the surface than would be predicted without relativistic calculations.

**QUICK QUIZ 9.4** You are packing for a trip to another star. During the journey, you will be traveling at $0.99c$. You are trying to decide whether you should buy smaller sizes of your clothing because you will be thinner on your trip as a result of length contraction. You are also considering saving money by reserving a smaller cabin to sleep in because you will be shorter when you lie down. Should you **(a)** buy smaller sizes of clothing, **(b)** reserve a smaller cabin, **(c)** do neither, or **(d)** do both?

### EXAMPLE 9.2 A Voyage to Sirius

An astronaut takes a trip to Sirius, located 8.00 ly from the Earth. The astronaut measures the time interval for the one-way journey to be 6.00 yr. If the spacecraft moves at a constant speed of $0.800c$, how can the 8.00-ly distance be reconciled with the 6.00-yr duration measured by the astronaut?

**Solution** The 8.00 ly represents the proper length (the distance from the Earth to Sirius) measured by an observer for whom both the Earth and Sirius are at rest. The astronaut measures Sirius to be approaching her at $0.800c$ but also measures the distance contracted to

$$L = \frac{L_p}{\gamma} = \frac{8.00 \text{ ly}}{\gamma} = (8.00 \text{ ly})\sqrt{1 - \frac{v^2}{c^2}}$$

$$= (8.00 \text{ ly})\sqrt{1 - \frac{(0.800c)^2}{c^2}} = 4.80 \text{ ly}$$

So the travel time measured on her clock is

$$\Delta t = \frac{L}{v} = \frac{4.80 \text{ ly}}{0.800c} = \frac{4.80 \text{ ly}}{0.800(1.00 \text{ ly/yr})} = 6.00 \text{ yr}$$

Notice that we have used the speed of light as $c = 1.00$ ly/yr to determine this last result.

### EXAMPLE 9.3 The Triangular Spacecraft

A spacecraft in the form of a triangle flies by an observer on the Earth with a speed of $0.95c$ along the $x$ direction. According to an observer on the spacecraft (Fig. 9.9a), the distances $L_p$ and $y$ are measured to be 52 m and 25 m, respectively. What are the dimensions of the spacecraft as measured by the Earth observer when the spacecraft is in motion along the direction shown in Figure 9.9b?

**Solution** In Figure 9.9a, we show the shape of the spacecraft as measured[6] by the observer on the spacecraft. The proper length along the direction of motion

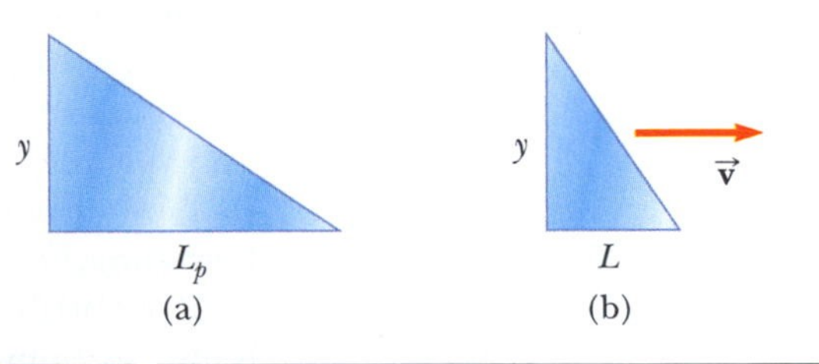


**FIGURE 9.9** (Example 9.3) (a) When the spacecraft is at rest, its shape is measured as shown. (b) The spacecraft is measured to have this shape when it moves to the right with a speed $v$. Note that only its $x$ dimension is contracted in this case.

[6]Notice that we are careful here to say the shape "as measured" by an observer rather than "as seen" by an observer. What an observer *sees* when looking at an object is the set of light rays entering the eye at a given instant. These rays left different parts of the object at different times because different parts of the object are at different distances from the eye. Figure 9.9b is a representation of the light rays leaving different parts of the object *simultaneously*. Viewing an object moving at high speed introduces additional changes to the object besides length contraction, including apparent rotations of the object.

is $L_p = 52$ m. The Earth observer watching the moving spacecraft measures the horizontal length of the spacecraft to be contracted to

$$L = L_p\sqrt{1 - \frac{v^2}{c^2}} = (52 \text{ m})\sqrt{1 - \frac{(0.95c)^2}{c^2}} = 16 \text{ m}$$

The 25-m vertical height is unchanged because it is perpendicular to the direction of relative motion between observer and spacecraft. Figure 9.9b represents the spacecraft's shape as measured by the Earth observer.

## 9.5 THE LORENTZ TRANSFORMATION EQUATIONS

Suppose an event that occurs at some point $P$ is reported by two observers: one at rest in a frame S and another in a frame S′ that is moving to the right with speed $v$ as in Figure 9.10. The observer in S reports the event with space–time coordinates $(x, y, z, t)$, and the observer in S′ reports the same event using the coordinates $(x', y', z', t')$. If two events occur at $P$ and $Q$ in Figure 9.10, Equation 9.1 predicts that $\Delta x = \Delta x'$; that is, the distance between the two points in space at which the events occur does not depend on the motion of the observer. Because this notion is contradictory to that of length contraction, the Galilean transformation is not valid when $v$ approaches the speed of light. In this section, we state the correct transformation equations that apply for all speeds in the range $0 \le v < c$.

S, S′, y, y′, $\vec{v}$, P (event), Q (event), vt, x′, Δx′, x, Δx, O, x, O′, x′

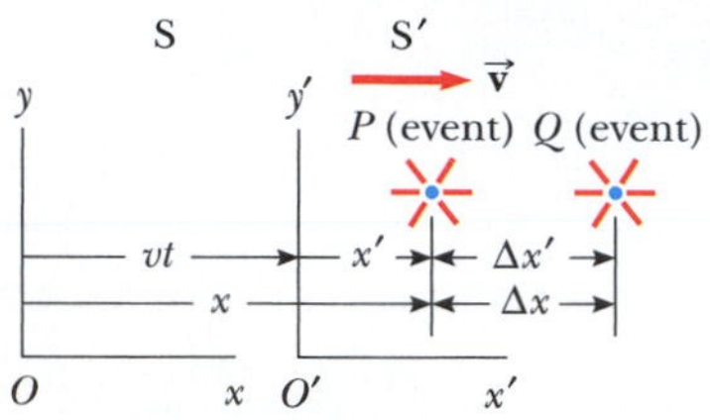

FIGURE 9.10 Events occur at points $P$ and $Q$ and are observed by an observer at rest in the S frame and another in the S′ frame, which is moving to the right with a speed $v$.

The equations that relate these measurements and enable us to transform coordinates from S to S′ are the **Lorentz transformation equations:**

■ Lorentz transformation for S→S′

$$\begin{aligned} x' &= \gamma(x - vt) \\ y' &= y \\ z' &= z \\ t' &= \gamma\left(t - \frac{v}{c^2}x\right) \end{aligned} \qquad [9.8]$$

These transformation equations were developed by Hendrik A. Lorentz (1853–1928) in 1890 in connection with electromagnetism. Einstein, however, recognized their physical significance and took the bold step of interpreting them within the framework of special relativity.

We see that the value for $t'$ assigned to an event by observer $O'$ depends both on the time $t$ and on the coordinate $x$ as measured by observer $O$. Therefore, in relativity, space and time are not separate concepts but rather are closely interwoven with each other in what we call **space–time.** This case is unlike that of the Galilean transformation in which $t = t'$.

If we wish to transform coordinates in the S′ frame to coordinates in the S frame, we simply replace $v$ by $-v$ and interchange the primed and unprimed coordinates in Equation 9.8:

■ Inverse Lorentz transformation for S′→S

$$\begin{aligned} x &= \gamma(x' + vt') \\ y &= y' \\ z &= z' \\ t &= \gamma\left(t' + \frac{v}{c^2}x'\right) \end{aligned} \qquad [9.9]$$

When $v \ll c$, the Lorentz transformation reduces to the Galilean transformation. To check, note that if $v \ll c$, $v^2/c^2 \ll 1$, so $\gamma$ approaches 1 and Equation 9.8 reduces in this limit to Equations 9.1 and 9.2:

$$x' = x - vt \qquad y' = y \qquad z' = z \qquad t' = t$$

## Lorentz Velocity Transformation

Let us now derive the **Lorentz velocity transformation,** which is the relativistic counterpart of the Galilean velocity transformation, Equation 9.3. Once again S′ is a frame of reference that moves at a speed $v$ relative to another frame S along the common $x$ and $x'$ axes. Suppose an object is measured in S′ to have an instantaneous velocity component $u'_x$ given by

$$u'_x = \frac{dx'}{dt'} \qquad [9.10]$$

Using Equations 9.8, we have

$$dx' = \gamma(dx - v\,dt) \qquad \text{and} \qquad dt' = \gamma\left(dt - \frac{v}{c^2}\,dx\right)$$

Substituting these values into Equation 9.10 gives

$$u'_x = \frac{dx'}{dt'} = \frac{dx - v\,dt}{dt - \dfrac{v}{c^2}\,dx} = \frac{\dfrac{dx}{dt} - v}{1 - \dfrac{v}{c^2}\dfrac{dx}{dt}}$$

Note, though, that $dx/dt$ is the velocity component $u_x$ of the object measured in S, so this expression becomes

$$u'_x = \frac{u_x - v}{1 - \dfrac{u_x v}{c^2}} \qquad [9.11]$$

▪ Lorentz velocity transformation for S → S′

Similarly, if the object has velocity components along $y$ and $z$, the components in S′ are

$$u'_y = \frac{u_y}{\gamma\left(1 - \dfrac{u_x v}{c^2}\right)} \qquad \text{and} \qquad u'_z = \frac{u_z}{\gamma\left(1 - \dfrac{u_x v}{c^2}\right)} \qquad [9.12]$$

When $u_x$ or $v$ is much smaller than $c$ (the nonrelativistic case), the denominator of Equation 9.11 approaches unity and so $u'_x \approx u_x - v$. This result corresponds to the Galilean velocity transformations. In the other extreme, when $u_x = c$, Equation 9.11 becomes

$$u'_x = \frac{c - v}{1 - \dfrac{cv}{c^2}} = \frac{c\left(1 - \dfrac{v}{c}\right)}{1 - \dfrac{v}{c}} = c$$

From this result, we see that an object whose speed approaches $c$ relative to an observer in S also has a speed approaching $c$ relative to an observer in S′, independent of the relative motion of S and S′. Note that this conclusion is consistent with Einstein's second postulate, namely, that the speed of light must be $c$ relative to all inertial frames of reference.

To obtain $u_x$ in terms of $u'_x$, we replace $v$ by $-v$ in Equation 9.11 and interchange the roles of primed and unprimed variables:

$$u_x = \frac{u'_x + v}{1 + \dfrac{u'_x v}{c^2}} \qquad [9.13]$$

▪ Inverse Lorentz velocity transformation for S′ → S

### PITFALL PREVENTION 9.3

**What can the observers agree on?** We have seen several measurements on which the two observers $O$ and $O'$ do not agree. These measurements include (1) the time interval between events that take place in the same position in one of the frames, (2) the distance between two points that remain fixed in one of their frames, (3) the velocity components of a moving particle, and (4) whether two events occurring at different locations in both frames are simultaneous. It is worth noting here what the two observers *can* agree on: (1) the relative speed $v$ with which they move with respect to each other, (2) the speed $c$ of any ray of light, and (3) the simultaneity of two events taking place at the same position and time in some frame.

**QUICK QUIZ 9.5** You are driving on a freeway at a relativistic speed. Straight ahead of you, a technician standing on the ground turns on a searchlight and a beam of light moves exactly vertically upward, as seen by the technician. As you observe the beam of light, you measure the magnitude of the vertical component of its velocity as **(a)** equal to $c$, **(b)** greater than $c$, or **(c)** less than $c$. If the technician aims the searchlight directly at you instead of upward, you measure the magnitude of the horizontal component of its velocity as **(d)** equal to $c$, **(e)** greater than $c$, or **(f)** less than $c$.

### EXAMPLE 9.4 Relative Velocity of Spacecraft

Two spacecraft A and B are moving directly toward each other as in Figure 9.11. An observer on the Earth measures the speed of A to be $0.750c$ and the speed of B to be $0.850c$. Find the velocity of B with respect to A.

**Solution** Conceptualize the problem by studying Figure 9.11. Note that the spacecraft are approaching each other, so the speed of one as measured by an observer in the other will be larger than the speed of either as measured by an observer on the Earth.

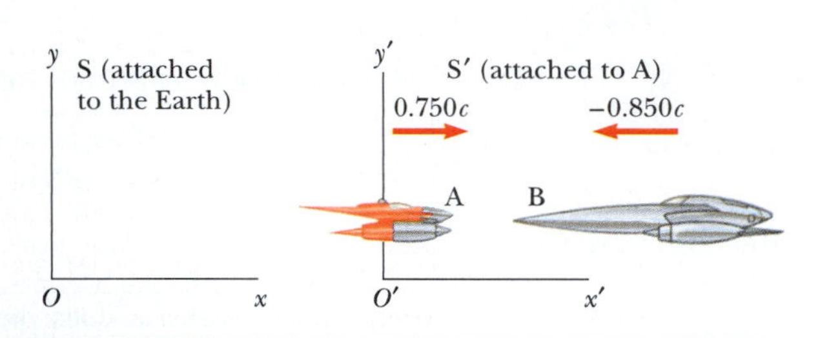


FIGURE 9.11 (Example 9.4) Two spacecraft A and B move in opposite directions.

Because the speeds of the spacecraft are large fractions of the speed of light, we categorize the problem as a relativistic one. We analyze the problem by taking the S′ frame as being attached to A so that $v = 0.750c$ relative to the Earth observer (in the S frame). Spacecraft B can be considered as an object moving with a velocity component $u_x = -0.850c$ relative to the Earth observer. Hence, the velocity of B with respect to A is the velocity of B as measured by the observer $O'$ on A, which can be obtained by using Equation 9.11:

$$u'_x = \frac{u_x - v}{1 - \dfrac{u_x v}{c^2}} = \frac{-0.850c - 0.750c}{1 - \dfrac{(-0.850c)(0.750c)}{c^2}}$$

$$= -0.980c$$

To finalize the problem, note that the negative sign in the result indicates that spacecraft B is moving in the negative $x$ direction as observed by A. Also note that the relative speed is larger than each of the individual speeds but is smaller than the speed of light, as it must be.

### INTERACTIVE EXAMPLE 9.5 Relativistic Leaders of the Pack

Two motorcycle pack leaders named David and Emily are racing at relativistic speeds along perpendicular paths as in Figure 9.12. How fast does Emily recede over David's right shoulder as seen by David?

**Solution** To determine Emily's speed of recession as seen by David, we take S′ to move in the $x$ direction along with David. Figure 9.12 represents the situation as seen by a police officer at rest in frame S who observes the following:

David: $v = 0.75c$

Emily: $u_x = 0$ $\quad u_y = -0.90c$

We calculate $u'_x$ and $u'_y$ for Emily using Equations 9.11 and 9.12:

$$u'_x = \frac{u_x - v}{1 - \dfrac{u_x v}{c^2}} = \frac{0 - 0.75c}{1 - \dfrac{(0)(0.75c)}{c^2}} = -0.75c$$

$$u'_y = \frac{u_y}{\gamma\left(1 - \dfrac{u_x v}{c^2}\right)} = \frac{\sqrt{1 - \dfrac{(0.75c)^2}{c^2}}\,(-0.90c)}{\left(1 - \dfrac{(0)(0.75c)}{c^2}\right)}$$

$$= -0.60c$$

Therefore, the speed of Emily as observed by David is

$$u' = \sqrt{(u'_x)^2 + (u'_y)^2} = \sqrt{(-0.75c)^2 + (-0.60c)^2}$$

$$= 0.96c$$

PhysicsNow™ Investigate this situation with various speeds of David and Emily by logging into PhysicsNow at **www.pop4e.com** and going to Interactive Example 9.5.

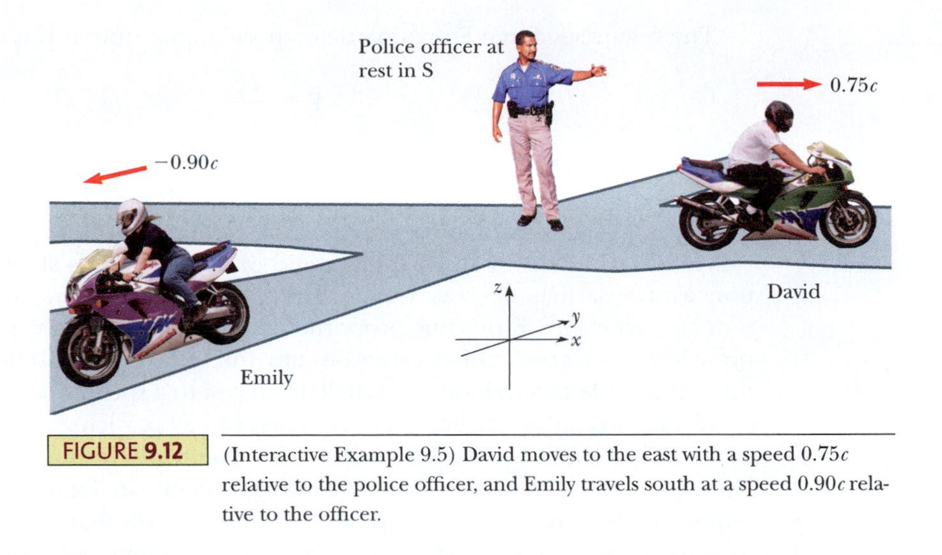


**FIGURE 9.12** (Interactive Example 9.5) David moves to the east with a speed $0.75c$ relative to the police officer, and Emily travels south at a speed $0.90c$ relative to the officer.

## 9.6 RELATIVISTIC MOMENTUM AND THE RELATIVISTIC FORM OF NEWTON'S LAWS

We have seen that to describe the motion of particles within the framework of special relativity properly, the Galilean transformation must be replaced by the Lorentz transformation. Because the laws of physics must remain unchanged under the Lorentz transformation, we must generalize Newton's laws and the definitions of momentum and energy to conform to the Lorentz transformation and the principle of relativity. These generalized definitions should reduce to the classical (nonrelativistic) definitions for $v \ll c$ or $u \ll c$. (As we have done previously, we will use $v$ for the speed of one reference frame relative to another and $u$ for the speed of a particle.)

First, recall that the total momentum of an isolated system of particles is conserved. Suppose a collision between two particles is described in a reference frame S in which the momentum of the system is measured to be conserved. If the velocities in a second reference frame S′ are calculated using the Lorentz transformation and the Newtonian definition of momentum, $\vec{\mathbf{p}} = m\vec{\mathbf{u}}$, is used, it is found that the momentum of the system is *not* measured to be conserved in the second reference frame. This finding violates one of Einstein's postulates: The laws of physics are the same in all inertial frames. Therefore, assuming the Lorentz transformation to be correct, we must modify the definition of momentum.

The relativistic equation for the momentum of a particle of mass $m$ that maintains the principle of conservation of momentum is

$$\vec{\mathbf{p}} \equiv \frac{m\vec{\mathbf{u}}}{\sqrt{1-\dfrac{u^2}{c^2}}} \qquad [9.14]$$

■ Definition of relativistic momentum

where $\vec{\mathbf{u}}$ is the velocity of the particle. When $u$ is much less than $c$, the denominator of Equation 9.14 approaches unity, so $\vec{\mathbf{p}}$ approaches $m\vec{\mathbf{u}}$. Therefore, the relativistic equation for $\vec{\mathbf{p}}$ reduces to the classical expression when $u$ is small compared with $c$. Equation 9.14 is often written in simpler form as

$$\vec{\mathbf{p}} = \gamma m\vec{\mathbf{u}} \qquad [9.15]$$

using our previously defined expression[7] for $\gamma$.

**PITFALL PREVENTION 9.4**

**WATCH OUT FOR "RELATIVISTIC MASS"** Some older treatments of relativity maintained the conservation of momentum principle at high speeds by using a model in which the mass of a particle increases with speed. You might still encounter this notion of "relativistic mass" in your outside reading, especially in older books. Be aware that this notion is no longer widely accepted; today, mass is considered as *invariant*, independent of speed. The mass of an object in all frames is considered to be the mass as measured by an observer at rest with respect to the object.

[7]We defined $\gamma$ previously in terms of the speed $v$ of one frame relative to another frame. The same symbol is also used for $(1 - u^2/c^2)^{-1/2}$, where $u$ is the speed of a particle.

The relativistic force $\vec{\mathbf{F}}$ on a particle whose momentum is $\vec{\mathbf{p}}$ is defined as

$$\vec{\mathbf{F}} \equiv \frac{d\vec{\mathbf{p}}}{dt} \qquad [9.16]$$

where $\vec{\mathbf{p}}$ is given by Equation 9.14. This expression preserves both classical mechanics in the limit of low velocities and conservation of momentum for an isolated system ($\Sigma \vec{\mathbf{F}}_{\text{ext}} = 0$) both relativistically and classically.

We leave it to Problem 9.57 at the end of the chapter to show that the acceleration $\vec{\mathbf{a}}$ of a particle decreases under the action of a constant force, in which case $a \propto (1 - u^2/c^2)^{3/2}$. From this proportionality, note that as the particle's speed approaches $c$, the acceleration caused by any finite force approaches zero. It is therefore impossible to accelerate a particle from rest to a speed $u \geq c$.

Hence, $c$ is an upper limit for the speed of any particle. In fact, it is possible to show that **no matter, energy, or information can travel through space faster than $c$.** Note that the relative speeds of the two spacecraft in Example 9.4 and the two motorcyclists in Interactive Example 9.5 were both less than $c$. If we had attempted to solve these examples with Galilean transformations, we would have obtained relative speeds larger than $c$ in both cases.

**EXAMPLE 9.6** Momentum of an Electron

An electron, which has a mass of $9.11 \times 10^{-31}$ kg, moves with a speed of $0.750c$. Find its relativistic momentum and compare it with the momentum calculated from the classical expression.

**Solution** Using Equation 9.14 with $u = 0.750c$, we have

$$p = \frac{m_e u}{\sqrt{1 - \frac{u^2}{c^2}}}$$

$$p = \frac{(9.11 \times 10^{-31}\text{ kg})(0.750 \times 3.00 \times 10^8\text{ m/s})}{\sqrt{1 - \frac{(0.750c)^2}{c^2}}}$$

$$= 3.10 \times 10^{-22}\text{ kg}\cdot\text{m/s}$$

The classical expression, if used (inappropriately) for this high-speed particle, gives

$$p = m_e u = 2.05 \times 10^{-22}\text{ kg}\cdot\text{m/s}$$

Hence, the correct relativistic result is more than 50% larger than the classical result!

## 9.7 RELATIVISTIC ENERGY

We have seen that the definition of momentum requires generalization to make it compatible with the principle of relativity. We find that the definition of kinetic energy must also be modified.

To derive the relativistic form of the work–kinetic energy theorem, let us start with the definition of the work done by a force of magnitude $F$ on a particle initially at rest. Recall from Chapter 6 that the work–kinetic energy theorem states that the work done by a net force acting on a particle equals the change in kinetic energy of the particle. Because the initial kinetic energy is zero, we conclude that the work $W$ done in accelerating a particle from rest is equivalent to the relativistic kinetic energy $K$ of the particle:

$$W = \Delta K = K - 0 = K = \int_{x_1}^{x_2} F\,dx = \int_{x_1}^{x_2} \frac{dp}{dt}\,dx \qquad [9.17]$$

where we are considering the special case of force and displacement vectors along the $x$ axis for simplicity. To perform this integration and find the relativistic kinetic energy as a function of $u$, we first evaluate $dp/dt$, using Equation 9.14:

$$\frac{dp}{dt} = \frac{d}{dt}\frac{mu}{\sqrt{1-\dfrac{u^2}{c^2}}} = \frac{m(du/dt)}{\left(1-\dfrac{u^2}{c^2}\right)^{3/2}}$$

Substituting this expression for $dp/dt$ and $dx = u\,dt$ into Equation 9.17 gives

$$K = \int_0^t \frac{m(du/dt)\,u\,dt}{\left(1-\dfrac{u^2}{c^2}\right)^{3/2}} = m\int_0^u \frac{u}{\left(1-\dfrac{u^2}{c^2}\right)^{3/2}}\,du$$

Evaluating the integral, we find that

$$K = \frac{mc^2}{\sqrt{1-\dfrac{u^2}{c^2}}} - mc^2 = \gamma mc^2 - mc^2 = (\gamma - 1)mc^2 \qquad [9.18]$$

■ **Relativistic kinetic energy**

At low speeds, where $u/c \ll 1$, Equation 9.18 should reduce to the classical expression $K = \frac{1}{2}mu^2$. We can show this reduction by using the binomial expansion $(1 - x^2)^{-1/2} \approx 1 + \frac{1}{2}x^2 + \cdots$ for $x \ll 1$, where the higher-order powers of $x$ are ignored in the expansion because they are so small. In our case, $x = u/c$, so

$$\gamma = \frac{1}{\sqrt{1-\dfrac{u^2}{c^2}}} = \left(1-\frac{u^2}{c^2}\right)^{-1/2} \approx 1 + \frac{1}{2}\frac{u^2}{c^2} + \cdots$$

Substituting into Equation 9.18 gives

$$K \approx \left(1 + \frac{1}{2}\frac{u^2}{c^2} + \cdots\right)mc^2 - mc^2 = \tfrac{1}{2}mu^2$$

which agrees with the classical result. Figure 9.13 shows a comparison of the speed–kinetic energy relationships for a particle using the nonrelativistic expression for $K$ (the blue curve) and the relativistic expression for $K$ (the brown curve). The curves are in good agreement at low speeds, but deviate at higher speeds. The nonrelativistic expression indicates a violation of special relativity because it suggests that sufficient energy can be added to the particle to accelerate it to a speed larger than $c$. In the relativistic case, the particle speed never exceeds $c$, regardless

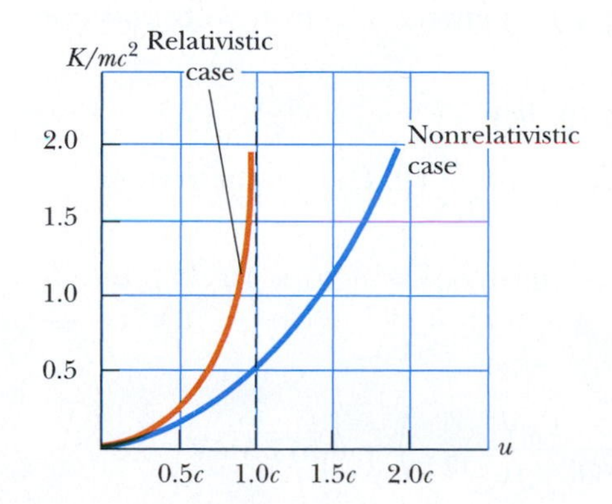


**FIGURE 9.13** A graph comparing relativistic and nonrelativistic kinetic energy of a particle. The energies are plotted as a function of speed $u$. In the relativistic case, $u$ is always less than $c$.

of the kinetic energy, which is consistent with experimental results. When an object's speed is less than one-tenth the speed of light, the classical kinetic energy equation differs by less than 1% from the relativistic equation (which is experimentally verified at all speeds). Therefore, for practical calculations it is valid to use the classical equation when the object's speed is less than $0.1c$.

The constant term $mc^2$ in Equation 9.18, which is independent of the speed, is called the **rest energy** $E_R$ of the particle:

■ Rest energy

$$E_R = mc^2 \tag{9.19}$$

The term $\gamma mc^2$ in Equation 9.18 depends on the particle speed and is the sum of the kinetic and rest energies. We define $\gamma mc^2$ to be the **total energy** $E$; that is, total energy = kinetic energy + rest energy:

$$E = \gamma mc^2 = K + mc^2 = K + E_R \tag{9.20}$$

or, when $\gamma$ is replaced by its equivalent,

■ Total energy of a relativistic particle

$$E = \frac{mc^2}{\sqrt{1 - \dfrac{u^2}{c^2}}} \tag{9.21}$$

The relation $E_R = mc^2$ shows that **mass is a manifestation of energy.** It also shows that a small mass corresponds to an enormous amount of energy. This concept is fundamental to much of the field of nuclear physics.

In many situations, the momentum or energy of a particle is measured rather than its speed. It is therefore useful to have an expression relating the total energy $E$ to the relativistic momentum $p$, which is accomplished by using the expressions $E = \gamma mc^2$ and $p = \gamma mu$. By squaring these equations and subtracting, we can eliminate $u$ (see Problem 9.37). The result, after some algebra, is

■ Energy–momentum relationship for a relativistic particle

$$E^2 = p^2c^2 + (mc^2)^2 \tag{9.22}$$

When the particle is at rest, $p = 0$, and so $E = E_R = mc^2$. That is, the total energy equals the rest energy.

For the case of particles that have zero mass, such as photons (massless, chargeless particles of light to be discussed further in Chapter 28), we set $m = 0$ in Equation 9.22 and see that

$$E = pc \tag{9.23}$$

This equation is an exact expression relating energy and momentum for photons, which always travel at the speed of light.

When dealing with subatomic particles, it is convenient to express their energy in a unit called an *electron volt* (eV). The equality between electron volts and our standard energy unit is

$$1 \text{ eV} = 1.60 \times 10^{-19} \text{ J}$$

For example, the mass of an electron is $9.11 \times 10^{-31}$ kg. Hence, the rest energy of the electron is

$$E_R = m_ec^2 = (9.11 \times 10^{-31} \text{ kg})(3.00 \times 10^8 \text{ m/s})^2 = 8.20 \times 10^{-14} \text{ J}$$

Converting to eV, we have

$$E_R = m_ec^2 = (8.20 \times 10^{-14} \text{ J})\left(\frac{1 \text{ eV}}{1.60 \times 10^{-19} \text{ J}}\right) = 0.511 \text{ MeV}$$

**QUICK QUIZ 9.6** The following *pairs* of energies represent the rest energy and total energy of three different particles: particle 1: $E$, $2E$; particle 2: $E$, $3E$; particle 3: $2E$, $4E$. Rank the particles, from greatest to least, according to their (**a**) mass, (**b**) kinetic energy, and (**c**) speed.

**EXAMPLE 9.7** **The Energy of a Speedy Proton**

Let us consider the relativistic motion of a proton.

**A** Find the proton's rest energy in electron volts.

**Solution** To find the rest energy, we use Equation 9.19,

$$E_R = m_pc^2 = (1.67 \times 10^{-27}\ \text{kg})(3.00 \times 10^8\ \text{m/s})^2$$

$$= (1.50 \times 10^{-10}\ \text{J})\left(\frac{1.00\ \text{eV}}{1.60 \times 10^{-19}\ \text{J}}\right)$$

$$= 938\ \text{MeV}$$

**B** The total energy of a proton is three times its rest energy. With what speed is the proton moving?

**Solution** Because the total energy $E$ is three times the rest energy, $E = \gamma mc^2$ (Eq. 9.20) gives

$$E = 3m_pc^2 = \frac{m_pc^2}{\sqrt{1 - \frac{u^2}{c^2}}}$$

$$3 = \frac{1}{\sqrt{1 - \frac{u^2}{c^2}}}$$

Solving for $u$ gives

$$\left(1 - \frac{u^2}{c^2}\right) = \frac{1}{9} \quad \text{or} \quad \frac{u^2}{c^2} = \frac{8}{9}$$

$$u = \frac{\sqrt{8}}{3}c = 2.83 \times 10^8\ \text{m/s}$$

**C** Determine the kinetic energy of the proton in part B in electron volts.

**Solution** We use Equation 9.20:

$$K = E - m_pc^2 = 3m_pc^2 - m_pc^2 = 2m_pc^2$$

Because $m_pc^2 = 938$ MeV, $K = 1.88$ GeV.

**D** What is the magnitude of the proton's momentum in part B?

**Solution** We can use Equation 9.22 to calculate the momentum with $E = 3m_pc^2$:

$$E^2 = p^2c^2 + (m_pc^2)^2 = (3m_pc^2)^2$$

$$p^2c^2 = 9(m_pc^2)^2 - (m_pc^2)^2 = 8(m_pc^2)^2$$

$$p = \sqrt{8}\frac{m_pc^2}{c} = \sqrt{8}\frac{(938\ \text{MeV})}{c}$$

$$= 2.65 \times 10^3\ \frac{\text{MeV}}{c}$$

The unit of momentum is written MeV/$c$, which is a momentum unit often used in particle studies.

## 9.8 MASS AND ENERGY

Equation 9.20, $E = \gamma mc^2$, which represents the total energy of a particle, suggests that even when a particle is at rest ($\gamma = 1$) it still possesses enormous energy through its mass. The clearest experimental proof of the equivalence of mass and energy occurs in nuclear and elementary particle interactions in which the conversion of mass into kinetic energy takes place. Hence, we cannot use the principle of conservation of energy in relativistic situations exactly as it is outlined in Chapter 7. We must include rest energy as another form of energy storage.

This concept is important in atomic and nuclear processes, in which the change in mass during the process is on the order of the initial mass. For example, in a conventional nuclear reactor, the uranium nucleus undergoes *fission,* a reaction that results in several lighter fragments having considerable kinetic energy. In the case of a $^{235}$U atom, which is used as fuel in nuclear power plants, the fragments are two lighter nuclei and a few neutrons. The total mass of the fragments is less than that of the $^{235}U$ by an amount $\Delta m$. The corresponding energy $\Delta mc^2$ associated with this mass difference is exactly equal to the total kinetic energy of the fragments. The kinetic energy is transferred by collisions with water molecules as the fragments move through water, raising the internal energy of the water. This internal energy is used to produce steam for the generation of electric power.

Next, consider a basic *fusion* reaction in which two deuterium atoms combine to form one helium atom. The decrease in mass that results from the creation of one helium atom from two deuterium atoms is $\Delta m = 4.25 \times 10^{-29}$ kg. Hence, the corresponding energy that results from one fusion reaction is calculated to be $\Delta mc^2 = 3.83 \times 10^{-12}\ \text{J} = 23.9$ MeV. To appreciate the magnitude of this result, consider that if 1 g of deuterium is converted to helium, the energy released is on the order of $10^{12}$ J! At the year 2006 cost of electric energy, this energy would be worth about $20 000. We will see more details of these nuclear processes in Chapter 30.

**EXAMPLE 9.8 Mass Change in a Radioactive Decay**

The $^{216}Po$ nucleus is unstable and exhibits radioactivity, which we will study in detail in Chapter 30. It decays to $^{212}Pb$ by emitting an alpha particle, which is a helium nucleus, $^4He$.

**A** Find the mass change in this decay.

**Solution** Using values in Table A.3, we see that the initial and final masses are

$$m_i = m(^{216}\text{Po}) = 216.001\,905\ \text{u}$$

$$m_f = m(^{212}\text{Pb}) + m(^4\text{He})$$
$$= 211.991\,888\ \text{u} + 4.002\,603\ \text{u}$$
$$= 215.994\,491\ \text{u}$$

Therefore, the mass change is

$$\Delta m = 216.001\,905\ \text{u} - 215.994\,491\ \text{u} = 0.007\,414\ \text{u}$$
$$= 1.23 \times 10^{-29}\ \text{kg}$$

**B** Find the energy that this mass change represents.

**Solution** The energy associated with this mass change is

$$E = \Delta mc^2 = (1.23 \times 10^{-29}\ \text{kg})(3.00 \times 10^8\ \text{m/s})^2$$
$$= 1.11 \times 10^{-12}\ \text{J} = 6.92\ \text{MeV}$$

This energy appears as the kinetic energies of the alpha particle and the $^{212}Pb$ nucleus after the decay.

## 9.9 GENERAL RELATIVITY

Up to this point, we have sidestepped a curious puzzle. Mass has two seemingly different properties: It determines a force of mutual gravitational attraction between two objects (Newton's law of universal gravitation), and it also represents the resistance of a single object to being accelerated (Newton's second law), regardless of the type of force producing the acceleration. How can one quantity have two such different properties? An answer to this question, which puzzled Newton and many other physicists over the years, was provided when Einstein published his theory of gravitation, known as *general relativity,* in 1916. Because it is a mathematically complex theory, we offer merely a hint of its elegance and insight.

In Einstein's view, the dual behavior of mass was evidence for a very intimate and basic connection between the two behaviors. He pointed out that no mechanical experiment (e.g., dropping an object) could distinguish between the two situations illustrated in Figures 9.14a and 9.14b. In Figure 9.14a, a person is standing in an elevator on the surface of a planet and feels pressed into the floor due to the gravitational force. If he releases his briefcase, he observes it moving toward the floor with acceleration $\vec{\mathbf{g}} = -g\hat{\mathbf{j}}$. In Figure 9.14b, the person is in an elevator in empty space accelerating upward with $\vec{\mathbf{a}}_{\text{el}} = +g\hat{\mathbf{j}}$. The person feels pressed into the floor with the same force as in Figure 9.14a. If he releases his briefcase, he observes it moving toward the floor with acceleration $g$, just as in the previous situation. In each case, an object released by the observer undergoes a downward

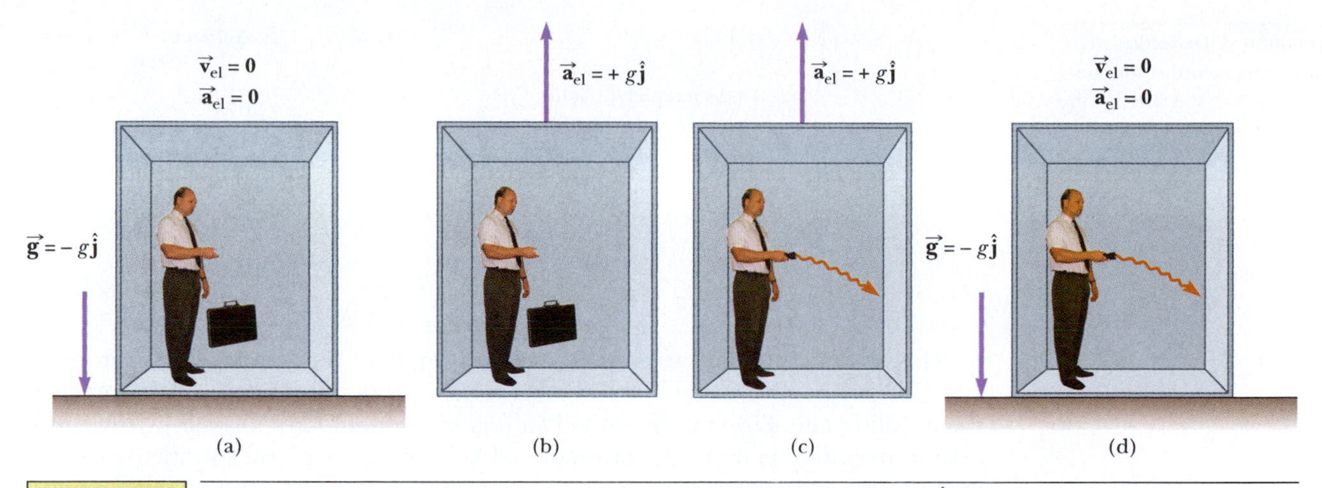


FIGURE 9.14 (a) The observer is at rest in an elevator in a uniform gravitational field $\vec{g} = -g\hat{j}$, directed downward. The observer drops his briefcase, which moves downward with acceleration $g$. (b) The observer is in a region where gravity is negligible, but the elevator moves upward with an acceleration $\vec{a}_{el} = +g\hat{j}$. The observer releases his briefcase, which moves downward (according to the observer) with acceleration $g$ relative to the floor of the elevator. According to Einstein, the frames of reference in parts (a) and (b) are equivalent in every way. No local experiment can distinguish any difference between the two frames. (c) In the accelerating frame, a ray of light would appear to bend downward due to the acceleration. (d) If parts (a) and (b) are truly equivalent, as Einstein proposed, part (c) suggests that a ray of light would bend downward in a gravitational field.

acceleration of magnitude $g$ relative to the floor. In Figure 9.14a, the person is at rest in an inertial frame in a gravitational field due to the planet. (A gravitational field exists around any object with mass, such as a planet. We will define the gravitational field formally in Chapter 11.) In Figure 9.14b, the person is in a noninertial frame accelerating in gravity-free space. Einstein's claim is that these two situations are completely equivalent.

Einstein carried this idea further and proposed that *no* experiment, mechanical or otherwise, could distinguish between the two cases. This extension to include all phenomena (not just mechanical ones) has interesting consequences. For example, suppose a light pulse is sent horizontally across the elevator as in Figure 9.14c, in which the elevator is accelerating upward in empty space. From the point of view of an observer in an inertial frame outside the elevator, the light travels in a straight line while the floor of the elevator accelerates upward. According to the observer on the elevator, however, the trajectory of the light pulse bends downward as the floor of the elevator (and the observer) accelerates upward. Therefore, based on the equality of parts (a) and (b) of the figure for all phenomena, Einstein proposed that **a beam of light should also be bent downward by a gravitational field,** as in Figure 9.14d.

The two postulates of Einstein's **general theory of relativity** are as follows:

■ Postulates of general relativity

- All the laws of nature have the same form for observers in any frame of reference, whether accelerated or not.
- In the vicinity of any given point, a gravitational field is equivalent to an accelerated frame of reference in the absence of gravitational effects. (This postulate is known as the **principle of equivalence.**)

One interesting effect predicted by general relativity is that the passage of time is altered by gravity. A clock in the presence of gravity runs more slowly than one for which gravity is negligible. Consequently, the frequencies of radiation emitted by atoms in the presence of a strong gravitational field are shifted to lower values compared with the same emissions in a weak field. This gravitational shift has been detected in light emitted by atoms in massive stars. It has also been verified on the Earth by comparing the frequencies of gamma rays (a high-energy form of electromagnetic radiation) emitted from nuclei separated vertically by about 20 m.

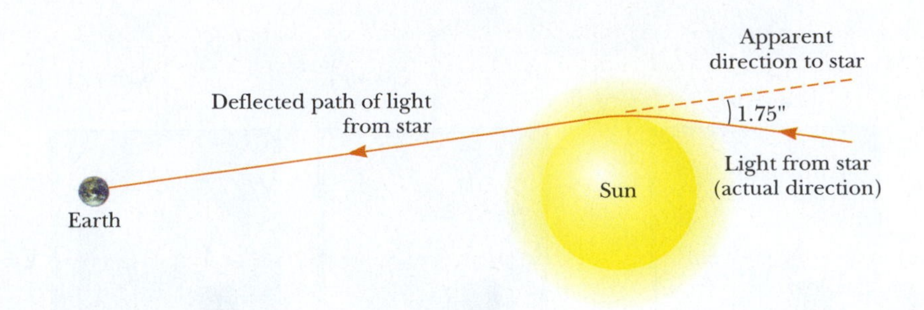


**FIGURE 9.15** Deflection of starlight passing near the Sun. Because of this effect, the Sun or other remote objects can act as a *gravitational lens.* In his general theory of relativity, Einstein calculated that starlight just grazing the Sun's surface should be deflected by an angle of 1.75 seconds of arc.

The second postulate suggests that a gravitational field may be "transformed away" at any point if we choose an appropriate accelerated frame of reference, a freely falling one. Einstein developed an ingenious method of describing the acceleration necessary to make the gravitational field "disappear." He specified a certain quantity, the *curvature of space–time,* that describes the gravitational effect of a mass. In fact, the curvature of space–time completely replaces Newton's gravitational theory. According to Einstein, there is no such thing as a gravitational force. Rather, the presence of a mass causes a curvature of space–time in the vicinity of the mass, and this curvature dictates the space-time path that all freely moving objects must follow.

One important test of general relativity is the prediction that a light ray passing near the Sun should be deflected by some angle. This prediction was confirmed by astronomers as the bending of starlight during a total solar eclipse shortly following World War I (Fig. 9.15).

As an example of the effects of curved space–time, imagine two travelers moving on parallel paths a few meters apart on the surface of the Earth and maintaining an exact northward heading along two longitude lines. As they observe each other near the equator, they will claim that their paths are exactly parallel. As they approach the North Pole, however, they will notice that they are moving closer together and that they will actually meet at the North Pole. Thus, they will claim that they moved along parallel paths, but moved toward each other, *as if there were an attractive force between them.* They will make this conclusion based on their everyday experience of moving on flat surfaces. From our mental representation, however, we realize that they are walking on a curved surface, and the geometry of the curved surface, rather than an attractive force, causes them to converge. In a similar way, general relativity replaces the notion of forces with the movement of objects through curved space–time.

If a concentration of mass in space becomes very great, as is believed to occur when a large star exhausts its nuclear fuel and collapses to a very small volume, a **black hole** may form. Here the curvature of space–time is so extreme that, within a certain distance from the center of the black hole, all matter and light become trapped. We will say more about black holes in Chapter 11.

## Thinking Physics 9.2

Atomic clocks are extremely accurate; in fact, an error of 1 s in 3 million years is typical. This error can be described as about 1 part in $10^{14}$. On the other hand, the atomic clock in Boulder, Colorado, near Denver, is often 15 ns faster than the one in Washington, D.C., after only one day. This error is one of about 1 part in $6 \times 10^{12}$, which is about 17 times larger than the previously expressed error. If atomic clocks are so accurate, why does a clock in Boulder not remain in synchronization with one in Washington, D.C.? (*Hint:* Denver is known as the Mile High City.)

**Reasoning** According to the general theory of relativity, the passage of time depends on gravity. Time is measured to run more slowly in strong gravitational fields. Washington, D.C., is at an elevation very close to sea level, but Boulder is about a mile higher in altitude. This difference results in a weaker gravitational field at Boulder than at Washington, D.C. As a result, time is measured to run more rapidly in Boulder than in Washington, D.C. ■

## 9.10 FROM MARS TO THE STARS

CONTEXT CONNECTION

In this chapter, we have discussed the strange effects of traveling at high speeds. Do we need to consider these effects in our planned mission to Mars?

To answer this question, let us consider a typical spacecraft speed necessary to travel from the Earth to Mars. This speed is on the order of $10^4$ m/s. Let us evaluate $\gamma$ for this speed:

$$\gamma = \frac{1}{\sqrt{1 - \frac{u^2}{c^2}}} = \frac{1}{\sqrt{1 - \frac{(10^4 \text{ m/s})^2}{(3.00 \times 10^8 \text{ m/s})^2}}} = 1.000\,000\,000\,6$$

where we have completely ignored the rules of significant figures so that we could find the first nonzero digit to the right of the decimal place!

It is clear from this result that relativistic considerations are not important for our trip to Mars. Yet what about deeper travels into space? Suppose we wish to travel to another star. This distance is several orders of magnitude larger. The nearest star is about 4.2 ly from the Earth. In comparison, Mars is $4.0 \times 10^{-5}$ ly at its farthest from the Earth. Therefore, we are talking about a distance to the nearest star that is five orders of magnitude larger than the distance to Mars. Very long travel times will be needed to reach even the nearest star. At the escape speed from the Sun, for example, assuming that this speed is maintained during the entire trip, the travel time is 30 000 years to the nearest star. This time period is clearly prohibitive, especially if we would like the people who leave the Earth to be the same people who arrive at the star!

We can use the principles of relativity to reduce this travel time significantly by traveling at very high speeds. Suppose our spacecraft travels at a constant speed of $0.99c$. The travel time as measured by an observer on the Earth then is

$$\Delta t = \frac{L_p}{u} = \frac{4.2 \text{ ly}}{0.99(1.0 \text{ ly/yr})} = 4.2 \text{ yr}$$

where the distance between the Earth and the destination star is the proper length $L_p$.

Because the spacecraft occupants see both the Earth and the destination star moving, the distance between them is measured to be shorter than that measured by observers on the Earth. We can use length contraction to calculate the distance from the Earth to the star as measured by the spacecraft occupants:

$$L = \frac{L_p}{\gamma} = L_p\sqrt{1 - \frac{u^2}{c^2}} = (4.2 \text{ ly})\sqrt{1 - \frac{(0.99c)^2}{c^2}} = 0.59 \text{ ly}$$

The time interval required to reach the star is now

$$\Delta t = \frac{L}{u} = \frac{0.59 \text{ ly}}{0.99(1.0 \text{ ly/yr})} = 0.60 \text{ yr}$$

which is clearly a reduction in travel time from the low-speed trip!

There are three major problems with this scenario, however. The first is the technological challenge of designing and building a spacecraft and rocket engine assembly that can attain a speed of $0.99c$. Second is the design of a safety system that will provide early warnings about running into asteroids, meteoroids, or other bits of matter while traveling at almost light speed through space. Even a small piece of rock could be disastrous if struck at $0.99c$. The third problem is related to the twin paradox discussed earlier in this chapter. During the trip to the star, 4.2 yr will pass on the Earth. If the travelers return to the Earth, another 4.2 yr will pass. Therefore, the travelers will have aged by only 2(0.6 yr) = 1.2 yr, but 8.4 yr will have passed on the Earth. For stars farther away than the nearest star, these effects could result in the personnel assisting with the liftoff from the Earth no longer being alive when the travelers return. In conclusion, we see that travel to the stars will be an enormous challenge!

## SUMMARY



The two basic postulates of **special relativity** are:

- All the laws of physics are the same in all inertial reference frames.
- The speed of light in vacuum has the same value, in all inertial frames, regardless of the velocity of the observer or the velocity of the source emitting the light.

Three consequences of special relativity are:

- Events that are simultaneous for one observer may not be simultaneous for another observer who is in motion relative to the first.
- Clocks in motion relative to an observer are measured to be slowed down by a factor $\gamma$. This phenomenon is known as **time dilation.**
- Lengths of objects in motion are measured to be shorter in the direction of motion. This phenomenon is known as **length contraction.**

To satisfy the postulates of special relativity, the Galilean transformations must be replaced by the **Lorentz transformation equations:**

$$x' = \gamma(x - vt)$$
$$y' = y \qquad [9.8]$$
$$z' = z$$
$$t' = \gamma\left(t - \frac{v}{c^2}x\right)$$

where $\gamma = (1 - v^2/c^2)^{-1/2}$.

The relativistic form of the **Lorentz velocity transformation** is

$$u'_x = \frac{u_x - v}{1 - \dfrac{u_x v}{c^2}} \qquad [9.11]$$

where $u_x$ is the speed of an object as measured in the S frame and $u'_x$ is its speed measured in the S′ frame.

The relativistic expression for the momentum of a particle moving with a velocity $\vec{\mathbf{u}}$ is

$$\vec{\mathbf{p}} \equiv \frac{m\vec{\mathbf{u}}}{\sqrt{1 - \dfrac{u^2}{c^2}}} = \gamma m\vec{\mathbf{u}} \qquad [9.14, 9.15]$$

The relativistic expression for the kinetic energy of a particle is

$$K = \gamma mc^2 - mc^2 = (\gamma - 1)mc^2 \qquad [9.18]$$

where $E_R = mc^2$ is the **rest energy** of the particle.

The **total energy** $E$ of a particle is given by the expression

$$E = \frac{mc^2}{\sqrt{1 - \dfrac{u^2}{c^2}}} \qquad [9.21]$$

The total energy of a particle is the sum of its rest energy and its kinetic energy: $E = E_R + K$.

The relativistic momentum of a particle is related to its total energy through the equation

$$E^2 = p^2c^2 + (mc^2)^2 \qquad [9.22]$$

The **general theory of relativity** claims that no experiment can distinguish between a gravitational field and an accelerating reference frame. It correctly predicts that the path of light is affected by a gravitational field.

## QUESTIONS

☐ = answer available in the *Student Solutions Manual and Study Guide*

1. On what two speed measurements do two observers in relative motion always agree?
2. A spacecraft with the shape of a sphere moves past an observer on Earth with a speed $0.5c$. What shape does the observer measure for the spacecraft as it moves past?
3. The speed of light in water is 230 Mm/s. Suppose an electron is moving through water at 250 Mm/s. Does that violate the principle of relativity?
4. Two identical clocks are synchronized. One is then put in orbit directed eastward around the Earth, and the other remains on the Earth. According to an observer on the Earth, which clock runs more slowly? When the moving clock returns to the Earth, are the two still synchronized?

**5.** Explain why it is necessary, when defining the length of a rod, to specify that the positions of the ends of the rod are to be measured simultaneously.

**6.** A train is approaching you at very high speed as you stand next to the tracks. Just as an observer on the train passes you, you both begin to play the same Beethoven symphony on portable compact disc players. (a) According to you, whose CD player finishes the symphony first? (b) According to the observer on the train, whose CD player finishes the symphony first? (c) Whose CD player really finishes the symphony first?

**7.** List some ways our day-to-day lives would change if the speed of light were only 50 m/s.

**8.** A particle is moving at a speed less than $c/2$. If the speed of the particle is doubled, what happens to its momentum?

**9.** Give a physical argument that shows that it is impossible to accelerate an object of mass $m$ to the speed of light, even with a continuous force acting on it.

**10.** The upper limit of the speed of an electron is the speed of light $c$. Does that mean that the momentum of the electron has an upper limit?

**11.** Because mass is a measure of energy, can we conclude that the mass of a compressed spring is greater than the mass of the same spring when it is not compressed?

**12.** It is said that Einstein, in his teenage years, asked the question, "What would I see in a mirror if I carried it in my hands and ran at the speed of light?" How would you answer this question?

**13.** Some distant astronomical objects, called quasars, are receding from us at half the speed of light (or greater). What is the speed of the light we receive from these quasars?

**14.** Photons of light have zero mass. How is it possible that they have momentum?

**15.** "Newtonian mechanics correctly describes objects moving at ordinary speeds, and relativistic mechanics correctly describes objects moving very fast." "Relativistic mechanics must make a smooth transition as it reduces to Newtonian mechanics in a case where the speed of an object becomes small compared with the speed of light." Argue for or against each of these two statements.

**16.** Two cards have straight edges. Suppose the top edge of one card crosses the bottom edge of another card at a small angle as shown in Figure Q9.16a. A person slides the cards together at a moderately high speed. In what direction does the intersection point of the edges move? Show that it can move at a speed greater than the speed of light.

A small flashlight is suspended in a horizontal plane and set into rapid rotation. Show that the spot of light it produces on a distant screen can move across the screen at a speed greater than the speed of light. (If you use a laser pointer, as shown in Fig. Q9.16b, make sure that the direct laser light cannot enter a person's eyes.) Argue that these experiments do not invalidate the principle that no material, no energy, and no information can move faster than light moves in a vacuum.

**17.** With regard to reference frames, how does general relativity differ from special relativity?

**18.** Two identical clocks are in the same house, one upstairs in a bedroom and the other downstairs in the kitchen. Which clock runs more slowly? Explain.

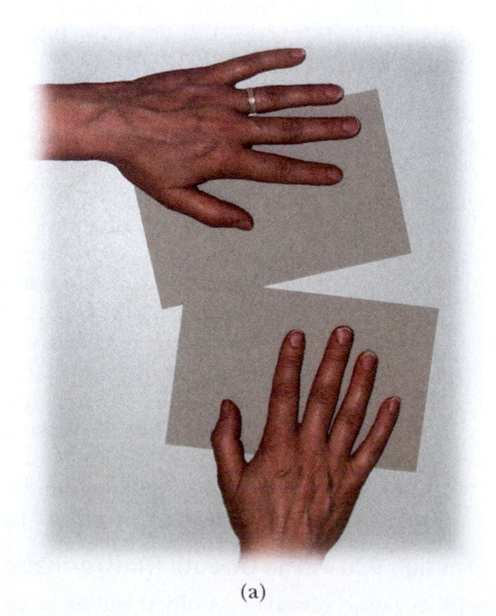

(a)

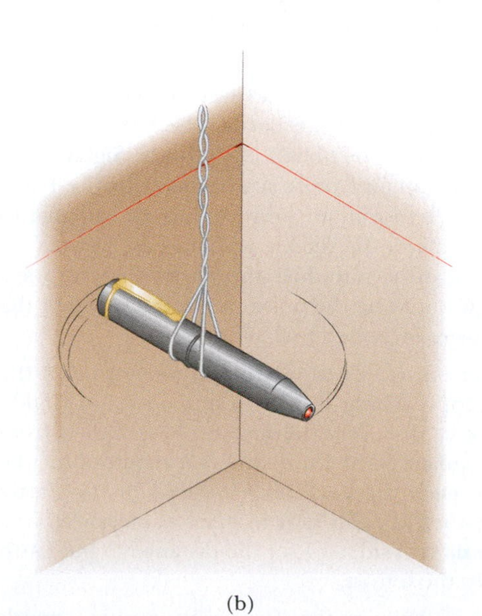

(b)

FIGURE Q9.16

## PROBLEMS

1, 2, 3 = straightforward, intermediate, challenging
☐ = full solution available in the *Student Solutions Manual and Study Guide*
PhysicsNow™ = coached problem with hints available at www.pop4e.com
= computer useful in solving problem
= paired numerical and symbolic problems
= biomedical application

### Section 9.1 ■ The Principle of Newtonian Relativity

1. In a laboratory frame of reference, an observer notes that Newton's second law is valid. Show that it is also valid for an observer moving at a constant speed, small compared with the speed of light, relative to the laboratory frame.

2. Show that Newton's second law is *not* valid in a reference frame moving past the laboratory frame of Problem 9.1 with a constant acceleration.

3. A 2 000-kg car moving at 20.0 m/s collides and locks together with a 1 500-kg car at rest at a stop sign. Show that momentum is conserved in a reference frame moving at 10.0 m/s in the direction of the moving car.

### Section 9.2 ■ The Michelson–Morley Experiment
### Section 9.3 ■ Einstein's Principle of Relativity
### Section 9.4 ■ Consequences of Special Relativity

Problem 3.36 in Chapter 3 can be assigned with this section.

4. How fast must a meter stick be moving if its length is measured to shrink to 0.500 m?

5. At what speed does a clock move if it is measured to run at a rate that is one-half the rate of a clock at rest with respect to an observer?

6. An astronaut is traveling in a space vehicle that has a speed of $0.500c$ relative to the Earth. The astronaut measures her pulse rate at 75.0 beats per minute. Signals generated by the astronaut's pulse are radioed to the Earth when the vehicle is moving in a direction perpendicular to the line that connects the vehicle with an observer on the Earth. (a) What pulse rate does the Earth observer measure? (b) What would be the pulse rate if the speed of the space vehicle were increased to $0.990c$?

7. An astronomer on the Earth observes a meteoroid in the southern sky approaching the Earth at a speed of $0.800c$. At the time of its discovery the meteoroid is 20.0 ly from the Earth. Calculate (a) the time interval required for the meteoroid to reach the Earth as measured by the Earth-bound astronomer, (b) this time interval as measured by a tourist on the meteoroid, and (c) the distance to the Earth as measured by the tourist.

8. A muon formed high in the Earth's atmosphere travels at speed $v = 0.990c$ for a distance of 4.60 km before it decays into an electron, a neutrino, and an antineutrino ($\mu^- \rightarrow e^- + \nu + \bar{\nu}$). (a) How long does the muon live, as measured in its reference frame? (b) How far does the Earth travel, as measured in the frame of the muon?

9. An atomic clock moves at 1 000 km/h for 1.00 h as measured by an identical clock on the Earth. How many nanoseconds slow will the moving clock be compared with the Earth clock at the end of the 1.00-h interval?

10. For what value of $v$ does $\gamma = 1.010\,0$? Observe that for speeds lower than this value, time dilation and length contraction are effects amounting to less than 1%.

11. PhysicsNow™ A spacecraft with a proper length of 300 m takes 0.750 $\mu$s to pass an Earth observer. Determine the speed of the spacecraft as measured by the Earth observer.

12. (a) An object of proper length $L_p$ takes a time interval $\Delta t$ to pass an Earth observer. Determine the speed of the object as measured by the Earth observer. (b) A column of tanks, 300 m long, takes 75.0 s to pass a child waiting at a street corner on her way to school. Determine the speed of the armored vehicles. (c) Show that the answer to part (a) includes the answer to Problem 9.11 as a special case and includes the answer to part (b) as another special case.

13. A friend passes by you in a spacecraft traveling at a high speed. He tells you that his craft is 20.0 m long and that the identically constructed craft you are sitting in is 19.0 m long. According to your observations, (a) how long is your spacecraft, (b) how long is your friend's craft, and (c) what is the speed of your friend's craft?

14. The identical twins Speedo and Goslo join a migration from the Earth to Planet X. It is 20.0 ly away in a reference frame in which both planets are at rest. The twins, of the same age, depart at the same time on different spacecraft. Speedo's craft travels steadily at $0.950c$, and Goslo's travels at $0.750c$. Calculate the age difference between the twins after Goslo's spacecraft lands on Planet X. Which twin is the older?

15. An interstellar space probe is launched from the Earth. After a brief period of acceleration it moves with a constant velocity, with a magnitude of 70.0% of the speed of light. Its nuclear-powered batteries supply the energy to keep its data transmitter active continuously. The batteries have a lifetime of 15.0 yr as measured in a rest frame. (a) How long do the batteries on the space probe last as measured by Mission Control on the Earth? (b) How far is the probe from the Earth when its batteries fail as measured by Mission Control? (c) How far is the probe from the Earth when its batteries fail as measured by its built-in trip odometer? (d) For what total time interval after launch are data received from the probe by Mission Control? Note that radio waves travel at the speed of light and fill the space between the probe and the Earth at the time of battery failure.

### Section 9.5 ■ The Lorentz Transformation Equations

16. Suzanne observes two light pulses to be emitted from the same location, but separated in time by 3.00 $\mu$s. Mark sees

the emission of the same two pulses separated in time by 9.00 $\mu$s. (a) How fast is Mark moving relative to Suzanne? (b) According to Mark, what is the separation in space of the two pulses?

**17.** A moving rod is measured to have a length of 2.00 m and to be oriented at an angle of 30.0° with respect to the direction of motion as shown in Figure P9.17. The rod has a speed of 0.995$c$. (a) What is the proper length of the rod? (b) What is the orientation angle in the proper frame?

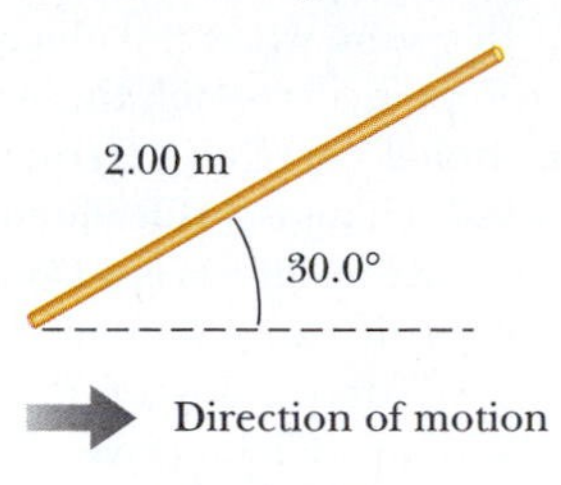


FIGURE **P9.17**

**18.** An observer in reference frame S measures two events as simultaneous. Event A occurs at the point (50.0 m, 0, 0) at the instant 9:00:00 Universal time on January 15, 2005. Event B occurs at the point (150 m, 0, 0) at the same moment. A second observer, moving past with a velocity of $0.800c\hat{\mathbf{i}}$, also observes the two events. In her reference frame S′, which event occurred first and what time interval elapsed between the events?

**19.** A red light flashes at position $x_R = 3.00$ m and time $t_R = 1.00 \times 10^{-9}$ s, and a blue light flashes at $x_B = 5.00$ m and $t_B = 9.00 \times 10^{-9}$ s, all measured in the S reference frame. Reference frame S′ has its origin at the same point as S at $t = t' = 0$; frame S′ moves uniformly to the right. Both flashes are observed to occur at the same place in S′. (a) Find the relative speed between S and S′. (b) Find the location of the two flashes in frame S′. (c) At what time does the red flash occur in the S′ frame?

**20.** A Klingon spacecraft moves away from the Earth at a speed of 0.800$c$ (Fig. P9.20). The starship *Enterprise* pursues at a speed of 0.900$c$ relative to the Earth. Observers on the Earth measure the *Enterprise* overtaking the Klingon craft at a relative speed of 0.100$c$. With what speed is the *Enterprise* overtaking the Klingon craft as measured by the crew of the *Enterprise*?

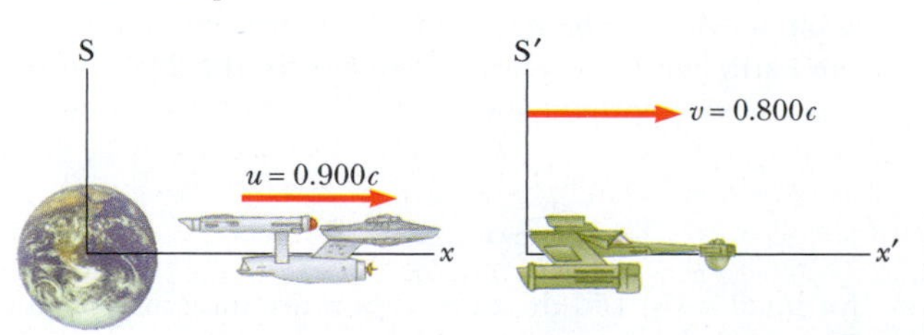


FIGURE **P9.20**

**21.** PhysicsNow™ Two jets of material from the center of a radio galaxy are ejected in opposite directions. Both jets move at 0.750$c$ relative to the galaxy. Determine the speed of one jet relative to the other.

**22.** A spacecraft is launched from the surface of the Earth with a velocity of 0.600$c$ at an angle of 50.0° above the horizontal positive $x$ axis. Another spacecraft is moving past with a velocity of 0.700$c$ in the negative $x$ direction. Determine the magnitude and direction of the velocity of the first spacecraft as measured by the pilot of the second spacecraft.

## Section 9.6 ▪ Relativistic Momentum and the Relativistic Form of Newton's Laws

**23.** Calculate the momentum of an electron moving with a speed of (a) 0.010 0$c$, (b) 0.500$c$, and (c) 0.900$c$.

**24.** The nonrelativistic expression for the momentum of a particle, $p = mu$, agrees with experimental results if $u \ll c$. For what speed does the use of this equation give an error in the momentum of (a) 1.00% and (b) 10.0%?

**25.** A golf ball travels with a speed of 90.0 m/s. By what fraction does its relativistic momentum magnitude $p$ differ from its classical value $mu$? That is, find the ratio $(p - mu)/mu$.

**26.** The speed limit on a certain roadway is 90.0 km/h. Suppose speeding fines are made proportional to the amount by which a vehicle's momentum exceeds the momentum it would have when traveling at the speed limit. The fine for driving at 190 km/h (that is, 100 km/h more than the speed limit) is $80.0. What, then, will be the fine for traveling at (a) 1 090 km/h and (b) 1 000 000 090 km/h?

**27.** PhysicsNow™ An unstable particle at rest breaks into two fragments of unequal mass. The mass of one fragment is $2.50 \times 10^{-28}$ kg and that of the other is $1.67 \times 10^{-27}$ kg. If the lighter fragment has a speed of 0.893$c$ after the breakup, what is the speed of the heavier fragment?

**28.** Show that the speed of an object having momentum of magnitude $p$ and mass $m$ is

$$u = \frac{c}{\sqrt{1 + (mc/p)^2}}$$

## Section 9.7 ▪ Relativistic Energy

**29.** Determine the energy required to accelerate an electron from (a) 0.500$c$ to 0.900$c$ and (b) 0.900$c$ to 0.990$c$.

**30.** Show that, for any object moving at less than one-tenth the speed of light, the relativistic kinetic energy agrees with the result of the classical equation $K = \frac{1}{2}mu^2$ to within less than 1%. Therefore, for most purposes the classical equation is good enough to describe these objects, whose motion we call *nonrelativistic*.

**31.** An electron has a kinetic energy five times greater than its rest energy. Find (a) its total energy and (b) its speed.

**32.** Find the kinetic energy of a 78.0-kg spacecraft launched out of the solar system with speed 106 km/s by using (a) the classical equation $K = \frac{1}{2}mu^2$ and (b) the relativistic equation.

**33.** PhysicsNow™ A proton moves at 0.950$c$. Calculate its (a) rest energy, (b) total energy, and (c) kinetic energy.

**34.** A cube of steel has a volume of 1.00 cm$^3$ and a mass of 8.00 g when at rest on the Earth. If this cube is now given a speed $u = 0.900c$, what is its density as measured by a stationary observer? Note that relativistic density is defined as $E_R/c^2V$.

**35.** The rest energy of an electron is 0.511 MeV. The rest energy of a proton is 938 MeV. Assume that both particles have kinetic energies of 2.00 MeV. Find the speed of (a) the electron and (b) the proton. (c) By how much does the speed of the electron exceed that of the proton? (d) Repeat the calculations assuming that both particles have kinetic energies of 2 000 MeV.

**36.** An unstable particle with a mass of $3.34 \times 10^{-27}$ kg is initially at rest. The particle decays into two fragments that fly off along the $x$ axis with velocity components $0.987c$ and $-0.868c$. Find the masses of the fragments. (*Suggestion:* Conserve both energy and momentum.)

**37.** Show that the energy–momentum relationship $E^2 = p^2c^2 + (mc^2)^2$ follows from the expressions $E = \gamma mc^2$ and $p = \gamma mu$.

**38.** An object having mass 900 kg and traveling at speed $0.850c$ collides with a stationary object having mass 1 400 kg. The two objects stick together. Find (a) the speed and (b) the mass of the composite object.

**39.** A pion at rest ($m_\pi = 273m_e$) decays to a muon ($m_\mu = 207m_e$) and an antineutrino ($m_{\bar{\nu}} \approx 0$). The reaction is written $\pi^- \rightarrow \mu^- + \bar{\nu}$. Find the kinetic energy of the muon and the energy of the antineutrino in electron volts. (*Suggestion:* Conserve both energy and momentum.)

## Section 9.8 ■ Mass and Energy

**40.** When 1.00 g of hydrogen combines with 8.00 g of oxygen, 9.00 g of water is formed. During this chemical reaction, $2.86 \times 10^5$ J of energy is released. How much mass do the constituents of this reaction lose? Is the loss of mass likely to be detectable?

**41.** The power output of the Sun is $3.85 \times 10^{26}$ W. How much mass is converted to energy in the Sun each second?

**42.** In a nuclear power plant, the fuel rods last 3 yr before they are replaced. If a plant with rated thermal power 1.00 GW operates at 80.0% capacity for the 3.00 yr, what is the loss of mass of the fuel?

**43.** A gamma ray (a high-energy photon) can produce an electron ($e^-$) and a positron ($e^+$) when it enters the electric field of a heavy nucleus: $\gamma \rightarrow e^+ + e^-$. What minimum gamma-ray energy is required to accomplish this task? (*Note:* The masses of the electron and the positron are equal.)

**44.** (a) In a crash test, two 1 500-kg cars, both moving at 20.0 m/s, collide head-on and stick together. Consider the whole quantity of wreckage before it loses any energy by such processes as thermal radiation. Is its mass greater or less than the original total mass of the two cars? By how much? (b) Repeat the problem for a relativistic crash test in which two 1 500-kg space vehicles, both moving at 200 Mm/s, meet head-on in a completely inelastic collision.

## Section 9.9 ■ General Relativity

**45.** An Earth satellite used in the Global Positioning System (GPS) moves in a circular orbit with radius $2.66 \times 10^7$ m and period 11 h 58 min. (a) Determine its speed. (b) The satellite contains an oscillator producing the principal nonmilitary GPS signal. Its frequency is 1 575.42 MHz in the reference frame of the satellite. When it is received on the Earth's surface, what is the fractional change in this frequency due to time dilation, as described by special relativity? (c) The gravitational "blueshift" of the frequency according to general relativity is a separate effect. It is called a blueshift to indicate a change to a higher frequency. The magnitude of that fractional change is given by

$$\frac{\Delta f}{f} = \frac{\Delta U_g}{mc^2}$$

where $\Delta U_g$ is the change in gravitational potential energy of an object–Earth system when the object of mass $m$ is moved between the two points at which the signal is observed. Calculate this fractional change in frequency. (d) What is the overall fractional change in frequency? Superposed on both of these relativistic effects is a Doppler shift that is generally much larger. It can either increase or decrease the frequency received, depending on the motion of a particular satellite relative to a GPS receiver (Fig. P9.45).

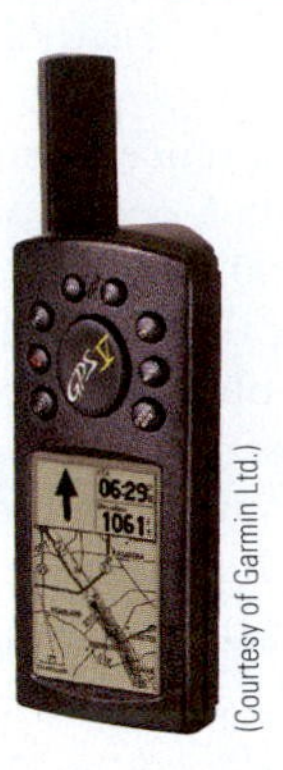


FIGURE **P9.45** This Global Positioning System (GPS) receiver incorporates relativistically corrected time calculations in its analysis of signals it receives from orbiting satellites, allowing the unit to determine its position on the Earth's surface to within a few meters. If these corrections were not made, the location error would be about 1 km.

## Section 9.10 ■ Context Connection—From Mars to the Stars

**46.** In 1963, Mercury astronaut Gordon Cooper orbited the Earth 22 times. The press stated that for each orbit he aged 2 millionths of a second less than he would have if he had remained on the Earth. (a) Assuming that he was 160 km above the Earth moving at 7.82 km/s in a circular orbit, determine the time difference between someone on the Earth and the orbiting astronaut for the 22 orbits. You may use the approximation

$$\frac{1}{\sqrt{1-x}} \approx 1 + \frac{x}{2}$$

for small $x$. (b) Did the press report accurate information? Explain.

**47.** An astronaut wishes to visit the Andromeda galaxy, making a one-way trip that will take 30.0 yr in the spacecraft's frame of reference. Assume that the galaxy is 2.00 million ly away and that the astronaut's speed is constant. (a) How fast must he travel relative to the Earth? (b) What will be the kinetic energy of his 1 000-metric-ton spacecraft? (c) What is the cost of this energy if it is purchased at a typical consumer price for energy from the electric company of \$0.130/kWh?

## Additional Problems

**48.** An electron has a speed of $0.750c$. (a) Find the speed of a proton that has the same kinetic energy as the electron. (b) Find the speed of a proton that has the same momentum as the electron.

**49.** **Physics Now™** The cosmic rays of highest energy are protons that have kinetic energy on the order of $10^{13}$ MeV. (a) How long would it take a proton of this energy to travel across the Milky Way galaxy, having a diameter on the order of $10^5$ ly, as measured in the proton's frame? (b) From the point of view of the proton, how many kilometers across is the galaxy?

**50.** Ted and Mary are playing a game of catch in frame S′, which is moving at $0.600c$ with respect to frame S, while Jim, at rest in frame S, watches the action (Fig. P9.50). Ted throws the ball to Mary at $0.800c$ (according to Ted), and their separation (measured in S′) is $1.80 \times 10^{12}$ m. (a) According to Mary, how fast is the ball moving? (b) According to Mary, how long does it take the ball to reach her? (c) According to Jim, how far apart are Ted and Mary, and how fast is the ball moving? (d) According to Jim, how long does it take the ball to reach Mary?

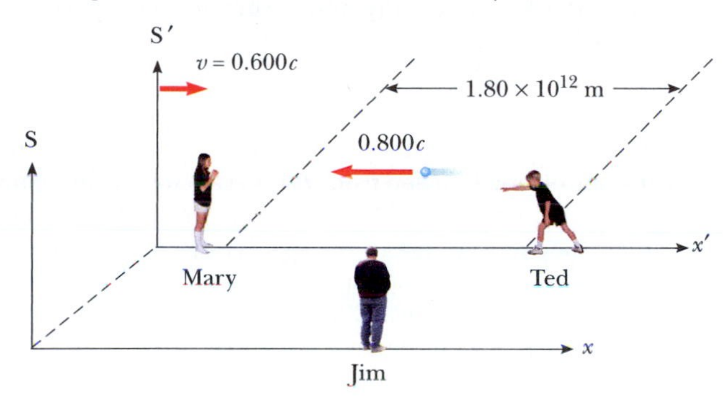


FIGURE **P9.50**

**51.** The net nuclear fusion reaction inside the Sun can be written as $4\,{}^1\text{H} \rightarrow {}^4\text{He} + E$. The rest energy of each hydrogen atom is 938.78 MeV and the rest energy of the helium-4 atom is 3 728.4 MeV. Calculate the percentage of the starting mass that is transformed to other forms of energy.

**52.** An object disintegrates into two fragments. One of the fragments has mass $1.00\ \text{MeV}/c^2$ and momentum $1.75\ \text{MeV}/c$ in the positive $x$ direction. The other fragment has mass $1.50\ \text{MeV}/c^2$ and momentum $2.00\ \text{MeV}/c$ in the positive $y$ direction. Find (a) the mass and (b) the speed of the original object.

**53.** An alien spaceship traveling at $0.600c$ toward the Earth launches a landing craft with an advance guard of purchasing agents and physics teachers. The lander travels in the same direction with a speed of $0.800c$ relative to the mother ship. As observed on the Earth, the spaceship is 0.200 ly from the Earth when the lander is launched. (a) What speed do the Earth observers measure for the approaching lander? (b) What is the distance to the Earth at the time of lander launch as observed by the aliens? (c) How long does it take the lander to reach the Earth as observed by the aliens on the mother ship? (d) If the lander has a mass of $4.00 \times 10^5$ kg, what is its kinetic energy as observed in the Earth reference frame?

**54.** A physics professor on the Earth gives an exam to her students, who are in a spacecraft traveling at speed $v$ relative to the Earth. The moment the craft passes the professor, she signals the start of the exam. She wishes her students to have a time interval $T_0$ (spacecraft time) to complete the exam. Show that she should wait a time interval (Earth time) of

$$T = T_0\sqrt{\frac{1 - v/c}{1 + v/c}}$$

before sending a light signal telling them to stop. (*Suggestion:* Remember that it takes some time for the second light signal to travel from the professor to the students.)

**55.** A supertrain (proper length 100 m) travels at a speed of $0.950c$ as it passes through a tunnel (proper length 50.0 m). As seen by a trackside observer, is the train ever completely within the tunnel? If so, with how much space to spare?

**56.** Energy reaches the upper atmosphere of the Earth from the Sun at the rate of $1.79 \times 10^{17}$ W. If all this energy were absorbed by the Earth and not re-emitted, how much would the mass of the Earth increase in 1.00 yr?

**57.** A particle with electric charge $q$ moves along a straight line in a uniform electric field $\vec{\mathbf{E}}$ with a speed of $u$. The electric force exerted on the charge is $q\vec{\mathbf{E}}$. The motion and the electric field are both in the $x$ direction. (a) Show that the acceleration of the particle in the $x$ direction is given by

$$a = \frac{du}{dt} = \frac{qE}{m}\left(1 - \frac{u^2}{c^2}\right)^{3/2}$$

(b) Discuss the significance of the dependence of the acceleration on the speed. (c) If the particle starts from rest at $x = 0$ at $t = 0$, how could you proceed to find the speed of the particle and its position at time $t$?

**58.** Imagine that the entire Sun collapses to a sphere of radius $R_g$ such that the work required to remove a small mass $m$ from the surface would be equal to its rest energy $mc^2$. This radius is called the *gravitational radius* for the Sun. Find $R_g$. (It is believed that the ultimate fate of very massive stars is to collapse beyond their gravitational radii into black holes.)

**59.** The creation and study of new elementary particles is an important part of contemporary physics. Especially interesting is the discovery of a very massive particle. To create a particle of mass $M$ requires an energy $Mc^2$. With enough energy, an exotic particle can be created by allowing a fast-moving particle of ordinary matter, such as a proton, to collide with a similar target particle. Let us consider a perfectly inelastic collision between two protons in which an incident proton with mass $m_p$, kinetic energy $K$, and momentum magnitude $p$ joins with an originally stationary target proton to form a single product particle of mass $M$. You might think that the creation of a new product particle, nine times more massive than in a previous experiment, would require just nine times more energy for the incident proton. Unfortunately, not all the kinetic energy of the incoming proton is available to create the product particle because conservation of momentum requires that after the collision the system as a whole still must have some kinetic energy. Only a fraction of the energy of the incident particle is thus available to create a new particle. Determine how the energy available for particle creation depends on the energy of the moving proton. In particular, show that the energy available to create a product particle is given by

$$Mc^2 = 2m_pc^2\sqrt{1 + \frac{K}{2m_pc^2}}$$

From this result, when the kinetic energy $K$ of the incident proton is large compared with its rest energy $m_pc^2$, we see that $M$ approaches $(2m_pK)^{1/2}/c$. Thus, if the energy of the incoming proton is increased by a factor of nine, the mass you can create increases only by a factor of three. This disappointing result is the main reason that most modern accelerators, such as those at CERN (in Europe), at Fermilab (near Chicago), at SLAC (at Stanford), and at DESY (in Germany), use *colliding beams.* Here the total momentum of a pair of interacting particles can be zero. The center of mass can be at rest after the collision, so in principle all the initial kinetic energy can be used for particle creation, according to

$$Mc^2 = 2mc^2 + K = 2mc^2\left(1 + \frac{K}{2mc^2}\right)$$

where $K$ is the total kinetic energy of two identical colliding particles. Here, if $K >> mc^2$, we have $M$ directly proportional to $K$, as we would desire. These machines are difficult to build and to operate, but they open new vistas in physics.

**60.** An observer in a coasting spacecraft moves toward a mirror at speed $v$ relative to the reference frame labeled by S in Figure P9.60. The mirror is stationary with respect to S. A light pulse emitted by the spacecraft travels toward the mirror and is reflected back to the craft. The front of the craft is a distance $d$ from the mirror (as measured by observers in S) at the moment the light pulse leaves the craft. What is the total travel time of the pulse as measured by observers in (a) the S frame and (b) the front of the spacecraft?

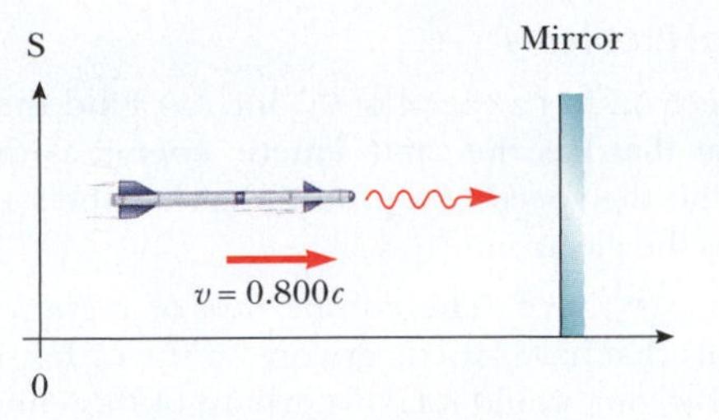


FIGURE P9.60

**61.** A rod of length $L_0$ moving with a speed $v$ along the horizontal direction makes an angle $\theta_0$ with respect to the $x'$ axis. (a) Show that the length of the rod as measured by a stationary observer is $L = L_0[1 - (v^2/c^2)\cos^2\theta_0]^{1/2}$. (b) Show that the angle that the rod makes with the $x$ axis is given by $\tan\theta = \gamma\tan\theta_0$. These results show that the rod is both contracted and rotated. (Take the lower end of the rod to be at the origin of the primed coordinate system.)

**62.** Prepare a graph of the relativistic kinetic energy and the classical kinetic energy, both as a function of speed, for an object with a mass of your choice. At what speed does the classical kinetic energy underestimate the experimental value by 1%, by 5%, and by 50%?

**63.** Suppose our Sun is about to explode. In an effort to escape, we depart in a spacecraft at $v = 0.800c$ and head toward the star Tau Ceti, 12.0 ly away. When we reach the midpoint of our journey from the Earth, we see our Sun explode and, unfortunately, at the same instant we see Tau Ceti explode as well. (a) In the spacecraft's frame of reference, should we conclude that the two explosions occurred simultaneously? If not, which occurred first? (b) In a frame of reference in which the Sun and Tau Ceti are at rest, did they explode simultaneously? If not, which exploded first?

## ANSWERS TO QUICK QUIZZES

**9.1** (d). The two events (the pulse leaving the flashlight and the pulse hitting the far wall) take place at different locations for both observers, so neither measures the proper time interval.

**9.2** (a). The two events are the beginning and the end of the movie, both of which take place at rest with respect to the spacecraft crew. Therefore, the crew measures the proper time interval of 2 h. Any observer in motion with respect to the spacecraft, which includes the observer on Earth, will measure a longer time interval because of time dilation.

**9.3** (a). If their on-duty time is based on clocks that remain on the Earth, they will have larger paychecks. A shorter time interval will have passed for the astronauts in their frame of reference than for their employer back on the Earth.

**9.4** (c). Both your body and your sleeping cabin are at rest in your reference frame; therefore, they will have their proper length according to you. There will be no change in measured lengths of objects, including yourself, within your spacecraft.

**9.5** (c), (d). Because of your motion toward the source of the light, the light beam has a horizontal component of velocity as measured by you. The magnitude of the vector sum of the horizontal and vertical component vectors must be equal to $c$, so the magnitude of the vertical component must be smaller than $c$. When the searchlight is aimed directly toward you, there is only a horizontal component of the velocity of the light and you must measure a speed of $c$.

**9.6** (a) $m_3 > m_2 = m_1$; the rest energy of particle 3 is $2E$, whereas it is $E$ for particles 1 and 2. (b) $K_3 = K_2 > K_1$; the kinetic energy is the difference between the total energy and the rest energy. The kinetic energy is $4E - 2E = 2E$ for particle 3, $3E - E = 2E$ for particle 2, and $2E - E = E$ for particle 1. (c) $u_2 > u_3 = u_1$; from Equation 9.21, $E = \gamma E_R$. Solving for the square of the particle speed $u$, we find that $u^2 = c^2[1 - (E_R/E)^2]$. Therefore, the particle with the smallest ratio of rest energy to total energy will have the largest speed. Particles 1 and 3 have the same ratio as each other, and the ratio of particle 2 is smaller.

# CHAPTER 10

# Rotational Motion

(Courtesy of Tourism Malaysia)

The Malaysian pastime of *gasing* involves the spinning of tops that can have masses up to 5 kg. Professional spinners can spin their tops so that they might rotate for 1 to 2 h before stopping. We will study the rotational motion of objects such as these tops in this chapter.

## CHAPTER OUTLINE


10.1 Angular Position, Speed, and Acceleration
10.2 Rotational Kinematics: The Rigid Object Under Constant Angular Acceleration
10.3 Relations Between Rotational and Translational Quantities
10.4 Rotational Kinetic Energy
10.5 Torque and the Vector Product
10.6 The Rigid Object in Equilibrium
10.7 The Rigid Object Under a Net Torque
10.8 Angular Momentum
10.9 Conservation of Angular Momentum
10.10 Precessional Motion of Gyroscopes
10.11 Rolling Motion of Rigid Objects
10.12 Context Connection — Turning the Spacecraft
SUMMARY


When an extended object, such as a wheel, rotates about its axis, the motion cannot be analyzed by treating the object as a particle because at any given time different parts of the object are moving with different speeds and in different directions. We can, however, analyze the motion by considering an extended object to be composed of a collection of moving particles.

In dealing with a rotating object, analysis is greatly simplified by assuming that the object is rigid. A **rigid object** is one that is nondeformable; that is, it is an object in which the relative locations of all particles of which the object is composed remain constant. All real objects are deformable to some extent; our rigid-object model, however, is useful in many situations in which deformation is negligible.

## 10.1 ANGULAR POSITION, SPEED, AND ACCELERATION

We began our study of translational motion in Chapter 2 by defining the terms *position, velocity,* and *acceleration.* For example, we locate a particle in one-dimensional space with the position variable $x$. In this chapter, we will insert the word *translational* before our previously studied kinematic variables to distinguish them from the analogous *rotational* variables that we will develop.

Let us think about a rotating object. How would we describe its position in its rotational motion? We do so by describing its *orientation* relative to some fixed reference direction. For example, imagine two soldiers performing a military about-face maneuver. They both begin by facing due north. One soldier, who has been practicing diligently, ends up after the maneuver with his body facing due south. The second, who has not been practicing, ends up facing southeast. We could describe their respective rotational positions after the maneuver by reporting the angle through which each turned from the original direction. The first soldier turned through 180°, but the second turned through only 135°. Thus, we can use an angle measured from a reference direction as a measure of **rotational position,** or **angular position,** which is our starting point for our description of rotational motion.

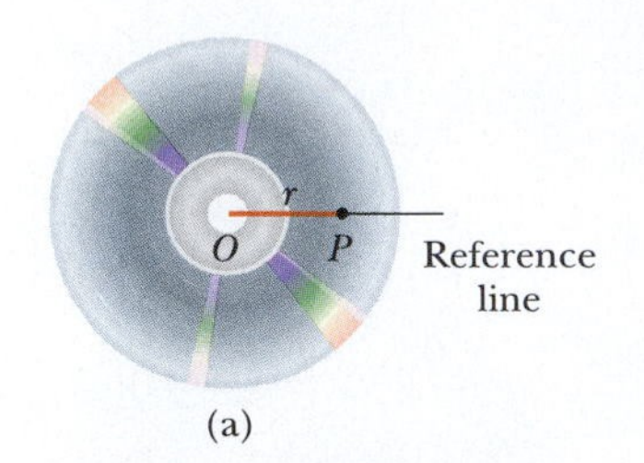


(a)

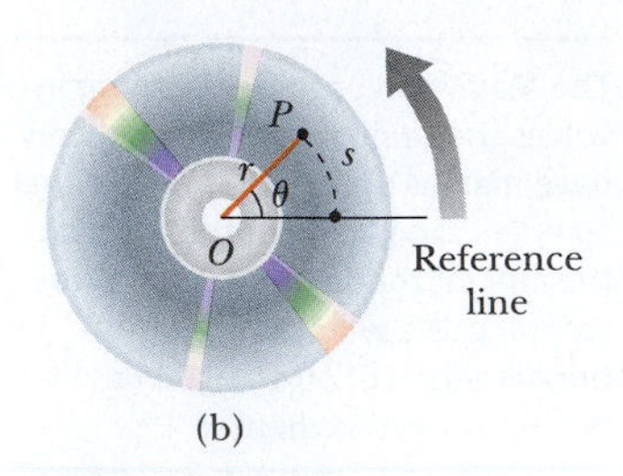


(b)

**FIGURE 10.1** A compact disc rotating about a fixed axis through $O$ perpendicular to the plane of the figure. (a) To define angular position for the disc, a fixed reference line is chosen. A particle at $P$ is located at a radial distance $r$ from the rotation axis at $O$. (b) As the disc rotates, point $P$ moves through an arc length $s$ on a circular path of radius $r$. The angular position of $P$ is $\theta$.

**PITFALL PREVENTION 10.1**

**REMEMBER THE RADIAN** Keep in mind that Equation 10.1b defines an angle expressed in *radians.* Don't fall into the trap of using this equation for angles measured in degrees. Also, be sure to set your calculator in radian mode when doing problems in rotation.

Figure 10.1 illustrates an overhead view of a rotating compact disc. The disc is rotating about a fixed axis through $O$. The axis is perpendicular to the plane of the figure. Let us investigate the motion of only one of the millions of "particles" making up the disc. A particle at $P$ is at a fixed distance $r$ from the origin and rotates about it in a circle of radius $r$. (In fact, *every* particle on the disc undergoes circular motion about $O$.) It is convenient to represent the position of $P$ with its polar coordinates $(r, \theta)$, where $r$ is the distance from the origin to $P$ and $\theta$ is measured *counterclockwise* from some reference line shown in Figure 10.1. In this representation, the only coordinate for the particle that changes in time is the angle $\theta$; $r$ remains constant. As the particle moves along the circle from the reference line ($\theta = 0$) to an angular position $\theta$, it moves through an arc of length $s$ as in Figure 10.1b. The arc length $s$ is related to the angle $\theta$ through the relationship

$$s = r\theta \qquad [10.1a]$$

$$\theta = \frac{s}{r} \qquad [10.1b]$$

It is important to note the units of $\theta$ in Equation 10.1b. Because $\theta$ is the ratio of an arc length and the radius of the circle, it is a pure number. Nonetheless, we commonly give $\theta$ the artificial unit **radian** (rad), where

■ The radian

one radian is the angle subtended by an arc length equal to the radius of the arc.

Because the circumference of a circle is $2\pi r$, it follows from Equation 10.1b that 360° corresponds to an angle of $(2\pi r/r)$ rad $= 2\pi$ rad. (Also note that $2\pi$ rad corresponds to one complete revolution.) Hence, 1 rad $= 360°/2\pi \approx 57.3°$. To convert an angle in degrees to an angle in radians, we use $\pi$ rad $= 180°$, so

$$\theta\ (\text{rad}) = \frac{\pi}{180°}\,\theta\ (\text{deg})$$

For example, 60° equals $\pi/3$ rad, and 45° equals $\pi/4$ rad.

Because the disc in Figure 10.1 is a rigid object, as the particle moves along the circle from the reference line every other particle on the object rotates through the same angle $\theta$. Therefore, **we can associate the angle $\theta$ with the entire rigid object as well as with an individual particle,** which allows us to define the angular position of a rigid object in its rotational motion. We choose a radial line on the object, such as a line connecting $O$ and a chosen particle on the object. The angular position of the rigid object is the angle $\theta$ between this radial line on the object and the fixed reference line in space, which is often chosen as the $x$ axis. This process is similar to the way we identify the position of an object in translational motion as the distance $x$ between the object and the reference position, which is the origin, $x = 0$.

As a particle on a rigid object travels from position Ⓐ to position Ⓑ in a time interval $\Delta t$ as in Figure 10.2, the radial line of length $r$ sweeps out an angle $\Delta\theta = \theta_f - \theta_i$. This quantity $\Delta\theta$ is defined as the **angular displacement** of the rigid object:

$$\Delta\theta \equiv \theta_f - \theta_i$$

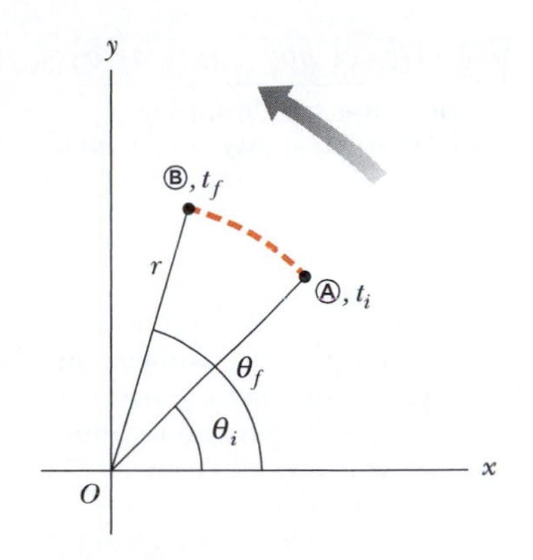


**FIGURE 10.2** A particle on a rotating rigid object moves from Ⓐ to Ⓑ along the arc of a circle. In the time interval $\Delta t = t_f - t_i$, the radial line of length $r$ sweeps out an angle $\Delta\theta = \theta_f - \theta_i$.

The rate at which this angular displacement occurs can vary. If the rigid object spins rapidly, this displacement can occur in a short time interval. If it rotates slowly, this displacement occurs in a longer time interval. These different rotation rates can be quantified by introducing *angular speed.* We define the **average angular speed** $\omega_{avg}$ as the ratio of the angular displacement of a rigid object to the time interval $\Delta t$ during which the displacement occurs:

▪ Average angular speed

$$\omega_{avg} \equiv \frac{\theta_f - \theta_i}{t_f - t_i} = \frac{\Delta\theta}{\Delta t} \quad [10.2]$$

In analogy to linear speed, the **instantaneous angular speed** $\omega$ is defined as the limit of the ratio $\Delta\theta/\Delta t$ as $\Delta t$ approaches zero:

▪ Instantaneous angular speed

$$\omega \equiv \lim_{\Delta t \to 0} \frac{\Delta\theta}{\Delta t} = \frac{d\theta}{dt} \quad [10.3]$$

Angular speed has units of rad/s (or $s^{-1}$ because radians are not dimensional). Let us adopt the convention that the fixed axis of rotation for an object is the $z$ axis, which is directed out of the page in Figures 10.1 and 10.2. We shall take $\omega$ to be positive when $\theta$ is increasing (counterclockwise motion in Figs. 10.1 and 10.2) and negative when $\theta$ is decreasing (clockwise motion).

If the instantaneous angular speed of a particle changes from $\omega_i$ to $\omega_f$ in the time interval $\Delta t$, the particle has an angular acceleration. The **average angular acceleration** $\alpha_{avg}$ of a particle moving in a circular path is defined as the ratio of the change in the angular speed to the time interval $\Delta t$:

▪ Average angular acceleration

$$\alpha_{avg} \equiv \frac{\omega_f - \omega_i}{t_f - t_i} = \frac{\Delta\omega}{\Delta t} \quad [10.4]$$

In analogy to linear acceleration, the **instantaneous angular acceleration** is defined as the limit of the ratio $\Delta\omega/\Delta t$ as $\Delta t$ approaches zero:

▪ Instantaneous angular acceleration

$$\alpha \equiv \lim_{\Delta t \to 0} \frac{\Delta\omega}{\Delta t} = \frac{d\omega}{dt} \quad [10.5]$$

Angular acceleration has units of rad/$s^2$ or $s^{-2}$.

As pointed out in the introduction, we will focus much of our attention in this chapter on rigid objects. Approximating a real object as a rigid object is a simplification model, similar to the particle model, which we call the **rigid object model.** If we were to imagine a rotating block of gelatin, which is *not* a rigid object, the motion is very complicated because of the combination of rotation of the particles

**PITFALL PREVENTION 10.2**

**Specify Your Axis** In solving rotation problems, you will need to specify an axis of rotation, a feature we did not see in our study of translational motion. The choice is arbitrary, but once you make it, you need to maintain that choice consistently throughout the problem. In some problems, a natural axis is suggested by the physical situation, such as the center of an automobile wheel. In other problems, the choice may not be obvious, and you will need to choose an axis.

and the movement of particles within the deformable block relative to one another. Another example of a nonrigid object is our own Sun; the region of the Sun near the solar equator is moving with a higher angular speed than the region near the poles. We shall not analyze such messy problems, however. Instead, our analyses will address only rigid objects. As we investigate rotational motion, we shall develop a number of analysis models for rigid objects that have analogs in our analysis models for particles.

With our simplification model of a rigid object, we can make a statement about the various particles in the rigid object: When a rigid object is rotating about a *fixed* axis, **every particle on the object rotates about that axis through the same angle in a given time interval and has the same angular speed and the same angular acceleration.** That is, the quantities $\theta$, $\omega$, and $\alpha$ characterize the rotational motion of the entire rigid object as well as individual particles in the object. Using these quantities, we can greatly simplify the analysis of rigid-object rotation.

The angular position $\theta$, angular speed $\omega$, and angular acceleration $\alpha$ of a rigid object are analogous to translational position $x$, translational speed $v$, and translational acceleration $a$, respectively, for the corresponding one-dimensional motion of a particle discussed in Chapter 2. The variables $\theta$, $\omega$, and $\alpha$ differ dimensionally from the variables $x$, $v$, and $a$ only by a length factor, as we shall see shortly.

We have not associated any direction with the angular speed and angular acceleration.[1] Strictly speaking, these variables are the magnitudes of the angular velocity and angular acceleration vectors $\vec{\omega}$ and $\vec{\alpha}$. Because we are considering rotation about a fixed axis, we can indicate the directions of these vectors by assigning a positive or negative sign to $\omega$ and $\alpha$, as discussed for $\omega$ after Equation 10.3. For rotation about a fixed axis, the only direction in space that uniquely specifies the rotational motion is the direction along the axis, but we still must specify one of the two directions along this axis as positive.

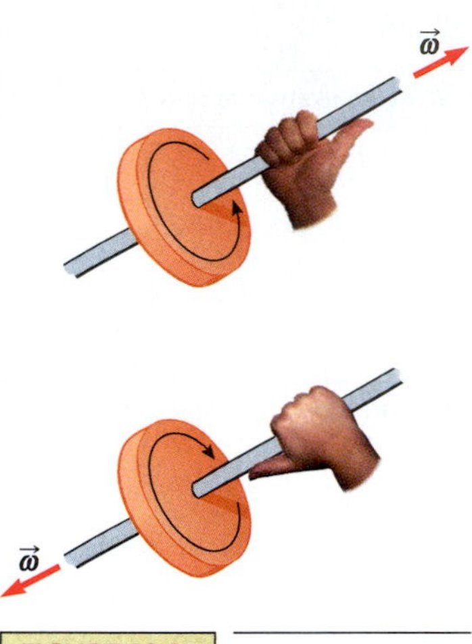


**FIGURE 10.3** The orange disk rotates in the directions indicated. The right-hand rule determines the direction of the angular velocity vector.

The direction of $\vec{\omega}$ is along the axis of rotation, which is the $z$ axis in Figure 10.1. By convention, we take the direction of $\vec{\omega}$ to be *out of* the plane of the diagram when the rotation is counterclockwise and *into* the plane of the diagram when the rotation is clockwise. To further illustrate this convention, it is convenient to use the **right-hand rule** illustrated by Figure 10.3. The four fingers of the right hand are wrapped in the direction of the rotation. The extended right thumb points in the direction of $\vec{\omega}$.

The direction of $\vec{\alpha}$ follows from its vector definition as $d\vec{\omega}/dt$. For rotation about a fixed axis, the direction of $\vec{\alpha}$ is the same as $\vec{\omega}$ if the angular speed (the magnitude of $\vec{\omega}$) is increasing in time and is antiparallel to $\vec{\omega}$ if the angular speed is decreasing in time.

The full vector treatment of rotational motion is beyond the scope of this book and not necessary for our level of understanding, so we will not use vector notation for most of this chapter.

**QUICK QUIZ 10.1** A rigid object is rotating in a counterclockwise sense around a fixed axis. Each of the following pairs of quantities represents an initial angular position and a final angular position of the rigid object. **(i)** Which of the sets can *only* occur if the rigid object rotates through more than 180°? **(a)** 3 rad, 6 rad **(b)** −1 rad, 1 rad **(c)** 1 rad, 5 rad **(ii)** If each of the displacements occurs during the same time interval, which choice represents the lowest average angular speed?

[1]Although we do not verify it here, the instantaneous angular velocity and instantaneous angular acceleration are vector quantities, but the corresponding average values are not because angular displacements do not add as vector quantities for finite rotations.

## 10.2 ROTATIONAL KINEMATICS: THE RIGID OBJECT UNDER CONSTANT ANGULAR ACCELERATION

In our study of one-dimensional motion, we found that the simplest accelerated motion to analyze is motion under constant translational acceleration (Chapter 2). Likewise, for rotational motion about a fixed axis, the simplest accelerated motion to analyze is motion of a rigid object under constant angular acceleration. We will identify this situation as an analysis model that can be used to solve a wide variety of rotational problems.

If we write Equation 10.5 in the form $d\omega = \alpha\, dt$ and let $\omega = \omega_i$ at $t_i = 0$, we can integrate this expression directly to find the final angular speed $\omega_f$ of the rigid object as a function of time:

$$\omega_f = \omega_i + \alpha t \qquad \text{(for constant } \alpha\text{)} \qquad [10.6]$$

Likewise, if we rewrite Equation 10.3 and substitute Equation 10.6, we can integrate once more (with $\theta = \theta_i$ at $t_i = 0$) to find the angular position of the rigid object as a function of time:

$$\theta_f = \theta_i + \omega_i t + \tfrac{1}{2}\alpha t^2 \qquad \text{(for constant } \alpha\text{)} \qquad [10.7]$$

If we eliminate $t$ from Equations 10.6 and 10.7, we obtain

$$\omega_f^2 = \omega_i^2 + 2\alpha(\theta_f - \theta_i) \qquad \text{(for constant } \alpha\text{)} \qquad [10.8]$$

If we eliminate $\alpha$, we find

$$\theta_f = \theta_i + \tfrac{1}{2}(\omega_i + \omega_f)t \qquad \text{(for constant } \alpha\text{)} \qquad [10.9]$$

Notice that these kinematic expressions for rotational motion of a rigid object under constant angular acceleration are of the *same mathematical form* as those for translational motion of a particle under constant acceleration, with the substitutions $x \rightarrow \theta$, $v \rightarrow \omega$, and $a \rightarrow \alpha$. The similarities between rotational and translational kinematic equations are shown in Table 10.1.

**QUICK QUIZ 10.2** Consider again the pairs of angular positions for the rigid object in Quick Quiz 10.1. If the object starts from rest at the initial angular position, moves counterclockwise with constant angular acceleration, and arrives at the final angular position with the same angular speed in all three cases, for which choice is the angular acceleration the highest?

**TABLE 10.1** A Comparison of Equations for Rotational and Translational Motion: Kinematic Equations

| Rotational Motion About a Fixed Axis with $\alpha$ = Constant Variables: $\theta_f$ and $\omega_f$ | Translational Motion with $a$ = Constant Variables: $x_f$ and $v_f$ |
|---|---|
| $\omega_f = \omega_i + \alpha t$ | $v_f = v_i + at$ |
| $\theta_f = \theta_i + \omega_i t + \frac{1}{2}\alpha t^2$ | $x_f = x_i + v_i t + \frac{1}{2}at^2$ |
| $\theta_f = \theta_i + \frac{1}{2}(\omega_i + \omega_f)t$ | $x_f = x_i + \frac{1}{2}(v_i + v_f)t$ |
| $\omega_f^2 = \omega_i^2 + 2\alpha(\theta_f - \theta_i)$ | $v_f^2 = v_i^2 + 2a(x_f - x_i)$ |

**PITFALL PREVENTION 10.3**

**JUST LIKE TRANSLATION?** Table 10.1 suggests that rotational kinematics is just like translational kinematics. That is almost true, but keep in mind two differences that you must address. (1) In rotational kinematics, as suggested in Pitfall Prevention 10.2, you need to specify a rotation axis. (2) In rotational motion, the object keeps returning to its original orientation; therefore, you may be asked for the number of revolutions made by a rigid object, a concept that has no meaning in translational motion.

**EXAMPLE 10.1** **Rotating Wheel**

A wheel rotates with a constant angular acceleration of 3.50 rad/s$^2$.

**A** If the angular speed of the wheel is 2.00 rad/s at $t = 0$, through what angle does the wheel rotate between $t = 0$ and $t = 2.00$ s?

**Solution** We assume that the wheel is perfectly rigid, so we can use the rigid object model. Because the angular acceleration in the problem is given as constant, we model the wheel as a rigid object under constant angular acceleration and use the rotational kinematic equations. We use Equation 10.7, setting $\theta_i = 0$, and obtain

$$\theta_f = \theta_i + \omega_i t + \tfrac{1}{2}\alpha t^2$$
$$= 0 + (2.00 \text{ rad/s})(2.00 \text{ s}) + \tfrac{1}{2}(3.50 \text{ rad/s}^2)(2.00 \text{ s})^2$$
$$= \boxed{11.0 \text{ rad}}$$

which is equivalent to $11.0 \text{ rad}/(2\pi \text{ rad/rev}) = 1.75$ rev.

**B** What is the angular speed of the wheel at $t = 2.00$ s?

**Solution** We use Equation 10.6:

$$\omega_f = \omega_i + \alpha t = 2.00 \text{ rad/s} + (3.50 \text{ rad/s}^2)(2.00 \text{ s})$$
$$= \boxed{9.00 \text{ rad/s}}$$

We could also obtain this result using Equation 10.8 and the results of part A. Try it!

## 10.3 RELATIONS BETWEEN ROTATIONAL AND TRANSLATIONAL QUANTITIES

In this section, we shall derive some useful relations between the angular speed and angular acceleration of a single particle on a rotating rigid object and its translational speed and translational acceleration. Keep in mind that when a rigid object rotates about a fixed axis, *every* particle of the object moves in a circle whose center is on the axis of rotation.

Consider a particle on a rotating rigid object, moving in a circle of radius $r$ about the $z$ axis, as in Active Figure 10.4. Because the particle moves along a circular path, its translational velocity vector $\vec{v}$ is always tangent to the path; hence, we often call this quantity **tangential velocity.** The magnitude of the tangential velocity of the particle is, by definition, the **tangential speed,** given by $v = ds/dt$, where $s$ is the distance traveled by the particle along the circular path. Recalling from Equation 10.1a that $s = r\theta$ and noting that $r$ is a constant, we have

$$v = \frac{ds}{dt} = r\frac{d\theta}{dt}$$

$$\boxed{v = r\omega} \qquad [10.10]$$

That is, the tangential speed of the particle equals the distance of the particle from the axis of rotation multiplied by the particle's angular speed.

We can relate the angular acceleration of the particle to its tangential acceleration $a_t$—which is the component of its acceleration tangent to the path of motion—by taking the time derivative of $v$:

$$a_t = \frac{dv}{dt} = r\frac{d\omega}{dt}$$

$$\boxed{a_t = r\alpha} \qquad [10.11]$$

That is, the tangential component of the translational acceleration of a particle undergoing circular motion equals the distance of the particle from the axis of rotation multiplied by the angular acceleration.

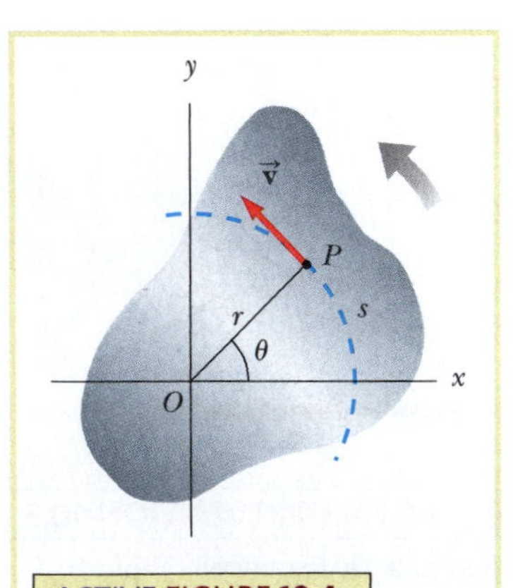


**ACTIVE FIGURE 10.4**
As a rigid object rotates about the fixed axis through $O$, the point $P$ has a tangential velocity $\vec{v}$ that is always tangent to the circular path of radius $r$.

**PhysicsNow™** Log into PhysicsNow at **www.pop4e.com** and go to Active Figure 10.4 to move point $P$ and see the change in the tangential velocity.

In Chapter 3, we found that a particle rotating in a circular path undergoes a centripetal, or radial, acceleration of magnitude $v^2/r$ directed toward the center of rotation (Fig. 10.5). Because $v = r\omega$, we can express the centripetal acceleration of the particle in terms of the angular speed as

$$a_c = \frac{v^2}{r} = r\omega^2 \qquad [10.12]$$

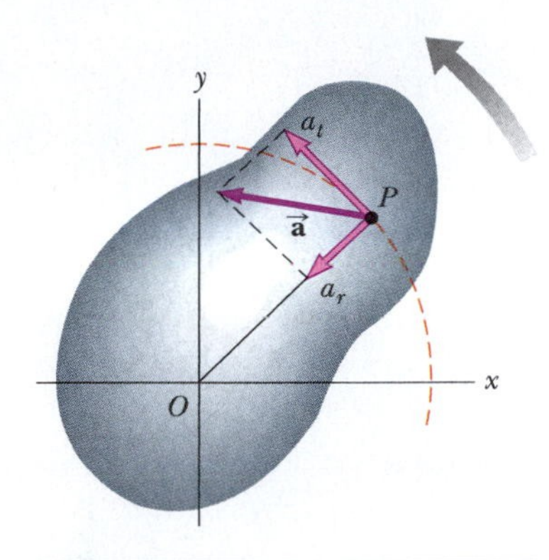


FIGURE 10.5 As a rigid object rotates about a fixed axis through $O$, a particle at point $P$ experiences a tangential component $a_t$ and a radial component $a_r$ of translational acceleration. The total translational acceleration of this particle is $\vec{\mathbf{a}} = \vec{\mathbf{a}}_t + \vec{\mathbf{a}}_r$, where $\vec{\mathbf{a}}_r = -a_c\hat{\mathbf{r}}$.

The *total translational acceleration* of the particle is $\vec{\mathbf{a}} = \vec{\mathbf{a}}_t + \vec{\mathbf{a}}_r$. The magnitude of the total translational acceleration of the particle is therefore

$$a = \sqrt{a_t^2 + a_r^2} = \sqrt{r^2\alpha^2 + r^2\omega^4} = r\sqrt{\alpha^2 + \omega^4} \qquad [10.13]$$

**QUICK QUIZ 10.3** Benjamin and Torrey are riding on a merry-go-round. Benjamin rides on a horse at the outer rim of the circular platform, twice as far from the center of the circular platform as Torrey, who rides on an inner horse. **(i)** When the merry-go-round is rotating at a constant angular speed, what is Benjamin's angular speed? **(a)** twice Torrey's **(b)** the same as Torrey's **(c)** half of Torrey's **(d)** impossible to determine. **(ii)** When the merry-go-round is rotating at a constant angular speed, what is Benjamin's tangential speed from the same list of choices?

## ■ Thinking Physics 10.1

A phonograph record (LP, for *long-playing*) rotates at a constant *angular* speed. A compact disc (CD) rotates so that the surface sweeps past the laser at a constant *tangential* speed. Consider two circular grooves of information on an LP, one near the outer edge and one near the inner edge. Suppose the outer groove "contains" 1.8 s of music. Does the inner groove also contain 1.8 s of music? And for the CD, do the inner and outer "grooves" contain the same time interval of music?

**Reasoning** On the LP the inner and outer grooves must both rotate once in the same time interval. Therefore, each groove, regardless of where it is on the record, contains the same time interval of information. Of course, on the inner grooves, this same information must be compressed into a smaller circumference. On a CD, the constant tangential speed requires that no such compression occur; the digital pits representing the information are spaced uniformly everywhere on the surface. Therefore, there is more information in an outer "groove," because of its larger circumference and, as a result, a longer time interval of music than in the inner "groove." ■

## ■ Thinking Physics 10.2

The launch area for the European Space Agency is not in Europe, but rather in South America. Why?

**Reasoning** Placing a satellite in Earth orbit requires providing a large tangential speed to the satellite, which is the task of the rocket propulsion system. Anything that reduces the requirements on the propulsion system is a welcome contribution. The surface of the Earth is already traveling toward the east at a high speed due to the rotation of the Earth. Therefore, if rockets are launched toward the east, the rotation of the Earth provides some initial tangential speed, reducing somewhat the requirements on the propulsion system. If rockets were launched from Europe, which is at a relatively large latitude, the contribution of the Earth's rotation is relatively small because the distance between Europe and the rotation axis of the Earth is relatively small. The ideal place for launching is at the equator, which is as far as one can be from the rotation axis of the Earth and still be on the surface of the Earth. This location results in the largest possible tangential speed due to the Earth's rotation. The European Space Agency exploits this advantage by launching from French Guiana, which is only a few degrees north of the equator.

A second advantage of this location is that launching toward the east takes the spacecraft over water. In the event of an accident or a failure, the wreckage will fall into the ocean rather than into populated areas as it would if launched to the east from Europe. Similarly, the United States launches spacecraft from Florida rather than California, despite the more favorable weather conditions in California. ■

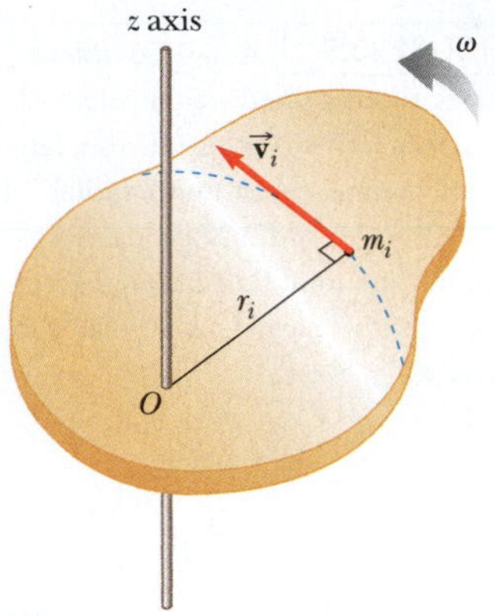

z axis

$\omega$

$\vec{\mathbf{v}}_i$

$m_i$

$r_i$

$O$

**FIGURE 10.6** A rigid object rotating about the $z$ axis with angular speed $\omega$. The kinetic energy of the particle of mass $m_i$ is $\frac{1}{2}m_i v_i^2$. The kinetic energy of the rigid object is called its rotational kinetic energy.

## 10.4 ROTATIONAL KINETIC ENERGY

Imagine that you begin a workout session on a stationary exercise bicycle. You apply a force with your feet on the pedals, moving them through a displacement; as a result, you have done work. The result of this work is the spinning of the wheel. This rotational motion represents kinetic energy because an object with mass is in motion. In this section, we will investigate this kinetic energy for rotating objects. In a later section, we will consider the work done in rotational motion and develop a rotational version of the work–kinetic energy theorem.

Let us consider a rigid object as a collection of particles and assume that it rotates about a fixed $z$ axis with an angular speed $\omega$ (Fig. 10.6). Each particle of the object is in motion and therefore has some kinetic energy, determined by its mass and tangential speed. If the mass of the $i$th particle is $m_i$ and its tangential speed is $v_i$, the kinetic energy of this particle is

$$K_i = \tfrac{1}{2}m_i v_i^2$$

### PITFALL PREVENTION 10.4

**NO SINGLE MOMENT OF INERTIA** We have pointed out that moment of inertia is analogous to mass, but there is one major difference. Mass is an inherent property of an object and has a single value. The moment of inertia of an object depends on your choice of rotation axis; therefore, an object has no single value of the moment of inertia. An object does have a minimum value of the moment of inertia, which is that calculated around an axis passing through the center of mass of the object.

We can express the *total* kinetic energy $K_R$ of the rotating rigid object as the sum of the kinetic energies of the individual particles. Therefore, incorporating Equation 10.10,

$$K_R = \sum_i K_i = \sum_i \tfrac{1}{2}m_i v_i^2 = \tfrac{1}{2}\sum_i m_i r_i^2\omega^2$$

$$= \tfrac{1}{2}\left(\sum_i m_i r_i^2\right)\omega^2$$

where we have factored $\omega^2$ from the sum because it is common to every particle in the object. The quantity in parentheses is called the **moment of inertia** $I$ of the rigid object:

▪ Moment of inertia for a system of particles

$$I = \sum_i m_i r_i^2 \qquad [10.14]$$

Therefore, we can express the kinetic energy of the rotating rigid object around the $z$ axis as

$$K_R = \frac{1}{2}I\omega^2 \qquad [10.15]$$

■ Kinetic energy of a rotating rigid object

From the definition of moment of inertia, we see that it has dimensions of $ML^2$ ($kg \cdot m^2$ in SI units). The moment of inertia is a measure of an object's *resistance to change in its angular speed.* Therefore, it plays a role in rotational motion identical to the role mass plays in translational motion. Notice that moment of inertia depends not only on the mass of the rigid object but also on *how the mass is distributed around the rotation axis.*

Although we shall commonly refer to the quantity $\frac{1}{2}I\omega^2$ as the **rotational kinetic energy,** it is not a new form of energy. It is ordinary kinetic energy because it was derived from a sum over individual kinetic energies of the particles contained in the rigid object. It is a new role for kinetic energy for us, however, because we have only considered kinetic energy associated with translation through space so far. **On the storage side of the continuity equation for energy (see Eq. 6.20), we should now consider that the kinetic energy term should be the sum of the changes in both translational and rotational kinetic energy.** Therefore, in energy versions of system models, we should keep in mind the possibility of rotational kinetic energy.

Equation 10.14 gives the moment of inertia of a collection of particles. For an extended, continuous object, we can calculate the moment of inertia by dividing the object into many small elements with mass $\Delta m_i$. Then, the moment of inertia is approximately $I \approx \sum r_i^2 \Delta m_i$, where $r_i$ is the perpendicular distance of the element of mass $\Delta m_i$ from the rotation axis. Now we take the limit as $\Delta m_i \rightarrow 0$, in which case the sum becomes an integral:

$$I = \lim_{\Delta m_i \to 0} \sum_i r_i^2 \Delta m_i = \int r^2\, dm \qquad [10.16]$$

It is usually easier to calculate moments of inertia in terms of the volume of the elements rather than their mass, and we can easily make this change by using Equation 1.1, $\rho = m/V$, where $\rho$ is the density of the object and $V$ is its volume. We can express the mass of an element by writing Equation 1.1 in differential form, $dm = \rho\, dV$. Using this form, Equation 10.16 becomes

$$I = \int \rho r^2\, dV \qquad [10.17]$$

■ Moment of inertia for an extended, continuous object

If the object is homogenous, the density $\rho$ is constant and the integral can be evaluated for a given geometry. If $\rho$ is not uniform over the volume of the object, its variation with position must be known in order to perform the integration.

For symmetric objects, the moment of inertia can be expressed in terms of the total mass of the object and one or more dimensions of the object. Table 10.2 shows the moments of inertia of various common symmetric objects.

**QUICK QUIZ 10.4** A section of hollow pipe and a solid cylinder have the same radius, mass, and length. They both rotate about their long central axes with the same angular speed. Which object has the higher rotational kinetic energy? **(a)** the pipe **(b)** the solid cylinder **(c)** they have the same rotational kinetic energy **(d)** impossible to determine

**TABLE 10.2 Moments of Inertia of Homogeneous Rigid Objects With Different Geometries**

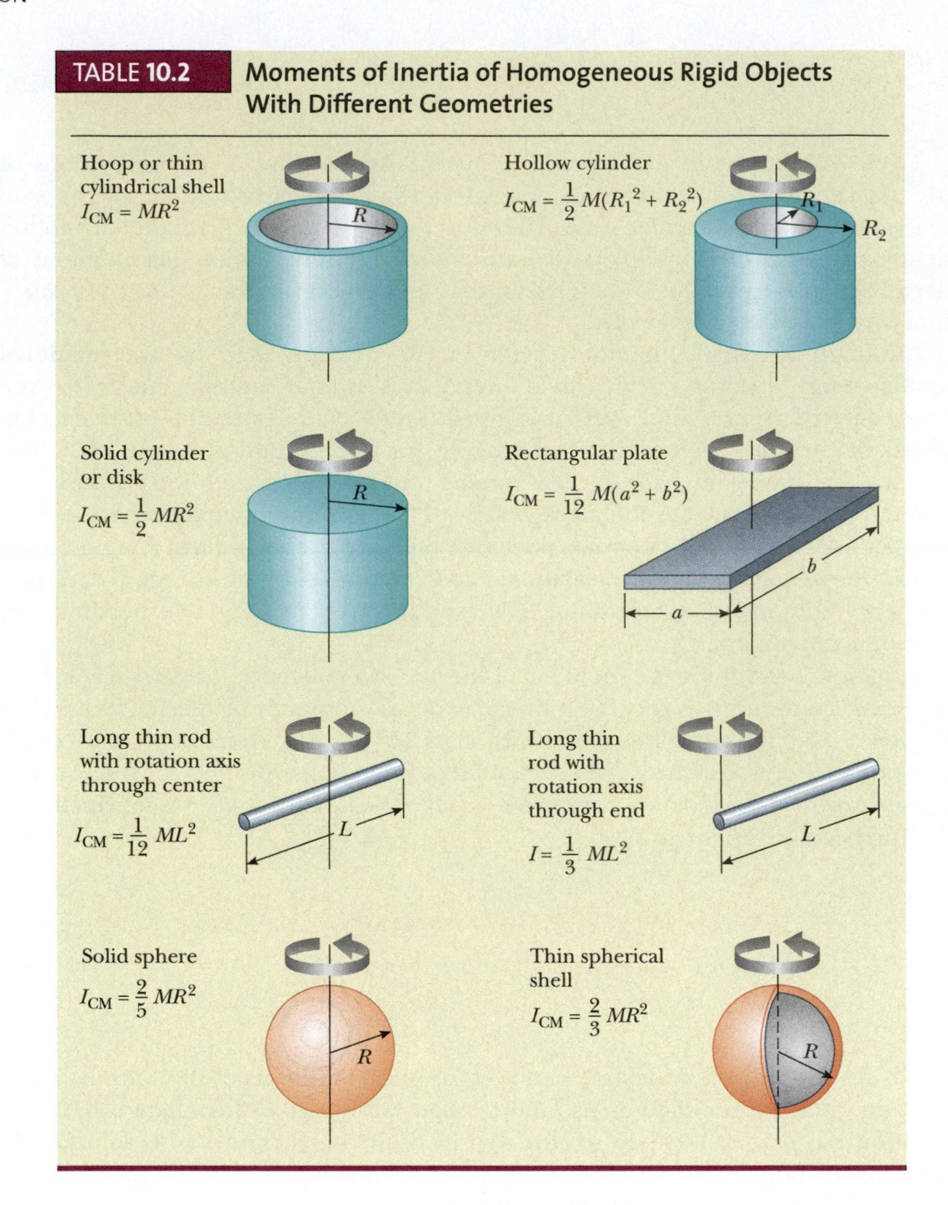


**EXAMPLE 10.2 The Oxygen Molecule**

Consider the diatomic oxygen molecule $O_2$, which is rotating in the *xy* plane about the *z* axis passing through its center, perpendicular to its length. The mass of each oxygen atom is $2.66 \times 10^{-26}$ kg, and at room temperature, the average separation between the two oxygen atoms is $d = 1.21 \times 10^{-10}$ m.

**A** Calculate the moment of inertia of the molecule about the *z* axis.

**Solution** We model the molecule as a rigid object, consisting of two particles (the two oxygen atoms), in rotation. Because the distance of each particle from the *z* axis is $d/2$, the moment of inertia about the *z* axis is

$$I = \sum_i m_i r_i^2 = m\left(\frac{d}{2}\right)^2 + m\left(\frac{d}{2}\right)^2 = \frac{md^2}{2}$$

$$= \frac{(2.66 \times 10^{-26}\ \text{kg})(1.21 \times 10^{-10}\ \text{m})^2}{2}$$

$$= \boxed{1.95 \times 10^{-46}\ \text{kg} \cdot \text{m}^2}$$

**B** A typical angular speed of a molecule is $4.60 \times 10^{12}$ rad/s. If the oxygen molecule is rotating with this angular speed about the *z* axis, what is its rotational kinetic energy?

**Solution** We use Equation 10.15:

$$K_R = \tfrac{1}{2}I\omega^2$$

$$= \tfrac{1}{2}(1.95 \times 10^{-46}\ \text{kg} \cdot \text{m}^2)(4.60 \times 10^{12}\ \text{rad/s})^2$$

$$= \boxed{2.06 \times 10^{-21}\ \text{J}}$$

## EXAMPLE 10.3 Four Rotating Objects

Four small spheres are fastened to the corners of a frame of negligible mass lying in the $xy$ plane (Fig. 10.7).

**A** If the rotation of the system occurs about the $y$ axis, as in Figure 10.7a, with an angular speed $\omega$, find the moment of inertia $I_y$ about the $y$ axis and the rotational kinetic energy about this axis.

**Solution** Because the spheres are small, we will model them as particles. First, note that the two spheres of mass $m$ that lie on the $y$ axis do not contribute to $I_y$. Because they are modeled as particles, $r_i = 0$ for these spheres about this axis. Applying Equation 10.14, we have for the two spheres on the $x$ axis

$$I_y = \sum_i m_i r_i^2 = Ma^2 + Ma^2 = 2Ma^2$$

Therefore, the rotational kinetic energy about the $y$ axis is

$$K_R = \tfrac{1}{2}I_y\omega^2 = \tfrac{1}{2}(2Ma^2)\omega^2 = Ma^2\omega^2$$

That the spheres of mass $m$ do not enter into this result makes sense because they have no motion about the chosen axis of rotation; hence, they have no kinetic energy.

**B** Suppose the system rotates in the $xy$ plane about an axis (the $z$ axis) through $O$ (Fig. 10.7b). Calculate the moment of inertia about the $z$ axis and the rotational energy about this axis.

**Solution** Because $r_i$ in Equation 10.14 is the perpendicular distance to the axis of rotation, we have

$$I_z = \sum_i m_i r_i^2 = Ma^2 + Ma^2 + mb^2 + mb^2 = 2Ma^2 + 2mb^2$$

$$K_R = \tfrac{1}{2}I_z\omega^2 = \tfrac{1}{2}(2Ma^2 + 2mb^2)\omega^2 = (Ma^2 + mb^2)\omega^2$$

Comparing the results for parts A and B, we see explicitly that the moment of inertia and therefore the rotational energy associated with a given angular speed depend on the axis of rotation. In part B, we expect the result to include all masses and distances because all four spheres are in motion for rotation in the $xy$ plane. Furthermore, that the rotational energy in part A is smaller than in part B indicates that there is less resistance to changes in rotational motion about the $y$ axis than about the $z$ axis.

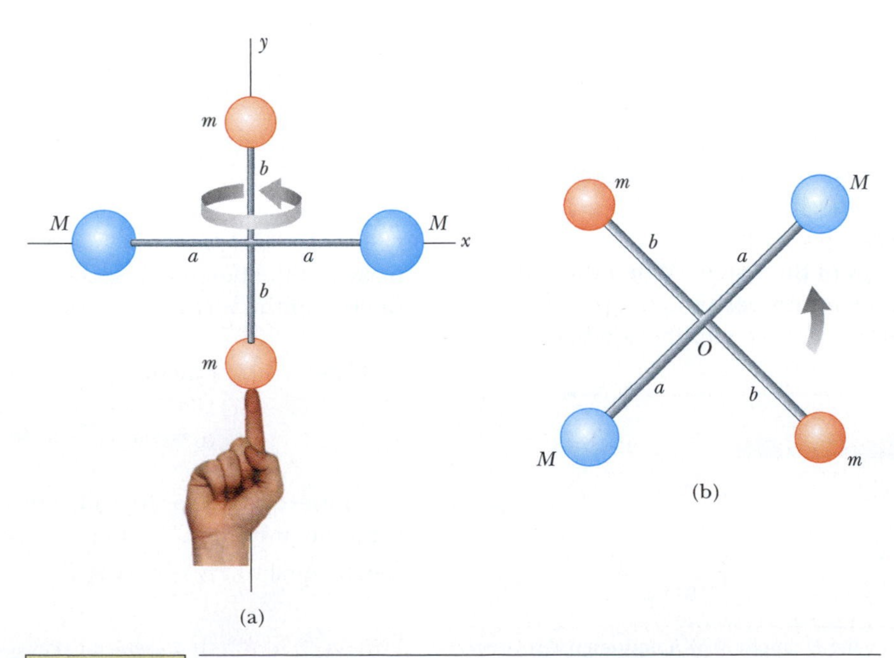


FIGURE 10.7 (Example 10.3) Four spheres form an unusual baton. (a) The baton is rotated about the $y$ axis. (b) The baton is rotated about the $z$ axis.

## EXAMPLE 10.4 Moment of Inertia of a Uniform Solid Cylinder

A uniform solid cylinder has a radius $R$, mass $M$, and length $L$. Calculate its moment of inertia about its central axis (the $z$ axis shown in Fig. 10.8).

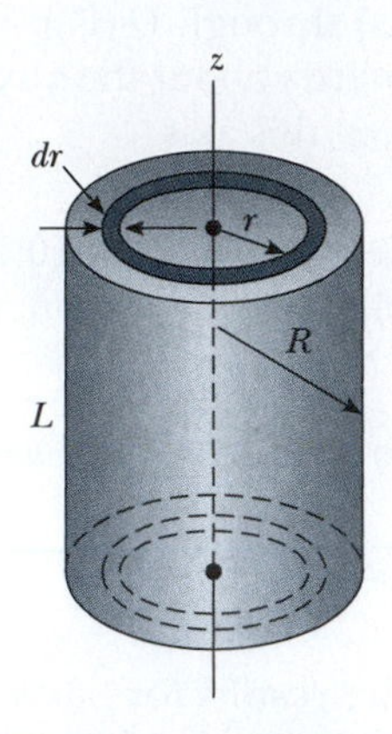


FIGURE 10.8 (Example 10.4) The geometry for calculating the moment of inertia about the central axis of a uniform solid cylinder.

**Solution** The integral in Equation 10.17 can be evaluated relatively simply by dividing the cylinder into many cylindrical shells of radius $r$, thickness $dr$, and length $L$ as shown in Figure 10.8. The volume $dV$ of a shell is its cross-sectional area multiplied by its length: $dV = (dA)L = (2\pi r\, dr)L$. Equation 10.17 gives the moment of inertia:

$$I = \int \rho r^2\, dV = \int_0^R \rho r^2 (2\pi r L)\, dr$$

$$= 2\pi\rho L \int_0^R r^3\, dr = \tfrac{1}{2}\pi\rho L R^4$$

The volume of the entire cylinder is $\pi R^2 L$, so the density is $\rho = M/V = M/\pi R^2 L$. Substituting this value of $\rho$ in the above result gives

$$I = \tfrac{1}{2}\pi\left(\frac{M}{\pi R^2 L}\right) L R^4 = \boxed{\tfrac{1}{2}MR^2}$$

Note that this result, which appears in Table 10.2, does not depend on $L$. Therefore, it applies equally well to a long cylinder and a flat disk.

## INTERACTIVE EXAMPLE 10.5 Rotating Rod

A uniform rod of length $L$ and mass $M$ is free to rotate on a frictionless pin through one end (Fig. 10.9). The rod is released from rest in the horizontal position.

**A** What is the angular speed of the rod at its lowest position?

**Solution** We consider the rod and the Earth as an isolated system and use the energy version of the isolated system model. Consider the mechanical energy of the system. When the rod is horizontal, as in Figure 10.9, it has no rotational kinetic energy. Let us also define this position of the rod as representing the zero of gravitational potential energy of the system. When the rod's center of mass is at the lowest position, the potential energy of the system is $-MgL/2$ and the rod has rotational kinetic energy $\frac{1}{2}I\omega^2$, where $I$ is the moment of inertia about the pivot.

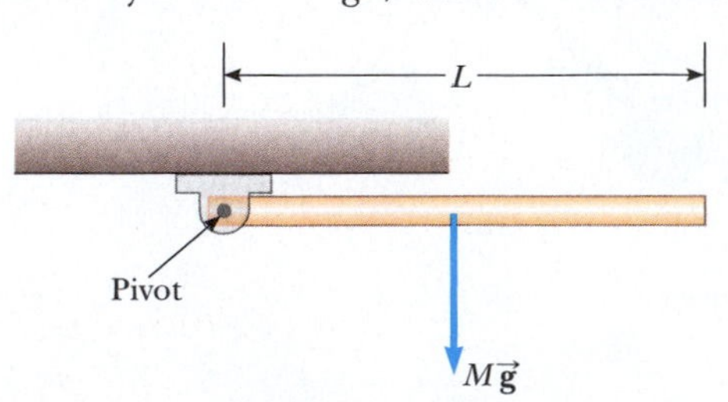


FIGURE 10.9 (Interactive Example 10.5) A uniform rod rotates freely under the influence of gravity around a pivot at the left end.

Because $I = \frac{1}{3}ML^2$ (see Table 10.2) for a geometric model of a long, thin rod and because mechanical energy of the isolated system is conserved, we have

$$K_i + U_i = K_f + U_f$$

$$0 + 0 = -\tfrac{1}{2}MgL + \tfrac{1}{2}I\omega^2 = -\tfrac{1}{2}MgL + \tfrac{1}{2}(\tfrac{1}{3}ML^2)\omega^2$$

$$\omega = \boxed{\sqrt{\frac{3g}{L}}}$$

**B** Determine the tangential speed of the center of mass and the tangential speed of the lowest point on the rod in the vertical position.

**Solution** Using Equation 10.10, we have

$$v_{CM} = r\omega = \frac{L}{2}\,\omega = \boxed{\tfrac{1}{2}\sqrt{3gL}}$$

The lowest point on the rod, because it is twice as far from the pivot as the center of mass, has a tangential speed equal to $2v_{CM} = \boxed{\sqrt{3gL}}$.

**PhysicsNow™** By logging into PhysicsNow at **www.pop4e.com** and going to Interactive Example 10.5 you can alter the mass and length of the rod and see the effect on the velocity at the lowest point.

## 10.5 TORQUE AND THE VECTOR PRODUCT

Recall our stationary exercise bicycle from the preceding section. We caused the rotational motion of the wheel by applying forces to the pedals. When a net force is exerted on a rigid object pivoted about some axis and the line of action[2] of the force does not pass through the pivot, the object tends to rotate about that axis. For example, when you push on a door, the door rotates about an axis through the hinges. The tendency of a force to rotate an object about some axis is measured by a vector quantity called **torque.** Torque is the cause of changes in rotational motion and is analogous to force, which causes changes in translational motion. Consider the wrench pivoted about the axis through $O$ in Figure 10.10. The applied force $\vec{\mathbf{F}}$ generally can act at an angle $\phi$ with respect to the position vector $\vec{\mathbf{r}}$ locating the point of application of the force. We define the torque $\tau$ resulting from the force $\vec{\mathbf{F}}$ with the expression[3]

$$\tau \equiv rF\sin\phi \qquad [10.18]$$

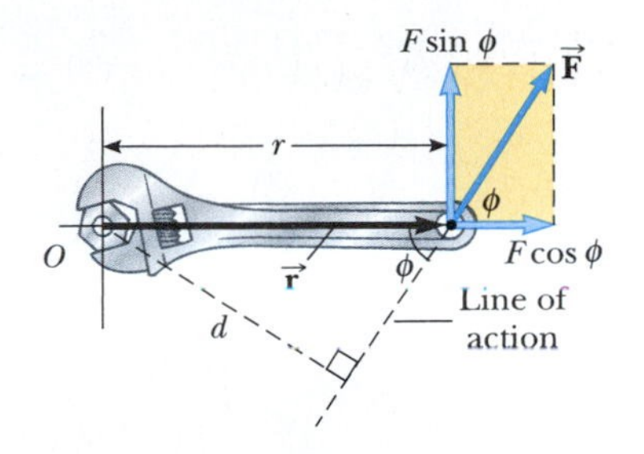


FIGURE 10.10 A force $\vec{\mathbf{F}}$ is applied to a wrench in an effort to loosen a bolt. The force has a greater rotating tendency about $O$ as $F$ increases and as the moment arm $d$ increases. The component $F\sin\phi$ tends to rotate the system about $O$.

It is very important to recognize that **torque is defined only when a reference axis is specified,** from which the distance $r$ is determined. We can interpret Equation 10.18 in two different ways. Looking at the force components in Figure 10.10, we see that the component $F\cos\phi$ parallel to $\vec{\mathbf{r}}$ will not cause a rotation of the wrench around the pivot point because its line of action passes right through the pivot point. Similarly, you cannot open a door by pushing on the hinges! Therefore, only the perpendicular component $F\sin\phi$ causes a rotation of the wrench about the pivot. In this case, we can write Equation 10.18 as

$$\tau = r(F\sin\phi)$$

so that the torque is the product of the distance to the point of application of the force and the perpendicular component of the force. In some problems, this method is the easiest way to interpret the calculation of the torque.

The second way to interpret Equation 10.18 is to associate the sine function with the distance $r$ so that we can write

$$\tau = F(r\sin\phi) = Fd$$

The quantity $d = r\sin\phi$, called the **moment arm** (or *lever arm*) of the force $\vec{\mathbf{F}}$, represents the perpendicular distance from the rotation axis to the line of action of $\vec{\mathbf{F}}$. In some problems, this approach to the calculation of the torque is easier than that of resolving the force into components.

If two or more forces are acting on a rigid object, as in Active Figure 10.11, each has a tendency to produce a rotation about the pivot at $O$. For example, if the object is initially at rest, $\vec{\mathbf{F}}_2$ tends to rotate the object clockwise and $\vec{\mathbf{F}}_1$ tends to rotate the object counterclockwise. We shall use the convention that the sign of the torque resulting from a force is positive if its turning tendency is counterclockwise around the rotation axis and negative if its turning tendency is clockwise. For example, in Active Figure 10.11, the torque resulting from $\vec{\mathbf{F}}_1$, which has a moment arm of $d_1$, is *positive* and equal to $+F_1d_1$; the torque from $\vec{\mathbf{F}}_2$ is *negative* and equal to $-F_2d_2$. Hence, the *net* torque acting on the rigid object about an axis through $O$ is

$$\tau_{\text{net}} = \tau_1 + \tau_2 = F_1d_1 - F_2d_2$$

**PITFALL PREVENTION 10.5**

**TORQUE DEPENDS ON YOUR CHOICE** Like moment of inertia, torque has no unique value. Its value depends on your choice of rotation axis.

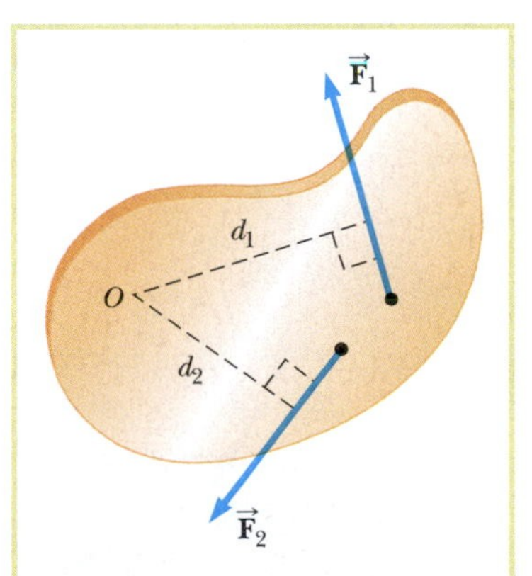


ACTIVE FIGURE 10.11 The force $\vec{\mathbf{F}}_1$ tends to rotate the object counterclockwise about an axis through $O$, and $\vec{\mathbf{F}}_2$ tends to rotate the object clockwise.

**PhysicsNow™** Change the magnitudes, directions, and points of application of forces $\vec{\mathbf{F}}_1$ and $\vec{\mathbf{F}}_2$ to see how the object accelerates under the action of the two forces by logging into PhysicsNow at www.pop4e.com and going to Active Figure 10.11.

[2] The line of action of a force is an imaginary line colinear with the force vector and extending to infinity in both directions.

[3] In general, torque is a vector. For rotation about a fixed axis, however, we will use italic, nonbold notation and specify the direction with a positive or a negative sign as we did for angular speed and acceleration in Section 10.1. We will treat the vector nature of torque briefly in a short while.

From the definition of torque, we see that the rotating tendency increases as $F$ increases and as $d$ increases. For example, we cause more rotation of a door if (a) we push harder or (b) we push at the doorknob rather than at a point close to the hinges. **Torque should not be confused with force.** Torque *depends* on force, but it also depends on *where the force is applied.* Torque has units of force times length, or newton · meters (N · m) in SI units.[4]

So far, we have not discussed the vector nature of torque aside from assigning a positive or negative value to $\tau$. Consider a force $\vec{\mathbf{F}}$ acting on a particle of a rigid object located at the vector position $\vec{\mathbf{r}}$ (Active Fig. 10.12). The *magnitude* of the torque due to this force relative to an axis through the origin is $|rF \sin \phi|$, where $\phi$ is the angle between $\vec{\mathbf{r}}$ and $\vec{\mathbf{F}}$. The axis about which $\vec{\mathbf{F}}$ would tend to produce rotation of the object is perpendicular to the plane formed by $\vec{\mathbf{r}}$ and $\vec{\mathbf{F}}$. If the force lies in the $xy$ plane, as in Active Figure 10.12, the torque is represented by a vector parallel to the $z$ axis. The force in Active Figure 10.12 creates a torque that tends to rotate the object counterclockwise when we are looking down the $z$ axis. We define the direction of torque such that the vector $\vec{\boldsymbol{\tau}}$ is in the positive $z$ direction (i.e., coming toward your eyes). If we reverse the direction of $\vec{\mathbf{F}}$ in Active Figure 10.12, $\vec{\boldsymbol{\tau}}$ is in the negative $z$ direction. With this choice, the torque vector can be defined to be equal to the **vector product,** or **cross product,** of $\vec{\mathbf{r}}$ and $\vec{\mathbf{F}}$:

■ Definition of torque using the cross product

$$\vec{\boldsymbol{\tau}} \equiv \vec{\mathbf{r}} \times \vec{\mathbf{F}} \qquad [10.19]$$

We now give a formal definition of the vector product, first introduced in Section 1.8. Given any two vectors $\vec{\mathbf{A}}$ and $\vec{\mathbf{B}}$, the vector product $\vec{\mathbf{A}} \times \vec{\mathbf{B}}$ is defined as a third vector $\vec{\mathbf{C}}$, the *magnitude* of which is $AB \sin \theta$, where $\theta$ is the angle between $\vec{\mathbf{A}}$ and $\vec{\mathbf{B}}$:

$$\vec{\mathbf{C}} = \vec{\mathbf{A}} \times \vec{\mathbf{B}} \qquad [10.20]$$

$$C = |\vec{\mathbf{C}}| \equiv AB \sin \theta \qquad [10.21]$$

Note that the quantity $AB \sin \theta$ is equal to the area of the parallelogram formed by $\vec{\mathbf{A}}$ and $\vec{\mathbf{B}}$, as shown in Figure 10.13. The *direction* of $\vec{\mathbf{A}} \times \vec{\mathbf{B}}$ is perpendicular to the plane formed by $\vec{\mathbf{A}}$ and $\vec{\mathbf{B}}$ and is determined by the right-hand rule illustrated in Figure 10.13. The four fingers of the right hand are pointed along $\vec{\mathbf{A}}$ and then "wrapped" into $\vec{\mathbf{B}}$ through the angle $\theta$. The direction of the upright thumb is the direction of $\vec{\mathbf{A}} \times \vec{\mathbf{B}}$. Because of the notation, $\vec{\mathbf{A}} \times \vec{\mathbf{B}}$ is often read "$\vec{\mathbf{A}}$ cross $\vec{\mathbf{B}}$," hence the term *cross product.*

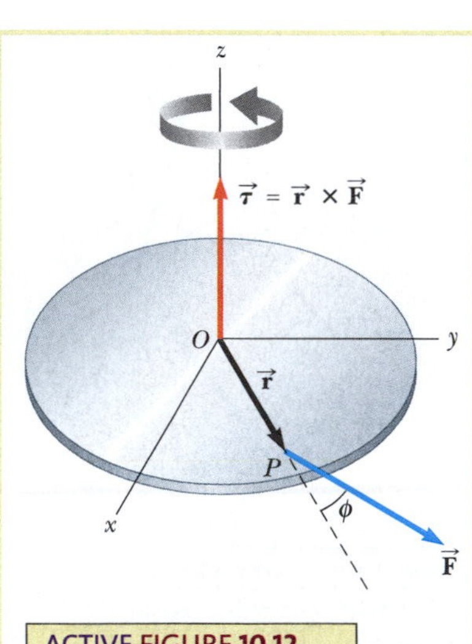


**ACTIVE FIGURE 10.12**
The torque vector $\vec{\boldsymbol{\tau}}$ lies in a direction perpendicular to the plane formed by the position vector $\vec{\mathbf{r}}$ and the applied force vector $\vec{\mathbf{F}}$.

**Physics Now™** By logging into PhysicsNow at **www.pop4e.com** and going to Active Figure 10.12 you can move point $P$ and change the force vector $\vec{\mathbf{F}}$ to see the effect on the torque vector.

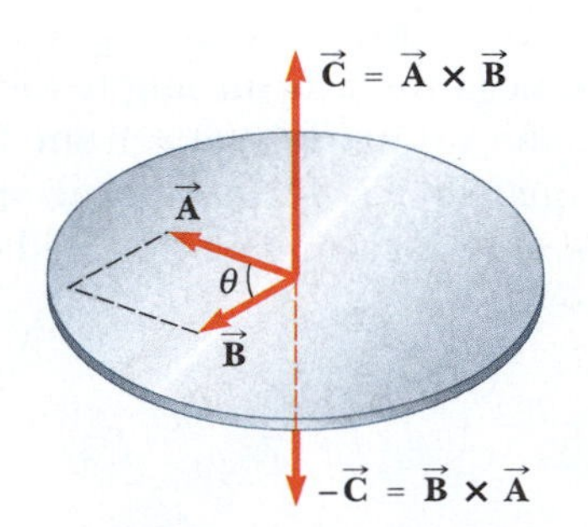

**FIGURE 10.13** The vector product $\vec{\mathbf{A}} \times \vec{\mathbf{B}}$ is a third vector $\vec{\mathbf{C}}$ having a magnitude $AB \sin \theta$ equal to the area of the parallelogram shown. The vector $\vec{\mathbf{C}}$ is perpendicular to the plane formed by $\vec{\mathbf{A}}$ and $\vec{\mathbf{B}}$, and its direction is determined by the right-hand rule.

[4]In Chapter 6, we saw the product of newtons and meters when we defined work and called this product a *joule.* We do not use this term here because the joule is only to be used when discussing energy. For torque, the unit is simply the newton · meter, or N · m.

Some properties of the vector product follow from its definition:

- Unlike the case of the scalar product, the vector product is not commutative; in fact,

$$\vec{\mathbf{A}} \times \vec{\mathbf{B}} = -\vec{\mathbf{B}} \times \vec{\mathbf{A}} \tag{10.22}$$

  Therefore, if you change the order of the vector product, you must change the sign. One can easily verify this relation with the right-hand rule (see Fig. 10.13).
- If $\vec{\mathbf{A}}$ is parallel to $\vec{\mathbf{B}}$ ($\theta = 0°$ or $180°$), then $\vec{\mathbf{A}} \times \vec{\mathbf{B}} = 0$; therefore, it follows that $\vec{\mathbf{A}} \times \vec{\mathbf{A}} = 0$.
- If $\vec{\mathbf{A}}$ is perpendicular to $\vec{\mathbf{B}}$, then $|\vec{\mathbf{A}} \times \vec{\mathbf{B}}| = AB$. It is left to Problem 10.25 to show, from Equations 10.20 and 10.21 and the definition of unit vectors, that the vector products of the unit vectors $\hat{\mathbf{i}}$, $\hat{\mathbf{j}}$, and $\hat{\mathbf{k}}$ obey the following expressions:

$$\begin{aligned} \hat{\mathbf{i}} \times \hat{\mathbf{i}} &= \hat{\mathbf{j}} \times \hat{\mathbf{j}} = \hat{\mathbf{k}} \times \hat{\mathbf{k}} = 0 \\ \hat{\mathbf{i}} \times \hat{\mathbf{j}} &= -\hat{\mathbf{j}} \times \hat{\mathbf{i}} = \hat{\mathbf{k}} \\ \hat{\mathbf{j}} \times \hat{\mathbf{k}} &= -\hat{\mathbf{k}} \times \hat{\mathbf{j}} = \hat{\mathbf{i}} \\ \hat{\mathbf{k}} \times \hat{\mathbf{i}} &= -\hat{\mathbf{i}} \times \hat{\mathbf{k}} = \hat{\mathbf{j}} \end{aligned} \tag{10.23}$$

  Signs are interchangeable. For example, $\hat{\mathbf{i}} \times (-\hat{\mathbf{j}}) = -\hat{\mathbf{i}} \times \hat{\mathbf{j}} = -\hat{\mathbf{k}}$.

**QUICK QUIZ 10.5** If you are trying to loosen a stubborn screw from a piece of wood with a screwdriver and fail, should you find a screwdriver for which the handle is **(a)** longer or **(b)** fatter? If you are trying to loosen a stubborn bolt from a piece of metal with a wrench and fail, should you find a wrench for which the handle is **(c)** longer or **(d)** fatter?

## EXAMPLE 10.6 The Net Torque on a Cylinder

A one-piece cylinder is shaped as in Figure 10.14, with a core section protruding from the larger drum. The cylinder is free to rotate around the central axis shown in the drawing. A rope wrapped around the drum, of radius $R_1$, exerts a force $\vec{\mathbf{T}}_1$ to the right on the cylinder. A rope wrapped around the core, of radius $R_2$, exerts a force $\vec{\mathbf{T}}_2$ downward on the cylinder.

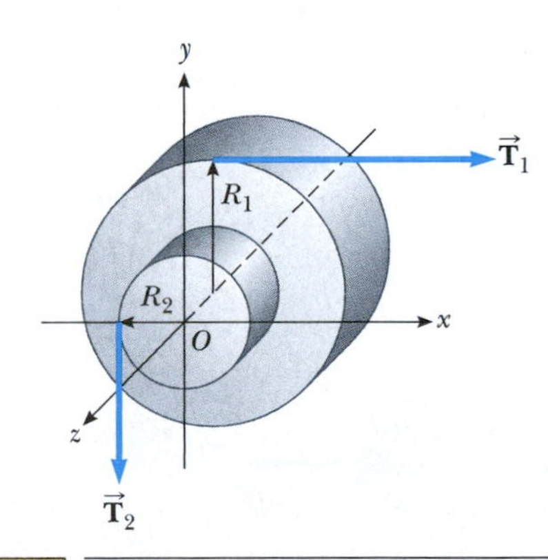


**FIGURE 10.14** (Example 10.6) A solid cylinder pivoted about the $z$ axis through $O$. The moment arm of $\vec{\mathbf{T}}_1$ is $R_1$, and the moment arm of $\vec{\mathbf{T}}_2$ is $R_2$.

**A** What is the net torque acting on the cylinder about the rotation axis (which is the $z$ axis in Fig. 10.14)?

**Solution** The torque due to $\vec{\mathbf{T}}_1$ is $-R_1T_1$. It is negative because it tends to produce a clockwise rotation from the point of view in Figure 10.14. The torque due to $\vec{\mathbf{T}}_2$ is $+R_2T_2$ and is positive because it tends to produce a counterclockwise rotation. Therefore, the net torque about the rotation axis is

$$\tau_{\text{net}} = \tau_1 + \tau_2 = R_2T_2 - R_1T_1$$

**B** Suppose $T_1 = 5.0$ N, $R_1 = 1.0$ m, $T_2 = 6.0$ N, and $R_2 = 0.50$ m. What is the net torque about the rotation axis and which way does the cylinder rotate if it starts from rest?

**Solution** We substitute numerical values in the result from part A:

$$\tau_{\text{net}} = (6.0 \text{ N})(0.50 \text{ m}) - (5.0 \text{ N})(1.0 \text{ m}) = -2.0 \text{ N} \cdot \text{m}$$

Because the net torque is negative, the cylinder rotates clockwise from rest.

**EXAMPLE 10.7** **The Vector Product**

Two vectors lying in the $xy$ plane are given by the equations $\vec{\mathbf{A}} = 2\hat{\mathbf{i}} + 3\hat{\mathbf{j}}$ and $\vec{\mathbf{B}} = -\hat{\mathbf{i}} + 2\hat{\mathbf{j}}$. Find $\vec{\mathbf{A}} \times \vec{\mathbf{B}}$ and verify explicitly that $\vec{\mathbf{A}} \times \vec{\mathbf{B}} = -\vec{\mathbf{B}} \times \vec{\mathbf{A}}$.

**Solution** Using Equation 10.23 for the vector product of unit vectors gives

$$\vec{\mathbf{A}} \times \vec{\mathbf{B}} = (2\hat{\mathbf{i}} + 3\hat{\mathbf{j}}) \times (-\hat{\mathbf{i}} + 2\hat{\mathbf{j}})$$
$$= 2\hat{\mathbf{i}} \times 2\hat{\mathbf{j}} + 3\hat{\mathbf{j}} \times (-\hat{\mathbf{i}}) = 4\hat{\mathbf{k}} + 3\hat{\mathbf{k}} = 7\hat{\mathbf{k}}$$

(We have omitted the terms containing $\hat{\mathbf{i}} \times \hat{\mathbf{i}}$ and $\hat{\mathbf{j}} \times \hat{\mathbf{j}}$ because, as Equation 10.23 shows, they are equal to zero.)

We can show that $\vec{\mathbf{A}} \times \vec{\mathbf{B}} = -\vec{\mathbf{B}} \times \vec{\mathbf{A}}$, because

$$\vec{\mathbf{B}} \times \vec{\mathbf{A}} = (-\hat{\mathbf{i}} + 2\hat{\mathbf{j}}) \times (2\hat{\mathbf{i}} + 3\hat{\mathbf{j}})$$
$$= -\hat{\mathbf{i}} \times 3\hat{\mathbf{j}} + 2\hat{\mathbf{j}} \times 2\hat{\mathbf{i}} = -3\hat{\mathbf{k}} - 4\hat{\mathbf{k}} = -7\hat{\mathbf{k}}$$

Therefore, $\vec{\mathbf{A}} \times \vec{\mathbf{B}} = -\vec{\mathbf{B}} \times \vec{\mathbf{A}}$.

## 10.6 THE RIGID OBJECT IN EQUILIBRIUM

We have defined a rigid object and have discussed torque as the cause of changes in rotational motion of a rigid object. We can now establish models for a rigid object subject to torques that are analogous to those for a particle subject to forces. We begin by imagining a rigid object with balanced torques, which will give us an analysis model that we call the rigid object in equilibrium.

Consider two forces of equal magnitude and opposite direction applied to an object as shown in Figure 10.15a. The force directed to the right tends to rotate the object clockwise about an axis perpendicular to the diagram through $O$, whereas the force directed to the left tends to rotate it counterclockwise about that axis. Because the forces are of equal magnitude and act at the same perpendicular distance from $O$, their torques are equal in magnitude. Therefore, the net torque on the rigid object is zero. The situation shown in Figure 10.15b is another case in which the net torque about $O$ is zero (although the net *force* on the object is not zero), and we can devise many more cases.

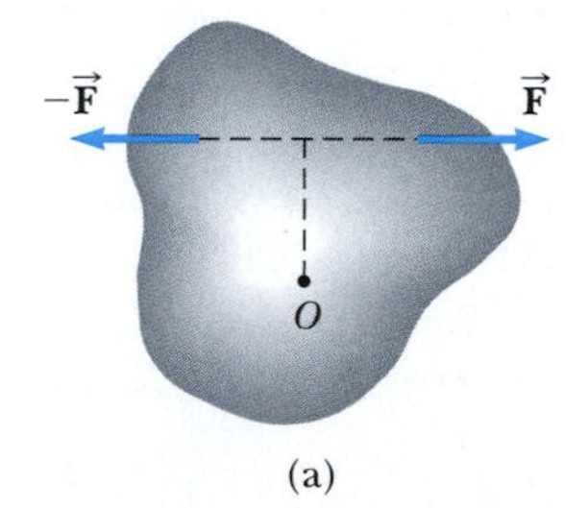


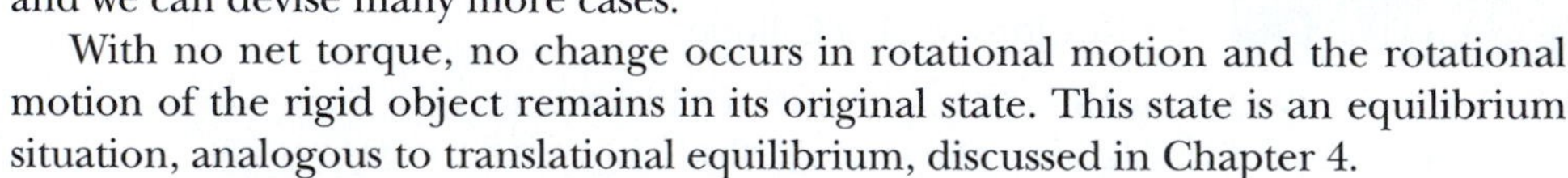

With no net torque, no change occurs in rotational motion and the rotational motion of the rigid object remains in its original state. This state is an equilibrium situation, analogous to translational equilibrium, discussed in Chapter 4.

We now have **two conditions for complete equilibrium of an object,** which can be stated as follows:

- The net external force must equal zero:

$$\sum \vec{\mathbf{F}} = 0 \qquad [10.24]$$

- The net external torque must be zero about *any* axis:

$$\sum \vec{\boldsymbol{\tau}} = 0 \qquad [10.25]$$

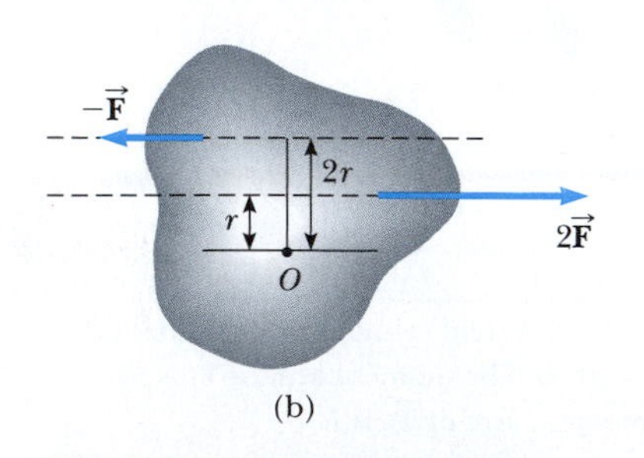


**FIGURE 10.15** (a) The two forces acting on the object are equal in magnitude and opposite in direction. Because they also act along the same line of action, the net torque is zero and the object is in equilibrium. (b) Another situation in which two forces act on an object to produce zero net torque about $O$ (but *not* zero net force).

The first condition is a statement of translational equilibrium. The second condition is a statement of rotational equilibrium. In the special case of **static equilibrium,** the object is at rest, so it has no translational or angular speed (i.e., $v_{CM} = 0$ and $\omega = 0$).

The two vector expressions given by Equations 10.24 and 10.25 are equivalent, in general, to six scalar equations: three from the first condition of equilibrium and three from the second (corresponding to $x$, $y$, and $z$ components). Hence, in a complex system involving several forces acting in various directions, you would be faced with solving a set of equations with many unknowns. Here, we restrict our discussion to situations in which all the forces on an object lie in the $xy$ plane. (Forces whose vector representations are in the same plane are said to be *coplanar.*) With this restriction, we need to deal with only three scalar equations. Two of them come from balancing the forces on the object in the $x$ and $y$ directions. The third comes from the torque equation, namely, that the net torque on the object about an axis through *any* point in the $xy$ plane must be zero. Hence, the two conditions of equilibrium provide the equations

$$\sum F_x = 0 \qquad \sum F_y = 0 \qquad \sum \tau_z = 0 \qquad [10.26]$$

where the axis of the torque equation is arbitrary.

In working static equilibrium problems, it is important to recognize all external forces acting on the object. Failure to do so will result in an incorrect analysis. The following procedure is recommended when analyzing an object in equilibrium under the action of several external forces:

### PROBLEM-SOLVING STRATEGY Rigid Object in Equilibrium

1. **Conceptualize** Think about the object that is in equilibrium and identify the forces on it. Imagine what effect each force would have on the rotation of the object if it were the only force acting.
2. **Categorize** Confirm that the object under consideration is indeed a rigid object in equilibrium.
3. **Analyze** Draw a free-body diagram and label all external forces acting on the object. Try to guess the correct direction for each force.

   Resolve all forces into rectangular components, choosing a convenient coordinate system. Then apply the first condition for equilibrium, Equation 10.24. Remember to keep track of the signs of the various force components.

   Choose a convenient axis for calculating the net torque on the rigid object. Remember that the choice of the axis for the torque equation is arbitrary; therefore, choose an axis that will simplify your calculation as much as possible. Usually, the most convenient axis for calculating torques is one through a point at which several forces act, so their torques around this axis are zero. If you don't know a force or don't need to know a force, it is often beneficial to choose an axis through the point at which this force acts. Apply the second condition for equilibrium, Equation 10.25.

   Solve the simultaneous equations for the unknowns in terms of the known quantities.
4. **Finalize** Make sure your results are consistent with the free-body diagram. If you selected a direction that leads to a negative sign in your solution for a force, do not be alarmed; it merely means that the direction of the force is the opposite of what you guessed. Add up the vertical and horizontal forces on the object and confirm that each set of components adds to zero. Add up the torques on the object and confirm that the sum equals zero.

### INTERACTIVE EXAMPLE 10.8 Standing on a Horizontal Beam

A uniform horizontal beam of length 8.00 m and weight 200 N is attached to a wall by a pin connection. Its far end is supported by a cable that makes an angle of 53.0° with the horizontal (Fig. 10.16a). If a 600-N man stands 2.00 m from the wall, find the tension in the cable and the force exerted by the wall on the beam at the pivot.

**Solution** The beam–man system is at rest and remains at rest, so it is clearly in static equilibrium. First, we must identify all the external forces acting on the system, which we do in the free-body diagram in Figure 10.16b. These forces are the gravitational forces on the beam and the man, the force $\vec{\mathbf{T}}$ exerted by the cable, and the force $\vec{\mathbf{R}}$ exerted by the wall at the pivot (the direction of this force is unknown). (The force between the man and the beam is internal to the system, so it is not included in the free-body diagram.) Notice that we have imagined the gravitational force on the beam as acting at its center of gravity. Because the beam is uniform, the center of gravity is at the geometric center. If we resolve $\vec{\mathbf{T}}$ and $\vec{\mathbf{R}}$ into horizontal and vertical components (Fig. 10.16c) and apply the first condition for equilibrium for the beam, we have

$$(1) \quad \sum F_x = R\cos\theta - T\cos 53.0° = 0$$

$$(2) \quad \sum F_y = R\sin\theta + T\sin 53.0° - 600\text{ N} - 200\text{ N} = 0$$

Because we have three unknowns—$R$, $T$, and $\theta$—we cannot obtain a solution from these two expressions alone.

To generate a third expression, let us invoke the condition for rotational equilibrium because the beam can be modeled as a rigid object in equilibrium. A convenient axis to choose for our torque equation is the one that passes through the pivot at the wall. The feature that makes this point so convenient is that the

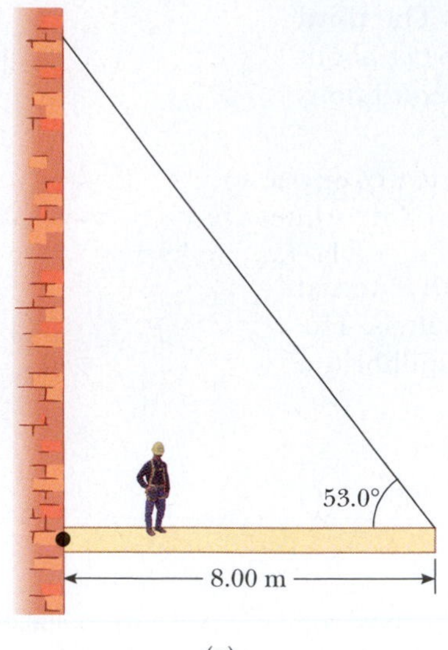


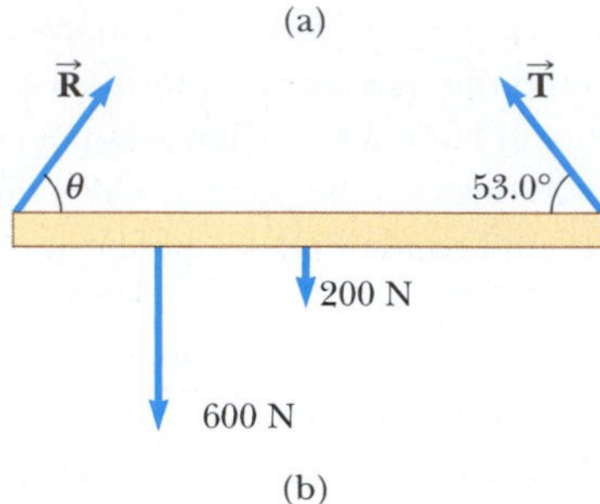


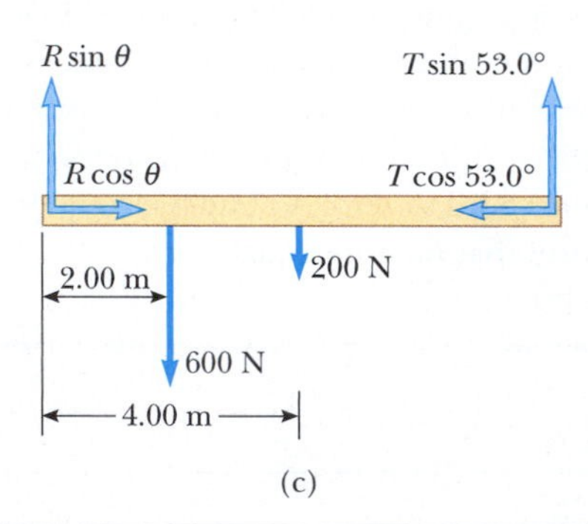


**FIGURE 10.16** (Interactive Example 10.8) (a) A uniform beam supported by a cable. A man walks out on the beam. (b) The free-body diagram for the beam–man system. (c) The free-body diagram with forces resolved into horizontal and vertical components.

force $\vec{\mathbf{R}}$ and the horizontal component of $\vec{\mathbf{T}}$ both have a lever arm of zero, and hence zero torque, about this pivot. Recalling our convention for the sign of the torque about an axis and noting that the lever arms of the 600-N, 200-N, and $T\sin 53°$ forces are 2.00 m, 4.00 m, and 8.00 m, respectively, we have

$$\sum \tau = (T\sin 53.0°)(8.00\text{ m}) - (600\text{ N})(2.00\text{ m}) - (200\text{ N})(4.00\text{ m}) = 0$$

$$T = 313\text{ N}$$

The torque equation gives us one of the unknowns directly, thanks to our judicious choice of the axis! This value is substituted into (1) and (2) to give

$$R\cos\theta = 188\text{ N}$$

$$R\sin\theta = 550\text{ N}$$

We divide these two equations to find

$$\tan\theta = \frac{550\text{ N}}{188\text{ N}} = 2.93$$

$$\theta = 71.1°$$

Finally,

$$R = \frac{188\text{ N}}{\cos\theta} = \frac{188\text{ N}}{\cos 71.1°} = 581\text{ N}$$

If we had selected some other axis for the torque equation, the results would have been the same although the details of the solution would be somewhat different. For example, if we had chosen to have the axis pass through the center of gravity of the beam, the torque equation would involve both $T$ and $R$. This equation, coupled with (1) and (2), however could still be solved for the unknowns $T$, $R$, and $\theta$, yielding the same results. Try it!

**PhysicsNow™** Adjust the position of the person and observe the effect on the forces by logging into PhysicsNow at **www.pop4e.com** and going to Interactive Example 10.8.

## INTERACTIVE EXAMPLE 10.9 The Leaning Ladder

A uniform ladder of length $\ell$ and mass $m$ rests against a smooth, vertical wall (Fig. 10.17a). If the coefficient of static friction between ladder and ground is $\mu_s = 0.40$, find the minimum angle $\theta_{\min}$ such that the ladder does not slip.

**Solution** The ladder is at rest and remains at rest, so we model it as a rigid object in equilibrium. The free-body diagram showing all the external forces acting on the ladder is illustrated in Figure 10.17b. The force exerted by the ground on the ladder is the vector sum of a normal force $\vec{\mathbf{n}}$ and the force of static friction $\vec{\mathbf{f}}_s$. The reaction force $\vec{\mathbf{P}}$ exerted by the wall on the ladder is horizontal because the wall is smooth, meaning that it is frictionless. Therefore, $\vec{\mathbf{P}}$ is simply the normal force on the ladder from the wall.

From the first condition of equilibrium applied to the ladder, we have

$$\sum F_x = f_s - P = 0$$
$$\sum F_y = n - mg = 0$$

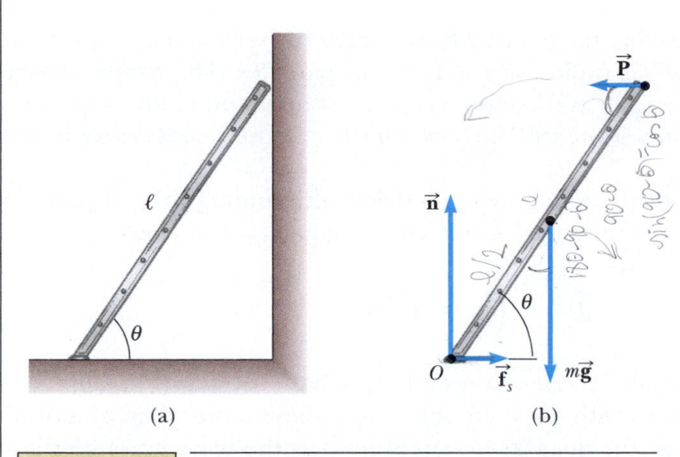


FIGURE 10.17 (Interactive Example 10.9) (a) A uniform ladder at rest, leaning against a frictionless wall. (b) The free-body diagram for the ladder.

We see from the second equation that $n = mg$. Furthermore, when the ladder is on the verge of slipping, the force of static friction must be a maximum, given by $f_{s,\,\max} = \mu_s n$.

To find $\theta$, we use the second condition of equilibrium. When the torques are taken about the origin $O$ at the bottom of the ladder, we have

$$\sum \tau_O = P\ell \sin\theta - mg\frac{\ell}{2}\cos\theta = 0$$

This expression gives

$$\tan\theta_{\min} = \frac{mg}{2P} = \frac{n}{2f_{s,\,\max}} = \frac{n}{2(\mu_s n)} = \frac{1}{2(0.40)} = 1.25$$

$$\theta_{\min} = 51°$$

It is interesting that the result does not depend on $\ell$ or $m$. The answer depends only on $\mu_s$.

**PhysicsNow™** Adjust the angle of the ladder and watch what happens when it is released by logging into PhysicsNow at **www.pop4e.com** and going to Interactive Example 10.9.

## 10.7 THE RIGID OBJECT UNDER A NET TORQUE

In the preceding section, we investigated the equilibrium situation in which the net torque on a rigid object is zero. What if the net torque on a rigid object is not zero? In analogy with Newton's second law for translational motion, we should expect the angular speed of the rigid object to change. The net torque will cause angular acceleration of the rigid object. We describe this situation with a new analysis model, the rigid object under a net torque, and investigate this model in this section.

Let us imagine a rotating rigid object again as a collection of particles. The rigid object will be subject to a number of forces applied at various locations on the rigid object at which individual particles will be located. Therefore, we can imagine that the forces on the rigid object are exerted on individual particles of the rigid object. We will calculate the net torque on the object due to the torques resulting from these forces around the rotation axis of the rotating object. Any applied force can be represented by its radial component and its tangential component. The radial component of an applied force provides no torque because its line of action goes through the rotation axis. Therefore, only the tangential component of an applied force contributes to the torque.

On any given particle, described by index variable $i$, within the rigid object, we can use Newton's second law to describe the tangential acceleration of the particle:

$$F_{ti} = m_i a_{ti}$$

where the $t$ subscript refers to tangential components. Let us multiply both sides of this expression by $r_i$, the distance of the particle from the rotation axis:

$$r_i F_{ti} = r_i m_i a_{ti}$$

Using Equation 10.11 and recognizing the definition of torque ($\tau = rF\sin\phi = rF_t$ in this case), we can rewrite this expression as

$$\tau_i = m_i r_i^{\,2}\alpha_i$$

Now, let us add up the torques on all particles of the rigid object:

$$\sum_i \tau_i = \sum_i m_i r_i^2 \alpha_i$$

The left side is the net torque on all particles of the rigid object. The net torque associated with *internal* forces is zero, however. To understand why, recall that Newton's third law tells us that the internal forces occur in equal and opposite pairs that lie along the line of separation of each pair of particles. The torque due to each action–reaction force pair is therefore zero. On summation of all torques, we see that the *net internal torque vanishes.* The term on the left, then, reduces to the net *external* torque.

On the right, we adopt the rigid object model by demanding that all particles have the same angular acceleration $\alpha$. Therefore, this equation becomes

$$\sum \tau = \left(\sum_i m_i r_i^2\right) \alpha$$

where the torque and angular acceleration no longer have subscripts because they refer to quantities associated with the rigid object as a whole rather than to individual particles. We recognize the quantity in parentheses as the moment of inertia $I$. Therefore,

■ Rotational analog to Newton's second law

$$\sum \tau = I\alpha \qquad [10.27]$$

That is, **the net torque acting on the rigid object is proportional to its angular acceleration,** and the proportionality constant is the moment of inertia. It is important to note that $\Sigma\, \tau = I\alpha$ is the rotational analog of Newton's second law of motion, $\Sigma\, F = ma$.

**QUICK QUIZ 10.6** You turn off your electric drill and find that the time interval for the rotating bit to come to rest due to frictional torque in the drill is $\Delta t$. You replace the drill bit with a larger one that results in a doubling of the moment of inertia of the drill's entire rotating mechanism. When this larger bit is rotated at the same angular speed as the first and turned off, the frictional torque remains the same as that for the previous situation. What is the time interval for this second bit to come to rest? **(a)** $4\Delta t$ **(b)** $2\Delta t$ **(c)** $\Delta t$ **(d)** $0.5\Delta t$ **(e)** $0.25\Delta t$ **(f)** impossible to determine

## EXAMPLE 10.10 An Atwood Machine with a Massive Pulley

In Example 4.4, we analyzed an Atwood machine in which two objects with unequal masses hang from a string that passes over a light, frictionless pulley. Suppose the pulley, which is modeled as a disk, has mass $M$ and radius $R$, and suppose the pulley surface is not frictionless so that the string does not slide on the pulley (Fig. 10.18a). We will assume that the frictional torque acting at the bearing of the pulley is negligible. Calculate the magnitude of the acceleration of the two objects.

**Solution** Conceptualize the problem by imagining the motion of the two objects in Figure 10.18, as we did in Example 4.4. The difference here is that the pulley is not considered to be massless. Because this problem involves a massive object in rotation as well as two other objects in translational motion, we categorize it as one involving the rigid object under a net torque model (for the pulley) and the particle under a net force model (for the hanging objects).

To analyze the problem, we first set up a coordinate system. Because counterclockwise angular acceleration of the pulley is defined as positive, we define the positive directions for $m_1$ and $m_2$ as shown in Figure 10.18 so that all accelerations, translational and rotational, are positive if $m_1$ accelerates downward. If the pulley has mass and friction, the tensions $T_1$ and $T_2$ in the string on either side of the pulley are not equal in magnitude as they are in Example 4.4. Indeed, it is the

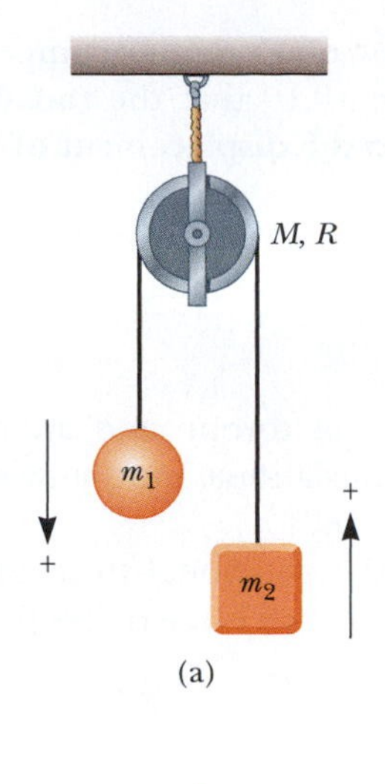


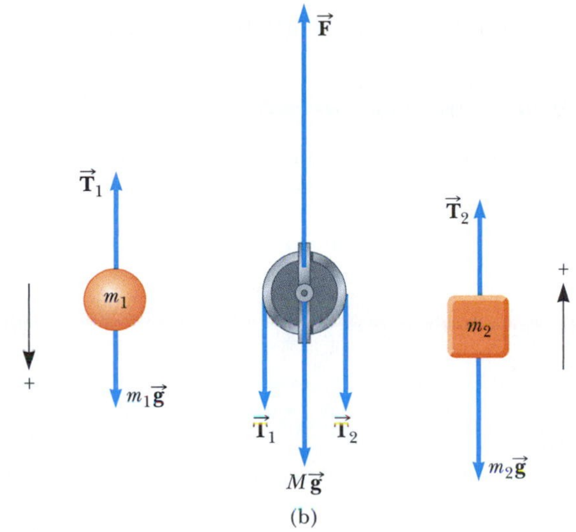


FIGURE 10.18 (Example 10.10) (a) An Atwood machine with a massive pulley. The pulley is modeled as a disk. (b) Free-body diagrams for the two hanging objects and the pulley.

difference in torque due to these different tensions that provides the net torque to cause the angular acceleration of the pulley (Fig. 10.18b). Consequently, some of what we do here will look similar to Example 4.4, with the exception of incorporating $T_1$ and $T_2$ in our mathematical representation instead of just a single $T$. The forces $M\vec{\mathbf{g}}$ and $\vec{\mathbf{F}}$ (the force supporting the pulley) both act through the pulley axle, so these forces do not contribute to the torque on the pulley.

With the help of the free-body diagrams in Figure 10.18b, we apply Newton's second law to $m_1$ so that

$$(1)\quad \sum F_y = m_1 g - T_1 = m_1 a$$

and for $m_2$,

$$(2)\quad \sum F_y = T_2 - m_2 g = m_2 a$$

We cannot solve these two equations for $a$, as is done in Example 4.4, because we have three unknowns: $a$, $T_1$, and $T_2$. We can find a third equation by applying Equation 10.27 to the pulley (Fig. 10.18b):

$$\sum \tau = T_1 R - T_2 R = I\alpha = (\tfrac{1}{2}MR^2)\left(\frac{a}{R}\right) = \tfrac{1}{2}MRa$$

$$(3)\quad T_1 - T_2 = \tfrac{1}{2}Ma$$

We substitute into (3) expressions for $T_1$ and $T_2$ from (1) and (2):

$$(m_1 g - m_1 a) - (m_2 a + m_2 g) = \tfrac{1}{2}Ma$$

$$a = \left(\frac{m_1 - m_2}{m_1 + m_2 + \frac{1}{2}M}\right) g$$

To finalize this problem, notice that this result differs from the result for Example 4.4 only in the extra term $\frac{1}{2}M$ in the denominator. If the pulley mass $M \rightarrow 0$, this expression reduces to that in Example 4.4.

## Work and Energy in Rotational Motion

In translational motion, we found energy concepts, and in particular the reduction of the continuity equation for energy called the work–kinetic energy theorem, to be extremely useful in describing the motion of a system. Energy concepts can be equally useful in simplifying the analysis of rotational motion, as we saw in the isolated system analysis in Interactive Example 10.5. From the continuity equation for energy, we expect that for rotation of an object about a fixed axis, the work done by external forces on the object will equal the change in the rotational kinetic energy as long as energy is not stored by any other means. To show that this case is in fact true, we begin by finding an expression for the work done by a torque.

Consider a rigid object pivoted at the point $O$ in Figure 10.19. Suppose a single external force $\vec{\mathbf{F}}$ is applied at the point $P$ and $d\vec{\mathbf{s}}$ is the displacement of the point of application of the force. The small amount of work $dW$ done on the object by $\vec{\mathbf{F}}$ as the point of application rotates through an infinitesimal distance $ds = r\,d\theta$ in a time interval $dt$ is

$$dW = \vec{\mathbf{F}} \cdot d\vec{\mathbf{s}} = (F \sin \phi) r\, d\theta$$

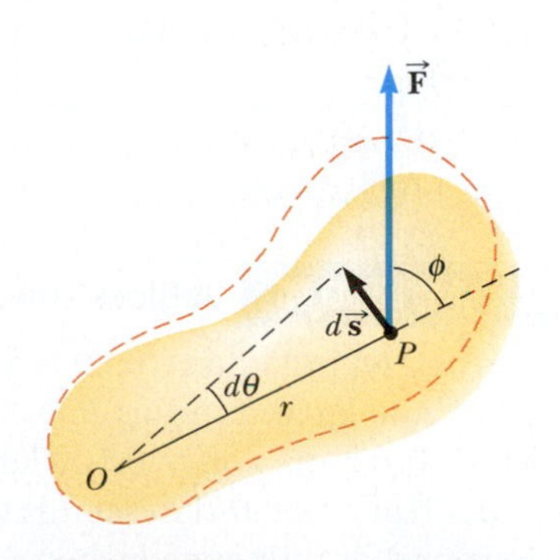


FIGURE 10.19 A rigid object rotates about an axis through $O$ under the action of an external force $\vec{\mathbf{F}}$ applied at $P$.

where $F \sin \phi$ is the tangential component of $\vec{\mathbf{F}}$, or the component of the force along the displacement. Note from Figure 10.19 that the **radial component of $\vec{\mathbf{F}}$ does no work because it is perpendicular to the displacement of the point of application of the force.**

Because the magnitude of the torque due to $\vec{\mathbf{F}}$ about the origin is defined as $rF \sin \phi$, we can write the work done for the infinitesimal rotation in the form

$$dW = \tau \, d\theta \tag{10.28}$$

Notice that this expression is the product of torque and angular displacement, making it analogous to the work done in translational motion, which is the product of force and translational displacement (Eq. 6.2).

Now, we will combine this result with the rotational form of Newton's second law, $\tau = I\alpha$. Using the chain rule from calculus, we can express the torque as

$$\tau = I\alpha = I\frac{d\omega}{dt} = I\frac{d\omega}{d\theta}\frac{d\theta}{dt} = I\frac{d\omega}{d\theta}\,\omega$$

Rearranging this expression and noting that $\tau \, d\theta = dW$ from Equation 10.28, we have

$$\tau \, d\theta = dW = I\omega \, d\omega$$

Integrating this expression, we find the total work done by the torque:

$$W = \int_{\theta_i}^{\theta_f} \tau \, d\theta = \int_{\omega_i}^{\omega_f} I\omega \, d\omega$$

■ Work–kinetic energy theorem for pure rotation

$$W = \tfrac{1}{2} I\omega_f^2 - \tfrac{1}{2} I\omega_i^2 = \Delta K_R \tag{10.29}$$

Notice that this equation has exactly the same mathematical form as the work–kinetic energy theorem for translation. If a system consists of components that are both translating and rotating, the work–kinetic energy theorem generalizes to $W = \Delta K + \Delta K_R$.

We finish this discussion of energy concepts for rotation by investigating the *rate* at which work is being done by $\vec{\mathbf{F}}$ on an object rotating about a fixed axis. This rate is obtained by dividing the left and right sides of Equation 10.28 by $dt$:

$$\frac{dW}{dt} = \tau \frac{d\theta}{dt} \tag{10.30}$$

The quantity $dW/dt$ is, by definition, the instantaneous power $\mathcal{P}$ delivered by the force. Furthermore, because $d\theta/dt = \omega$, Equation 10.30 reduces to

■ Power delivered to a rotating object

$$\mathcal{P} = \tau\omega \tag{10.31}$$

This expression is analogous to $\mathcal{P} = Fv$ in the case of translational motion.

## EXAMPLE 10.11 A Block Unwinding from a Wheel

The wheel in Figure 10.20 is a solid disk of mass $M = 2.00$ kg and radius $R = 30.0$ cm. The suspended block has a mass $m = 0.500$ kg. If the suspended block starts from rest and descends to a position 1.00 m lower, what is its speed when it is at this position?

**Solution** The work done on the system of the block and the wheel is due to the gravitational force $m\vec{\mathbf{g}}$ acting on the hanging block. From the continuity equation for energy, the work must be equal to the change in kinetic energy of the system:

$$W = \Delta K + \Delta K_R$$

where $K$ is the translational kinetic energy of the block and $K_R$ is the rotational kinetic energy of the wheel.

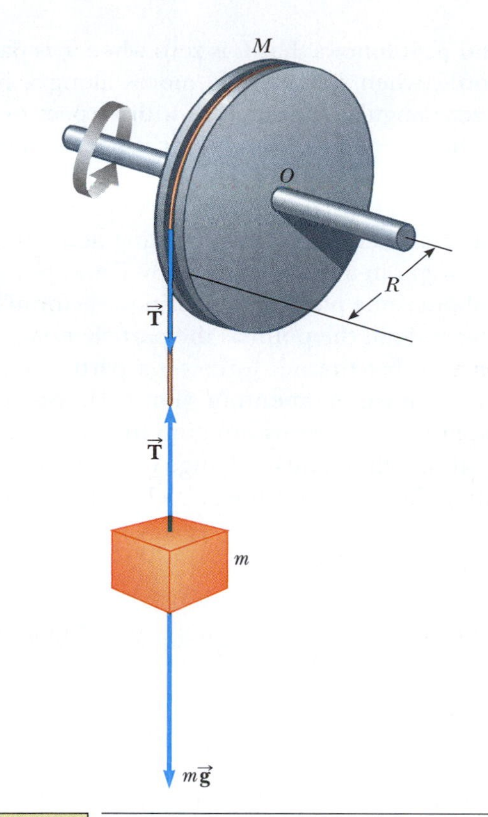


FIGURE 10.20 (Example 10.11) An object hangs from a cord wrapped around a wheel. The tension in the cord produces a torque about the axle passing through $O$.

The system begins from rest, so we can write this expression as

$$W = \vec{F} \cdot \Delta\vec{r} = (-mg\hat{j}) \cdot (-\Delta y\hat{j}) = mg(\Delta y) = \tfrac{1}{2}mv^2 + \tfrac{1}{2}I\omega^2$$

where $v$ is the speed of the block at its final position. It is also the speed of the string at this instant as well as the speed of a point on the rim of the wheel at this instant. Therefore, $\omega = v/R$. In addition, because the wheel is a solid disk, its moment of inertia is $I = \frac{1}{2}MR^2$. Consequently,

$$mg(\Delta y) = \tfrac{1}{2}mv^2 + \tfrac{1}{2}(\tfrac{1}{2}MR^2)\left(\frac{v}{R}\right)^2 = \tfrac{1}{2}mv^2 + \tfrac{1}{4}Mv^2$$

Solving for $v$, we find that

$$v = \sqrt{\frac{mg(\Delta y)}{\frac{1}{2}m + \frac{1}{4}M}}$$

$$= \sqrt{\frac{(0.500\text{ kg})(9.80\text{ m/s}^2)(1.00\text{ m})}{\frac{1}{2}(0.500\text{ kg}) + \frac{1}{4}(2.00\text{ kg})}} = \boxed{2.56\text{ m/s}}$$

## 10.8 ANGULAR MOMENTUM

Imagine an object rotating in space with no motion of its center of mass. Each particle in the object is moving in a circular path, so momentum is associated with the motion of each particle. Although the object has no linear momentum (its center of mass is not moving through space), a "quantity of motion" is associated with its rotation. We will investigate the **angular momentum** that the object has in this section.

Let us first consider a particle of mass $m$, situated at the vector position $\vec{r}$ and moving with a momentum $\vec{p}$, as shown in Active Figure 10.21. For now, we don't consider it as a particle on a rigid object; it is any particle moving with momentum $\vec{p}$. We will apply the result to a rotating rigid object shortly. The **instantaneous angular momentum** $\vec{L}$ of the particle relative to the origin $O$ is defined by the vector product of its instantaneous position vector $\vec{r}$ and the instantaneous linear momentum $\vec{p}$:

$$\boxed{\vec{L} = \vec{r} \times \vec{p}} \qquad [10.32]$$

The SI units of angular momentum are $kg \cdot m^2/s$. Note that both the magnitude and the direction of $\vec{L}$ depend on the choice of origin. The direction of $\vec{L}$ is perpendicular to the plane formed by $\vec{r}$ and $\vec{p}$, and the sense of $\vec{L}$ is governed by the right-hand rule. For example, in Active Figure 10.21, $\vec{r}$ and $\vec{p}$ are assumed to be in the $xy$ plane and $\vec{L}$ points in the $z$ direction. Because $\vec{p} = m\vec{v}$, the magnitude of $\vec{L}$ is

$$L = mvr\sin\phi \qquad [10.33]$$

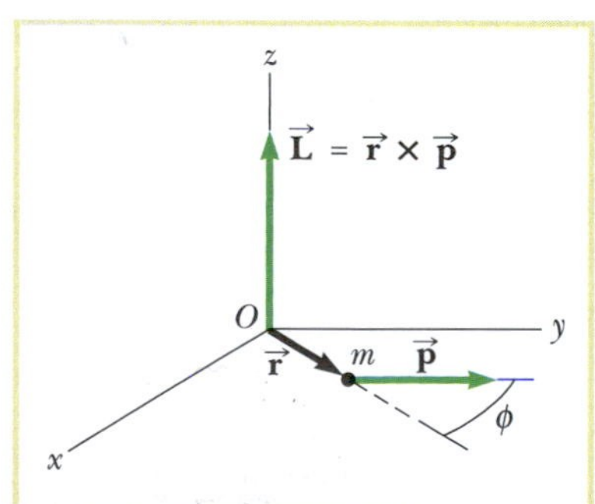


ACTIVE FIGURE 10.21
The angular momentum $\vec{L}$ of a particle of mass $m$ and linear momentum $\vec{p}$ located at the position $\vec{r}$ is given by $\vec{L} = \vec{r} \times \vec{p}$. The value of $\vec{L}$ depends on the origin about which it is measured and is a vector perpendicular to both $\vec{r}$ and $\vec{p}$.

**PhysicsNow™** By logging into PhysicsNow at **www.pop4e.com** and going to Active Figure 10.21 you can change the position vector $\vec{r}$ and the momentum vector $\vec{p}$ to see the effect on the angular momentum vector.

**PITFALL PREVENTION 10.6**

**IS ROTATION NECESSARY FOR ANGULAR MOMENTUM?** We can define angular momentum even if the particle is not moving in a circular path. Even a particle moving in a straight line has angular momentum about any axis displaced from the path of the particle.

where $\phi$ is the angle between $\vec{\mathbf{r}}$ and $\vec{\mathbf{p}}$. It follows that $\vec{\mathbf{L}}$ is zero when $\vec{\mathbf{r}}$ is parallel to $\vec{\mathbf{p}}$ ($\phi = 0°$ or $180°$). In other words, when the particle moves along a line that passes through the origin, it has zero angular momentum with respect to the origin, which is equivalent to stating that the momentum vector is not tangent to *any* circle drawn about the origin. On the other hand, if $\vec{\mathbf{r}}$ is perpendicular to $\vec{\mathbf{p}}$ ($\phi = 90°$), $L$ is a maximum and equal to $mvr$. In fact, at that instant the particle moves exactly as though it were on the rim of a wheel of radius $r$ rotating at angular speed $\omega = v/r$ about an axis through the origin in a plane defined by $\vec{\mathbf{r}}$ and $\vec{\mathbf{p}}$. A particle has nonzero angular momentum about some point if the position vector of the particle measured from that point rotates about the point as the particle moves.

For translational motion, we found that the net force on a particle equals the time rate of change of the particle's linear momentum (Eq. 8.4). We shall now show that Newton's second law implies an analogous situation for rotation: that the net torque acting on a particle equals the time rate of change of the particle's angular momentum. Let us start by writing the torque on the particle in the form

$$\vec{\boldsymbol{\tau}} = \vec{\mathbf{r}} \times \vec{\mathbf{F}} = \vec{\mathbf{r}} \times \frac{d\vec{\mathbf{p}}}{dt} \qquad [10.34]$$

where we have used $\vec{\mathbf{F}} = d\vec{\mathbf{p}}/dt$ (Eq. 8.4). Now let us differentiate Equation 10.32 with respect to time, using the product rule for differentiation:

$$\frac{d\vec{\mathbf{L}}}{dt} = \frac{d}{dt}(\vec{\mathbf{r}} \times \vec{\mathbf{p}}) = \vec{\mathbf{r}} \times \frac{d\vec{\mathbf{p}}}{dt} + \frac{d\vec{\mathbf{r}}}{dt} \times \vec{\mathbf{p}}$$

It is important to adhere to the order of factors in the vector product because the vector product is not commutative.

The last term on the right in the preceding equation is zero because $\vec{\mathbf{v}} = d\vec{\mathbf{r}}/dt$ is parallel to $\vec{\mathbf{p}}$. Therefore,

$$\frac{d\vec{\mathbf{L}}}{dt} = \vec{\mathbf{r}} \times \frac{d\vec{\mathbf{p}}}{dt} \qquad [10.35]$$

Comparing Equations 10.34 and 10.35, we see that

■ Torque on a particle equals time rate of change of angular momentum of the particle

$$\vec{\boldsymbol{\tau}} = \frac{d\vec{\mathbf{L}}}{dt} \qquad [10.36]$$

This result is the rotational analog of Newton's second law, $\vec{\mathbf{F}} = d\vec{\mathbf{p}}/dt$. Equation 10.36 says that the torque acting on a particle is equal to the time rate of change of the particle's angular momentum. Note that Equation 10.36 is valid only if the axes used to define $\vec{\boldsymbol{\tau}}$ and $\vec{\mathbf{L}}$ are the *same*. Equation 10.36 is also valid when several forces are acting on the particle, in which case $\vec{\boldsymbol{\tau}}$ is the *net* torque on the particle. Of course, the same origin must be used in calculating all torques as well as the angular momentum.

Now, let us apply these ideas to a system of particles. The total angular momentum $\vec{\mathbf{L}}$ of the system of particles about some point is defined as the vector sum of the angular momenta of the individual particles:

$$\vec{\mathbf{L}} = \vec{\mathbf{L}}_1 + \vec{\mathbf{L}}_2 + \cdots + \vec{\mathbf{L}}_n = \sum_i \vec{\mathbf{L}}_i$$

where the vector sum is over all the $n$ particles in the system.

Because the individual angular momenta of the particles may change in time, the total angular momentum may also vary in time. In fact, from Equations 10.34 and 10.35 we find that the time rate of change of the total angular momentum of the system equals the vector sum of *all* torques, including those associated with internal forces between particles and those associated with external forces.

As we found in our discussion of the rigid object under a net torque, however, the sum of the internal torques is zero. Therefore, we conclude that the total angular momentum can vary with time *only* if there is a net *external* torque on the system, so that we have

$$\sum \vec{\boldsymbol{\tau}}_{\text{ext}} = \sum_i \frac{d\vec{\mathbf{L}}_i}{dt} = \frac{d}{dt}\sum_i \vec{\mathbf{L}}_i = \frac{d\vec{\mathbf{L}}_{\text{tot}}}{dt} \qquad [10.37]$$

■ **Net external torque on a system equals time rate of change of angular momentum of the system**

That is, the time rate of change of the total angular momentum of the system about some origin in an inertial frame equals the net external torque acting on the system about that origin. Note that Equation 10.37 is the rotational analog of $\sum \vec{\mathbf{F}}_{\text{ext}} = d\vec{\mathbf{p}}_{\text{tot}}/dt$ (Eq. 8.40) for a system of particles.

This result is valid for a system of particles that change their positions with respect to one another, that is, a nonrigid object. In this discussion of angular momentum of a system of particles, notice that we never imposed the rigid-object condition.

Equation 10.37 is the primary equation in the angular momentum version of the nonisolated system model. The system's angular momentum changes in response to an interaction with the environment, described by means of the net torque on the system.

One final result can be obtained for angular momentum, which will serve as an analog to the definition of linear momentum. Let us imagine a rigid object rotating about an axis. Each particle of mass $m_i$ in the rigid object moves in a circular path of radius $r_i$, with a tangential speed $v_i$. Therefore, the total angular momentum of the rigid object is

$$L = \sum_i m_i v_i r_i$$

Let us now replace the tangential speed with the product of the radial distance and the angular speed (Eq. 10.10):

$$L = \sum_i m_i v_i r_i = \sum_i m_i (r_i \omega) r_i = \left(\sum_i m_i {r_i}^2\right)\omega$$

We recognize the combination in the parentheses as the moment of inertia, so we can write the angular momentum of the rigid object as

$$L = I\omega$$

■ **Angular momentum of an object with moment of inertia $I$**

which is the rotational analog to $p = mv$. Table 10.3 is a continuation of Table 10.1, with additional translational and rotational analogs that we have developed in the past few sections.

**TABLE 10.3 A Comparison of Equations for Rotational and Translational Motion: Dynamic Equations**

| | Rotational Motion About a Fixed Axis | Translational Motion |
|---|---|---|
| Kinetic energy | $K_R = \frac{1}{2}I\omega^2$ | $K = \frac{1}{2}mv^2$ |
| Equilibrium | $\sum \vec{\boldsymbol{\tau}} = 0$ | $\sum \vec{\mathbf{F}} = 0$ |
| Newton's second law | $\sum \tau = I\alpha$ | $\sum \vec{\mathbf{F}} = m\vec{\mathbf{a}}$ |
| Newton's second law | $\vec{\boldsymbol{\tau}} = \frac{d\vec{\mathbf{L}}}{dt}$ | $\vec{\mathbf{F}} = \frac{d\vec{\mathbf{p}}}{dt}$ |
| Momentum | $L = I\omega$ | $\vec{\mathbf{p}} = m\vec{\mathbf{v}}$ |
| Conservation principle | $\vec{\mathbf{L}}_i = \vec{\mathbf{L}}_f$ | $\vec{\mathbf{p}}_i = \vec{\mathbf{p}}_f$ |
| Power | $\mathcal{P} = \tau\omega$ | $\mathcal{P} = Fv$ |

*Note:* Equations in translation motion expressed in terms of vectors have rotational analogs in terms of vectors. Because the full vector treatment of rotation is beyond the scope of this book, however, some rotational equations are given in nonvector form.

**QUICK QUIZ 10.7** A solid sphere and a hollow sphere have the same mass and radius. They are rotating with the same angular speed. Which is the one with the higher angular momentum? **(a)** the solid sphere **(b)** the hollow sphere **(c)** both have the same angular momentum **(d)** impossible to determine

**EXAMPLE 10.12 The Atwood Machine Once Again**

Consider again the Atwood machine with the massive pulley in Example 10.10. Determine the acceleration of the two objects using an angular momentum approach.

**Solution** This example is of a nonrigid object experiencing a net torque, so we use the nonisolated system model. We will evaluate the angular momentum of the system at any time and then differentiate the angular momentum, setting it equal to the net external torque. We will solve the resulting expression for the acceleration of the objects.

Let us first calculate the angular momentum of the system, which consists of the two objects plus the pulley. At the instant $m_1$ and $m_2$ have a speed $v$, the angular momentum of $m_1$ around the axle of the pulley is $m_1vR$ and that of $m_2$ is $m_2vR$. At the same instant, the angular momentum of the pulley around its center is $L = I\omega = Iv/R$. Therefore, the total angular momentum of the system is

$$(1)\quad L = m_1vR + m_2vR + I\frac{v}{R}$$

$$= m_1vR + m_2vR + (\tfrac{1}{2}MR^2)\frac{v}{R}$$

$$= (m_1 + m_2 + \tfrac{1}{2}M)vR$$

Now let us evaluate the total external torque on the system about the axle. The weight of the pulley and the force of the axle upward on the pulley have zero moment arm around the center of the pulley, so they do not contribute to the torque. The external forces on the system that produce torques about the axle are $m_1\vec{\mathbf{g}}$, with a torque of $m_1gR$, and $m_2\vec{\mathbf{g}}$, with a torque of $-m_2gR$. Combining the net external torque with (1) and Equation 10.37 gives us

$$\tau_{\text{ext}} = \frac{dL}{dt}$$

$$m_1gR - m_2gR = \frac{d}{dt}[(m_1 + m_2 + \tfrac{1}{2}M)vR]$$

$$(2)\quad m_1gR - m_2gR = (m_1 + m_2 + \tfrac{1}{2}M)R\frac{dv}{dt}$$

Because $dv/dt = a$, we can solve Equation (2) for $a$ to find

$$a = \left(\frac{m_1 - m_2}{m_1 + m_2 + \frac{1}{2}M}\right)g$$

which is the same result as that obtained in Example 10.10. You may wonder why we did not include the tension forces that the cord exerts on the objects in evaluating the net torque about the axle. The reason is that these forces are *internal* to the system under consideration. Only the *external* torques contribute to the change in angular momentum.

## 10.9 CONSERVATION OF ANGULAR MOMENTUM

In Chapter 8, we found that the total linear momentum of a system of particles is conserved when the net external force acting on the system is zero. In rotational motion, we have an analogous conservation law that states that **the total angular momentum of a system is conserved if the net external torque acting on the system is zero.**

Because the net external torque acting on the system equals the time rate of change of the system's angular momentum, we see from Equation 10.37 that if

$$\sum \vec{\boldsymbol{\tau}}_{\text{ext}} = \frac{d\vec{\mathbf{L}}_{\text{tot}}}{dt} = 0 \qquad [10.38]$$

then

▪ Conservation of angular momentum for an isolated system

$$\vec{\mathbf{L}}_{\text{tot}} = \text{constant} \rightarrow \vec{\mathbf{L}}_{\text{tot},\,i} = \vec{\mathbf{L}}_{\text{tot},\,f} \qquad [10.39]$$

Equation 10.39 represents a third conservation law to add to our list of fundamental conservation principles. We can now state that **the total energy, linear**

**momentum, and angular momentum of an isolated system are all conserved.** We have focused our attention in this chapter on rigid objects; the conservation of angular momentum principle, however, is a general result of the isolated system model. Therefore, **the angular momentum of an isolated system is conserved whether the system is a rigid object or not.**

At any instant of time, the angular momentum of a system of particles about a fixed axis has a magnitude given by $L = I\omega$, where $I$ is the moment of inertia of the system about the axis. In this case, if the net external torque on the system is zero, we can express the conservation of angular momentum principle as $I\omega$ = constant. Imagine a situation in which a rotating system undergoes a change in moment of inertia. Because of the principle of conservation of angular momentum, there must be a corresponding change in the angular speed.

Many examples can be used to demonstrate this effect; some of them should be familiar to you. You may have observed a figure skater spinning (Fig. 10.22). The angular speed of the skater is large when his hands and feet are close to the trunk of his body. Ignoring friction between skater and ice, we see that there are no external torques on the skater. The moment of inertia of his body increases as his hands and feet are moved away from his body at the finish of the spin. According to the principle of conservation of angular momentum, his angular speed must decrease.

An interesting astrophysical example of conservation of angular momentum occurs when, at the end of its lifetime, a massive star uses up all its fuel and collapses under the influence of gravitational forces, causing a gigantic outburst of energy called a supernova explosion. The best-studied example of a remnant of a supernova explosion is the Crab Nebula, a chaotic, expanding mass of gas (Fig. 10.23). In a supernova, part of the star's mass is released into space, where it eventually condenses into new stars and planets. Most of what is left behind typically collapses into a **neutron star,** an extremely dense sphere of matter with a diameter of about 10 km in comparison with the $10^6$-km diameter of the original star and containing a large fraction of the star's original mass. As the moment of inertia of the system decreases during the collapse, the star's rotational speed increases, similar to the change in speed of the skater in Figure 10.22. More than 700 rapidly rotating neutron stars have been identified since the first discovery of such astronomical bodies in 1967, with periods of rotation ranging from a millisecond to several seconds. The neutron star—an object with a mass greater than the Sun, rotating about its axis many times each second—is a most dramatic system!

(© Stuart Franklin/Getty Images)

FIGURE 10.22 Angular momentum is conserved as Russian figure skater Evgeni Plushenko performs during the 2004 World Figure Skating Championships. When his arms and legs are close to his body, his moment of inertia is small and his angular speed is large. To slow down for the finish of his spin, he moves his arms and legs outward, increasing his moment of inertia.

(David Malin, Anglo-Australian Observatory)

FIGURE 10.23 The Crab Nebula, in the constellation Taurus. This nebula is the remnant of a supernova explosion, which was seen on Earth in the year A.D. 1054. It is located some 6 300 lightyears away and is approximately 6 lightyears in diameter, still expanding outward.

**QUICK QUIZ 10.8** (i) A competitive diver leaves the diving board and falls toward the water with her body straight and rotating slowly. She pulls her arms and legs into a tight tuck position. What happens to her angular speed? (a) It increases. (b) It decreases. (c) It stays the same. (d) It is impossible to determine. (ii) From the same list of choices, what happens to the rotational kinetic energy of her body?

## EXAMPLE 10.13 A Revolving Puck on a Horizontal, Frictionless Surface

A puck of mass $m$ on a horizontal, frictionless table is connected to a string that passes through a small hole in the table. The puck is set into circular motion of radius $R$, at which time its speed is $v_i$ (Fig. 10.24).

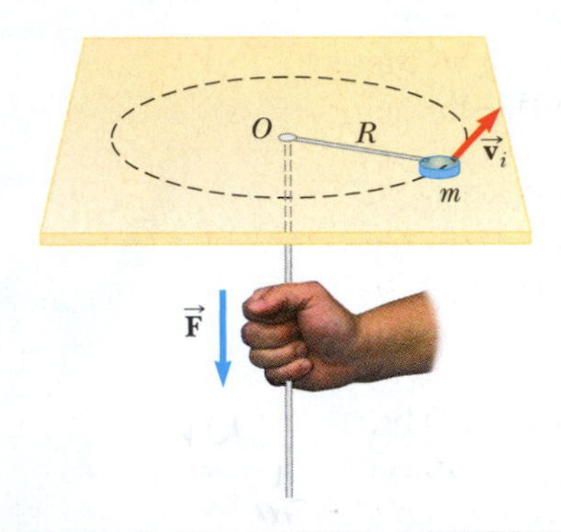


**FIGURE 10.24** (Example 10.13) When the string is pulled downward, the speed of the puck changes.

**A** If the string is pulled from the bottom so that the radius of the circular path is decreased to $r$, what is the final speed $v_f$ of the puck?

**Solution** We identify the system as the puck. We will calculate torque about the center of rotation $O$. Note that the gravitational force acting on the puck is balanced by the upward normal force, so these forces cancel, resulting in zero net torque from these forces. The force $\vec{F}$ of the string on the puck acts toward the center of rotation, and the vector position $\vec{r}$ is directed away from $O$. Therefore, we see that $\vec{\tau} = \vec{r} \times \vec{F} = 0$, so no torque is applied on the puck due to this force. Three forces are acting on the puck, but zero net torque occurs. Therefore, $\vec{L}$ is a constant of the motion. The puck can be modeled as a particle moving in a circular path, so $L = mv_iR = mv_fr$, or

$$v_f = \frac{v_iR}{r}$$

From this result we see that as $r$ decreases, the speed $v$ increases.

**B** Is the kinetic energy of the puck conserved in this process?

**Solution** We set up the ratio of the final kinetic energy to the initial kinetic energy:

$$\frac{K_f}{K_i} = \frac{\frac{1}{2}mv_f^2}{\frac{1}{2}mv_i^2} = \frac{1}{v_i^2}\left(\frac{v_iR}{r}\right)^2 = \frac{R^2}{r^2}$$

Because this ratio is not equal to 1, kinetic energy is not conserved. Furthermore, because $R > r$, the kinetic energy of the puck has increased. This increase corresponds to energy entering the system of the puck by means of the work done by the person pulling the string.

## EXAMPLE 10.14 Rotation Period of a Neutron Star

A star undergoes a supernova explosion. The material left behind forms a sphere of radius $8.0 \times 10^6$ m just after the explosion with a rotation period of 15 h. This remaining material collapses into a neutron star of radius 8.0 km. What is the rotation period $T$ of the neutron star?

**Solution** We model the star as an isolated system. As the moment of inertia of the stellar core decreases during its collapse, the angular speed increases. From conservation of angular momentum,

$$I_i\omega_i = I_f\omega_f$$

We have no information about the variation in the density of material with radius in either the initial star or the neutron star, so we choose a simplification model in which the density is uniform. Our result will most likely not be entirely accurate, but we can consider it an estimate. Using the moment of inertia of a sphere of uniform density (see Table 10.2),

$$\left(\tfrac{2}{5}MR_i^{\,2}\right)\omega_i = \left(\tfrac{2}{5}MR_f^{\,2}\right)\omega_f \quad \rightarrow \quad \omega_f = \left(\frac{R_i}{R_f}\right)^2 \omega_i$$

Now, because $\omega = 2\pi/T$, we have

$$\frac{2\pi}{T_f} = \left(\frac{R_i}{R_f}\right)^2 \frac{2\pi}{T_i} \quad \rightarrow \quad T_f = \left(\frac{R_f}{R_i}\right)^2 T_i$$

Substituting numerical values, we have

$$T_f = \left(\frac{8.0 \times 10^3 \text{ m}}{8.0 \times 10^6 \text{ m}}\right)^2 (15 \text{ h}) = 1.5 \times 10^{-5} \text{ h} = 0.054 \text{ s}$$

## 10.10 PRECESSIONAL MOTION OF GYROSCOPES

Angular momentum is the basis of the operation of a **gyroscope,** which is a spinning object used to control or maintain the orientation in space of the object or a system containing the object. As an example, consider a quarterback passing a football. If he imparts no spin to the ball, there is no angular momentum to be conserved and forces from the air might cause the ball to tumble as it moves through its trajectory. If a spin is imparted to the ball along the long axis of the football, however, the angular momentum vector stays fixed in direction and the football maintains its orientation throughout the trajectory, resulting in much less air resistance and a longer pass. In this application, the football is acting as a gyroscope to maintain its own orientation in space.

An unusual and fascinating type of motion you probably have observed is that of a top spinning rapidly about its axis of symmetry, as shown in Figure 10.25a. The top is acting as a gyroscope and one might expect the orientation to remain fixed in space. If the top is leaning over, however, it is observed that the symmetry axis rotates about the $z$ axis, sweeping out a cone (see Fig. 10.25b). This phenomenon is called **precessional motion.** The angular speed of the symmetry axis about the vertical is usually slow relative to the angular speed of the top about the symmetry axis.

It is quite natural to wonder why the top does not maintain its direction of spin. Because the center of mass of the top is not directly above the pivot point $O$, a net torque is acting on the top about $O$, a torque resulting from the gravitational force $M\vec{\mathbf{g}}$. The top would certainly fall over if it were not spinning. Because it is spinning, however, it has an angular momentum $\vec{\mathbf{L}}$ directed along its symmetry axis. As we shall show, the motion of this symmetry axis about the $z$ axis (the precessional motion) occurs because the torque produces a change in the *direction* of the symmetry axis. This motion is an excellent example of the importance of the directional nature of angular momentum.

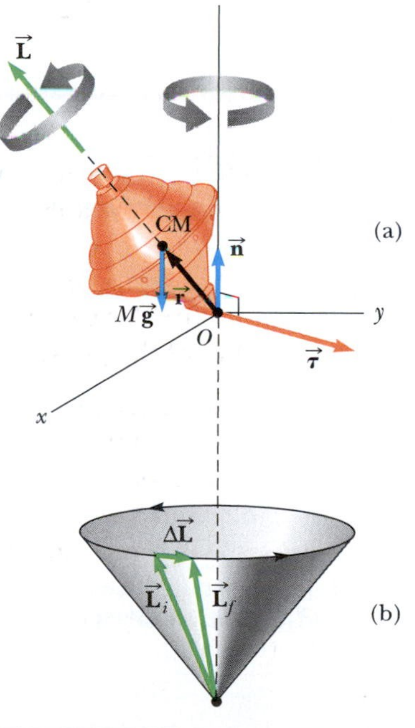


**FIGURE 10.25** Precessional motion of a top spinning about its symmetry axis. (a) The only external forces acting on the top are the normal force $\vec{\mathbf{n}}$ and the gravitational force $M\vec{\mathbf{g}}$. The direction of the angular momentum $\vec{\mathbf{L}}$ is along the axis of symmetry. The right-hand rule indicates that $\vec{\boldsymbol{\tau}} = \vec{\mathbf{r}} \times \vec{\mathbf{F}} = \vec{\mathbf{r}} \times M\vec{\mathbf{g}}$ is in the $xy$ plane. (b). The direction of $\Delta\vec{\mathbf{L}}$ is parallel to that of $\vec{\boldsymbol{\tau}}$ in part (a). That $\vec{\mathbf{L}}_f = \vec{\mathbf{L}}_i + \Delta\vec{\mathbf{L}}$ indicates that the top precesses about the $z$ axis.

The two forces acting on the top are the downward gravitational force $M\vec{\mathbf{g}}$ and the normal force $\vec{\mathbf{n}}$ acting upward at the pivot point $O$. The normal force produces no torque about the pivot because its moment arm through that point is zero. The gravitational force, however, produces a torque $\vec{\boldsymbol{\tau}} = \vec{\mathbf{r}} \times M\vec{\mathbf{g}}$ about $O$, where the direction of $\vec{\boldsymbol{\tau}}$ is perpendicular to the plane formed by $\vec{\mathbf{r}}$ and $M\vec{\mathbf{g}}$. By necessity, the vector $\vec{\boldsymbol{\tau}}$ lies in a horizontal plane perpendicular to the angular momentum vector. The net torque and angular momentum of the top are related through Equation 10.37:

$$\sum \vec{\boldsymbol{\tau}} = \frac{d\vec{\mathbf{L}}}{dt}$$

From this expression we see that the nonzero torque produces a change in angular momentum $d\vec{\mathbf{L}}$, a change that is in the same direction as $\sum \vec{\boldsymbol{\tau}}$. Therefore, like the torque vector, $d\vec{\mathbf{L}}$ must also be perpendicular to $\vec{\mathbf{L}}$. Figure 10.25b illustrates the resulting precessional motion of the symmetry axis of the top. In a time interval $\Delta t$, the change in angular momentum is $\Delta\vec{\mathbf{L}} = \vec{\mathbf{L}}_f - \vec{\mathbf{L}}_i = \sum \vec{\boldsymbol{\tau}}\,\Delta t$. Because $\Delta\vec{\mathbf{L}}$ is perpendicular to $\vec{\mathbf{L}}$, the magnitude of $\vec{\mathbf{L}}$ does not change ($|\vec{\mathbf{L}}_i| = |\vec{\mathbf{L}}_f|$). Rather, what is changing is the *direction* of $\vec{\mathbf{L}}$. Because the change in angular momentum $\Delta\vec{\mathbf{L}}$ is in the direction of $\Sigma\,\vec{\boldsymbol{\tau}}$, which lies in the *xy* plane, the top undergoes precessional motion.

With careful manufacturing tolerances, precession due to gravitational torque can be made very small and gyroscopes can be used for guidance systems in vehicles, whereby a change in the direction of the velocity of a vehicle is detected as a change between the direction of the angular momentum of the gyroscope and a reference direction attached to the vehicle. With proper electronic feedback, the deviation from the desired direction of motion can be removed, bringing the angular momentum back in line with the reference direction. Precession rates for highly specialized military gyroscopes are as low as 0.02° per day.

## 10.11 ROLLING MOTION OF RIGID OBJECTS

In this section, we shall investigate the special case of rotational motion in which a round object rolls on a surface. Many everyday examples exist for such motion, including automobile tires rolling on roads and bowling balls rolling toward the pins.

Suppose a cylinder is rolling on a straight path as in Figure 10.26. The center of mass moves in a straight line, but a point on the rim moves in a more complex path called a *cycloid.* Let us further assume that the cylinder of radius $R$ is uniform and rolls on a surface with friction. We make a rather odd, but valid, simplification model here for rolling objects. The surfaces must exert friction forces on each other; otherwise, the cylinder would simply slide rather than roll. If the friction force on the cylinder is large enough, the cylinder rolls without slipping. In this situation, the friction force is static rather than kinetic because the contact point of the cylinder with the surface is at rest relative to the surface at any instant. The static friction force acts through no displacement, so it does no work on the cylinder and causes no decrease in mechanical energy of the cylinder. In real rolling objects, deformations of the surfaces result in some rolling resistance. If both surfaces are hard, however, they will deform very little, and rolling resistance can be negligibly small. Therefore, we can model the rolling motion as maintaining constant mechanical energy. The wheel was a great invention!

As the cylinder rotates through an angle $\theta$, its center of mass moves a distance of $s = r\theta$. Therefore, the speed and acceleration of the center of mass for **pure rolling motion** are

■ **Relations between translational and rotational variables for a rolling object**

$$v_{\text{CM}} = \frac{ds}{dt} = R\frac{d\theta}{dt} = R\omega \qquad [10.40]$$

$$a_{\text{CM}} = \frac{dv_{\text{CM}}}{dt} = R\frac{d\omega}{dt} = R\alpha \qquad [10.41]$$

(Courtesy of Henry Leap and Jim Lehman)

**FIGURE 10.26** Light sources at the center and rim of a rolling cylinder illustrate the different paths these points take. The center moves in a straight line *(green line)*, whereas a point on the rim moves in the path of a cycloid *(red curve)*.

The translational velocities of various points on the rolling cylinder are illustrated in Figure 10.27. Note that the translational velocity of any point is in a direction perpendicular to the line from that point to the contact point. At any instant, the point $P$ is at rest relative to the surface because sliding does not occur.

We can express the **total kinetic energy** of a rolling object of mass $M$ and moment of inertia $I$ as the combination of the rotational kinetic energy around the center of mass plus the translational kinetic energy of the center of mass:

$$K = \frac{1}{2}I_{CM}\omega^2 + \frac{1}{2}Mv_{CM}{}^2 \qquad [10.42]$$

■ Total kinetic energy of a rolling object

A useful theorem called the **parallel axis theorem** enables us to express this energy in terms of the moment of inertia $I_p$ through any axis parallel to the axis through the center of mass of an object. This theorem states that

$$I_p = I_{CM} + MD^2 \qquad [10.43]$$

where $D$ is the distance from the center-of-mass axis to the parallel axis and $M$ is the total mass of the object. Let us use this theorem to express the moment of inertia around an axis passing through the contact point $P$ between the rolling object and the surface. The distance from this point to the center of mass of the symmetric object is its radius, so

$$I_p = I_{CM} + MR^2$$

If we write the translational speed of the center of mass of the object in Equation 10.42 in terms of the angular speed, we have

$$K = \frac{1}{2}I_{CM}\omega^2 + \frac{1}{2}MR^2\omega^2 = \frac{1}{2}(I_{CM} + MR^2)\,\omega^2 = \frac{1}{2}I_p\omega^2 \qquad [10.44]$$

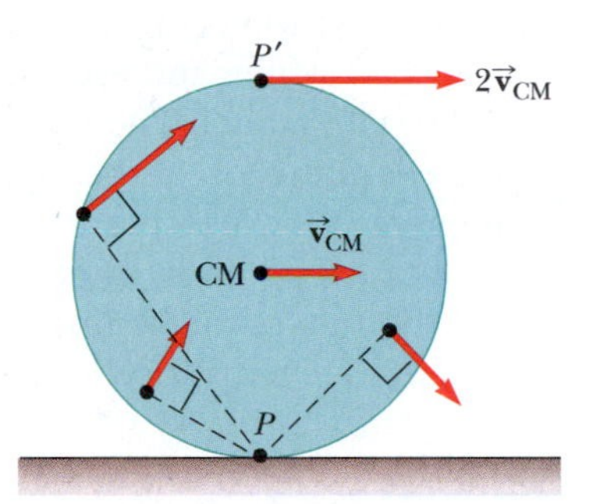


**FIGURE 10.27** All points on a rolling object move in a direction perpendicular to a line through the instantaneous point of contact $P$. The center of the object moves with a velocity $\vec{v}_{CM}$, whereas the point $P'$ moves with a velocity $2\vec{v}_{CM}$.

Therefore, the kinetic energy of the rolling object can be considered as equivalent to a purely rotational kinetic energy of the object rotating around its contact point.

We can use the energy version of the isolated system model to treat a class of problems concerning the rolling motion of a rigid object down a rough incline. In these types of problems, gravitational potential energy of the object–Earth system decreases as the rotational and translational kinetic energies of the object increase. For example, consider a sphere rolling without slipping after being released from rest at the top of an incline. Note that accelerated rolling motion is possible only if a friction force is present between the sphere and the incline to produce a net torque about the center of mass. Despite the presence of friction, no loss of mechanical energy occurs because the contact point is at rest relative to the surface at any instant. (On the other hand, if the sphere were to slip, mechanical energy of the sphere–incline–Earth system would be lost due to the nonconservative force of kinetic friction.)

Using $v_{CM} = R\omega$ for pure rolling motion, we can express Equation 10.42 as

$$K = \frac{1}{2}I_{CM}\left(\frac{v_{CM}}{R}\right)^2 + \frac{1}{2}Mv_{CM}{}^2$$

$$K = \frac{1}{2}\left(\frac{I_{CM}}{R^2} + M\right)v_{CM}{}^2 \qquad [10.45]$$

For the system of the sphere and the Earth, we define the zero configuration of gravitational potential energy to be when the sphere is at the bottom of the incline. Therefore, conservation of mechanical energy gives us

$$K_f + U_f = K_i + U_i$$

$$\frac{1}{2}\left(\frac{I_{CM}}{R^2} + M\right)v_{CM}{}^2 + 0 = 0 + Mgh$$

$$v_{CM} = \left(\frac{2gh}{1 + I_{CM}/MR^2}\right)^{1/2} \qquad [10.46]$$

**QUICK QUIZ 10.9** Two items A and B are placed at the top of an incline and released from rest. For *each* of the three pairs of items in (i), (ii), and (iii), which item arrives at the bottom of the incline first? **(i)** a ball A rolling without slipping and a box B sliding on a frictionless portion of the incline **(ii)** a sphere A that has twice the mass and twice the radius of a sphere B, where both roll without slipping **(iii)** a sphere A that has the same mass and radius as a sphere B, but sphere A is solid while sphere B is hollow and both roll without slipping. Choose from the following list for each of the three pairs of items. **(a)** item A **(b)** item B **(c)** items A and B arrive at the same time **(d)** impossible to determine

## EXAMPLE 10.15 Sphere Rolling Down an Incline

**A** If the object in Active Figure 10.28 is a solid sphere, calculate the speed of its center of mass at the bottom.

**Solution** We shall consider the sphere and the Earth as an isolated system and use the energy version of the isolated system model. The energy of the system when the sphere is at the top of the incline is gravitational potential energy only. We choose the zero configuration of gravitational potential energy to be when the sphere is at the bottom of the incline. Therefore, conservation of mechanical energy for the system gives us

$$K_f + U_f = K_i + U_i$$

$$(\tfrac{1}{2}Mv_{\text{CM},f}^2 + \tfrac{1}{2}I_{\text{CM}}\omega_f^2) + 0 = 0 + Mgh$$

Using Equation 10.40 to relate the translational and angular speeds, and substituting the moment of inertia for a sphere, we have

$$\tfrac{1}{2}Mv_{\text{CM},f}^2 + \tfrac{1}{2}(\tfrac{2}{5}MR^2)\frac{v_{\text{CM},f}^2}{R^2} = Mgh$$

$$\tfrac{1}{2}Mv_{\text{CM},f}^2 + \tfrac{1}{5}Mv_{\text{CM},f}^2 = \tfrac{7}{10}Mv_{\text{CM},f}^2 = Mgh$$

$$v_{\text{CM},f} = \sqrt{\tfrac{10}{7}gh}$$

**B** Determine the magnitude of the translational acceleration of the center of mass.

**Solution** To find the acceleration, let us recognize that the constant gravitational force should cause a constant acceleration of the center of mass of the sphere. From Equation 2.13,

$$v_{\text{CM},f}^2 = v_{\text{CM},i}^2 + 2a_{\text{CM}}(x_{\text{CM},f} - x_{\text{CM},i})$$

we can solve for the acceleration

$$a_{\text{CM}} = \frac{v_{\text{CM},f}^2 - v_{\text{CM},i}^2}{2(x_{\text{CM},f} - x_{\text{CM},i})} = \frac{\tfrac{10}{7}gh - 0}{2\left(\dfrac{h}{\sin\theta}\right)} = \tfrac{5}{7}g\sin\theta$$

These results are quite interesting in that both the speed and the acceleration of the center of mass are independent of the mass and radius of the sphere. That is, *all* homogeneous solid spheres experience the same speed and acceleration on a given incline!

If we repeated the calculations for a hollow sphere, a solid cylinder, or a hoop, we would obtain similar results with different numerical factors appearing in the expressions for $v_{\text{CM},f}$ and $a_{\text{CM}}$. These factors depend only on the moment of inertia about the center of mass for the specific object. In all cases, the acceleration of the center of mass is less than $g\sin\theta$, the value it would have if the plane were frictionless and no rolling occurred.

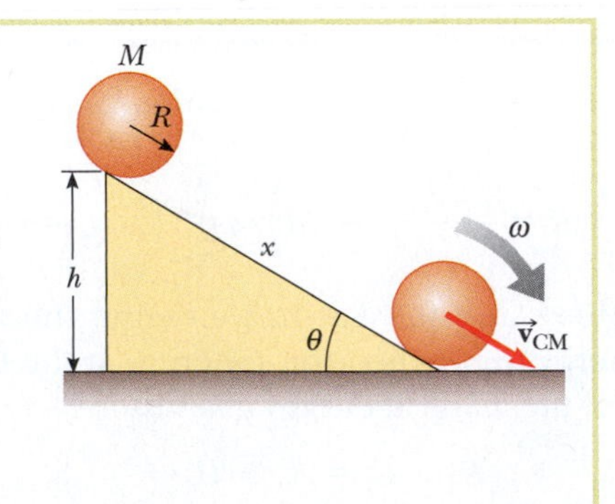


**ACTIVE FIGURE 10.28**
(Example 10.15) A round object rolling down an incline. Mechanical energy of the object–surface–Earth system is conserved if no slipping occurs and there is no rolling resistance.

**PhysicsNow™** Roll several objects down the hill and see the effect on the final speed by logging into PhysicsNow at **www.pop4e.com** and going to Active Figure 10.28.

## 10.12 TURNING THE SPACECRAFT

CONTEXT CONNECTION

In the Context Connection of Chapter 8, we discussed how to make a spacecraft move in empty space by firing its rocket engines. Let us know consider how to make the spacecraft turn in empty space.

One way to change the orientation of a spacecraft is to have small rocket engines that fire perpendicularly out the side of the spacecraft, providing a torque around its center of mass. This torque causes an angular acceleration around the center of mass of the spacecraft and therefore an angular speed. This rotation can be stopped to give the spacecraft the desired final orientation by firing the sideward-mounted rocket engines in the opposite direction. This option is desirable, and many spacecraft have such sideward-mounted rocket engines. An undesirable feature of this technique is that it consumes nonrenewable fuel on the spacecraft, both to initiate and to stop the rotation.

Let us consider another possibility related to angular momentum. Suppose the spacecraft carries a gyroscope that is not rotating, as in Figure 10.29a. In this case, the angular momentum of the spacecraft about its center of mass is zero. Suppose the gyroscope is set into rotation. Now, it would appear that the spacecraft system has a nonzero angular momentum because of the rotation of the gyroscope. Yet there is no external torque on the system, so the angular momentum of the isolated system must remain zero according to the principle of conservation of angular momentum. This principle can be satisfied by realizing that the spacecraft will turn in the direction opposite to that of the gyroscope so that the angular momentum vectors of the gyroscope and the spacecraft cancel, resulting in no angular momentum of the system. The result of rotating the gyroscope, as in Figure 10.29b, is that the spacecraft turns! By including three gyroscopes with mutually perpendicular axles, any desired rotation in space can be achieved. Once the desired orientation is achieved, the rotation of the gyroscope is halted.

This effect occurred in an undesirable situation with the *Voyager 2* spacecraft during its flight. The spacecraft carried a tape recorder whose reels rotated at high speeds. Each time the tape recorder was turned on, the reels acted as gyroscopes and the spacecraft started an undesirable rotation in the opposite direction. This rotation had to be counteracted by Mission Control by using the sideward-firing jets to stop the rotation!

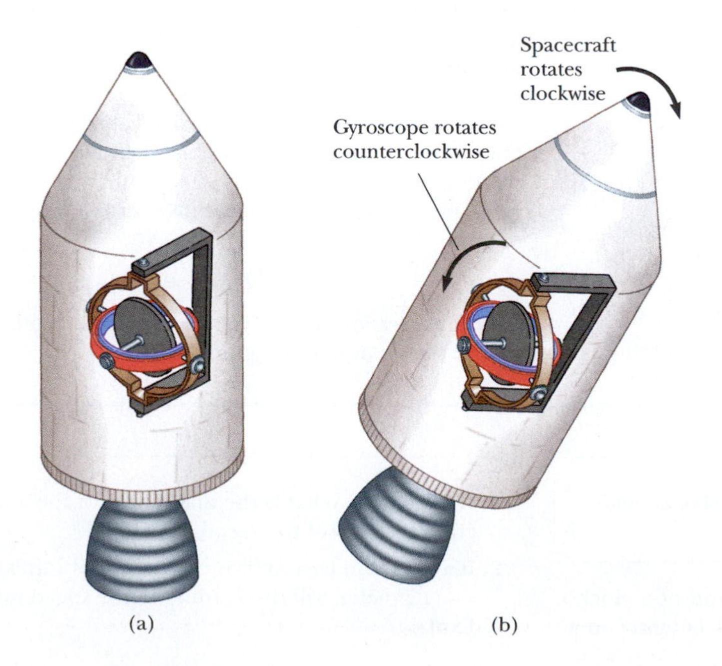


**FIGURE 10.29** (a) A spacecraft carries a gyroscope that is not spinning. (b) When the gyroscope is set into rotation, the spacecraft turns the other way so that the angular momentum of the system is conserved.

## SUMMARY

**Physics Now™** Take a practice test by logging into PhysicsNow at **www.pop4e.com** and clicking on the Pre-Test link for this chapter.

The **instantaneous angular speed** of a particle rotating in a circle or of a rigid object rotating about a fixed axis is

$$\omega \equiv \frac{d\theta}{dt} \qquad [10.3]$$

where $\omega$ is in rad/s or $s^{-1}$.

The **instantaneous angular acceleration** of a particle rotating in a circle or of a rigid object rotating about a fixed axis is

$$\alpha \equiv \frac{d\omega}{dt} \qquad [10.5]$$

and has units of rad/$s^2$ or $s^{-2}$.

When a rigid object rotates about a fixed axis, every part of the object has the same angular speed and the same angular acceleration. Different parts of the object, in general, have different translational speeds and different translational accelerations, however.

If a particle (or object) undergoes rotational motion about a fixed axis under constant angular acceleration $\alpha$, one can apply equations of kinematics by analogy with kinematic equations for translational motion with constant translational acceleration:

$$\omega_f = \omega_i + \alpha t \qquad [10.6]$$

$$\theta_f = \theta_i + \omega_i t + \tfrac{1}{2}\alpha t^2 \qquad [10.7]$$

$$\omega_f^2 = \omega_i^2 + 2\alpha(\theta_f - \theta_i) \qquad [10.8]$$

$$\theta_f = \theta_i + \tfrac{1}{2}(\omega_i + \omega_f)t \qquad [10.9]$$

When a particle rotates about a fixed axis, the angular position, the angular speed, and the angular acceleration are related to the tangential position, the tangential speed, and the tangential acceleration through the relationships

$$s = r\theta \qquad [10.1a]$$

$$v = r\omega \qquad [10.10]$$

$$a_t = r\alpha \qquad [10.11]$$

The **moment of inertia** of a system of particles is

$$I = \sum_i m_i r_i^2 \qquad [10.14]$$

If a rigid object rotates about a fixed axis with angular speed $\omega$, its **rotational kinetic energy** can be written

$$K_R = \tfrac{1}{2}I\omega^2 \qquad [10.15]$$

where $I$ is the moment of inertia about the axis of rotation.

The moment of inertia of a continuous object of density $\rho$ is

$$I = \int \rho r^2 dV \qquad [10.17]$$

The **torque** $\vec{\tau}$ due to a force $\vec{\mathbf{F}}$ about an origin in an inertial frame is defined to be

$$\vec{\tau} \equiv \vec{\mathbf{r}} \times \vec{\mathbf{F}} \qquad [10.19]$$

where $\vec{\mathbf{r}}$ is the position vector of the point of application of the force.

Given two vectors $\vec{\mathbf{A}}$ and $\vec{\mathbf{B}}$, their **vector product** or **cross product** $\vec{\mathbf{A}} \times \vec{\mathbf{B}}$ is a vector $\vec{\mathbf{C}}$ having the magnitude

$$C \equiv AB \sin\theta \qquad [10.21]$$

where $\theta$ is the angle between $\vec{\mathbf{A}}$ and $\vec{\mathbf{B}}$. The direction of $\vec{\mathbf{C}}$ is perpendicular to the plane formed by $\vec{\mathbf{A}}$ and $\vec{\mathbf{B}}$, and is determined by the right-hand rule.

The net torque acting on an object is proportional to the angular acceleration of the object, and the proportionality constant is the moment of inertia $I$:

$$\sum \tau = I\alpha \qquad [10.27]$$

The **angular momentum** $\vec{\mathbf{L}}$ of a particle with linear momentum $\vec{\mathbf{p}} = m\vec{\mathbf{v}}$ is

$$\vec{\mathbf{L}} \equiv \vec{\mathbf{r}} \times \vec{\mathbf{p}} \qquad [10.32]$$

where $\vec{\mathbf{r}}$ is the vector position of the particle relative to the origin. If $\phi$ is the angle between $\vec{\mathbf{r}}$ and $\vec{\mathbf{p}}$, the magnitude of $\vec{\mathbf{L}}$ is

$$L = mvr \sin\phi \qquad [10.33]$$

The net external torque acting on a system is equal to the time rate of change of its angular momentum:

$$\sum \vec{\tau}_{\text{ext}} = \frac{d\vec{\mathbf{L}}_{\text{tot}}}{dt} \qquad [10.37]$$

The law of conservation of angular momentum states that the total angular momentum of a system remains constant if the net external torque acting on the system is zero:

$$\vec{\mathbf{L}}_{\text{tot},i} = \vec{\mathbf{L}}_{\text{tot},f} \qquad [10.39]$$

The **total kinetic energy** of a rigid object, such as a cylinder, that is rolling on a rough surface without slipping equals the rotational kinetic energy $\frac{1}{2}I_{\text{CM}}\omega^2$ about the object's center of mass plus the translational kinetic energy $\frac{1}{2}Mv_{\text{CM}}^2$ of the center of mass:

$$K = \tfrac{1}{2}I_{\text{CM}}\omega^2 + \tfrac{1}{2}Mv_{\text{CM}}^2 \qquad [10.42]$$

In this expression, $v_{\text{CM}}$ is the speed of the center of mass and $v_{\text{CM}} = R\omega$ for pure rolling motion.

## QUESTIONS

☐ = answer available in the *Student Solutions Manual and Study Guide*

1. ☐ What is the angular speed of the second hand of a clock? What is the direction of $\vec{\omega}$ as you view a clock hanging on a vertical wall? What is the magnitude of the angular acceleration vector $\vec{\alpha}$ of the second hand?

2. If a car's standard tires are replaced with tires of larger outside diameter, will the reading of the speedometer change? Explain.

**3.** Suppose just two external forces act on a stationary rigid object and the two forces are equal in magnitude and opposite in direction. Under what condition does the object start to rotate?

**4.** Suppose you remove two eggs from the refrigerator, one hard-boiled and the other uncooked. You wish to determine which is the hard-boiled egg without breaking the eggs. You can do so by spinning the two eggs on the floor and comparing the rotational motions. Which egg spins faster? Which rotates more uniformly? Explain.

**5.** If you see an object rotating, is there necessarily a net torque acting on it?

**6.** Which of the entries in Table 10.2 applies to finding the moment of inertia of a long, straight sewer pipe rotating about its axis of symmetry? of an embroidery hoop rotating about an axis through its center and perpendicular to its plane? of a uniform door turning on its hinges? of a coin turning about an axis through its center and perpendicular to its faces?

**7.** In a tape recorder, the tape is pulled past the read-and-write heads at a constant speed by the drive mechanism. Consider the reel from which the tape is pulled. As the tape is pulled from it, the radius of the roll of remaining tape decreases. How does the torque on the reel change with time? How does the angular speed of the reel change in time? If the drive mechanism is switched on so that the tape is suddenly jerked with a large force, is the tape more likely to break when it is being pulled from a nearly full reel or from a nearly empty reel?

**8.** Vector $\vec{\mathbf{A}}$ is in the negative $y$ direction and vector $\vec{\mathbf{B}}$ is in the negative $x$ direction. What are the directions of (a) $\vec{\mathbf{A}} \times \vec{\mathbf{B}}$ and (b) $\vec{\mathbf{B}} \times \vec{\mathbf{A}}$?

**9.** For a helicopter to be stable as it flies, it must have at least two propellers. Why?

**10.** Often when a high diver wants to turn a flip in midair, she draws her legs up against her chest. Why does this movement make her rotate faster? What should she do when she wants to come out of her flip?

**11.** Why does a long pole help a tightrope walker stay balanced?

**12.** In some motorcycle races, the riders drive over small hills and the motorcycle becomes airborne for a short time. If the motorcycle racer keeps the throttle open while leaving the hill and going into the air, the motorcycle tends to nose upward. Why?

**13.** If global warming continues over the next one hundred years, it is likely that some polar ice will melt and the water will be distributed closer to the Equator. How would that change the moment of inertia of the Earth? Would the length of the day (one revolution) increase or decrease?

**14.** Two uniform solid spheres, a large, massive sphere and a small sphere with low mass, are rolled down a hill. Which one reaches the bottom of the hill first? Next, we roll a large, low-density sphere and a small high-density sphere of equal mass. Which one wins in this case?

**15.** In a soapbox derby race, the cars have no engines; they simply coast down a hill to race with one another. Suppose you are designing a car for a coasting race. Do you want to use large wheels or small wheels? Do you want to use solid disk-like wheels or hoop-like wheels? Should the wheels be heavy or light?

**16.** Stand with your back against a wall. Why can't you put your heels firmly against the wall and then bend forward without falling?

**17.** A ladder stands on the ground, leaning against a wall. Would you feel safer climbing up the ladder if you were told that the ground is frictionless but the wall is rough, or that the wall is frictionless but the ground is rough? Justify your answer.

**18.** (a) Give an example in which the net force acting on an object is zero and yet the net torque is nonzero. (b) Give an example in which the net torque acting on an object is zero and yet the net force is nonzero.

# PROBLEMS

1, 2, 3 = straightforward, intermediate, challenging
☐ = full solution available in the *Student Solutions Manual and Study Guide*
**PhysicsNow**™ = coached problem with hints available at www.pop4e.com
= computer useful in solving problem
= paired numerical and symbolic problems
= biomedical application

## Section 10.1 ▪ Angular Position, Speed, and Acceleration

**1.** During a certain period of time, the angular position of a swinging door is described by $\theta = 5.00 + 10.0t + 2.00t^2$, where $\theta$ is in radians and $t$ is in seconds. Determine the angular position, angular speed, and angular acceleration of the door (a) at $t = 0$ and (b) at $t = 3.00$ s.

## Section 10.2 ▪ Rotational Kinematics: The Rigid Object Under Constant Angular Acceleration

**2.** A dentist's drill starts from rest. After 3.20 s of constant angular acceleration, it turns at a rate of $2.51 \times 10^4$ rev/min. (a) Find the drill's angular acceleration. (b) Determine the angle (in radians) through which the drill rotates during this period.

**3.** **PhysicsNow**™ An electric motor rotating a grinding wheel at 100 rev/min is switched off. The wheel then moves with a constant negative angular acceleration of magnitude 2.00 rad/s$^2$. (a) During what time interval does the wheel come to rest? (b) Through how many radians does it turn while it is slowing down?

**4.** A centrifuge in a medical laboratory rotates at an angular speed of 3 600 rev/min. When switched off, it rotates through 50.0 revolutions before coming to rest. Find the constant angular acceleration of the centrifuge.

**5.** The tub of a washer goes into its spin cycle, starting from rest and gaining angular speed steadily for 8.00 s, at which time it is turning at 5.00 rev/s. At this point, the person doing the laundry opens the lid and a safety switch turns off the washer. The tub smoothly slows to rest in 12.0 s. Through how many revolutions does the tub turn while it is in motion?

**6.** A rotating wheel requires 3.00 s to rotate through 37.0 rev. Its angular speed at the end of the 3.00-s interval is 98.0 rad/s. What is the constant angular acceleration of the wheel?

**7.** (a) Find the angular speed of the Earth's rotation on its axis. As the Earth turns toward the east, we see the sky turning toward the west at this same rate.

(b) *The rainy Pleiads wester*
*And seek beyond the sea*
*The head that I shall dream of*
*That shall not dream of me.*
—A. E. Housman (© Robert E. Symons)

Cambridge, England, is at longitude 0° and Saskatoon, Saskatchewan, is at longitude 107° west. How much time elapses after the Pleiades set in Cambridge until these stars fall below the western horizon in Saskatoon?

## Section 10.3 ■ Relations Between Rotational and Translational Quantities

**8.** Make an order-of-magnitude estimate of the number of revolutions through which a typical automobile tire turns in 1 yr. State the quantities you measure or estimate and their values.

**9.** **PhysicsNow™** A disk 8.00 cm in radius rotates at a constant rate of 1 200 rev/min about its central axis. Determine (a) its angular speed, (b) the tangential speed at a point 3.00 cm from its center, (c) the radial acceleration of a point on the rim, and (d) the total distance a point on the rim moves in 2.00 s.

**10.** A wheel 2.00 m in diameter lies in a vertical plane and rotates with a constant angular acceleration of 4.00 rad/s$^2$. The wheel starts at rest at $t = 0$, and the radius vector of a certain point $P$ on the rim makes an angle of 57.3° with the horizontal at this time. At $t = 2.00$ s, find (a) the angular speed of the wheel, (b) the tangential speed and the total acceleration of the point $P$, and (c) the angular position of the point $P$.

**11.** Figure P10.11 shows the drivetrain of a bicycle that has wheels 67.3 cm in diameter and pedal cranks 17.5 cm long. The cyclist pedals at a steady cadence of 76.0 rev/min. The chain engages with a front sprocket 15.2 cm in diameter and a rear sprocket 7.00 cm in diameter. (a) Calculate the speed of a link of the chain relative to the bicycle frame. (b) Calculate the angular speed of the bicycle wheels. (c) Calculate the speed of the bicycle relative to the road. (d) What pieces of data, if any, are not necessary for the calculations?

**12.** A digital audio compact disc carries data, each bit of which occupies 0.6 $\mu$m along a continuous spiral track from the inner circumference of the disc to the outside edge. A CD player turns the disc to carry the track counterclockwise above a lens at a constant speed of 1.30 m/s. Find the required angular speed (a) at the beginning of the recording, where the spiral has a radius of 2.30 cm, and (b) at the end of the recording, where the spiral has a radius of 5.80 cm. (c) A full-length recording lasts for 74 min 33 s. Find the average angular acceleration of the disc. (d) Assuming that the acceleration is constant, find the total angular displacement of the disc as it plays. (e) Find the total length of the track.

FIGURE P10.11

**13.** A car traveling on a flat (unbanked) circular track accelerates uniformly from rest with a tangential acceleration of 1.70 m/s$^2$. The car makes it one fourth of the way around the circle before it skids off the track. Determine the coefficient of static friction between the car and track from these data.

## Section 10.4 ■ Rotational Kinetic Energy

**14.** Rigid rods of negligible mass lying along the $y$ axis connect three particles (Fig. P10.14). The system rotates about the $x$ axis with an angular speed of 2.00 rad/s. Find (a) the moment of inertia about the $x$ axis and the total rotational kinetic energy evaluated from $\frac{1}{2}I\omega^2$ and (b) the tangential speed of each particle and the total kinetic energy evaluated from $\Sigma \frac{1}{2}m_iv_i^2$.

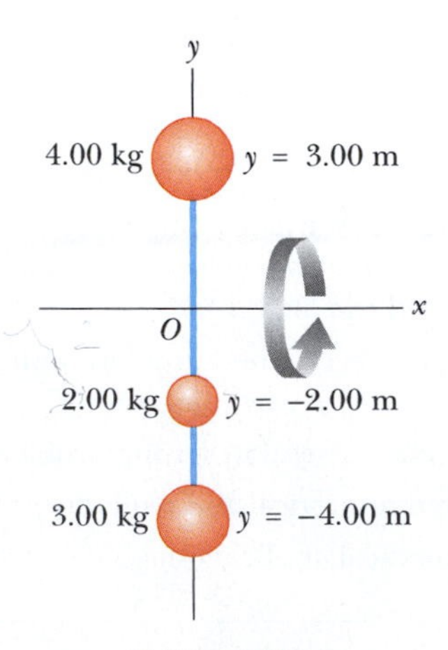


FIGURE P10.14

**15.** This problem describes one experimental method for determining the moment of inertia of an irregularly shaped object such as the payload for a satellite. Figure P10.15 shows a counterweight of mass $m$ suspended by a cord wound around a spool of radius $r$, forming part of a turntable supporting the object. The turntable can rotate without friction. When the counterweight is released from rest, it descends through a distance $h$, acquiring a speed $v$. Show that the moment of inertia $I$ of the rotating apparatus (including the turntable) is $mr^2(2gh/v^2 - 1)$.

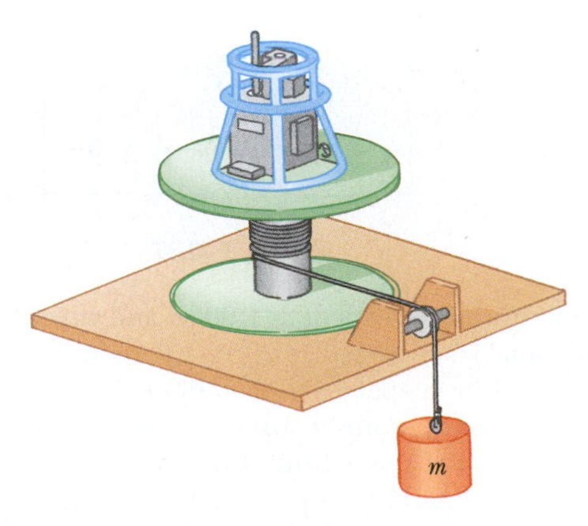


FIGURE **P10.15**

16. Big Ben, the Parliament tower clock in London, has an hour hand 2.70 m long with a mass of 60.0 kg and a minute hand 4.50 m long with a mass of 100 kg (Fig. P10.16). Calculate the total rotational kinetic energy of the two hands about the axis of rotation. (You may model the hands as uniform long, thin rods.)

FIGURE **P10.16** Problems 10.16, 10.42, and 10.64.

17. Consider two objects with $m_1 > m_2$ connected by a light string that passes over a pulley having a moment of inertia of $I$ about its axis of rotation as shown in Figure P10.17. The string does not slip on the pulley or stretch. The pulley turns without friction. The two objects are released from rest separated by a vertical distance $2h$. (a) Use the principle of conservation of energy to find the translational speeds of the objects as they pass each other. (b) Find the angular speed of the pulley at this time.

18. As a gasoline engine operates, a flywheel turning with the crankshaft stores energy after each fuel explosion, providing the energy required to compress the next charge of fuel and air. For the engine of a certain lawn tractor, suppose a flywheel must be no more than 18.0 cm in diameter. Its thickness, measured along its axis of rotation, must be no larger than 8.00 cm. The flywheel must release energy 60.0 J when its angular speed drops from 800 rev/min to 600 rev/min. Design a sturdy, steel flywheel to meet these requirements with the smallest mass that you can reasonably attain. Assume that the material has the density listed for iron in Table 15.1. Specify the shape and mass of the flywheel.

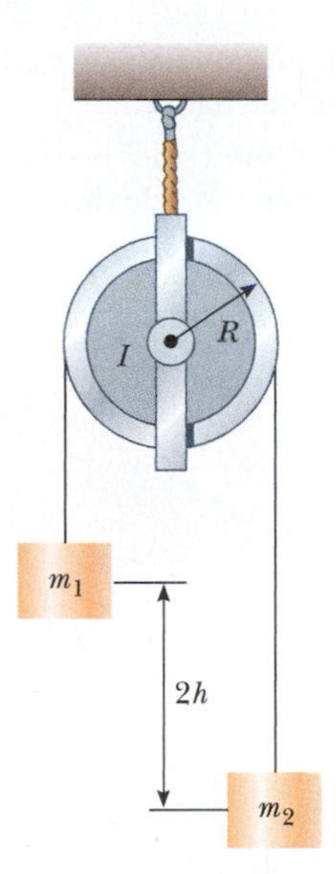


FIGURE **P10.17**

19. A *war-wolf* or *trebuchet* is a device used during the Middle Ages to throw rocks at castles and sometimes now used to fling pianos as a sport. A simple trebuchet is shown in Figure P10.19. Model it as a stiff rod of negligible mass, 3.00 m long, joining particles of mass 60.0 kg and 0.120 kg at its ends. It can turn on a frictionless horizontal axle perpendicular to the rod and 14.0 cm from the large-mass particle. The rod is released from rest in a horizontal orientation. Find the maximum speed that the small-mass object attains.

FIGURE **P10.19**

## Section 10.5 ▪ Torque and the Vector Product

20. The fishing pole in Figure P10.20 makes an angle of 20.0° with the horizontal. What is the torque exerted by the fish

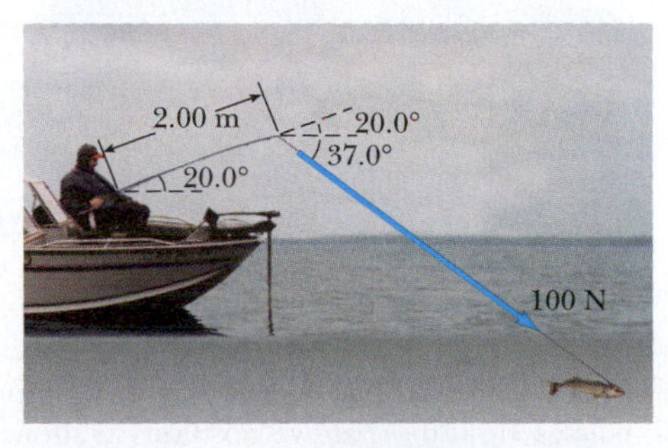


FIGURE **P10.20**

about an axis perpendicular to the page and passing through the angler's hand?

**21.** PhysicsNow™ Find the net torque on the wheel in Figure P10.21 about the axle through $O$, taking $a = 10.0$ cm and $b = 25.0$ cm.

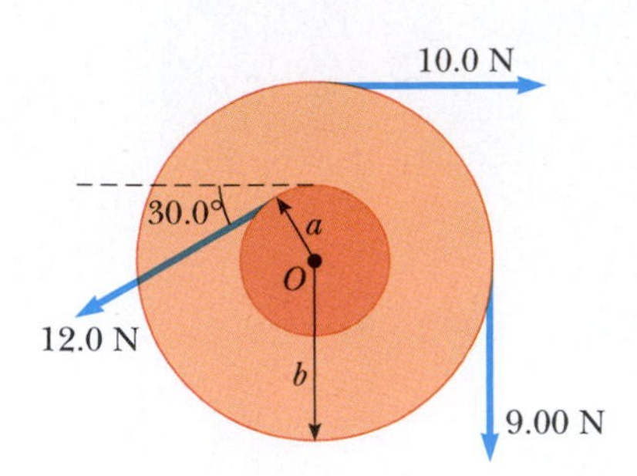


FIGURE **P10.21**

**22.** Given $\vec{\mathbf{M}} = 6\hat{\mathbf{i}} + 2\hat{\mathbf{j}} - \hat{\mathbf{k}}$ and $\vec{\mathbf{N}} = 2\hat{\mathbf{i}} - \hat{\mathbf{j}} - 3\hat{\mathbf{k}}$, calculate the vector product $\vec{\mathbf{M}} \times \vec{\mathbf{N}}$.

**23.** A force of $\vec{\mathbf{F}} = (2.00\hat{\mathbf{i}} + 3.00\hat{\mathbf{j}})$ N is applied to an object that is pivoted about a fixed axle aligned along the $z$ coordinate axis. The force is applied at the point $\vec{\mathbf{r}} = (4.00\hat{\mathbf{i}} + 5.00\hat{\mathbf{j}})$ m. Find (a) the magnitude of the net torque about the $z$ axis and (b) the direction of the torque vector $\vec{\boldsymbol{\tau}}$.

**24.** Two vectors are given by $\vec{\mathbf{A}} = -3\hat{\mathbf{i}} + 7\hat{\mathbf{j}} - 4\hat{\mathbf{k}}$ and $\vec{\mathbf{B}} = 6\hat{\mathbf{i}} - 10\hat{\mathbf{j}} + 9\hat{\mathbf{k}}$. Evaluate the following quantities: (a) $\cos^{-1}[\vec{\mathbf{A}} \cdot \vec{\mathbf{B}}/AB]$ and (b) $\sin^{-1}[|\vec{\mathbf{A}} \times \vec{\mathbf{B}}|/AB]$. (c) Which give(s) the angle between the vectors?

**25.** Use the definition of the vector product and the definitions of the unit vectors $\hat{\mathbf{i}}$, $\hat{\mathbf{j}}$, and $\hat{\mathbf{k}}$ to prove Equations 10.23. You may assume that the $x$ axis points to the right, the $y$ axis up, and the $z$ axis toward you (not away from you). This choice is said to make the coordinate system *right-handed.*

## Section 10.6 ▪ The Rigid Object in Equilibrium

**26.** In exercise physiology studies, it is sometimes important to determine the location of a person's center of mass, which can be done with the arrangement shown in Figure P10.26. A light plank rests on two scales, which read $F_{g1} = 380$ N and $F_{g2} = 320$ N. A distance of 2.00 m separates the scales. How far from the woman's feet is her center of mass?

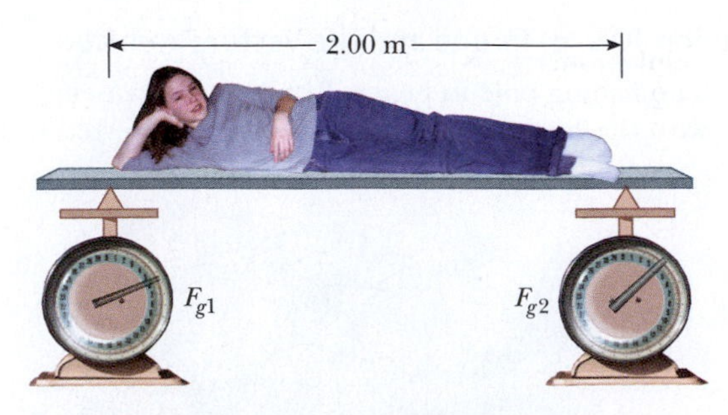


FIGURE **P10.26**

**27.** A uniform beam of mass $m_b$ and length $\ell$ supports blocks with masses $m_1$ and $m_2$ at two positions as shown in Figure P10.27. The beam rests on two knife edges. For what value of $x$ will the beam be balanced at $P$ such that the normal force at $O$ is zero?

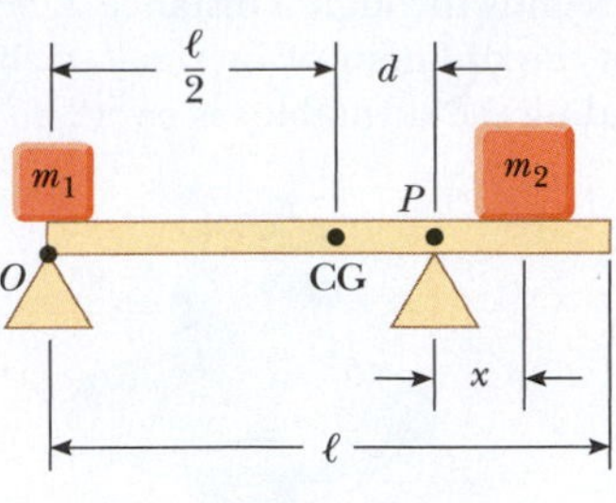


FIGURE **P10.27**

**28.** A uniform plank of length 6.00 m and mass 30.0 kg rests horizontally across two horizontal bars of a scaffold. The bars are 4.50 m apart, and 1.50 m of the plank hangs over one side of the scaffold. Draw a free-body diagram of the plank. How far can a painter of mass 70.0 kg walk on the overhanging part of the plank before it tips?

**29.** Figure P10.29 shows a claw hammer as it is being used to pull a nail out of a horizontal board. A force of 150 N is exerted horizontally as shown. Find (a) the force exerted by the hammer claws on the nail and (b) the force exerted by the surface on the point of contact with the hammer head. Assume that the force the hammer exerts on the nail is parallel to the nail.

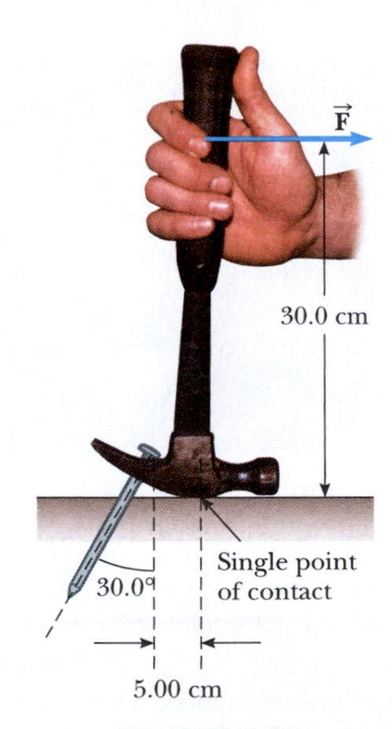


FIGURE **P10.29**

**30.** A uniform ladder of length $L$ and mass $m_1$ rests against a frictionless wall. The ladder makes an angle $\theta$ with the horizontal. (a) Find the horizontal and vertical forces the ground exerts on the base of the ladder when a firefighter of mass $m_2$ is a distance $x$ from the bottom. (b) If the ladder is just on the verge of slipping when the firefighter is a distance $d$ from the bottom, what is the coefficient of static friction between ladder and ground?

**31.** PhysicsNow™ A uniform sign of weight $F_g$ and width $2L$ hangs from a light, horizontal beam hinged at the wall and supported by a cable (Fig. P10.31). Determine (a) the tension in the cable and (b) the components of the reaction force exerted by the wall on the beam, in terms of $F_g$, $d$, $L$, and $\theta$.

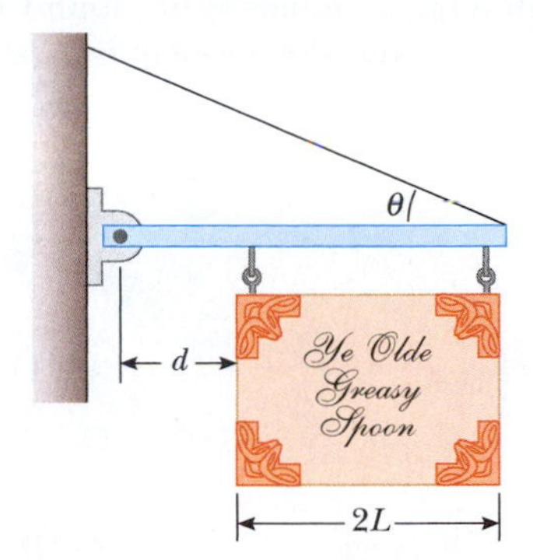


FIGURE P10.31

32. A crane of mass 3 000 kg supports a load of 10 000 kg as shown in Figure P10.32. The crane is pivoted with a frictionless pin at $A$ and rests against a smooth support at $B$. Find the reaction forces at $A$ and $B$.

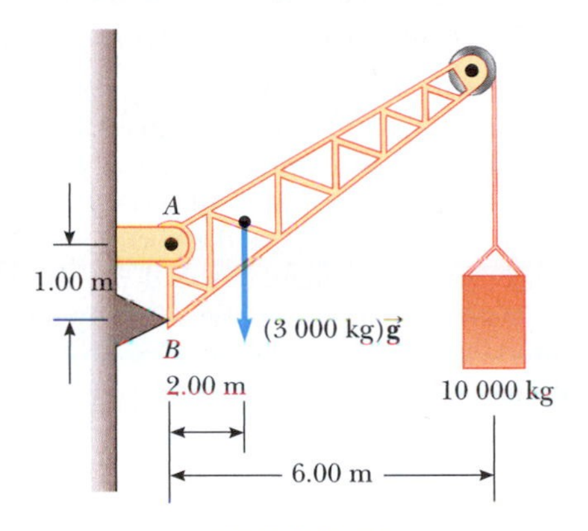


FIGURE P10.32

## Section 10.7 ■ The Rigid Object Under a Net Torque

33. The combination of an applied force and a friction force produces a constant total torque of 36.0 N·m on a wheel rotating about a fixed axis. The applied force acts for 6.00 s. During this time the angular speed of the wheel increases from 0 to 10.0 rad/s. The applied force is then removed, and the wheel comes to rest in 60.0 s. Find (a) the moment of inertia of the wheel, (b) the magnitude of the frictional torque, and (c) the total number of revolutions of the wheel.

34. A potter's wheel—a thick stone disk of radius 0.500 m and mass 100 kg—is freely rotating at 50.0 rev/min. The potter can stop the wheel in 6.00 s by pressing a wet rag against the rim and exerting a radially inward force of 70.0 N. Find the effective coefficient of kinetic friction between wheel and rag.

35. An electric motor turns a flywheel through a drive belt that joins a pulley on the motor and a pulley that is rigidly attached to the flywheel, as shown in Figure P10.35. The flywheel is a solid disk with a mass of 80.0 kg and a diameter of 1.25 m. It turns on a frictionless axle. Its pulley has much smaller mass and a radius of 0.230 m. The tension in the upper (taut) segment of the belt is 135 N, and the flywheel has a clockwise angular acceleration of 1.67 rad/s$^2$. Find the tension in the lower (slack) segment of the belt.

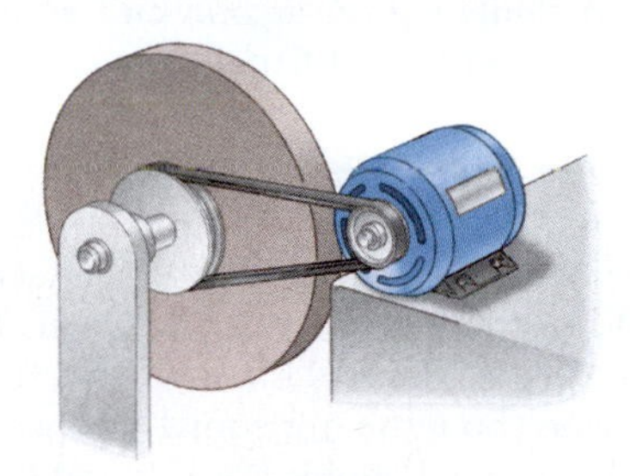

FIGURE P10.35

36. In Figure P10.36, the sliding block has a mass of 0.850 kg, the counterweight has a mass of 0.420 kg, and the pulley is a hollow cylinder with a mass of 0.350 kg, an inner radius of 0.020 0 m, and an outer radius of 0.030 0 m. The coefficient of kinetic friction between the block and the horizontal surface is 0.250. The pulley turns without friction on its axle. The light cord does not stretch and does not slip on the pulley. The block has a velocity of 0.820 m/s toward the pulley when it passes through a photogate. (a) Use energy methods to predict its speed after it has moved to a second photogate, 0.700 m away. (b) Find the angular speed of the pulley at the same moment.

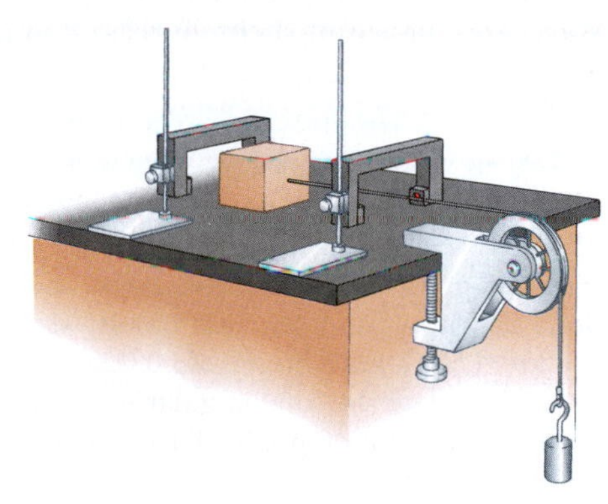

FIGURE P10.36

37. Two blocks, as shown in Figure P10.37, are connected by a string of negligible mass passing over a pulley of radius 0.250 m and moment of inertia $I$. The block on the frictionless incline is moving up with a constant acceleration of 2.00 m/s$^2$. (a) Determine $T_1$ and $T_2$, the tensions in the

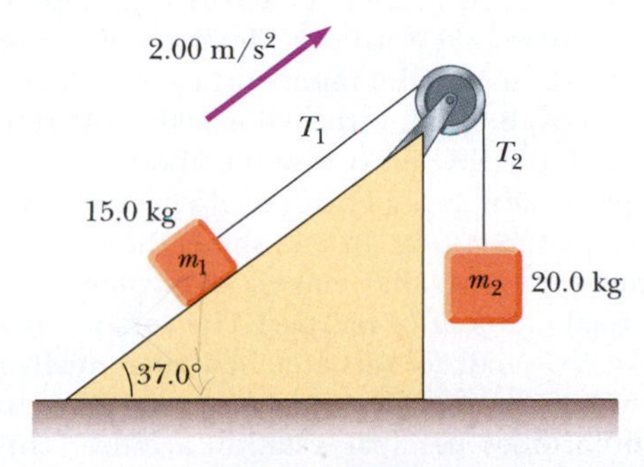


FIGURE P10.37

two parts of the string. (b) Find the moment of inertia of the pulley.

**38.** A uniform rod of length $L$ and mass $M$ is free to rotate about a frictionless pivot at one end as shown in Figure 10.9. The rod is released from rest in the horizontal position. What are the *initial* angular acceleration of the rod and the *initial* translational acceleration of the right end of the rod?

**39.** An object with a weight of 50.0 N is attached to the free end of a light string wrapped around a reel of radius 0.250 m and mass 3.00 kg. The reel is a solid disk, free to rotate in a vertical plane about the horizontal axis passing through its center. The suspended object is released 6.00 m above the floor. (a) Determine the tension in the string, the acceleration of the object, and the speed with which the object hits the floor. (b) Verify your last answer by using the principle of conservation of energy to find the speed with which the object hits the floor.

## Section 10.8 ■ Angular Momentum

**40.** Heading straight toward the summit of Pikes Peak, an airplane of mass 12 000 kg flies over the plains of Kansas at nearly constant altitude 4.30 km with constant velocity 175 m/s west. (a) What is the airplane's vector angular momentum relative to a wheat farmer on the ground directly below the airplane? (b) Does this value change as the airplane continues its motion along a straight line? (c) What is its angular momentum relative to the summit of Pikes Peak?

**41.** PhysicsNow™ The position vector of a particle of mass 2.00 kg is given as a function of time by $\vec{r} = (6.00\hat{i} + 5.00t\hat{j})$ m. Determine the angular momentum of the particle about the origin as a function of time.

**42.** Big Ben (Fig. P10.16), the Parliament tower clock in London, has hour and minute hands with lengths of 2.70 m and 4.50 m and masses of 60.0 kg and 100 kg, respectively. Calculate the total angular momentum of these hands about the center point. Treat the hands as long, thin, uniform rods.

**43.** A particle of mass 0.400 kg is attached to the 100-cm mark of a meter stick of mass 0.100 kg. The meter stick rotates on a horizontal, frictionless table with an angular speed of 4.00 rad/s. Calculate the angular momentum of the system when the stick is pivoted about an axis (a) perpendicular to the table through the 50.0-cm mark and (b) perpendicular to the table through the 0-cm mark.

**44.** A space station is constructed in the shape of a hollow ring of mass $5.00 \times 10^4$ kg. Members of the crew walk on a deck formed by the inner surface of the outer cylindrical wall of the ring, with radius 100 m. At rest when constructed, the ring is set rotating about its axis so that the people inside experience an effective free-fall acceleration equal to $g$. (Fig. P10.44 shows the ring together with some other parts that make a negligible contribution to the total moment of inertia.) The rotation is achieved by firing two small rockets attached tangentially to opposite points on the outside of the ring. (a) What angular momentum does the space station acquire? (b) How long must the rockets be fired if each exerts a thrust of 125 N? (c) Prove that the total torque on the ring, multiplied by the time interval found in part (b), is equal to the change in angular momentum, found in part (a). This equality represents the *angular impulse–angular momentum theorem.*

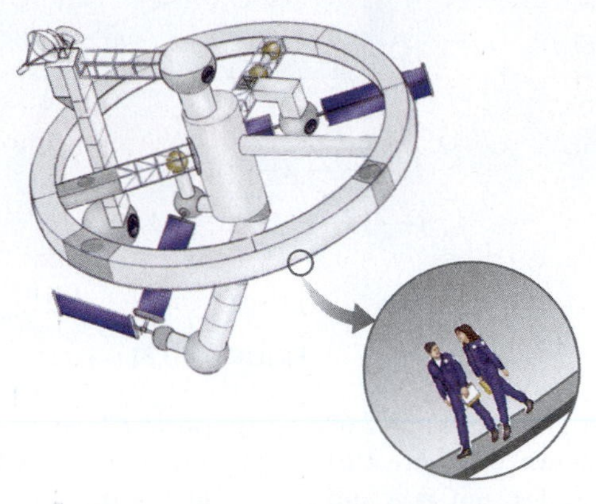

FIGURE P10.44 Problems 10.44 and 10.50.

## Section 10.9 ■ Conservation of Angular Momentum

**45.** A cylinder with moment of inertia $I_1$ rotates about a vertical, frictionless axle with angular speed $\omega_i$. A second cylinder, this one having moment of inertia $I_2$ and initially not rotating, drops onto the first cylinder (Fig. P10.45). Because of friction between the surfaces, the two eventually reach the same angular speed $\omega_f$. (a) Calculate $\omega_f$. (b) Show that the kinetic energy of the system decreases in this interaction and calculate the ratio of the final to the initial rotational energy.

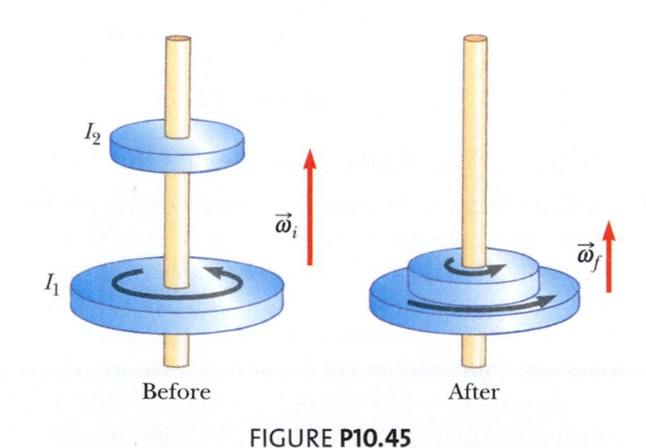


FIGURE P10.45

**46.** A playground merry-go-round of radius $R = 2.00$ m has a moment of inertia $I = 250$ kg·m$^2$ and is rotating at 10.0 rev/min about a frictionless vertical axle. Facing the axle, a 25.0-kg child hops onto the merry-go-round and manages to sit down on the edge. What is the new angular speed of the merry-go-round?

**47.** A 60.0-kg woman stands at the rim of a horizontal turntable having a moment of inertia of 500 kg·m$^2$ and a radius of 2.00 m. The turntable is initially at rest and is free to rotate about a frictionless, vertical axle through its center. The woman then starts walking around the rim clockwise (as viewed from above the system) at a constant speed of 1.50 m/s relative to the Earth. (a) In what direction and with what angular speed does the turntable rotate? (b) How much work does the woman do to set herself and the turntable into motion?

**48.** A student sits on a freely rotating stool holding two weights, each of mass 3.00 kg (Fig. P10.48). When his arms are extended horizontally, the weights are 1.00 m from the axis of rotation and he rotates with an angular speed of 0.750 rad/s. The moment of inertia of the student plus stool is 3.00 kg · m$^2$ and is assumed to be constant. The student pulls the weights inward horizontally to a position 0.300 m from the rotation axis. (a) Find the new angular speed of the student. (b) Find the kinetic energy of the rotating system before and after he pulls the weights inward.

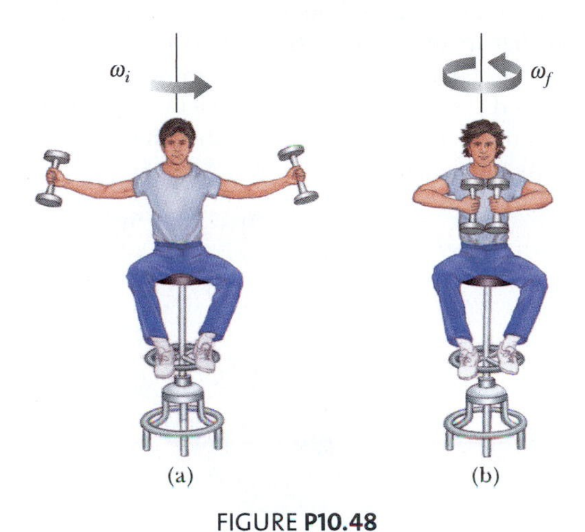


FIGURE **P10.48**

**49.** A puck of mass 80.0 g and radius 4.00 cm slides along an air table at a speed of 1.50 m/s as shown in Figure P10.49a. It makes a glancing collision with a second puck of radius 6.00 cm and mass 120 g (initially at rest) such that their rims just touch. Because their rims are coated with instant-acting glue, the pucks stick together and spin after the collision (Fig. P10.49b). (a) What is the angular momentum of the system relative to the center of mass? (b) What is the angular speed about the center of mass?

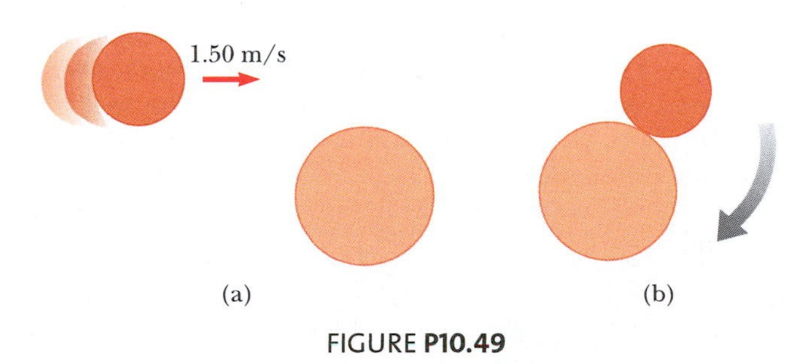


FIGURE **P10.49**

**50.** A space station shaped like a giant wheel has a radius of 100 m and a moment of inertia of $5.00 \times 10^8$ kg·m$^2$. A crew of 150 is living on the rim, and the station's rotation causes the crew to experience an apparent free-fall acceleration of $g$ (Fig. P10.44). When 100 people move to the center of the station for a union meeting, the angular speed changes. Assume that the average mass for each inhabitant is 65.0 kg. What apparent free-fall acceleration is experienced by the managers remaining at the rim?

**51.** The puck in Figure 10.24 has a mass of 0.120 kg. The distance of the puck from the center of rotation is originally 40.0 cm, and the puck is sliding with a speed of 80.0 cm/s. The string is pulled downward 15.0 cm through the hole in the frictionless table. Determine the work done on the puck. (*Suggestion:* Consider the change of kinetic energy.)

## Section 10.10 ■ Precessional Motion of Gyroscopes

**52.** The angular momentum vector of a precessing gyroscope sweeps out a cone as shown in Figure 10.25b. Its angular speed, called its precessional frequency, is given by $\omega_p = \tau/L$, where $\tau$ is the magnitude of the torque on the gyroscope and $L$ is the magnitude of its angular momentum. In the motion called *precession of the equinoxes,* represented in Figure P10.52, the Earth's axis of rotation precesses about the perpendicular to its orbital plane with a period of $2.58 \times 10^4$ yr. Model the Earth as a uniform sphere and calculate the torque on the Earth that is causing this precession.

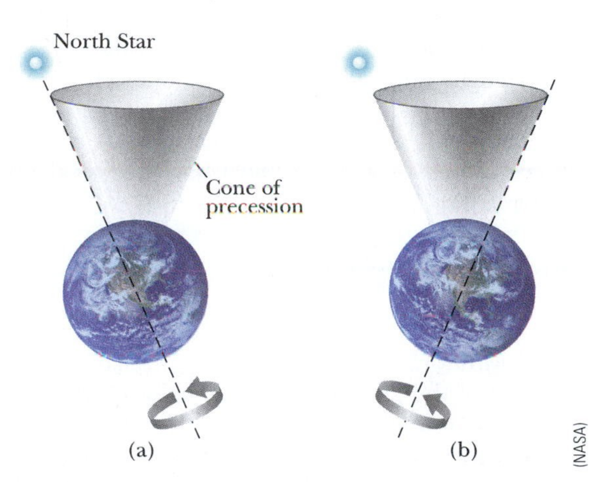


FIGURE **P10.52** (a) At present, the spin axis of the Earth points toward the North Star. (b) Torque on the spinning Earth will cause it to precess, so the spin axis will no longer be pointing in this direction in the future.

## Section 10.11 ■ Rolling Motion of Rigid Objects

**53.** A cylinder of mass 10.0 kg rolls without slipping on a horizontal surface. At a certain instant its center of mass has a speed of 10.0 m/s. Determine (a) the translational kinetic energy of its center of mass, (b) the rotational kinetic energy about its center of mass, and (c) its total energy.

**54.** A uniform solid disk and a uniform hoop are placed side by side at the top of an incline of height $h$. If they are released from rest and roll without slipping, which object reaches the bottom first? Verify your answer by calculating their speeds when they reach the bottom in terms of $h$.

**55.** A tennis ball is a hollow sphere with a thin wall. It is set rolling without slipping at 4.03 m/s on a horizontal section of a track as shown in Figure P10.55. It rolls around the inside of a vertical circular loop 90.0 cm in diameter and finally leaves the track at a point 20.0 cm below the horizontal section. (a) Find the speed of the ball at the top of the loop. Demonstrate that it will not fall from the track. (b) Find its speed as it leaves the track. (c) Suppose static friction between ball and track were negligible so that the ball slid instead of rolling. Would its speed then be higher, lower, or the same at the top of the loop? Explain.

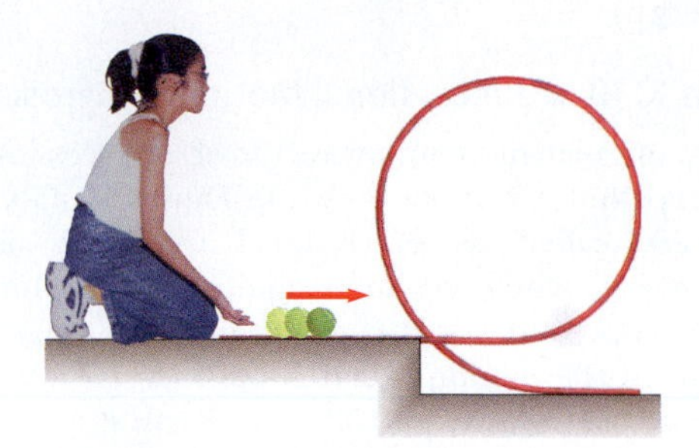

FIGURE **P10.55**

**56.** A metal can containing condensed mushroom soup has mass 215 g, height 10.8 cm, and diameter 6.38 cm. It is placed at rest on its side at the top of a 3.00-m-long incline that is at 25.0° to the horizontal and is then released to roll straight down. It takes 1.50 s to reach the bottom of the incline. Assuming mechanical energy conservation, calculate the moment of inertia of the can. Which pieces of data, if any, are unnecessary in calculating the solution?

## Section 10.12 ■ Context Connection—Turning the Spacecraft

**57.** A spacecraft is in empty space. It carries on board a gyroscope with a moment of inertia of $I_g = 20.0 \text{ kg}\cdot\text{m}^2$ about the axis of the gyroscope. The moment of inertia of the spacecraft around the same axis is $I_s = 5.00 \times 10^5 \text{ kg}\cdot\text{m}^2$. Neither the spacecraft nor the gyroscope is originally rotating. The gyroscope can be powered up in a negligible period of time to an angular speed of $100 \text{ s}^{-1}$. If the orientation of the spacecraft is to be changed by 30.0°, for how long should the gyroscope be operated?

## Additional Problems

**58.** **Review problem.** A mixing beater consists of three thin rods, each 10.0 cm long. The rods diverge from a central hub, separated from each other by 120°, and all turn in the same plane. A ball is attached to the end of each rod. Each ball has cross-sectional area $4.00 \text{ cm}^2$ and is so shaped that it has a drag coefficient of 0.600. Calculate the power input required to spin the beater at 1 000 rev/min (a) in air and (b) in water.

**59.** A long uniform rod of length $L$ and mass $M$ is pivoted about a horizontal, frictionless pin through one end. The rod is released from rest in a vertical position as shown in Figure P10.59. At the instant the rod is horizontal, find (a) its angular speed, (b) the magnitude of its angular acceleration, (c) the $x$ and $y$ components of the acceleration of its center of mass, and (d) the components of the reaction force at the pivot.

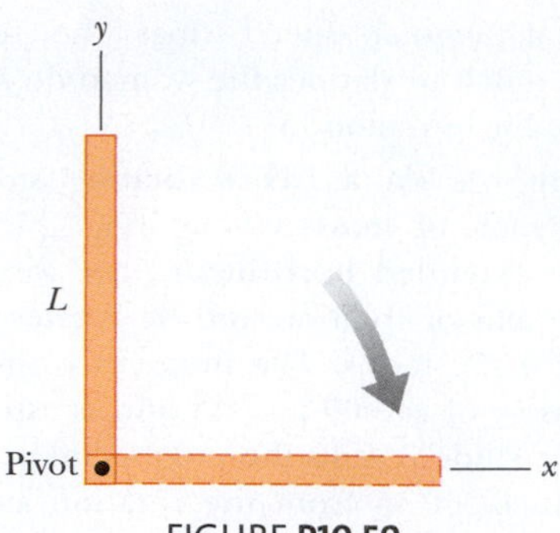


FIGURE **P10.59**

**60.** A uniform, hollow, cylindrical spool has inside radius $R/2$, outside radius $R$, and mass $M$ (Fig. P10.60). It is mounted so that it rotates on a fixed, horizontal axle. A counterweight of mass $m$ is connected to the end of a string wound around the spool. The counterweight falls from rest at $t = 0$ to a position $y$ at time $t$. Show that the torque due to the friction forces between spool and axle is

$$\tau_f = R\left[m\left(g - \frac{2y}{t^2}\right) - M\frac{5y}{4t^2}\right]$$

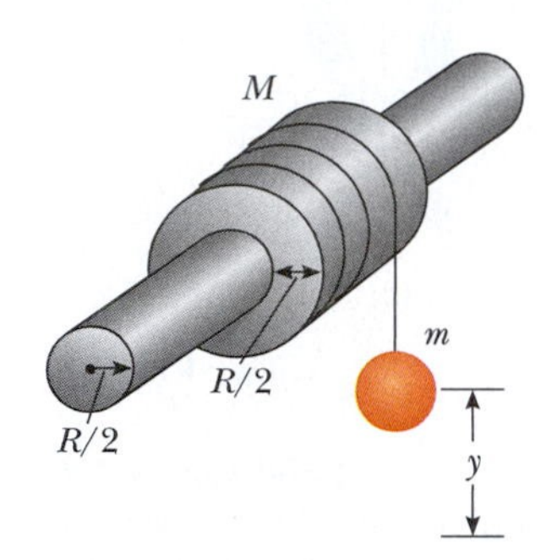


FIGURE **P10.60**

**61.** The reel shown in Figure P10.61 has radius $R$ and moment of inertia $I$. One end of the block of mass $m$ is connected to a spring of force constant $k$, and the other end is fastened to a cord wrapped around the reel. The reel axle and the incline are frictionless. The reel is wound counterclockwise so that the spring stretches a distance $d$ from its unstretched position and is then released from rest. (a) Find the angular speed of the reel when the spring is again unstretched. (b) Evaluate the angular speed numerically at this point taking $I = 1.00 \text{ kg}\cdot\text{m}^2$, $R = 0.300$ m, $k = 50.0$ N/m, $m = 0.500$ kg, $d = 0.200$ m, and $\theta = 37.0°$.

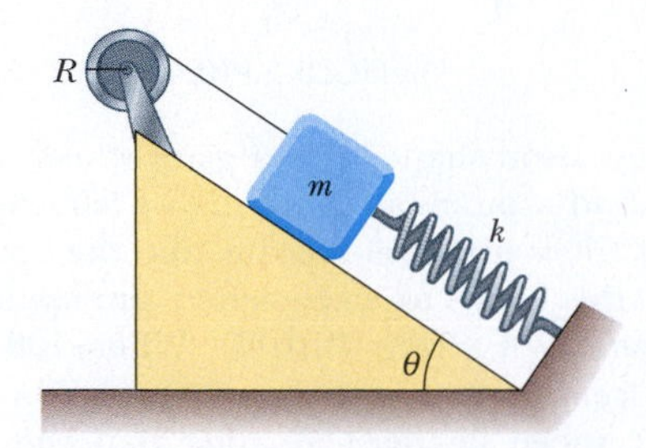


FIGURE **P10.61**

**62.** A block of mass $m_1 = 2.00$ kg and a block of mass $m_2 = 6.00$ kg are connected by a massless string over a pulley in the shape of a solid disk having radius $R = 0.250$ m and mass $M = 10.0$ kg. These blocks are allowed to move on a fixed block-wedge of angle $\theta = 30.0°$ as shown in Figure P10.62. The coefficient of kinetic friction is 0.360 for both blocks. Draw free-body diagrams of both blocks and of the pulley. Determine (a) the acceleration of the two blocks and (b) the tensions in the string on both sides of the pulley.

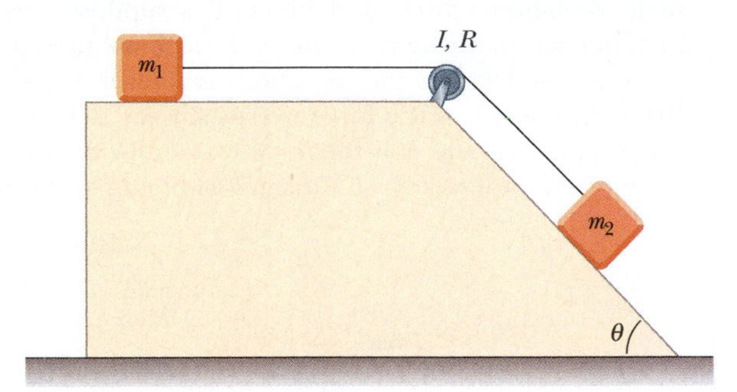


FIGURE **P10.62**

**63.** A common demonstration, illustrated in Figure P10.63, consists of a ball resting at one end of a uniform board of length $\ell$, hinged at the other end, and elevated at an angle $\theta$. A light cup is attached to the board at $r_c$ so that it will catch the ball when the support stick is suddenly removed. (a) Show that the ball will lag behind the falling board when $\theta$ is less than 35.3°. (b) Assume that the board is 1.00 m long and is supported at this limiting angle. Show that the cup must be 18.4 cm from the moving end.

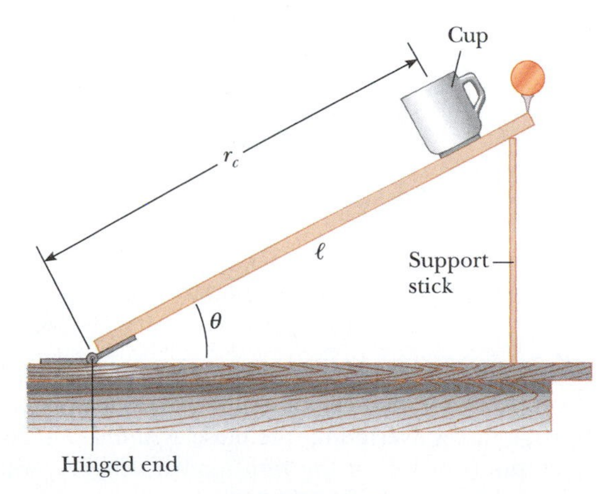


FIGURE **P10.63**

**64.** The hour hand and the minute hand of Big Ben, the Parliament tower clock in London, are 2.70 m and 4.50 m long and have masses of 60.0 kg and 100 kg, respectively (see Fig. P10.16). (a) Determine the total torque due to the weight of these hands about the axis of rotation when the time reads (i) 3:00, (ii) 5:15, (iii) 6:00, (iv) 8:20, and (v) 9:45. (You may model the hands as long, thin, uniform rods.) (b) Determine all times when the total torque about the axis of rotation is zero. Determine the times to the nearest second, solving a transcendental equation numerically.

**65.** A string is wound around a uniform disk of radius $R$ and mass $M$. The disk is released from rest with the string vertical and its top end tied to a fixed bar (Fig. P10.65). Show that (a) the tension in the string is one-third the weight of the disk, (b) the magnitude of the acceleration of the center of mass is $2g/3$, and (c) the speed of the center of mass is $(4gh/3)^{1/2}$ after the disk has descended through distance $h$. Verify your answer to (c) using the energy approach.

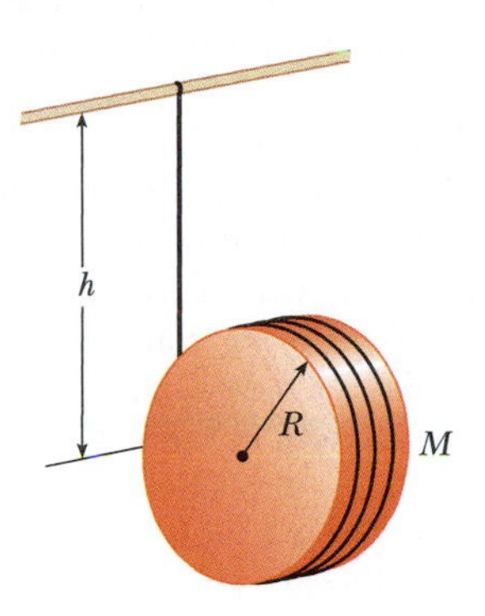


FIGURE **P10.65**

**66.** A new General Electric stove has a mass of 68.0 kg and the dimensions shown in Figure P10.66. The stove comes with a warning that it can tip forward if a person stands or sits on the oven door when it is open. What can you conclude about the weight of such a person? Could it be a child? List the assumptions you make in solving this problem. The stove is supplied with a wall bracket to prevent the accident.

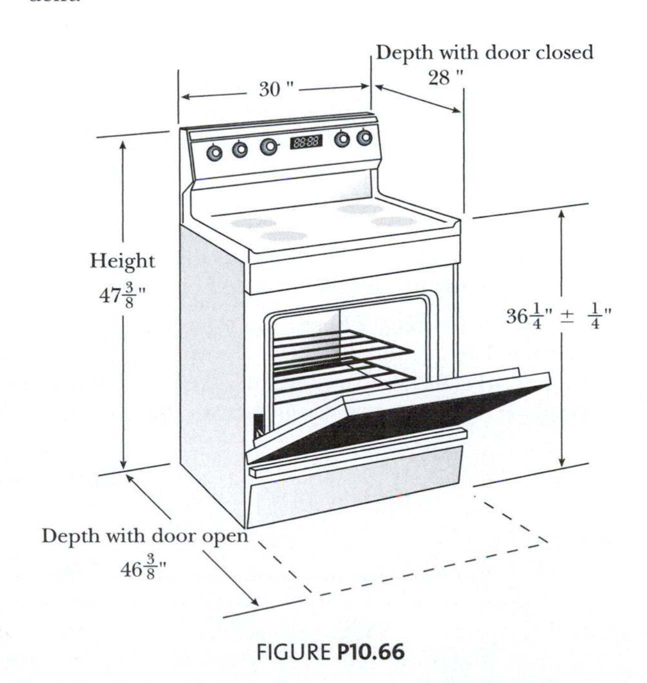


FIGURE **P10.66**

**67.** (a) Without the wheels, a bicycle frame has a mass of 8.44 kg. Each of the wheels can be roughly modeled as a uniform solid disk with a mass of 0.820 kg and a radius of 0.343 m. Find the kinetic energy of the whole bicycle when it is moving forward at 3.35 m/s. (b) Before the invention of a wheel turning on an axle, ancient people moved heavy loads by placing rollers under them. (Modern people use rollers, too. Any hardware store will sell you a roller bearing for a lazy Susan.) A stone block of mass 844 kg moves forward at 0.335 m/s, supported by two uniform cylindrical tree trunks each of mass 82.0 kg and radius 0.343 m. No slipping occurs between the block and the rollers or between the rollers and the ground. Find the total kinetic energy of the moving objects.

**68.** A skateboarder with his board can be modeled as a particle of mass 76.0 kg, located at his center of mass. As shown in Figure P7.59 on page 218, the skateboarder starts from rest in a crouching position at one lip of a half-pipe (point Ⓐ). The half-pipe forms one half of a cylinder of radius 6.80 m with its axis horizontal. On his descent, the skateboarder moves without friction and maintains his crouch so that his center of mass moves through one quarter of a circle of radius 6.30 m. (a) Find his speed at the bottom of the half-pipe (point Ⓑ). (b) Find his angular momentum about the center of curvature. (c) Immediately after passing point Ⓑ, he stands up and raises his arms, lifting his center of gravity from 0.500 m to 0.950 m above the concrete (point Ⓒ). Explain why his angular momentum is constant in this maneuver, whereas his linear momentum and his mechanical energy are not constant. (d) Find his speed immediately after he stands up, when his center of mass is moving in a quarter circle of radius 5.85 m. (e) What work did the skateboarder's legs do on his body as he stood up? Next, the skateboarder glides upward with his center of mass moving in a quarter circle of radius 5.85 m. His body is horizontal when he passes point Ⓓ, the far lip of the half-pipe. (f) Find his speed at this location. At last he goes ballistic, twisting around while his center of mass moves vertically. (g) How high above point Ⓓ does he rise? (h) Over what time interval is he airborne before he touches down, facing downward and again in a crouch, 2.34 m below the level of point Ⓓ? (i) Compare the solution to this problem with the solution to Problem 7.59. Which is more accurate? Why? (*Caution:* Do not try this maneuver yourself without the required skill and protective equipment, or in a drainage channel to which you do not have legal access.)

**69.** Two astronauts (Fig. P10.69), each having a mass $M$, are connected by a rope of length $d$ having negligible mass. They are isolated in space, orbiting their center of mass at speeds $v$. Treating the astronauts as particles, calculate (a) the magnitude of the angular momentum of the system and (b) the rotational energy of the system. By pulling on the rope, one of the astronauts shortens the distance between them to $d/2$. (c) What is the new angular momentum of the system? (d) What are the astronauts' new speeds? (e) What is the new rotational energy of the system? (f) How much work does the astronaut do in shortening the rope?

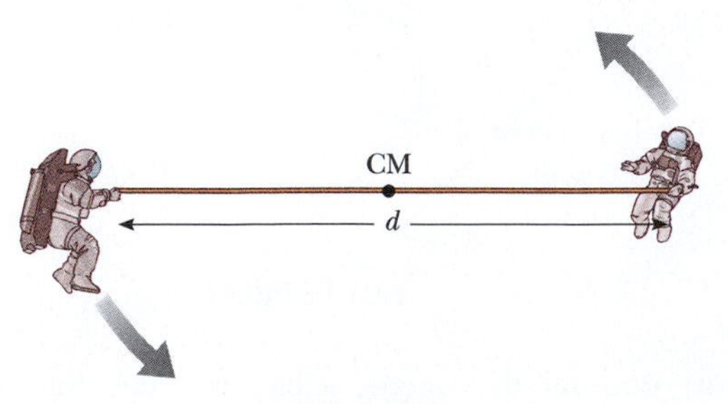


FIGURE P10.69

**70.** When a person stands on tiptoe (a strenuous position), the position of the foot is as shown in Figure P10.70a. The total gravitational force on the body $\vec{\mathbf{F}}_g$ is supported by the force $\vec{\mathbf{n}}$ exerted by the floor on the toes of one foot. A mechanical model for the situation is shown in Figure P10.70b, where $\vec{\mathbf{T}}$ is the force exerted by the Achilles tendon on the foot and $\vec{\mathbf{R}}$ is the force exerted by the tibia on the foot. Find the values of $T$, $R$, and $\theta$ when $F_g = 700$ N.

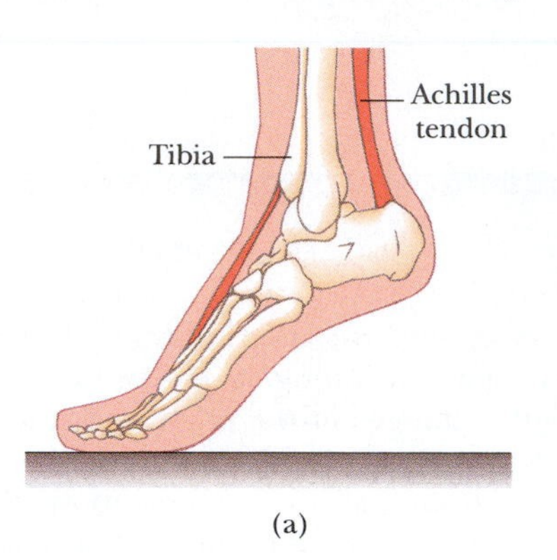


(a)

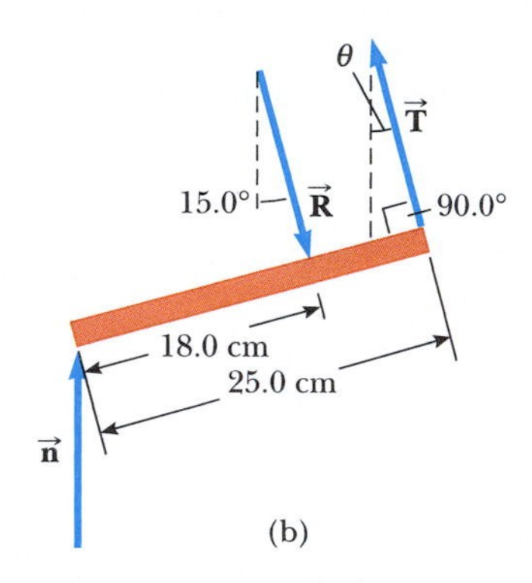


(b)

FIGURE P10.70

**71.** A person bending forward to lift a load "with his back" (Fig. P10.71a) rather than "with his knees" can be injured by large forces exerted on the muscles and vertebrae. The spine pivots mainly at the fifth lumbar vertebra, with the principal supporting force provided by the erector spinalis muscle in the back. To see the magnitude of the forces involved and to understand why back problems are common among humans, consider the model shown in Fig. P10.71b for a person bending forward to lift a 200-N object. The spine and upper body are represented as a uniform horizontal rod of weight 350 N, pivoted at the base of the spine. The erector spinalis muscle, attached at a point two thirds of the way up the spine, maintains the position of the back. The angle between the spine and this muscle is 12.0°. Find the tension in the back muscle and the compressional force in the spine.

FIGURE **P10.71**

**72.** A wad of sticky clay with mass $m$ and velocity $\vec{v}_i$ is fired at a solid cylinder of mass $M$ and radius $R$ (Fig. P10.72). The cylinder is initially at rest and is mounted on a fixed horizontal axle that runs through its center of mass. The line of motion of the projectile is perpendicular to the axle and at a distance $d < R$ from the center. (a) Find the angular speed of the system just after the clay strikes and sticks to the surface of the cylinder. (b) Is mechanical energy of the clay–cylinder system conserved in this process? Explain your answer.

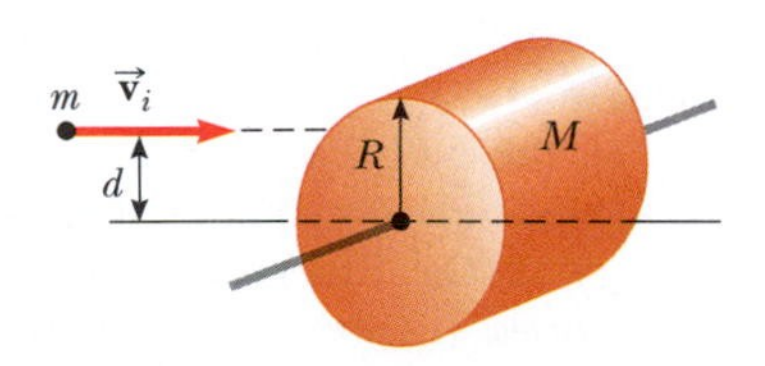


FIGURE **P10.72**

**73.** A force acts on a rectangular cabinet weighing 400 N as shown in Figure P10.73. (a) Assuming that the cabinet slides with constant speed when $F = 200$ N and $h = 0.400$ m, find the coefficient of kinetic friction and the position of the resultant normal force. (b) Taking $F = 300$ N, find the value of $h$ for which the cabinet just begins to tip.

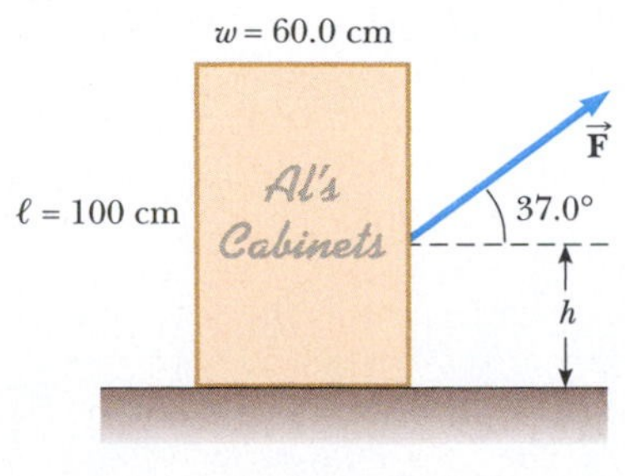


FIGURE **P10.73**

**74.** The following equations are obtained from a free-body diagram of a rectangular farm gate, supported by two hinges on the left-hand side. A bucket of grain is hanging from the latch.

$$-A + C = 0$$

$$+B - 392\text{ N} - 50.0\text{ N} = 0$$

$$A(0) + B(0) + C(1.80\text{ m}) - 392\text{ N}(1.50\text{ m}) - 50.0\text{ N}(3.00\text{ m}) = 0$$

(a) Draw the free-body diagram and complete the statement of the problem, specifying the unknowns. (b) Determine the values of the unknowns and state the physical meaning of each.

**75.** A stepladder of negligible weight is constructed as shown in Figure P10.75. A painter of mass 70.0 kg stands on the ladder 3.00 m from the bottom. Assuming that the floor is frictionless, find (a) the tension in the horizontal bar connecting the two halves of the ladder, (b) the normal forces at $A$ and $B$, and (c) the components of the reaction force at the single hinge $C$ that the left half of the ladder exerts on the right half. (*Suggestion:* Treat the ladder as a single object, but also treat each half of the ladder separately.)

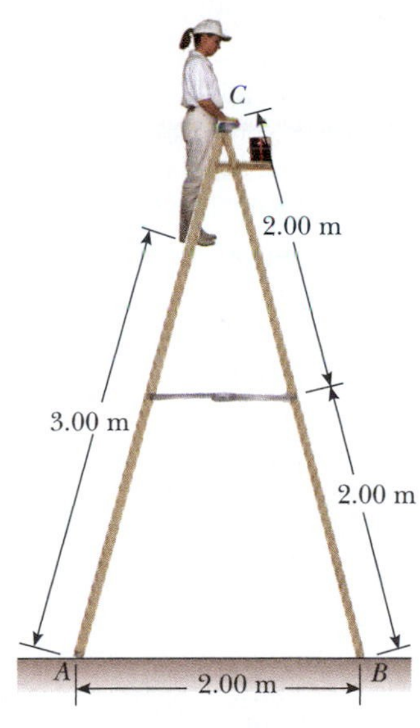


FIGURE **P10.75**

**76.** A solid sphere of mass $m$ and radius $r$ rolls without slipping along the track shown in Figure P10.76. It starts from rest with the lowest point of the sphere at height $h$ above the

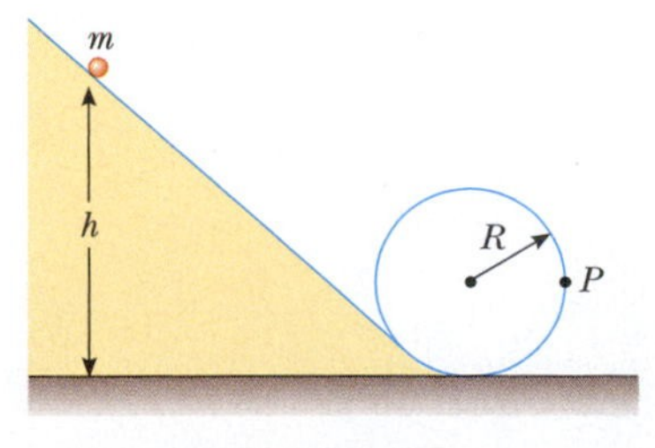


FIGURE **P10.76**

bottom of the loop of radius $R$, much larger than $r$. (a) What is the minimum value of $h$ (in terms of $R$) such that the sphere completes the loop? (b) What are the force components on the sphere at the point $P$ if $h = 3R$?

**77.** Figure P10.77 shows a vertical force applied tangentially to a uniform cylinder of weight $F_g$. The coefficient of static friction between the cylinder and all surfaces is 0.500. In terms of $F_g$, find the maximum force $P$ that can be applied that does not cause the cylinder to rotate. (*Suggestion:* When the cylinder is on the verge of slipping, both friction forces are at their maximum values. Why?)

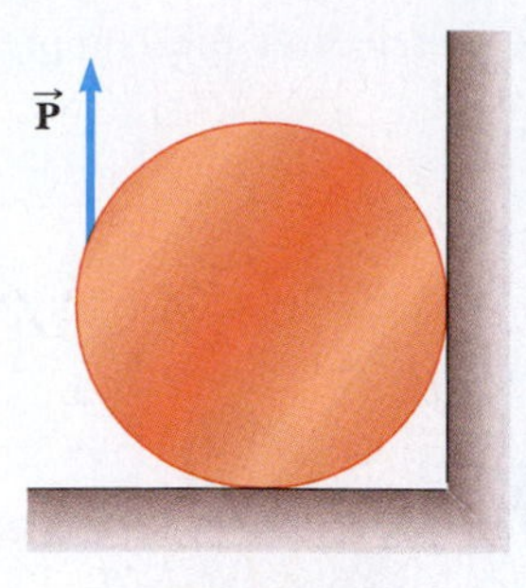


FIGURE **P10.77**

## ANSWERS TO QUICK QUIZZES

**10.1** (i), (c). For a rotation of more than 180°, the angular displacement must be larger than $\pi = 3.14$ rad. The angular displacements in the three choices are (a) 6 rad − 3 rad = 3 rad, (b) 1 rad − (−1) rad = 2 rad, and (c) 5 rad − 1 rad = 4 rad. (ii), (b). Because all angular displacements occur in the same time interval, the displacement with the lowest value will be associated with the lowest average angular speed.

**10.2** (b). In Equation 10.8, both the initial and final angular speeds are the same in all three cases. As a result, the angular acceleration is inversely proportional to the angular displacement. Therefore, the highest angular acceleration is associated with the lowest angular displacement.

**10.3** (i), (b). The system of the platform, Benjamin, and Torrey is a rigid object, so all points on the rigid object have the same angular speed. (ii), (a). The tangential speed is proportional to the radial distance from the rotation axis.

**10.4** (a). Almost all the mass of the pipe is at the same distance from the rotation axis, so it has a larger moment of inertia than the solid cylinder.

**10.5** (b), (c). The fatter handle of the screwdriver gives you a larger moment arm and increases the torque that you can apply with a given force from your hand. The longer handle of the wrench gives you a larger moment arm and increases the torque that you can apply with a given force from your hand.

**10.6** (b). With twice the moment of inertia and the same frictional torque, there is half the angular acceleration. With half the angular acceleration, it will require twice as long to change the speed to zero.

**10.7** (b). The hollow sphere has a larger moment of inertia than the solid sphere because much of its mass is far from the rotation axis. Because $L = I\omega$ and $\omega$ is the same for both objects, the hollow sphere has a larger angular momentum.

**10.8** (i), (a). The diver is an isolated system, so the product $I\omega$ remains constant. As the moment of inertia of the diver decreases, the angular speed increases by the same factor. For example, if $I$ goes down by a factor of 2, $\omega$ goes up by a factor of 2. (ii), (a). The rotational kinetic energy varies as the square of $\omega$. If $I$ is halved, $\omega^2$ increases by a factor of 4 and the energy increases by a factor of 2.

**10.9** (i), (b). All the gravitational potential energy of the box–Earth system is transformed to kinetic energy of translation. For the ball, some of the gravitational potential energy of the ball–Earth system is transformed to rotational kinetic energy, leaving less for translational kinetic energy, so the ball moves downhill more slowly than the box does. (ii), (c). In Equation 10.46, $I_{CM}$ for a sphere is $\frac{2}{5}MR^2$. Therefore, $MR^2$ will cancel and the remaining expression on the right-hand side of the equation is independent of mass and radius. (iii), (a). The moment of inertia of the hollow sphere B is larger than that of sphere A. As a result, Equation 10.46 tells us that sphere B will have a smaller speed of the center of mass, so sphere A should arrive first.

CHAPTER 11

# Gravity, Planetary Orbits, and the Hydrogen Atom

The Rosette Nebula is a region of gas and dust surrounding an open cluster of stars. Bundles of matter in the Universe such as this one interact with other bundles of matter by means of the gravitational force. The red color is due to hydrogen atoms, excited by light from the stars in the cluster, making transitions from the $n = 3$ quantum state to the $n = 2$ state. In this chapter, we will study both the gravitational force and the origin of the red color in the hydrogen atoms.

## CHAPTER OUTLINE

11.1 Newton's Law of Universal Gravitation Revisited
11.2 Structural Models
11.3 Kepler's Laws
11.4 Energy Considerations in Planetary and Satellite Motion
11.5 Atomic Spectra and the Bohr Theory of Hydrogen
11.6 Context Connection—Changing From a Circular to an Elliptical Orbit
SUMMARY

At the beginning of our discussion of mechanics in Chapter 1, we introduced the notion of modeling and defined four categories of models: geometric, simplification, analysis, and structural. We can apply our analysis models to two very common *structural models.* In this chapter we shall discuss a structural model for a large system—the Solar System—and a structural model for a small system—the hydrogen atom.

We return to Newton's law of universal gravitation—one of the fundamental force laws in nature—and show how it, together with our analysis models, enables us to understand the motions of planets, moons, and artificial Earth satellites.

We conclude this chapter with a discussion of Niels Bohr's model of the hydrogen atom, which represents an interesting mixture of classical and nonclassical physics. Despite the hybrid nature of the model, some of its predictions agree with experimental measurements made on hydrogen atoms. This discussion will be our first major venture into the area of *quantum physics,* which we will continue in Chapter 28.

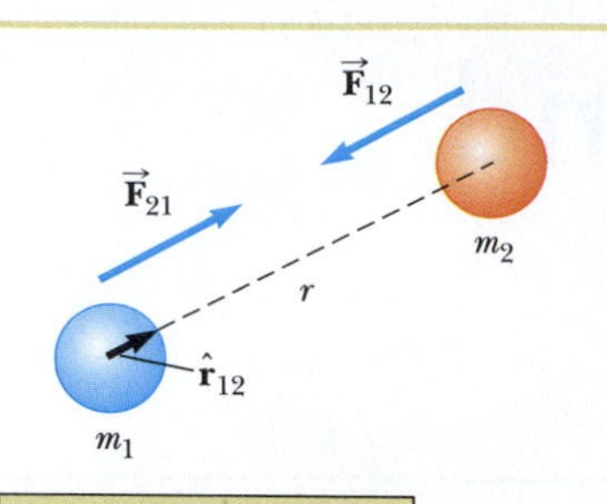


ACTIVE FIGURE 11.1 The gravitational force between two particles is attractive. The unit vector $\hat{\mathbf{r}}_{12}$ is directed from particle 1 toward particle 2. Note that $\vec{\mathbf{F}}_{21} = -\vec{\mathbf{F}}_{12}$.

**Physics Now™** By logging into PhysicsNow at **www.pop4e.com** and going to Active Figure 11.1 you can change the separation distance between the particles to see the effect on the gravitational force.

## 11.1 NEWTON'S LAW OF UNIVERSAL GRAVITATION REVISITED

Prior to 1686, many data had been collected on the motions of the Moon and the planets, but a clear understanding of the forces involved with the motions was not yet attainable. In that year, Isaac Newton provided the key that unlocked the secrets of the heavens. He knew, from the first law of motion, that a net force had to be acting on the Moon. If not, the Moon would move in a straight-line path rather than in its almost circular orbit. Newton reasoned that this force between the Moon and the Earth was an attractive force. He also concluded that there could be nothing special about the Earth–Moon system or the Sun and its planets that would cause gravitational forces to act on them alone.

As you should recall from Chapter 5, every particle in the Universe attracts every other particle with a force that is directly proportional to the product of their masses and inversely proportional to the square of the distance between them. If two particles have masses $m_1$ and $m_2$ and are separated by a distance $r$, the magnitude of the gravitational force between them is

$$F_g = G\frac{m_1 m_2}{r^2} \qquad [11.1]$$

where $G$ is the **universal gravitational constant** whose value in SI units is

$$G = 6.673 \times 10^{-11}\ \text{N}\cdot\text{m}^2/\text{kg}^2 \qquad [11.2]$$

**PITFALL PREVENTION 11.1**

**Be clear on $g$ and $G$** Be sure you understand the difference between $g$ and $G$. The symbol $g$ represents the magnitude of the free-fall acceleration near a planet. At the surface of the Earth, $g$ has the value $9.80\ \text{m/s}^2$. On the other hand, $G$ is a universal constant that has the same value everywhere in the Universe.

The force law given by Equation 11.1 is often referred to as an **inverse-square law** because the magnitude of the force varies as the inverse square of the separation of the particles. We can express this attractive force in vector form by defining a unit vector $\hat{\mathbf{r}}_{12}$ directed from $m_1$ toward $m_2$ as shown in Active Figure 11.1. The force exerted by $m_1$ on $m_2$ is

$$\vec{\mathbf{F}}_{12} = -G\frac{m_1 m_2}{r^2}\hat{\mathbf{r}}_{12} \qquad [11.3]$$

where the negative sign indicates that particle 1 is attracted toward particle 2. Likewise, by Newton's third law, the force exerted by $m_2$ on $m_1$, designated $\vec{\mathbf{F}}_{21}$, is equal in magnitude to $\vec{\mathbf{F}}_{12}$ and in the opposite direction. That is, these forces form an action–reaction pair, and $\vec{\mathbf{F}}_{21} = -\vec{\mathbf{F}}_{12}$.

As Newton demonstrated, **the gravitational force exerted by a finite-sized, spherically symmetric mass distribution on a particle outside the distribution is the same as if the entire mass of the distribution were concentrated at its center.** For example, the force on a particle of mass $m$ at the Earth's surface has the magnitude

$$F_g = G\frac{M_E m}{R_E^2}$$

where $M_E$ is the Earth's mass and $R_E$ is the Earth's radius. This force is directed toward the center of the Earth.

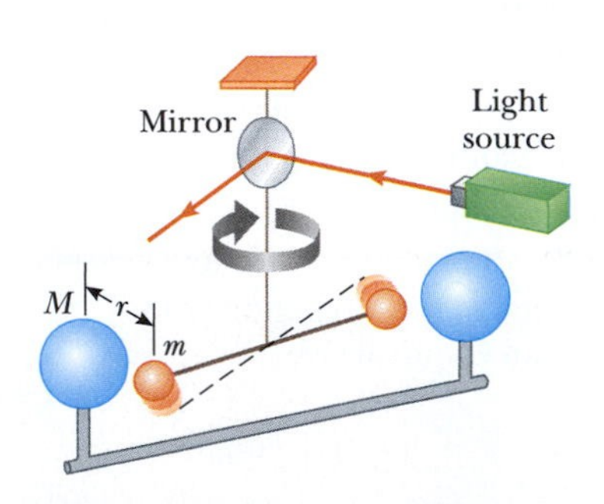


FIGURE 11.2 Schematic diagram of the Cavendish apparatus for measuring $G$. As the small spheres of mass $m$ are attracted to the large spheres of mass $M$, the rod rotates through a small angle. A light beam reflected from a mirror on the rotating apparatus measures the angle of rotation. The dashed line represents the original position of the rod. (In reality, the length of wire above the mirror is much larger than that below it.)

### Measurement of the Gravitational Constant

The universal gravitational constant $G$ was first measured in an important experiment by Sir Henry Cavendish in 1798. The apparatus he used consists of two small spheres, each of mass $m$, fixed to the ends of a light horizontal rod suspended by a thin wire as in Figure 11.2. Two large spheres, each of mass $M$, are then placed near the smaller spheres. The attractive force between the smaller and larger spheres causes the rod to rotate and twist the wire. If the system is oriented as shown in Figure 11.2, the rod rotates clockwise when viewed from the top. The angle through which it rotates is measured by the deflection of a light beam that is reflected from a mirror attached to the wire. The experiment is carefully repeated with different masses at various separations. In addition to providing a value for $G$, the results confirm that the force is attractive, proportional to the product $mM$, and inversely proportional to the square of the distance $r$.

It is interesting that $G$ is the least well known of the fundamental constants, with a percentage uncertainty thousands of times larger than those for other constants such as the speed of light $c$ and the fundamental electric charge $e$. Several measurements of $G$ made in the 1990s varied significantly from the previous value and from one another! The search for a more precise value of $G$ continues to be an area of active research.

**QUICK QUIZ 11.1** A planet has two moons of equal mass. Moon 1 is in a circular orbit of radius $r$. Moon 2 is in a circular orbit of radius $2r$. What is the magnitude of the gravitational force exerted by the planet on Moon 2? **(a)** four times as large as that on Moon 1 **(b)** twice as large as that on Moon 1 **(c)** equal to that on Moon 1 **(d)** half as large as that on Moon 1 **(e)** one-fourth as large as that on Moon 1

## Thinking Physics 11.1

The novel *Icebound*, by Dean Koontz (Bantam Books, 2000), is a story of a group of scientists trapped on a floating iceberg near the North Pole. One of the devices the scientists have with them is a transmitter with which they can fix their position with "the aid of a geosynchronous polar satellite." Can a satellite in a *polar* orbit be *geosynchronous*?

**Reasoning** A geosynchronous satellite is one that stays over one location on the Earth's surface at all times. Therefore, an antenna on the surface that receives signals from the satellite, such as a television dish, can stay pointed in a fixed direction toward the sky. The satellite must be in an orbit with the correct radius such that its orbital period is the same as that of the Earth's rotation. This orbit results in the satellite appearing to have no east–west motion relative to the observer at the chosen location. Another requirement is that a geosynchronous satellite *must be in orbit over the equator*. Otherwise it would appear to undergo a north–south oscillation during one orbit. Therefore, it would be impossible to have a geosynchronous satellite in a *polar* orbit. Even if such a satellite were at the proper distance from the Earth, it would be moving rapidly in the north–south direction, resulting in the necessity of accurate tracking equipment. What's more, it would be below the horizon for long periods of time, making it useless for determining one's position. ■

## The Gravitational Field

When Newton first published his theory of gravitation, his contemporaries found it difficult to accept the concept of a force that one object could exert on another without anything happening in the space between them. They asked how it was possible for two objects with mass to interact even though they were not in contact with each other. Although Newton himself could not answer this question, his theory was considered a success because it satisfactorily explained the motions of the planets.

An alternative mental representation of the gravitational force is to think of the gravitational interaction as a two-step process involving a *field*, as discussed in Section 4.1. First, one object (a *source mass*) creates a **gravitational field** $\vec{\mathbf{g}}$ throughout the space around it. Then, a second object (a *test mass*) of mass $m$ residing in this field experiences a force $\vec{\mathbf{F}}_g = m\vec{\mathbf{g}}$. In other words, we model the *field* as exerting a force on the test mass rather than the source mass exerting the force directly. The gravitational field is defined by

$$\vec{\mathbf{g}} \equiv \frac{\vec{\mathbf{F}}_g}{m} \qquad [11.4]$$

■ Gravitational field

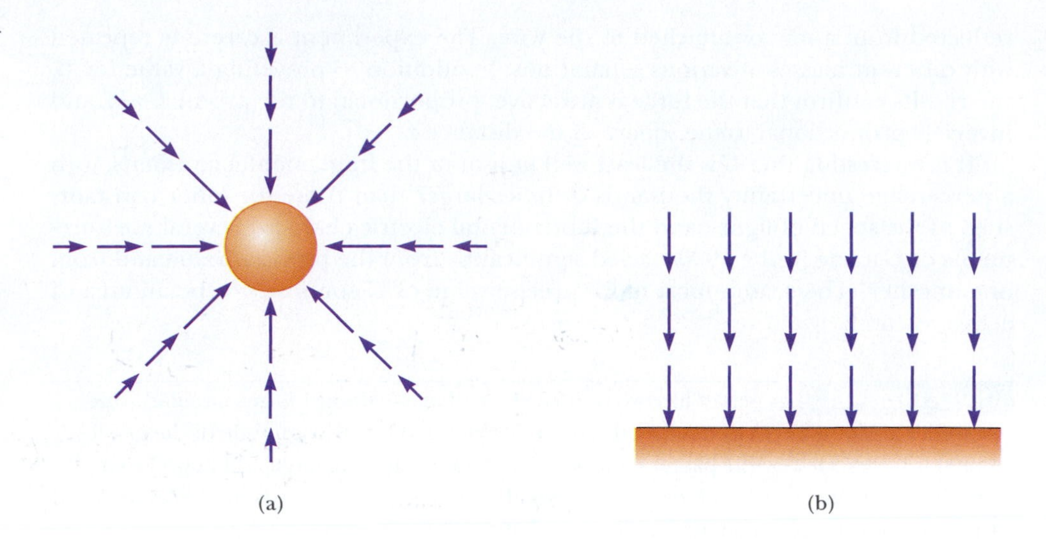


**FIGURE 11.3** (a) The gravitational field vectors in the vicinity of a uniform spherical mass vary in both direction and magnitude. (b) The gravitational field vectors in a small region near the Earth's surface are uniform; that is, they all have the same direction and magnitude.

That is, the gravitational field at a point in space equals the gravitational force that a test mass $m$ experiences at that point divided by the mass. Consequently, if $\vec{\mathbf{g}}$ is known at some point in space, a particle of mass $m$ experiences a gravitational force $\vec{\mathbf{F}}_g = m\vec{\mathbf{g}}$ when placed at that point. We will also see the model of a particle in a field for electricity and magnetism in later chapters, where it plays a much larger role than it does for gravity.

As an example, consider an object of mass $m$ near the Earth's surface. The gravitational force on the object is directed toward the center of the Earth and has a magnitude $mg$. Therefore, we see that the gravitational field experienced by the object at some point has a magnitude equal to the free-fall acceleration at that point. Because the gravitational force on the object has a magnitude $GM_Em/r^2$ (where $M_E$ is the mass of the Earth), the field $\vec{\mathbf{g}}$ at a distance $r$ from the center of the Earth is given by

$$\vec{\mathbf{g}} = \frac{\vec{\mathbf{F}}_g}{m} = -\frac{GM_E}{r^2}\hat{\mathbf{r}} \qquad [11.5]$$

where $\hat{\mathbf{r}}$ is a unit vector pointing radially outward from the Earth and the negative sign indicates that the field vector points toward the center of the Earth as shown in Figure 11.3a. Note that the field vectors at different points surrounding the spherical mass vary in both direction and magnitude. In a small region near the Earth's surface, $\vec{\mathbf{g}}$ is approximately constant and the downward field is uniform as indicated in Figure 11.3b. Equation 11.5 is valid at all points outside the Earth's surface, assuming that the Earth is spherical and that rotation can be neglected. At the Earth's surface, where $r = R_E$, $\vec{\mathbf{g}}$ has a magnitude of 9.80 m/s$^2$.

### INTERACTIVE EXAMPLE 11.1 An Earth Satellite

A satellite of mass $m$ moves in a circular orbit about the Earth with a constant speed $v$ and at a height of $h$ = 1 000 km above the Earth's surface as in Figure 11.4. (For clarity, this figure is not drawn to scale.) Find the orbital speed of the satellite.

**Solution** The only external force on the satellite is the gravitational force exerted by the Earth. This force is directed toward the center of the satellite's circular path. We apply Newton's second law to the satellite modeled as a particle in uniform circular motion. Because the magnitude of the gravitational force between the Earth and the satellite is $GM_Em/r^2$ we find that

$$F_g = G\frac{M_Em}{r^2} = m\frac{v^2}{r}$$

$$v = \sqrt{\frac{GM_E}{r}}$$

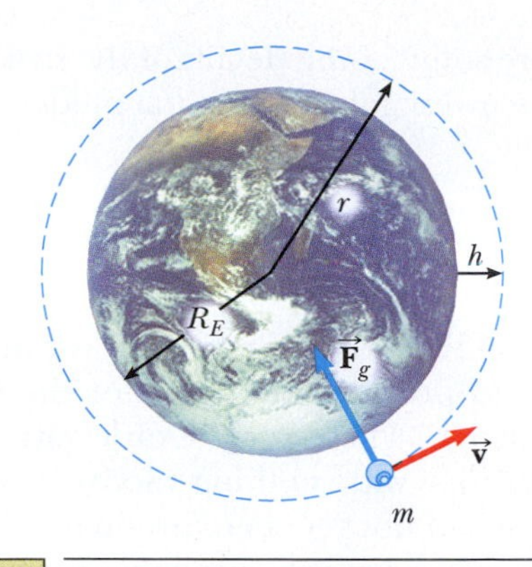


FIGURE 11.4 (Example 11.1) A satellite of mass $m$ moving around the Earth in a circular orbit of radius $r$ with constant speed $v$. The only force acting on the satellite is the gravitational force $\vec{\mathbf{F}}_g$. (Not drawn to scale.)

In this expression, the distance $r$ is the Earth's radius plus the height of the satellite; that is, $r = R_E + h = 6.37 \times 10^6 + 1.00 \times 10^6 = 7.37 \times 10^6$ m, so that

$$v = \sqrt{\frac{(6.67 \times 10^{-11}\ \text{N}\cdot\text{m}^2/\text{kg}^2)(5.98 \times 10^{24}\ \text{kg})}{7.37 \times 10^6\ \text{m}}}$$

$$= 7.36 \times 10^3\ \text{m/s} \approx \boxed{16\ 400\ \text{mi/h}}$$

Note that $v$ is independent of the mass of the satellite!

**PhysicsNow™** You can adjust the altitude of the satellite and observe the orbit by logging into PhysicsNow at **www.pop4e.com** and going to Interactive Example 11.1.

## 11.2 STRUCTURAL MODELS

In Chapter 1, we mentioned that we would discuss four categories of models. The fourth category is **structural models.** In these models, we propose theoretical structures in an attempt to understand the behavior of a system with which we cannot interact directly because it is far different in scale—either much smaller or much larger—from our macroscopic world.

One of the earliest structural models to be explored was that of the place of the Earth in the Universe. The movements of the planets, stars, and other celestial bodies have been observed by people for thousands of years. Early in history, scientists regarded the Earth as the center of the Universe because it appeared that objects in the sky moved around the Earth. This organization of the Earth and other objects is a structural model for the Universe called the *geocentric model.* It was elaborated and formalized by the Greek astronomer Claudius Ptolemy in the second century A.D. and was accepted for the next 1400 years. In 1543, Polish astronomer Nicolaus Copernicus (1473–1543) offered a different structural model in which the Earth is part of a local Solar System, suggesting that the Earth and the other planets revolve in perfectly circular orbits about the Sun (the *heliocentric model*).

In general, a structural model contains the following features:

■ Features of structural models

1. A description of the physical components of the system; in the heliocentric model, the components are the planets and the Sun.
2. A description of where the components are located relative to one another and how they interact; in the heliocentric model, the planets are in orbit around the Sun and they interact via the gravitational force.
3. A description of the time evolution of the system; the heliocentric model assumes a steady-state Solar System, with planets revolving in orbits around the Sun with fixed periods.
4. A description of the agreement between predictions of the model and actual observations and, possibly, predictions of new effects that have not yet been observed; the heliocentric model predicts Earth-based observations of Mars that are in agreement with historical and present measurements. The geocentric model was also able to find agreement between predictions and observations, but only at the expense of a very complicated structural model in which the planets moved in circles built on other circles. The heliocentric model, along with Newton's law of universal gravitation, predicted that a spacecraft could be sent from the Earth to Mars long before it was actually first done in the 1970s.

In Sections 11.3 and 11.4, we explore some of the details of the structural model of the Solar System. In Section 11.5, we investigate a structural model of the hydrogen atom.

(Art Resource)

**JOHANNES KEPLER** (1571–1630)

Kepler, a German astronomer, is best known for developing the laws of planetary motion based on the careful observations of Tycho Brahe.

## 11.3 KEPLER'S LAWS

Danish astronomer Tycho Brahe (1546–1601) made accurate astronomical measurements over a period of 20 years and provided the basis for the currently accepted structural model of the Solar System. These precise observations, made on the planets and 777 stars, were carried out with nothing more elaborate than a large sextant and compass; the telescope had not yet been invented.

German astronomer Johannes Kepler, who was Brahe's assistant, acquired Brahe's astronomical data and spent about 16 years trying to deduce a mathematical model for the motions of the planets. After many laborious calculations, he found that Brahe's precise data on the revolution of Mars about the Sun provided the answer. Kepler's analysis first showed that the concept of circular orbits about the Sun in the heliocentric model had to be abandoned. He discovered that the orbit of Mars could be accurately described by a curve called an *ellipse.* He then generalized this analysis to include the motions of all planets. The complete analysis is summarized in three statements, known as **Kepler's laws of planetary motion,** each of which is discussed in the following sections.

Newton demonstrated that these laws are consequences of the gravitational force that exists between any two masses. Newton's law of universal gravitation, together with his laws of motion, provides the basis for a full mathematical representation of the motion of planets and satellites.

### Kepler's First Law

We are familiar with circular orbits of objects around gravitational force centers from Interactive Example 11.1. Kepler's first law indicates that the circular orbit is a very special case and that elliptical orbits are the general situation:[1]

Each planet in the Solar System moves in an elliptical orbit with the Sun at one focus.

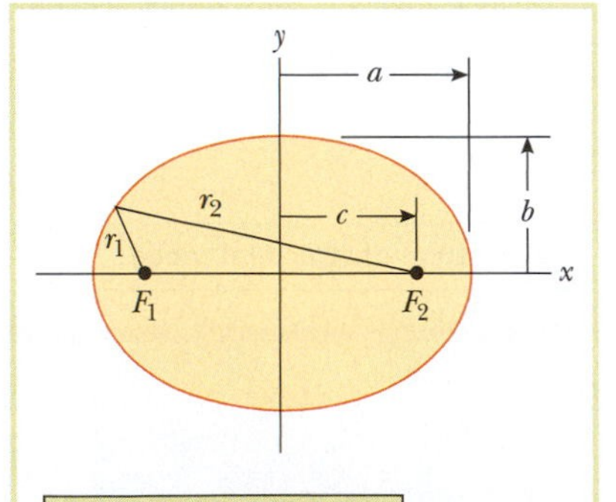


ACTIVE FIGURE 11.5
Plot of an ellipse. The semimajor axis has length $a$, and the semiminor axis has length $b$. A focus is located at a distance $c$ from the center on each side of the center.

**PhysicsNow™** Log into PhysicsNow at **www.pop4e.com** and go to Active Figure 11.5 to move the focal points or enter values for $a$, $b$, $c$, and $e$ and see the resulting elliptical shape.

Active Figure 11.5 shows the geometry of an ellipse, which serves as our geometric model for the elliptical orbit of a planet.[2] An ellipse is mathematically defined by choosing two points, $F_1$ and $F_2$, each of which is a called a **focus,** and then drawing a curve through points for which the sum of the distances $r_1$ and $r_2$ from $F_1$ and $F_2$ is a constant. The longest distance through the center between points on the ellipse (and passing through both foci) is called the **major axis,** and this distance is $2a$. In Active Figure 11.5, the major axis is drawn along the $x$ direction. The distance $a$ is called the **semimajor axis.** Similarly, the shortest distance through the center between points on the ellipse is called the **minor axis** of length $2b$, where the distance $b$ is the **semiminor axis.** Either focus of the ellipse is located at a distance $c$ from the center of the ellipse, where $a^2 = b^2 + c^2$. In the elliptical orbit of a planet around the Sun, the Sun is at one focus of the ellipse. Nothing is at the other focus.

[1] We choose a simplification model in which a body of mass $m$ is in orbit around a body of mass $M$, with $M \gg m$. In this way, we model the body of mass $M$ to be stationary. In reality, that is not true; both $M$ and $m$ move around the center of mass of the system of two objects. That is how we indirectly detect planets around other stars; we see the "wobbling" motion of the star as the planet and the star rotate about the center of mass.

[2] Actual orbits show perturbations due to moons in orbit around the planet and passages of the planet near other planets. We will ignore these perturbations and adopt a simplification model in which the planet follows a perfectly elliptical orbit.

The **eccentricity** of an ellipse is defined as $e \equiv c/a$ and describes the general shape of the ellipse. For a circle, $c = 0$ and the eccentricity is therefore zero. The smaller $b$ is than $a$, the shorter the ellipse is along the $y$ direction compared with its extent in the $x$ direction in Active Figure 11.5. As $b$ decreases, $c$ increases and the eccentricity $e$ increases. Therefore, higher values of eccentricity correspond to longer and thinner ellipses. The range of values of the eccentricity for an ellipse is $0 < e < 1$. Eccentricities higher than 1 correspond to hyperbolas.

Eccentricities for planetary orbits vary widely in the Solar System. The eccentricity of the Earth's orbit is 0.017, which makes it nearly circular. On the other hand, the eccentricity of Pluto's orbit is 0.25, the highest of all the nine planets. Figure 11.6a shows an ellipse with the eccentricity of that of Pluto's orbit. Notice that even this highest eccentricity orbit is difficult to distinguish from a circle, which is why Kepler's first law is an admirable accomplishment.

The eccentricity of the orbit of Comet Halley is 0.97, describing an orbit whose major axis is much longer than its minor axis as shown in Figure 11.6b. As a result, Comet Halley spends much of its 76-year period far from the Sun and invisible from the Earth. It is only visible to the naked eye during a small part of its orbit when it is near the Sun.

Let us imagine now a planet in an elliptical orbit such as that shown in Active Figure 11.5 with the Sun at focus $F_2$. When the planet is at the far left in the diagram, the distance between the planet and the Sun is $a + c$. This point is called the *aphelion*, where the planet is the farthest away from the Sun that it can be in the orbit (for an object in orbit around the Earth, this point is called the *apogee*). Conversely, when the planet is at the right end of the ellipse, the point is called the *perihelion* (for an Earth orbit, the *perigee*), and the distance between the planet and the Sun is $a - c$.

Kepler's first law is a direct result of the inverse-square nature of the gravitational force. We have discussed circular and elliptical orbits, which are the allowed shapes of orbits for objects that are *bound* to the gravitational force center. These objects include planets, asteroids, and comets that move repeatedly around the Sun, as well as moons orbiting a planet. *Unbound* objects might also occur, such as a meteoroid from deep space that might pass by the Sun once and then never return. The gravitational force between the Sun and these objects also varies as the inverse square of the separation distance, and the allowed paths for these objects are parabolas and hyperbolas.

**PITFALL PREVENTION 11.2**

**WHERE IS THE SUN?** The Sun is located at one focus of the elliptical orbit of a planet. It is *not* located at the center of the ellipse.

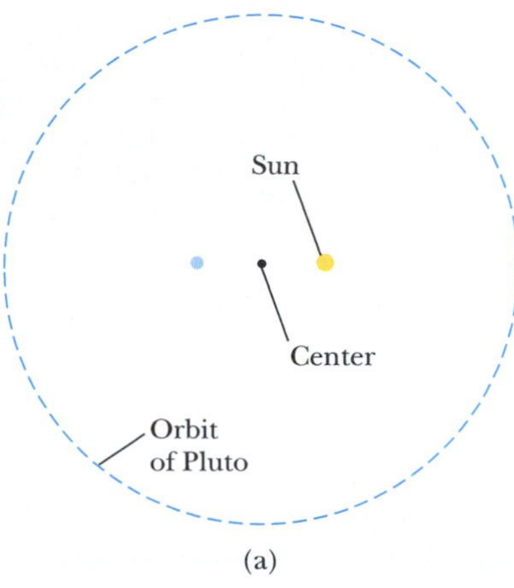


(a)

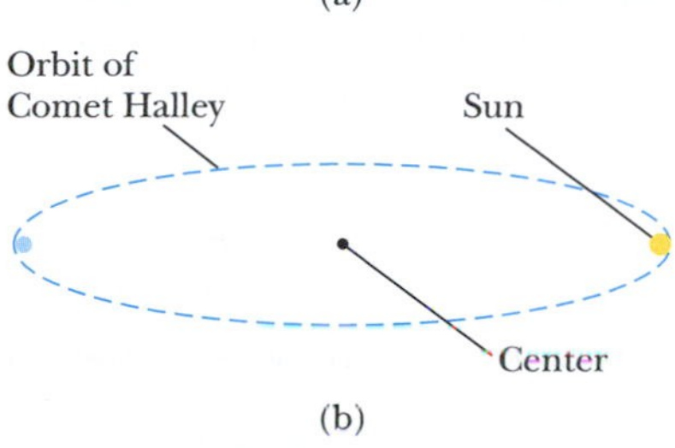


(b)

**FIGURE 11.6** (a) The shape of the orbit of Pluto, which has the highest eccentricity ($e = 0.25$) among the planets in the Solar System. The Sun is located at the large yellow dot, which is a focus of the ellipse. Nothing physical is located at the center of the orbit (the small dot) or the other focus (the blue dot). (b) The shape of the orbit of Comet Halley.

## Kepler's Second Law

Let us now look at the second of Kepler's laws:

▮ Kepler's second law

> The radius vector drawn from the Sun to any planet sweeps out equal areas in equal time intervals.

This law can be shown to be a consequence of angular momentum conservation as follows. Consider a planet of mass $M_p$ moving about the Sun in an elliptical orbit (Active Fig. 11.7a). Let us consider the planet as a system. We shall assume that the Sun is much more massive than the planet, so the Sun does not move. The gravitational force acting on the planet is a central force, that is, a force that is always directed along the radius vector. Therefore, the force on the planet is directed toward the Sun. The torque on the planet due to this central force is zero because $\vec{\mathbf{F}}$ is parallel to $\vec{\mathbf{r}}$. That is,

$$\vec{\boldsymbol{\tau}} \equiv \vec{\mathbf{r}} \times \vec{\mathbf{F}} = \vec{\mathbf{r}} \times F(r)\hat{\mathbf{r}} = 0$$

Recall that the external net torque on a system equals the time rate of change of angular momentum of the system; that is, $\vec{\boldsymbol{\tau}} = d\vec{\mathbf{L}}/dt$. Therefore, because $\vec{\boldsymbol{\tau}} = 0$ for the planet, the angular momentum $\vec{\mathbf{L}}$ of the planet is a constant of the motion:

$$\vec{\mathbf{L}} = \vec{\mathbf{r}} \times \vec{\mathbf{p}} = M_p \vec{\mathbf{r}} \times \vec{\mathbf{v}} = \text{constant}$$

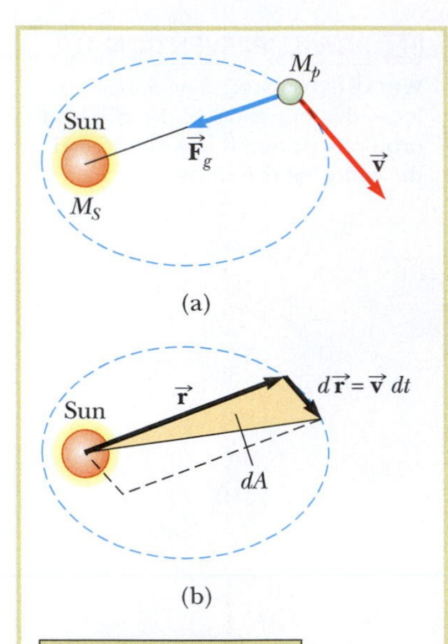


ACTIVE FIGURE 11.7
(a) The gravitational force acting on a planet acts toward the Sun, along the radius vector. (b) As a planet orbits the Sun, the area swept out by the radius vector in a time interval $dt$ is equal to one-half the area of the parallelogram formed by the vectors $\vec{r}$ and $d\vec{r} = \vec{v}\,dt$.

**Physics Now™** By logging into PhysicsNow at www.pop4e.com and going to Active Figure 11.7 you can assign a value of the eccentricity and see the resulting motion of the planet around the Sun.

We can relate this result to the following geometric consideration. In a time interval $dt$, the radius vector $\vec{r}$ in Active Figure 11.7b sweeps out the area $dA$, which equals one-half the area $|\vec{r} \times d\vec{r}|$ of the parallelogram formed by the vectors $\vec{r}$ and $d\vec{r}$. Because the displacement of the planet in the time interval $dt$ is given by $d\vec{r} = \vec{v}\,dt$, we have

$$dA = \tfrac{1}{2}|\vec{r} \times d\vec{r}| = \tfrac{1}{2}|\vec{r} \times \vec{v}\,dt| = \frac{L}{2M_p}\,dt$$

$$\frac{dA}{dt} = \frac{L}{2M_p} = \text{constant} \qquad [11.6]$$

where $L$ and $M_p$ are both constants. Therefore, we conclude that the radius vector from the Sun to any planet sweeps out equal areas in equal times.

It is important to recognize that this result is a consequence of the gravitational force being a central force, which in turn implies that angular momentum of the planet is constant. Therefore, the law applies to *any* situation that involves a central force, whether inverse-square or not.

### ■ Thinking Physics 11.2

The Earth is closer to the Sun when it is winter in the Northern Hemisphere than when it is summer. July and January both have 31 days. In which month, if either, does the Earth move through a longer distance in its orbit?

**Reasoning** The Earth is in a slightly elliptical orbit around the Sun. Because of angular momentum conservation, the Earth moves more rapidly when it is close to the Sun and more slowly when it is farther away. Therefore, because it is closer to the Sun in January, it is moving faster and will cover more distance in its orbit than it will in July. ■

## Kepler's Third Law

Kepler's third law reads as follows:

> The square of the orbital period of any planet is proportional to the cube of the semimajor axis of the elliptical orbit.

This law can be shown easily for circular orbits. Consider a planet of mass $M_p$ that is assumed to be moving about the Sun (mass $M_S$) in a circular orbit as in Figure 11.8. Because the gravitational force provides the centripetal acceleration of the planet as it moves in a circle, we use Newton's second law for a particle in uniform circular motion:

$$\frac{GM_S M_p}{r^2} = \frac{M_p v^2}{r}$$

The orbital speed of the planet is $2\pi r/T$, where $T$ is the period; therefore, the preceding expression becomes

$$\frac{GM_S}{r^2} = \frac{(2\pi r/T)^2}{r}$$

$$T^2 = \left(\frac{4\pi^2}{GM_S}\right) r^3 = K_S r^3$$

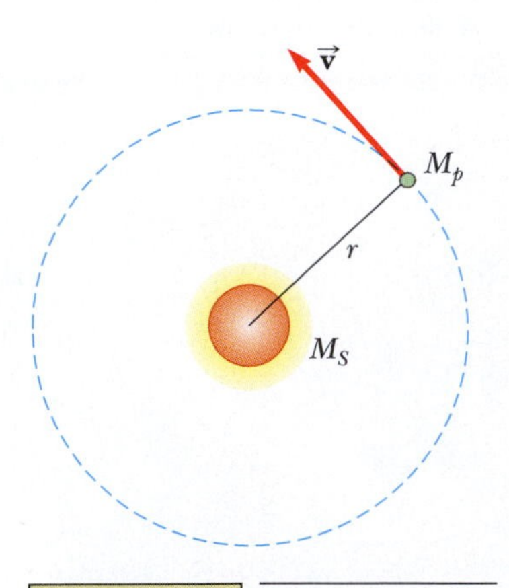


FIGURE 11.8 A planet of mass $M_p$ moving in a circular orbit about the Sun. Kepler's third law relates the period of the orbit to the radius.

**TABLE 11.1 Useful Planetary Data**

| Body | Mass (kg) | Mean Radius (m) | Period (s) | Average Distance from Sun (m) | $\frac{T^2}{a^3}$ ($s^2/m^3$) |
|---|---|---|---|---|---|
| Mercury | $3.18 \times 10^{23}$ | $2.43 \times 10^{6}$ | $7.60 \times 10^{6}$ | $5.79 \times 10^{10}$ | $2.97 \times 10^{-19}$ |
| Venus | $4.88 \times 10^{24}$ | $6.06 \times 10^{6}$ | $1.94 \times 10^{7}$ | $1.08 \times 10^{11}$ | $2.99 \times 10^{-19}$ |
| Earth | $5.98 \times 10^{24}$ | $6.37 \times 10^{6}$ | $3.156 \times 10^{7}$ | $1.496 \times 10^{11}$ | $2.97 \times 10^{-19}$ |
| Mars | $6.42 \times 10^{23}$ | $3.37 \times 10^{6}$ | $5.94 \times 10^{7}$ | $2.28 \times 10^{11}$ | $2.98 \times 10^{-19}$ |
| Jupiter | $1.90 \times 10^{27}$ | $6.99 \times 10^{7}$ | $3.74 \times 10^{8}$ | $7.78 \times 10^{11}$ | $2.97 \times 10^{-19}$ |
| Saturn | $5.68 \times 10^{26}$ | $5.85 \times 10^{7}$ | $9.35 \times 10^{8}$ | $1.43 \times 10^{12}$ | $2.99 \times 10^{-19}$ |
| Uranus | $8.68 \times 10^{25}$ | $2.33 \times 10^{7}$ | $2.64 \times 10^{9}$ | $2.87 \times 10^{12}$ | $2.95 \times 10^{-19}$ |
| Neptune | $1.03 \times 10^{26}$ | $2.21 \times 10^{7}$ | $5.22 \times 10^{9}$ | $4.50 \times 10^{12}$ | $2.99 \times 10^{-19}$ |
| Pluto | $\approx 1.4 \times 10^{22}$ | $\approx 1.5 \times 10^{6}$ | $7.82 \times 10^{9}$ | $5.91 \times 10^{12}$ | $2.96 \times 10^{-19}$ |
| Moon | $7.36 \times 10^{22}$ | $1.74 \times 10^{6}$ | — | — | — |
| Sun | $1.991 \times 10^{30}$ | $6.96 \times 10^{8}$ | — | — | — |

*Note:* For a more complete set of data, see, for example, the *Handbook of Chemistry and Physics* (Boca Raton, FL: CRC Press, published annually).

where $K_S$ is a constant given by

$$K_S = \frac{4\pi^2}{GM_S} = 2.97 \times 10^{-19}\ s^2/m^3$$

For elliptical orbits, Kepler's third law is expressed by starting with $T^2 = K_S r^3$ and replacing $r$ with the length $a$ of the semimajor axis (see Fig. 11.5):

$$T^2 = \left(\frac{4\pi^2}{GM_S}\right) a^3 = K_S a^3 \qquad [11.7]$$

■ Kepler's third law

Equation 11.7 is Kepler's third law. Because the semimajor axis of a circular orbit is its radius, Equation 11.7 is valid for both circular and elliptical orbits. Note that the constant of proportionality $K_S$ is independent of the mass of the planet. Equation 11.7 is therefore valid for *any* planet. If we were to consider the orbit of a satellite about the Earth, such as the Moon, the constant would have a different value, with the Sun's mass replaced by the Earth's mass; that is, $K_E = 4\pi^2/GM_E$.

Table 11.1 is a collection of useful planetary data. The last column verifies that the ratio $T^2/a^3$ is constant. The small variations in the values in this column are because of uncertainties in the data measured for the periods and semimajor axes of the planets.

**QUICK QUIZ 11.2** A comet is in a highly elliptical orbit around the Sun. The period of the comet's orbit is 90 days. Which of the following statements is true about the possibility of a collision between this comet and the Earth? **(a)** Collision is not possible. **(b)** Collision is possible. **(c)** Not enough information is available to determine whether a collision is possible.

## 11.4 ENERGY CONSIDERATIONS IN PLANETARY AND SATELLITE MOTION

So far we have approached orbital mechanics from the point of view of forces and angular momentum. Let us now investigate the motion of planets in orbit from the *energy* point of view.

Consider an object of mass $m$ moving with a speed $v$ in the vicinity of a massive object of mass $M \gg m$. This two-object system might be a planet moving around the Sun, a satellite orbiting the Earth, or a comet making a one-time flyby past the Sun. We will treat the two objects of mass $m$ and $M$ as an isolated system. If we assume that $M$ is at rest in an inertial reference frame (because $M \gg m$), the total mechanical energy $E$ of the two-object system is the sum of the kinetic energy of the object of mass $m$ and the gravitational potential energy of the system:

$$E = K + U_g$$

Recall from Chapter 7 that the gravitational potential energy $U_g$ associated with *any pair* of particles of masses $m_1$ and $m_2$ separated by a distance $r$ is given by

$$U_g = -\frac{Gm_1m_2}{r}$$

where we have defined $U_g \rightarrow 0$ as $r \rightarrow \infty$; therefore, in our case, the mechanical energy of the system of $m$ and $M$ is

$$E = \tfrac{1}{2}mv^2 - \frac{GMm}{r} \qquad [11.8]$$

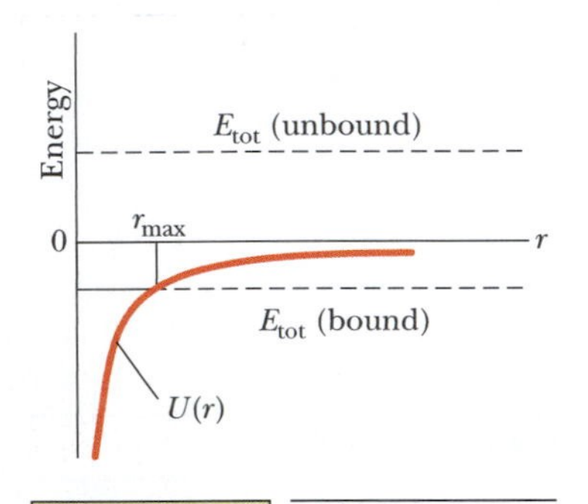


**FIGURE 11.9** The lower total energy line represents a bound system. The separation distance $r$ between the two gravitationally bound objects never exceeds $r_{max}$. The upper total energy line represents an unbound system of two objects interacting gravitationally. The separation distance $r$ between the two objects can have any value.

Equation 11.8 shows that $E$ may be positive, negative, or zero, depending on the value of $v$ at a particular separation distance $r$. If we consider the energy diagram method of Section 7.7, we can show the potential and total energies of the system as a function of $r$ as in Figure 11.9. A planet moving around the Sun and a satellite in orbit around the Earth are *bound* systems, such as those we discussed in Section 11.3; the Earth will always stay near the Sun and the satellite near the Earth. In Figure 11.9, these systems are represented by a total energy that is negative. The point at which the total energy line intersects the potential energy curve is a turning point, the maximum separation distance $r_{max}$ between the two bound objects.

A one-time meteoroid flyby represents an unbound system. The meteoroid interacts with the Sun but is not bound to it. Therefore, the meteoroid can in theory move infinitely far away from the Sun as represented in Figure 11.9 by a total energy line in the positive region of the graph. This line never intersects the potential energy curve, so all values of $r$ are possible.

For a bound system, such as the Earth and Sun, $E$ is necessarily less than zero because we have chosen the convention that $U_g \rightarrow 0$ as $r \rightarrow \infty$. We can easily establish that $E < 0$ for the system consisting of an object of mass $m$ moving in a circular orbit about an object of mass $M \gg m$. Applying Newton's second law to the object of mass $m$ in uniform circular motion gives

$$\sum F = ma \quad \rightarrow \quad \frac{GMm}{r^2} = \frac{mv^2}{r}$$

Many artificial satellites have been placed in orbit about the Earth. This diagram shows a plot of all known unclassified satellites and satellite debris larger in size than a baseball as of 1995. Note the large number of geosynchronous satellites (see Thinking Physics 11.1) that form a visible circle above the Earth's equator.

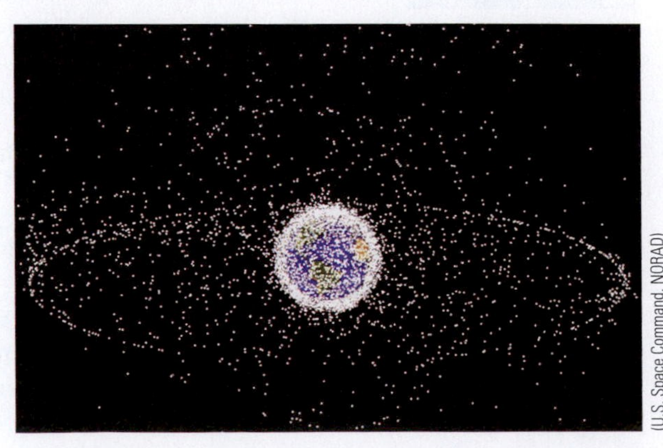

(U.S. Space Command, NORAD)

Multiplying both sides by $r$ and dividing by 2 gives

$$\frac{1}{2}mv^2 = \frac{GMm}{2r} \quad [11.9]$$

Substituting this result into Equation 11.8, we obtain

$$E = \frac{GMm}{2r} - \frac{GMm}{r}$$

$$E = -\frac{GMm}{2r} \quad [11.10]$$

This result clearly shows that **the total mechanical energy must be negative in the case of circular orbits.** Furthermore, Equation 11.9 shows that **the kinetic energy of an object in a circular orbit is equal to one-half the magnitude of the potential energy of the system** (when the potential energy is chosen to be zero at infinite separation).

The total mechanical energy is also negative in the case of elliptical orbits. The expression for $E$ for elliptical orbits is the same as Equation 11.10, with $r$ replaced by the semimajor axis $a$:

$$E = -\frac{GMm}{2a} \quad [11.11]$$

■ Total energy of a planet–star system

The total energy, the total angular momentum, and the total linear momentum of a planet–star system are constants of the motion, according to the isolated system model.

**QUICK QUIZ 11.3** A comet moves in an elliptical orbit around the Sun. Which point in its orbit represents the highest value of **(a)** the speed of the comet, **(b)** the potential energy of the comet–Sun system, **(c)** the kinetic energy of the comet, and **(d)** the total energy of the comet–Sun system?

## EXAMPLE 11.2 A Satellite in an Elliptical Orbit

A satellite moves in an elliptical orbit about the Earth as in Figure 11.10. The minimum and maximum distances from the surface of the Earth are 400 km and 3 000 km, respectively. Find the speeds of the satellite at apogee and perigee.

**Solution** Figure 11.10 helps conceptualize the motion of the satellite. Gravity is a central force, so there is zero torque exerted on the satellite. There is also zero torque on the Earth for the same reason. Consequently, we categorize the problem as one involving an isolated system for which the angular momentum is conserved. Because the mass of the satellite is negligible compared with the Earth's mass, we take the center of mass of the Earth to be at rest. Therefore, we only need to consider the angular momentum of the satellite. To analyze the problem, we assign subscripts $a$ and $p$ for the apogee and perigee positions, and apply the principle of conservation of angular momentum for the satellite at these two positions. The result is $L_p = L_a$, or

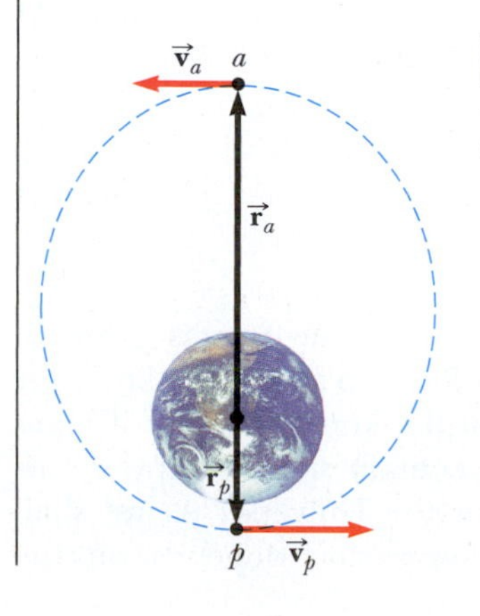


**FIGURE 11.10** (Example 11.2) A satellite in an elliptical orbit about the Earth.

$$mv_pr_p = mv_ar_a$$

$$v_pr_p = v_ar_a$$

Using the Earth's radius of $6.37 \times 10^6$ m and the given data, we find that $r_a = 9.37 \times 10^6$ m and $r_p = 6.77 \times 10^6$ m. Therefore,

$$(1) \quad \frac{v_p}{v_a} = \frac{r_a}{r_p} = \frac{9.37 \times 10^6 \text{ m}}{6.77 \times 10^6 \text{ m}} = 1.38$$

Because the satellite and the Earth form an isolated system, we can apply conservation of energy for the system and obtain $E_p = E_a$, or

$$U_p + K_p = U_a + K_a$$

$$-G\frac{M_E m}{r_p} + \frac{1}{2}m{v_p}^2 = -G\frac{M_E m}{r_a} + \frac{1}{2}m{v_a}^2$$

$$(2) \quad 2GM_E\left(\frac{1}{r_a} - \frac{1}{r_p}\right) = ({v_a}^2 - {v_p}^2)$$

Because we know the numerical values of $G$, $M_E$, $r_p$, and $r_a$, we can use Equations (1) and (2) to determine the two unknowns $v_p$ and $v_a$. Solving the equations simultaneously, we obtain

$$v_p = \boxed{8.27 \text{ km/s}} \qquad v_a = \boxed{5.98 \text{ km/s}}$$

To finalize the problem, note that $v_p > v_a$, as we would expect.

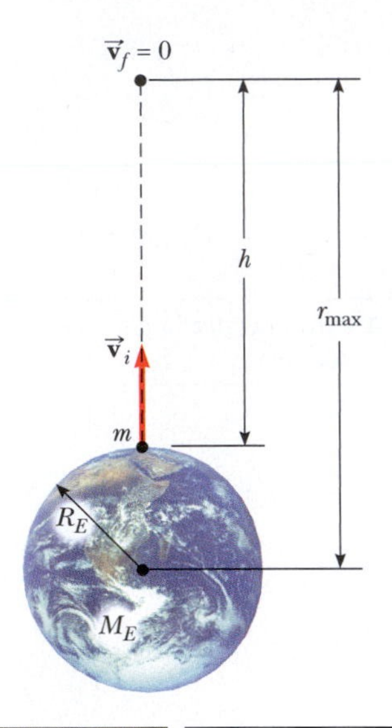


FIGURE 11.11 An object of mass $m$ projected upward from the Earth's surface with an initial speed $v_i$ reaches a maximum altitude $h = r_{max} - R_E$.

## Escape Speed

Suppose an object of mass $m$ is projected vertically from the Earth's surface with an initial speed $v_i$ as in Figure 11.11. We can use energy considerations to find the minimum value of the initial speed such that the object will continue to move away from the Earth forever. Equation 11.8 gives the total energy of the object–Earth system at any point when the speed of the object and its distance from the center of the Earth are known. At the surface of the Earth, $r_i = R_E$. When the object reaches its maximum altitude, $v_f = 0$ and $r_f = r_{max}$. Because the total energy of the system is conserved, substitution of these conditions into Equation 11.8 gives

$$\frac{1}{2}mv_i^2 - \frac{GM_E m}{R_E} = -\frac{GM_E m}{r_{max}}$$

Solving for $v_i^2$ gives

$$v_i^2 = 2GM_E\left(\frac{1}{R_E} - \frac{1}{r_{max}}\right) \qquad [11.12]$$

If the initial speed is known, this expression can therefore be used to calculate the maximum altitude $h$ because we know that $h = r_{max} - R_E$.

We are now in a position to calculate the minimum speed the object must have at the Earth's surface to continue to move away forever. This **escape speed** $v_{esc}$ results in the speed asymptotically approaching *zero*. Letting $r_{max} \rightarrow \infty$ in Equation 11.12 and setting $v_i = v_{esc}$, we have

$$v_{esc} = \sqrt{\frac{2GM_E}{R_E}} \qquad [11.13]$$

Note that this expression for $v_{esc}$ is independent of the mass of the object projected from the Earth. For example, a spacecraft has the same escape speed as a molecule. Furthermore, the result is independent of the *direction* of the velocity.

Note also that Equations 11.12 and 11.13 can be applied to objects projected from *any* planet. That is, in general, the escape speed from any planet of mass $M$ and radius $R$ is

$$v_{esc} = \sqrt{\frac{2GM}{R}} \qquad [11.14]$$

A list of escape speeds for the planets, the Moon, and the Sun is given in Table 11.2. Note that the values vary from 1.1 km/s for Pluto to about 618 km/s for the Sun. These results, together with some ideas from the kinetic theory of gases (Chapter 16), explain why our atmosphere does not contain significant amounts of hydrogen, which is the most abundant element in the Universe. As we shall see later, gas molecules have an average kinetic energy that depends on the

**PITFALL PREVENTION 11.3**

**You can't really escape** Although Equation 11.13 provides the escape speed, remember that this conventional name is misleading. It is impossible to escape *completely* from the Earth's gravitational influence because the gravitational force is of infinite range. No matter how far away you are, you will always feel some gravitational force due to the Earth. In practice, however, this force will be much smaller than forces due to other astronomical objects closer to you, so the gravitational force from the Earth can be ignored.

**TABLE 11.2 Escape Speeds from the Surfaces of the Planets, the Moon, and the Sun**

| Planet | $v_{esc}$ (km/s) |
|---|---|
| Mercury | 4.3 |
| Venus | 10.3 |
| Earth | 11.2 |
| Mars | 5.0 |
| Jupiter | 60 |
| Saturn | 36 |
| Uranus | 22 |
| Neptune | 24 |
| Pluto | 1.1 |
| Moon | 2.3 |
| Sun | 618 |

temperature of the gas. Lighter molecules in an atmosphere have translational speeds that are closer to the escape speed than more massive molecules, so they have a higher probability of escaping from the planet and the lighter molecules diffuse into space. This mechanism explains why the Earth does not retain hydrogen molecules and helium atoms in its atmosphere but does retain much heavier molecules, such as oxygen and nitrogen. On the other hand, Jupiter has a very large escape speed (60 km/s), which enables it to retain hydrogen, the primary constituent of its atmosphere.

## Black Holes

In Chapter 10, we briefly described a rare event called a supernova, the catastrophic explosion of a very massive star. The material that remains in the central core of such an object continues to collapse, and the core's ultimate fate depends on its mass. If the core has a mass less than 1.4 times the mass of our Sun, it gradually cools down and ends its life as a white dwarf star. If, however, the core's mass is greater than that, it may collapse further due to gravitational forces. What remains is a neutron star, discussed in Chapter 10, in which the mass of a star is compressed to a radius of about 10 km. (On the Earth, a teaspoon of this material would weigh about 5 billion tons!)

An even more unusual star death may occur when the core has a mass greater than about three solar masses. The collapse may continue until the star becomes a very small object in space, commonly referred to as a **black hole.** In effect, black holes are the remains of stars that have collapsed under their own gravitational force. If an object such as a spacecraft comes close to a black hole, it experiences an extremely strong gravitational force and is trapped forever.

The escape speed from any spherical body depends on the mass and radius of the body. The escape speed for a black hole is very high because of the concentration of the star's mass into a sphere of very small radius. If the escape speed exceeds the speed of light $c$, radiation from the body (e.g. visible light) cannot escape and the body appears to be black, hence the origin of the term *black hole.* The critical radius $R_S$ at which the escape speed is $c$ is called the **Schwarzschild radius** (Fig. 11.12). The imaginary surface of a sphere of this radius surrounding the black hole is called the **event horizon,** which is the limit of how close you can approach the black hole and hope to be able to escape.

Although light from a black hole cannot escape, light from events taking place near the black hole should be visible. For example, it is possible for a binary star system to consist of one normal star and one black hole. Material surrounding the ordinary star can be pulled into the black hole, forming an **accretion disk** around the

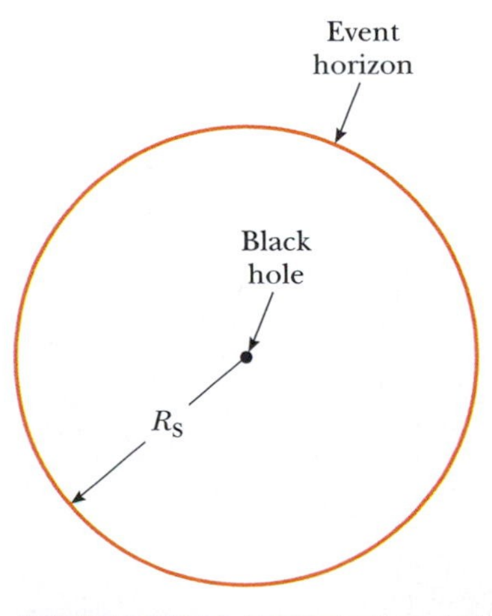


FIGURE 11.12 A black hole. The distance $R_S$ equals the Schwarzschild radius. Any event occurring within the boundary of radius $R_S$, called the event horizon, is invisible to an outside observer.

FIGURE 11.13 A binary star system consisting of an ordinary star on the left and a black hole on the right. Matter pulled from the ordinary star forms an accretion disk around the black hole, in which matter is raised to very high temperatures, resulting in the emission of x-rays.

black hole as suggested in Figure 11.13. Friction among particles in the accretion disk results in transformation of mechanical energy into internal energy. As a result, the orbital height of the material above the event horizon decreases and the temperature rises. This high-temperature material emits a large amount of radiation, extending well into the x-ray region of the electromagnetic spectrum. These x-rays are characteristic of a black hole. Several possible candidates for black holes have been identified by observation of these x-rays.

Evidence also supports the existence of supermassive black holes at the centers of galaxies, with masses very much larger than the Sun. (The evidence is strong for a supermassive black hole of mass 2 to 3 million solar masses at the center of our galaxy.) Theoretical models for these bizarre objects predict that jets of material should be evident along the rotation axis of the black hole. Figure 11.14 shows a Hubble Space Telescope photograph of the galaxy M87. The jet of material coming from this galaxy is believed to be evidence for a supermassive black hole at the center of the galaxy.

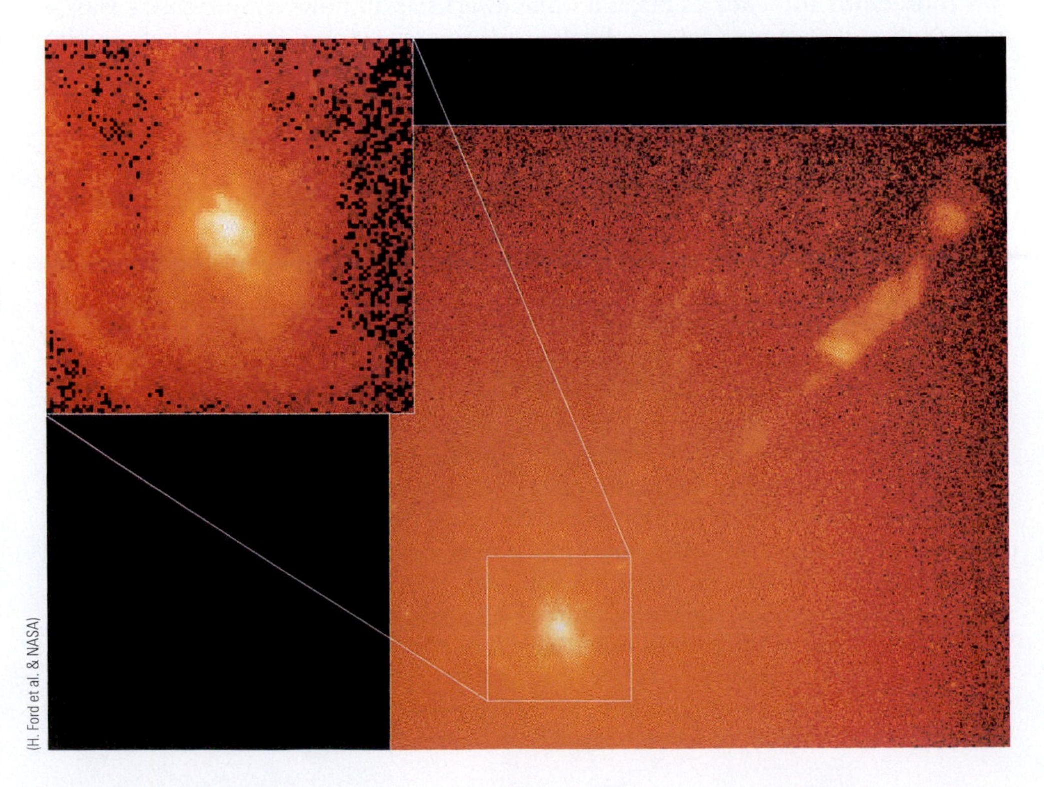

(H. Ford et al. & NASA)

FIGURE 11.14 Hubble Space Telescope images of the galaxy M87. The inset shows the center of the galaxy. The wider view shows a jet of material moving away from the center of the galaxy toward the upper right of the figure at about one-tenth the speed of light. Such jets are believed to be evidence of a supermassive black hole at the galaxy's center.

Black holes are of considerable interest to those searching for **gravity waves,** which are ripples in space–time caused by changes in a gravitational system. These ripples can be caused by a star collapsing into a black hole, a binary star consisting of a black hole and a visible companion, and supermassive black holes at a galaxy center. A gravity wave detector, the Laser Interferometer Gravitational Wave Observatory (LIGO), is currently being built and tested in the United States, and hopes are high for detecting gravitational waves with this instrument.

## 11.5 ATOMIC SPECTRA AND THE BOHR THEORY OF HYDROGEN

In the preceding sections, we described a structural model for a large-scale system, the Solar System. Let us now do the same for a very small-scale system, the hydrogen atom. We shall find that a Solar System model of the atom, with a few extra features, provides explanations for some of the experimental observations made on the hydrogen atom.

As you may have already learned in a chemistry course, the hydrogen atom is the simplest known atomic system and an especially important one to understand. Much of what is learned about the hydrogen atom (which consists of one proton and one electron) can be extended to single-electron ions such as $He^+$ and $Li^{2+}$. Furthermore, a thorough understanding of the physics underlying the hydrogen atom can then be used to describe more complex atoms and the periodic table of the elements.

Atomic systems can be investigated by observing *electromagnetic waves* emitted from the atom. Our eyes are sensitive to visible light, one type of electromagnetic wave. The wave will be one of our four simplification models around which we will identify analysis models, as we have done for a particle, a system, and a rigid object. A common form of periodic wave is the sinusoidal wave, whose shape is depicted in Figure 11.15. If this graph represents an electromagnetic wave, the vertical axis represents the magnitude of the electric field. (We will study electric fields in Chapter 19.) The horizontal axis is position in the direction of travel of the wave. The distance between two consecutive crests of the wave is called the **wavelength** $\lambda$. As the wave travels to the right with a speed $v$, any point on the wave travels a distance of one wavelength in a time interval of one period $T$ (the time interval for one cycle), so the wave speed is given by $v = \lambda/T$. The inverse of the period, $1/T$, is called the **frequency** $f$ of the wave; it represents the number of cycles per second. Therefore, the speed of the wave is often written as $v = \lambda f$. In this section, because we shall deal with electromagnetic waves—which travel at the speed of light $c$—the appropriate relation is

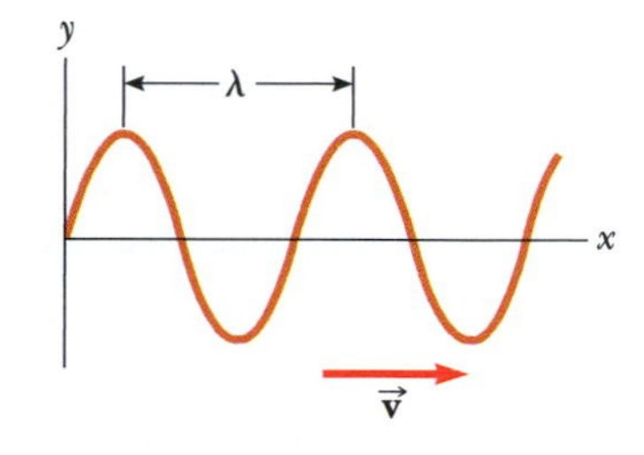


FIGURE 11.15 A sinusoidal wave traveling to the right with wave speed $v$. Any point on the wave moves a distance of one wavelength $\lambda$ in a time interval equal to the period $T$ of the wave.

$$c = \lambda f \tag{11.15}$$

■ Relation between wavelength, frequency, and wave speed

Suppose an evacuated glass tube is filled with hydrogen (or some other gas). If a voltage applied between metal electrodes in the tube is great enough to produce an electric current in the gas, the tube emits light with colors that are characteristic of the gas. (That is how a neon sign works.) When the emitted light is analyzed with a device called a spectroscope, in which the light passes through a narrow slit, a series of discrete **spectral lines** is observed, each line corresponding to a different wavelength, or color, of light. Such a series of spectral lines is commonly referred to as an **emission spectrum.** The wavelengths contained in a given spectrum are characteristic of the element emitting the light. Figure 11.16 is a semigraphical representation of the spectra of various elements. It is semigraphical because the horizontal axis is linear in wavelength, but the vertical axis has no significance. Because no two elements emit the same line spectrum, this phenomenon represents a marvelous and reliable technique for identifying elements in a substance.

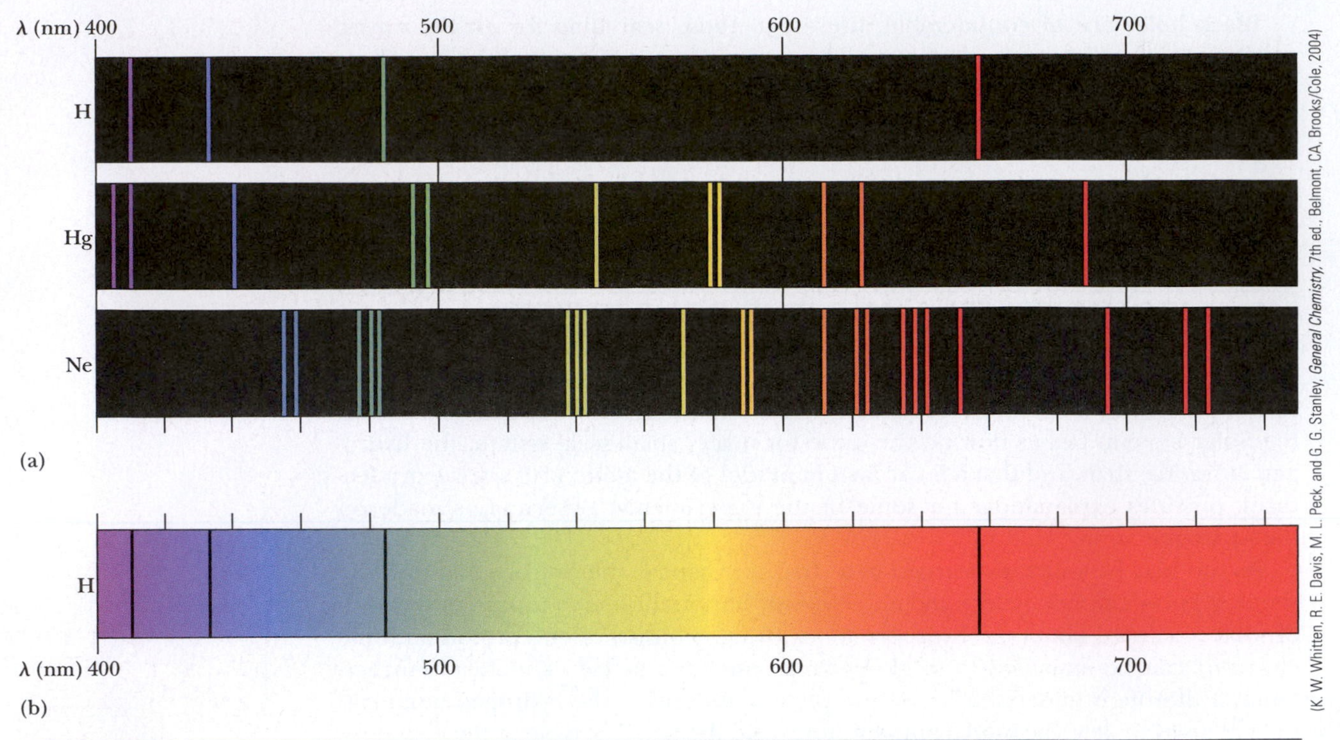


FIGURE 11.16 Visible spectra. (a) Line spectra produced by emission in the visible range for the elements hydrogen, mercury, and neon. (b) The absorption spectrum for hydrogen. The dark absorption lines occur at the same wavelengths as the emission lines for hydrogen shown in (a).

In addition to emitting light at specific wavelengths, an element can also absorb light at specific wavelengths. The spectral lines corresponding to this process form what is known as an **absorption spectrum.** An absorption spectrum can be obtained by passing a continuous radiation spectrum (one containing all wavelengths) through a vapor of the element being analyzed. The absorption spectrum consists of a series of dark lines superimposed on the otherwise continuous spectrum (Fig. 11.16b).

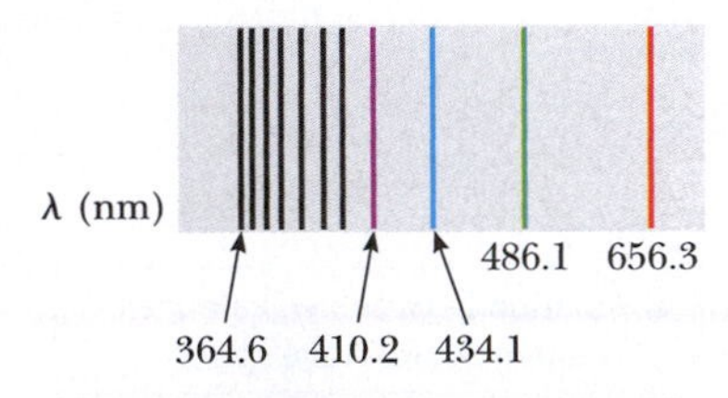


FIGURE 11.17 A series of spectral lines for atomic hydrogen. The prominent lines labeled are part of the Balmer series.

The emission spectrum of hydrogen shown in Figure 11.17 includes four prominent lines that occur at wavelengths of 656.3 nm, 486.1 nm, 434.1 nm, and 410.2 nm. In 1885, Johann Balmer (1825–1898) found that the wavelengths of these and less prominent lines can be described by the following simple empirical equation:

$$\lambda = 364.56 \frac{n^2}{n^2 - 4} \qquad n = 3, 4, 5, \ldots$$

in which $n$ is an integer starting at 3 and the wavelengths given by this expression are in nanometers. These spectral lines are called the **Balmer series.** The first line in the Balmer series, at 656.3 nm, corresponds to $n = 3$, the line at 486.1 nm corresponds to $n = 4$, and so on. At the time this equation was formulated, it had no valid theoretical basis; it simply predicted the wavelengths correctly. Therefore, this equation is not based on a model but is simply a trial-and-error equation that happens to work. A few years later, Johannes Rydberg (1854–1919) recast the equation in the following form:

■ **Rydberg equation**

$$\frac{1}{\lambda} = R_{\mathrm{H}}\left(\frac{1}{2^2} - \frac{1}{n^2}\right) \qquad n = 3, 4, 5, \ldots \qquad [11.16]$$

where $n$ may have integral values of 3, 4, 5, . . . and $R_H$ is a constant, now called the **Rydberg constant,** which has the value

$$R_H = 1.097\,373\,2 \times 10^7 \text{ m}^{-1}$$

Equation 11.16 is no more based on a model than is Balmer's equation. In this form, however, we can compare it with the predictions of a structural model of the hydrogen atom that is described below.

At the beginning of the 20th century, scientists were perplexed by the failure of classical physics to explain the characteristics of atomic spectra. Why did atoms of a given element emit only certain wavelengths of radiation so that the emission spectrum displayed discrete lines? Furthermore, why did the atoms absorb many of the same wavelengths that they emitted? In 1913, Niels Bohr provided an explanation of atomic spectra that includes some features of the currently accepted theory. Using the simplest atom, hydrogen, Bohr described a structural model for the atom. His model of the hydrogen atom contains some classical features that can be related to our analysis models as well as some revolutionary postulates that could not be justified within the framework of classical physics. The basic assumptions of the Bohr model as it applies to the hydrogen atom are as follows:

(Photo courtesy of AIP Niels Bohr Library, Margarethe Bohr Collection)

**NIELS BOHR** (1885–1962)

Bohr, a Danish physicist, was an active participant in the early development of quantum mechanics and provided much of its philosophical framework. During the 1920s and 1930s, Bohr headed the Institute for Advanced Studies in Copenhagen. The institute was a magnet for many of the world's best physicists and provided a forum for the exchange of ideas. Bohr was awarded the 1922 Nobel Prize in Physics for his investigation of the structure of atoms and of the radiation emanating from them.

1. The electron moves in a circular orbit about the proton under the influence of the electric force of attraction as in Figure 11.18. This notion is purely classical and is very similar to our previous discussion of planets in orbit around the Sun in our structural model of the Solar System.
2. Only certain electron orbits are stable, and they are the only orbits in which we find the electron. In these orbits, the hydrogen atom does not emit energy in the form of radiation. Hence, the total energy of the atom remains constant, and classical mechanics can be used to describe the electron's motion. This restriction to certain orbits is a new idea that is not consistent with classical physics. As we shall see in Chapter 24, an accelerating electron should emit energy by electromagnetic radiation. Therefore, according to the continuity equation for energy, the emission of radiation from the atom should result in a decrease in the energy of the atom. Bohr's postulate boldly claims that this radiation simply does not happen.
3. Radiation is emitted by the hydrogen atom when the atom makes a transition from a more energetic initial state to a lower state. The transition cannot be visualized or treated classically. In particular, the frequency $f$ of the radiation emitted in the transition is related to the change in the atom's energy. The frequency of the emitted radiation is found from

$$E_i - E_f = hf \qquad [11.17]$$

where $E_i$ is the energy of the initial state, $E_f$ is the energy of the final state, $h$ is **Planck's constant** ($h = 6.63 \times 10^{-34}$ J·s; we will see Planck's constant extensively in our studies of modern physics), and $E_i > E_f$. The notion of energy being emitted only when a transition occurs is nonclassical. Given this notion, however, Equation 11.17 is simply the continuity equation for energy, $\Delta E = \Sigma T \rightarrow E_f - E_i = -hf$. On the left is the change in energy of the system—the atom—and on the right is the energy transferred out of the system by electromagnetic radiation.

4. The size of the allowed electron orbits is determined by a condition imposed on the electron's orbital angular momentum. The allowed orbits are those for which the electron's orbital angular momentum about the nucleus is an integral multiple of $\hbar \equiv h/2\pi$:

$$m_e vr = n\hbar \qquad n = 1, 2, 3, \ldots \qquad [11.18]$$

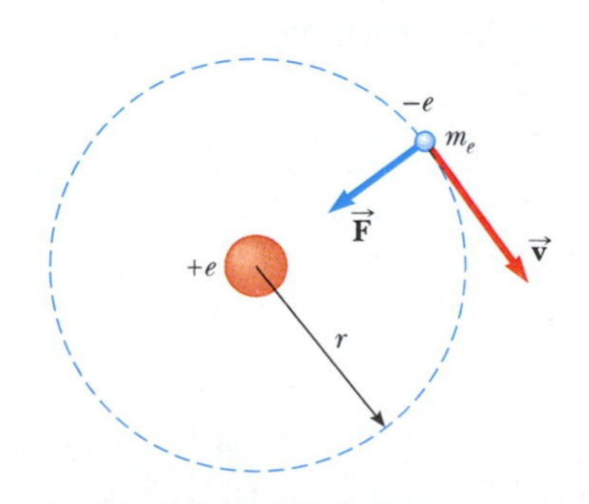


FIGURE 11.18 A pictorial representation of Bohr's model of the hydrogen atom, in which the electron is in a circular orbit about the proton.

This new idea cannot be related to any of the models we have developed so far. It can be related, however, to a model that will be developed in later chapters, and we shall return to this idea at that time to see how it is predicted by the model. This concept is our first introduction to a notion from **quantum mechanics,** which describes the behavior of microscopic particles.

Using these four assumptions, Bohr built a structural model that explains the emission wavelengths of the hydrogen atom. The electric potential energy of the system shown in Figure 11.18 is given by Equation 7.23, $U_e = -k_e e^2/r$, where $k_e$ is the electric constant, $e$ is the charge on the electron, and $r$ is the electron–proton separation. Therefore, the total energy of the atom, which contains both kinetic and potential energy terms, is

$$E = K + U_e = \tfrac{1}{2}m_e v^2 - k_e \frac{e^2}{r} \qquad [11.19]$$

According to assumption 2, the energy of the system remains constant; the system is isolated because the structural model does not allow for electromagnetic radiation for a given orbit.

Applying Newton's second law to this system, we see that the magnitude of the attractive electric force on the electron, $k_e e^2/r^2$ (Eq. 5.15), is equal to the product of its mass and its centripetal acceleration ($a_c = v^2/r$):

$$\frac{k_e e^2}{r^2} = \frac{m_e v^2}{r}$$

From this expression, the kinetic energy of the electron is found to be

$$K = \tfrac{1}{2}m_e v^2 = \frac{k_e e^2}{2r} \qquad [11.20]$$

Substituting this value of $K$ into Equation 11.19 gives the following expression for the total energy $E$ of the hydrogen atom:

■ Total energy of the hydrogen atom

$$E = -\frac{k_e e^2}{2r} \qquad [11.21]$$

Note that the total energy is negative,[3] indicating a bound electron–proton system. Therefore, energy in the amount of $k_e e^2/2r$ must be added to the atom just to separate the electron and proton by an infinite distance and make the total energy zero.[4] An expression for $r$, the radius of the allowed orbits, can be obtained by eliminating $v$ by substitution between Equations 11.18 and 11.20:

■ Radii of Bohr orbits in hydrogen

$$r_n = \frac{n^2\hbar^2}{m_e k_e e^2} \qquad n = 1, 2, 3 \ldots \qquad [11.22]$$

This result shows that the radii have discrete values, or are *quantized.* The integer $n$ is called a **quantum number** and specifies the particular allowed **quantum state** of the atomic system.

[3]Compare this expression with Equation 11.11 for a gravitational system.

[4]This process is called *ionizing* the atom. In theory, ionization requires separating the electron and proton by an infinite distance. In reality, however, the electron and proton are in an environment with huge numbers of other particles. Therefore, ionization means separating the electron and proton by a distance large enough so that the interaction of these particles with other entities in their environment is larger than the remaining interaction between them.

The orbit for which $n = 1$ has the smallest radius; it is called the **Bohr radius** $a_0$ and has the value

$$a_0 = \frac{\hbar^2}{m_e k_e e^2} = 0.052\,9 \text{ nm} \qquad [11.23]$$

■ The Bohr radius

The first three Bohr orbits are shown to scale in Active Figure 11.19.

The quantization of the orbit radii immediately leads to quantization of the energy of the atom, which can be seen by substituting $r_n = n^2 a_0$ into Equation 11.21. The allowed energies of the atom are

$$E_n = -\frac{k_e e^2}{2a_0}\left(\frac{1}{n^2}\right) \qquad n = 1, 2, 3, \ldots \qquad [11.24]$$

Insertion of numerical values into Equation 11.24 gives

$$E_n = -\frac{13.606 \text{ eV}}{n^2} \qquad n = 1, 2, 3, \ldots \qquad [11.25]$$

■ Energies of quantum states of the hydrogen atom

(Recall from Section 9.7 that 1 eV $= 1.60 \times 10^{-19}$ J.) The lowest quantum state, corresponding to $n = 1$, is called the **ground state** and has an energy of $E_1 = -13.606$ eV. The next state, the **first excited state,** has $n = 2$ and an energy of $E_2 = E_1/2^2 = -3.401$ eV. Active Figure 11.20 is an **energy level diagram** showing the energies of these discrete energy states and the corresponding quantum numbers. This diagram is another semigraphical representation. The vertical axis is linear in energy, but the horizontal axis has no significance. The horizontal lines correspond to the allowed energies. The atomic system cannot have any energies other than those represented by the lines. The vertical lines with arrowheads represent transitions between states, during which energy is emitted.

The upper limit of the quantized levels, corresponding to $n \rightarrow \infty$(or $r \rightarrow \infty$) and $E \rightarrow 0$, represents the state for which the electron is removed from the atom.[5] Above this energy is a continuum of available states for the ionized atom. The minimum energy required to ionize the atom is called the **ionization energy.** As can be seen from Active Figure 11.20, the ionization energy for hydrogen, based on Bohr's calculation, is 13.6 eV. This finding constituted a major achievement for the Bohr theory because the ionization energy for hydrogen had already been measured to be 13.6 eV.

Active Figure 11.20 also shows various transitions of the atom from one state to a lower state, as referred to in Bohr's assumption 3. As the energy of the atom decreases in a transition, the difference in energy between the states is carried away by electromagnetic radiation. Those transitions ending on $n = 2$ are shown in color, corresponding to the color of the light they represent. The transitions ending on $n = 2$ form the Balmer series of spectral lines, the wavelengths of which are correctly predicted by the Rydberg equation (see Eq. 11.16). Active Figure 11.20 also shows other spectral series (the Lyman series and the Paschen series) that were found after Balmer's discovery.

Equation 11.24, together with Bohr's third postulate, can be used to calculate the frequency of the radiation that is emitted when the atom makes a transition[6] from a high-energy state to a low-energy state:

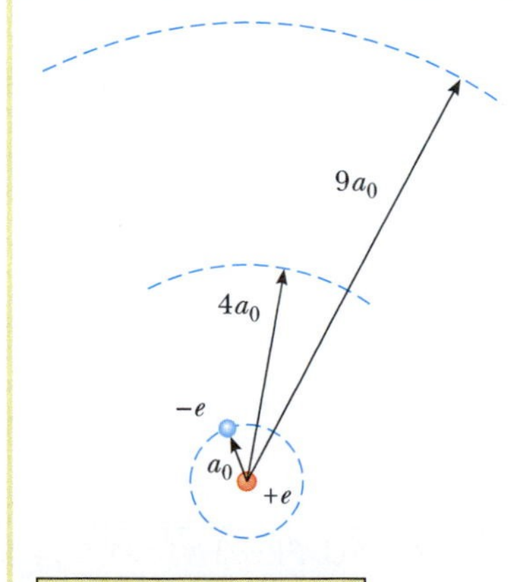


**ACTIVE FIGURE 11.19**
The first three circular orbits predicted by the Bohr model for hydrogen.

**Physics Now™** Log into PhysicsNow at **www.pop4e.com** and go to Active Figure 11.19 to choose the initial and final states of the hydrogen atom and observe the transitions in this figure and in Active Figure 11.20.

[5]The phrase "the electron is removed from the atom" is very commonly used, but, of course, we realize that we mean that the electron and proton are separated *from each other.*

[6]The phrase "the electron makes a transition" is also commonly used, but we will use "the atom makes a transition" to emphasize that the energy belongs to the system of the atom, not just to the electron. This wording is similar to our discussion in Chapter 7 of gravitational potential energy belonging to the system of an object and the Earth, not to the object alone.

**ACTIVE FIGURE 11.20**

An energy level diagram for hydrogen. The discrete allowed energies are plotted on the vertical axis. Nothing is plotted on the horizontal axis, but the horizontal extent of the diagram is made large enough to show allowed transitions. Quantum numbers are given on the left and energies (in electron volts) on the right. Vertical arrows represent the four lowest-energy transitions in each of the spectral series shown. The colored arrows for the Balmer series indicate that this series results in visible light.

**Physics Now™** Log into PhysicsNow at www.pop4e.com and go to Active Figure 11.20 to choose the initial and final states of the hydrogen atom and observe the transitions in this figure and in Active Figure 11.19.

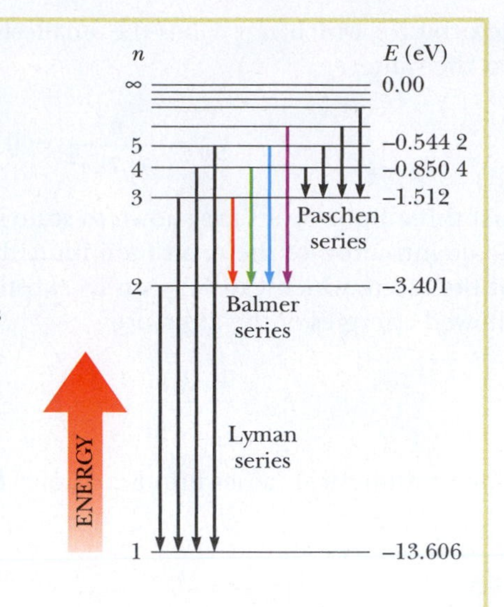


■ Frequency of radiation emitted from hydrogen

$$f = \frac{E_i - E_f}{h} = \frac{k_e e^2}{2a_0 h}\left(\frac{1}{n_f^2} - \frac{1}{n_i^2}\right) \quad [11.26]$$

Because the quantity expressed in the Rydberg equation is wavelength, it is convenient to convert frequency to wavelength, using $c = f\lambda$, to obtain

■ Emission wavelengths of hydrogen

$$\frac{1}{\lambda} = \frac{f}{c} = \frac{k_e e^2}{2a_0 hc}\left(\frac{1}{n_f^2} - \frac{1}{n_i^2}\right) \quad [11.27]$$

Notice that the *theoretical* expression, Equation 11.27, is identical to the *empirical* Rydberg equation (Equation 11.16), provided that the combination of constants $k_e e^2/2a_0 hc$ is equal to the experimentally determined Rydberg constant and that $n_f = 2$. After Bohr demonstrated the agreement of the constants in these two equations to a precision of about 1%, it was soon recognized as the crowning achievement of his structural model of the atom.

One question remains: What is the significance of $n_f = 2$? Its importance is simply because those transitions ending on $n_f = 2$ result in radiation that happens to lie in the visible; therefore, they were easily observed! As seen in Active Figure 11.20, other series of lines end on other final states. These lines lie in regions of the spectrum not visible to the eye, the infrared and ultraviolet. The generalized Rydberg equation for any initial and final states is

$$\frac{1}{\lambda} = R_{\mathrm{H}}\left(\frac{1}{n_f^2} - \frac{1}{n_i^2}\right) \quad [11.28]$$

In this equation, different series correspond to different values of $n_f$ and different lines within a series correspond to varying values of $n_i$.

Bohr immediately extended his structural model for hydrogen to other elements in which all but one electron had been removed. Ionized elements such as $He^+$, $Li^{2+}$, and $Be^{3+}$ were suspected to exist in hot stellar atmospheres, where frequent atomic collisions occur with enough energy to completely remove one or

**PITFALL PREVENTION 11.4**

**THE BOHR MODEL IS GREAT, BUT . . .** The Bohr model correctly predicts the ionization energy for hydrogen, but it cannot account for the spectra of more complex atoms and is unable to predict many subtle spectral details of hydrogen and other simple atoms. Despite its cultural popularity, the notion of electrons in well-defined orbits about the nucleus is *not* consistent with current models of the atom.

more atomic electrons. Bohr showed that many mysterious lines observed in the Sun and several stars could not be due to hydrogen, but were correctly predicted by his theory if attributed to singly ionized helium.

**QUICK QUIZ 11.4** A hydrogen atom makes a transition from the $n = 3$ level to the $n = 2$ level. It then makes a transition from the $n = 2$ level to the $n = 1$ level. Which transition results in emission of the longest-wavelength photon? **(a)** the first transition **(b)** the second transition **(c)** neither, because the wavelengths are the same for both transitions

**INTERACTIVE EXAMPLE 11.3 An Electronic Transition in Hydrogen**

A hydrogen atom makes a transition from the $n = 2$ state to the ground state (corresponding to $n = 1$). Find the wavelength and frequency of the emitted radiation.

**Solution** We can use Equation 11.28 directly to obtain $\lambda$, with $n_i = 2$ and $n_f = 1$:

$$\frac{1}{\lambda} = R_H\left(\frac{1}{n_f^2} - \frac{1}{n_i^2}\right) = R_H\left(\frac{1}{1^2} - \frac{1}{2^2}\right) = \frac{3R_H}{4}$$

$$\lambda = \frac{4}{3R_H} = \frac{4}{3(1.097 \times 10^{-7}\ \text{m}^{-1})}$$

$$= 1.215 \times 10^{-7}\ \text{m} = 121.5\ \text{nm (ultraviolet)}$$

Because $c = f\lambda$, the frequency of the radiation is

$$f = \frac{c}{\lambda} = \frac{3.00 \times 10^8\ \text{m/s}}{1.215 \times 10^{-7}\ \text{m}} = 2.47 \times 10^{15}\ \text{s}^{-1}$$

**PhysicsNow™** Investigate transitions of the atom between states by logging into PhysicsNow at **www.pop4e.com** and going to Interactive Example 11.3.

## 11.6 CHANGING FROM A CIRCULAR TO AN ELLIPTICAL ORBIT

CONTEXT CONNECTION

In Interactive Example 11.1, we discussed a spacecraft in a circular orbit around the Earth. From our studies of Kepler's laws in this chapter, we are also aware that an elliptical orbit is possible for our spacecraft. Let us investigate how the motion of our spacecraft can be changed from a circular to an elliptical orbit, which will set us up for the conclusion to our *Mission to Mars* Context.

Let us identify the system as the spacecraft and the Earth, *but not the portion of the fuel in the spacecraft that we use to change the orbit.* In a given orbit, the mechanical energy of the spacecraft–Earth system is given by Equation 11.10,

$$E = -\frac{GMm}{2r}$$

This energy includes the kinetic energy of the spacecraft and the potential energy associated with the gravitational force between the spacecraft and the Earth. If the rocket engines are fired, the exhausted fuel can be seen as doing work on the spacecraft–Earth system because the thrust force moves through a displacement. As a result, the mechanical energy of the spacecraft–Earth system increases.

The spacecraft has a new, higher energy but is constrained to be in an orbit that includes the original starting point. It cannot be in a higher-energy circular orbit having a larger radius because this orbit would not contain the starting point. The only possibility is that the orbit is elliptical. Figure 11.21 shows the change from the original circular orbit to the new elliptical orbit for our spacecraft.

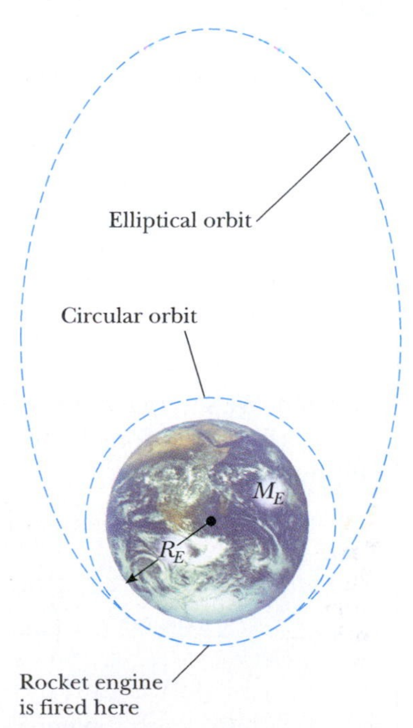


**FIGURE 11.21** A spacecraft, originally in a circular orbit about the Earth, fires its engines and enters an elliptical orbit about the Earth.

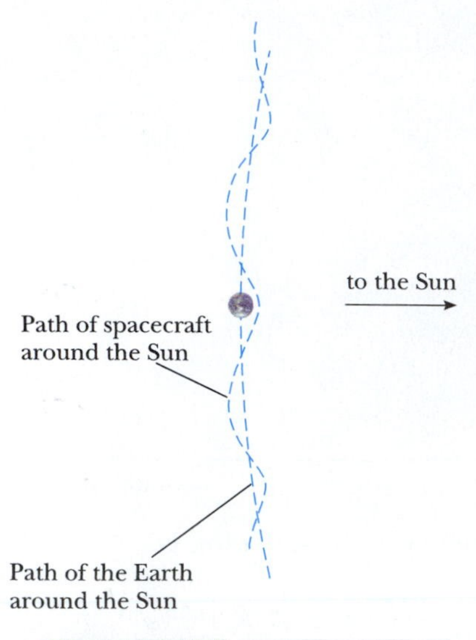


**FIGURE 11.22** A spacecraft in orbit about the Earth can be modeled as one in a circular orbit about the Sun, with its orbit about the Earth appearing as small perturbations from the circular orbit.

Equation 11.11 gives the energy of the spacecraft–Earth system for an elliptical orbit. Therefore, if we know the new energy of the orbit, we can find the semimajor axis of the elliptical orbit. Conversely, if we know the semimajor axis of an elliptical orbit we would like to achieve, we can calculate how much additional energy is required from the rocket engines. This information can then be converted to a required burn time for the rockets.

Larger amounts of energy increase supplied by the rocket engines will move the spacecraft into elliptical orbits with larger semimajor axes. What happens if the burn time of the engines is so long that the total mechanical energy of the spacecraft–Earth system becomes positive? A positive energy refers to an *unbound* system. Therefore, in this case, the spacecraft will *escape* from the Earth, going into a hyperbolic path that would not bring it back to the Earth.

This process is the essence of what must be done to transfer to Mars. Our rocket engines must be fired to leave the circular parking orbit and escape the Earth. At this point, our thinking must shift to a spacecraft–Sun system rather than a spacecraft–Earth system. From this point of view, the spacecraft in orbit around the Earth can also be considered to be in a circular orbit around the Sun, moving along with the Earth, as shown in Figure 11.22. The orbit is not a perfect circle because there are perturbations corresponding to its extra motion around the Earth, but these perturbations are small compared with the radius of the orbit around the Sun. When our engines are fired to escape from the Earth, our orbit around the Sun changes from a circular orbit (ignoring the perturbations) to an elliptical one with the Sun at one focus. We shall choose the semimajor axis of our elliptical orbit so that it intersects the orbit of Mars! In the Context 2 Conclusion, we shall look at more details of this process.

## EXAMPLE 11.4 How High Do We Go?

Imagine that you are in a spacecraft in circular orbit around the Earth, at a height of 300 km from the surface. You fire your rocket engines, and as a result the magnitude of the mechanical energy of the spacecraft–Earth system decreases by 10.0%. (Because the mechanical energy is negative, a decrease in magnitude is an increase in energy.) What is the greatest height of your spacecraft above the surface of the Earth in your new orbit?

**Solution** We set up a ratio of the energies of the two orbits, using Equations 11.10 and 11.11 for circular and elliptical orbits:

$$\frac{E_{\text{elliptical}}}{E_{\text{circular}}} = \frac{\left(-\dfrac{GMm}{2a}\right)}{\left(-\dfrac{GMm}{2r}\right)} = \frac{r}{a}$$

The ratio on the left is 0.900 because of the 10.0% decrease in magnitude of the mechanical energy. Therefore,

$$0.900 = \frac{r}{a} \quad \rightarrow \quad a = \frac{r}{0.900} = 1.11r$$

From this equation, we can find the semimajor axis for the elliptical orbit:

$$a = 1.11r = 1.11(6.37 \times 10^3 \text{ km} + 3.00 \times 10^2 \text{ km})$$
$$= 7.40 \times 10^3 \text{ km}$$

The maximum distance from the center of the Earth will occur when the spacecraft is at apogee and is given by

$$r_{\max} = 2a - r = 2(7.40 \times 10^3 \text{ km}) - (6.67 \times 10^3 \text{ km})$$
$$= 8.14 \times 10^3 \text{ km}$$

Now, if we subtract the radius of the Earth, we have the maximum height above the surface:

$$h_{\max} = r_{\max} - R_E = 8.14 \times 10^3 \text{ km} - 6.37 \times 10^3 \text{ km}$$
$$= 1.77 \times 10^3 \text{ km}$$

## SUMMARY

 Take a practice test by logging into PhysicsNow at www.pop4e.com and clicking on the Pre-Test link for this chapter.

**Newton's law of universal gravitation** states that the gravitational force of attraction between any two particles of masses $m_1$ and $m_2$ separated by a distance $r$ has the magnitude

$$F_g = G\frac{m_1 m_2}{r^2} \qquad [11.1]$$

where $G$ is the **universal gravitational constant** whose value is $6.673 \times 10^{-11}\ \text{N}\cdot\text{m}^2/\text{kg}^2$.

Rather than considering the gravitational force as a direct interaction between two objects, we can imagine that one object sets up a **gravitational field** in space:

$$\vec{\mathbf{g}} \equiv \frac{\vec{\mathbf{F}}_g}{m} \qquad [11.4]$$

A second object in this field experiences a force $\vec{\mathbf{F}}_g = m\vec{\mathbf{g}}$ when placed in this field.

**Kepler's laws of planetary motion** state the following:

1. Each planet in the Solar System moves in an elliptical orbit with the Sun at one focus.
2. The radius vector drawn from the Sun to any planet sweeps out equal areas in equal time intervals.
3. The square of the orbital period of any planet is proportional to the cube of the semimajor axis of the elliptical orbit.

**Kepler's first law** is a consequence of the inverse-square nature of the law of universal gravitation. The **semimajor axis** of an ellipse is $a$, where $2a$ is the longest dimension of the ellipse. The **semiminor axis** of the ellipse is $b$, where $2b$ is the shortest dimension of the ellipse. The **eccentricity** of the ellipse is $e = c/a$, where $c$ is the distance between the center and a focus and $a^2 = b^2 + c^2$.

**Kepler's second law** is a consequence of the gravitational force being a central force. For a central force, the angular momentum of the planet is conserved.

**Kepler's third law** is a consequence of the inverse-square nature of the universal law of gravitation. Newton's second law, together with the force law given by Equation 11.1, verifies that the period $T$ and semimajor axis $a$ of the orbit of a planet about the Sun are related by

$$T^2 = \left(\frac{4\pi^2}{GM_S}\right)a^3 \qquad [11.7]$$

where $M_S$ is the mass of the Sun.

If an isolated system consists of a particle of mass $m$ moving with a speed $v$ in the vicinity of a massive body of mass $M$, the *total energy* of the system is constant and is

$$E = \tfrac{1}{2}mv^2 - \frac{GMm}{r} \qquad [11.8]$$

If $m$ moves in an elliptical orbit of major axis $2a$ about $M$, where $M \gg m$, the total energy of the system is

$$E = -\frac{GMm}{2a} \qquad [11.11]$$

The total energy is negative for any bound system, that is, one in which the orbit is closed, such as a circular or an elliptical orbit.

The Bohr model of the atom successfully describes the spectra of atomic hydrogen and hydrogen-like ions. One basic assumption of this structural model is that the electron can exist only in discrete orbits such that the angular momentum $m_e vr$ is an integral multiple of $\hbar \equiv h/2\pi$. Assuming circular orbits and a simple electrical attraction between the electron and proton, the energies of the quantum states for hydrogen are calculated to be

$$E_n = -\frac{k_e e^2}{2a_0}\left(\frac{1}{n^2}\right) \qquad n = 1, 2, 3, \ldots \qquad [11.24]$$

where $k_e$ is the Coulomb constant, $e$ is the fundamental electric charge, $n$ is a positive integer called a **quantum number,** and $a_0 = 0.052\,9$ nm is the **Bohr radius.**

If the hydrogen atom makes a transition from a state whose quantum number is $n_i$ to one whose quantum number is $n_f$, where $n_f < n_i$, the frequency of the radiation emitted by the atom is

$$f = \frac{k_e e^2}{2a_0 h}\left(\frac{1}{n_f^2} - \frac{1}{n_i^2}\right) \qquad [11.26]$$

Using $E_i - E_f = hf = hc/\lambda$, one can calculate the wavelengths of the radiation for various transitions. The calculated wavelengths are in excellent agreement with those in observed atomic spectra.

## QUESTIONS

☐ = answer available in the *Student Solutions Manual and Study Guide*

1. If the gravitational force on an object is directly proportional to its mass, why don't objects with large masses fall with greater acceleration than small ones?

2. The gravitational force exerted by the Sun on you is downward into the Earth at night and upward into the sky during the day. If you had a sensitive enough bathroom scale, would you expect to weigh more at night than during the day? Note also that you are farther away from the Sun at night than during the day. Would you expect to weigh less?

3. The gravitational force that the Sun exerts on the Moon is about twice as great as the gravitational force that the Earth exerts on the Moon. Why doesn't the Sun pull the Moon away from the Earth during a total eclipse of the Sun?

4. A satellite in orbit is not truly traveling through a vacuum. It is moving through very thin air. Does the resulting air friction cause the satellite to slow down?

5. Explain why it takes more fuel for a spacecraft to travel from the Earth to the Moon than for the return trip. Estimate the difference.

6. Explain why no work is done on a planet as it moves in a circular orbit around the Sun, even though a gravitational force is acting on the planet. What is the net work done on a planet during each revolution as it moves around the Sun in an elliptical orbit?

7. Why don't we put a geosynchronous weather satellite in orbit around the 45th parallel? Wouldn't that be more useful in the United States than one in orbit around the equator?

8. If a hole could be dug to the center of the Earth, would the force on an object of mass $m$ still obey Equation 11.1 there? What do you think the force on $m$ would be at the center of the Earth?

9. At what position in its elliptical orbit is the speed of a planet a maximum? At what position is the speed a minimum?

10. Each *Voyager* spacecraft was accelerated toward escape speed from the Sun by Jupiter's gravitational force exerted on the spacecraft. How is that possible?

11. In his 1798 experiment, Cavendish was said to have "weighed the Earth." Explain this statement.

12. The *Apollo 13* spacecraft developed trouble in the oxygen system about halfway to the Moon. Why did the mission continue on around the Moon and then return home, rather than immediately turn back to the Earth?

13. Suppose the system of a hydrogen atom obeyed classical mechanics rather than quantum mechanics. Why should such a hypothetical atom emit a continuous spectrum rather than the observed line spectrum?

14. Can the electron in the ground state of hydrogen absorb a photon of energy (a) less than 13.6 eV and (b) greater than 13.6 eV?

15. Explain why, in the Bohr model, the total energy of the atom is negative.

16. Let $-E$ represent the energy of a hydrogen atom. What is the kinetic energy of the electron? What is the potential energy of the atom?

# PROBLEMS

1, 2, 3 = straightforward, intermediate, challenging
☐ = full solution available in the *Student Solutions Manual and Study Guide*
**Physics Now™** = coached problem with hints available at www.pop4e.com
= computer useful in solving problem
= paired numerical and symbolic problems
= biomedical application

## Section 11.1 ▪ Newton's Law of Universal Gravitation Revisited

Problems 5.31 through 5.33 in Chapter 5 can be assigned with this section.

1. Two ocean liners, each with a mass of 40 000 metric tons, are moving on parallel courses, 100 m apart. What is the magnitude of the acceleration of one of the liners toward the other due to their mutual gravitational attraction? Treat the ships as particles.

2. A 200-kg object and a 500-kg object are separated by 0.400 m. (a) Find the net gravitational force exerted by these objects on a 50.0-kg object placed midway between them. (b) At what position (other than an infinitely remote one) can the 50.0-kg object be placed so as to experience a net force of zero?

3. **Physics Now™** In introductory physics laboratories, a typical Cavendish balance for measuring the gravitational constant $G$ uses lead spheres with masses of 1.50 kg and 15.0 g whose centers are separated by about 4.50 cm. Calculate the gravitational force between these spheres, treating each as a particle located at the center of the sphere.

4. Three uniform spheres of mass 2.00 kg, 4.00 kg, and 6.00 kg are placed at the corners of a right triangle as shown in Figure P11.4. Calculate the resultant gravitational force on the 4.00-kg object, assuming that the spheres are isolated from the rest of the Universe.

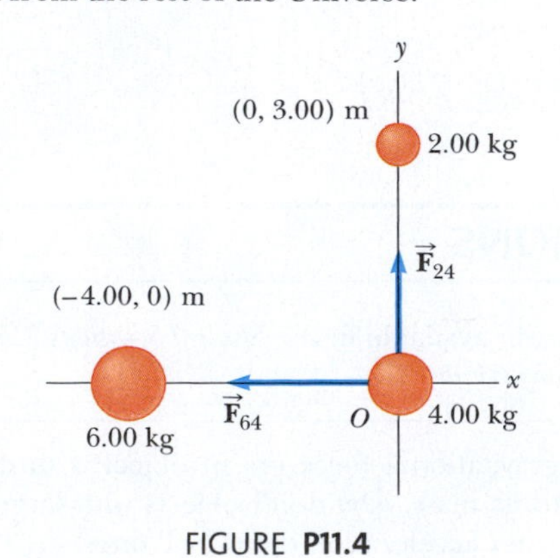


FIGURE **P11.4**

5. During a solar eclipse, the Moon, the Earth, and the Sun all lie on the same line, with the Moon between the Earth and the Sun. (a) What force is exerted by the Sun on the Moon? (b) What force is exerted by the Earth on the Moon? (c) What force is exerted by the Sun on the Earth?

6. A student proposes to measure the gravitational constant $G$ by suspending two spherical objects from the ceiling of a tall cathedral and measuring the deflection of the cables from the vertical. Draw a free-body diagram of one of the objects. If two 100.0-kg objects are suspended at the lower ends of cables 45.00 m long and the cables are attached to the ceiling 1.000 m apart, what is the separation of the objects?

7. The free-fall acceleration on the surface of the Moon is about one-sixth that on the surface of the Earth. Assuming that the radius of the Moon is about $0.250R_E$, find the ratio of their average densities, $\rho_{\text{Moon}}/\rho_{\text{Earth}}$.

8. On the way to the Moon, the *Apollo* astronauts passed a point after which the Moon's gravitational pull became stronger than the Earth's. (a) Determine the distance of this point from the center of the Earth. (b) What is the acceleration due to the Earth's gravitation at this point?

9. **Review problem.** Miranda, a satellite of Uranus, is shown in Figure P11.9a. It can be modeled as a sphere of radius 242 km and mass $6.68 \times 10^{19}$ kg. (a) Find the free-fall acceleration on its surface. (b) A cliff on Miranda is 5 000 m high. It appears on the limb at the 11 o'clock position in Figure P11.9a and is magnified in Figure P11.9b. If a devotee of extreme sports runs horizontally off the top of the cliff at 8.50 m/s, for what time interval will he be in flight? (Or will he be in orbit?) (c) How far from the base of the vertical cliff will he strike the icy surface of Miranda? (d) What will be his vector impact velocity?

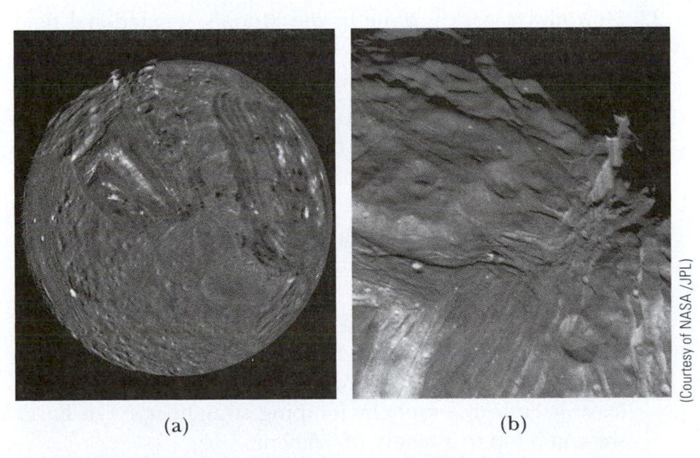


FIGURE P11.9 (a) Miranda, a moon of Uranus. (b) A magnified image of a 5 000-m cliff on Miranda

10. A spacecraft in the shape of a long cylinder has a length of 100 m, and its mass with occupants is 1 000 kg. It has strayed too close to a black hole having a mass 100 times that of the Sun (Fig. P11.10). The nose of the spacecraft points toward the black hole, and the distance between the nose and the center of the black hole is 10.0 km. (a) Determine the total force on the spacecraft. (b) What is the difference in the gravitational fields acting on the occupants in the nose of the ship and on those in the rear of the ship, farthest from the black hole? This difference in accelerations grows rapidly as the ship approaches the black hole. It puts the body of the ship under extreme tension and eventually tears it apart.

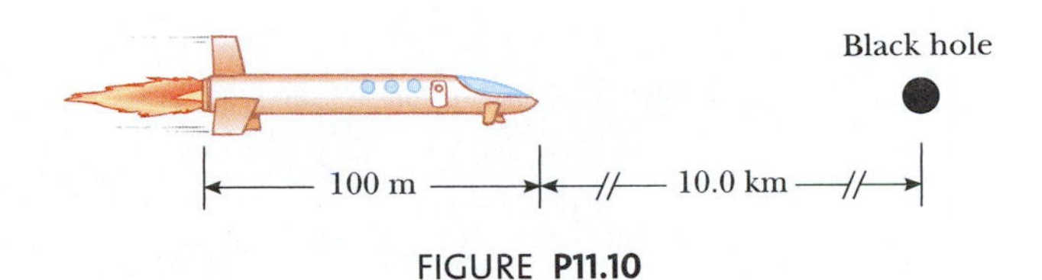


FIGURE P11.10

11. Compute the magnitude and direction of the gravitational field at a point $P$ on the perpendicular bisector of the line joining two objects of equal mass separated by a distance $2a$, as shown in Figure P11.11.

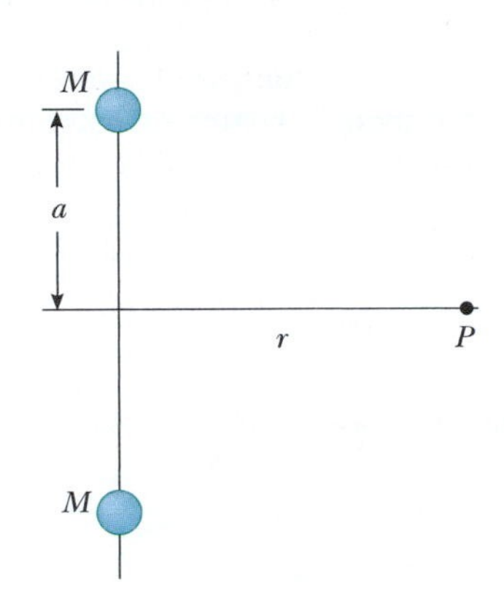


FIGURE P11.11

12. A satellite of mass 300 kg is in a circular orbit around the Earth at an altitude equal to the Earth's mean radius. Find (a) the satellite's orbital speed, (b) the period of its revolution, and (c) the gravitational force acting on it.

## Section 11.3 ■ Kepler's Laws

13. A communication satellite in geosynchronous orbit remains above a single point on the Earth's equator as the planet rotates on its axis. (a) Calculate the radius of its orbit. (b) The satellite relays a radio signal from a transmitter near the North Pole to a receiver, also near the North Pole. Traveling at the speed of light, how long is the radio wave in transit?

14. The *Explorer VIII* satellite, placed into orbit November 3, 1960, to investigate the ionosphere, had the following orbit parameters: perigee, 459 km; apogee, 2 289 km (both distances above the Earth's surface); period, 112.7 min. Find the ratio $v_p/v_a$ of the speed at perigee to that at apogee.

15. **Physics Now™** Io, a satellite of Jupiter, has an orbital period of 1.77 days and an orbital radius of $4.22 \times 10^5$ km. From these data, determine the mass of Jupiter.

16. Comet Halley approaches the Sun to within 0.570 AU (Fig. P11.16), and its orbital period is 75.6 yr. (AU is the symbol for astronomical unit, where 1 AU = $1.50 \times 10^{11}$ m is the mean Earth–Sun distance.) How far from the Sun will this comet travel before it starts its return journey?

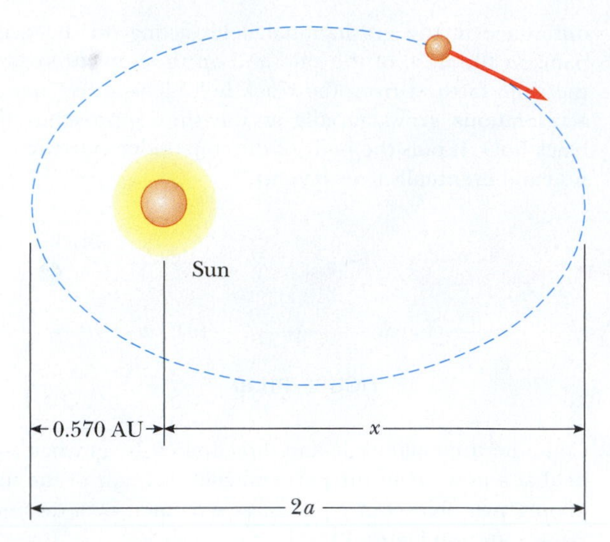


FIGURE **P11.16** The elliptical orbit of Comet Halley (not to scale).

**17.** Plaskett's binary system consists of two stars that revolve in a circular orbit about a center of mass midway between them. Therefore, the masses of the two stars are equal (Fig. P11.17). Assume that the orbital speed of each star is 220 km/s and that the orbital period of each is 14.4 days. Find the mass $M$ of each star. (For comparison, the mass of our Sun is $1.99 \times 10^{30}$ kg.)

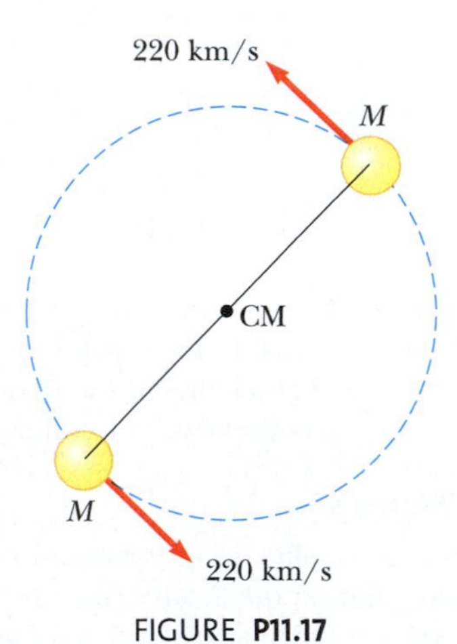


FIGURE **P11.17**

**18.** Two planets X and Y travel counterclockwise in circular orbits about a star as shown in Figure P11.18. The radii of their orbits are in the ratio 3:1. At some time, they are aligned as shown in Figure P11.18a, making a straight line with the star. During the next five years, the angular displacement of planet X is 90.0°, as shown in Figure P11.18b. Where is planet Y at this time?

**19.** Suppose the Sun's gravity were switched off. The planets would leave their nearly circular orbits and fly away in straight lines, as described by Newton's first law. Would Mercury ever be farther from the Sun than Pluto? If so, find how long it would take for Mercury to achieve this passage. If not, give a convincing argument that Pluto is always farther from the Sun than is Mercury.

**20.** As thermonuclear fusion proceeds in its core, the Sun loses mass at a rate of $3.64 \times 10^9$ kg/s. During the 5 000-yr period of recorded history, by how much has the length of the year changed due to the loss of mass from the Sun? (*Suggestions:* Assume that the Earth's orbit is circular. No external torque acts on the Earth–Sun system, so angular momentum is conserved. If $x$ is small compared to 1, then $(1 + x)^n$ is nearly equal to $1 + nx$.)

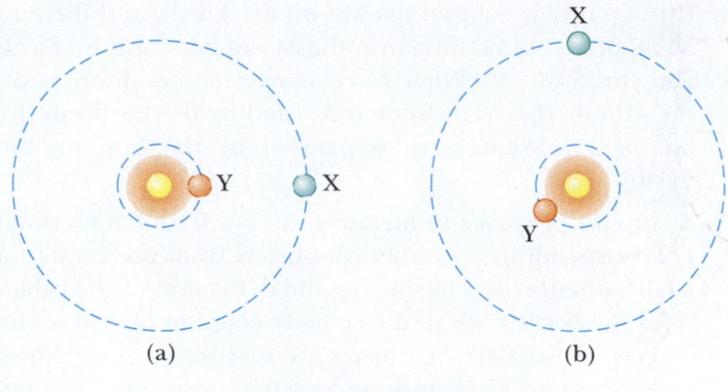


FIGURE **P11.18**

## Section 11.4 ■ Energy Considerations in Planetary and Satellite Motion

Problems 7.35 through 7.38 in Chapter 7 can be assigned with this section.

**21.** After our Sun exhausts its nuclear fuel, its ultimate fate may be to collapse to a *white dwarf* state, in which it has approximately the same mass as it has now but a radius equal to the radius of the Earth. Calculate (a) the average density of the white dwarf, (b) the free-fall acceleration, and (c) the gravitational potential energy associated with a 1.00-kg object at its surface.

**22.** How much work is done by the Moon's gravitational field as a 1 000-kg meteor comes in from outer space and impacts on the Moon's surface?

**23.** An asteroid is on a collision course with Earth. An astronaut lands on the rock to bury explosive charges that will blow the asteroid apart. Most of the small fragments will miss the Earth, and those that fall into the atmosphere will produce only a beautiful meteor shower. The astronaut finds that the density of the spherical asteroid is equal to the average density of the Earth. To ensure its pulverization, she incorporates into the explosives the rocket fuel and oxidizer intended for her return journey. What maximum radius can the asteroid have for her to be able to leave it entirely simply by jumping straight up? On Earth she can jump to a height of 0.500 m.

**24.** (a) Determine the amount of work that must be done on a 100-kg payload to elevate it to a height of 1 000 km above the Earth's surface. (b) Determine the amount of additional work that is required to put the payload into circular orbit at this elevation.

**25.** **Physics Now™** A space probe is fired up from the Earth's surface with an initial speed of $2.00 \times 10^4$ m/s. What will its speed be when it is very far from the Earth? Ignore friction and the rotation of the Earth.

By permission of John Hart FLP, and Creators Syndicate, Inc.

FIGURE **P11.27**

**26.** (a) What is the minimum speed, relative to the Sun, necessary for a spacecraft to escape the solar system if it starts at the Earth's orbit? (b) *Voyager 1* achieved a maximum speed of 125 000 km/h on its way to photograph Jupiter. Beyond what distance from the Sun is this speed sufficient to escape the solar system?

**27.** A "treetop satellite" (Fig. P11.27) moves in a circular orbit just above the surface of a planet, assumed to offer no air resistance. Show that its orbital speed $v$ and the escape speed from the planet are related by the expression $v_{\text{esc}} = \sqrt{2}v$.

**28.** An object is released from rest at an altitude $h$ above the surface of the Earth. (a) Show that its speed at a distance $r$ from the Earth's center, where $R_E \leq r \leq R_E + h$, is given by

$$v = \sqrt{2GM_E\left(\frac{1}{r} - \frac{1}{R_E + h}\right)}$$

(b) Assume that the release altitude is 500 km. Perform the integral

$$\Delta t = \int_i^f dt = \int_i^f -\frac{dr}{v}$$

to find the time of fall during which the object moves from the release point to the Earth's surface. The negative sign appears because the object is moving opposite to the radial direction, so its speed is $v = -dr/dt$. Perform the integral numerically.

**29.** A 500-kg satellite is in a circular orbit at an altitude of 500 km above the Earth's surface. Because of air friction, the satellite eventually falls to the Earth's surface, where it hits the ground with a speed of 2.00 km/s. How much energy was transformed into internal energy by means of air friction?

**30.** A satellite of mass $m$, originally on the surface of the Earth, is placed into Earth orbit at an altitude $h$. (a) With a circular orbit, how long does the satellite take to complete one orbit? (b) What is the satellite's speed? (c) What is the minimum energy input necessary to place this satellite in orbit? Ignore air resistance but include the effect of the planet's daily rotation. At what location on the Earth's surface and in what direction should the satellite be launched to minimize the required energy investment? Represent the mass and radius of the Earth as $M_E$ and $R_E$.

**31.** An object is fired vertically upward from the surface of the Earth (of radius $R_E$) with an initial speed $v_i$ that is comparable to but less than the escape speed $v_{\text{esc}}$. (a) Show that the object attains a maximum height $h$ given by

$$h = \frac{R_E v_i^2}{v_{\text{esc}}^2 - v_i^2}$$

(b) A space vehicle is launched vertically upward from the Earth's surface with an initial speed of 8.76 km/s, which is less than the escape speed of 11.2 km/s. What maximum height does it attain? (c) A meteorite falls toward the Earth. It is essentially at rest with respect to the Earth when it is at a height of $2.51 \times 10^7$ m. With what speed does the meteorite strike the Earth? (d) Assume that a baseball is tossed up with an initial speed that is very small compared with the escape speed. Show that the equation from part (a) is consistent with Equation 3.15.

**32.** Derive an expression for the work required to move an Earth satellite of mass $m$ from a circular orbit of radius $2R_E$ to one of radius $3R_E$.

**33.** A comet of mass $1.20 \times 10^{10}$ kg moves in an elliptical orbit around the Sun. Its distance from the Sun ranges between 0.500 AU and 50.0 AU. (a) What is the eccentricity of its orbit? (b) What is its period? (c) At aphelion, what is the potential energy of the comet–Sun system? (*Note:* 1 AU = one astronomical unit = the average distance from the Sun to the Earth = $1.496 \times 10^{11}$ m.)

## Section 11.5 ▪ Atomic Spectra and the Bohr Theory of Hydrogen

**34.** Within the Rosette Nebula shown in the photograph opening this chapter, a hydrogen atom emits light as it undergoes a transition from the $n = 3$ state to the $n = 2$ state. Calculate (a) the energy, (b) the wavelength, and (c) the frequency of the radiation.

**35.** (a) What value of $n_i$ is associated with the 94.96-nm spectral line in the Lyman series of hydrogen? (b) Could this wavelength be associated with the Paschen series or the Balmer series?

**36.** For a hydrogen atom in its ground state, use the Bohr model to compute (a) the orbital speed of the electron, (b) the kinetic energy of the electron, and (c) the electric potential energy of the atom.

**37.** Four possible transitions for a hydrogen atom are as follows:

(i) $n_i = 2;\ n_f = 5$ (ii) $n_i = 5;\ n_f = 3$

(iii) $n_i = 7;\ n_f = 4$ (iv) $n_i = 4;\ n_f = 7$

(a) In which transition is light of the shortest wavelength emitted? (b) In which transition does the atom gain the most energy? (c) In which transition(s) does the atom lose energy?

**38.** How much energy is required to ionize hydrogen (a) when it is in the ground state and (b) when it is in the state for which $n = 3$?

**39.** **PhysicsNow™** A hydrogen atom is in its first excited state ($n = 2$). Using the Bohr theory of the atom, calculate (a) the radius of the orbit, (b) the linear momentum of the electron, (c) the angular momentum of the electron, (d) the kinetic energy, (e) the potential energy, and (f) the total energy.

**40.** Show that the speed of the electron in the $n$th Bohr orbit in hydrogen is given by

$$v_n = \frac{k_e e^2}{n\hbar}$$

**41.** Two hydrogen atoms collide head-on and end up with zero kinetic energy. Each atom then emits light with a wavelength of 121.6 nm ($n = 2$ to $n = 1$ transition). At what speed were the atoms moving before the collision?

## Section 11.6 ▪ Context Connection—Changing From a Circular to an Elliptical Orbit

**42.** A spacecraft of mass $1.00 \times 10^4$ kg is in a circular orbit at an altitude of 500 km above the Earth's surface. Mission Control wants to fire the engines so as to put the spacecraft in an elliptical orbit around the Earth with an apogee of $2.00 \times 10^4$ km. How much energy must be used from the fuel to achieve this orbit? (Assume that all the fuel energy goes into increasing the orbital energy. This model will give a lower limit to the required energy because some of the energy from the fuel will appear as internal energy in the hot exhaust gases and engine parts.)

**43.** A spacecraft is approaching Mars after a long trip from the Earth. Its velocity is such that it is traveling along a parabolic trajectory under the influence of the gravitational force from Mars. The distance of closest approach will be 300 km above the Martian surface. At this point of closest approach, the engines will be fired to slow down the spacecraft and place it in a circular orbit 300 km above the surface. (a) By what percentage must the speed of the spacecraft be reduced to achieve the desired orbit? (b) How would the answer to part (a) change if the distance of closest approach and the desired circular orbit altitude were 600 km instead of 300 km? (*Note:* The energy of the spacecraft–Mars system for a parabolic orbit is $E = 0$.)

## Additional Problems

**44.** The Solar and Heliospheric Observatory (SOHO) spacecraft has a special orbit, chosen so that its view of the Sun is never eclipsed and it is always close enough to the Earth to transmit data easily. It moves in a near-circle around the Sun that is smaller than the Earth's circular orbit. Its period, however, is not less than 1 yr, but just equal to 1 yr. It is always located between the Earth and the Sun along the line joining them. Both objects exert gravitational forces on the observatory. Show that the spacecraft's distance from the Earth must be between $1.47 \times 10^9$ m and $1.48 \times 10^9$ m. In 1772, Joseph Louis Lagrange determined theoretically the special location allowing this orbit. The SOHO spacecraft took this position on February 14, 1996. (*Suggestions:* Use data that are precise to four digits. The mass of the Earth is $5.983 \times 10^{24}$ kg.)

**45.** Let $\Delta g_M$ represent the difference in the gravitational fields produced by the Moon at the points on the Earth's surface nearest to and farthest from the Moon. Find the fraction $\Delta g_M/g$, where $g$ is the Earth's gravitational field. (This difference is responsible for the occurrence of the *lunar tides* on the Earth.)

**46.** **Review problem.** Two identical hard spheres, each of mass $m$ and radius $r$, are released from rest in otherwise empty space with their centers separated by the distance $R$. They are allowed to collide under the influence of their gravitational attraction. (a) Show that the magnitude of the impulse received by each sphere before they make contact is given by $[Gm^3(1/2r - 1/R)]^{1/2}$. (b) Find the magnitude of the impulse each receives if they collide elastically.

**47.** (a) Show that the rate of change of the free-fall acceleration with distance above the Earth's surface is

$$\frac{dg}{dr} = -\frac{2GM_E}{R_E^3}$$

This rate of change over distance is called a *gradient.* (b) Assuming that $h$ is small in comparison to the radius of the Earth, show that the difference in free-fall acceleration between two points separated by vertical distance $h$ is

$$|\Delta g| = \frac{2GM_E h}{R_E^3}$$

(c) Evaluate this difference for $h = 6.00$ m, a typical height for a two-story building.

**48.** A ring of matter is a familiar structure in planetary and stellar astronomy. Examples include Saturn's rings and a ring nebula. Consider a uniform ring of mass $2.36 \times 10^{20}$ kg and radius $1.00 \times 10^{8}$ m. An object of mass 1 000 kg is placed at a point $A$ on the axis of the ring, $2.00 \times 10^{8}$ m from the center of the ring (Fig. P11.48). When the object is released, the attraction of the ring makes the object move along the axis toward the center of the ring (point $B$). (a) Calculate the gravitational potential energy of the object–ring system when the object is at $A$. (b) Calculate the gravitational potential energy of the system when the object is at $B$. (c) Calculate the speed of the object as it passes through $B$.

(NASA)

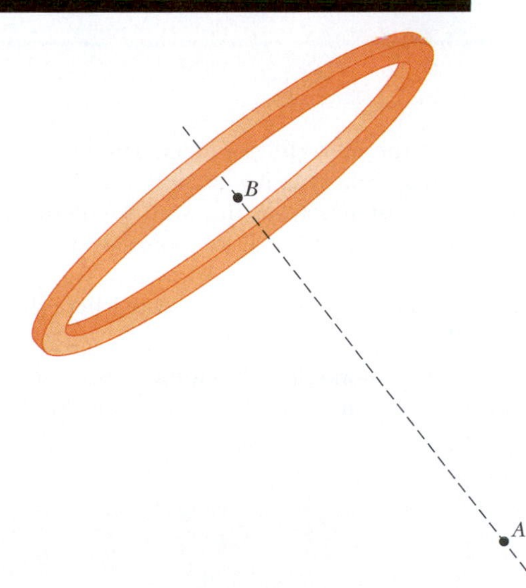


FIGURE **P11.48**

**49.** *Voyager 1* and *Voyager 2* surveyed the surface of Jupiter's moon Io and photographed active volcanoes spewing liquid sulfur to heights of 70 km above the surface of this moon. Find the speed with which the liquid sulfur left the volcano. Io's mass is $8.9 \times 10^{22}$ kg, and its radius is 1 820 km.

**50.** As an astronaut, you observe a small planet to be spherical. After landing on the planet, you set off, walking always straight ahead, and find yourself returning to your spacecraft from the opposite side after completing a lap of 25.0 km. You hold a hammer and a falcon feather at a height of 1.40 m, release them, and observe that they fall together to the surface in 29.2 s. Determine the mass of the planet.

**51.** Many people assume that air resistance acting on a moving object will always make the object slow down. It can actually be responsible for making the object speed up. Consider a 100-kg Earth satellite in a circular orbit at an altitude of 200 km. A small force of air resistance makes the satellite drop into a circular orbit with an altitude of 100 km. (a) Calculate its initial speed. (b) Calculate its final speed in this process. (c) Calculate the initial energy of the satellite–Earth system. (d) Calculate the final energy of the system. (e) Show that the system has lost mechanical energy and find the amount of the loss due to friction. (f) What force makes the satellite's speed increase? You will find a free-body diagram useful in explaining your answer.

**52.** The maximum distance from the Earth to the Sun (at our aphelion) is $1.521 \times 10^{11}$ m, and the distance of closest approach (at perihelion) is $1.471 \times 10^{11}$ m. The Earth's orbital speed at perihelion is $3.027 \times 10^{4}$ m/s. Determine (a) the Earth's orbital speed at aphelion, (b) the kinetic and potential energies of the Earth–Sun system at perihelion, and (c) the kinetic and potential energies at aphelion. Is the total energy constant? (Ignore the effect of the Moon and other planets.)

**53.** **PhysicsNow™** Two hypothetical planets of masses $m_1$ and $m_2$ and radii $r_1$ and $r_2$, respectively, are nearly at rest when they are an infinite distance apart. Because of their gravitational attraction, they head toward each other on a collision course. (a) When their center-to-center separation is $d$, find expressions for the speed of each planet and for their relative speed. (b) Find the kinetic energy of each planet just before they collide, taking $m_1 = 2.00 \times 10^{24}$ kg, $m_2 = 8.00 \times 10^{24}$ kg, $r_1 = 3.00 \times 10^{6}$ m, and $r_2 = 5.00 \times 10^{6}$ m. (*Note:* Both energy and momentum of the system are conserved.)

**54.** Assume that you are agile enough to run across a horizontal surface at 8.50 m/s, independently of the value of the gravitational field. What would be (a) the radius and (b) the mass of an airless spherical asteroid of uniform density $1.10 \times 10^{3}$ kg/m$^3$ on which you could launch yourself into orbit by running? (c) What would be your period? (d) Take your mass as 90.0 kg. If the asteroid were originally stationary, your running would set it into rotation with what period?

**55.** Studies of the relationship of the Sun to its galaxy—the Milky Way—have revealed that the Sun is located near the outer edge of the galactic disc, about 30 000 ly from the center. The Sun has an orbital speed of approximately 250 km/s around the galactic center. (a) What is the period of the Sun's galactic motion? (b) What is the order of magnitude of the mass of the Milky Way galaxy? Suppose the galaxy is made mostly of stars of which the Sun is typical. What is the order of magnitude of the number of stars in the Milky Way?

**56.** The oldest artificial satellite in orbit is *Vanguard I*, launched March 3, 1958. Its mass is 1.60 kg. In its initial orbit, its minimum distance from the center of the Earth was 7.02 Mm and its speed at this perigee point was 8.23 km/s. (a) Find the total energy of the satellite–Earth system. (b) Find the magnitude of the angular momentum of the satellite. (c) Find its speed at apogee and its maximum (apogee) distance from the center of the Earth. (d) Find the semimajor axis of its orbit. (e) Determine its period.

**57.** Astronomers detect a distant meteoroid moving along a straight line that, if extended, would pass at a distance $3R_E$ from the center of the Earth, where $R_E$ is the radius of the Earth. What minimum speed must the meteoroid have if the Earth's gravitation is not to deflect the meteoroid to make it strike the Earth?

**58.** A spherical planet has uniform density $\rho$. Show that the minimum period for a satellite in orbit around it is

$$T_{\min} = \sqrt{\frac{3\pi}{G\rho}}$$

independent of the radius of the planet.

**59.** Two stars of masses $M$ and $m$, separated by a distance $d$, revolve in circular orbits about their center of mass (Fig. P11.59). Show that each star has a period given by

$$T^2 = \frac{4\pi^2 d^3}{G(M+m)}$$

Proceed as follows: Apply Newton's second law to each star. Note that the center-of-mass condition requires that $Mr_2 = mr_1$, where $r_1 + r_2 = d$.

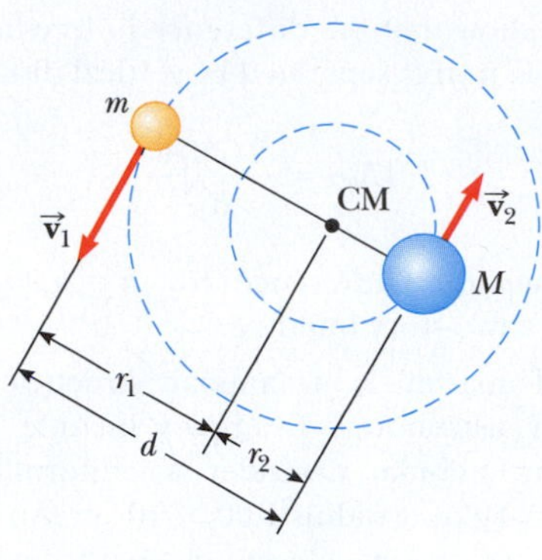


FIGURE **P11.59**

**60.** (a) A 5.00-kg object is released $1.20 \times 10^7$ m from the center of the Earth. It moves with what acceleration relative to the Earth? (b) A $2.00 \times 10^{24}$ kg object is released $1.20 \times 10^7$ m from the center of the Earth. It moves with what acceleration relative to the Earth? Assume that the objects behave as pairs of particles, isolated from the rest of the Universe.

**61.** The positron is the antiparticle to the electron. It has the same mass and a positive electric charge of the same magnitude as the electron charge. Positronium is a hydrogen-like atom consisting of a positron and an electron revolving around each other. Using the Bohr model, find the allowed distances between the two particles and the allowed energies of the system.

## ANSWERS TO QUICK QUIZZES

**11.1** (e). The gravitational force follows an inverse-square behavior, so doubling the distance causes the force to be one-fourth as large.

**11.2** (a). From Kepler's third law and the given period, the major axis of the asteroid can be calculated. It is found to be $1.2 \times 10^{11}$ m. Because this axis is smaller than the Earth–Sun distance, the asteroid cannot possibly collide with the Earth.

**11.3** (a) Perihelion. Because of conservation of angular momentum, the speed of the comet is highest at its closest position to the Sun. (b) Aphelion. The potential energy of the comet–Sun system is highest when the comet is at its farthest distance from the Sun. (c) Perihelion. The kinetic energy is highest at the point at which the speed of the comet is highest. (d) All points. The total energy of the system is the same regardless of where the comet is in its orbit.

**11.4** (a). The longest-wavelength photon is associated with the lowest-energy transition, which is $n = 3$ to $n = 2$.

## A Successful Mission Plan

Now that we have explored the physics of classical mechanics, let us return to our central question for the *Mission to Mars* Context:

***How can we undertake a successful transfer of a spacecraft from the Earth to Mars?***

We make use of the physical principles that we now understand and apply them to our journey from the Earth to Mars.

Let us start with a more modest proposition. Suppose a spacecraft is in a circular orbit around the Earth and you are a passenger on the spacecraft. If you toss a wrench in the direction of travel, tangent to the circular path, what orbital path will the wrench follow?

Let us adopt a simplification model in which the spacecraft is much more massive than the wrench. Conservation of momentum for the isolated system of the wrench and the spacecraft tells us that the spacecraft must slow down slightly once the wrench is thrown. Because of the mass difference between the wrench and spacecraft, however, we can ignore the small change in the spacecraft's speed. The wrench now enters a new orbit, from its perigee position, and the wrench–Earth system has more energy than it had when the wrench was in the circular orbit. Because the orbital energy is related to the major axis, the wrench is injected into an elliptical orbit as discussed in the Context Connection of Chapter 11 and as shown in Figure 1. Therefore, the path of the wrench is changed from a circular orbit to an elliptical orbit by providing the wrench–Earth system with extra energy. The energy is provided by the force you apply to the wrench tangent to the circular orbit because you have done work on the system. The elliptical orbit will take the wrench farther from the Earth than the circular orbit. If there were another spacecraft in a higher circular orbit than your spacecraft, you could throw the wrench so that it transfers from one spacecraft to another as shown in Figure 2. For that to occur, the elliptical orbit of the wrench must intersect with the higher spacecraft orbit. Furthermore, the wrench and the second spacecraft must arrive at the same point at the same time.

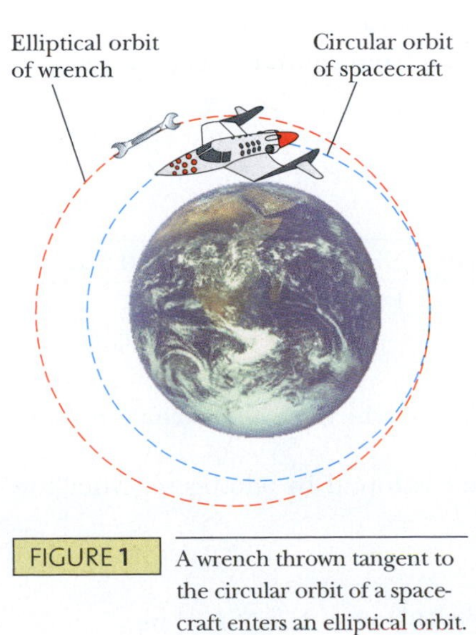


FIGURE 1 A wrench thrown tangent to the circular orbit of a spacecraft enters an elliptical orbit.

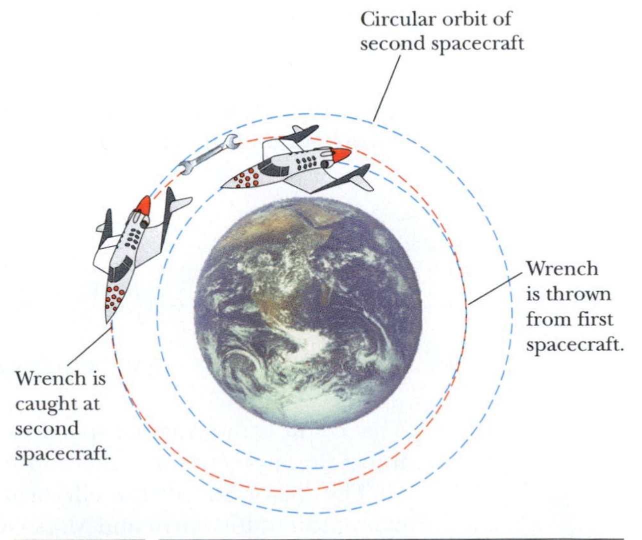


FIGURE 2 If a second spacecraft were in a higher circular orbit, the wrench could be carefully thrown so as to be transferred from one spacecraft to the other.

This scenario is the essence of our planned mission from the Earth to Mars. Rather than transferring a wrench between two spacecraft in orbit around the Earth, we will transfer a spacecraft between two planets in orbit around the Sun. Kinetic energy is added to the wrench–Earth system by throwing the wrench. Kinetic energy is added to the spacecraft–Sun system by firing the engines.

What if you were to throw the wrench harder and harder in the previous example? The wrench would be placed in a larger and larger elliptical orbit around the Earth. As you increased the launch velocity, you could inject the wrench into a *hyperbolic* escape orbit, relative to the Earth, and into an *elliptical* orbit around the *Sun*. This approach is the one we will take for the trip from the Earth to Mars; we will break free from a circular *parking* orbit around the Earth and move into an elliptical *transfer* orbit around the Sun. The spacecraft will then continue on its journey to Mars, where it will enter a new parking orbit.

Now let us focus our attention on the transfer orbit part of the journey. One simple transfer orbit is called a **Hohmann transfer,** the type of transfer imparted to the wrench shown in Figure 2. The Hohmann transfer involves the least energy expenditure and thus requires the smallest amount of fuel. As might be expected for a lowest-energy transfer, the transfer time for a Hohmann transfer is longer than for other types of orbits. We shall investigate the Hohmann transfer because of its simplicity and its general usefulness in planetary transfers.

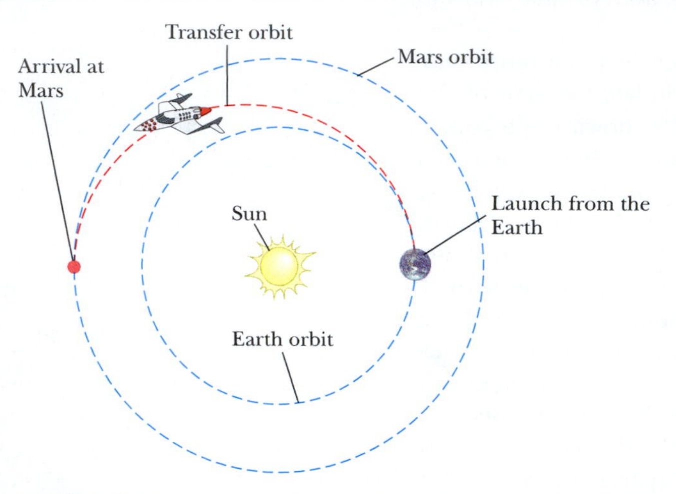


FIGURE 3 The Hohmann transfer orbit from the Earth to Mars. It is similar to transferring the wrench from one spacecraft to another in Figure 2, but here we are transferring a spacecraft from one planet to another.

The rocket engine on the spacecraft is fired from the parking orbit such that the spacecraft enters an elliptical orbit around the Sun at its perihelion and encounters the planet at the spacecraft's aphelion. Therefore, the spacecraft makes exactly one half of a revolution about its elliptical path during the transfer as shown in Figure 3.

This process is energy efficient because fuel is expended only at the beginning and the end. The movement between parking orbits around the Earth and Mars is free; the spacecraft simply follows Kepler's laws while in an elliptical orbit around the Sun.

Let us perform a simple numerical calculation to see how to apply the mechanical laws to this process. We assume that the spacecraft is in a parking orbit above the Earth's surface. Notice also that the spacecraft is in orbit around the Sun, with a perturbation in its orbit caused by the Earth. Therefore, if we calculate the tangential speed of the Earth about the Sun, we can let this speed represent the average speed of the spacecraft around the Sun. This speed is calculated from Newton's second law for a particle in uniform circular motion:

$$F = ma \quad \rightarrow \quad G\frac{M_{\text{Sun}}\, m_{\text{Earth}}}{r^2} = m_{\text{Earth}}\frac{v^2}{r}$$

$$\rightarrow \quad v = \sqrt{\frac{GM_{\text{Sun}}}{r}} = \sqrt{\frac{(6.67 \times 10^{-11}\,\text{N}\cdot\text{m}^2/\text{kg}^2)(1.99 \times 10^{30}\,\text{kg})}{1.50 \times 10^{11}\,\text{m}}}$$

$$= 2.97 \times 10^4\ \text{m/s}$$

This result is the original speed of the spacecraft, to which we add a change $\Delta v$ to inject the spacecraft into the transfer orbit.

The major axis of the elliptical transfer orbit is found by adding together the orbit radii of the Earth and Mars (see Fig. 3):

$$\text{Major axis} = 2a = r_{\text{Earth}} + r_{\text{Mars}}$$

$$= 1.50 \times 10^{11}\ \text{m} + 2.28 \times 10^{11}\ \text{m} = 3.78 \times 10^{11}\ \text{m}$$

Therefore

$$a = 1.89 \times 10^{11}\ \text{m}$$

From this value, Kepler's third law is used to find the travel time, which is one half of the period of the orbit:

$$\begin{aligned}\Delta t_{\text{travel}} &= \tfrac{1}{2}T = \tfrac{1}{2}\sqrt{\frac{4\pi^2}{GM_{\text{Sun}}}\,a^3}\\ &= \tfrac{1}{2}\sqrt{\frac{4\pi^2}{(6.67 \times 10^{-11}\ \text{N}\cdot\text{m}^2/\text{kg}^2)(1.99 \times 10^{30}\ \text{kg})}(1.89 \times 10^{11}\ \text{m})^3}\\ &= 2.24 \times 10^7\ \text{s} = 0.711\ \text{yr} = 260\ \text{d}\end{aligned}$$

Therefore, the journey to Mars will require 260 Earth days. We can also determine where in their orbits Mars and the Earth must be so that the planet will be there when the spacecraft arrives. Mars has an orbital period of 687 Earth days. During the transfer time, the angular position *change* of Mars is

$$\Delta\theta_{\text{Mars}} = \frac{260\ \text{d}}{687\ \text{d}}(2\pi) = 2.38\ \text{rad} = 136°$$

Therefore, for the spacecraft and Mars to arrive at the same point at the same time, the spacecraft must be launched when Mars is 180° − 136° = 44° ahead of the Earth in its orbit. This geometry is shown in Figure 4.

With relatively simple mathematics, that is as far as we can go in describing the details of a trip to Mars. We have found the desired path, the time for the trip, and the position of Mars at launch time. Another important issue for the spacecraft captain would be that of the amount of fuel required for the trip. This question is related to the speed changes necessary to put us into a transfer orbit. These types of calculations involve energy considerations and are explored in Problem 3.

Although many considerations for a successful mission to Mars have not been addressed, we have successfully designed a transfer orbit from the Earth to Mars that is consistent with the laws of mechanics. We consequently declare success for our endeavor and bring this *Mission to Mars* Context to a close.

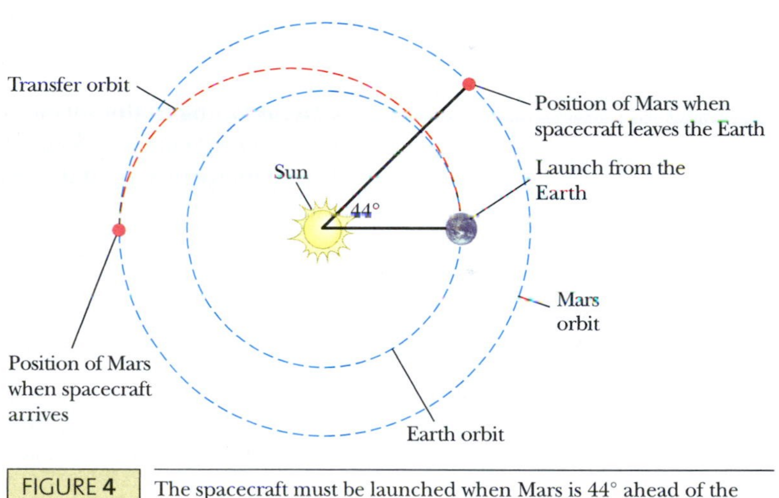


FIGURE 4 The spacecraft must be launched when Mars is 44° ahead of the Earth in its orbit.

## Questions

1. Some science fiction stories describe a twin planet to the Earth. It is exactly 180° ahead of us in the same orbit as the Earth, so we will never see it because it is on the other side of the Sun. Assuming you are in a spacecraft in orbit around the Earth, describe conceptually how you could visit this planet by altering your orbit.
2. You are in an orbiting spacecraft. Another spacecraft is in precisely the same orbit but is 1 km ahead of you, moving in the same direction around the circle. Through an oversight, your food supplies have been exhausted, but there is more than enough food in the other spacecraft. The commander of the other spacecraft is going to throw, from her spacecraft to yours, a picnic basket full of sandwiches. Give a qualitative description of how she should throw it.

## Problems

1. Consider a Hohmann transfer from the Earth to Venus. (a) How long will this transfer take? (b) Should Venus be ahead of or behind the Earth in its orbit

when the spacecraft leaves the Earth on its way to the rendezvous? How many degrees is Venus ahead or behind the Earth?

2. You are on a space station in a circular orbit 500 km above the surface of the Earth. Your passenger and guest is a large, strong, intelligent extraterrestrial. You cannot answer her penetrating questions about bigotry and war, so you try to teach her to play golf. Walking on the space station surface with magnetic shoes, you demonstrate a drive. The alien tees up a golf ball and hits it with incredible power, sending it off with speed $\Delta v$, relative to the space station, in a direction parallel to the instantaneous velocity vector of the space station. You notice that after you then complete precisely 2.00 orbits of the Earth, the golf ball also returns to the same location, so you reach up and catch the ball as it is passing the space station. With what speed $\Delta v$ was the golf ball hit?

3. Investigate what the engine has to do to make a spacecraft follow the Hohmann transfer orbit from the Earth to Mars described in the text. Short-duration burns of our rocket engine are required to change the speed of our spacecraft whenever we alter our orbit. There are no brakes in space, so fuel is required both to increase and to decrease the speed of the spacecraft. First, ignore the gravitational attraction between the spacecraft and the planets. (a) Calculate the speed change required for switching the craft from a circular orbit around the Sun at the Earth's distance to the transfer orbit to Mars. (b) Calculate the speed change required for switching from the transfer orbit to a circular orbit around the Sun at the distance of Mars. Now consider the effects of the two planets' gravity. (c) Calculate the speed change required to carry the craft from the Earth's surface to its own independent orbit around the Sun. You may suppose the craft is launched from the Earth's equator toward the east. (d) Model the craft as falling to the surface of Mars from solar orbit. Calculate the magnitude of the speed change required to make a soft landing on Mars at the end of the fall. Mars rotates on its axis with a period of 24.6 h.

# CONTEXT 3

## Earthquakes

Earthquakes result in massive movement of the ground, as evidenced by the accompanying photograph of railroad tracks in Mexico, damaged severely by an earthquake in 1985. Anyone who has experienced a serious earthquake can attest to the violent shaking it produces. In this Context, we shall focus on earthquakes as an application of our study of the physics of vibrations and waves.

(Guillermo Aldana Espinosa/National Geographic Image Collection)

FIGURE 1 In this Context, we shall study the physics of vibrations and waves by investigating earthquakes. These deformed railroad tracks suggest the large amounts of energy released in an earthquake.

The cause of an earthquake is a release of energy within the Earth at a point called the *focus,* or *hypocenter,* of the earthquake. The point on the Earth's surface radially above the focus is called the *epicenter.* As the energy from the focus reaches the surface, it spreads out along the surface of the Earth. We might expect that the risk of damage in an earthquake decreases as one moves farther from the epicenter, and over long distances that assumption is correct. For example, structures in Kansas are not affected by earthquakes in California. In regions close to the earthquake, however, the notion of decrease in risk with distance is not consistent. Consider, for example, the following quotations describing damage in two different earthquakes.

After the Northridge, California, earthquake, January 17, 1994:[1]

> Although the city [Santa Monica] sits 25 kilometers from the shock's epicenter, it suffered more [damage] than did other areas less than one-third that distance from the jolt.

After the Michoacán earthquake, September 19, 1985:[2]

> An earthquake rattled the coast of Mexico in the state of Michoacán, about 400 kilometers west of Mexico City. Near the coast, the shaking of the ground was mild and caused little damage. As the seismic waves raced

(Joseph Sohm/ChromoSohm Inc./CORBIS)

FIGURE 2 The Northridge earthquake in California in 1994 caused billions of dollars in damage.

[1] *Science News,* Volume 145, 1994, p. 287.

[2] *American Scientist,* November–December 1992, p. 566.

inland, the ground shook even less, and by the time the waves were 100 kilometers from Mexico City, the shaking had nearly subsided. Nevertheless, the seismic waves induced severe shaking in the city, and some areas continued to shake for several minutes after the seismic waves had passed. Some 300 buildings collapsed and more than 20,000 people died.

(T. Campion/SYGMA)

FIGURE 3 Severe damage occurred in localized regions of Mexico City in 1985, even though the epicenter of the Michoacán earthquake was hundreds of kilometers away.

It is clear from these quotations that the notion of a simple decrease in risk with distance is misleading. We will use these quotations as motivation in our study of the physics of vibrations and waves so that we can better analyze the risk of damage to structures in an earthquake. Our study here will also be important when we investigate electromagnetic waves in Chapters 24 through 27. In this Context, we shall address the central question:

**How can we choose locations and build structures to minimize the risk of damage in an earthquake?**

CHAPTER 12

# Oscillatory Motion

To reduce swaying in tall buildings because of wind, tuned dampers are placed near the top of the building. These mechanisms include an object of large mass that oscillates under computer control at the same frequency as the building, reducing the swaying. The large sphere in the photograph on the left is part of the tuned damper system of the building in the photograph on the right, called Taipei 101, in Taiwan. The building, also called the Taipei Financial Center, was completed in 2004, at which time it held the record for the world's tallest building.

(Courtesy of Motioneering Inc.)

(© Simon Kwong/Reuters/CORBIS)

## CHAPTER OUTLINE


12.1 Motion of a Particle Attached to a Spring

12.2 Mathematical Representation of Simple Harmonic Motion

12.3 Energy Considerations in Simple Harmonic Motion

12.4 The Simple Pendulum

12.5 The Physical Pendulum

12.6 Damped Oscillations

12.7 Forced Oscillations

12.8 Context Connection — Resonance in Structures

SUMMARY


You are most likely familiar with several examples of *periodic* motion, such as the oscillations of an object on a spring, the motion of a pendulum, and the vibrations of a stringed musical instrument. Numerous other systems exhibit periodic behavior. For example, the molecules in a solid oscillate about their equilibrium positions; electromagnetic waves, such as light waves, radar, and radio waves, are characterized by oscillating electric and magnetic field vectors; in alternating-current circuits, such as in your household electrical service, voltage and current vary periodically with time. In this chapter, we will investigate mechanical systems that exhibit periodic motion.

We have experienced a number of situations in which the net force on a particle is constant. In these situations, the acceleration of the particle is also constant and we can describe the motion of the particle using the kinematic equations of Chapter 2. If a force acting on a particle varies in time, the acceleration of the particle also changes with time and so the kinematic equations cannot be used.

A special kind of periodic motion occurs when the force that acts on a particle is always directed toward an equilibrium position and is proportional to the position of the particle relative to the

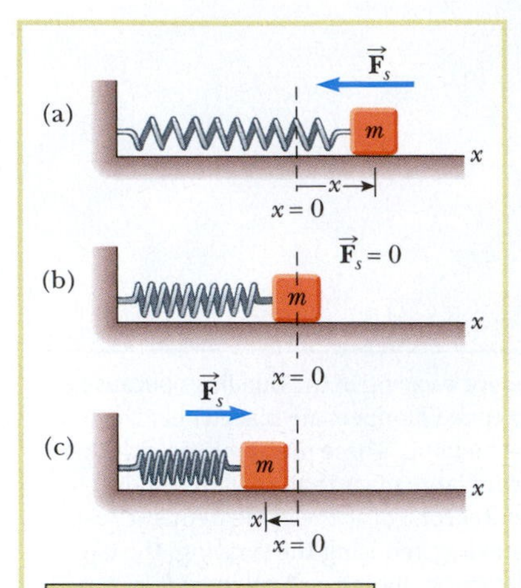


ACTIVE FIGURE 12.1

A block attached to a spring on a frictionless track moves in simple harmonic motion. (a) When the block is displaced to the right of equilibrium, the position is positive and the force and acceleration are negative. (b) At the equilibrium position $x = 0$, the force and acceleration of the block are zero but the speed is a maximum. (c) When the position is negative, the force and acceleration of the block are positive.

**Physics Now™** By logging into PhysicsNow at www.pop4e.com and going to Active Figure 12.1 you can choose the spring constant and the initial position and velocities of the block to see the resulting simple harmonic motion.

### PITFALL PREVENTION 12.1

**THE ORIENTATION OF THE SPRING** Active Figure 12.1 shows the pictorial representation we will use to study the behavior of systems with springs: a *horizontal* spring with an attached block sliding on a frictionless surface. Another possible pictorial representation is a block hanging from a *vertical* spring. All the results that we discuss for the horizontal spring will be the same for the vertical spring, except for one difference. When the block is placed on the vertical spring, its weight will cause the spring to extend. If the position of the block at which it hangs at rest on the spring is defined as $x = 0$, the results of this chapter will apply to this system also.

equilibrium position. We shall study this special type of varying force in this chapter. When this type of force acts on a particle, the particle exhibits *simple harmonic motion,* which will serve as an analysis model for a large class of oscillation problems.

## 12.1 MOTION OF A PARTICLE ATTACHED TO A SPRING

At this point in your study of physics, you have probably started to develop a set of mental models associated with the analysis models we have developed. By mental model, we mean a typical physical situation that comes to your mind each time you identify the analysis model to be used in a problem. For example, a rock falling in the absence of air resistance is a possible mental model for the analysis model of a particle under constant acceleration. Collisions between two billiard balls represent a mental model to help understand the momentum version of the isolated system model. A bicycle wheel might offer a mental model for the rigid object under a net torque model.

In the case of oscillatory motion, a useful mental model is an object of mass $m$ attached to a horizontal spring as in Active Figure 12.1. If the spring is unstretched, the object is at rest on a frictionless surface at its **equilibrium position,** which is defined as $x = 0$ (Active Fig. 12.1b). If the object is pulled to the side to position $x$ and released, it will oscillate back and forth as we discussed in Section 6.4. We will use the particle model to ignore the object's size and analyze a particle on a spring as our system.

Recall from Chapter 6 that when the particle attached to an idealized massless spring is located at a position $x$, the spring exerts a force $F_s$ on it given by **Hooke's law,**

$$F_s = -kx \qquad [12.1]$$

where $k$ is the **force constant** or the **spring constant** of the spring. We call the force in Equation 12.1 a **linear restoring force** because it is proportional to the position of the particle relative to the equilibrium position and is always directed toward the equilibrium position. That is, when the particle is displaced to the right in Active Figure 12.1a, $x$ is positive and the spring force is negative, to the left. When the particle is displaced to the left of $x = 0$ (Active Fig. 12.1c), $x$ is negative and the spring force is positive, to the right. When a particle is under the effect of a linear restoring force, the motion it follows is a special type of oscillatory motion called **simple harmonic motion.** You can test whether or not a particle will undergo simple harmonic motion by seeing if the force on the particle is linear in $x$. A system undergoing simple harmonic motion is called a **simple harmonic oscillator,** and we describe the motion of the particle with an analysis model called the particle in simple harmonic motion.

Let us imagine a particle subject to a linear restoring force such as that given by Equation 12.1. Applying Newton's second law in the $x$ direction to the particle gives us

$$\sum F = F_s = ma \quad \rightarrow \quad -kx = ma$$

$$a = -\frac{k}{m}x \qquad [12.2]$$

That is, **the acceleration of a particle in simple harmonic motion is proportional to the position of the particle relative to the equilibrium position and is in the opposite direction.** If the particle is released from rest at position $x = A$, its *initial* acceleration is $-kA/m$. When the particle passes through the equilibrium position $x = 0$, its acceleration is zero. At this instant, its speed is a maximum because the acceleration changes sign. The particle then continues to travel to the left of the equilibrium position with a positive acceleration and finally reaches $x = -A$, at which time its acceleration is $+kA/m$ and its speed is again zero. The particle completes a full cycle of its motion by returning to the original position, again passing through $x = 0$ with maximum speed. Thus, we see that the particle oscillates between the turning points $x = \pm A$. In the absence of friction, this motion will continue forever because the force exerted by the spring is conservative. Real systems are generally subject to friction and so cannot oscillate forever. We explore the details of the situation with friction in Section 12.6.

**QUICK QUIZ 12.1** A block on the end of a spring is pulled to position $x = A$ and released. In one full cycle of its motion, through what total distance does it travel? **(a)** $A/2$ **(b)** $A$ **(c)** $2A$ **(d)** $4A$

## 12.2 MATHEMATICAL REPRESENTATION OF SIMPLE HARMONIC MOTION

Let us now develop a mathematical representation of the motion we described in the preceding section. Recall that, by definition, $a = dv/dt = d^2x/dt^2$, so we can express Equation 12.2 as

$$\frac{d^2x}{dt^2} = -\frac{k}{m}x \qquad [12.3]$$

We denote the ratio $k/m$ with the symbol $\omega^2$ (we choose $\omega^2$ rather than $\omega$ to make the solution simpler in form),

$$\omega^2 = \frac{k}{m} \qquad [12.4]$$

and Equation 12.3 can be written in the form

$$\frac{d^2x}{dt^2} = -\omega^2 x \qquad [12.5]$$

What we now require is a mathematical solution to Equation 12.5, that is, a function $x(t)$ that satisfies this second-order differential equation. This function will be a mathematical representation of the particle's position as a function of time. We seek a function $x(t)$ such that the second derivative of the function is the same as the original function with a negative sign and multiplied by $\omega^2$. The trigonometric functions sine and cosine exhibit this behavior, so we can build a solution around one or both of them. The following cosine function is a solution to the differential equation:

$$x(t) = A\cos(\omega t + \phi) \qquad [12.6]$$

■ Position of a particle in simple harmonic motion

where $A$, $\omega$, and $\phi$ are constants.[1] To see explicitly that this expression is a solution to Equation 12.5, note that

$$\frac{dx}{dt} = A\frac{d}{dt}\cos(\omega t + \phi) = -\omega A\sin(\omega t + \phi) \qquad [12.7]$$

$$\frac{d^2x}{dt^2} = -\omega A\frac{d}{dt}\sin(\omega t + \phi) = -\omega^2 A\cos(\omega t + \phi) \qquad [12.8]$$

**PITFALL PREVENTION 12.2**

**A NONCONSTANT ACCELERATION**
Notice that the acceleration of the particle in simple harmonic motion is not constant; Equation 12.2 shows that it varies with position $x$. Therefore, as pointed out in the introduction to the chapter, we *cannot* apply the kinematic equations of Chapter 2 in this situation. We now explore the correct approach in Section 12.2.

[1] In earlier chapters, we saw many examples in which we evaluated a trigonometric function of an angle. The argument of a trigonometric function, such as sine or cosine, *must* be a pure number. The radian is a pure number because it is a ratio of lengths. Degrees are pure simply because the degree is a completely artificial "unit"; it is not related to measurements of lengths. The notion of requiring a pure number for a trigonometric function is important in Equation 12.6, where the angle is expressed in terms of other measurements. Therefore, $\omega$ *must* be expressed in radians per second (and not, for example, in revolutions per second) if $t$ is expressed in seconds. Furthermore, the argument of other types of functions must also be pure numbers, including logarithms and exponential functions.

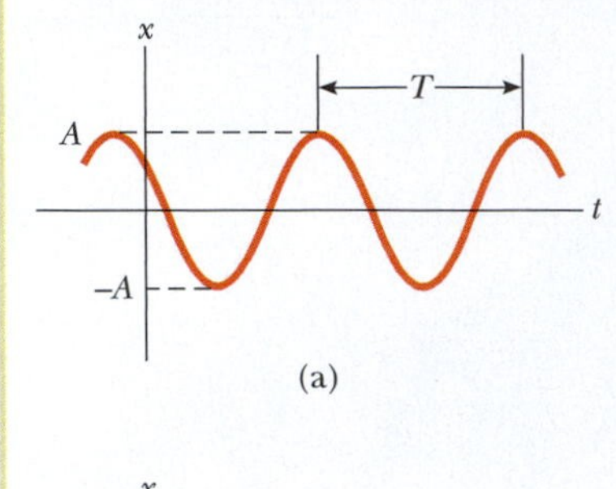


(a)

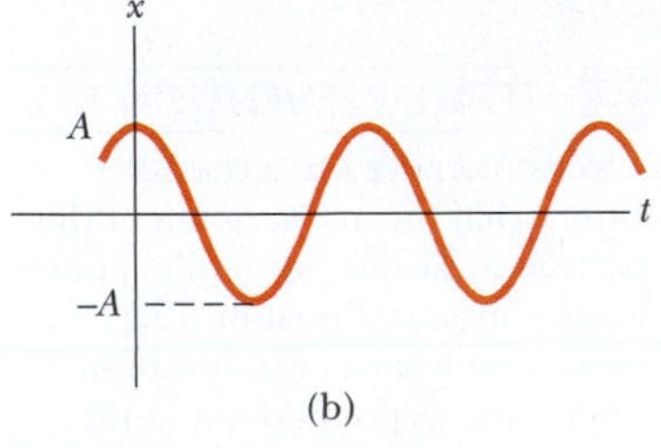


(b)

**ACTIVE FIGURE 12.2**

(a) A graphical representation (position versus time) for the system in Active Figure 12.1, a particle in simple harmonic motion. The amplitude of the motion is $A$ and the period is $T$. (b) The $x$-$t$ curve in the special case in which $x = A$ and $v = 0$ at $t = 0$.

**Physics Now™** By logging into PhysicsNow at www.pop4e.com and going to Active Figure 12.2 you can adjust the graphical representation and see the resulting simple harmonic motion of the block in Active Figure 12.1.

Comparing Equations 12.6 and 12.8, we see that $d^2x/dt^2 = -\omega^2 x$ and that Equation 12.5 is satisfied.

The parameters $A$, $\omega$, and $\phi$ are constants of the motion. To give physical significance to these constants, it is convenient to form a graphical representation of the motion by plotting $x$ as a function of $t$ as in Active Figure 12.2a. First, we note that $A$, called the **amplitude** of the motion, is simply the **maximum value of the position of the particle in either the positive or negative $x$ direction.** The constant $\omega$ is called the **angular frequency** and has units of radians per second (rad/s). From Equation 12.4, the angular frequency is

$$\omega = \sqrt{\frac{k}{m}} \qquad [12.9]$$

The constant angle $\phi$ is called the **phase constant** (or phase angle) and, along with the amplitude $A$, is determined uniquely by the position and velocity of the particle at $t = 0$. If the particle is at its maximum position $x = A$ at $t = 0$, the phase constant is $\phi = 0$ and the graphical representation of the motion is shown in Active Figure 12.2b. The quantity $(\omega t + \phi)$ is called the **phase** of the motion. Note that the function $x(t)$ is periodic and that its value is the same each time $\omega t$ increases by $2\pi$ rad.

Equations 12.1, 12.5, and 12.6 form the basis for the analysis model of the particle in simple harmonic motion. We can be assured that a particle is undergoing simple harmonic motion if (1) we analyze the situation and find that the force on the particle is of the mathematical form of Equation 12.1, (2) we analyze the situation and find that it is described by a differential equation of the form of Equation 12.5, or (3) we analyze the situation and find that the position of the particle is described by Equation 12.6.

Let us investigate further the mathematical description of the motion. The **period** $T$ of the motion is the time interval required for the particle to go through one full cycle of its motion (see Active Fig. 12.2a). That is, the values of $x$ and $v$ for the particle at time $t$ equal the values of $x$ and $v$ at time $t + T$. We can relate the period to the angular frequency by noting that the phase increases by $2\pi$ rad in a time interval of $T$:

$$[\omega(t + T) + \phi] - (\omega t + \phi) = 2\pi$$

Simplifying this expression, we see that $\omega T = 2\pi$, or

▪ Relation of period to angular frequency

$$T = \frac{2\pi}{\omega} \qquad [12.10]$$

The inverse of the period is called the **frequency** $f$ of the motion. Whereas the period is the time interval per oscillation, the frequency represents **the number of oscillations the particle makes per unit time interval:**

$$f = \frac{1}{T} = \frac{\omega}{2\pi} \qquad [12.11]$$

The units of $f$ are cycles per second, or **hertz** (Hz). Rearranging Equation 12.11 gives

$$\omega = 2\pi f = \frac{2\pi}{T} \qquad [12.12]$$

We can use Equations 12.9, 12.10, and 12.11 to express the period and frequency of the motion for the particle–spring system in terms of the characteristics $m$ and $k$ of the system as

**PITFALL PREVENTION 12.3**

**Two kinds of frequency** We identify two kinds of frequency for a simple harmonic oscillator: $f$, called simply the *frequency*, is measured in hertz, and $\omega$, the *angular frequency*, is measured in radians per second. Be sure you are clear about which frequency is being discussed or requested in a given problem. Equations 12.11 and 12.12 show the relationship between the two frequencies.

$$T = \frac{2\pi}{\omega} = 2\pi\sqrt{\frac{m}{k}} \qquad [12.13]$$

■ Period in terms of system parameters

$$f = \frac{1}{T} = \frac{1}{2\pi}\sqrt{\frac{k}{m}} \qquad [12.14]$$

■ Frequency in terms of system parameters

That is, the period and frequency depend *only* on the mass of the particle and the force constant of the spring and *not* on the parameters of the motion, such as $A$ or $\phi$. As we might expect, the frequency is larger for a stiffer spring (larger value of $k$) and decreases with increasing mass of the particle.

We can obtain the velocity and acceleration[2] of a particle undergoing simple harmonic motion from Equations 12.7 and 12.8:

$$v = \frac{dx}{dt} = -\omega A \sin(\omega t + \phi) \qquad [12.15]$$

■ Velocity of a particle in simple harmonic motion

$$a = \frac{d^2x}{dt^2} = -\omega^2 A \cos(\omega t + \phi) \qquad [12.16]$$

■ Acceleration of a particle in simple harmonic motion

From Equation 12.15 we see that because the sine and cosine functions oscillate between $\pm 1$, the extreme values of $v$ are $\pm \omega A$. Likewise, Equation 12.16 tells us that the extreme values of the acceleration are $\pm \omega^2 A$. Therefore, the *maximum* values of the magnitudes of the speed and acceleration are

$$v_{\max} = \omega A = \sqrt{\frac{k}{m}}\,A \qquad [12.17]$$

$$a_{\max} = \omega^2 A = \frac{k}{m}\,A \qquad [12.18]$$

■ Maximum values of speed and acceleration of a particle in simple harmonic motion

Figure 12.3a plots position versus time for an arbitrary value of the phase constant. The associated velocity–time and acceleration–time curves are illustrated in Figures 12.3b and 12.3c. They show that the phase of the velocity differs from the

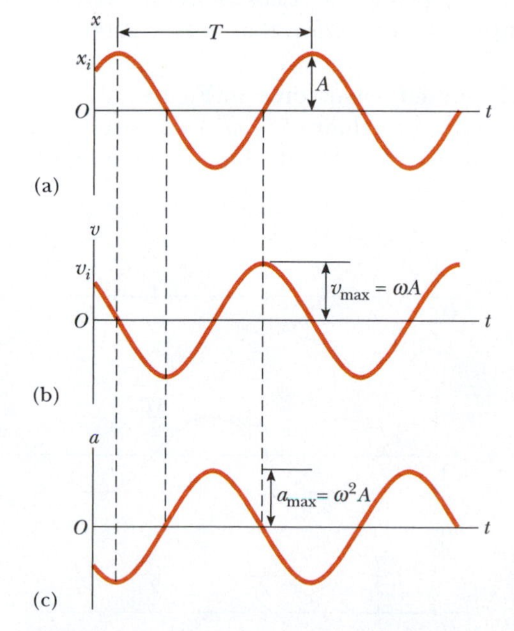


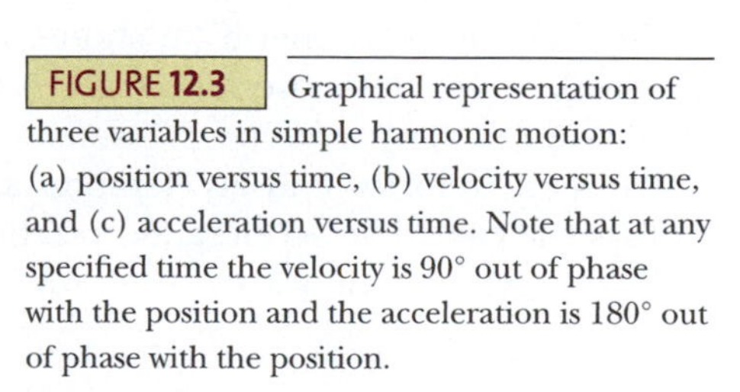

**FIGURE 12.3** Graphical representation of three variables in simple harmonic motion: (a) position versus time, (b) velocity versus time, and (c) acceleration versus time. Note that at any specified time the velocity is 90° out of phase with the position and the acceleration is 180° out of phase with the position.

[2]Because the motion of a simple harmonic oscillator takes place in one dimension, we will denote velocity as $v$ and acceleration as $a$, with the direction indicated by a positive or negative sign, as in Chapter 2.

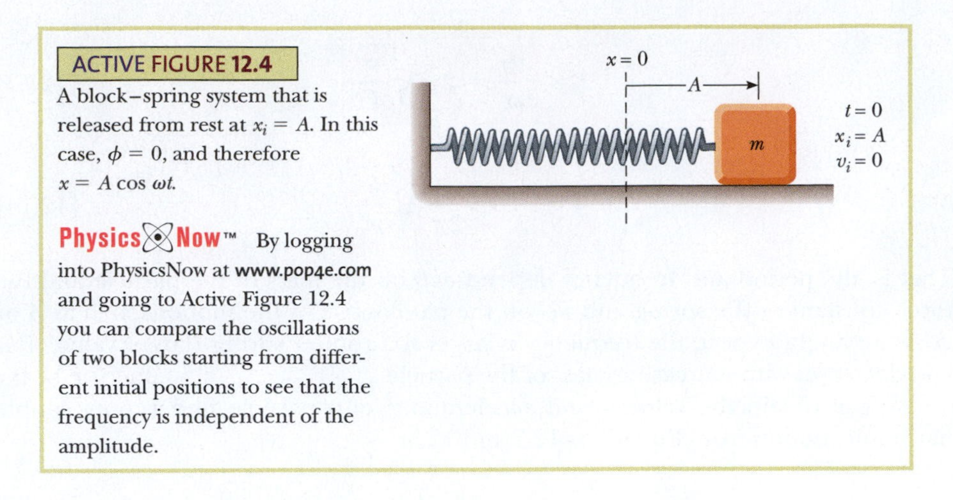


ACTIVE FIGURE 12.4 A block–spring system that is released from rest at $x_i = A$. In this case, $\phi = 0$, and therefore $x = A \cos \omega t$.

**Physics Now™** By logging into PhysicsNow at www.pop4e.com and going to Active Figure 12.4 you can compare the oscillations of two blocks starting from different initial positions to see that the frequency is independent of the amplitude.

phase of the position by $\pi/2$ rad, or 90°. That is, when $x$ is a maximum or a minimum, the velocity is zero. Likewise, when $x$ is zero, the speed is a maximum. Furthermore, note that the phase of the acceleration differs from the phase of the position by $\pi$ rad, or 180°. For example, when $x$ is a maximum, $a$ has a maximum magnitude in the opposite direction.

Equation 12.6 describes simple harmonic motion of a particle in general. Let us now see how to evaluate the constants of the motion. The angular frequency $\omega$ is evaluated using Equation 12.9. The constants $A$ and $\phi$ are evaluated from the initial conditions, that is, the state of the oscillator at $t = 0$.

Suppose we initiate the motion by pulling the particle from equilibrium by a distance $A$ and releasing it from rest at $t = 0$ as in Active Figure 12.4. We must then require that our solutions for $x(t)$ and $v(t)$ (Eqs. 12.6 and 12.15) obey the initial conditions that $x(0) = A$ and $v(0) = 0$:

$$x(0) = A \cos \phi = A$$

$$v(0) = -\omega A \sin \phi = 0$$

These conditions are met if we choose $\phi = 0$, giving $x = A \cos \omega t$ as our solution. To check this solution, we note that it satisfies the condition that $x(0) = A$ because $\cos 0 = 1$.

Position, velocity, and acceleration are plotted versus time in Figure 12.5a for this special case. The acceleration reaches extreme values of $\mp \omega^2 A$ when the position has extreme values of $\pm A$. Furthermore, the velocity has extreme values

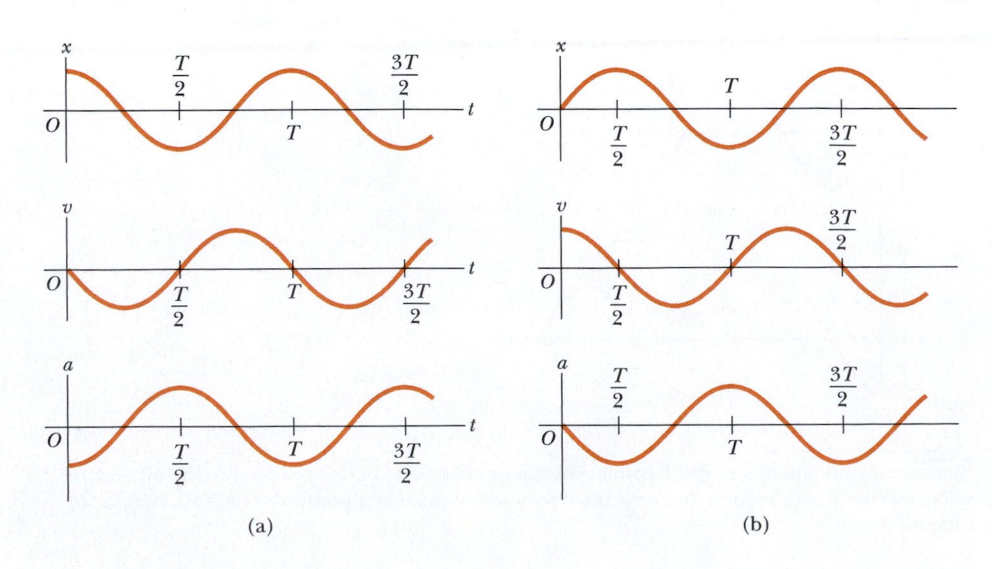


FIGURE 12.5 (a) Position, velocity, and acceleration versus time for a block undergoing simple harmonic motion under the initial conditions that at $t = 0$, $x(0) = A$ and $v(0) = 0$. (b) Position, velocity, and acceleration versus time for a block undergoing simple harmonic motion under the initial conditions that at $t = 0$, $x(0) = 0$ and $v(0) = v_i$.

of $\pm \omega A$, which both occur at $x = 0$. Hence, the quantitative solution agrees with our qualitative description of this system.

Let us consider another possibility. Suppose the system is oscillating and we define $t = 0$ as the instant that the particle passes through the unstretched position of the spring while moving to the right (Active Fig. 12.6) with speed $v_i$. We must then require that our solutions for $x(t)$ and $v(t)$ obey the initial conditions that $x(0) = 0$ and $v(0) = v_i$:

$$x(0) = A \cos \phi = 0$$

$$v(0) = -\omega A \sin \phi = v_i$$

The first of these conditions tells us that $\phi = -\pi/2$. With this value for $\phi$, the second condition tells us that $A = v_i/\omega$. Hence, the solution is given by

$$x = \frac{v_i}{\omega} \cos\left(\omega t - \frac{\pi}{2}\right)$$

Figure 12.5b shows the graphs of position, velocity, and acceleration versus time for this choice of $t = 0$. Note that these curves are the same as those in Figure 12.5a, but shifted to the right by one fourth of a cycle. This shift is described mathematically by the phase constant $\phi = -\pi/2$, which is one fourth of a full cycle of $2\pi$.

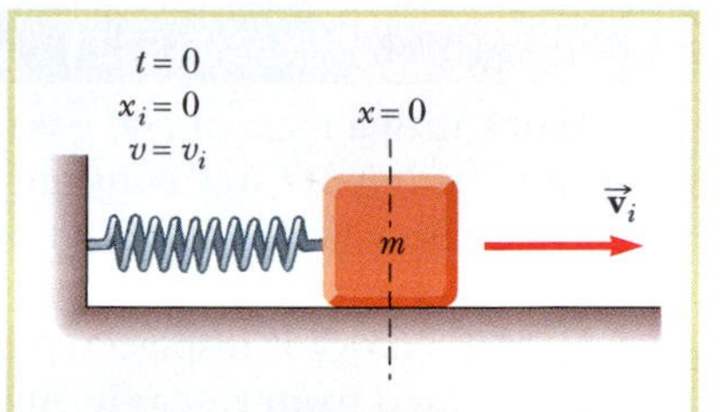


**ACTIVE FIGURE 12.6** The block–spring system is undergoing oscillation, and $t = 0$ is defined at an instant when the block passes through the equilibrium position $x = 0$ and is moving to the right with speed $v_i$.

**Physics Now™** By logging into PhysicsNow at www.pop4e.com and going to Active Figure 12.6 you can compare the oscillations of two blocks with different velocities at $t = 0$ to see that the frequency is independent of the amplitude.

**QUICK QUIZ 12.2** Consider a graphical representation (Fig. 12.7) of simple harmonic motion as described mathematically in Equation 12.6. **(i)** When the object is at point Ⓐ on the graph, what are, respectively, its position and velocity? **(a)** both positive **(b)** both negative **(c)** positive and zero **(d)** negative and zero **(e)** positive and negative **(f)** negative and positive **(ii)** From the same list of choices, what are the respective signs of the velocity and acceleration when the object is at position Ⓐ on the graph?

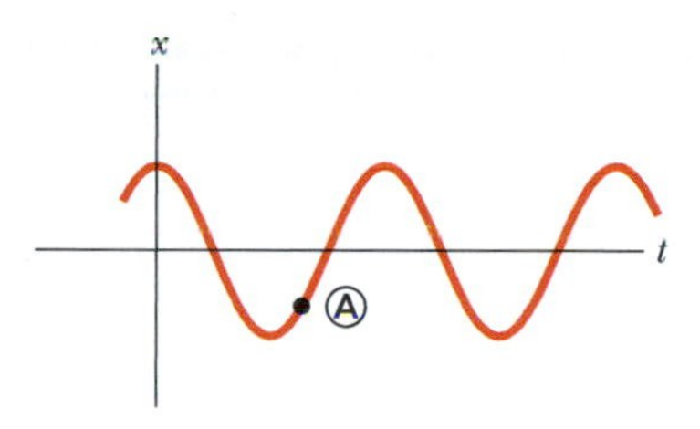


**FIGURE 12.7** (Quick Quiz 12.2) An $x$-$t$ graph for an object undergoing simple harmonic motion. At a particular time, the object's position is indicated by Ⓐ in the diagram.

## Thinking Physics 12.1

We know that the period of oscillation of an object attached to a spring is proportional to the square root of the mass of the object (Eq. 12.13). Therefore, if we perform an experiment in which we place objects with a range of masses on the end of a spring and measure the period of oscillation of each object–spring system, a graph of the square of the period versus the mass will result in a straight line as suggested in Figure 12.8. We find, however, that the line does not go through the origin. Why not?

**Reasoning** The line does not go through the origin because the spring itself has mass. Therefore, the resistance to changes in motion of the system is a combination of the mass of the object on the end of the spring and the mass of the oscillating spring coils. The entire mass of the spring is not oscillating in the same way, however. The coil of the spring attached to the object is oscillating over the same amplitude as the object, but the coil at the fixed end of the spring is not oscillating at all. For a cylindrical spring, energy arguments can be used to show that the effective additional mass representing the oscillations of the spring is one third of the mass of the spring. The square of the period is proportional to the total oscillating mass, but the graph in Figure 12.8 shows the square of the period versus only the mass of the object on the spring. A graph of period squared versus total mass (mass of the object on the spring plus the effective oscillating mass of the spring) would pass through the origin. ■

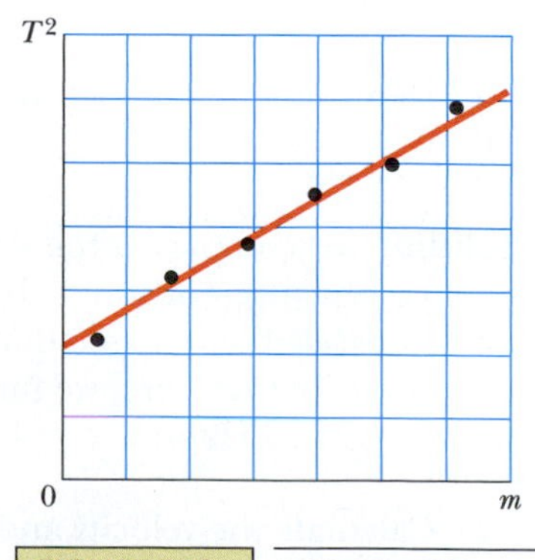


**FIGURE 12.8** (Thinking Physics 12.1) A graph of experimental data: the square of the period versus mass of a block in a block–spring system.

### INTERACTIVE EXAMPLE 12.1 A Block–Spring System

A block with a mass of 200 g is connected to a light horizontal spring of force constant 5.00 N/m and is free to oscillate on a horizontal, frictionless surface.

**A** If the block is displaced 5.00 cm from equilibrium and released from rest as in Active Figure 12.4, find the period of its motion.

**Solution** The situation (we assume an ideal spring) tells us to use the simple harmonic motion model. Using Equation 12.13,

$$T = 2\pi\sqrt{\frac{m}{k}} = 2\pi\sqrt{\frac{200 \times 10^{-3}\ \text{kg}}{5.00\ \text{N/m}}} = 1.26\ \text{s}$$

**B** Determine the maximum speed and maximum acceleration of the block.

**Solution** Using Equations 12.17 and 12.18, with $A = 5.00 \times 10^{-2}$ m, we have

$$v_{\text{max}} = \omega A = \frac{2\pi}{T} A = \left(\frac{2\pi}{1.26\ \text{s}}\right)(5.00 \times 10^{-2}\ \text{m})$$

$$= 0.250\ \text{m/s}$$

$$a_{\text{max}} = \omega^2 A = \left(\frac{2\pi}{T}\right)^2 A = \left(\frac{2\pi}{1.26\ \text{s}}\right)^2 (5.00 \times 10^{-2}\ \text{m})$$

$$= 1.25\ \text{m/s}^2$$

**C** Express the position, velocity, and acceleration of this object as functions of time, assuming that $\phi = 0$.

**Solution** From Equations 12.6, 12.15, and 12.16,

$$x = A\cos\omega t = (0.050\,0\ \text{m})\cos 5.00t$$

$$v = -\omega A \sin\omega t = -(0.250\ \text{m/s})\sin 5.00t$$

$$a = -\omega^2 A\cos\omega t = -(1.25\ \text{m/s}^2)\cos 5.00t$$

**Physics Now™** You can adjust the mass of the object, the force constant of the spring, and the starting position by logging into PhysicsNow at **www.pop4e.com** and going to Interactive Example 12.1.

### EXAMPLE 12.2 An Oscillating Particle

A particle oscillates with simple harmonic motion along the $x$ axis. Its position varies with time according to the equation

$$x = (4.00\ \text{m})\cos\left(\pi t + \frac{\pi}{4}\right)$$

where $t$ is in seconds.

**A** Determine the amplitude, frequency, and period of the motion.

**Solution** By comparing this equation with the general equation for simple harmonic motion, $x = A\cos(\omega t + \phi)$, we see that $A = 4.00$ m and $\omega = \pi$ rad/s; therefore, we find that $f = \omega/2\pi = \pi/2\pi =$ 0.500 Hz and $T = 1/f =$ 2.00 s.

**B** Calculate the velocity and acceleration of the particle at any time $t$.

**Solution** Using Equations 12.15 and 12.16,

$$v = \frac{dx}{dt} = -(4.00\pi\ \text{m/s})\sin\left(\pi t + \frac{\pi}{4}\right)$$

$$a = \frac{dv}{dt} = -(4.00\pi^2\ \text{m/s}^2)\cos\left(\pi t + \frac{\pi}{4}\right)$$

**C** What are the position and the velocity of the particle at time $t = 0$?

**Solution** The position function is given in the text of the problem. Evaluating this expression at $t = 0$ gives us

$$x = (4.00\ \text{m})\cos\left(\frac{\pi}{4}\right) = 2.83\ \text{m}$$

From part B, we evaluate the velocity function at $t = 0$:

$$v = -(4.00\pi\ \text{m/s})\sin\left(\frac{\pi}{4}\right)$$

$$= -8.89\ \text{m/s}$$

**EXAMPLE 12.3 Initial Conditions**

Suppose the initial position $x_i$ and initial velocity $v_i$ of a harmonic oscillator of known angular frequency are given; that is, $x(0) = x_i$ and $v(0) = v_i$. Find general expressions for the amplitude and the phase constant in terms of these initial parameters.

**Solution** With these initial conditions, Equations 12.6 and 12.15 give us at $t = 0$

$$x_i = A\cos\phi \quad \text{and} \quad v_i = -\omega A\sin\phi$$

Dividing these two equations eliminates $A$, giving $v_i/x_i = -\omega\tan\phi$, or

$$\tan\phi = -\frac{v_i}{\omega x_i}$$

Furthermore, if we take the sum $x_i^2 + (v_i/\omega)^2 = A^2\cos^2\phi + A^2\sin^2\phi = A^2$ (where we have used Eqs. 12.6 and 12.15) and solve for $A$, we find that

$$A = \sqrt{x_i^2 + \left(\frac{v_i}{\omega}\right)^2}$$

## 12.3 ENERGY CONSIDERATIONS IN SIMPLE HARMONIC MOTION

If an object attached to a spring slides on a frictionless surface, we can consider the combination of the spring and the attached object to be an isolated system. As a result, we can apply the energy version of the isolated system model to the system. Let us examine the mechanical energy of the system described in Active Figure 12.1. Because the surface is frictionless, the total mechanical energy of the system is constant. We model the object as a particle. The kinetic energy, in the simplification model in which the spring is massless, is associated only with the motion of the particle of mass $m$. We use Equation 12.15 to express the kinetic energy as

$$K = \tfrac{1}{2}mv^2 = \tfrac{1}{2}m\omega^2A^2\sin^2(\omega t + \phi) \qquad [12.19]$$

■ Kinetic energy of a simple harmonic oscillator

Elastic potential energy in this system is associated with the spring and for any position $x$ of the particle is, as we found in Chapter 7, $U = \frac{1}{2}kx^2$. Using Equation 12.6, we have

$$U = \tfrac{1}{2}kx^2 = \tfrac{1}{2}kA^2\cos^2(\omega t + \phi) \qquad [12.20]$$

■ Potential energy of a simple harmonic oscillator

We see that $K$ and $U$ are always positive quantities and that each varies with time. We can express the **total energy** of the simple harmonic oscillator as

$$E = K + U = \tfrac{1}{2}m\omega^2A^2\sin^2(\omega t + \phi) + \tfrac{1}{2}kA^2\cos^2(\omega t + \phi)$$

Because $\omega^2 = k/m$, we can write this expression as

$$E = \tfrac{1}{2}kA^2[\sin^2(\omega t + \phi) + \cos^2(\omega t + \phi)]$$

Because $\sin^2\theta + \cos^2\theta = 1$ for any angle $\theta$, this equation reduces to

$$E = \tfrac{1}{2}kA^2 \qquad [12.21]$$

■ Total energy of a simple harmonic oscillator

That is, the total energy of an isolated simple harmonic oscillator is a constant of the motion and proportional to the square of the amplitude. In fact, the total energy is just equal to the maximum potential energy stored in the spring when $x = \pm A$. At these points, $v = 0$ and there is no kinetic energy. At the equilibrium position, $x = 0$ and $U = 0$, so the total energy is all in the form of kinetic energy of the particle,

$$K_{\text{max}} = \tfrac{1}{2}mv_{\text{max}}^2 = \tfrac{1}{2}kA^2$$

These results are appropriate for the simplification model in which we consider the spring to be massless. In the real situation in which the spring has mass, additional kinetic energy is associated with the motion of the spring. In numerical problems in this book, we will consider only massless springs unless otherwise noted.

(a) Kinetic energy and potential energy versus time for a simple harmonic oscillator with $\phi = 0$. (b) Kinetic energy and potential energy versus position for a simple harmonic oscillator. In either plot, note that $K + U =$ constant.

**PhysicsNow™** By logging into PhysicsNow at www.pop4e.com and going to Active Figure 12.9 you can compare the physical oscillation of a block with energy graphs in this figure as well as with energy bar graphs.

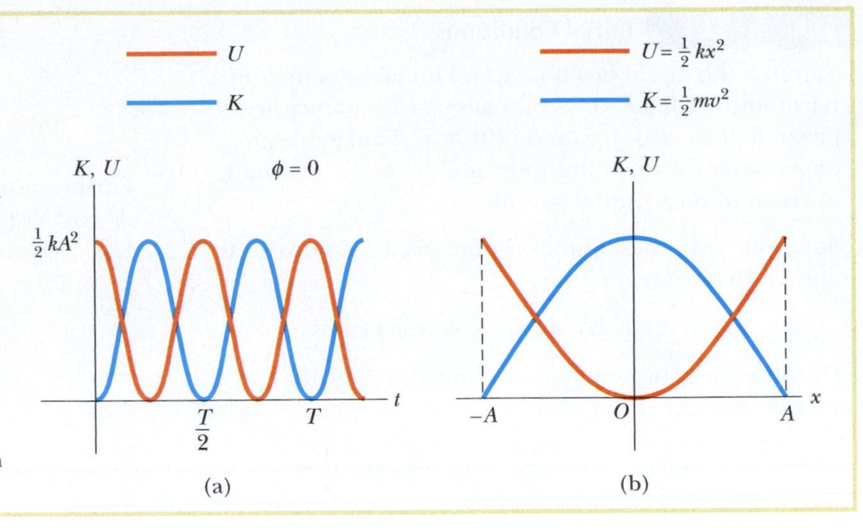


Graphical representations of the kinetic and potential energies versus time for the system of a particle on a massless spring are shown in Active Figure 12.9a, where $\phi = 0$. In this situation, the sum of the kinetic and potential energies at all times is a constant equal to $\frac{1}{2}kA^2$, the total energy of the system. The variations of $K$ and $U$ with position are plotted in Active Figure 12.9b. Energy in the system is continuously being transformed between potential energy (in the spring) and kinetic energy (of the object attached to the spring). Active Figure 12.10 illustrates the position, velocity, acceleration, kinetic energy, and potential energy of the particle–spring system for one full period of the motion. Most of the ideas discussed so far for simple harmonic motion are incorporated in this important figure. We suggest that you study it carefully.

Finally, we can use conservation of mechanical energy for an isolated system to obtain the velocity for an arbitrary position $x$ of the particle, expressing the total energy as

$$E = K + U = \tfrac{1}{2}mv^2 + \tfrac{1}{2}kx^2 = \tfrac{1}{2}kA^2$$

▮ Velocity as a function of position for a particle in simple harmonic motion

$$v = \pm\sqrt{\frac{k}{m}\,(A^2 - x^2)} = \pm\,\omega\sqrt{A^2 - x^2} \qquad [12.22]$$

This expression confirms that the speed is a maximum at $x = 0$ and is zero at the turning points $x = \pm A$.

## ▮ Thinking Physics 12.2

An object oscillating on the end of a horizontal spring slides back and forth over a frictionless surface. During one oscillation, you set an identical object at the maximum displacement point, with instant-acting glue on its surface. Just as the oscillating object reaches its largest displacement and is momentarily at rest, it adheres to the new object by means of the glue and the two objects continue the oscillation together. Does the period of the oscillation change? Does the amplitude of oscillation change? Does the energy of the oscillation change?

**Reasoning** The period of oscillation changes because the period depends on the mass that is oscillating (Eq. 12.13). The amplitude does not change. Because the new object was added under the special condition that the original object was at rest, the combined objects are at rest at this point also, defining the amplitude as the same as in the original oscillation. The energy does not change either. At the maximum displacement point, the energy is all potential energy stored in the spring, which depends only on the force constant and the amplitude, not on

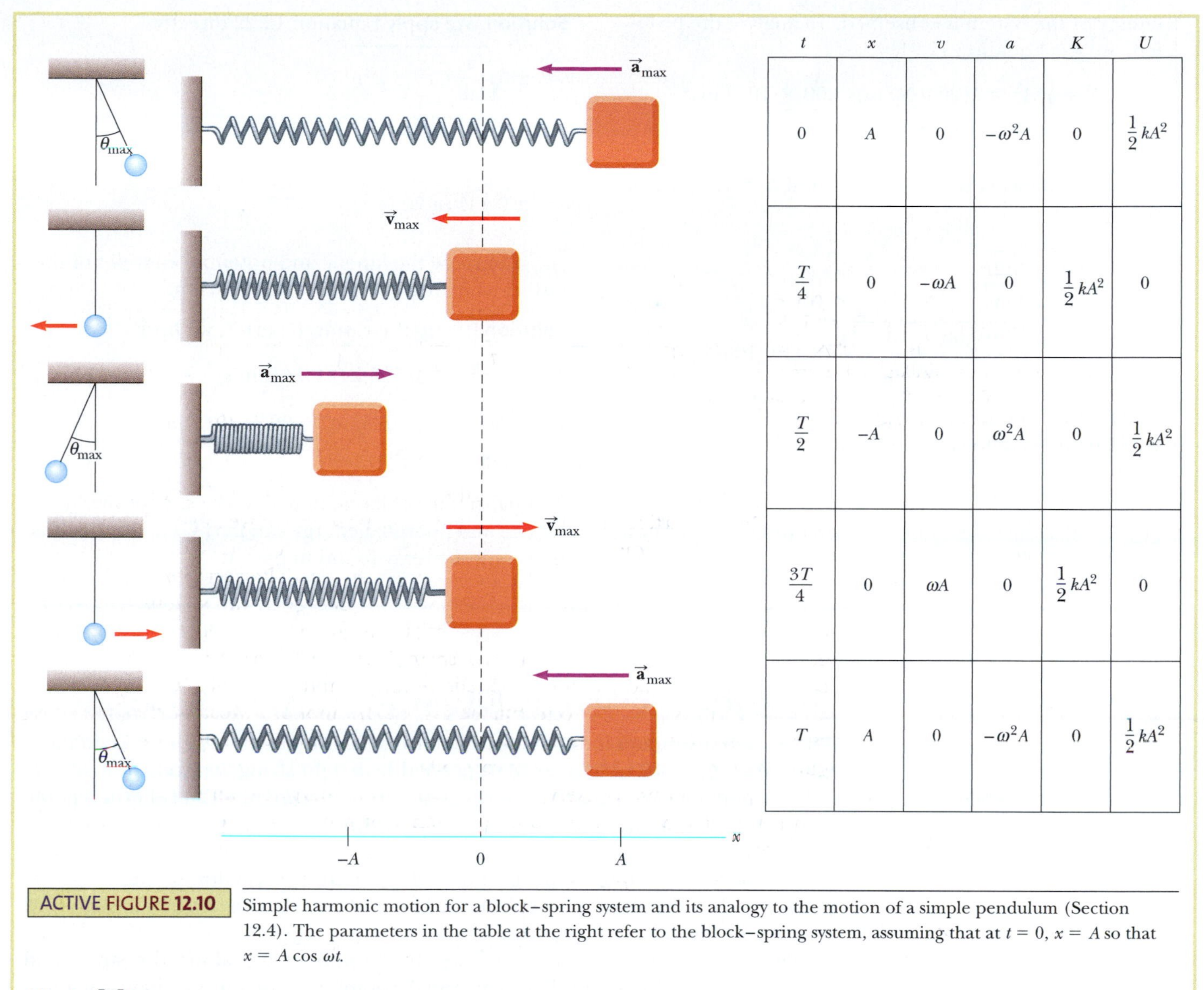


| $t$ | $x$ | $v$ | $a$ | $K$ | $U$ |
|---|---|---|---|---|---|
| 0 | $A$ | 0 | $-\omega^2 A$ | 0 | $\frac{1}{2}kA^2$ |
| $\frac{T}{4}$ | 0 | $-\omega A$ | 0 | $\frac{1}{2}kA^2$ | 0 |
| $\frac{T}{2}$ | $-A$ | 0 | $\omega^2 A$ | 0 | $\frac{1}{2}kA^2$ |
| $\frac{3T}{4}$ | 0 | $\omega A$ | 0 | $\frac{1}{2}kA^2$ | 0 |
| $T$ | $A$ | 0 | $-\omega^2 A$ | 0 | $\frac{1}{2}kA^2$ |

**ACTIVE FIGURE 12.10** Simple harmonic motion for a block–spring system and its analogy to the motion of a simple pendulum (Section 12.4). The parameters in the table at the right refer to the block–spring system, assuming that at $t = 0$, $x = A$ so that $x = A \cos \omega t$.

**PhysicsNow™** By logging into PhysicsNow at **www.pop4e.com** and going to Active Figure 12.10 you can set the initial position of the block and see the block–spring system and the analogous pendulum in motion.

the mass of the object. The object of increased mass will pass through the equilibrium point with lower speed than in the original oscillation but with the same kinetic energy. Another approach is to think about how energy could be transferred into the oscillating system. No work was done on the system (nor did any other form of energy transfer occur), so the energy in the system cannot change. ■

## EXAMPLE 12.4 Oscillations on a Horizontal Surface

A 0.500-kg object connected to a massless spring of force constant 20.0 N/m oscillates on a horizontal, frictionless track.

**A** Calculate the total energy of the system and the maximum velocity of the object if the amplitude of the motion is 3.00 cm.

**Solution** Conceptualize the problem by studying the block–spring system in Active Figure 12.10. Because the object slides on a frictionless surface, we can categorize the problem as one involving an isolated system of the object and the spring. Because only conservative forces are acting within the system, the mechanical energy of the system is conserved. To analyze the problem, we use Equation 12.21:

$$E = \tfrac{1}{2}kA^2 = \tfrac{1}{2}(20.0\ \text{N/m})(3.00 \times 10^{-2}\ \text{m})^2$$
$$= 9.00 \times 10^{-3}\ \text{J}$$

When the object is at $x = 0$, $U = 0$ and $E = \frac{1}{2}mv_{\text{max}}^2$; therefore,

$$v_{\text{max}} = \sqrt{\frac{2E}{m}}$$
$$= \sqrt{\frac{2(9.00 \times 10^{-3}\ \text{J})}{0.500\ \text{kg}}} = \pm 0.190\ \text{m/s}$$

The positive and negative signs indicate that the object could be moving to either the right or the left at this instant.

**B** What is the velocity of the object when the position is equal to 2.00 cm?

**Solution** We apply Equation 12.22 directly:

$$v = \pm\sqrt{\frac{k}{m}(A^2 - x^2)}$$
$$= \pm\sqrt{\frac{20.0\ \text{N/m}}{0.50\ \text{kg}}[(3.00 \times 10^{-2}\ \text{m})^2 - (2.00 \times 10^{-2}\ \text{m})^2]}$$
$$= \pm 0.141\ \text{m/s}$$

**C** Compute the kinetic and potential energies of the system when the position equals 2.00 cm.

**Solution** Using the result to part B, we find

$$K = \tfrac{1}{2}mv^2 = \tfrac{1}{2}(0.500\ \text{kg})(0.141\ \text{m/s})^2 = 5.00 \times 10^{-3}\ \text{J}$$

$$U = \tfrac{1}{2}kx^2 = \tfrac{1}{2}(20.0\ \text{N/m})(2.00 \times 10^{-2}\ \text{m})^2$$
$$= 4.00 \times 10^{-3}\ \text{J}$$

To finalize the problem, note that the sum of the kinetic and potential energies in part C equals the total mechanical energy found in part A.

## 12.4 THE SIMPLE PENDULUM

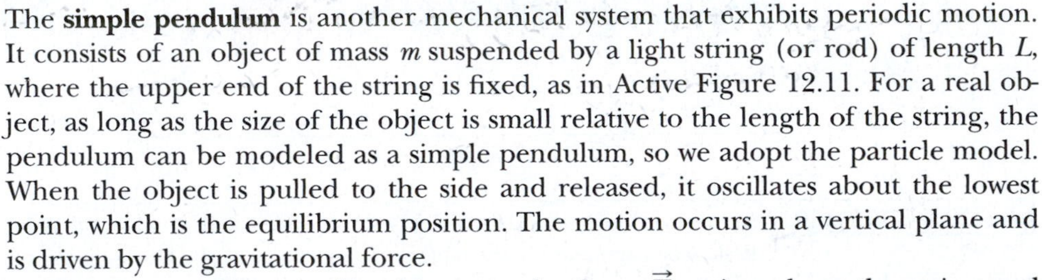

The **simple pendulum** is another mechanical system that exhibits periodic motion. It consists of an object of mass $m$ suspended by a light string (or rod) of length $L$, where the upper end of the string is fixed, as in Active Figure 12.11. For a real object, as long as the size of the object is small relative to the length of the string, the pendulum can be modeled as a simple pendulum, so we adopt the particle model. When the object is pulled to the side and released, it oscillates about the lowest point, which is the equilibrium position. The motion occurs in a vertical plane and is driven by the gravitational force.

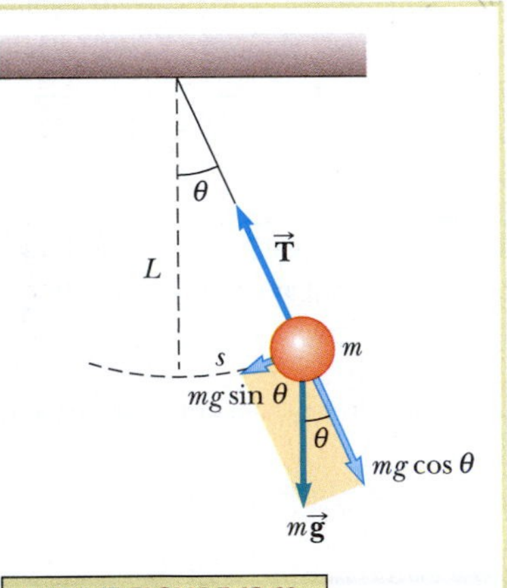


ACTIVE FIGURE 12.11
When $\theta$ is small, the oscillation of the simple pendulum can be modeled as simple harmonic motion about the equilibrium position ($\theta = 0$). The restoring force is $mg \sin\theta$, the component of the gravitational force tangent to the arc.

**PhysicsNow™** By logging into PhysicsNow at **www.pop4e.com** and going to Active Figure 12.11 you can adjust the mass of the bob, the length of the string, and the initial angle and see the resulting oscillation of the pendulum.

The forces acting on the object are the force $\vec{T}$ acting along the string and the gravitational force $m\vec{g}$. The tangential component of the gravitational force $mg\sin\theta$ always acts toward $\theta = 0$, opposite the displacement. The gravitational force is therefore a restoring force, and we can use Newton's second law to write the equation of motion in the tangential direction as

$$F_t = ma_t \quad \rightarrow \quad -mg\sin\theta = m\frac{d^2s}{dt^2}$$

where $s$ is the position measured along the circular arc in Active Figure 12.11 and the negative sign indicates that $F_t$ acts toward the equilibrium position. Because $s = L\theta$ (Eq. 10.1a) and $L$ is constant, this equation reduces to

$$\frac{d^2\theta}{dt^2} = -\frac{g}{L}\sin\theta$$

Compare this equation to Equation 12.5, which is of a similar, but not identical, mathematical form. The right side is proportional to $\sin\theta$ rather than to $\theta$; hence, we conclude that the motion is *not* simple harmonic motion because the equation describing the motion is not of the form of Equation 12.5. If we assume that $\theta$ is *small* (less than about 10° or 0.2 rad), however, we can use a simplification model called the **small angle approximation,** in which $\sin\theta \approx \theta$, where $\theta$ is measured in radians. Table 12.1 shows angles, in degrees and radians, and the sines of these

**TABLE 12.1 Angles and Sines of Angles**

| Angle in Degrees | Angle in Radians | Sine of Angle | Percent Difference |
|---|---|---|---|
| 0° | 0.000 0 | 0.000 0 | 0.0% |
| 1° | 0.017 5 | 0.017 5 | 0.0% |
| 2° | 0.034 9 | 0.034 9 | 0.0% |
| 3° | 0.052 4 | 0.052 3 | 0.0% |
| 5° | 0.087 3 | 0.087 2 | 0.1% |
| 10° | 0.174 5 | 0.173 6 | 0.5% |
| 15° | 0.261 8 | 0.258 8 | 1.2% |
| 20° | 0.349 1 | 0.342 0 | 2.1% |
| 30° | 0.523 6 | 0.500 0 | 4.7% |

angles. As long as $\theta$ is less than about 10°, the angle in radians and its sine are the same, at least to within an accuracy of less than 1.0%.

Therefore, for small angles, the equation of motion becomes

$$\frac{d^2\theta}{dt^2} = -\frac{g}{L}\theta \qquad [12.23]$$

Now we have an expression with exactly the same mathematical form as Equation 12.5, with $\omega^2 = g/L$, and so we conclude that the motion is approximately simple harmonic motion for small amplitudes. Modeling the solution after Equation 12.6, $\theta$ can therefore be written as $\theta = \theta_{max}\cos(\omega t + \phi)$, where $\theta_{max}$ is the **maximum angular position** and the angular frequency $\omega$ is

$$\omega = \sqrt{\frac{g}{L}} \qquad [12.24]$$

▪ Angular frequency for a simple pendulum

The period of the motion is

$$T = \frac{2\pi}{\omega} = 2\pi\sqrt{\frac{L}{g}} \qquad [12.25]$$

▪ Period for a simple pendulum

In other words, **the period and frequency of a simple pendulum oscillating at small angles depend only on the length of the string and the free-fall acceleration.** Because the period is *independent* of the mass, we conclude that *all* simple pendulums of equal length at the same location oscillate with equal periods. Experiments show that this conclusion is correct. The analogy between the motion of a simple pendulum and the particle–spring system is illustrated in Active Figure 12.10.

**PITFALL PREVENTION 12.4**

**NOT TRUE SIMPLE HARMONIC MOTION** Remember that the pendulum *does not* exhibit true simple harmonic motion for *any* angle. If the angle is less than about 10°, the motion can be *modeled* as simple harmonic.

**QUICK QUIZ 12.3** A grandfather clock depends on the period of a pendulum to keep correct time. **(i)** Suppose a grandfather clock is calibrated correctly and then a mischievous child slides the bob of the pendulum downward on the oscillating rod. Does the grandfather clock run **(a)** slow, **(b)** fast, or **(c)** correctly? **(ii)** Suppose the grandfather clock is calibrated correctly at sea level and is then taken to the top of a very tall mountain. Does the grandfather clock run **(a)** slow, **(b)** fast, or **(c)** correctly?

## Thinking Physics 12.3

You set up two oscillating systems: a simple pendulum and a block hanging from a vertical spring. You carefully adjust the length of the pendulum so that both oscillators have the same period. You now take the two oscillators to the Moon. Will they still have the same period as each other? What happens if you observe the two oscillators in an orbiting spacecraft? (Assume that the spring is one with open space between the coils when it is unstretched, so the spring can be both stretched and compressed.)

**Reasoning** The block hanging from the spring will have the same period on the Moon that it had on the Earth because the period depends on the mass of the block and the force constant of the spring, neither of which have changed. The pendulum's period on the Moon will be different from its period on the Earth because the period of the pendulum depends on the value of $g$. Because $g$ is smaller on the Moon than on the Earth, the pendulum will oscillate with a longer period.

In the orbiting spacecraft, the block–spring system will oscillate with the same period as on the Earth when it is set into motion because the period does not depend on gravity. The pendulum will not oscillate at all; if you pull it to the side from a direction you define as "vertical" and release it, it stays there. Because the spacecraft is in free-fall while in orbit around the Earth, the effective gravity is zero and there is no restoring force on the pendulum. ■

**EXAMPLE 12.5** **A Measure of Height**

A man enters a tall tower, needing to know its height. He notes that a long pendulum extends from the ceiling almost to the floor and that its period is 12.0 s. How tall is the tower?

**Solution** We adopt a simplification model in which the height of the tower is equal to the length of the pendulum. If we use $T = 2\pi\sqrt{L/g}$ and solve for $L$, we have

$$L = \frac{gT^2}{4\pi^2} = \frac{(9.80\ \text{m/s}^2)(12.0\ \text{s})^2}{4\pi^2} = 35.7\ \text{m}$$

## 12.5 THE PHYSICAL PENDULUM

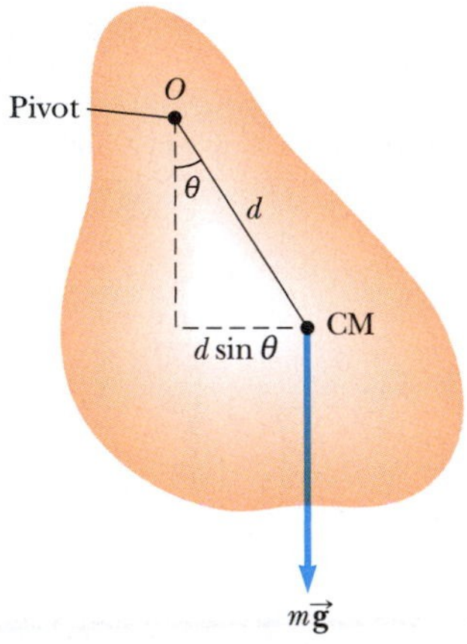

**FIGURE 12.12** The physical pendulum consists of a rigid object pivoted at the point $O$, which is not at the center of mass.

If a hanging object that cannot be modeled as a particle oscillates about a fixed axis that does not pass through its center of mass, it must be treated as a **physical, or compound, pendulum.** For the simple pendulum we generated Equation 12.23 from the particle under a net force model. For the physical pendulum we will need to use the rigid object under a net torque model from Chapter 10. Consider a rigid object pivoted at a point $O$ that is a distance $d$ from the center of mass (Fig. 12.12). The torque about $O$ is provided by the gravitational force, and its magnitude is $mgd \sin \theta$. Using Newton's second law for rotation, $\sum \tau = I\alpha$ (Eq. 10.27), where $I$ is the moment of inertia of the object about the axis through $O$, we have

$$-mgd \sin \theta = I\frac{d^2\theta}{dt^2}$$

The negative sign on the left indicates that the torque about $O$ tends to decrease $\theta$. That is, the gravitational force produces a restoring torque.

If we again assume that $\theta$ is small, the small angle approximation $\sin \theta \approx \theta$ is valid and the equation of motion reduces to

$$\frac{d^2\theta}{dt^2} = -\left(\frac{mgd}{I}\right)\theta \qquad [12.26]$$

Note that this equation has the same mathematical form as Equation 12.5, with $\omega^2 = mgd/I$, and so the motion of the object is approximately simple harmonic motion for small amplitudes. That is, the solution of Equation 12.26 is $\theta = \theta_{max} \cos(\omega t + \phi)$, where $\theta_{max}$ is the maximum angular position and

■ Angular frequency for a physical pendulum

$$\omega = \sqrt{\frac{mgd}{I}}$$

The period is

$$T = \frac{2\pi}{\omega} = 2\pi\sqrt{\frac{I}{mgd}} \qquad [12.27]$$

■ Period for a physical pendulum

One can use this result to measure the moment of inertia of a rigid object. If the location of the center of mass and hence the distance $d$ are known, the moment of inertia can be obtained through a measurement of the period. Finally, note that Equation 12.27 becomes the equation for the period of a simple pendulum (Eq. 12.25) when $I = md^2$—that is, when all the mass is concentrated at a point—and the physical pendulum reduces to the simple pendulum.

**QUICK QUIZ 12.4** Two students, Alex and Brian, are in a museum watching the swinging of a pendulum with a large bob. Alex says, "I'm going to sneak past the fence and stick some chewing gum on the top of the pendulum bob to change its period of oscillation." Brian says, "That won't change the period. The period of a pendulum is independent of mass." Which student is correct? **(a)** Alex **(b)** Brian

**EXAMPLE 12.6 A Swinging Sign**

A circular sign of mass $M$ and radius $R$ is hung on a nail from a small loop located at one edge (Fig. 12.13).

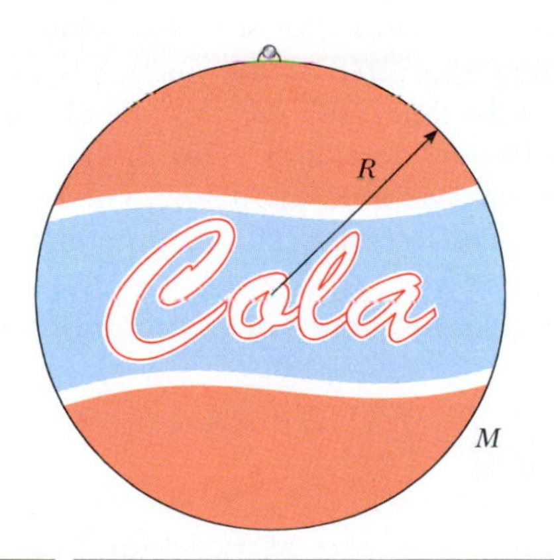


FIGURE 12.13 (Example 12.6) A circular sign oscillating about a pivot as a physical pendulum.

After it is placed on the nail, the sign oscillates in a vertical plane. Find the period of oscillation if the amplitude of the motion is small.

**Solution** The moment of inertia of a disk about an axis through the center is $\frac{1}{2}MR^2$ (see Table 10.2). The pivot point for the sign is through a point on the rim, so we use the parallel axis theorem (see Eq. 10.43) to find the moment of inertia about the pivot:

$$I_p = \tfrac{1}{2}MR^2 + MR^2 = \tfrac{3}{2}MR^2$$

The distance $d$ from the pivot to the center of mass is the radius $R$. Substituting these quantities into Equation 12.27 gives

$$T = 2\pi\sqrt{\frac{\frac{3}{2}MR^2}{MgR}} = 2\pi\sqrt{\frac{3R}{2g}}$$

## 12.6 DAMPED OSCILLATIONS

The oscillatory motions we have considered so far have occurred under the simplification model of an ideal frictionless system, that is, one that oscillates indefinitely under the action of only a linear restoring force. In many realistic systems, resistive forces, such as friction, are present and retard the motion of the system. Consequently, the mechanical energy of the system diminishes in time, and the motion is described as a **damped oscillation.**

Consider an object moving through a medium such as a liquid or a gas. One common type of resistive force on the object, which we discussed in Chapter 5, is proportional to the velocity of the object and acts in the direction opposite that of the object's velocity relative to the medium. This type of force is often observed when an object is oscillating slowly in air, for instance. Because the resistive force can be expressed as $\vec{\mathbf{R}} = -b\vec{\mathbf{v}}$, where $b$ is a constant related to the strength of the resistive force, and the restoring force exerted on the system is $-kx$, Newton's second law gives us

$$\sum F_x = -kx - bv = ma_x$$

$$-kx - b\frac{dx}{dt} = m\frac{d^2x}{dt^2} \qquad [12.28]$$

The solution of this differential equation requires mathematics that may not yet be familiar to you, so it will simply be stated without proof. When the parameters of the system are such that $b < \sqrt{4mk}$ so that the resistive force is small, the solution to Equation 12.28 is

$$x = (Ae^{-(b/2m)t})\cos(\omega t + \phi) \qquad [12.29]$$

where the angular frequency of the motion is

$$\omega = \sqrt{\frac{k}{m} - \left(\frac{b}{2m}\right)^2} \qquad [12.30]$$

This result can be verified by substituting Equation 12.29 into Equation 12.28. Notice that Equation 12.29 is similar to Equation 12.6, with the new feature that the amplitude (in the parentheses before the cosine function) depends on time. In Active Figure 12.14a, we see one example of a damped system. The object suspended from the spring experiences both a force from the spring and a resistive force from the surrounding liquid. Active Figure 12.14b shows the position as a function of time for such a damped oscillator. We see that **when the resistive force is relatively small, the oscillatory character of the motion is preserved but the amplitude of vibration decreases in time** and the motion ultimately ceases. This system is known as an **underdamped oscillator.** The dashed blue lines in Active Figure 12.14b, which form the *envelope* of the oscillatory curve, represent the exponential factor that appears in Equation 12.29. The exponential factor shows that the *amplitude decays exponentially with time.*

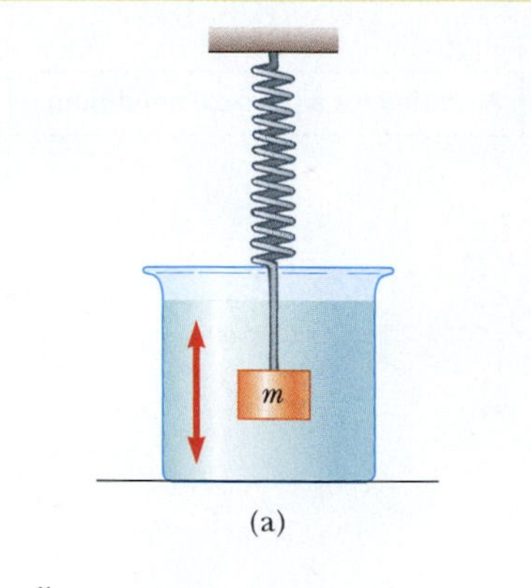


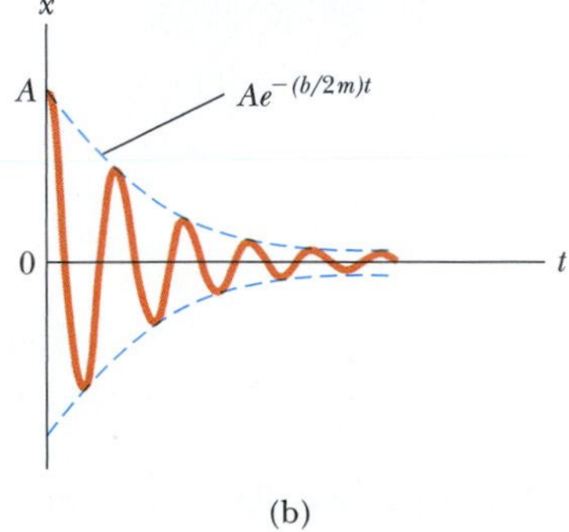


**ACTIVE FIGURE 12.14**
(a) One example of a damped oscillator is an object attached to a spring and submerged in a viscous liquid. (b) Graph of the position versus time for a damped oscillator with small damping. Note the decrease in amplitude with time.

**PhysicsNow™** By logging into PhysicsNow at **www.pop4e.com** and going to Active Figure 12.14 you can adjust the spring constant, the mass of the object, and the damping constant and see the resulting damped oscillation of the object.

It is convenient to express the angular frequency of vibration of a damped system (Eq. 12.30) in the form

$$\omega = \sqrt{\omega_0^2 - \left(\frac{b}{2m}\right)^2}$$

where $\omega_0 = \sqrt{k/m}$ represents the angular frequency of oscillation in the absence of a resistive force (the undamped oscillator). In other words, when $b = 0$, the resistive force is zero and the system oscillates with angular frequency $\omega_0$, called the **natural frequency.**[3] As the magnitude of the resistive force increases, the oscillations dampen more rapidly. When $b$ reaches a critical value $b_c$, so that $b_c/2m = \omega_0$, the system does not oscillate and is said to be **critically damped.** In this case, it returns to equilibrium in an exponential manner with time, as in Figure 12.15.

If the medium is highly viscous and the parameters meet the condition that $b/2m > \omega_0$, the system is **overdamped.** Again, the displaced system does not oscillate but simply returns to its equilibrium position. As the damping increases, the time interval required for the particle to approach the equilibrium position also increases, as indicated in Figure 12.15. In any case, when a resistive force is present, the mechanical energy of the oscillator eventually falls to zero. The mechanical energy is transformed into internal energy in the oscillating system and the resistive medium.

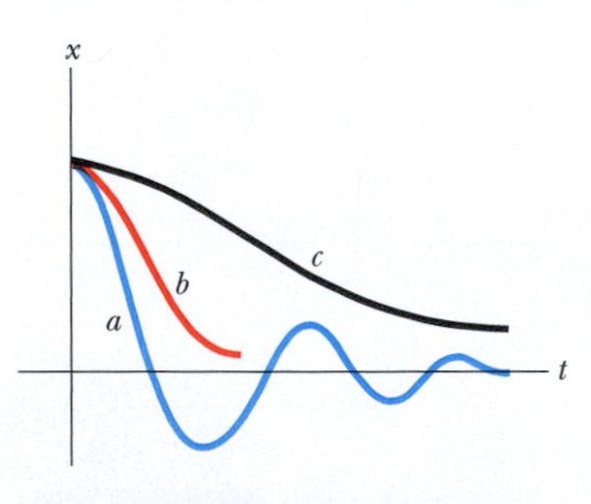


**FIGURE 12.15** Plots of position versus time for an underdamped oscillator (a), a critically damped oscillator (b), and an overdamped oscillator (c).

[3] In practice, both $\omega_0$ and $f_0 = \omega_0/2\pi$ are described as the natural frequency. The context of the discussion will help you determine which frequency is being discussed.

## 12.7 FORCED OSCILLATIONS

We have seen that the mechanical energy of a damped oscillator decreases in time as a result of the resistive force. It is possible to compensate for this energy decrease by applying an external force that does positive work on the system. Such an oscillator then undergoes **forced oscillations.** At any instant, energy can be transferred into the system by an applied force that acts in the direction of motion of the oscillator. For example, a child on a swing can be kept in motion by appropriately timed "pushes." The amplitude of motion remains constant if the energy input per cycle of motion exactly equals the decrease in mechanical energy in each cycle that results from resistive forces.

A common example of a forced oscillator is a damped oscillator driven by an external force that varies periodically, such as $F(t) = F_0 \sin \omega t$, where $\omega$ is the angular frequency of the driving force and $F_0$ is a constant. In general, the frequency $\omega$ of the driving force is different from the natural frequency $\omega_0$ of the oscillator. Newton's second law in this situation gives

$$\sum F_x = ma_x \quad \rightarrow \quad F_0 \sin \omega t - b\frac{dx}{dt} - kx = m\frac{d^2x}{dt^2} \qquad [12.31]$$

Again, the solution of this equation is rather lengthy and will not be presented. After the driving force on an initially stationary object begins to act, the amplitude of the oscillation will increase. After a sufficiently long time interval, when the energy input per cycle from the driving force equals the amount of mechanical energy transformed to internal energy for each cycle, a steady-state condition is reached in which the oscillations proceed with constant amplitude. In this case, Equation 12.31 has the solution

$$x = A\cos(\omega t + \phi) \qquad [12.32]$$

where

$$A = \frac{F_0/m}{\sqrt{\left(\omega^2 - \omega_0^{\,2}\right)^2 + \left(\dfrac{b\omega}{m}\right)^2}} \qquad [12.33]$$

and where $\omega_0 = \sqrt{k/m}$ is the natural frequency of the undamped oscillator ($b = 0$).

Equation 12.33 shows that the amplitude of the forced oscillator is constant for a given driving force because it is being driven in steady state by an external force. For small damping the amplitude becomes large when the frequency of the driving force is near the natural frequency of oscillation, or when $\omega \approx \omega_0$ as can be seen in Equation 12.33. The dramatic increase in amplitude near the natural frequency is called **resonance,** and the natural frequency $\omega_0$ is called the **resonance frequency** of the system.

Figure 12.16 is a graph of amplitude as a function of frequency for the forced oscillator, with varying resistive forces. Note that the amplitude increases with decreasing damping ($b \rightarrow 0$) and that the resonance curve flattens as the damping increases. In the absence of a damping force ($b = 0$), we see from Equation 12.33 that the steady-state amplitude approaches infinity as $\omega \rightarrow \omega_0$. In other words, if there are no resistive forces in the system and we continue to drive an oscillator with a sinusoidal force at the resonance frequency, the amplitude of motion will build up without limit. This situation does not occur in practice because some damping is always present in real oscillators.

Resonance appears in many areas of physics. For example, certain electric circuits have resonance frequencies. This fact is exploited in radio tuners, which allow you to select the station you wish to hear. Vibrating strings and columns of air also

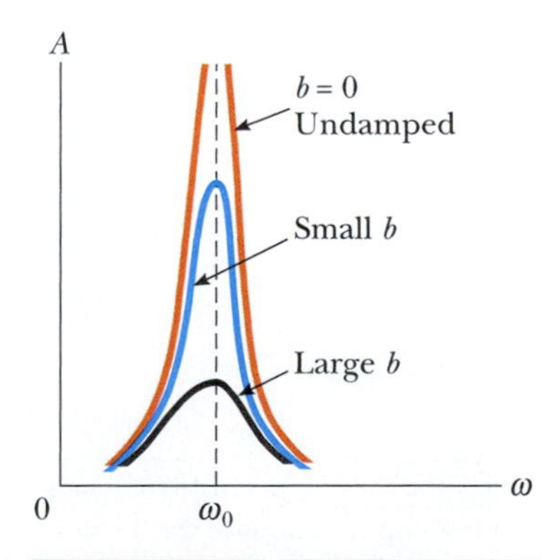


**FIGURE 12.16** Graph of amplitude versus frequency for a damped oscillator when a periodic driving force is present. When the frequency of the driving force equals the natural frequency $\omega_0$, resonance occurs. Note that the shape of the resonance curve depends on the size of the damping coefficient $b$.

have resonance frequencies, which allow them to be used for musical instruments, which we shall discuss in Chapter 14.

## 12.8 RESONANCE IN STRUCTURES

CONTEXT CONNECTION

In the preceding section, we investigated the phenomenon of resonance in which an oscillating system exhibits its maximum response to a periodic driving force when the frequency of the driving force matches the oscillator's natural frequency. We now apply this understanding to the interaction between the shaking of the ground during an earthquake and structures attached to the ground. The structure is the oscillator. It has a set of natural frequencies, determined by its stiffness, its mass, and the details of its construction. The periodic driving force is supplied by the shaking of the ground.

A disastrous result can occur if a natural frequency of the building matches a frequency contained in the ground shaking. In this case, the resonance vibrations of the building can build to a very large amplitude, large enough to damage or destroy the building. This result can be avoided in two ways. The first involves designing the structure so that natural frequencies of the building lie outside the range of earthquake frequencies. (A typical range of earthquake frequencies is 0–15 Hz.) Such a building can be designed by varying its size or mass structure. The second method involves incorporating sufficient damping in the building. This method may not change the resonance frequency significantly, but it will lower the response to the natural frequency as in Figure 12.16. It will also flatten the resonance curve, so the building will respond to a wide range of frequencies but with relatively small amplitude at any given frequency.

We now describe two examples involving resonance excitations in bridge structures. Soldiers are commanded to break step when marching across a bridge. This command takes into account resonance; if the marching frequency of the soldiers matches that of the bridge, the bridge could be set into resonance oscillation. If the amplitude becomes large enough, the bridge could actually collapse. Just such a situation occurred on April 14, 1831, when the Broughton suspension bridge in England collapsed while troops marched over it. Investigations after the accident showed that the bridge was near failure, and the resonance vibration induced by the marching soldiers caused it to fail sooner than it otherwise might have.

The second example of such a structural resonance occurred in 1940, when the Tacoma Narrows Bridge in Washington State was destroyed by resonant vibrations (Fig. 12.17). The winds were not particularly strong on that occasion, but the bridge still collapsed because vortices (turbulences) generated by the wind blowing through the bridge occurred at a frequency that matched a natural frequency of the bridge. The flapping of this wind across the roadway (similar to the flapping of

(Special Collections Division, Univ. of Wash. Libraries, Photo by Farquharson)

(a) (b)

**FIGURE 12.17** (a) In 1940, steady winds set up vibrations in the Tacoma Narrows Bridge, causing it to oscillate at a frequency near one of the natural frequencies of the bridge structure. (b) Once established, this resonance condition led to the bridge's collapse.

a flag in a strong breeze) provided the periodic driving force that brought the bridge down into the river.

Resonance gives us our first clue to responding to the central question for this Context. Suppose a building is far from the epicenter of an earthquake so that the ground shaking is small. If the shaking frequency matches a natural frequency of the building, a very effective energy coupling occurs between the ground and the building. Therefore, even for relatively small shaking, the ground, by resonance, can feed energy into the building efficiently enough to cause the failure of the structure. The structure must be carefully designed so as to reduce the resonance response.

## SUMMARY



The particle in simple harmonic motion model is used for a particle experiencing a linear restoring force, expressed by **Hooke's law,**

$$F_s = -kx \quad [12.1]$$

where $k$ is the **force constant** of the spring. The motion caused by such a force is called **simple harmonic motion,** and the system is called a **simple harmonic oscillator.** The position of a particle in simple harmonic motion varies periodically in time according to the relation

$$x(t) = A\cos(\omega t + \phi) \quad [12.6]$$

where $A$ is the **amplitude** of the motion, $\omega$ is the **angular frequency,** and $\phi$ is the **phase constant.** The values of $A$ and $\phi$ depend on the initial position and velocity of the particle.

The time for one complete oscillation is called the **period** $T$ of the motion. The inverse of the period is the **frequency** $f$ of the motion, which equals the number of oscillations per second:

$$f = \frac{1}{T} = \frac{\omega}{2\pi} \quad [12.11]$$

A particle–spring system oscillating without friction exhibits simple harmonic motion with the period

$$T = \frac{2\pi}{\omega} = 2\pi\sqrt{\frac{m}{k}} \quad [12.13]$$

where $m$ is the mass of the particle attached to the spring of force constant $k$.

The velocity and acceleration of a particle in simple harmonic oscillation are

$$v = \frac{dx}{dt} = -\omega A\sin(\omega t + \phi) \quad [12.15]$$

$$a = \frac{d^2x}{dt^2} = -\omega^2 A\cos(\omega t + \phi) \quad [12.16]$$

Therefore, the maximum speed of the particle is $\omega A$ and its maximum acceleration is of magnitude $\omega^2 A$. The speed is zero when the particle is at its turning points, $x = \pm A$, and the speed is a maximum at the equilibrium position, $x = 0$. The magnitude of the acceleration is a maximum at the turning points and is zero at the equilibrium position.

The kinetic energy and potential energy of a simple harmonic oscillator vary with time and are given by

$$K = \tfrac{1}{2}mv^2 = \tfrac{1}{2}m\omega^2A^2\sin^2(\omega t + \phi) \quad [12.19]$$

$$U = \tfrac{1}{2}kx^2 = \tfrac{1}{2}kA^2\cos^2(\omega t + \phi) \quad [12.20]$$

The **total energy** of a simple harmonic oscillator is a constant of the motion and is

$$E = \tfrac{1}{2}kA^2 \quad [12.21]$$

The potential energy of a simple harmonic oscillator is a maximum when the particle is at its turning points (maximum displacement from equilibrium) and is zero at the equilibrium position. The kinetic energy is zero at the turning points and is a maximum at the equilibrium position.

A **simple pendulum** of length $L$ exhibits simple harmonic motion for small angular displacements from the vertical, with a period of

$$T = 2\pi\sqrt{\frac{L}{g}} \quad [12.25]$$

The period of a simple pendulum is independent of the mass of the suspended object.

A **physical pendulum** exhibits simple harmonic motion for small angular displacements from equilibrium about a pivot that does not go through the center of mass. The period of this motion is

$$T = 2\pi\sqrt{\frac{I}{mgd}} \quad [12.27]$$

where $I$ is the moment of inertia about an axis through the pivot and $d$ is the distance from the pivot to the center of mass.

**Damped oscillations** occur in a system in which a resistive force opposes the motion of the oscillating object. If such a system is set in motion and then left to itself, its mechanical energy decreases in time because of the presence of the nonconservative resistive force. It is possible to compensate for this transformation of energy by driving the system with an external periodic force. The oscillator in this case is undergoing **forced oscillations.** When the frequency of the driving force matches the natural frequency of the *undamped* oscillator, energy is efficiently transferred to the oscillator and its steady-state amplitude is a maximum. This situation is called **resonance.**

## QUESTIONS

☐ = answer available in the *Student Solutions Manual and Study Guide*

1. Is a bouncing ball an example of simple harmonic motion? Is the daily movement of a student from home to school and back simple harmonic motion? Why or why not?

2. Does the displacement of an oscillating particle between $t = 0$ and a later time $t$ necessarily equal the position of the particle at time $t$? Explain.

3. If the position of a particle varies as $x = -A \cos \omega t$, what is the phase constant in Equation 12.6? At what position is the particle at $t = 0$?

4. Can the amplitude $A$ and phase constant $\phi$ be determined for an oscillator if only the position is specified at $t = 0$? Explain.

5. Determine whether or not the following quantities can be in the same direction for a simple harmonic oscillator: (a) position and velocity, (b) velocity and acceleration, (c) position and acceleration.

6. A block is hung on a spring and the frequency $f$ of the oscillation of the system is measured. The block, a second identical block, and the spring are carried into space in a spacecraft. The two blocks are attached to opposite ends of the spring, and the system is taken out into space on a space walk. The spring is extended, and the system is released to oscillate while floating in space. What is the frequency of oscillation for this system in terms of $f$?

7. A block–spring system undergoes simple harmonic motion with amplitude $A$. Does the total energy change if the mass is doubled but the amplitude is not changed? Do the kinetic and potential energies depend on the mass? Explain.

8. The equations listed in Table 2.2 give position as a function of time, velocity as a function of time, and velocity as function of position for an object moving in a straight line with constant acceleration. The quantity $v_{xi}$ appears in every equation. Do any of these equations apply to an object moving in a straight line with simple harmonic motion? Using a similar format, make a table of equations describing simple harmonic motion. Include equations giving acceleration as a function of time and acceleration as a function of position. State the equations in such a form that they apply equally to a block–spring system, to a pendulum, and to other vibrating systems. What quantity appears in every equation?

9. What happens to the period of a simple pendulum if the pendulum's length is doubled? What happens to the period if the mass of the suspended bob is doubled?

10. If a grandfather clock were running slow, how could we adjust the pendulum's length to correct the time?

11. Will damped oscillations occur for any values of $b$ and $k$? Explain.

12. You stand on the end of a diving board and bounce to set it into oscillation. You find a maximum response, in terms of the amplitude of oscillation of the end of the board, when you bounce at frequency $f$. You now move to the middle of the board and repeat the experiment. Is the resonance frequency for forced oscillations at this point higher, lower, or the same as $f$? Why?

13. Is it possible to have damped oscillations when a system is at resonance? Explain.

14. You are looking at a small tree. You do not notice any breeze, and most of the leaves on the tree are motionless. One leaf, however, is fluttering back and forth wildly. After you wait a while, that leaf stops moving and you notice a different leaf moving much more than all the others. Explain what could cause the large motion of one particular leaf.

## PROBLEMS

1, 2, 3 = straightforward, intermediate, challenging
☐ = full solution available in the *Student Solutions Manual and Study Guide*
Physics Now™ = coached problem with hints available at www.pop4e.com
= computer useful in solving problem
= paired numerical and symbolic problems
= biomedical application

*Note:* Ignore the mass of every spring except in problems 12.54 and 12.56.

### Section 12.1 ▪ Motion of a Particle Attached to a Spring

Problems 6.13, 6.15, 6.17, 6.18, 6.30, and 6.56 in Chapter 6 can also be assigned with this section.

1. A ball dropped from a height of 4.00 m makes an elastic collision with the ground. Assuming that no mechanical energy is lost due to air resistance, (a) show that the ensuing motion is periodic and (b) determine the period of the motion. (c) Is the motion simple harmonic? Explain.

### Section 12.2 ▪ Mathematical Representation of Simple Harmonic Motion

2. In an engine, a piston oscillates with simple harmonic motion so that its position varies according to the expression

$$x = (5.00 \text{ cm}) \cos(2t + \pi/6)$$

where $x$ is in centimeters and $t$ is in seconds. At $t = 0$, find (a) the position of the piston, (b) its velocity, and (c) its acceleration. (d) Find the period and amplitude of the motion.

3. The position of a particle is given by the expression $x = (4.00 \text{ m}) \cos(3.00\pi t + \pi)$, where $x$ is in meters and $t$ is in

seconds. Determine (a) the frequency and period of the motion, (b) the amplitude of the motion, (c) the phase constant, and (d) the position of the particle at $t = 0.250$ s.

4. A particle moves in simple harmonic motion with a frequency of 3.00 Hz and an amplitude of 5.00 cm. (a) Through what total distance does the particle move during one cycle of its motion? (b) What is its maximum speed? Where does this maximum speed occur? (c) Find the maximum acceleration of the particle. Where in the motion does the maximum acceleration occur?

5. **Physics Now™** A particle moving along the $x$ axis in simple harmonic motion starts from its equilibrium position, the origin, at $t = 0$ and moves to the right. The amplitude of its motion is 2.00 cm and the frequency is 1.50 Hz. (a) Show that the position of the particle is given by

$$x = (2.00 \text{ cm}) \sin(3.00\pi t)$$

Determine (b) the maximum speed and the earliest time ($t > 0$) at which the particle has this speed, (c) the maximum acceleration and the earliest time ($t > 0$) at which the particle has this acceleration, and (d) the total distance traveled between $t = 0$ and $t = 1.00$ s.

6. **Review problem.** A particle moves along the $x$ axis. It is initially at the position 0.270 m, moving with velocity 0.140 m/s and acceleration $-0.320$ m/s$^2$. First, assume that it moves with constant acceleration for 4.50 s. Find (a) its position and (b) its velocity at the end of this time interval. Next, assume that it moves with simple harmonic motion for 4.50 s and that $x = 0$ is its equilibrium position. Find (c) its position and (d) its velocity at the end of this time interval.

7. The initial position, velocity, and acceleration of an object moving in simple harmonic motion are $x_i$, $v_i$, and $a_i$; the angular frequency of oscillation is $\omega$. (a) Show that the position and velocity of the object for all time can be written as

$$x(t) = x_i \cos \omega t + \left(\frac{v_i}{\omega}\right) \sin \omega t$$

$$v(t) = -x_i \omega \sin \omega t + v_i \cos \omega t$$

(b) Using $A$ to represent the amplitude of the motion, show that

$$v^2 - ax = v_i^2 - a_i x_i = \omega^2 A^2$$

8. A simple harmonic oscillator takes 12.0 s to undergo five complete vibrations. Find (a) the period of its motion, (b) the frequency in hertz, and (c) the angular frequency in radians per second.

9. A 7.00 kg object is hung from the bottom end of a vertical spring fastened to an overhead beam. The object is set into vertical oscillations having a period of 2.60 s. Find the force constant of the spring.

10. A vibration sensor, used in testing a washing machine, consists of a cube of aluminum 1.50 cm on edge mounted on one end of a strip of spring steel (like a hacksaw blade) that lies in a vertical plane. The strip's mass is small compared with that of the cube, but the strip's length is large compared with the size of the cube. The other end of the strip is clamped to the frame of the washing machine that is not operating. A horizontal force of 1.43 N applied to the cube is required to hold it 2.75 cm away from its equilibrium position. If it is released, what is its frequency of vibration?

11. A 0.500-kg object attached to a spring with a force constant of 8.00 N/m vibrates in simple harmonic motion with an amplitude of 10.0 cm. Calculate (a) the maximum value of its speed and acceleration, (b) the speed and acceleration when the object is 6.00 cm from the equilibrium position, and (c) the time interval required for the object to move from $x = 0$ to $x = 8.00$ cm.

12. A 1.00-kg glider attached to a spring with a force constant of 25.0 N/m oscillates on a horizontal, frictionless air track. At $t = 0$, the glider is released from rest at $x = -3.00$ cm (that is, the spring is compressed by 3.00 cm). Find (a) the period of its motion, (b) the maximum values of its speed and acceleration, and (c) the position, velocity, and acceleration as functions of time.

## Section 12.3 ■ Energy Considerations in Simple Harmonic Motion

13. **Physics Now™** An automobile having a mass of 1 000 kg is driven into a brick wall in a safety test. The bumper behaves like a spring of force constant $5.00 \times 10^6$ N/m and compresses 3.16 cm as the car is brought to rest. What was the speed of the car before impact, assuming that no mechanical energy is lost during impact with the wall?

14. A 200-g block is attached to a horizontal spring and executes simple harmonic motion with a period of 0.250 s. The total energy of the system is 2.00 J. Find (a) the force constant of the spring and (b) the amplitude of the motion.

15. A block of unknown mass is attached to a spring with a spring constant of 6.50 N/m and undergoes simple harmonic motion with an amplitude of 10.0 cm. When the block is halfway between its equilibrium position and the end point, its speed is measured to be 30.0 cm/s. Calculate (a) the mass of the block, (b) the period of the motion, and (c) the maximum acceleration of the block.

16. A block–spring system oscillates with an amplitude of 3.50 cm. The spring constant is 250 N/m and the mass of the block is 0.500 kg. Determine (a) the mechanical energy of the system, (b) the maximum speed of the block, and (c) the maximum acceleration.

17. A 50.0-g object connected to a spring with a force constant of 35.0 N/m oscillates on a horizontal, frictionless surface with an amplitude of 4.00 cm. Find (a) the total energy of the system and (b) the speed of the object when the position is 1.00 cm. Find (c) the kinetic energy and (d) the potential energy when the position is 3.00 cm.

18. A 2.00-kg object is attached to a spring and placed on a horizontal, smooth surface. A horizontal force of 20.0 N is required to hold the object at rest when it is pulled 0.200 m from its equilibrium position (the origin of the $x$ axis). The object is now released from rest with an initial position of $x_i = 0.200$ m, and it subsequently undergoes simple harmonic oscillations. Find (a) the force constant of the spring, (b) the frequency of the oscillations, and (c) the maximum speed of the object. Where does this

maximum speed occur? (d) Find the maximum acceleration of the object. Where does it occur? (e) Find the total energy of the oscillating system. Find (f) the speed and (g) the acceleration of the object when its position is equal to one third of the maximum value.

**19.** The amplitude of a system moving in simple harmonic motion is doubled. Determine the change in (a) the total energy, (b) the maximum speed, (c) the maximum acceleration, and (d) the period.

**20.** A 65.0-kg bungee jumper steps off a bridge with a light bungee cord tied to her and to the bridge (Figure P12.20). The unstretched length of the cord is 11.0 m. She reaches the bottom of her motion 36.0 m below the bridge before bouncing back. Her motion can be separated into an 11.0-m free-fall and a 25.0-m section of simple harmonic oscillation. (a) For what time interval is she in free-fall? (b) Use the principle of conservation of energy to find the spring constant of the bungee cord. (c) What is the location of the equilibrium point where the spring force balances the gravitational force acting on the jumper? Note that this point is taken as the origin in our mathematical description of simple harmonic oscillation. (d) What is the angular frequency of the oscillation? (e) What time interval is required for the cord to stretch by 25.0 m? (f) What is the total time interval for the entire 36.0-m drop?

FIGURE **P12.20** Problems 12.20 and 12.44.

**21.** A particle executes simple harmonic motion with an amplitude of 3.00 cm. At what position does its speed equal half of its maximum speed?

## Section 12.4 ▪ The Simple Pendulum
## Section 12.5 ▪ The Physical Pendulum

Problem 1.62 in Chapter 1 can also be assigned with this section.

**22.** A "seconds pendulum" is one that moves through its equilibrium position once each second. (The period of the pendulum is precisely 2 s.) The length of a seconds pendulum is 0.992 7 m at Tokyo, Japan, and 0.994 2 m at Cambridge, England. What is the ratio of the free-fall accelerations at these two locations?

**23.** **PhysicsNow™** A simple pendulum has a mass of 0.250 kg and a length of 1.00 m. It is displaced through an angle of 15.0° and then released. What are (a) the maximum speed, (b) the maximum angular acceleration, and (c) the maximum restoring force? Solve this problem once by using the simple harmonic motion model for the motion of the pendulum, and then solve the problem more precisely by using more general principles.

**24.** The angular position of a pendulum is represented by the equation $\theta = (0.032\ 0 \text{ rad}) \cos \omega t$, where $\theta$ is in radians and $\omega = 4.43$ rad/s. Determine the period and length of the pendulum.

**25.** A particle of mass $m$ slides without friction inside a hemispherical bowl of radius $R$. Show that if it starts from rest with a small displacement from equilibrium, the particle moves in simple harmonic motion with an angular frequency equal to that of a simple pendulum of length $R$ (that is, $\omega = \sqrt{g/R}$).

**26.** A small object is attached to the end of a string to form a simple pendulum. The period of its harmonic motion is measured for small angular displacements and three lengths, each time by clocking the motion with a stopwatch for 50 oscillations. For lengths of 1.000 m, 0.750 m, and 0.500 m, total time intervals of 99.8 s, 86.6 s, and 71.1 s are measured for 50 oscillations. (a) Determine the period of motion for each length. (b) Determine the mean value of $g$ obtained from these three independent measurements, and compare it with the accepted value. (c) Plot $T^2$ versus $L$, and obtain a value for $g$ from the slope of your best-fit straight-line graph. Compare this value with that obtained in part (b).

**27.** A physical pendulum in the form of a planar object moves in simple harmonic motion with a frequency of 0.450 Hz. The pendulum has a mass of 2.20 kg, and the pivot is located 0.350 m from the center of mass. Determine the moment of inertia of the pendulum about the pivot point.

**28.** A very light rigid rod with a length of 0.500 m extends straight out from one end of a meter stick. The stick is suspended from a pivot at the far end of the rod and is set into oscillation. (a) Determine the period of oscillation. (*Suggestion:* Use the parallel-axis theorem from Section 10.11.) (b) By what percentage does the period differ from the period of a simple pendulum 1.00 m long?

**29.** Consider the physical pendulum of Figure 12.12. (a) Representing its moment of inertia about an axis passing through its center of mass and parallel to the axis passing through its pivot point as $I_{CM}$, show that its period is

$$T = 2\pi\sqrt{\frac{I_{CM} + md^2}{mgd}}$$

where $d$ is the distance between the pivot point and center of mass. (b) Show that the period has a minimum value when $d$ satisfies $md^2 = I_{CM}$.

## Section 12.6 ■ Damped Oscillations

**30.** Show that the time rate of change of mechanical energy for a damped, undriven oscillator is given by $dE/dt = -bv^2$ and hence is always negative. (*Suggestion:* Differentiate the expression for the mechanical energy of an oscillator, $E = \frac{1}{2}mv^2 + \frac{1}{2}kx^2$, and use Equation 12.28.)

**31.** A pendulum with a length of 1.00 m is released from an initial angle of 15.0°. After 1 000 s, its amplitude has been reduced by friction to 5.50°. What is the value of $b/2m$?

**32.** Show that Equation 12.29 is a solution of Equation 12.28 provided that $b^2 < 4mk$.

## Section 12.7 ■ Forced Oscillations

**33.** A 2.00-kg object attached to a spring moves without friction and is driven by an external force $F = (3.00\text{ N})\sin(2\pi t)$. Assuming that the force constant of the spring is 20.0 N/m, determine (a) the period and (b) the amplitude of the motion.

**34.** The front of her sleeper wet from teething, a baby rejoices in the day by crowing and bouncing up and down in her crib. Her mass is 12.5 kg, and the crib mattress can be modeled as a light spring with force constant 4.30 kN/m. (a) The baby soon learns to bounce with maximum amplitude and minimum effort by bending her knees at what frequency? (b) She learns to use the mattress as a trampoline—losing contact with it for part of each cycle—when her amplitude exceeds what value?

**35.** Considering an undamped, forced oscillator ($b = 0$), show that Equation 12.32 is a solution of Equation 12.31, with an amplitude given by Equation 12.33.

**36.** Damping is negligible for a 0.150-kg object hanging from a light 6.30-N/m spring. A sinusoidal force with an amplitude of 1.70 N drives the system. At what frequency will the force make the object vibrate with an amplitude of 0.440 m?

**37.** You are a research biologist. You take your emergency pager along to a fine restaurant even though its batteries are getting low. You switch the small pager to vibrate instead of beep, and you put it into a side pocket of your suit coat. The arm of your chair presses the light cloth against your body at one spot. Fabric with a length of 8.21 cm hangs freely below that spot, with the pager at the bottom. A coworker urgently needs instructions and calls you from your laboratory. The pager's motion makes the hanging part of your coat swing back and forth with remarkably large amplitude. The waiter, maître d', wine steward, and nearby diners notice immediately and fall silent. Your daughter pipes up and says, "Daddy, look! Your cockroaches must have gotten out again!" Find the frequency at which your pager vibrates.

## Section 12.8 ■ Context Connection—Resonance in Structures

**38.** Four people, each with a mass of 72.4 kg, are in a car with a mass of 1 130 kg. An earthquake strikes. The vertical oscillations of the ground surface make the car bounce up and down on its suspension springs, but the driver manages to pull off the road and stop. When the frequency of the shaking is 1.80 Hz, the car exhibits a maximum amplitude of vibration. The earthquake ends and the four people leave the car as fast as they can. By what distance does the car's undamaged suspension lift the car's body as the people get out?

**39.** People who ride motorcycles and bicycles learn to look out for bumps in the road and especially for *washboarding,* a condition in which many equally spaced ridges are worn into the road. What is so bad about washboarding? A motorcycle has several springs and shock absorbers in its suspension, but you can model it as a single spring supporting a block. You can estimate the force constant by thinking about how far the spring compresses when a large biker sits down on the seat. A motorcyclist traveling at highway speed must be particularly careful of washboard bumps that are a certain distance apart. What is the order of magnitude of their separation distance? State the quantities you take as data and the values you measure or estimate for them.

## Additional Problems

**40.** An object of mass $m_1 = 9.00$ kg is in equilibrium while connected to a light spring of constant $k = 100$ N/m that is fastened to a wall as shown in Figure P12.40a. A second object, $m_2 = 7.00$ kg, is slowly pushed up against $m_1$, compressing the spring by the amount $A = 0.200$ m (see Fig. P12.40b). The system is then released and both objects start moving to the right on the frictionless surface. (a) When $m_1$ reaches the equilibrium point, $m_2$ loses contact with $m_1$ (see Fig. P12.40c) and moves to the right with speed $v$. Determine the value of $v$. (b) How far apart are the objects when the spring is fully stretched for the first time ($D$ in Fig. P12.40d)? (*Suggestion:* First determine the period of oscillation and the amplitude of the $m_1$–spring system after $m_2$ loses contact with $m_1$.)

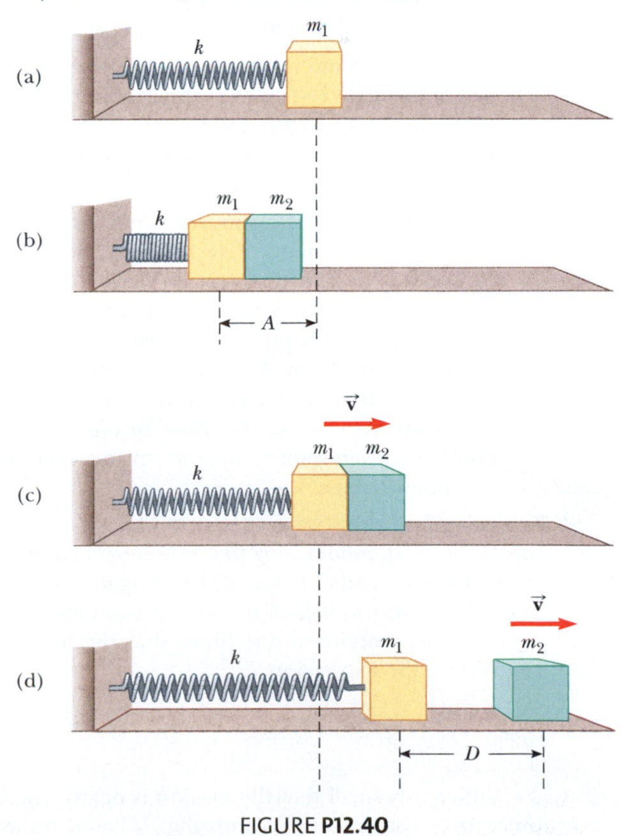


FIGURE P12.40

**41.** **Physics Now™** A large block $P$ executes horizontal simple harmonic motion as it slides across a frictionless surface with a frequency $f = 1.50$ Hz. Block $B$ rests on it, as shown in Figure P12.41, and the coefficient of static friction between the two is $\mu_s = 0.600$. What maximum amplitude of oscillation can the system have if block $B$ is not to slip?

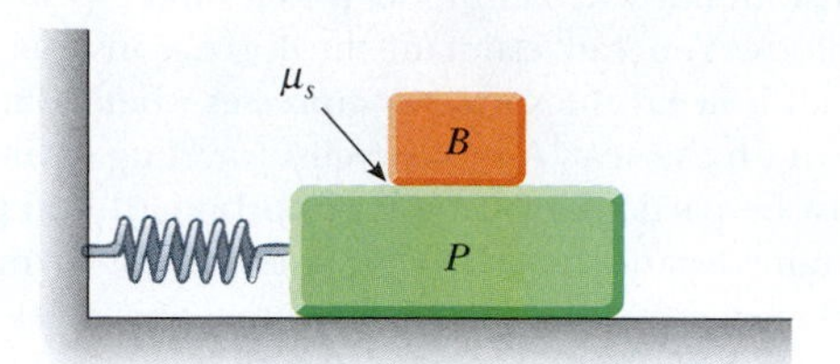


FIGURE **P12.41**

**42.** (a) A hanging spring stretches by 35.0 cm when an object of mass 450 g is hung on it at rest. In this situation, we define its position as $x = 0$. The object is pulled down an additional 18.0 cm and released from rest to oscillate without friction. What is its position $x$ at a time 84.4 s later? (b) A hanging spring stretches by 35.5 cm when an object of mass 440 g is hung on it at rest. We define this new position as $x = 0$. This object is also pulled down an additional 18.0 cm and released from rest to oscillate without friction. Find its position 84.4 s later. (c) Why are the answers to parts (a) and (b) different by such a large percentage when the data are so similar? Does this circumstance reveal a fundamental difficulty in calculating the future? (d) Find the distance traveled by the vibrating object in part (a). (e) Find the distance traveled by the object in part (b).

**43.** The mass of the deuterium molecule ($D_2$) is twice that of the hydrogen molecule ($H_2$). If the vibrational frequency of $H_2$ is $1.30 \times 10^{14}$ Hz, what is the vibrational frequency of $D_2$? Assume that the "spring constant" of attracting forces is the same for the two molecules.

**44.** After a thrilling plunge, bungee jumpers bounce freely on the bungee cord through many cycles (Fig. P12.20). After the first few cycles, the cord does not go slack. Your little brother can make a pest of himself by figuring out the mass of each person, using a proportion that you set up by solving the following problem. An object of mass $m$ is oscillating freely on a light vertical spring with a period $T$. An object of unknown mass $m'$ on the same spring oscillates with a period $T'$. Determine (a) the spring constant and (b) the unknown mass.

**45.** To account for the walking speed of a bipedal or quadrupedal animal, model a leg that is not contacting the ground as a uniform rod of length $\ell$, swinging as a physical pendulum through one half of a cycle, in resonance. Let $\theta_{max}$ represent its amplitude. (a) Show that the animal's speed is given by the expression

$$\frac{\sqrt{6g\ell}\,\sin\theta_{max}}{\pi}$$

if $\theta_{max}$ is sufficiently small that the motion is nearly simple harmonic. An empirical relationship that is based on the same model and applies over a wider range of angles is

$$\frac{\sqrt{6g\ell\,\cos(\theta_{max}/2)}\,\sin\theta_{max}}{\pi}$$

(b) Evaluate the walking speed of a human with leg length 0.850 m and leg-swing amplitude 28.0°. (c) What leg length would give twice the speed for the same angular amplitude?

**46.** **Review problem.** The problem extends the reasoning of Problem 8.46 in Chapter 8. Two gliders are set in motion on an air track. Glider 1 has mass $m_1 = 0.240$ kg and velocity $0.740\hat{\mathbf{i}}$ m/s. It will have a rear-end collision with glider 2, of mass $m_2 = 0.360$ kg, which has original velocity $0.120\hat{\mathbf{i}}$ m/s. A light spring of force constant 45.0 N/m is attached to the back end of glider 2 as shown in Figure P8.46. When glider 1 touches the spring, superglue instantly and permanently makes it stick to its end of the spring. (a) Find the common velocity the two gliders have when the spring compression is a maximum. (b) Find the maximum spring compression distance. (c) Argue that the motion after the gliders become attached consists of the center of mass of the two-glider system moving with the constant velocity found in part (a) while both gliders oscillate in simple harmonic motion relative to the center of mass. (d) Find the energy of the center-of-mass motion. (e) Find the energy of the oscillation.

**47.** A pendulum of length $L$ and mass $M$ has a spring of force constant $k$ connected to it at a distance $h$ below its point of suspension (Fig. P12.47). Find the frequency of vibration of the system for small values of the amplitude (small $\theta$). Assume that the vertical suspension of length $L$ is rigid, but ignore its mass.

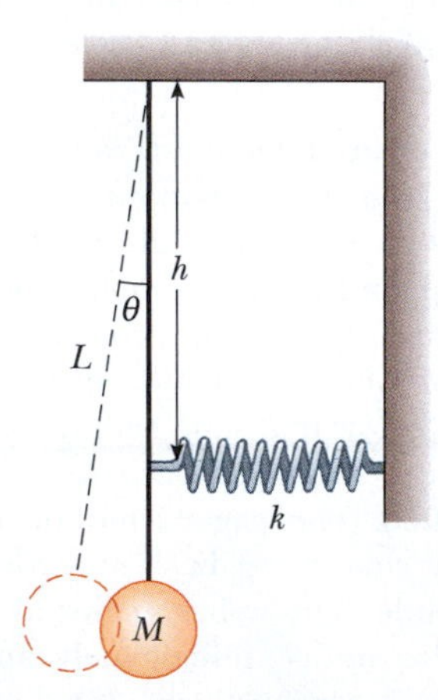


FIGURE **P12.47**

**48.** A particle with a mass of 0.500 kg is attached to a spring with a force constant of 50.0 N/m. At time $t = 0$ the particle has its maximum speed of 20.0 m/s and is moving to the left. (a) Determine the particle's equation of motion, specifying its position as a function of time. (b) Where in the motion is the potential energy three times the kinetic energy? (c) Find the length of a simple pendulum with the same period. (d) Find the minimum time interval required for the particle to move from $x = 0$ to $x = 1.00$ m.

**49.** A horizontal plank of mass $m$ and length $L$ is pivoted at one end. The plank's other end is supported by a spring of force constant $k$ (Fig. P12.49). The moment of inertia of the plank about the pivot is $\frac{1}{3}mL^2$. The plank is displaced by a small angle $\theta$ from its horizontal equilibrium position and released. (a) Show that it moves with simple harmonic motion with an angular frequency $\omega = \sqrt{3k/m}$. (b) Evaluate the frequency, assuming that the mass is 5.00 kg and that the spring has a force constant of 100 N/m.

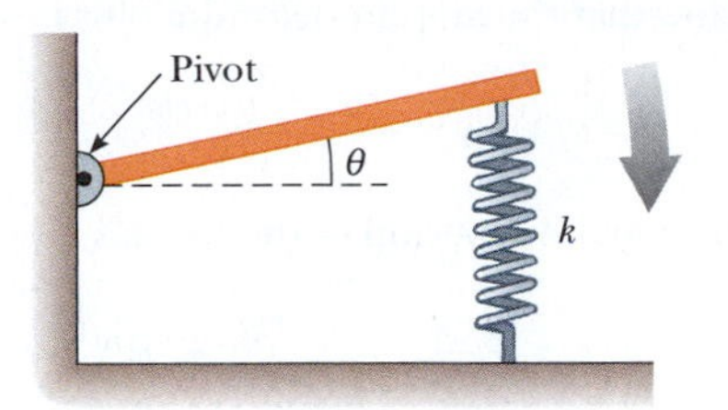


FIGURE **P12.49**

**50.** **Review problem.** A particle of mass 4.00 kg is attached to a spring with a force constant of 100 N/m. It is oscillating on a horizontal, frictionless surface with an amplitude of 2.00 m. A 6.00-kg object is dropped vertically on top of the 4.00-kg object as it passes through its equilibrium point. The two objects stick together. (a) By how much does the amplitude of the vibrating system change as a result of the collision? (b) By how much does the period change? (c) By how much does the mechanical energy of the system change? (d) Account for the change in energy.

**51.** A simple pendulum with a length of 2.23 m and a mass of 6.74 kg is given an initial speed of 2.06 m/s at its equilibrium position. Assume that it undergoes simple harmonic motion and determine its (a) period, (b) total energy, and (c) maximum angular displacement.

**52.** **Review problem.** One end of a light spring with force constant 100 N/m is attached to a vertical wall. A light string is tied to the other end of the horizontal spring. The string changes from horizontal to vertical as it passes over a solid pulley of diameter 4.00 cm. The pulley is free to turn on a fixed, smooth axle. The vertical section of the string supports a 200-g object. The string does not slip at its contact with the pulley. Find the frequency at which the object oscillates if the mass of the pulley is (a) negligible, (b) 250 g, and (c) 750 g.

**53.** **PhysicsNow™** A ball of mass $m$ is connected to two rubber bands of length $L$, each under tension $T$, as shown in Figure P12.53. The ball is displaced by a small distance $y$ perpendicular to the length of the rubber bands. Assuming that the tension does not change, show that (a) the restoring force is $-(2T/L)y$ and (b) the system exhibits simple harmonic motion with an angular frequency $\omega = \sqrt{2T/mL}$.

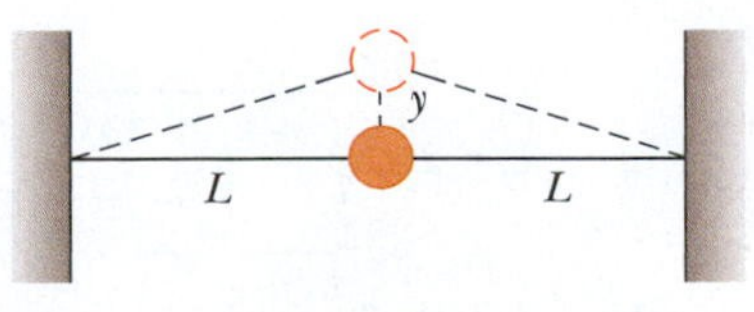


FIGURE **P12.53**

**54.** A block of mass $M$ is connected to a spring of mass $m$ and oscillates in simple harmonic motion on a horizontal, frictionless track (Fig. P12.54). The force constant of the spring is $k$ and the equilibrium length is $\ell$. Assume that all portions of the spring oscillate in phase and that the velocity of a segment $dx$ is proportional to the distance $x$ from the fixed end; that is, $v_x = (x/\ell)v$. Also, note that the mass of a segment of the spring is $dm = (m/\ell)\,dx$, Find (a) the kinetic energy of the system when the block has a speed $v$ and (b) the period of oscillation.

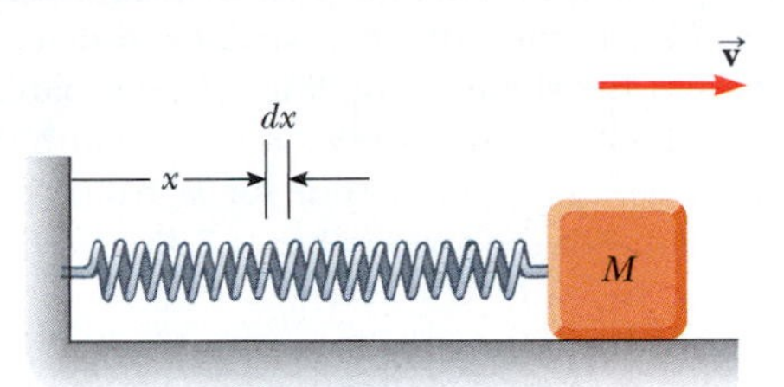


FIGURE **P12.54**

**55.** A smaller disk of radius $r$ and mass $m$ is attached rigidly to the face of a second larger disk of radius $R$ and mass $M$ as shown in Figure P12.55. The center of the small disk is located at the edge of the large disk. The large disk is mounted at its center on a frictionless axle. The assembly is rotated through a small angle $\theta$ from its equilibrium position and released. (a) Show that the speed of the center of the small disk as it passes through the equilibrium position is

$$v = 2\left[\frac{Rg(1 - \cos\theta)}{(M/m) + (r/R)^2 + 2}\right]^{1/2}$$

(b) Show that the period of the motion is

$$T = 2\pi\left[\frac{(M + 2m)R^2 + mr^2}{2mgR}\right]^{1/2}$$

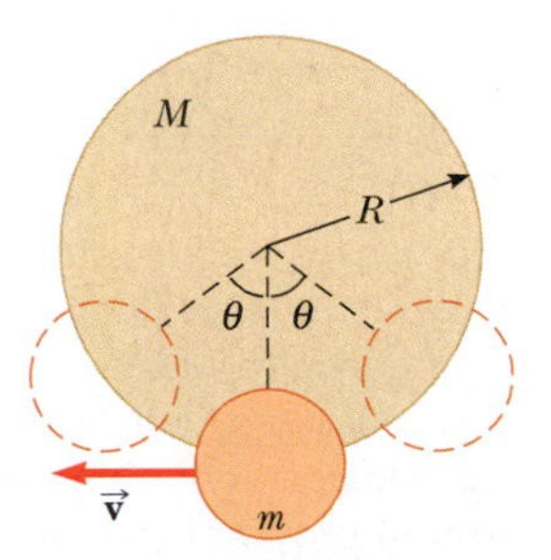


FIGURE **P12.55**

**56.** When a block of mass $M$, connected to the end of a spring of mass $m_s = 7.40$ g and force constant $k$, is set into simple harmonic motion, the period of its motion is

$$T = 2\pi\sqrt{\frac{M + (m_s/3)}{k}}$$

A two-part experiment is conducted with the use of blocks of various masses suspended vertically from the spring as shown in Figure P12.56. (a) Static extensions of 17.0, 29.3, 35.3, 41.3, 47.1, and 49.3 cm are measured for $M$ values of

20.0, 40.0, 50.0, 60.0, 70.0, and 80.0 g, respectively. Construct a graph of $Mg$ versus $x$ and perform a linear least-squares fit to the data. From the slope of your graph, determine a value for $k$ for this spring. (b) The system is now set into simple harmonic motion, and periods are measured with a stopwatch. With $M = 80.0$ g, the total time interval for ten oscillations is measured to be 13.41 s. The experiment is repeated with $M$ values of 70.0, 60.0, 50.0, 40.0, and 20.0 g, with corresponding time intervals for ten oscillations of 12.52, 11.67, 10.67, 9.62, and 7.03 s. Compute the experimental value for $T$ from each of these measurements. Plot a graph of $T^2$ versus $M$ and determine a value for $k$ from the slope of the linear least-squares fit through the data points. Compare this value of $k$ with that obtained in part (a). (c) Obtain a value for $m_s$ from your graph and compare it with the given value of 7.40 g.

FIGURE P12.56

57. An object is hung from a spring, set into vertical vibration, and immersed in a beaker of oil. Its motion is graphed in Active Figure 12.14b. The object has mass 375 g, the spring has force constant 100 N/m, and the damping coefficient is $b = 0.100$ N · s/m. (a) How long does it take for the amplitude to drop to half its initial value? (b) How long does it take for the mechanical energy to drop to half its initial value? (c) Show that, in general, the fractional rate at which the amplitude decreases in a damped harmonic oscillator is one-half the fractional rate at which the mechanical energy decreases.

58. Your thumb squeaks on a plate you have just washed. Your sneakers squeak on the gym floor. Car tires squeal when you start or stop abruptly. Mortise joints groan in an old barn. The concertmaster's violin sings out over a full orchestra. You can make a goblet sing by wiping your moistened finger around its rim. As you slide it across the table, a Styrofoam cup may not make much sound, but it makes the surface of some water inside it dance in a complicated resonance vibration. When chalk squeaks on a blackboard, you can see that it makes a row of regularly spaced dashes. As these examples suggest, vibration commonly results when friction acts on a moving elastic object. The oscillation is not simple harmonic motion, but is called *stick-and-slip*. This problem models stick-and-slip motion.

A block of mass $m$ is attached to a fixed support by a horizontal spring with force constant $k$ and negligible mass (Fig. P12.58). Hooke's law describes the spring both in extension and in compression. The block sits on a long horizontal board with which it has coefficient of static friction $\mu_s$ and a smaller coefficient of kinetic friction $\mu_k$. The board moves to the right at constant speed $v$. Assume that the block spends most of its time sticking to the board and moving to the right, so the speed $v$ is small in comparison to the average speed the block has as it slips back toward the left. (a) Show that the maximum extension of the spring from its unstressed position is very nearly given by $\mu_s mg/k$. (b) Show that the block oscillates around an equilibrium position at which the spring is stretched by $\mu_k mg/k$. (c) Graph the block's position versus time. (d) Show that the amplitude of the block's motion is

$$A = \frac{(\mu_s - \mu_k)\,mg}{k}$$

(e) Show that the period of the block's motion is

$$T = \frac{2(\mu_s - \mu_k)\,mg}{vk} + \pi\sqrt{\frac{m}{k}}$$

(f) Evaluate the frequency of the motion assuming that $\mu_s = 0.400$, $\mu_k = 0.250$, $m = 0.300$ kg, $k = 12.0$ N/m, and $v = 2.40$ cm/s. (g) What happens to the frequency if the mass increases? (h) What happens if the spring constant increases? (i) What happens if the speed of the board increases? (j) What happens if the coefficient of static friction increases relative to the coefficient of kinetic friction? Note that it is the excess of static over kinetic friction that is important for the vibration. "The squeaky wheel gets the grease" because even a viscous fluid cannot exert a force of static friction.

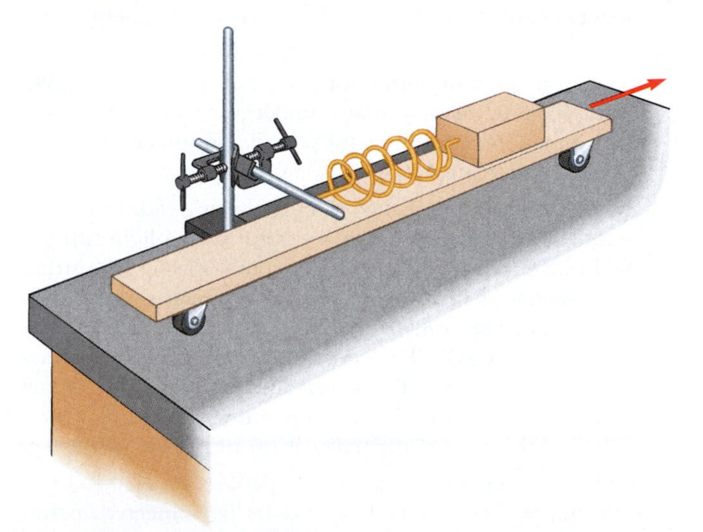

FIGURE P12.58

59. A block of mass $m$ is connected to two springs of force constants $k_1$ and $k_2$ in two ways as shown in Figure P12.59. In both cases, the block moves on a frictionless table after it is displaced from equilibrium and released. Show that in the two cases the block exhibits simple harmonic motion with periods

(a) $$T = 2\pi\sqrt{\frac{m(k_1 + k_2)}{k_1 k_2}}$$

(b) $$T = 2\pi\sqrt{\frac{m}{k_1 + k_2}}$$

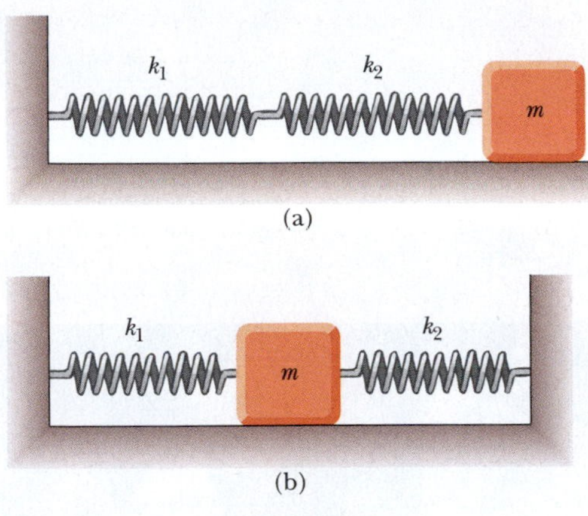


FIGURE **P12.59**

**60. Review problem.** Imagine that a hole is drilled through the center of the Earth to the other side. An object of mass $m$ at a distance $r$ from the center of the Earth is pulled toward the center of the Earth only by the mass within the sphere of radius $r$ (the reddish region in Fig. P12.60). (a) Write Newton's law of gravitation for an object at the distance $r$ from the center of the Earth and show that the force on it is of Hooke's law form, $F = -kr$, where the effective force constant is $k = \frac{4}{3}\pi\rho G m$. Here $\rho$ is the density of the Earth, assumed uniform, and $G$ is the gravitational constant. (b) Show that a sack of mail dropped into the hole will execute simple harmonic motion if it moves without friction. How long does it take to arrive at the other side of the Earth? (c) At the same time as the sack of mail is dropped in the hole, a golf ball is struck so that it becomes a treetop satellite. (See Problem 11.27 in Chapter 11.) Which object, sack of mail or golf ball, arrives first at the end of the hole halfway around the Earth?

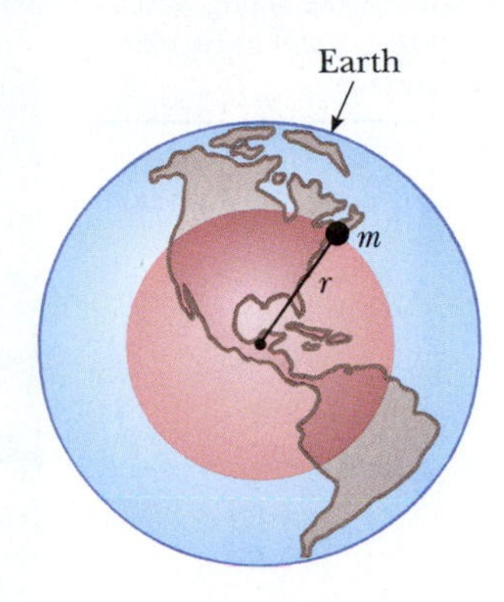


FIGURE **P12.60**

## ANSWERS TO QUICK QUIZZES

**12.1** (d). From its maximum positive position to the equilibrium position, the block travels a distance $A$. It then goes an equal distance past the equilibrium position to its maximum negative position. It then repeats these two motions in the reverse direction to return to its original position and complete one cycle.

**12.2** (i), (f). The object is in the region $x < 0$, so the position is negative. Because the object is moving back toward the origin in this region, the velocity is positive. (ii), (a). The velocity is positive, as in (i). Because the spring is pulling the object toward equilibrium from the negative $x$ region, the acceleration is also positive.

**12.3** (i), (a). With a longer length, the period of the pendulum will increase. Therefore, it will take slightly longer to execute each swing, so each second according to the clock will take longer than an actual second and the clock will run *slow.* (ii), (a). At the top of the mountain, the value of $g$ is less than that at sea level. As a result, the period of the pendulum will increase slightly and the clock will run slow.

**12.4** (a). Although changing the mass of a simple pendulum does not change the frequency, because the bob is large means that we must model the pendulum as a physical pendulum rather than a simple pendulum. When the gum is placed on top of the bob, the moment of inertia and the center of mass of the physical pendulum are altered slightly. According to Equation 12.27, these changes alter the period of the pendulum.

# Mechanical Waves

Drops of water fall from a leaf into a pond. The disturbance caused by the falling water moves away from the drop point as circular ripples on the water surface.

(Don Bonsey/Stone Getty)

## CHAPTER OUTLINE


13.1 Propagation of a Disturbance
13.2 The Wave Model
13.3 The Traveling Wave
13.4 The Speed of Transverse Waves on Strings
13.5 Reflection and Transmission of Waves
13.6 Rate of Energy Transfer by Sinusoidal Waves on Strings
13.7 Sound Waves
13.8 The Doppler Effect
13.9 Context Connection — Seismic Waves
SUMMARY


Most of us experienced waves as children when we dropped pebbles into a pond. The disturbance created by a pebble manifests itself as ripples that move outward from the point at which the pebble lands in the water, like the ripples from the falling water drops in the opening photograph. If you were to carefully examine the motion of a leaf floating near the point where the pebble enters the water, you would see that the leaf moves up and down and back and forth about its original position but does not undergo any net displacement away from or toward the source of the disturbance. The *disturbance* in the water moves over a long distance, but a given small *element of the water* oscillates only over a very small distance. This behavior is the essence of wave motion.

The world is full of other kinds of waves, including sound waves, waves on strings, seismic waves, radio waves, and x-rays. Most waves can be placed in one of two categories. **Mechanical**

**waves** are waves that disturb and propagate through a medium; the ripple in the water because of the pebble and a sound wave, for which air is the medium, are examples of mechanical waves. **Electromagnetic waves** are a special class of waves that do not require a medium to propagate, as discussed with regard to the absence of the ether in Section 9.2; light waves and radio waves are two familiar examples. In this chapter, we shall confine our attention to the study of mechanical waves, deferring our study of electromagnetic waves to Chapter 24.

## 13.1 PROPAGATION OF A DISTURBANCE

In the introduction, we alluded to the essence of wave motion: the transfer of a *disturbance* through space without the accompanying transfer of *matter*. The propagation of the disturbance also represents a transfer of energy; thus, we can view waves as a means of energy transfer. In the list of energy transfer mechanisms in Section 6.6, we see two entries that depend on waves: mechanical waves and electromagnetic radiation. These entries are to be contrasted with another entry—matter transfer—in which the energy transfer is accompanied by a movement of matter through space.

All waves carry energy, but the amount of energy transmitted through a medium and the mechanism responsible for the energy transport differ from case to case. For instance, the power of ocean waves during a storm is much greater than that of sound waves generated by a musical instrument.

**All mechanical waves require (1) some source of disturbance, (2) a medium that can be disturbed, and (3) some physical mechanism through which elements of the medium can influence one another.** This final requirement assures that a disturbance to one element will cause a disturbance to the next so that the disturbance will indeed propagate through the medium.

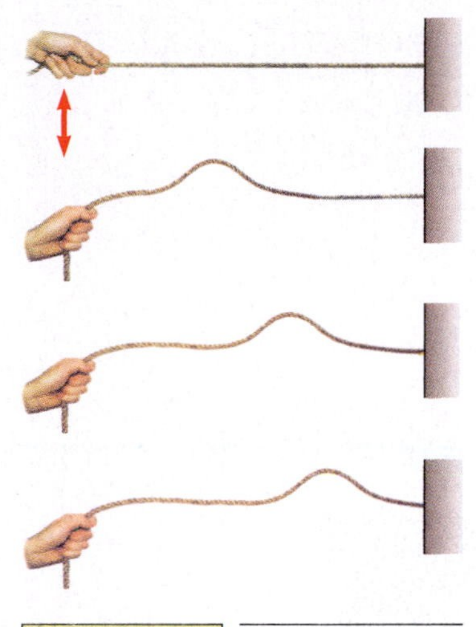

FIGURE 13.1 A pulse traveling down a stretched rope. The shape of the pulse is approximately unchanged as it travels along the rope.

One way to demonstrate wave motion is to flip the free end of a long rope that is under tension and has its opposite end fixed as in Figure 13.1. In this manner, a single **pulse** is formed and travels (to the right in Fig. 13.1) with a definite speed. The rope is the medium through which the pulse travels. Figure 13.1 represents consecutive "snapshots" of the traveling pulse. The shape of the pulse changes very little as it travels along the rope.

**As the pulse travels, each rope element that is disturbed moves in a direction perpendicular to the direction of propagation.** Figure 13.2 illustrates this point for a particular element, labeled *P*. Note that there is no motion of any part of the rope that is in the direction of the wave. A disturbance such as this one in which the elements of the disturbed medium move perpendicularly to the direction of propagation is called a **transverse wave.**

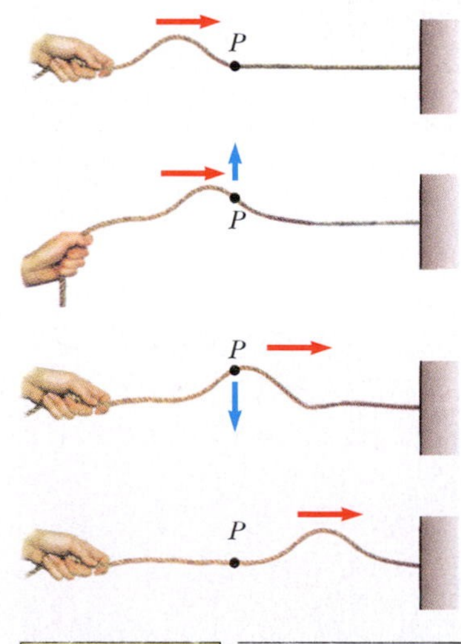


FIGURE 13.2 A pulse traveling on a stretched rope is a transverse disturbance. That is, any element of the rope, such as that at *P*, moves *(blue arrows)* in a direction perpendicular to the propagation of the pulse *(red arrows)*.

In another class of waves, called **longitudinal waves,** the elements of the medium undergo displacements *parallel* to the direction of propagation. Sound waves in air, for instance, are longitudinal. Their disturbance corresponds to a series of high- and low-pressure regions that may travel through air or through any material medium with a certain speed. A longitudinal pulse can be easily produced in a stretched spring as in Figure 13.3. A group of coils at the free end is pushed forward and pulled back. This action produces a pulse in the form of a compressed region of coils that travels along the spring.

So far, we have provided pictorial representations of a traveling pulse and hope you have begun to develop a mental representation of such a pulse. Let us now

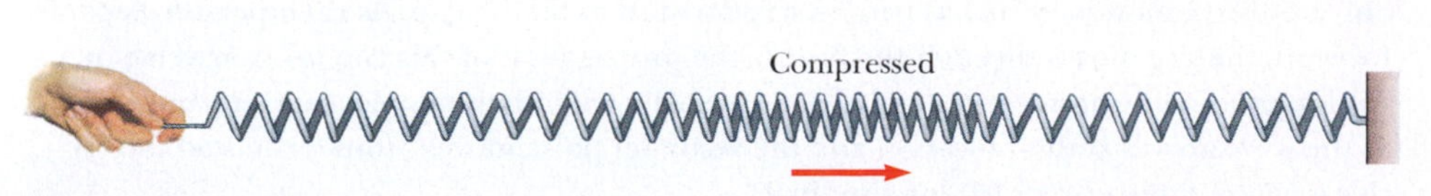


FIGURE 13.3 A longitudinal pulse along a stretched spring. The displacement of the coils is in the direction of the wave motion. The compressed region moves to the right along the spring.

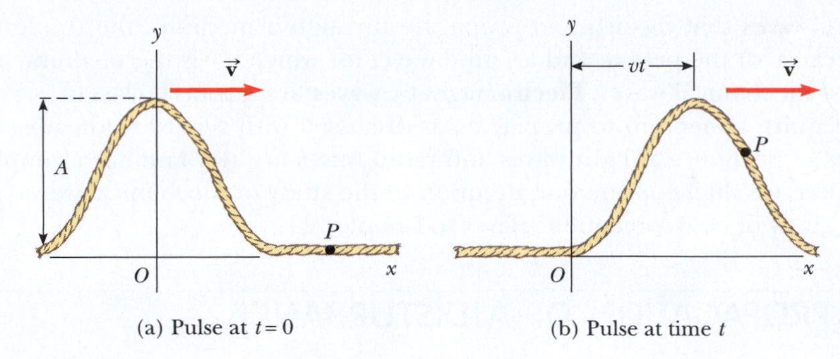


**FIGURE 13.4** A one-dimensional pulse traveling to the right with a speed $v$. (a) At $t = 0$, the shape of the pulse is given by $y = f(x)$. (b) At some later time $t$, the shape remains unchanged and the vertical position of any element of the medium is given by $y = f(x - vt)$.

develop a mathematical representation for the propagation of this pulse. Consider a pulse traveling to the right with constant speed $v$ on a long, stretched string as in Figure 13.4. The pulse moves along the $x$ axis (the axis of the string), and the transverse (up-and-down) displacement of the elements of the string is described by means of the position $y$.

Figure 13.4a represents the shape and position of the pulse at time $t = 0$. At this time, the shape of the pulse, whatever it may be, can be represented by some mathematical function that we will write as $y(x, 0) = f(x)$. This function describes the vertical position $y$ of the element of the string located at each value of $x$ at time $t = 0$. Because the speed of the pulse is $v$, the pulse has traveled to the right a distance $vt$ at time $t$ (Fig. 13.4b). We adopt a simplification model in which the shape of the pulse does not change with time.[1] Therefore, at time $t$, the shape of the pulse is the same as it was at time $t = 0$, as in Figure 13.4a. Consequently, an element of the string at $x$ at this time has the same $y$ position as an element located at $x - vt$ had at time $t = 0$:

$$y(x, t) = y(x - vt, 0)$$

In general, then, we can represent the position $y$ for all values of $x$ and $t$, measured in a stationary frame with the origin at $O$, as

■ Pulse traveling to the right

$$y(x, t) = f(x - vt) \qquad [13.1a]$$

If the pulse travels to the left, the position of an element of the string is described by

■ Pulse traveling to the left

$$y(x, t) = f(x + vt) \qquad [13.1b]$$

The function $y$, sometimes called the **wave function,** depends on the two variables $x$ and $t$. For this reason, it is often written $y(x, t)$, which is read "$y$ as a function of $x$ and $t$."

It is important to understand the meaning of $y$. Consider a point $P$ on the string, identified by a particular value of its $x$ coordinate as in Figure 13.4. As the pulse passes through $P$, the $y$ coordinate of this point increases, reaches a maximum, and then decreases to zero. **The wave function $y(x, t)$ represents the $y$ position of any element of string located at position $x$ at any time $t$.** Furthermore, if $t$ is fixed (e.g., in the case of taking a snapshot of the pulse), the wave function $y$ as a function of $x$, sometimes called the **waveform,** defines a curve representing the actual geometric shape of the pulse at that time.

**QUICK QUIZ 13.1** In a long line of people waiting to buy tickets, the first person leaves and a pulse of motion occurs as people step forward to fill the gap. As each person steps forward, the gap moves through the line. Is the propagation of this gap **(a)** transverse or **(b)** longitudinal? Consider "the wave" at a baseball game when people stand up and shout as the wave arrives at their location and the resultant pulse moves around the stadium. Is this wave **(c)** transverse or **(d)** longitudinal?

[1] In reality, the pulse changes its shape and gradually spreads out during the motion. This effect, called *dispersion,* is common to many mechanical waves, but we adopt a simplification model that ignores this effect.

**EXAMPLE 13.1** **A Pulse Moving to the Right**

A pulse moving to the right along the $x$ axis is represented by the wave function

$$y(x, t) = \frac{2.0}{(x - 3.0t)^2 + 1}$$

where $x$ and $y$ are measured in centimeters and $t$ is in seconds. Let us plot the waveform at $t = 0$, $t = 1.0$ s, and $t = 2.0$ s.

**Solution** First, note that this function is of the form $y = f(x - vt)$. By inspection, we see that the speed of the wave is $v = 3.0$ cm/s. The location of the peak of the pulse occurs at the value of $x$ for which the denominator is a minimum, that is, where $(x - 3.0t) = 0$. Therefore, the peaks occur at $x = 0.0$ cm at $t = 0$, at $x = 3.0$ cm at $t = 1.0$ s, and $x = 6.0$ cm at $t = 2.0$ s.

At times $t = 0$, $t = 1.0$ s, and $t = 2.0$ s, the wave function expressions are

$$y(x, 0) = \frac{2.0}{x^2 + 1} \quad \text{at } t = 0$$

$$y(x, 1.0) = \frac{2.0}{(x - 3.0)^2 + 1} \quad \text{at } t = 1.0 \text{ s}$$

$$y(x, 2.0) = \frac{2.0}{(x - 6.0)^2 + 1} \quad \text{at } t = 2.0 \text{ s}$$

We can now use these expressions to plot the wave function versus $x$ at these times. For example, let us evaluate $y(x, 0)$ at $x = 0.50$ cm:

$$y(0.50, 0) = \frac{2.0}{(0.50)^2 + 1} = 1.6 \text{ cm}$$

Likewise, $y(1.0, 0) = 1.0$ cm, $y(2.0, 0) = 0.40$ cm, and so on. A continuation of this procedure for other values of $x$ yields the waveform shown in Figure 13.5a. In a similar manner, one obtains the graphs of $y(x, 1.0)$ and $y(x, 2.0)$, shown in Figures 13.5b and 13.5c, respectively. These snapshots show that the pulse moves to the right without changing its shape and has a constant speed of 3.0 cm/s.

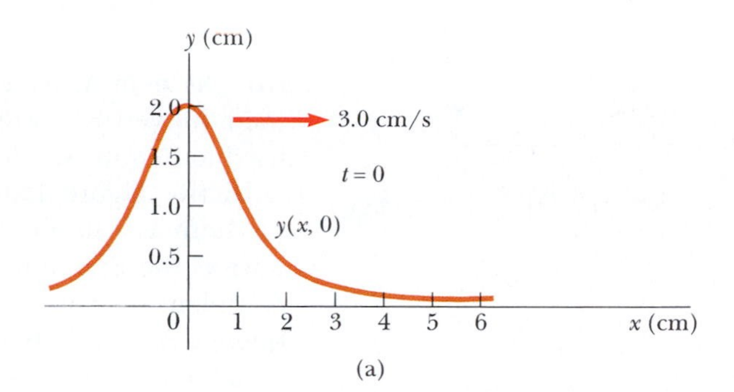


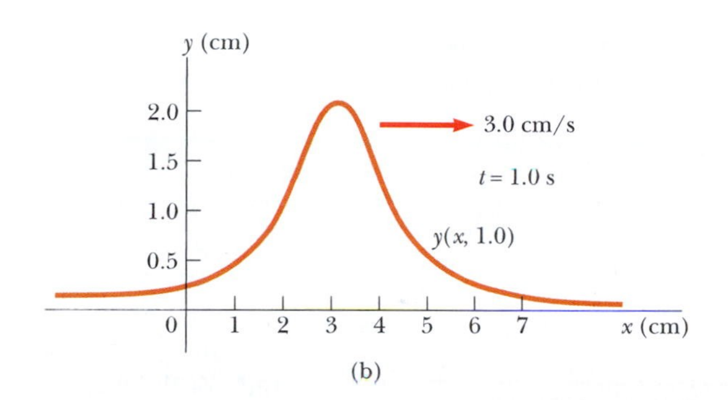


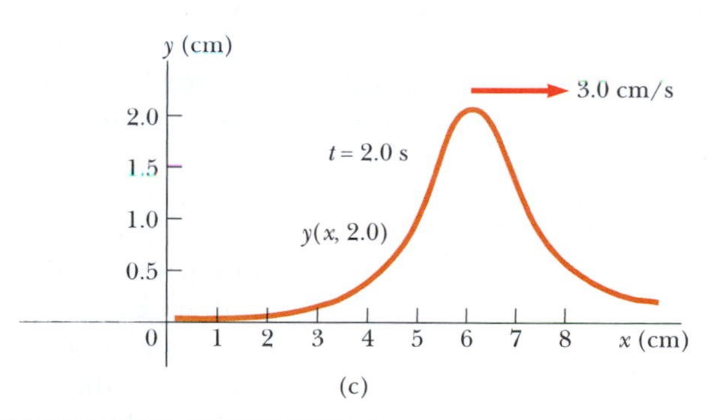


**FIGURE 13.5** (Example 13.1) Graphs of the function $y(x, t) = 2.0/[(x - 3.0t)^2 + 1]$ at (a) $t = 0$, (b) $t = 1.0$ s, and (c) $t = 2.0$ s.

## 13.2 THE WAVE MODEL

We have discussed creating a disturbance moving through a medium such as a stretched string by a simple up-and-down displacement of the end of the string. This action results in a pulse moving along the medium. A continuous wave is created by shaking the end of the string in simple harmonic motion, which we studied in Chapter 12. If we do that, the string will take on the shape shown by the curve in the graph in Active Figure 13.6a, with this shape remaining the same but moving toward the right. This shape is what we call a **sinusoidal** wave because the waveform in Active Figure 13.6a is that of a sine wave. The point with the largest positive displacement of the string is called the **crest** of the wave. The lowest point is called the **trough.** The crest and trough move along with the wave, and a particular point on

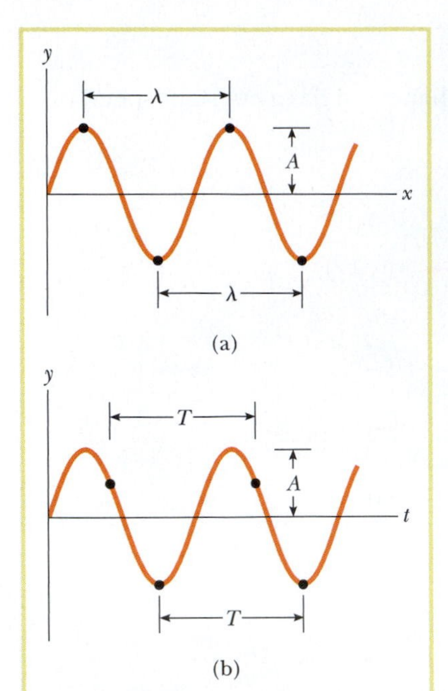


ACTIVE FIGURE 13.6
(a) A graph of the $y$ position of elements of a medium versus $x$ position, measured along the length of the medium. The wavelength $\lambda$ of a wave is the distance between adjacent crests or adjacent troughs. (b) A graph of the $y$ position of *one* element of the medium as a function of time. The period $T$ of the wave is the same as the time interval required for the element to complete one oscillation.

**PhysicsNow™** By logging into PhysicsNow at **www.pop4e.com** and going to Active Figure 13.6 you can change the parameters to see the effect on the wave function.

the string alternates between locations on a crest and a trough. In idealized wave motion in an idealized medium, each element of the medium undergoes simple harmonic motion around its equilibrium position.

Three physical characteristics are important in describing a sinusoidal wave: **wavelength, frequency,** and **wave speed. One wavelength is the minimum distance between any two identical points on a wave** such as adjacent crests or adjacent troughs as in Active Figure 13.6a, which is a graph of $y$ position of elements of the medium versus $x$ position for a sinusoidal wave at a specific time. The symbol $\lambda$ is used to denote wavelength.

Active Figure 13.6b shows $y$ position versus time for a single element of the medium as a sinusoidal wave is passing through its position $x$. The **period** $T$ of the wave is the time interval required for an element of the medium to undergo one complete oscillation. The **frequency** $f$ of sinusoidal waves is the same as the frequency of simple harmonic motion of an element of the medium. The period is equal to the inverse of the frequency:

$$T = \frac{1}{f} \quad [13.2]$$

Waves travel through the medium with a specific **wave speed,** which depends on the properties of the medium being disturbed. For instance, sound waves travel through air at 20°C with a speed of about 343 m/s, whereas the speed of sound in most solids is higher than 343 m/s. We will learn more about wavelength, frequency, and wave speed in the next section.

Another important parameter for the wave in Active Figure 13.6 is the **amplitude** of the wave. Amplitude is the maximum position of an element of the medium relative to the equilibrium position. It is denoted by $A$ and is the same as the amplitude of the simple harmonic motion of the elements of the medium.

One method of producing a traveling sinusoidal wave on a very long string is shown in Active Figure 13.7. One end of the string is connected to a blade that is set vibrating. As the blade oscillates vertically with simple harmonic motion, a traveling wave moving to the right is set up on the string. Active Figure 13.7 represents snapshots of the wave at intervals of one quarter of a period. **Each element of the string, such as that at $P$, oscillates vertically in the $y$ direction with simple harmonic motion.** Every element of the string can therefore be treated as a simple harmonic oscillator vibrating with a frequency equal to the frequency of vibration of the blade that drives the string. Although each element oscillates in the $y$ direction, the wave (or disturbance) travels in the $x$ direction with a speed $v$. Of course, this situation is the definition of a transverse wave. In this case, the energy carried by the traveling wave is supplied by the vibrating blade.

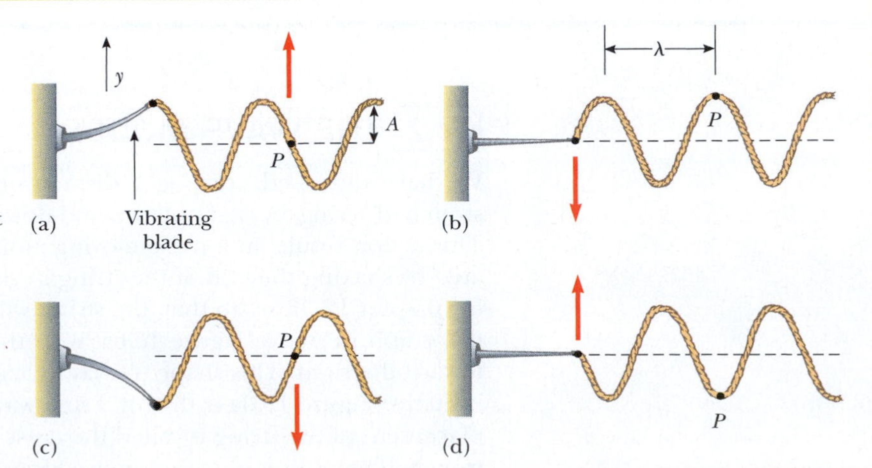


ACTIVE FIGURE 13.7
One method for producing a sinusoidal wave on a continuous string. The left end of the string is connected to a blade that is set into vibration. Every element of the string, such as the one at point $P$, oscillates with simple harmonic motion in the vertical direction.

**PhysicsNow™** Log into PhysicsNow at **www.pop4e.com** and go to Active Figure 13.7 to adjust the frequency of the blade.

In the early chapters of this book, we developed several analysis models based on the particle model. With our introduction to waves, we can develop a new simplification model, the **wave model,** which will allow us to explore more analysis models for solving problems. An ideal particle has zero size. We can build physical objects with nonzero size as combinations of particles. Thus, the particle can be considered a basic building block. An ideal wave has a single frequency and is infinitely long; that is, the wave exists throughout the Universe. (It is beyond the mathematical scope of this text at this point to prove this fact, but a wave of finite length must necessarily have a mixture of frequencies.) We will find that we can combine ideal waves, just as we combined particles, and thus the ideal wave can be considered as a basic building block. We will explore this concept in Section 14.7.

The wave model starts with an ideal wave having a single frequency, wavelength, wave speed, and amplitude. From this beginning, we describe waves in a variety of situations that serve as analysis models to help us solve problems. In the next section, we further develop our mathematical representation for the wave model.

### PITFALL PREVENTION 13.1

**WHAT'S THE DIFFERENCE BETWEEN ACTIVE FIGURES 13.6a AND 13.6b?** The curves in both parts of the figure are the same, but (a) is a graph of vertical position versus horizontal position and (b) is vertical position versus time. Part (a) can also be interpreted as a pictorial representation of the wave *for a series of elements of the medium,* which is what you would see at an instant of time. Part (b) is a graphical representation of the position of *one element of the medium* as a function of time.

## 13.3 THE TRAVELING WAVE

Let us investigate further the mathematics of a sinusoidal wave (Active Fig. 13.8). The brown curve represents a snapshot of a sinusoidal wave at $t = 0$, and the blue curve represents a snapshot of the wave at some later time $t$. In what follows, we will develop the principal features and mathematical representations of the model of a **traveling wave.** This analysis model is used in situations in which a wave moves through space without interacting with any other waves or particles.

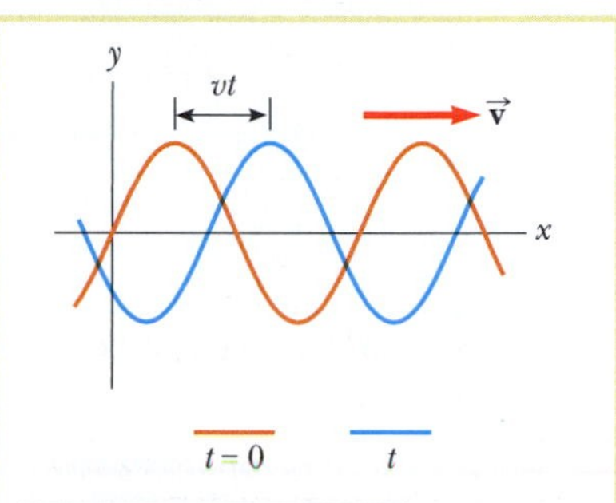


ACTIVE FIGURE 13.8
A one-dimensional sinusoidal wave traveling to the right with a speed $v$. The brown curve represents a snapshot of the wave at $t = 0$, and the blue curve represents a snapshot at some later time $t$.

**PhysicsNow™** By logging into PhysicsNow at **www.pop4e.com** and going to Active Figure 13.8 you can watch the wave move and take snapshots of it at various times.

At $t = 0$, the brown curve in Active Figure 13.8 can be described mathematically as

$$y = A \sin\left(\frac{2\pi}{\lambda} x\right) \qquad [13.3]$$

where the amplitude $A$, as usual, represents the maximum value of the position of an element relative to the equilibrium position and $\lambda$ is the wavelength as defined in Active Figure 13.6a. Therefore, we see that the value of $y$ is the same when $x$ is increased by an integral multiple of $\lambda$. If the wave moves to the right with a speed of $v$, the wave function at some later time $t$ is

$$y = A \sin\left[\frac{2\pi}{\lambda}(x - vt)\right] \qquad [13.4]$$

That is, the sinusoidal wave has moved to the right a distance of $vt$ at time $t$ as in Active Figure 13.8. Note that the wave function has the form $f(x - vt)$ and represents a wave traveling to the right. If the wave were traveling to the left, the quantity $x - vt$ would be replaced by $x + vt$, just as in the case of the traveling pulse described by Equations 13.1a and 13.1b.

Because the period $T$ is the time interval required for the wave to travel a distance of one wavelength, the speed, wavelength, and period are related by

$$v = \frac{\lambda}{T} \qquad [13.5]$$

Substituting Equation 13.5 into Equation 13.4, we find that

$$y = A \sin\left[2\pi\left(\frac{x}{\lambda} - \frac{t}{T}\right)\right] \qquad [13.6]$$

This form of the wave function shows the periodic nature of $y$ in both space and time. That is, at any given time $t$ (a snapshot of the wave), $y$ has the same value at the positions $x$, $x + \lambda$, $x + 2\lambda$, and so on. Furthermore, at any given position $x$ (at which a single element of the medium is undergoing simple harmonic motion), the values of $y$ at times $t$, $t + T$, $t + 2T$, and so on are the same.

We can express the sinusoidal wave function in a compact form by defining two other quantities: **angular wave number** $k$ (often called simply the **wave number**) and **angular frequency** $\omega$:

■ Angular wave number

$$k \equiv \frac{2\pi}{\lambda} \qquad [13.7]$$

■ Angular frequency

$$\omega \equiv \frac{2\pi}{T} = 2\pi f \qquad [13.8]$$

Note that in Equation 13.8 we use the definition of frequency, $f = 1/T$. Using these definitions, Equation 13.6 can be written in the more compact form

■ Wave function for a sinusoidal wave

$$y = A \sin(kx - \omega t) \qquad [13.9]$$

We shall use this form most frequently.

Using Equations 13.7 and 13.8, we can express the wave speed $v$ (Eq. 13.5) in the alternative forms

■ Speed of a traveling sinusoidal wave

$$v = \frac{\omega}{k} \qquad [13.10]$$

$$v = \lambda f \qquad [13.11]$$

The wave function given by Equation 13.9 assumes that the position $y$ is zero at $x = 0$ and $t = 0$, but that need not be the case. If the transverse position of an element is not zero at $x = 0$ and $t = 0$, we generally express the wave function in the form

$$y = A \sin(kx - \omega t + \phi) \qquad [13.12]$$

where $\phi$ is called the **phase constant** and can be determined from the initial conditions.

**QUICK QUIZ 13.2** A sinusoidal wave of frequency $f$ is traveling along a stretched string. The string is brought to rest and a second traveling wave of frequency $2f$ is established on the same string. **(i)** What is the wave speed of the second wave? **(a)** twice that of the first wave **(b)** half that of the first wave **(c)** the same as that of the first wave **(d)** impossible to determine **(ii)** What is the wavelength of the second wave? **(a)** twice that of the first wave **(b)** half that of the first wave **(c)** the same as that of the first wave **(d)** impossible to determine **(iii)** What is the amplitude of the second wave? **(a)** twice that of the first wave **(b)** half that of the first wave **(c)** the same as that of the first wave **(d)** impossible to determine

**EXAMPLE 13.2** A Traveling Sinusoidal Wave

A sinusoidal wave traveling in the positive $x$ direction has an amplitude of 15.0 cm, a wavelength of 40.0 cm, and a frequency of 8.00 Hz. The vertical position of an element of the medium at $t = 0$ and $x = 0$ is also 15.0 cm as shown in Figure 13.9.

**A** Find the angular wave number, period, angular frequency, and speed of the wave.

**Solution** This problem is a simple one in which we apply the traveling wave model. Using Equations 13.7,

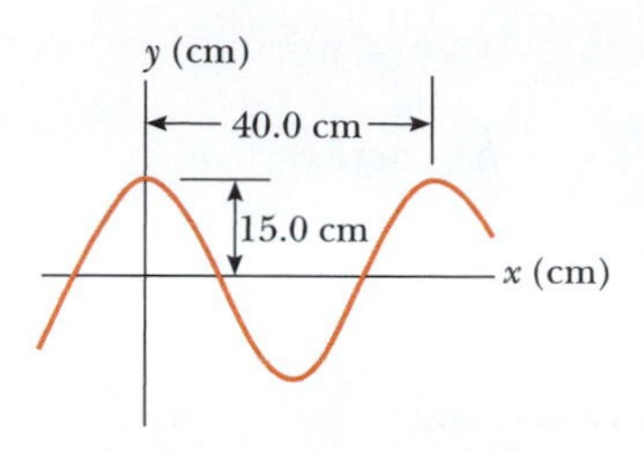


FIGURE 13.9 (Example 13.2) A sinusoidal wave of wavelength $\lambda = 40.0$ cm and amplitude $A = 15.0$ cm. The wave function can be written in the form $y = A\cos(kx - \omega t)$.

13.8, and 13.11, we find the following:

$$k = \frac{2\pi}{\lambda} = \frac{2\pi \text{ rad}}{40.0 \text{ cm}} = \boxed{0.157 \text{ rad/cm}}$$

$$T = \frac{1}{f} = \frac{1}{8.00 \text{ s}^{-1}} = \boxed{0.125 \text{ s}}$$

$$\omega = 2\pi f = 2\pi(8.00 \text{ s}^{-1}) = \boxed{50.3 \text{ rad/s}}$$

$$v = f\lambda = (8.00 \text{ s}^{-1})(40.0 \text{ cm}) = \boxed{320 \text{ cm/s}}$$

B Determine the phase constant $\phi$ and write a general expression for the wave function.

**Solution** Because $A = 15.0$ cm and because it is given that $y = 15.0$ cm at $x = 0$ and $t = 0$, substitution into Equation 13.12 gives

$$15.0 = 15.0 \sin\phi \qquad \text{or} \qquad \sin\phi = 1$$

We see that $\phi = \boxed{\pi/2 \text{ rad}}$ (or 90°). Hence, the wave function is of the form

$$y = A\sin\left(kx - \omega t + \frac{\pi}{2}\right) = A\cos(kx - \omega t)$$

As we can see by inspection, the wave function must have this form because the cosine argument is displaced by 90° from the sine function. Substituting the values for $A$, $k$, and $\omega$ into this expression gives

$$y = \boxed{(15.0 \text{ cm})\cos(0.157x - 50.3t)}$$

## The Linear Wave Equation

If the waveform at $t = 0$ is as described in Active Figure 13.7b, the wave function can be written

$$y = A\sin(kx - \omega t)$$

We can use this expression to describe the motion of any element of the string. The element at point $P$ (or any other point on the string) moves only vertically, so its $x$ coordinate *remains constant.* The **transverse velocity** $v_y$ of the element and its **transverse acceleration** $a_y$ are therefore

$$v_y = \left.\frac{dy}{dt}\right|_{x=\text{constant}} = \frac{\partial y}{\partial t} = -\omega A\cos(kx - \omega t) \qquad [13.13]$$

$$a_y = \left.\frac{dv_y}{dt}\right|_{x=\text{constant}} = \frac{\partial v_y}{\partial t} = \frac{\partial^2 y}{\partial t^2} = -\omega^2 A\sin(kx - \omega t) \qquad [13.14]$$

The maximum values of these quantities are simply the absolute values of the coefficients of the cosine and sine functions:

$$v_{y,\,\max} = \omega A \qquad [13.15]$$

$$a_{y,\,\max} = \omega^2 A \qquad [13.16]$$

You should recognize from Equations 13.13 and 13.14 that the transverse velocity and transverse acceleration of any element of the string do not reach their maximum values simultaneously. In fact, the transverse velocity reaches its maximum value ($\omega A$) when position $y = 0$, whereas the transverse acceleration reaches its maximum magnitude ($\omega^2 A$) when $y = \pm A$. These relationships are due to the sine and cosine functions differing by a phase constant of $\pi/2$. Finally, note that Equations 13.15 and 13.16 are identical to the corresponding equations for simple harmonic motion (Eq. 12.17 and Eq. 12.18).

Let us take derivatives of our wave function with respect to position at a fixed time, similar to the process by which we took derivatives with respect to time in

**PITFALL PREVENTION 13.2**

**TWO KINDS OF SPEED/VELOCITY** Be sure to differentiate between $v$, the speed of the wave as it propagates through the medium, and $v_y$, the transverse velocity of an element of the string. The speed $v$ is constant for a uniform medium, whereas $v_y$ varies sinusoidally.

Equations 13.13 and 13.14:

$$\left.\frac{dy}{dx}\right|_{t=\text{constant}} = \frac{\partial y}{\partial x} = -kA\cos(kx - \omega t) \qquad [13.17]$$

$$\left.\frac{d^2y}{dx^2}\right|_{t=\text{constant}} = \frac{\partial^2 y}{\partial x^2} = -k^2 A\sin(kx - \omega t) \qquad [13.18]$$

Comparing Equations 13.14 and 13.18, we see that

$$A\sin(kx - \omega t) = -\frac{1}{k^2}\frac{\partial^2 y}{\partial x^2} = -\frac{1}{\omega^2}\frac{\partial^2 y}{\partial t^2} \rightarrow \frac{\partial^2 y}{\partial x^2} = \frac{k^2}{\omega^2}\frac{\partial^2 y}{\partial t^2}$$

Using Equation 13.10, we can rewrite this expression as

▪ Linear wave equation

$$\frac{\partial^2 y}{\partial x^2} = \frac{1}{v^2}\frac{\partial^2 y}{\partial t^2} \qquad [13.19]$$

which is known as the **linear wave equation.** If we analyze a situation and find this kind of relationship between derivatives of a function describing the situation, wave motion is occurring. Equation 13.19 is a differential equation representation of the traveling wave model. The solutions to the equation describe **linear mechanical waves.** We have developed the linear wave equation from a sinusoidal mechanical wave traveling through a medium, but it is much more general. The linear wave equation successfully describes waves on strings, sound waves, and also electromagnetic waves.[2] What's more, although the sinusoidal wave that we have studied is a solution to Equation 13.19, the general solution to the equation is *any* function of the form $y(x, t) = f(x \pm vt)$ as discussed in Section 13.1.

Nonlinear waves are more difficult to analyze, but they are an important area of current research, especially in optics. An example of a nonlinear mechanical wave is one for which the amplitude is not small compared with the wavelength.

**EXAMPLE 13.3 A Solution to the Linear Wave Equation**

Verify that the wave function presented in Example 13.1 is a solution to the linear wave equation.

**Solution** The wave function is

$$y(x, t) = \frac{2.0}{(x - 3.0t)^2 + 1}$$

By taking partial derivatives of this function with respect to $x$ and to $t$, we find that

$$\frac{\partial^2 y}{\partial x^2} = \frac{12(x - 3.0t)^2 - 4.0}{[(x - 3.0t)^2 + 1]^3}$$

$$\frac{\partial^2 y}{\partial t^2} = \frac{108(x - 3.0t)^2 - 36}{[(x - 3.0t)^2 + 1]^3} = 9.0\frac{[12(x - 3.0t)^2 - 4.0]}{[(x - 3.0t)^2 + 1]^3}$$

Comparing these two expressions, we see that

$$\frac{\partial^2 y}{\partial x^2} = \frac{1}{9.0}\frac{\partial^2 y}{\partial t^2}$$

Comparing this result with Equation 13.19, we see that the wave function is a solution to the linear wave equation if the speed at which the pulse moves is 3.0 cm/s. We have already determined in Example 13.1 that this speed is indeed the speed of the pulse, so we have proven what we set out to do.

## 13.4 THE SPEED OF TRANSVERSE WAVES ON STRINGS

An aspect of the behavior of linear mechanical waves is that **the wave speed depends only on the properties of the medium through which the wave travels.** Waves for which the amplitude $A$ is small relative to the wavelength $\lambda$ are well represented

[2]In the case of electromagnetic waves, $y$ is interpreted to represent an electric field, which we will study in Chapter 24.

as linear waves. In this section, we determine the speed of a transverse wave traveling on a stretched string.

Let us use a mechanical analysis to derive the expression for the speed of a pulse traveling on a stretched string under tension $T$. Consider a pulse moving to the right with a uniform speed $v$, measured relative to a stationary (with respect to the Earth) inertial reference frame. Recall from Chapter 9 that Newton's laws are valid in any inertial reference frame. Therefore, let us view this pulse from a different inertial reference frame, one that moves along with the pulse at the same speed so that the pulse appears to be at rest in the frame as in Figure 13.10a. In this reference frame, the pulse remains fixed and each element of the string moves to the left through the pulse shape.

A short element of the string, of length $\Delta s$, forms an arc with a radius of curvature $R$ as shown in Figure 13.10a and magnified in Figure 13.10b. We use a simplification model in which this arc is an arc of a perfect circle. In our moving frame of reference, the element of the string moves to the left with speed $v$ through the arc. As it travels through the arc, we can model the element as a particle in uniform circular motion. This element has a centripetal acceleration of $v^2/R$, which is supplied by components of the force $\vec{\mathbf{T}}$ in the string at each end of the element. The force $\vec{\mathbf{T}}$ acts on each side of the element, tangent to the arc, as in Figure 13.10b. The horizontal components of $\vec{\mathbf{T}}$ cancel, and each vertical component $T \sin \theta$ acts radially inward toward the center of the arc. Hence, the magnitude of the total radial force on the element is $2T \sin \theta$. Because the element is small, $\theta$ is small and we can use the small-angle approximation $\sin \theta \approx \theta$. Therefore, the magnitude of the total radial force can be expressed as

$$F_r = 2T \sin \theta \approx 2T\theta$$

The element has mass $m = \mu \Delta s$, where $\mu$ is the mass per unit length of the string. Because the element forms part of a circle and subtends an angle of $2\theta$ at the center, $\Delta s = R(2\theta)$, and hence

$$m = \mu \, \Delta s = 2\mu R\theta$$

The radial component of Newton's second law applied to the element gives

$$F_r = \frac{mv^2}{R} \quad \rightarrow \quad 2T\theta = \frac{2\mu R\theta v^2}{R} \quad \rightarrow \quad T = \mu v^2$$

where $F_r$ is the force that supplies the centripetal acceleration of the element. Solving for $v$ gives

$$v = \sqrt{\frac{T}{\mu}} \qquad [13.20]$$

■ Speed of a wave on a stretched string

Notice that this derivation is based on the linear wave assumption that the pulse height is small relative to the length of the pulse. Using this assumption, we were able to use the approximation $\sin \theta \approx \theta$. Furthermore, the model assumes that the tension $T$ is not affected by the presence of the pulse, so $T$ is the same at all points on the string. Finally, this proof does *not* assume any particular shape for the pulse. We therefore conclude that a pulse of *any shape* will travel on the string with speed $v = \sqrt{T/\mu}$, without changing its shape.

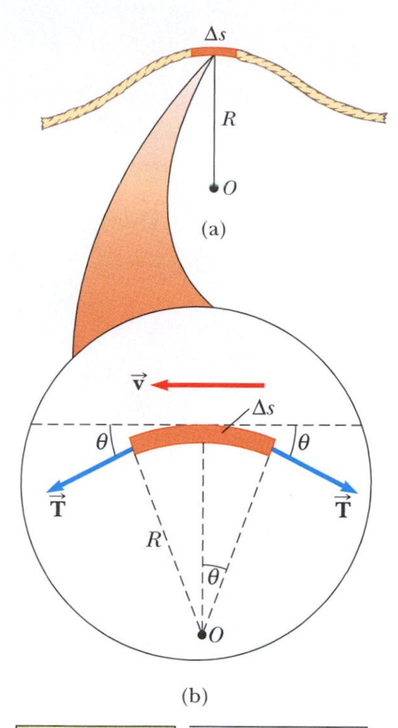


**FIGURE 13.10** (a) To obtain the speed $v$ of a wave on a stretched string, it is convenient to describe the motion of a small element of the string in a moving frame of reference. (b) The horizontal components of the force $\vec{\mathbf{T}}$ on a small element of length $\Delta s$ cancel. The radial components add, so there is a net force in the radial direction.

**PITFALL PREVENTION 13.3**

**MULTIPLE $T$'S** Be careful not to confuse the $T$ for the magnitude of the tension in this discussion with the $T$ we are using in this chapter for the period of a wave. The context of the equation should help you to identify which one it is. The alphabet simply doesn't have enough letters to allow us to assign a unique letter to each variable!

**QUICK QUIZ 13.3** Suppose you create a pulse by moving the free end of a taut string up and down once with your hand. The string is attached at its other end to a distant wall. The pulse reaches the wall in a time interval $\Delta t$. Which of the following actions, taken by itself, decreases the time interval required for the pulse to reach the wall? More than one choice may be correct. **(a)** Moving your hand more quickly, but still only up and down once by the same amount. **(b)** Moving your hand more slowly, but still only up and down once by the same amount. **(c)** Moving your hand a greater distance up and down in the same amount of time. **(d)** Moving your hand a smaller distance up and down in the same amount of time. **(e)** Using a heavier string of the same length and under the same tension. **(f)** Using a lighter string of the same length and under the same tension. **(g)** Using a string of the same linear mass density but under decreased tension. **(h)** Using a string of the same linear mass density but under increased tension.

## Thinking Physics 13.1

A secret agent is trapped in a building on top of an elevator car at a lower floor. He attempts to signal a fellow agent on the roof by tapping a message in Morse code on the elevator cable so that transverse pulses move upward on the cable. As the pulses move up the cable toward the accomplice, does the speed with which they move stay the same, increase, or decrease? If the pulses are sent 1 s apart, are they received 1 s apart by the agent on the roof?

**Reasoning** The elevator cable can be modeled as a vertical string. The speed of waves on the cable is a function of the tension in the cable. As the waves move higher on the cable, they encounter increased tension because each higher point on the cable must support the weight of all the cable below it (and the elevator). Therefore, the speed of the pulses increases as they move higher on the cable. The frequency of the pulses will not be affected because each pulse takes the same time interval to reach the top. They will still arrive at the top of the cable at intervals of 1 s. ■

### EXAMPLE 13.4 The Speed of a Pulse on a Cord

A uniform cord has a mass of 0.300 kg and a total length of 6.00 m. Tension is maintained in the cord by suspending an object of mass 2.00 kg from one end (Fig. 13.11). Find the speed of a pulse on this cord.

**Solution** Because the suspended object can be modeled

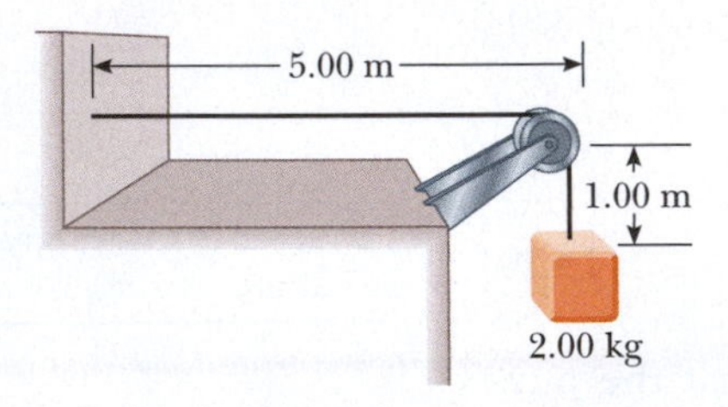


FIGURE 13.11 (Example 13.4) The tension $T$ in the cord is maintained by the suspended object. The wave speed is given by the expression $v = \sqrt{T/\mu}$.

as a particle in equilibrium, the tension $T$ in the cord is equal to the weight of the suspended 2.00-kg object:

$$T = mg = (2.00 \text{ kg})(9.80 \text{ m/s}^2) = 19.6 \text{ N}$$

(This calculation of the tension neglects the small mass of the cord. Strictly speaking, the horizontal portion of the cord can never be exactly straight—it will sag slightly—and therefore the tension is not uniform.)

The mass per unit length $\mu$ is

$$\mu = \frac{m}{\ell} = \frac{0.300 \text{ kg}}{6.00 \text{ m}} = 0.050\,0 \text{ kg/m}$$

Therefore, the wave speed is

$$v = \sqrt{\frac{T}{\mu}} = \sqrt{\frac{19.6 \text{ N}}{0.050\,0 \text{ kg/m}}} = 19.8 \text{ m/s}$$

### INTERACTIVE EXAMPLE 13.5 Rescuing the Hiker

An 80.0-kg hiker is trapped on a mountain ledge following a storm. A helicopter rescues the hiker by hovering above him and lowering a cable to him. The mass of the cable is 8.00 kg and its length is 15.0 m. A chair of mass 70.0 kg is attached to the end of the cable. The hiker attaches himself to the chair and the helicopter then accelerates upward. Terrified by hanging from the cable in midair, the hiker tries to signal the pilot by sending transverse pulses up the cable. A pulse takes 0.250 s to travel the length of the cable. What is the acceleration of the helicopter?

**Solution** To conceptualize this problem, imagine the effect of the helicopter's acceleration on the cable. The higher the upward acceleration, the larger the tension in the cable. In turn, the larger the tension, the higher the speed of pulses on the cable. Therefore, we categorize this problem as a combination of one involving Newton's laws and one involving the speed of pulses on a string. To analyze the problem, we use the time interval for the pulse to travel from the hiker to the helicopter to find the speed of the pulses on the cable:

$$v = \frac{\Delta x}{\Delta t} = \frac{15.0 \text{ m}}{0.250 \text{ s}} = 60.0 \text{ m/s}$$

The speed of pulses on the cable is given by Equation 13.20, which allows us to find the tension in the cable:

$$v = \sqrt{\frac{T}{\mu}} \rightarrow T = \mu v^2 = \left(\frac{8.00 \text{ kg}}{15.0 \text{ m}}\right)(60.0 \text{ m/s})^2$$

$$= 1.92 \times 10^3 \text{ N}$$

Newton's second law relates the tension in the cable to the acceleration of the hiker and the chair, which is the same as the acceleration of the helicopter (we ignore the mass of the cable relative to that of the hiker and the chair):

$$\sum F = ma \rightarrow T - mg = ma$$

$$a = \frac{T}{m} - g = \frac{1.92 \times 10^3 \text{ N}}{150.0 \text{ kg}} - 9.80 \text{ m/s}^2$$

$$= 3.00 \text{ m/s}^2$$

To finalize this problem, note that a real cable has stiffness in addition to tension. Stiffness tends to return a cable or a wire to its original straight-line shape even when it is not under tension. For example, a piano wire, which has stiffness, will straighten if released from a curved shape, whereas normal package wrapping string will not.

Stiffness represents a restoring force in addition to tension, which tends to increase the speed of waves on the cable over that due to tension alone. Consequently, for a real cable, the speed of 60.0 m/s that we determined is most likely associated with a tension lower than $1.92 \times 10^3$ N and a correspondingly smaller acceleration of the helicopter.

**PhysicsNow™** Investigate the rescue situation by logging into PhysicsNow at **www.pop4e.com** and going to Interactive Example 13.5.

## 13.5 REFLECTION AND TRANSMISSION OF WAVES

So far, we have only considered a wave traveling through a medium with no changes in the medium and no interactions with anything other than the elements of the medium. This model is the traveling wave model. This situation is similar to a particle traveling through empty space and obeying Newton's first law. Although these situations demonstrate important physics, things become more interesting when particles and waves interact with something. Let us see what happens when a wave encounters a boundary between two media.

For simplicity, consider a single pulse once again. When a traveling pulse reaches a boundary, part or all of the pulse is *reflected.* Any part not reflected is said to be *transmitted* through the boundary. Suppose a pulse travels on a string that is fixed at one end (Fig. 13.12). When the pulse reaches the fixed boundary, it is reflected. In the simplification model in which the support attaching the string to the wall is rigid, none of the pulse is transmitted through the fixed end.

Note that the reflected pulse (Figs. 13.12d and 13.12e) has exactly the same amplitude as the incoming pulse but is inverted. The inversion can be explained as follows. The pulse is created initially with an upward and then downward force on the free end of the string. As the pulse arrives at the fixed end of the string, the string first produces an upward force on the support. By Newton's third law, the support exerts a reaction force in the opposite direction on the string. Therefore, the positive shape of the pulse results in a downward and then upward force on the string as the entirety of the pulse encounters the rigid end. This situation is equivalent to a person replacing the fixed support and applying a downward and then an upward force to the string. Therefore, reflection at a rigid end causes the pulse to invert on reflection.

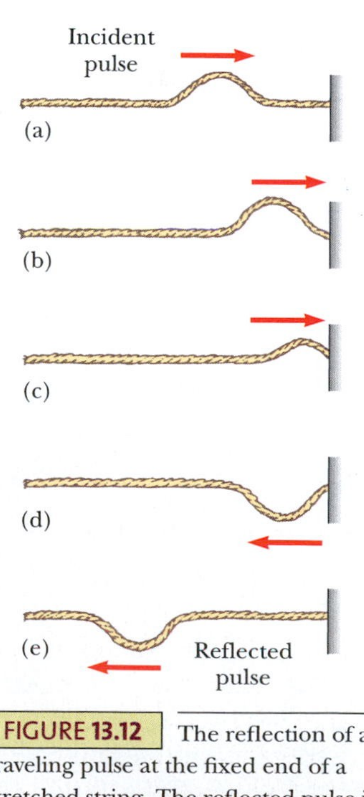


FIGURE 13.12 The reflection of a traveling pulse at the fixed end of a stretched string. The reflected pulse is inverted, but its shape remains the same.

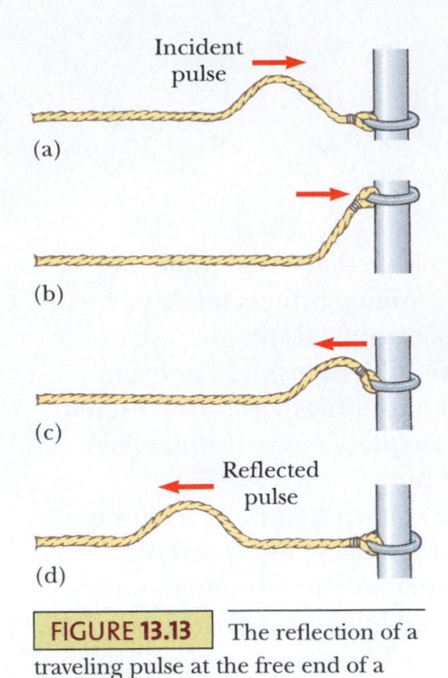


**FIGURE 13.13** The reflection of a traveling pulse at the free end of a stretched string. In this case, the reflected pulse is not inverted.

Now consider a second idealized situation in which reflection is total and transmission is zero. In this simplification model, the pulse arrives at the end of a string that is perfectly free to move vertically, as in Figure 13.13. The tension at the free end is maintained by tying the string to a ring of negligible mass that is free to slide vertically on a frictionless post. Again, the pulse is reflected, but this time it is not inverted. As the pulse reaches the post, it exerts a force on the free end, causing the ring to accelerate upward. In the process, the ring reaches the top of its motion and is then returned to its original position by the downward component of the tension force. Therefore, the ring experiences the same motion as if it were raised and lowered by hand. This motion produces a reflected pulse that is not inverted and whose amplitude is the same as that of the incoming pulse.

Finally, in some situations the boundary is intermediate between these two extreme cases; that is, it is neither completely rigid nor completely free. In that case, part of the wave is transmitted and part is reflected. For instance, suppose a string is attached to a denser string as in Active Figure 13.14. When a pulse traveling on the first string reaches the boundary between the two strings, part of the pulse is reflected and inverted and part is transmitted to the denser string. Both the reflected and transmitted pulses have smaller amplitude than the incident pulse. The inversion in the reflected pulse is similar to the behavior of a pulse meeting a rigid boundary. As the pulse travels from the initial string to the denser string, the junction acts more like a rigid end than a free end. Therefore, the reflected pulse is inverted.

When a pulse traveling on a dense string strikes the boundary of a less dense string, as in Active Figure 13.15, again part is reflected and part transmitted. This time, however, the reflected pulse is not inverted. As the pulse travels from the dense string to the less dense one, the junction acts more like a free end than a rigid end.

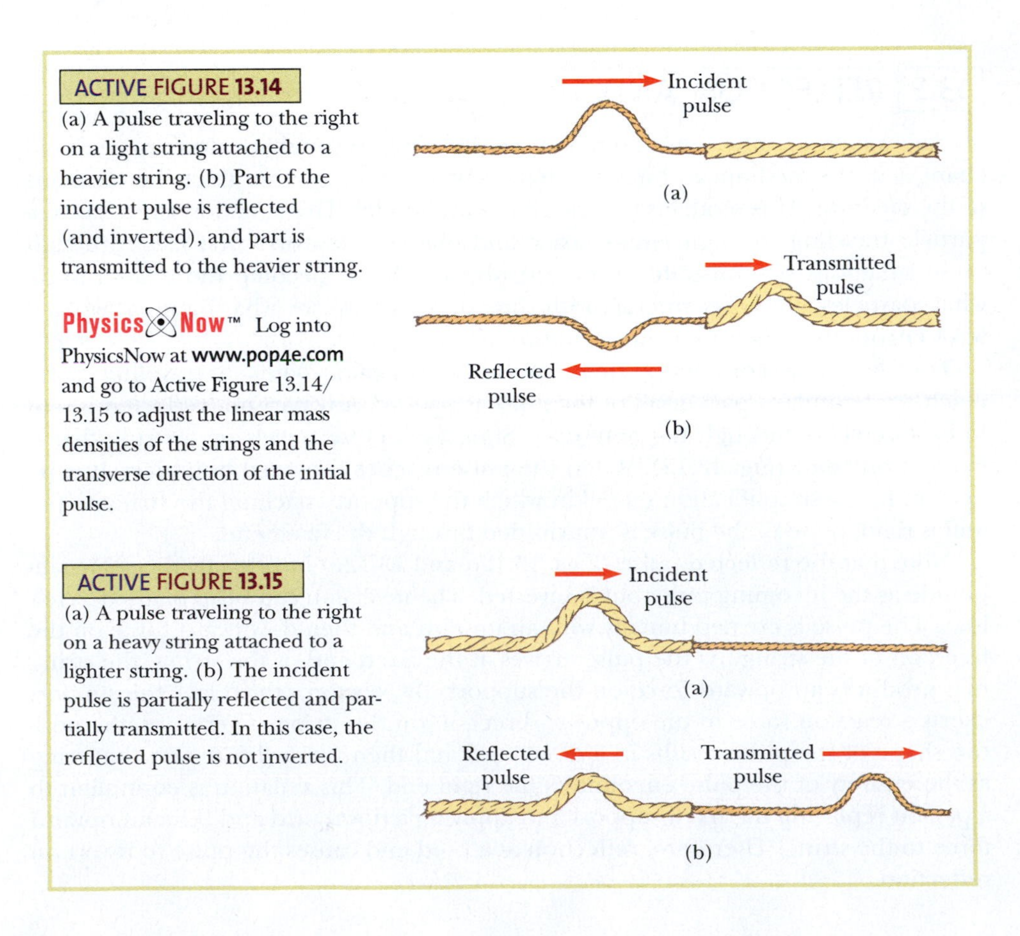


**ACTIVE FIGURE 13.14** (a) A pulse traveling to the right on a light string attached to a heavier string. (b) Part of the incident pulse is reflected (and inverted), and part is transmitted to the heavier string.

**Physics Now™** Log into PhysicsNow at **www.pop4e.com** and go to Active Figure 13.14/13.15 to adjust the linear mass densities of the strings and the transverse direction of the initial pulse.

**ACTIVE FIGURE 13.15** (a) A pulse traveling to the right on a heavy string attached to a lighter string. (b) The incident pulse is partially reflected and partially transmitted. In this case, the reflected pulse is not inverted.

The limiting value between the two cases in Figures 13.14 and 13.15 would be that in which both strings have the same linear mass density. In this case, no boundary exists between the two media. Both strings are identical. As a result, no reflection occurs and transmission is total.

In the preceding section, we found that the speed of a wave on a string increases as the mass per unit length of the string decreases. In other words, a pulse travels more slowly on a dense string than on a less dense one if both are under the same tension. This comparison is illustrated by the lengths of the red velocity vectors in Figures 13.14 and 13.15.

This discussion has focused on pulses arriving at a boundary. If a sinusoidal wave on a string arrives at a rigid end, the inversion of the waveform is equivalent to shifting the entire wave by half a wavelength. This equivalence can be seen by looking back at Active Figure 13.7. The wave in Active Figure 13.7d is the inversion of the wave in Active Figure 13.7b. Notice, however, that Active Figure 13.7b would look like Active Figure 13.7d if the wave were shifted to the right or left by half a wavelength. Therefore, because a full wavelength can be associated with an angle of 360°, we describe the inversion of a wave at a rigid end as a **180° phase shift.** We will see this effect again in Chapter 27 when we discuss reflection of light waves from materials.

## 13.6 RATE OF ENERGY TRANSFER BY SINUSOIDAL WAVES ON STRINGS

As waves propagate through a medium, they transport energy. This fact is easily demonstrated by hanging an object on a stretched string and sending a pulse down the string as in Figure 13.16. When the pulse meets the suspended object, the object is momentarily displaced as in Figure 13.16b. In the process, energy is transferred to the object because work must be done in moving it upward. This section examines the rate at which energy is transferred along a string. We shall assume a one-dimensional sinusoidal wave in the calculation of the energy transferred.

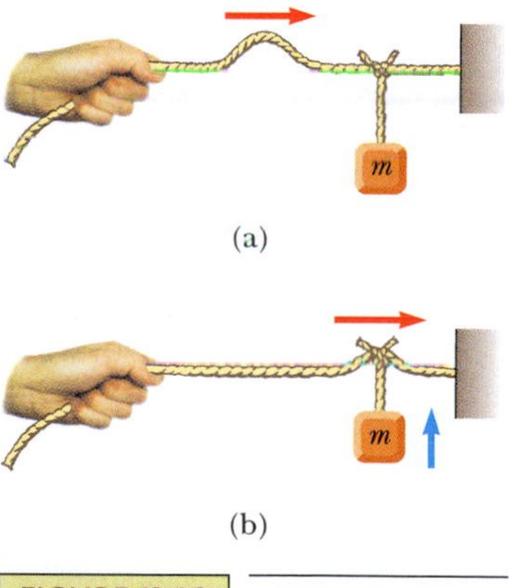


FIGURE 13.16 (a) A pulse traveling to the right on a stretched string on which an object has been suspended. (b) Energy is transmitted to the suspended object when the pulse arrives.

Consider a sinusoidal wave traveling on a string (Fig. 13.17). The source of the energy is some external agent at the left end of the string, which does work in producing the oscillations. We can consider the string to be a nonisolated system. As the external agent performs work on the end of the string, moving it up and down, energy enters the system of the string and propagates along its length. Let us focus our attention on an element of the string of length $\Delta x$ and mass $\Delta m$. Each such element moves vertically with simple harmonic motion. Therefore, we can model each element of the string as a simple harmonic oscillator, with the oscillation in the $y$ direction. All elements have the same angular frequency $\omega$ and the same amplitude $A$. The kinetic energy $K$ associated with a particle in simple harmonic motion is $K = \frac{1}{2}mv^2$, where $v$ varies sinusoidally during the oscillation. If we apply this equation to an element of length $\Delta x$, we see that the kinetic energy $\Delta K$ of this element is

$$\Delta K = \tfrac{1}{2}(\Delta m)\,v_y^{\,2}$$

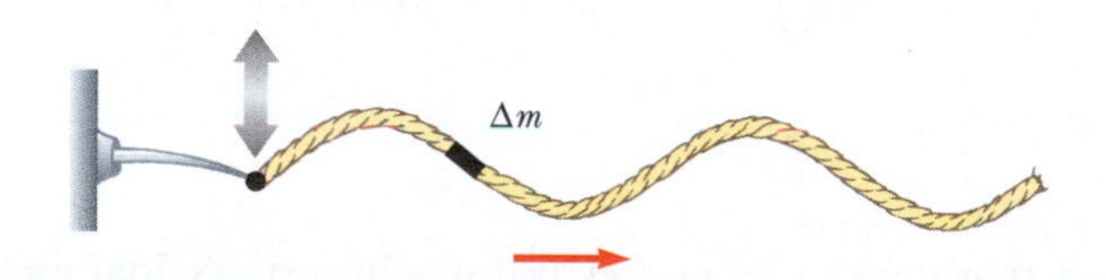


FIGURE 13.17 A sinusoidal wave traveling along the $x$ axis on a stretched string. Every element, such as the one labeled with its mass $\Delta m$, moves vertically, and each element has the same total energy. The average power transmitted by the wave equals the energy contained in one wavelength divided by the period of the wave.

If $\mu$ is the mass per unit length of the string, the element of length $\Delta x$ has a mass $\Delta m$ that is equal to $\mu\,\Delta x$. Hence, we can express the kinetic energy of an element of the string as

$$\Delta K = \tfrac{1}{2}(\mu\,\Delta x)v_y^2 \qquad [13.21]$$

As the length of the element of the string shrinks to zero, this expression becomes a differential relationship:

$$dK = \tfrac{1}{2}(\mu\,dx)v_y^2$$

We substitute for the general velocity of an element of the string using Equation 13.13:

$$\begin{aligned} dK &= \tfrac{1}{2}\mu[\omega A\cos(kx - \omega t)]^2\,dx \\ &= \tfrac{1}{2}\mu\omega^2A^2\cos^2(kx - \omega t)\,dx \end{aligned}$$

If we take a snapshot of the wave at time $t = 0$, the kinetic energy of a given element is

$$dK = \tfrac{1}{2}\mu\omega^2A^2\cos^2 kx\,dx$$

Let us integrate this expression over all the string elements in a wavelength of the wave, which will give us the kinetic energy in one wavelength:

$$\begin{aligned} K_\lambda &= \int_0^\lambda \tfrac{1}{2}\mu\omega^2A^2\cos^2 kx\,dx = \tfrac{1}{2}\mu\omega^2A^2\int_0^\lambda \cos^2 kx\,dx \\ &= \tfrac{1}{2}\mu\omega^2A^2\left[\tfrac{1}{2}x + \frac{1}{4k}\sin 2kx\right]_0^\lambda = \tfrac{1}{2}\mu\omega^2A^2\left[\tfrac{1}{2}\lambda\right] = \tfrac{1}{4}\mu\omega^2A^2\lambda \end{aligned}$$

In addition to this kinetic energy, there is potential energy associated with each element of the string due to its displacement from the equilibrium position. A similar analysis as that above for the total potential energy in a wavelength gives the same result:

$$U_\lambda = \tfrac{1}{4}\mu\omega^2A^2\lambda$$

The total energy in one wavelength of the wave is the sum of the kinetic and potential energies:

$$E_\lambda = K_\lambda + U_\lambda = \tfrac{1}{2}\mu\omega^2A^2\lambda \qquad [13.22]$$

As the wave moves along the string, this amount of energy passes by a given point on the string during one period of the oscillation. Therefore, the **power,** or rate of energy transfer, associated with the wave is

$$\mathcal{P} = \frac{E_\lambda}{\Delta t} = \frac{\tfrac{1}{2}\mu\omega^2A^2\lambda}{T} = \tfrac{1}{2}\mu\omega^2A^2\left(\frac{\lambda}{T}\right)$$

■ Rate of energy transfer for a wave

$$\mathcal{P} = \tfrac{1}{2}\mu\omega^2A^2v \qquad [13.23]$$

This result shows that the rate of energy transfer by a sinusoidal wave on a string is proportional to (a) the square of the angular frequency, (b) the square of the amplitude, and (c) the wave speed. In fact, *all* sinusoidal waves have the following general property: **The rate of energy transfer in any sinusoidal wave is proportional to the square of the angular frequency and to the square of the amplitude.**

**QUICK QUIZ 13.4** Which of the following, taken by itself, would be most effective in increasing the rate at which energy is transferred by a wave traveling along a string? **(a)** reducing the linear mass density of the string by one half **(b)** doubling the wavelength of the wave **(c)** doubling the tension in the string **(d)** doubling the amplitude of the wave

**EXAMPLE 13.6 Power Supplied to a Vibrating String**

A string having a linear mass density of $\mu = 5.00 \times 10^{-2}$ kg/m is under a tension of 80.0 N. How much power must be supplied to the string to generate sinusoidal waves at a frequency of 60.0 Hz and an amplitude of 6.00 cm?

**Solution** The wave speed on the string is

$$v = \sqrt{\frac{T}{\mu}} = \left(\frac{80.0 \text{ N}}{5.00 \times 10^{-2} \text{ kg/m}}\right)^{1/2} = 40.0 \text{ m/s}$$

Because $f = 60.0$ Hz, the angular frequency $\omega$ of the sinusoidal waves on the string has the value

$$\omega = 2\pi f = 2\pi(60.0 \text{ Hz}) = 377 \text{ s}^{-1}$$

Using these values in Equation 13.23 for the power, with $A = 6.00 \times 10^{-2}$ m, gives

$$\begin{aligned}\mathcal{P} &= \tfrac{1}{2}\mu\omega^2 A^2 v \\ &= \tfrac{1}{2}(5.00 \times 10^{-2} \text{ kg/m})(377 \text{ s}^{-1})^2 \\ &\quad \times (6.00 \times 10^{-2} \text{ m})^2(40.0 \text{ m/s}) = 512 \text{ W}\end{aligned}$$

## 13.7 SOUND WAVES

Let us turn our attention from transverse waves to longitudinal waves. As stated in Section 13.2, for longitudinal waves the elements of the medium undergo displacements parallel to the direction of wave motion. Sound waves in air are the most important examples of longitudinal waves. Sound waves can travel through any material medium, however, and their speed depends on the properties of that medium. Table 13.1 provides examples of the speed of sound in different media.

The displacements accompanying a sound wave in air are longitudinal displacements of small elements of air from their equilibrium positions. Such displacements result if the source of the waves, such as the diaphragm of a loudspeaker, oscillates in air. If the oscillation of the diaphragm is described by simple harmonic motion, a sinusoidal sound wave propagates away from the loudspeaker. For instance, a one-dimensional sound wave can be produced in a long, narrow tube containing a gas by means of a vibrating piston at one end, as in Figure 13.18.

It is difficult to draw a pictorial representation of longitudinal waves because the displacements of the elements of the medium are in the same direction as that of the propagation of the wave. Figure 13.18 is one way to represent these types of waves. The darker color in the figure represents a region where the gas is compressed; consequently, the density and pressure are *above* their equilibrium values. Such a compressed region of gas, called a **compression,** is formed when the piston is being pushed into the tube. The compression moves along the tube, continuously compressing the layers in front of it. When the piston is withdrawn from the tube, the gas in front of it expands, and consequently the pressure and density in this region fall below their equilibrium values. These low-pressure regions, called **rarefactions,** are represented by the lighter areas in Figure 13.18. The rarefactions also propagate along the tube, following the compressions. Both regions move with a speed equal to the speed of sound in that medium.

As the piston oscillates back and forth in a sinusoidal fashion, regions of compression and rarefaction are continuously set up. The distance between two successive compressions (or two successive rarefactions) equals the wavelength $\lambda$. As these regions travel along the tube, any small element of the medium moves with simple harmonic motion parallel to the direction of the wave (in other words,

**TABLE 13.1**

**Speed of Sound in Various Media**

| Medium | $v$ (m/s) |
|---|---|
| Gases | |
| Hydrogen (0°C) | 1 286 |
| Helium (0°C) | 972 |
| Air (20°C) | 343 |
| Air (0°C) | 331 |
| Oxygen (0°C) | 317 |
| Liquids at 25°C | |
| Glycerol | 1 904 |
| Sea water | 1 533 |
| Water | 1 493 |
| Mercury | 1 450 |
| Kerosene | 1 324 |
| Methyl alcohol | 1 143 |
| Carbon tetrachloride | 926 |
| Solids[a] | |
| Pyrex glass | 5 640 |
| Iron | 5 950 |
| Aluminum | 6 420 |
| Brass | 4 700 |
| Copper | 5 010 |
| Gold | 3 240 |
| Lucite | 2 680 |
| Lead | 1 960 |
| Rubber | 1 600 |

[a]Values given are for propagation of longitudinal waves in bulk media. Speeds for longitudinal waves in thin rods are smaller, and speeds of transverse waves in bulk are smaller yet.

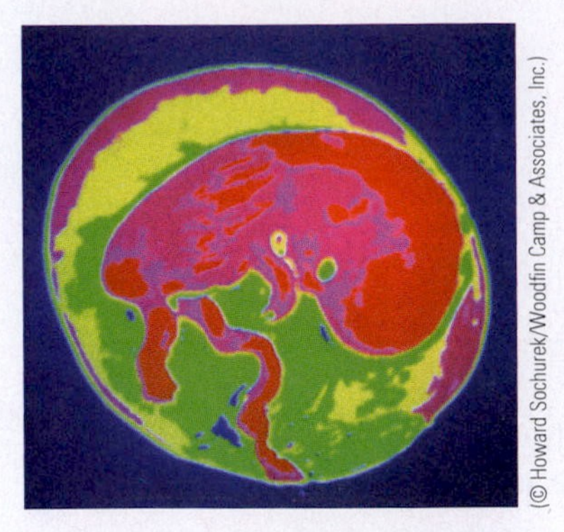



An ultrasound image showing a young human fetus and umbilical cord. Ultrasound refers to sound waves that are higher in frequency than those audible to humans. The sound waves transmit through the body of the mother and reflect from the skin of the fetus. The reflected sound waves are organized by the electronics of the ultrasonic imaging system into a visual image of the fetus. ■

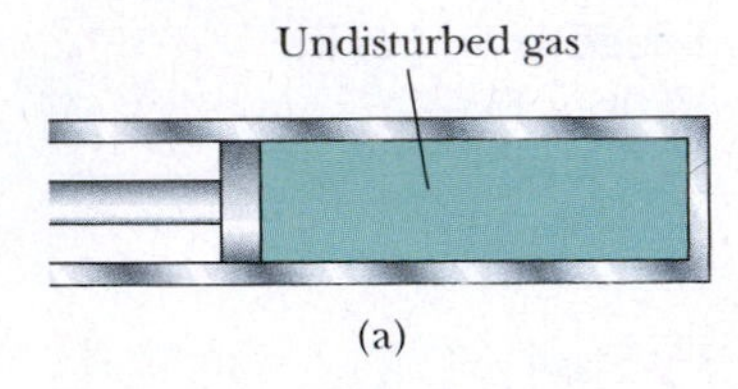


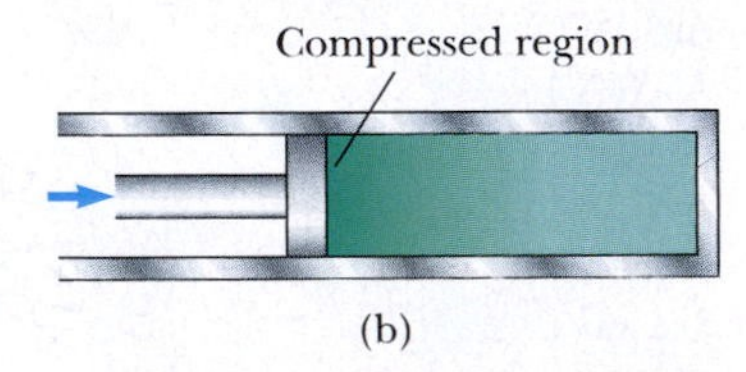


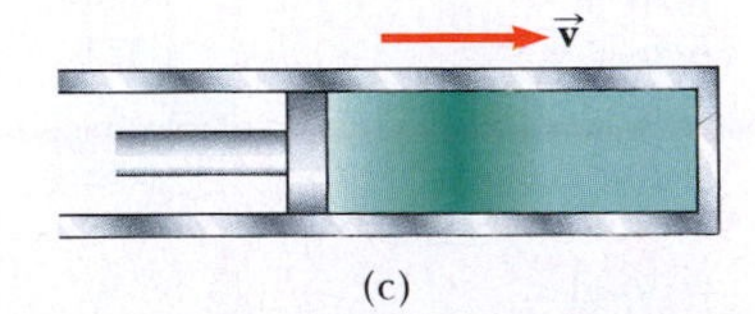


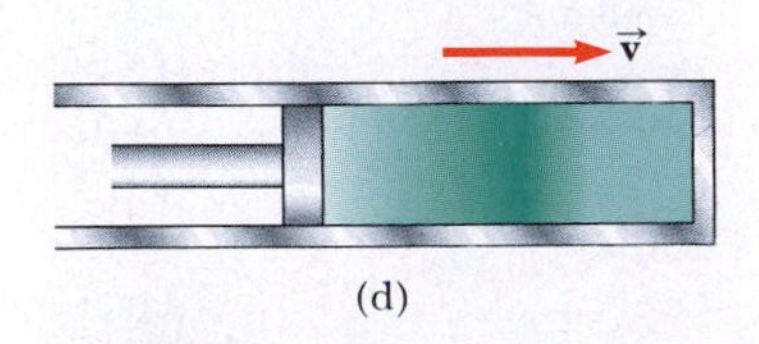


FIGURE 13.18 A longitudinal wave propagating along a tube filled with a compressible gas. The source of the wave is a vibrating piston at the left. The high- and low-pressure regions are dark and light, respectively.

longitudinally). If $s(x, t)$ is the position of a small element measured relative to its equilibrium position, we can express this position function as

$$s(x, t) = s_{\max} \sin(kx - \omega t) \tag{13.24}$$

where $s_{\max}$ is the **maximum position relative to equilibrium,** often called the **displacement amplitude.** Equation 13.24 represents the **displacement wave,** where $k$ is the wave number and $\omega$ is the angular frequency of the piston. The variation $\Delta P$ in the pressure[3] of the gas measured from its equilibrium value is also sinusoidal; it is given by

$$\Delta P = \Delta P_{\max} \cos(kx - \omega t) \tag{13.25}$$

**The pressure amplitude $\Delta P_{\max}$ is the maximum change in pressure from the equilibrium value,** and Equation 13.25 represents the **pressure wave.** The pressure amplitude is proportional to the displacement amplitude $s_{\max}$:

$$\Delta P_{\max} = \rho v \omega s_{\max} \tag{13.26}$$

where $\rho$ is the density of the medium, $v$ is the wave speed, and $\omega s_{\max}$ is the maximum longitudinal speed of an element of the medium. It is these pressure variations in a sound wave that result in an oscillating force on the eardrum, leading to the sensation of hearing.

Therefore, we see that a sound wave may be considered as either a displacement wave or a pressure wave. A comparison of Equations 13.24 and 13.25 shows that **the pressure wave is 90° out of phase with the displacement wave.** Graphs of these functions are shown in Figure 13.19. Note that the change in pressure from equilibrium is a maximum when the displacement is zero, whereas the displacement is a maximum when the pressure change is zero.

Note that Figure 13.19 presents two graphical representations of the longitudinal wave: one for position of the elements of the medium and the other for pressure variation. They are *not* pictorial representations for longitudinal waves, however. For transverse waves, the element displacement is perpendicular to the direction of propagation and the pictorial and graphical representations look the same because the perpendicularity of the oscillations and propagation is matched by the perpendicularity of $x$ and $y$ axes. For longitudinal waves, the oscillations and propagation exhibit no perpendicularity, so those pictorial representations look like Figure 13.18.

The speed of a sound wave in air depends only on the temperature of the air. For a small range of temperatures around room temperature, the speed of sound is described by

$$v = 331 \text{ m/s} + (0.6 \text{ m/s}\cdot{}^\circ\text{C})\, T_C \tag{13.27}$$

where $T_C$ is the temperature in degrees Celsius and the speed of sound at 0°C is 331 m/s.

## ■ Thinking Physics 13.2

Why does thunder produce an extended "rolling" sound when its source, a lightning strike, occurs in a fraction of a second? How does lightning produce thunder in the first place?

**Reasoning** Let us assume that we are at ground level and ignore ground reflections. When cloud-to-ground lightning strikes, a channel of ionized air carries a very large electric current from the cloud to the ground. (We will study electric

[3]We will formally introduce pressure in Chapter 15. In the case of longitudinal waves in a gas, each compressed area is a region of higher-than-average pressure and density, and each stretched region is a region of lower-than-average pressure and density.

current in Chapter 21.) The result is a very rapid temperature increase of this channel of air as it carries the current. The temperature increase causes a sudden expansion of the air. This expansion is so sudden and so intense that a tremendous disturbance is produced in the air: thunder. The thunder rolls because the lightning channel is a long, extended source; the entire length of the channel produces the sound at essentially the same instant of time. Sound produced at the end of the channel nearest you reaches you first, but sounds from progressively farther portions of the channel reach you shortly thereafter. If the lightning channel were a perfectly straight line, the resulting sound might be a steady roar, but the zigzagged shape of the path results in the rolling variation in loudness. ■

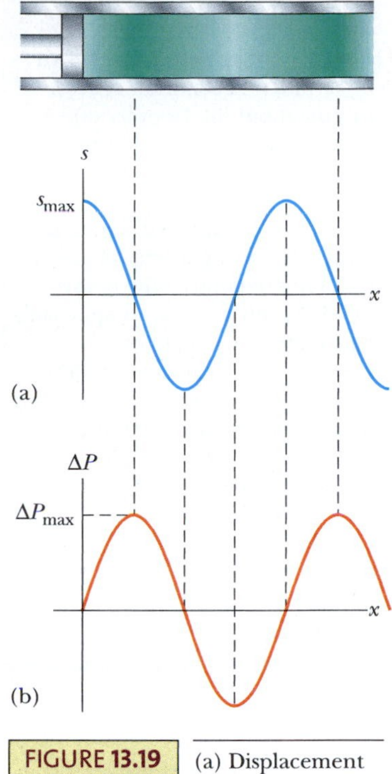


FIGURE 13.19 (a) Displacement versus position and (b) pressure versus position for a sinusoidal longitudinal wave. The displacement wave is 90° out of phase with the pressure wave.

## 13.8 THE DOPPLER EFFECT

When someone honks the horn of a vehicle as it travels along a highway, the frequency of the sound you hear is higher as the vehicle approaches you than it is as the vehicle moves away from you. This change is one example of the **Doppler effect,** named after Christian Johann Doppler (1803–1853), an Austrian physicist.

The Doppler effect for sound is experienced whenever there is relative motion between the source of sound and the observer. Motion of the source or observer toward the other results in the observer's hearing a frequency that is higher than the true frequency of the source. Motion of the source or observer away from the other results in the observer hearing a frequency that is lower than the true frequency of the source.

Although we shall restrict our attention to the Doppler effect for sound waves, it is associated with waves of all types. The Doppler effect for electromagnetic waves is used in police radar systems to measure the speeds of motor vehicles. Likewise, astronomers use the effect to determine the relative motions of stars, galaxies, and other celestial objects. In 1842, Doppler first reported the frequency shift in connection with light emitted by two stars revolving about each other in double-star systems. In the early 20th century, the Doppler effect for light from galaxies was used to argue for the expansion of the Universe, which led to the Big Bang theory, discussed in Chapter 31.

To see what causes this apparent frequency change, imagine you are in a boat lying at anchor on a gentle sea where the waves have a period of $T = 2.0$ s. Thus, every 2.0 s a crest hits your boat. Figure 13.20a shows this situation with the water waves moving toward the left. If you start a stopwatch at $t = 0$ just as one crest hits, the stopwatch reads 2.0 s when the next crest hits, 4.0 s when the third crest hits, and so on. From these observations you conclude that the wave frequency is $f = 1/T = 0.50$ Hz. Now suppose you start your motor and head directly into the oncoming waves as shown in Figure 13.20b. Again you set your stopwatch to $t = 0$ as a crest hits the bow of your boat. This time, however, because you are moving toward the next wave crest as it moves toward you, it hits you less than 2.0 s after the first hit. In other words, the period you observe is shorter than the 2.0-s period you observed when you were stationary. Because $f = 1/T$, you observe a higher wave frequency than when you were at rest.

If you turn around and move in the same direction as the waves (Fig. 13.20c), you observe the opposite effect. You set your watch to $t = 0$ as a crest hits the stern of the boat. Because you are now moving away from the next crest, more than 2.0 s has elapsed on your watch by the time that crest catches you. Therefore, you observe a lower frequency than when you were at rest.

These effects occur because the relative speed between your boat and the crest of a wave depends on the direction of travel and on the speed of your boat. When you are moving toward the right in Figure 13.20b, this relative speed is higher than that of the wave speed, which leads to the observation of an increased frequency. When you turn around and move to the left, the relative speed is lower, as is the observed frequency of the water waves.

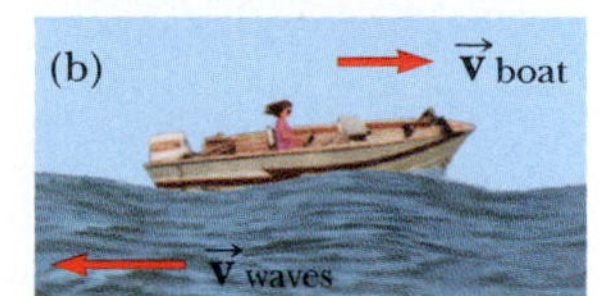


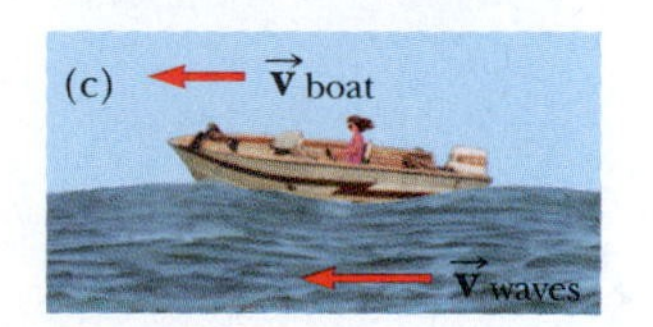


FIGURE 13.20 (a) Waves moving toward a stationary boat. The waves travel to the left and their source is far to the right of the boat, out of the frame of the drawing. (b) The boat moving toward the wave source. (c) The boat moving away from the wave source.

**PITFALL PREVENTION 13.4**

**DOPPLER EFFECT DOES NOT DEPEND ON DISTANCE** A common misconception about the Doppler effect is that it depends on the distance between the source and the observer. Although the intensity of a sound will vary as the distance changes, the apparent frequency will not; the frequency depends only on the speed. As you listen to an approaching source, you will detect increasing intensity but constant frequency. As the source passes, you will hear the frequency suddenly drop to a new constant value and the intensity begin to decrease.

Let us now examine an analogous situation with sound waves in which we replace the water waves with sound waves, the water surface becomes the air, and the person on the boat becomes an observer listening to the sound. In this case, an observer $O$ is moving with a speed of $v_O$ and a sound source $S$ is stationary. For simplicity, we assume that the air is also stationary and that the observer moves directly toward the source.

The red lines in Active Figure 13.21 represent circles connecting the crests of sound waves moving away from the source. Therefore, the radial distance between adjacent red lines is one wavelength. We shall take the frequency of the source to be $f$, the wavelength to be $\lambda$, and the speed of sound to be $v$. A stationary observer would detect a frequency $f$, where $f = v/\lambda$ (i.e., when the source and observer are both at rest, the observed frequency must equal the true frequency of the source). If the observer moves toward the source with the speed $v_O$, however, the relative speed of sound experienced by the observer is higher than the speed of sound in air. Using our relative speed discussion of Section 3.6, if the sound is coming toward the observer at $v$ and the observer is moving toward the sound at $v_O$, the relative speed of sound as measured by the observer is

$$v_{\text{rel}} = v + v_O$$

The frequency of sound heard by the observer is based on this apparent speed of sound:

$$f' = \frac{v_{\text{rel}}}{\lambda} = \frac{v + v_O}{\lambda} = f\left(\frac{v + v_O}{v}\right) \quad \text{(observer moving toward source)} \quad [13.28]$$

Now consider the situation in which the source moves with a speed of $v_S$ relative to the medium and the observer is at rest. Active Figure 13.22a shows this situation. Because the source is moving, the crest of each new wave is emitted from the source to the right of the position of the emission of the previous crest a distance $v_S T$, where $T$ is the period of the wave being generated by the source. Therefore, the center of each colored circle (indicated by the identically colored dot) in Active Figure 13.22a is shifted to the right by this distance relative to the circle representing the previous crest. If the source moves directly toward observer A in Active Figure 13.22a, the crests detected by the observer along a line between the source and observer are closer to one another than they would be if the source were at rest. As a result, the wavelength $\lambda'$ measured by observer A is shorter than the true wavelength $\lambda$ of the source. The wavelength is *shortened* by the distance $v_S T$, and the observed wavelength has the value $\lambda' = \lambda - v_S/f$. Because $\lambda = v/f$, the frequency heard by observer A is

$$f' = \frac{v}{\lambda'} = f\left(\frac{v}{v - v_S}\right) \quad \text{(source moving toward observer)} \quad [13.29]$$

**ACTIVE FIGURE 13.21**

An observer $O$ (the cyclist) moving with a speed $v_O$ toward a stationary point source $S$, the horn of a parked car. The observer hears a frequency $f'$ that is greater than the source frequency.

**Physics Now™** Log into PhysicsNow at **www.pop4e.com** and go to Active Figure 13.21 to adjust the speed of the observer.

That is, the frequency is *increased* when the source moves toward the observer. In a similar manner, if the source moves away from observer B at rest, the sign of $v_S$ is reversed in Equation 13.29 and the frequency is lower.

In Equation 13.29, notice that the denominator approaches zero when the speed of the source approaches the speed of sound, resulting in the frequency $f'$ approaching infinity. Such a situation results in waves that cannot escape from the source in the direction of motion of the source. This concentration of energy in front of the source results in a *shock wave.* Such a disturbance is noted when a jet aircraft flying at a speed equal to or greater than the speed of sound produces a *sonic boom.*

(Courtesy of U.S. Navy. Photo by Ensign John Gay)

A shock wave due to a jet traveling at the speed of sound is made visible as a fog of water vapor. The large pressure variation in the shock wave causes the water in the air to condense into water droplets. ■

Finally, if both the source and the observer are in motion, the following general equation for the observed frequency is found:

$$f' = f\left(\frac{v + v_O}{v - v_S}\right) \qquad [13.30]$$

In this expression, the signs for the values substituted for $v_O$ and $v_S$ depend on the direction of the velocity. A positive value is used for motion of the observer or the source *toward* the other, and a negative sign is used for motion of one *away from* the other.

When working with any Doppler effect problem, remember the following rule concerning signs: The word *toward* is associated with an *increase* in the observed frequency, and the words *away from* are associated with a *decrease* in the observed frequency.

Doppler measurements of blood flow

The Doppler effect is used in medicine to measure the speed of blood flow. In ultrasound Doppler procedures, an ultrasonic sound wave is sent into the skin from a transducer. The sound waves reflect from moving blood cells, undergoing a frequency shift based on the speed of the cells. The instrumentation detects the reflected sound waves and converts the frequency information to a speed of flow of the blood. It is also possible to use the Doppler shift of light to measure the speed of blood flow. That can be done for blood vessels just under the skin by shining light from a low power laser onto the skin surface and monitoring the reflected light. The procedure can be performed on internal blood vessels by means of optical fibers (see Section 25.8) entering the body through natural openings or small incisions.

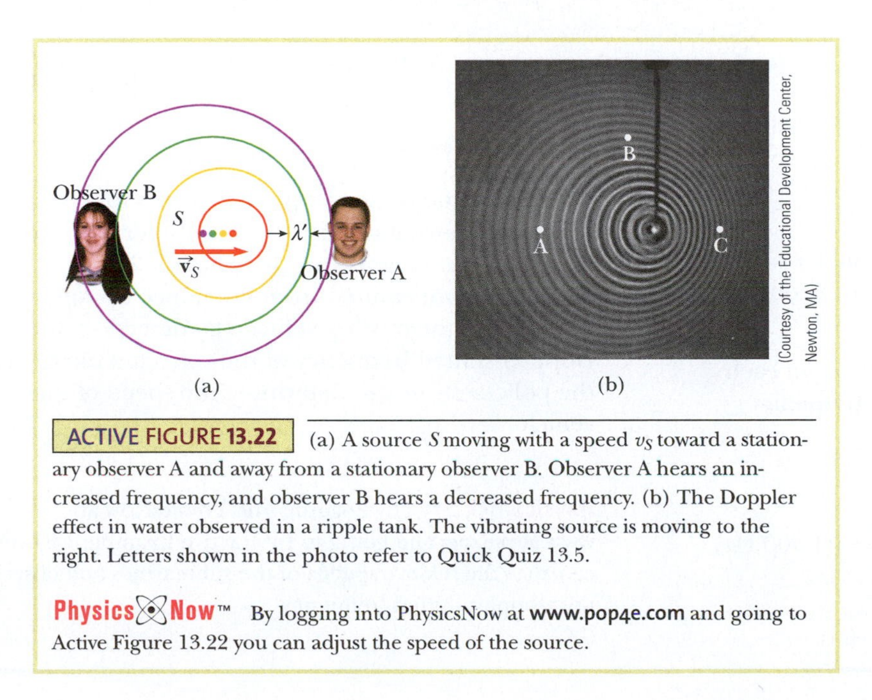


**ACTIVE FIGURE 13.22** (a) A source $S$ moving with a speed $v_S$ toward a stationary observer A and away from a stationary observer B. Observer A hears an increased frequency, and observer B hears a decreased frequency. (b) The Doppler effect in water observed in a ripple tank. The vibrating source is moving to the right. Letters shown in the photo refer to Quick Quiz 13.5.

**Physics Now™** By logging into PhysicsNow at **www.pop4e.com** and going to Active Figure 13.22 you can adjust the speed of the source.

**QUICK QUIZ 13.5** Consider detectors of water waves at three locations A, B, and C in Active Figure 13.22b. Which of the following statements is true? **(a)** The wave speed is highest at location A. **(b)** The wave speed is highest at location C. **(c)** The detected wavelength is largest at location B. **(d)** The detected wavelength is largest at location C. **(e)** The detected frequency is highest at location C. **(f)** The detected frequency is highest at location A.

**QUICK QUIZ 13.6** You stand on a platform at a train station and listen to a train approaching the station at a constant velocity. While the train approaches, but before it arrives, you hear **(a)** the intensity and the frequency of the sound both increasing, **(b)** the intensity and the frequency of the sound both decreasing, **(c)** the intensity increasing and the frequency decreasing, **(d)** the intensity decreasing and the frequency increasing, **(e)** the intensity increasing and the frequency remaining the same, or **(f)** the intensity decreasing and the frequency remaining the same.

### INTERACTIVE EXAMPLE 13.7 Doppler Submarines

Submarines A and B are traveling toward each other under water. Sub A travels through the water at a speed of 8.00 m/s, emitting a sonar wave at a frequency of 1 400 Hz. Sub B travels through the water at a speed of 9.00 m/s. The speed of sound in the water is 1 533 m/s.

**A** What frequency is detected by an observer riding on sub B as the subs approach each other?

**Solution** We use Equation 13.30 to find the Doppler-shifted frequency. As the two submarines approach each other, the observer in sub B hears the frequency

$$f' = \left(\frac{v + v_O}{v - v_S}\right) f$$

$$= \left(\frac{1\,533 \text{ m/s} + (+9.00 \text{ m/s})}{1\,533 \text{ m/s} - (+8.00 \text{ m/s})}\right)(1\,400 \text{ Hz})$$

$$= 1\,416 \text{ Hz}$$

**B** The subs barely miss each other and pass. What frequency is detected by an observer riding on sub B as the subs recede from each other?

**Solution** As the two submarines recede from each other, the observer in sub B hears the frequency

$$f' = \left(\frac{v + v_O}{v - v_S}\right) f$$

$$= \left(\frac{1\,533 \text{ m/s} + (-9.00 \text{ m/s})}{1\,533 \text{ m/s} - (-8.00 \text{ m/s})}\right)(1\,400 \text{ Hz})$$

$$= 1\,385 \text{ Hz}$$

**C** While the subs are approaching each other, some of the sound from sub A will reflect from sub B and return to sub A. If this sound were to be detected by an observer on sub A, what is its frequency?

**Solution** The sound of apparent frequency 1 416 Hz found in part A will be reflected from a moving source (sub B) and then detected by a moving observer (sub A). Therefore, the frequency detected by sub A is

$$f'' = \left(\frac{v + v_O}{v - v_S}\right) f'$$

$$= \left(\frac{1\,533 \text{ m/s} + (+8.00 \text{ m/s})}{1\,533 \text{ m/s} - (+9.00 \text{ m/s})}\right)(1\,416 \text{ Hz})$$

$$= 1\,432 \text{ Hz}$$

This technique is used by police officers to measure the speed of a moving car, using the Doppler effect for electromagnetic radiation (see Section 24.3). Microwaves are emitted from the police car and reflected by the moving vehicle. By detecting the Doppler-shifted frequency of the reflected microwaves, the police officer can determine the speed of the vehicle.

**PhysicsNow™** By Logging into PhysicsNow at www.pop4e.com and going to Interactive Example 13.7 you can alter the relative speeds of the submarines and observe the Doppler-shifted frequency.

## 13.9 SEISMIC WAVES

CONTEXT CONNECTION

When an earthquake occurs, a sudden release of energy takes place at a location called the **focus** or **hypocenter** of the earthquake. The **epicenter** is the point on the Earth's surface radially above the hypocenter. The released energy will propagate away from the focus of the earthquake by means of **seismic waves.** Seismic waves are like the sound waves that we have studied in the later sections of this chapter in that they are mechanical disturbances moving through a medium.

In discussing mechanical waves in this chapter, we identified two types: transverse and longitudinal. In the case of mechanical waves moving through air, we have only a longitudinal possibility. For mechanical waves moving through a solid, however, both possibilities are available because of the strong interatomic forces between elements of the solid. Therefore, in the case of seismic waves, energy propagates away from the focus both by longitudinal and transverse waves.

In the language used in earthquake studies, these two types of waves are named according to the order of their arrival at a seismograph. The longitudinal wave travels at a higher speed than the transverse wave. As a result, the longitudinal wave arrives at a seismograph first and is thus called the ***P* wave,** where *P* stands for *primary.* The slower moving transverse wave arrives next, so it is called the ***S* wave,** or *secondary* wave.

Let us see why longitudinal waves travel faster than transverse waves. The speed of *all* mechanical waves follows an expression of the general form

$$v = \sqrt{\frac{\text{elastic property}}{\text{inertial property}}} \qquad [13.31]$$

For a wave traveling on a string, we have seen the speed given by Equation 13.20:

$$v = \sqrt{\frac{T}{\mu}}$$

where the elastic property is the tension in the string. It is the tension in the string that returns a displaced element of the string to equilibrium. The appropriate inertial property is the linear mass density of the string.

For a transverse wave moving in a bulk solid, the elastic property is the *shear modulus S* of the material.[4] The shear modulus is a parameter that measures the deformation of a solid to a shear force, a force in the sideways direction. For example, lay your textbook down on a table and place your hand flat on the cover. Now, move your hand in a direction away from the book spine. The book will deform so that its cross-section changes from a rectangle to a parallelogram. The amount by which the book deforms under a given force from your hand is related to the shear modulus of the book. The speed of a transverse wave (an $S$ wave) in a bulk solid is

$$v_S = \sqrt{\frac{S}{\rho}} \qquad [13.32]$$

where $\rho$ is the density and $S$ is the shear modulus of the material.

For a longitudinal wave moving in a gas or liquid, the elastic property in Equation 13.31 is the *bulk modulus B* of the material. The bulk modulus is a parameter that measures the change in volume of a sample of material due to a force compressing it that is uniform over a surface area. The speed of sound in a gas is given by

$$v = \sqrt{\frac{B}{\rho}} \qquad [13.33]$$

where $B$ is the bulk modulus of the gas and $\rho$ is the gas density.

[4]For details on various elastic moduli for materials see R. A. Serway and J. W. Jewett Jr., *Physics for Scientists and Engineers,* 6th ed. (Brooks-Cole, Belmont, CA: 2004), Section 12.4.

Now we consider longitudinal waves moving through a bulk solid. As a wave passes through a sample of the material, the material is compressed, so the wave speed should depend on the bulk modulus. As the material is compressed along the direction of travel of the wave, however, it is also distorted in the perpendicular direction. (Imagine a partially inflated balloon that is pressed downward against a table. It spreads out in the direction parallel to the table.) The result is a shear distortion of the sample of material. Therefore, the wave speed should depend on both the bulk modulus and the shear modulus! Careful analysis shows that this wave speed is

$$v_P = \sqrt{\frac{B + \frac{4}{3}S}{\rho}} \qquad [13.34]$$

Notice that this equation for the speed of a *P* wave gives a value that is larger than that for the *S* wave in Equation 13.32.

The wave speed for a seismic wave depends on the medium through which it travels. Typical values are 8 km/s for a *P* wave and 5 km/s for an *S* wave. Figure 13.23 shows a typical seismograph trace of a distant earthquake, with the *S* wave clearly arriving after the *P* wave.

The *P* and *S* waves move through the body of the Earth and can be detected by seismographs at various locations around the planet. Once these waves reach the surface, the energy can propagate by additional types of waves along the surface. In a *Rayleigh wave,* the motion of the elements of the medium at the surface is a combination of longitudinal and transverse displacements so that the net motion of a point on the surface is circular or elliptical. This motion is similar to the path followed by elements of water on the ocean surface as a wave passes by, as in Active Figure 13.24. The *Love wave* is a transverse surface wave in which the transverse oscillations are parallel to the surface. Therefore, no vertical displacement of the surface occurs in a Love wave.

It is possible to use the *P* and *S* waves traveling through the body of the Earth to gain information about the structure of the Earth's interior. Measurements of a given earthquake by seismographs at various locations on the surface indicate that the Earth has an interior region that allows the passage of *P* waves but not *S* waves. This fact can be understood if this particular region is modeled as having liquid

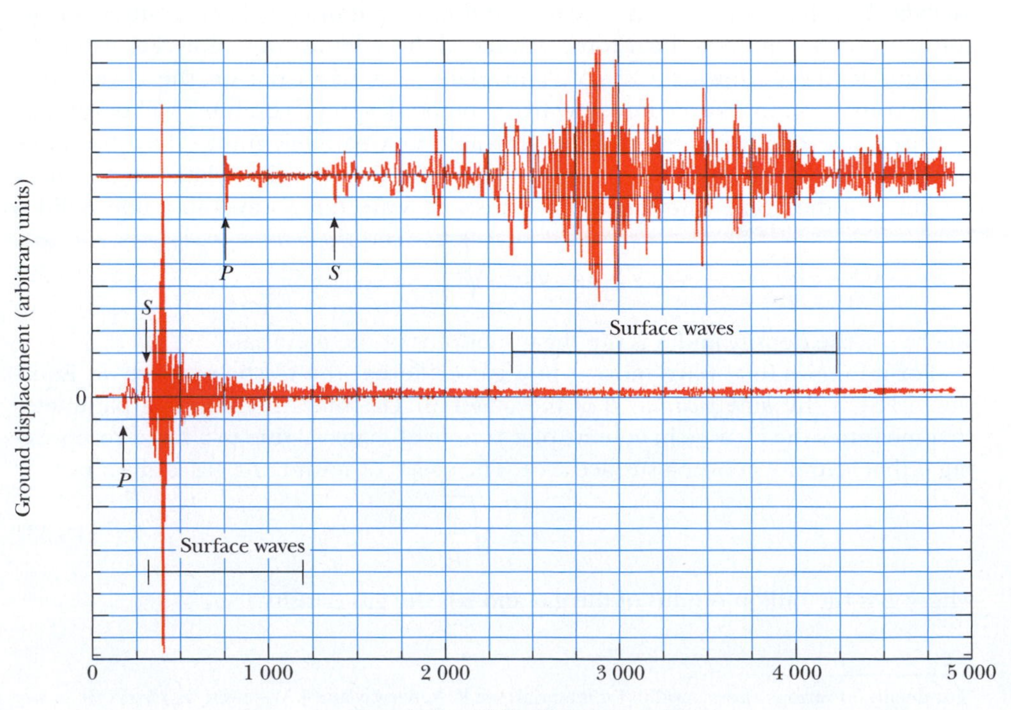


**FIGURE 13.23** A seismograph trace, showing the arrival of *P* and *S* waves from the Northridge, California, earthquake at San Pablo, Spain (*top trace*) and Albuquerque, New Mexico (*bottom trace*). The *P* wave arrives first because it travels the fastest, followed by the slower moving *S* wave. The farther the seismograph station is from the epicenter, the longer the time interval between the arrivals of the *P* and *S* waves.

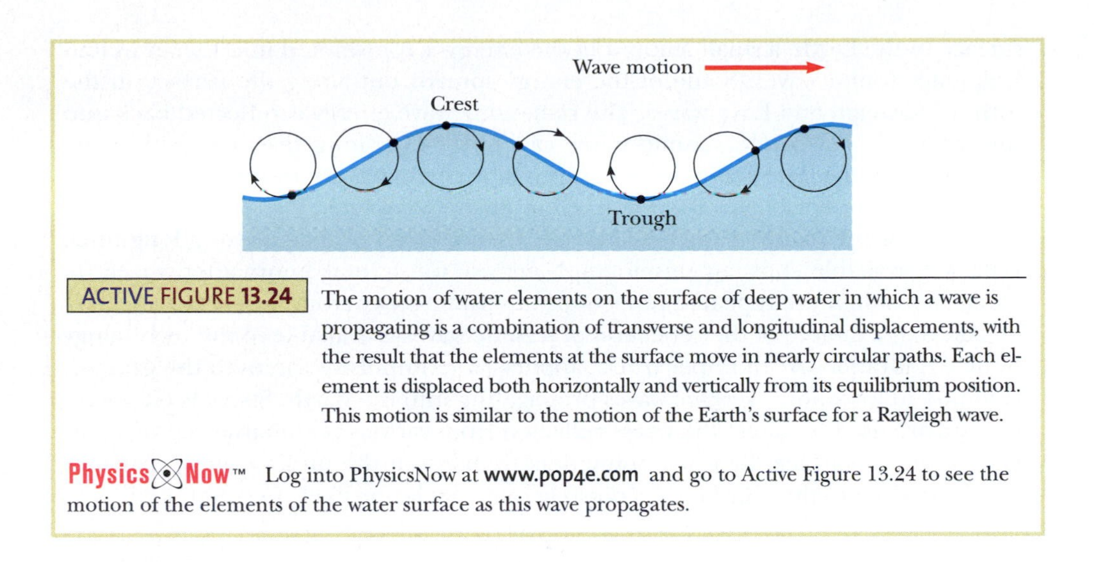


ACTIVE FIGURE 13.24 The motion of water elements on the surface of deep water in which a wave is propagating is a combination of transverse and longitudinal displacements, with the result that the elements at the surface move in nearly circular paths. Each element is displaced both horizontally and vertically from its equilibrium position. This motion is similar to the motion of the Earth's surface for a Rayleigh wave.

**PhysicsNow™** Log into PhysicsNow at **www.pop4e.com** and go to Active Figure 13.24 to see the motion of the elements of the water surface as this wave propagates.

characteristics. Similar to a gas, a liquid cannot sustain a transverse force. Therefore, the transverse *S* waves cannot pass through this region. This information leads us to a structural model in which the Earth has a **liquid core** between radii of approximately $1.2 \times 10^3$ km and $3.5 \times 10^3$ km.

Other measurements of seismic waves allow additional interpretations of layers within the interior of the Earth, including a **solid core** at the center, a rocky region called the **mantle,** and a relatively thin outer layer called the **crust.** Figure 13.25 shows this structure. Using x-rays or ultrasound in medicine to provide information about the interior of the human body is somewhat similar to using seismic waves to provide information about the interior of the Earth.

As *P* and *S* waves propagate in the interior of the Earth, they will encounter variations in the medium. At each boundary at which the properties of the medium change, reflection and transmission occur. When the seismic wave arrives at the

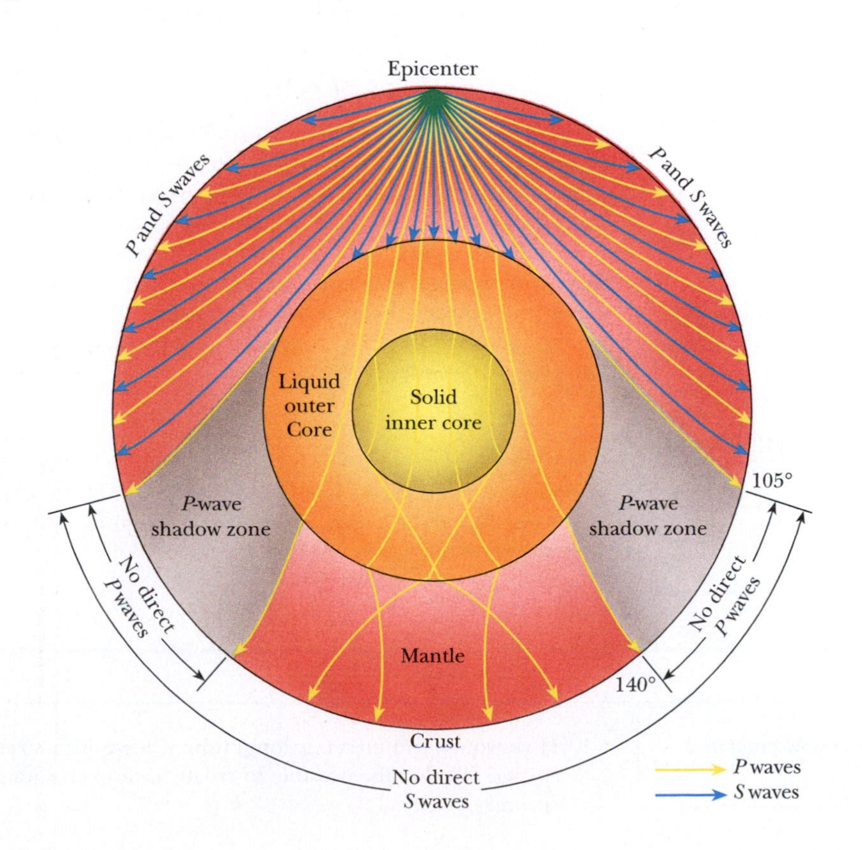


FIGURE 13.25 Cross-section of the Earth showing paths of waves produced by an earthquake. Only *P* waves (yellow) can propagate in the liquid core. The *S* waves (blue) do not enter the liquid core. When the *P* waves transmit from one region to another, such as from the mantle to the liquid core, they experience *refraction,* a change in the direction of propagation. We will study refraction for light in Chapter 25. Because of the refraction for seismic waves, there is a "shadow" zone between 105° and 140° from the epicenter in which no waves following a direct path (i.e., a path with no reflections) arrive.

surface of the Earth, a small amount of the energy is transmitted into the air as low-frequency sound waves. Some of the energy spreads out along the surface in the form of Rayleigh and Love waves. The remaining wave energy is reflected back into the interior. As a result, seismic waves can travel over long distances within the Earth and can be detected at seismographs at many locations around the globe. In addition, because a relatively large fraction of the wave energy continues to be reflected at each encounter with the surface, the wave can propagate for a long time. Data are available showing seismograph activity for several hours after an earthquake, a result of the repeated reflections of seismic waves from the surface.

Another example of the reflection of seismic waves is available in the technology of oil exploration. A "thumper truck" applies large impulsive forces to the ground, resulting in low-energy seismic waves propagating into the Earth. Specialized microphones are used to detect the waves reflected from various boundaries between layers under the surface. By using computers to map out the underground structure corresponding to these layers, it is possible to detect layers likely to contain oil.

## SUMMARY

**PhysicsNow™** Take a practice test by logging into PhysicsNow at www.pop4e.com and clicking on the Pre-Test link for this chapter.

A **transverse wave** is a wave in which the elements of the medium move in a direction perpendicular to the direction of the wave velocity. An example is a wave moving along a stretched string.

**Longitudinal waves** are waves in which the elements of the medium move back and forth parallel to the direction of the wave velocity. Sound waves in air are longitudinal.

Any one-dimensional wave traveling with a speed of $v$ in the positive $x$ direction can be represented by a **wave function** of the form $y = f(x - vt)$. Likewise, the wave function for a wave traveling in the negative $x$ direction has the form $y = f(x + vt)$.

The wave function for a one-dimensional sinusoidal wave traveling to the right can be expressed as

$$y = A\sin\left[\frac{2\pi}{\lambda}(x - vt)\right] = A\sin(kx - \omega t) \quad [13.4, 13.9]$$

where $A$ is the **amplitude,** $\lambda$ is the **wavelength,** $k$ is the **angular wave number,** and $\omega$ is the **angular frequency.** If $T$ is the **period** and $f$ is the **frequency,** $v$, $k$, and $\omega$ can be written as

$$v = \frac{\lambda}{T} = \lambda f \quad [13.5, 13.11]$$

$$k \equiv \frac{2\pi}{\lambda} \quad [13.7]$$

$$\omega \equiv \frac{2\pi}{T} = 2\pi f \quad [13.8]$$

The speed of a transverse wave traveling on a stretched string of mass per unit length $\mu$ and tension $T$ is

$$v = \sqrt{\frac{T}{\mu}} \quad [13.20]$$

When a pulse traveling on a string meets a fixed end, the pulse is reflected and inverted. If the pulse reaches a free end, it is reflected but not inverted.

The **power** transmitted by a sinusoidal wave on a stretched string is

$$\mathcal{P} = \tfrac{1}{2}\mu\omega^2 A^2 v \quad [13.23]$$

The change in frequency of a sound wave heard by an observer whenever there is relative motion between a wave source and the observer is called the **Doppler effect.** When the source and observer are moving toward each other, the observer hears a higher frequency than the true frequency of the source. When the source and observer are moving away from each other, the observer hears a lower frequency than the true frequency of the source. The following general equation provides the observed frequency:

$$f' = f\left(\frac{v + v_O}{v - v_S}\right) \quad [13.30]$$

A positive value is used for $v_O$ or $v_S$ for motion of the observer or source *toward* the other, and a negative sign is used for motion *away from* the other.

## QUESTIONS

☐ = answer available in the *Student Solutions Manual and Study Guide*

1. How would you create a longitudinal wave in a stretched spring? Would it be possible to create a transverse wave in a spring?

2. By what factor would you have to multiply the tension in a stretched string so as to double the wave speed?
3. When a pulse travels on a taut string, does it always invert upon reflection? Explain.
4. Consider a wave traveling on a taut rope. What is the difference, if any, between the speed of the wave and the speed of a small element of the rope?
5. What happens to the wavelength of a wave on a string when the frequency is doubled? Assume that the tension in the string remains constant.
6. What happens to the speed of a wave on a taut string when the frequency is doubled? Assume that the tension in the string remains constant.
7. If you stretch a rubber hose and pluck it, you can observe a pulse traveling up and down the hose. What happens to the speed of the pulse if you stretch the hose more tightly? What happens to the speed if you fill the hose with water?
8. If one end of a heavy rope is attached to one end of a light rope, the speed of a wave will change as the wave goes from the heavy rope to the light one. Will it increase or decrease? What happens to the frequency? What happens to the wavelength?
9. A vibrating source generates a sinusoidal wave on a string under constant tension. If the power delivered to the string is doubled, by what factor does the amplitude change? Does the wave speed change under these circumstances?
10. Why are sound waves characterized as longitudinal?
11. If an alarm clock is placed in a good vacuum and then activated, no sound is heard. Explain.
12. If the wavelength of sound is reduced by a factor of 2, what happens to its frequency? What happens to its speed?
13. By listening to a band or orchestra, how can you determine that the speed of sound is the same for all frequencies?
14. *The Tunguska event.* On June 30, 1908, a meteor burned up and exploded in the atmosphere above the Tunguska River valley in Siberia. It knocked down trees over thousands of square kilometers and started a forest fire, but apparently caused no human casualties. A witness sitting on his doorstep outside the zone of falling trees recalled events in the following sequence. He saw a moving light in the sky, brighter than the sun and descending at a low angle to the horizon. He felt his face become warm. He felt the ground shake. An invisible agent picked him up and immediately dropped him about a meter farther away from where the light had been. He heard a very loud protracted rumbling. Suggest an explanation for these observations and for the order in which they happened.
15. How can an object move with respect to an observer so that the sound from it is not shifted in frequency?
16. Suppose the wind blows. Does that cause a Doppler effect for sound propagating through the air? Is it like a moving source or a moving observer?
17. In an earthquake, both $S$ (transverse) and $P$ (longitudinal) waves propagate from the focus of the earthquake. The focus is in the ground below the epicenter on the surface. Assume that the waves move in straight lines through uniform material. The $S$ waves travel through the Earth more slowly than the $P$ waves (at about 5 km/s versus 8 km/s). By detecting the time of arrival of the waves, how can one determine the distance to the focus of the quake? How many detection stations are necessary to locate the focus unambiguously?

## PROBLEMS

1, 2, 3 = straightforward, intermediate, challenging
☐ = full solution available in the *Student Solutions Manual and Study Guide*
**PhysicsNow™** = coached problem with hints available at www.pop4e.com
= computer useful in solving problem
= paired numerical and symbolic problems
= biomedical application

### Section 13.1 ▪ Propagation of a Disturbance

1. At $t = 0$, a transverse pulse in a wire is described by the function

$$y = \frac{6}{x^2 + 3}$$

where $x$ and $y$ are in meters. Write the function $y(x, t)$ that describes this pulse if it is traveling in the positive $x$ direction with a speed of 4.50 m/s.

2. Ocean waves with a crest-to-crest distance of 10.0 m can be described by the wave function

$$y(x, t) = (0.800 \text{ m}) \sin[0.628(x - vt)]$$

where $v = 1.20$ m/s. (a) Sketch $y(x, t)$ at $t = 0$. (b) Sketch $y(x, t)$ at $t = 2.00$ s. Note that the entire wave form has shifted 2.40 m in the positive $x$ direction in this time interval.

### Section 13.2 ▪ The Wave Model

### Section 13.3 ▪ The Traveling Wave

3. A sinusoidal wave is traveling along a rope. The oscillator that generates the wave completes 40.0 vibrations in 30.0 s. Also, a given maximum travels 425 cm along the rope in 10.0 s. What is the wavelength?
4. For a certain transverse wave, the distance between two successive crests is 1.20 m and eight crests pass a given point along the direction of travel every 12.0 s. Calculate the wave speed.

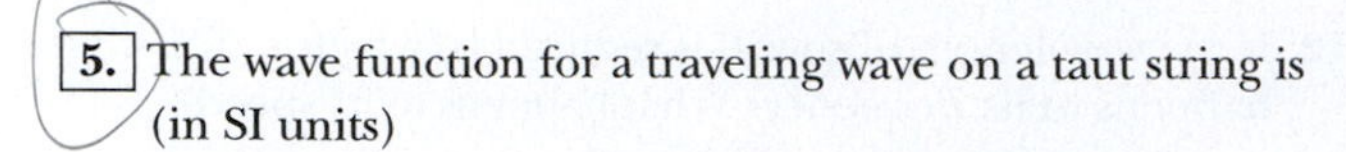

5. The wave function for a traveling wave on a taut string is (in SI units)

$$y(x, t) = (0.350 \text{ m}) \sin(10\pi t - 3\pi x + \pi/4)$$

(a) What are the speed and direction of travel of the wave? (b) What is the vertical position of an element of the string at $t = 0$, $x = 0.100$ m? (c) What are the wavelength and frequency of the wave? (d) What is the maximum transverse speed of an element of the string?

6. A wave is described by $y = (2.00 \text{ cm}) \sin(kx - \omega t)$, where $k = 2.11$ rad/m, $\omega = 3.62$ rad/s, $x$ is in meters, and $t$ is in seconds. Determine the amplitude, wavelength, frequency, and speed of the wave.

7. The string shown in Active Figure 13.8 is driven at a frequency of 5.00 Hz. The amplitude of the motion is 12.0 cm and the wave speed is 20.0 m/s. Furthermore, the wave is such that $y = 0$ at $x = 0$ and $t = 0$. Determine (a) the angular frequency and (b) wave number for this wave. (c) Write an expression for the wave function. Calculate (d) the maximum transverse speed and (e) the maximum transverse acceleration of an element of the string.

8. Consider the sinusoidal wave of Example 13.2, with the wave function

$$y = (15.0 \text{ cm}) \cos(0.157x - 50.3t)$$

At a certain instant, let point $A$ be at the origin and point $B$ be the first point along the $x$ axis where the wave is 60.0° out of phase with point $A$. What is the coordinate of point $B$?

9. **PhysicsNow™** (a) Write the expression for $y$ as a function of $x$ and $t$ for a sinusoidal wave traveling along a rope in the *negative x* direction with the following characteristics: $A = 8.00$ cm, $\lambda = 80.0$ cm, $f = 3.00$ Hz, and $y(0, t) = 0$ at $t = 0$. (b) Write the expression for $y$ as a function of $x$ and $t$ for the wave in part (a) assuming that $y(x, 0) = 0$ at the point $x = 10.0$ cm.

10. A transverse wave on a string is described by the wave function

$$y = (0.120 \text{ m}) \sin\left(\frac{\pi}{8}x + 4\pi t\right)$$

(a) Determine the transverse speed and acceleration of an element of the string at $t = 0.200$ s for the point on the string located at $x = 1.60$ m. (b) What are the wavelength, period, and speed of propagation of this wave?

11. A transverse sinusoidal wave on a string has a period $T = 25.0$ ms and travels in the negative $x$ direction with a speed of 30.0 m/s. At $t = 0$, an element of the string at $x = 0$ has a transverse position of 2.00 cm and is traveling downward with a speed of 2.00 m/s. (a) What is the amplitude of the wave? (b) What is the initial phase angle? (c) What is the maximum transverse speed of an element of the string? (d) Write the wave function for the wave.

12. Show that the wave function $y = e^{b(x - vt)}$ is a solution of the linear wave equation (Eq. 13.19), where $b$ is a constant.

## Section 13.4 ■ The Speed of Transverse Waves on Strings

13. A telephone cord is 4.00 m long and has a mass of 0.200 kg. A transverse pulse is produced by plucking one end of the taut cord. The pulse makes four trips down and back along the cord in 0.800 s. What is the tension in the cord?

14. An astronaut on the Moon wishes to measure the local value of the free-fall acceleration by timing pulses traveling down a wire that has an object of large mass suspended from it. Assume that a wire has a mass of 4.00 g and a length of 1.60 m and that a 3.00-kg object is suspended from it. A pulse requires 36.1 ms to traverse the length of the wire. Calculate $g_{\text{Moon}}$ from these data. (You may ignore the mass of the wire when calculating the tension in it.)

15. Transverse waves travel with a speed of 20.0 m/s in a string under a tension of 6.00 N. What tension is required for a wave speed of 30.0 m/s in the same string?

16. **Review problem.** A light string with a mass per unit length of 8.00 g/m has its ends tied to two walls separated by a distance equal to three-fourths the length of the string (Fig. P13.16). An object of mass $m$ is suspended from the center of the string, putting a tension in the string. (a) Find an expression for the transverse wave speed in the string as a function of the mass of the hanging object. (b) What should be the mass of the object suspended from the string so as to produce a wave speed of 60.0 m/s?

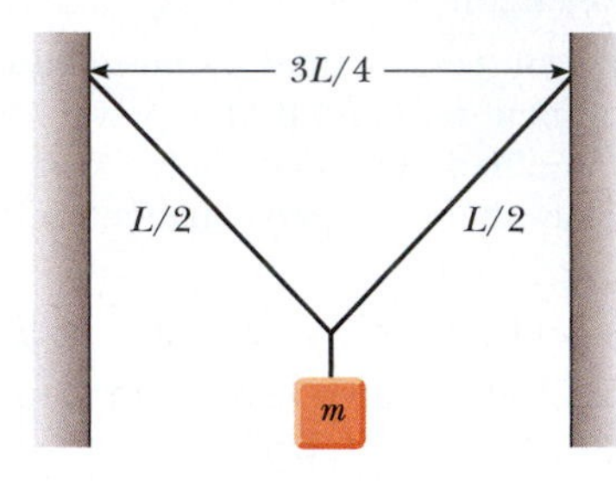


FIGURE **P13.16**

17. **PhysicsNow™** A 30.0-m steel wire and a 20.0-m copper wire, both with 1.00-mm diameters, are connected end to end and stretched to a tension of 150 N. How long does it take a transvers e wave to travel the entire length of the two wires?

## Section 13.5 ■ Reflection and Transmission of Waves

18. A series of pulses, each of amplitude 0.150 m, are sent down a string that is attached to a post at one end. The pulses are reflected at the post and travel back along the string without loss of amplitude. When two waves are present on the same string, the net displacement of a particular element of the string is the sum of the displacements of the individual waves at that point. What is the net displacement of an element at a point on the string where two pulses are crossing (a) if the string is rigidly attached to the post and (b) if the end at which reflection occurs is free to slide up and down?

## Section 13.6 ■ Rate of Energy Transfer by Sinusoidal Waves on Strings

19. A taut rope has a mass of 0.180 kg and a length of 3.60 m. What power must be supplied to the rope so as to generate sinusoidal waves having an amplitude of 0.100 m and

a wavelength of 0.500 m and traveling with a speed of 30.0 m/s?

**20.** It is found that a 6.00-m segment of a long string contains four complete waves and has a mass of 180 g. The string is vibrating sinusoidally with a frequency of 50.0 Hz and a peak-to-valley displacement of 15.0 cm. (The peak-to-valley distance is the vertical distance from the farthest positive position to the farthest negative position.) (a) Write the function that describes this wave traveling in the positive $x$ direction. (b) Determine the power being supplied to the string.

**21.** **PhysicsNow™** Sinusoidal waves 5.00 cm in amplitude are to be transmitted along a string that has a linear mass density of $4.00 \times 10^{-2}$ kg/m. If the source can deliver a maximum power of 300 W and the string is under a tension of 100 N, what is the highest frequency at which the source can operate?

**22.** A horizontal string can transmit a maximum power $\mathcal{P}_0$ (without breaking) if a wave with amplitude $A$ and angular frequency $\omega$ is traveling along it. To increase this maximum power, a student folds the string and uses this "double string" as a medium. Determine the maximum power that can be transmitted along the "double string," assuming that the tension is constant.

## Section 13.7 ▪ Sound Waves

*Note:* Use the following values as needed unless otherwise specified. The equilibrium density of air at 20°C is $\rho = 1.20$ kg/m$^3$. The speed of sound in air is $v = 343$ m/s at 20°C. Pressure variations $\Delta P$ are measured relative to atmospheric pressure, $1.013 \times 10^5$ N/m$^2$.

Problem 2.55 in Chapter 2 can also be assigned with this section.

**23.** Suppose you hear a clap of thunder 16.2 s after seeing the associated lightning stroke. The speed of sound waves in air is 343 m/s and the speed of light in air is $3.00 \times 10^8$ m/s. How far are you from the lightning stroke?

**24.** A dolphin in sea water at a temperature of 25°C emits sound directed toward the bottom of the ocean 150 m below. How much time passes before it hears an echo?

**25.** Many artists sing very high notes in *ad lib* ornaments and cadenzas. The highest note written for a singer in a published score was F-sharp above high C, 1.480 kHz, for Zerbinetta in the original version of Richard Strauss's opera *Ariadne auf Naxos.* (a) Find the wavelength of this sound in air. (b) In response to complaints, Strauss later transposed the note down to F above high C, 1.397 kHz. By what increment did the wavelength change?

**26.** A bat (Fig. P13.26) can detect very small objects, such as an insect whose length is approximately equal to one wavelength of the sound the bat makes. If a bat emits chirps at a frequency of 60.0 kHz and the speed of sound in air is 340 m/s, what is the smallest insect the bat can detect?

**27.** An ultrasonic tape measure uses frequencies above 20 MHz to determine dimensions of structures such as buildings. It does so by emitting a pulse of ultrasound into air and then measuring the time interval for an echo to return from a reflecting surface whose distance away is to be measured. The distance is displayed as a digital readout. For a tape measure that emits a pulse of ultrasound with a frequency of 22.0 MHz, (a) what is the distance to an object from which the echo pulse returns after 24.0 ms when the air temperature is 26°C? (b) What should be the duration of the emitted pulse if it is to include ten cycles of the ultrasonic wave? (c) What is the spatial length of such a pulse?

(Joe McDonald/Visuals Unlimited)

FIGURE **P13.26** Problems 13.26 and 13.59.

**28.** Ultrasound is used in medicine both for diagnostic imaging and for therapy. For diagnosis, short pulses of ultrasound are passed through the patient's body. An echo reflected from a structure of interest is recorded, and from the time interval for the return of the echo the distance to the structure can be determined. A single transducer emits and detects the ultrasound. An image of the structure is obtained by reducing the data with a computer. With sound of low intensity, this technique is noninvasive and harmless. It is used to examine fetuses, tumors, aneurysms, gallstones, hearts, and many other structures. To reveal detail, the wavelength of the reflected ultrasound must be small compared with the size of the object reflecting the wave. (a) What is the wavelength of ultrasound with a frequency of 2.40 MHz, used in echo cardiography to map the beating heart? (b) In the whole set of imaging techniques, frequencies in the range 1.00 to 20.0 MHz are used. What is the range of wavelengths corresponding to this range of frequencies? The speed of ultrasound in human tissue is about 1 500 m/s (nearly the same as the speed of sound in water).

**29.** **PhysicsNow™** An experimenter wishes to generate in air a sound wave that has a displacement amplitude of $5.50 \times 10^{-6}$ m. The pressure amplitude is to be limited to 0.840 N/m$^2$. What is the minimum wavelength the sound wave can have?

**30.** A sinusoidal sound wave is described by the displacement wave function

$$s(x, t) = (2.00\ \mu\text{m}) \cos[(15.7\ \text{m}^{-1})x - (858\ \text{s}^{-1})t]$$

(a) Find the amplitude, wavelength, and speed of this wave. (b) Determine the instantaneous displacement from equilibrium of the elements of the medium at the position $x = 0.050\ 0$ m at $t = 3.00$ ms. (c) Determine the maximum speed of the element's oscillatory motion.

**31.** Write an expression that describes the pressure variation as a function of position and time for a sinusoidal sound wave in air, taking $\lambda = 0.100$ m and $\Delta P_{max} = 0.200$ N/m$^2$.

**32.** Calculate the pressure amplitude of a 2.00-kHz sound wave in air, assuming that the displacement amplitude is equal to $2.00 \times 10^{-8}$ m.

## Section 13.8 ▪ The Doppler Effect

**33.** A driver travels northbound on a highway at a speed of 25.0 m/s. A police car, traveling southbound at a speed of 40.0 m/s, approaches with its siren producing sound at a frequency of 2 500 Hz. (a) What frequency does the driver observe as the police car approaches? (b) What frequency does the driver detect after the police car passes him? (c) Repeat parts (a) and (b) for the case when the police car is traveling northbound.

**34.** Expectant parents are thrilled to hear their unborn baby's heartbeat, revealed by an ultrasonic motion detector. Suppose the fetus's ventricular wall moves in simple harmonic motion with an amplitude of 1.80 mm and a frequency of 115 per minute. (a) Find the maximum linear speed of the heart wall. Suppose the motion detector in contact with the mother's abdomen produces sound at 2 000 000.0 Hz that travels through tissue at 1.50 km/s. (b) Find the maximum frequency at which sound arrives at the wall of the baby's heart. (c) Find the maximum frequency at which reflected sound is received by the motion detector. By electronically "listening" for echoes at a frequency different from the broadcast frequency, the motion detector can produce beeps of audible sound in synchronization with the fetal heartbeat.

**35.** **PhysicsNow™** Standing at a crosswalk, you hear a frequency of 560 Hz from the siren of an approaching ambulance. After the ambulance passes, the observed frequency of the siren is 480 Hz. Determine the ambulance's speed from these observations.

**36.** A block with a speaker bolted to it is connected to a spring having spring constant $k = 20.0$ N/m as shown in Figure P13.36. The total mass of the block and speaker is 5.00 kg, and the amplitude of this unit's motion is 0.500 m. The speaker emits sound waves of frequency 440 Hz. Determine the highest and lowest frequencies heard by the person to the right of the speaker. Assume that the speed of sound is 343 m/s.

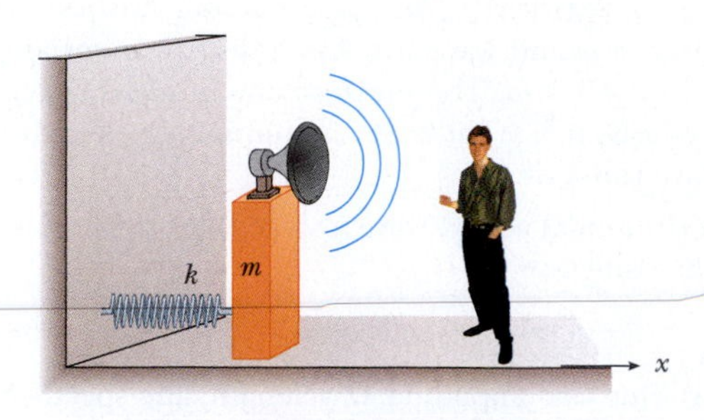


FIGURE P13.36

**37.** A tuning fork vibrating at 512 Hz falls from rest and accelerates at 9.80 m/s$^2$. How far below the point of release is the tuning fork when waves of frequency 485 Hz reach the release point? Take the speed of sound in air to be 340 m/s.

**38.** At the Winter Olympics, an athlete rides her luge down the track while a bell just above the wall of the chute rings continuously. When her sled passes the bell, she hears the frequency of the bell fall by the musical interval called a minor third. That is, the frequency she hears drops to five sixths of its original value. (a) Find the speed of sound in air at the ambient temperature −10.0°C. (b) Find the speed of the athlete.

**39.** A siren mounted on the roof of a firehouse emits sound at a frequency of 900 Hz. A steady wind is blowing with a speed of 15.0 m/s. Taking the speed of sound in calm air to be 343 m/s, find the wavelength of the sound (a) upwind of the siren and (b) downwind of the siren. Firefighters are approaching the siren from various directions at 15.0 m/s. What frequency does a firefighter hear (c) if she is approaching from an upwind position so that she is moving in the direction in which the wind is blowing and (d) if she is approaching from a downwind position and moving against the wind?

## Section 13.9 ▪ Context Connection—Seismic Waves

**40.** Two points *A* and *B* on the surface of the Earth are at the same longitude and 60.0° apart in latitude. Suppose an earthquake at point *A* creates a *P* wave that reaches point *B* by traveling straight through the body of the Earth at a constant speed of 7.80 km/s. The earthquake also radiates a Rayleigh wave, which travels along the surface of the Earth at 4.50 km/s. (a) Which of these two seismic waves arrives at *B* first? (b) What is the time difference between the arrivals of the two waves at *B*? Take the radius of the Earth to be 6 370 km.

**41.** A seismographic station receives *S* and *P* waves from an earthquake, 17.3 s apart. Assume that the waves have traveled over the same path at speeds of 4.50 km/s and 7.80 km/s. Find the distance from the seismograph to the hypocenter of the quake.

## Additional Problems

**42.** "The wave" is a particular type of pulse that can propagate through a large crowd gathered at a sports arena (Fig. P13.42). The elements of the medium are the spectators, with zero position corresponding to their being seated and maximum position corresponding to their standing and raising their arms. When a large fraction of the spectators participate in the wave motion, a somewhat stable pulse shape can develop. The wave speed depends on people's reaction time, which is typically on the order of 0.1 s. Estimate the order of magnitude, in minutes, of the time interval required for such a pulse to make one circuit around a large sports stadium. State the quantities you measure or estimate and their values.

**43.** **Review problem.** A block of mass *M*, supported by a string, rests on a frictionless incline making an angle $\theta$ with the horizontal (Fig. P13.43). The length of the string is *L* and its mass is $m \ll M$. Derive an expression for the time interval required for a transverse wave to travel from one end of the string to the other.

FIGURE **P13.42**

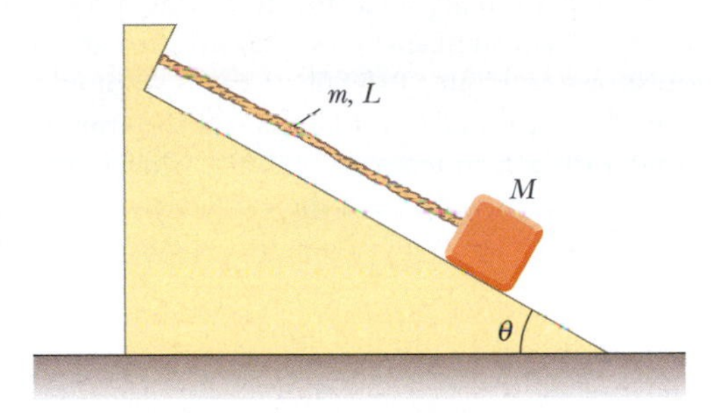

FIGURE **P13.43**

44. **Review problem.** A block of mass $M$ hangs from a rubber cord. The block is supported so that the cord is not stretched. The unstretched length of the cord is $L_0$ and its mass is $m$, much less than $M$. The "spring constant" for the cord is $k$. The block is released and stops at the lowest point. (a) Determine the tension in the cord when the block is at this lowest point. (b) What is the length of the cord in this "stretched" position? (c) Find the speed of a transverse wave in the cord, assuming that the block is held in this lowest position.

45. **Review problem.** A block of mass 0.450 kg is attached to one end of a cord of mass 0.003 20 kg; the other end of the cord is attached to a fixed point. The block rotates with constant angular speed in a circle on a horizontal, frictionless table. Through what angle does the block rotate in the time interval required for a transverse wave to travel along the string from the center of the circle to the block?

46. (a) Show that the speed of longitudinal waves along a spring of force constant $k$ is $v = \sqrt{kL/\mu}$, where $L$ is the unstretched length of the spring and $\mu$ is the mass per unit length. (b) A spring with a mass of 0.400 kg has an unstretched length of 2.00 m and a force constant of 100 N/m. Using the result you obtained in part (a), determine the speed of longitudinal waves along this spring.

47. A rope of total mass $m$ and length $L$ is suspended vertically. Show that a transverse pulse travels the length of the rope in a time interval $\Delta t = 2\sqrt{L/g}$. (*Suggestion:* First find an expression for the wave speed at any point a distance $x$ from the lower end by considering the tension in the rope as resulting from the weight of the segment below that point.)

48. Assume that an object of mass $M$ is suspended from the bottom of the rope in Problem 13.47. (a) Show that the time interval for a transverse pulse to travel the length of the rope is

$$\Delta t = 2\sqrt{\frac{L}{mg}}\left(\sqrt{M+m} - \sqrt{M}\right)$$

(b) Show that this expression reduces to the result of Problem 13.47 when $M = 0$. (c) Show that for $m \ll M$, the expression in part (a) reduces to

$$\Delta t = \sqrt{\frac{mL}{Mg}}$$

49. A pulse traveling along a string of linear mass density $\mu$ is described by the wave function

$$y = [A_0e^{-bx}]\sin(kx - \omega t)$$

where the factor in brackets is said to be the amplitude. (a) What is the power $\mathcal{P}(x)$ carried by this wave at a point $x$? (b) What is the power carried by this wave at the origin? (c) Compute the ratio $\mathcal{P}(x)/\mathcal{P}(0)$.

50. An earthquake on the ocean floor in the Gulf of Alaska produces a *tsunami* (sometimes incorrectly called a "tidal wave") that reaches Hilo, Hawaii, 4 450 km away, in a time interval of 9 h 30 min. Tsunamis have enormous wavelengths (100 to 200 km), and the propagation speed for these waves is $v \approx \sqrt{g\bar{d}}$, where $\bar{d}$ is the average depth of the water. From the information given, find the average wave speed and the average ocean depth between Alaska and Hawaii. (This method was used in 1856 to estimate the average depth of the Pacific Ocean long before soundings were made to give a direct determination.)

51. A string on a musical instrument is held under tension $T$ and extends from the point $x = 0$ to the point $x = L$. The string is overwound with wire in such a way that its mass per unit length $\mu(x)$ increases uniformly from $\mu_0$ at $x = 0$ to $\mu_L$ at $x = L$. (a) Find an expression for $\mu(x)$ as a function of $x$ over the range $0 \le x \le L$. (b) Show that the time interval required for a transverse pulse to travel the length of the string is given by

$$\Delta t = \frac{2L(\mu_L + \mu_0 + \sqrt{\mu_L\mu_0})}{3\sqrt{T}(\sqrt{\mu_L} + \sqrt{\mu_0})}$$

52. A flowerpot is knocked off a balcony 20.0 m above the sidewalk and falls toward an unsuspecting 1.75-m-tall man who is standing below. How close to the sidewalk can the

flowerpot fall before it is too late for a warning shouted from the balcony to reach the man in time? Assume that the man below requires 0.300 s to respond to the warning.

**53.** A sound wave in a cylinder is described by Equations 13.24 through 13.26. Show that $\Delta P = \pm \rho v \omega \sqrt{s_{\max}^2 - s^2}$.

**54.** On a Saturday morning, pickup trucks and sport utility vehicles carrying garbage to the town landfill form a nearly steady procession on a country road, all traveling at 19.7 m/s. From one direction, two trucks arrive at the dump every 3 min. A bicyclist is also traveling toward the landfill, at 4.47 m/s. (a) With what frequency do the trucks pass him? (b) A hill does not slow down the trucks, but makes the out-of-shape cyclist's speed drop to 1.56 m/s. How often do noisy, smelly, inefficient, garbage-dripping, roadhogging trucks whiz past him now?

**55.** The ocean floor is underlain by a layer of basalt that constitutes the crust, or uppermost layer, of the Earth in that region. Below this crust is found denser periodotite rock that forms the Earth's mantle. The boundary between these two layers is called the Mohorovicic discontinuity ("Moho" for short). If an explosive charge is set off at the surface of the basalt, it generates a seismic wave that is reflected back out at the Moho. If the speed of this wave in basalt is 6.50 km/s and the two-way travel time is 1.85 s, what is the thickness of this oceanic crust?

**56.** A train whistle ($f$ = 400 Hz) sounds higher or lower in frequency depending on whether it approaches or recedes. (a) Prove that the difference in frequency between the approaching and receding train whistle is

$$\Delta f = \frac{2u/v}{1 - u^2/v^2} f$$

where $u$ is the speed of the train and $v$ is the speed of sound. (b) Calculate this difference for a train moving at a speed of 130 km/h. Take the speed of sound in air to be 340 m/s.

**57.** To permit measurement of her speed, a sky diver carries a buzzer emitting a steady tone at 1 800 Hz. A friend on the ground at the landing site directly below listens to the amplified sound he receives. Assume that the air is calm and that the sound speed is 343 m/s, independent of altitude. While the sky diver is falling at terminal speed, her friend on the ground receives waves of frequency 2 150 Hz. (a) What is the sky diver's speed of descent? (b) Suppose the sky diver can hear the sound of the buzzer reflected from the ground. What frequency does she receive?

**58.** A police car is traveling east at 40.0 m/s along a straight road, overtaking a car ahead of it moving east at 30.0 m/s. The police car has a malfunctioning siren that is stuck at 1 000 Hz. (a) Sketch the appearance of the wave fronts of the sound produced by the siren. Show the wave fronts both to the east and to the west of the police car. (b) What would be the wavelength in air of the siren sound if the police car were at rest? (c) What is the wavelength in front of the car? (d) What is it behind the police car? (e) What is the frequency heard by the driver being chased?

**59.** A bat, moving at 5.00 m/s, is chasing a flying insect (Fig. P13.26). If the bat emits a 40.0-kHz chirp and receives back an echo at 40.4 kHz, at what speed is the insect moving toward or away from the bat? Take the speed of sound in air to be $v$ = 340 m/s.

**60.** The Doppler Equation 13.30 is valid when the motion between the observer and the source occurs on a straight line so that the source and observer are moving either directly toward or directly away from each other. If this restriction is relaxed, one must use the more general Doppler equation

$$f' = \left(\frac{v + v_O \cos \theta_O}{v - v_S \cos \theta_S}\right) f$$

where $\theta_O$ and $\theta_S$ are defined in Figure P13.60a. (a) Show that if the observer and source are moving away from each other, the preceding equation reduces to Equation 13.30 with negative values for both $v_O$ and $v_S$. (b) Use the preceding equation to solve the following problem. A train moves at a constant speed of 25.0 m/s toward the intersection shown in Figure P13.60b. A car is stopped near the intersection, 30.0 m from the tracks. If the train's horn emits sound with a frequency of 500 Hz, what is the frequency heard by the passengers in the car when the train is 40.0 m from the intersection? Take the speed of sound to be 343 m/s.

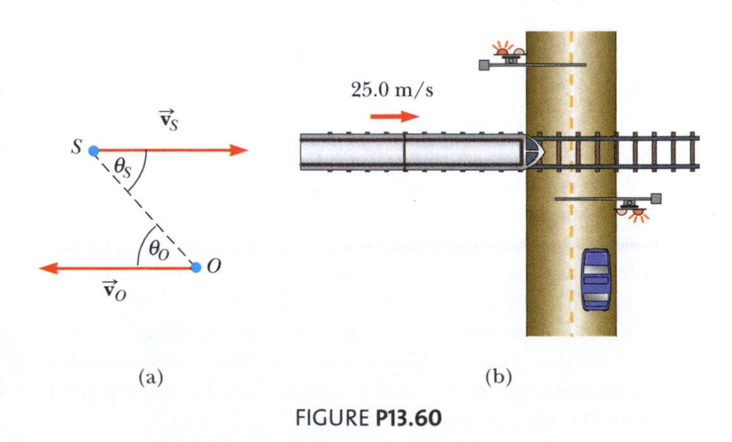


FIGURE P13.60

## ANSWERS TO QUICK QUIZZES

**13.1** (b), (c). The movement of the people in the line is parallel to the direction of propagation of the gap. The fans participating in the "wave" stand up vertically as the wave sweeps past them horizontally.

**13.2** (i), (c). The wave speed is determined by the medium, so it is unaffected by changing the frequency. (ii), (b). Because the wave speed remains the same, the result of doubling the frequency is that the wavelength is half as large. (iii), (d). The amplitude of a wave is unrelated to the frequency, so we cannot determine the new amplitude without further information.

**13.3** Only choices (f) and (h) are correct. Choices (a) and (b) affect the transverse speed of an element of the string but not the wave speed along the string. Choices (c) and (d) change the amplitude. Choices (e) and (g) increase the time interval by decreasing the wave speed.

**13.4** (d). Doubling the amplitude of the wave causes the power to be larger by a factor of 4. In (a), halving the linear mass density of the string causes the power to change by a factor of 0.71; the rate decreases. In (b), doubling the wavelength of the wave halves the frequency and causes the power to change by a factor of 0.25; the rate decreases. In (c), doubling the tension in the string changes the wave speed and causes the power to change by a factor of 1.4, which is not as large as in part (d).

**13.5** (e). The wave speed cannot be changed by moving the source, so (a) and (b) are incorrect. The detected wavelength is largest at A, so (c) and (d) are incorrect. Choice (f) is incorrect because the detected frequency is lowest at location A. Choice (e) is correct because at location C the wavelength is the smallest, so the frequency must be the largest.

**13.6** (e). The intensity of the sound increases because the train is moving closer to you. Because the train moves at a constant velocity, the Doppler-shifted frequency remains fixed.

CHAPTER 14

# Superposition and Standing Waves

The rich sound of a piano is due to standing waves on strings under tension. Many such strings can be seen in this photograph. Waves also travel on the soundboard, which is visible below the strings.

(Kathy Ferguson Johnson/PhotoEdit/PictureQuest)

## CHAPTER OUTLINE


14.1 The Principle of Superposition
14.2 Interference of Waves
14.3 Standing Waves
14.4 Standing Waves in Strings
14.5 Standing Waves in Air Columns
14.6 Beats: Interference in Time
14.7 Nonsinusoidal Wave Patterns
14.8 Context Connection—Building on Antinodes
SUMMARY


In Chapter 13, we introduced the wave model. We have seen that waves are very different from particles. An ideal particle is of zero size, but an ideal wave is of infinite length. Another important difference between waves and particles is that we can explore the possibility of two or more waves combining at one point in the same medium. We can combine particles to form extended objects, but the particles must be at different locations. In contrast, two waves can both be present at a given location, and the ramifications of this possibility are explored in this chapter.

One ramification of the combination of waves is that only certain allowed frequencies can exist on systems with boundary conditions; that is, the frequencies are *quantized.* In Chapter 11, we learned about quantized energies of the hydrogen atom. Quantization is at the heart of quantum mechanics, a subject that is introduced formally in Chapter 28. We shall see that waves under boundary conditions explain many of the quantum phenomena. For our present purposes in this chapter, quantization enables us to understand the behavior of the wide array of musical instruments that are based on strings and air columns.

## 14.1 THE PRINCIPLE OF SUPERPOSITION

Many interesting wave phenomena in nature cannot be described by a single wave. Instead, one must analyze complex waveforms in terms of a combination of traveling waves. To analyze such wave combinations, we make use of the **principle of superposition:**

> If two or more traveling waves are moving through a medium and combine at a given point, the resultant position of the element of the medium at that point is the sum of the positions due to the individual waves.

■ Principle of superposition

This rather striking property is exhibited by many waves in nature, including waves on strings, sound waves, and surface water waves. It is also exhibited by electromagnetic waves, for which the electric fields of the combined waves are added. Waves that obey this principle are called *linear waves.* In general, linear waves have an amplitude that is small relative to their wavelength. Waves that violate the superposition principle are called *nonlinear waves* and, as mentioned in Chapter 13, are often characterized by large amplitudes. In this book, we shall deal only with linear waves.

A simple pictorial representation of the superposition principle is obtained by considering two pulses traveling in opposite directions on a stretched string as in Active Figure 14.1. The wave function for the pulse moving to the right is $y_1$, and the wave function for the pulse moving to the left is $y_2$. The pulses have the same speed but different shapes. Each pulse is assumed to be symmetric (although that is not a necessary condition), and in both cases displacements of the elements of the string in the vertical direction are taken to be positive. When the waves overlap, the resulting waveform is given by $y_1 + y_2$. After the time interval during which the pulses combine, they separate and continue moving in their original directions (Active Fig. 14.1d). Note that the final waveforms remain unchanged as if the two

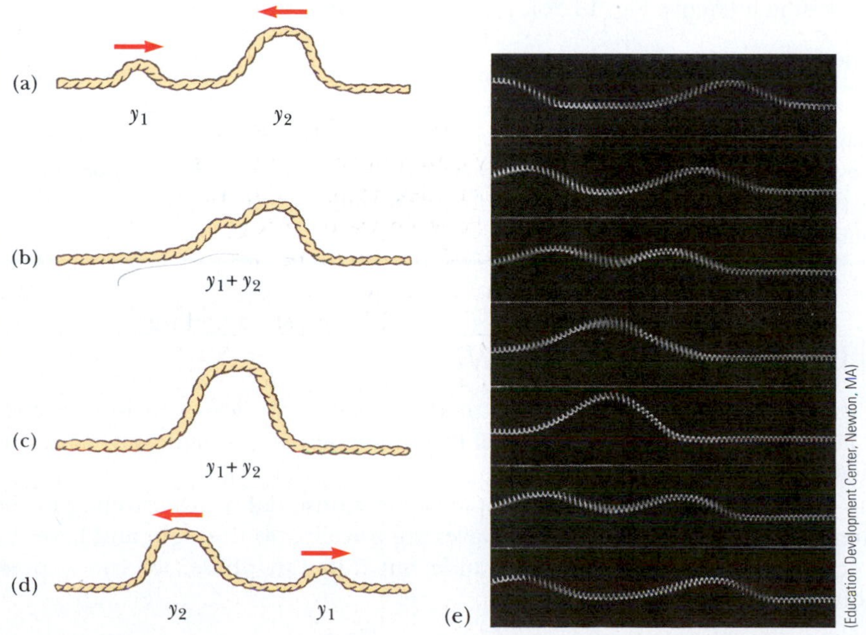

(a) $y_1$ $y_2$

(b) $y_1 + y_2$

(c) $y_1 + y_2$

(d) $y_2$ $y_1$

(e)

(Education Development Center, Newton, MA)

**ACTIVE FIGURE 14.1** (*Left*) Two pulses traveling on a stretched string in opposite directions pass through each other. When the pulses overlap, as in (b) and (c), the net displacement of each element of the string equals the sum of the displacements produced by each pulse. Because each pulse produces positive displacements of the string, we refer to their superposition as *constructive interference.* (*Right*) Photograph of the superposition of two equal and symmetric pulses traveling in opposite directions on a stretched spring.

**Physics Now™** By logging into PhysicsNow at **www.pop4e.com** and going to Active Figure 14.1 you can choose the amplitude and orientation of each of the pulses and observe the interference as they pass each other.

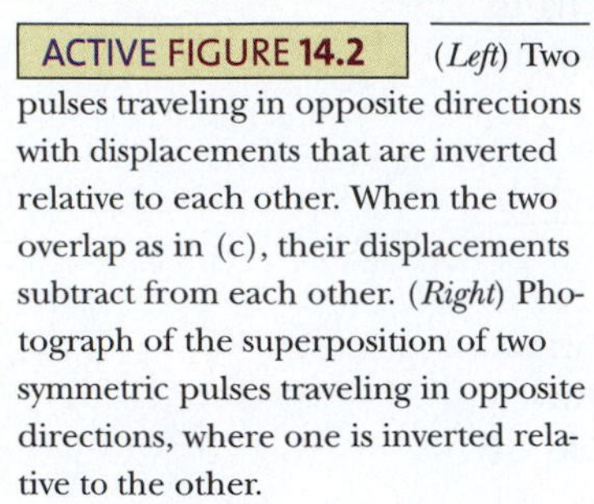

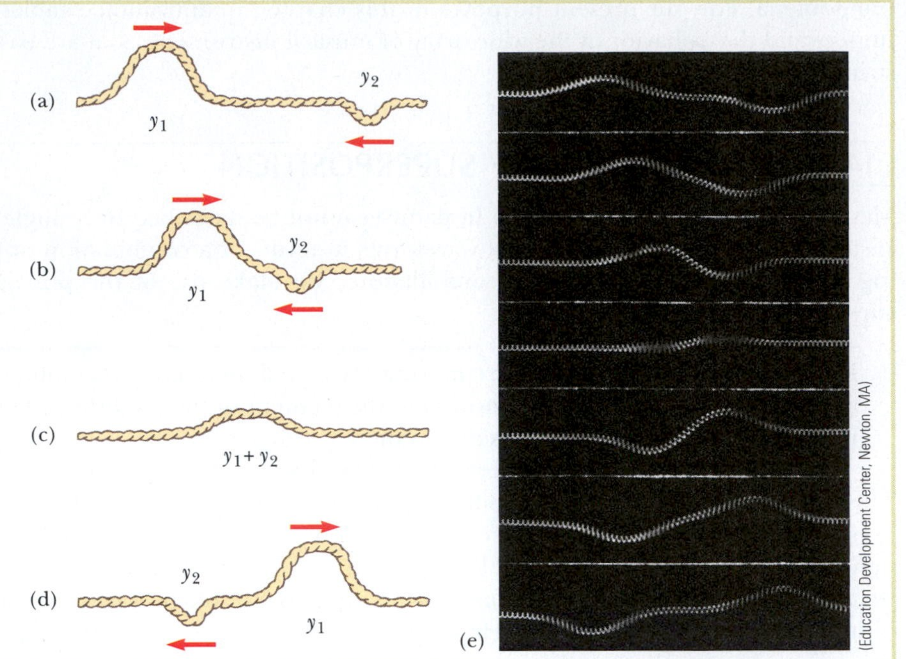


**ACTIVE FIGURE 14.2** (*Left*) Two pulses traveling in opposite directions with displacements that are inverted relative to each other. When the two overlap as in (c), their displacements subtract from each other. (*Right*) Photograph of the superposition of two symmetric pulses traveling in opposite directions, where one is inverted relative to the other.

**PhysicsNow™** By logging into PhysicsNow at **www.pop4e.com** and going to Active Figure 14.2 you can choose the amplitude and orientation of each of the pulses and observe the interference as they pass each other.

pulses had never met! The combination of separate waves in the same region of space to produce a resultant wave is called **interference.** Notice that the interference exists only while the waves are in the same region of space, and there is no permanent effect on the pulses after they separate.

For the two pulses shown in Active Figure 14.1, the vertical displacements are in the same direction and so the resultant waveform (when the pulses overlap) exhibits an amplitude greater than those of the individual pulses. Now consider two identical pulses, again traveling in opposite directions on a stretched string, but this time one pulse is inverted relative to the other as in Active Figure 14.2. In this case, when the pulses begin to overlap, the resultant waveform is the sum of the two separate waveforms again, but one of the displacements is negative. Again, the two pulses pass through each other. When they exactly overlap, they partially *cancel* each other. At this time (Active Fig. 14.2c), the resultant amplitude is small.

**PITFALL PREVENTION 14.1**

**DO WAVES REALLY *INTERFERE*?** In popular usage, the term *interfere* implies that an agent affects a situation in some way so as to preclude something from happening. For example, in American football, *pass interference* means that a defending player has affected the receiver so that he is unable to catch the ball. This usage is very different from that in physics, in which waves pass through one another and interfere but do not affect one another in any way. In physics, we will consider interference to be similar to the notion of *combination,* as described in this chapter, as opposed to the popular usage.

**QUICK QUIZ 14.1** Two pulses move in opposite directions on a string and are identical in shape except that one has positive displacements of the elements of the string and the other has negative displacements. What happens at the moment that the two pulses completely overlap on the string? **(a)** The energy associated with the pulses has disappeared. **(b)** The string is not moving. **(c)** The string forms a straight line. **(d)** The pulses have vanished and will not reappear.

## 14.2 INTERFERENCE OF WAVES

In this section, we shall investigate the mathematics of the waves in interference analysis model. Additional applications of this model applied to light waves are presented in Chapter 27.

Let us apply the superposition principle to two sinusoidal waves traveling in the same direction in a medium. If the two waves are traveling to the right and have the same frequency, wavelength, and amplitude but differ in phase, we can express their individual wave functions as

$$y_1 = A\sin(kx - \omega t) \qquad \text{and} \qquad y_2 = A\sin(kx - \omega t + \phi)$$

where $\phi$ is the phase difference between the two waves. Let us imagine that these two waves coincide in the medium. For example, these expressions might represent two waves traveling along the same string. In this situation, the resultant wave function $y$ is, according to the principle of superposition,

$$y = y_1 + y_2 = A[\sin(kx - \omega t) + \sin(kx - \omega t + \phi)]$$

To simplify this expression, it is convenient to use the trigonometric identity

$$\sin a + \sin b = 2 \cos\left(\frac{a-b}{2}\right) \sin\left(\frac{a+b}{2}\right)$$

If we let $a = kx - \omega t$ and $b = kx - \omega t + \phi$, the resultant wave function $y$ reduces to

$$y = \left(2A \cos\frac{\phi}{2}\right) \sin\left(kx - \omega t + \frac{\phi}{2}\right) \qquad [14.1]$$

This mathematical representation of the resultant wave has several important features. The resultant wave function $y$ is also a sinusoidal wave and has the *same* frequency and wavelength as the individual waves. The amplitude of the resultant wave is $2A \cos(\phi/2)$ and the phase angle is $\phi/2$. If the phase angle $\phi$ equals 0, $\cos(\phi/2) = \cos 0 = 1$ and the amplitude of the resultant wave is $2A$. In other words, the amplitude of the resultant wave is twice the amplitude of either individual wave. In this case, the waves are said to be everywhere *in phase* ($\phi = 0$) and to **interfere constructively.** That is, the crests of the individual waves occur at the same positions, as is shown by the blue line in Active Figure 14.3a. In general, constructive interference occurs when $\cos(\phi/2) = \pm 1$ or when $\phi = 0, 2\pi, 4\pi, \ldots$ .

■ Constructive interference

On the other hand, if $\phi$ is equal to $\pi$ radians or to any *odd* multiple of $\pi$, $\cos(\phi/2) = \cos(\pi/2) = 0$ and the resultant wave has *zero* amplitude everywhere. In this case, the two waves **interfere destructively.** That is, the crest of one wave coincides with the trough of the second (Active Fig. 14.3b) and their displacements cancel at every point. Finally, when the phase constant has a value between 0 and $\pi$, as in Active Figure 14.3c, the resultant wave has an amplitude whose value is somewhere between 0 and $2A$.

■ Destructive interference

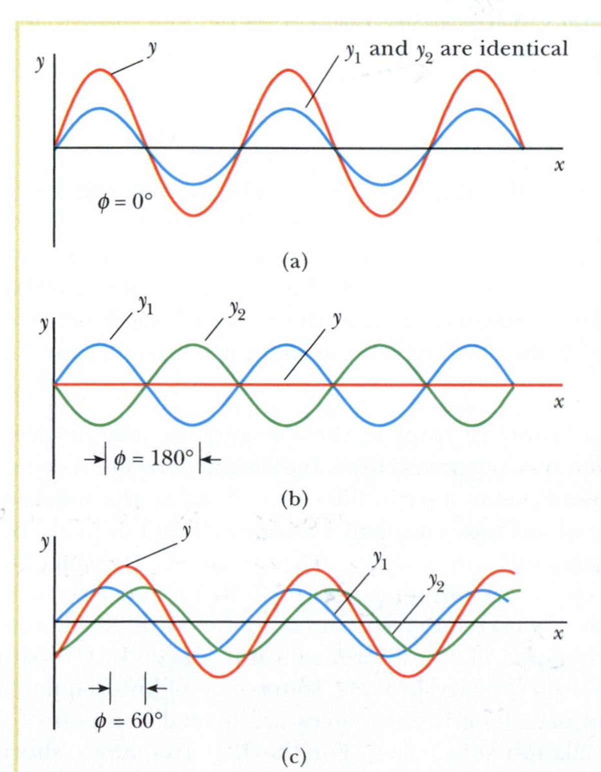


ACTIVE FIGURE 14.3
The superposition of two identical waves $y_1$ and $y_2$. (a) When the two waves are in phase, the result is constructive interference. (b) When the two waves are $\pi$ rad out of phase, the result is destructive interference. (c) When the phase angle has a value other than 0 or $\pi$ rad, the resultant wave $y$ falls somewhere between the extremes shown in (a) and (b).

**PhysicsNow™** Log into PhysicsNow at **www.pop4e.com** and go to Active Figure 14.3 to change the phase relationship between the waves and observe the wave representing the superposition.

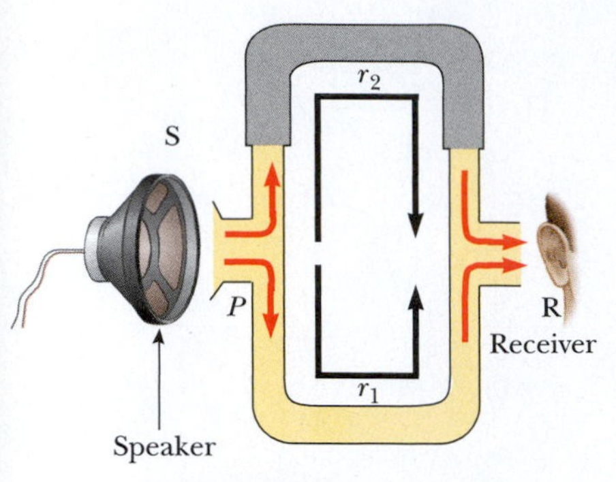


**FIGURE 14.4** An acoustical system for demonstrating interference of sound waves. Sound waves from the speaker propagate into the tube and the energy splits into two parts at point $P$. The waves from the two paths, which combine at the opposite side, are detected at the receiver $R$. The upper path length $r_2$ can be varied by sliding the upper section.

Although we have used waves having the same amplitude in the preceding discussion, waves of differing amplitudes will interfere in a similar way. If the waves are in phase, the combined amplitude is the sum of the individual amplitudes. If they are 180° out of phase, the combined amplitude is the difference between the individual amplitudes.

Figure 14.4 shows a simple device for demonstrating interference of sound waves. Sound from speaker S is sent into a tube at $P$, where there is a T-shaped junction. Half the sound energy travels in one direction and half in the opposite direction. Therefore, the sound waves that reach receiver R at the other side can travel along either of two paths. The total distance from speaker to receiver is called the **path length** $r$. The length of the lower path is fixed at $r_1$. The upper path length $r_2$ can be varied by sliding the U-shaped tube (similar to that on a slide trombone). When the difference in the path lengths $\Delta r = |r_2 - r_1|$ is either zero or some integral multiple of the wavelength $\lambda$, the two waves reaching the receiver are in phase and interfere constructively as in Active Figure 14.3a. In this case, a maximum in the sound intensity is detected at the receiver. If path length $r_2$ is adjusted so that $\Delta r$ is $\lambda/2, 3\lambda/2, \ldots, n\lambda/2$ (for $n$ odd), the two waves are exactly 180° out of phase at the receiver and hence cancel each other. In this case of completely destructive interference, no sound is detected at the receiver. This simple experiment is a striking illustration of interference. In addition, it demonstrates that a phase difference may arise between two waves generated by the same source when they travel along paths of unequal lengths.

It is often useful to express a path difference in terms of the phase angle $\phi$ between the two waves. Because a path difference of one wavelength corresponds to a phase angle of $2\pi$ rad, we obtain the ratio $\phi/2\pi = \Delta r/\lambda$ or

▪ Relationship between path difference and phase angle

$$\Delta r = \frac{\phi}{2\pi}\lambda \qquad [14.2]$$

Therefore, for example, a phase difference of 180° or $\pi$ rad corresponds to a shift of $\lambda/2$. Conversely, a one-quarter-wavelength shift corresponds to a 90° phase difference.

Nature provides many examples of interference phenomena. Later in the text we shall describe several interesting interference effects involving light waves.

## ▪ Thinking Physics 14.1

If stereo speakers are connected to the amplifier "out of phase," one speaker is moving outward when the other is moving inward. The result is a weakness in the bass notes, which can be corrected by reversing the wires on one of the speaker connections. Why are only the bass notes affected in this case and not the treble notes? For help in answering this question, note that the range of wavelengths of sound from a standard piano is from 0.082 m for the highest C to 13 m for the lowest A.

**Reasoning** Imagine that you are sitting in front of the speakers, midway between them. Then, the sound from each speaker travels the same distance to you, so there is no phase difference in the sound due to a path difference. Because the speakers are connected out of phase, the sound waves are half a wavelength out of phase on leaving the speaker and, consequently, on arriving at your ear. As a result, the sound for all frequencies cancels in the simplification model of a zero-size head located exactly on the midpoint between the speakers. If the ideal head were moved off the centerline, an additional phase difference is introduced by the path length difference for the sound from the two speakers. In the case of low-frequency, long-wavelength bass notes, the path length differences are a small fraction of a wavelength, so significant cancellation still occurs. For the high-frequency, short-wavelength treble notes, a small movement of the ideal head results in a much larger fraction of a wavelength in path length difference or even multiple wavelengths. Therefore, the treble notes could be in phase with this head movement. If we now add that the head is not of zero size and that it has two ears, we can see that complete cancellation is not possible and, with even small movements of the head, one or both ears will be at or near maxima for the treble notes. The size of the head is much smaller than bass wavelengths, however, so the bass notes are significantly weakened over much of the region in front of the speakers. ■

### INTERACTIVE EXAMPLE 14.1 Two Speakers Driven by the Same Source

Two speakers placed 3.00 m apart are driven in phase by the same oscillator (Fig. 14.5). A listener is originally at point *O*, which is located 8.00 m from the center of the line connecting the two speakers. The listener then moves to point *P*, which is a perpendicular distance 0.350 m from *O*, at which the first cancellation of waves occurs, resulting in a minimum in sound intensity. What is the frequency of the oscillator?

**Solution** The first cancellation occurs when the two waves reaching the listener at *P* are 180° out of phase or, in other words, when their path difference equals $\lambda/2$. To calculate the path difference, we must first find the path lengths $r_1$ and $r_2$. Consider the two geometric model triangles shaded in Figure 14.5. Making use of these triangles, we find the path lengths to be

$$r_1 = \sqrt{(8.00\text{ m})^2 + (1.15\text{ m})^2} = 8.08\text{ m}$$

$$r_2 = \sqrt{(8.00\text{ m})^2 + (1.85\text{ m})^2} = 8.21\text{ m}$$

Hence, the path difference is $r_2 - r_1 = 0.13$ m. Because we require that this path difference be equal to $\lambda/2$ for the first minimum, we find that $\lambda = 0.26$ m.

To obtain the oscillator frequency, we use $v = \lambda f$, where $v$ is the speed of sound in air, 343 m/s:

$$f = \frac{v}{\lambda} = \frac{343\text{ m/s}}{0.26\text{ m}} = 1.3\text{ kHz}$$

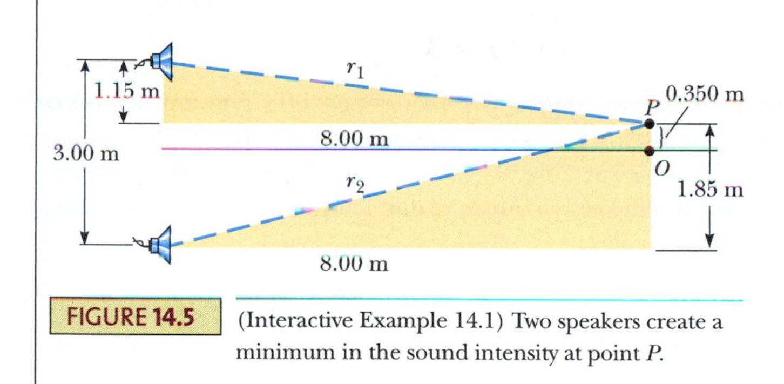


FIGURE 14.5 (Interactive Example 14.1) Two speakers create a minimum in the sound intensity at point *P*.

**PhysicsNow™** By logging into PhysicsNow at www.pop4e.com and going to Interactive Example 14.1 you can vary the point *P* at which the first minimum occurs to determine the frequency of the sound waves.

## 14.3 STANDING WAVES

The sound waves from the speakers in Interactive Example 14.1 leave the speakers in the forward direction, and we considered interference at a point in front of the speakers. Suppose we turn the speakers so that they face each other as in Figure 14.6 and then have them emit sound of the same frequency and amplitude. In this situation, two identical waves travel in opposite directions in the same medium. These waves combine in accordance with the superposition principle.

We can analyze such a situation by considering wave functions for two transverse sinusoidal waves having the same amplitude, frequency, and wavelength but traveling in opposite directions in the same medium:

$$y_1 = A\sin(kx - \omega t) \quad \text{and} \quad y_2 = A\sin(kx + \omega t)$$

where $y_1$ represents a wave traveling in the $+x$ direction and $y_2$ represents a wave traveling in the $-x$ direction. According to the principle of superposition, adding these two functions gives the resultant wave function $y$:

$$y = y_1 + y_2 = A\sin(kx - \omega t) + A\sin(kx + \omega t)$$

FIGURE 14.6 Two speakers emit sound waves toward each other. Between the speakers, identical waves traveling in opposite directions combine to form standing waves.

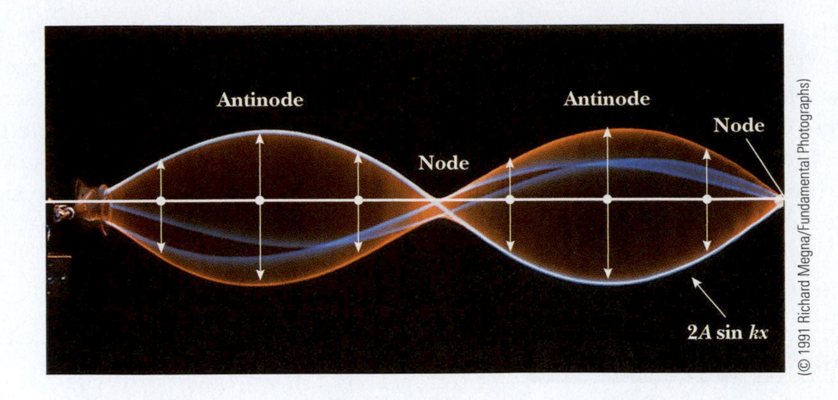


**FIGURE 14.7** Multiflash photograph of a standing wave on a string. The vertical displacement from equilibrium of an individual element of the string is proportional to $\cos \omega t$. That is, each element vibrates at an angular frequency $\omega$. The amplitude of the vertical oscillation of any element on the string depends on the horizontal position of the element. Each element vibrates within the confines of the envelope function $2A \sin kx$.

## PITFALL PREVENTION 14.2

**THREE TYPES OF AMPLITUDE** We need to distinguish carefully here between the **amplitude of the individual waves,** which is $A$, and the **amplitude of the simple harmonic motion of the elements of the medium,** which is $2A \sin kx$. A given element in a standing wave vibrates within the constraints of the *envelope* function $2A \sin kx$, where $x$ is that element's position in the medium. That vibration is in contrast to traveling sinusoidal waves, in which all elements oscillate with the same amplitude and the same frequency and the amplitude $A$ of the wave is the same as the amplitude $A$ of the simple harmonic motion of the elements. Furthermore, we can identify the **amplitude of the standing wave** as $2A$.

Using the trigonometric identity $\sin(a \pm b) = \sin a \cos b \pm \cos a \sin b$, this expression reduces to

$$y = (2A \sin kx) \cos \omega t \qquad [14.3]$$

Notice that this function does not look mathematically like a traveling wave because there is no function of $kx - \omega t$. Equation 14.3 represents the wave function of a **standing wave** such as that shown in Figure 14.7. A standing wave is an oscillation pattern that results from two waves traveling in opposite directions. Mathematically, this equation looks more like simple harmonic motion than wave motion for traveling waves. Every element of the medium vibrates in simple harmonic motion with the same angular frequency $\omega$ (according to the factor $\cos \omega t$). The amplitude of motion of a given element (the factor $2A \sin kx$), however, depends on its position along the medium, described by the variable $x$. From this result, we see that the simple harmonic motion of every element has an angular frequency of $\omega$ and a position-dependent amplitude of $2A \sin kx$.

Because the amplitude of the simple harmonic motion of an element at any value of $x$ is equal to $2A \sin kx$, we see that the *maximum* amplitude of the simple harmonic motion has the value $2A$. This maximum amplitude is described as the amplitude of the standing wave. It occurs when the coordinate $x$ for an element satisfies the condition $\sin kx = 1$, or when

$$kx = \frac{\pi}{2}, \frac{3\pi}{2}, \frac{5\pi}{2}, \ldots$$

Because $k = 2\pi/\lambda$, the positions of maximum amplitude, called **antinodes,** are

▮ Positions of antinodes

$$x = \frac{\lambda}{4}, \frac{3\lambda}{4}, \frac{5\lambda}{4}, \ldots = \frac{n\lambda}{4} \qquad [14.4]$$

where $n = 1, 3, 5, \ldots$. Note that **adjacent antinodes are separated by a distance $\lambda/2$.**

Similarly, the simple harmonic motion has a *minimum* amplitude of zero when $x$ satisfies the condition $\sin kx = 0$, or when $kx = \pi, 2\pi, 3\pi, \ldots$, giving

▮ Positions of nodes

$$x = \frac{\lambda}{2}, \lambda, \frac{3\lambda}{2}, \ldots = \frac{n\lambda}{2} \qquad [14.5]$$

where $n = 1, 2, 3, \ldots$. **These points of zero amplitude, called nodes, are also spaced by $\lambda/2$.** The distance between a node and an adjacent antinode is $\lambda/4$. The standing wave patterns produced at various times by two waves traveling in opposite directions are represented graphically in Active Figure 14.8. The upper part of each figure represents the individual traveling waves and the lower part represents the standing wave patterns. The nodes of the standing wave are labeled N and the

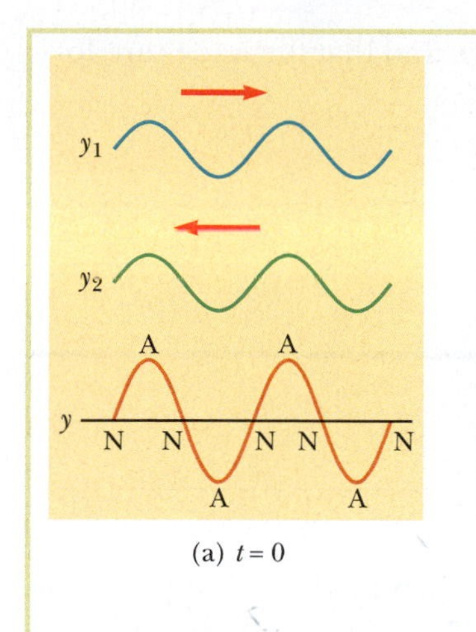

(a) $t = 0$

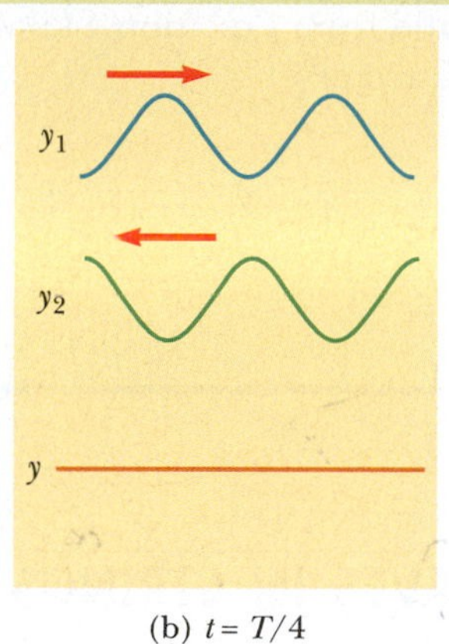

(b) $t = T/4$

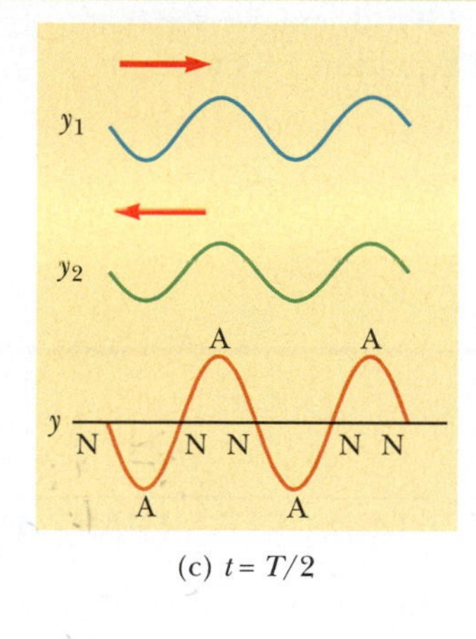

(c) $t = T/2$

**ACTIVE FIGURE 14.8**

Standing wave patterns at various times produced by two waves of equal amplitude traveling in opposite directions. For the resultant wave $y$, the nodes (N) are points of zero displacement and the antinodes (A) are points of maximum displacement.

**PhysicsNow™** Log into PhysicsNow at **www.pop4e.com** and go to Active Figure 14.8 to choose the wavelength of the waves and see the standing wave that results.

antinodes are labeled A. At $t = 0$ (Active Fig. 14.8a), the two waves are in phase, giving a wave pattern with amplitude $2A$. One quarter of a period later, at $t = T/4$ (Active Fig. 14.8b), the individual waves have moved one quarter of a wavelength (one to the right and the other to the left). At this time, the waves are 180° out of phase. The individual displacements of the elements of the medium from their equilibrium positions are of equal magnitude and opposite direction for all values of $x$; hence, the resultant wave has zero displacement everywhere. At $t = T/2$ (Active Fig. 14.8c), the individual waves are again in phase, producing a wave pattern that is inverted relative to the $t = 0$ pattern.

**QUICK QUIZ 14.2** Consider Active Figure 14.8 as representing a standing wave on a string. Define the velocity of elements of the string as positive if they are moving upward in the figure. **(i)** At the moment the string has the shape shown at the bottom of Active Figure 14.8a, the instantaneous velocity of elements along the string **(a)** is zero for all elements, **(b)** is positive for all elements, **(c)** is negative for all elements, or **(d)** varies with the position of the element. **(ii)** From the same set of choices, choose the best answer at the moment the string has the shape shown at the bottom of Active Figure 14.8b.

## EXAMPLE 14.2 Formation of a Standing Wave

Two transverse waves traveling in opposite directions produce a standing wave. The individual wave functions are

$$y_1 = (4.0 \text{ cm}) \sin(3.0x - 2.0t)$$

$$y_2 = (4.0 \text{ cm}) \sin(3.0x + 2.0t)$$

where $x$ and $y$ are in centimeters.

**A** Find the maximum transverse position of an element of the medium at $x = 2.3$ cm.

**Solution** When the two waves are summed, the result is a standing wave whose mathematical representation is given by Equation 14.3, with $A = 4.0$ cm and $k = 3.0$ rad/cm:

$$y = (2A \sin kx) \cos \omega t = [(8.0 \text{ cm}) \sin 3.0x] \cos 2.0t$$

Therefore, the maximum transverse position of an element at the position $x = 2.3$ cm is

$$y_{\text{max}} = [(8.0 \text{ cm}) \sin 3.0x]_{x=2.3 \text{ cm}}$$
$$= (8.0 \text{ cm}) \sin(6.9 \text{ rad}) = 4.6 \text{ cm}$$

**B** Find the positions of the nodes and antinodes.

**Solution** Because $k = 2\pi/\lambda = 3.0 \text{ rad/cm}$, we see that $\lambda = 2\pi/3$ cm. Therefore, from Equation 14.4 we find that the antinodes are located at

$$x = n\left(\frac{\pi}{6.0}\right) \text{ cm} \qquad (n = 1, 3, 5, \ldots)$$

and from Equation 14.5 we find that the nodes are located at

$$x = n\frac{\lambda}{2} = n\left(\frac{\pi}{3.0}\right) \text{ cm} \qquad (n = 1, 2, 3, \ldots)$$

## 14.4 STANDING WAVES IN STRINGS

In the preceding section, we discussed standing waves formed by identical waves moving in opposite directions in the same medium. One way to establish a standing wave on a string is to combine incoming and reflected waves from a rigid end. If a string is stretched between *two* rigid supports (Active Fig. 14.9a) and waves are established on the string, standing waves will be set up in the string by the continuous superposition of the waves incident on and reflected from the ends. This physical system is a model for the source of sound in any stringed instrument, such as the guitar, the violin, and the piano. The string has a number of natural patterns of vibration, called **normal modes,** and each mode has a characteristic frequency.

This discussion is our first introduction to an important analysis model, the **wave under boundary conditions.** When boundary conditions are applied to a wave, we find very interesting behavior that has no analog in the physics of particles. The most prominent aspect of this behavior is **quantization.** We shall find that only certain waves—those that satisfy the boundary conditions—are allowed. The notion of quantization was introduced in Chapter 11 when we discussed the Bohr model of the atom. In that model, angular momentum was quantized. As we shall see in Chapter 29, this quantization is just an application of the wave under boundary conditions model.

In the standing wave pattern on a stretched string, the ends of the string must be nodes because these points are fixed, establishing the boundary condition on the waves. The rest of the pattern can be built from this boundary condition along with the requirement that nodes and antinodes are equally spaced and separated by one fourth of a wavelength. The simplest pattern that satisfies these conditions

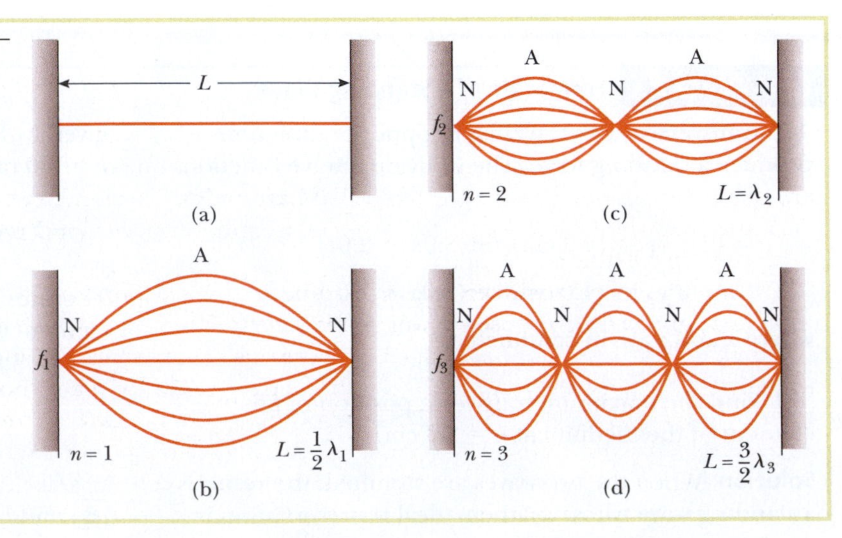


**ACTIVE FIGURE 14.9** (a) A string of length $L$ fixed at both ends. The normal modes of vibration form a harmonic series. In each case, the shape of the string is shown at several instants within one period: (b) the fundamental frequency, or first harmonic; (c) the second harmonic; and (d) the third harmonic.

**PhysicsNow™** By logging into PhysicsNow at **www.pop4e.com** and going to Active Figure 14.9 you can choose the mode number and observe the corresponding standing wave.

has the required nodes at the ends of the string and an antinode at the center point (Active Fig. 14.9b). For this normal mode, the length of the string equals $\lambda/2$ (the distance between adjacent nodes):

$$L = \frac{\lambda_1}{2} \qquad \text{or} \qquad \lambda_1 = 2L$$

The next normal mode, of wavelength $\lambda_2$ (Active Fig. 14.9c), occurs when the length of the string equals one wavelength, that is, when $\lambda_2 = L$. In this mode, the two halves of the string are moving in opposite directions at a given instant, and we sometimes say that two *loops* occur. The third normal mode (Active Fig. 14.9d) corresponds to the case when the length equals $3\lambda/2$; therefore, $\lambda_3 = 2L/3$. In general, the wavelengths of the various normal modes can be conveniently expressed as

$$\lambda_n = \frac{2L}{n} \qquad (n = 1, 2, 3, \ldots) \qquad [14.6]$$

▪ Wavelengths of normal modes

where the index $n$ refers to the $n$th mode of vibration. The natural frequencies associated with these modes are obtained from the relationship $f = v/\lambda$, where the wave speed $v$ is determined by the tension $T$ and linear mass density $\mu$ of the string and therefore is the same for all frequencies. Using Equation 14.6, we find that the frequencies of the normal modes are

$$f_n = \frac{v}{\lambda_n} = \frac{n}{2L}v \qquad (n = 1, 2, 3, \ldots) \qquad [14.7]$$

▪ Frequencies of normal modes as functions of wave speed and length of string

Because $v = \sqrt{T/\mu}$ (Equation 13.20), we can express the natural frequencies of a stretched string as

$$f_n = \frac{n}{2L}\sqrt{\frac{T}{\mu}} \qquad (n = 1, 2, 3, \ldots) \qquad [14.8]$$

▪ Frequencies of normal modes as functions of string tension and linear mass density

Equation 14.8 demonstrates the quantization that we mentioned as a feature of the wave under boundary conditions model. The frequencies are quantized because only certain frequencies of waves satisfy the boundary conditions and can exist on the string. The lowest frequency, corresponding to $n = 1$, is called the **fundamental frequency** $f_1$ and is

$$f_1 = \frac{1}{2L}\sqrt{\frac{T}{\mu}} \qquad [14.9]$$

▪ Fundamental frequency of a stretched string

Equation 14.8 shows that the frequencies of the higher modes are integral multiples of the fundamental frequency, that is, $2f_1$, $3f_1$, $4f_1$, and so on. These higher natural frequencies, together with the fundamental frequency, form a **harmonic series** and the various frequencies are called **harmonics.** The fundamental $f_1$ is the first harmonic; the frequency $f_2 = 2f_1$ is the second harmonic; and the frequency $f_n$ is the $n$th harmonic.

If a stretched string is distorted to a shape that corresponds to any one of its harmonics, after being released it will vibrate at the frequency of that harmonic. If the string is plucked, bowed, or struck, however, as occurs when playing a stringed instrument, the resulting vibration will include frequencies of many modes, including the fundamental. In effect, the string "selects" a mixture of normal-mode frequencies when disturbed by a finger or a bow. The frequency of the combination is that of the fundamental because that is the rate at which the waveform repeats; the frequency associated with the string by a listener is that of the fundamental.

The frequency of a given string on a stringed instrument can be changed either by varying the string's tension $T$ or by changing the length $L$ of the vibrating portion of the string. For example, the tension in the strings of guitars and violins is adjusted by a screw mechanism or by tuning pegs on the neck of the instrument. As the tension increases, the frequencies of the normal modes increase according to Equation 14.8. Once the instrument is "tuned," the player varies the frequency by moving his or her fingers along the neck, thereby changing the length of the vibrating portion of the string. As this length is reduced, the frequency increases because the normal-mode frequencies are inversely proportional to the length of the vibrating portion of the string.

Imagine that we have several strings of the same length under the same tension but varying linear mass density $\mu$. The strings will have different wave speeds and therefore different fundamental frequencies. The linear mass density can be changed either by varying the diameter of the string or by wrapping extra mass around the string. Both of these possibilities can be seen on the guitar, on which the higher-frequency strings vary in diameter and the lower-frequency strings have additional wire wrapped around them.

**QUICK QUIZ 14.3** Which of the following is true when a standing wave is set up on a string fixed at both ends? **(a)** The number of nodes is equal to the number of antinodes. **(b)** The wavelength is equal to the length of the string divided by an integer. **(c)** The frequency is equal to the number of nodes times the fundamental frequency. **(d)** The center of the string is either a node or an antinode.

## INTERACTIVE EXAMPLE 14.3 Give Me a C Note

A middle C string on a piano has a fundamental frequency of 262 Hz, and the A note has a fundamental frequency of 440 Hz.

**A** Calculate the frequencies of the next two harmonics of the C string.

**Solution** Because the higher frequencies are integer multiples of the fundamental frequency,

$$f_2 = 2f_1 = 524 \text{ Hz}$$

$$f_3 = 3f_1 = 786 \text{ Hz}$$

**B** If the strings for the A and C notes are assumed to have the same mass per unit length and the same length, determine the ratio of tensions in the two strings.

**Solution** Using Equation 14.9 for the two strings vibrating at their fundamental frequencies gives

$$\left.\begin{aligned} f_{1\text{A}} &= \frac{1}{2L}\sqrt{\frac{T_\text{A}}{\mu}} \\ f_{1\text{C}} &= \frac{1}{2L}\sqrt{\frac{T_\text{C}}{\mu}} \end{aligned}\right\} \rightarrow \frac{f_{1\text{A}}}{f_{1\text{C}}} = \sqrt{\frac{T_\text{A}}{T_\text{C}}}$$

$$\frac{T_\text{A}}{T_\text{C}} = \left(\frac{f_{1\text{A}}}{f_{1\text{C}}}\right)^2 = \left(\frac{440 \text{ Hz}}{262 \text{ Hz}}\right)^2 = 2.82$$

**C** In a real piano, the assumption we made in part B is only partially true. The string densities are equal, but the A string is 64% as long as the C string. What is the ratio of their tensions?

**Solution** We start from the same point as in part B, but the string lengths do not cancel in the ratio:

$$\left.\begin{aligned} f_{1\text{A}} &= \frac{1}{2L_\text{A}}\sqrt{\frac{T_\text{A}}{\mu}} \\ f_{1\text{C}} &= \frac{1}{2L_\text{C}}\sqrt{\frac{T_\text{C}}{\mu}} \end{aligned}\right\} \rightarrow \frac{f_{1\text{A}}}{f_{1\text{C}}} = \frac{L_\text{C}}{L_\text{A}}\sqrt{\frac{T_\text{A}}{T_\text{C}}}$$

$$\frac{T_\text{A}}{T_\text{C}} = \left(\frac{L_\text{A}}{L_\text{C}}\right)^2\left(\frac{f_{1\text{A}}}{f_{1\text{C}}}\right)^2 = (0.64)^2\left(\frac{440 \text{ Hz}}{262 \text{ Hz}}\right)^2$$

$$= 1.16$$

**Physics Now™** Investigate this situation for different combinations of notes by logging into PhysicsNow at **www.pop4e.com** and going to Interactive Example 14.3.

## 14.5 STANDING WAVES IN AIR COLUMNS

We have discussed musical instruments that use strings, which include guitars, violins, and pianos. What about instruments classified as brasses or woodwinds? These instruments produce music using a column of air. Standing longitudinal waves can be set up in an air column, such as an organ pipe or a clarinet, as the result of interference between longitudinal sound waves traveling in opposite directions. Whether a node or an antinode occurs at the end of an air column depends on whether that end is open or closed. **The closed end of an air column is a displacement node,** just as the fixed end of a vibrating string is a displacement node. Furthermore, because the pressure wave is 90° out of phase with the displacement wave (Section 13.7), **the closed end of an air column corresponds to a pressure antinode** (i.e., a point of maximum pressure variation). On the other hand, **the open end of an air column is approximately a displacement antinode and a pressure node.**

You may wonder how a sound wave can reflect from an open end because there may not appear to be a change in the medium at this point. It is indeed true that the medium through which the sound wave moves is air both inside and outside the pipe. Sound is a pressure wave, however, and a compression region of the sound wave is constrained by the sides of the pipe as long as the region is inside the pipe. As the compression region exits at the open end of the pipe, the constraint of the pipe is removed and the compressed air is free to expand into the atmosphere. Therefore, there is a change in the *character* of the medium between the inside of the pipe and the outside even though there is no change in the *material* of the medium. This change in character is sufficient to allow some reflection.

Strictly speaking, the open end of an air column is not exactly an antinode. A compression in the sound wave does not reach full expansion until it passes somewhat beyond the open end. Therefore, to calculate frequencies of the normal modes accurately, an **end correction** must be added to the length of the air column at each open end. For a thin-walled tube of circular cross-section, this end correction is about $0.6R$, where $R$ is the tube's radius. Hence, the effective acoustical length of the tube is somewhat greater than the physical length $L$.

We can determine the modes of vibration of an air column by applying the appropriate boundary condition at the end of the column, along with the requirement that nodes and antinodes be separated by one fourth of a wavelength. We shall find that the frequency for sound waves in air columns is quantized, similar to the results found for waves on strings under boundary conditions.

The first three modes of vibration of an air column that is open at both ends are shown in Figure 14.10a. Note that the ends are displacement antinodes (approximately). In the fundamental mode, the wavelength is twice the length of the air column; hence, the frequency of the fundamental $f_1$ is $v/2L$. Similarly, the frequencies of the higher harmonics are $2f_1$, $3f_1$, . . . . Therefore, in an air column that is open at both ends, the natural frequencies of vibration form a harmonic series; that is, the higher harmonics are integral multiples of the fundamental frequency. Because all harmonics are present, we can express the natural frequencies of vibration as

$$f_n = n\frac{v}{2L} \qquad (n = 1, 2, 3, \ldots) \qquad [14.10]$$

■ Natural frequencies of an air column open at both ends

where $v$ is the speed of sound in air.

If an air column is closed at one end and open at the other, the closed end is a displacement node and the open end is a displacement antinode (Fig. 14.10b). In this case, the wavelength for the fundamental mode is four times the length of the column. Hence, the fundamental frequency $f_1$ is equal to $v/4L$, and the frequencies

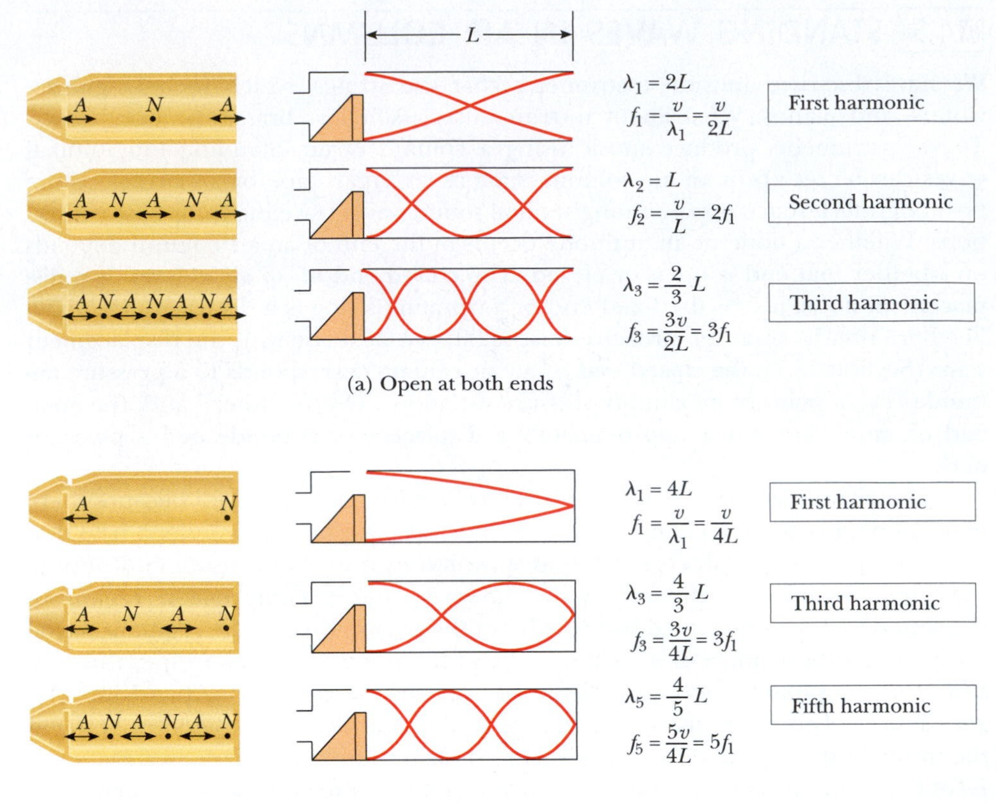


**FIGURE 14.10** Motion of elements of air in standing longitudinal waves in an air column, along with graphical representations of the displacements of the elements. (a) In an air column open at both ends, the harmonic series created consists of all integer multiples of the fundamental frequency: $f_1, 2f_1, 3f_1, \ldots$. (b) In an air column closed at one end and open at the other, the harmonic series consists of only odd-integer multiples of the fundamental frequency: $f_1, 3f_1, 5f_1, \ldots$.

of the higher harmonics are equal to $3f_1, 5f_1, \ldots$. That is, **in an air column that is closed at one end, only odd harmonics are present,** and the frequencies are

■ Natural frequencies of an air column closed at one end and open at the other

$$f_n = n\frac{v}{4L} \qquad (n = 1, 3, 5, \ldots) \qquad [14.11]$$

Standing waves in air columns are the primary sources of the sounds produced by wind instruments. In a woodwind instrument, a key is pressed, which opens a hole in the side of the column. This hole defines the end of the vibrating column of air (because the hole acts as an open end at which pressure can be released), so that the column is effectively shortened and the fundamental frequency rises. In a brass instrument, the length of the air column is changed by an adjustable section, as in a trombone, or by adding segments of tubing, as is done in a trumpet when a valve is pressed.

**PITFALL PREVENTION 14.3**

**SOUND WAVES IN AIR ARE NOT TRANSVERSE** Note that the standing longitudinal waves are drawn as transverse waves in Figure 14.10. It is difficult to draw longitudinal displacements because they are in the same direction as the propagation. Therefore, it is best to interpret the curves in Figure 14.10 as a graphical representation of the waves (our diagrams of string waves are pictorial representations), with the vertical axis representing horizontal position of the elements of the medium.

**QUICK QUIZ 14.4** Standing waves in a pipe open at both ends are excited at a fundamental frequency $f_{open}$. When one end is closed and the pipe is again excited, the fundamental frequency is $f_{closed}$. Which of the following expressions describes how these two frequencies that are heard compare? **(a)** $f_{closed} = f_{open}$ **(b)** $f_{closed} = \frac{1}{2}f_{open}$ **(c)** $f_{closed} = 2f_{open}$ **(d)** $f_{closed} = \frac{3}{2}f_{open}$

**QUICK QUIZ 14.5** Balboa Park in San Diego has an outdoor organ. When the air temperature increases, what happens to the fundamental frequency of one of the organ pipes? **(a)** It stays the same. **(b)** It goes down. **(c)** It goes up. **(d)** It is impossible to determine.

## ■ Thinking Physics 14.2

A bugle has no valves, keys, slides, or finger holes. How can it play a song?

**Reasoning** Songs for the bugle are limited to harmonics of the fundamental frequency because the bugle has no control over frequencies by means of valves, keys, slides, or finger holes. The player obtains different notes by changing the tension in the lips as the bugle is played to excite different harmonics. The normal playing range of a bugle is among the third, fourth, fifth, and sixth harmonics of the fundamental. As examples, "Reveille" is played with just the three notes D (294 Hz), G (392 Hz), and B (490 Hz), and "Taps" is played with these same three notes and the D one octave above the lower D (588 Hz). Note that the frequencies of these four notes are, respectively, three, four, five, and six times the fundamental of 98 Hz. ■

## ■ Thinking Physics 14.3

If an orchestra doesn't warm up before a performance, the strings go flat and the wind instruments go sharp during the performance. Why?

**Reasoning** Without warming up, all the instruments will be at room temperature at the beginning of the concert. As the wind instruments are played, they fill with warm air from the player's exhalation. The increase in temperature of the air in the instrument causes an increase in the speed of sound, which raises the fundamental frequencies of the air columns. As a result, the wind instruments go sharp. The strings on the stringed instruments also increase in temperature due to the friction of rubbing with the bow. This increase in temperature results in thermal expansion, which causes a decrease in the tension in the strings. (We will study thermal expansion in Chapter 16.) With a decrease in tension, the wave speed on the strings drops and the fundamental frequencies decrease. Therefore, the stringed instruments go flat. ■

## EXAMPLE 14.4 Harmonics in a Pipe

A pipe has a length of 1.23 m.

**A** Determine the frequencies of the first three harmonics if the pipe is open at each end. Take $v = 343$ m/s as the speed of sound in air.

**Solution** The first harmonic of a pipe open at both ends is

$$f_1 = \frac{v}{2L} = \frac{343 \text{ m/s}}{2(1.23 \text{ m})} = \boxed{139 \text{ Hz}}$$

Because all harmonics are possible for a pipe open at both ends, the second and third harmonics are $f_2 = 2f_1 = \boxed{278 \text{ Hz}}$ and $f_3 = 3f_1 = \boxed{417 \text{ Hz}}$.

**B** What are the three frequencies requested in part A if the pipe is closed at one end?

**Solution** The fundamental frequency of a pipe closed at one end is

$$f_1 = \frac{v}{4L} = \frac{343 \text{ m/s}}{4(1.23 \text{ m})} = \boxed{69.7 \text{ Hz}}$$

In this case, only odd harmonics are present, so the next two harmonics have frequencies $f_3 = 3f_1 = \boxed{209 \text{ Hz}}$ and $f_5 = 5f_1 = \boxed{349 \text{ Hz}}$.

**C** For the pipe open at both ends, how many harmonics are present in the normal human hearing range (20–20 000 Hz)?

**Solution** Because all harmonics are present, $f_n = nf_1$. For $f_n = 20\,000$ Hz, we have $n = 20\,000/139 = 144$, so $\boxed{144}$ harmonics are present in the audible range. Actually, only the first few harmonics have sufficient amplitude to be heard.

### EXAMPLE 14.5 Measuring the Frequency of a Tuning Fork

A simple apparatus for demonstrating standing waves in a tube is described in Figure 14.11a. A long, vertical tube open at both ends is partially submerged in a beaker of water, and a vibrating tuning fork of unknown frequency is placed near the top. The length $L$ of the air column is adjusted by moving the tube vertically. The sound waves generated by the fork excite a resonance response in the air column when its length equals that associated with one of the harmonic frequencies of the tube.

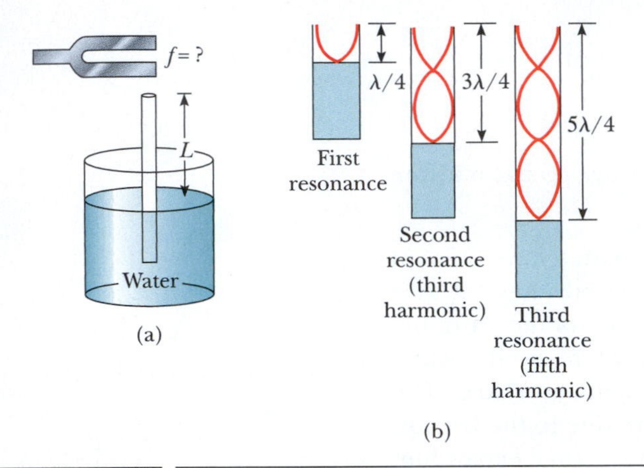


FIGURE 14.11 (Example 14.5) (a) Apparatus for demonstrating the resonance of sound waves in a tube closed at one end. The length $L$ of the air column is varied by moving the tube vertically while it is partially submerged in water. (b) The first three normal modes of the system shown in (a).

For a certain tube, the smallest value of $L$ for which a peak occurs in the sound intensity is 9.00 cm. From this measurement, determine the frequency of the tuning fork and the value of $L$ for the next two resonant modes.

**Solution** To conceptualize the problem, we realize that although the tube is open at the bottom end to allow the water in, the water surface acts like a rigid barrier at that end. This setup can therefore be categorized as an air column closed at one end, and the fundamental has frequency $v/4L$, where $L$ is the length of the tube from the open end to the water surface (Fig. 14.11b). To analyze the problem, we take $v = 343$ m/s for the speed of sound in air and $L = 0.090\,0$ m. Then, from Equation 14.11, we have

$$f_1 = \frac{v}{4L} = \frac{343 \text{ m/s}}{4(0.090\,0 \text{ m})} = 953 \text{ Hz}$$

From the information about the fundamental mode, we see that the wavelength is $\lambda = 4L = 0.360$ m. Because the frequency of the source is constant, the next two natural modes (Fig. 14.11b) correspond to lengths of $3\lambda/4 = 0.270$ m and $5\lambda/4 = 0.450$ m.

To finalize the problem, note in Figure 14.11 that the wavelength of the sound remains fixed because it is determined by the tuning fork. A resonance response occurs whenever the level of the water coincides with a node of the standing wave. In this condition, the driving frequency of the tuning fork matches the natural frequency of the air column, and the amplitude of the sound increases.

## 14.6 BEATS: INTERFERENCE IN TIME

The interference phenomena that we have discussed so far involve the superposition of two or more waves with the same frequency. Because the resultant displacement of an element in the medium in this case depends on the position of the element, we can refer to the phenomenon as **spatial interference.** Standing waves in strings and air columns are common examples of spatial interference.

We now consider another type of interference effect, one that results from the superposition of two waves with slightly *different* frequencies. In this case, when the two waves of amplitudes $A_1$ and $A_2$ are observed at a given point, they are alternately in and out of phase. We refer to this phenomenon as **interference in time** or **temporal interference.** When the waves are in phase, the combined amplitude is $A_1 + A_2$. When they are out of phase, the combined amplitude is $|A_1 - A_2|$. The combination therefore varies between small and large amplitudes, resulting in what we call **beats.**

Although beats occur for all types of waves, they are particularly noticeable for sound waves. For example, if two tuning forks of slightly different frequencies are struck, you hear a sound of pulsating intensity.

The number of beats you hear per second, the **beat frequency,** equals the difference in frequency between the two sources. The maximum beat frequency that the human ear can detect is about 20 beats/s. When the beat frequency exceeds this value, it blends with the sounds producing the beats.

One can use beats to tune a stringed instrument, such as a piano, by beating a note against a reference tone of known frequency. The frequency of the string can then be adjusted to equal the frequency of the reference by changing the string's tension until the beats disappear; the two frequencies are then the same.

Let us look at the mathematical representation of beats. Consider two waves with equal amplitudes traveling through a medium with slightly different frequencies $f_1$ and $f_2$. We can represent the position of an element of the medium associated with each wave at a fixed point, which we choose as $x = 0$, as

$$y_1 = A \cos 2\pi f_1 t \qquad \text{and} \qquad y_2 = A \cos 2\pi f_2 t$$

Using the superposition principle, we find that the resultant position at that point is given by

$$y = y_1 + y_2 = A(\cos 2\pi f_1 t + \cos 2\pi f_2 t)$$

It is convenient to write this expression in a form that uses the trigonometric identity

$$\cos a + \cos b = 2 \cos\left(\frac{a-b}{2}\right) \cos\left(\frac{a+b}{2}\right)$$

Letting $a = 2\pi f_1 t$ and $b = 2\pi f_2 t$, we find that

$$y = \left[2A \cos 2\pi \left(\frac{f_1 - f_2}{2}\right) t\right] \cos 2\pi \left(\frac{f_1 + f_2}{2}\right) t \qquad [14.12]$$

Graphs demonstrating the individual waves as well as the resultant wave are shown in Active Figure 14.12. From the factors in Equation 14.12, we see that the resultant wave has an effective frequency equal to the average frequency $(f_1 + f_2)/2$ and an amplitude of

$$A_{x=0} = 2A \cos 2\pi \left(\frac{f_1 - f_2}{2}\right) t \qquad [14.13]$$

That is, the *amplitude varies in time* with a frequency of $(f_1 - f_2)/2$. When $f_1$ is close to $f_2$, this amplitude variation is slow compared with the frequency of the individual waves, as illustrated by the envelope (broken line) of the resultant wave in Active Figure 14.12b.

Note that a maximum in amplitude will be detected whenever

$$\cos 2\pi \left(\frac{f_1 - f_2}{2}\right) t = \pm 1$$

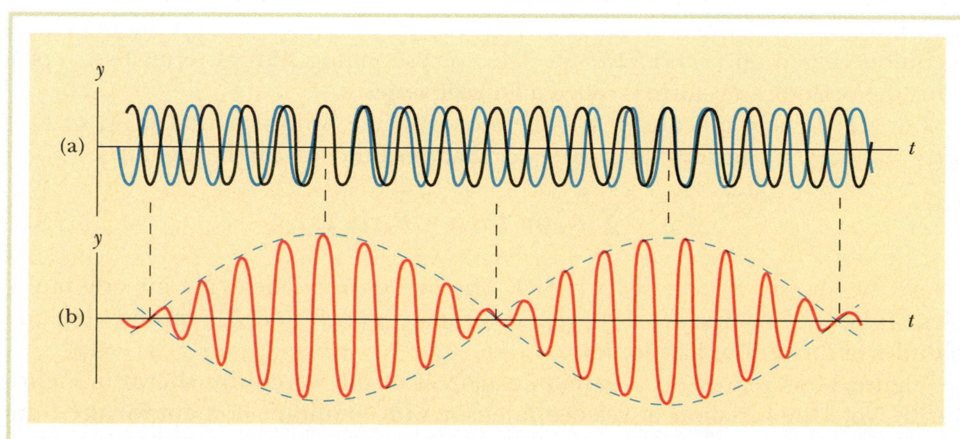


**ACTIVE FIGURE 14.12** Beats are formed by the combination of two waves of slightly different frequencies. (a) The blue and black curves represent the individual waves. (b) The combined wave has an amplitude (broken line) that oscillates in time.

**PhysicsNow™** By logging into PhysicsNow at **www.pop4e.com** and going to Active Figure 14.12 you can choose the two frequencies and observe the corresponding beats.

That is, the amplitude maximizes twice in each cycle of the function on the left in the preceding expression. Therefore, the number of beats per second, or the beat frequency $f_b$, is twice the frequency of this function:

**Beat frequency**

$$f_b = |f_1 - f_2| \quad [14.14]$$

For instance, if two tuning forks vibrate individually at frequencies of 438 Hz and 442 Hz, respectively, the resultant sound wave of the combination has a frequency of $(f_1 + f_2)/2 = 440$ Hz (the musical note A) and a beat frequency of $|f_1 - f_2| = 4$ Hz. That is, the listener hears the 440-Hz sound wave go through an intensity maximum four times every second.

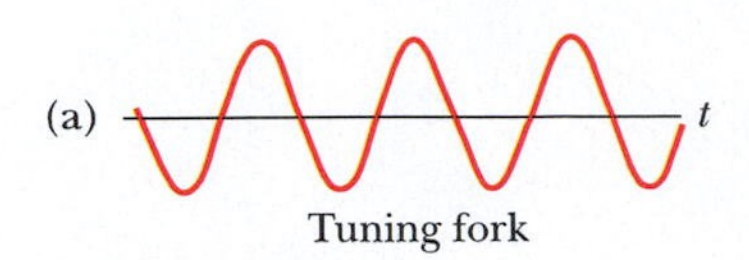


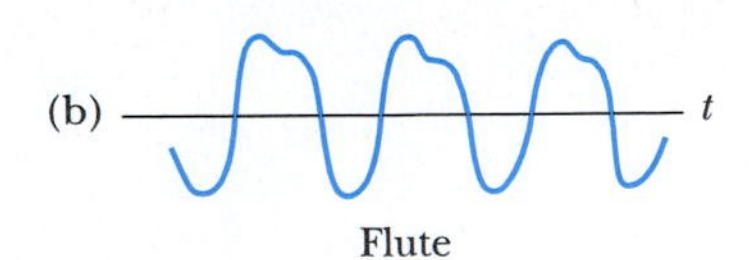


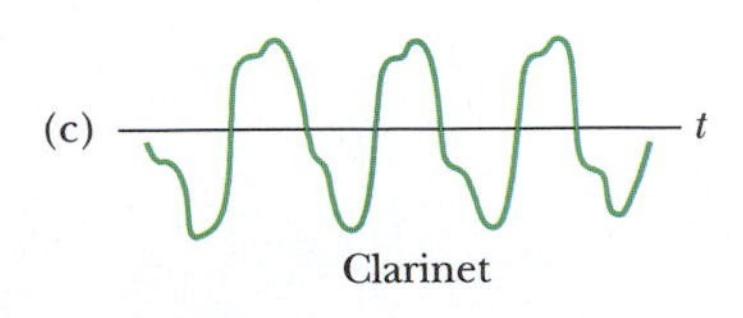


**FIGURE 14.13** Waveforms of sound produced by (a) a tuning fork, (b) a flute, and (c) a clarinet, each at approximately the same frequency.

**QUICK QUIZ 14.6** You are tuning a guitar by comparing the sound of the string with that of a standard tuning fork. You notice a beat frequency of 5 Hz when both sounds are present. You tighten the guitar string and the beat frequency rises to 8 Hz. To tune the string exactly to the tuning fork, what should you do? **(a)** Continue to tighten the string. **(b)** Loosen the string. **(c)** It is impossible to determine.

## 14.7 NONSINUSOIDAL WAVE PATTERNS

The sound wave patterns produced by most instruments are not sinusoidal. Some characteristic waveforms produced by a tuning fork, a flute, and a clarinet are shown in Figure 14.13. Although each instrument has its own characteristic pattern, Figure 14.13 shows that all three waveforms are periodic. A struck tuning fork produces primarily one harmonic (the fundamental), whereas the flute and clarinet produce many frequencies, which include the fundamental and various harmonics. The nonsinusoidal waveforms produced by a violin or clarinet, and the corresponding richness of musical tones, are the result of the superposition of various harmonics.

This phenomenon is in contrast to a percussive musical instrument, such as the drum, in which the combination of frequencies does not form a harmonic series. When frequencies that are integer multiples of a fundamental frequency are combined, the result is a *musical* sound. A listener can assign a pitch to the sound based on the fundamental frequency. Pitch is a psychological reaction to a sound that allows the listener to place the sound on a scale of low to high (bass to treble). Combinations of frequencies that are not integer multiples of a fundamental result in a *noise* rather than a musical sound. It is much harder for a listener to assign a pitch to a noise than to a musical sound.

### PITFALL PREVENTION 14.4

**Pitch versus Frequency** A very common mistake made in speech when talking about sound is to use the term *pitch* when one means *frequency*. Frequency is the physical measurement of the number of oscillations per second, as we have defined. Pitch is a psychological reaction of humans to sound that enables a human to place the sound on a scale from high to low or from treble to bass. Therefore, frequency is the stimulus and pitch is the response. Although pitch is related mostly (but not completely) to frequency, they are not the same. A phrase such as "the pitch of the sound" is incorrect because pitch is not a physical property of the sound.

Analysis of nonsinusoidal waveforms appears at first sight to be a formidable task. If the waveform is periodic, however, it can be represented with arbitrary precision by the combination of a sufficiently large number of sinusoidal waves that form a harmonic series. In fact, one can represent any periodic function or any function over a finite interval as a series of sine and cosine terms by using a mathematical technique based on *Fourier's theorem.* The corresponding sum of terms that represents the periodic waveform is called a **Fourier series.**

Let $y(t)$ be any function that is periodic in time, with a period of $T$, so that $y(t + T) = y(t)$. **Fourier's theorem** states that this function can be written

**Fourier's theorem**

$$y(t) = \sum_n (A_n \sin 2\pi f_n t + B_n \cos 2\pi f_n t) \quad [14.15]$$

where the lowest frequency is $f_1 = 1/T$. The higher frequencies are integral multiples of the fundamental, so $f_n = nf_1$. The coefficients $A_n$ and $B_n$ represent the amplitudes of the various harmonics.

Figure 14.14 represents a harmonic analysis of the waveforms shown in Figure 14.13. Note the variation of relative intensity with harmonic content for the flute and the clarinet. In general, any musical sound contains components that are members of a harmonic series with varying relative intensities.

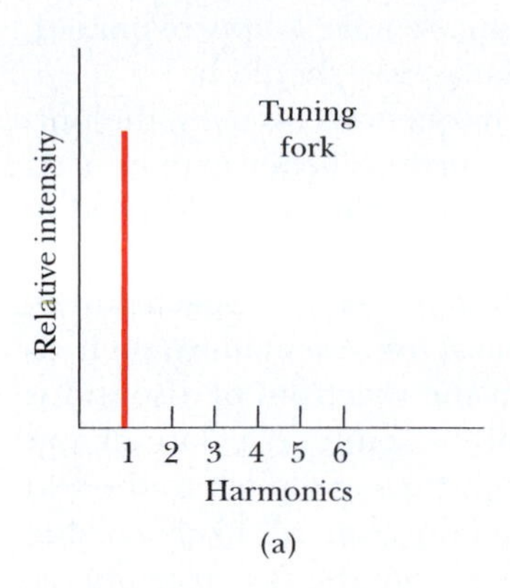


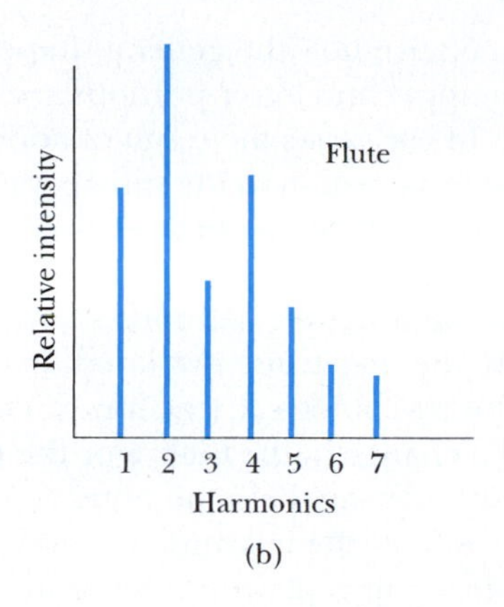


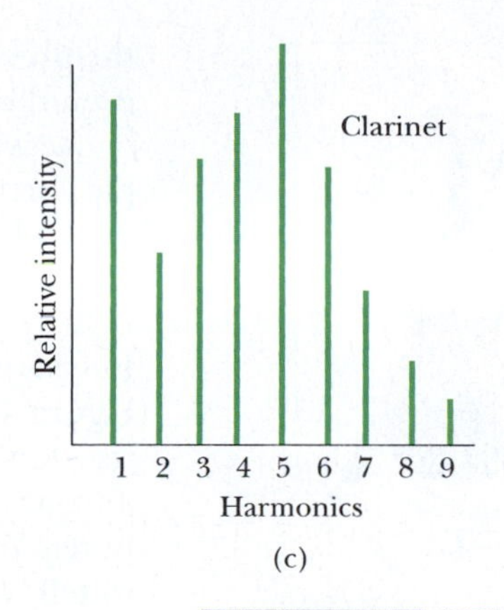


**FIGURE 14.14** Harmonics of the waveforms shown in Figure 14.13. Note the variations in intensity of the various harmonics.

We have discussed the *analysis* of a wave pattern using Fourier's theorem. The analysis involves determining the coefficients of the trigonometric functions in Equation 14.15 from a knowledge of the wave pattern. We can also perform the reverse process, *Fourier synthesis.* In this process, the various harmonics are added together to form a resultant wave pattern. As an example of Fourier synthesis, consider the building of a square wave as shown in Active Figure 14.15. The symmetry of the square wave results in only odd multiples of the fundamental combining in the synthesis. In Active Figure 14.15a, the brown curve shows the combination of $f$ and $3f$. In Active Figure 14.15b, we have added $5f$ to the combination and obtained

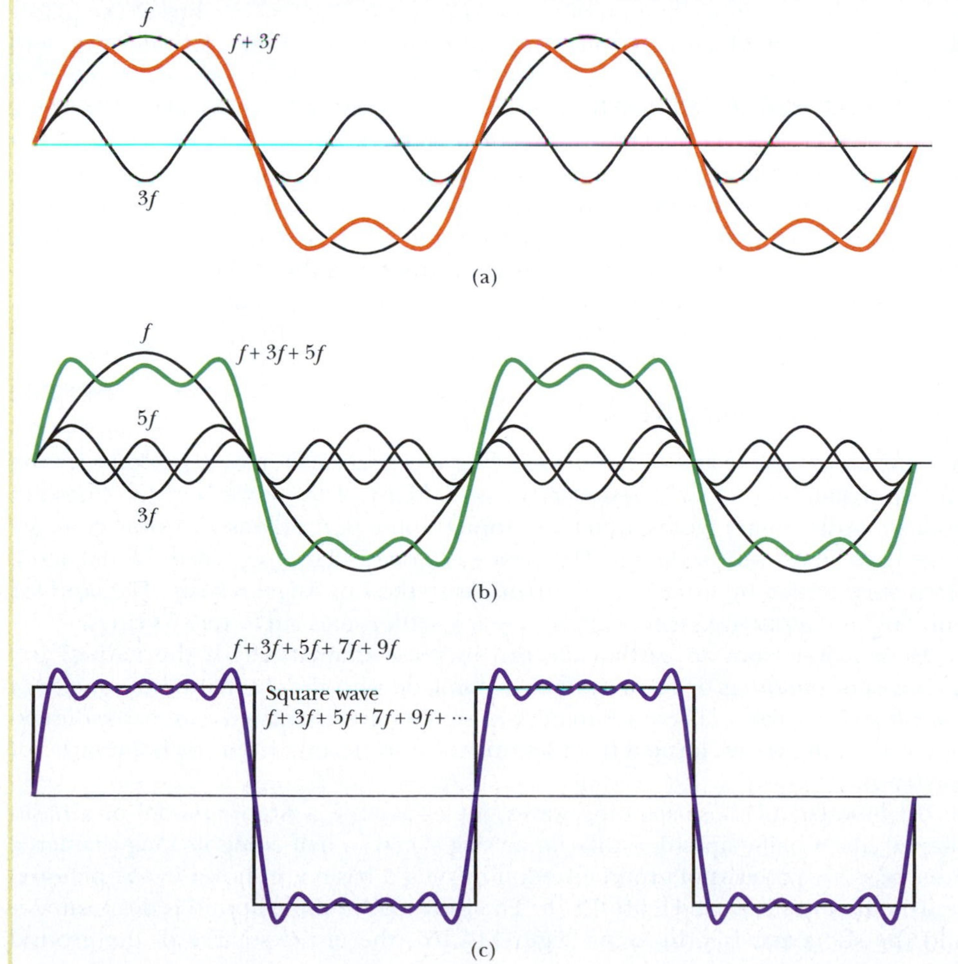


**ACTIVE FIGURE 14.15** Fourier synthesis of a square wave represented by the sum of odd multiples of the first harmonic, which has frequency $f$. (a) Waves of frequency $f$ and $3f$ are added. (b) One more odd frequency of $5f$ is added. (c) The synthesis curve approaches the square wave when odd frequencies up to $9f$ are combined.

**Physics Now™** By logging into PhysicsNow at **www.pop4e.com** and going to Active Figure 14.15 you can add in harmonics with frequencies higher than $9f$ to try to synthesize a square wave.

(a)

(b)

(c)

(Photographs courtesy of (a) and (b) Getty Images; (c) Photodisc Green/Getty Images)

Each musical instrument has its own characteristic sound and mixture of harmonics. Instruments shown are (a) the violin, (b) the saxophone, and (c) the trumpet. ■

the green curve. Notice how the general shape of the square wave is approximated, even though the upper and lower portions are not as flat as they should be.

Active Figure 14.15c shows the result of adding odd frequencies up to $9f$, the purple curve. This approximation to the square wave (black curve) is better than in parts (a) and (b). To approximate the square wave as closely as possible, we would need to add all odd multiples of the fundamental frequency up to infinite frequency.

The physical mixture of harmonics can be described as the **spectrum** of the sound, with the spectrum displayed in a graphical representation such as Figure 14.14. The psychological reaction to changes in the spectrum of a sound is the detection of a change in the **timbre** or the **quality** of the sound. If a clarinet and a trumpet are both playing the same note, you will assign the same pitch to the two notes. Yet if only one of the instruments then plays the note, you will likely be able to tell which instrument is playing. The sounds you hear from the two instruments differ in timbre because of a different physical mixture of harmonics. For example, the timbre due to the sound of a trumpet is different from that of a clarinet. You have probably developed words to describe timbres of various instruments, such as "brassy," "mellow," and "tinny."

Fourier's theorem allows us to understand the excitation process of musical instruments. In a stringed instrument that is plucked, such as a guitar, the string is pulled aside and released. After release, the string oscillates almost freely; a small damping causes the amplitude to decay to zero eventually. The mixture of harmonic frequencies depends on the length of the string, its linear mass density, and the plucking point.

On the other hand, a bowed stringed instrument, such as a violin, or a wind instrument is a forced oscillator. In the case of the violin, the alternate sticking and slipping of the bow on the string provides the periodic driving force. In the case of a wind instrument, the vibration of a reed (in a woodwind), of the lips of the player (in a brass), or the blowing of air across an edge (as in a flute) provides the periodic driving force. According to Fourier's theorem, these periodic driving forces contain a mixture of harmonic frequencies. The violin string or the air column in a wind instrument is therefore driven with a wide variety of frequencies. The frequency actually played is determined by *resonance,* which we studied in Chapter 12. The maximum response of the instrument will be to those frequencies that match or are very close to the harmonic frequencies of the instrument. The spectrum of the instrument therefore depends heavily on the strengths of the various harmonics in the initial periodic driving force.

## 14.8 BUILDING ON ANTINODES

CONTEXT CONNECTION

As an example of the application of standing waves to earthquakes, we consider the effects of standing waves in *sedimentary basins.* Many of the world's major cities are built on sedimentary basins, which are topographic depressions that over geologic time have filled with sediment. These areas provide large expanses of flat land, often surrounded by attractive mountains, as in the Los Angeles basin. Flat land for building and attractive scenery attracted early settlers and led to today's cities.

Destruction from an earthquake can increase dramatically if the natural frequencies of buildings or other structures coincide with the resonant frequencies of the underlying basin. These resonant frequencies are associated with three-dimensional standing waves, formed from seismic waves reflecting from the boundaries of the basin.

To understand these standing waves, let us assume a simple model of a basin shaped like a half-ellipsoid, similar to an egg sliced in half along its long diameter. Four possible patterns of ground motion in such a basin are shown in the pictorial representation in Active Figure 14.16. The long axis of the ellipsoid is designated $x$ and the short axis is $y$. In Active Figure 14.16a, the entire surface of the ground

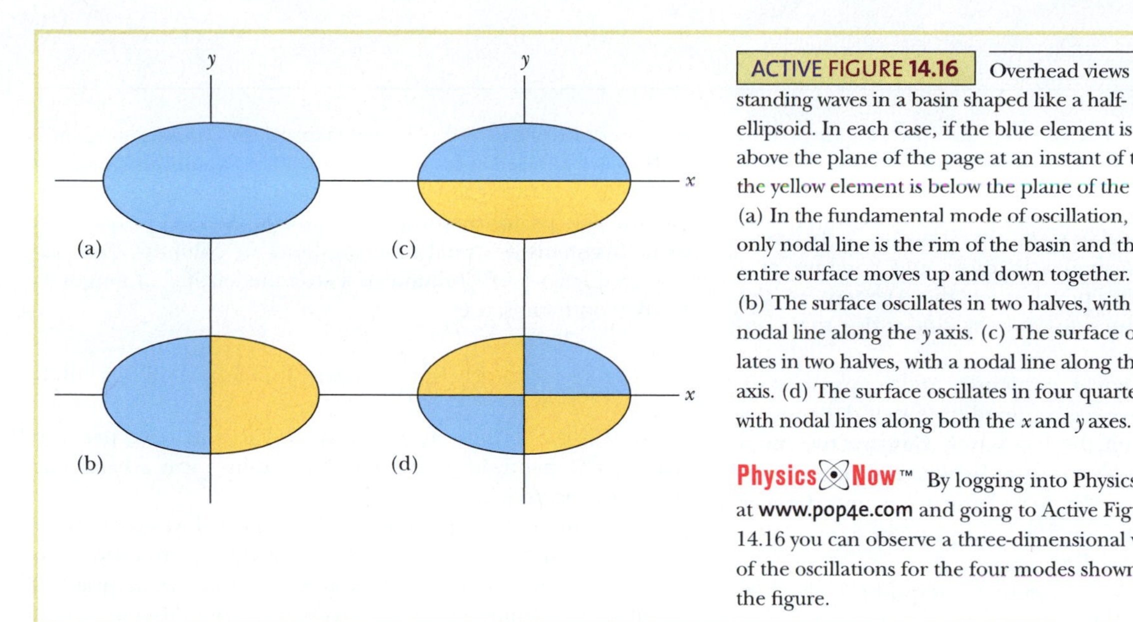


**ACTIVE FIGURE 14.16** Overhead views of standing waves in a basin shaped like a half-ellipsoid. In each case, if the blue element is above the plane of the page at an instant of time, the yellow element is below the plane of the page. (a) In the fundamental mode of oscillation, the only nodal line is the rim of the basin and the entire surface moves up and down together. (b) The surface oscillates in two halves, with a nodal line along the $y$ axis. (c) The surface oscillates in two halves, with a nodal line along the $x$ axis. (d) The surface oscillates in four quarters, with nodal lines along both the $x$ and $y$ axes.

**Physics Now™** By logging into PhysicsNow at **www.pop4e.com** and going to Active Figure 14.16 you can observe a three-dimensional view of the oscillations for the four modes shown in the figure.

moves up and down (that is, in and out of the page) except at a nodal curve running around the edge of the basin.

In Figures 14.16b and 14.16c, half the ground surface lies above and half lies below the equilibrium position, and each half oscillates up and down on either side of a nodal line. The nodal line is along the $y$ axis in Active Figure 14.16b and along the $x$ axis in Active Figure 14.16c. In Active Figure 14.16d, nodal lines occur along both the $x$ and $y$ axes and the surface oscillates in four segments, with two above the equilibrium position at any time and the other two below.

The standing wave patterns in a basin arise from seismic waves traveling horizontally between the boundaries of the basin. For structures built on sedimentary basins, the degree of seismic risk will depend on the standing wave modes excited by the interference of seismic waves trapped in the basin. It is clear that structures built on regions of maximum ground motion (i.e., the antinodes) will suffer maximum shaking, whereas structures residing near nodes will experience relatively mild ground motion. These considerations appear to have played an important role in the selective destruction that occurred in Mexico City in the Michoacán earthquake in 1985 and in the 1989 Loma Prieta earthquake, which caused the collapse of a section of the Nimitz Freeway in Oakland, California.

A similar effect occurs in bounded bodies of water, such as harbors and bays. A standing wave pattern established in such a body of water is called a **seiche.** This wave pattern can result in variations in the water level that exhibit a period of several minutes, superposed on the longer-period tidal variations. Seiches can be caused by earthquakes, tsunamis, winds, or weather disturbances. You can create a seiche in your bathtub by sliding back and forth at just the right frequency such that the water sloshes back and forth at such a large amplitude that much of it spills out onto the floor.

During the Northridge earthquake of 1994, swimming pools throughout southern California overflowed as a result of seiches set up by the shaking of the ground. In a more dramatic example, seismic waves from the 1964 Alaska earthquake caused severe seiches in the bays and bayous of Louisiana, some causing the water level to shift by 2 m.

We have now considered the role of standing waves in the damage caused by an earthquake. In the Context Conclusion, we will gather together the principles of vibrations and waves that we have learned to respond more fully to the central question of this Context. ■

# SUMMARY



The **principle of superposition** states that if two or more traveling waves are moving through a medium and combine at a given point, the resultant position of the element of the medium at that point is the sum of the positions due to the individual waves.

When two waves with equal amplitudes and frequencies superpose, the resultant wave has an amplitude that depends on the phase angle $\phi$ between the two waves. **Constructive interference** occurs when the two waves are *in phase* everywhere, corresponding to $\phi = 0, 2\pi, 4\pi, \ldots$. **Destructive interference** occurs when the two waves are 180° out of phase everywhere, corresponding to $\phi = \pi, 3\pi, 5\pi, \ldots$.

**Standing waves** are formed from the superposition of two sinusoidal waves that have the same frequency, amplitude, and wavelength but are traveling in *opposite* directions. The resultant standing wave is described by the wave function

$$y = (2A \sin kx) \cos \omega t \qquad [14.3]$$

The maximum amplitude points (called **antinodes**) are separated by a distance $\lambda/2$. Halfway between antinodes are points of zero amplitude (called **nodes**).

The **wave under boundary conditions** model tells us that when boundary conditions are applied to a wave, we find that only certain waves—those that satisfy the boundary conditions—are allowed. This restriction leads to **quantization** of the frequencies of the system.

One can set up standing waves with quantized frequencies in such systems as stretched strings and air columns. The natural frequencies of vibration of a stretched string of length $L$, fixed at both ends, are

$$f_n = \frac{n}{2L}\sqrt{\frac{T}{\mu}} \qquad (n = 1, 2, 3, \ldots) \qquad [14.8]$$

where $T$ is the tension in the string and $\mu$ is its mass per unit length. The natural frequencies of vibration form a **harmonic series,** that is, $f_1, 2f_1, 3f_1, \ldots$.

The standing wave patterns for longitudinal waves in an air column depend on whether the ends of the column are open or closed. If the column is open at both ends, the natural frequencies of vibration form a harmonic series. If one end is closed, only odd harmonics of the fundamental are present.

The phenomenon of **beats** occurs as a result of the superposition of two traveling waves of slightly different frequencies. For sound waves at a given point, one hears an alternation in sound intensity with time.

Any periodic waveform can be represented by the combination of sinusoidal waves that form a harmonic series. The process is based on **Fourier's theorem.**

# QUESTIONS

☐ = answer available in the *Student Solutions Manual and Study Guide*

1. Does the phenomenon of wave interference apply only to sinusoidal waves?
2. Can two pulses traveling in opposite directions on the same string reflect from each other? Explain.
3. When two waves interfere, can the amplitude of the resultant wave be greater than either of the two original waves? If so, under what conditions can that happen?
4. For certain positions of the movable section shown in Figure 14.4, no sound is detected at the receiver, a situation corresponding to destructive interference. This situation suggests that energy is somehow lost. What happens to the energy transmitted by the speaker?
5. When two waves interfere constructively or destructively, is there any gain or loss in energy? Explain.
6. What limits the amplitude of motion of a real vibrating system that is driven at one of its resonant frequencies?
7. Explain why your voice seems to sound better than usual when you sing in the shower.
8. What is the purpose of the slide on a trombone or of the valves on a trumpet?
9. Why does a vibrating guitar string sound louder when placed on the instrument than it would if allowed to vibrate in air while off the instrument?
10. Explain why all harmonics are present in an organ pipe open at both ends, but only the odd harmonics are present in a pipe closed at one end.
11. An archer shoots an arrow from a bow. Does the string of the bow exhibit standing waves after the arrow leaves? If so, and if the bow is perfectly symmetric so that the arrow leaves from the center of the string, what harmonics are excited?
12. Explain how a musical instrument such as a piano may be tuned by using the phenomenon of beats.
13. An airplane mechanic notices that the sound from a twin-engine aircraft rapidly varies in loudness when both engines are running. What could be causing this variation from loud to soft?
14. Despite a reasonably steady hand, a person often spills his coffee when carrying it to his seat. Discuss resonance as a possible cause of this difficulty and devise a means for solving the problem.
15. You have a standard tuning fork whose frequency is 262 Hz and a second tuning fork with an unknown frequency. When you tap both of them on the heel of one of your sneakers, you hear beats with a frequency of 4 per second.

Thoughtfully chewing your gum, you wonder whether the unknown frequency is 258 Hz or 266 Hz. How can you decide?

**16.** When the base of a vibrating tuning fork is placed against a chalkboard, the sound that it emits becomes louder because the vibrations of the tuning fork are transmitted to the chalkboard. Because it has a larger area than the tuning fork, the vibrating chalkboard sets more air into vibration. Therefore, the chalkboard is a better radiator of sound than the tuning fork. How does that affect the time interval during which the fork vibrates? Does that agree with the principle of conservation of energy?

## PROBLEMS

1, 2, 3 = straightforward, intermediate, challenging
☐ = full solution available in the *Student Solutions Manual and Study Guide*
PhysicsNow = coached problem with hints available at www.pop4e.com
= computer useful in solving problem
= paired numerical and symbolic problems
= biomedical application

### Section 14.1 ■ The Principle of Superposition

**1.** Two waves in one string are described by the wave functions

$$y_1 = 3.0 \cos(4.0x - 1.6t) \quad \text{and} \quad y_2 = 4.0 \sin(5.0x - 2.0t)$$

where $y$ and $x$ are in centimeters and $t$ is in seconds. Find the superposition of the waves $y_1 + y_2$ at the points (a) $x = 1.00$, $t = 1.00$; (b) $x = 1.00$, $t = 0.500$; and (c) $x = 0.500$, $t = 0$. (Remember that the arguments of the trigonometric functions are in radians.)

**2.** Two pulses A and B are moving in opposite directions along a taut string with a speed of 2.00 cm/s. The amplitude of A is twice the amplitude of B. The pulses are shown in Figure P14.2 at $t = 0$. Sketch the shape of the string at $t = 1$, 1.5, 2, 2.5, and 3 s.

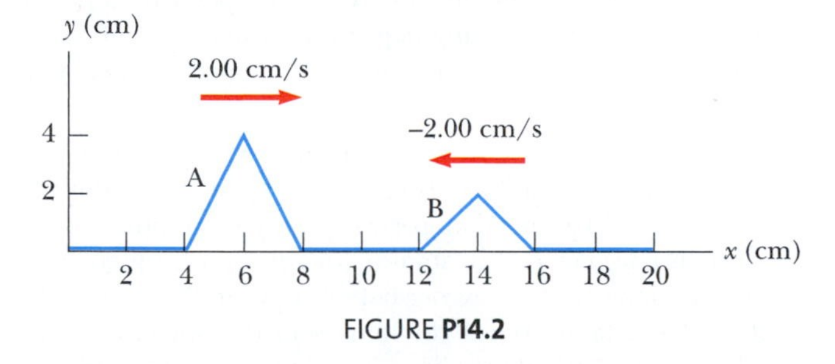


FIGURE **P14.2**

**3.** Two pulses traveling on the same string are described by

$$y_1 = \frac{5}{(3x - 4t)^2 + 2} \quad \text{and} \quad y_2 = \frac{-5}{(3x + 4t - 6)^2 + 2}$$

(a) In which direction does each pulse travel? (b) At what time do the two cancel everywhere? (c) At what point do the two pulses always cancel?

### Section 14.2 ■ Interference of Waves

**4.** Two waves are traveling in the same direction along a stretched string. The waves are 90.0° out of phase. Each wave has an amplitude of 4.00 cm. Find the amplitude of the resultant wave.

**5.** PhysicsNow™ Two traveling sinusoidal waves are described by the wave functions

$$y_1 = (5.00 \text{ m}) \sin[\pi(4.00x - 1\,200t)]$$

and

$$y_2 = (5.00 \text{ m}) \sin[\pi(4.00x - 1\,200t - 0.250)]$$

where $x$, $y_1$, and $y_2$ are in meters and $t$ is in seconds. (a) What is the amplitude of the resultant wave? (b) What is the frequency of the resultant wave?

**6.** Two identical sinusoidal waves with wavelengths of 3.00 m travel in the same direction at a speed of 2.00 m/s. The second wave originates from the same point as the first, but at a later time. The amplitude of the resultant wave is the same as that of each of the two initial waves. Determine the minimum possible time interval between the starting moments of the two waves.

**7.** A tuning fork generates sound waves with a frequency of 246 Hz. The waves travel in opposite directions along a hallway, are reflected by end walls, and return. The hallway is 47.0 m long and the tuning fork is located 14.0 m from one end. What is the phase difference between the reflected waves when they meet at the tuning fork? The speed of sound in air is 343 m/s.

**8.** Two loudspeakers are placed on a wall 2.00 m apart. A listener stands 3.00 m from the wall directly in front of one of the speakers. A single oscillator is driving the speakers at a frequency of 300 Hz. (a) What is the phase difference between the two waves when they reach the observer? (b) What is the frequency closest to 300 Hz to which the oscillator may be adjusted such that the observer hears minimal sound?

**9.** Two sinusoidal waves in a string are defined by the functions

$$y_1 = (2.00 \text{ cm}) \sin(20.0x - 32.0t)$$

and

$$y_2 = (2.00 \text{ cm}) \sin(25.0x - 40.0t)$$

where $y$ and $x$ are in centimeters and $t$ is in seconds. (a) What is the phase difference between these two waves at the point $x = 5.00$ cm at $t = 2.00$ s? (b) What is the positive $x$ value closest to the origin for which the two phases differ by $\pm\pi$ at $t = 2.00$ s? (This location is where the two waves add to zero.)

**10.** Two speakers are driven by the same oscillator of frequency $f$. They are located a distance $d$ from each other on a vertical pole. A man walks straight toward the lower speaker in a direction perpendicular to the pole as shown in Figure P14.10. (a) How many times will he hear a minimum in sound intensity? (b) How far is he from the pole at these moments? Let $v$ represent the speed of sound and assume that the ground does not reflect sound.

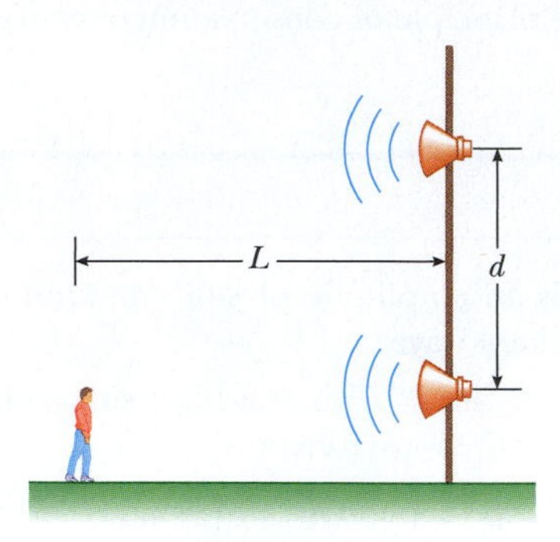


FIGURE **P14.10**

**11.** Two identical speakers 10.0 m apart are driven by the same oscillator with a frequency of $f = 21.5$ Hz (Fig. P14.11). (a) Explain why a receiver at point $A$ records a minimum in sound intensity from the two speakers. (b) If the receiver is moved in the plane of the speakers, what path should it take so that the intensity remains at a minimum? That is, determine the relationship between $x$ and $y$ (the coordinates of the receiver) that causes the receiver to record a minimum in sound intensity. Take the speed of sound to be 344 m/s.

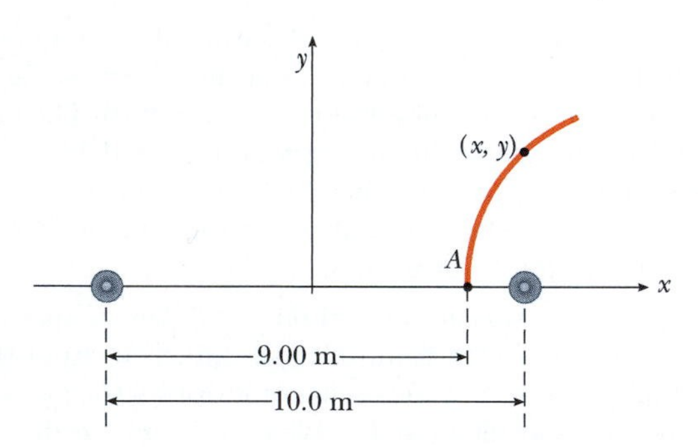


FIGURE **P14.11**

## Section 14.3 ■ Standing Waves

**12.** Two sinusoidal waves traveling in opposite directions interfere to produce a standing wave with the wave function

$$y = (1.50 \text{ m}) \sin(0.400x) \cos(200t)$$

where $x$ is in meters and $t$ is in seconds. Determine the wavelength, frequency, and speed of the interfering waves.

**13.** **Physics Now™** Two speakers are driven in phase by a common oscillator at 800 Hz and face each other at a distance of 1.25 m. Locate the points along a line joining the two speakers where relative minima of sound pressure amplitude would be expected. (Use $v = 343$ m/s.)

**14.** Verify by direct substitution that the wave function for a standing wave given in Equation 14.3,

$$y = (2A \sin kx) \cos \omega t$$

is a solution of the general linear wave equation, Equation 13.19:

$$\frac{\partial^2 y}{\partial x^2} = \frac{1}{v^2}\frac{\partial^2 y}{\partial t^2}$$

**15.** Two sinusoidal waves combining in a medium are described by the wave functions

$$y_1 = (3.0 \text{ cm}) \sin \pi(x + 0.60t)$$

and

$$y_2 = (3.0 \text{ cm}) \sin \pi(x - 0.60t)$$

where $x$ is in centimeters and $t$ is in seconds. Determine the maximum transverse position of an element of the medium at (a) $x = 0.250$ cm, (b) $x = 0.500$ cm, and (c) $x = 1.50$ cm. (d) Find the three smallest values of $x$ corresponding to antinodes.

**16.** Two waves simultaneously present in a long string are given by the wave functions

$$y_1 = A \sin(kx - \omega t + \phi) \quad \text{and} \quad y_2 = A \sin(kx + \omega t)$$

In the case $\phi = 0$, the chapter text shows that they add to a standing wave. Demonstrate (a) that the addition of the arbitrary phase constant $\phi$ changes only the position of the nodes and, in particular, (b) that the distance between nodes is still one half the wavelength.

## Section 14.4 ■ Standing Waves in Strings

**17.** Find the fundamental frequency and the next three frequencies that could cause standing wave patterns on a string that is 30.0 m long, has a mass per length of $9.00 \times 10^{-3}$ kg/m, and is stretched to a tension of 20.0 N.

**18.** A standing wave is established in a 120-cm-long string fixed at both ends. The string vibrates in four segments when driven at 120 Hz. (a) Determine the wavelength. (b) What is the fundamental frequency of the string?

**19.** A string with a mass of 8.00 g and a length of 5.00 m has one end attached to a wall; the other end is draped over a pulley and attached to a hanging object with a mass of 4.00 kg. If the string is plucked, what is the fundamental frequency of vibration?

**20.** In the arrangement shown in Figure P14.20, an object can be hung from a string (with linear mass density $\mu = 0.002\,00$ kg/m) that passes over a light pulley. The string is connected to a vibrator (of constant frequency $f$), and the length of the string between point $P$ and the pulley is $L = 2.00$ m. When the mass $m$ of the object is either 16.0 kg or 25.0 kg, standing waves are observed, but no standing waves are observed with any mass between these

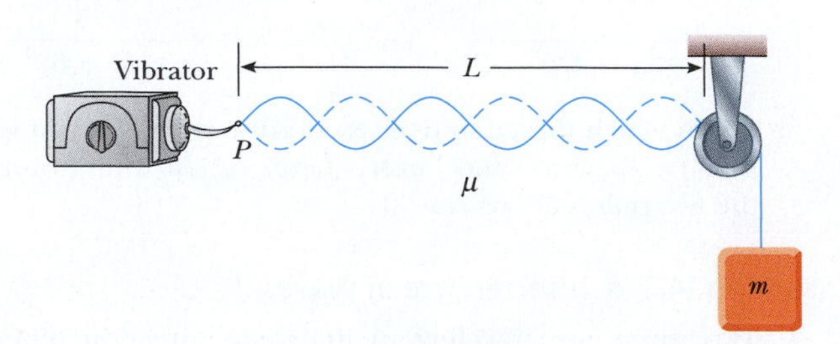


FIGURE **P14.20**

values. (a) What is the frequency of the vibrator? (*Note:* The greater the tension in the string, the smaller the number of nodes in the standing wave.) (b) What is the largest object mass for which standing waves could be observed?

**21.** A string of length $L$, mass per unit length $\mu$, and tension $T$ is vibrating at its fundamental frequency. What effect will the following have on the fundamental frequency? (a) The length of the string is doubled, with all other factors held constant. (b) The mass per unit length is doubled, with all other factors held constant. (c) The tension is doubled, with all other factors held constant.

**22.** The top string of a guitar has a fundamental frequency of 330 Hz when it is allowed to vibrate as a whole, along all its 64.0-cm length from the neck to the bridge. A fret is provided for limiting vibration to just the lower two thirds of the string. (a) If the string is pressed down at this fret and plucked, what is the new fundamental frequency? (b) The guitarist can play a "natural harmonic" by gently touching the string at the location of this fret and plucking the string at about one sixth of the way along its length from the bridge. What frequency will be heard then?

**23.** The A string on a cello vibrates in its first normal mode with a frequency of 220 Hz. The vibrating segment is 70.0 cm long and has a mass of 1.20 g. (a) Find the tension in the string. (b) Determine the frequency of vibration when the string vibrates in three segments.

**24.** A violin string has a length of 0.350 m and is tuned to concert G, with $f_G = 392$ Hz. Where must the violinist place her finger to play concert A, with $f_A = 440$ Hz? If this position is to remain correct to one-half the width of a finger (that is, to within 0.600 cm), what is the maximum allowable percentage change in the string tension?

**25. Review problem.** A sphere of mass $M$ is supported by a string that passes over a light horizontal rod of length $L$ (Fig. P14.25). Given that the angle is $\theta$ and that $f$ represents the fundamental frequency of standing waves in the portion of the string above the rod, determine the mass of this portion of the string.

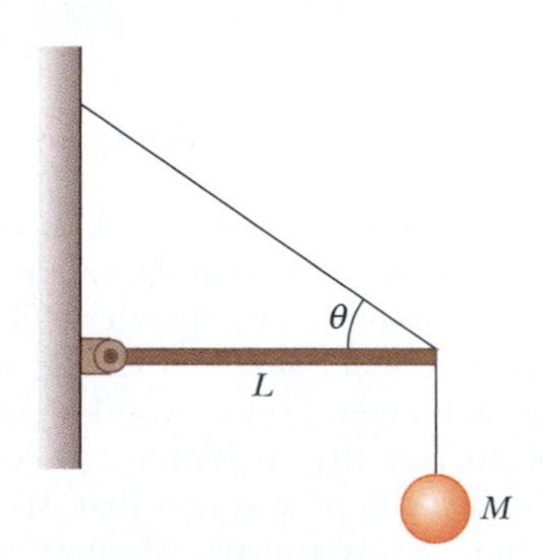


FIGURE **P14.25**

**26.** A standing wave pattern is observed in a thin wire with a length of 3.00 m. The equation of the wave is

$$y = (0.002 \text{ m}) \sin(\pi x) \cos(100\pi t)$$

where $x$ is in meters and $t$ is in seconds. (a) How many loops does this pattern exhibit? (b) What is the fundamental frequency of vibration of the wire? (c) If the original frequency is held constant and the tension in the wire is increased by a factor of nine, how many loops are present in the new pattern?

## Section 14.5 ▪ Standing Waves in Air Columns

*Note:* Unless otherwise specified, assume that the speed of sound in air is 343 m/s at 20°C and is described by

$$v = 331 \text{ m/s} + (0.6 \text{ m/s}\cdot{}^\circ\text{C})T_C$$

at any Celsius temperature $T_C$.

**27.** Calculate the length of a pipe that has a fundamental frequency of 240 Hz if the pipe is (a) closed at one end and (b) open at both ends.

**28.** The overall length of a piccolo is 32.0 cm. The resonating air column vibrates as in a pipe open at both ends. (a) Find the frequency of the lowest note that a piccolo can play, assuming that the speed of sound in air is 340 m/s. (b) Opening holes in the side effectively shortens the length of the resonant column. Assuming that the highest note a piccolo can sound is 4 000 Hz, find the distance between adjacent antinodes for this mode of vibration.

**29.** The windpipe of one typical whooping crane is 5.00 feet long. What is the fundamental resonant frequency of the bird's trachea, modeled as a narrow pipe closed at one end? Assume a temperature of 37°C.

**30.** The fundamental frequency of an open organ pipe corresponds to middle C (261.6 Hz on the chromatic musical scale). The third resonance of a closed organ pipe has the same frequency. What is the length of each of the two pipes?

**31.** **PhysicsNow™** A shower stall has dimensions 86.0 cm × 86.0 cm × 210 cm. If you were singing in this shower, which frequencies would sound the richest (because of resonance)? Assume that the shower stall acts as a pipe closed at both ends, with nodes at opposite sides, and that the voices of various singers range from 130 Hz to 2 000 Hz. Let the speed of sound in the hot air be 355 m/s.

**32.** Do not stick anything into your ear! Estimate the length of your ear canal, from its opening at the external ear to the eardrum. If you regard the canal as a narrow tube that is open at one end and closed at the other, at approximately what fundamental frequency would you expect your hearing to be most sensitive? Explain why you can hear especially soft sounds just around this frequency.

**33.** **PhysicsNow™** Two adjacent natural frequencies of an organ pipe are determined to be 550 Hz and 650 Hz. Calculate the fundamental frequency and length of this pipe. (Use $v = 340$ m/s.)

**34.** As shown in Figure P14.34, water is pumped into a tall vertical cylinder at a volume flow rate $R$. The radius of the cylinder is $r$, and at the open top of the cylinder a tuning fork is vibrating with a frequency $f$. As the water rises, how much time elapses between successive resonances?

**35.** A glass tube (open at both ends) of length $L$ is positioned near an audio speaker of frequency $f = 680$ Hz. For what values of $L$ will the tube resonate with the speaker?

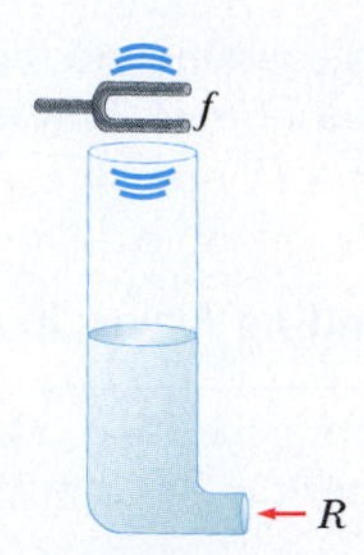


FIGURE P14.34

**36.** A tuning fork with a frequency of 512 Hz is placed near the top of the tube shown in Figure 14.11a. The water level is lowered so that the length $L$ slowly increases from an initial value of 20.0 cm. Determine the next two values of $L$ that correspond to resonant modes.

**37.** An air column in a glass tube is open at one end and closed at the other by a movable piston. The air in the tube is warmed above room temperature and a 384-Hz tuning fork is held at the open end. Resonance is heard when the piston is 22.8 cm from the open end and again when it is 68.3 cm from the open end. (a) What speed of sound is implied by these data? (b) How far from the open end will the piston be when the next resonance is heard?

**38.** With a particular fingering, a flute plays a note with frequency 880 Hz at 20.0°C. The flute is open at both ends. (a) Find the air column length. (b) Find the frequency it produces at the beginning of the halftime performance at a late-season American football game, when the ambient temperature is $-5.00$°C and the player has not had a chance to warm up the flute.

## Section 14.6 ■ Beats: Interference in Time

**39.** **PhysicsNow™** In certain ranges of a piano keyboard, more than one string is tuned to the same note to provide extra loudness. For example, the note at 110 Hz has two strings at this frequency. If one string slips from its normal tension of 600 N to 540 N, what beat frequency is heard when the hammer strikes the two strings simultaneously?

**40.** While attempting to tune the note C at 523 Hz, a piano tuner hears 2.00 beats/s between a reference oscillator and the string. (a) What are the possible frequencies of the string? (b) When she tightens the string slightly, she hears 3.00 beats/s. What is the frequency of the string now? (c) By what percentage should the piano tuner now change the tension in the string to bring it into tune?

**41.** A student holds a tuning fork oscillating at 256 Hz. He walks toward a wall at a constant speed of 1.33 m/s. (a) What beat frequency does he observe between the tuning fork and its echo? (b) How fast must he walk away from the wall to observe a beat frequency of 5.00 Hz?

## Section 14.7 Nonsinusoidal Wave Patterns

**42.** Suppose a flutist plays a 523-Hz C note with first harmonic displacement amplitude $A_1 = 100$ nm. From Figure 14.14b read, by proportion, the displacement amplitudes of harmonics 2 through 7. Take these values as the coefficients $A_2$ through $A_7$ in the Fourier analysis of the sound and assume that $B_1 = B_2 = \cdots = B_7 = 0$. Construct a graph of the waveform of the sound. Your waveform will not look exactly like the flute waveform in Figure 14.13b because you simplify by ignoring cosine terms; nevertheless, it produces the same sensation to human hearing.

**43.** An A-major chord consists of the notes called A, C$^{\#}$, and E. It can be played on a piano by simultaneously striking strings with fundamental frequencies of 440.00 Hz, 554.37 Hz, and 659.26 Hz. The rich consonance of the chord is associated with near equality of the frequencies of some of the higher harmonics of the three tones. Consider the first five harmonics of each string and determine which harmonics show near equality.

## Section 14.8 ■ Context Connection—Building on Antinodes

**44.** An earthquake can produce a seiche in a lake in which the water sloshes back and forth from end to end with remarkably large amplitude and long period. Consider a seiche produced in a rectangular farm pond as shown in the cross-sectional view of Figure P14.44. (The figure is not drawn to scale.) Suppose the pond is 9.15 m long and of uniform width and depth. You measure that a pulse produced at one end reaches the other end in 2.50 s. (a) What is the wave speed? (b) To produce the seiche, several people stand on the bank at one end and paddle together with snow shovels, moving them in simple harmonic motion. What should be the frequency of this motion?

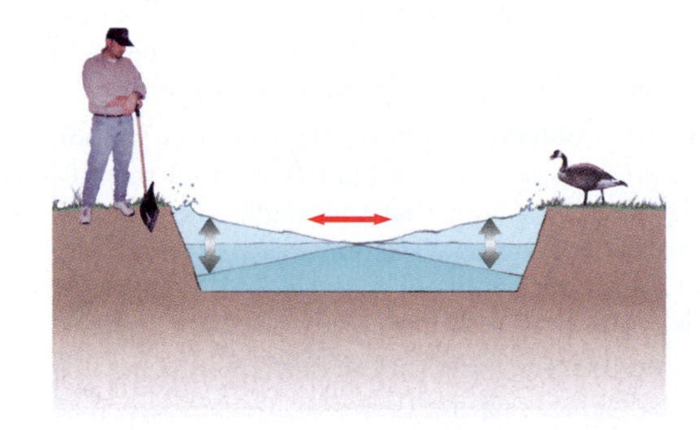

FIGURE P14.44

**45.** The Bay of Fundy, Nova Scotia, has the highest tides in the world. Assume that in midocean and at the mouth of the bay, the Moon's gravity gradient and the Earth's rotation make the water surface oscillate with an amplitude of a few centimeters and a period of 12 h 24 min. At the head of the bay, the amplitude is several meters. Argue for or against the proposition that the tide is magnified by standing wave resonance. Assume that the bay has a length of 210 km and a uniform depth of 36.1 m. The speed of long-wavelength water waves is given by $\sqrt{gd}$, where $d$ is the water's depth.

## Additional Problems

**46.** Figure P14.46a is a photograph of a vibrating wine glass. A special technique makes black and white stripes appear where the glass is moving, with closer spacing where the amplitude is larger. Six nodes and six antinodes alternate around the rim of the glass in the vibration photographed, but consider instead the case of a standing wave vibration with four nodes and four antinodes equally spaced around the 20.0-cm circumference of the rim of a goblet. If transverse waves move around the glass at 900 m/s, an opera singer would have to produce a high harmonic with what frequency to shatter the glass with a resonant vibration as shown in Figure P14.46b?

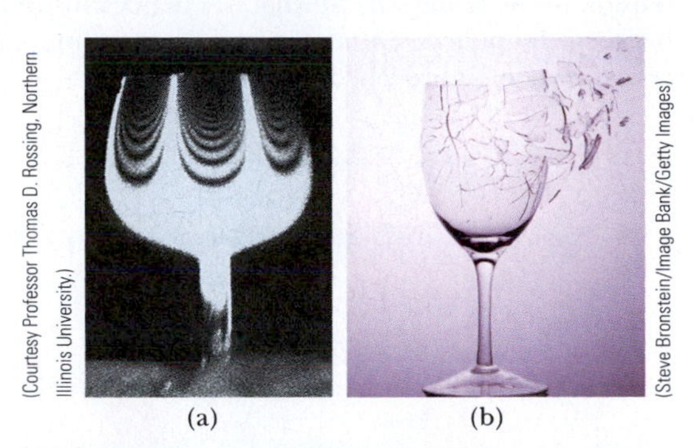

(Courtesy Professor Thomas D. Rossing, Northern Illinois University.) (Steve Bronstein/Image Bank/Getty Images)

(a) (b)

**FIGURE P14.46** (a) Nodes (in white) and antinodes (where the stripes converge to black) alternate around the rim of a vibrating wine glass. (b) A glass shatters when vibrating with large amplitude.

**47.** On a marimba (Fig. P14.47), the wooden bar that sounds a tone when struck vibrates in a transverse standing wave having three antinodes and two nodes. The lowest-frequency note is 87.0 Hz, produced by a bar 40.0 cm long. (a) Find the speed of transverse waves on the bar. (b) A resonant pipe suspended vertically below the center of the bar enhances the loudness of the emitted sound. If the pipe is open at the top end only and the speed of sound in air is 340 m/s, what is the length of the pipe required to resonate with the bar in part (a)?

(Murray Greenberg)

**FIGURE P14.47** Marimba players in Mexico City.

**48.** A nylon string has mass 5.50 g and length 86.0 cm. One end is tied to the floor and the other end to a small magnet, with a mass negligible compared with the string. A magnetic field (which we will study in Chapter 22) exerts an upward force of 1.30 N on the magnet, wherever the magnet is located. At equilibrium, the string is vertical and motionless, with the magnet at the top. When it is carrying a small-amplitude wave, you may assume the string is always under uniform tension 1.30 N. (a) Find the speed of transverse waves on the string. (b) The string's vibration possibilities are a set of standing wave states, each with a node at the fixed bottom end and an antinode at the free top end. Find the node–antinode distances for each one of the three simplest states. (c) Find the frequency of each of these states.

**49.** Two train whistles have identical frequencies of 180 Hz. When one train is at rest in the station and the other is moving nearby, a commuter standing on the station platform hears beats with a frequency of 2.00 beats/s when the whistles sound at the same time. What are the two possible speeds and directions that the moving train can have?

**50.** A loudspeaker at the front of a room and an identical loudspeaker at the rear of the room are being driven by the same oscillator at 456 Hz. A student walks at a uniform rate of 1.50 m/s along the length of the room. She hears a single tone repeatedly becoming louder and softer. (a) Model these variations as beats between the Doppler-shifted sounds the student receives. Calculate the number of beats the student hears each second. (b) Model the two speakers as producing a standing wave in the room and the student as walking between antinodes. Calculate the number of intensity maxima the student hears each second.

**51.** A student uses an audio oscillator of adjustable frequency to measure the depth of a water well. The student hears two successive resonances at 51.5 Hz and 60.0 Hz. How deep is the well?

**52.** A string fixed at both ends and having a mass of 4.80 g, a length of 2.00 m, and a tension of 48.0 N vibrates in its second ($n = 2$) normal mode. What is the wavelength in air of the sound emitted by this vibrating string?

**53.** Two wires are welded together end to end. The wires are made of the same material, but the diameter of one is twice that of the other. They are subjected to a tension of 4.60 N. The thin wire has a length of 40.0 cm and a linear mass density of 2.00 g/m. The combination is fixed at both ends and vibrated in such a way that two antinodes are present, with the node between them being right at the weld. (a) What is the frequency of vibration? (b) How long is the thick wire?

**54.** A string is 0.400 m long and has a mass per unit length of $9.00 \times 10^{-3}$ kg/m. What must be the tension in the string if its second harmonic has the same frequency as the second resonance mode of a 1.75-m-long pipe open at one end?

**55.** A standing wave is set up in a string of variable length and tension by a vibrator of variable frequency. Both ends of the string are fixed. When the vibrator has a frequency $f$, in a string of length $L$ and under tension $T$, $n$ antinodes are set up in the string. (a) If the length of the string is doubled, by what factor should the frequency be changed so that the same number of antinodes is produced? (b) If the frequency and length are held constant, what tension will produce $n + 1$ antinodes? (c) If the frequency is tripled and the length of the string is halved, by what factor should the tension be changed so that twice as many antinodes are produced?

**56.** **Review problem.** For the arrangement shown in Figure P14.56, $\theta = 30.0°$, the inclined plane and the small pulley are frictionless, the string supports the object of mass $M$ at the bottom of the plane, and the string has mass $m$ that is small compared with $M$. The system is in equilibrium and the vertical part of the string has a length $h$. Standing waves are set up in the vertical section of the string. (a) Find the tension in the string. (b) Model the shape of the string as one leg and the hypotenuse of a right triangle. Find the whole length of the string. (c) Find the mass per unit length of the string. (d) Find the speed of waves on the string. (e) Find the lowest frequency for a standing wave. (f) Find the period of the standing wave having three nodes. (g) Find the wavelength of the standing wave having three nodes. (h) Find the frequency of the beats resulting from the interference of the sound wave of lowest frequency generated by the string with another sound wave having a frequency that is 2.00% greater.

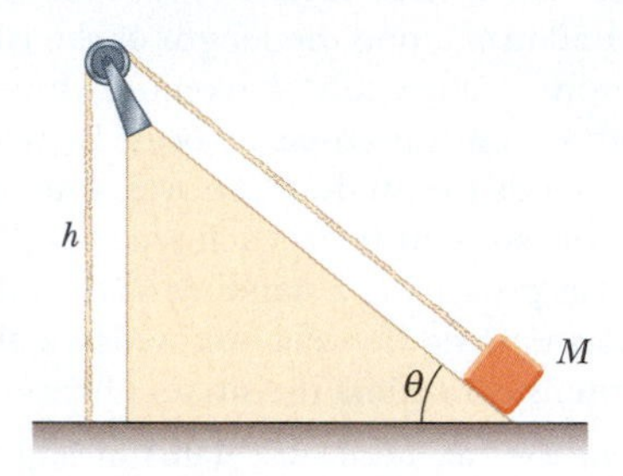


FIGURE **P14.56**

**57.** Two waves are described by the wave functions

$$y_1(x, t) = 5.0 \sin(2.0x - 10t)$$

$$y_2(x, t) = 10 \cos(2.0x - 10t)$$

where $y_1$, $y_2$, and $x$ are in meters and $t$ is in seconds. Show that the wave resulting from their superposition is sinusoidal. Determine the amplitude and phase of this sinusoidal wave.

**58.** A 0.010 0-kg wire, 2.00 m long, is fixed at both ends and vibrates in its simplest mode under a tension of 200 N. When a vibrating tuning fork is placed near the wire, a beat frequency of 5.00 Hz is heard. (a) What could be the frequency of the tuning fork? (b) What should the tension in the wire be if the beats are to disappear?

**59. Review problem.** A 12.0-kg object hangs in equilibrium from a string with a total length of $L = 5.00$ m and a linear mass density of $\mu = 0.001\,00$ kg/m. The string is wrapped around two light, frictionless pulleys that are separated by a distance of $d = 2.00$ m (Fig. P14.59a). (a) Determine the tension in the string. (b) At what frequency must the string between the pulleys vibrate to form the standing wave pattern shown in Figure P14.59b?

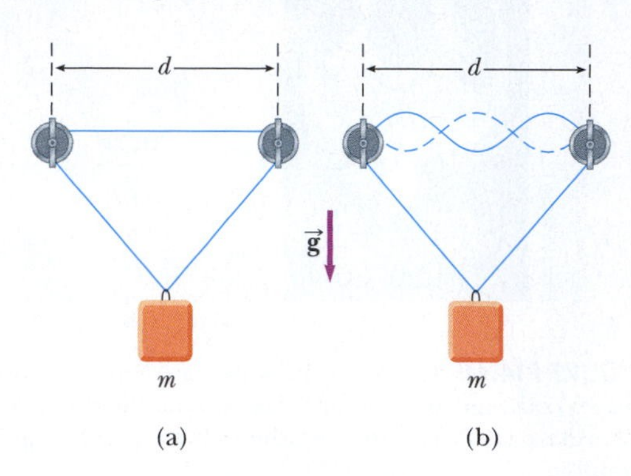


FIGURE **P14.59**

**60.** A quartz watch contains a crystal oscillator in the form of a block of quartz that vibrates by contracting and expanding. Two opposite faces of the block, 7.05 mm apart, are antinodes, moving alternately toward each other and away from each other. The plane halfway between these two faces is a node of the vibration. The speed of sound in quartz is 3.70 km/s. Find the frequency of the vibration. An oscillating electric voltage accompanies the mechanical oscillation; the quartz is described as *piezoelectric*. An electric circuit feeds in energy to maintain the oscillation and also counts the voltage pulses to keep time.

## ANSWERS TO QUICK QUIZZES

**14.1** (c). The pulses completely cancel each other in terms of displacement of elements of the string from equilibrium, but the string is still moving. A short time later, the string will be displaced again and the pulses will have passed each other.

**14.2** (i), (a). The pattern shown at the bottom of Active Figure 14.8a corresponds to the extreme position of the string. All elements of the string have momentarily come to rest. Notice that the time derivative of Equation 14.3 gives the transverse velocity of each element of the string: $v(t) = dy/dt = -(2\omega A \sin kx) \sin \omega t$. Whenever $\omega t = n\pi$, the velocity of every element of the string is equal to zero. (ii), (d). Near a nodal point, elements on one side of the point are moving upward at this instant and elements on the other side are moving downward.

**14.3** (d). Choice (a) is incorrect because the number of nodes is one greater than the number of antinodes. Choice (b) is only true for half of the modes; it is not true for any odd-numbered mode. Choice (c) would be correct if we replace the word *nodes* with *antinodes*.

**14.4** (b). With both ends open, the pipe has a fundamental frequency given by Equation 14.10: $f_{\text{open}} = v/2L$. With one end closed, the pipe has a fundamental frequency given by Equation 14.11:

$$f_{\text{closed}} = \frac{v}{4L} = \frac{1}{2}\left(\frac{v}{2L}\right) = \frac{1}{2} f_{\text{open}}$$

**14.5** (c). The increase in temperature causes the speed of sound to go up. According to Equation 14.10, this effect will result in an increase in the fundamental frequency of a given organ pipe.

**14.6** (b). Tightening the string has caused the frequencies to be farther apart based on the increase in the beat frequency, so you want to loosen the string.

## Minimizing the Risk

We have explored the physics of vibrations and waves. Let us now return to our central question for this *Earthquakes* Context:

***How can we choose locations and build structures to minimize the risk of damage in an earthquake?***

To answer this question, we shall use the physical principles that we now understand more clearly and apply them to our choices of locations and structural design.

In our discussion of simple harmonic oscillation, we learned about resonance. Designers of structures in earthquake-prone areas need to pay careful attention to resonance vibrations from shaking of the ground. The design features to be considered include ensuring that the resonance frequencies of the building do not match typical earthquake frequencies. In addition, the structural details should include sufficient damping to ensure that the amplitude of resonance vibration does not destroy the structure.

In Chapter 13, we discussed the role of the medium in the propagation of a wave. For seismic waves moving across the surface of the Earth, the soil on the surface is the medium. Because soil varies from one location to another, the speed of seismic waves will vary at different locations. A particularly dangerous situation exists for structures built on loose soil or mudfill. In these types of media, the interparticle forces are much weaker than in a more solid foundation such as granite bedrock. As a result, the wave speed is less in loose soil than in bedrock.

Consider Equation 13.23, which provides an expression for the rate of energy transfer by waves. This equation was derived for waves on strings, but the proportionality to the square of the amplitude and the speed is general. Because of conservation of energy, the rate of energy transfer for a wave must remain constant regardless of the medium. Thus, according to Equation 13.23, **if the wave speed decreases, as it does for seismic waves moving from rock into loose soil, the amplitude must increase.** As a result, the shaking of structures built on loose soil is of larger magnitude than for those built on solid bedrock.

This factor contributed to the collapse of the Nimitz Freeway during the Loma Prieta earthquake, near San Francisco, in 1989. Figure 1 shows the results of the earthquake on the freeway. The portion of the freeway that collapsed was built on mudfill, but the surviving portion was built on bedrock. The amplitude of oscillation in the portion built on mudfill was more than five times as large as the amplitude of other portions.

(Paul X. Scott/SYGMA)

FIGURE 1 Portions of the double-decked Nimitz Freeway in Oakland, California, collapsed during the Loma Prieta earthquake of 1989.

Another danger for structures on loose soil is the possibility of

(Photo by M. Hamada; used by permission from the Multidisciplinary Center for Earthquake Engineering Research, Univ. of Buffalo)

FIGURE 2 The collapse of this crane during the Kobe, Japan, earthquake of 1995 was caused by liquefaction of the underlying soil.

**liquefaction** of the soil. When soil is shaken, the elements of soil can move with respect to one another and the soil tends to act like a liquid rather than a solid. It is possible for the structure to sink into the soil during an earthquake. If the liquefaction is not uniform over the foundation of the structure, the structure can tip over, as seen in Figure 2. As a result, even if the earthquake is not sufficient to damage the structure, it will be unusable in its tipped-over orientation.

As discussed in Chapter 14, building structures where standing seismic waves can be established is dangerous. Such construction was a factor in the Michoacán earthquake of 1985. The shape of the bedrock under Mexico City resulted in standing waves, with severe damage to buildings located at antinodes.

In summary, to minimize risk of damage in an earthquake, architects and engineers must design structures to prevent destructive resonances, avoid building on loose soil, and pay attention to the underground rock formations so as to be aware of possible standing wave patterns. Other precautions can also be taken. For example, buildings can be constructed with **seismic isolation** from the ground. This method involves mounting the structure on **isolation dampers,** heavy-duty bearings that dampen the oscillations of the building, resulting in reduced amplitude of vibration.

We have not addressed many other considerations for earthquake safety in structures, but we have been able to apply many of our concepts from oscillations and waves so as to understand some aspects of logical choices in locating and designing structures.

## Problems

1. For seismic waves spreading out from a point (the epicenter) on the surface of the Earth, the intensity of the waves decreases with distance according to an inverse proportionality to distance. That is, the wave intensity is proportional to $1/r$, where $r$ is the distance from the epicenter to the observation point. This rule applies if the medium is uniform. The intensity of the wave is proportional to the rate of energy transfer for the wave. Furthermore, we have shown that the energy of vibration of an oscillator is proportional to the square of the amplitude of the vibration. Assume that a particular earthquake causes ground shaking with an amplitude of 5.0 cm at a distance of 10 km from the epicenter. If the medium is uniform, what is the amplitude of the ground shaking at a point 20 km from the epicenter?
2. As mentioned in the text, the amplitude of oscillation during the Loma Prieta earthquake of 1989 was five times greater in areas of mudfill than in areas of bedrock. From this information, find the factor by which the seismic wave speed changed as the waves moved from the bedrock to the mudfill. Ignore any reflection of wave energy and any change in density between the two media.

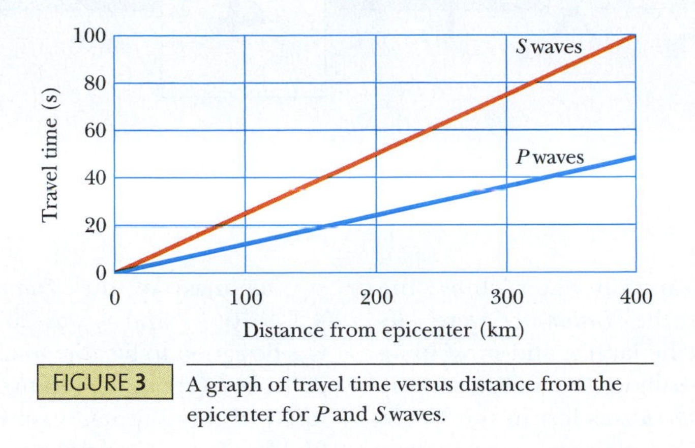


FIGURE 3 A graph of travel time versus distance from the epicenter for *P* and *S* waves.

3. Figure 3 is a graphical representation of the travel time for *P* and *S* waves from the epicenter of an earthquake to a seismograph as a function of the distance of travel. The following table shows the measured times of day for arrival of *P* waves from a particular earthquake at three seismograph locations. In the last column, fill in the times of day for the arrival of the *S* waves at the three seismograph locations.

| Seismograph Station | Distance from Epicenter (km) | *P* Wave Arrival Time | *S* Wave Arrival Time |
|---|---|---|---|
| #1 | 200 | 15:46:06 | |
| #2 | 160 | 15:46:01 | |
| #3 | 105 | 15:45:54 | |

# CONTEXT 4

## Search for the *Titanic*

The *Titanic* and its sister ships, the *Olympic* and the *Britannic,* were designed to be the largest and most luxurious liners sailing the ocean at the time. The *Titanic* was lost in the North Atlantic on its maiden voyage from Southampton, England, to New York, on April 15, 1912. It was claimed by many (although not by the White Star Line, which operated the *Titanic*) that the ship was unsinkable. This claim was proven to be untrue when the *Titanic* struck an iceberg at 11:40 P.M. on April 14 and sank less than 3 h later. This event was one of the worst maritime disasters of all time, with more than 1 500 lives lost because of a severe shortage of lifeboats. Amazingly, British ships of the time were not required to carry enough lifeboats for all passengers on board. It is fortunate that the *Titanic* was not completely full on its maiden voyage, as there would have been only enough lifeboats for a third of its passengers. As it was, there were enough lifeboats for a little more than half of the 2 200 passengers on board, but only 705 were saved due to the partial filling of the boats, which occurred for a number of reasons.

FIGURE 1 The *Titanic* on its way from Southampton to New York.

The mass of the *Titanic* was over $4.2 \times 10^7$ kg and it was 269 m long. It was designed to be able to travel at 24 to 25 knots (about 12–13 m/s), and the safety of the ship was ensured by lateral bulkheads at several places across the ship with electrically operated

FIGURE 2 During the sinking process in the early morning of April 15, 1912, the bow of the *Titanic* is under water and the stern is lifted out of the water. The huge forces necessary to hold the stern aloft caused the *Titanic* to split in the middle before sinking.

watertight doors. It is ironic to note that the design of the bulkheads actually worked against the safety of the ship, and the closing of the watertight doors, according to some experts after the disaster, caused the ship to sink more rapidly than if they had been left open.

The accidental sinking of the *Titanic* resulted from a remarkable confluence of bad luck, complacency, and poor policy. Several events occurred during the last day of its voyage that would not have led to the foundering of the ship if they had happened in just a slightly different way.

The *Titanic* has captured the interest of the public for many years. Many theatrical movies related to the ship have

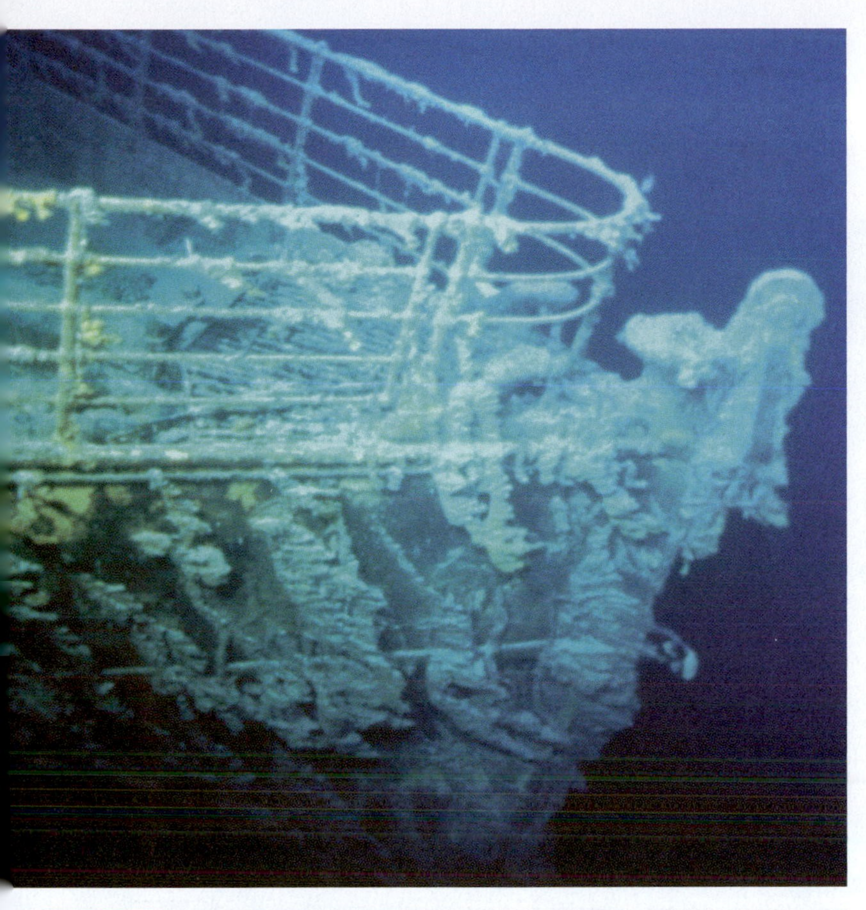

FIGURE 3 The bow of the *Titanic* as it rests on the ocean floor 80 years after it sank in the North Atlantic.

(Courtesy of RMS Titanic, Inc.)

FIGURE 4 In 1998, a 20-ton section of the *Titanic*'s hull was raised to the surface and is now part of a touring exhibit of *Titanic* artifacts. Called the "Big Piece," the hull section contains four portholes, three of which still contain the original glass.

been produced, including *Titanic* (1953), *A Night to Remember* (1958), *Raise the Titanic* (1980), *Titanic* (1997), and *Ghosts of the Abyss* (2003). A musical play, *Titanic* (1997), has been produced, and the *Titanic* plays a role in the Broadway and movie musical *The Unsinkable Molly Brown* (1964). A large number of books have been written on the disaster, many of which were reissued when the *Titanic* became wildly popular in response to the 1997 film.

Finding the *Titanic* on the ocean floor was a dream of many individuals ever since the ship was lost. The wreck of the *Titanic* was discovered in 1985 by a research team from Woods Hole Oceanographic Institute, led by Dr. Robert Ballard. This discovery led to interesting ethical questions because subsequent expeditions to the site salvaged items from the wreck, making them available for exhibition and sale to the public. We will use the *Titanic,* the search for its wreckage, and the underwater visits to its gravesite in this short Context on fluids as we address the central question:

**How can we safely visit the wreck of the *Titanic*?**

CHAPTER 15

# Fluid Mechanics

Icebergs float in the cold waters of the North Atlantic. Although the visible portion of an iceberg may tower over a passing ship, only about 11% of the iceberg is above water.

(Norman Tobley/Taxi/Getty Images)

## CHAPTER OUTLINE


15.1 Pressure
15.2 Variation of Pressure with Depth
15.3 Pressure Measurements
15.4 Buoyant Forces and Archimedes's Principle
15.5 Fluid Dynamics
15.6 Streamlines and the Continuity Equation for Fluids
15.7 Bernoulli's Equation
15.8 Other Applications of Fluid Dynamics
15.9 Context Connection—A Near Miss Even Before Leaving Southampton
SUMMARY


Matter is normally classified as being in one of three states: solid, liquid, or gas. Everyday experience tells us that a solid has a definite volume and shape. A brick maintains its familiar shape and size over a long time. We also know that a liquid has a definite volume but no definite shape. For example, a cup of liquid water has a fixed volume but assumes the shape of its container. Finally, an unconfined gas has neither definite volume nor definite shape. For example, if there is a leak in the natural gas supply in your home, the escaping gas continues to expand into the surrounding atmosphere. These definitions help us picture the states of matter, but they are somewhat artificial. For example, asphalt, glass, and plastics are normally considered solids, but over a long time interval they tend to flow like liquids. Likewise, most substances can be a solid, liquid, or gas (or combinations of these states), depending on the temperature and pressure. In general, the time interval required for a particular substance to change its shape in response to an external force determines whether we treat the substance as a solid, liquid, or gas.

A **fluid** is a collection of molecules that are randomly arranged and held together by weak cohesive forces between molecules and forces exerted by the walls of a container. Both liquids and gases are fluids. In our treatment of the mechanics of fluids, we shall see that no new physical principles are needed to explain such effects as the buoyant force on a submerged object and vascular flutter in an artery. In this chapter, we shall apply a number of familiar analysis models to the physics of fluids.

## 15.1 PRESSURE

Our first task in understanding the physics of fluids is to define a new quantity to describe fluids. Imagine applying a force to the surface of an object, with the force having components both parallel to and perpendicular to the surface. If the object is a solid at rest on a table, the force component perpendicular to the surface may cause the object to flatten, depending on how hard the object is. Assuming that the object does not slide on the table, the component of the force parallel to the surface of the object will cause the object to distort. As an example, suppose you place your physics book flat on a table and apply a force with your hand parallel to the front cover and perpendicular to the spine. The book will distort, with the bottom pages staying fixed at their original location and the top pages shifting horizontally by some distance. The cross-section of the book changes from a rectangle to a parallelogram. This kind of force parallel to the surface is called a *shearing force.*

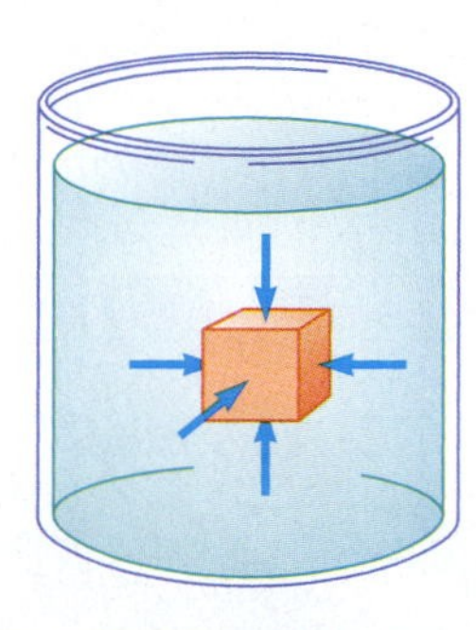

FIGURE 15.1 The force exerted by the fluid on a submerged object at any point is perpendicular to the surface of the object. The force exerted by the fluid on the walls of the container is perpendicular to the walls at all points.

We shall adopt a simplification model in which the fluids we study will be nonviscous; that is, no friction exists between adjacent layers of the fluid. Nonviscous fluids and static fluids do not sustain shearing forces. If you imagine placing your hand on a water surface and pushing parallel to the surface, your hand simply slides over the water; you cannot distort the water as you did the book. This phenomenon occurs because the interatomic forces in a fluid are not strong enough to lock atoms in place with respect to one another. The fluid cannot be modeled as a rigid object as in Chapter 10. If we try to apply a shearing force, the molecules of the fluid simply slide past one another.

Therefore, the only type of force that can exist in a fluid is one that is perpendicular to a surface. For example, the forces exerted by the fluid on the object in Figure 15.1 are everywhere perpendicular to the surfaces of the object.

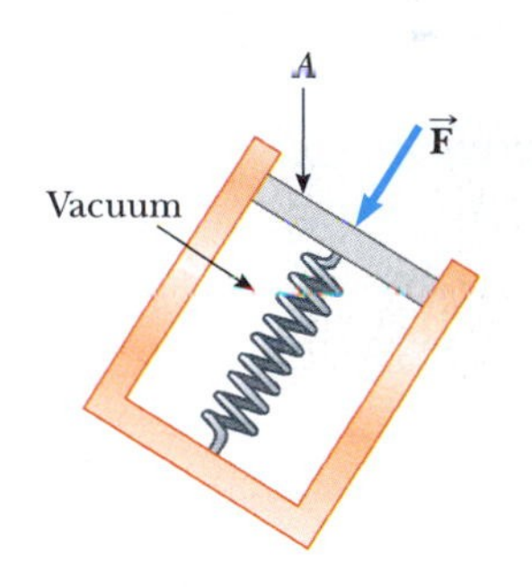


FIGURE 15.2 A simple device for measuring pressure in a fluid.

The force that a fluid exerts on a surface originates in the collisions of molecules of the fluid with the surface. Each collision results in the reversal of the component of the velocity vector of the molecule perpendicular to the surface. By the impulse–momentum theorem and Newton's third law, each collision results in a force on the surface. A huge number of these impulsive forces occur every second, resulting in a constant macroscopic force on the surface. This force is spread out over the area of the surface and is related to a new quantity called **pressure.**

### PITFALL PREVENTION 15.1

**FORCE AND PRESSURE** Equation 15.1 makes a clear distinction between force and pressure. Another important distinction is that *force is a vector* and *pressure is a scalar.* No direction is associated with pressure, but the direction of the force associated with the pressure is perpendicular to the surface of interest.

The pressure at a specific point in a fluid can be measured with the device pictured in Figure 15.2. The device consists of an evacuated cylinder enclosing a light piston connected to a spring. As the device is submerged in a fluid, the fluid presses in on the top of the piston and compresses the spring until the inward force of the fluid is balanced by the outward force of the spring. The force exerted on the piston by the fluid can be measured if the spring is calibrated in advance.

If $F$ is the magnitude of the force exerted by the fluid on the piston and $A$ is the surface area of the piston, **the pressure $P$ of the fluid at the level to which the device has been submerged is defined as the ratio of force to area:**

▪ Definition of pressure

$$P \equiv \frac{F}{A} \qquad [15.1]$$

Although we have defined pressure in terms of our device in Figure 15.2, the definition is general. Because pressure is force per unit area, it has units of newtons per square meter in the SI system. Another name for the SI unit of pressure is the **pascal** (Pa):

■ The pascal

$$1 \text{ Pa} \equiv 1 \text{ N/m}^2 \qquad [15.2]$$

Hypodermic needles

Notice that pressure and force are different quantities. We can have a very large pressure from a relatively small force by making the area over which the force is applied small. Such is the case with hypodermic needles. The area of the tip of the needle is very small, so a small force pushing on the needle is sufficient to cause a pressure large enough to puncture the skin. We can also create a small pressure from a large force by enlarging the area over which the force acts. Such is the principle behind the design of snowshoes. If a person were to walk on deep snow with regular shoes, it is possible for his or her feet to break through the snow and sink. Snowshoes, however, allow the force on the snow due to the weight of the person to spread out over a larger area, reducing the pressure enough so that the snow surface is not broken (Fig. 15.3).

(Royalty-free/CORBIS)

**FIGURE 15.3** Snowshoes keep you from sinking into soft snow because they spread the downward force you exert on the snow over a large area, reducing the pressure on the snow surface.

The atmosphere exerts a pressure on the surface of the Earth and all objects at the surface. This pressure is responsible for the action of suction cups, drinking straws, vacuum cleaners, and many other devices. In our calculations and end-of-chapter problems, we usually take atmospheric pressure to be

$$P_0 = 1.00 \text{ atm} \approx 1.013 \times 10^5 \text{ Pa} \qquad [15.3]$$

**QUICK QUIZ 15.1** Suppose you are standing directly behind someone who steps back and accidentally stomps on your foot with the heel of one shoe. Would you be better off if that person were **(a)** a large male professional basketball player wearing sneakers or **(b)** a petite woman wearing spike-heeled shoes?

### ■ Thinking Physics 15.1

Suction cups can be used to hold objects onto surfaces. Why don't astronauts use suction cups to hold onto the outside surface of an orbiting spacecraft?

**Reasoning** A suction cup works because air is pushed out from under the cup when it is pressed against a surface. When the cup is released, it tends to spring back a bit, causing the trapped air under the cup to expand. This expansion causes a reduced pressure inside the cup. Thus, the difference between the atmospheric pressure on the outside of the cup and the reduced pressure inside provides a net force pushing the cup against the surface. For astronauts in orbit around the Earth, almost no air exists outside the surface of the spacecraft. Therefore, if a suction cup were to be pressed against the outside surface of the spacecraft, the pressure differential needed to press the cup to the surface is not present. ■

## 15.2 VARIATION OF PRESSURE WITH DEPTH

The study of fluid mechanics involves the density of a substance, defined in Equation 1.1 as the mass per unit volume for the substance. Table 15.1 lists the densities of various substances. These values vary slightly with temperature because the volume of a substance is temperature-dependent (as we shall see in Chapter 16). Note that under standard conditions (0°C and atmospheric pressure) the densities of gases are on the order of 1/1 000 the densities of solids and liquids. This difference implies that the average molecular spacing in a gas under these conditions is about ten times greater in each dimension than in a solid or liquid.

As divers know well, the pressure in the sea or a lake increases as they dive to greater depths. Likewise, atmospheric pressure decreases with increasing altitude.

**TABLE 15.1** Densities of Some Common Substances at Standard Temperature (0°C) and Pressure (Atmospheric)

| Substance | $\rho$ (kg/m$^3$) | Substance | $\rho$ (kg/m$^3$) |
|---|---|---|---|
| Air | 1.29 | Ice | $0.917 \times 10^3$ |
| Aluminum | $2.70 \times 10^3$ | Iron | $7.86 \times 10^3$ |
| Benzene | $0.879 \times 10^3$ | Lead | $11.3 \times 10^3$ |
| Copper | $8.92 \times 10^3$ | Mercury | $13.6 \times 10^3$ |
| Ethyl alcohol | $0.806 \times 10^3$ | Oak | $0.710 \times 10^3$ |
| Fresh water | $1.00 \times 10^3$ | Oxygen gas | 1.43 |
| Glycerin | $1.26 \times 10^3$ | Pine | $0.373 \times 10^3$ |
| Gold | $19.3 \times 10^3$ | Platinum | $21.4 \times 10^3$ |
| Helium gas | $1.79 \times 10^{-1}$ | Sea water | $1.03 \times 10^3$ |
| Hydrogen gas | $8.99 \times 10^{-2}$ | Silver | $10.5 \times 10^3$ |

For this reason, aircraft flying at high altitudes must have pressurized cabins to provide sufficient oxygen for the passengers.

We now show mathematically how the pressure in a liquid increases with depth. Consider a liquid of density $\rho$ at rest as in Figure 15.4. Let us select a sample of the liquid contained within an imaginary cylinder of cross-sectional area $A$ extending from depth $d$ to depth $d + h$. This sample of liquid is in equilibrium and at rest. Therefore, according to Newton's second law, the net force on the sample must be equal to zero. We will investigate the forces on the sample related to the pressure on it.

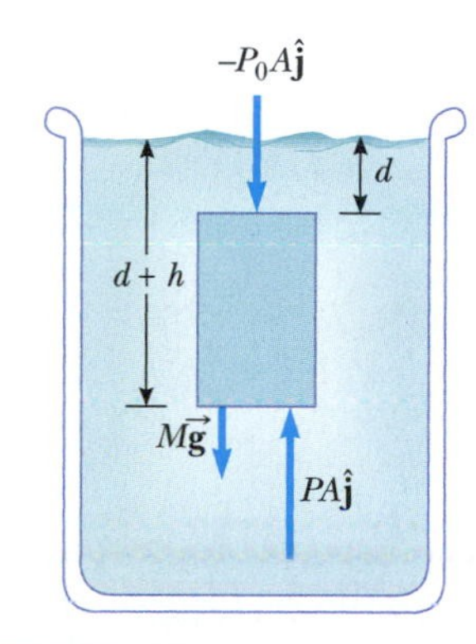


**FIGURE 15.4** The net force on the sample of liquid within the darker region must be zero because the sample is in equilibrium.

The liquid external to our sample exerts forces at all points on the sample's surface, perpendicular to it. On the sides of the sample of liquid in Figure 15.4, forces due to the pressure act horizontally and cancel in pairs on opposite sides of the sample for a net horizontal force of zero. The pressure exerted by the liquid on the sample's bottom face is $P$ and the pressure on the top face is $P_0$. Therefore, from Equation 15.1, the magnitude of the upward force exerted by the liquid on the bottom of the sample is $PA$, and the magnitude of the downward force exerted by the liquid on the top is $P_0A$. In addition, a gravitational force is exerted on the sample. Because the sample is in equilibrium, the net force in the vertical direction must be zero:

$$\sum F_y = 0 \quad \rightarrow \quad PA - P_0A - Mg = 0$$

Because the mass of liquid in the sample is $M = \rho V = \rho Ah$, the gravitational force on the liquid in the sample is $Mg = \rho gAh$. Therefore,

$$PA = P_0A + \rho gAh$$

or

$$P = P_0 + \rho gh \qquad [15.4]$$

■ Variation of pressure with depth in a liquid

If the top surface of our sample is at $d = 0$ so that it is open to the atmosphere, $P_0$ is atmospheric pressure. Equation 15.4 indicates that the pressure in a liquid depends only on the depth $h$ within the liquid. The pressure is therefore the same at all points having the same depth, independent of the shape of the container.

In view of Equation 15.4, any increase in pressure at the surface must be transmitted to every point in the liquid. This behavior was first recognized by French scientist Blaise Pascal (1623–1662) and is called **Pascal's law: A change in the pressure applied to an enclosed fluid is transmitted undiminished to every point of the fluid and to the walls of the container.** You use Pascal's law when you squeeze the sides of your toothpaste tube. The increase in pressure on the sides of the tube increases the pressure everywhere, which pushes a stream of toothpaste out of the opening.

■ Pascal's law

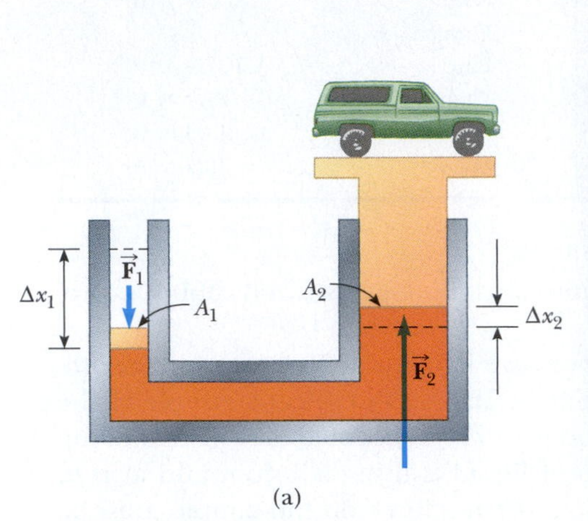

(a)

(b)

(David Frazier)

**FIGURE 15.5** (a) Diagram of a hydraulic press. Because the increase in pressure is the same at the left and right sides, a small force $\vec{F}_1$ at the left produces a much larger force $\vec{F}_2$ at the right. (b) A vehicle under repair is supported by a hydraulic lift in a garage.

An important application of Pascal's law is the hydraulic press illustrated by Figure 15.5. A force $\vec{F}_1$ is applied to a small piston of area $A_1$. The pressure is transmitted through a liquid to a larger piston of area $A_2$, and force $\vec{F}_2$ is exerted by the liquid on this piston. Because the pressure is the same at both pistons, we see that $P = F_1/A_1 = F_2/A_2$. The force magnitude $F_2$ is therefore larger than $F_1$ by the multiplying factor $A_2/A_1$. Hydraulic brakes, car lifts, hydraulic jacks, and forklifts all make use of this principle.

**QUICK QUIZ 15.2** The pressure at the bottom of a filled glass of water ($\rho = 1\,000$ kg/m$^3$) is $P$. The water is poured out and the glass is filled with ethyl alcohol ($\rho = 806$ kg/m$^3$). What is the pressure at the bottom of the glass? **(a)** smaller than $P$ **(b)** equal to $P$ **(c)** larger than $P$ **(d)** indeterminate

## ■ Thinking Physics 15.2

 Measuring blood pressure

Blood pressure is normally measured with the cuff of the sphygmomanometer around the arm. Suppose the blood pressure were measured with the cuff around the calf of the leg of a standing person. Would the reading of the blood pressure be the same here as it is for the arm?

**Reasoning** The blood pressure measured at the calf would be larger than that measured at the arm. If we imagine the vascular system of the body to be a vessel containing a liquid (blood), the pressure in the liquid will increase with depth. The blood at the calf is deeper in the liquid than that at the arm and is at a higher pressure.

Blood pressures are normally taken at the arm because it is at approximately the same height as the heart. If blood pressures at the calf were used as a standard, adjustments would need to be made for the height of the person and the blood pressure would be different if the person were lying down. ■

## EXAMPLE 15.1 The Car Lift

In a car lift used in a service station, compressed air exerts a force on a small piston of circular cross-section having a radius of 5.00 cm. This pressure is transmitted by an incompressible liquid to a second piston of radius 15.0 cm.

**A** What force must the compressed air exert to lift a car weighing 13 300 N?

**Solution** Because the pressure exerted by the compressed air is transmitted undiminished throughout the liquid, we have

$$F_1 = \left(\frac{A_1}{A_2}\right)F_2 = \frac{\pi(5.00 \times 10^{-2}\text{ m})^2}{\pi(15.0 \times 10^{-2}\text{ m})^2}(1.33 \times 10^4\text{ N})$$
$$= 1.48 \times 10^3\text{ N}$$

**B** What air pressure will produce this force?

**Solution** The air pressure that will produce this force is

$$P = \frac{F_1}{A_1} = \frac{1.48 \times 10^3\text{ N}}{\pi(5.00 \times 10^{-2}\text{ m})^2} = 1.88 \times 10^5\text{ Pa}$$

This pressure is approximately twice atmospheric pressure.

**C** Consider the lift as a nonisolated system and show that the input energy transfer is equal in magnitude to the output energy transfer.

**Solution** The energy input and output are by means of work done by the forces as the pistons move. To determine the work done, we must find the magnitude of the displacement through which each force acts. Because the liquid is modeled to be incompressible, the volume of the cylinder through which the input piston moves must equal that through which the output piston moves. The lengths of these cylinders are the magnitudes $\Delta x_1$ and $\Delta x_2$ of the displacements of the forces (see Fig. 15.5a). Setting the volumes equal, we have

$$V_1 = V_2 \quad \rightarrow \quad A_1\Delta x_1 = A_2\Delta x_2$$
$$\frac{A_1}{A_2} = \frac{\Delta x_2}{\Delta x_1}$$

Evaluating the ratio of the input work to the output work, we find that

$$\frac{W_1}{W_2} = \frac{F_1\,\Delta x_1}{F_2\,\Delta x_2} = \left(\frac{F_1}{F_2}\right)\left(\frac{\Delta x_1}{\Delta x_2}\right) = \left(\frac{A_1}{A_2}\right)\left(\frac{A_2}{A_1}\right) = 1$$

which verifies that the work input and output are the same, as they must be to conserve energy.

## EXAMPLE 15.2 The Force on a Dam

Water is filled to a height $H$ behind a dam of width $w$ (Fig. 15.6). Determine the resultant force on the dam.

**Solution** We cannot calculate the force on the dam by simply multiplying the area by the pressure because the pressure varies with depth. The problem can be solved by finding the force $dF$ on a narrow horizontal strip at depth $h$ and then integrating the expression over the height of the dam to find the total force.

The pressure at the depth $h$ beneath the surface at the red strip in Figure 15.6 is

$$P = \rho gh = \rho g(H - y)$$

(We have not included atmospheric pressure in our calculation because it acts on both sides of the dam, resulting in a net contribution of zero to the total force.) From Equation 15.1, we find the force on the red strip of area $dA$:

$$F = PA \quad \rightarrow \quad dF = P\,dA$$

Because $dA = w\,dy$, we have

$$dF = P\,dA = \rho g(H - y)w\,dy$$

Therefore, the total force on the dam is

$$F = \int_0^H \rho g(H - y)w\,dy = \tfrac{1}{2}\rho g w H^2$$

Note that because the pressure increases with depth, the dam is designed such that its thickness increases with depth as in Figure 15.6.

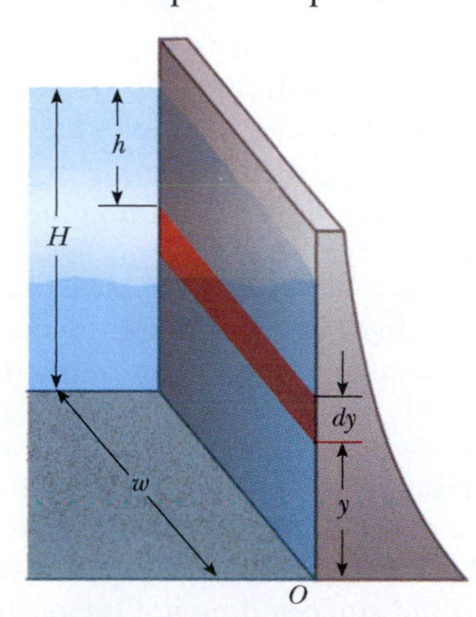


FIGURE 15.6 (Example 15.2) The total force on a dam is obtained from the expression $F = \int P\,dA$, where $dA$ is the area of the red strip.

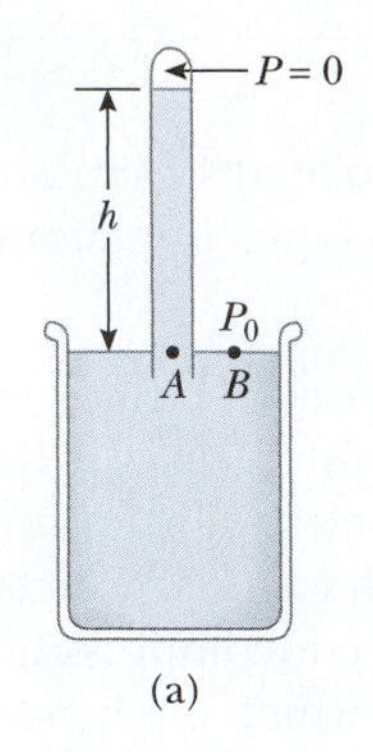


(a)

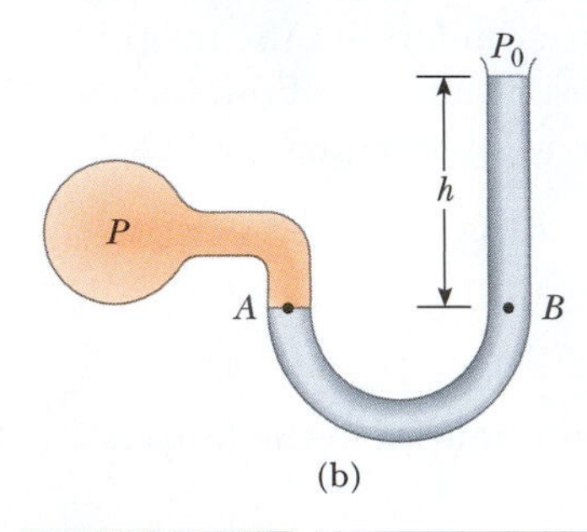


(b)

**FIGURE 15.7** Two devices for measuring pressure: (a) a mercury barometer and (b) an open-tube manometer.

## 15.3 PRESSURE MEASUREMENTS

During the weather report on a television news program, the *barometric pressure* is often provided. Barometric pressure is the current pressure of the atmosphere, which varies over a small range from the standard value provided in Equation 15.3. How is this pressure measured?

One instrument used to measure atmospheric pressure is the common barometer, invented by Evangelista Torricelli (1608–1647). A long tube closed at one end is filled with mercury and then inverted into a dish of mercury (Fig. 15.7a). The closed end of the tube is nearly a vacuum, so the pressure at the top of the mercury column can be taken as zero. In Figure 15.7a, the pressure at point $A$ due to the column of mercury must equal the pressure at point $B$ due to the atmosphere. If that were not the case, a net force would move mercury from one point to the other until equilibrium was established. It therefore follows that $P_0 = \rho_{\text{Hg}} gh$, where $\rho_{\text{Hg}}$ is the density of the mercury and $h$ is the height of the mercury column. As atmospheric pressure varies, the height of the mercury column varies, so the height can be calibrated to measure atmospheric pressure. Let us determine the height of a mercury column for one atmosphere of pressure, $P_0 = 1 \text{ atm} = 1.013 \times 10^5$ Pa:

$$P_0 = \rho_{\text{Hg}} gh \quad \rightarrow \quad h = \frac{P_0}{\rho_{\text{Hg}} g} = \frac{1.013 \times 10^5 \text{ Pa}}{(13.6 \times 10^3 \text{ kg/m}^3)(9.80 \text{ m/s}^2)} = 0.760 \text{ m}$$

Based on a calculation such as this one, one atmosphere of pressure is defined as the pressure equivalent of a column of mercury that is exactly 0.760 0 m in height at 0°C.

The open-tube manometer illustrated in Figure 15.7b is a device for measuring the pressure of a gas contained in a vessel. One end of a U-shaped tube containing a liquid is open to the atmosphere, and the other end is connected to a system of unknown pressure $P$. The pressures at points $A$ and $B$ must be the same (otherwise, the curved portion of the liquid would experience a net force and would accelerate), and the pressure at $A$ is the unknown pressure of the gas. Therefore, equating the unknown pressure $P$ to the pressure at point $B$, we see that $P = P_0 + \rho gh$. The difference in pressure $P - P_0$ is equal to $\rho gh$. Pressure $P$ is called the **absolute pressure,** and the difference $P - P_0$ is called the **gauge pressure.** For example, the pressure you measure in your bicycle tire is gauge pressure.

## 15.4 BUOYANT FORCES AND ARCHIMEDES'S PRINCIPLE

In this section, we investigate the origin of a **buoyant force,** which is **an upward force exerted on an object by the surrounding fluid.** Buoyant forces are evident in many situations. Anyone who has ridden in a boat, for example, has experienced a buoyant force. Another common example is the relative ease with which you can lift someone in a swimming pool compared with lifting that same individual on dry land. According to **Archimedes's principle:**

> Any object completely or partially submerged in a fluid experiences an upward buoyant force whose magnitude is equal to the weight of the fluid displaced by the object.

Archimedes's principle can be verified in the following manner. Suppose we focus our attention on a small parcel of a larger fluid such as the indicated cube of fluid in the container of Figure 15.8. This cube of fluid is in equilibrium under the action of the forces exerted on it by the fluid surrounding it. One of these forces in the vertical direction is the gravitational force. Because the cube is in equilibrium, the net force on it in the vertical direction must be zero. What cancels the downward gravitational force so that the cube remains in equilibrium? Apparently, the rest of the fluid inside the container is applying an upward force, the buoyant

**ARCHIMEDES** (ca. 287–212 B.C.)

Archimedes, a Greek mathematician, physicist, and engineer, was perhaps the greatest scientist of antiquity. He was the first to compute accurately the ratio of a circle's circumference to its diameter, and he showed how to calculate the volume and surface area of spheres, cylinders, and other geometric shapes. He is well known for discovering the nature of the buoyant force.

force. Therefore, the magnitude $B$ of the buoyant force must be exactly equal to the weight of the fluid inside the cube:

$$\sum F_y = 0 \quad \rightarrow \quad B - F_g = 0 \quad \rightarrow \quad B = Mg$$

where $M$ is the mass of the fluid in the cube.

Now imagine that the cube of fluid is replaced by a cube of steel of the same dimensions. What is the buoyant force on the steel? The fluid surrounding a cube behaves in the same way whether it is exerting pressure on a cube of fluid or a cube of steel; therefore, **the buoyant force acting on the steel is the same as the buoyant force acting on a cube of fluid of the same dimensions.** This result applies for a submerged object of any shape, size, or density.

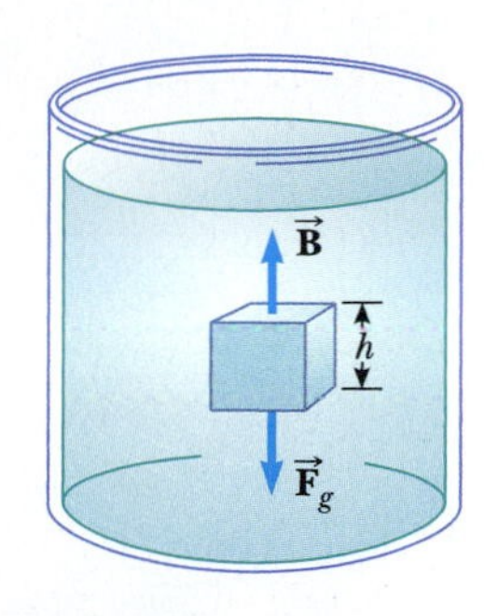


FIGURE 15.8 The external forces on the cube of liquid are the gravitational force $\vec{F}_g$ and the buoyant force $\vec{B}$. Under equilibrium conditions, $B = F_g$.

Let us now show more explicitly that the magnitude of the buoyant force is equal to the weight of the displaced fluid. Although that is true for both liquids and gases, we will perform the derivation for a liquid. On the sides of the cube of liquid in Figure 15.8, forces due to the pressure act horizontally and cancel in pairs on opposite sides of the cube for a net horizontal force of zero. In a liquid, the pressure at the bottom of the cube is greater than the pressure at the top by an amount $\rho_f gh$, where $\rho_f$ is the density of the liquid and $h$ is the height of the cube. Therefore, the upward force $F_{\text{bot}}$ on the bottom is greater than the downward force $F_{\text{top}}$ on the top of the cube. The net vertical force *exerted by the liquid* (we are ignoring the gravitational force for now) is

$$\sum F_{\text{liquid}} = B = F_{\text{bot}} - F_{\text{top}}$$

Expressing the forces in terms of pressure gives us

$$B = P_{\text{bot}}A - P_{\text{top}}A = \Delta PA = \rho_f ghA$$

$$B = \rho_f gV \qquad [15.5]$$

■ Archimedes's principle

where $V = hA$ is the volume of the cube. Because the mass of the liquid in the cube is $M = \rho_f V$, we see that

$$B = Mg$$

which is the weight of the displaced liquid.

Before proceeding with a few examples, it is instructive to compare two common cases: the buoyant force acting on a totally submerged object and that acting on a floating object.

**PITFALL PREVENTION 15.2**

**BUOYANT FORCE IS EXERTED BY THE FLUID** Notice the important point in this discussion of the buoyant force: **it is exerted by the fluid.** It is not determined by properties of the object except for the amount of fluid displaced by the object. Therefore, if several objects of different densities but the same volume are immersed in a fluid, they will all experience the same buoyant force. Whether they sink or float will be determined by the relationship between the buoyant force and the weight.

## Case I: A Totally Submerged Object

When an object is totally submerged in a liquid of density $\rho_f$, the magnitude of the upward buoyant force is $B = \rho_f gV$, where $V$ is the volume of the liquid displaced by the object. Because the object is totally submerged, the volume $V_O$ of the object and the volume $V$ of liquid displaced by the object are the same, $V = V_O$. If the object has a density $\rho_O$, its weight is $Mg = \rho_O V_O g$. Therefore, the net force on it is $\Sigma F = B - Mg = (\rho_f - \rho_O) V_O g$. We see that if the density of the object is less than the density of the liquid as in Active Figure 15.9a, the net force is positive and the unsupported object accelerates upward. If the density of the object is greater than the density of the liquid as in Active Figure 15.9b, the net force is negative and the unsupported object sinks.

The same behavior is exhibited by an object immersed in a gas, such as the air in the atmosphere.[1] If the object is less dense than air, like a helium-filled balloon, the object floats upward. If it is denser, like a rock, it falls downward.

[1]The general behavior is the same, but the buoyant force varies with height in the atmosphere due to the variation in density of the air.

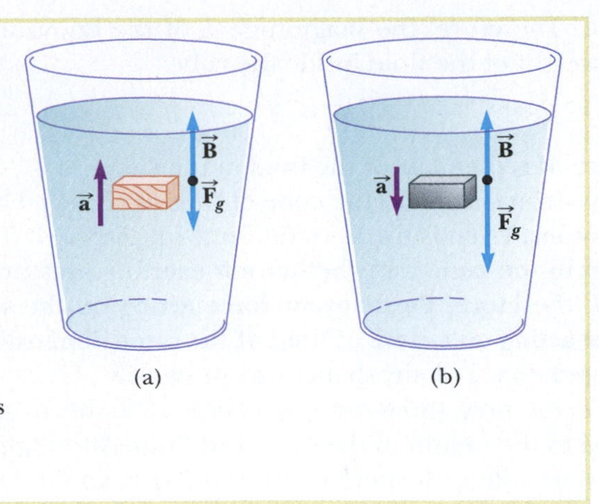


ACTIVE FIGURE 15.9 (a) A totally submerged object that is less dense than the fluid in which it is submerged experiences a net upward force. (b) A totally submerged object that is denser than the fluid sinks.

**Physics Now™** Log into PhysicsNow at **www.pop4e.com** and go to Active Figure 15.9 to move the object to new positions as well as change the density of the object to see the results.

ACTIVE FIGURE 15.10 An object floating on the surface of a liquid experiences two forces, the gravitational force $\vec{\mathbf{F}}_g$ and the buoyant force $\vec{\mathbf{B}}$. Because the object floats in equilibrium, $B = F_g$.

**Physics Now™** By logging into PhysicsNow at **www.pop4e.com** and going to Active Figure 15.10 you can change the densities of the object and the liquid.

## Case II: A Floating Object

Now consider an object in static equilibrium floating on the surface of a liquid, that is, an object that is only partially submerged such as the ice cube floating in water in Active Figure 15.10. Because it is only partially submerged, the volume $V$ of liquid displaced by the object is only a fraction of the total volume $V_O$ of the object. The volume of the liquid displaced by the object corresponds to that volume of the object beneath the liquid surface. Because the object is in equilibrium, the upward buoyant force is balanced by the downward gravitational force exerted on the object. The buoyant force has a magnitude $B = \rho_f gV$. Because the weight of the object is $Mg = \rho_O V_O g$ and because Newton's second law tells us that $\sum F = 0$ in the vertical direction, $Mg = B$. We see that $\rho_f gV = \rho_O V_O g$, or

$$\frac{\rho_O}{\rho_f} = \frac{V}{V_O} \qquad [15.6]$$

Therefore, the fraction of the volume of the object under the liquid surface is equal to the ratio of the object density to the liquid density.

Let us consider examples of both cases. Under normal conditions, the average density of a fish is slightly greater than the density of water. That being the case, a fish would sink if it did not have some mechanism to counteract the net downward force. The fish does so by internally regulating the size of its swim bladder, a gas-filled cavity within the fish's body. Increasing its size increases the amount of water displaced, which increases the buoyant force. In this manner, fish are able to swim to various depths. Because the fish is totally submerged in the water, this example illustrates Case I.

As an example of Case II, imagine a large cargo ship. When the ship is at rest, the upward buoyant force from the water balances the weight so that the ship is in equilibrium. Only part of the volume of the ship is under water. If the ship is loaded with heavy cargo, it sinks deeper into the water. The increased weight of the ship due to the cargo is balanced by the extra buoyant force related to the extra volume of the ship that is now beneath the water surface.

(Richard Megna/Fundamental Photographs)

These hot-air balloons float on air because they are filled with air at high temperature. The buoyant force on a balloon due to the surrounding air is equal to the weight of the balloon, resulting in a net force of zero. ■

QUICK QUIZ 15.3 An apple is held completely submerged just below the surface of a container of water. The apple is then moved to a deeper point in the water. Compared with the force needed to hold the apple just below the surface, what is the force needed to hold it at a deeper point? **(a)** larger **(b)** the same **(c)** smaller **(d)** impossible to determine

**QUICK QUIZ 15.4** You are shipwrecked and floating in the middle of the ocean on a raft. Your cargo on the raft includes a treasure chest full of gold that you found before your ship sank and the raft is just barely afloat. To keep you floating as high as possible in the water, should you **(a)** leave the treasure chest on top of the raft, **(b)** secure the treasure chest to the underside of the raft, or **(c)** hang the treasure chest in the water with a rope attached to the raft? (Assume that throwing the treasure chest overboard is not an option you wish to consider!)

## Thinking Physics 15.3

A florist delivery person is delivering a flower basket to a home. The basket includes an attached helium-filled balloon, which suddenly comes loose from the basket and begins to accelerate upward toward the sky. Startled by the release of the balloon, the delivery person drops the flower basket. As the basket falls, the basket–Earth system experiences an increase in kinetic energy and a decrease in gravitational potential energy, consistent with conservation of mechanical energy. The balloon–Earth system, however, experiences an increase in *both* gravitational potential energy and kinetic energy. Is that consistent with the principle of conservation of mechanical energy? If not, from where is the extra energy coming?

**Reasoning** In the case of the system of the flower basket and the Earth, a good approximation to the motion of the basket can be made by ignoring the effects of the air. Therefore, the basket–Earth system can be analyzed with the isolated system model and mechanical energy is conserved. For the balloon–Earth system, we cannot ignore the effects of the air because it is the buoyant force of the air that causes the balloon to rise. Therefore, the balloon–Earth system is analyzed with the nonisolated system model. The buoyant force of the air does work across the boundary of the system, and that work results in an increase in both the kinetic and gravitational potential energies of the system. ■

### EXAMPLE 15.3 Eureka!

Archimedes supposedly was asked to determine whether a crown made for the king consisted of pure gold. Legend has it that Archimedes solved this problem by weighing the crown first in air and then in water as shown in Figure 15.11. Suppose the scale reads 7.84 N in air and 6.84 N in water. What should Archimedes have told the king?

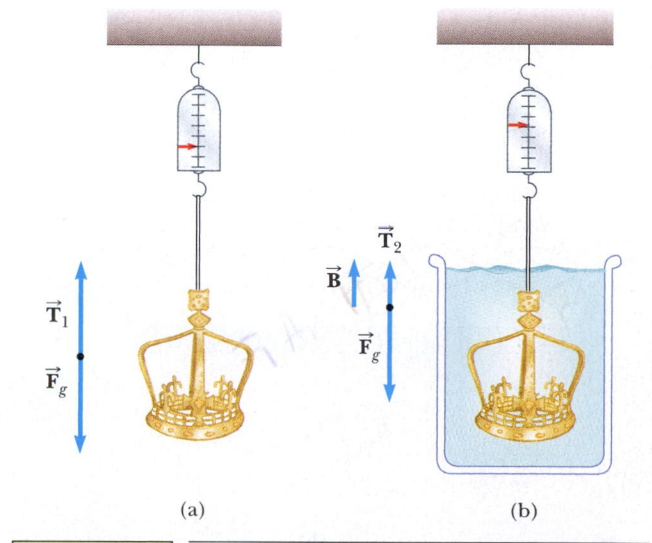


FIGURE 15.11 (Example 15.3) (a) When the crown is suspended in air, the scale reads its true weight because $T_1 = F_g$ (the buoyancy due to air is negligible). (b) When the crown is immersed in water, the buoyant force $\vec{\mathbf{B}}$ reduces the scale reading to $T_2 = F_g - B$.

**Solution** Our strategy will be based on determining the density of the crown and comparing it with the density of gold. Figure 15.11 helps us conceptualize the problem. Because of our understanding of the buoyant force, we realize that the scale reading will be smaller in Figure 15.11b than in Figure 15.11a. The scale reading is a measure of one of the forces on the crown and we recognize that the crown is stationary. Therefore, we can categorize this problem as one in which we model the crown as a particle in equilibrium. To analyze the problem, note that when the crown is suspended in air, the scale reads the true weight $T_1 = F_g$ (neglecting the buoyancy of air). When it is immersed in water, the buoyant force $\vec{\mathbf{B}}$ reduces the scale reading to an *apparent* weight of $T_2 = F_g - B$. Because the crown is in equilibrium, the net force on it is zero. When the crown is in water, then,

$$\sum F = B + T_2 - F_g = 0$$

so that

$$B = F_g - T_2 = 7.84 \text{ N} - 6.84 \text{ N} = 1.00 \text{ N}$$

Because this buoyant force is equal in magnitude to the weight of the displaced water, we have $\rho_w g V_w = 1.00$ N, where $V_w$ is the volume of the displaced water and $\rho_w$ is its density. Also, the volume of the crown $V_c$ is equal to the volume of the displaced water because the crown is completely submerged. Therefore,

$$V_c = V_w = \frac{1.00 \text{ N}}{\rho_w g} = \frac{1.00 \text{ N}}{(1\,000 \text{ kg/m}^3)(9.80 \text{ m/s}^2)}$$
$$= 1.02 \times 10^{-4} \text{ m}^3$$

Finally, the density of the crown is

$$\rho_c = \frac{m_c}{V_c} = \frac{m_c g}{V_c g} = \frac{7.84 \text{ N}}{(1.02 \times 10^{-4} \text{ m}^3)(9.80 \text{ m/s}^2)}$$
$$= 7.84 \times 10^3 \text{ kg/m}^3$$

To finalize the problem, from Table 15.1 we see that the density of gold is $19.3 \times 10^3$ kg/m$^3$. Therefore, Archimedes should have told the king that he had been cheated. Either the crown was hollow or it was not made of pure gold.

## INTERACTIVE EXAMPLE 15.4 Changing String Vibration with Water

One end of a horizontal string is attached to a vibrating blade and the other end passes over a pulley as in Figure 15.12a. A sphere of mass 2.00 kg hangs on the end of the string. The string is vibrating in its second harmonic. A container of water is raised under the sphere so that the sphere is completely submerged. After that is done, the string vibrates in its fifth harmonic as shown in Figure 15.12b. What is the radius of the sphere?

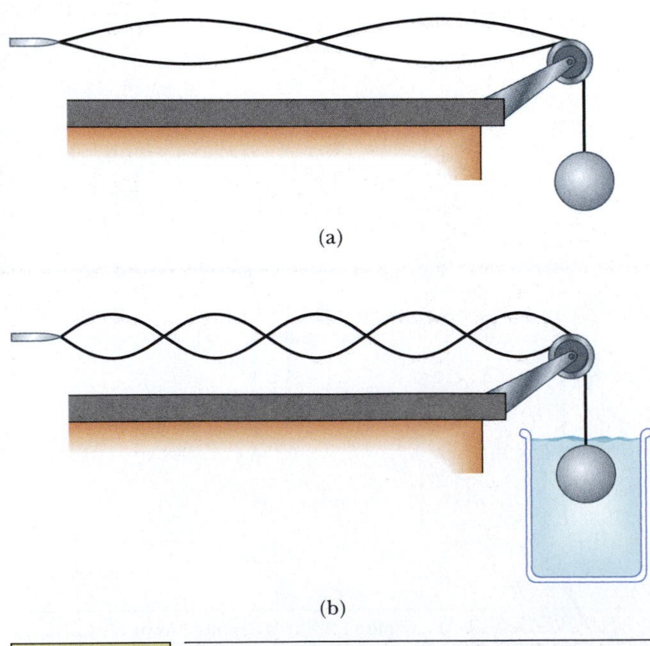


FIGURE 15.12 (Interactive Example 15.4) (a) When the sphere hangs in air, the string vibrates in its second harmonic. (b) When the sphere is immersed in water, the string vibrates in its fifth harmonic.

**Solution** In Figure 15.12a, Newton's second law applied to the sphere tells us that the initial tension $T_i$ in the string is equal to the weight of the sphere:

$$T_i - mg = 0 \quad \rightarrow \quad T_i = mg$$

$$T_i = (2.00 \text{ kg})(9.80 \text{ m/s}^2) = 19.6 \text{ N}$$

where the subscript $i$ is used to indicate initial variables before we immerse the sphere in water. Once the sphere is immersed in water, the tension in the string will decrease to $T_f$. Applying Newton's second law to the sphere again in this situation, we have

$$(1) \quad T_f + B - mg = 0 \quad \rightarrow \quad B = mg - T_f$$

The desired quantity, the radius of the sphere, will appear in the expression for the buoyant force $B$. Before we can proceed in this direction, however, we need to evaluate $T_f$. We do so from the standing wave information. We write the equation for the frequency of a standing wave on a string (Equation 14.8) twice: once before we immerse the sphere and once after, and divide the equations:

$$(2) \quad \left.\begin{aligned} f &= \frac{n_i}{2L}\sqrt{\frac{T_i}{\mu}} \\ f &= \frac{n_f}{2L}\sqrt{\frac{T_f}{\mu}} \end{aligned}\right\} \quad \rightarrow \quad 1 = \frac{n_i}{n_f}\sqrt{\frac{T_i}{T_f}}$$

where the frequency $f$ is the same in both cases because it is determined by the vibrating blade. In addition, the linear mass density $\mu$ and the length $L$ of the vibrating portion of the string are the same in both cases. Solving (2) for $T_f$ gives

$$T_f = \left(\frac{n_i}{n_f}\right)^2 T_i = \left(\frac{2}{5}\right)^2 (19.6\text{ N}) = 3.14\text{ N}$$

Substituting this value into equation (1), we can evaluate the buoyant force on the sphere:

$$B = mg - T_f = 19.6\text{ N} - 3.14\text{ N} = 16.5\text{ N}$$

Finally, expressing the buoyant force in terms of the radius of the sphere, we solve for the radius,

$$B = \rho_{\text{water}} g V_{\text{sphere}} = \rho_{\text{water}} g \left(\tfrac{4}{3}\pi r^3\right)$$

$$r = \left(\frac{3B}{4\pi\rho_{\text{water}} g}\right)^{1/3}$$

$$= \left(\frac{3(16.5\text{ N})}{4\pi(1\,000\text{ kg/m}^3)(9.80\text{ m/s}^2)}\right)^{1/3}$$

$$= 7.38 \times 10^{-2}\text{ m} = 7.38\text{ cm}$$

**PhysicsNow™** You can adjust the mass of the sphere by logging into PhysicsNow at **www.pop4e.com** and going to Interactive Example 15.4.

## 15.5 FLUID DYNAMICS

Thus far, our study of fluids has been restricted to fluids at rest, or **fluid statics.** We now turn our attention to **fluid dynamics,** the study of fluids in motion. Instead of trying to study the motion of each particle of the fluid as a function of time, we describe the properties of the fluid as a whole.

### Flow Characteristics

When fluid is in motion, its flow is of one of two main types. The flow is said to be **steady,** or **laminar,** if each particle of the fluid follows a smooth path so that the paths of different particles never cross each other as in Figure 15.13. Therefore, in steady flow, the velocity of the fluid at any point remains constant in time.

Above a certain critical speed, fluid flow becomes **turbulent.** Turbulent flow is an irregular flow characterized by small, whirlpool-like regions as in Figure 15.14. As an example, the flow of water in a river becomes turbulent in regions where rocks and other obstructions are encountered, often forming "white-water" rapids.

(Andy Sacks/Tony Stone Images/Getty Images)

**FIGURE 15.13** An illustration of steady flow around an automobile in a test wind tunnel. The streamlines in the airflow are made visible by smoke particles.

**FIGURE 15.14** Hot gases from a cigarette made visible by smoke particles. The smoke first moves in laminar flow at the bottom and then in turbulent flow above.

The term **viscosity** is commonly used in fluid flow to characterize the degree of internal friction in the fluid. This internal friction, or viscous force, is associated with the resistance of two adjacent layers of the fluid against moving relative to each other. Because viscosity represents a nonconservative force, part of a fluid's kinetic energy is converted to internal energy when layers of fluid slide past one another. This conversion is similar to the mechanism by which an object sliding on a rough horizontal surface experiences a transformation of kinetic energy to internal energy.

Because the motion of a real fluid is very complex and not yet fully understood, we adopt a simplification model. As we shall see, many features of real fluids in motion can be understood by considering the behavior of an ideal fluid. In our simplification model, we make the following four assumptions:

1. *Nonviscous fluid.* In a nonviscous fluid, internal friction is ignored. An object moving through the fluid experiences no viscous force.
2. *Incompressible fluid.* The density of the fluid is assumed to remain constant regardless of the pressure in the fluid.
3. *Steady flow.* In steady flow, we assume that the velocity of the fluid at each point remains constant in time.
4. *Irrotational flow.* Fluid flow is irrotational if the fluid has no angular momentum about any point. If a small paddle wheel placed anywhere in the fluid does not rotate about the wheel's center of mass, the flow is irrotational. (If the wheel were to rotate, as it would if turbulence were present, the flow would be rotational.)

The first two assumptions in our simplification model are properties of our ideal fluid. The last two are descriptions of the way that the fluid flows.

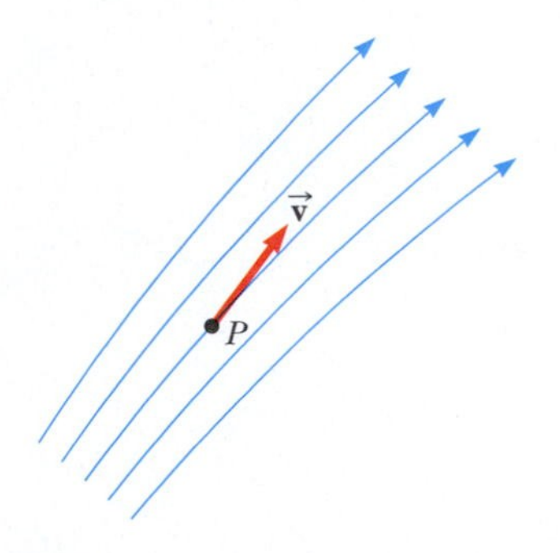


**FIGURE 15.15** This diagram represents a set of streamlines (*blue lines*). A particle at $P$ follows one of these streamlines, and its velocity is tangent to the streamline at each point along its path.

## 15.6 STREAMLINES AND THE CONTINUITY EQUATION FOR FLUIDS

If you are watering your garden and your garden hose is too short, you might do one of two things to help you reach the garden with the water (before you look for a longer hose!). You might attach a nozzle to the end of the hose, or, in the absence of a nozzle, you might place your thumb over the end of the hose, allowing the water to come out of a narrower opening. Why does either of these techniques cause the water to come out faster so that it can be projected over a longer range? We shall see the answer to this question in this section.

The path taken by a particle of the fluid under steady flow is called a **streamline.** The velocity of the particle is always tangent to the streamline as shown in Figure 15.15. No two streamlines can cross each other; if they did, a particle could move either way at the crossover point and then the flow would not be steady.

Consider an ideal fluid flowing through a pipe of nonuniform size as in Figure 15.16. The particles in the fluid move along the streamlines in steady flow. Let us analyze this situation using the nonisolated system in steady-state model. We have seen this model used for energy in Chapter 7, but we mentioned at that time that the model can be used for any conserved quantity. The volume of an incompressible fluid is a conserved quantity. Assuming no leaks in our pipe, we can neither create nor destroy fluid, just as we could not create nor destroy energy in Chapters 6 and 7.

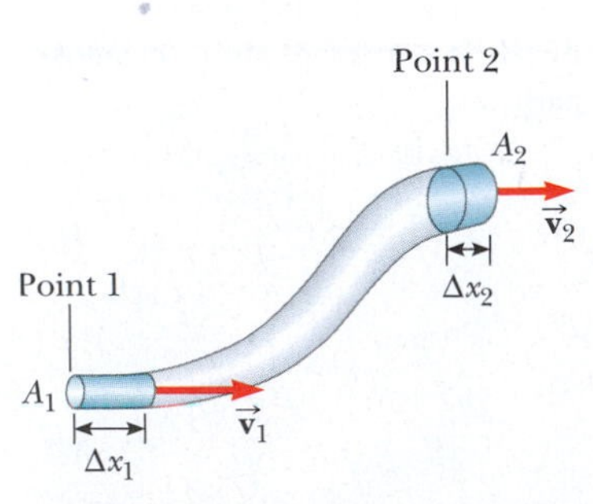


**FIGURE 15.16** A fluid moving with steady flow through a pipe of varying cross-sectional area. The volume of fluid flowing through $A_1$ in a time interval $\Delta t$ must equal the volume flowing through $A_2$ in the same time interval.

We choose as our system the region of space in the pipe from point 1 to point 2 in Figure 15.16. Let us assume that this region is filled with fluid at all times. As the fluid flows in the pipe, fluid enters the system at point 1 and leaves the system at point 2. Imagine that the fluid moves through a displacement $\Delta x_1$ at point 1 and moves through a displacement $\Delta x_2$ at point 2 as it leaves the system. The volume of fluid entering the system at point 1 is $A_1 \Delta x_1$ and the volume leaving at point 2 is $A_2 \Delta x_2$. Because the volume of an incompressible fluid is a conserved quantity, these two volumes must be equal for the system to be in steady state. If that were not true, the volume of fluid in the system would be changing. Therefore,

$$A_1 \Delta x_1 = A_2 \Delta x_2$$

Let us divide this equation by the time interval during which the fluid moves:

$$\frac{A_1 \Delta x_1}{\Delta t} = \frac{A_2 \Delta x_2}{\Delta t}$$

In the limit as the time interval shrinks to zero, the ratio of the displacement of the fluid to the time interval is the instantaneous speed of the fluid, so we can write this expression as

$$A_1 v_1 = A_2 v_2 \quad [15.7]$$

Continuity equation for fluids

The product $Av$, which has the dimensions of volume per time, is called the **volume flow rate.** Equation 15.7, called the **continuity equation for fluids,** says that **the product of the area and the fluid speed at all points along the pipe is a constant.** Therefore, the speed is high where the tube is constricted and low where the tube is wide. Hence, a nozzle or your thumb over the garden hose allows you to project the water farther. By reducing the area through which the water flows, you increase its speed. Therefore, you project the water from the hose with a high initial velocity, resulting in a large value of the range, as discussed for projectiles in Chapter 3.

**QUICK QUIZ 15.5** You tape two different sodas straws together end to end to make a longer straw with no leaks. The two straws have radii of 3 mm and 5 mm. You drink a soda through your combination straw. In which straw is the speed of the liquid higher? **(a)** It is higher in whichever one is nearest your mouth. **(b)** It is higher in the one of radius 3 mm. **(c)** It is higher in the one of radius 5 mm. **(d)** Neither, because the speed is the same in both straws.

### EXAMPLE 15.5 Watering a Garden

A water hose 2.50 cm in diameter is used by a gardener to fill a 30.0-L bucket. The gardener notes that it takes 1.00 min to fill the bucket. A nozzle with an opening of cross-sectional area 0.500 cm$^2$ is then attached to the hose. The nozzle is held so that water is projected horizontally from a point 1.00 m above the ground. Over what horizontal distance can the water be projected?

**Solution** We identify point 1 within the hose and point 2 at the exit of the nozzle. We first find the speed of the water in the hose from the bucket-filling information. The cross-sectional area of the hose is

$$A_1 = \pi r^2 = \pi \frac{d^2}{4} = \pi \left[\frac{(2.50 \text{ cm})^2}{4}\right] = 4.91 \text{ cm}^2$$

According to the data given, the volume flow rate is equal to 30.0 L/min:

$$A_1 v_1 = 30.0 \text{ L/min} = \frac{30.0 \times 10^3 \text{ cm}^3}{60.0 \text{ s}} = 500 \text{ cm}^3/\text{s}$$

$$v_1 = \frac{500 \text{ cm}^3/\text{s}}{4.91 \text{ cm}^2} = 102 \text{ cm/s} = 1.02 \text{ m/s}$$

Now we use the continuity equation for fluids to find the speed $v_2 = v_{xi}$ with which the water exits the nozzle. The subscript $i$ anticipates that this speed will be the *initial* velocity component of the water projected from the hose, and the subscript $x$ indicates that the initial velocity vector of the projected water is in the horizontal direction. So,

$$A_1 v_1 = A_2 v_2 = A_2 v_{xi}$$

$$v_{xi} = \frac{A_1}{A_2} v_1 = \frac{4.91 \text{ cm}^2}{0.500 \text{ cm}^2}(1.02 \text{ m/s}) = 10.0 \text{ m/s}$$

We now shift our thinking away from fluids and to projectile motion because the water is in free-fall once it exits the nozzle. An element of the water is modeled as a particle under constant acceleration as it falls through a vertical distance of 1.00 m starting from rest at $t = 0$. We find the time at which the water strikes the ground. From Equation 3.13,

$$y_f = y_i + v_{yi}t - \tfrac{1}{2}gt^2$$

$$-1.00 \text{ m} = 0 + 0 - \tfrac{1}{2}(9.80 \text{ m/s}^2)t^2$$

$$t = \sqrt{\frac{2(1.00 \text{ m})}{9.80 \text{ m/s}^2}} = 0.452 \text{ s}$$

In the horizontal direction, the element of water is modeled as a particle under constant velocity. We apply Equation 3.12 to find the horizontal position as the water strikes the ground:

$$x_f = x_i + v_{xi}t = 0 + (10.0 \text{ m/s})(0.452 \text{ s}) = 4.52 \text{ m}$$

## 15.7 BERNOULLI'S EQUATION

(© Bettmann/CORBIS)

**DANIEL BERNOULLI** (1700–1782)

Bernoulli, a Swiss physicist and mathematician, made important discoveries in fluid dynamics. His most famous work, *Hydrodynamica*, published in 1738, is both a theoretical and a practical study of equilibrium, pressure, and speed in fluids. In this publication, Bernoulli also attempted the first explanation of the behavior of gases with changing pressure and temperature; this effort was the beginning of the kinetic theory of gases, which we will study in Chapter 16.

You have probably had the experience of driving on a highway and having a large truck pass by you at high speed. In that situation, you may have had the frightening feeling that your car was being pulled in toward the truck as it passed. We will see the origin for this effect in this section.

As a fluid moves through a region where its speed, elevation above the Earth's surface, or both change, the pressure in the fluid varies with these changes. The relationship between fluid speed, pressure, and elevation was first derived in 1738 by Swiss physicist Daniel Bernoulli. Consider the flow of a segment of an ideal fluid through a nonuniform pipe in a time interval $\Delta t$ as illustrated in Figure 15.17. At the beginning of the time interval, the segment of fluid consists of the blue shaded portion (portion 1) at the left and the unshaded portion. During the time interval, the left end of the segment moves to the right through a displacement $\Delta x_1$, which is the length of the blue shaded portion at the left. Meanwhile, the right end of the segment moves to the right through a displacement $\Delta x_2$, which is the length of the blue shaded portion (portion 2) at the upper right of Figure 15.17. Therefore, at the end of the time interval, the segment of fluid consists of the unshaded portion and the blue shaded portion at the upper right.

Now consider forces exerted on this segment by fluid to the left and the right of the segment. The force exerted by the fluid on the left end has a magnitude $P_1A_1$. The work done by this force on the segment in a time interval $\Delta t$ is $W_1 = F_1\,\Delta x_1 = P_1A_1\,\Delta x_1 = P_1V$, where $V$ is the volume of portion 1. In a similar manner, the work done by the fluid to the right of the segment in the same time interval $\Delta t$ is $W_2 = -P_2A_2\,\Delta x_2 = -P_2V$. (The volume of portion 1 equals the volume of portion 2.) This work is negative because the force on the segment of fluid is to the left and the displacement of the point of application of the force is to the right. Therefore, the net work done on the segment by these forces in the time interval $\Delta t$ is

$$W = (P_1 - P_2)V \qquad [15.8]$$

Part of this work goes into changing the kinetic energy of the segment of fluid and part goes into changing the gravitational potential energy of the segment–Earth system. Because we are assuming streamline flow, the kinetic energy of the unshaded portion of the segment in Figure 15.17 is unchanged during the time interval. The only change is that before the time interval we have portion 1 traveling at $v_1$, whereas after the time interval we have portion 2 traveling at $v_2$. Therefore, the change in the kinetic energy of the segment of fluid is

$$\Delta K = \tfrac{1}{2}mv_2^2 - \tfrac{1}{2}mv_1^2 \qquad [15.9]$$

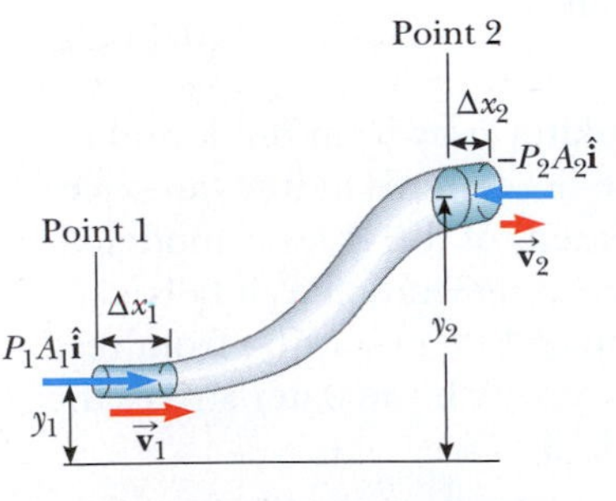



**FIGURE 15.17** A fluid in laminar flow through a constricted pipe. The volume of the shaded portion on the left is equal to the volume of the shaded portion on the right.

where $m$ is the mass of either portion 1 or portion 2. Because the volumes of both portions are the same, they also have the same mass.

Considering the gravitational potential energy of the segment–Earth system, once again there is no change during the time interval for the unshaded portion of the fluid. The net change is that the mass of the fluid in portion 1 has effectively been moved to the location of portion 2. Consequently, the change in gravitational potential energy of the system is

$$\Delta U = mgy_2 - mgy_1 \qquad [15.10]$$

The total work done on the segment–Earth system by the fluid outside the segment is equal to the change in mechanical energy of the system: $W = \Delta K + \Delta U$. Substituting for each of these terms gives

$$(P_1 - P_2)V = \tfrac{1}{2}mv_2^2 - \tfrac{1}{2}mv_1^2 + mgy_2 - mgy_1 \qquad [15.11]$$

If we divide each term by the portion volume $V$ and recall that $\rho = m/V$, this expression reduces to

$$P_1 - P_2 = \tfrac{1}{2}\rho v_2^2 - \tfrac{1}{2}\rho v_1^2 + \rho g y_2 - \rho g y_1$$

Rearranging terms, we obtain

$$P_1 + \tfrac{1}{2}\rho v_1^2 + \rho g y_1 = P_2 + \tfrac{1}{2}\rho v_2^2 + \rho g y_2 \qquad [15.12]$$

■ **Bernoulli's equation**

which is **Bernoulli's equation** applied to an ideal fluid. It is often expressed as

$$P + \tfrac{1}{2}\rho v^2 + \rho g y = \text{constant} \qquad [15.13]$$

Bernoulli's equation says that the sum of the pressure $P$, the kinetic energy per unit volume $\frac{1}{2}\rho v^2$, and gravitational potential energy per unit volume $\rho g y$ has the same value at all points along a streamline.

When the fluid is at rest, $v_1 = v_2 = 0$ and Equation 15.12 becomes

$$P_1 - P_2 = \rho g(y_2 - y_1) = \rho g h$$

which agrees with Equation 15.4.

Although Equation 15.13 was derived for an incompressible fluid, the general behavior of pressure with speed is true even for gases: as the speed increases, the pressure decreases. This *Bernoulli effect* explains the experience with the truck on the highway at the opening of this section. As air passes between your car and the truck, it must pass through a relatively narrow channel. According to the continuity equation, the speed of the air is higher. According to the Bernoulli effect, this higher-speed air exerts less pressure on your car than the slower-moving air on the other side of your car. Thus, there is a net force pushing you toward the truck.

**QUICK QUIZ 15.6** You observe two helium balloons floating next to each other at the ends of strings secured to a table. The facing surfaces of the balloons are separated by 1 to 2 cm. You blow through the opening between the balloons. What happens to the balloons? **(a)** They move toward each other. **(b)** They move away from each other. **(c)** They are unaffected.

### EXAMPLE 15.6 Sinking the Cruise Ship

A scuba diver is hunting for fish with a spear gun. He accidentally fires the gun so that a spear punctures the side of a cruise ship. The hole is located at a depth of 10.0 m below the water surface. With what speed does the water enter the cruise ship through the hole?

**Solution** We identify point 1 as the water surface outside the ship, which we will assign as $y = 0$. At this point, the water is static, so $v_1 = 0$. We identify point 2 as a point just inside the hole in the interior of the ship because that is the point at which we wish to evaluate the speed of the water. This point is at a depth $y = -h = -10.0$ m below the water surface. We use Bernoulli's equation to compare these two points. At both points, the water is open to atmospheric pressure, so $P_1 = P_2 = P_0$.

Based on this argument, Bernoulli's equation becomes

$$P_0 + \tfrac{1}{2}\rho(0)^2 + \rho g(0) = P_0 + \tfrac{1}{2}\rho v_2^2 + \rho g(-h) \rightarrow$$

$$v_2 = \sqrt{2gh} = \sqrt{2(9.80\text{ m/s}^2)(10.0\text{ m})} = 14\text{ m/s}$$

## INTERACTIVE EXAMPLE 15.7 Torricelli's Law

An enclosed tank containing a liquid of density $\rho$ has a hole in its side at a distance $y_1$ from the tank's bottom (Fig. 15.18). The hole is open to the atmosphere, and its diameter is much smaller than the diameter of the tank. The air above the liquid is maintained at a pressure $P$.

**A** Determine the speed of the liquid as it leaves the hole when the liquid's level is a distance $h$ above the hole.

**Solution** Because $A_2 \gg A_1$, the liquid is approximately at rest at the top of the tank, where the pressure is $P$. Applying Bernoulli's equation to points 1 and 2 and noting that at the hole $P_1$ is equal to atmospheric pressure $P_0$, we find that

$$P_0 + \tfrac{1}{2}\rho v_1^2 + \rho g y_1 = P + \rho g y_2$$

In this case, $y_2 - y_1 = h$; therefore, this expression reduces to

$$v_1 = \sqrt{\frac{2(P - P_0)}{\rho} + 2gh}$$

When $P$ is much greater than $P_0$ and $P/\rho \gg 2gh$ (so that the term $2gh$ can be neglected), the exit speed of the water is mainly a function of $P$. If the tank is open to the atmosphere, $P = P_0$ and $v_1 = \sqrt{2gh}$. In other words, for an open tank the speed of liquid coming out through a hole a distance $h$ below the surface is equal to that acquired by an object falling freely through a vertical distance $h$. This phenomenon is known as **Torricelli's law.**

**B** Suppose the position of the hole in Figure 15.18 could be adjusted vertically. If the tank is open to the atmosphere and sitting on a table, what position of the hole would cause the water to land on the table at the farthest distance from the tank?

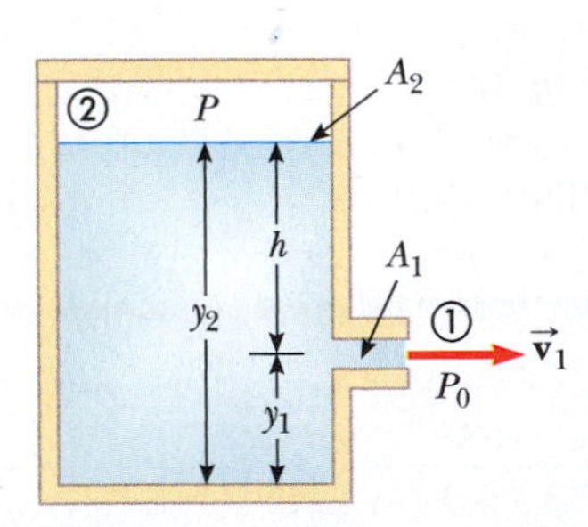


**FIGURE 15.18** (Interactive Example 15.7) A liquid leaves a hole in a tank at speed $v_1$.

**Solution** Because the tank is open to the atmosphere, the pressure at both points 1 and 2 is atmospheric pressure. Therefore, Bernoulli's equation becomes

$$P_0 + \tfrac{1}{2}\rho v_1^2 + \rho g y_1 = P_0 + \rho g y_2 \quad \rightarrow \quad v_1 = \sqrt{2g(y_2 - y_1)}$$

We model a parcel of water exiting the hole as a projectile. We find the time at which the parcel strikes the table from a hole at an arbitrary position:

$$y_f = y_i + v_{yi}t - \tfrac{1}{2}gt^2$$

$$0 = y_1 + 0 - \tfrac{1}{2}gt^2$$

$$t = \sqrt{\frac{2y_1}{g}}$$

Therefore, the horizontal position of the parcel at the time it strikes the table is

$$x_f = x_i + v_{xi}t = 0 + \sqrt{2g(y_2 - y_1)}\sqrt{\frac{2y_1}{g}} = 2\sqrt{(y_2 y_1 - y_1^2)}$$

Now we maximize the horizontal position by taking the derivative of $x_f$ with respect to $y_1$ (because $y_1$, the height of the hole, is the variable that can be adjusted) and setting it equal to zero:

$$\frac{dx_f}{dy_1} = \tfrac{1}{2}(2)[(y_2 y_1 - y_1^2)]^{-1/2}(y_2 - 2y_1) = 0$$

This expression is satisfied if

$$y_1 = \tfrac{1}{2}y_2$$

Therefore, the hole should be halfway between the bottom of the tank and the upper surface of the water to maximize the horizontal distance. Below this location, the water is projected at a higher speed but falls for a short time interval, reducing the horizontal range. Above this point, the water spends more time in the air but is projected with a smaller horizontal speed.

**PhysicsNow™** By logging into PhysicsNow at www.pop4e.com and going to Interactive Example 15.7 you can move the hole vertically to see where the water lands.

## 15.8 OTHER APPLICATIONS OF FLUID DYNAMICS

Consider the streamlines that flow around an airplane wing as shown in Figure 15.19. Let us assume that the airstream approaches the wing horizontally from the right. The tilt of the wing causes the airstream to be deflected downward. Because the airstream is deflected by the wing, the wing must exert a force on the airstream. According to Newton's third law, the airstream must exert an equal and opposite force $\vec{\mathbf{F}}$ on the wing. This force has a vertical component called the **lift** (or aerodynamic lift) and a horizontal component called **drag.** The lift depends on several factors, such as the speed of the airplane, the area of the wing, its curvature, and the angle between the wing and the horizontal. As this angle increases, turbulent flow can set in above the wing to reduce the lift.

In general, an object experiences lift by any effect that causes the fluid to change its direction as it flows past the object. Some factors that influence lift are the shape of the object, its orientation with respect to the fluid flow, spinning motion (for example, a curve ball thrown in a baseball game due to the spinning of the baseball), and the texture of the object's surface.

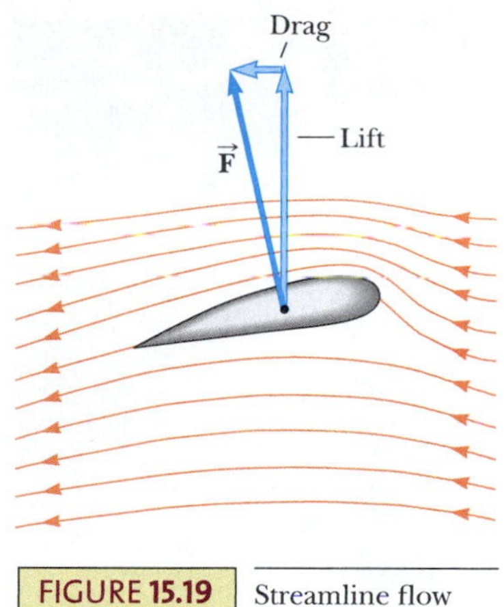


**FIGURE 15.19** Streamline flow around a moving airplane wing. The air approaching from the right is deflected downward by the wing.

A number of devices operate in a manner similar to the *atomizer* in Figure 15.20. A stream of air passing over an open tube reduces the pressure above the tube. This reduction in pressure causes the liquid to rise into the air stream. The liquid is then dispersed into a fine spray of droplets. This type of system is used in perfume bottles and paint sprayers.

Bernoulli's principle explains one symptom of advanced arteriosclerosis called *vascular flutter.* The artery is constricted as a result of an accumulation of plaque on its inner walls (Fig. 15.21). Plaque is a combination of fat, cell debris, connective tissue, and sometimes calcium that forms a flat patch inside a blood vessel. The blood speed through the constriction is higher than elsewhere according to the continuity equation for fluids. According to Bernoulli's principle, the pressure in the constriction is lower than elsewhere. If the blood speed is sufficiently high in the constricted region, the artery may collapse under the larger external pressure, causing a momentary interruption in blood flow. At this point, the speed of the blood goes to zero, its pressure rises again, and the vessel reopens. As the blood rushes through the constricted artery, the internal pressure drops and again the artery closes. Such variations in blood flow can be heard with a stethoscope.

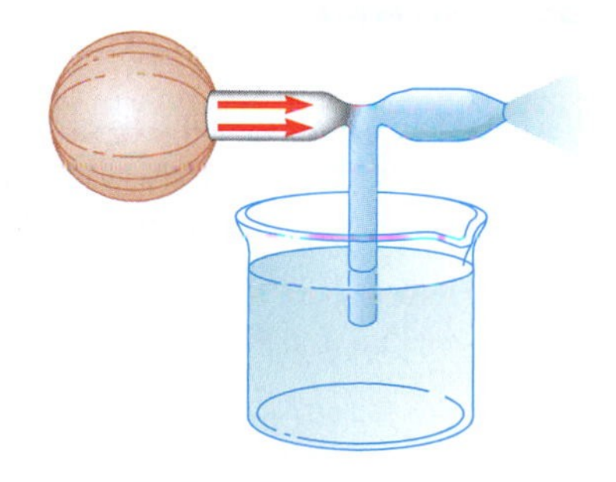

**FIGURE 15.20** A stream of air passing over a tube dipped into a liquid will cause the liquid to rise in the tube.

## 15.9 A NEAR MISS EVEN BEFORE LEAVING SOUTHAMPTON

CONTEXT CONNECTION

As the *Titanic* began its maiden voyage, it experienced a potentially disastrous incident before it left Southampton harbor. (If this disaster had occurred, however, it is highly likely that it would have been much less disastrous in terms of lives lost than the incident that actually occurred later in the voyage.) The *Titanic* passed closely by the *New York,* which was tied securely next to the *Oceanic* at the dock, with the keels of the two ships parallel. As the *Titanic* passed by, the *New York* was forced toward it, the *New York*'s mooring ropes snapped, and its stern swung out toward the *Titanic.* It was only quick thinking by the harbor pilot on the *Titanic,* who reversed the engines, causing the *Titanic* to slow and allow the *New York* to pass by safely, that saved the two ships from a collision. As it was, a collision was averted by only a few feet, and the *Titanic* was delayed by over an hour in her departure. Figure 15.22 is a photograph taken from the *Titanic,* showing how close the ships came to colliding.

It is ironic that the captain of the *Titanic,* E. J. Smith, who watched the near miss from the bridge, was captain on one of the *Titanic*'s sister ships, the *Olympic,* when a similar incident occurred seven months before the *New York* incident. In this case, the cruiser *Hawke* was pulled toward the *Olympic* and a collision was not averted. The *Hawke*'s bow was seriously damaged in the collision, and the hull of the *Olympic* was punctured above and below the waterline. Both ships were able to return to port but needed extensive repairs.

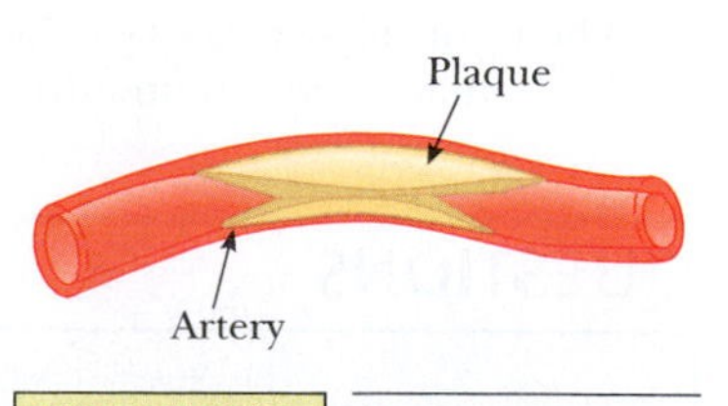


**FIGURE 15.21** Blood must travel faster than normal through a constricted region of an artery.

FIGURE 15.22 While leaving Southampton harbor, the *Titanic (left)* experienced a near miss with the *New York (right)* due to Bernoulli's principle. If this accident had actually occurred, it may have changed the timing enough that the *Titanic,* once under way, might not have been sunk by an iceberg.

Why did these events occur? The answer lies in Bernoulli's principle. As ships move through the water, they push water out of the way and the water moves around the sides of the ship. Imagine now that a ship such as the *Titanic* passes near another ship, such as the *New York,* with their keels parallel. The water moving around the side of the *Titanic* toward the *New York* is forced into a narrow channel between the ships. Because water is incompressible, its volume remains constant when it is squeezed into a narrow channel. The initial tendency is for the compressed water to rise into the air between the ships because the air above the water offers little resistance to being compressed. As soon as the water level between the ships rises, however, the water will begin to flow in a direction parallel to the keels toward the lower-level water near the bow and stern of the ships. Therefore, the *water between the ships is moving at a higher speed than the water on the opposite sides of the ships.* According to Bernoulli's principle, this rapidly moving water exerts less pressure on the sides of the ships than the slower moving water on the outer sides. The result is a *net force pushing the two ships toward each other.*

Therefore, captains of boats and ships are advised not to pass too close by other boats moving in a parallel direction. If that does occur, the boats could be pushed into each other. This effect occurs for air, explaining the effect of the passing truck in Section 15.7.

In this Context Connection section, we investigated an application of Bernoulli's principle. In the Context Conclusion, we shall explore the difficulties in visiting the *Titanic* because of its great depth under the ocean surface. ■

## SUMMARY

**Physics Now™** Take a practice test by logging into PhysicsNow at **www.pop4e.com** and clicking on the Pre-Test link for this chapter.

The **pressure** $P$ in a fluid is the force per unit area that the fluid exerts on a surface:

$$P \equiv \frac{F}{A} \qquad [15.1]$$

In the SI system, pressure has units of newtons per square meter, and $1\ \text{N/m}^2 = 1$ pascal (Pa).

The pressure in a liquid varies with depth $h$ according to the expression

$$P = P_0 + \rho g h \qquad [15.4]$$

where $P_0$ is the pressure at the surface of the liquid and $\rho$ is the density of the liquid, assumed uniform.

**Pascal's law** states that when a change in pressure is applied to a fluid, the change in pressure is transmitted undiminished to every point in the fluid and to every point on the walls of the container.

When an object is partially or fully submerged in a fluid, the fluid exerts an upward force on the object called the **buoyant force.** According to **Archimedes's principle,** the buoyant force is equal to the weight of the fluid displaced by the object.

Various aspects of fluid dynamics can be understood by adopting a simplification model in which the fluid is nonviscous and incompressible and the fluid motion is a steady flow with no turbulence.

Using this model, two important results regarding fluid flow through a pipe of nonuniform size can be obtained:

1. The flow rate through the pipe is a constant, which is equivalent to stating that the product of the cross-sectional area $A$ and the speed $v$ at any point is a constant. This behavior is described by the **continuity equation for fluids:**

$$A_1 v_1 = A_2 v_2 = \text{constant} \qquad [15.7]$$

2. The sum of the pressure, kinetic energy per unit volume, and gravitational potential energy per unit volume has the same value at all points along a streamline. This behavior is described by **Bernoulli's equation:**

$$P_1 + \tfrac{1}{2}\rho v_1^2 + \rho g y_1 = P_2 + \tfrac{1}{2}\rho v_2^2 + \rho g y_2 \qquad [15.12]$$

## QUESTIONS

☐ = answer available in the *Student Solutions Manual and Study Guide*

1. Two drinking glasses having equal weights but different shapes and different cross-sectional areas are filled to the same level with water. According to the expression

$P = P_0 + \rho gh$, the pressure is the same at the bottom of both glasses. In view of this fact, why does one weigh more than the other?

2. Figure Q15.2 shows aerial views from directly above two dams. Both dams are equally wide (the vertical dimension in the diagram) and equally high (into the page in the diagram). The dam on the left holds back a very large lake, whereas the dam on the right holds back a narrow river. Which dam has to be built stronger?

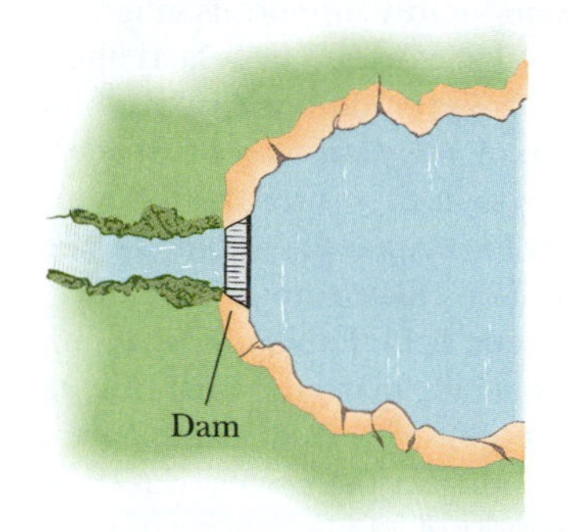


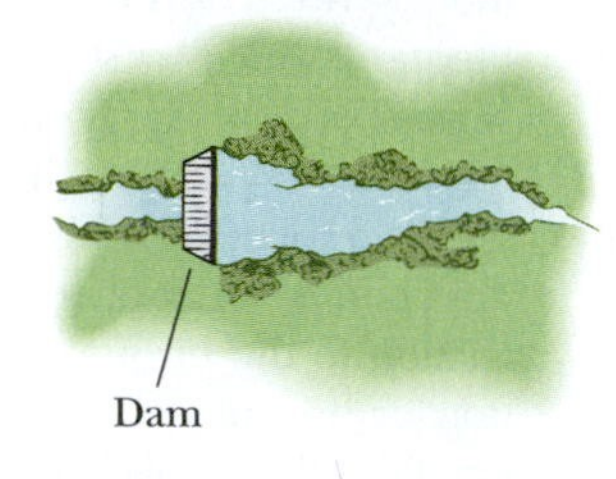


FIGURE Q15.2

3. Some physics students attach a long tube to the opening of a hot water bottle made of strong rubber. Leaving the hot water bottle on the ground, they hoist the other end of the tube to the roof of a multistory campus building. Students at the top of the building pour water into the tube. The students on the ground watch the bottle fill with water. On the roof, the students are surprised to see that the tube never seems to fill up: they can continue to pour more and more water down the tube. On the ground, the hot water bottle swells up like a balloon and bursts, drenching the students. Explain these observations.

4. Suppose a damaged ship can just barely keep afloat in the ocean. It is towed toward shore and into a river, heading toward a dry dock for repair. As it is pulled up the river, it sinks. Why?

5. A fish rests on the bottom of a bucket of water while the bucket is being weighed on a scale. When the fish begins to swim around, does the scale reading change?

6. Lead has a greater density than iron, and both are denser than water. Is the buoyant force on a lead object greater than, less than, or equal to the buoyant force on an iron object of the same volume?

7. Is the buoyant force a conservative force? Is a potential energy associated with it? Explain your answers.

8. If the air stream from a hair dryer is directed over a Ping-Pong ball, the ball can be levitated. Explain.

9. The water supply for a city is often provided from reservoirs built on high ground. Water flows from the reservoir, through pipes, and into your home when you turn the tap on your faucet. Why is the water flow more rapid out of a faucet on the first floor of a building than in an apartment on a higher floor?

10. When ski jumpers are airborne (Fig. Q15.10), why do they bend their bodies forward and keep their hands at their sides?

(© TempSport/CORBIS)

FIGURE Q15.10

11. Explain why a sealed bottle partially filled with a liquid can float in a basin of the same liquid.

12. When is the buoyant force on a swimmer greater, after exhaling or after inhaling?

13. A barge is carrying a load of gravel along a river. As it approaches a low bridge the captain realizes that the top of the pile of gravel is not going to make it under the bridge. The captain orders the crew to shovel gravel quickly from the pile into the water. Is that a good decision?

14. A person in a boat floating in a small pond throws an anchor overboard. Does the level of the pond rise, fall, or remain the same?

15. An empty metal soap dish barely floats in water. A bar of Ivory soap floats in water. When the soap is stuck in the soap dish, the combination sinks. Explain why.

16. A piece of unpainted porous wood barely floats in a container partly filled with water. If the container is sealed and pressurized above atmospheric pressure, does the wood rise, fall, or remain at the same level?

17. Because atmospheric pressure is about $10^5$ N/m$^2$ and the area of a person's chest is about 0.13 m$^2$, the force of the atmosphere on one's chest is around 13 000 N. In view of this enormous force, why don't our bodies collapse?

18. A small piece of steel is tied to a block of wood. When the wood is placed in a tub of water with the steel on top, half of the block is submerged. If the block is inverted so that the steel is under water, does the amount of the block submerged increase, decrease, or remain the same? What happens to the water level in the tub when the block is inverted?

19. An unopened can of diet cola floats when placed in a tank of water, whereas a can of regular cola of the same brand sinks in the tank. What do you suppose could explain this behavior?

20. Prairie dogs (Fig. Q15.20) ventilate their burrows by building a mound around one entrance, which is open to a stream of air when wind blows from any direction. A second entrance at ground level is open to almost stagnant air. How does this construction create an airflow through the burrow?

(Pamela Zilly/The Image Bank/Getty)

FIGURE Q15.20

21. You are a passenger on a spacecraft. For your survival and comfort, the interior contains air just like that at the surface of the Earth. The craft is coasting through a very empty region of space. That is, a nearly perfect vacuum exists just outside the wall. Suddenly, a meteoroid pokes a hole, about the size of a large coin, right through the wall next to your seat. What will happen? Is there anything you can or should do about it?

22. Consider a stationary fluid in contact with a solid surface. If the force exerted by the fluid is entirely characterized by a pressure, the force must be perpendicular to the surface. That is, a bulk fluid exerts a normal force but cannot exert a force of static friction. In contrast, a *moving* fluid can have the effect of exerting kinetic friction if it possesses viscosity; for example, think of the force that molasses exerts on a stirring spoon. In Chapter 5, we modeled this drag force as possibly proportional to the first or to the second power of the speed of an object moving through the fluid. Now we are saying that the drag force must go to zero when the speed approaches zero. A *thin film,* as opposed to a bulk fluid, can exert a force parallel to a solid surface. For example, a droplet or a thin film of water can temporarily support its weight by adhering to a vertical surface that it wets. These facts about moving fluids and thin films, however, do not invalidate the theorem, which we state again: *A bulk fluid cannot exert a force of static friction.* Apply this theorem to answer the following questions. (a) As we will study in Chapter 22, a compass needle on a frictionless pivot would oscillate forever around the direction of an applied magnetic field. Explain how adding friction to the pivot would generally make the final orientation of the needle inaccurate, but with fluid damping the needle will approach rest pointing in the correct direction. (b) A carpenter's level can consist of a bubble of air in water within a tube forming an arc of a circle. Explain how, after a quick calibration, it can be a very accurate level. (c) Assume that just after you step out of a shower the whole bathtub and also the bar of soap are thoroughly wet, covered with more than a thin film of water. If you drop the soap into the tub, will it come to rest? If so, where? Explain.

# PROBLEMS

1, 2, 3 = straightforward, intermediate, challenging
☐ = full solution available in the *Student Solutions Manual and Study Guide*
**Physics Now™** = coached problem with hints available at www.pop4e.com
(computer icon) = computer useful in solving problem
(shaded box) = paired numerical and symbolic problems
(biomedical icon) = biomedical application

## Section 15.1 ▪ Pressure

1. Calculate the mass of a solid iron sphere that has a diameter of 3.00 cm.

2. The four tires of an automobile are inflated to a gauge pressure of 200 kPa. Each tire has an area of 0.024 0 $m^2$ in contact with the ground. Determine the weight of the automobile.

3. A 50.0-kg woman balances on one heel of a pair of high-heeled shoes. If the heel is circular and has a radius of 0.500 cm, what pressure does she exert on the floor?

4. What is the total mass of the Earth's atmosphere? (The radius of the Earth is $6.37 \times 10^6$ m, and atmospheric pressure at the surface is $1.013 \times 10^5$ N/$m^2$.)

## Section 15.2 ▪ Variation of Pressure with Depth

5. The spring of the pressure gauge shown in Figure 15.2 has a force constant of 1 000 N/m and the piston has a diameter of 2.00 cm. As the gauge is lowered into water, what change in depth causes the piston to move in by 0.500 cm?

6. (a) Calculate the absolute pressure at an ocean depth of 1 000 m. Assume that the density of sea water is 1 024 kg/$m^3$ and that the air above exerts a pressure of 101.3 kPa. (b) At this depth, what force must the frame around a circular submarine porthole having a diameter of 30.0 cm exert to counterbalance the force exerted by the water?

7. **Physics Now™** What must be the contact area between a suction cup (completely exhausted) and a ceiling if the cup is to support the weight of an 80.0-kg student?

8. (a) A very powerful vacuum cleaner has a hose 2.86 cm in diameter (Fig. P15.8a). With no nozzle on the hose, what is the weight of the heaviest brick that the cleaner can lift? (b) A very powerful octopus uses one sucker of diameter 2.86 cm on each of the two shells of a clam in an attempt to pull the shells apart (Fig. P15.8b). Find the greatest force the octopus can exert in salt water 32.3 m deep. (*Caution:* Experimental verification can be interesting, but do not drop a brick on your foot. Do not overheat the motor of a vacuum cleaner. Do not get an octopus mad at you.)

(a)

(b)

FIGURE **P15.8**

**9.** For the cellar of a new house, a hole is dug in the ground, with vertical sides going down 2.40 m. A concrete foundation wall is built all the way across the 9.60-m width of the excavation. This foundation wall is 0.183 m away from the front of the cellar hole. During a rainstorm, drainage from the street fills up the space in front of the concrete wall, but not the cellar behind the wall. The water does not soak into the clay soil. Find the force the water causes on the foundation wall. For comparison, the weight of the water is given by $2.40 \text{ m} \times 9.60 \text{ m} \times 0.183 \text{ m} \times 1\,000 \text{ kg/m}^3 \times 9.80 \text{ m/s}^2 = 41.3 \text{ kN}$.

**10.** A swimming pool has dimensions $30.0 \text{ m} \times 10.0 \text{ m}$ and a flat bottom. When the pool is filled to a depth of 2.00 m with fresh water, what is the force caused by the water on the bottom? On each end? On each side?

**11.** **Review problem.** Piston ① in Figure P15.11 has a diameter of 0.250 in. Piston ② has a diameter of 1.50 in. Determine the magnitude $F$ of the force necessary to support the 500 lb load in the absence of friction.

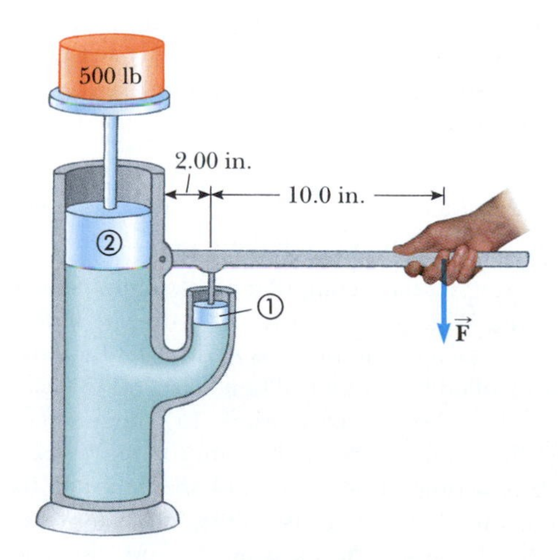


FIGURE **P15.11**

## Section 15.3 ▪ Pressure Measurements

**12.** Figure P15.12 shows Superman attempting to drink water through a very long straw. With his great strength he achieves maximum possible suction. The walls of the tubular straw do not collapse. (a) Find the maximum height through which he can lift the water. (b) Still thirsty, the Man of Steel repeats his attempt on the Moon, which has no atmosphere. Find the difference between the water levels inside and outside the straw.

FIGURE **P15.12**

**13.** **Physics Now™** Blaise Pascal duplicated Torricelli's barometer using a red Bordeaux wine, of density $984 \text{ kg/m}^3$, as the working liquid (Fig. P15.13). What was the height $h$ of the wine column for normal atmospheric pressure? Would you expect the vacuum above the column to be as good as for mercury?

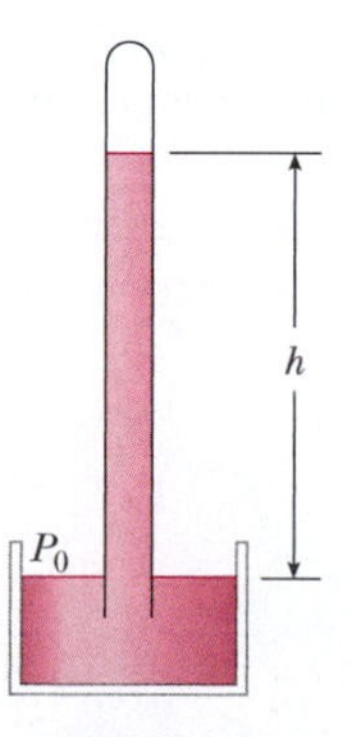


FIGURE **P15.13**

14. Mercury is poured into a U-tube as shown in Figure P15.14a. The left arm of the tube has cross-sectional area $A_1$ of 10.0 $cm^2$ and the right arm has a cross-sectional area $A_2$ of 5.00 $cm^2$. One hundred grams of water are then poured into the right arm as shown in Figure P15.14b. (a) Determine the length of the water column in the right arm of the U-tube. (b) Given that the density of mercury is 13.6 $g/cm^3$, what distance $h$ does the mercury rise in the left arm?

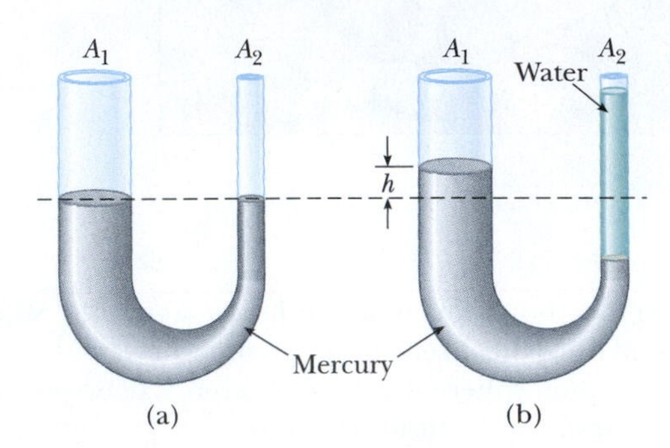


FIGURE **P15.14**

15. Normal atmospheric pressure is $1.013 \times 10^5$ Pa. The approach of a storm causes the height of a mercury barometer to drop by 20.0 mm from the normal height. What is the atmospheric pressure? (The density of mercury is 13.59 $g/cm^3$.)

16. The human brain and spinal cord are immersed in the cerebrospinal fluid. The fluid is normally continuous between the cranial and spinal cavities. It normally exerts a pressure of 100 to 200 mm of $H_2O$ above the prevailing atmospheric pressure. In medical work, pressures are often measured in millimeters of $H_2O$ because body fluids, including the cerebrospinal fluid, typically have the same density as water. The pressure of the cerebrospinal fluid can be measured by means of a *spinal tap* as illustrated in Figure P15.16. A hollow tube is inserted into the spinal column, and the height to which the fluid rises is observed. If the fluid rises to a height of 160 mm, we write its gauge pressure as 160 mm $H_2O$. (a) Express this pressure in pascals, in atmospheres, and in millimeters of mercury. (b) Sometimes it is necessary to determine whether an accident victim has suffered a crushed vertebra that is blocking flow of the cerebrospinal fluid in the spinal column. In other cases, a physician may suspect that a tumor or other growth is blocking the spinal column and inhibiting flow of cerebrospinal fluid. Such conditions can be investigated by means of *Queckenstedt's test.* In this procedure, the veins in the patient's neck are compressed so as to make the blood pressure rise in the brain. The increase in pressure in the blood vessels is transmitted to the cerebrospinal fluid. What should be the normal effect on the height of the fluid in the spinal tap? (c) Suppose compressing the veins had no effect on the fluid level. What might account for that?

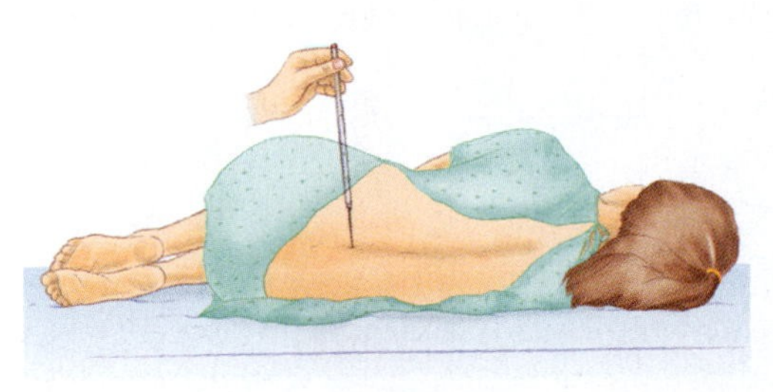

FIGURE **P15.16**

17. Mercury is a poison. The liquid and mercury vapor can enter the body through the skin and mucous membranes. Do not carry out the procedure described here without appropriate safety precautions. Assume that a barometer is constructed as follows. A rigid, thin-walled plastic tube, closed at one end, has mass 480 g, inner diameter 2.10 cm, and height 160 cm. Mercury is poured into it to fill the tube. (a) Find the mass of the metal. With the open end covered, the tube is inverted and held just above the flat bottom of an originally empty pan (mass 320 g) on a table. The tube is hung from a cord attached between its closed end and the ceiling. Next, the bottom end of the tube is uncovered. It is found that more than half, but not all, of the mercury runs out of the tube into the pan. The flow stops when the open end of the tube is below the level of the mercury in the pan and the level of mercury in the tube is 76.0 cm higher than the level in the pan. We say that the barometer reading is 76.0 cm. (b) Find the tension in the cord. (c) Find the normal force exerted by the table on the pan. (d) If another barometer were placed into the space within the tube above the mercury, what would it read? (This question was discussed by town philosophical societies in colonial America.) (e) A slow leak at the closed end of the tube allows air to enter over a period of days. What happens to the tension in the cord? What happens to the normal force exerted by the table?

## Section 15.4 ■ Buoyant Forces and Archimedes's Principle

18. A Styrofoam slab has thickness $h$ and density $\rho_s$. When a swimmer of mass $m$ is resting on it, the slab floats in fresh water with its top at the same level as the water surface. Find the area of the slab.

19. A Ping-Pong ball has a diameter of 3.80 cm and average density of 0.084 0 $g/cm^3$. What force is required to hold it completely submerged under water?

20. The weight of a rectangular block of low-density material is 15.0 N. With a thin string, the center of the horizontal bottom face of the block is tied to the bottom of a beaker partly filled with water. When 25.0% of the block's volume is submerged, the tension in the string is 10.0 N. (a) Sketch a free-body diagram for the block, showing all forces acting on it. (b) Find the buoyant force on the block. (c) Oil of density 800 $kg/m^3$ is now steadily added to the beaker, forming a layer above the water and surrounding the block. The oil exerts forces on each of the four side walls of the block that the oil touches. What are the directions of these forces? (d) What happens to the string tension as the oil is added? Explain how the oil has this effect on the string tension. (e) The string breaks when its tension reaches 60.0 N. At this moment, 25.0% of the block's volume is still below the waterline. What addi-

tional fraction of the block's volume is below the top surface of the oil? (f) After the string breaks, the block comes to a new equilibrium position in the beaker. It is now in contact only with the oil. What fraction of the block's volume is submerged?

21. A 10.0-kg block of metal measuring 12.0 cm × 10.0 cm × 10.0 cm is suspended from a scale and immersed in water as shown in Figure P15.21. The 12.0-cm dimension is vertical and the top of the block is 5.00 cm below the surface of the water. (a) What are the forces acting on the top and on the bottom of the block? (Use $P_0 = 1.013\,0 \times 10^5\ \text{N/m}^2$.) (b) What is the reading of the spring scale? (c) Show that the buoyant force equals the difference between the forces at the top and the bottom of the block.

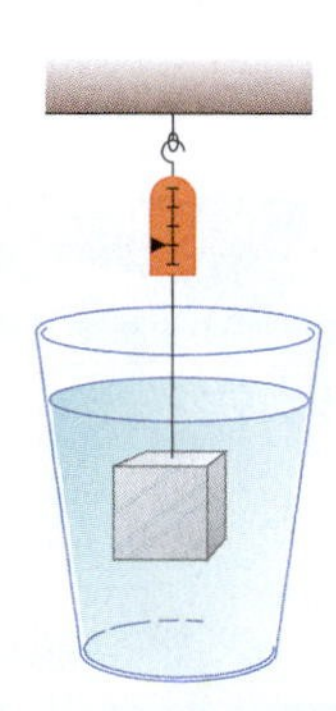

FIGURE P15.21

22. To an order of magnitude, how many helium-filled toy balloons would be required to lift you? Because helium is an irreplaceable resource, develop a theoretical answer rather than an experimental answer. In your solution, state what physical quantities you take as data and the values you measure or estimate for them.

23. **Physics Now™** A cube of wood having an edge dimension of 20.0 cm and a density of 650 kg/m$^3$ floats on water. (a) What is the distance from the horizontal top surface of the cube to the water level? (b) What mass of lead should be placed on top of the cube so that its top will be just level with the water?

24. Determination of the density of a fluid has many important applications. A car battery contains sulfuric acid, for which density is a measure of concentration. For the battery to function properly, the density must be within a range specified by the manufacturer. Similarly, the effectiveness of antifreeze in your car's engine coolant depends on the density of the mixture (usually ethylene glycol and water). When you donate blood to a blood bank, its screening includes determination of the density of your blood because higher density correlates with higher hemoglobin content. A *hydrometer* is an instrument used to determine liquid density. A simple one is sketched in Figure P15.24. The bulb of a syringe is squeezed and released to let the atmosphere lift a sample of the liquid of interest into a tube containing a calibrated rod of known density. The rod, of length $L$ and average density $\rho_0$, floats partially immersed in the liquid of density $\rho$. A length $h$ of the rod protrudes above the surface of the liquid. Show that the density of the liquid is given by

$$\rho = \frac{\rho_0 L}{L - h}$$

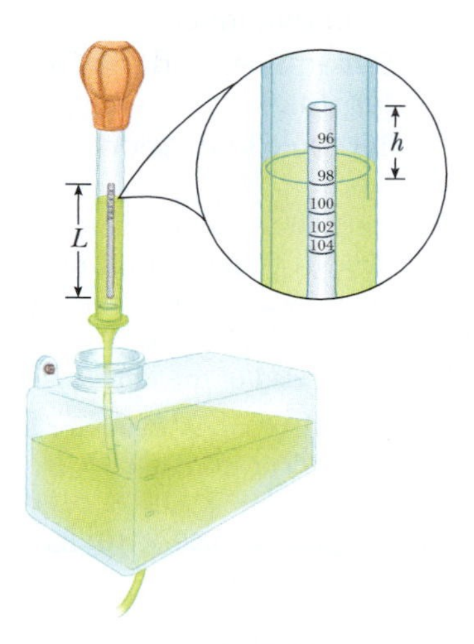


FIGURE P15.24

25. How many cubic meters of helium are required to lift a balloon with a 400-kg payload to a height of 8 000 m? (Take $\rho_{\text{He}} = 0.180\ \text{kg/m}^3$.) Assume that the balloon maintains a constant volume and that the density of air decreases with the altitude $z$ according to the expression $\rho_{\text{air}} = \rho_0 e^{-z/8\,000}$, where $z$ is in meters and the density of air at sea level is $\rho_0 = 1.25\ \text{kg/m}^3$.

26. A bathysphere used for deep-sea exploration has a radius of 1.50 m and a mass of $1.20 \times 10^4$ kg. To dive, this submarine takes on mass in the form of sea water. Determine the amount of mass the submarine must take on if it is to descend at a constant speed of 1.20 m/s, when the resistive force on it is 1 100 N in the upward direction. The density of sea water is $1.03 \times 10^3\ \text{kg/m}^3$.

27. A plastic sphere floats in water with 50.0% of its volume submerged. This same sphere floats in glycerin with 40.0% of its volume submerged. Determine the densities of the glycerin and the sphere.

28. **Review problem.** A long, cylindrical rod of radius $r$ is weighted on one end so that it floats upright in a fluid having a density $\rho$. It is pushed down a distance $x$ from its equilibrium position and released. Show that the rod will execute simple harmonic motion if the resistive effects of the fluid are negligible and determine the period of the oscillations.

29. Decades ago, it was thought that huge herbivorous dinosaurs such as *Apatosaurus* and *Brachiosaurus* habitually walked on the bottom of lakes, extending their long necks up to the surface to breathe. *Brachiosaurus* had its nostrils on the top of its head. In 1977, Knut Schmidt-Nielsen pointed out that breathing would be too much work for such a creature. For a simple model, consider a sample consisting of 10.0 L of air at absolute pressure 2.00 atm,

with density 2.40 kg/m$^3$, located at the surface of a freshwater lake. Find the work required to transport it to a depth of 10.3 m, with its temperature, volume, and pressure remaining constant. This energy investment is greater than the energy that can be obtained by metabolism of food with the oxygen in that quantity of air.

## Section 15.5 ■ Fluid Dynamics
## Section 15.6 ■ Streamlines and the Continuity Equation for Fluids
## Section 15.7 ■ Bernoulli's Equation

**30.** (a) A water hose 2.00 cm in diameter is used to fill a 20.0-L bucket. If it takes 1.00 min to fill the bucket, what is the speed $v$ at which water moves through the hose? (*Note:* 1 L = 1 000 cm$^3$.) (b) The hose has a nozzle 1.00 cm in diameter. Find the speed of the water at the nozzle.

**31.** A horizontal pipe 10.0 cm in diameter has a smooth reduction to a pipe 5.00 cm in diameter. If the pressure of the water in the larger pipe is $8.00 \times 10^4$ Pa and the pressure in the smaller pipe is $6.00 \times 10^4$ Pa, at what rate does water flow through the pipes?

**32.** Water flows through a fire hose of diameter 6.35 cm at a rate of 0.012 0 m$^3$/s. The fire hose ends in a nozzle of inner diameter 2.20 cm. What is the speed with which the water exits the nozzle?

**33.** **PhysicsNow™** A large storage tank with an open top is filled to a height $h_0$. The tank is punctured at a height $h$ above the bottom of the tank (Fig. P15.33). Find an expression for how far from the tank the exiting stream lands.

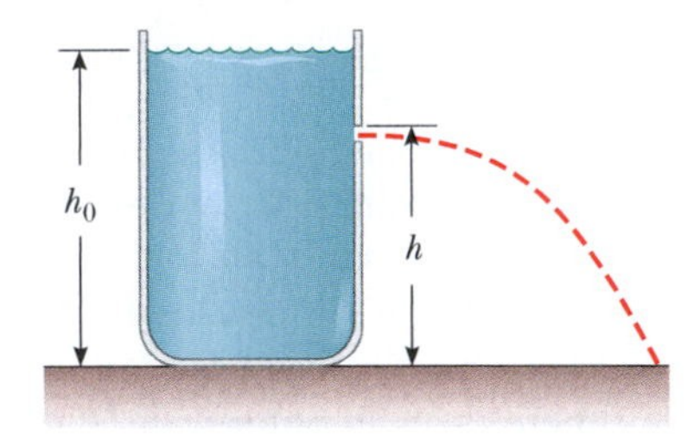


FIGURE **P15.33** Problems 15.33 and 15.34.

**34.** A large storage tank, open at the top and filled with water, develops a small hole in its side (Fig. P15.33) at a point 16.0 m below the water level. If the rate of flow from the leak is $2.50 \times 10^{-3}$ m$^3$/min, determine (a) the speed at which the water leaves the hole and (b) the diameter of the hole.

**35.** A village maintains a large tank with an open top, containing water for emergencies. The water can drain from the tank through a hose of diameter 6.60 cm. The hose ends with a nozzle of diameter 2.20 cm. A rubber stopper is inserted into the nozzle. The water level in the tank is kept 7.50 m above the nozzle. (a) Calculate the friction force exerted by the nozzle on the stopper. (b) The stopper is removed. What mass of water flows from the nozzle in 2.00 h? (c) Calculate the gauge pressure of the flowing water in the hose just behind the nozzle.

**36.** Water falls over a dam of height $h$ with a mass flow rate of $R$, in kilograms per second. (a) Show that the power available from the water is

$$\mathcal{P} = Rgh$$

where $g$ is the free-fall acceleration. (b) Each hydroelectric unit at the Grand Coulee Dam takes in water at a rate of $8.50 \times 10^5$ kg/s from a height of 87.0 m. The power developed by the falling water is converted to electric power with an efficiency of 85.0%. How much electric power does each hydroelectric unit produce?

**37.** Figure P15.37 shows a stream of water in steady flow from a kitchen faucet. At the faucet the diameter of the stream is 0.960 cm. The stream fills a 125-cm$^3$ container in 16.3 s. Find the diameter of the stream 13.0 cm below the opening of the faucet.

(George Semple)

FIGURE **P15.37**

**38.** A legendary Dutch boy saved Holland by plugging a hole in a dike with his finger, 1.20 cm in diameter. If the hole was 2.00 m below the surface of the North Sea (density 1 030 kg/m$^3$), (a) what was the force on his finger? (b) If he pulled his finger out of the hole, during what time interval would the released water fill 1 acre of land to a depth of 1 ft? Assume that the hole remained constant in size. (A typical U.S. family of four uses 1 acre-foot of water, 1 234 m$^3$, in 1 year.)

**39.** Water is pumped up from the Colorado River to supply Grand Canyon Village, located on the rim of the canyon. The river is at an elevation of 564 m and the village is at an elevation of 2 096 m. Imagine that water is pumped through a single, long pipe 15.0 cm in diameter, driven by a single pump at the bottom end. (a) What is the minimum pressure at which the water must be pumped if it is to arrive at the village? (b) If 4 500 m$^3$ of water are pumped per day, what is the speed of the water in the pipe? (c) What additional pressure is necessary to deliver this flow? (*Note:* You may assume that the free-fall acceleration and the density of air are constant over this range of elevations. The pressures you calculate are too high for an ordinary pipe. In fact, the water is lifted in stages by several pumps through shorter pipes.)

**40.** Old Faithful Geyser in Yellowstone Park (Fig. P15.40) erupts at approximately 1-h intervals and the height of the water column reaches 40.0 m. (a) Model the rising stream as a series of separate drops. Analyze the free-fall motion of one of the drops to determine the speed at which the water leaves the ground. (b) Model the rising stream as an ideal fluid in streamline flow. Use Bernoulli's equation to determine the speed of the water as it leaves ground level. (c) What is the pressure (above atmospheric) in the heated underground chamber if its depth is 175 m? You may assume that the chamber is large compared with the geyser's vent.

FIGURE **P15.40**

**41.** An airplane is cruising at altitude 10 km. The pressure outside the craft is 0.287 atm; within the passenger compartment the pressure is 1.00 atm and the temperature is 20°C. A small leak occurs in one of the window seals in the passenger compartment. Model the air as an ideal fluid to find the speed of the stream of air flowing through the leak.

## Section 15.8 ▪ Other Applications of Fluid Dynamics

**42.** An airplane has a mass of $1.60 \times 10^4$ kg and each wing has an area of 40.0 $m^2$. During level flight, the pressure on the lower wing surface is $7.00 \times 10^4$ Pa. Determine the pressure on the upper wing surface.

**43.** A siphon is used to drain water from a tank as illustrated in Figure P15.43. The siphon has a uniform diameter. Assume steady flow without friction. (a) Assuming that the distance $h = 1.00$ m, find the speed of outflow at the end of the siphon. (b) What is the limitation on the height of the top of the siphon above the water surface? (For the flow of the liquid to be continuous, the pressure must not drop below the vapor pressure of the liquid.)

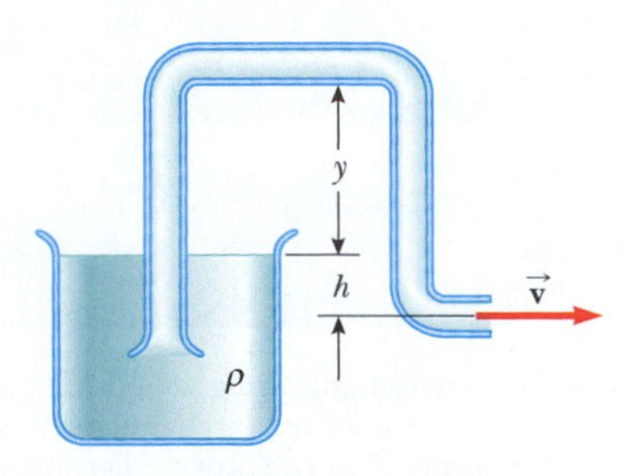


FIGURE **P15.43**

**44.** The Bernoulli effect can have important consequences for the design of buildings. For example, wind can blow around a skyscraper at remarkably high speed, creating low pressure. The higher atmospheric pressure in the still air inside the buildings can cause windows to pop out. As originally constructed, the John Hancock Building in Boston popped window panes, which fell many stories to the sidewalk below. (a) Suppose a horizontal wind blows in streamline flow with a speed of 11.2 m/s outside a large pane of plate glass with dimensions 4.00 m × 1.50 m. Assume that the density of the air is 1.30 $kg/m^3$. The air inside the building is at atmospheric pressure. What is the total force exerted by air on the window pane? (b) If a second skyscraper is built nearby, the air speed can be especially high where wind passes through the narrow separation between the buildings. Solve part (a) again, this time taking the wind speed as 22.4 m/s, twice as high.

**45.** A hypodermic syringe contains a medicine with the density of water (Fig. P15.45). The barrel of the syringe has a cross-sectional area $A = 2.50 \times 10^{-5}$ $m^2$ and the needle has a cross-sectional area $a = 1.00 \times 10^{-8}$ $m^2$. In the absence of a force on the plunger, the pressure everywhere is 1 atm. A force $\vec{F}$ of magnitude 2.00 N acts on the plunger, making medicine squirt horizontally from the needle. Determine the speed of the medicine as it leaves the needle's tip.

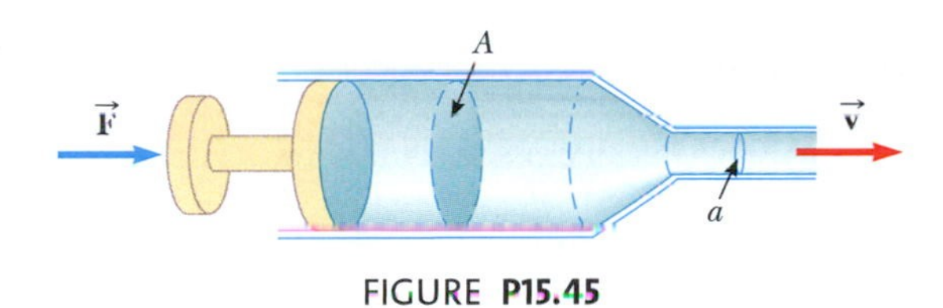


FIGURE **P15.45**

## Section 15.9 ▪ Context Connection—A Near Miss Even Before Leaving Southampton

**46.** According to the caption of the chapter-opening photograph, about 11% of an iceberg is above water. (a) Confirm this value mathematically. (b) Suppose an iceberg were floating in fresh water rather than in sea water. Would a larger or smaller percentage be above the waterline? Calculate this percentage.

**47.** The *Titanic* is docked in Southampton harbor just before boarding. You, as a ticket agent for White Star Lines, notice where the water level is on a scale of numbers marked on the side of the vessel. Because your ticket-collecting job is so boring that you need something to occupy your mind, you make a note of the water level and check it again after everyone boards. Over an interval of 2 h, 2 205 passengers, of average mass 75.0 kg, board the *Titanic.* You notice that the ship has sunk 1.00 cm deeper in the water with the passengers on board. What is the horizontal area enclosed by the waterline of the *Titanic*?

**48.** **Review problem.** Assume that the *Titanic* is drifting in Southampton harbor before its fateful journey and the captain wishes to stop the drift by dropping an anchor. The iron anchor has a mass of 2 000 kg. It is attached to a massless rope. The rope is wrapped around a reel in the form of a solid disk of radius 0.250 m and mass 300 kg that rotates on a frictionless axle. (a) Find the angular displacement of the reel when the anchor moves down 15.0 m. (b) Find the acceleration of the anchor as it falls through the air, which offers negligible resistance. (c) While the anchor continues to drop through the water, the water exerts a drag force of 2 500 N on it. With what acceleration does the anchor move through the water? (d) While the anchor drops through the water, what torque is exerted on the reel?

## Additional Problems

**49.** The true weight of an object can be measured in a vacuum, where buoyant forces are absent. An object of volume $V$ is weighed in air on an equal-arm balance with the use of counterweights of density $\rho$. Let the density of air be $\rho_{air}$ and the balance reading be $F'_g$. Show that the true weight $F_g$ is

$$F_g = F'_g + \left(V - \frac{F'_g}{\rho g}\right)\rho_{air} g$$

**50.** Water is forced out of a fire extinguisher by air pressure as shown in Figure P15.50. How much gauge air pressure in the tank (above atmospheric) is required for the water jet to have a speed of 30.0 m/s when the water level is 0.500 m below the nozzle?

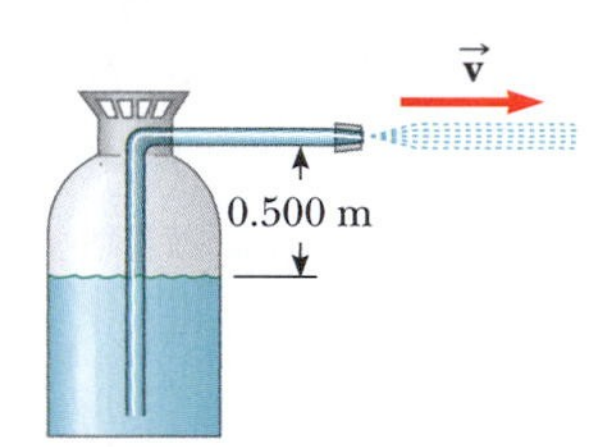


FIGURE **P15.50**

**51.** A light spring of constant $k = 90.0$ N/m is attached vertically to a table (Fig. P15.51a). A 2.00-g balloon is filled with helium (density $= 0.180$ kg/m$^3$) to a volume of 5.00 m$^3$ and is then connected to the spring, causing it to stretch as shown in Figure P15.51b. Determine the extension distance $L$ when the balloon is in equilibrium.

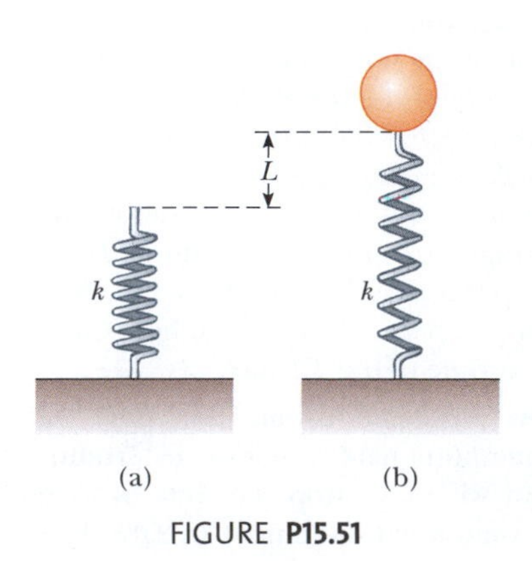


FIGURE **P15.51**

**52.** Evangelista Torricelli was the first person to realize that we live at the bottom of an ocean of air. He correctly surmised that the pressure of our atmosphere is attributable to the weight of the air. The density of air at 0°C at the Earth's surface is 1.29 kg/m$^3$. The density decreases with increasing altitude (as the atmosphere thins). On the other hand, if we assume that the density is constant at 1.29 kg/m$^3$ up to some altitude $h$ and is zero above that altitude, $h$ would represent the depth of the ocean of air. Use this model to determine the value of $h$ that gives a pressure of 1.00 atm at the surface of the Earth. Would the peak of Mount Everest rise above the surface of such an atmosphere?

**53.** **Physics Now™** **Review problem.** With reference to Figure 15.6, show that the total torque exerted by the water behind the dam about a horizontal axis through $O$ is $\frac{1}{6}\rho g w H^3$. Show that the effective line of action of the total force exerted by the water is at a distance $\frac{1}{3}H$ above $O$.

**54.** In about 1657, Otto von Guericke, inventor of the air pump, evacuated a sphere made of two brass hemispheres. Two teams of eight horses each could pull the hemispheres apart only on some trials, and then "with greatest difficulty," with the resulting sound likened to a cannon firing (Fig. P15.54). (a) Show that the force $F$ required to pull the evacuated hemispheres apart is $\pi R^2(P_0 - P)$, where $R$ is the radius of the hemispheres and $P$ is the pressure inside the hemispheres, which is much less than $P_0$. (b) Determine the force, taking $P = 0.100P_0$ and $R = 0.300$ m.

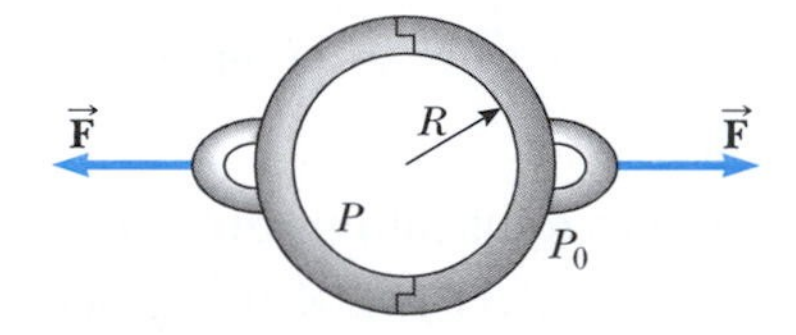

(The Granger Collection)

FIGURE **P15.54** The colored engraving, dated 1672, illustrates Otto von Guericke's demonstration of the force due to air pressure as it might have been performed before Emperor Ferdinand III in about 1657.

55. A beaker of mass $m_b$ containing oil of mass $m_0$ and density $\rho_0$ rests on a scale. A block of iron of mass $m_{Fe}$ is suspended from a spring scale and completely submerged in the oil as shown in Figure P15.55. Determine the equilibrium readings of both scales.

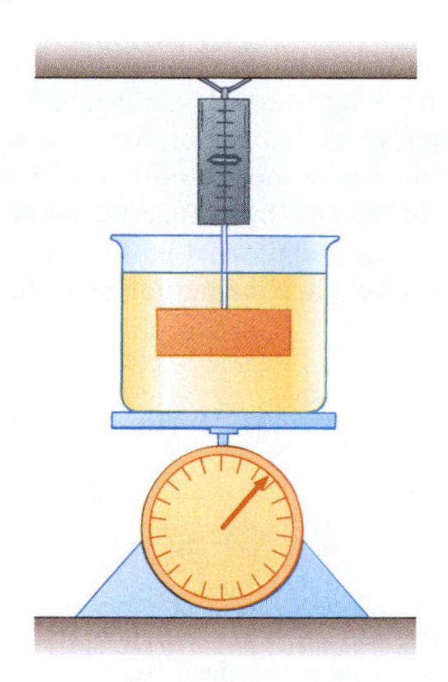

FIGURE **P15.55**

56. **Review problem.** A copper cylinder hangs at the bottom of a steel wire of negligible mass. The top end of the wire is fixed. When the wire is struck, it emits sound with a fundamental frequency of 300 Hz. The copper cylinder is then submerged in water so that half its volume is below the waterline. Determine the new fundamental frequency.

57. **Review problem.** This problem extends the reasoning of Problem 5.54 in Chapter 5 on sedimentation and centrifugation. According to Stokes's law, water exerts on a slowly moving immersed spherical object a resistive force described by

$$\vec{\mathbf{R}} = (I){-0.018\,8\ \mathrm{N\cdot s/m^2}}\ r\vec{\mathbf{v}}$$

where $r$ is the radius of the sphere and $\vec{\mathbf{v}}$ is its velocity. (a) Spherical cells of average density $1.02 \times 10^3$ kg/m$^3$ and radius 8.00 $\mu$m are suspended in water. Find the terminal speed with which the cells drift down. (b) Over what time interval will all the cells settle out of a tube 8.00 cm high? (c) The sedimentation rate can be greatly increased by the use of a centrifuge. Assume that it spins the tube at 3 000 rev/min in a horizontal plane, with the middle of the tube at 9.00 cm from the axis of rotation. Find the acceleration of the middle of the tube. (d) This acceleration has the effect of an enhanced free-fall acceleration. Model it as uniform over the length of the tube. Over what time interval will all the suspended cells settle out the water in this situation?

58. Show that the variation of atmospheric pressure with altitude is given by $P = P_0e^{-\alpha y}$, where $\alpha = \rho_0 g/P_0$, $P_0$ is atmospheric pressure at some reference level $y = 0$, and $\rho_0$ is the atmospheric density at this level. Assume that the decrease in atmospheric pressure over an infinitesimal change in altitude (so that the density is approximately uniform) is given by $dP = -\rho g\, dy$ and that the density of air is proportional to the pressure.

59. An incompressible, nonviscous fluid is initially at rest in the vertical portion of the pipe shown in Figure P15.59a, where $L = 2.00$ m. When the valve is opened, the fluid flows into the horizontal section of the pipe. What is the speed of the fluid when it is all in the horizontal section as shown in Figure P15.59b? Assume that the cross-sectional area of the entire pipe is constant.

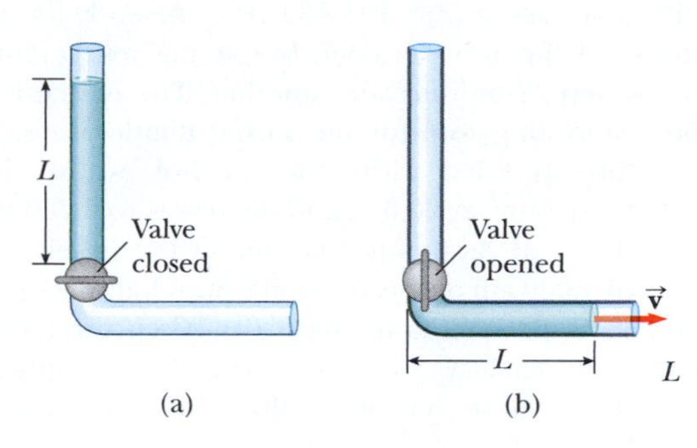


FIGURE **P15.59**

60. A cube of ice whose edges measure 20.0 mm is floating in a glass of ice-cold water with one of its faces parallel to the water's surface. (a) How far below the water surface is the bottom face of the block? (b) Ice-cold ethyl alcohol is gently poured onto the water surface to form a layer 5.00 mm thick above the water. The alcohol does not mix with the water. When the ice cube again attains hydrostatic equilibrium, what will be the distance from the top of the water to the bottom face of the block? (c) Additional cold ethyl alcohol is poured onto the water's surface until the top surface of the alcohol coincides with the top surface of the ice cube (in hydrostatic equilibrium). How thick is the required layer of ethyl alcohol?

61. A U-tube open at both ends is partially filled with water (Fig. P15.61a). Oil having a density 750 kg/m$^3$ is then poured into the right arm and forms a column $L = 5.00$ cm high (Fig. P15.61b). (a) Determine the difference $h$ in the heights of the two liquid surfaces. (b) The right arm is then shielded from any air motion while air is blown across the top of the left arm until the surfaces of the two liquids are at the same height (Fig. P15.61c). Determine the speed of the air being blown across the left arm. Take the density of air as 1.29 kg/m$^3$.

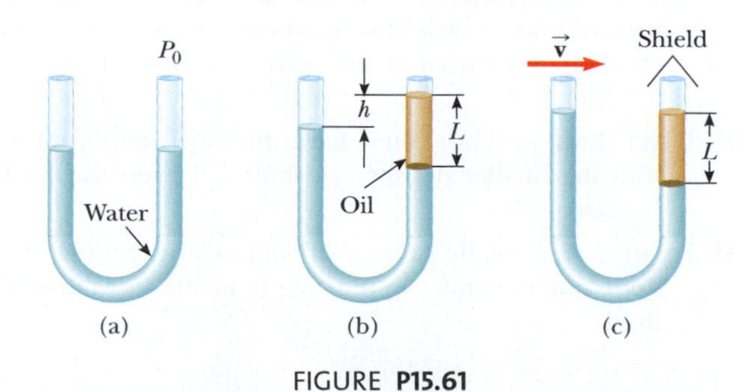


FIGURE **P15.61**

62. The water supply of a building is fed through a main pipe 6.00 cm in diameter. A 2.00-cm-diameter faucet tap, located 2.00 m above the main pipe, is observed to fill a

25.0 L container in 30.0 s. (a) What is the speed at which the water leaves the faucet? (b) What is the gauge pressure in the 6-cm main pipe? (Assume that the faucet is the only "leak" in the building.)

**63.** The *spirit-in-glass thermometer,* invented in Florence, Italy, around 1654, consists of a tube of liquid (the spirit) containing a number of submerged glass spheres with slightly different masses (Fig. P15.63). At sufficiently low temperatures, all the spheres float, but as the temperature rises, the spheres sink one after another. The device is a crude but interesting tool for measuring temperature. Suppose the tube is filled with ethyl alcohol, whose density is 0.789 45 g/cm$^3$ at 20.0°C and decreases to 0.780 97 g/cm$^3$ at 30.0°C. (a) Assuming that one of the spheres has a radius of 1.000 cm and is in equilibrium halfway up the tube at 20.0°C, determine its mass. (b) When the temperature increases to 30.0°C, what mass must a second sphere of the same radius have to be in equilibrium at the halfway point? (c) At 30.0°C, the first sphere has fallen to the bottom of the tube. What upward force does the bottom of the tube exert on this sphere?

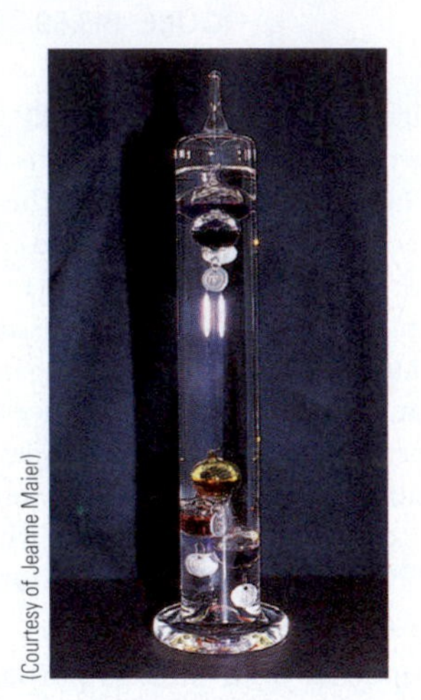


FIGURE **P15.63**

**64.** The hull of an experimental boat is to be lifted above the water by a hydrofoil mounted below its keel as shown in Figure P15.64. The hydrofoil has a shape like that of an airplane wing. Its area projected onto a horizontal surface is $A$. When the boat is towed at sufficiently high speed, water of density $\rho$ moves in streamline flow so that its average speed at the top of the hydrofoil is $n$ times larger than its speed $v_b$ below the hydrofoil. (a) Ignoring the buoyant force, show that the upward lift force exerted by the water on the hydrofoil has a magnitude given by

$$F \approx \tfrac{1}{2}(n^2 - 1)\rho v_b{}^2 A$$

(b) The boat has mass $M$. Show that the liftoff speed is given by

$$v \approx \sqrt{\frac{2Mg}{(n^2 - 1)A\rho}}$$

(c) Assume that an 800-kg boat is to lift off at 9.50 m/s. Evaluate the area $A$ required for the hydrofoil if its design yields $n = 1.05$.

FIGURE **P15.64**

## ANSWERS TO QUICK QUIZZES

**15.1** (a). Because the basketball player's weight is distributed over the larger surface area of the shoe, the pressure ($F/A$) that he applies is relatively small. The woman's lesser weight is distributed over the very small cross-sectional area of the spiked heel, so the pressure is high.

**15.2** (a). Because both fluids have the same depth, the one with the smaller density (alcohol) will exert the smaller pressure.

**15.3** (b). For a totally submerged object, the buoyant force does not depend on the depth in an incompressible fluid.

**15.4** (b) or (c). In all three cases, the weight of the treasure chest causes a downward force on the raft that makes it sink into the water. In (b) and (c), however, the treasure chest also displaces water, which provides a buoyant force in the upward direction, reducing the effect of the weight of the chest on the raft.

**15.5** (b). The liquid moves at the highest speed in the straw with the smallest cross sectional area.

**15.6** (a). The high-speed air between the balloons results in low pressure in this region. The higher pressure on the outer surfaces of the balloons pushes them toward each other.

# Finding and Visiting the *Titanic*

We have now investigated the physics of fluids and can respond to our central question for the *Search for the Titanic* Context:

***How can we safely visit the wreck of the Titanic?***

Many individuals believed that the *Titanic* was unsinkable. One factor in this belief was the series of watertight bulkheads that divided the hull of the ship into several watertight compartments. Even if the hull were breached so that a compartment became flooded, the incoming water could be isolated to that compartment by closing watertight doors in the bulkhead.

According to the design of the ship, the *Titanic* could be kept afloat if its four forwardmost compartments were flooded. Unfortunately, the collision with the iceberg caused a breach in the first five compartments. As these forward compartments filled, the extra weight of the water in the bow of the ship resulted in the bow sinking into the water and the stern lifting out of the water (Fig. 1).

Despite the shipbuilders' pride in their watertight compartments, they were not watertight at the top. The bulkheads only went up to a certain height in the ship and then ended. Therefore, as the *Titanic* tilted forward, water from one compartment simply spilled over the top of the bulkhead into the next compartment and the compartments filled one by one.

Some experts after the disaster claimed that opening the watertight doors in the bulkheads would have kept the *Titanic* afloat longer, with an increased possibility of another ship arriving in time to save those who were not able to leave in the lifeboats. According to this hypothesis, if the water entering the forward compartments had been allowed to distribute evenly along the ship by passing through the doors in the bulkheads, the ship would not have tilted so that water could spill over the tops of the bulkheads. The sinking of a ship is a complicated event, however, and this hypothesis is not universally accepted.

FIGURE 1 The *Titanic* struck the iceberg near the bow, so the forward compartments filled with water and sank, lifting the stern of the ship above the water.

In the region of the sinking of the *Titanic,* the depth of the ocean is about 4 km. When the wreckage was located and visited, the depth was measured to be 3 784 m.

Let us use Equation 15.4 to calculate the pressure at this depth of sea water:

$$P = P_0 + \rho g h$$
$$= 1.013 \times 10^5 \text{ Pa} + (1.03 \times 10^3 \text{ kg/m}^3)(9.80 \text{ m/s}^2)(3\,784 \text{ m})$$
$$= 3.83 \times 10^7 \text{ Pa} = 378 \text{ atm}$$

Therefore, the pressure is 378 times that at the surface! A human being could not survive at this pressure.

Plans for finding and possibly salvaging the *Titanic* began immediately after it sank. Families of some of the wealthy victims contacted salvage companies with requests for a salvage operation. One plan suggested filling the *Titanic* with Ping-Pong balls so that its overall density would be less than that of water and the ship would float to the surface! This plan, of course, ignores the obvious problems of the Ping-Pong balls' failure to withstand the tremendous pressure at that depth.

An early expedition to find the *Titanic* occurred in 1980 and met with failure. After unsuccessful searches were carried out by a number of teams, Dr. Robert Ballard of Woods Hole Oceanographic Institute discovered the wreck in 1985, in cooperation with a team from IFREMER, the French National Institute of Oceanography. The search began by towing a sonar device, which emitted sound waves through the water and analyzed the reflection of the waves from solid objects such as the hull of the *Titanic.* The search pattern was a tedious back-and-forth sweeping of the area near the reported location of the sinking of the ship, looking for sonar reflections and checking possible sites with a magnetometer for the presence of an iron hull.

After failing to find the *Titanic* with the sonar system, Ballard switched to a visual search using an underwater video system called Argo. After three more grueling weeks with no reward, the searchers saw one of the *Titanic*'s boilers in the early morning of September 1, 1985. After this first evidence of the wreck was found, the remainder was located quickly.

The visual evidence indicated clearly that the *Titanic* had split in two as had been reported by some of the survivors in 1912. As it tilted steeply in the water due to the sinking of the bow, the midsection was subjected to forces that it was not designed to sustain. After the break occurred, but while the two sections were still connected, the stern section settled back into the water, with the bow section hanging from it underwater. As more water entered the bow section, it pulled the stern section into a vertical orientation and then broke free, beginning its trip to the bottom. The stern bobbed for a while as it filled with water and then sank into the ocean.

FIGURE 2 The bow section of the *Titanic* rests on the ocean floor relatively intact.

The two sections of the *Titanic* lie about 600 m apart on the ocean floor. The bow section (Fig. 2) is fairly intact, but the stern section (Fig. 3) is tremendously damaged. As the bow section sank, it was already filled with water. As the pressure of the water outside the bow section increased during the plummet to the bottom, the pressure inside the section increased. On the other hand, the stern section spent its time in the air before sinking. Therefore, as it sank, a significant volume of air was still trapped inside the stern section. As the water pressure increased while the stern section sank, the air pressure inside could not increase along with the external water pressure because many air pockets existed in the relatively sealed sections of the structure. Therefore, some areas of the hull of the stern section experienced very large pressure on the outside surface, with relatively low pressure on the inside surface. This extreme imbalance in pressures possibly caused an implosion of the stern section at some depth, causing severe destruction of the structure. Further damage was caused by the sudden impact of hitting the bottom. With little structural integrity left, the decks pancaked downward as the stern hit the ocean floor.

**FIGURE 3** The stern section of the *Titanic* is heavily damaged and lies in pieces on the ocean floor.

The *Titanic* has been visited by a number of teams for purposes of research, salvage, and even filmmaking, by James Cameron, the director of the 1997 version of the film *Titanic* and the 2003 IMAX film *Ghosts of the Abyss.* What is necessary to travel to such depths? Our calculation of the pressure at the location of the *Titanic* indicates that special submarines must be used that can withstand such high pressure while maintaining normal atmospheric pressure inside for the human occupants. That was first done by Ballard in the summer of 1986 using a deep-sea submersible called *Alvin* with a remote-controlled robot named *Jason Junior.* Figure 4 shows the structure of *Alvin.*

*Alvin* has a titanium alloy hull that can withstand the pressure at the depth of the *Titanic.* The submersible has room for three occupants, although they are quite cramped. A number of air tanks on the craft can be flooded with water. Blocks of iron can also be jettisoned. When the tanks are filled with air and the iron blocks are attached, *Alvin* floats on water. When the air tanks are flooded with water, the submersible sinks. In visiting the *Titanic,* the first step is to fill the air tanks with water and then wait 2.5 h to sink to the bottom. After visiting the wreckage, the iron blocks are jettisoned. After doing so, the buoyant force on *Alvin* is larger than its weight and it starts upward on another long journey to the surface. The descent and ascent of *Alvin* are effective examples of applications of the physics described in Case I in Section 15.4.

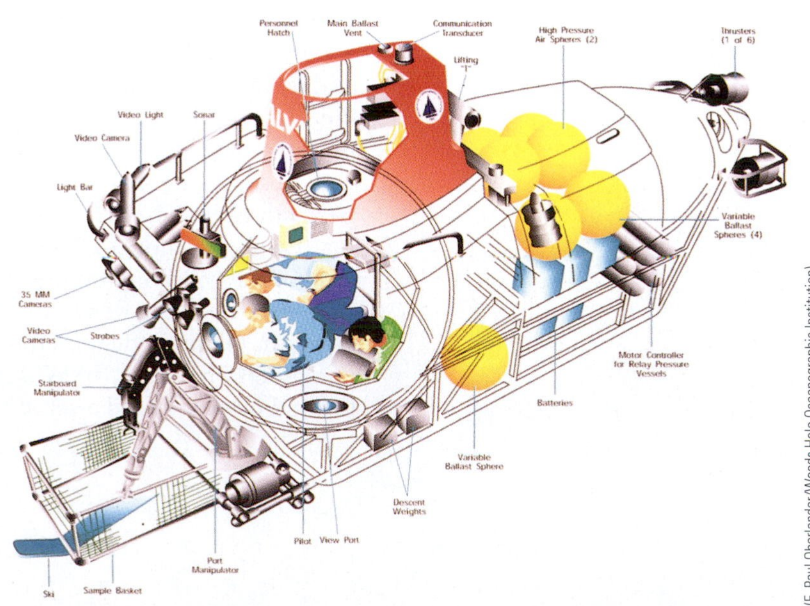




**FIGURE 4** The submersible *Alvin,* which can carry three scientists to the great depths at which the *Titanic* currently lies.

Once *Alvin* reaches the *Titanic,* the only means of viewing it are by visual inspection through the portholes or by video, using *Jason Junior.* It is impossible to don scuba gear and exit the submersible because of the tremendous pressure. There is much more to the story of the *Titanic,* but we need to return to our investigations into physics.

## Problems

1. When the *Titanic* is in its normal sailing position, the torque due to the gravitational force about a horizontal axis through the midpoint of the ship is zero. Imagine now that the bow of the ship is under water and the stern is in the air above the water during the sinking process as shown in Figure 1. The waterline is at the midpoint of the ship and the keel makes a 45.0° angle with the horizontal. The net torque is zero here also because the *Titanic* as a whole is in equilibrium. If we take as our system the stern half of the ship, the torque counteracting the weight of the stern section must be applied by the framework of the ship at the midpoint. Calculate the torque about the midpoint of the *Titanic* required to hold the stern section in the air. Model the ship as a uniform rod of length 269 m and mass $4.2 \times 10^7$ kg. This torque caused the *Titanic* to split near the middle of the ship during the sinking process.
2. The *Titanic* had two almost identical sister ships, the *Britannic* and the *Olympic.* The *Britannic* sank in 1916 off the coast of Athens, Greece, possibly due to a mine planted during World War I. It now sits below 119 m of sea water. (a) What is the pressure at the location of the *Britannic*? (b) Look up a practical limit for scuba diving and determine whether a person can visit the *Britannic* by scuba diving.
3. The submersible *Alvin* requires 2.5 h to sink to the location of the *Titanic.* (a) What is the average speed during descent? (b) Assume that the speed of *Alvin* remains constant during the entire descent. Is the average density of *Alvin* greater than, less than, or equal to the density of sea water? (*Note:* Include in your analysis the resistive force on *Alvin* as it moves through the water.)
4. The *Titanic* is only one of many maritime disasters. In 1956, despite the use of radar, which was invented after the time of the *Titanic,* a collision occurred between the Italian luxury liner *Andrea Doria* and the Swedish liner *Stockholm.* Deaths were relatively few because of the long time interval between the collision and the sinking of the *Andrea Doria,* but a remarkable event occurred. A 14-year-old girl, asleep in her bed on the *Andrea Doria* before the collision, awoke on the bow of the *Stockholm.* The bow of the latter ship pierced the hull of the *Andrea Doria* at the location of the girl's berth and miraculously scooped her up with comparatively minor injuries.

   Let us imagine that the collision between the *Andrea Doria* and the *Stockholm* is perfectly inelastic. (In fact, the *Stockholm* drew away after the collision, but ours is a reasonable model.) The weight of the *Andrea Doria* is 29 100 tons. At the time of the collision, it is traveling at full speed of 23 knots at 15° south of west. The *Stockholm* has a weight of 12 165 tons and is traveling at 18 knots at 30° east of south. (a) Immediately after the collision, what is the velocity, in knots, of the entangled wreckage? (b) What fraction of the initial kinetic energy was transformed or transferred away in the collision?

# Tables

## TABLE A.1 Conversion Factors

**Length**

| | m | cm | km | in. | ft | mi |
|---|---|---|---|---|---|---|
| 1 meter | 1 | $10^2$ | $10^{-3}$ | 39.37 | 3.281 | $6.214 \times 10^{-4}$ |
| 1 centimeter | $10^{-2}$ | 1 | $10^{-5}$ | 0.393 7 | $3.281 \times 10^{-2}$ | $6.214 \times 10^{-6}$ |
| 1 kilometer | $10^3$ | $10^5$ | 1 | $3.937 \times 10^4$ | $3.281 \times 10^3$ | 0.621 4 |
| 1 inch | $2.540 \times 10^{-2}$ | 2.540 | $2.540 \times 10^{-5}$ | 1 | $8.333 \times 10^{-2}$ | $1.578 \times 10^{-5}$ |
| 1 foot | 0.304 8 | 30.48 | $3.048 \times 10^{-4}$ | 12 | 1 | $1.894 \times 10^{-4}$ |
| 1 mile | 1 609 | $1.609 \times 10^5$ | 1.609 | $6.336 \times 10^4$ | 5 280 | 1 |

**Mass**

| | kg | g | slug | u |
|---|---|---|---|---|
| 1 kilogram | 1 | $10^3$ | $6.852 \times 10^{-2}$ | $6.024 \times 10^{26}$ |
| 1 gram | $10^{-3}$ | 1 | $6.852 \times 10^{-5}$ | $6.024 \times 10^{23}$ |
| 1 slug | 14.59 | $1.459 \times 10^4$ | 1 | $8.789 \times 10^{27}$ |
| 1 atomic mass unit | $1.660 \times 10^{-27}$ | $1.660 \times 10^{-24}$ | $1.137 \times 10^{-28}$ | 1 |

*Note:* 1 metric ton = 1 000 kg.

**Time**

| | s | min | h | day | yr |
|---|---|---|---|---|---|
| 1 second | 1 | $1.667 \times 10^{-2}$ | $2.778 \times 10^{-4}$ | $1.157 \times 10^{-5}$ | $3.169 \times 10^{-8}$ |
| 1 minute | 60 | 1 | $1.667 \times 10^{-2}$ | $6.994 \times 10^{-4}$ | $1.901 \times 10^{-6}$ |
| 1 hour | 3 600 | 60 | 1 | $4.167 \times 10^{-2}$ | $1.141 \times 10^{-4}$ |
| 1 day | $8.640 \times 10^4$ | 1 440 | 24 | 1 | $2.738 \times 10^{-5}$ |
| 1 year | $3.156 \times 10^7$ | $5.259 \times 10^5$ | $8.766 \times 10^3$ | 365.2 | 1 |

**Speed**

| | m/s | cm/s | ft/s | mi/h |
|---|---|---|---|---|
| 1 meter per second | 1 | $10^2$ | 3.281 | 2.237 |
| 1 centimeter per second | $10^{-2}$ | 1 | $3.281 \times 10^{-2}$ | $2.237 \times 10^{-2}$ |
| 1 foot per second | 0.304 8 | 30.48 | 1 | 0.681 8 |
| 1 mile per hour | 0.447 0 | 44.70 | 1.467 | 1 |

*Note*: 1 mi/min = 60 mi/h = 88 ft/s.

**Force**

| | N | lb |
|---|---|---|
| 1 newton | 1 | 0.224 8 |
| 1 pound | 4.448 | 1 |

*(Continued)*

## TABLE A.1 Conversion Factors *(Continued)*

**Work, Energy, Heat**

| | J | ft · lb | eV |
|---|---|---|---|
| 1 joule | 1 | 0.737 6 | $6.242 \times 10^{18}$ |
| 1 foot-pound | 1.356 | 1 | $8.464 \times 10^{18}$ |
| 1 electron volt | $1.602 \times 10^{-19}$ | $1.182 \times 10^{-19}$ | 1 |
| 1 calorie | 4.186 | 3.087 | $2.613 \times 10^{19}$ |
| 1 British thermal unit | $1.055 \times 10^{3}$ | $7.779 \times 10^{2}$ | $6.585 \times 10^{21}$ |
| 1 kilowatt-hour | $3.600 \times 10^{6}$ | $2.655 \times 10^{6}$ | $2.247 \times 10^{25}$ |

| | cal | Btu | kWh |
|---|---|---|---|
| 1 joule | 0.238 9 | $9.481 \times 10^{-4}$ | $2.778 \times 10^{-7}$ |
| 1 foot-pound | 0.323 9 | $1.285 \times 10^{-3}$ | $3.766 \times 10^{-7}$ |
| 1 electron volt | $3.827 \times 10^{-20}$ | $1.519 \times 10^{-22}$ | $4.450 \times 10^{-26}$ |
| 1 calorie | 1 | $3.968 \times 10^{-3}$ | $1.163 \times 10^{-6}$ |
| 1 British thermal unit | $2.520 \times 10^{2}$ | 1 | $2.930 \times 10^{-4}$ |
| 1 kilowatt-hour | $8.601 \times 10^{5}$ | $3.413 \times 10^{2}$ | 1 |

**Pressure**

| | Pa | atm |
|---|---|---|
| 1 pascal | 1 | $9.869 \times 10^{-6}$ |
| 1 atmosphere | $1.013 \times 10^{5}$ | 1 |
| 1 centimeter mercury[a] | $1.333 \times 10^{3}$ | $1.316 \times 10^{-2}$ |
| 1 pound per square inch | $6.895 \times 10^{3}$ | $6.805 \times 10^{-2}$ |
| 1 pound per square foot | 47.88 | $4.725 \times 10^{-4}$ |

| | cm Hg | lb/in.$^2$ | lb/ft$^2$ |
|---|---|---|---|
| 1 pascal | $7.501 \times 10^{-4}$ | $1.450 \times 10^{-4}$ | $2.089 \times 10^{-2}$ |
| 1 atmosphere | 76 | 14.70 | $2.116 \times 10^{3}$ |
| 1 centimeter mercury[a] | 1 | 0.194 3 | 27.85 |
| 1 pound per square inch | 5.171 | 1 | 144 |
| 1 pound per square foot | $3.591 \times 10^{-2}$ | $6.944 \times 10^{-3}$ | 1 |

[a]At 0°C and at a location where the free-fall acceleration has its "standard" value, 9.806 65 m/s$^2$.

## TABLE A.2 Symbols, Dimensions, and Units of Physical Quantities

| Quantity | Common Symbol | Unit[a] | Dimensions[b] | Unit in Terms of Base SI Units |
|---|---|---|---|---|
| Acceleration | $\vec{a}$ | m/s$^2$ | L/T$^2$ | m/s$^2$ |
| Amount of substance | $n$ | MOLE | | mol |
| Angle | $\theta, \phi$ | radian (rad) | 1 | |
| Angular acceleration | $\vec{\alpha}$ | rad/s$^2$ | T$^{-2}$ | s$^{-2}$ |
| Angular frequency | $\omega$ | rad/s | T$^{-1}$ | s$^{-1}$ |
| Angular momentum | $\vec{L}$ | kg · m$^2$/s | ML$^2$/T | kg · m$^2$/s |
| Angular velocity | $\vec{\omega}$ | rad/s | T$^{-1}$ | s$^{-1}$ |
| Area | $A$ | m$^2$ | L$^2$ | m$^2$ |
| Atomic number | $Z$ | | | |

*(Continued)*

TABLE A.2 **Symbols, Dimensions, and Units of Physical Quantities** *(Continued)*

| Quantity | Common Symbol | Unit[a] | Dimensions[b] | Unit in Terms of Base SI Units |
|---|---|---|---|---|
| Capacitance | $C$ | farad (F) | $Q^2T^2/ML^2$ | $A^2 \cdot s^4/kg \cdot m^2$ |
| Charge | $q, Q, e$ | coulomb (C) | Q | $A \cdot s$ |
| Charge density | | | | |
| Line | $\lambda$ | C/m | Q/L | $A \cdot s/m$ |
| Surface | $\sigma$ | $C/m^2$ | $Q/L^2$ | $A \cdot s/m^2$ |
| Volume | $\rho$ | $C/m^3$ | $Q/L^3$ | $A \cdot s/m^3$ |
| Conductivity | $\sigma$ | $1/\Omega \cdot m$ | $Q^2T/ML^3$ | $A^2 \cdot s^3/kg \cdot m^3$ |
| Current | $I$ | AMPERE | Q/T | A |
| Current density | $\vec{\mathbf{J}}$ | $A/m^2$ | $Q/TL^2$ | $A/m^2$ |
| Density | $\rho$ | $kg/m^3$ | $M/L^3$ | $kg/m^3$ |
| Dielectric constant | $\kappa$ | | | |
| Electric dipole moment | $\vec{\mathbf{p}}$ | $C \cdot m$ | QL | $A \cdot s \cdot m$ |
| Electric field | $\vec{\mathbf{E}}$ | V/m | $ML/QT^2$ | $kg \cdot m/A \cdot s^3$ |
| Electric flux | $\Phi_E$ | $V \cdot m$ | $ML^3/QT^2$ | $kg \cdot m^3/A \cdot s^3$ |
| Electromotive force | $\mathcal{E}$ | volt (V) | $ML^2/QT^2$ | $kg \cdot m^2/A \cdot s^3$ |
| Energy | $E, U, K$ | joule (J) | $ML^2/T^2$ | $kg \cdot m^2/s^2$ |
| Entropy | $S$ | J/K | $ML^2/T^2 \cdot K$ | $kg \cdot m^2/s^2 \cdot K$ |
| Force | $\vec{\mathbf{F}}$ | newton (N) | $ML/T^2$ | $kg \cdot m/s^2$ |
| Frequency | $f$ | hertz (Hz) | $T^{-1}$ | $s^{-1}$ |
| Heat | $Q$ | joule (J) | $ML^2/T^2$ | $kg \cdot m^2/s^2$ |
| Inductance | $L$ | henry (H) | $ML^2/Q^2$ | $kg \cdot m^2/A^2 \cdot s^2$ |
| Length | $\ell, L$ | METER | L | m |
| Displacement | $\Delta x, \Delta\vec{\mathbf{r}}$ | | | |
| Distance | $d, h$ | | | |
| Position | $x, y, z, \vec{\mathbf{r}}$ | | | |
| Magnetic dipole moment | $\vec{\boldsymbol{\mu}}$ | $N \cdot m/T$ | $QL^2/T$ | $A \cdot m^2$ |
| Magnetic field | $\vec{\mathbf{B}}$ | tesla (T) (= $Wb/m^2$) | M/QT | $kg/A \cdot s^2$ |
| Magnetic flux | $\Phi_B$ | weber (Wb) | $ML^2/QT$ | $kg \cdot m^2/A \cdot s^2$ |
| Mass | $m, M$ | KILOGRAM | M | kg |
| Molar specific heat | $C$ | $J/mol \cdot K$ | | $kg \cdot m^2/s^2 \cdot mol \cdot K$ |
| Moment of inertia | $I$ | $kg \cdot m^2$ | $ML^2$ | $kg \cdot m^2$ |
| Momentum | $\vec{\mathbf{p}}$ | $kg \cdot m/s$ | ML/T | $kg \cdot m/s$ |
| Period | $T$ | s | T | s |
| Permeability of free space | $\mu_0$ | $N/A^2$ (= H/m) | $ML/Q^2$ | $kg \cdot m/A^2 \cdot s^2$ |
| Permittivity of free space | $\epsilon_0$ | $C^2/N \cdot m^2$ (= F/m) | $Q^2T^2/ML^3$ | $A^2 \cdot s^4/kg \cdot m^3$ |
| Potential | $V$ | volt (V) (= J/C) | $ML^2/QT^2$ | $kg \cdot m^2/A \cdot s^3$ |
| Power | $\mathcal{P}$ | watt (W) (= J/s) | $ML^2/T^3$ | $kg \cdot m^2/s^3$ |
| Pressure | $P$ | pascal (Pa) (= $N/m^2$) | $M/LT^2$ | $kg/m \cdot s^2$ |
| Resistance | $R$ | ohm ($\Omega$) (= V/A) | $ML^2/Q^2T$ | $kg \cdot m^2/A^2 \cdot s^3$ |
| Specific heat | $c$ | $J/kg \cdot K$ | $L^2/T^2 \cdot K$ | $m^2/s^2 \cdot K$ |
| Speed | $v$ | m/s | L/T | m/s |
| Temperature | $T$ | KELVIN | K | K |
| Time | $t$ | SECOND | T | s |
| Torque | $\vec{\boldsymbol{\tau}}$ | $N \cdot m$ | $ML^2/T^2$ | $kg \cdot m^2/s^2$ |
| Velocity | $\vec{\mathbf{v}}$ | m/s | L/T | m/s |
| Volume | $V$ | $m^3$ | $L^3$ | $m^3$ |
| Wavelength | $\lambda$ | m | L | m |
| Work | $W$ | joule (J) (= $N \cdot m$) | $ML^2/T^2$ | $kg \cdot m^2/s^2$ |

[a]The base SI units are given in uppercase letters.

[b]The symbols M, L, T, and Q denote mass, length, time, and charge, respectively.

TABLE A.3 Table of Atomic Masses

| Atomic Number $Z$ | Element | Symbol | Chemical Atomic Mass (u) | Mass Number (*indicates radioactive) $A$ | Atomic Mass (u) | Percent Abundance | Half-Life (if radioactive) $T_{1/2}$ |
|---|---|---|---|---|---|---|---|
| 0 | (Neutron) | n | | 1* | 1.008 665 | | 10.4 min |
| 1 | Hydrogen | H | 1.007 94 | 1 | 1.007 825 | 99.988 5 | |
| | Deuterium | D | | 2 | 2.014 102 | 0.011 5 | |
| | Tritium | T | | 3* | 3.016 049 | | 12.33 yr |
| 2 | Helium | He | 4.002 602 | 3 | 3.016 029 | 0.000 137 | |
| | | | | 4 | 4.002 603 | 99.999 863 | |
| | | | | 6* | 6.018 888 | | 0.81 s |
| 3 | Lithium | Li | 6.941 | 6 | 6.015 122 | 7.5 | |
| | | | | 7 | 7.016 004 | 92.5 | |
| | | | | 8* | 8.022 487 | | 0.84 s |
| 4 | Beryllium | Be | 9.012 182 | 7* | 7.016 929 | | 53.3 days |
| | | | | 9 | 9.012 182 | 100 | |
| | | | | 10* | 10.013 534 | | $1.5 \times 10^6$ yr |
| 5 | Boron | B | 10.811 | 10 | 10.012 937 | 19.9 | |
| | | | | 11 | 11.009 306 | 80.1 | |
| | | | | 12* | 12.014 352 | | 0.020 2 s |
| 6 | Carbon | C | 12.010 7 | 10* | 10.016 853 | | 19.3 s |
| | | | | 11* | 11.011 434 | | 20.4 min |
| | | | | 12 | 12.000 000 | 98.93 | |
| | | | | 13 | 13.003 355 | 1.07 | |
| | | | | 14* | 14.003 242 | | 5 730 yr |
| | | | | 15* | 15.010 599 | | 2.45 s |
| 7 | Nitrogen | N | 14.006 7 | 12* | 12.018 613 | | 0.011 0 s |
| | | | | 13* | 13.005 739 | | 9.96 min |
| | | | | 14 | 14.003 074 | 99.632 | |
| | | | | 15 | 15.000 109 | 0.368 | |
| | | | | 16* | 16.006 101 | | 7.13 s |
| | | | | 17* | 17.008 450 | | 4.17 s |
| 8 | Oxygen | O | 15.999 4 | 14* | 14.008 595 | | 70.6 s |
| | | | | 15* | 15.003 065 | | 122 s |
| | | | | 16 | 15.994 915 | 99.757 | |
| | | | | 17 | 16.999 132 | 0.038 | |
| | | | | 18 | 17.999 160 | 0.205 | |
| | | | | 19* | 19.003 579 | | 26.9 s |
| 9 | Fluorine | F | 18.998 403 2 | 17* | 17.002 095 | | 64.5 s |
| | | | | 18* | 18.000 938 | | 109.8 min |
| | | | | 19 | 18.998 403 | 100 | |
| | | | | 20* | 19.999 981 | | 11.0 s |
| | | | | 21* | 20.999 949 | | 4.2 s |
| 10 | Neon | Ne | 20.179 7 | 18* | 18.005 697 | | 1.67 s |
| | | | | 19* | 19.001 880 | | 17.2 s |
| | | | | 20 | 19.992 440 | 90.48 | |
| | | | | 21 | 20.993 847 | 0.27 | |
| | | | | 22 | 21.991 385 | 9.25 | |
| | | | | 23* | 22.994 467 | | 37.2 s |
| 11 | Sodium | Na | 22.989 77 | 21* | 20.997 655 | | 22.5 s |
| | | | | 22* | 21.994 437 | | 2.61 yr |
| | | | | 23 | 22.989 770 | 100 | |
| | | | | 24* | 23.990 963 | | 14.96 h |
| 12 | Magnesium | Mg | 24.305 0 | 23* | 22.994 125 | | 11.3 s |
| | | | | 24 | 23.985 042 | 78.99 | |
| | | | | 25 | 24.985 837 | 10.00 | |

*(Continued)*

TABLE A.3 **Table of Atomic Masses** *(Continued)*

| Atomic Number Z | Element | Symbol | Chemical Atomic Mass (u) | Mass Number (*indicates radioactive) A | Atomic Mass (u) | Percent Abundance | Half-Life (if radioactive) $T_{1/2}$ |
|---|---|---|---|---|---|---|---|
| (12) | Magnesium | | | 26 | 25.982 593 | 11.01 | |
| | | | | 27* | 26.984 341 | | 9.46 min |
| 13 | Aluminum | Al | 26.981 538 | 26* | 25.986 892 | | $7.4 \times 10^5$ yr |
| | | | | 27 | 26.981 539 | 100 | |
| | | | | 28* | 27.981 910 | | 2.24 min |
| 14 | Silicon | Si | 28.085 5 | 28 | 27.976 926 | 92.229 7 | |
| | | | | 29 | 28.976 495 | 4.683 2 | |
| | | | | 30 | 29.973 770 | 3.087 2 | |
| | | | | 31* | 30.975 363 | | 2.62 h |
| | | | | 32* | 31.974 148 | | 172 yr |
| 15 | Phosphorus | P | 30.973 761 | 30* | 29.978 314 | | 2.50 min |
| | | | | 31 | 30.973 762 | 100 | |
| | | | | 32* | 31.973 907 | | 14.26 days |
| | | | | 33* | 32.971 725 | | 25.3 days |
| 16 | Sulfur | S | 32.066 | 32 | 31.972 071 | 94.93 | |
| | | | | 33 | 32.971 458 | 0.76 | |
| | | | | 34 | 33.967 869 | 4.29 | |
| | | | | 35* | 34.969 032 | | 87.5 days |
| | | | | 36 | 35.967 081 | 0.02 | |
| 17 | Chlorine | Cl | 35.452 7 | 35 | 34.968 853 | 75.78 | |
| | | | | 36* | 35.968 307 | | $3.0 \times 10^5$ yr |
| | | | | 37 | 36.965 903 | 24.22 | |
| 18 | Argon | Ar | 39.948 | 36 | 35.967 546 | 0.336 5 | |
| | | | | 37* | 36.966 776 | | 35.04 days |
| | | | | 38 | 37.962 732 | 0.063 2 | |
| | | | | 39* | 38.964 313 | | 269 yr |
| | | | | 40 | 39.962 383 | 99.600 3 | |
| | | | | 42* | 41.963 046 | | 33 yr |
| 19 | Potassium | K | 39.098 3 | 39 | 38.963 707 | 93.258 1 | |
| | | | | 40* | 39.963 999 | 0.011 7 | $1.28 \times 10^9$ yr |
| | | | | 41 | 40.961 826 | 6.730 2 | |
| 20 | Calcium | Ca | 40.078 | 40 | 39.962 591 | 96.941 | |
| | | | | 41* | 40.962 278 | | $1.0 \times 10^5$ yr |
| | | | | 42 | 41.958 618 | 0.647 | |
| | | | | 43 | 42.958 767 | 0.135 | |
| | | | | 44 | 43.955 481 | 2.086 | |
| | | | | 46 | 45.953 693 | 0.004 | |
| | | | | 48 | 47.952 534 | 0.187 | |
| 21 | Scandium | Sc | 44.955 910 | 41* | 40.969 251 | | 0.596 s |
| | | | | 45 | 44.955 910 | 100 | |
| 22 | Titanium | Ti | 47.867 | 44* | 43.959 690 | | 49 yr |
| | | | | 46 | 45.952 630 | 8.25 | |
| | | | | 47 | 46.951 764 | 7.44 | |
| | | | | 48 | 47.947 947 | 73.72 | |
| | | | | 49 | 48.947 871 | 5.41 | |
| | | | | 50 | 49.944 792 | 5.18 | |
| 23 | Vanadium | V | 50.941 5 | 48* | 47.952 254 | | 15.97 days |
| | | | | 50* | 49.947 163 | 0.250 | $1.5 \times 10^{17}$ yr |
| | | | | 51 | 50.943 964 | 99.750 | |
| 24 | Chromium | Cr | 51.996 1 | 48* | 47.954 036 | | 21.6 h |
| | | | | 50 | 49.946 050 | 4.345 | |

*(Continued)*

TABLE A.3 **Table of Atomic Masses** *(Continued)*

| Atomic Number $Z$ | Element | Symbol | Chemical Atomic Mass (u) | Mass Number (*indicates radioactive) $A$ | Atomic Mass (u) | Percent Abundance | Half-Life (if radioactive) $T_{1/2}$ |
|---|---|---|---|---|---|---|---|
| (24) | Chromium | | | 52 | 51.940 512 | 83.789 | |
| | | | | 53 | 52.940 654 | 9.501 | |
| | | | | 54 | 53.938 885 | 2.365 | |
| 25 | Manganese | Mn | 54.938 049 | 54* | 53.940 363 | | 312.1 days |
| | | | | 55 | 54.938 050 | 100 | |
| 26 | Iron | Fe | 55.845 | 54 | 53.939 615 | 5.845 | |
| | | | | 55* | 54.938 298 | | 2.7 yr |
| | | | | 56 | 55.934 942 | 91.754 | |
| | | | | 57 | 56.935 399 | 2.119 | |
| | | | | 58 | 57.933 280 | 0.282 | |
| | | | | 60* | 59.934 077 | | $1.5 \times 10^6$ yr |
| 27 | Cobalt | Co | 58.933 200 | 59 | 58.933 200 | 100 | |
| | | | | 60* | 59.933 822 | | 5.27 yr |
| 28 | Nickel | Ni | 58.693 4 | 58 | 57.935 348 | 68.076 9 | |
| | | | | 59* | 58.934 351 | | $7.5 \times 10^4$ yr |
| | | | | 60 | 59.930 790 | 26.223 1 | |
| | | | | 61 | 60.931 060 | 1.139 9 | |
| | | | | 62 | 61.928 349 | 3.634 5 | |
| | | | | 63* | 62.929 673 | | 100 yr |
| | | | | 64 | 63.927 970 | 0.925 6 | |
| 29 | Copper | Cu | 63.546 | 63 | 62.929 601 | 69.17 | |
| | | | | 65 | 64.927 794 | 30.83 | |
| 30 | Zinc | Zn | 65.39 | 64 | 63.929 147 | 48.63 | |
| | | | | 66 | 65.926 037 | 27.90 | |
| | | | | 67 | 66.927 131 | 4.10 | |
| | | | | 68 | 67.924 848 | 18.75 | |
| | | | | 70 | 69.925 325 | 0.62 | |
| 31 | Gallium | Ga | 69.723 | 69 | 68.925 581 | 60.108 | |
| | | | | 71 | 70.924 705 | 39.892 | |
| 32 | Germanium | Ge | 72.61 | 70 | 69.924 250 | 20.84 | |
| | | | | 72 | 71.922 076 | 27.54 | |
| | | | | 73 | 72.923 459 | 7.73 | |
| | | | | 74 | 73.921 178 | 36.28 | |
| | | | | 76 | 75.921 403 | 7.61 | |
| 33 | Arsenic | As | 74.921 60 | 75 | 74.921 596 | 100 | |
| 34 | Selenium | Se | 78.96 | 74 | 73.922 477 | 0.89 | |
| | | | | 76 | 75.919 214 | 9.37 | |
| | | | | 77 | 76.919 915 | 7.63 | |
| | | | | 78 | 77.917 310 | 23.77 | |
| | | | | 79* | 78.918 500 | | $\leq 6.5 \times 10^4$ yr |
| | | | | 80 | 79.916 522 | 49.61 | |
| | | | | 82* | 81.916 700 | 8.73 | $1.4 \times 10^{20}$ yr |
| 35 | Bromine | Br | 79.904 | 79 | 78.918 338 | 50.69 | |
| | | | | 81 | 80.916 291 | 49.31 | |
| 36 | Krypton | Kr | 83.80 | 78 | 77.920 386 | 0.35 | |
| | | | | 80 | 79.916 378 | 2.28 | |
| | | | | 81* | 80.916 592 | | $2.1 \times 10^5$ yr |
| | | | | 82 | 81.913 485 | 11.58 | |
| | | | | 83 | 82.914 136 | 11.49 | |
| | | | | 84 | 83.911 507 | 57.00 | |
| | | | | 85* | 84.912 527 | | 10.76 yr |
| | | | | 86 | 85.910 610 | 17.30 | |

*(Continued)*

TABLE A.3 **Table of Atomic Masses** *(Continued)*

| Atomic Number $Z$ | Element | Symbol | Chemical Atomic Mass (u) | Mass Number (*indicates radioactive) $A$ | Atomic Mass (u) | Percent Abundance | Half-Life (if radioactive) $T_{1/2}$ |
|---|---|---|---|---|---|---|---|
| 37 | Rubidium | Rb | 85.467 8 | 85 | 84.911 789 | 72.17 | |
| | | | | 87* | 86.909 184 | 27.83 | $4.75 \times 10^{10}$ yr |
| 38 | Strontium | Sr | 87.62 | 84 | 83.913 425 | 0.56 | |
| | | | | 86 | 85.909 262 | 9.86 | |
| | | | | 87 | 86.908 880 | 7.00 | |
| | | | | 88 | 87.905 614 | 82.58 | |
| | | | | 90* | 89.907 738 | | 29.1 yr |
| 39 | Yttrium | Y | 88.905 85 | 89 | 88.905 848 | 100 | |
| 40 | Zirconium | Zr | 91.224 | 90 | 89.904 704 | 51.45 | |
| | | | | 91 | 90.905 645 | 11.22 | |
| | | | | 92 | 91.905 040 | 17.15 | |
| | | | | 93* | 92.906 476 | | $1.5 \times 10^{6}$ yr |
| | | | | 94 | 93.906 316 | 17.38 | |
| | | | | 96 | 95.908 276 | 2.80 | |
| 41 | Niobium | Nb | 92.906 38 | 91* | 90.906 990 | | $6.8 \times 10^{2}$ yr |
| | | | | 92* | 91.907 193 | | $3.5 \times 10^{7}$ yr |
| | | | | 93 | 92.906 378 | 100 | |
| | | | | 94* | 93.907 284 | | $2 \times 10^{4}$ yr |
| 42 | Molybdenum | Mo | 95.94 | 92 | 91.906 810 | 14.84 | |
| | | | | 93* | 92.906 812 | | $3.5 \times 10^{3}$ yr |
| | | | | 94 | 93.905 088 | 9.25 | |
| | | | | 95 | 94.905 842 | 15.92 | |
| | | | | 96 | 95.904 679 | 16.68 | |
| | | | | 97 | 96.906 021 | 9.55 | |
| | | | | 98 | 97.905 408 | 24.13 | |
| | | | | 100 | 99.907 477 | 9.63 | |
| 43 | Technetium | Tc | | 97* | 96.906 365 | | $2.6 \times 10^{6}$ yr |
| | | | | 98* | 97.907 216 | | $4.2 \times 10^{6}$ yr |
| | | | | 99* | 98.906 255 | | $2.1 \times 10^{5}$ yr |
| 44 | Ruthenium | Ru | 101.07 | 96 | 95.907 598 | 5.54 | |
| | | | | 98 | 97.905 287 | 1.87 | |
| | | | | 99 | 98.905 939 | 12.76 | |
| | | | | 100 | 99.904 220 | 12.60 | |
| | | | | 101 | 100.905 582 | 17.06 | |
| | | | | 102 | 101.904 350 | 31.55 | |
| | | | | 104 | 103.905 430 | 18.62 | |
| 45 | Rhodium | Rh | 102.905 50 | 103 | 102.905 504 | 100 | |
| 46 | Palladium | Pd | 106.42 | 102 | 101.905 608 | 1.02 | |
| | | | | 104 | 103.904 035 | 11.14 | |
| | | | | 105 | 104.905 084 | 22.33 | |
| | | | | 106 | 105.903 483 | 27.33 | |
| | | | | 107* | 106.905 128 | | $6.5 \times 10^{6}$ yr |
| | | | | 108 | 107.903 894 | 26.46 | |
| | | | | 110 | 109.905 152 | 11.72 | |
| 47 | Silver | Ag | 107.868 2 | 107 | 106.905 093 | 51.839 | |
| | | | | 109 | 108.904 756 | 48.161 | |
| 48 | Cadmium | Cd | 112.411 | 106 | 105.906 458 | 1.25 | |
| | | | | 108 | 107.904 183 | 0.89 | |
| | | | | 109* | 108.904 986 | | 462 days |
| | | | | 110 | 109.903 006 | 12.49 | |
| | | | | 111 | 110.904 182 | 12.80 | |

*(Continued)*

TABLE A.3 **Table of Atomic Masses** *(Continued)*

| Atomic Number $Z$ | Element | Symbol | Chemical Atomic Mass (u) | Mass Number (*indicates radioactive) $A$ | Atomic Mass (u) | Percent Abundance | Half-Life (if radioactive) $T_{1/2}$ |
|---|---|---|---|---|---|---|---|
| (48) | Cadmium | | | 112 | 111.902 757 | 24.13 | |
| | | | | 113* | 112.904 401 | 12.22 | $9.3 \times 10^{15}$ yr |
| | | | | 114 | 113.903 358 | 28.73 | |
| | | | | 116 | 115.904 755 | 7.49 | |
| 49 | Indium | In | 114.818 | 113 | 112.904 061 | 4.29 | |
| | | | | 115* | 114.903 878 | 95.71 | $4.4 \times 10^{14}$ yr |
| 50 | Tin | Sn | 118.710 | 112 | 111.904 821 | 0.97 | |
| | | | | 114 | 113.902 782 | 0.66 | |
| | | | | 115 | 114.903 346 | 0.34 | |
| | | | | 116 | 115.901 744 | 14.54 | |
| | | | | 117 | 116.902 954 | 7.68 | |
| | | | | 118 | 117.901 606 | 24.22 | |
| | | | | 119 | 118.903 309 | 8.59 | |
| | | | | 120 | 119.902 197 | 32.58 | |
| | | | | 121* | 120.904 237 | | 55 yr |
| | | | | 122 | 121.903 440 | 4.63 | |
| | | | | 124 | 123.905 275 | 5.79 | |
| 51 | Antimony | Sb | 121.760 | 121 | 120.903 818 | 57.21 | |
| | | | | 123 | 122.904 216 | 42.79 | |
| | | | | 125* | 124.905 248 | | 2.7 yr |
| 52 | Tellurium | Te | 127.60 | 120 | 119.904 020 | 0.09 | |
| | | | | 122 | 121.903 047 | 2.55 | |
| | | | | 123* | 122.904 273 | 0.89 | $1.3 \times 10^{13}$ yr |
| | | | | 124 | 123.902 820 | 4.74 | |
| | | | | 125 | 124.904 425 | 7.07 | |
| | | | | 126 | 125.903 306 | 18.84 | |
| | | | | 128* | 127.904 461 | 31.74 | $>8 \times 10^{24}$ yr |
| | | | | 130* | 129.906 223 | 34.08 | $\leq 1.25 \times 10^{21}$ yr |
| 53 | Iodine | I | 126.904 47 | 127 | 126.904 468 | 100 | |
| | | | | 129* | 128.904 988 | | $1.6 \times 10^{7}$ yr |
| 54 | Xenon | Xe | 131.29 | 124 | 123.905 896 | 0.09 | |
| | | | | 126 | 125.904 269 | 0.09 | |
| | | | | 128 | 127.903 530 | 1.92 | |
| | | | | 129 | 128.904 780 | 26.44 | |
| | | | | 130 | 129.903 508 | 4.08 | |
| | | | | 131 | 130.905 082 | 21.18 | |
| | | | | 132 | 131.904 145 | 26.89 | |
| | | | | 134 | 133.905 394 | 10.44 | |
| | | | | 136* | 135.907 220 | 8.87 | $\geq 2.36 \times 10^{21}$ yr |
| 55 | Cesium | Cs | 132.905 45 | 133 | 132.905 447 | 100 | |
| | | | | 134* | 133.906 713 | | 2.1 yr |
| | | | | 135* | 134.905 972 | | $2 \times 10^{6}$ yr |
| | | | | 137* | 136.907 074 | | 30 yr |
| 56 | Barium | Ba | 137.327 | 130 | 129.906 310 | 0.106 | |
| | | | | 132 | 131.905 056 | 0.101 | |
| | | | | 133* | 132.906 002 | | 10.5 yr |
| | | | | 134 | 133.904 503 | 2.417 | |
| | | | | 135 | 134.905 683 | 6.592 | |
| | | | | 136 | 135.904 570 | 7.854 | |
| | | | | 137 | 136.905 821 | 11.232 | |
| | | | | 138 | 137.905 241 | 71.698 | |
| 57 | Lanthanum | La | 138.905 5 | 137* | 136.906 466 | | $6 \times 10^{4}$ yr |
| | | | | 138* | 137.907 107 | 0.090 | $1.05 \times 10^{11}$ yr |

*(Continued)*

TABLE A.3 **Table of Atomic Masses** *(Continued)*

| Atomic Number Z | Element | Symbol | Chemical Atomic Mass (u) | Mass Number (*indicates radioactive) A | Atomic Mass (u) | Percent Abundance | Half-Life (if radioactive) $T_{1/2}$ |
|---|---|---|---|---|---|---|---|
| (57) | Lanthanum | | | 139 | 138.906 349 | 99.910 | |
| 58 | Cerium | Ce | 140.116 | 136 | 135.907 144 | 0.185 | |
| | | | | 138 | 137.905 986 | 0.251 | |
| | | | | 140 | 139.905 434 | 88.450 | |
| | | | | 142* | 141.909 240 | 11.114 | $>5 \times 10^{16}$ yr |
| 59 | Praseodymium | Pr | 140.907 65 | 141 | 140.907 648 | 100 | |
| 60 | Neodymium | Nd | 144.24 | 142 | 141.907 719 | 27.2 | |
| | | | | 143 | 142.909 810 | 12.2 | |
| | | | | 144* | 143.910 083 | 23.8 | $2.3 \times 10^{15}$ yr |
| | | | | 145 | 144.912 569 | 8.3 | |
| | | | | 146 | 145.913 112 | 17.2 | |
| | | | | 148 | 147.916 888 | 5.7 | |
| | | | | 150* | 149.920 887 | 5.6 | $>1 \times 10^{18}$ yr |
| 61 | Promethium | Pm | | 143* | 142.910 928 | | 265 days |
| | | | | 145* | 144.912 744 | | 17.7 yr |
| | | | | 146* | 145.914 692 | | 5.5 yr |
| | | | | 147* | 146.915 134 | | 2.623 yr |
| 62 | Samarium | Sm | 150.36 | 144 | 143.911 995 | 3.07 | |
| | | | | 146* | 145.913 037 | | $1.0 \times 10^{8}$ yr |
| | | | | 147* | 146.914 893 | 14.99 | $1.06 \times 10^{11}$ yr |
| | | | | 148* | 147.914 818 | 11.24 | $7 \times 10^{15}$ yr |
| | | | | 149* | 148.917 180 | 13.82 | $>2 \times 10^{15}$ yr |
| | | | | 150 | 149.917 272 | 7.38 | |
| | | | | 151* | 150.919 928 | | 90 yr |
| | | | | 152 | 151.919 728 | 26.75 | |
| | | | | 154 | 153.922 205 | 22.75 | |
| 63 | Europium | Eu | 151.964 | 151 | 150.919 846 | 47.81 | |
| | | | | 152* | 151.921 740 | | 13.5 yr |
| | | | | 153 | 152.921 226 | 52.19 | |
| | | | | 154* | 153.922 975 | | 8.59 yr |
| | | | | 155* | 154.922 889 | | 4.7 yr |
| 64 | Gadolinium | Gd | 157.25 | 148* | 147.918 110 | | 75 yr |
| | | | | 150* | 149.918 656 | | $1.8 \times 10^{6}$ yr |
| | | | | 152* | 151.919 788 | 0.20 | $1.1 \times 10^{14}$ yr |
| | | | | 154 | 153.920 862 | 2.18 | |
| | | | | 155 | 154.922 619 | 14.80 | |
| | | | | 156 | 155.922 120 | 20.47 | |
| | | | | 157 | 156.923 957 | 15.65 | |
| | | | | 158 | 157.924 100 | 24.84 | |
| | | | | 160 | 159.927 051 | 21.86 | |
| 65 | Terbium | Tb | 158.925 34 | 159 | 158.925 343 | 100 | |
| 66 | Dysprosium | Dy | 162.50 | 156 | 155.924 278 | 0.06 | |
| | | | | 158 | 157.924 405 | 0.10 | |
| | | | | 160 | 159.925 194 | 2.34 | |
| | | | | 161 | 160.926 930 | 18.91 | |
| | | | | 162 | 161.926 795 | 25.51 | |
| | | | | 163 | 162.928 728 | 24.90 | |
| | | | | 164 | 163.929 171 | 28.18 | |
| 67 | Holmium | Ho | 164.930 32 | 165 | 164.930 320 | 100 | |
| | | | | 166* | 165.932 281 | | $1.2 \times 10^{3}$ yr |
| 68 | Erbium | Er | 167.6 | 162 | 161.928 775 | 0.14 | |
| | | | | 164 | 163.929 197 | 1.61 | |

*(Continued)*

TABLE A.3 Table of Atomic Masses *(Continued)*

| Atomic Number $Z$ | Element | Symbol | Chemical Atomic Mass (u) | Mass Number (*indicates radioactive) $A$ | Atomic Mass (u) | Percent Abundance | Half-Life (if radioactive) $T_{1/2}$ |
|---|---|---|---|---|---|---|---|
| (68) | Erbium | | | 166 | 165.930 290 | 33.61 | |
| | | | | 167 | 166.932 045 | 22.93 | |
| | | | | 168 | 167.932 368 | 26.78 | |
| | | | | 170 | 169.935 460 | 14.93 | |
| 69 | Thulium | Tm | 168.934 21 | 169 | 168.934 211 | 100 | |
| | | | | 171* | 170.936 426 | | 1.92 yr |
| 70 | Ytterbium | Yb | 173.04 | 168 | 167.933 894 | 0.13 | |
| | | | | 170 | 169.934 759 | 3.04 | |
| | | | | 171 | 170.936 322 | 14.28 | |
| | | | | 172 | 171.936 378 | 21.83 | |
| | | | | 173 | 172.938 207 | 16.13 | |
| | | | | 174 | 173.938 858 | 31.83 | |
| | | | | 176 | 175.942 568 | 12.76 | |
| 71 | Lutecium | Lu | 174.967 | 173* | 172.938 927 | | 1.37 yr |
| | | | | 175 | 174.940 768 | 97.41 | |
| | | | | 176* | 175.942 682 | 2.59 | $3.78 \times 10^{10}$ yr |
| 72 | Hafnium | Hf | 178.49 | 174* | 173.940 040 | 0.16 | $2.0 \times 10^{15}$ yr |
| | | | | 176 | 175.941 402 | 5.26 | |
| | | | | 177 | 176.943 220 | 18.60 | |
| | | | | 178 | 177.943 698 | 27.28 | |
| | | | | 179 | 178.945 815 | 13.62 | |
| | | | | 180 | 179.946 549 | 35.08 | |
| 73 | Tantalum | Ta | 180.947 9 | 180 | 179.947 466 | 0.012 | |
| | | | | 181 | 180.947 996 | 99.988 | |
| 74 | Tungsten (Wolfram) | W | 183.84 | 180 | 179.946 706 | 0.12 | |
| | | | | 182 | 181.948 206 | 26.50 | |
| | | | | 183 | 182.950 224 | 14.31 | |
| | | | | 184 | 183.950 933 | 30.64 | |
| | | | | 186 | 185.954 362 | 28.43 | |
| 75 | Rhenium | Re | 186.207 | 185 | 184.952 956 | 37.40 | |
| | | | | 187* | 186.955 751 | 62.60 | $4.4 \times 10^{10}$ yr |
| 76 | Osmium | Os | 190.23 | 184 | 183.952 491 | 0.02 | |
| | | | | 186* | 185.953 838 | 1.59 | $2.0 \times 10^{15}$ yr |
| | | | | 187 | 186.955 748 | 1.96 | |
| | | | | 188 | 187.955 836 | 13.24 | |
| | | | | 189 | 188.958 145 | 16.15 | |
| | | | | 190 | 189.958 445 | 26.26 | |
| | | | | 192 | 191.961 479 | 40.78 | |
| | | | | 194* | 193.965 179 | | 6.0 yr |
| 77 | Iridium | Ir | 192.217 | 191 | 190.960 591 | 37.3 | |
| | | | | 193 | 192.962 924 | 62.7 | |
| 78 | Platinum | Pt | 195.078 | 190* | 189.959 930 | 0.014 | $6.5 \times 10^{11}$ yr |
| | | | | 192 | 191.961 035 | 0.782 | |
| | | | | 194 | 193.962 664 | 32.967 | |
| | | | | 195 | 194.964 774 | 33.832 | |
| | | | | 196 | 195.964 935 | 25.242 | |
| | | | | 198 | 197.967 876 | 7.163 | |
| 79 | Gold | Au | 196.966 55 | 197 | 196.966 552 | 100 | |
| 80 | Mercury | Hg | 200.59 | 196 | 195.965 815 | 0.15 | |
| | | | | 198 | 197.966 752 | 9.97 | |
| | | | | 199 | 198.968 262 | 16.87 | |
| | | | | 200 | 199.968 309 | 23.10 | |
| | | | | 201 | 200.970 285 | 13.18 | |

*(Continued)*

**TABLE A.3** Table of Atomic Masses *(Continued)*

| Atomic Number Z | Element | Symbol | Chemical Atomic Mass (u) | Mass Number (*indicates radioactive) $A$ | Atomic Mass (u) | Percent Abundance | Half-Life (if radioactive) $T_{1/2}$ |
|---|---|---|---|---|---|---|---|
| (80) | Mercury | | | 202 | 201.970 626 | 29.86 | |
| | | | | 204 | 203.973 476 | 6.87 | |
| 81 | Thallium | Tl | 204.383 3 | 203 | 202.972 329 | 29.524 | |
| | | | | 204* | 203.973 849 | | 3.78 yr |
| | | | | 205 | 204.974 412 | 70.476 | |
| | | (Ra E″) | | 206* | 205.976 095 | | 4.2 min |
| | | (Ac C″) | | 207* | 206.977 408 | | 4.77 min |
| | | (Th C″) | | 208* | 207.982 005 | | 3.053 min |
| | | (Ra C″) | | 210* | 209.990 066 | | 1.30 min |
| 82 | Lead | Pb | 207.2 | 202* | 201.972 144 | | $5 \times 10^4$ yr |
| | | | | 204* | 203.973 029 | 1.4 | $\geq 1.4 \times 10^{17}$ yr |
| | | | | 205* | 204.974 467 | | $1.5 \times 10^7$ yr |
| | | | | 206 | 205.974 449 | 24.1 | |
| | | | | 207 | 206.975 881 | 22.1 | |
| | | | | 208 | 207.976 636 | 52.4 | |
| | | (Ra D) | | 210* | 209.984 173 | | 22.3 yr |
| | | (Ac B) | | 211* | 210.988 732 | | 36.1 min |
| | | (Th B) | | 212* | 211.991 888 | | 10.64 h |
| | | (Ra B) | | 214* | 213.999 798 | | 26.8 min |
| 83 | Bismuth | Bi | 208.980 38 | 207* | 206.978 455 | | 32.2 yr |
| | | | | 208* | 207.979 727 | | $3.7 \times 10^5$ yr |
| | | | | 209 | 208.980 383 | 100 | |
| | | (Ra E) | | 210* | 209.984 105 | | 5.01 days |
| | | (Th C) | | 211* | 210.987 258 | | 2.14 min |
| | | | | 212* | 211.991 272 | | 60.6 min |
| | | (Ra C) | | 214* | 213.998 699 | | 19.9 min |
| | | | | 215* | 215.001 832 | | 7.4 min |
| 84 | Polonium | Po | | 209* | 208.982 416 | | 102 yr |
| | | (Ra F) | | 210* | 209.982 857 | | 138.38 days |
| | | (Ac C′) | | 211* | 210.986 637 | | 0.52 s |
| | | (Th C′) | | 212* | 211.988 852 | | 0.30 $\mu$s |
| | | (Ra C′) | | 214* | 213.995 186 | | 164 $\mu$s |
| | | (Ac A) | | 215* | 214.999 415 | | 0.001 8 s |
| | | (Th A) | | 216* | 216.001 905 | | 0.145 s |
| | | (Ra A) | | 218* | 218.008 966 | | 3.10 min |
| 85 | Astatine | At | | 215* | 214.998 641 | | $\approx 100$ $\mu$s |
| | | | | 218* | 218.008 682 | | 1.6 s |
| | | | | 219* | 219.011 297 | | 0.9 min |
| 86 | Radon | Rn | | | | | |
| | | (An) | | 219* | 219.009 475 | | 3.96 s |
| | | (Tn) | | 220* | 220.011 384 | | 55.6 s |
| | | (Rn) | | 222* | 222.017 570 | | 3.823 days |
| 87 | Francium | Fr | | | | | |
| | | (Ac K) | | 223* | 223.019 731 | | 22 min |
| 88 | Radium | Ra | | | | | |
| | | (Ac X) | | 223* | 223.018 497 | | 11.43 days |
| | | (Th X) | | 224* | 224.020 202 | | 3.66 days |
| | | (Ra) | | 226* | 226.025 403 | | 1 600 yr |
| | | (Ms $Th_1$) | | 228* | 228.031 064 | | 5.75 yr |
| 89 | Actinium | Ac | | 227* | 227.027 747 | | 21.77 yr |
| | | (Ms $Th_2$) | | 228* | 228.031 015 | | 6.15 h |
| 90 | Thorium | Th | 232.038 1 | | | | |

*(Continued)*

## TABLE A.3 Table of Atomic Masses *(Continued)*

| Atomic Number $Z$ | Element | Symbol | Chemical Atomic Mass (u) | Mass Number (*indicates radioactive) $A$ | Atomic Mass (u) | Percent Abundance | Half-Life (if radioactive) $T_{1/2}$ |
|---|---|---|---|---|---|---|---|
| (90) | Thorium | (Rd Ac) | | 227* | 227.027 699 | | 18.72 days |
| | | (Rd Th) | | 228* | 228.028 731 | | 1.913 yr |
| | | | | 229* | 229.031 755 | | 7 300 yr |
| | | (Io) | | 230* | 230.033 127 | | 75.000 yr |
| | | (UY) | | 231* | 231.036 297 | | 25.52 h |
| | | (Th) | | 232* | 232.038 050 | 100 | $1.40 \times 10^{10}$ yr |
| | | $(UX_1)$ | | 234* | 234.043 596 | | 24.1 days |
| 91 | Protactinium | Pa | 231.035 88 | 231* | 231.035 879 | | 32.760 yr |
| | | (Uz) | | 234* | 234.043 302 | | 6.7 h |
| 92 | Uranium | U | 238.028 9 | 232* | 232.037 146 | | 69 yr |
| | | | | 233* | 233.039 628 | | $1.59 \times 10^5$ yr |
| | | | | 234* | 234.040 946 | 0.005 5 | $2.45 \times 10^5$ yr |
| | | (Ac U) | | 235* | 235.043 923 | 0.720 0 | $7.04 \times 10^8$ yr |
| | | | | 236* | 236.045 562 | | $2.34 \times 10^7$ yr |
| | | (UI) | | 238* | 238.050 783 | 99.274 5 | $4.47 \times 10^9$ yr |
| 93 | Neptunium | Np | | 235* | 235.044 056 | | 396 days |
| | | | | 236* | 236.046 560 | | $1.15 \times 10^5$ yr |
| | | | | 237* | 237.048 167 | | $2.14 \times 10^6$ yr |
| 94 | Plutonium | Pu | | 236* | 236.046 048 | | 2.87 yr |
| | | | | 238* | 238.049 553 | | 87.7 yr |
| | | | | 239* | 239.052 156 | | $2.412 \times 10^4$ yr |
| | | | | 240* | 240.053 808 | | 6 560 yr |
| | | | | 241* | 241.056 845 | | 14.4 yr |
| | | | | 242* | 242.058 737 | | $3.73 \times 10^6$ yr |
| | | | | 244* | 244.064 198 | | $8.1 \times 10^7$ yr |

*Sources:* Chemical atomic masses are from T. B. Coplen, "Atomic Weights of the Elements 1999," a technical report to the International Union of Pure and Applied Chemistry, and published in *Pure and Applied Chemistry* 73(4), 667–683, 2001. Atomic masses of the isotopes are from G. Audi and A. H. Wapstra, "The 1995 Update to the Atomic Mass Evaluation," *Nuclear Physics* A595, vol. 4, 409–480, December 25, 1995. Percent abundance values are from K. J. R. Rosman and P. D. P. Taylor, "Isotopic Compositions of the Elements 1999," a technical report to the International Union of Pure and Applied Chemistry, and published in *Pure and Applied Chemistry* 70(1), 217–236, 1998.

# Mathematics Review

This appendix in mathematics is intended as a brief review of operations and methods. Early in this course, you should be totally familiar with basic algebraic techniques, analytic geometry, and trigonometry. The sections on differential and integral calculus are more detailed and are intended for those students who have difficulty applying calculus concepts to physical situations.

## B.1 SCIENTIFIC NOTATION

Many quantities that scientists deal with often have very large or very small values. The speed of light, for example, is about 300 000 000 m/s, and the ink required to make the dot over an *i* in this textbook has a mass of about 0.000 000 001 kg. Obviously, it is very cumbersome to read, write, and keep track of such numbers. We avoid this problem by using a method dealing with powers of the number ten:

$$10^0 = 1$$

$$10^1 = 10$$

$$10^2 = 10 \times 10 = 100$$

$$10^3 = 10 \times 10 \times 10 = 1\,000$$

$$10^4 = 10 \times 10 \times 10 \times 10 = 10\,000$$

$$10^5 = 10 \times 10 \times 10 \times 10 \times 10 = 100\,000$$

and so on. The number of zeros corresponds to the power to which ten is raised, called the **exponent** of ten. For example, the speed of light, 300 000 000 m/s, can be expressed as $3 \times 10^8$ m/s.

In this method, some representative numbers smaller than unity are the following:

$$10^{-1} = \frac{1}{10} = 0.1$$

$$10^{-2} = \frac{1}{10 \times 10} = 0.01$$

$$10^{-3} = \frac{1}{10 \times 10 \times 10} = 0.001$$

$$10^{-4} = \frac{1}{10 \times 10 \times 10 \times 10} = 0.000\,1$$

$$10^{-5} = \frac{1}{10 \times 10 \times 10 \times 10 \times 10} = 0.000\,01$$

In these cases, the number of places the decimal point is to the left of the digit 1 equals the value of the (negative) exponent. Numbers expressed as some power of ten multiplied by another number between one and ten are said to be in **scientific notation.** For example, the scientific notation for 5 943 000 000 is $5.943 \times 10^9$ and that for 0.000 083 2 is $8.32 \times 10^{-5}$.

When numbers expressed in scientific notation are being multiplied, the following general rule is very useful:

$$10^n \times 10^m = 10^{n+m} \quad \text{[B.1]}$$

where $n$ and $m$ can be *any* numbers (not necessarily integers). For example, $10^2 \times 10^5 = 10^7$. The rule also applies if one of the exponents is negative: $10^3 \times 10^{-8} = 10^{-5}$.

When dividing numbers expressed in scientific notation, note that

$$\frac{10^n}{10^m} = 10^n \times 10^{-m} = 10^{n-m} \quad \text{[B.2]}$$

## Exercises

With help from the preceding rules, verify the answers to the following equations.

1. $86\,400 = 8.64 \times 10^4$
2. $9\,816\,762.5 = 9.816\,762\,5 \times 10^6$
3. $0.000\,000\,039\,8 = 3.98 \times 10^{-8}$
4. $(4.0 \times 10^8)(9.0 \times 10^9) = 3.6 \times 10^{18}$
5. $(3.0 \times 10^7)(6.0 \times 10^{-12}) = 1.8 \times 10^{-4}$
6. $\dfrac{75 \times 10^{-11}}{5.0 \times 10^{-3}} = 1.5 \times 10^{-7}$
7. $\dfrac{(3 \times 10^6)(8 \times 10^{-2})}{(2 \times 10^{17})(6 \times 10^5)} = 2 \times 10^{-18}$

# B.2 ALGEBRA

## Some Basic Rules

When algebraic operations are performed, the laws of arithmetic apply. Symbols such as $x$, $y$, and $z$ are usually used to represent unspecified quantities, called the **unknowns.**

First, consider the equation

$$8x = 32$$

If we wish to solve for $x$, we can divide (or multiply) each side of the equation by the same factor without destroying the equality. In this case, if we divide both sides by 8, we have

$$\frac{8x}{8} = \frac{32}{8}$$

$$x = 4$$

Next consider the equation

$$x + 2 = 8$$

In this type of expression, we can add or subtract the same quantity from each side. If we subtract 2 from each side, we have

$$x + 2 - 2 = 8 - 2$$

$$x = 6$$

In general, if $x + a = b$, then $x = b - a$.

Now consider the equation

$$\frac{x}{5} = 9$$

If we multiply each side by 5, we are left with $x$ on the left by itself and 45 on the right:

$$\left(\frac{x}{5}\right)(5) = 9 \times 5$$

$$x = 45$$

In all cases, *whatever operation is performed on the left side of the equality must also be performed on the right side.*

The following rules for multiplying, dividing, adding, and subtracting fractions should be recalled, where $a$, $b$, $c$, and $d$ are four numbers:

| | Rule | Example |
|---|---|---|
| **Multiplying** | $\left(\frac{a}{b}\right)\left(\frac{c}{d}\right) = \frac{ac}{bd}$ | $\left(\frac{2}{3}\right)\left(\frac{4}{5}\right) = \frac{8}{15}$ |
| **Dividing** | $\frac{(a/b)}{(c/d)} = \frac{ad}{bc}$ | $\frac{2/3}{4/5} = \frac{(2)(5)}{(4)(3)} = \frac{10}{12}$ |
| **Adding** | $\frac{a}{b} \pm \frac{c}{d} = \frac{ad \pm bc}{bd}$ | $\frac{2}{3} - \frac{4}{5} = \frac{(2)(5) - (4)(3)}{(3)(5)} = -\frac{2}{15}$ |

## Exercises

In the following exercises, solve for $x$:

| | | Answers |
|---|---|---|
| 1. | $a = \frac{1}{1 + x}$ | $x = \frac{1 - a}{a}$ |
| 2. | $3x - 5 = 13$ | $x = 6$ |
| 3. | $ax - 5 = bx + 2$ | $x = \frac{7}{a - b}$ |
| 4. | $\frac{5}{2x + 6} = \frac{3}{4x + 8}$ | $x = -\frac{11}{7}$ |

## Powers

When powers of a given quantity $x$ are multiplied, the following rule applies:

$$x^n x^m = x^{n+m} \qquad \text{[B.3]}$$

For example, $x^2 x^4 = x^{2+4} = x^6$.

When dividing the powers of a given quantity, the rule is

$$\frac{x^n}{x^m} = x^{n-m} \qquad \text{[B.4]}$$

For example, $x^8/x^2 = x^{8-2} = x^6$.

A power that is a fraction, such as $\frac{1}{3}$, corresponds to a root as follows:

$$x^{1/n} = \sqrt[n]{x} \qquad \text{[B.5]}$$

For example, $4^{1/3} = \sqrt[3]{4} = 1.5874$. (A scientific calculator is useful for such calculations.)

Finally, any quantity $x^n$ raised to the $m$th power is

$$(x^n)^m = x^{nm} \tag{B.6}$$

Table B.1 summarizes the rules of exponents.

**TABLE B.1 Rules of Exponents**

| |
|---|
| $x^0 = 1$ |
| $x^1 = x$ |
| $x^n x^m = x^{n+m}$ |
| $x^n/x^m = x^{n-m}$ |
| $x^{1/n} = \sqrt[n]{x}$ |
| $(x^n)^m = x^{nm}$ |

## Exercises

Verify the following equations.

1. $3^2 \times 3^3 = 243$
2. $x^5 x^{-8} = x^{-3}$
3. $x^{10}/x^{-5} = x^{15}$
4. $5^{1/3} = 1.709\ 975$ (Use your calculator.)
5. $60^{1/4} = 2.783\ 158$ (Use your calculator.)
6. $(x^4)^3 = x^{12}$

## Factoring

Some useful formulas for factoring an equation are the following:

$$ax + ay + az = a(x + y + x) \qquad \text{common factor}$$

$$a^2 + 2ab + b^2 = (a + b)^2 \qquad \text{perfect square}$$

$$a^2 - b^2 = (a + b)(a - b) \qquad \text{differences of squares}$$

## Quadratic Equations

The general form of a quadratic equation is

$$ax^2 + bx + c = 0 \tag{B.7}$$

where $x$ is the unknown quantity and $a$, $b$, and $c$ are numerical factors referred to as **coefficients** of the equation. This equation has two roots, given by

$$x = \frac{-b \pm \sqrt{b^2 - 4ac}}{2a} \tag{B.8}$$

If $b^2 \geq 4ac$, the roots are real.

**EXAMPLE B.1**

The equation $x^2 + 5x + 4 = 0$ has the following roots corresponding to the two signs of the square-root term:

$$x = \frac{-5 \pm \sqrt{5^2 - (4)(1)(4)}}{2(1)} = \frac{-5 \pm \sqrt{9}}{2} = \frac{-5 \pm 3}{2}$$

$$x_+ = \frac{-5 + 3}{2} = -1 \qquad x_- = \frac{-5 - 3}{2} = -4$$

where $x_+$ refers to the root corresponding to the positive sign and $x_-$ refers to the root corresponding to the negative sign.

## Exercises

Solve the following quadratic equations.

| | | Answers | |
|---|---|---|---|
| 1. | $x^2 + 2x - 3 = 0$ | $x_+ = 1$ | $x_- = -3$ |
| 2. | $2x^2 - 5x + 2 = 0$ | $x_+ = 2$ | $x_- = \frac{1}{2}$ |
| 3. | $2x^2 - 4x - 9 = 0$ | $x_+ = 1 + \sqrt{22}/2$ | $x_- = 1 - \sqrt{22}/2$ |

## Linear Equations

A linear equation has the general form

$$y = mx + b \quad \text{[B.9]}$$

where $m$ and $b$ are constants. This equation is referred to as being linear because the graph of $y$ versus $x$ is a straight line as shown in Figure B.1. The constant $b$, called the **y-intercept,** represents the value of $y$ at which the straight line intersects the $y$ axis. The constant $m$ is equal to the **slope** of the straight line. If any two points on the straight line are specified by the coordinates $(x_1, y_1)$ and $(x_2, y_2)$ as in Figure B.1, the slope of the straight line can be expressed as

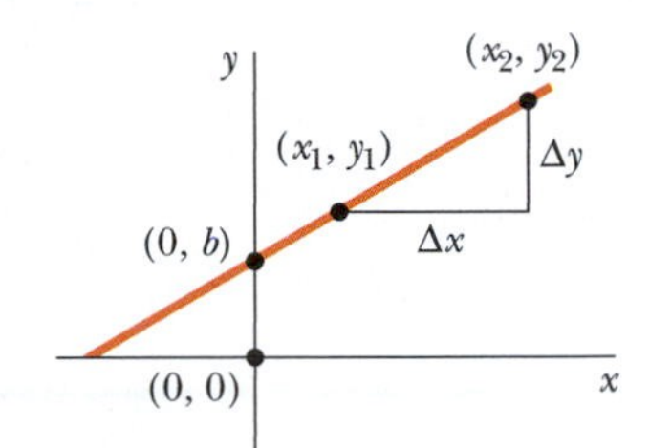


FIGURE B.1 A straight line graphed on an $x$-$y$ coordinate system. The slope of the line is the ratio of $\Delta y$ to $\Delta x$.

$$\text{Slope} = \frac{y_2 - y_1}{x_2 - x_1} = \frac{\Delta y}{\Delta x} \quad \text{[B.10]}$$

Note that $m$ and $b$ can have either positive or negative values. If $m > 0$, the straight line has a *positive* slope as in Figure B.1. If $m < 0$, the straight line has a *negative* slope. In Figure B.1, both $m$ and $b$ are positive. Three other possible situations are shown in Figure B.2.

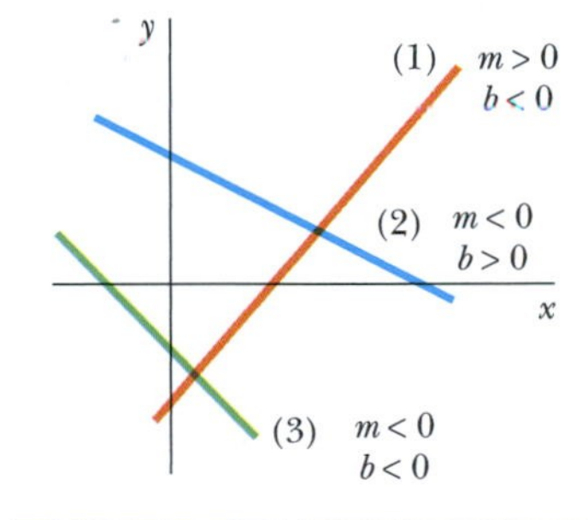


FIGURE B.2 The brown line has a positive slope and a negative $y$-intercept. The blue line has a negative slope and a positive $y$-intercept. The green line has a negative slope and a negative $y$-intercept.

## Exercises

1. Draw graphs of the following straight lines: (a) $y = 5x + 3$, (b) $y = -2x + 4$, (c) $y = -3x - 6$.
2. Find the slopes of the straight lines described in Exercise 1.

**Answers** (a) 5 (b) $-2$ (c) $-3$

3. Find the slopes of the straight lines that pass through the following sets of points: (a) $(0, -4)$ and $(4, 2)$, (b) $(0, 0)$ and $(2, -5)$, (c) $(-5, 2)$ and $(4, -2)$.

**Answers** (a) $\frac{3}{2}$ (b) $-\frac{5}{2}$ (c) $-\frac{4}{9}$

## Solving Simultaneous Linear Equations

Consider the equation $3x + 5y = 15$, which has two unknowns, $x$ and $y$. Such an equation does not have a unique solution. For example, note that $(x = 0, y = 3)$, $(x = 5, y = 0)$, and $\left(x = 2, y = \frac{9}{5}\right)$ are all solutions to this equation.

If a problem has two unknowns, a unique solution is possible only if we have *two* equations. In general, if a problem has $n$ unknowns, its solution requires $n$ equations. To solve two simultaneous equations involving two unknowns, $x$ and $y$, we solve one of the equations for $x$ in terms of $y$ and substitute this expression into the other equation.

**EXAMPLE B.2**

Solve the two simultaneous equations

$$(1) \quad 5x + y = -8$$

$$(2) \quad 2x - 2y = 4$$

**Solution** From (2), $x = y + 2$. Substitution of this equation into (1) gives

$$5(y + 2) + y = -8$$

$$6y = -18$$

$$y = -3$$

$$x = y + 2 = -1$$

**Alternative Solution** Multiply each term in (1) by the factor 2 and add the result to (2):

$$10x + 2y = -16$$

$$2x - 2y = 4$$

$$12x = -12$$

$$x = -1$$

$$y = x - 2 = -3$$

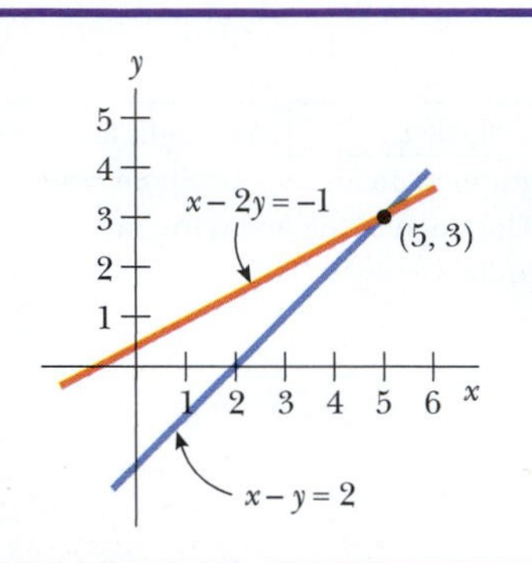


**FIGURE B.3** A graphical solution for two linear equations.

Two linear equations containing two unknowns can also be solved by a graphical method. If the straight lines corresponding to the two equations are plotted in a conventional coordinate system, the intersection of the two lines represents the solution. For example, consider the two equations

$$x - y = 2$$

$$x - 2y = -1$$

These equations are plotted in Figure B.3. The intersection of the two lines has the coordinates $x = 5$ and $y = 3$, which represents the solution to the equations. You should check this solution by the analytical technique discussed earlier.

## Exercises

Solve the following pairs of simultaneous equations involving two unknowns.

| | | Answers |
|---|---|---|
| **1.** | $x + y = 8$<br>$x - y = 2$ | $x = 5, y = 3$ |
| **2.** | $98 - T = 10a$<br>$T - 49 = 5a$ | $T = 65, a = 3.27$ |
| **3.** | $6x + 2y = 6$<br>$8x - 4y = 28$ | $x = 2, y = -3$ |

## Logarithms

Suppose a quantity $x$ is expressed as a power of some quantity $a$:

$$x = a^y \qquad [B.11]$$

The number $a$ is called the **base** number. The **logarithm** of $x$ with respect to the base $a$ is equal to the exponent to which the base must be raised to satisfy the expression $x = a^y$:

$$y = \log_a x \qquad [B.12]$$

Conversely, the **antilogarithm** of $y$ is the number $x$:

$$x = \text{antilog}_a\, y \qquad [B.13]$$

In practice, the two bases most often used are base 10, called the *common* logarithm base, and base $e = 2.718\,282$, called Euler's constant or the *natural* logarithm base. When common logarithms are used,

$$y = \log_{10} x \qquad (\text{or } x = 10^y) \tag{B.14}$$

When natural logarithms are used,

$$y = \ln x \qquad (\text{or } x = e^y) \tag{B.15}$$

For example, $\log_{10} 52 = 1.716$, so $\text{antilog}_{10}\, 1.716 = 10^{1.716} = 52$. Likewise, $\ln 52 = 3.951$, so $\text{antiln}\, 3.951 = e^{3.951} = 52$.

In general, you can convert between base 10 and base $e$ with the equality

$$\ln x = (2.302\,585) \log_{10} x \tag{B.16}$$

Finally, some useful properties of logarithms are the following:

$$\left.\begin{aligned} \log(ab) &= \log a + \log b \\ \log(a/b) &= \log a - \log b \\ \log(a^n) &= n \log a \end{aligned}\right\} \text{ any base}$$

$$\ln e = 1$$

$$\ln e^a = a$$

$$\ln\left(\frac{1}{a}\right) = -\ln a$$

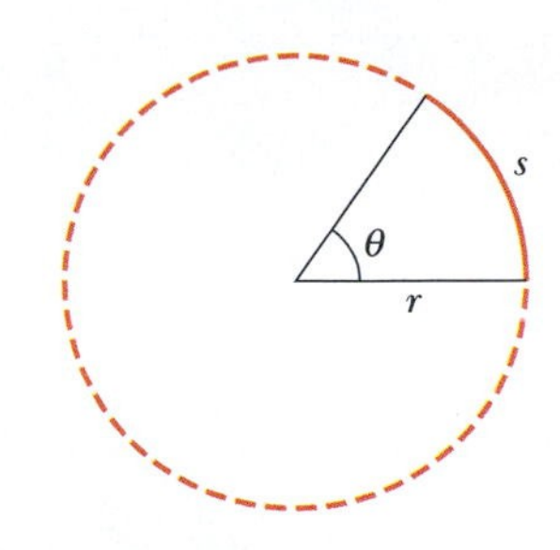


FIGURE B.4 The angle $\theta$ in radians is the ratio of the arc length $s$ to the radius $r$ of the circle.

## B.3 GEOMETRY

The **distance** $d$ between two points having coordinates $(x_1, y_1)$ and $(x_2, y_2)$ is

$$d = \sqrt{(x_2 - x_1)^2 + (y_2 - y_1)^2} \tag{B.17}$$

**Radian measure:** The arc length $s$ of a circular arc (Fig. B.4) is proportional to the radius $r$ for a fixed value of $\theta$ (in radians):

$$s = r\theta \qquad \theta = \frac{s}{r} \tag{B.18}$$

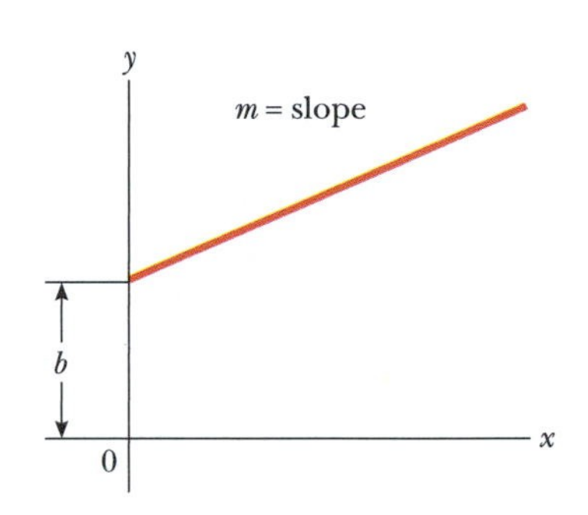


FIGURE B.5 A straight line with a slope of $m$ and a $y$-intercept of $b$.

Table B.2 gives the **areas** and **volumes** for several geometric shapes used throughout this text.

The equation of a **straight line** (Fig. B.5) is

$$y = mx + b \tag{B.19}$$

where $b$ is the $y$-intercept and $m$ is the slope of the line.

The equation of a **circle** of radius $R$ centered at the origin is

$$x^2 + y^2 = R^2 \tag{B.20}$$

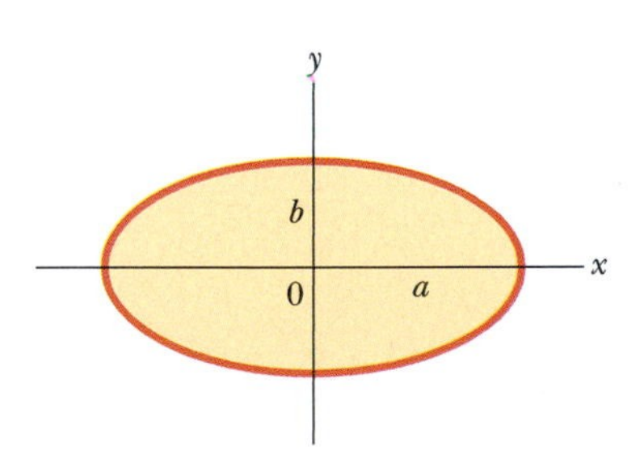


FIGURE B.6 An ellipse with semimajor axis $a$ and semiminor axis $b$.

The equation of an **ellipse** having the origin at its center (Fig. B.6) is

$$\frac{x^2}{a^2} + \frac{y^2}{b^2} = 1 \tag{B.21}$$

where $a$ is the length of the semimajor axis (the longer one) and $b$ is the length of the semiminor axis (the shorter one).

The equation of a **parabola** the vertex of which is at $y = b$ (Fig. B.7) is

$$y = ax^2 + b \tag{B.22}$$

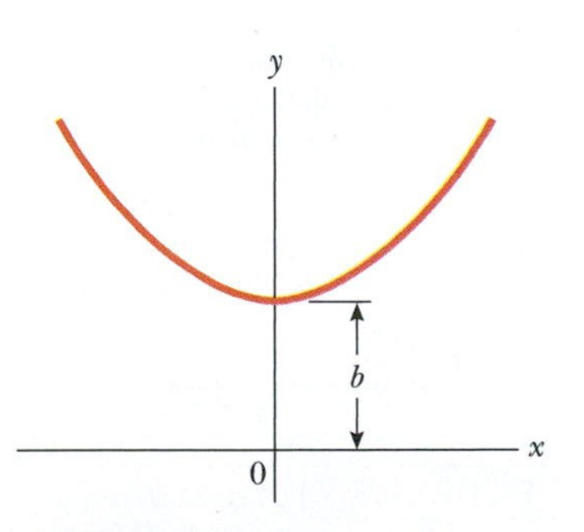


FIGURE B.7 A parabola.

**TABLE B.2 Useful Information for Geometry**

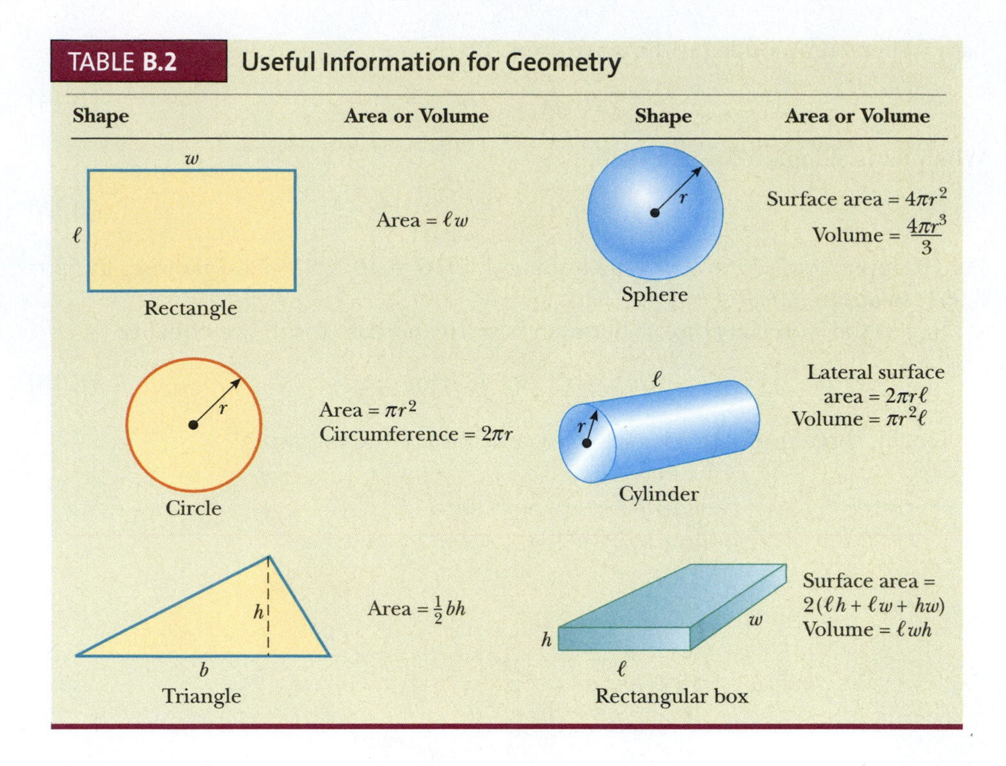

| Shape | Area or Volume | Shape | Area or Volume |
|---|---|---|---|
| Rectangle | Area = $\ell w$ | Sphere | Surface area = $4\pi r^2$<br>Volume = $\frac{4\pi r^3}{3}$ |
| Circle | Area = $\pi r^2$<br>Circumference = $2\pi r$ | Cylinder | Lateral surface area = $2\pi r\ell$<br>Volume = $\pi r^2\ell$ |
| Triangle | Area = $\frac{1}{2}bh$ | Rectangular box | Surface area = $2(\ell h + \ell w + hw)$<br>Volume = $\ell wh$ |

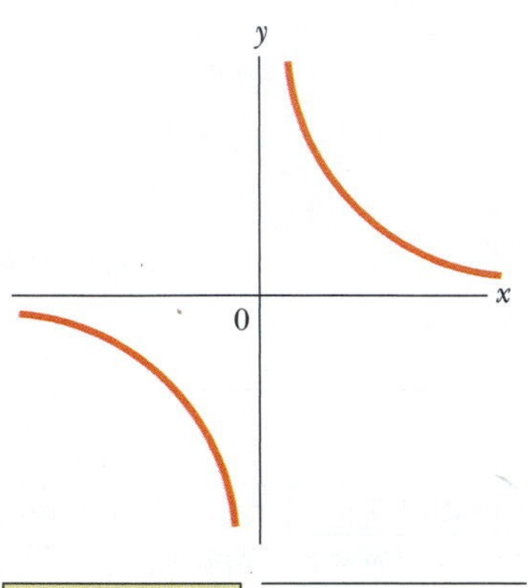


**FIGURE B.8** A hyperbola.

The equation of a **rectangular hyperbola** (Fig. B.8) is

$$xy = \text{constant} \qquad \text{[B.23]}$$

## B.4 TRIGONOMETRY

That portion of mathematics based on the special properties of the right triangle is called trigonometry. By definition, a right triangle is a triangle containing a 90° angle. Consider the right triangle shown in Figure B.9, where side $a$ is opposite the angle $\theta$, side $b$ is adjacent to the angle $\theta$, and side $c$ is the hypotenuse of the triangle. The three basic trigonometric functions defined by such a triangle are the sine (sin), cosine (cos), and tangent (tan) functions. In terms of the angle $\theta$, these functions are defined by

$$\sin\theta = \frac{\text{side opposite } \theta}{\text{hypotenuse}} = \frac{a}{c} \qquad \text{[B.24]}$$

$$\cos\theta = \frac{\text{side adjacent to } \theta}{\text{hypotenuse}} = \frac{b}{c} \qquad \text{[B.25]}$$

$$\tan\theta = \frac{\text{side opposite } \theta}{\text{side adjacent to } \theta} = \frac{a}{b} \qquad \text{[B.26]}$$

The Pythagorean theorem provides the following relationship among the sides of a right triangle:

$$c^2 = a^2 + b^2 \qquad \text{[B.27]}$$

From the preceding definitions and the Pythagorean theorem, it follows that

$$\sin^2\theta + \cos^2\theta = 1$$

$$\tan\theta = \frac{\sin\theta}{\cos\theta}$$

$a$ = opposite side
$b$ = adjacent side
$c$ = hypotenuse

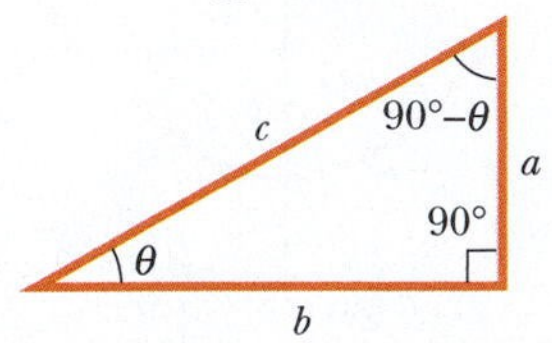


**FIGURE B.9** A right triangle, used to define the basic functions of trigonometry.

**TABLE B.3 Some Trigonometric Identities**

| | |
|---|---|
| $\sin^2\theta + \cos^2\theta = 1$ | $\csc^2\theta = 1 + \cot^2\theta$ |
| $\sec^2\theta = 1 + \tan^2\theta$ | $\sin^2\dfrac{\theta}{2} = \frac{1}{2}(1 - \cos\theta)$ |
| $\sin 2\theta = 2\sin\theta\cos\theta$ | $\cos^2\dfrac{\theta}{2} = \frac{1}{2}(1 + \cos\theta)$ |
| $\cos 2\theta = \cos^2\theta - \sin^2\theta$ | $1 - \cos\theta = 2\sin^2\dfrac{\theta}{2}$ |
| $\tan 2\theta = \dfrac{2\tan\theta}{1 - \tan^2\theta}$ | $\tan\dfrac{\theta}{2} = \sqrt{\dfrac{1 - \cos\theta}{1 + \cos\theta}}$ |

$\sin(A \pm B) = \sin A\cos B \pm \cos A\sin B$

$\cos(A \pm B) = \cos A\cos B \mp \sin A\sin B$

$\sin A \pm \sin B = 2\sin[\frac{1}{2}(A \pm B)]\cos[\frac{1}{2}(A \mp B)]$

$\cos A + \cos B = 2\cos[\frac{1}{2}(A + B)]\cos[\frac{1}{2}(A - B)]$

$\cos A - \cos B = 2\sin[\frac{1}{2}(A + B)]\sin[\frac{1}{2}(B - A)]$

The cosecant, secant, and cotangent functions are defined by

$$\csc\theta = \frac{1}{\sin\theta} \qquad \sec\theta = \frac{1}{\cos\theta} \qquad \cot\theta = \frac{1}{\tan\theta}$$

The following relationships are derived directly from the right triangle shown in Figure B.9:

$$\sin\theta = \cos(90° - \theta)$$

$$\cos\theta = \sin(90° - \theta)$$

$$\cot\theta = \tan(90° - \theta)$$

Some properties of trigonometric functions are

$$\sin(-\theta) = -\sin\theta$$

$$\cos(-\theta) = \cos\theta$$

$$\tan(-\theta) = -\tan\theta$$

The following relationships apply to *any* triangle, as shown in Figure B.10:

$$\alpha + \beta + \gamma = 180°$$

$$\text{Law of cosines} \quad \begin{cases} a^2 = b^2 + c^2 - 2bc\cos\alpha \\ b^2 = a^2 + c^2 - 2ac\cos\beta \\ c^2 = a^2 + b^2 - 2ab\cos\gamma \end{cases}$$

$$\text{Law of sines} \quad \frac{a}{\sin\alpha} = \frac{b}{\sin\beta} = \frac{c}{\sin\gamma}$$

Table B.3 lists a number of useful trigonometric identities.

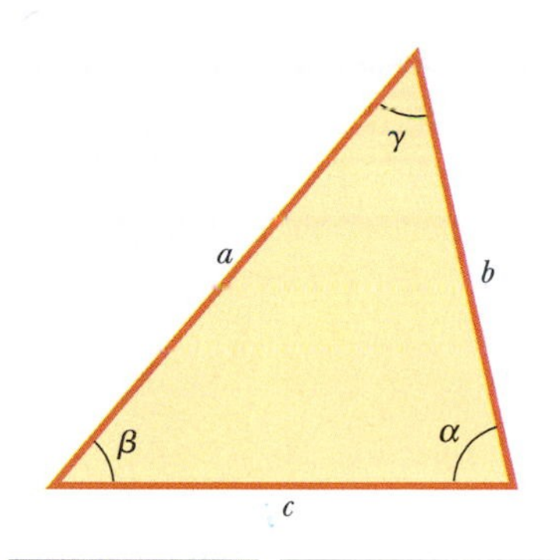

FIGURE B.10 An arbitrary, non-right triangle.

**EXAMPLE B.3**

Consider the right triangle in Figure B.11 in which $a = 2.00$, $b = 5.00$, and $c$ is unknown. From the Pythagorean theorem we have

$$c^2 = a^2 + b^2 = 2.00^2 + 5.00^2 = 4.00 + 25.0 = 29.0$$

$$c = \sqrt{29.0} = 5.39$$

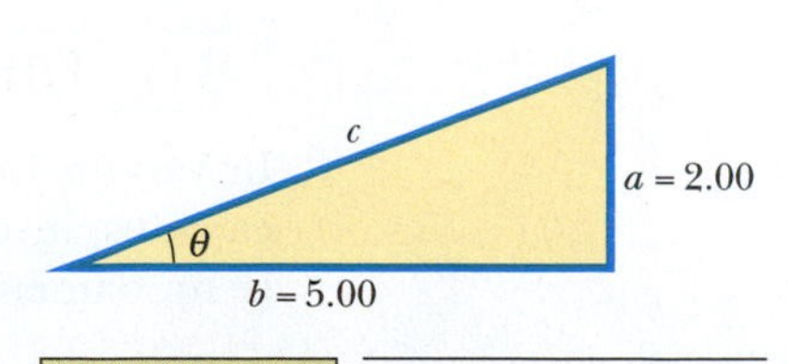

FIGURE B.11 (Example B.3)

To find the angle $\theta$, note that

$$\tan\theta = \frac{a}{b} = \frac{2.00}{5.00} = 0.400$$

Using a calculator, we find that

$$\theta = \tan^{-1}(0.400) = 21.8°$$

where $\tan^{-1}(0.400)$ is the notation for "angle whose tangent is 0.400," sometimes written as arctan(0.400).

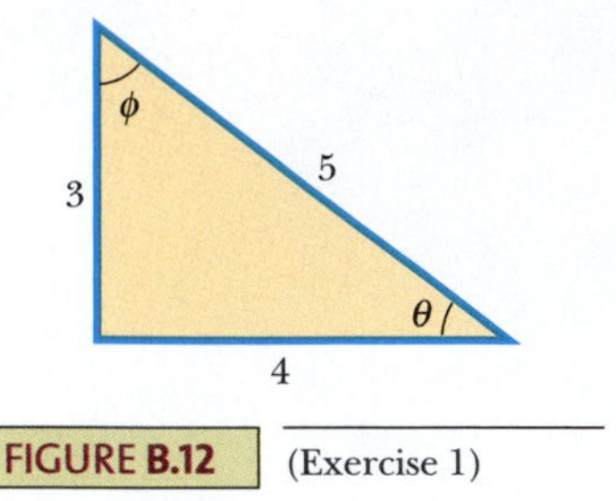


FIGURE B.12 (Exercise 1)

## Exercises

1. In Figure B.12, identify (a) the side opposite $\theta$ and (b) the side adjacent to $\phi$ and then find (c) $\cos\theta$, (d) $\sin\phi$, and (e) $\tan\phi$.

**Answers** (a) 3 (b) 3 (c) $\frac{4}{5}$ (d) $\frac{4}{5}$ (e) $\frac{4}{3}$

2. In a certain right triangle, the two sides that are perpendicular to each other are 5.00 m and 7.00 m long. What is the length of the third side?

**Answer** 8.60 m

3. A right triangle has a hypotenuse of length 3.0 m, and one of its angles is 30°. (a) What is the length of the side opposite the 30° angle? (b) What is the length of the side adjacent to the 30° angle?

**Answers** (a) 1.5 m (b) 2.6 m

## B.5 SERIES EXPANSIONS

$$(a+b)^n = a^n + \frac{n}{1!}a^{n-1}b + \frac{n(n-1)}{2!}a^{n-2}b^2 + \cdots$$

$$(1+x)^n = 1 + nx + \frac{n(n-1)}{2!}x^2 + \cdots$$

$$e^x = 1 + x + \frac{x^2}{2!} + \frac{x^3}{3!} + \cdots$$

$$\ln(1 \pm x) = \pm x - \tfrac{1}{2}x^2 \pm \tfrac{1}{3}x^3 - \cdots$$

$$\left.\begin{aligned} \sin x &= x - \frac{x^3}{3!} + \frac{x^5}{5!} - \cdots \\ \cos x &= 1 - \frac{x^2}{2!} + \frac{x^4}{4!} - \cdots \\ \tan x &= x + \frac{x^3}{3} + \frac{2x^5}{15} + \cdots \qquad |x| < \frac{\pi}{2} \end{aligned}\right\} x \text{ in radians}$$

For $x \ll 1$, the following approximations can be used:[1]

$$(1+x)^n \approx 1 + nx \qquad \sin x \approx x$$

$$e^x \approx 1 + x \qquad \cos x \approx 1$$

$$\ln(1 \pm x) \approx \pm x \qquad \tan x \approx x$$

## B.6 DIFFERENTIAL CALCULUS

In various branches of science, it is sometimes necessary to use the basic tools of calculus, invented by Newton, to describe physical phenomena. The use of calculus is fundamental in the treatment of various problems in Newtonian mechanics,

[1]The approximations for the functions $\sin x$, $\cos x$, and $\tan x$ are for $x \leq 0.1$ rad.

electricity, and magnetism. In this section, we simply state some basic properties and "rules of thumb" that should be a useful review to the student.

First, a **function** must be specified that relates one variable to another (e.g., a coordinate as a function of time). Suppose one of the variables is called $y$ (the dependent variable) and the other $x$ (the independent variable). We might have a function relationship such as

$$y(x) = ax^3 + bx^2 + cx + d$$

If $a$, $b$, $c$, and $d$ are specified constants, $y$ can be calculated for any value of $x$. We usually deal with continuous functions, that is, those for which $y$ varies "smoothly" with $x$.

The **derivative** of $y$ with respect to $x$ is defined as the limit, as $\Delta x$ approaches zero, of the slopes of chords drawn between two points on the $y$ versus $x$ curve. Mathematically, we write this definition as

$$\frac{dy}{dx} = \lim_{\Delta x \to 0} \frac{\Delta y}{\Delta x} = \lim_{\Delta x \to 0} \frac{y(x + \Delta x) - y(x)}{\Delta x} \quad \text{[B.28]}$$

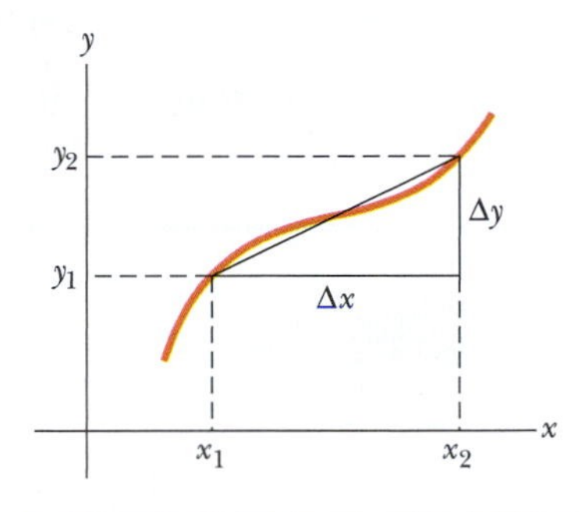


**FIGURE B.13** The lengths $\Delta x$ and $\Delta y$ are used to define the derivative of this function at a point.

where $\Delta y$ and $\Delta x$ are defined as $\Delta x = x_2 - x_1$ and $\Delta y = y_2 - y_1$ (Fig. B.13). It is important to note that $dy/dx$ *does not* mean $dy$ divided by $dx$, but rather is simply a notation of the limiting process of the derivative as defined by Equation B.28.

A useful expression to remember when $y(x) = ax^n$, where $a$ is a *constant* and $n$ is *any* positive or negative number (integer or fraction), is

$$\frac{dy}{dx} = nax^{n-1} \quad \text{[B.29]}$$

If $y(x)$ is a polynomial or algebraic function of $x$, we apply Equation B.29 to *each* term in the polynomial and take $d[\text{constant}]/dx = 0$. In Examples 4 through 7, we evaluate the derivatives of several functions.

## Special Properties of the Derivative

**A. Derivative of the product of two functions** If a function $f(x)$ is given by the product of two functions—say, $g(x)$ and $h(x)$—the derivative of $f(x)$ is defined as

$$\frac{d}{dx} f(x) = \frac{d}{dx}[g(x)h(x)] = g\frac{dh}{dx} + h\frac{dg}{dx} \quad \text{[B.30]}$$

**B. Derivative of the sum of two functions** If a function $f(x)$ is equal to the sum of two functions, the derivative of the sum is equal to the sum of the derivatives:

$$\frac{d}{dx} f(x) = \frac{d}{dx}[g(x) + h(x)] = \frac{dg}{dx} + \frac{dh}{dx} \quad \text{[B.31]}$$

**C. Chain rule of differential calculus** If $y = f(x)$ and $x = g(z)$, then $dy/dz$ can be written as the product of two derivatives:

$$\frac{dy}{dz} = \frac{dy}{dx}\frac{dx}{dz} \quad \text{[B.32]}$$

**D. The second derivative** The second derivative of $y$ with respect to $x$ is defined as the derivative of the function $dy/dx$ (the derivative of the derivative). It is usually written as

$$\frac{d^2y}{dx^2} = \frac{d}{dx}\left(\frac{dy}{dx}\right) \quad \text{[B.33]}$$

**EXAMPLE B.4**

Suppose $y(x)$ (that is, $y$ as a function of $x$) is given by

$$y(x) = ax^3 + bx + c$$

where $a$ and $b$ are constants. Then it follows that

$$y(x + \Delta x) = a(x + \Delta x)^3 + b(x + \Delta x) + c$$
$$= a(x^3 + 3x^2 \Delta x + 3x \Delta x^2 + \Delta x^3) + b(x + \Delta x) + c$$

so

$$\Delta y = y(x + \Delta x) - y(x)$$
$$= a(3x^2 \Delta x + 3x \Delta x^2 + \Delta x^3) + b \Delta x$$

Substituting this equation into Equation B.28 gives

$$\frac{dy}{dx} = \lim_{\Delta x \to 0} \frac{\Delta y}{\Delta x} = \lim_{\Delta x \to 0} a[3x^2 + 3x \Delta x + \Delta x^2] + b$$

$$\frac{dy}{dx} = 3ax^2 + b$$

**EXAMPLE B.5**

Find the derivative of

$$y(x) = 8x^5 + 4x^3 + 2x + 7$$

**Solution** By applying Equation B.29 to each term independently and remembering that $d/dx$ (constant) $= 0$, we have

$$\frac{dy}{dx} = 8(5)x^4 + 4(3)x^2 + 2(1)x^0 + 0$$

$$\frac{dy}{dx} = 40x^4 + 12x^2 + 2$$

**EXAMPLE B.6**

Find the derivative of $y(x) = x^3/(x + 1)^2$ with respect to $x$.

**Solution** We can rewrite this function as $y(x) = x^3(x + 1)^{-2}$ and apply Equation B.30:

$$\frac{dy}{dx} = (x + 1)^{-2} \frac{d}{dx}(x^3) + x^3 \frac{d}{dx}(x + 1)^{-2}$$
$$= (x + 1)^{-2}\, 3x^2 + x^3(-2)(x + 1)^{-3}$$

$$\frac{dy}{dx} = \frac{3x^2}{(x + 1)^2} - \frac{2x^3}{(x + 1)^3}$$

**EXAMPLE B.7**

A useful formula that follows from Equation B.30 is the derivative of the quotient of two functions. Show that

$$\frac{d}{dx}\left[\frac{g(x)}{h(x)}\right] = \frac{h \dfrac{dg}{dx} - g \dfrac{dh}{dx}}{h^2}$$

**Solution** We can write the quotient as $gh^{-1}$ and then apply Equations B.29 and B.30:

$$\frac{d}{dx}\left(\frac{g}{h}\right) = \frac{d}{dx}(gh^{-1}) = g\frac{d}{dx}(h^{-1}) + h^{-1}\frac{d}{dx}(g)$$
$$= -gh^{-2}\frac{dh}{dx} + h^{-1}\frac{dg}{dx}$$
$$= \frac{h \dfrac{dg}{dx} - g \dfrac{dh}{dx}}{h^2}$$

Some of the more commonly used derivatives of functions are listed in Table B.4.

## B.7 INTEGRAL CALCULUS

We think of integration as the inverse of differentiation. As an example, consider the expression

$$f(x) = \frac{dy}{dx} = 3ax^2 + b \qquad [B.34]$$

which was the result of differentiating the function

$$y(x) = ax^3 + bx + c$$

in Example 4. We can write Equation B.34 as $dy = f(x)\,dx = (3ax^2 + b)\,dx$ and obtain $y(x)$ by "summing" over all values of $x$. Mathematically, we write this inverse operation

$$y(x) = \int f(x)\,dx$$

For the function $f(x)$ given by Equation B.34, we have

$$y(x) = \int (3ax^2 + b)\,dx = ax^3 + bx + c$$

where $c$ is a constant of the integration. This type of integral is called an *indefinite integral* because its value depends on the choice of $c$.

A general **indefinite integral** $I(x)$ is defined as

$$I(x) = \int f(x)\,dx \qquad \text{[B.35]}$$

where $f(x)$ is called the *integrand* and $f(x) = dI(x)/dx$.

For a *general continuous* function $f(x)$, the integral can be described as the area under the curve bounded by $f(x)$ and the $x$ axis, between two specified values of $x$, say, $x_1$ and $x_2$, as in Figure B.14.

The area of the blue element in Figure B.14 is approximately $f(x_i)\,\Delta x_i$. If we sum all these area elements between $x_1$ and $x_2$ and take the limit of this sum as $\Delta x_i \rightarrow 0$, we obtain the *true* area under the curve bounded by $f(x)$ and the $x$ axis, between the limits $x_1$ and $x_2$:

$$\text{Area} = \lim_{\Delta x_i \rightarrow 0} \sum_i f(x_i)\Delta x_i = \int_{x_1}^{x_2} f(x)\,dx \qquad \text{[B.36]}$$

Integrals of the type defined by Equation B.36 are called **definite integrals.**

One common integral that arises in practical situations has the form

$$\int x^n\,dx = \frac{x^{n+1}}{n+1} + c \qquad (n \neq -1) \qquad \text{[B.37]}$$

This result is obvious because differentiation of the right-hand side with respect to $x$ gives $f(x) = x^n$ directly. If the limits of the integration are known, this integral becomes a *definite integral* and is written

$$\int_{x_1}^{x_2} x^n\,dx = \frac{x^{n+1}}{n+1}\bigg|_{x_1}^{x_2} = \frac{x_2^{\,n+1} - x_1^{\,n+1}}{n+1} \qquad (n \neq -1) \qquad \text{[B.38]}$$

**TABLE B.4** Derivatives for Several Functions

| |
|---|
| $\dfrac{d}{dx}(a) = 0$ |
| $\dfrac{d}{dx}(ax^n) = nax^{n-1}$ |
| $\dfrac{d}{dx}(e^{ax}) = ae^{ax}$ |
| $\dfrac{d}{dx}(\sin ax) = a \cos ax$ |
| $\dfrac{d}{dx}(\cos ax) = -a \sin ax$ |
| $\dfrac{d}{dx}(\tan ax) = a \sec^2 ax$ |
| $\dfrac{d}{dx}(\cot ax) = -a \csc^2 ax$ |
| $\dfrac{d}{dx}(\sec x) = \tan x \sec x$ |
| $\dfrac{d}{dx}(\csc x) = -\cot x \csc x$ |
| $\dfrac{d}{dx}(\ln ax) = \dfrac{1}{x}$ |

*Note:* The letters $a$ and $n$ are constants.

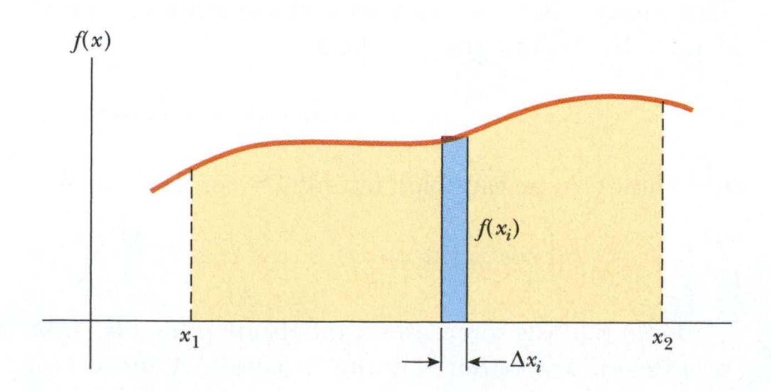


**FIGURE B.14** The definite integral of a function is the area under the curve of the function between the limits $x_1$ and $x_2$.

EXAMPLES

1. $\int_0^a x^2\,dx = \frac{x^3}{3}\Big]_0^a = \frac{a^3}{3}$

2. $\int_0^b x^{3/2}\,dx = \frac{x^{5/2}}{5/2}\Big]_0^b = \frac{2}{5}b^{5/2}$

3. $\int_3^5 x\,dx = \frac{x^2}{2}\Big]_3^5 = \frac{5^2 - 3^2}{2} = 8$

## Partial Integration

Sometimes it is useful to apply the method of *partial integration* (also called "integrating by parts") to evaluate certain integrals. The method uses the property that

$$\int u\,dv = uv - \int v\,du \qquad \text{[B.39]}$$

where $u$ and $v$ are *carefully* chosen so as to reduce a complex integral to a simpler one. In many cases, several reductions have to be made. Consider the function

$$I(x) = \int x^2 e^x\,dx$$

which can be evaluated by integrating by parts twice. First, if we choose $u = x^2$, $v = e^x$, we obtain

$$\int x^2 e^x\,dx = \int x^2\,d(e^x) = x^2 e^x - 2\int e^x x\,dx + c_1$$

Now, in the second term, choose $u = x$, $v = e^x$, which gives

$$\int x^2 e^x\,dx = x^2 e^x - 2xe^x + 2\int e^x\,dx + c_1$$

or

$$\int x^2 e^x\,dx = x^2 e^x - 2xe^x + 2e^x + c_2$$

## The Perfect Differential

Another useful method to remember is the use of the *perfect differential,* in which we look for a change of variable such that the differential of the function is the differential of the independent variable appearing in the integrand. For example, consider the integral

$$I(x) = \int \cos^2 x \sin x\,dx$$

This integral becomes easy to evaluate if we rewrite the differential as $d(\cos x) = -\sin x\,dx$. The integral then becomes

$$\int \cos^2 x \sin x\,dx = -\int \cos^2 x\,d(\cos x)$$

If we now change variables, letting $y = \cos x$, we obtain

$$\int \cos^2 x \sin x\,dx = -\int y^2 dy = -\frac{y^3}{3} + c = -\frac{\cos^3 x}{3} + c$$

Table B.5 lists some useful indefinite integrals. Table B.6 gives Gauss's probability integral and other definite integrals. A more complete list can be found in

**TABLE B.5 Some Indefinite Integrals (An arbitrary constant should be added to each of these integrals.)**

$$\int x^n\,dx = \frac{x^{n+1}}{n+1} \quad (\text{provided } n \neq -1)$$

$$\int \frac{dx}{x} = \int x^{-1}dx = \ln x$$

$$\int \frac{dx}{a+bx} = \frac{1}{b}\ln(a+bx)$$

$$\int \frac{x\,dx}{a+bx} = \frac{x}{b} - \frac{a}{b^2}\ln(a+bx)$$

$$\int \frac{dx}{x(x+a)} = -\frac{1}{a}\ln\frac{x+a}{x}$$

$$\int \frac{dx}{(a+bx)^2} = -\frac{1}{b(a+bx)}$$

$$\int \frac{dx}{a^2+x^2} = \frac{1}{a}\tan^{-1}\frac{x}{a}$$

$$\int \frac{dx}{a^2-x^2} = \frac{1}{2a}\ln\frac{a+x}{a-x} \quad (a^2 - x^2 > 0)$$

$$\int \frac{dx}{x^2-a^2} = \frac{1}{2a}\ln\frac{x-a}{x+a} \quad (x^2 - a^2 > 0)$$

$$\int \frac{x\,dx}{a^2 \pm x^2} = \pm\tfrac{1}{2}\ln(a^2 \pm x^2)$$

$$\int \frac{dx}{\sqrt{a^2-x^2}} = \sin^{-1}\frac{x}{a} = -\cos^{-1}\frac{x}{a} \quad (a^2 - x^2 > 0)$$

$$\int \frac{dx}{\sqrt{x^2 \pm a^2}} = \ln\left(x + \sqrt{x^2 \pm a^2}\right)$$

$$\int \frac{x\,dx}{\sqrt{a^2-x^2}} = -\sqrt{a^2-x^2}$$

$$\int \frac{x\,dx}{\sqrt{x^2 \pm a^2}} = \sqrt{x^2 \pm a^2}$$

$$\int \sqrt{a^2-x^2}\,dx = \tfrac{1}{2}\left(x\sqrt{a^2-x^2} + a^2\sin^{-1}\frac{x}{a}\right)$$

$$\int x\sqrt{a^2-x^2}\,dx = -\tfrac{1}{3}(a^2-x^2)^{3/2}$$

$$\int \sqrt{x^2 \pm a^2}\,dx = \tfrac{1}{2}\left[x\sqrt{x^2 \pm a^2} \pm a^2\ln\left(x + \sqrt{x^2 \pm a^2}\right)\right]$$

$$\int x\left(\sqrt{x^2 \pm a^2}\right)dx = \tfrac{1}{3}(x^2 \pm a^2)^{3/2}$$

$$\int e^{ax}\,dx = \frac{1}{a}e^{ax}$$

$$\int \ln ax\,dx = (x\ln ax) - x$$

$$\int xe^{ax}\,dx = \frac{e^{ax}}{a^2}(ax - 1)$$

$$\int \frac{dx}{a+be^{cx}} = \frac{x}{a} - \frac{1}{ac}\ln(a+be^{cx})$$

$$\int \sin ax\,dx = -\frac{1}{a}\cos ax$$

$$\int \cos ax\,dx = \frac{1}{a}\sin ax$$

$$\int \tan ax\,dx = -\frac{1}{a}\ln(\cos ax) = \frac{1}{a}\ln(\sec ax)$$

$$\int \cot ax\,dx = \frac{1}{a}\ln(\sin ax)$$

$$\int \sec ax\,dx = \frac{1}{a}\ln(\sec ax + \tan ax) = \frac{1}{a}\ln\left[\tan\left(\frac{ax}{2} + \frac{\pi}{4}\right)\right]$$

$$\int \csc ax\,dx = \frac{1}{a}\ln(\csc ax - \cot ax) = \frac{1}{a}\ln\left(\tan\frac{ax}{2}\right)$$

$$\int \sin^2 ax\,dx = \frac{x}{2} - \frac{\sin 2ax}{4a}$$

$$\int \cos^2 ax\,dx = \frac{x}{2} + \frac{\sin 2ax}{4a}$$

$$\int \frac{dx}{\sin^2 ax} = -\frac{1}{a}\cot ax$$

$$\int \frac{dx}{\cos^2 ax} = \frac{1}{a}\tan ax$$

$$\int \tan^2 ax\,dx = \frac{1}{a}(\tan ax) - x$$

$$\int \cot^2 ax\,dx = -\frac{1}{a}(\cot ax) - x$$

$$\int \sin^{-1} ax\,dx = x(\sin^{-1} ax) + \frac{\sqrt{1-a^2x^2}}{a}$$

$$\int \cos^{-1} ax\,dx = x(\cos^{-1} ax) - \frac{\sqrt{1-a^2x^2}}{a}$$

$$\int \frac{dx}{(x^2+a^2)^{3/2}} = \frac{x}{a^2\sqrt{x^2+a^2}}$$

$$\int \frac{x\,dx}{(x^2+a^2)^{3/2}} = -\frac{1}{\sqrt{x^2+a^2}}$$

various handbooks, such as *The Handbook of Chemistry and Physics* (Boca Raton, FL: CRC Press, published annually).

## B.8 PROPAGATION OF UNCERTAINTY

In laboratory experiments, a common activity is to take measurements that act as raw data. These measurements are of several types—length, time interval, temperature, voltage, and so on—and are taken by a variety of instruments. Regardless of the measurement and the quality of the instrumentation, **there is always uncertainty associated with a physical measurement.** This uncertainty is a combination of that associated with the instrument and that related to the system being measured.

**TABLE B.6 Gauss's Probability Integral and Other Definite Integrals**

$$\int_0^\infty x^n e^{-ax}\,dx = \frac{n!}{a^{n+1}}$$

$$I_0 = \int_0^\infty e^{-ax^2}\,dx = \frac{1}{2}\sqrt{\frac{\pi}{a}} \qquad \text{(Gauss's probability integral)}$$

$$I_1 = \int_0^\infty xe^{-ax^2}\,dx = \frac{1}{2a}$$

$$I_2 = \int_0^\infty x^2e^{-ax^2}\,dx = -\frac{dI_0}{da} = \frac{1}{4}\sqrt{\frac{\pi}{a^3}}$$

$$I_3 = \int_0^\infty x^3e^{-ax^2}\,dx = -\frac{dI_1}{da} = \frac{1}{2a^2}$$

$$I_4 = \int_0^\infty x^4e^{-ax^2}\,dx = \frac{d^2I_0}{da^2} = \frac{3}{8}\sqrt{\frac{\pi}{a^5}}$$

$$I_5 = \int_0^\infty x^5e^{-ax^2}\,dx = \frac{d^2I_1}{da^2} = \frac{1}{a^3}$$

$$\vdots$$

$$I_{2n} = (-1)^n\frac{d^n}{da^n}I_0$$

$$I_{2n+1} = (-1)^n\frac{d^n}{da^n}I_1$$

An example of the former is the inability to determine exactly the position of a length measurement between the lines on a meter stick. An example of uncertainty related to the system being measured is the variation of temperature within a sample of water so that a single temperature for the sample is difficult to determine.

Uncertainties can be expressed in two ways. **Absolute uncertainty** refers to an uncertainty expressed in the same units as the measurement. Thus, the length of a computer disk label might be expressed as (5.5 ± 0.1) cm. The uncertainty of ± 0.1 cm by itself is not descriptive enough for some purposes, however. This uncertainty is large if the measurement is 1.0 cm, but it is small if the measurement is 100 m. To give a more descriptive account of the uncertainty, **fractional uncertainty** or **percent uncertainty** is used. In this type of description, the uncertainty is divided by the actual measurement. Therefore, the length of the computer disk label could be expressed as

$$\ell = 5.5\,\text{cm} \pm \frac{0.1\,\text{cm}}{5.5\,\text{cm}} = 5.5\,\text{cm} \pm 0.018 \qquad \text{(fractional uncertainty)}$$

or as

$$\ell = 5.5\,\text{cm} \pm 1.8\% \qquad \text{(percent uncertainty)}$$

When combining measurements in a calculation, the percent uncertainty in the final result is generally larger than the uncertainty in the individual measurements. This **propagation of uncertainty** is one of the challenges of experimental physics.

Some simple rules can provide a reasonable estimate of the uncertainty in a calculated result.

**Multiplication and division:** When measurements with uncertainties are multiplied or divided, add the *percent uncertainties* to obtain the percent uncertainty in the result.

Example: The Area of a Rectangular Plate

$$A = \ell w = (5.5 \text{ cm} \pm 1.8\%) \times (6.4 \text{ cm} \pm 1.6\%) = 35 \text{ cm}^2 \pm 3.4\%$$
$$= (35 \pm 1) \text{ cm}^2$$

**Addition and subtraction:** When measurements with uncertainties are added or subtracted, add the *absolute uncertainties* to obtain the absolute uncertainty in the result.

Example: A Change in Temperature

$$\Delta T = T_2 - T_1 = (99.2 \pm 1.5)°\text{C} - (27.6 \pm 1.5)°\text{C} = (71.6 \pm 3.0)°\text{C}$$
$$= 71.6°\text{C} \pm 4.2\%$$

**Powers:** If a measurement is taken to a power, the percent uncertainty is multiplied by that power to obtain the percent uncertainty in the result.

Example: The Volume of a Sphere

$$V = \tfrac{4}{3}\pi r^3 = \tfrac{4}{3}\pi(6.20 \text{ cm} \pm 2.0\%)^3 = 998 \text{ cm}^3 \pm 6.0\%$$
$$= (998 \pm 60) \text{ cm}^3$$

For complicated calculations, many uncertainties are added together. This can cause the uncertainty in the final result to be undesirably large. Experiments should be designed such that calculations are as simple as possible.

Notice that uncertainties in a calculation always add. As a result, an experiment involving a subtraction should be avoided if possible, especially if the measurements being subtracted are close together. The result of such a calculation is a small difference in the measurements and uncertainties that add together. It is possible that the uncertainty in the result could be larger than the result itself!

# Periodic Table of the Elements

Symbol — **Ca** 20 — Atomic number
Atomic mass † — 40.078
$4s^2$ — Electron configuration

| Group I | Group II | Transition elements | | | | | | |
|---|---|---|---|---|---|---|---|---|
| **H** 1<br>1.007 9<br>$1s$ | | | | | | | | |
| **Li** 3<br>6.941<br>$2s^1$ | **Be** 4<br>9.0122<br>$2s^2$ | | | | | | | |
| **Na** 11<br>22.990<br>$3s^1$ | **Mg** 12<br>24.305<br>$3s^2$ | | | | | | | |
| **K** 19<br>39.098<br>$4s^1$ | **Ca** 20<br>40.078<br>$4s^2$ | **Sc** 21<br>44.956<br>$3d^14s^2$ | **Ti** 22<br>47.867<br>$3d^24s^2$ | **V** 23<br>50.942<br>$3d^34s^2$ | **Cr** 24<br>51.996<br>$3d^54s^1$ | **Mn** 25<br>54.938<br>$3d^54s^2$ | **Fe** 26<br>55.845<br>$3d^64s^2$ | **Co** 27<br>58.933<br>$3d^74s^2$ |
| **Rb** 37<br>85.468<br>$5s^1$ | **Sr** 38<br>87.62<br>$5s^2$ | **Y** 39<br>88.906<br>$4d^15s^2$ | **Zr** 40<br>91.224<br>$4d^25s^2$ | **Nb** 41<br>92.906<br>$4d^45s^1$ | **Mo** 42<br>95.94<br>$4d^55s^1$ | **Tc** 43<br>(98)<br>$4d^55s^2$ | **Ru** 44<br>101.07<br>$4d^75s^1$ | **Rh** 45<br>102.91<br>$4d^85s^1$ |
| **Cs** 55<br>132.91<br>$6s^1$ | **Ba** 56<br>137.33<br>$6s^2$ | 57–71* | **Hf** 72<br>178.49<br>$5d^26s^2$ | **Ta** 73<br>180.95<br>$5d^36s^2$ | **W** 74<br>183.84<br>$5d^46s^2$ | **Re** 75<br>186.21<br>$5d^56s^2$ | **Os** 76<br>190.23<br>$5d^66s^2$ | **Ir** 77<br>192.2<br>$5d^76s^2$ |
| **Fr** 87<br>(223)<br>$7s^1$ | **Ra** 88<br>(226)<br>$7s^2$ | 89–103** | **Rf** 104<br>(261)<br>$6d^27s^2$ | **Db** 105<br>(262)<br>$6d^37s^2$ | **Sg** 106<br>(266) | **Bh** 107<br>(264) | **Hs** 108<br>(269) | **Mt** 109<br>(268) |

| | | | | | | |
|---|---|---|---|---|---|---|
| *Lanthanide series | **La** 57<br>138.91<br>$5d^16s^2$ | **Ce** 58<br>140.12<br>$5d^14f^16s^2$ | **Pr** 59<br>140.91<br>$4f^36s^2$ | **Nd** 60<br>144.24<br>$4f^46s^2$ | **Pm** 61<br>(145)<br>$4f^56s^2$ | **Sm** 62<br>150.36<br>$4f^66s^2$ |
| **Actinide series | **Ac** 89<br>(227)<br>$6d^17s^2$ | **Th** 90<br>232.04<br>$6d^27s^2$ | **Pa** 91<br>231.04<br>$5f^26d^17s^2$ | **U** 92<br>238.03<br>$5f^36d^17s^2$ | **Np** 93<br>(237)<br>$5f^46d^17s^2$ | **Pu** 94<br>(244)<br>$5f^66d^07s^2$ |

*Note:* Atomic mass values given are averaged over isotopes in the percentages in which they exist in nature.
†For an unstable element, mass number of the most stable known isotope is given in parentheses.
††Elements 111, 112, and 114 have not yet been named
†††For a description of the atomic data, visit **physics.nist.gov/atomic**.

| | | | Group III | Group IV | Group V | Group VI | Group VII | Group 0 |
|---|---|---|---|---|---|---|---|---|
| | | | | | | | H 1<br>1.007 9<br>$1s^1$ | He 2<br>4.002 6<br>$1s^2$ |
| | | | B 5<br>10.811<br>$2p^1$ | C 6<br>12.011<br>$2p^2$ | N 7<br>14.007<br>$2p^3$ | O 8<br>15.999<br>$2p^4$ | F 9<br>18.998<br>$2p^5$ | Ne 10<br>20.180<br>$2p^6$ |
| | | | Al 13<br>26.982<br>$3p^1$ | Si 14<br>28.086<br>$3p^2$ | P 15<br>30.974<br>$3p^3$ | S 16<br>32.066<br>$3p^4$ | Cl 17<br>35.453<br>$3p^5$ | Ar 18<br>39.948<br>$3p^6$ |
| Ni 28<br>58.693<br>$3d^84s^2$ | Cu 29<br>63.546<br>$3d^{10}4s^1$ | Zn 30<br>65.39<br>$3d^{10}4s^2$ | Ga 31<br>69.723<br>$4p^1$ | Ge 32<br>72.61<br>$4p^2$ | As 33<br>74.922<br>$4p^3$ | Se 34<br>78.96<br>$4p^4$ | Br 35<br>79.904<br>$4p^5$ | Kr 36<br>83.80<br>$4p^6$ |
| Pd 46<br>106.42<br>$4d^{10}$ | Ag 47<br>107.87<br>$4d^{10}5s^1$ | Cd 48<br>112.41<br>$4d^{10}5s^2$ | In 49<br>114.82<br>$5p^1$ | Sn 50<br>118.71<br>$5p^2$ | Sb 51<br>121.76<br>$5p^3$ | Te 52<br>127.60<br>$5p^4$ | I 53<br>126.90<br>$5p^5$ | Xc 54<br>131.29<br>$5p^6$ |
| Pt 78<br>195.08<br>$5d^96s^1$ | Au 79<br>196.97<br>$5d^{10}6s^1$ | Hg 80<br>200.59<br>$5d^{10}6s^2$ | Tl 81<br>204.38<br>$6p^1$ | Pb 82<br>207.2<br>$6p^2$ | Bi 83<br>208.98<br>$6p^3$ | Po 84<br>(209)<br>$6p^4$ | At 85<br>(210)<br>$6p^5$ | Rn 86<br>(222)<br>$6p^6$ |
| Ds 110<br>(271) | 111†† <br>(272) | 112††<br>(285) | | 114††<br>(289) | | | | |

| | | | | | | | | |
|---|---|---|---|---|---|---|---|---|
| Eu 63<br>151.96<br>$4f^76s^2$ | Gd 64<br>157.25<br>$5d^14f^76s^2$ | Tb 65<br>158.93<br>$5d^14f^86s^2$ | Dy 66<br>162.50<br>$4f^{10}6s^2$ | Ho 67<br>164.93<br>$4f^{11}6s^2$ | Er 68<br>167.26<br>$4f^{12}6s^2$ | Tm 69<br>168.93<br>$4f^{13}6s^2$ | Yb 70<br>173.04<br>$4f^{14}6s^2$ | Lu 71<br>174.97<br>$5d^14f^{14}6s^2$ |
| Am 95<br>(243)<br>$5f^76d^07s^2$ | Cm 96<br>(247)<br>$5f^76d^17s^2$ | Bk 97<br>(247)<br>$5f^86d^17s^2$ | Cf 98<br>(251)<br>$5f^{10}6d^07s^2$ | Es 99<br>(252)<br>$5f^{11}6d^07s^2$ | Fm 100<br>(257)<br>$5f^{12}6d^07s^2$ | Md 101<br>(258)<br>$5f^{13}6d^07s^2$ | No 102<br>(259)<br>$6d^07s^2$ | Lr 103<br>(262)<br>$6d^17s^2$ |

APPENDIX D

# SI Units

**TABLE D.1** SI Base Units

| Base Quantity | SI Base Unit Name | Symbol |
|---|---|---|
| Length | meter | m |
| Mass | kilogram | kg |
| Time | second | s |
| Electric current | ampere | A |
| Temperature | kelvin | K |
| Amount of substance | mole | mol |
| Luminous intensity | candela | cd |

**TABLE D.2** Some Derived SI Units

| Quantity | Name | Symbol | Expression in Terms of Base Units | Expression in Terms of Other SI Units |
|---|---|---|---|---|
| Plane angle | radian | rad | m/m | |
| Frequency | hertz | Hz | $s^{-1}$ | |
| Force | newton | N | $kg \cdot m/s^2$ | J/m |
| Pressure | pascal | Pa | $kg/m \cdot s^2$ | $N/m^2$ |
| Energy; work | joule | J | $kg \cdot m^2/s^2$ | $N \cdot m$ |
| Power | watt | W | $kg \cdot m^2/s^3$ | J/s |
| Electric charge | coulomb | C | $A \cdot s$ | |
| Electric potential | volt | V | $kg \cdot m^2/A \cdot s^3$ | W/A |
| Capacitance | farad | F | $A^2 \cdot s^4/kg \cdot m^2$ | C/V |
| Electric resistance | ohm | w | $kg \cdot m^2/A^2 \cdot s^3$ | V/A |
| Magnetic flux | weber | Wb | $kg \cdot m^2/A \cdot s^2$ | $V \cdot s$ |
| Magnetic field | tesla | T | $kg/A \cdot s^2$ | |
| Inductance | henry | H | $kg \cdot m^2/A^2 \cdot s^2$ | $T \cdot m^2/A$ |

# Nobel Prizes

All Nobel Prizes in Physics are listed (and marked with a P), as well as relevant Nobel Prizes in Chemistry (C). The key dates for some of the scientific work are supplied; they often antedate the prize considerably.

**1901** (P) *Wilhelm Roentgen* for discovering x-rays (1895).

**1902** (P) *Hendrik A. Lorentz* for predicting the Zeeman effect and *Pieter Zeeman* for discovering the Zeeman effect, the splitting of spectral lines in magnetic fields.

**1903** (P) *Antoine-Henri Becquerel* for discovering radioactivity (1896) and *Pierre Curie* and *Marie Curie* for studying radioactivity.

**1904** (P) *Lord Rayleigh* for studying the density of gases and discovering argon.

(C) *William Ramsay* for discovering the inert gas elements helium, neon, xenon, and krypton, and placing them in the periodic table.

**1905** (P) *Philipp Lenard* for studying cathode rays, electrons (1898–1899).

**1906** (P) *J. J. Thomson* for studying electrical discharge through gases and discovering the electron (1897).

**1907** (P) *Albert A. Michelson* for inventing optical instruments and measuring the speed of light (1880s).

**1908** (P) *Gabriel Lippmann* for making the first color photographic plate, using interference methods (1891).

(C) *Ernest Rutherford* for discovering that atoms can be broken apart by alpha rays and for studying radioactivity.

**1909** (P) *Guglielmo Marconi* and *Carl Ferdinand Braun* for developing wireless telegraphy.

**1910** (P) *Johannes D. van der Waals* for studying the equation of state for gases and liquids (1881).

**1911** (P) *Wilhelm Wien* for discovering Wien's law giving the peak of a blackbody spectrum (1893).

(C) *Marie Curie* for discovering radium and polonium (1898) and isolating radium.

**1912** (P) *Nils Dalén* for inventing automatic gas regulators for lighthouses.

**1913** (P) *Heike Kamerlingh Onnes* for the discovery of superconductivity and liquefying helium (1908).

**1914** (P) *Max T. F. von Laue* for studying x-rays from their diffraction by crystals, showing that x-rays are electromagnetic waves (1912).

(C) *Theodore W. Richards* for determining the atomic weights of 60 elements, indicating the existence of isotopes.

**1915** (P) *William Henry Bragg* and *William Lawrence Bragg*, his son, for studying the diffraction of x-rays in crystals.

**1917** (P) *Charles Barkla* for studying atoms by x-ray scattering (1906).

**1918** (P) *Max Planck* for discovering energy quanta (1900).

**1919** (P) *Johannes Stark* for discovering the Stark effect, the splitting of spectral lines in electric fields (1913).

**1920** (P) *Charles-Édouard Guillaume* for discovering invar, a nickel–steel alloy with low coefficient of expansion.

(C) *Walther Nernst* for studying heat changes in chemical reactions and formulating the third law of thermodynamics (1918).

**1921** (P) *Albert Einstein* for explaining the photoelectric effect and for his services to theoretical physics (1905).

(C) *Frederick Soddy* for studying the chemistry of radioactive substances and discovering isotopes (1912).

**1922** (P) *Niels Bohr* for his model of the atom and its radiation (1913).

(C) *Francis W. Aston* for using the mass spectrograph to study atomic weights, thus discovering 212 of the 287 naturally occurring isotopes.

**1923** (P) *Robert A. Millikan* for measuring the charge on an electron (1911) and for studying the photoelectric effect experimentally (1914).

**1924** (P) *Karl M. G. Siegbahn* for his work in x-ray spectroscopy.

**1925** (P) *James Franck* and *Gustav Hertz* for discovering the Franck–Hertz effect in electron–atom collisions.

**1926** (P) *Jean-Baptiste Perrin* for studying Brownian motion to validate the discontinuous structure of matter and measure the size of atoms.

**1927** (P) *Arthur Holly Compton* for discovering the Compton effect on x-rays, their change in wavelength when they collide with matter (1922), and *Charles T. R. Wilson* for inventing the cloud chamber, used to study charged particles (1906).

**1928** (P) *Owen W. Richardson* for studying the thermionic effect and electrons emitted by hot metals (1911).

**1929** (P) *Louis Victor de Broglie* for discovering the wave nature of electrons (1923).

**1930** (P) *Chandrasekhara Venkata Raman* for studying Raman scattering, the scattering of light by atoms and molecules with a change in wavelength (1928).

**1932** (P) *Werner Heisenberg* for creating quantum mechanics (1925).

**1933** (P) *Erwin Schrödinger* and *Paul A. M. Dirac* for developing wave mechanics (1925) and relativistic quantum mechanics (1927).

(C) *Harold Urey* for discovering heavy hydrogen, deuterium (1931).

**1935** (P) *James Chadwick* for discovering the neutron (1932).

(C) *Irène Joliot-Curie* and *Frédéric Joliot-Curie* for synthesizing new radioactive elements.

**1936** (P) *Carl D. Anderson* for discovering the positron in particular and antimatter in general (1932) and *Victor F. Hess* for discovering cosmic rays.

(C) *Peter J. W. Debye* for studying dipole moments and diffraction of x-rays and electrons in gases.

**1937** (P) *Clinton Davisson* and *George Thomson* for discovering the diffraction of electrons by crystals, confirming de Broglie's hypothesis (1927).

**1938** (P) *Enrico Fermi* for producing the transuranic radioactive elements by neutron irradiation (1934–1937).

**1939** (P) *Ernest O. Lawrence* for inventing the cyclotron.

**1943** (P) *Otto Stern* for developing molecular-beam studies (1923) and using them to discover the magnetic moment of the proton (1933).

**1944** (P) *Isidor I. Rabi* for discovering nuclear magnetic resonance in atomic and molecular beams.

(C) *Otto Hahn* for discovering nuclear fission (1938).

**1945** (P) *Wolfgang Pauli* for discovering the exclusion principle (1924).

**1946** (P) *Percy W. Bridgman* for studying physics at high pressures.

**1947** (P) *Edward V. Appleton* for studying the ionosphere.

**1948** (P) *Patrick M. S. Blackett* for studying nuclear physics with cloud-chamber photographs of cosmic-ray interactions.

**1949** (P) *Hideki Yukawa* for predicting the existence of mesons (1935).

**1950** (P) *Cecil F. Powell* for developing the method of studying cosmic rays with photographic emulsions and discovering new mesons.

**1951** (P) *John D. Cockcroft* and *Ernest T. S. Walton* for transmuting nuclei in an accelerator (1932).

(C) *Edwin M. McMillan* for producing neptunium (1940) and *Glenn T. Seaborg* for producing plutonium (1941) and further transuranic elements.

**1952** (P) *Felix Bloch* and *Edward Mills Purcell* for discovering nuclear magnetic resonance in liquids and gases (1946).

**1953** (P) *Frits Zernike* for inventing the phase-contrast microscope, which uses interference to provide high contrast.

**1954** (P) *Max Born* for interpreting the wave function as a probability (1926) and other quantum-mechanical discoveries and *Walther Bothe* for developing the coincidence method to study subatomic particles (1930–1931), producing, in particular, the particle interpreted by Chadwick as the neutron.

**1955** (P) *Willis E. Lamb Jr.*, for discovering the Lamb shift in the hydrogen spectrum (1947) and *Polykarp Kusch* for determining the magnetic moment of the electron (1947).

**1956** (P) *John Bardeen, Walter H. Brattain,* and *William Shockley* for inventing the transistor (1956).

**1957** (P) *T.-D. Lee* and *C.-N. Yang* for predicting that parity is not conserved in beta decay (1956).

**1958** (P) *Pavel A. Čerenkov* for discovering Čerenkov radiation (1935) and *Ilya M. Frank* and *Igor Tamm* for interpreting it (1937).

**1959** (P) *Emilio G. Segrè* and *Owen Chamberlain* for discovering the antiproton (1955).

**1960** (P) *Donald A. Glaser* for inventing the bubble chamber to study elementary particles (1952).

(C) *Willard Libby* for developing radiocarbon dating (1947).

**1961** (P) *Robert Hofstadter* for discovering internal structure in protons and neutrons and *Rudolf L. Mössbauer* for discovering the Mössbauer effect of recoilless gamma-ray emission (1957).

**1962** (P) *Lev Davidovich Landau* for studying liquid helium and other condensed matter theoretically.

**1963** (P) *Eugene P. Wigner* for applying symmetry principles to elementary-particle theory and *Maria Goeppert Mayer* and *J. Hans D. Jensen* for studying the shell model of nuclei (1947).

**1964** (P) *Charles H. Townes, Nikolai G. Basov,* and *Alexandr M. Prokhorov* for developing masers (1951–1952) and lasers.

**1965** (P) *Sin-itiro Tomonaga, Julian S. Schwinger,* and *Richard P. Feynman* for developing quantum electrodynamics (1948).

**1966** (P) *Alfred Kastler* for his optical methods of studying atomic energy levels.

**1967** (P) *Hans Albrecht Bethe* for discovering the routes of energy production in stars (1939).

**1968** (P) *Luis W. Alvarez* for discovering resonance states of elementary particles.

**1969** (P) *Murray Gell-Mann* for classifying elementary particles (1963).

**1970** (P) *Hannes Alfvén* for developing magnetohydrodynamic theory and *Louis Eugène Félix Néel* for discovering antiferromagnetism and ferrimagnetism (1930s).

**1971** (P) *Dennis Gabor* for developing holography (1947).

(C) *Gerhard Herzberg* for studying the structure of molecules spectroscopically.

**1972** (P) *John Bardeen, Leon N. Cooper,* and *John Robert Schrieffer* for explaining superconductivity (1957).

**1973** (P) *Leo Esaki* for discovering tunneling in semiconductors, *Ivar Giaever* for discovering tunneling in superconductors, and *Brian D. Josephson* for predicting the Josephson effect, which involves tunneling of paired electrons (1958–1962).

**1974** (P) *Anthony Hewish* for discovering pulsars and *Martin Ryle* for developing radio interferometry.

**1975** (P) *Aage N. Bohr, Ben R. Mottelson,* and *James Rainwater* for discovering why some nuclei take asymmetric shapes.

**1976** (P) *Burton Richter* and *Samuel C. C. Ting* for discovering the J/psi particle, the first charmed particle (1974).

**1977** (P) *John H. Van Vleck, Nevill F. Mott,* and *Philip W. Anderson* for studying solids quantum-mechanically.

(C) *Ilya Prigogine* for extending thermodynamics to show how life could arise in the face of the second law.

**1978** (P) *Arno A. Penzias* and *Robert W. Wilson* for discovering the cosmic background radiation (1965) and *Pyotr Kapitsa* for his studies of liquid helium.

**1979** (P) *Sheldon L. Glashow, Abdus Salam,* and *Steven Weinberg* for developing the theory that unified the weak and electromagnetic forces (1958–1971).

**1980** (P) *Val Fitch* and *James W. Cronin* for discovering CP (charge-parity) violation (1964), which possibly explains the cosmological dominance of matter over antimatter.

**1981** (P) *Nicolaas Bloembergen* and *Arthur L. Schawlow* for developing laser spectroscopy and *Kai M. Siegbahn* for developing high-resolution electron spectroscopy (1958).

**1982** (P) *Kenneth G. Wilson* for developing a method of constructing theories of phase transitions to analyze critical phenomena.

**1983** (P) *William A. Fowler* for theoretical studies of astrophysical nucleosynthesis and *Subramanyan Chandrasekhar* for studying physical processes of importance to stellar structure and evolution, including the prediction of white dwarf stars (1930).

**1984** (P) *Carlo Rubbia* for discovering the W and Z particles verifying the electroweak unification, and *Simon van der Meer* for developing the method of stochastic cooling of the CERN beam that allowed the discovery (1982–1983).

**1985** (P) *Klaus von Klitzing* for the quantized Hall effect, relating to conductivity in the presence of a magnetic field (1980).

**1986** (P) *Ernst Ruska* for inventing the electron microscope (1931) and *Gerd Binnig* and *Heinrich Rohrer* for inventing the scanning-tunneling electron microscope (1981).

**1987** (P) *J. Georg Bednorz* and *Karl Alex Müller* for the discovery of high-temperature superconductivity (1986).

**1988** (P) *Leon M. Lederman, Melvin Schwartz,* and *Jack Steinberger* for a collaborative experiment that led to the development of a new tool for studying the weak nuclear force, which affects the radioactive decay of atoms.

**1989** (P) *Norman Ramsay* for various techniques in atomic physics and *Hans Dehmelt* and *Wolfgang Paul* for the development of techniques for trapping single-charge particles.

**1990** (P) *Jerome Friedman, Henry Kendall,* and *Richard Taylor* for experiments important to the development of the quark model.

**1991** (P) *Pierre-Gilles de Gennes* for discovering that methods developed for studying order phenomena in simple systems can be generalized to more complex forms of matter, in particular to liquid crystals and polymers.

**1992** (P) *George Charpak* for developing detectors that trace the paths of evanescent subatomic particles produced in particle accelerators.

**1993** (P) *Russell Hulse* and *Joseph Taylor* for discovering evidence of gravitational waves.

**1994** (P) *Bertram N. Brockhouse* and *Clifford G. Shull* for pioneering work in neutron scattering.

**1995** (P) *Martin L. Perl* and *Frederick Reines* for discovering the tau particle and the neutrino, respectively.

**1996** (P) *David M. Lee, Douglas C. Osheroff,* and *Robert C. Richardson* for developing a superfluid using helium-3.

**1997** (P) *Steven Chu, Claude Cohen-Tannoudji,* and *William D. Phillips* for developing methods to cool and trap atoms with laser light.

**1998** (P) *Robert B. Laughlin, Horst L. Störmer,* and *Daniel C. Tsui* for discovering a new form of quantum fluid with fractionally charged excitations.

**1999** (P) *Gerardus 'T Hooft* and *Martinus J. G. Veltman* for studies in the quantum structure of electroweak interactions in physics.

**2000** (P) *Zhores I. Alferov* and *Herbert Kroemer* for developing semiconductor heterostructures used in high-speed electronics and optoelectronics and *Jack St. Clair Kilby* for participating in the invention of the integrated circuit.

**2001** (P) *Eric A. Cornell, Wolfgang Ketterle,* and *Carl E. Wieman* for the achievement of Bose–Einstein condensation in dilute gases of alkali atoms.

**2002** (P) *Raymond Davis Jr.* and *Masatoshi Koshiba* for the detection of cosmic neutrinos and *Riccardo Giacconi* for contributions to astrophysics that led to the discovery of cosmic x-ray sources.

**2003** *Alexei A. Abrikosov, Vitaly L. Ginzburg,* and *Anthony J. Leggett* for pioneering contributions to the theory of superconductors and superfluids.

**2004** *David J. Gross, H. David Politzer,* and *Frank Wilczeck* for the discovery of asymptotic freedom in the theory of the strong interaction.

# Answers to Odd-Numbered Problems

## Chapter 1

1. $5.52 \times 10^3$ kg/m$^3$, between the density of aluminum and that of iron and greater than the densities of typical surface rocks
3. $4\pi\rho(r_2^3 - r_1^3)/3$
5. No
7. (b) only
9. (a) 0.071 4 gal/s (b) $2.70 \times 10^{-4}$ m$^3$/s (c) 1.03 h
11. 667 lb/s
13. 151 $\mu$m
15. 2.86 cm
17. (a) 2.07 mm (b) $8.62 \times 10^{13}$ times as large
19. $\sim 10^6$ balls
21. $\sim 10^2$ tuners
23. (a) 3 (b) 4 (c) 3 (d) 2
25. 31 556 926.0 s
27. 5.2 m$^3$, 3%
29. 108° and 288°
31. 3.46 or $-$ 3.46
33. (a) 2.24 m (b) 2.24 m at 26.6°
35. (a) $r$, $180° - \theta$ (b) $2r$, $180° + \theta$ (c) $3r$, $-\theta$
37. (a) 10.0 m (b) 15.7 m (c) 0
39. Approximately 420 ft at $-$ 3°
41. 47.2 units at 122°
43. 196 cm at 345°
45. (a) $2.00\hat{\mathbf{i}} - 6.00\hat{\mathbf{j}}$ (b) $4.00\hat{\mathbf{i}} + 2.00\hat{\mathbf{j}}$ (c) 6.32 (d) 4.47 (e) 288°; 26.6°
47. 240 m at 237°
49. (a) 10.4 cm (b) 35.5°
51. (a) $8.00\hat{\mathbf{i}} + 12.0\hat{\mathbf{j}} - 4.00\hat{\mathbf{k}}$ (b) $2.00\hat{\mathbf{i}} + 3.00\hat{\mathbf{j}} - 1.00\hat{\mathbf{k}}$ (c) $-24.0\hat{\mathbf{i}} - 36.0\hat{\mathbf{j}} + 12.0\hat{\mathbf{k}}$
53. (a) $49.5\hat{\mathbf{i}} + 27.1\hat{\mathbf{j}}$ (b) 56.4 units at 28.7°
55. 70.0 m
57. 0.141 nm
59. 4.50 m$^2$
61. 0.449%
63. (a) 0.529 cm/s (b) 11.5 cm/s
65. $\sim 10^{11}$ stars
67. (a) 185 N at 77.8° from the $+x$ axis (b) $(-39.3\hat{\mathbf{i}} - 181\hat{\mathbf{j}})$ N
69. (a) (10.0 m, 16.0 m)
71. (a) $\vec{\mathbf{R}}_1 = a\hat{\mathbf{i}} + b\hat{\mathbf{j}}$; $R_1 = \sqrt{a^2 + b^2}$ (b) $\vec{\mathbf{R}}_2 = a\hat{\mathbf{i}} + b\hat{\mathbf{j}} + c\hat{\mathbf{k}}$; $R_2 = \sqrt{a^2 + b^2 + c^2}$

## Chapter 2

1. (a) 2.30 m/s (b) 16.1 m/s (c) 11.5 m/s
3. (a) 5 m/s (b) 1.2 m/s (c) $-$ 2.5 m/s (d) $-$3.3 m/s (e) 0
5. (a) $-$ 2.4 m/s (b) $-$ 3.8 m/s (c) 4.0 s
7. (b) $v_{t\,=\,5.0\text{ s}} = 23$ m/s, $v_{t\,=\,4.0\text{ s}} = 18$ m/s, $v_{t\,=\,3.0\text{ s}} = 14$ m/s, $v_{t\,=\,2.0\text{ s}} = 9.0$ m/s (c) 4.6 m/s$^2$ (d) 0
9. 5.00 m
11. (a) 20.0 m/s, 5.00 m/s (b) 262 m
13. (a) 2.00 m (b) $-$ 3.00 m/s (c) $-$ 2.00 m/s$^2$
15. (a) 1.3 m/s$^2$ (b) 2.0 m/s$^2$ at 3 s (c) at $t = 6$ s and for $t > 10$ s (d) $-$ 1.5 m/s$^2$ at 8 s
17. (a) 6.61 m/s (b) $-$ 0.448 m/s$^2$
19. $-$ 16.0 cm/s$^2$
21. (a) 20.0 s (b) no
23. 3.10 m/s
25. (a) 35.0 s (b) 15.7 m/s
27. yes; 212 m, 11.4 s
29. (a) 29.4 m/s (b) 44.1 m
31. (a) 10.0 m/s up (b) 4.68 m/s down
33. (a) 7.82 m (b) 0.782 s
35. (b) 7.4 m/s$^2$ and 2.1 m/s$^2$ (c) 48 m and 170 m (d) 2.74 s
37. (a) 70.0 mi/h·s = 31.3 m/s$^2$ = 3.19$g$ (b) 321 ft = 97.8 m
39. (a) $-$ 202 m/s$^2$ (b) 198 m
41. $2.74 \times 10^5$ m/s$^2$, which is $2.79 \times 10^4$ $g$
43. (a) 3.00 m/s (b) 6.00 s (c) $-$ 0.300 m/s$^2$ (d) 2.05 m/s
45. 1.60 m/s$^2$
47. (a) 41.0 s (b) 1.73 km (c) $-$ 184 m/s
49. (a) 5.43 m/s$^2$ and 3.83 m/s$^2$ (b) 10.9 m/s and 11.5 m/s (c) Maggie by 2.62 m

**51.** (a) 3.00 s (b) − 15.3 m/s (c) 31.4 m/s down and 34.8 m/s down

**53.** (c) $v_{\text{boy}}^2/h$, 0 (d) $v_{\text{boy}}$, 0

**55.** (a) 26.4 m (b) 6.82%

**57.** $0.577v$

## Chapter 3

**1.** (a) 4.87 km at 209° from east (b) 23.3 m/s (c) 13.5 m/s at 209°

**3.** (a) $(0.800\hat{\mathbf{i}} - 0.300\hat{\mathbf{j}})\text{m/s}^2$ (b) 339° (c) $(360\hat{\mathbf{i}} - 72.7\hat{\mathbf{j}})$m, − 15.2°

**5.** (a) $\vec{\mathbf{r}} = (5.00t\,\hat{\mathbf{i}} + 1.50t^2\,\hat{\mathbf{j}})\text{m}$, $\vec{\mathbf{v}} = (5.00\hat{\mathbf{i}} + 3.00t\,\hat{\mathbf{j}})\text{m/s}$ (b) (10.0 m, 6.00 m), 7.81 m/s

**7.** (a) $3.34\hat{\mathbf{i}}$ m/s (b) − 50.9°

**9.** 12.0 m/s

**11.** 22.4° or 89.4°

**13.** 67.8°

**15.** (a) The ball clears by 0.889 m while (b) descending

**17.** (a) 18.1 m/s (b) 1.13 m (c) 2.79 m

**19.** 9.91 m/s

**21.** $\tan^{-1}[(2gh)^{1/2}/v]$

**23.** 377 m/s$^2$

**25.** 10.5 m/s, 219 m/s$^2$ inward

**27.** $7.58 \times 10^3$ m/s, $5.80 \times 10^3$ s

**29.** 1.48 m/s$^2$ inward and 29.9° backward

**31.** (a) 13.0 m/s$^2$ (b) 5.70 m/s (c) 7.50 m/s$^2$

**33.** $2.02 \times 10^3$ s; 21.0% longer

**35.** 153 km/h at 11.3° north of west

**37.** 15.3 m

**39.** $0.975g$

**41.** (a) 101 m/s (b) 32 700 ft (c) 20.6 s (d) 180 m/s

**43.** 54.4 m/s$^2$

**45.** (a) 41.7 m/s (b) 3.81 s (c) $(34.1\hat{\mathbf{i}} - 13.4\hat{\mathbf{j}})$m/s; 36.7 m/s

**47.** 10.7 m/s

**49.** (a) 6.80 km (b) 3.00 km vertically above the impact point (c) 66.2°

**51.** (a) 20.0 m/s, 5.00 s (b) $(16.0\hat{\mathbf{i}} - 27.1\hat{\mathbf{j}})$m/s (c) 6.53 s (d) $24.5\hat{\mathbf{i}}$ m

**53.** (a) 22.9 m/s (b) 360 m from the base of the cliff (c) $\vec{\mathbf{v}} = (114\hat{\mathbf{i}} - 44.3\hat{\mathbf{j}})$m/s

**55.** (a) 1.52 km (b) 36.1 s (c) 4.05 km

**57.** (a) 43.2 m (b) $(9.66\hat{\mathbf{i}} - 25.6\hat{\mathbf{j}})$m/s

**59.** 4.00 km/h

**61.** Safe distances are less than 270 m and greater than $3.48 \times 10^3$ m from the western shore.

## Chapter 4

**1.** (a) 1/3 (b) 0.750 m/s$^2$

**3.** $(6.00\hat{\mathbf{i}} + 15.0\hat{\mathbf{j}})$ N; 16.2 N

**5.** (a) $(2.50\hat{\mathbf{i}} + 5.00\hat{\mathbf{j}})$ N (b) 5.59 N

**7.** (a) 5.00 m/s$^2$ at 36.9° (b) 6.08 m/s$^2$ at 25.3°

**9.** (a) 534 N down (b) 54.5 kg

**11.** 2.55 N for an 88.7-kg person

**13.** (a) $3.64 \times 10^{-18}$ N (b) $8.93 \times 10^{-30}$ N is 408 billion times smaller

**15.** (a) $\sim 10^{-22}$ m/s$^2$ (b) $\sim 10^{-23}$ m

**17.** (a) 15.0 lb up (b) 5.00 lb up (c) 0

**21.** (a) From a free-body diagram of the forces on the bit of string touching the weight hanger we have $\Sigma F_y = 0$: $-F_g + T\sin\theta = 0$, so $T = F_g/\sin\theta$. The force the child feels gets smaller, changing from $T$ to $T\cos\theta$ when the counterweight hangs from the string. On the other hand, the kite does not notice what you are doing and the tension in the main part of the string stays constant. You do not need a level because you learned in physics lab to sight to a horizontal line in a building. Share with the parents your estimate of the experimental uncertainty, which you made by thinking critically about the measurement, repeating trials, practicing in advance, and looking for variations and improvements in technique, including using other observers. You will then be glad to have the parents themselves repeat your measurements.
(b) 1.79 N

**23.** (a) $a = g\tan\theta$ (b) 4.16 m/s$^2$

**25.** 100 N and 204 N

**27.** 8.66 N east

**29.** 3.73 m

**31.** A is in compression 3.83 kN and B is in tension 3.37 kN

**33.** 950 N

**35.** (a) $F_x > 19.6$ N (b) $F_x \leq -78.4$ N

(c)

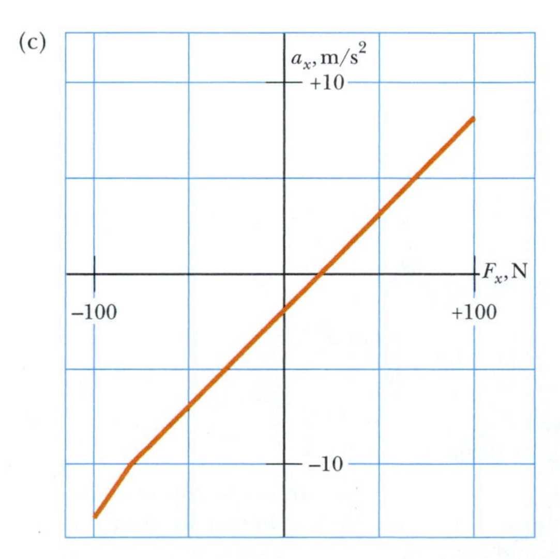


**37.** (a) 706 N (b) 814 N (c) 706 N (d) 648 N

**39.** (a) Removing mass (b) 13.7 mi/h·s

**41.** (a)

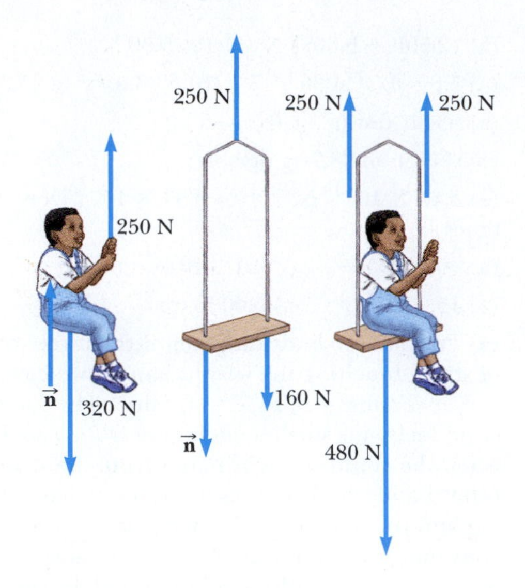


(b) $0.408 \text{ m/s}^2$ (c) 83.3 N

**43.** (a) $2.00 \text{ m/s}^2$ forward (b) 4.00 N forward on 2 kg, 6.00 N forward on 3 kg, 8.00 N forward on 4 kg (c) 14.0 N between 2 kg and 3 kg, 8.00 N between 4 kg and 3 kg (d) The 3-kg block models the heavy block of wood. The contact force on your back is represented by $Q$, which is much less than $F$. The difference between $F$ and $Q$ is the net force causing acceleration of the 5-kg pair of objects. The acceleration is real and nonzero but lasts for so short a time interval that it is never associated with a large velocity. The frame of the building and your legs exert forces, small compared with the hammer blow, to bring the partition, block, and you to rest again over a time interval large compared with the duration of the hammer blow.

**45.** (a) $Mg/2$, $Mg/2$, $Mg/2$, $3Mg/2$, $Mg$ (b) $Mg/2$

**47.** $(M + m_1 + m_2)(m_2 g/m_1)$

**49.** (c) 3.56 N

**51.** 1.16 cm

**53.** (a) 30.7° (b) 0.843 N

**55.** $mg \sin\theta \cos\theta \hat{\mathbf{i}} + (M + m\cos^2\theta) g\hat{\mathbf{j}}$

**57.** (a) $T_1 = \dfrac{2mg}{\sin\theta_1}$, $T_2 = \dfrac{mg}{\sin\theta_2} = \dfrac{mg}{\sin[\tan^{-1}(\frac{1}{2}\tan\theta_1)]}$,

$$T_3 = \frac{2\,mg}{\tan\theta_1}$$

(b) $\theta_2 = \tan^{-1}\left(\dfrac{\tan\theta_1}{2}\right)$

## Chapter 5

**1.** $\mu_s = 0.306$; $\mu_k = 0.245$

**3.** (a) 3.34 (b) The car would flip over backwards; or the wheels would skid, spinning in place, and the time would increase.

**5.** (a) 1.11 s (b) 0.875 s

**7.** $\mu_s = 0.727$, $\mu_k = 0.577$

**9.** (a) $1.78 \text{ m/s}^2$ (b) 0.368 (c) 9.37 N (d) 2.67 m/s

**11.** (a)

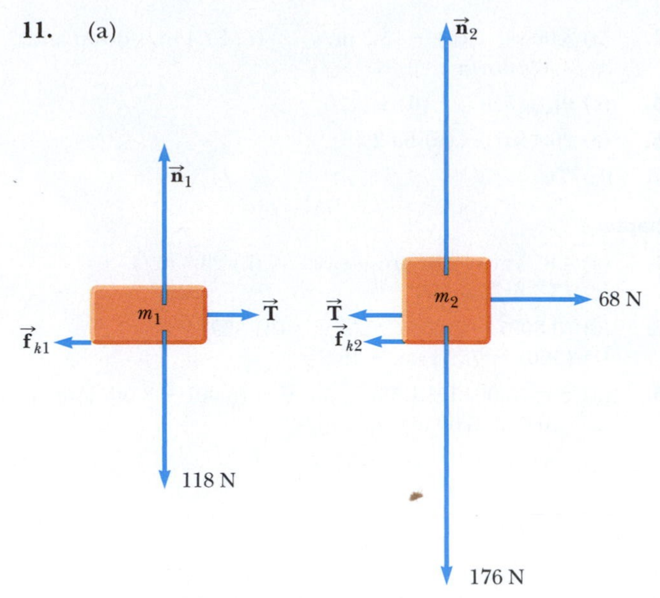


(b) 27.2 N, $1.29 \text{ m/s}^2$

**13.** any value between 31.7 N and 48.6 N

**15.** any speed up to 8.08 m/s

**17.** $v \leq 14.3$ m/s

**19.** (a) 68.6 N toward the center of the circle and 784 N up (b) $0.857 \text{ m/s}^2$

**21.** No. The jungle-lord needs a vine of tensile strength 1.38 kN.

**23.** 3.13 m/s

**25.** (a) $32.7 \text{ s}^{-1}$ (b) $9.80 \text{ m/s}^2$ down (c) $4.90 \text{ m/s}^2$ down

**27.** (a) $1.47 \text{ N}\cdot\text{s/m}$ (b) $2.04 \times 10^{-3}$ s (c) $2.94 \times 10^{-2}$ N

**29.** (a) $0.034\,7 \text{ s}^{-1}$ (b) 2.50 m/s (c) $a = -cv$

**31.** 2.97 nN

**33.** $0.613 \text{ m/s}^2$ toward the Earth

**35.** $-0.212 \text{ m/s}^2$

**37.** (a) $M = 3m\sin\theta$ (b) $T_1 = 2mg\sin\theta$, $T_2 = 3mg\sin\theta$

(c) $a = \dfrac{g\sin\theta}{1 + 2\sin\theta}$

(d) $T_1 = 4mg\sin\theta\left(\dfrac{1+\sin\theta}{1+2\sin\theta}\right)$

$T_2 = 6mg\sin\theta\left(\dfrac{1+\sin\theta}{1+2\sin\theta}\right)$

(e) $M_{\max} = 3m(\sin\theta + \mu_s\cos\theta)$

(f) $M_{\min} = 3m(\sin\theta - \mu_s\cos\theta)$

(g) $T_{2,\max} - T_{2,\min} = (M_{\max} - M_{\min})g = 6\mu_s mg\cos\theta$

**39.** (b)

| $\theta$ | 0 | 15° | 30° | 45° | 60° |
|---|---|---|---|---|---|
| **P (N)** | 40.0 | 46.4 | 60.1 | 94.3 | 260 |

**41.** (a) 0.087 1 (b) 27.4 N

**43.** (a) 2.13 s (b) 1.67 m

**45.** (a)

$$v_{\min} = \sqrt{\frac{Rg(\tan\theta - \mu_s)}{1 + \mu_s\tan\theta}}$$

$$v_{\max} = \sqrt{\frac{Rg(\tan\theta + \mu_s)}{1 - \mu_s\tan\theta}}$$

(b) $\mu_s = \tan\theta$
(c) $8.57 \text{ m/s} \le v \le 16.6 \text{ m/s}$

**47.** 0.835 rev/s

**49.** (b) 732 N down at the equator and 735 N down at the poles

**51.** (a) $1.58 \text{ m/s}^2$ (b) 455 N (c) 329 N
(d) 397 N upward and 9.15° inward

**53.** 2.14 rev/min

**55.** (b) 2.54 s; 23.6 rev/min

**57.** (a) 0.013 2 m/s (b) 1.03 m/s (c) 6.87 m/s

**59.** 12.8 N

## Chapter 6

**1.** (a) 31.9 J (b) 0 (c) 0 (d) 31.9 J

**3.** − 4.70 kJ

**5.** 5.33 W

**7.** (a) 16.0 J (b) 36.9°

**9.** (a) 11.3° (b) 156° (c) 82.3°

**11.** (a) 7.50 J (b) 15.0 J (c) 7.50 J (d) 30.0 J

**13.** (a) 0.938 cm (b) 1.25 J

**15.** (a) 575 N/m (b) 46.0 J

**17.** 12.0 J

**19.** (b) $mgR$

**21.** (a) 1.20 J (b) 5.00 m/s (c) 6.30 J

**23.** (a) 60.0 J (b) 60.0 J

**25.** 878 kN up

**27.** 0.116 m

**29.** (a) 650 J (b) 588 J (c) 0 (d) 0 (e) 62.0 J
(f) 1.76 m/s

**31.** (a) − 168 J (b) 184 J (c) 500 J (d) 148 J
(e) 5.65 m/s

**33.** 2.04 m

**35.** 875 W

**37.** $46.2

**39.** (a) 423 mi/gal (b) 776 mi/gal

**41.** 830 N

**43.** 2.92 m/s

**45.** (a) $(2 + 24t^2 + 72t^4)$ J (b) $12t \text{ m/s}^2$; $48t$ N
(c) $(48t + 288t^3)$ W (d) 1 250 J

**47.** (a) $\dfrac{mgnhh_s}{v + nh_s}$ (b) $\dfrac{mgvh}{v + nh_s}$

**49.** 7.37 N/m

**51.** (b) 240 W

**53.** (a) 4.12 m (b) 3.35 m

**55.** 1.68 m/s

**57.** $-1.37 \times 10^{-21}$ J

**59.** 0.799 J

**61.** (a) 2.17 kW (b) 58.6 kW

## Chapter 7

**1.** (a) 259 kJ, 0, − 259 kJ (b) 0, − 259 kJ, − 259 kJ

**3.** 22.0 kW

**5.** (a) $v = (3gR)^{1/2}$ (b) 0.098 0 N down

**7.** 1.84 m

**9.** (a) 4.43 m/s (b) 5.00 m

**11.** (b) 60.0°

**13.** (a) 1.24 kW (b) 20.9%

**15.** (a) 125 J (b) 50.0 J (c) 66.7 J
(d) Nonconservative; the work done depends on the path.

**17.** 10.2 m

**19.** (a) 22.0 J, 40.0 J (b) Yes; the total mechanical energy changes.

**21.** 26.5 m/s

**23.** 3.74 m/s

**25.** (a) − 160 J (b) 73.5 J (c) 28.8 N (d) 0.679

**27.** (a) 1.40 m/s (b) 4.60 cm after release
(c) 1.79 m/s

**29.** (a) 0.381 m (b) 0.143 m (c) 0.371 m

**31.** (a) 40.0 J (b) −40.0 J (c) 62.5 J

**33.** $(A/r^2)$ away from the other particle

**35.** (a) $-4.77 \times 10^9$ J (b) 569 N (c) 569 N up

**37.** $2.52 \times 10^7$ m

**39.** (a) + at Ⓑ, − at Ⓓ, 0 at Ⓐ, Ⓒ, and Ⓔ (b) Ⓒ stable;
Ⓐ and Ⓔ unstable
(c)

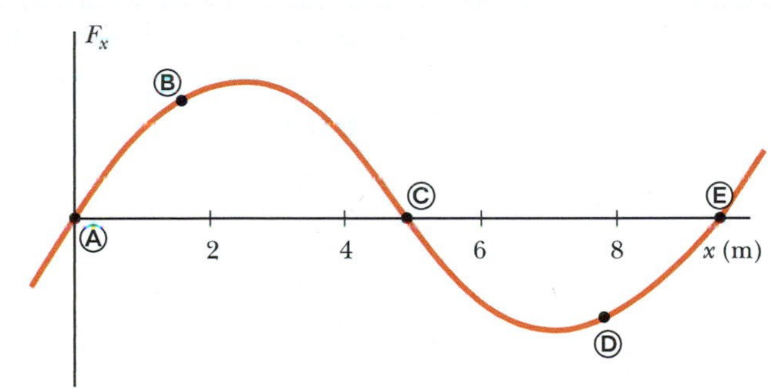


**41.** (b)

Equilibrium at $x = 0$. (c) 0.823 m/s

**43.** 0.27 MJ/kg for a battery. 17 MJ/kg for hay is 63 times larger. 44 MJ/kg for gasoline is 2.6 times larger still. 142 MJ/kg for hydrogen is 3.2 times larger than that.

**45.** $\sim 10^3$ W peak or $\sim 10^2$ W sustainable

**47.** $(8gh/15)^{1/2}$

**49.** (a) 0.225 J (b) $\Delta E_{\text{mech}} = -0.363$ J (c) No; the normal force changes in a complicated way.

**51.** 0.328

**53.** 1.24 m/s

**55.** (a) 0.400 m (b) 4.10 m/s (c) The block stays on the track.

**57.** (a) 6.15 m/s (b) 9.87 m/s

**59.** (a) 11.1 m/s (b) 19.6 m/s² upward (c) $2.23 \times 10^3$ N upward (d) $1.01 \times 10^3$ J (e) 5.14 m/s (f) 1.35 m (g) 1.39 s

**63.** (a) 14.1 m/s (b) −7.90 kJ (c) 800 N (d) 771 N (e) 1.57 kN up

## Context 1 Conclusion

**1.** (a) 315 kJ (b) 220 kJ (c) 187 kJ (d) 127 kJ (e) 14.0 m/s (f) 40.5% (g) 187 kJ

## Chapter 8

**1.** (a) $(9.00\hat{\mathbf{i}} - 12.0\hat{\mathbf{j}})$ kg·m/s (b) 15.0 kg·m/s at 307°

**3.** 40.5 g

**5.** (a) 6.00 m/s toward the left (b) 8.40 J

**7.** (a) 13.5 N·s (b) 9.00 kN (c) 18.0 kN

**9.** 260 N normal to the wall

**11.** 15.0 N in the direction of the initial velocity of the exiting water stream

**13.** (a) 2.50 m/s (b) 37.5 kJ

**15.** (a) $v_{gx} = 1.15$ m/s (b) $v_{px} = -0.346$ m/s

**17.** force on truck driver = $1.78 \times 10^3$ N; force on car driver = $8.89 \times 10^3$ N in the opposite direction

**19.** (a) 0.284 (b) 115 fJ and 45.4 fJ

**21.** 91.2 m/s

**23.** (a) 4.85 m/s (b) 8.41 m

**25.** orange: $v_i \cos\theta$; yellow: $v_i \sin\theta$

**27.** 2.50 m/s at −60.0°

**29.** $(3.00\hat{\mathbf{i}} - 1.20\hat{\mathbf{j}})$ m/s

**31.** (a) $(-9.33\hat{\mathbf{i}} - 8.33\hat{\mathbf{j}})$ Mm/s (b) 439 fJ

**33.** $\vec{\mathbf{r}}_{CM} = (11.7\hat{\mathbf{i}} + 13.3\hat{\mathbf{j}})$ cm

**35.** (b) $3.57 \times 10^8$ J

**37.** (a) $(1.40\hat{\mathbf{i}} + 2.40\hat{\mathbf{j}})$ m/s (b) $(7.00\hat{\mathbf{i}} + 12.0\hat{\mathbf{j}})$ kg·m/s

**39.** 0.700 m

**41.** (a) 39.0 MN (b) 3.20 m/s² up

**43.** (a) 442 metric tons (b) 19.2 metric tons

**45.** 4.41 kg

**47.** (a) $1.33\hat{\mathbf{i}}$ m/s (b) $-235\hat{\mathbf{i}}$ N (c) 0.680 s (d) $-160\hat{\mathbf{i}}$ N·s and $+160\hat{\mathbf{i}}$ N·s (e) 1.81 m (f) 0.454 m (g) −427 J (h) +107 J (i) Equal friction forces act through different distances on person and cart to do different amounts of work on them. The total work on both together, −320 J, becomes +320 J of extra internal energy in this perfectly inelastic collision.

**49.** (a) 2.07 m/s² (b) 3.88 m/s

**51.** (a) −0.667 m/s (b) 0.952 m

**53.** (a) $-0.256\hat{\mathbf{i}}$ m/s and $0.128\hat{\mathbf{i}}$ m/s (b) $-0.064\,2\hat{\mathbf{i}}$ m/s and 0 (c) 0 and 0

**55.** $2v_i$ and 0

**57.** (a) $m/M = 0.403$ (b) no changes; no difference

**59.** (a) 3.75 kg·m/s² to the right (b) 3.75 N to the right (c) 3.75 N (d) 2.81 J (e) 1.41 J (f) Friction between sand and belt causes half of the input work to appear as extra internal energy.

## Chapter 9

**5.** $0.866c$

**7.** (a) 25.0 yr (b) 15.0 yr (c) 12.0 ly

**9.** 1.54 ns

**11.** $0.800c$

**13.** (a) 20.0 m (b) 19.0 m (c) $0.312c$

**15.** (a) 21.0 yr (b) 14.7 ly (c) 10.5 ly (d) 35.7 yr

**17.** (a) 17.4 m (b) 3.30°

**19.** (a) $2.50 \times 10^8$ m/s (b) 4.97 m (c) $-1.33 \times 10^{-8}$ s

**21.** $0.960c$

**23.** (a) $2.73 \times 10^{-24}$ kg·m/s (b) $1.58 \times 10^{-22}$ kg·m/s (c) $5.64 \times 10^{-22}$ kg·m/s

**25.** $4.50 \times 10^{-14}$

**27.** $0.285c$

**29.** (a) 0.582 MeV (b) 2.45 MeV

**31.** (a) 3.07 MeV (b) $0.986c$

**33.** (a) 938 MeV (b) 3.00 GeV (c) 2.07 GeV

**35.** (a) $0.979c$ (b) $0.065\,2c$ (c) $0.914c = 274$ Mm/s (d) $0.999\,999\,97c$; $0.948c$; $0.052\,3c = 15.7$ Mm/s

**39.** 4.08 MeV and 29.6 MeV

**41.** $4.28 \times 10^9$ kg/s

**43.** 1.02 MeV

**45.** (a) 3.87 km/s (b) $-8.36 \times 10^{-11}$ (c) $5.29 \times 10^{-10}$ (d) $+4.46 \times 10^{-10}$

**47.** (a) $v/c = 1 - 1.12 \times 10^{-10}$ (b) $6.00 \times 10^{27}$ J (c) \$$2.17 \times 10^{20}$

**49.** (a) a few hundred seconds (b) $\sim 10^8$ km

**51.** 0.712%

**53.** (a) $0.946c$ (b) 0.160 ly (c) 0.114 yr (d) $7.50 \times 10^{22}$ J

**55.** yes, with 18.8 m to spare

**57.** (b) For $u$ small compared to $c$, the relativistic expression agrees with the classical expression. As $u$ approaches $c$, the acceleration approaches zero, so the object can never reach or surpass the speed of light.
(c) Perform $\int (1 - u^2/c^2)^{-3/2} du = (qE/m)\int dt$ to obtain $u = qEct(m^2c^2 + q^2E^2t^2)^{-1/2}$ and then $\int dx = \int qEct(m^2c^2 + q^2E^2t^2)^{-1/2} dt$ to obtain $x = (c/qE)[(m^2c^2 + q^2E^2t^2)^{1/2} - mc]$.

**63.** (a) The refugees conclude that Tau Ceti exploded 16.0 yr before the Sun.
(b) A stationary observer at the midpoint concludes that they exploded simultaneously.

## Chapter 10

**1.** (a) 5.00 rad, 10.0 rad/s, 4.00 rad/s$^2$
(b) 53.0 rad, 22.0 rad/s, 4.00 rad/s$^2$

**3.** (a) 5.24 s (b) 27.4 rad

**5.** 50.0 rev

**7.** (a) $7.27 \times 10^{-5}$ rad/s (b) $2.57 \times 10^4$ s = 428 min

**9.** (a) 126 rad/s (b) 3.77 m/s (c) 1.26 km/s$^2$
(d) 20.1 m

**11.** (a) 0.605 m/s (b) 17.3 rad/s (c) 5.82 m/s
(d) The crank length is unnecessary.

**13.** 0.572

**17.** (a) $\sqrt{\frac{2(m_1 - m_2)gh}{m_1 + m_2 + I/R^2}}$ (b) $\sqrt{\frac{2(m_1 - m_2)gh}{m_1R^2 + m_2R^2 + I}}$

**19.** 24.5 m/s

**21.** $-3.55$ N·m

**23.** $\vec{\tau} = (2.00\hat{\mathbf{k}})$ N·m

**27.** $[(m_1 + m_b)d + m_1\ell/2]/m_2$

**29.** (a) 1.04 kN at 60.0° (b) $(370\hat{\mathbf{i}} + 900\hat{\mathbf{j}})$ N

**31.** (a) $T = F_g(L + d)/\sin\theta(2L + d)$
(b) $R_x = F_g(L + d)\cot\theta/(2L + d)$; $R_y = F_gL/(2L + d)$

**33.** (a) 21.6 kg·m$^2$ (b) 3.60 N·m (c) 52.4 rev

**35.** 21.5 N

**37.** (a) 118 N and 156 N (b) 1.17 kg·m$^2$

**39.** (a) 11.4 N, 7.57 m/s$^2$, 9.53 m/s down (b) 9.53 m/s

**41.** $(60.0\hat{\mathbf{k}})$ kg·m$^2$/s

**43.** (a) 0.433 kg·m$^2$/s (b) 1.73 kg·m$^2$/s

**45.** (a) $\omega_f = \omega_i I_1/(I_1 + I_2)$ (b) $I_1/(I_1 + I_2)$

**47.** (a) 0.360 rad/s counterclockwise (b) 99.9 J

**49.** (a) $7.20 \times 10^{-3}$ kg·m$^2$/s (b) 9.47 rad/s

**51.** $5.99 \times 10^{-2}$ J

**53.** (a) 500 J (b) 250 J (c) 750 J

**55.** (a) 2.38 m/s. Its weight is insufficient to provide the centripetal acceleration. (b) 4.31 m/s
(c) The ball does not reach the top of the loop.

**57.** 131 s

**59.** (a) $(3g/L)^{1/2}$ (b) $3g/2L$ (c) $-\frac{3}{2}g\hat{\mathbf{i}} - \frac{3}{4}g\hat{\mathbf{j}}$
(d) $-\frac{3}{2}Mg\hat{\mathbf{i}} + \frac{1}{4}Mg\hat{\mathbf{j}}$

**61.** (a) $\sqrt{\frac{2mgd\sin\theta + kd^2}{I + mR^2}}$ (b) 1.74 rad/s

**67.** (a) 61.2 J (b) 50.8 J

**69.** (a) $Mvd$ (b) $Mv^2$ (c) $Mvd$ (d) $2v$
(e) $4Mv^2$ (f) $3Mv^2$

**71.** $T = 2.71$ kN, $R_x = 2.65$ kN

**73.** (a) 20.1 cm to the left of the front edge; $\mu_k = 0.571$
(b) 0.501 m

**75.** (a) 133 N (b) $n_A = 429$ N and $n_B = 257$ N
(c) $R_x = 133$ N and $R_y = -257$ N

**77.** $\frac{3}{8}F_g$

## Chapter 11

**1.** $2.67 \times 10^{-7}$ m/s$^2$

**3.** $7.41 \times 10^{-10}$ N

**5.** (a) $4.39 \times 10^{20}$ N toward the Sun (b) $1.99 \times 10^{20}$ N away from the Sun (c) $3.55 \times 10^{22}$ N toward the Sun

**7.** $\rho_M/\rho_E = 2/3$

**9.** (a) 7.61 cm/s$^2$ (b) 363 s (c) 3.08 km
(d) 28.9 m/s at 72.9° below the horizontal

**11.** $\vec{\mathbf{g}} = 2MGr(r^2 + a^2)^{-3/2}$ toward the center of mass

**13.** (a) $4.23 \times 10^7$ m (b) 0.285 s

**15.** $1.90 \times 10^{27}$ kg

**17.** $1.26 \times 10^{32}$ kg

**19.** After 3.93 yr, Mercury would be farther from the Sun than Pluto.

**21.** (a) $1.84 \times 10^9$ kg/m$^3$ (b) $3.27 \times 10^6$ m/s$^2$
(c) $-2.08 \times 10^{13}$ J

**23.** 1.78 km

**25.** $1.66 \times 10^4$ m/s

**29.** $1.58 \times 10^{10}$ J

**31.** (b) $1.00 \times 10^7$ m (c) $1.00 \times 10^4$ m/s

**33.** (a) 0.980 (b) 127 yr (c) $-2.13 \times 10^{17}$ J

**35.** (a) 5 (b) no; no

**37.** (a) ii (b) i (c) ii and iii

**39.** (a) 0.212 nm (b) $9.95 \times 10^{-25}$ kg·m/s
(c) $2.11 \times 10^{-34}$ kg·m$^2$/s (d) 3.40 eV
(e) $-6.80$ eV (f) $-3.40$ eV

**41.** $4.42 \times 10^4$ m/s

**43.** (a) 29.3% (b) no change

**45.** $2.26 \times 10^{-7}$

**47.** (c) $1.85 \times 10^{-5}$ m/s$^2$

**49.** $v = 492$ m/s

**51.** (a) 7.79 km/s (b) 7.85 km/s (c) $-3.04$ GJ
(d) $-3.08$ GJ (e) loss = 46.9 MJ (f) A component of the Earth's gravity pulls forward on the satellite on its downward-banking trajectory.

**53.** (a) $m_2(2G/d)^{1/2}(m_1 + m_2)^{-1/2}$ and $m_1(2G/d)^{1/2}(m_1 + m_2)^{-1/2}$; relative speed $(2G/d)^{1/2}(m_1 + m_2)^{1/2}$
(b) $1.07 \times 10^{32}$ J and $2.67 \times 10^{31}$ J

**55.** (a) 200 Myr (b) $\sim 10^{41}$ kg; $\sim 10^{11}$ stars

**57.** $(GM_E/4R_E)^{1/2}$

**61.** $r_n = (0.106\text{ nm})n^2$, $E_n = -6.80\text{ eV}/n^2$, for $n = 1, 2, 3, \ldots$

## Context 2 Conclusion

**1.** (a) 146 d (b) Venus 53.9° behind the Earth

**3.** (a) 2.95 km/s (b) 2.65 km/s (c) 10.7 km/s
(d) 4.80 km/s

## Chapter 12

**1.** (a) The motion repeats precisely. (b) 1.81 s
(c) No, the force is not in the form of Hooke's law.

**3.** (a) 1.50 Hz, 0.667 s (b) 4.00 m (c) $\pi$ rad
(d) 2.83 m

**5.** (b) 18.8 cm/s, 0.333 s (c) 178 cm/s$^2$, 0.500 s
(d) 12.0 cm

**9.** 40.9 N/m

**11.** (a) 40.0 cm/s, 160 cm/s$^2$ (b) 32.0 cm/s, – 96.0 cm/s$^2$ (c) 0.232 s

**13.** 2.23 m/s

**15.** (a) 0.542 kg (b) 1.81 s (c) 1.20 m/s$^2$

**17.** (a) 28.0 mJ (b) 1.02 m/s (c) 12.2 mJ (d) 15.8 mJ

**19.** (a) $E$ increases by a factor of 4 (b) $v_{max}$ is doubled. (c) $a_{max}$ is doubled. (d) The period is unchanged.

**21.** 2.60 cm and – 2.60 cm

**23.** Assume simple harmonic motion: (a) 0.820 m/s (b) 2.57 rad/s$^2$ (c) 0.641 N More precisely: (a) 0.817 m/s (b) 2.54 rad/s$^2$ (c) 0.634 N

**27.** 0.944 kg · m$^2$

**31.** $1.00 \times 10^{-3}$ s$^{-1}$

**33.** (a) 1.00 s (b) 5.09 cm

**37.** 1.74 Hz

**39.** If the cyclist goes over them at one certain speed, the washboard bumps can excite a resonance vibration of the bike, so large in amplitude as to make the rider lose control. ~$10^1$ m

**41.** 6.62 cm

**43.** $9.19 \times 10^{13}$ Hz

**45.** (b) 1.04 m/s (c) four times larger, 3.40 m

**47.** $f = (2\pi L)^{-1}(gL + kh^2/M)^{1/2}$

**49.** (b) 1.23 Hz

**51.** (a) 3.00 s (b) 14.3 J (c) 25.5°

**57.** (a) 5.20 s (b) 2.60 s

## Chapter 13

**1.** $y = 6\,[(x - 4.5t)^2 + 3]^{-1}$

**3.** 0.319 m

**5.** (a) $(3.33\hat{\mathbf{i}})$ m/s (b) −5.48 cm (c) 0.667 m, 5.00 Hz (d) 11.0 m/s

**7.** (a) 31.4 rad/s (b) 1.57 rad/m (c) $y = (0.120 \text{ m}) \sin(1.57x - 31.4t)$ (d) 3.77 m/s (e) 118 m/s$^2$

**9.** (a) $y = (8.00 \text{ cm}) \sin(7.85x + 6\pi t)$ (b) $y = (8.00 \text{ cm}) \sin(7.85x + 6\pi t - 0.785)$

**11.** (a) 0.021 5 m (b) 1.95 rad (c) 5.41 m/s (d) $y(x, t) = (0.021\,5 \text{ m}) \sin(8.38x + 80.0\pi t + 1.95)$

**13.** 80.0 N

**15.** 13.5 N

**17.** 0.329 s

**19.** 1.07 kW

**21.** 55.1 Hz

**23.** 5.56 km

**25.** (a) 23.2 cm (b) 1.38 cm

**27.** (a) 4.16 m (b) 0.455 μs (c) 0.158 mm

**29.** 5.81 m

**31.** $\Delta P = (0.200 \text{ N/m}^2) \sin(62.8x/\text{m} - 2.16 \times 10^4 t/\text{s})$

**33.** (a) 3.04 kHz (b) 2.08 kHz (c) 2.62 kHz; 2.40 kHz

**35.** 26.4 m/s

**37.** 19.3 m

**39.** (a) 0.364 m (b) 0.398 m (c) 941 Hz (d) 938 Hz

**41.** 184 km

**43.** $(Lm/Mg \sin\theta)^{1/2}$

**45.** 0.084 3 rad

**49.** (a) $\dfrac{\mu\omega^3}{2k} A_0^2 e^{-2bx}$ (b) $\dfrac{\mu\omega^3}{2k} A_0^2$ (c) $e^{-2bx}$

**51.** (a) $\mu_0 + (\mu_L - \mu_0)x/L$

**55.** 6.01 km

**57.** (a) 55.8 m/s (b) 2 500 Hz

**59.** The gap between bat and insect is closing at 1.69 m/s.

## Chapter 14

**1.** (a) – 1.65 cm (b) – 6.02 cm (c) 1.15 cm

**3.** (a) $+x$, $-x$ (b) 0.750 s (c) 1.00 m

**5.** (a) 9.24 m (b) 600 Hz

**7.** 91.3°

**9.** (a) 156° (b) 0.058 4 cm

**11.** (a) To reach the receiver, waves from the more distant source must travel an extra distance $\Delta r = \lambda/2$ and interfere destructively with waves from the closer source. (b) It should move along the hyperbola represented by $9.00x^2 - 16.0y^2 = 144$.

**13.** at 0.089 1 m, 0.303 m, 0.518 m, 0.732 m, 0.947 m, and 1.16 m from one speaker

**15.** (a) 4.24 cm (b) 6.00 cm (c) 6.00 cm (d) 0.500 cm, 1.50 cm, 2.50 cm

**17.** 0.786 Hz, 1.57 Hz, 2.36 Hz, 3.14 Hz

**19.** 15.7 Hz

**21.** (a) reduced by $\frac{1}{2}$ (b) reduced by $1/\sqrt{2}$ (c) increased by $\sqrt{2}$

**23.** (a) 163 N (b) 660 Hz

**25.** $\dfrac{Mg}{4Lf^2 \tan\theta}$

**27.** (a) 0.357 m (b) 0.715 m

**29.** 57.9 Hz

**31.** $n$(206 Hz) for $n = 1$ to 9 and $n$(84.5 Hz) for $n = 2$ to 23

**33.** 50.0 Hz, 1.70 m

**35.** $n$(0.252 m) with $n = 1, 2, 3, \ldots$

**37.** (a) 350 m/s (b) 1.14 m

**39.** 5.64 beats/s

**41.** (a) 1.99 beats/s (b) 3.38 m/s

**43.** The second harmonic of E is close to the third harmonic of A, and the fourth harmonic of C# is close to the fifth harmonic of A.

**45.** The condition for resonance is satisfied because the 12 h 24 min period of free oscillation agrees precisely with the period of the lunar excitation.

**47.** (a) 34.8 m/s (b) 0.977 m

**49.** 3.85 m/s away from the station and 3.77 m/s toward the station

**51.** 21.5 m

**53.** (a) 59.9 Hz (b) 20.0 cm

**55.** (a) $\frac{1}{2}$ (b) $[n/(n+1)]^2T$ (c) $\frac{9}{16}$

**57.** $y_1 + y_2 = 11.2 \sin(2.00x - 10.0t + 63.4°)$

**59.** (a) 78.9 N (b) 211 Hz

## Context 3 Conclusion

**1.** 3.5 cm

**2.** The speed decreases by a factor of 25.

**3.** Station 1: 15:46:32; Station 2: 15:46:22; Station 3: 15:46:08, all with uncertainties of ±1 s

## Chapter 15

**1.** 0.111 kg

**3.** 6.24 MPa

**5.** 1.62 m

**7.** $7.74 \times 10^{-3}$ m$^2$

**9.** 271 kN horizontally backward

**11.** 2.31 lb

**13.** 10.5 m; no because some alcohol and water evaporate

**15.** 98.6 kPa

**17.** (a) 7.54 kg (b) 39.8 N (c) 41.9 N up (d) zero (e) The tension decreases and the normal force increases.

**19.** 0.258 N

**21.** (a) $1.017\,9 \times 10^3$ N down, $1.029\,7 \times 10^3$ N up (b) 86.2 N (c) 11.8 N

**23.** (a) 7.00 cm (b) 2.80 kg

**25.** 1 430 m$^3$

**27.** 1 250 kg/m$^3$ and 500 kg/m$^3$

**29.** 1.01 kJ

**31.** 12.8 kg/s

**33.** $2\sqrt{h(h_0 - h)}$

**35.** (a) 27.9 N (b) $3.32 \times 10^4$ kg (c) $7.26 \times 10^4$ Pa

**37.** 0.247 cm

**39.** (a) 1 atm + 15.0 MPa (b) 2.95 m/s (c) 4.34 kPa

**41.** 347 m/s

**43.** (a) 4.43 m/s (b) The siphon can be no higher than 10.3 m.

**45.** 12.6 m/s

**47.** $1.61 \times 10^4$ m$^2$

**51.** 0.604 m

**55.** The top scale reads $(1 - \rho_0/\rho_{Fe})m_{Fe}g$. The bottom scale reads $[m_b + m_0 + \rho_0 m_{Fe}/\rho_{Fe}]g$.

**57.** (a) 2.79 μm/s (b) 7.95 h (c) $8.88 \times 10^3$ m/s$^2$ (d) 31.6 s

**59.** 4.43 m/s

**61.** (a) 1.25 cm (b) 13.8 m/s

**63.** (a) 3.307 g (b) 3.271 g (c) $3.48 \times 10^{-4}$ N

## Context 4 Conclusion

**1.** $9.8 \times 10^9$ N·m

**2.** (a) 1.30 MPa (b) yes, but only with specialized equipment and techniques

**3.** (a) 0.42 m/s (b) greater

**4.** (a) 16 knots at 56° west of south (b) 47%

# Credits

## Photographs



**Invitation.** **1:** © Stockbyte **3:** © David Parker Photo Researchers, Inc.

**Chapter 1.** **4:** Mark Wagner/Stone/Getty Images **6:** Courtesy of National Institute of Standards and Technology, U.S. Department of Commerce **10:** Phil Boorman/Getty Images **14:** Mack Henley/Visuals Unlimited

**Contcxt 1.** **34:** Courtesy of The Exhibition Alliance, Hamil ton, N.Y. **35:** top left, © Bettmann/CORBIS; bottom right, © Martin Bond/Photo Researchers, Inc. **36:** © Mehau Kulyk/ Photo Researchers, Inc.

**Chapter 2.** **37:** Jean Y. Ruszniewski/Getty Images **56:** top left, North Wind Picture Archive; bottom, © 1993 James Sugar/Black Star **57:** George Semple **60:** George Lepp/ Stone/Getty **66:** top left, Courtesy of the U.S. Air Force; top right, Photri, Inc., **67:** Courtesy Amtrak NEC Media Relations

**Chapter 3.** **69:** © Arndt/Premium Stock/PictureQuest **74:** © The Telegraph Colour Library/Getty Images **79:** Tony Duffy/Getty Images **89:** Frederick McKinney/FPG/Getty **90:** top left, Jed Jacobsohn/Getty Images; middle left, Bill Lee/Dembinsky Photo Associates; bottom left, Sam Sargent/ Liaison International **91:** Courtesy of NASA **92:** Courtesy of NASA

**Chapter 4.** **96:** Steve Raymer/CORBIS **99:** Giraudon/Art Source **104:** NASA **105:** top right, John Gillmoure, The Stock Market **116:** Roger Viollet, Mill Valley, CA, University Science Books, 1982 **118:** © Tony Arruza/CORBIS

**Chapter 5.** **125:** Paul Hardy/CORBIS **134:** top, Robin Smith/Getty Images; bottom left, © Tom Carroll/Index Stock Imagery/Picture Quest **143:** Jump Run Productions/Image Bank **146:** a) Courtesy of GM; b) Courtesy of GM-Hummer **148:** Mike Powell/Getty Images **150:** Frank Cezus/FPG International

**Chapter 6.** **156:** Billy Hustace/Getty Images **170:** a–c, e, f) George Semple; d) Digital Vision/Getty Images **172:** Sinclair Stammers/Science Photo Library/Photo Researchers, Inc. **185:** Ron Chapple/FPG/Getty

**Chapter 7.** **188:** © Harold E. Edgerton/Courtesy of Palm Press, Inc. **213:** Gamma **219:** Engraving from Scientific American, July 1888

**Context Conclusion 1.** **220:** Courtesy of Honda Motor Co., Inc **222:** top left, Photo by Brent Romans/www.Edmunds.com; bottom left, © Adam Hart-Davis/Photo Researchers, Inc.

**Context 2.** **223:** Japanese Aerospace Exploration Agency (JAXA) **224:** top, courtesy of NASA/JPL; bottom, Courtesy of NASA/JPL/Cornell **225:** Pierre Mion/National Geographic Image Collection

**Chapter 8.** **226:** © Harold and Esther Edgerton Foundation 2002, courtesy of Palm Press, Inc. **232:** Courtesy of Saab **233:** Tim Wright/CORBIS **246:** Richard Megna, Fundamental Photographs **251:** © Bill Stormont/The Stock Market

**Chapter 9.** **259:** Emily Serway **262:** AIP Emilio Segré Visual Archives, Michelson Collection **264:** AIP Niels Bohr Library **288:** Courtesy of Garmin Ltd.

**Chapter 10.** **291:** Courtesy of Tourism Malaysia **317:** top left, © Stuart Franklin/Getty Images; bottom, David Malin, Anglo-Australian Observatory **320:** Courtesy of Henry Leap and Jim Lehman **327:** John Lawrence/Stone/Getty Images

**Chapter 11.** **337:** © 1987 Royal Observatory/Anglo-Australian Observatory, by David F. Mahlin from U.K. Schmidt plates **342:** Art Resource **346:** U.S. Space Command, NORAD **350:** bottom, H. Ford et al. & NASA **352:** K. W. Whitten, R. E. Davis, M. L. Peck, and G. G. Stanley, General Chemistry, 7th ed., Belmont, CA, Brooks/Cole, 2004 **353:** Photo courtesy of AIP Niels Bohr Library, Margarethe Bohr Collection **361:** Courtesy of NASA/JPL **365:** NASA

**Context Conclusion 3.** **371:** top right, Guillermo Aldana Espinosa/National Geographic Image Collection; bottom right, Joseph Sohm/ChromoSohm Inc./CORBIS **372:** T. Campion/SYGMA

**Chapter 12.** **373:** © Simon Kwong/Reuters/CORBIS **390:** Special Collections Division, Univ. of Wash. Libraries, Photo by Farquharson **394:** Telegraph Colour Library/FPG International

**Chapter 13.** **400:** Don Bonsey/Stone Getty **416:** © Howard Sochurek/Woodfin Camp & Associates, Inc. **419:** top right, Courtesy of U.S. Navy. Photo by Ensign John Gay; bottom right, Courtesy of the Educational Development Center, Newton, MA **427:** Joe McDonald/Visuals Unlimited **429:** Gregg Adams/Stone

**Chapter 14.** **432:** Kathy Ferguson Johnson/PhotoEdit/PictureQuest **433:** bottom right, Education Development Center, Newton, MA **434:** top right, Education Development Center, Newton, MA **438:** © 1991 Richard Megna/Fundamental Photographs **450:** Photographs courtesy of (a) and (b) Getty Images; (c) Photodisc Green/Getty Images **457:** top left, Courtesy Professor Thomas D. Rossing, Northern Illinois University; top right, Steve Bronstein/Image Bank/Getty Images; bottom left, Murray Greenberg

**Context Conclusion 3.** **459:** Paul X. Scott/SYGMA **460:** Photo by M. Hamada; used by permission from the Multidisciplinary Center for Earthquake Engineering Research, Univ. of Buffalo

**Context 4.** **462:** bottom left, Illustrations by Ken Marschall © 1992 from *Titanic: An Illustrated History,* a Hyperion/Madison Press Book); top right, Illustrations by Ken Marschall © 1992 from *Titanic: An Illustrated History,* a Hyperion/Madison Press Book **463:** top left, Courtesy of RMS Titanic, Inc.; top right, Courtesy of RMS Titanic, Inc.

**Chapter 15.** **464:** Norman Tobley/Taxi/Getty Images **466:** Royalty-free/CORBIS **468:** top right, David Frazier **470:** © Hulton Deutsch Collection/CORBIS **472:** Richard Megna/Fundamental Photographs **475:** Andy Sacks/Tony Stone Images/Getty Images **476:** Werner Wolff/Black Star **478:** © Bettmann/CORBIS **482:** Courtesy of The Father Browne S.J. Collection **483:** © TempSport/CORBIS **484:** Pamela Zilly/The Image Bank/Getty **488:** George Semple **489:** Stan Osolinski/Dembinsky Photo Associates **490:** bottom right, The Granger Collection **492:** Courtesy of Jeanne Maier

**Context Conclusion 4.** **493:** Illustrations by Ken Marschall © 1992 from *Titanic: An Illustrated History,* a Hyperion/Madison Press Book **494:** Illustrations by Ken Marschall © 1992 from *Titanic: An Illustrated History,* a Hyperion/Madison Press Book **495:** top, Illustrations by Ken Marschall © 1992 from *Titanic: An Illustrated History,* a Hyperion/Madison Press Book; bottom, E. Paul Oberlander/Woods Hole Oceanographic Institution

## Tables and Illustrations

This page constitutes an extension of the copyright page. We have made every effort to trace the ownership of all copyrighted material and to secure permission from copyright holders. In the event of any question arising as to the use of any material, we will be pleased to make the necessary corrections in future printings. Thanks are due to the following authors, publishers, and agents for permission to use the material indicated.

**Chapter 1.** **33:** By permission of John L. Hart FLP, and Creators Syndicate, Inc.

**Chapter 11.** **363:** By permission of John L. Hart FLP, and Creators Syndicate, Inc.

**Chapter 14.** **452:** Adapted from C. A. Culver, Musical Acoustic, 4th Edition, New York, McGraw-Hill, 1956, p. 128. **453:** Adapted from C. A. Culver, Musical Acoustic, 4th Edition, New York, McGraw-Hill, 1956.

# Index

Page numbers in *italics* indicate figures; page numbers followed by "n" indicate footnotes; page numbers followed by "t" indicate tables.


**A**

Absolute frame of reference, 262
Absolute pressure, 470
Absorption spectrum, 352
Acceleration, 1, 47–49, 292. *See also* Angular acceleration; Average acceleration; Centripetal acceleration; Translational acceleration
  average, 45, 61
  centripetal, 87, 132–133
  constant, 51–53, 86
    two-dimensional motion with, 71–73
  force proportional to, 48–49
  free-fall, 55–59
  in simple harmonic motion, 378–379
  in uniform circular motion, 79–81
  instantaneous angular, 293, 324
  instantaneous, 45, 49, 61
    definition of, 45, 71
  lateral, 86
  mass and, 115
  negative, 48, 50
  nonconstant, 375
  of center of mass, 246
  positive, 50
  radial, 82–83
  required by consumers, 59–61
  tangential, 82–83
  uniform, 54
  units of, 8t, 102
  zero, 50
Acceleration-time curves, 45–46, 377–378
Acceleration vectors, 69–71
Accretion disk, 349–350
Action force, 104–106
Addition of vectors
  associative law of, 16
  commutative law of, 16
  triangle method of, 16
Aging, effect of relativity on, 268–269
Air bags, value of, 232
Air columns, standing waves in, 443–446
Airplane wing, streamline flow around, 480–481
Airstream flow, 480–481
Algebraic expressions, 8
Alternating-current circuits, oscillating waves in, 373
Alternative representations, 24–25
Alternative-fuel vehicles, 3, 34–36
  acceleration in, 59–61
  fuel cell, 221–222
  hybrid, 220–221
Altitude, atmospheric pressure and, 466–467
*Alvin* submarine, 495–496
Amplitude, 391, 404, 424
  energy transfer rate and, 414
  maximum point of, 452
  of oscillating motion, 376
  of seismic waves, 459–460
  types of, 438
  varying in time with frequency, 447–448
  zero, 438–439, 452
Analysis models, 23
Analytical procedures, 4
Analyze, in problem-solving strategy, 25
Angular acceleration, 293
  constant, 295–296
  counterclockwise, 310–311
  net torque of rigid object and, 310
  of rigid object, 294
  tangential component of, 296
Angular frequency, 376, 391, 424
  energy transfer rate and, 414
  of physical pendulum, 386
  of simple pendulum, 385
  of traveling wave, 406
Angular momentum, 313–316, 324
  conservation of, 316–319, 324
  quantized, 440
Angular position, 292
  of rigid object, 294
Angular speed, 293
  average, 293
  constant, 297
  of rigid object, 294
  resistance to change in, 299
Angular wave number, 406, 424
Antinodes, 452
  building on, 450–451
  on string, 440–441
  positions of, 438–439
Aphelion, 343
Apogee, 343
Archimedes, 470
Archimedes's principle, 470–471, 473–474, 482
Area, 6
  calculation of, 12
  units of, 8t
Argo video system, 494
Atmosphere, hydrogen in, 348–349
Atmospheric pressure, 482
  altitude and, 466–467
  measurements of, 470
Atomic spectra, hydrogen, 351–357
Atomizer, mechanics of, 481
Atwood machine, 111–112
  angular momentum of, 316
  with massive pulley, 310–311
Automobiles
  air bags in, 232
  bumpers of, 233
  drag coefficients of, 145–147

Automobiles *(Continued)*
forces on, 114–115
horsepower ratings of, 179–180
lateral acceleration of, 86
maximum speed of, 136
performance, 179
traditional, 179
Average acceleration, 49, 61
angular, 293
definition of, 45, 70–71
equation for, 80
Average position, 242
Average power, 177, 181
Axis
major, 342
minor, 342
of rotation, 294
semimajor, 342, 344–345, 359
semiminor, 342, 359

**B**

Ballard, Dr. Robert, 463, 494
Balloon, helium-filled, 473
Balmer, Johann, 352
Balmer series, 352–353, 355
Banked roadway, 137
Barometric pressure, 470
*Beagle 2,* 224
Beat frequency, 446–447, *448*
Beats, 446–448, 452
Benz, Karl, 34
Bernoulli, Daniel, 478
Bernoulli effect, 479
Bernoulli's equation, 478–480, 482
Bernoulli's principle, 481, 482
Big Bang theory, 145
Binary star system, *350,* 351
Biodiesel fuel, potential energy in, 208
Bioluminescence, 172–173
Black holes, 349–351
supermassive, 350
Block-spring collision, 199–200
Block-spring system, 176–177
oscillating motion in, 379–380
simple harmonic motion for, *383*
Blood flow
Bernoulli's principle in, 481
Doppler measurements of, *419*
Blood pressure, measuring, 468
Bohr, Niels, 337, 353
Bohr radius, 354–355, 359
Bohr theory of hydrogen, 351–357
Bound system, 346
Boundary conditions, 432–433
at end of air column, 443
wave under, 440
Bridges, collapse of, 390–391
Broughton suspension bridge, collapse of, 390
Bulk modulus, 421
Buoyant force, 141n, 482
Archimedes's principle and, 470–475
on floating object, 472–473
on totally submerged object, 471–472

**C**

C string, harmonics of, 442
Calculus, 2
Cameron, James, 495
Carbon dioxide, 207
Carbon monoxide
from propane fuel, 209
in combustion, 207
Cartesian coordinate system, 13
designation of points in, *13*
Categorize, in problem-solving strategy, 25
Cavendish apparatus, *338*
Center of gravity, 243–244
Center of mass, 242–245
acceleration of, 246
determination of, 244
motion of, 247
of right triangle, 245
of three particles, 244
velocity of, 245, 250
versus center of gravity, 243–244
Centrifugal force, 134
Centripetal acceleration, 87, 132–133
in loop-the-loop maneuver, 137–138
magnitude of, 79–80
nonconstant, 81
of automobiles, 86
of Earth, 81
Centripetal force, 133
Circular motion
nonuniform, 138–140
uniform, 79–81
Circular orbit
mechanical energy of, 346–347
parking, 368
to elliptical orbit, 357–358
Cohesive forces, between liquid molecules, 464–465
Collisions, 227
block-spring, 199–200
definition of, 233–234
elastic, 234, 235–236, 250
inelastic, 234, 236–237, 250
one-dimensional, 234, 236–237
perfectly, 234, 236–237, 239, 250
momenta before and after, 232
one-dimensional, 234–239, 235–236
slowing neutrons in, 237–238
two-dimensional, 239–242
value of air bags in, 232
value of bumpers in, 233
Color force, 145
Comet Halley, motion of, 343
Commutative scalar product, 161
Component vectors, 17–18
Components, 17
Compressed air lift, 469
Compression, 415
Conceptualize, in problem-solving strategy, 25
Conduction, 171
Conical pendulum, 135–136
Conservation of angular momentum, 316–319
Conservation of energy, 171
Conservation of momentum, 229–231
Conservative forces, 195–196, 209
potential energy and, 200–201
Constant force, 157–160
Contact forces, 97
Contextual approach, 2–3
Continuity equation, 476–478
for energy, 171–172, 180–181
for fluids, 482
Convection, 171n
Conversion factor, 9
Coordinate systems, 12–14
Coordinates, Galilean transformation of, 261–262
Copernican theory, 134–135
Coplanar forces, 306
Coulomb constant, 144
Coulomb's law, 144
Crab Nebula, *317*
Crests, 403–404
in Doppler effect, 418–419
Cross product, 304–306, 324
Crude oil, discovery of, 35
Crust, Earth, 423
Cugnot, Nicholas Joseph, 34

Curved track, motion on, 199
Cycloid, 320
Cylinder
  moment of inertia of, 302
  net torque on, 305
  rolling motion of, 320–321

**D**

da Vinci, Leonardo, 34
Daimler, Gottlieb, 34
Dam, force on, 469
Damped oscillations, 387–388, 391
Deceleration, 48
Definite integral, 162–163
Deformation, of rotating object, 291
Density, 26
  definition of, 7
  of common substances at standard temperature and pressure, 467t
  of fluids, 466
Derivative, 42
Derived quantities, 6
Diesel fuel, potential energy in, 207–208
Differential equation, 141
Dimension, 8
Dimensional analysis, 8–9, 26
Dispersion, 402n
Displacement, 14–15
  angular, 293
  arbitrary, 165
  average velocity and, 38–39
  cause of, 159
  direction of, 158–159
  negative, 40–41
  small, 162–163
Displacement amplitude, 416
Displacement antinode, 443
Displacement node, 443
Displacement vectors, 21
  magnitude of, 15
Displacement wave, 416
Distance, 15, 38–39
Distributive law of multiplication, 161
Disturbance, propagation of, 401–403
Doppler effect, 417–420, 424
  misconception about, 417
Doppler-shifted frequency, 420
Dot product, 160–161, 180
Drag, 481
Drag coefficients, 142
  of automobiles, 145–147
Dynamics, 96

**E**

Earth
  centripetal acceleration of, 81
  cross-section of, *423*
  crust of, 423
  kinetic energy of, 229–230
  liquid core of, 423
  mantle of, 423
  mass and radius of, 338
  place of in Universe, 341
  satellite of, 340–341
  solid core of, 423
Earth-atmosphere system, energy transfer mechanisms in, 202–203
Earthquakes, 371–372
  epicenter of, 421
  focus or hypocenter of, 421
  minimizing damage in, 459–461
  paths of waves in, 423
  seismic waves in, 421–424
  standing waves in, 450–451
Eccentricity, 343, 344, 359
Einstein, Albert
  general relativity theory of, 280–283
  relativity theory of, 2, 260, 263–272
Elastic collision. *See* Collisions, elastic
Elastic modulus, 421
Elastic potential energy, 195–196, 209
Electric force, 205
  potential energy for, 205
Electric potential energy, 210
Electric vehicles, 34, 209
  hybrid, 220–221
Electrical transmission, 171, 173, 181
Electromagnetic force, 144, 147
Electromagnetic radiation, 171, 173, 181
  for hydrogen atom orbit, 354
  from atomic transitions, 355–356
Electromagnetic waves, 351, 401
  Doppler effect for, 417
  frequency of, 351
  superposition of, 433
  wavelength of, 351
Electromagnetism, unified theory of, 2
Electron-proton system, bound, 354
Electron volt, 278
Electrons
  acceleration of, 54
  momentum of, 276
  transition of, 355, 357
Electrostatic force, 144, 145
Electroweak force, 145
Elliptical orbit, 343, 368
  circular orbit changing to, 357–358
  mechanical energy of, 346–347
  of satellite, 347–348
  semimajor axis of, 344–345
Elliptical transfer orbit, 368
  major axis of, 368–369
Emission spectrum, 351
  of hydrogen, 352
Emission wavelengths, hydrogen, 356
End correction, 443
Energy. *See also* Energy transfer; Kinetic energy; Mechanical energy; Potential energy
  conservation of, 171–172, 181, 197
  forms of, 156–157
  in planetary and satellite motion, 345–351
  in rotational motion, 311–313
  in simple harmonic motion, 381–384, 391
  internal, 169–170, 173, 175, 181, 188–190
    increase in, 175
    kinetic energy transformed to, 204
  ionization, 355
  mass and, 279–280
  of hydrogen atom, 354
  of proton, 279
  potential, 188–210
  quantized, 432–433
  relativistic, 276–279
  sources of, developing, 35–36
  total, 284
    in simple harmonic motion, 381, 391
  total relativistic, 278
  transformation of, 191–192
  transition, 355–356
Energy level diagrams, 355, 356
  stability of equilibrium and, 206–207
Energy-momentum relationship, 278

Energy transfer, 156–157, 169, 173, 197
  mechanisms of, 170–173, 181
    in Earth-atmosphere system, 202–203
    in home, 202
    in human body, 203
    in nonisolated systems, 202–203
  rate of, by sinusoidal waves on string, 413–415
Environment, 157, 180
Epicenter, 421
  of earthquake, 371
  seismic waves spreading from, 460
Equilibrium
  maximum position relative to, 416
  neutral, 207, 210
  particle in, 107–108
  rigid object in, 306–309
  stability of, 206–207, 210
  unstable, 207, 210
Equilibrium position, 374
Equivalence principle, 281
Escape speed, 348–349
Ethanol, potential energy of, 208
European Space Agency, 223–224
Event horizon, 349
Events, 261, 284

**F**

Falling objects, 55–59
Feynman, Richard, 134–135
Field, gravitational force and, 339–340
Field forces, 97–98
Finalize, in problem-solving strategy, 25
Fission, 279–280
Floating object, buoyant forces on, 472–473
Flow characteristics, 475–476
Flow rate, 482
Fluid. *See also* Gas; Liquid
  buoyant force from, 470–475
  continuity equation for, 476–478
  definition of, 464–465
  dynamics of, 475–476, 482
    in airplanes, 480–481
    in blood flow, 481
  incompressible, 476
  irrotational flow, 476
  mechanics of, 464–482
  nonviscous, 476
  pressure and depth of, 493–494
  pressure of, 465–466
Fluid *(Continued)*
  pressure variation with depth of, 466–469
  statics of, 475
  steady flow, 476
  viscosity of, 476
Focus, 342
  of earthquake, 371
Force. *See also specific forces*
  action, 104–106
  centrifugal, 134
  centripetal, 133
  concept of, 97–98
  conservative, 195–196, 209
  constant, 157–160
    work done by, 157–160
  fundamental, 143–145, 147
  gravitational, 103–104
  in equilibrium, 306–309
  nonconservative, 196–200, 209
  normal, 105–106
  of friction, 126–132
  on automobiles, 114–115
  point of application of, 169
  pressure and, 465
  proportional to acceleration, 48–49
  reaction, 104–106
  relativistic, 276
  sum of, 101
  unit of, 102
  varying, 162–166, 180
  vector nature of, 98
  versus torque, 304
Forced oscillations, 389–390, 391
Ford, Henry, 35
Fourier series, 448
Fourier synthesis, 449
Fourier's theorem, 448–450, 452
Free-body diagram, 106
Free-fall
  ball in, 193
  for projectile motion, 78–79
  magnitude of acceleration of, 115
Freely falling objects, 55–59
Frequency, 351, 376, 377, 391, 424
  *See also* Angular frequency
  beat, 446–447, *448*
  of air column, 443
  of musical instruments, 450
  of normal modes, 441
  of stringed instruments, 442
  of tuning fork, 446
  versus pitch, 448
Friction
  coefficients of, 127, 130
  forces of, 126–132
  kinetic, 147, 173–177
  static, 127–131, 147
Friction force
  and internal energy, 175
  direction of, 127
  on automobiles, 114–115
  point of application of, 196–197
Frictionless pulley, 131
Frictionless surface, 107
  acceleration on, 110, 112–113
  block pulled on, 167–168
  revolving object on, 318
Fuel cell vehicles
  advantages of, 221
  disadvantages of, 221–222
Fuels, potential energy in, 207–209
Fundamental frequency, 441
  of air column, 443–444
  of string, 441
Fundamental quantities, 6
Fusion reaction, 280

**G**

Galaxy M87, *350*
Galilean transformation, 272
  of coordinates, 261–262
  of velocities, 261–262
Galileo Galilei, 1, 99
  free-fall acceleration theory of, 55–56
Gas, 464
Gasing, *291*
Gasoline
  in hybrid electric vehicles, 221
  potential energy in, 207–209
Gasoline engines, electric starter for, 35
Gasoline-powered car, 34–35
Gauge pressure, 470
General Motors EV1, 209, 220
  acceleration of, 59–60
  drag coefficient on, 146
  fuel in, 209
General relativity
  postulates of, 281
  theory of, 284
Geocentric model, 341
Geometric model, 22–23
Geosynchronous orbit, 340
Glancing collision, 239–240
Global warming, 207
Goddard, Robert, 249
Graphical representation, 24–25, 38

Graphs, slopes of, 40
Gravitation, universal, law of, 280, 337, 338–341, 359
Gravitational constant, 143–144, 338–339, 359
Gravitational field, 339–340, 359
  light beam and, 281–282
Gravitational force, 103–104, 143–144, 147, 338
  acting on planets, 343–344
  average position of, 243–244
  buoyant force in equilibrium with, 471
  friction forces and, 133
  in precessional motion, 319
  on floating object, 472
  on liquid, 467
  on spacecraft, 223
  potential energy for, 203–205
  weight and, 113–114
  work done by, 159, 191–192
Gravitational lens, 282
Gravitational mass, 104
Gravitational potential energy, 189–190, 198, 209, 210, 478–479, 482
  of planet-star system, 346
  stored in system, 190–191
Gravity, 337
  center of, 243–244
Gravity waves, 351
Greenhouse effect, 207
Ground state, 355
Gyroscopes, 319–320
  in spacecraft, 323

**H**

Hafele, J. C., 267
Harmonic motion
  amplitude of, 438
  simple, 374–375, 391
    acceleration in, 377
    energy considerations in, 381–384
    mathematical representation of, 375–381
    of oscillating particle, 380
    position, velocity, and acceleration in, 378–379
    velocity in, 377
Harmonic series, 441, 443, 452
Harmonics, 441, 452
  Fourier synthesis of, 449
  in pipe, 445
  natural frequencies in, 443–444
  odd, 444
Harmonics *(Continued)*
  of C string, 442
  physical mixture of, 450
  superposition of, 448
Heat, 171, 173, 181
  of combustion, 207
Heliocentric model, 341
Hertz (Hz), 376
Hohmann transfer, 369–370
Hohmann transfer orbit, 368
Home system, energy transfer mechanisms in, 202
Honda Civic, 220
Honda Insight, 220
  acceleration of, 60
  forces on, 115
  lateral acceleration of, 86
Hooke's law, 163–164, 165–166, 195, 201, 374, 391
Horizontal motion, of projectile, 77–78, 86
Horizontal range, of projectile, 75–76
Horsepower (hp), 177–178
  ratings of vehicles, 179–180
Hose, water flow through, 477–478
Hot-air balloons, floating, 472
Hubble Space Telescope images, *350*
Huygens, Christian, 34
Human body, energy transfer mechanisms in, 203
Hybrid electric vehicles, 60, 220–221
  gas mileage for, 220
  parallel and series, 220
Hydraulic press, diagram of, *468*
Hydrocarbons, oxidation of, 207
Hydrogen
  atomic spectra of, 351–357
  electronic transition in, 357
  emission spectrum of, 352
  emission wavelengths of, 356
  frequency of radiation from, 356
  in atmosphere, 348–349
Hydrogen atom
  Bohr model of, 353–357, 359
  Bohr orbits in, 354
  energy level diagram for, *356*
  in Rosette Nebula, *337*
  total energy of, 354
Hydrogen fuel cell, 221
Hyperbolas, 343
Hyperbolic escape orbit, 368
Hypocenter, 421
  of earthquake, 371

**I**

Icebergs, visible portion of, *464*
Impulse
  approximation, 232
  of net force, 231, 250
Impulse-momentum theorem, 231–233, 250
Impulsive force, 232
Incline, object sliding down, 132
Indefinite integral, 162–163
Inertia, 100
  moment of, 298–299, 324
    and angular momentum, 315
    of homogenous rigid objects, *300*
    of uniform solid cylinder, 302
Inertial frame of reference, 99, 115
  wave speed in, 409
Instantaneous power, 177, 181
Interference
  constructive, 435, 452
  definition of, 434
  destructive, 435, 436, 452
  in acoustical system, 436
  of waves, 434–437
  spatial, 446
  temporal, 446–448
Interferometer, 262–263
Internal combustion engine, 34
  chemical reaction in, 207
  mass production of, 35
International Space Station, 223
Inverse tangent function, 85
Ionization, 354n
Ionization energy, 355
Ionized elements, 356–357
Irrotational flow, 476
Isolated system, 190–194, 209
  total energy of, 197–200
  total momentum of, 229
Isolation dampers, 460

**J**

Japanese Aerospace Exploration Agency (JAXA), 223–224
Javelin, free-fall model for travel of, 78–79
Joule, 158

**K**

Kaon
  decay of, 230–231
  neutral, 230
Keating, R. E., 267
Kepler, Johannes, 1, 342

Kepler's laws of planetary motion, 342, 368
first, 342–343, 359
second, 343–344, 359
third, 344–345, 359, 369
Kilogram, 26
definition of, 5
Kilowatt-hour (kWh), 178
Kinematic equation, 53, 61
Kinematics, 37
rotational, 295–296
variables in rotational motion, 292
Kinetic energy, 180, 181
conservation of in collisions, 234, 237–238
conserved in proton-proton collision, 240–241
in atomic and nuclear processes, 279–280
in balloon-Earth system, 473
in Bernoulli's equation, 482
in collisions, 235
in perfectly inelastic collision, 236–237
in simple harmonic motion, 381–382, 391
in two-dimensional collisions, 239–240
of hydrogen atom, 354
of rolling object, 321
of rotating object, 318
of sinusoidal wave on string, 414
of system, 188–189
of viscous fluid, 476
potential energy transformed into, 204
relativistic, 276–277, 284
rotational, 298–302, 324
total, 324
translational, 312
work-kinetic energy theorem and, 166–168
Kinetic friction, 181
coefficient of, 127, 147
force of, 126–127, 132, 147
situations involving, 173–177
Kobe, Japan, earthquake, *460*

**L**

Laminar flow, through constricted pipe, *478*
Laser Interferometer Gravitational Wave Observatory (LIGO), 351
Length, 5, 26
approximate values of, 6–7, 7t
contraction in special relativity, 269–272, 284
Lift, 481
Light beam, bending, 281–282
Light clock, 267
Light waves, 263
oscillating, 373
Limiting process, 43–44
Linear mass density, 421
Linear momentum, 227–231, 313, 316–317, 324
conservation of, 229–231, 250
definition of, 228, 315
for system of particles, 246–247
Linear wave equation, 407–408
Linear waves, 433
speed of, 409
Liquefaction, 459–460
Liquid
characteristics of, 464
pressure in, 482
Loma Prieta earthquake, 459
amplitude of oscillations during, 460
destruction caused by, 451
Longitudinal wave, 401, 424
oscillations and propagation in, 416
propagating along tube, 416
sound, 415–417
Loop-the-loop maneuver, 137–138
Lorentz transformation equations, 272–275, 284
Lorentz velocity transformation, 273–274, 284
Love wave, 422, 424
Luminiferous ether, 262
Lyman series, 355

**M**

Manometer, 470
Mantle, 423
Mars Climate Orbiter, 223
*Mars Express* orbiter, 224
Mars Global Surveyor, 223
Mars Pathfinder, 223
Mars Polar Lander, 223
Mass, 5, 100–101. *See also* Center of mass
acceleration and, 115
approximate values of, 6–7
as manifestation of energy, 278
center of, 242–245, 250
Mass *(Continued)*
changing in radioactive decay, 280
continuous distribution of, 243, 245
definition of, 100
dual behavior of, 280–281
energy and, 279–280
of various objects, 7t
per unit volume, 7
relativistic, 275
spherically symmetric distributions of, 204
units of, 102
weight and, 101, 103–104
Mass production, 35
Mathematical representation, 25
Matter, states of, 464
Matter transfer, 171, 181
Maximum height, projectile, 75–76
Maxwell, James Clerk, 2
Measurement standards, 5–6
Mechanical energy
conservation of, 191, 192, 199–200, 209
in collisions, 238–239
total, 191, 209
transformation of, 195
Mechanical waves, 170–171, 181, 400–401
disturbance in, 401–403
linear, 408
reflection and transmission of, 411–413
traveling, 405–408
types of, 401–425
Mechanics, classical, 1–2
Mental representation, 24
Mercury barometer, 470
Meteoroid
interacting with Sun, 346
motion of, 343
Meter, 26
definition of, 5
Methane, 208
Mexico City, earthquake destruction in, 451
Michelson, Albert A., 262
Michelson-Morley experiment, 262–263
Michoacán earthquake, 371–372, 460
destruction caused by, 451
Microwaves, 420
Models, 22–26
categories of, 22, 337

Moment arm, 303
Momentum, 226–227. *See also* Angular momentum; Linear momentum
  conservation of, 229–231
    in collisions, 234
    in proton-proton collision, 240–241
    in rocket propulsion, 248–250
    in two-dimensional collisions, 240–242
  impulse and, 231–233
  in elastic collisions, 234, 235–236
  in isolated system, 229
  in perfectly inelastic collision, 237
  instantaneous angular, 313–314
  linear, 227–231
  of electron, 276
  of system of particles, 245, 246–247
  relativistic, 275–276, 284
  time rate of change of, 228–229
Moon, escape speeds from surface of, 349t
Motion
  direction of, in length contraction, 270
  frequency of oscillatory, 376, 377
  in one dimension, 37–61
  in plane, 73
  in presence of velocity-dependent resistive forces, 140–143
  in two dimensions, 69–87
  laws of, 1, 96–115
  Newton's first law of, 98–100
  Newton's second law of, 101–103
  Newton's third law of, 104–106
  nonuniform circular, 138–140
  of system of particles, 245–247
  oscillatory, 373–391
  period of oscillatory, 376, 377
  projectile, 73–79, 86
  quantity of, 228
  rotational, 291–324
  uniform circular, 132–138
Motion diagrams, 50–51
Multiplication, distributive law of, 161
Muons, 267
  time dilation for, 267
Musical instruments
  characteristic sounds of, 450
  standing waves produced by, 443–446
Musical sound, nonsinusoidal wave patterns of, 448–450

**N**

National Aerodynamics and Space Administration (NASA), Mars mission of, 223–224
Natural frequency, 388, 443–444
  of oscillation, 389, 391
Natural gas, potential energy in, 208
Nature, fundamental forces of, 143–145
Net force
  impulse of, 231, 250
  particle motion under, 101–103
  particle under, 108–109
  variable, work done by, 163
Net torque
  external, 324
    angular momentum and, 315
  internal, 310
    angular momentum and, 314
  on cylinder, 305
  rigid object under, 309–313
Net work, 163
Neutron star, 317
  rotation period of, 318–319
Neutrons, slowed by collisions, 237–238
Newton, 102
  per square meter, 465–466
Newton, Isaac, 1–2
  relativity principle of, 259–262, 263
  self-propelled vehicle of, 34
Newton · meter, 158
Newton's law of universal gravitation, 143–144, 337, 338–341, 359
Newton's laws of motion, 55, 96, 98–115
  applications of, 107–114, 125–147
  first, 98–100, 115, 229
  in inertial reference frame, 409
  relativistic form of, 275–276
  second, 101–103, 115, 147, 179, 227, 368, 411
    application of, 130–138
    applied to satellites, 340–341
    for floating objects, 472
    for particle, 228–229
Newton's laws of motion *(Continued)*
    for system of particles, 246–247
    for translational motion, 309
    for uniform circular motion, 344
    rotational analog to, 309–310, 314
    with velocity-dependent resistive forces, 142–143
  third, 104–106, 115, 226, 227
Nimitz Freeway collapse, 451, 459
Nodal lines, in sedimentary basin, 451
Nodes, 452
  on string, 440–441
  positions of, 438–439
Nonconservative force, 209
Nonisolated system, 169–173
Nonlinear waves, 408, 433
Normal force, 105–106
Northridge, California, earthquake, 371, 451
*Nozomi* orbiter, 223, 224
Nuclear processes, mass and energy in, 279–280
Nucleons, 144–145

**O**

Oil crisis, 209
One-dimensional motion. *See* Motion, in one dimension
*Opportunity* rover, 224–225
Orbital period, 344–345
Orbiting object, escape speed of, 348–349
Order-of-magnitude calculations, 10
Oscillating particle, 380
Oscillations
  damped, 387–388, 391
  forced, 389–390, 391
  number of per unit time interval, 376
  of simple pendulum, 384–386
  simple harmonic, 374, 378–379, 391, 459
    energy of, 381–384
Oscillatory motion, 373–391
  of particle attached to spring, 374–375
Oxygen molecule, rotation of, 300

**P**

*P* waves, 421–424
  in earthquake, 461
Parabolas, 343
Parabolic path, of projectile, 74–78, 247

Parabolic trajectory, 73–74
Parallel axis theorem, 321
Particle-Earth system, potential energy of, 203–204
Particle models, 22, 37, 45–47, 51–53, 61
Particle under constant acceleration model, 51–53, 61
Particle under constant speed model, 45–47
Particles
  in equilibrium, 107–108
  in uniform circular motion, 79–81, 132–138
  motion of system of, 245–247
  under net force, 108–109
Pascal (Pa), 465–466
Pascal's law, 467–468, 482
Paschen series, 355
Path length, 436, 437
Pendulum
  conical, 135–136
  physical, 386–387, 391
  simple, 384–386, 391
    motion of, *383*
Perigee, 343
Perihelion, 343
Period, 377, 391, 424
  of particle in uniform circular motion, 81
  of physical pendulum, 387
  of simple pendulum, 385
Periodic motion, 373
Phase, 376
Phase constant, 376, 391, 405, 406
Phase shift, 180°, 413
Physics
  classical, 1–2
  contextual approach to, 2–3
  goal of, 4
  modern, 2
Pictorial representation, 24, 39
  simplified, 24
Pions, negative and positive, 230–231
Pipe, harmonics in, 445
Pitch, versus frequency, 448
Planck's constant, 353
Plane, motion in, 73
Plane polar coordinates, 13–14
Planetary motion, 1
  Copernican theory of, 134–135
  energy considerations in, 345–351
  Kepler's laws of, 342–345, 359, 368
  two objects in, 346
Planetary orbits, 337
  eccentricities of, 343
Planets
  escape speeds from surface of, 349t
  useful data on, 345t
Polar satellite, 340
Position, 292
  as function of time, 52
  as function of velocity and time, 52
  change of, 14
  of particle in simple harmonic motion, 375
Position-time graph, 38, *42*, 43–44
Position vector, 69–71
  as function of time, 72
  of projectile, 75
Potential energy
  as property of system, 189
  change in, 204–205
  conservative forces and, 200–201
  elastic, 195–196, 209
  electric, 210
  for gravitational and electric forces, 203–205
  gravitational, 189–190, 198, 209, 210
  in fuels, 207–209
  in simple harmonic motion, 381–382
  negative, 204
  of sinusoidal wave on string, 414
  of system, 188–190
Power, 181, 414
  associated with sinusoidal wave, 414
  delivered by elevator motor, 178–179
  delivered to rotating object, 312
  general expression for, 177–178
  instantaneous, 177
Powers of ten, prefixes for, 7t
Precessional motion, 319–320
Pressure, 482. *See also* Atmospheric pressure; Blood pressure
  definition of, 465
  depth of fluid and, 493–494
  in compressed air lift, 469
  measurements of, 470
  units of, 465–466
  variation of with depth, 466–469
Pressure amplitude, 416
Pressure differential, 466
Pressure node, 443
Pressure wave, 416
Problem-solving strategy, 25
  analyze, 25
Problem-solving strategy *(Continued)*
  categorize, 25
  conceptualize, 25
  finalize, 25
  for determining projectile motion, 76
  for particle under constant acceleration, 53–54
Projectile
  exploding, 247
  horizontal motion of, 77–78
  horizontal range and maximum height of, 75–76
  motion of, 73–79, 86
Projections, of a vector, 17
Propagation, of disturbance, 401–403
Propane, potential energy in, 208–209
Proper time interval, 266–267
Proton-proton collision, 240–241
Protons
  energy of, 279
  repulsive electrostatic force of, 144–145
Pulses, 401, 409
  at boundary, 413
  propagation of, 401–402
  reflection of, 411, *412*
  shape of, 409
  speed of on cord, 410
  superposition of, 433–434
  transmission of, 412–413
Pure rolling motion, 320–321
Pythagorean theorem, 85

**Q**

Quantization, 432–433, 440, 452
Quantum mechanics, 2, 354
Quantum number, 354, 359
Quantum physics, 337
Quantum state, 354
  energies of, 355
Quark model, 145
Quarks, 144
Quasi-steady state, 202

**R**

Radar, 373
Radial acceleration, 82–83, 138–139
Radian (rad), 292
Radiation, 171n
Radio waves, oscillating, 373
Radioactive decay, 280
Ramp, crate sliding down, 198
Rarefactions, 415

Rayleigh wave, 422, 424
Reaction force, 104–106, 107, 226
Reasonable values, 6
Rectangular coordinate system, 13
Reference clock, 6
Reference frames
  absolute, 262
  inertial, 99, 115
  noninertial, 99
  time measurement and, 264
Reflection, mechanical wave, 411–413
Relative velocity, 83–85, 87
Relativistic energy, 276–279
Relativistic momentum, 275–276
Relativistic particle, total energy of, 278
Relativity, 259–284. *See also* General relativity; Special relativity
  Einstein's principle of, 263–264
  general theory of, 280–283, 284
  in space travel, 283–284
  length contraction and, 269–272
  Lorentz transformation equations and, 272–275
  Michelson-Morley experiment and, 262–263
  Newtonian, 259–263
  special theory of, 260, 284
    consequences of, 264–272
  theory of, 2
  time dilation and, 265–269
  twin paradox and, 268–269
Repulsion, 205
Resistive force, 147
  magnitude of, 140
  mathematical representation of, 141
  proportional to object speed, 140–142
  proportional to object speed squared, 142–143
  velocity-dependent, 140–143
Resonance, 389, 391, 450
  in structures, 390–391
Resonance frequency, 389–390
Rest energy, 278, 284
Restoring force, 163
Resultant displacement, 21
Resultant force, 101
Resultant vector, 16
Right-hand rule, 294
Rigid object
  angular position, speed, and acceleration of, 292–294
  definition of, 291
Rigid object *(Continued)*
  in equilibrium, 306–309
  modeling of, 23, 293–294
  moment of inertia of, 298–300
  rolling motion of, 320–423
  rotational kinetic energy of, 298–302
  under constant angular acceleration, 295–296
  under net torque, 309–313
Roadways
  switchbacks on, 159–160
Rocket propulsion, 248–250
Rod, rotating, 302
Rolling motion, 320–423, 324
Rolling object, translational and rotational variables of, 320–321
Rolling sound of thunder, 416–417
Rosette Nebula, *337*
Rotating ball, 139–140
Rotational equilibrium, 306
Rotational motion, 291, 324
  angular momentum and, 313–316
  angular position, speed, and acceleration in, 292–294
  conservation of angular momentum in, 316–319
  dynamic equations of, 315t
  equilibrium and, 306–309
  kinematic equations of, 295–296
  kinetic energy of, 298–302, 324
  net torque in, 309–313
  of gyroscopes, 319–320
  of spacecraft, 323
  rolling as special case of, 320–322
  torque and vector product in, 303–306
  translational quantities and, 296–298
  work and energy in, 311–313
Rotational motion equations, 295t
Rotational position, 292
Rotational variables, 292
Rydberg, Johannes, 352
Rydberg constant, 353
Rydberg equation, 352–353, 355–356
  empirical, 356
  generalized, 356

**S**

*S* waves, 421–424
  in earthquake, 461
Satellite
  artificial, *346*
Satellite *(Continued)*
  energy considerations in motion of, 345–351
  gravitational force applied to, 340–341
  in elliptical orbit, 347–348
Scalar product, 17, 180
  integral of, 174
  of two vectors, 160–162
Scalars, 14–15, 26
  multiplication of vector by, 17
Schwarzschild radius, 349
Scientific notation, 12
Scott, David, 56
Second, 26
  definition of, 6
Sedimentary basins, standing waves in, 450–451
Seiche, 451
Seismic isolation, 460
Seismic waves, 371–372, 421–424
  speed of, 422
  spread of, 459–460
Seismograph trace, *422*
Self-propelled vehicles, 34–36
Shear modulus, 421
Shearing force, 465
Shock wave, 419
SI system, 5, 6
  units of, 8t
Significant figures, 11–12
Simple harmonic motion, 374–384
Simple pendulum, 384–386
Simplification models, 23, 37
  for trajectory, 73–74
Simultaneity, 264
Sinusoidal waves, *351,* 403
  at boundary, 413
  energy transfer rate of on string, 413–415
  in harmonic series, 452
  one-dimensional, *405*
  physical characteristics of, 404
  traveling, 405–408, 424
Small angle approximation, 384–385
Smith, E. J., 481
Snowshoes, design of, 466
Soil, liquefaction of, 459–460
Solar day, 5–6
Solar system
  Copernican theory of, 134–135
  Kepler's laws of planetary motion in, 342–345
  structural model of, 337, 341–342
Solid, 464

Sonic boom, 419
Sound
  spectrum of, 450
  timbre of, 450
Sound waves, 415–417
  Doppler effect for, 417–420
  interference of, 436–437
  longitudinal, 401
  nonsinusoidal patterns of, 448–450
  resonance of in tube, 446
  speed of in various media, 415t
  standing, in air column, 437–440
Space-time, 272
  curvature of, 282
  ripples in, 351
Space travel
  length contraction in, 269–272
  physics of, 223–225
  relativity in, 283–284
Spacecraft
  circular to elliptical orbit of, 357–358
  gravitational forces on, 223
  relative velocity of, 274
  rotational motion of, 323
  successful mission of, 367–370
Spatial interference, 446
Special relativity
  consequences of, 264–272
  postulates of, 263, 284
  second postulate of, 265–266
Spectral lines, 351–356
  absorption, 351
  Balmer series of, 352, 355
  emission, 351
Spectroscope, 351
Speed, 6. *See also* Angular speed; Velocity
  average, 61
    definition of, 38
    versus average velocity, 38
  constant
    in loop-the-loop maneuver, 137–138
  instantaneous, 42–43, 61
  instantaneous angular, 293, 324
  of light, constancy, 263–264
  of sound in various media, 415t
  of transverse waves on strings, 408–411
Sphere, rolling down incline, 322
Sphygmomanometer, 468
Spinning top, 319–320
*Spirit* rover, 224
Spring
  compression of in collision, 238–239
  dropping block onto, 168
  motion of particle attached to, 374–375
  work done by, 163–166, 195
  work required to stretch, 165–166
Spring-drive car, 34
Stable equilibrium, 206–207, 210
Standing waves, 437–440, 452
  amplitude of, 438
  formation of, 439–440
  in air columns, 443–446
  in earthquakes, 450–451
  in strings, 440–442
  multiflash photograph of, *438*
Static equilibrium, 306
Static friction
  coefficient of, 127, 147
  force of, 126, 136
  maximum force of, 147
Steady flow, 476
Steady-state, 389
  flow in, 476–477
  nonisolated system in, 202–203, 210
Steam-driven vehicles, 34
Steam engine, 35
Stereo speakers, out of phase, 436–437
Stiffness, in cable tension, 410–411
Streamline flow, 478–479
  around airplane wing, 480–481
Streamlines, 476–478
Stringed instrument, tuning, 447
Strings
  frequencies of standing waves on, 441
  pulses overlapping on, 433–434
  standing waves in, 440–442
  tension and linear mass density of, 441
Stroboscopic photograph, definition of, 50
Strong force, 144–145, 147
Structural models, 23–24, 337, 341–342
  features of, 341
Structures, resonance in, 390–391
Submarines, Doppler effect in, 420
Submerged object, buoyant forces on, 471–472
Sun
  bound and unbound systems in relation to, 346
  escape speeds from surface of, 349t
  formation of, 204
  position of, in elliptical orbit of planets, 343
  radius vector from planet to, 343–344
Supernova, remnant of, *317*
Superposition
  of harmonics, 448
  principle of, 433–434, 437, 452
Switchbacks, 159–160
Symmetric object, moment of inertia of, 299
System, 157, 180
  center of mass of, 242–245
  identifying, 157
  increasing internal energy of, 175
  isolated, 172, 190–194, 197–200, 209
  motion of, 245–247
  nonisolated, 169–173, 180–181, 210
  nonisolated in steady state, 202–203
  potential energy of, 188–190, 200–201
  total mechanical energy of, 209
System boundary, 157, 180

**T**

Tabular representation, 25, 38
Tacoma Narrows Bridge, destruction of, 390–391
Tailpipe emissions
  carcinogens in, 208
  from propane-fueled vehicles, 209
Tangent function, 18
Tangential acceleration, 82–83, 138–139, 324
Tangential position, 324
Tangential speed, 296, 297–298, 324
Tangential velocity, 296
Temporal interference, 446–448
Tension, 107
  in cable, 410–411
Terminal speed, 141, 147
  calculation of, 143
Thermal conduction, 171
Thrust, 248–249
Thunder, rolling sound of, 416–417
Timbre, 450

Time, 5–6
  dilation in special relativity, 265–269
  simultaneity and relativity of, 264
Time constant, 141
Time dilation, 265–269, 284
  twin paradox and, 268–269
Time intervals, 266–267
  approximate values of, 7t
  in calculating average velocity, 39–40
  proper length and, 370–371
*Titanic*
  design of, 493
  maiden voyage of, 481–482
  mass of, 462
  salvaging, 494–496
  search for, 462–463
  sinking of, 493–494
Torque, 324. *See also* Net torque
  angular momentum and, 314–315
  definition of, 303
  in rotational motion, 303–306
  versus force, 304
Torricelli, Evangelista, 470
Torricelli's law, 480
Toyota Prius, 220, *221*
  acceleration of, 60
  forces on, 115
Trajectory
  of projectile, 73–74
Transformations, energy, 156–157
Translational acceleration, 324
  of rolling object, 322
  tangential component of, 296
  total, 297
Translational equilibrium, 306
Translational kinematic variables, 292
Translational motion
  angular momentum and, 314
  dynamic equations of, 315t
  kinematic equations of, 295t
Translational quantities, 296–298
Translational speed, 321
Transmission, of mechanical waves, 411–413
Transverse acceleration, 407
Transverse velocity, 407
Transverse wave, 401, 424
  speed of on strings, 408–411
Trigonometric function, 13, 375n
Trough, 403–404
Tuning forks
  frequencies of, 448
  measuring frequency of, 446
Turbulent flow, 475–476
Turning points, 206–207
Twin paradox, 268–269
Two-dimensional motion, 69–87
  with constant acceleration, 71–73

**U**
Ultrasound image, *416*
Unbalanced force, 101
Unbound system, 346
Uniform circular motion, 79–81
  Newton's second law applied to, 132–138
Uniform gravitational field, *281*
Unit vectors, 19–20
Units of measure, 5–7
  conversion of, 9–10
Universal gravitational constant, 144, 338–339, 359
Universal gravitation, Newton's law of, 280, 338–341, 359
Universe
  energy forms in, 156–157
U.S. customary system, 6
  units of, 8t

**V**
Vacuum, rocket in, 249–250
Varying force, work done by, 162–166
Vascular flutter, 481
Vector product, 17, 324
  definition of, 304
  properties of, 305
  torque vector and, 304–306
Vector sum, 16, 19, 20
Vectors, 14–15, 26
  addition of, 16, 19–20
  components of, 17–20
  direction of, 18
  equality of, 15–16
  magnitude of, 18
  multiplication of, 17
  multiplication of by scalar, 17
  negative of, 16
  position, velocity, and acceleration, 69–71
  scalar product of, 160–162
  subtraction of, 17
Velocity, 292
  as function of position, 52
  as function of time, 51–53
  average, 38–41, 44–45, 61
    calculation of, 39–41, 44–45
    definition of, 38–39, 70
    for particle under constant acceleration, 52
Velocity *(Continued)*
  changing in rocket propulsion, 248
  constant
    analysis of particle under, 45–47
  dimensions of, 9
  in simple harmonic motion, 378–379, 382
  instantaneous, 41–45, 44–45, 61
    calculating, 44–45
    definition of, 70
  magnitude and direction of, 79
  of center of mass, 245, 250
  positive, 50
  relative, 83–85, 87
  units of, 8t
Velocity-dependent forces, 145
Velocity-dependent resistive forces, 140–143
Velocity-time graphs, 39–40, 48–49
  for simple harmonic oscillator, 377–378
Velocity transformation, 273–274, 284
  Galilean, 261–262
  Lorentz, 273
Velocity vectors, 69–71
  as function of time, 71–72
  change in direction of, 82–83
Vertical motion, projectile, 86
Vibrating strings
  power supplied to, 415
  resonance of, 389–390
Viking Project, 223
Viscosity, in fluid flow, 476
Volcanic eruption, 69
Volta, Alessandro, 34
Volume, units of, 8t
*Vostok* spacecraft, 223
*Voyager 2* spacecraft, rotation of, 323

**W**
Water
  disturbance of, 400
  waves, 423
Watt (W), 177
Wave function, 402, 424
  for sinusoidal wave, 405, 406
Wave model, 403–405
Wave number, 406, 424
Wave speed, 351, 404
  frequency and, 420
  normal mode frequencies as functions of, 441

Wave speed *(Continued)*
of transverse wave on strings, 408–411
of traveling sinusoidal wave, 405, 406
Wave under boundary conditions, 452
Waveform, 402
Wavelength, 351, 424
of normal modes, 441
Waves
combination of, 432–433
crest of, 403–404
interference of, 434–437
nonsinusoidal patterns of, 448–450
standing, 437–440
in air columns, 443–446
in strings, 440–442
superposition of, 433–434

Waves *(Continued)*
transfer of disturbance in, 400–401
traveling, 405–408
trough of, 403–404
types of, 400–401
Weak force, 145, 147
Weight
gravitational force and, 103–104, 113–114
mass and, 115
versus mass, 5, 101
Wheel, block unwinding from, 312–313
Wind instruments, standing waves produced by, 444
Wind tunnel test, *475*
Work, 181
as method of energy transfer, 170
definition of, 158

Work *(Continued)*
done by constant force, 157–160
done by gravitational force, 191–192
done by spring, 163–166
done by varying force, 162–166
in rotational motion, 311–313
SI units for, 158
Work-kinetic energy theorem, 166–168, 172, 176–177, 180, 191
for rotational motion, 311–313
relativistic, 276–279

**Z**
Zero acceleration, 50
Zero amplitude, 438–439, 452
Zero net force, 100

## Standard Abbreviations and Symbols for Units

| Symbol | Unit | Symbol | Unit |
|---|---|---|---|
| A | ampere | K | kelvin |
| u | atomic mass unit | kg | kilogram |
| atm | atmosphere | kmol | kilomole |
| Btu | British thermal unit | L | liter |
| C | coulomb | lb | pound |
| °C | degree Celsius | ly | lightyear |
| cal | calorie | m | meter |
| d | day | min | minute |
| eV | electron volt | mol | mole |
| °F | degree Fahrenheit | N | newton |
| F | farad | Pa | pascal |
| ft | foot | rad | radian |
| G | gauss | rev | revolution |
| g | gram | s | second |
| H | henry | T | tesla |
| h | hour | V | volt |
| hp | horsepower | W | watt |
| Hz | hertz | Wb | weber |
| in. | inch | yr | year |
| J | joule | Ω | ohm |

## Mathematical Symbols Used in the Text and Their Meaning

| Symbol | Meaning |
|---|---|
| $=$ | is equal to |
| $\equiv$ | is defined as |
| $\neq$ | is not equal to |
| $\propto$ | is proportional to |
| $\sim$ | is on the order of |
| $>$ | is greater than |
| $<$ | is less than |
| $\gg (\ll)$ | is much greater (less) than |
| $\approx$ | is approximately equal to |
| $\Delta x$ | the change in $x$ |
| $\sum_{i=1}^{N} x_i$ | the sum of all quantities $x_i$ from $i = 1$ to $i = N$ |
| $\lvert x \rvert$ | the magnitude of $x$ (always a nonnegative quantity) |
| $\Delta x \rightarrow 0$ | $\Delta x$ approaches zero |
| $\frac{dx}{dt}$ | the derivative of $x$ with respect to $t$ |
| $\frac{\partial x}{\partial t}$ | the partial derivative of $x$ with respect to $t$ |
| $\int$ | integral |